Springer Polar Sciences

Springer Polar Sciences

Springer Polar Sciences is an interdisciplinary book series that is dedicated to research in the Arctic, sub-Arctic regions, and the Antarctic. In recent years, the polar regions have received increased scientific and public interest. Both the Arctic and Antarctic have been recognized as key regions in the regulation of the global climate, and polar ecosystems have been identified to be particularly susceptible to the ongoing environmental changes. Consequently, the international efforts in polar research have been enhanced considerably, and a wealth of new findings is being produced at a growing rate by the international community of polar researchers.

Springer Polar Sciences aims to present a broad platform that will include state-of-the-art research, bringing together both science and humanities to facilitate an exchange of knowledge between the various polar science communities. The Series offers an outlet to publish contributions, monographs, edited works, conference proceedings, etc. Topics and perspectives will be broad and will include, but not be limited to: climate change impacts, environmental change, polar ecology, governance, health, economics, indigenous populations, tourism and resource extraction activities. Books published in the series will appeal to scientists, students, polar researchers and policy makers.

More information about this series at http://www.springer.com/series/15180

Alexander Kokhanovsky • Claudio Tomasi
Editors

Physics and Chemistry of the Arctic Atmosphere

Springer

Editors
Alexander Kokhanovsky
Vitrociset Belgium SPRL
Darmstadt, Germany

Claudio Tomasi
Institute of Atmospheric Sciences
and Climate (ISAC)
Italian National Research Council (CNR)
Bologna, Italy

ISSN 2510-0475 ISSN 2510-0483 (electronic)
Springer Polar Sciences
ISBN 978-3-030-33565-6 ISBN 978-3-030-33566-3 (eBook)
https://doi.org/10.1007/978-3-030-33566-3

This Springer imprint is published by the registered company Springer Nature Switzerland AG.
The registered company address is: Gewerbestrasse 11, 6330 Cham, Switzerland

This book is dedicated to the memory of
Prof. Wolfgang von Hoyningen-Huene
(1947–2019)

Preface

Our planet experiences significant climatic change due to increasing anthropogenic influences and also due to natural variations. The changes have accelerated in recent decades, especially in the polar regions, where substantial warming of both the atmosphere and the underlying surface has been reported. This leads to impacts on terrestrial and freshwater species, communities, and ecosystems. Climate change in the Arctic will continue, with implications for biological resources and globally important feedbacks to climate.

With this in mind, a survey of modern research related to properties of the Arctic atmosphere has been conducted and is presented in this book. The book consists of eleven chapters that survey the dynamics and thermodynamics of the Arctic atmosphere, Arctic aerosols, fog, clouds (tropospheric, stratospheric, and mesospheric), radiation, and trace gases. In addition, recent results from remote sensing of Arctic atmospheric aerosols are outlined, and observed and projected changes in the Arctic climate are described. We believe that this book, aimed at a better understanding of various processes and trends characteristic of the Arctic atmosphere, will be useful both for scientists dealing with various aspects of Arctic atmospheric research and for graduate and undergraduate students.

August 30, 2019

Contents

Contributors

Gerd Baumgarten Leibniz-Institute of Atmospheric Physics (IAP), Kühlungsborn, Germany

Silvia Becagli Department of Chemistry "Ugo Schiff", University of Florence, Florence, Italy

Simon Bélanger Département de biologie, chimie et géographie, Université du Québec à Rimouski, Rimouski, QC, Canada

Kristof Bognar Department of Physics, University of Toronto, Toronto, ON, Canada

Francesco Cairo Institute of Atmospheric Sciences and Climate, National Research Council of Italy, Rome, Italy

Thomas Carlund Swedish Meteorological and Hydrological Institute (SMHI), Norrköping, Sweden

Dmitry Chechin Alfred Wegener Institute for Polar and Marine Research, Bremerhaven, Germany
A.M. Obukhov Institute of Atmospheric Physics of the Russian Academy of Sciences, Moscow, Russia

Tiziana Colavitto State Scientific High School "Giuseppe Peano", Rome, Italy

Abhay Devasthale Atmospheric remote sensing unit, Research and development department, Swedish Meteorological and Hydrological Institute (SMHI), Norrköping, Sweden

Oxana Drofa Institute of Atmospheric Sciences and Climate (ISAC), Italian National Research Council (CNR), Bologna, Italy

André Ehrlich Leipzig Institute for Meteorology (LIM), University of Leipzig, Leipzig, Germany

Martin Gallagher Faculty of Natural Sciences, Department of Earth & Environmental Sciences, University of Manchester, Manchester, UK

Ismail Gultepe Environment and Climate Change Canada, Meteorological Research Division, Toronto, ON, Canada
Faculty of Engineering and Applied Science, Ontario Technical University, Oshawa, ON, Canada

Andreas Herber Alfred Wegener Institute Helmholtz Centre for Polar and Marine Research, Bremerhaven, Germany

Andrew J. Heymsfield NCAR, Boulder, CO, USA

Brent Holben Biospheric Sciences Laboratory, NASA Goddard Space Flight Center, Greenbelt, MD, USA

Marius O. Jonassen The University Centre in Svalbard, Longyearbyen, Norway
Geophysical Institute, University of Bergen, Bergen, Norway

Alexey Karpechko Finnish Meteorological Institute, Helsinki, Finland

Jeff Key National Oceanic and Atmospheric Administration, Madison, WI, USA

Stefan Kinne Max-Planck Institute for Meteorology, Hamburg, Germany

Torben Koenigk Swedish Meteorological and Hydrological Institute, Norrköping, Sweden

Alexander Kokhanovsky Vitrociset Belgium SPRL, Darmstadt, Germany

Christian Lanconelli Joint Research Centre, Directorate D – Sustainable Resources, Unit D6 – Knowledge for Sustainable Development and Food Security, European Commission, Ispra (Varese), Italy

Pierre Larouche Institut Maurice-Lamontagne, Pêches et Océans Canada, Mont-Joli, QC, Canada

Rodica Lindenmaier Department of Physics, University of Toronto, Toronto, ON, Canada

Franz-Josef Lübken Leibniz-Institute of Atmospheric Physics (IAP), Kühlungsborn, Germany

Angelo Lupi Institute of Atmospheric Sciences and Climate (ISAC), Italian National Research Council (CNR), Bologna, Italy

Christof Lüpkes Alfred Wegener Institute for Polar and Marine Research, Bremerhaven, Germany

Mauro Mazzola Institute of Atmospheric Sciences and Climate (ISAC), Italian National Research Council (CNR), Bologna, Italy

Roland Neuber Alfred Wegener Institute Helmholtz Centre for Polar and Marine Research, Potsdam, Germany

Boyan H. Petkov Institute of Atmospheric Sciences and Climate (ISAC), Italian National Research Council (CNR), Bologna, Italy

Patricia K. Quinn NOAA PMEL, Seattle, WA, USA

Vladimir Radionov Arctic and Antarctic Research Institute, Saint Petersburg, Russia

Christoph Ritter Alfred Wegener Institute Helmholtz Centre for Polar and Marine Research, Potsdam, Germany

Sébastien Roche Department of Physics, University of Toronto, Toronto, ON, Canada

Joseph Sedlar Cooperative Institute for Research in Environmental Sciences (CIRES), University of Colorado Boulder, Boulder, CO, USA

William R. Simpson Department of Chemistry, Biochemistry, and Geophysical Institute, University of Alaska Fairbanks, Fairbanks, AK, USA

Alexander Smirnov Science Systems and Applications, Inc., Lanham, MD, USA
Biospheric Sciences Laboratory, NASA Goddard Space Flight Center, Greenbelt, MD, USA

Thomas Spengler Geophysical Institute, University of Bergen, Bergen, Norway

Kimberly Strong Department of Physics, University of Toronto, Toronto, ON, Canada

Jason L. Tackett Science Systems and Applications, Inc., Hampton, VA, USA

Annick Tepstra Geophysical Institute, University of Bergen, Bergen, Norway

Michael Tjernström Department of Meteorology, University of Stockholm (MISU), Stockholm, Sweden
Bert Bolin Center for Climate Research, University of Stockholm, Stockholm, Sweden

Carlos Toledano Faculty of Science, University of Valladolid, Valladolid, Spain

Claudio Tomasi Institute of Atmospheric Sciences and Climate (ISAC), Italian National Research Council (CNR), Bologna, Italy

Rita Traversi Department of Chemistry "Ugo Schiff", University of Florence, Florence, Italy

Roberto Udisti Department of Chemistry "Ugo Schiff", University of Florence, Florence, Italy
Institute of Atmospheric Sciences and Climate (ISAC), National Research Council (CNR), Bologna, Italy

Timo Vihma Finnish Meteorological Institute, Helsinki, Finland

Vito Vitale Institute of Atmospheric Sciences and Climate (ISAC), Italian National Research Council (CNR), Bologna, Italy

Christian von Savigny Institute of Physics, University of Greifswald, Greifswald, Germany

Manfred Wendisch Leipzig Institute for Meteorology (LIM), University of Leipzig, Leipzig, Germany

David M. Winker NASA Langley Research Center, Hampton, VA, USA

Xiangdong Zhang University of Alaska, Fairbanks, AK, USA

Tymon Zielinski Institute of Oceanology, Polish Academy of Sciences, Sopot, Poland

Chapter 1
Dynamical Processes in the Arctic Atmosphere

Marius O. Jonassen, Dmitry Chechin, Alexey Karpechko, Christof Lüpkes, Thomas Spengler, Annick Tepstra, Timo Vihma, and Xiangdong Zhang

Abstract The scales of dynamical processes in the Arctic atmosphere range from turbulence in the atmospheric boundary layer (ABL) via interactive mesoscale processes, such as orographic flows and Polar lows, to synoptic-scale cyclones, and further to hemispherical-scale circulations characterized by the Polar front jet stream and planetary waves. Specific boundary conditions for tropospheric dynamics in the Arctic include (a) sea ice and snow, which strongly affect the surface energy budget, (b) large transports of heat and moisture from lower-latitudes, and (c) the wintertime stratospheric Polar vortex, which has a large impact on tropospheric large-scale circulation and synoptic-scale cyclones. Knowledge on dynamics of the Arctic atmosphere is advancing but, compared to mid- and low-latitudes, still limited due to lack of process-level observations from the Arctic. The dynamics of the Arctic atmosphere poses a challenge for numerical weather prediction (NWP) and climate models, in particular in the case of ABL, orographic flows, Polar lows, and troposphere-stratosphere coupling. More research is also needed to

M. O. Jonassen (✉)
The University Centre in Svalbard, Longyearbyen, Norway

Geophysical Institute, University of Bergen, Bergen, Norway
e-mail: Marius.jonassen@unis.no

D. Chechin
Alfred Wegener Institute for Polar and Marine Research, Bremerhaven, Germany

A.M. Obukhov Institute of Atmospheric Physics of the Russian Academy of Sciences, Moscow, Russia

A. Karpechko · T. Vihma
Finnish Meteorological Institute, Helsinki, Finland

C. Lüpkes
Alfred Wegener Institute for Polar and Marine Research, Bremerhaven, Germany

T. Spengler · A. Tepstra
Geophysical Institute, University of Bergen, Bergen, Norway

X. Zhang
University of Alaska, Fairbanks, AK, USA

A. Kokhanovsky, C. Tomasi (eds.), *Physics and Chemistry of the Arctic Atmosphere*, Springer Polar Sciences, https://doi.org/10.1007/978-3-030-33566-3_1

better understand how the atmospheric dynamics affects and is affected by climate warming.

Keywords Atmospheric boundary layer · Cold-air outbreak · Downslope wind · Orographic flow · Planetary waves · Polar low · Polar front jet stream · Polar vortex · Synoptic-scale cyclones · Troposphere-stratosphere interactions

1.1 Introduction

Dynamical processes in the Arctic atmosphere cover a broad range of spatial and temporal scales. The hemispherical-scale circulation is characterized by the stratospheric Polar vortex, the tropospheric Polar front jet stream, the Arctic Oscillation, and planetary waves (Sect. 1.2). The synoptic-scale dynamics are dominated by transient cyclones (Sect. 1.3) and anticyclones, including high-pressure blockings. The synoptic-scale flow often results in marine cold-air outbreaks, which in winter are favourable conditions for generation of Polar lows – the most vigorous mesoscale dynamical features in the Arctic (Sect. 1.4). Various orographic effects on the flow include patterns in synoptic spatial scale, such as the Greenland and Ural Blockings, as well as meso- and local-scale features, such as katabatic and other downslope winds, barrier winds, tip jets, gap flows, and downslope wind storms (Sect. 1.5). In all spatial scales, the flow features occurring in the lower troposphere are affected by atmospheric boundary-layer (ABL) dynamics, which control the exchange of momentum, heat, and moisture between the atmosphere and Earth surface (Sect. 1.6). This couples the Arctic atmosphere, both dynamically and thermodynamically, with the open ocean, sea ice, land ice, and terrestrial surfaces.

The broad range of spatial scales of Arctic atmospheric dynamics is associated with temporal variability in scales of seconds (ABL turbulence), hours (some orographic flow features), days (transient cyclones), weeks (jet stream patterns, planetary waves, high-pressure blockings), seasonal (troposphere – stratosphere interactions), and inter-annual/decadal (large-scale coupling between the atmosphere and ocean). In addition to the natural variability, the Arctic climate system is rapidly changing. A major challenge is to better understand how the atmospheric dynamics will respond to the Arctic Amplification of climate warming.

Our knowledge on dynamical processes in the Arctic atmosphere is based on theoretical understanding of atmospheric dynamics, observations, and model experiments. Theories on atmospheric dynamics have mostly been developed on the basis of observations from mid and low latitudes, and the applicability of these theories for the Arctic environment has been evaluated on the basis of limited observations from the Arctic. The regular network of in situ observations is limited to coastal and archipelago stations, most of them only equipped for near-surface observations. In addition there are stations for regular radiosonde soundings; a total of approximately 36 stations are in operation north of 65 °N, but there are no regular soundings north of 80 °N. These regular observations are complemented by field expeditions based on research vessels, drifting ice stations or aircraft. Research vessel and aircraft expeditions are mostly carried out during summer and spring, allowing process-level observations among others on dynamics of the atmospheric boundary layer and its interaction with cloud physics, radiative transfer, and the ocean and sea ice surface. Year-round drifting ice stations have been operated in the Arctic Ocean regularly from 1950 to 1991 and from 2003 to 2012, mostly by Russian colleagues. In addition, data relevant for atmospheric dynamics, such as surface pressure and near-surface air temperature, are collected by automatic stations deployed on sea ice, above all by the International Arctic Buoy Program since 1979. The data collected at regular stations and field expeditions are mostly point observations, except of research aircraft observations. Even point observations are useful for understanding ABL dynamics but not enough from the point of view of meso- and synoptic-scale dynamics. Satellite data are typically better than in situ observations with respect to spatial and temporal resolution, but usually worse with respect to accuracy and vertical resolution. Satellite observations are also hampered by the Polar night and high latitudes (geostationary satellite data not available).

Dynamics of the Arctic atmosphere poses a challenge for numerical weather prediction (NWP) and climate models, in particular in the case of ABL (Sect. 1.6), mesoscale processes (Sects. 1.4 and 1.5), and troposphere-stratosphere coupling (Sect. 1.2). Operational forecasting of Polar lows is a challenge due to their fast development and small spatial scale, and in climate models Polar lows are often not resolved at all, although they sometimes have effects that are also important for climate, e.g., when generating deep convection in the ocean (Sect. 1.4). In high latitudes, NWP and climate models as well as atmospheric reanalyses typically have their largest temperature, humidity and wind errors in the ABL (Sect. 1.6). These errors are critical, as improved modelling capabilities are needed to better understand the recent rapid Arctic amplification of climate warming and to reliably project the future. In addition, increasing navigation, aviation, and offshore activities in the Arctic calls for more accurate weather forecasts for the Arctic. To improve modelling of the Arctic troposphere-stratosphere coupling, ABL and mesoscale processes, we need improved understanding on the relevant dynamical processes.

1.2 Large-Scale Circulation in the Troposphere and Stratosphere

1.2.1 Stratospheric Circulation

The stratosphere is a region in the atmosphere extending approximately from 10 km to 50 km above the surface and mostly known for being home to Earth's ozone layer. The stratosphere is also characterised by a distinct circulation which is capable of affecting weather and climate down to the surface. In this sub-chapter we first describe general characteristics of the Arctic stratosphere and factors affecting its variability and then describe the variability and recent changes in the stratosphere.

1.2.1.1 General Characteristics of the Arctic Stratosphere and Its Circulation

In the stratosphere air temperature increases, or remains almost constant with altitude. The temperature profile is mainly controlled by radiative balance between heating due to absorption of solar ultraviolet (UV) radiation by ozone and cooling due to radiative emission mostly by carbon dioxide. Such thermal stratification means that the stratosphere is very stable, i.e. vertical motion is restricted. As a result stratospheric dynamics differs considerably from that of the troposphere where convection plays significant role in the dynamics.

Outside of the equatorial belt the stratospheric characteristics experience pronounced seasonal cycle. In summer, solar irradiance peaks over the North Pole and the Arctic stratosphere warms with respect to lower latitudes. This leads to a formation of a pole-centred summer anticyclone (Fig. 1.1b). During summer, the Arctic stratosphere remains very close to radiative equilibrium with little interannual and interdecadal variability.

During autumn the polar air starts to cool faster than that in mid-latitudes and the summertime anticyclone is replaced by a wintertime pole-centred cyclone, called polar vortex (Fig. 1.1a). The edge of the polar vortex is surrounded by strong westerly winds, called polar night jet, with typical wind speed in the upper stratosphere (10–1 hPa) of 30–40 m/s and occasionally up to 60–70 m/s during mid-winter. Unlike in summer, stratospheric winds and temperatures in winter experience large variability at synoptic, intraseasonal and interannual time scales. The variability is linked to interaction between the zonal mean circulation and planetary waves, and is driven by variability in the strength of planetary wave sources as well as variability in the transmitting properties of zonal mean stratospheric circulation, as discussed in more detail below.

The difference between summer and winter circulation arises from the fact that planetary-scale stationary waves generated in the lower atmosphere by orography and ocean-land contrasts can only propagate their energy upward when background zonal winds are westerly (Charney and Drazin 1961). Thus, as zonal wind direction changes from summer easterlies to winter westerlies the stratosphere becomes supportive of planetary waves propagating from below. Charney and Drazin (1961) showed that the wave propagation is strongly controlled by zonal mean zonal winds: upward wave propagation is only possible when zonal winds are positive (i.e. westerly) and weaker than a critical wind speed, which is inversely proportional to zonal wave number. Under typical wind speed of the polar night jet, only longest waves with zonal wave numbers one and two can propagate to the stratosphere.

Propagation of the planetary waves in the atmosphere is best described by the Eliassen-Palm flux (Edmon et al. 1980) which is, under realistic approximations, proportional to wave's group velocity multiplied by a scalar called wave activity,

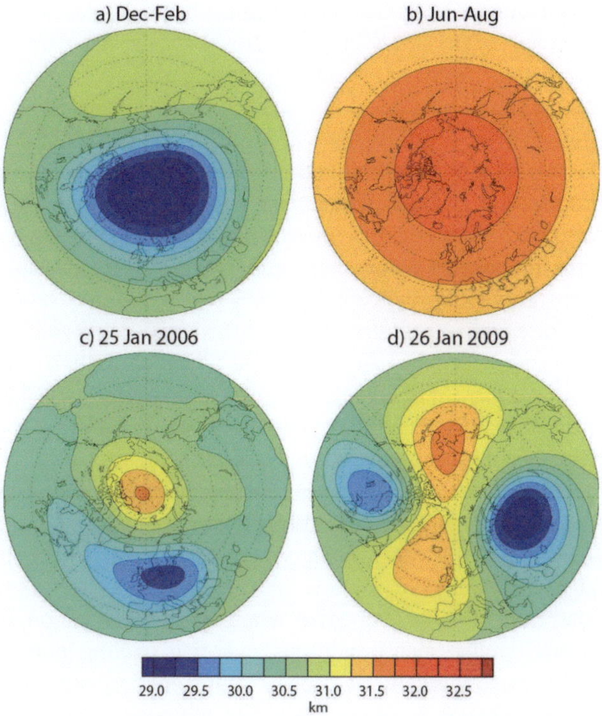

Fig. 1.1 Geopotential height at 10 hPa with four panels: (**a**) climatological winter mean, (**b**) climatological summer mean, (**c**) example of polar vortex displacement during an SSW; (**d**) example of polar vortex split during an SSW

which is quadratic in wave's amplitude. Thus upward wave propagation is associated with transfer of wave activity to the stratosphere. Wave activity can either be dissipated in the stratosphere or reflected back to the troposphere. The dissipation characterised by a convergence of the EP flux is the most typical situation. The dissipation is often associated with planetary wave breaking, which happens when wave approaches zero wind line, acting as a critical line for the stationary waves with zero phase speed. While details of wave breaking and dissipation are not well understood (Andrews et al. 1987), their effect on the zonal mean circulation is a deceleration of westerly winds. The deceleration destroys the thermal wind balance. In order to restore the balance, the vertical motion is induced in both polar and low latitudes. The descending motion in the polar region warms the polar atmosphere through adiabatic compression, while the ascending motion in the low latitudes cools the atmosphere through adiabatic expansion. The adiabatic heating results in a reduced meridional temperature gradient, which again satisfies the thermal wind balance. The fact that stratospheric temperatures are affected by dynamical heating also means that the winter stratosphere is not in a radiative equilibrium. Comparison of observed temperatures with those calculated under the assumption of no dynamical heating reveals that the dynamics heats the Arctic stratosphere by 30 K (lower stratosphere) to 100 K (upper stratosphere) in mid-winter.

While upward wave propagation and deceleration of the westerly winds is a permanent process, the intensity of the wave forcing varies in time because it depends on strength of wave sources and transmitting properties of the atmosphere. As a result, there is strong variability in the wintertime stratosphere. In an extreme case, wave propagation and dissipation in the stratosphere can lead to a strong warming of the polar atmosphere in a process called Sudden Stratospheric Warming (SSW) when the meridional temperature gradient between mid-latitudes and the pole reverses its sign. During such episodes westerly winds may reverse to easterly and the polar vortex becomes strongly disturbed and relocated. Typically, the vortex either becomes displaced from the pole or split into two smaller vortices (Fig. 1.1c, d). If such an event happens during mid-winter, an SSW is called a major SSW. It may take from a few days to a few weeks before a combination of dynamical processes and radiative relaxation towards radiative equilibrium restores the polar vortex. When such event occurs near the end of winter, or during spring, the polar vortex may not restore at all, and a transition towards the summer type of the circulation occurs. In this case, which is a typical ending of stratospheric winter, the SSW is called the final SSW.

The opposite extreme case can arise when wave propagation from the troposphere weakens. Under such conditions radiative cooling dominates thermal balance, polar vortex strengthens, and lower stratospheric polar temperatures can decrease down to 185–190 K. When such low temperatures are reached, water vapour and nitrous acid present in small amount in the stratospheric air start to freeze forming Polar Stratospheric Clouds (PSCs). PSCs play a key role in chemical ozone depletion (see Chap. 8).

Zonal wind deceleration and warming of the Arctic stratosphere are not the only effects of planetary wave dissipation in the stratosphere. Vertical motion generated to maintain thermal wind balance requires existence of meridional circulation to ensure mass conservation, with poleward motion in the stratosphere and returning flow beneath. This overturning meridional circulation is slow, with typical velocities of several centimetres per second; however it plays an important role in transport of chemicals. In particular, atmospheric ozone mostly produced in the tropical atmosphere is transported to extratropics, including the Arctic, by this circulation. The meridional circulation responsible for the transport of chemicals in the stratosphere is known as Brewer-Dobson circulation. Two branches of the Brewer-Dobson circulation are usually distinguished: a deep branch affecting the whole depth of the stratosphere, which is driven by the dissipation of the planetary waves propagating upward along the polar night jet, and a shallow branch, which extends up to ~ the 50 hPa level and is driven by planetary and synoptic waves propagating to the lower stratosphere in sub-tropics along the sub-tropical jet stream. While both branches contribute to the ascending motion in the tropics, it is the deep branch that is associated with the polar downwelling and adiabatic warming of the Arctic stratosphere.

1.2.1.2 Arctic Stratosphere Variability and Change

There is strong variability in the characteristics of the Arctic polar vortex on all timescales, from synoptic to decadal. This variability is mostly driven by the interaction between planetary waves and zonal mean flow. Sources of the variability are associated with factors affecting generation of the waves as well as modulating propagation of the waves within the stratosphere. These factors are considered below.

Blocking in the tropospheric circulation is a remarkable example of amplification of wave amplitude at synoptic scale. There is strong statistical connection between occurrence of blocking and SSWs (Martius et al. 2009). Although blocking is associated with synoptic-scale waves, which are not able to propagate to the stratosphere, blocking can enhance the amplitude of planetary waves through constructive interference (Garfinkel et al. 2010) and thus affect planetary wave flux to the stratosphere.

The Madden-Julian Oscillation (MJO) in the tropical atmosphere-ocean system can modulate the stratospheric variability at monthly scales. There is a significant connection between phases of MJO and state of the Arctic stratospheric vortex, which is attributed to MJO's ability to generate planetary waves that propagate poleward and consequently upward to the stratosphere (Garfinkel et al. 2012a). For example, enhanced convection over the western Pacific associated with MJO triggers enhanced planetary wave flux to the stratosphere, which is followed by a weakening of the polar vortex approximately 10 days later.

Another tropical phenomenon, El Niño, which is known to affect weather and climate globally, also affects the Arctic stratosphere. During El Niño winters, there is increased wave activity flux due to wavenumber one to the stratosphere. Composite analysis shows that weakening of stratospheric winds and warming of the Arctic stratosphere is observed during January–February of El Niño winters (Cagnazzo and Manzini 2009). El Niño variability also modulates frequency of SSWs, but in a somewhat unexpected way. The frequency of SSWs is enhanced during both El Niño and La Niña winters, while SSWs are rather rare during neutral winters (Butler et al. 2011).

Anomalies in the sea ice and terrestrial snow can also affect generation of planetary waves and thus stratospheric circulation. It has been reported that large-scale cooling associated with enhanced snow cover over Eurasia during boreal autumn is followed by an enhanced upward propagation of planetary waves and weakening of the polar vortex (Cohen et al. 2014). A similar connection was found between polar vortex weakening and sea ice loss over Barents and Kara sea region (Kim et al. 2014). The later effect is of interest because the rapid sea ice decline observed since the last decade of the twentieth century can potentially lead to long-term changes in the Arctic stratospheric circulation.

The Quasi-Biennial Oscillation (QBO) in the stratospheric equatorial winds is known to affect the strength of the polar vortex at inter-annual scale (Baldwin et al. 2001). During the easterly phase of QBO, when zonal winds in the equatorial lower stratosphere at about 30–50 hPa blow west, the polar vortex is weaker than during the westerly QBO phase. It has been hypothesized (Holton and Tan 1982) that the shift of the zero wind line to the Northern Hemisphere sub-tropical stratosphere during the easterly phase makes planetary waves to break further north and thus decelerate the polar night jet more effectively than during the westerly phase, when the zero wind line moves to the southern hemisphere allowing planetary waves to propagate deeper to the tropics and less affect the polar vortex. It has also been suggested (Garfinkel et al. 2012b) that other factors affecting planetary wave propagation can play a more important role than the zero wind line shift in modulating planetary wave propagation by QBO.

The main source of variability on the decadal scale is the 11-year solar cycle. The modulation of the polar vortex by the solar cycle is linked to differential absorption of solar radiation between the tropical and high latitudes. Larger solar UV radiation during years of solar maximum leads to a stronger meridional temperature gradient and consequently, via thermal wind balance, to a stronger polar vortex (Gray et al. 2010). Weaker UV radiation during years of solar minimum results in an opposite situation. The other potential sources of the decadal Arctic stratosphere variability are the long modes of ocean circulation variability, such as Atlantic Multidecadal Oscillation (Omrani et al. 2014) and Pacific Decadal Oscillation (Woo et al. 2015); however, the exact mechanisms of influence and importance of these factors remain poorly understood.

In addition to the regular and quasi-regular factors outlined above, explosive eruptions of tropical volcanoes can also affect the atmospheric circulation (Timmreck 2012). The aerosol formed in the tropical stratosphere from erupted gases absorbs the terrestrial infrared radiation and thus warms the lower tropical stratosphere. The absorption increases the meridional temperature gradient and strengthens the polar vortex, i.e. the mechanism is similar to that occurring during solar maximum.

Like the rest of the climate system, the Arctic stratosphere is expected to change in response to ongoing anthropogenic emission of greenhouse gases and associated global warming. Increased concentration of CO_2 leads to increased longwave radiation emission to space, which is expected to cool the stratosphere globally. However, the response of the Arctic lower stratosphere, whose temperatures are strongly controlled by the dynamics, will strongly depend on the response of the atmospheric circulation. Climate models consistently predict strengthening of the Brewer-Dobson circulation under global warming scenarios. While stronger Brewer-Dobson circulation is often associated with a weaker polar vortex and warmer Arctic stratosphere, the response will depend on whether or not the anomalous downwelling associated with strengthened Brewer-Dobson circulation will extend to the Arctic stratosphere. Since theory predicting the response of atmospheric circulation to global warming is currently lacking, there is a large uncertainty in the future of the Arctic stratosphere with climate models simulating both strengthening and weakening of the polar vortex as a response to climate change (Manzini et al. 2014).

1.2.2 Stratosphere-Troposphere Coupling

For a long time it was assumed that stratospheric dynamical variability is driven by tropospheric variability, while the stratosphere has little influence on the processes below. Such a view is intuitively understandable since the mass of the stratosphere is much smaller, only about 15% of that of the troposphere. Furthermore there are no obvious sources of variability in the stratosphere, such as dynamical instabilities, which are present in the troposphere. However, it gradually became widely accepted that the stratosphere plays an active role in the stratosphere-troposphere dynamical coupling. Experiments with simple atmospheric models revealed that the stratosphere behaves like a chaotic dissipative dynamical system, and may develop its own variability even if tropospheric wave forcing is constant (Holton and Mass 1976; Yoden 1987). Moreover, it was demonstrated that the atmospheric adjustment to a mechanical force applied within the stratosphere (which in real world can result, for example, from convergence of planetary or gravity wave activity) involves a meridional circulation extending downward to the boundary layer, the effect known as a downward control principle (Haynes et al. 1991). Research during the last decade of the twentieth century has confirmed that large stratospheric anomalies, such as SSWs or polar vortex strengthening usually propagate downward in time and sometimes affect tropospheric circulation (e.g. Baldwin and Dunkerton 1999). For example, an SSW and associated weakening of the stratospheric zonal winds can induce a southward shift of the tropospheric jet and lead to cold anomalies over northern Eurasia and eastern North America. Composites of temperature and precipitation anomalies following downward propagation of geopotential height anomalies are illustrated in Fig. 1.2. Although the downward control principle predicts weakening of the tropospheric zonal winds following SSWs, it does not explain all important details of the observed response, such as the shift of the tropospheric jet position. A feedback from the tropospheric synoptic-scale eddies is understood to play an important role, but the details of this feedback remain a subject of ongoing research.

The other mechanism of downward effects is the reflection of planetary waves from the stratosphere. The reflection does not affect the zonal mean circulation in the way described above. Rather, observations suggest that it can amplify tropospheric stationary waves, sometimes resulting in blocking.

Presently, there is a strong interest in stratosphere-troposphere coupling associated with the fact that weather patterns following the stratospheric anomalies are remarkably persistent, and therefore potentially predictable, at scales up to several weeks, i.e., beyond the reach of state-of-art numerical weather prediction models (Sigmond et al. 2013; Scaife et al. 2016). The opportunity to improve climate predictability at sub-seasonal to seasonal scales attracts considerable research efforts, which means that our understanding of the stratosphere-troposphere coupling will likely advance in the years to come.

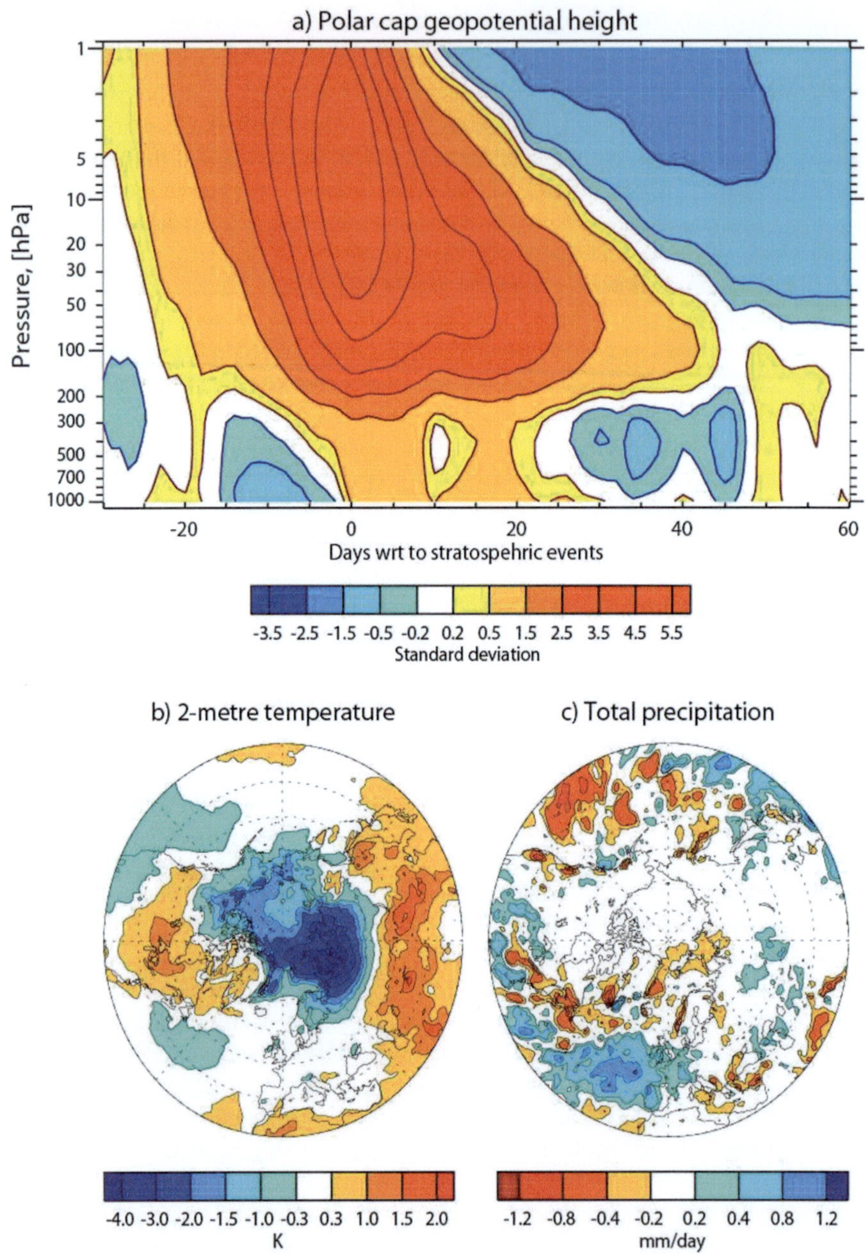

Fig. 1.2 (**a**) Composite of downward propagation of polar cap geopotential height anomalies; (**b**) composite of temperature anomalies during days 11–60 following stratospheric events; (**c**) composite of precipitation anomalies during days 11–60 following stratospheric events

1.2.3 Tropospheric Large-Scale Circulation

The state and variations of the Arctic troposphere strongly depend on the large-scale circulation in high- and mid-latitudes of the Northern Hemisphere. Important features of the circulation include the Polar front jet stream in the mid- and upper-troposphere, the Arctic Oscillation, planetary waves, and high-pressure blockings. These features interact with synoptic-scale cyclones (Sect. 1.3) and affect the large-scale boundary conditions for the occurrence of Polar lows (Sect. 1.4).

The Polar front jet stream guides the cyclone tracks due to the strong wind shear below the jet core. Particularly important properties of the jet stream are its latitude, speed, meandering, and the occurrence of events when the jet is split in two branches (Hall et al. 2015; Vavrus 2018). The split jet occurs typically in summer and the events are often associated with quasi-resonance of planetary waves with thermal and orographic forcing (Coumou et al. 2017). Hence, the split jet favours persistency of weather patterns, often resulting in floods, droughts, and heat waves in mid-latitudes. Arctic effects on the jet stream properties have recently received a lot of attention (e.g., Francis and Vavrus 2015), but also the jet stream has a strong influence on weather conditions in the Arctic. A meandering jet stream allows advection of warm, moist air masses from mid-latitudes to the Arctic, often resulting in extreme anomalies in weather and sea ice conditions (Vihma 2017; Rinke et al. 2017).

The Arctic Oscillation Index (AOI) characterizes the strength of the circumpolar zonal circulation. The AOI is calculated on the basis of the loading pattern of the first principal component of the mean-sea-level pressure (MSLP) or 1000 hPa geopotential height in the Northern Hemisphere north of 20 °N. The North Atlantic Oscillation (NAO) represents the role of the Atlantic sector on the circumpolar Arctic Oscillation. The NAO index (NAOI) has several definitions, but is always associated with a north–south- dipole structure in the MSLP (or 1000 hPa geopotential) field over the Atlantic, approximately between the Icelandic low and the Azores high (Ambaum et al. 2001). Both AOI and NAOI vary irregularly in time. Large positive (negative) values of AOI are related to strong (weak) westerly winds, and high (low) winter temperatures in mid-latitudes, but low (high) winter temperatures in the Arctic (Fig. 1.3). The effects of NAO in the Atlantic sector and Europe are qualitatively similar to those of AO. NAO has is often regarded as a cause of weather variations, but can also be seen as a consequence of variations in the jet stream and storm tracks (Vallis and Gerber 2008). NAOI is indeed significantly correlated with the jet stream latitude, a negative NAOI being associated with southward displacement of the jet stream (Woollings and Blackburn 2012). Several studies have suggested that the recent Arctic Amplification of climate warming has favoured a negative NAO (Kim et al. 2014; Nakamura et al. 2015; Francis and Skific 2015; Deser et al. 2016), but different conclusions have also been presented (Singarayer et al. 2006; Screen et al. 2014; Smith et al. 2017).

However, AO and NAO alone do not characterize all important aspects of large-scale circulation in northern mid- and high-latitudes. Several indices have been used

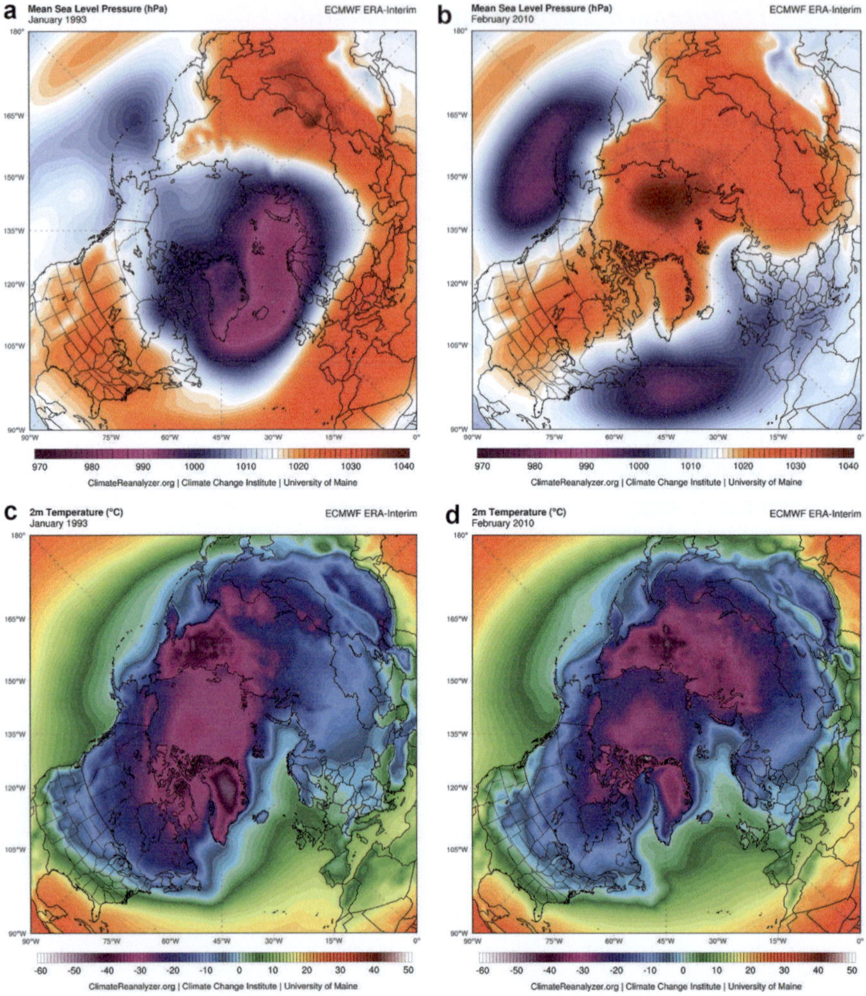

Fig. 1.3 Monthly mean fields of mean-sea-level pressure (in hPa, **a**, **b**) and 2-m air temperature (in °C, **c**, **d**) in January 1993 (left panels), when the Arctic Oscillation Index was highest in the record during 1900–2018, and in February 2010 (right panels), when the index was lowest in the record

to quantify the meridionality of the circulation. These include the Arctic Dipole (AD, Overland and Wang 2005), the closely related Dipole Anomaly (DA, Wu et al. 2006; Wang et al. 2009), the Central Arctic Index (CAI, Vihma et al. 2012), and the Meridional Circulation Index (Francis and Vavrus 2015). Although the exact definitions of AD, DA, and CAI differ, they all are related to the MSLP difference between the North American and Eurasian parts of the Arctic Ocean, and are therefore used to characterize the transport of marine air masses from the Pacific or Atlantic to the central Arctic and the atmospheric forcing on sea ice drift along the Transpolar Drift Stream. Unlike, AD, DA and CAI, MCI is calculated separately for each location, as MCI = $v|v| / (v^2 + u^2)$, where u and v are the zonal and meridional wind components, respectively. Hence, MCI is applicable in quantifying the waviness of the jet stream. In general, strong meridionality of the flow is related to negative values of AOI and NAOI. Further, important regional features of the circulation include the Siberian High, Ural Blocking, Greenland Blocking, the Scandinavian Pattern, and Atlantic Ridge.

The Siberian High (SH), usually centered approximately around Lake Baikal, is the most prominent feature of atmospheric circulation over Eurasia from September until April (Gong and Ho 2002). The SH affects weather patterns over Eurasia and farther in the Northern Hemisphere. The local effects in the region under the high pressure pattern are related to clear skies and weak winds, favouring extremely low winter temperatures, whereas remote effects in winter are felt via advection of cold northerly air masses to Southeast Asia (Wu et al. 2011) and advection of cold, continental air masses to Europe. The recent strengthening of SH has been associated with a negative trend in winter air temperatures over large parts of Eurasia (Cohen et al. 2014). Some studies have suggested that this is at least partly forced by the sea ice decline in the Barents and Kara seas via troposphere-stratosphere coupling (Kim et al. 2014; Jaiser et al. 2016; Furtado et al. 2016), whereas some studies suggest that this is due to natural variability (McCusker et al. 2016).

The Ural Blocking (UB) is a quasi-stationary anticyclonic circulation feature, which occurs around the Ural Mountains. It is considered a part of the negative phase of the East Atlantic / West Russia wave train, which is often associated with the positive phase of NAO (Luo et al. 2016a). There is a positive feedback between the sea ice loss in the Barents and Kara seas and UB (Gong and Luo 2017). When UB occurs, the associated clockwise circulation advects warm air masses to over the Barents and Kara seas, resulting in or enhancing the sea ice loss. The sea ice loss and related atmospheric warming regionally reduces the north-south temperature gradient in the sector around 60°E. This reduces the zonal flow, and favours the occurrence and strengthening of UB. Comparing periods 1979–1999 and 2000–2013, UB events have indeed become more persistent, contributing to the Warm Arctic – Cold Eurasia pattern in winter (Luo et al. 2016a, b). When UB merges with the Siberian High, anomalously cold winters occur over particularly large areas in central Asia.

The Scandinavian (SCA) pattern is seen, e.g., in the 700 hPa height anomalies. The pattern has the primary centre of action around Scandinavia, with two other centres of action with the opposite sign, one over the north-eastern Atlantic and the other over central Siberia. The positive phase of SCA is characterized by prominent anticyclonic anomalies around Scandinavia, in winter giving rise to below-normal temperatures across central Russia and western Europe, above-normal precipitation across southern Europe, and dry conditions over Scandinavia (Bueh and Nakamura 2007). A closely related patters, characterized by a positive pressure anomaly over the north-eastern Atlantic Ocean and a negative one over Scandinavia, is often called as the Atlantic Ridge (Barrier et al. 2014). Accordingly, the Atlantic Ridge favours north-westerly winds in Scandinavia, and is also associated with remarkable sea surface temperature anomalies.

The Aleutian low pressure center is a dominant atmospheric pattern over the North Pacific in winter. The Aleutian low appears in temporally averaged pressure fields, but the instantaneous conditions in the region are characterized by transient cyclones (Sect. 1.3) and a background stationary wave pattern. The wintertime variability of the Aleutian Low is associated with the Pacific – North American Pattern and the AO (Overland et al. 1999).

Due to its high orography, orientation perpendicular to westerly winds, and large extent in north-south direction, Greenland ice sheet tends to block the lower-tropospheric westerly flow. The strength of the blocking depends on both thermodynamic and dynamic factors (Rajewicz and Marshall 2014). The Greenland Blocking Index (GBI, Fang 2004; Hanna et al. 2013, 2016) is defined as the mean 500 hPa geopotential height for the 60–80 °N, 20–80 °W region, but also slightly different indices can be used to characterize the Greenland Blocking (Scherrer et al. 2006; Davini et al. 2012). Greenland Blocking has recently varied a lot with a lot of extreme values since 2001 (Hanna et al. 2016, 2018), probably related to the strongly meandering Polar front jet stream (Overland et al. 2015). The recent rapid climate warming over Greenland has contributed to thermodynamic forcing of Greenland Blocking, with warmer air expanding and raising geopotential heights (Hanna et al. 2016). According to Hanna et al. (2016), decadal variations of the GBI are related to variations in NAO and the Atlantic Multidecadal Oscillation (AMO).

1.3 Synoptic-Scale Cyclones

1.3.1 Theoretical Background

Extratropical synoptic-scale cyclones are fundamental atmospheric circulation system occurring on daily basis in the mid and high latitudes, including the Arctic region. Genesis, development, dissipation, and spatial structure of cyclones in midlatitudes have been extensively investigated (e.g., Bluestein 1993; Hoskins et al. 1985; Davis and Emanuel 1991; Lackmann 2002). However, studies particularly on cyclone dynamics over the Arctic have been very limited (e.g., Aizawa et al. 2014; Tao et al. 2017b). Based on the existing statistics, cyclones generally take 3–6 days to develop, reaching a sea level pressure (SLP) about 970–1000 hPa at cyclone centers, and 3–6 days to dissipate, and move eastward or northeastward at a speed of about 50 km/h. The cyclone diameter is roughly 1000–2500 km on average. Cyclones play predominant role in driving variation of weather elements or occurrence of weather events along their travelling pathways and adjacent areas. Intense cyclones can cause wind gusts, heavy rainfall or snowfall, and large fluctuations of air temperatures, possibly resulting in extreme events such as ocean wave surge, flooding, costal erosion, blizzard, and cold air outbreak. All of these extreme events certainly impact daily life and may cause severe consequences including destruction of infrastructures and loss of properties and even life.

Theoretical study of extratropical cyclones can be traced back to the First World War. A comprehensive understanding about the structure and developing mechanisms of cyclones in the mid-latitudes have been established along with development of atmospheric dynamics and observational network (e.g., Bluestein 1993; Holton 2004). The genesis of cyclones is fundamentally driven by baroclinic instability, i.e., perturbed thermal contrast between warm and cold air masses and resulting conversion of available potential energy to kinetic energy by secondary atmospheric circulation. The coupling between upper level waves and surface cyclones enhances the secondary circulation, in particular when the left exit of upper-level jet ahead of wave trough is located above the surface cyclone. This increases warm and cold temperature advection and, in turn, strengthens baroclinicity, leading to an increased conversion of available potential energy to kinetic energy and intensifying cyclone development. Once the upper level waves catch the surface cyclone and in particular when the surface cyclone is relocated beneath the left entrance of the jet behind the wave trough, the secondary circulation and baroclinicity weaken and then the cyclone decays. Due to the nature of the driving mechanism, cyclones in the mid-latitudes are predominantly characterized by baroclinic structures over their majority of lifetime.

To quantitatively describe details of the cyclone structures and driving mechanisms mentioned above, various theories have been developed, such as the Quasi-Geostrophic (QG) theory (e.g., Bluestein 1993). During recent decades, the potential vorticity (PV) has been increasingly employed to diagnostically analyze and interpret mid-latitude cyclone development (e.g., Hoskins et al. 1985; Davis and Emanuel 1991; Hirschberg and Fritsch 1991; Lackmann 2002; Liberato 2014). In particular, the PV perspective provides an insight into impacts of upper troposphere - lower stratosphere planetary waves on surface cyclones. Associated with deepening and southward amplification of the waves, positive PV anomaly occurs in mid-latitudes. The PV anomaly intensifies surface cyclone through enhancing troposphere baroclinicity ahead of upper-level PV anomaly or wave trough.

Although cyclones in the Arctic, in particular those originating from the mid-latitudes, can share many common features with the their counterparts in the midlatitudes, recent studies found that at least some intense and long-lived Arctic cyclones can be very different in their structures and driving mechanisms (Tanaka et al. 2012; Aizawa et al. 2014; Aizawa and Tanaka 2016; Tao et al. 2017a, b). Tao et al. (2017a) investigated a long-lived Arctic cyclone in September 2010. This cyclone demonstrates barotropic structures almost throughout its lifetime (Fig. 1.4), rather than baroclinic structures found in mid-latitude cyclones as discussed above. The baroclinic instability only occurred near the surface and in the low troposphere at the initial state of the surface cyclone genesis and development. After the generation of the surface cyclone, a synoptic-scale, symmetric stratospheric vortex quickly moved to be just above the surface cyclone, showing a barotropic structure with a warm core in the upper troposphere-lower stratosphere and a cold core in the low-mid troposphere. The positive PV anomaly induced by the downward intrusion

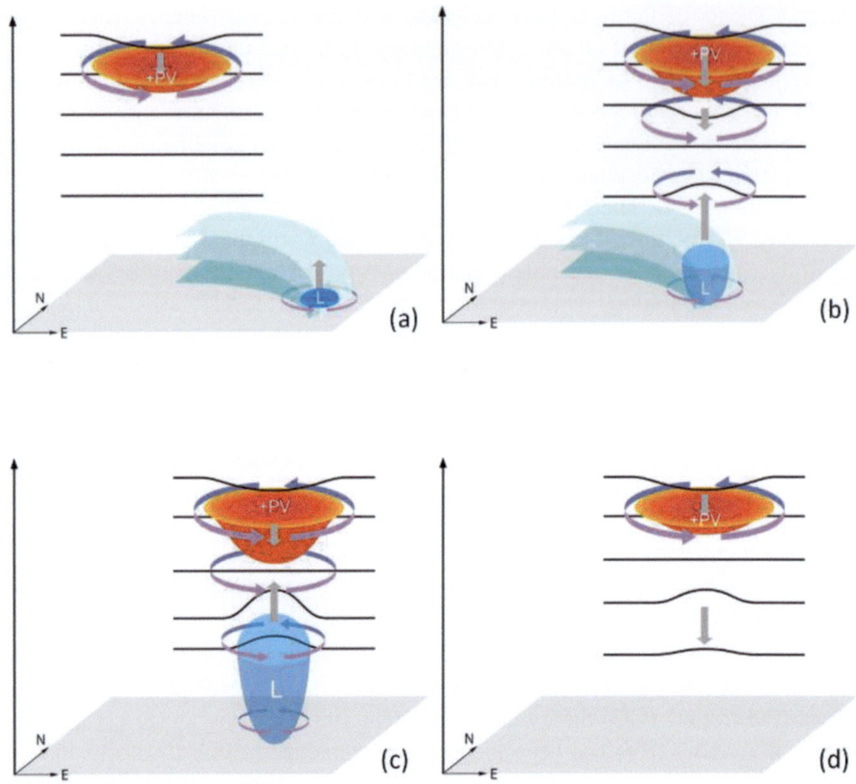

Fig. 1.4 The concept model showing Arctic cyclone structures and driving mechanisms in (**a**) Pre-generation phase; (**b**) development and intensification phase; (**c**) mature and persistence phase; and (**d**) decay and dissipation phase. The red shading represents the lower stratosphere positive PV anomaly associated with the downward intrusion of the stratospheric vortex; the blue shading the troposphere cyclone; the black solid lines the air temperature contours from the mid-troposphere to the lower stratosphere; and the light blue surfaces the air temperature isosurface associated with a surface and low-troposphere front. The vertical and horizontal curved vectors indicate the vertical motions and the horizontal circulations. (Adapted from Tao et al. 2017a, © American Meteorological Society. Used with permission)

of stratospheric vortex played a decisive role in intensifying the surface cyclone and maintaining its intensity over an extended time period. The strengthening and persistence of the PV anomaly resulted from the out-of-phase occurrence of the upper level warm core and the lower level cold core.

1.3.2 Observed Arctic Cyclones

Due to the nature of their driving mechanisms, extratropical cyclones are mainly generated over specific areas and moving along preferred tracks. Cyclones occurring in the Arctic region can originate from the mid-latitudes or be generated locally. There are two major climatological cyclones tracks in the Northern Hemisphere (Fig. 1.5; Zhang et al. 2004). One ranges from the northwest Pacific coastal area and shelf seas to the Bering Sea and Aleutian Islands and the other from the northeast coast of the North American continent to the Icelandic Sea and Barents Sea. The cyclone tracks show obvious seasonality, being more confined over the North Pacific and North Atlantic oceans during winter. During summer, there are more numerous cyclones occurring over the landmasses. Cyclones also have a lower central SLP,

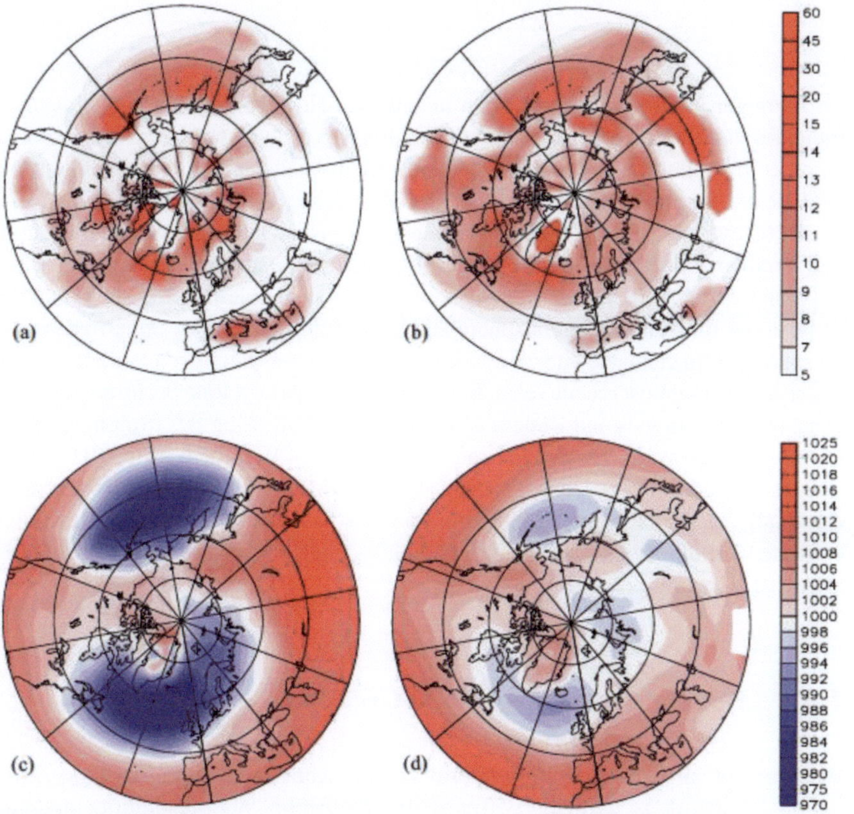

Fig. 1.5 The climatological count (unit: count per 10^5 km^2) of cyclone centers in (**a**) winter and (**b**) summer; and the composite climatological mean sea level pressure (unit: hPa) of cyclone centers in (**c**) winter and (**d**) summer. (Zhang et al. 2004; their Fig. 2; © American Meteorological Society. Used with permission)

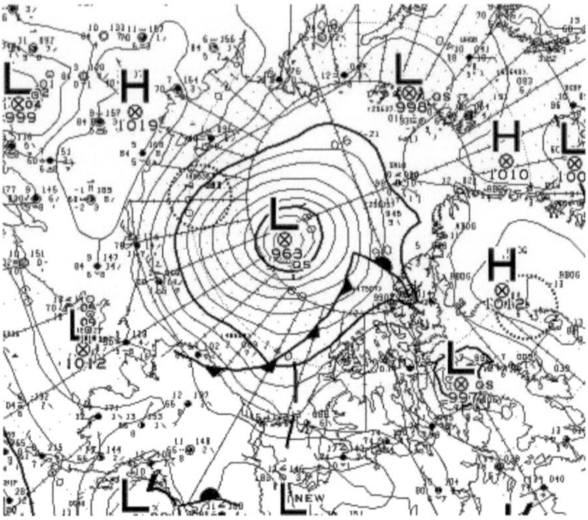

Fig. 1.6 A super cyclone occurring in early August 2012, which was followed by a record minimum of sea ice extent. (Courtesy of Environment Canada)

i.e., stronger intensity, in winter than in summer. However, climatological analysis indicates that there are more numerous and long-lived cyclones in the Arctic during summer than during winter, just opposite to the seasonal cycle of mid-latitude cyclone counts and lifetime (Zhang et al. 2004; Serreze and Barrett 2008).

Arctic cyclone activity has shown intensification since the mid-twentieth century and Arctic cyclones tend to have longer lifetime during summer (Zhang et al. 2004; Sepp and Jaagus 2011). These changes would enhance poleward transient transport of heat and moisture and atmosphere-sea ice-ocean interactions. Superimposed on these long-term changes, intense and long-lived cyclones have been more frequently observed during most recent years. For example, a super cyclone occurred in August 2012 (e.g., Simmonds and Rudeva 2012; Fig. 1.6). Following this cyclone, a record minimum of sea ice extent occurred in September 2012. In winter 2015/16, the intense cyclone "Frank" propagated from the North Atlantic into the Arctic. It drastically increased heat and moisture transport into the Arctic Ocean, causing an extreme warming event with an increase in the polar cap surface air temperature by about 10 °C within a few days (Kim et al. 2017).

1.3.3 Modeling Studies of Arctic Cyclones

Because not all details of atmospheric systems can be measured and the observational network is particularly sparse in the Arctic, high-resolution modeling experiment is an effective tool to reveal three-dimensional cyclone physical structures and driving mechanisms of cyclone genesis, development, and dissipation. Although modeling studies on Arctic cyclones have been few, several groups have employed different models to investigate physical processes (Aizawa et al. 2014;

Tao et al. 2017a, b) and to examine impacts of Arctic radiosonde observations on predicting Arctic cyclones (Yamazaki et al. 2015).

By employing a physically optimized high-resolution Weather Research and Forecasting (WRF) model for the Arctic region (Zhang et al. 2016), Tao et al. (2017a, b) realistically simulated a long-lived Arctic cyclone in September 2010 and the super cyclone in August 2012. Specifically, compared with the in-situ observations conducted during the R/V Miria cruises (Inoue and Hori 2011), the model accurately captured the time evolution of the observed profiles of air temperature, relative humidity, and wind from the radiosonde observations during the genesis and development period of the cyclone in September 2010. Based on the realistic simulation of the cyclone, physical mechanisms were quantitatively investigated.

As described above, the downward intrusion of positive PV anomaly plays critical role in the intensification and persistence of the surface cyclone. To understand temporal evolution of the PV anomaly, Tao et al. (2017a) quantitatively analyzed the simulated thermodynamic processes (Fig. 1.7). They found that adiabatic heating and cooling induced by vertical motions are the predominant mechanism responsible for the formation of the out-of-phase occurrence of the upper troposphere-lower stratosphere warm core and the low-mid troposphere cold core. This out-of-phase occurrence of the maximum temperature anomalies maintains strong static stability and, in turn, strong PV anomaly over an extended time period.

1.4 Polar Lows

1.4.1 Introduction

Polar lows are a subcategory of maritime Polar mesoscale cyclones (PMCs). They constitute the most severe PMCs and can be found in both hemispheres poleward of the main baroclinic zone during the winter season. Hence, in the Northern Hemisphere the majority of polar lows develop between November and April (e.g., Wilhelmsen 1985; Blechschmidt 2008; Noer et al. 2011). Polar lows usually last for about a day and horizontal scales, estimated from satellite imagery, range between 100–1000 km in diameter, with scales of $\sim$300 km being the most dominant (Condron and Bigg 2006). Due to their sudden occurrence, strong winds, and heavy snowfall, polar lows can have significant socio-economic impact on coastal communities, fisheries, transportation, tourism, and other marine activities, such as oil and gas exploitation. Besides their direct impact on human activities, polar lows potentially also play a role on the large-scale climate-system through modulation of heat and momentum exchange between the ocean and atmosphere in regions of deep-water formation (e.g., Condron and Renfrew 2012; Papritz and Pfahl 2016).

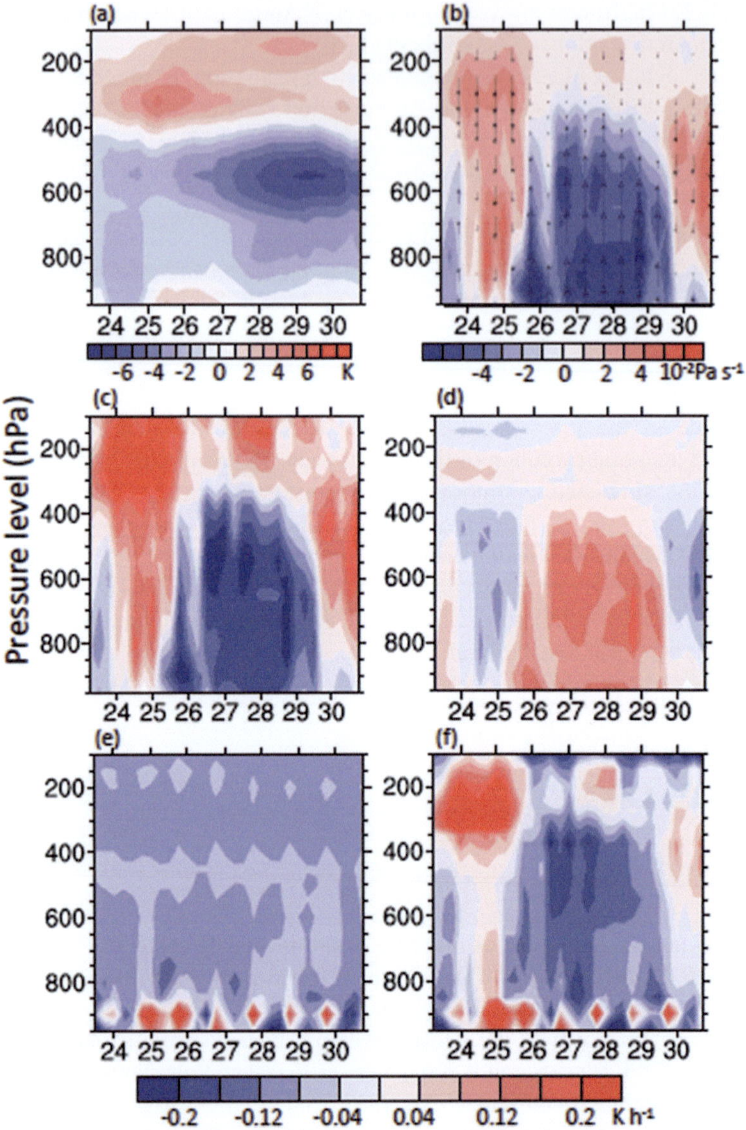

Fig. 1.7 (**a**) air temperature anomalies; (**b**) vertical motions (vectors and magnitude in color); (**c**) adiabatic heating rates due to vertical motion; (**d**) vertical temperature advection; (**e**) diabatic heating rates; and (**f**) the sum of (**c**)–(**e**) along the pressure-time vertical section during 18 UTC 23 September - 18 UTC 30 September 2010. All values were averaged over an area of 200 km radius around the cyclone center at 300 hPa (Tao et al. 2017a). (Reproduced with permission from the Royal Meteorological Society)

1.4.2 In-Situ Observations of Polar Lows

Airborne in-situ observations of polar lows are rare and only available towards the end of the polar low season due to polar night conditions, that restrict the time-window for aircraft operations at high latitudes. Despite this temporal bias in observations, insights into the structure of polar lows are still invaluable due to the limited availability of observations in these data sparse regions. The targeted polar lows (e.g., Shapiro et al. 1987; Bond and Shapiro 1991; Douglas et al. 1995; Brümmer et al. 2009; Kristjansson et al. 2011; Føre et al. 2011; Sergeev et al. 2016) were relatively shallow with cloud-top heights around 3 km, horizontal scales of 300–500 km, and maximum wind speeds around 20–35 m s^{-1}. Structurally, these polar lows featured a core that was 3–5 K warmer than their environment. In most cases frontal structures similar to mid-latitude cyclones accompanied their development (e.g., Shapiro et al. 1987; Bond and Shapiro 1991; Wagner et al. 2011). Furthermore, it is common for the targeted polar lows to developed along an existing confluence or a zone of intense horizontal shear (e.g., Bond and Shapiro 1991; Føre et al. 2011; Sergeev et al. 2016). Low-level flights also collected sensible and latent heat fluxes, which ranged between 200–500 W m^{-2}, with a Bowen ratio close to unity.

1.4.3 Ambient Environments for Polar Low Genesis

1.4.3.1 Synoptic Scale Configuration

The synoptic scale conditions during polar low genesis generally feature an upper-level cold trough and an extra-tropical cyclone to the east of the genesis location (Businger 1985; Yanase et al. 2016; Watanabe et al. 2017). The associated low-level flow enhances cold northerlies over the open water leading to an increase of surface heat fluxes (Papritz and Spengler 2017). Thus, polar lows are associated with synoptic-scale excursions of cold air masses, originating from ice-covered regions, over the adjacent ice-free ocean. Such events, referred to as cold-air outbreaks (CAOs), last several days and play a crucial role for air-sea interactions in these regions (e.g., Papritz and Spengler 2017). Polar lows typically develop either along the outer flanks of CAOs (for an example, see Fig. 1.8) or along convergence or shear zones embedded in the cold air mass. These genesis regions are characterized by significant baroclinicity and shear, as well as moisture contrasts. The synoptic and geographic settings determine the structure of the outer flanks of CAOs to a large extent, whereas the internal convergence zones are hypothesized to be associated with orography or inhomogeneities in the coastline or sea-ice edge from which the CAO originates (e.g., Nagata 1987, 1992).

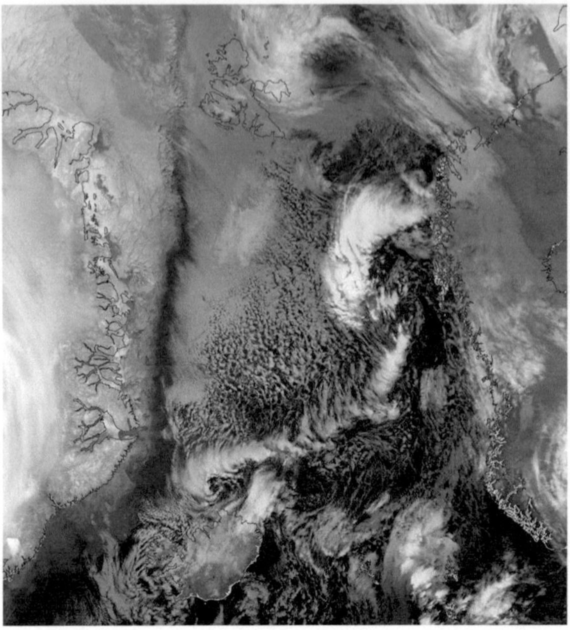

Fig. 1.8 Satellite image of mesoscale cyclones and developing polar lows over the Nordic Seas along the outer flanks of a cold air outbreak on 19 November 2015 at 11:19 UTC. (Source: Dundee Satellite Receiving Station)

1.4.3.2 Forward and Reverse Shear Classification

Polar low genesis environments can be categorized as forward or reverse shear (Duncan 1978), where forward (reverse) shear conditions are prevalent if the mean wind direction over the lower troposphere is in the same (opposite) direction as the wind shear vector. For both shear conditions, the temperature gradient near the surface features a baroclinic zone (Fig. 1.9). However, during forward shear conditions, this baroclinic zone extends throughout the entire troposphere with a strong jet at 300 hPa, whereas during reverse shear conditions the baroclinic zone is confined to the lower troposphere featuring an intense low-level jet (see Fig. 1.9; Terpstra et al. 2016; Michel et al. 2018). Polar lows forming in forward shear conditions are present over the entire Nordic Seas and propagate mainly eastward, whereas polar lows forming in reverse shear conditions occur more frequently over the Greenland and Norwegian seas and tend to propagate southwestward (Terpstra et al. 2016; Michel et al. 2018). The dominant direction of propagation for forward and reverse shear polar lows is similar in the Sea of Japan (Yanase et al. 2016).

Forward and reverse shear genesis environments distinguish between two different dynamical pathways for polar low formation. Forward shear conditions are similar to the classical mid-latitude baroclinic development and reverse shear conditions resemble the paradigm of frontal instability along a strong low-level shear zone (e.g., Schär and Davies 1990; Joly and Thorpe 1989). Overall, polar lows forming in reverse shear conditions become more intense in terms of wind and vorticity during their lifetime compared to forward shear polar lows (Michel et al.

Due to the limited representation of polar lows in coarse-resolution global reanalysis and model data, climatological studies of polar lows often resorted to diagnosing environmental conditions that are potentially conducive to polar low development (e.g., Claud et al. 2007; Kolstad 2006; Kolstad and Bracegirdle 2008), as opposed to detecting actual polar low development. These favorable conditions for polar low development have been associated with weather regimes, such as the North Atlantic Oscillation (NAO) and Scandinavian Blocking (SB) (e.g., Claud et al. 2007; Mallet et al. 2013). However, when considering actual cyclone tracks in the Nordic Seas region, the 500-hPa geopotential anomalies for polar low genesis have anomaly dipoles that are linked to neither the NAO nor SB, but rather feature a mix of both patterns (Michel et al. 2018).

1.4.4.2 Selection Criteria

Commonly adopted criteria to distinguish polar lows from PMCs include a low-level wind speed minimum of 15 m s^{-1} as well as low static stability, often identified by a threshold on the difference between temperatures at 500 hPa (T500) and the sea surface temperature (SST). The latter threshold varies regionally. For example, polar lows in the Pacific region tend to be more baroclinic and develop in more statically stable conditions (Businger 1985, 1987; Chen and von Storch 2013) than their Atlantic counterparts, rendering generic thresholds unsuitable for creating global polar low climatologies. In addition, the stability criterion can bias polar low climatologies towards certain categories of polar lows. For example, reverse shear genesis environments feature more statically unstable conditions than forward shear environments (Terpstra et al. 2016). Thus, studies applying the stability criteria are likely biased towards reverse shear polar lows. This bias imposes caveats for studies that try to investigate polar low occurrence in future climates (e.g., Kolstad and Bracegirdle 2008; Zahn and von Storch 2010), because it is not clear whether a potential reduction in reverse shear polar lows might be compensated by an increase in forward shear polar lows.

1.4.5 Dynamical Mechanisms

When polar lows were first identified in satellite images, they were mostly considered high-latitude versions of mid-latitude baroclinic cyclones, albeit with smaller time- and length scales. Subsequently, research indicated that surface fluxes and convection might play a prominent role, resulting in the hypothesis that polar lows might have a formation pathway similar to hurricanes. These two paradigms – baroclinic and symmetric convective instability – represent two ends of a spectrum for dynamic evolution of these type of systems, where baroclinicity is detrimental to the hurricane type evolution. Yanase and Niino (2007) used an idealized framework to map this parameter space for the development of polar lows under the influence of surface fluxes together with different intensities of environmental baroclinicity. With baroclinicity absent in the simulations, the cyclone develops hurricane-like in a symmetric manor with a spiraliform cloud structure fueled by surface sensible and latent heat fluxes. Increasing baroclinicity leads to more and more asymmetric development culminating in the more extratropical cyclone-like development with a comma shaped cloud structure and fully developed fronts.

In satellite images, polar lows only occasionally exhibit spiraliform cloud-structures and a cloud-free 'eye' during their mature stage. The majority of polar lows, however, have comma-shaped cloud structures (Mokhov et al. 2007), indicative that most polar low genesis occurs in a baroclinic environment. Furthermore, there is also a significant fraction of polar lows that develop along low-level frontal zones in reverse shear conditions that resemble frontal instability (e.g., Terpstra et al. 2016; Michel et al. 2018). While the initial development for this dynamic pathway is driven by horizontal shear (Schär and Davies 1990; Joly and Thorpe 1989), the development becomes more baroclinic as the cyclones deepen, where also diabatic processes can significantly modulate this type development (e.g., Hoskins and Berrisford 1988; Reed et al. 1993). The role of this process for polar low genesis remains, however, unclear.

1.4.5.1 Diabatic Processes

Latent heating was proposed as a primary ingredient for PL development (e.g., Sardie and Warner 1983; Forbes and Lottes 1985; Nordeng 1987), where mechanisms such as CISK (Convective Instability of the Second Kind) or WISHE (Wind-Induced Surface Heat Exchange) have been hypothesized as plausible paradigms for polar low development (Rasmussen 1979; Emanuel and Rotunno 1989; Craig and Gray 1996). However, subsequent studies highlighted the unfavorable conditions for CISK and WISHE during polar low development, such as the baroclinic nature of polar low genesis environments (e.g. Grønås and Kvamstø 1995; Claud et al. 2004; Brümmer et al. 2009; Føre et al. 2011, 2012), the lack of CAPE in high-latitude environments (Linders and Saetra 2010), and the dominance of asymmetric cloud structures (Mokhov et al. 2007). Thus, CISK and WISHE seem unlikely to play a prominent role during polar low development.

Henceforth, several other paradigms emphasizing the role for latent heating have been put forward for polar low development. Montgomery and Farrell (1992) hypothesized that classical baroclinic development, enhanced by latent heating, would be sufficient to generate polar lows. This pathway is similar to Type C cyclogenesis (Deveson et al. 2002), the diabatic extension of the classical Type A/B cyclogenesis (Petterssen and Smebye 1971), where Bracegirdle and Gray (2008) found that about 30% of polar lows fulfill the criteria for this dynamical pathway.

Another development mechanism for polar lows are diabatic Rossby waves, a concept introduced by Snyder and Lindzen (1991) and Parker and Thorpe (1995) and subsequently termed Diabatic Rossby Vortex (DRV; Moore and Montgomery 2005). A DRV comprises a low-level cyclone situated in a baroclinic zone, resulting in slantwise ascent and mid-tropospheric latent heat release associated with condensation ahead of the incipient low. In a DRV, diabatic processes are the dominant contributor to the potential voriticity structure without requiring upper-level support of a potential vorticity anomaly, which is fundamentally different from conventional baroclinic instability.

Despite the low absolute moisture content at high-latitudes, Terpstra et al. (2015) found that polar low development can feature DRV-like structures with diabatic potential vorticity anomalies of similar amplitude compared to mid-latitude cyclones. This is due to the proportionality of the amplitude of the diabatic potential vorticity anomalies to the vertical gradient of the heating. Hence, the reduced vertical extent of cyclones at high latitudes can compensate for the absolute heating rates being significantly lower compared to mid-latitudes (Terpstra et al. 2016).

Another diabatic contribution to polar low development is associated with the provision of sensible and latent heat via surface fluxes. However, there are contradicting results for the role of surface fluxes on the initial developing stage of polar lows, claiming a dominant (e.g., Kolstad et al. 2016; Kolstad and Bracegirdle 2017) and not so dominant role (e.g., Føre et al. 2012) for intensification. Furthermore, the exact processes and mechanisms that potentially contribute to an intensification of the development associated with surface fluxes are still debated and more work is needed to clarify their role in polar low intensification.

1.4.6 Climate-Scale Interaction of Polar Lows with the Ocean and Sea-Ice

Based on numerical climate simulations, it was hypothesized that the decline in Arctic sea ice extent plays a significant role in reducing polar low occurrence in future climates (e.g., Zahn and von Storch 2010; Kolstad and Bracegirdle 2008). However, despite the decline of Arctic sea ice extent over the last decades (e.g., Stroeve et al. 2007) there is no long-term trend in the number of polar lows for the period 1979–2014 (Zahn and von Storch 2008; Michel et al. 2018). Furthermore, the cyclogenesis pattern does not seem to be related to the sea ice seasonal cycle and inter-annual variations of its extent for the period 1979–2014, casting some doubt on possible inferences on future polar low formation based on changes in sea ice extent (Michel et al. 2018). It is likely that other drivers are more important for determining the occurrence of polar lows, such as the synoptic situation leading to possible changes in CAO intensity and location, or the availability of moisture.

Polar lows occur in areas crucial for deep water formation, such as the Nordic and Labrador seas. In these areas the energy and momentum exchange between the ocean and atmosphere can have a significant impact on deep oceanic convection as well as on the oceanic gyre circulation by changing the mean wind stress curl. Using an ocean model forced by reanalysis data in conjunction with a boosting of wind speed associated with PMCs, Condron et al. (2008) and Condron and Renfrew

(2012) showed that underestimation of wind-intensity associated with PMCs in climate models can have an influence on oceanic deep convection and hence on the Atlantic meridional overturning circulation.

While CAOs play a significant role for the heat extraction from the ocean in the Nordic seas (Papritz and Spengler 2017), polar lows can also play a significant role in eroding the CAO airmass via their increase in latent heat release leading to warming of the cold airmass (Papritz and Pfahl 2016). Hence, under-estimating polar lows in climate simulations might enhance the duration of CAOs resulting in overestimating the heat extraction from the ocean in areas crucial for deep water formation. However, it remains an open question how much PMCs actually affect the climate and ocean-circulation.

1.4.7 Forecasting of Polar Lows

Model resolution for regional numerical weather prediction (NWP) has continuously increased and is now commonly at around 2 km with convection permitting grid spacing (e.g., AROME Arctic, Müller et al. 2017). However, there is still a significant lack in polar low predictability, which crucially depends on the location and size of the regional model domain as well as the adequateness of the parameterization of the physical processes (Kristiansen et al. 2011). The sensitivity to size and location of the model domain point to the need for well constrained initial as well as lateral boundary conditions for polar low forecasting, though accurate forecasts of up to 2 days seem feasible.

Due to polar low genesis occurring in convergence zones or along shear zones, it appears especially important to appropriately initialize and predict these mesoscale features. However, polar orbiting satellites with a higher horizontal resolution cover the area only at a lower temporal frequency, and NWP in the polar regions thus often lacks information on these scales.

In addition, crucial physical processes associated with the energy and momentum exchange between the atmosphere and ocean as well as high-latitude clouds are inadequately represented in our current NWP systems (e.g., Bourassa et al. 2013).

Overall, the community has not sufficiently pinpointed the origin of the ongoing challenges with polar low forecasting, and it is unclear which of the aforementioned deficiencies in our NWP systems contribute most to the lack of our forecasting potential. However, with recent increase in satellite missions and polar field programs, there is room for future improvements of our forecasting capabilities on the mesoscale.

1.5 Orographic Effects on Dynamics of the Arctic Atmosphere

1.5.1 Theoretical Background

This section presents a short review of a selection of weather phenomena that are connected to orographic effects, with particular focus on the Arctic region and strong winds. Greenland, representing the arguably most dominant orographic feature in the Arctic, frequently modifies and distorts atmospheric flow over and around it, inducing phenomena such as barrier flows (Moore et al. 2005; Petersen et al. 2009; Harden et al. 2011; DuVivier et al. 2017), tip jets (Pickart et al. 2003; Doyle and Shapiro 1999; Renfrew et al. 2009; Moore et al. 2005; Outten et al. 2009) and katabatic winds/ downslope wind storms (Bromwich et al. 2001; Oltmanns et al. 2014; Doyle and Shapiro 1999). Orographic flow effects contribute to making the Arctic and the coast off Greenland to the home of some of the windiest areas on Earth (Sampe et al. 2007; Moore et al. 2008). Several of these flow phenomena are also found around other Arctic islands and archipelagos such as Svalbard (Skeie and Grønås 2000; Sandvik et al. 2002; Reeve and Kolstad 2011) and Novaya Zemlya (Moore 2013; Lydolph 1977). Other orographic phenomena taking place in the area include gap flows, such as documented in Alaska (Mass et al. 1995; Loescher et al. 2006).

Several reviews have been written on a wide range of aspects of mountain flows. Queney (1948) is one of the earliest, covering theoretical models of inviscid flow over hills and mountains. This, together with Scorer (1949), which deduced analytical solutions for trapped mountain lee waves, sparked an increased interest in the field. Much progress in the theory of mountain flow was made in the early years by refining and making Scorer's analysis more realistic (Wallington and Portnall 1958; Sawyer 1960; Foldvik 1962). More recent reviews of mountain flows include those of e.g., Smith (1979), Blumen (1990), Kaimal and Finnigan (1994), Belcher and Hunt (1998), and Wood (2000).

Idealised studies of mountain flows on a non-rotating plane commonly assume a hydrostatic flow that is inviscid and Boussinesq. Scale analysis shows that for these conditions the flow is governed by the mountain shape, its aspect ratio, the atmospheric profile and the non-dimensional mountain height (also known as the inverse Froude number) $\hat{h} = Nh/U$ (e.g. Smith and Grønås 1993), where h is the mountain height, U is the background flow velocity, and N is the Brunt-Väisälä frequency. $\hat{h}$ can be considered a measure of the non-linearity of the flow. For a non-rotating plane, Smith (1989) identified three flow regimes in frictionless, stratified and linear flow over a mountain. (1) For low values of $\hat{h}$, more flow passes over the mountain than around it, no upstream stagnation (or upstream "blocking") occurs, and typically gentle gravity waves form aloft, in the lee of the mountain. Linear mountain wave theory describes the flow response well within this range (Gill 1982). (2) As $\hat{h}$ approaches unity, the flow enters a high-drag state (large pressure perturbations) and wave breaking might take place. Flow in this regime typically contains a stagnation point above the downwind mountain slope, coinciding with the first steepening level of the streamlines over the slope. Such downwind stagnation may trigger wave breaking and or strong down-slope winds, as described by e.g. Peltier et al. (1979). (3) High values of $\hat{h}$ are typically associated with windward flow stagnation, allowing for splitting, i.e. the flow is diverted to either side of the mountain on its upwind side. Weak gravity waves may form and there might also be lee-side vortices (e.g. Smolarkiewicz et al. 1989) as for example observed behind isolated mountain ridges such as Beerenberg on Jan Mayen (e.g. Berger and Wille 1972). The pressure perturbations set up by the mountains are small within this regime as there is only weak flow lifting.

For mountains of a sufficiently large horizontal scale, such as found in Alaska, Greenland, Spitsbergen and Novaya Zemlya, the rotation of the Earth changes the flow morphology and the Rossby number enters as an additional parameter governing the flow. The Rossby number is defined as $R_0 = U/fL$, where U is the upstream flow velocity, f is the Coriolis parameter and L is a length scale, typically defined as the half-length of the mountain in the direction of the flow. In situations where air under the influence of Earth's rotation is forced to ascend a large mountain, it is stretched (and cooled adiabatically), and a positive pressure perturbation builds up underneath on the mountain's windward side. Air in a geostrophic balance entering this positive pressure perturbation is decelerated. As a result, the Coriolis force will weaken, and the air will accelerate down the large-scale pressure gradient, i.e. towards the left in the northern hemisphere, giving an upstream left-right flow asymmetry (Pierrehumbert et al. 1985) and also delaying the onset of upstream blocking (Thorsteinsson and Sigurdsson 1996). Through deflection of the low-level flow, this mechanism is believed to play a role in the formation of tip jets (e.g. Doyle and Shapiro 1999). Tip jets are narrow low-level jets that periodically develop around the vertex of large elliptical barriers, such as the southern tip of Greenland and Spitsbergen. A second mechanism argued to contribute to the formation of such jets is the conservation of the Bernoulli function (the sum of the enthalpy, kinetic energy and potential energy per unit mass (Schär 1993)), for example when air parcels descend at high velocity down the lee slope of Greenland.

In situations with cold and stably stratified air impinging on a mountain and with sufficiently high $\hat{h}$, the air will be unable to go over the mountain and will instead be blocked, and upstream damming of the air against the mountain occurs. As a result, a pressure gradient develops perpendicular to the mountain and there is a formation of an approximate geostrophic flow along the mountain frequently referred to as barrier wind (e.g. Schwerdtfeger 1975; Parish 1983). Barrier winds are known to occur in the Arctic e.g. along the coasts of Alaska (Loescher et al. 2006) and Greenland (Petersen et al. 2009; Harden et al. 2011; Harden and Renfrew 2012), and are often associated with very high wind speeds.

1.5.2 Observations of Orographic Flow in the Arctic

Flow over and around Greenland has been the focus of numerous studies. As pointed out by Doyle and Shapiro (1999), studies of orographic flow over Greenland are made difficult for several reasons, including its extreme orography, the proximity to the semi-permanent Icelandic low, and high latitude air-sea interaction processes. Thus, the literature on these phenomena around Greenland, and also for other locations in the Arctic, largely bases its results on numerical modelling. Though, there are a handful of studies based at least partly on observational data, and these are the topic of this section. Tip jets are dominating features at Cape Farewell in Greenland. The first to thoroughly address these jets by observational means were Pickart et al. (2003), who proposed the tip jets to be important in contributing to deep water formation in the Irminger Sea through an elevated surface heat flux and a strong wind stress curl. They used several decades of data from a network of automatic weather stations in Greenland to construct a climatology of tip-jet events. Their findings showed that tip-jet events were best captured by a station at Prins Christian Sund, where they are easily recognizable as short-lived episodes with strong winds, followed by relatively low atmospheric pressure and temperature.

Moore and Renfrew (2005) used QuickSCAT data (e.g. Hoffman et al. 2005) to develop a wintertime climatology for high wind speed flows along the coast of Greenland. Amongst their findings were high wind speed features, such as the previously identified (westerly) tip jets, and they also found tip jets with an easterly flow direction, the so-called reversed tip jet, as first described by Moore (2003). In addition, they identified barrier flows along two regions of the southeast coast of Greenland, which they called the Denmark Strait South and the Denmark Strait North. Maximum surface winds were estimated at over 50 m/s, with wind speeds above 25 m/s occurring around 10–15% of the time at each of these locations. These types of satellite-based results, in addition to numerical modelling studies, demonstrated the presence of until then little-understood orographic flows associated with high winds and strong atmospheric forcing of the ocean around Greenland. The meteorological stations in the area are few, however, and the satellite-based data are not available for larger portions of the Greenland coast due to sea ice. Also, based on scattering from ocean waves, satellite winds give no information about the vertical structure of the atmosphere. It was with this background the Greenland Flow Distortion Experiment (GFDex, Renfrew et al. 2008) was designed. It was the first field campaign to provide a detailed dataset of *in-situ* observations on mesoscale weather phenomena in the Greenland-Iceland region using, among others, aircraft, dropsondes and radiosondes. The project resulted in several papers, including Renfrew et al. (2009) investigating an easterly tip jet event off Cape Farewell, and Petersen et al. (2009) providing an overview of barrier winds off southeastern Greenland. In addition, McInnes et al. (2009) and Kristjánsson (2009) presented the first comprehensive observations of a lee-side cyclogenesis event. Another set of observations from this region was provided by two airborne field campaigns collectively called "Polar Winds", including data from a Doppler

Aerosol Wind Lidar. Using, among others, observations from these campaigns, which also included dropsonde data, DuVivier et al. (2017) investigated barrier winds in the Denmark Strait. One of their main findings centered on a case of 21 May, 2015, when they found a barrier wind in a shallow layer (~250–400 m) and had wind speeds between 23 and 28 m/s.

Another flow phenomenon occurring in Greenland that has received attention is mountain wave breaking and associated downslope windstorms. Doyle et al. (2005) investigated a wave breaking event that took place during deep southwesterly flow over Greenland during the Fronts and Atlantic Storm-Track Experiment (FASTEX), using aircraft, dropsonde and satellite data. Their upper-air data showed, among others, a large-amplitude wave with deep convectively unstable layers and a vertical velocity maximum of almost 10 m/s in the stratosphere, and the satellite-based surface data indicated the presence of high upstream wind speeds and a large-scale wake in the lee.

Other geographical areas in the Arctic with frequent occurrence of orographic flows have generally received less attention, but do include for example the Svalbard archipelago and surrounding waters. Reeve and Kolstad (2011) investigated tip jets forming at the southern tip of Spitsbergen, reminiscent in nature to those forming off Cape Farewell, using, among others, QuickSCAT data. These data showed that the jet is present 8% of the time throughout the year. Another study that used satellite data to investigate orographic flow in Svalbard is that of Sandvik and Furevik (2002), who employed synthetic aperture radar data together with surface-based observations and numerical modelling to investigate an orographically induced jet forming in Hinlopenstredet, the strait between Nordaustlandet and Spitsbergen. Their findings showed an enhancement of the wind in the core of the jet by the surrounding topography by a factor three.

1.5.3 Numerical Simulations of Orographic Flows in the Arctic

Naturally, numerical data of orographic flows are more readily available than observational data and a fair amount of studies have been conducted using numerical models to study orographic flows in the Arctic. These studies have either been supported by observations, as is the case for several of the above-cited studies, or they are based solely on numerical simulations. One of the earlier numerical studies of orographic flows in the Arctic is that by Doyle and Shapiro (1999), who ran a series of numerical simulations of idealised and observed flows in order to study the Greenland tip jet. They found the jet to be the most sensitive to the non-dimensional mountain height and the Rossby number, underlining the importance of flow deflection and downslope flow acceleration. Skeie and Grønås (2000) ran a set of idealised simulations focusing on strongly stratified easterly flows over Svalbard, with the non-dimensional mountain height varying between 0.67 and 2. They found strong downslope flows over the major mountains on central Spitsbergen at the lower end of $\hat{h}$, which they attributed to strong gravity wave activity aloft.

Furthermore, they found downstream jets emanating from the three larger fjords on the west side of the island: Isfjorden, Van Mijenfjorden and Hornsund. The strongest winds in their simulations were found in the southern part of Spitsbergen, which they explain in terms of gravity wave activity, but also with contributions from a left-right flow asymmetry induced by planetary rotation. An interesting aspect pointed out by Skeie and Grønås (2000) is that, due to the predominantly stably stratified atmosphere to the east of Spitsbergen, the non-linearity of orographic flows there is comparable to flows over far higher orography elsewhere in the world, e.g. over the Alps.

The availability of high-resolution regional reanalysis datasets has made it easier to compile climatologies of mountain flows, also over land where satellites do not provide wind data. The Arctic System Reanalysis Version 2 (ASRV2, Bromwich et al. 2017) is one such reanalysis dataset. As demonstrated by e.g. Oltmanns et al. (2014) and Moore et al. (2016) in their studies of downslope flows and barrier winds, it is widely recognized that relatively high horizontal grid resolution is necessary to appropriately resolve mountain flows. At a horizontal resolution of 15 km, ASRV2 resolves major orographic flow features around the larger mountains and mountain ranges in the Arctic. Moore (2013) used data from ASRV1 (30 km horizontal resolution) to study frequent occurrences of strong winds in Novaya Zemlya. He found these wind events to bear resemblance with Bora systems as well as with downslope windstorms. An interesting aspect of his study is the focus on air-sea interactions, which are intense over the largely ice free Barents Sea to the west of Novaya Zemlya. High-wind speed events together with the generally large air-sea temperature contrast in the Arctic winter, frequently induce strong surface heat fluxes, especially in the winter. This linkage has been studied amongst others in the context of the Greenland tip jet. Pickart et al. (2003) ran an idealised ocean model forced by atmospheric conditions representative for Greenland tip jet events and argued that the most likely cause of observed convection in the Irminger Sea is the Greenland tip jet.

Similar to the climatologies compiled by e.g. Moore (2013) of winds in Novaya Zemlya using ASRV1 data, by Harden et al. (2011) for winds around Greenland, and Reeve and Kolstad (2011) for the south Spitsbergen tip jet, we here present in Fig. 1.11 wind climatologies for Greenland and Spitsbergen using ASRV2 data. The Figure shows the annual frequency of occurrence of winds exceeding 20 m/s for Greenland and those exceeding 16 m/s for Svalbard. Easily recognizable are several of the orographic features discussed herein. We can identify strong winds off Cape Farewell in Greenland; this is the Greenland tip jet. There are strong winds also along the southeast coast of Greenland, in the vicinity of the Denmark strait; these are barrier winds and they have also been identified in a similar way by e.g. Harden et al. (2011) using data from ECMWF ERA Interim reanalysis (Dee et al. 2011). In addition, we see enhanced winds over the southeastern slope of Greenland, to the north of Cape Farewell; this could be related both to gravity wave activity and to katabatic winds as discussed by e.g. Oltmanns et al. (2014).

In Svalbard, we recognize a maximum in the frequency of high wind speeds to the south of Spitsbergen. This is the Spitsbergen tip jet investigated by Reeve

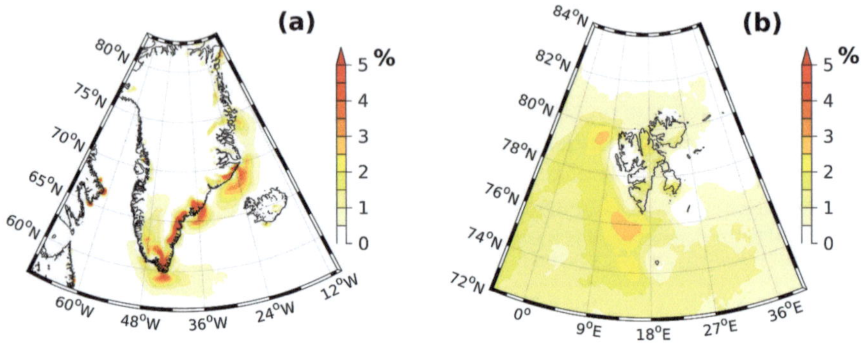

Fig. 1.11 (**a**) Percentage of time the 10 m wind speed is higher than a threshold of 20 m/s from the ASRV2 for Greenland and surrounding waters, and (**b**) as in (**a**), but for Svalbard and with a threshold of 16 m/s

and Kolstad (2011). Also seen in the flow around Spitsbergen is the presence of a maximum in the northwestern corner. This maximum is also found by Skeie and Grønås (2000).

1.6 Atmospheric Boundary-Layer Dynamics

1.6.1 Theoretical Background

The atmospheric boundary layer (ABL) is generally defined as the layer, which is influenced by the presence of the Earth's surface (e.g. Stull 1988). Surface friction, heating and cooling lead to turbulent motions in the ABL with eddy sizes on a scale in the order of the boundary layer depth h (Garratt 1992). This depth is, however, not uniquely specified and there are numerous definitions in the literature. h can be defined by characteristics of the vertical profiles of meteorological variables, such as the potential temperature, humidity and wind, but also the profiles of the turbulent fluxes or turbulent kinetic energy can be considered. So, h is sometimes defined as the nearest height to the surface in which the vertical profile of heat fluxes attain a minimum. In many applications h is, however, given as the bottom of the capping temperature inversion, which is always observed although its strength can vary considerably. Small values of h combined with strong capping inversions are one of the important characteristics of the polar ABL. Surface-based inversions occur often, but more often elevated inversions occur with h in the range between 50 m and 200 m height over large regions. The actual values depend strongly on the external forcing, e.g., by wind speed and horizontal advection. Especially the surface properties, namely the surface roughness and the surface temperature distribution have a large impact because they influence the turbulent eddies, which in turn shape the profiles of wind, temperature and humidity all over the ABL.

All surface parameters are finally related to the sea ice cover, its concentration, thickness, and snow pack on top of ice.

One of the most important global objectives for boundary layer research is the determination of the fluxes of energy and momentum in the ABL and at the Earth surface. The fluxes (radiation and turbulence) depend on both cloud processes and on the surface characteristics. For turbulence, processes in the lowest layers (roughly 10% of h) are important. In this commonly-called surface layer the turbulent fluxes can be determined by a combination of observations and theory, which is the Monin-Obukhov similarity theory (MOST, Monin and Obukhov 1954). It is based on several assumptions, the most important ones being horizontal homogeneity, height-constant turbulent fluxes in the surface layer, and the assumption that the vertical dimensionless gradients of, e.g., wind and temperature can be formulated as universal functions of the nondimensional parameter z/L, where z is the thickness of the near-surface layer and L is the Obukhov length (Obukhov 1946). It depends on the turbulent fluxes expressed by scaling parameters, such as the friction velocity and characteristic temperature scale. Combining this information the turbulent fluxes of momentum and heat follow as functions of wind speed and temperature, and of stability via correction functions depending on z/L, the surface roughness length z_0 and on the scalar roughness z_t. Since a numerical model provides wind speed and temperature at the lowermost atmospheric level, the fluxes can be determined either by an iterative method or by a non-iterative Louis-type (Louis 1979) method, in which the bulk Richardson number, instead of z/L, is used as a stability parameter. For Arctic conditions over sea ice, an iterative scheme by Andreas et al. (2010a, b) using stability correction functions (Grachev et al. 2007) based on more than a year of observations over sea ice (SHEBA drift, Uttal et al. 2002) is available. Also based on the functions by Grachev et al. (2007), Gryanik and Lüpkes (2018) provided recently a non-iterative scheme for the flux determination.

As mentioned above, the fluxes depend strongly on the surface roughness parameters z_0 and z_t which are connected with the transfer coefficients for momentum C_{dn} and heat C_{hn} in neutral stratification via

$$C_{dn} = [\kappa / \ln(z/z_0)]^2 \text{and } C_{hn} = \kappa^2 / [\ln(z/z_0)\ \ln(z/z_t)] \quad (1.1)$$

where κ is the von Karman constant. The turbulent fluxes can be formulated also in terms of these coefficients. For example, momentum flux F_m and heat flux F_h are given as

$$F_m = \rho\, C_{dn} f_m U^2; \quad F_h = -\rho\, c_p C_{hn} f_h U\ (\theta - \theta s) \quad (1.2)$$

where ρ is air density, c_p is the specific heat of air, θ is air potential temperature at z and the surface (subscript s). f_m and f_h are normalized stability dependent transfer coefficients which follow from MOST (see details in Garratt 1992 and Gryanik and Lüpkes 2018).

Although sea ice is smooth compared with land surfaces its topography has a large impact on the surface roughness and thus on the neutral transfer coefficients.

Important roughness elements are the sea ice pressure ridges forming due to the inhomogeneous drift of sea ice forced by wind and ocean currents, but also snow drift forms an uneven surface, e.g., in the form of so-called sastrugi. Although ridges range up to metres in height, the topography is on average much shallower with typical topographic elements around 50 cm high and distances in the range of 50 m to each other (see. e.g. Castellani et al. 2014; Tsamados et al. 2014). Moreover, there are the edges of drifting ice floes, leads and melt ponds, which all influence the roughness parameters. A possible parameterization approach for the determination of the surface roughness is based on a partitioning of the total surface drag into skin drag of the flat ice surface and form drag due to the obstacles (e.g. Hansen-Bauer and Gjessing 1988; Andreas et al. 2010a, b; Lüpkes and Gryanik 2015).

1.6.2 Observed ABL in the Arctic

The structure and processes of the Arctic ABL depend on the region, season, synoptic-scale weather conditions, and the surface type (Persson and Vihma 2017). Sea ice provides particular boundary conditions for the ABL, which often change rapidly due to sea ice drift and deformation. In conditions of free drift (not restricted by internal resistance of the ice field), ice drift speed is approximately 3% of the 10-m wind speed. Hence, the ice margin may be displaced by tens of km in a day, dramatically affecting the lower boundary conditions for the ABL, in particular in winter when the surface temperatures of sea ice and open water may differ by more than 30 K. Divergence of ice drift results in opening of leads, and convergence results in closing of leads as well as rafting or ridging of the ice floes. Both processes change the surface temperature and aerodynamic roughness length. The lower boundary conditions are naturally also affected by thermodynamic ice growth and melt, which are the dominating processes in seasonal time scales, but in synoptic time scales sea ice dynamics generates much faster changes in the surface type and properties. Due to the low heat conductivity of sea ice and, in particular, snow, the surface temperature may, however, respond rapidly to changes in the surface energy budget.

The surface energy balance is affected by incoming and reflected solar shortwave radiation, incoming and outgoing thermal longwave radiation, turbulent fluxes of sensible and latent heat, and the conductive heat flux, typically directed upwards from the ocean through ice and snow. In the absence of solar radiation in winter, the longwave radiation emitted by the snow/ice surface almost always exceeds the downward longwave radiation from the atmosphere, resulting in surface cooling. The energy balance is reached when the net longwave cooling is compensated by the downward turbulent heat flux (mostly sensible heat) and upward conductive heat flux from the ice and snow. This, together with the fact that the heat advection from lower latitudes to the Arctic peaks above the ABL typically results in a temperature inversion, which in the central Arctic may reach the height of 1.0–1.2 km with a temperature increase of 10–12 K (Serreze et al. 1992). Also specific humidity inversions are very common in the Arctic, and are partly linked to temperature inversions, as the saturation specific humidity depends on temperature, and partly generated by the vertical profile of moisture advection (Naakka et al. 2018). Even in winter, strong winds may mechanically mix the near-surface air, destroying both temperature and specific humidity inversions. In summer, when solar radiation is present and the surface net radiation is often close to zero, a well-mixed layer some 300–400 m thick is a typical situation in the central Arctic. The inversion strength is typically only 4–5 K (Serreze et al. 1992). It should be noted that the numbers above result from the decades with year-round radiosonde sounding data from drifting stations available, before the Arctic amplification. Recent data sets suggest generally weaker and thinner temperature inversions, at least in the Atlantic sector of the Arctic (Lüpkes et al. 2012).

In summer the thermal differences between sea ice and the open ocean are very small, and the ABL structure is not particularly sensitive to the type of the surface. In other seasons, however, the ABL over leads and polynyas is typically convective (Andreas 1980). As most leads are narrow, the localized convection does not reach high altitudes (Lüpkes et al. 2012), and part of the heat is recaptured by the sea ice downwind of the lead. In the vicinity (approximately 200 km distance) of the sea ice margin, the structure and properties of the ABL strongly depend on the wind direction with respect to sea ice edge. During off-ice flows in cold seasons, a cold Arctic air mass is advected to over the open ocean, which results to large upward fluxes of sensible and latent heat. A convective ABL grows and air temperatures increase with fetch over the open sea (Lüpkes and Schlünzen 1996). Due to the wind forcing on ice drift, such cases are typically associated with a diffuse ice margin (Vihma and Brümmer 2002). Accordingly, the ABL receives heat and moisture from the open leads already before the ultimate ice edge. During on-ice flows, warm, moist marine air masses are advected over the sea ice, and a stable boundary layer (SBL) grows downwind of the ice edge, which is typically distinct. The SBL growth is a gradual process (Vihma et al. 2003), as the downward turbulent heat flux has a magnitude much smaller than the upward flux in cases of off-ice flows. If low clouds or fog is formed over sea ice, the cloud radiative effects may have a dominating role for the SBL growth (Brümmer and Thiemann 2002). In cases of a flow parallel to the ice edge, the ABLs over the open sea and sea ice are close to a balance with the local surface: a SBL over sea ice and a CBL over the open sea. A strong horizontal temperature gradient may occur perpendicular to the ice edge.

A low-level jet (LLJ) is a common feature in the Arctic ABL. The occurrence of LLJs depends on the baroclinicity (horizontal temperature gradient), ABL stratification, and the flow direction relative to the ice edge (Chechin et al. 2013; Chechin and Lüpkes 2017). LLJs occur in convective boundary layer during cold-air outbreaks, and also interact with convective plumes over leads (Tetzlaff et al. 2015). Jakobson et al. (2013) observed that over sea ice in summer half of the LLJs observed were related to a frontal passage. Baroclinicity due to transient cyclones was the most important generation mechanism for LLJs, but also LLJs related to inertial oscillations and gusts were detected. In high-latitudes, with a small diurnal cycle, the LLJs generated by inertial oscillations are not classical nocturnal LLJs, but analogous to them in space (stabilization of the ABL due to warm-air advection over cold surface, as in Vihma et al. (2003)) or time (stabilization of the ABL after a passage of a storm, as in Andreas et al. (2000)). Challenges remain in understanding of LLJs associated to Arctic clouds and LLJs interacting with gravity waves.

The structure of and processes in the Arctic ABL strongly interact with clouds, fog, and radiative transfer (Solomon et al. 2011). Accroding to Tjernström (2007), in summer most of ABL turbulence in the Arctic is generated by clouds instead of wind shear or surface heating convection. Due to cloud-top radiative cooling and resulting top-down overturning, the cloud layer is often well mixed. If the in-cloud turbulence is strong and ABL stratification is weak or moderate, the turbulent eddies generated in the cloud may penetrate to the ice/ocean surface and contribute to the surface fluxes of momentum, heat, and moisture. However, if the in-cloud turbulence is weaker and ABL stratification strong, the mixing does not reach sub-cloud layers and the cloud layer becomes decoupled from the sub-cloud layer (Shupe et al. 2013). In such conditions, the ice/ocean surface cannot act as a moisture source for clouds, but a specific humidity inversion may provide the moisture source (Sedlar et al. 2012). The decoupling is further enhanced by sublimation from the ice crystals falling from the cloud, which cools the sub-cloud layer. The variations of ABL temperatures in synoptic time scales are principally controlled by the cloud cover, wind speed, and heat advection.

The structure of wintertime Arctic ABL varies between two basic states: a well-mixed ABL under cloudy skies and surface-based inversion under clear skies or optically thin clouds (Tjernström and Graversen 2009). In winter, the 2-m air temperature responds very fast to changes in cloud cover and wind speed, the two forcing factors being approximately equally strong (Walsh and Chapman 1998), whereas the heat advection has a cumulative effect with a longer response time of 2-m air temperature (Vihma and Pirazzini 2005). Challenges remain in improving our understanding of Arctic cloud physics (including the coupling of clouds, aerosols, radiative transfer, ABL turbulence, and cloud-generated turbulence).

Our main focus was in the marine Arctic. ABL over terrestrial Arctic is different due to (a) effects of complex orography (Sect. 1.5) and (b) lack of heat source from the warmer ocean below. In any case, the ABL over terrestrial Arctic includes many of the features observed over the marine Arctic, such the common occurrence of very stable stratification (Grachev et al. 2017) as well as temperature and humidity inversions (Kahl et al. 1992; Vihma et al. 2011).

1.6.3 ABL Modelling

Several studies have suggested that ABL schemes yield excessive turbulent fluxes in conditions of stable stratification (Cuxart et al. 2006; Tjernström et al. 2005), which often results in near-surface warm bias (Lüpkes et al. 2010; Atlaskin and Vihma 2012; Jakobson et al. 2012; de Boer et al. 2013). Problems directly related to ABL are decoupling of the snow/ice surface and air in cases of very stable stratification

(Derbyshire 1999) and limited vertical resolution applied (Byrkjedal et al. 2007). However, many of the model biases in ABL are not necessarily related to turbulence schemes, but to errors in the surface energy balance, which often originate from errors in modelling of radiative transfer and clouds (Tjernström et al. 2008), as well as in parameterizing the effects of heterogeneity of the sea ice surface (Burk et al. 1997; Persson and Vihma 2017).

A prerequisite to improve the performance of NWP and climate models in the Arctic ABL is better understanding and parameterization of subgrid-scale physical processes, such as interaction of turbulence, gravity waves, radiative transfer, cloud physics, and thermodynamics of sea ice and snow (Vihma 2014). This requires (a) more systematic utilization of existing Arctic observations together with coordinated model experiments applying a broad range of modelling approaches (Large-Eddy Simulation, coupled column modelling, as well as NWP and climate model experiments) and (b) new field campaigns to collect year-round process-level observations from the Arctic sea ice zone. For the latter, strong expectations are laid on the MOSAiC (Multidisciplinary drifting Observatory for Studies of Arctic Climate) field campaign from autumn 2019 to autumn 2020.

Acknowledgements The work on Chap. 1 has been supported by the Academy of Finland (contracts 286298 and 317999) and by the German Research Foundation (Deutsche Forschungsgemeinschaft; DFG) (Transregional Collaborative Research Center "ArctiC Amplification: Climate Relevant Atmospheric and SurfaCeProcesses, and Feedback Mechanisms $(AC)^3$" (TRR 172, project no. 268020496)). This work was supported by the US National Science Foundation (Grant #ARC-1023592, ARC-1107509, and PLR-1304684).

References

Aizawa, T., Tanaka, H. L., & Satoh, M. (2014). Rapid development of arctic cyclone in June 2008 simulated by the cloud resolving global model NICAM. *Meteorology and Atmospheric Physics, 126*, 105–117. https://doi.org/10.1007/s00703-013-0272-6.

Aizawa, T., & Tanaka, H. L. (2016). Axisymmetric structure of the long lasting summer Arctic cyclones. *Polar Science*. https://doi.org/10.1016/j.polar.2016.02.002.

Ambaum, M. H. P., Hoskins, B. J., & Stephenson, V. B. (2001). Arctic oscillation or North Atlantic oscillation? *Journal of Climate, 14*, 3495–3507.

Andreas, E. L. (1980). Estimation of heat and mass fluxes over Arctic leads. *Monthly Weather Review, 108*, 2057–2063.

Andreas, E. L., Claffy, K. J., & Makshtas, A. P. (2000). Low-level atmospheric jets and inversions over the western Weddell Sea. *Boundary-Layer Meteorology, 97*(3), 459–486.

Andreas, E. L., Horst, T. W., Grachev, A. A., Persson, P. O. G., Fairall, C. W., Guest, P. S., & Jordan, R. E. (2010a). Parameterising turbulent exchange over summer sea ice and the marginal ice zone. Quarterly journal of the royal meteorology. *Society, 136B*, 927–943.

Andreas, E. L., Persson, P. O. G., Jordan, R. E., Horst, T. W., Guest, P. S., Grachev, A. A., & Fairall, C. W. (2010b). Parameterizing turbulent exchange over sea ice in winter. *Journal of Hydrometeorology, 11*, 87–104.

Andrews, D. G., Holton, J. R., & Leovy, C. B. (1987). *Middle atmosphere dynamics*. Academic Press. 489 pp.

Atlaskin, E., & Vihma, T. (2012). Evaluation of NWP results for wintertime nocturnal boundary-layer temperatures over Europe and Finland. *Quarterly Journal of the Royal Meteorological Society, 138*, 1440–1451. https://doi.org/10.1002/qj.1885.

Baldwin, M. P., & Dunkerton, T. J. (1999). Propagation of the Arctic oscillation from the stratosphere to the troposphere. *Journal of Geophysical Research, 104*, 30 937–30 946.

Baldwin, M. P., Gray, L. J., Dunkerton, T. J., Hamilton, K., Haynes, P. H., Randel, W. J., Holton, J. R., Alexander, M. J., Hirota, I., Horinouchi, T., & Jones, D. B. A. (2001). The quasi-biennial oscillation. *Reviews of Geophysics, 39*(2), 179–229.

Barrier, N., Cassou, C., Deshayes, J., & Treguier, A. M. (2014). Response of North Atlantic Ocean circulation to atmospheric weather regimes. *Journal of Physical Oceanography, 44*(1), 179–201.

Belcher, S. E., & Hunt, J. C. R. (1998). Turbulent flow over hills and waves. *Annual Review of Fluid Mechanics, 30*, 507–538. https://doi.org/10.1146/annurev.fluid.30.1.507.

Berger, E., & Wille, R. (1972). Periodic flow phenomena. *Annual Review of Fluid Mechanics, 4*, 313–340. https://doi.org/10.1146/annurev.fl.04.010172.001525.

Blechschmidt, A. M. (2008). A 2-year climatology of polar low events over the Nordic seas from satellite remote sensing. *Geophysical Research Letters, 35*(L09), 815. https://doi.org/10.1029/2008GL033706.

Bluestein, H. B. (1993). *Synoptic-dynamic meteorology in midlatitudes, observations and theory of weather systems* (Vol. II). Oxford University Press. 594 pp.

Blumen, W. (1990). Mountain meteorology. In *Atmospheric processes over complex terrain* (pp. 1–4). Boston, MA: American Meteorological Society. http://link.springer.com/10.1007/978-1-935704-25-6_1.

Bond, N. A., & Shapiro, M. (1991). Polar lows over the Gulf of Alaska in conditions of reverse shear. *Monthly Weather Review, 119*, 551–572.

Bourassa, M. A., Gille, S. T., Bitz, C., Carlson, D., Cerovecki, I., Clayson, C. A., Cronin, M. F., Drennan, W. M., Fairall, C. W., Hofmann, R. N., Magnusdottir, G., Pinker, R. T., Renfrew, I. A., Serreze, M., Speer, K., Talley, L. D., & Wick, G. A. (2013). High-Latitude Ocean and sea ice surface fluxes: Challenges for climate research. *Bulletin of the American Meteorological Society*, 403–423.

Bracegirdle, T. J., & Gray, S. L. (2008). An objective climatology of the dynamical forcing of polar lows in the Nordic seas. *International Journal of Climatology, 28*, 1903–1919.

Bromwich, D. H., & Coauthors. (2001). Mesoscale modeling of katabatic winds over Greenland with the polar MM5$_{*}$. *Monthly Weather Review, 129*, 2290–2309. https://doi.org/10.1175/1520-0493(2001)129<2290:MMOKWO>2.0.CO;2.

Bromwich, D. H., & Coauthors. (2017). The arctic system reanalysis version 2. *Bulletin of the American Meteorological Society.*, BAMS-D-16-0215.1. https://doi.org/10.1175/BAMS-D-16-0215.1.

Brümmer, B., & Thiemann, S. (2002). The atmospheric boundary layer in an Arctic wintertime on-ice flow. *Boundary-Layer Meteorology, 104*, 53–72.

Brümmer, B., Müller, G., & Noer, G. (2009). A polar low pair over the Norwegian sea. *Monthly Weather Review, 137*(8), 2559–2575.

Bueh, C., & Nakamura, H. (2007). Scandinavian pattern and its climatic impact. *Quarterly Journal of the Royal Meteorological Society: A journal of the atmospheric sciences, applied meteorology and physical oceanography, 133*(629), 2117–2131.

Burk, S. D., Fett, R. W., & Englebretson, R. E. (1997). Numerical simulation of cloud plumes emanating from Arctic leads. *Journal of Geophysical Research, 102*. https://doi.org/10.1029/97JD00339.

Businger, S. (1985). The synoptic climatology of polar low outbreaks. *Tellus, 37A*, 419–432.

Businger, S. (1987). The synoptic climatology of polar low outbreaks over the Gulf of Alaska and the Bering Sea. *Tellus, 39A*, 307–325.

Butler, A. H., & Polvani, L. M. (2011). El Niño, La Niña, and stratospheric sudden warmings: A reevaluation in light of the observational record. *Geophysical Research Letters, 38*(13).

Byrkjedal, Ø., Esau, I. N., & Kvamstø, N. G. (2007). Sensitivity of simulated wintertime Arctic atmosphere to vertical resolution in the ARPEGE/IFS model. *Climate Dynamics, 30*, 687–701. https://doi.org/10.1007/s00382-007-0316-z.

Cagnazzo, C., & Manzini, E. (2009). Impact of the stratosphere on the winter tropospheric teleconnections between ENSO and the North Atlantic and European region. *Journal of Climate, 22*(5), 1223–1238.

Castellani, G., Lüpkes, C., Hendricks, S., & Gerdes, R. (2014). Variability of Arctic Sea-ice topography and its impact on the atmospheric surface drag. *Journal of Geophysical Research, Oceans, 119*, 6743–6762. https://doi.org/10.1002/2013JC009712.

Charney, J. G., & Drazin, P. G. (1961). Propagation of planetaryscale disturbances from the lower into the upper atmosphere. *Journal of Geophysical Research, 66*, 83–109.

Chechin, D., & Lüpkes, C. (2017). Boundary-layer development and low-level Baroclinicity during high-latitude cold-air outbreaks: A simple model. *Boundary-Layer Meteorology, 162*(91–116), 1–26. https://doi.org/10.1007/s10546-016-0193-2.

Chechin, D. G., Lüpkes, C., Repina, I. A., & Gryanik, V. M. (2013). Idealized dry quasi 2-D mesoscale simulations of cold-air outbreaks over the marginal sea ice zone with fine and coarse resolution. *Journal of Geophysical Research: Atmospheres, 118*(16), 8787–8813. https://doi.org/10.1002/jgrd.50679.

Chen, F., & von Storch, H. (2013). Trends and variability of North Pacific polar lows. *Advances in Meteorology*. https://doi.org/10.1155/2013/170387.

Claud, C., Heinemann, G., Raustein, E., & McMurdie, L. (2004). Polar low 'le Cygne': Satellite observations and numerical simulations. *Quarterly Journal of the Royal Meteorological Society, 130*, 1075–1102. https://doi.org/10.1256/qj.03.72.

Claud, C., Duchiron, B., & Terray, P. (2007). Associations between large-scale atmospheric circulation and polar low developments over the North Atlantic curing winter. *Journal of Geophysical Research, 112*. https://doi.org/10.1029/2006JD008251.

Cohen, J., Screen, J. A., Furtado, J. C., Barlow, M., Whittleston, D., Coumou, D., Francis, J., Dethloff, K., Entekhabi, D., Overland, J., & Jones, J. (2014). Recent Arctic amplification and extreme mid-latitude weather. *Nature Geoscience, 7*(9), 627.

Condron, A., & Bigg, G. R. (2006). Polar mesoscale cyclones in the Northeast Atlantic: Comparing climatologies from ERA-40 and satellite imagery. *Monthly Weather Review, 134*, 1518–1533.

Condron, A., Bigg, G. R., & Renfrew, I. A. (2008). Modeling the impact of polar mesocyclones on ocean circulation. *Journal of Geophysical Research: Oceans, 113*(C10).

Condron, A., & Renfrew, I. (2012). The impact of polar mesoscale storms on Northeast Atlantic Ocean circulation. *Nature Geoscience*.

Coumou, D., Kornhuber, K., Lehmann, J., & Petoukhov, V. (2017). Weakened flow, persistent circulation, and prolonged weather extremes in boreal summer. *Climate Extremes: Patterns and Mechanisms*, 61–73.

Craig, G. C., & Gray, S. L. (1996). CISK or WISHE as the mechanism for tropical cyclone intensification. *Journal of the Atmospheric Sciences, 53*, 3528–3540.

Cuxart, J., Holtslag, A. A. M., Beare, R., Beljaars, A., Cheng, A., Conangla, L., Ek, M., Freedman, F., Hamdi, R., Kerstein, A., Kitagawa, H., Lenderik, G., Lewellen, D., Mailhot, J., Mauritsen, T., Perov, V., Schayes, G., Steeneveld, G.-J., Svensson, G., Taylor, P., Wunsch, S., Weng, W., & Xu, K.-M. (2006). Single-column intercomparison for a stably stratified atmospheric boundary layer. *Boundary-Layer Meteorology, 118*, 273–303.

Davini, P., Cagnazzo, C., Neale, R., & Tribbia, J. (2012). Coupling between Greenland blocking and the North Atlantic oscillation pattern. *Geophysical Research Letters, 39*, L14701. https://doi.org/10.1029/2012GL052315.

Davis, C. A., & Emanuel, K. A. (1991). Potential corticity diagnostics of cyclogenesis. *Monthly Weather Review, 119*, 1929–1953. https://doi.org/10.1175/1520-0493(1991)119<1929:PVDOC>2.0.CO;2.

de Boer, G., Shupe, M. D., Caldwell, P. M., et al. (2013). Near-surface meteorology during the Arctic summer Cloud Ocean study (ASCOS): Evaluation of reanalyses and global climate models. *Atmospheric Chemistry and Physics, 14*, 427–445.

Dee, D. P., & Coauthors. (2011). The ERA-Interim reanalysis: configuration and performance of the data assimilation system. *Quarterly Journal of the Royal Meteorological Society, 137*, 553–597. https://doi.org/10.1002/qj.828.

Derbyshire, S. (1999). Boundary-layer decoupling over cold surfaces as a physical boundary-instability. *Boundary-Layer Meteorology, 90*, 297–325. https://doi.org/10.1023/A:1001710014316.

Deser, C., Sun, L., Tomas, R. A., & Screen, J. (2016). Does ocean coupling matter for the northern extratropical response to projected Arctic Sea ice loss? *Geophysical Research Letters, 43*(5), 2149–2157.

Deveson, A. C. L., Browning, K. A., & Hewson, T. D. (2002). A classification of FASTEX cyclones using height-attributable quasi-geostrophic vertical motion diagnostic. *Quarterly Journal of the Royal Meteorological Society, 128*, 93–117.

Douglas, M. W., Shapiro, M., Fedor, L., & Saukkonen, L. (1995). Research aircraft observatons of a polar low at the East Greenland ice edge. *Monthly Weather Review, 123*, 5–15.

Doyle, J. D., & Shapiro, M. A. (1999). Flow response to large-scale topography: The Greenland tip jet. *Tellus A: Dynamic Meteorology and Oceanography, 51*, 728–748. https://doi.org/10.3402/tellusa.v51i5.14471.

Doyle, J. D., Shapiro, M. A., Jiang, Q., Bartels, D. L., Doyle, J. D., Shapiro, M. A., Jiang, Q., & Bartels, D. L. (2005). Large-Amplitude Mountain wave breaking over Greenland. *Journal of the Atmospheric Sciences, 62*, 3106–3126. https://doi.org/10.1175/JAS3528.1.

Duncan, C. (1978). Baroclinic instability in a reversed shear flow. *Meteorite Magazine, 107*, 17–23.

DuVivier, A. K., Cassano, J. J., Greco, S., Emmitt, G. D., DuVivier, A. K., Cassano, J. J., Greco, S., & Emmitt, G. D. (2017). A case study of observed and modeled barrier flow in the Denmark Strait in may 2015. *Monthly Weather Review, 145*, 2385–2404. https://doi.org/10.1175/MWR-D-16-0386.1.

Emanuel, K. A., & Rotunno, R. (1989). Polar lows as Arctic hurricanes. *Tellus, 41A*, 1–17.

Edmon, H. J., Hoskins, B. J., & McIntyre, M. E. (1980). Eliassen–Palm cross-sections for the troposphere. *Journal of the Atmospheric Sciences, 37*, 2600–2616. Corrigendum, 38, 1115.

Ese, T., Kanestrom, I., & Pedersen, K. (1988). Climotology of polar lows over the Norwegian and Barents seas. *Tellus, 40A*, 248–255.

Fang, Z. F. (2004). Statistical relationship between the northern hemisphere sea ice and atmospheric circulation during wintertime. In *Observation, theory and modeling of atmospheric variability: Selected Papers of Nanjing Institute of Meteorology Alumni in Commemoration of Professor Jijia Zhang* (pp. 131–141).

Foldvik, A. (1962). Two-dimensional mountain waves – A method for the rapid computation of lee wavelengths and vertical velocities. *Quarterly Journal of the Royal Meteorological Society, 88*, 271–285. https://doi.org/10.1002/qj.49708837707.

Forbes, G. S., & Lottes, W. D. (1985). Classification of mesoscale vortices in polar airstreams and the influence of the large-scale environment on their evolutions. *Tellus, 37A*, 132–155.

Francis, J., & Skific, N. (2015). Evidence linking rapid Arctic warming to mid-latitude weather patterns. *Philosophical Transactions of the Royal Society A: Mathematical, Physical and Engineering Sciences, 373*(2045), 20140170.

Francis, J. A., & Vavrus, S. J. (2015). Evidence for a wavier jet stream in response to rapid Arctic warming. *Environmental Research Letters, 10*(1), 014005.

Føre, I., Kristjansson, J. E., Kolstad, E. W., Bracegirdle, T. J., Saetre, Ø., & Røsting, B. (2012). A 'hurricane-like' polar low fuelled by sensible heat flux: High-resolution numerical simulations. *Quarterly Journal of the Royal Meteorological Society*. https://doi.org/10.1002/qj.1876.

Føre, I., Kristjansson, J. E., Saetre, Ø., Breivik, Ø., Røsting, B., & Shapiro, M. (2011). The full life cycle of a polar low over the Norwegian sea observed by three research aircraft flights. *Quarterly Journal of the Royal Meteorological Society, 137*, 1659–1673.

Furtado, J. C., Cohen, J. L., & Tziperman, E. (2016). The combined influences of autumnal snow and sea ice on northern hemisphere winters. *Geophysical Research Letters, 43*, 3478–3485.

Garfinkel, C. I., Hartmann, D. L., & Sassi, F. (2010). Tropospheric precursors of anomalous northern hemisphere stratospheric polar vortices. *Journal of Climate, 23*, 3282–3299. https://doi.org/10.1175/2010JCLI3010.1.

Garfinkel, C. I., Feldstein, S. B., Waugh, D. W., Yoo, C., & Lee, S. (2012a). Observed Connection between Stratospheric Sudden Warmings and the Madden-Julian Oscillation. *Geophysical Research Letters, 39*. https://doi.org/10.1029/2012GL053144.

Garfinkel, C. I., Shaw, T. A., Hartmann, D. L., & Waugh, D. W. (2012b). Does the Holton-Tan Mechanism Explain How the Quasi-Biennial Oscillation Modulates the Arctic Polar Vortex?*Journal of the Atmospheric Sciences, 69*. https://doi.org/10.1175/JAS-D-11-0209.1.

Garratt, J. R. (1992). *The atmospheric boundary layer*. Cambridge University Press. 316 p.

Gill, A. E. (1982). *Atmosphere-ocean dynamics*. Academic Press. 662 pp.

Gong, D. Y., & Ho, C.-H. (2002). The Siberian high and climate change over middle to high latitude Asia. *Theoretical and Applied Climatology, 72*, 1–9.

Gong, T., & Luo, D. (2017). Ural blocking as an amplifier of the Arctic Sea ice decline in winter. *Journal of Climate, 30*, 2639–2654.

Grachev, A. A., Andreas, E. L., Fairall, C. W., Guest, P. S., & Persson, P. O. G. (2007). SHEBA flux-profile relationships in the stable atmospheric boundary layer. *Boundary-Layer Meteorology, 124*, 315–333.

Grachev, A. A., Persson, P. O. G., Uttal, T., Akish, E. A., Cox, C. J., Morris, S. M., Fairall, C. W., Stone, R. S., Lesins, G., Makshtas, A. P., & Repina, I. A. (2017). Seasonal and latitudinal variations of surface fluxes at two Arctic terrestrial sites. *Climate Dynamics*.https://doi.org/10.1007/s00382-017-3983-4.

Gray, L. J., et al. (2010). Solar influences on climate. *Reviews of Geophysics, 48*, RG4001. https://doi.org/10.1029/2009RG000282.

Grønås, S., & Kvamstø, N. G. (1995). Numerical simulations of the synoptic conditions and development of Arctic outbreak polar lows. *Tellus, 47A*, 797–814.

Gryanik, V. and Lüpkes, C. (2018) A new bulk parametrization of turbulent fluxes in the stably stratified surface layer over sea ice for climate models, European Geosciences Union General Assembly 2018, Vienna, 9–13 April 2018.

Hall, R., Erdelyi, R., Hanna, E., Jones, J. M., & Scaife, A. A. (2015). Drivers of North Atlantic polar front jet stream variability. *International Journal of Climatology, 35*(8), 1697–1720.

Hanna, E., Cropper, T. E., Hall, R. J., & Cappelen, J. (2016). Greenland blocking index 1851–2015: A regional climate change signal. *International Journal of Climatology, 36*(15), 4847–4861.

Hanna, E., Hall, R. J., Cropper, T. E., Ballinger, T. J., Wake, L., Mote, T., & Cappelen, J. (2018). Greenland blocking index daily series 1851-2015: Analysis of changes in extremes and links with North Atlantic and UK climate variability and change. *International Journal of Climatology, 38*, 3546–3564.

Hanna, E., Jones, J. M., Cappelen, J., Mernild, S. H., Wood, L., Steffen, K., & Huybrechts, P. (2013). The influence of North Atlantic atmospheric and oceanic forcing effects on 1900–2010 Greenland summer climate and ice melt/runoff. *International Journal of Climatology, 38*, 862–880.

Hanley, D., & Richards W. G. (1991) Polar lows in Atlantic Canadian waters 1977–1989. Report: MAES 2–91. Scientific Services Division, Atlantic Region, Atmospheric Environment Service.

Hansen-Bauer, I., & Gjessing, Y. T. (1988). Observations and model calculations of aerodynamic drag on sea ice in the Fram Strait. *Tellus, Service. A, 40*, 151–161.

Harden, B. E., & Renfrew, I. A. (2012). On the spatial distribution of high winds off Southeast Greenland. *Geophysical Research Letters, 39*. https://doi.org/10.1029/2012GL052245.

Harden, B. E., Renfrew, I. A., & Petersen, G. N. (2011). A climatology of wintertime barrier winds off Southeast Greenland. *Journal of Climate, 24*(17), 4701–4717. https://doi.org/10.1175/2011JCLI4113.1.

Harold, J. M., Bigg, G. R., & Turner, J. (1999). Mesocyclone activity over the North-East Atlantic. Part I: Vortex distribution and variability. *International Journal of Climatology, 19*, 1187–1204.

Haynes, P. H., McIntyre, M. E., Shepherd, T. G., Marks, C. J., & Shine, K. P. (1991). On the "downward control" of extratropical diabatic circulations by eddy-induced mean zonal forces. *Journal of the Atmospheric Sciences, 48*, 651–680.

Hirschberg, P. A., & Fritsch, J. M. (1991). Tropopause undulations and the development of Extratropical cyclones. Part II: Diagnostic analysis and conceptual model. *Monthly Weather Review, 119*, 518–550.

Hoffman, R. N., Leidner, S. M., Hoffman, R. N., & Leidner, S. M. (2005). An introduction to the near–real–time QuikSCAT data. *Weather and Forecasting, 20*, 476–493. https://doi.org/10.1175/WAF841.1.

Holton, J. R. (2004). *An introduction to dynamic meteorology*. Elsevier Academic Press. pp 517.

Holton, J. R., & Mass, C. (1976). Stratospheric vacillation cycles. *Journal of the Atmospheric Sciences, 33*, 2218–2225.

Holton, J. R., & Tan, H. C. (1982). The quasi-biennial oscillation in the northern hemisphere lower stratosphere. *Journal of the Meteorological Society of Japan. Ser. II, 60*(1), 140–148.

Hoskins, B. J., Mcintyre, M. E., & Robertson, A. W. (1985). On the use and significance of isentropic potential Vorticity maps. *Quarterly Journal of the Royal Meteorological Society, 111*, 877–946. https://doi.org/10.1002/qj.49711147002.

Hoskins B., & Berrisford P. (1988). A potential vorticity perspective of the storm of 15–16 October 1987, Weather, doi: https://doi.org/10.1002/j.1477-8696.1988.tb03890.

Inoue, J., & Hori, M. E. (2011). Arctic cyclogenesis at the marginal ice zone: A contributory mechanism for the temperature amplification? *Geophysical Research Letters, 38*, L12502. https://doi.org/10.11029/12011gl047696.

Jaiser, R., Nakamura, T., Handorf, D., Dethloff, K., Ukita, J., & Yamazaki, K. (2016). Atmospheric winter response to Arctic Sea ice changes in reanalysis data and model simulations. *Journal of Geophysical Research: Atmospheres, 121*(13), 7564–7577.

Jakobson, E., Vihma, T., Palo, T., Jakobson, L., Keernik, H., & Jaagus, J. (2012). Validation of atmospheric reanalyzes over the Central Arctic Ocean. *Geophysical Research Letters, 39*, L10802. https://doi.org/10.1029/2012GL051591.

Jakobson, L., Vihma, T., Jakobson, E., Palo, T., Männik, A., & Jaagus, J. (2013). Low-level jet characteristics over the Arctic Ocean in spring and summer. *Atmospheric Chemistry and Physics, 13*, 11089–11099.

Joly, A., & Thorpe, A. J. (1989). Warm and occluded fronts in twodimensional moist baroclinic instability. *Quarterly Journal of the Royal Meteorological Society, 115*, 513–534.

Kahl, J. D., Serreze, M. C., & Schnell, R. C. (1992). Tropospheric low-level temperature inversions in the Canadian Arctic. *Atmosphere-Ocean, 30*(4), 511–529. https://doi.org/10.1080/07055900.1992.9649453.

Kaimal, J. C. & Finnigan, J. J., (1994). Atmospheric boundary layer flows: Their structure and measurement. Oxford University Press, 289 pp. =y#v=onepage&q=Kaimal and Finnigan%2C 1994%3B&f=false. Accessed 22 Feb 2018.

Kim, B.-M., Son, S.-W., Min, S.-K., Jeong, J.-H., Kim, S.-J., Zhang, X., Shim, T., & Yoon, J.-H. (2014). Weakening of the stratospheric polar vortex by Arctic Sea-ice loss. *Nature Communications, 5*, 4646. https://doi.org/10.1038/ncomms5646.

Kim, B.-M., Hong, J.-Y., Jun, S.-Y., Zhang, X., Kwon, H., Kim, S.-J., Kim, J.-H., Kim, S.-W., & Kim, H.-K. (2017). Major cause of unprecedented Arctic warming in January 2016: Critical role of an Atlantic windstorm. *Scientific Reports, 7*, 40051. https://doi.org/10.1038/srep40051.

Kolstad, E. W. (2006). A new climatology of favourable conditions for reverse-shear polar lows. *Tellus, 58A*, 344–354.

Kolstad, E. W., & Bracegirdle, T. J. (2008). Marine cold-air outbreaks in the future: An assessment of IPCC AR4 model results for the northern hemisphere. *Climate Dynamics, 30*, 871–885.

Kolstad, E. W., Bracegirdle, T. J., & Zahn, M. (2016). Re-examining the roles of surface flux and latent heat release in a 'hurricane-like' polar low over the barents sea. *Journal of Geophysical Research – Atmospheres, 121*, 7853–7867. https://doi.org/10.1002/2015JD024633.

Kolstad, E. W., & Bracegirdle, T. J. (2017). Sensitivity of an apparently hurricane-like polar low to sea-surface temperature. *Quarterly Journal of the Royal Meteorological Society, 143*, 966–973.

Kristiansen, J., Sørland, S. L., Iversen, T., Bjørge, D., & Køltzow, M. Ø. (2011). High-resolution ensemble prediction of a polar low development. *Tellus A: Dynamic Meteorology and Oceanography, 63*(3), 585–604.

Kristjánsson, J. E., Thorsteinsson, S., & Røsting, B. (2009). Phase-locking of a rapidly developing extratropical cyclone by Greenland's orography. *Quarterly Journal of the Royal Meteorological Society, 135*, 1986–1998. https://doi.org/10.1002/qj.497.

Kristjansson, J., & Coauthors. (2011). The Norwegian IPY-THORPEX. Polar lows and Arctic fronts during the 2008 Andøya campaign. *Bulletin of the American Meteorological Society, 92*, 1443–1446.

Lackmann, G. M. (2002). Cold-frontal potential vorticity maxima, the low-level jet, and moisture transport in extratropical cyclones. *Monthly Weather Review, 130*, 59–74. https://doi.org/10.1175/1520-0493(2002)130<0059:CFPVMT>2.0.CO;2.

Liberato, M. L. R. (2014). The 19 January 2013 windstorm over the North Atlantic: Large-scale dynamics and impacts on Iberia. *Weather and Climate Extremes, 5-6*, 16–28. https://doi.org/10.1016/j.wace.2014.06.002.

Linders, T., & Saetra, Ø. (2010). Can CAPE maintain polar lows? *Journal of the Atmospheric Sciences, 67*(8), 2559–2571.

Loescher, K. A., Young, G. S., Colle, B. A., Winstead, N. S., Loescher, K. A., Young, G. S., Colle, B. A., & Winstead, N. S. (2006). Climatology of barrier jets along the Alaskan coast. Part I: Spatial and temporal distributions. *Monthly Weather Review, 134*, 437–453. https://doi.org/10.1175/MWR3037.1.

Louis, J. F. (1979). A parametric model of vertical eddy fluxes in the atmosphere. *Boundary-Layer Meteorology, 17*, 187–202, 1979.

Luo, D., Xiao, Y., Yao, Y., Dai, A., Simmonds, I., & Franzke, C. (2016a). Impact of Ural blockingon winter warm Arctic–cold Eurasian anomalies. Part I: Blocking-induced amplification. *Journal of Climate, 29*, 3925–3947. https://doi.org/10.1175/JCLI-D-15-0611.1.

Luo, D., Xiao, Y., Diao, Y., Dai, A., Franzke, C., & Simmonds, I. (2016b). Impact of Ural blocking on winter warm Arctic–cold Eurasian anomalies. Part II: The link to the North Atlantic oscillation. *Journal of Climate, 29*, 3949–3971. https://doi.org/10.1175/JCLI-D-15-0612.1.

Lüpkes, C., & Schlünzen, K. H. (1996). Modelling the arctic convective boundary-layer with different turbulence parameterisations. *Boundary-Layer Meteorology, 79*, 107–130.

Lüpkes, C., & Gryanik, V. M. (2015). A stability-dependent parametrization of transfer coefficients for momentum and heat over polar sea ice to be used in climate models. *Journal of Geophysical Research, 120*, 1–30.

Lüpkes, C., Vihma, T., Jakobson, E., König-Langlo, G., & Tetzlaff, A. (2010). Meteorological observations from ship cruises during summer to the Central Arctic: A comparison with reanalysis data. *Geophysical Research Letters, 37*, L09810. https://doi.org/10.1029/2010GL042724.

Lüpkes, C., Vihma, T., Birnbaum, G., et al. (2012). Mesoscale modelling of the Arctic atmospheric boundary layer and its interaction with sea ice. In P. Lemke & H.-W. Jacobi (Eds.), *Arctic climate change – The ACSYS decade and beyond* (Atmospheric and Oceanographic Sciences Library) (Vol. 43). https://doi.org/10.1007/978–94–007–2027–5.

Lydolph, P. E. (1977). *Climates of the Soviet Union: World survey of climatology* (Vol. 7). Amsterdam: Elsevier Science Publishing Company. 443 pp.

Mallet, P.-E., Claud, C., Cassou, C., Noer, G., & Kodera, K. (2013). Polar lows over the Nordic and Labrador seas: Synoptic circulation patterns and associations with North Atlantic-Europe wintertime weather regimes. *Journal of Geophysical Research, 118*, 2455–2472.

Manzini, E., & Coauthors. (2014). Northern winter climate change: Assessment of uncertainty in CMIP5 projections related to stratosphere–troposphere coupling. *Journal of Geophysical Research: Atmospheres, 119*, 7979–7998. https://doi.org/10.1002/2013JD021403.

Martius, O., Polvani, L. M., & Davies, H. C. (2009). Blocking precursors to stratospheric sudden warming events. *Geophysical Research Letters, 36*, L14806. https://doi.org/10.1029/2009GL038776.

Mass, C. F., Businger, S., Albright, M. D., Tucker, Z. A., Mass, C. F., Businger, S., Albright, M. D., & Tucker, Z. A. (1995). A windstorm in the Lee of a gap in a Coastal Mountain barrier. *Monthly Weather Review, 123*, 315–331. https://doi.org/10.1175/1520-0493(1995)123<0315:AWITLO>2.0.CO;2.

McCusker, K. E., Fyfe, J. C., & Sigmond, M. (2016). Twenty-five winters of unexpected Eurasian cooling unlikely due to Arctic Sea ice loss. *Nature Geoscience*, 838–842. https://doi.org/10.1038/ngeo2820.

McInnes, H., Kristjánsson, J. E., Schyberg, H., & Røsting, B. (2009). An assessment of a Greenland lee cyclone during the Greenland flow distortion experiment: An observational approach. *Quarterly Journal of the Royal Meteorological Society, 135*, 1968–1985. https://doi.org/10.1002/qj.524.

Michel, C., Terpstra, A., & Spengler, T. (2018). Polar mesoscale cyclone climatology for the nordic seas based on era-interim. *Journal of Climate, 31*, 2511–2532. https://doi.org/10.1175/JCLI-D-16-0890.1.

Mokhov, I. I., Akperov, M. G., Lagun, V. E., & Lutsenko, E. I. (2007). Intense arctic mesocyclones. *Izvestiya, Atmospheric and Oceanic Physics, 43*, 3.

Monin, A. S., & Obukhov, A. M. (1954). Basic laws of turbulent mixing in the surface layer of the atmosphere. *Trudy Geofiz. Instituta Akademii Nauk SSSR, 24*, 163–187.

Montgomery, M. T., & Farrell, B. F. (1992). Polar low dynamics. *Journal of the Atmospheric Sciences, 49*, 2484–2505.

Moore, G. W. K. (2003). Gale force winds over the Irminger Sea to the east of Cape farewell, Greenland. *Geophysical Research Letters, 30*. https://doi.org/10.1029/2003GL018012.

Moore, G. W. K. (2013). The Novaya Zemlya bora and its impact on Barents Sea air-sea interaction. *Geophysical Research Letters, 40*, 3462–3467. https://doi.org/10.1002/grl.50641.

Moore, R. W., & Montgomery, M. T. (2005). Analysis of an idealized, three-dimensional diabatic Rossby vortex: A coherent structure of the moist baroclinic atmosphere. *Journal of the Atmospheric Sciences, 62*, 2703–2725.

Moore, G. W. K., & Renfrew, I. A. (2005). Tip jets and barrier winds: A QuikSCAT climatology of high wind speed events around Greenland. *Journal of Climate, 18*(18), 3713–3725.

Moore, G. W. K., Renfrew, I. A., Moore, G. W. K., & Renfrew, I. A. (2005). Tip jets and barrier winds: A QuikSCAT climatology of high wind speed events around Greenland. *Journal of Climate, 18*, 3713–3725. https://doi.org/10.1175/JCLI3455.1.

Moore, G. W. K., Pickart, R. S., & Renfrew, I. A. (2008). Buoy observations from the windiest location in the world ocean, Cape farewell, Greenland. *Geophysical Research Letters, 35*, L18802. https://doi.org/10.1029/2008GL034845.

Moore, G. W. K., Bromwich, D. H., Wilson, A. B., Renfrew, I., & Bai, L. (2016). Arctic system reanalysis improvements in topographically forced winds near Greenland. *Quarterly Journal of the Royal Meteorological Society, 142*, 2033–2045. https://doi.org/10.1002/qj.2798.

Müller, M., Batrak, Y., Kristiansen, J., Køltzow, M. A. Ø., & Noer, G. (2017). Characteristics of a convection-scale weather forecasting system for the European Arctic. *Monthly Weather Review, 145*, 4771–4787.

Naakka, T., Nygård, T., & Vihma, T. (2018). Arctic humidity inversions: Climatology and processes. *Journal of Climate, 31*(10), 3765–3787.

Nagata, M. (1987). On the structure of a convergent cloud band over the Japan Sea in winter; a prediction experiments. *Journal of the Meteorological Society of Japan, 65*, 871–883.

Nagata, M. (1992). Modeling case study of the Japan Sea convergent cloud band in a varying large-scale environments; evolution and upscale effect. *Journal of the Meteorological Society of Japan, 70*, 649–671.

Nakamura, T., Yamazaki, K., Iwamoto, K., Honda, M., Miyoshi, Y., Ogawa, Y., & Ukita, J. (2015). A negative phase shift of the winter AO/NAO due to the recent Arctic Sea-ice reduction in late autumn. *Journal of Geophysical Research: Atmospheres, 120*(8), 3209–3227.

Noer, G., Saetre, Ø., Lien, T., & Gusdal, Y. (2011). A climatological study of polar lows in the Nordic seas. *Quarterly Journal of the Royal Meteorological Society*. https://doi.org/10.1002/qj.846.

Nordeng, T. E. (1987). The effect of vertical and slantwise convection on the simulation of polar lows. *Tellus, 39A*, 354–375.

Obukhov, A. M. (1946). Turbulence in an atmosphere with inhomogeneous temperature. *The Institute of Theoretical Geophysics of the Academy of Sciences of the USSR, 1*, 95–115.

Oltmanns, M., Straneo, F., Moore, G. W. K., Mernild, S. H., Oltmanns, M., Straneo, F., Moore, G. W. K., & Mernild, S. H. (2014). Strong downslope wind events in Ammassalik, Southeast Greenland. *Journal of Climate, 27*, 977–993. https://doi.org/10.1175/JCLI-D-13-00067.1.

Omrani, N. E., Keenlyside, N. S., Bader, J. R., & Manzini, E. (2014). Stratosphere key for wintertime atmospheric response to warm Atlantic decadal conditions. *Climate Dynamics, 42*, 649–663.

Outten, S. D., Renfrew, I. A., & Petersen, G. N. (2009). An easterly tip jet off Cape farewell, Greenland. II: Simulations and dynamics. *Quarterly Journal of the Royal Meteorological Society, 135*, 1934–1949. https://doi.org/10.1002/qj.531.

Overland, J. E., & Wang, M. (2005). The Arctic climate paradox: The recent decrease of the Arctic oscillation. *Geophysical Research Letters, 32*, L06701. https://doi.org/10.1029/2004GL021752.

Overland, J. E., Adams, J. M., & Bond, N. A. (1999). Decadal variability of the Aleutian low and its relation to high-latitude circulation. *Journal of Climate, 12*(5), 1542–1548.

Overland, J., Francis, J., Hall, R., Hanna, E., Kim, S.-J., & Vihma, T. (2015). The melting Arctic and mid-latitude weather patterns: Are they connected? *Journal of Climate, 28*, 7917–7932.

Papritz, L., & Pfahl, S. (2016). Importance of latent heating in mesocyclones for the decay of cold air outbreaks: A numerical process study from the Pacific sector of the Southern Ocean. *Monthly Weather Review, 144*, 315–336. https://doi.org/10.1175/MWR-D-15-0268.1.

Papritz, L., & Spengler, T. (2017). A lagrangian climatology of wintertime cold air outbreaks in the irminger and Nordic seas and their role in shaping air-sea heat fluxes. *Journal of Climate, 30*, 2717–2737. https://doi.org/10.1175/JCLI-D-16-0605.1.

Parish, T. R. (1983). The influence of the Antarctic peninsula on the wind field over the western Weddell Sea. *Journal of Geophysical Research, 88*, 2684. https://doi.org/10.1029/JC088iC04p02684.

Parker, D. J., & Thorpe, A. J. (1995). Conditional convective heating in a baroclinic atmosphere: A model of convective frontogenesis. *Journal of the Atmospheric Sciences, 52*(10), 1699–1711.

Peltier, W. R., Clark, T. L., Peltier, W. R., & Clark, T. L. (1979). The evolution and stability of finite-Amplitude Mountain waves. Part II: Surface Wave Drag and Severe Downslope Windstorms. *Journal of the Atmospheric Sciences, 36*, 1498–1529. https://doi.org/10.1175/1520-0469(1979)036<1498:TEASOF>2.0.CO;2.

Persson, P. O. G., & Vihma, T. (2017). Chapter 6: The atmosphere over sea ice. In D. N. Thomas (Ed.), *Sea ice* (3rd ed., pp. 160–196). London: Wiley-Blackwell.

Petersen, G. N., Renfrew, I. A., & Moore, G. W. K. (2009). An overview of barrier winds off southeastern Greenland during the Greenland flow distortion experiment. *Quarterly Journal of the Royal Meteorological Society, 135*, 1950–1967. https://doi.org/10.1002/qj.455.

Petterssen, S., & Smebye, S. (1971). On the development of extratropical cyclones. *Quarterly Journal of the Royal Meteorological Society, 97*, 457–482.

Pickart, R. S., Spall, M. A., Ribergaard, M. H., Moore, G. W. K., & Milliff, R. F. (2003). Deep convection in the Irminger Sea forced by the Greenland tip jet. *Nature, 424*, 152.

Pierrehumbert, R. T., Wyman, B., Pierrehumbert, R. T., & Wyman, B. (1985). Upstream effects of Mesoscale Mountains. *Journal of the Atmospheric Sciences, 42*, 977–1003. https://doi.org/10.1175/1520-0469(1985)042<0977:UEOMM>2.0.CO;2.

Queney, P. (1948). The problem of air flow over mountains: A summary of theoretical studies. *Bulletin of the American Meteorological Society, 29*, 16–26. https://doi.org/10.2307/26257649.

Rajewicz, J., & Marshall, S. J. (2014). Variability and trends in anticyclonic circulation over the Greenland ice sheet, 1948–2013. *Geophysical Research Letters, 41*(8), 2842–2850.

Rasmussen, E. (1979). The polar low as an extratropical CISK disturbance. *Quarterly Journal of the Royal Meteorological Society, 105*, 531–549.

Reed, R. J., Grell, G. A., & Kuo, Y.-H. (1993). The ERICA IOP 5 storm. Part II: Sensitivity tests and further diagnosis based on model output. *Monthly Weather Review, 121*, 1595–1612.

Reeve, M. A., & Kolstad, E. W. (2011). The Spitsbergen south Cape tip jet. *Quarterly Journal of the Royal Meteorological Society, 137*, 1739–1748. https://doi.org/10.1002/qj.876.

Renfrew, I. A., & Coauthors. (2008). The Greenland flow distortion experiment. *Bulletin of the American Meteorological Society, 89*, 1307–1324. https://doi.org/10.1175/2008BAMS2508.1.

Renfrew, I. A., Outten, S. D., & Moore, G. W. K. (2009). An easterly tip jet off Cape farewell, Greenland. I: Aircraft observations. *Quarterly Journal of the Royal Meteorological Society, 135*, 1919–1933. https://doi.org/10.1002/qj.513.

Rinke, A., Maturilli, M., Graham, R. M., Matthes, H., Handorf, D., Cohen, L., Hudson, S. R., & Moore, J. C. (2017). Extreme cyclone events in the Arctic: Wintertime variability and trends. *Environmental Research Letters, 12*, 094006.

Sampe, T., Xie, S.-P., Sampe, T., & Xie, S.-P. (2007). Mapping High Sea winds from space: A global climatology. *Bulletin of the American Meteorological Society, 88*, 1965–1978. https://doi.org/10.1175/BAMS-88-12-1965.

Sandvik, A. D., Furevik, B. R., Sandvik, A. D., & Furevik, B. R. (2002). Case study of a coastal jet at Spitsbergen—Comparison of SAR- and model-estimated wind. *Monthly Weather Review, 130*, 1040–1051. https://doi.org/10.1175/1520-0493(2002)130<1040:CSOACJ>2.0.CO;2.

Sandvik, A. D., & Furevik, B. R. (2002). Case study of a coastal jet at Spitsbergen—Comparison of SAR-and model-estimated wind. *Monthly Weather Review, 130*(4), 1040–1051.

Sardie, J. M., & Warner, T. T. (1983). On the mechanism for the development of polar lows. *Journal of the Atmospheric Sciences, 40*, 869–881.

Sawyer, J. S. (1960). Numerical calculation of the displacements of a stratified airstream crossing a ridge of small height. *Quarterly Journal of the Royal Meteorological Society, 86*, 326–345. https://doi.org/10.1002/qj.49708636905.

Scaife, A. A., Karpechko, A. Y., Baldwin, M. P., Brookshaw, A., Butler, A. H., Eade, R., Gordon, M., MacLachlan, C., Martin, N., Dunstone, N., & Smith, D. (2016). Seasonal winter forecasts and the stratosphere. *Atmospheric Science Letters, 17*(1), 51–56.

Schär, C., & Davies, H. C. (1990). An instability of mature cold fronts. *Journal of the Atmospheric Sciences, 47*, 929–950.

Schär, C. (1993). A generalization of Bernoulli's theorem. *Journal of the Atmospheric Sciences, 50*, 1437–1443. https://doi.org/10.1175/1520-0469(1993)050<1437:AGOBT>2.0.CO;2.

Scherrer, S., Croci-Maspoli, M., Schwierz, C., & Appenzeller, C. (2006). Two-dimensional indices of atmospheric blocking and their statistical relationship with winter climate patterns in the euro-Atlantic region. *International Journal of Climatology, 26*, 233–249.

Schwerdtfeger, W. (1975). The effect of the Antarctic peninsula on the temperature regime of the Weddell Sea. *Monthly Weather Review, 103*, 45–51. https://doi.org/10.1175/1520-0493(1975)103<0045:TEOTAP>2.0.CO;2.

Scorer, R. S. (1949). Theory of waves in the lee of mountains. *Quarterly Journal of the Royal Meteorological Society, 75*, 41–56. https://doi.org/10.1002/qj.49707532308.

Screen, J. A., Deser, C., Simmonds, I., & Tomas, R. (2014). Atmospheric impacts of Arctic Sea-ice loss, 1979–2009: Separating forced change from atmospheric internal variability. *Climate Dynamics, 43*(1–2), 333–344.

Sedlar, J., Shupe, M. D., & Tjernström, M. (2012). On the relationship between thermodynamic structure and cloud top, and its climate significance in the Arctic. *Journal of Climate, 25*(7), 2374–2393.

Sepp, M., & Jaagus, J. (2011). Changes in the activity and tracks of Arctic cyclones. *Climatic Change, 105*, 577–595. https://doi.org/10.1007/s10584-010-9893-7.

Sergeev, D., Renfrew, I., Spengler, T., & Dorling, S. (2016). Structure of a shear-line polar low. *Quarterly Journal of the Royal Meteorological Society*. https://doi.org/10.1002/qj.2911.

Serreze, M. C., & Barrett, A. P. (2008). The summer cyclone maximum over the Central Arctic Ocean. *Journal of Climate, 21*, 1048–1065. https://doi.org/10.1175/2007JCLI1810.1.

Serreze, M. C., Kahl, J. D., & Schnell, R. C. (1992). Low-level temperature inversions of the Eurasian Arctic and comparisons with soviet drifting station data. *Journal of Climate, 5*, 615–629.

Shapiro, M., Fedor, L., & Hampel, T. (1987). Research aircraft observations of a polar low over the Norwegian Sea. *Tellus, 39A*, 272–306.

Shupe, M. D., Persson, P. O. G., Brooks, I. M., Tjernström, M., Sedlar, J., Mauritsen, T., et al. (2013). Cloud and boundary layer interactions over the Arctic Sea ice in late summer. *Atmospheric Chemistry and Physics, 13*(18), 9379–9399.

Sigmond, M., Scinocca, J. F., Kharin, V. V., & Shepherd, T. G. (2013). Enhanced seasonal forecast skill following stratospheric sudden warmings. *Nature Geoscience, 6*, 98–102.

Simmonds, I., & Rudeva, I. (2012). The great Arctic cyclone of August 2012. *Geophysical Research Letters, 39*, L23709. https://doi.org/10.1029/2012GL054259.

Singarayer, J. S., Bamber, J. L., & Valdes, P. J. (2006). Twenty-first-century climate impacts from a declining Arctic Sea ice cover. *Journal of Climate, 19*(7), 1109–1125.

Skeie, P., & Grønås, S. (2000). Strongly stratified easterly flows across Spitsbergen. *Tellus A, 52*, 473–486. https://doi.org/10.1034/j.1600-0870.2000.01075.x.

Smith, R. B. (1979). The influence of mountains on the atmosphere. *Advances in Geophysics, 21*, 87–230. https://doi.org/10.1016/S0065-2687(08)60262-9.2018.

Smith, R. B. (1989). Mountain-induced stagnation points in hydrostatic flow. *Tellus A, 41A*, 270–274. https://doi.org/10.1111/j.1600-0870.1989.tb00381.x.

Smith, R. B., & Grønås, S. (1993). Stagnation points and bifurcation in 3-D mountain airflow. *Tellus A, 45*, 28–43. https://doi.org/10.1034/j.1600-0870.1993.00003.x.

Smith, D. M., Dunstone, N. J., Scaife, A. A., Fiedler, E. K., Copsey, D., & Hardiman, S. C. (2017). Atmospheric response to Arctic and Antarctic Sea ice: The importance of ocean–atmosphere coupling and the background state. *Journal of Climate, 30*(12), 4547–4565.

Smolarkiewicz, P. K., Rotunno, R., Smolarkiewicz, P. K., & Rotunno, R. (1989). Low Froude number flow past three-dimensional obstacles. Part I: Baroclinically Generated Lee Vortices. *Journal of the Atmospheric Sciences, 46*, 1154–1164. https://doi.org/10.1175/1520-0469(1989)046<1154:LFNFPT>2.0.CO;2.

Snyder, C., & Lindzen, R. S. (1991). Quasi-geostrophic wave-CISK in an unbounded baroclinic shear. *Journal of the Atmospheric Sciences, 48*(1), 76–86.

Solomon, A., Shupe, M. D., Persson, P. O. G., & Morrison, H. (2011). Moisture and dynamical interactions maintaining decoupled Arctic mixed-phase stratocumulus in the presence of a humidity inversion. *Atmospheric Chemistry and Physics, 11*(19), 10127–10148.

Stroeve, J., Holland, M. M., Meier, W., Scambos, T., & Serreze, M. (2007). Arctic Sea ice decline: Faster than forecast. *Geophysical Research Letters, 34*, L09501.

Stull, R. B. (1988). *An introduction to boundary layer meteorology*. Kluwer. 666 pp.

Tanaka, H. L., Yamagami, A., & Takahashi, S. (2012). The structure and behavior of the arctic cyclone in summer analyzed by the JRA-25/JCDAS data. *Polar Science, 6*, 44–69. https://doi.org/10.1016/j.polar.2012.03.001.

Tao, W., Zhang, J., & Zhang, X. (2017a). The role of stratosphere vortex downward intrusion in a long-lasting late-summer Arctic storm. *Quarterly Journal of the Royal Meteorological Society, 143*, 1953–1966. https://doi.org/10.1002/qj.3055.

Tao, W., Zhang, J., & Zhang, X. (2017b). Driving roles of tropospheric and stratospheric thermal anomalies in the intensification and persistence of 2012 Arctic Superstorm. *Geophysical Research Letters*. https://doi.org/10.1002/2017GL074778.

Terpstra, A., Michel, C., & Spengler, T. (2016). Forward and reverse shear environments during polar low genesis over the Northeast Atlantic. *Monthly Weather Review, 144*, 1341–1354. https://doi.org/10.1175/MWR-D-15-0314.1.

Terpstra, A., Spengler, T., & Moore, R. W. (2015). Idealised simulations of polar low development in an Arctic moist baroclinic environment. *Quarterly Journal of the Royal Meteorological Society*. https://doi.org/10.1002/qj.2507.

Tetzlaff, A., Lüpkes, C., & Hartmann, J. (2015). Aircraft-based observations of atmospheric boundary-layer modification over Arctic leads. *Quarterly Journal of the Royal Meteorological Society, 141*(692), 2839–2856.

Thorsteinsson, S., & Sigurdsson, S. (1996). Orogenic blocking and deflection of stratified air flow on an f-plane. *Tellus A, 48*, 572–583. https://doi.org/10.1034/j.1600-0870.1996.t01-3-00006.x.

Timmreck, C. (2012). Modeling the climatic effects of large explosive volcanic eruptions. *Wiley Interdisciplinary Reviews: Climate Change, 3*(6), 545–564.
Tjernström, M. (2007). Is there a diurnal cycle in the summer cloudcapped arctic boundary layer? *Journal of the Atmospheric Sciences, 64*, 3970–3986. https://doi.org/10.1175/2007jas2257.1.
Tjernström, M., & Graversen, R. G. (2009). The vertical structure of the lower Arctic troposphere analysed from observations and the ERA-40 reanalysis. *Quarterly Journal of the Royal Meteorological Society, 135*, 431–443. https://doi.org/10.1002/qj.380.
Tjernström, M., Zagar, M., Svensson, G., Cassano, J. J., Pfeifer, S., Rinke, A., Wyser, K., Dethloff, K., Jones, C., Semmler, T., & Shaw, M. (2005). Modelling the Arctic boundary layer: An evaluation of six ARCMIP regional-scale models using data from the SHEBA project. *Boundary-Layer Meteorology, 117*, 337–381.
Tjernström, M., Sedlar, J., & Shupe, M. D. (2008). How well do regional climate models reproduce radiation and clouds in the Arctic? An evaluation of ARCMIP simulations. *Journal of Applied Meteorology and Climatology, 47*, 2405–2422.
Tsamados, M., Feltham, D. L., Schroeder, D., Flocco, D., Farrell, S. L., Kurtz, N., Laxon, S. W., & Bacon, S. (2014). Impact of variable atmospheric and oceanic form drag on simulations of Arctic Sea ice. *Journal of Physical Oceanography, 44*(5), 1329–1353. https://doi.org/10.1175/JPO-D-13-0215.1.
Uttal, T., Curry, J. A., McPhee, M. G., et al. (2002). The surface heat budget of the Arctic. *Bulletin of the American Meteorological Society, 83*, 255–275.
Vallis, G. K., & Gerber, E. P. (2008). Local and hemispheric dynamics of the North Atlantic oscillation, annular patterns and the zonal index. *Dynamics of Atmospheres and Oceans, 44*, 184–212. https://doi.org/10.1016/j.dynatmoce.2007.04.003.
Vavrus, S. (2018). The influence of Arctic amplification on mid-latitude weather and climate. *Current Climate Change Reports*. https://doi.org/10.1007/s40641-018-0105-2.
Vihma, T. (2014). Effects of Arctic Sea ice decline on weather and climate: A review. *Surveys in Geophysics, 35*(5), 1175–1214.
Vihma, T. (2017). Weather extremes linked to interaction of the Arctic and mid-latitudes. In S.-Y. Wang et al. (Eds.), *Climate extremes: Mechanisms and potential prediction* (Geophysical monograph series, 226) (pp. 39–49). American Geophysical Union.
Vihma, T., & Brümmer, B. (2002). Observations and modelling of on-ice and off-ice flows in the northern Baltic Sea. *Boundary-Layer Meteorology, 103*, 1–27.
Vihma, T., & Pirazzini, R. (2005). On the factors controlling the snow surface and 2-m air temperatures over the Arctic Sea ice in winter. *Boundary-Layer Meteorology, 117*, 73–90.
Vihma, T., Hartmann, J., & Lüpkes, C. (2003). A case study of an on-ice air flow over the Arctic marginal sea ice zone. *Boundary-Layer Meteorology, 107*, 189–217, 2003.
Vihma, T., Kilpeläinen, T., Manninen, M., Sjöblom, A., Jakobson, E., Palo, T., Jaagus, J., & Maturilli, M. (2011). Characteristics of temperature and humidity inversions and low-level jets over Svalbard fjords in spring. *Advances in Meteorology, 2011*, 486807., 14 p. https://doi.org/10.1155/2011/486807.
Vihma, T., Tisler, P., & Uotila, P. (2012). Atmospheric forcing on the drift of Arctic Sea ice in 1989–2009. *Geophysical Research Letters, 39*, L02501. https://doi.org/10.1029/2011GL050118.
Walsh, J. E., & Chapman, W. L. (1998). Arctic cloud–radiation–temperature associations in observational data and atmospheric reanalyses. *Journal of Climate, 11*, 3030–3045.
Wang, J., Zhang, J., Watanabe, E., Ikeda, M., Mizobata, K., Walsh, J. E., Bai, X., & Wu, B. (2009). Is the dipole anomaly a major driver to record lows in Arctic summer sea ice extent? *Geophysical Research Letters, 36*, L05706. https://doi.org/10.1029/2008GL036706.
Wagner, J., Gohm, A., Dörnbrack, A., & Schäfler, A. (2011). The mesoscale structure of a polar low: Airborne lidar measurements and simulations. *Quarterly Journal of the Royal Meteorological Society, 137*, 1516–1531.
Wallington, C. E., & Portnall, J. (1958). A numerical study of the wavelength and amplitude of lee waves. *Quarterly Journal of the Royal Meteorological Society, 84*, 38–45. https://doi.org/10.1002/qj.49708435905.

Watanabe, S. I., Niino, H., & Yanase, W. (2016). Climatology of polar mesocyclones over the sea of Japan using a new objective tracking method. *Monthly Weather Review, 144*, 2503–2515. https://doi.org/10.1175/MWR-D-15-0349.1.

Watanabe, S. I. I., Niino, H., & Yanase, W. (2017). Structure and environment of polar mesocyclones over the northeastern part of the sea of Japan. *Monthly Weather Review, 145*(6), 2217–2233.

Wilhelmsen, K. (1985). Climatological study of gale-producing polar lows near Norway. *Tellus A: Dynamic Meteorology and Oceanography, 37*(5), 451–459.

Woo, S.-H., Sung, M.-K., Son, S.-W., & Kug, J.-S. (2015). Connection between weak stratospheric vortex events and the Pacific decadal oscillation. *Climate Dynamics, 45*, 3481–3492.

Wood, N. (2000). Wind flow over complex terrain: A historical perspective and the Prospect for large-Eddy modelling. *Boundary-Layer Meteorol., 96*, 11–32. https://doi.org/10.1023/A:1002017732694.

Woollings, T., & Blackburn, a M. (2012). The NorthAtlantic jet stream under climate change and its relation to the NAO and EA patterns. *Journal of Climate, 25*, 886–902. https://doi.org/10.1175/JCLI-D-11-00087.1.

Wu, B., Wang, J., & Walsh, J. E. (2006). Dipole anomaly in the winter Arctic atmosphere and its association with sea ice motion. *Journal of Climate, 19*, 210–225. https://doi.org/10.1175/JCLI3619.1.

Wu, B., Su, J., & Zhang, R. (2011). Effects of autumn-winter arctic sea ice on winter Siberian high. *Chinese Science Bulletin, 56*, 3220–3228. https://doi.org/10.1007/s11434-011-4696-4.

Yamazaki, A., Inoue, J., Dethloff, K., Maturilli, M., & Konig-Langlo, G. (2015). Impact of radiosonde observations on forecasting summertime Arctic cyclone formation. *Journal of Geophysical Research: Atmospheres, 120*, 3249–3273. https://doi.org/10.1002/2014JD022925.

Yanase, W., & Niino, H. (2007). Dependence of polar low development on baroclinicity and physical processes: An idealized high-resolution numerical experiment. *Journal of the Atmospheric Sciences, 64*, 3044–3067.

Yanase, W., Niino, H., Watanabe, S. I. I., Hodges, K., Zahn, M., Spengler, T., & Gurvich, I. A. (2016). Climatology of polar lows over the sea of Japan using the JRA-55 reanalysis. *Journal of Climate, 29*(2), 419–437.

Yoden, S. (1987). Bifurcation properties of a stratospheric vacillation model. *Journal of the Atmospheric Sciences, 44*, 1723–1733.

Zahn, M., & von Storch, H. (2008). A long-term climatology of North Atlantic polar lows. *Geophysical Research Letters, 35*, L22702. https://doi.org/10.1029/2008GL035769.

Zahn, M., & von Storch, H. (2010). Decreased frequency of North Atlantic polar lows associated with future climate warming. *Nature, 467*. https://doi.org/10.1038/nature09388.

Zappa, G., & Shaffrey, L. (2014). Can polar lows be objectively identified and tracked in the ECMWF operational analysis and the ERAinterim reanalysis? *Monthly Weather Review, 142*, 2596–2608.

Zhang, J., Liu, F., Tao, W., Krieger, J., Shulski, M., & Zhang, X. (2016). Mesoscale climatology and variability of surface winds over the Chukchi-Beaufort coastal areas. *Journal of Climate, 29*, 2721–2739. https://doi.org/10.1175/JCLI-D-15-0436.1.

Zhang, X., Walsh, J. E., Zhang, J., Bhatt, U. S., & Ikeda, M. (2004). Climatology and interannual variability of arctic cyclone activity: 1948-2002. *Journal of Climate, 17*, 2300–2317. https://doi.org/10.1175/1520-0442(2004)017<2300:CAIVOA>2.0.CO;2.

Chapter 2
Thermodynamics of the Arctic Atmosphere

Claudio Tomasi, Boyan H. Petkov, Oxana Drofa, and Mauro Mazzola

Abstract In a such changing environment like the Arctic, improving the understanding of the thermodynamic state and processes of the atmosphere is critical for the development of accurate prediction and climatic models. This is fundamental for example for studies on sea-ice development as well as on cloud formation. Taking into account the above remarks, it is very important to know the pressure, temperature and moisture conditions of the Arctic atmosphere throughout the year and over the whole tropospheric and stratospheric altitude range. Furthermore, knowing these data is necessary to realistically evaluate the radiative effects involving both the short-wave and long-wave radiation fluxes, which regulate the energy balance of the Arctic surface-atmosphere system.

In this contribution, continuous measurements of these parameters by means of radiosonde in the Arctic are reviewed, including correction algorithms, in order to obtain detailed climatologies in terms of seasonal, inter-annual and vertical behaviour.

Keywords Thermodynamic · Temperature · Humidity · Pressure · Radiosonde · Troposphere · Stratosphere · Sea-ice · Clouds

2.1 Introduction

The air pressure, temperature and moisture conditions of the Arctic atmosphere are in general subject to largely vary from winter to summer. Examining a set of 1320 radiosounding (hereinafter referred to more briefly as RS) measurements performed at the Arctic sites of Cambridge Bay, Resolute, Danmarkshavn, Eureka, Alert and Ny-Ålesund from January 2000 to December 2003, Tomasi et al. (2010) showed that air pressure p_o at the surface-level varied between 981 and 1045 hPa during

C. Tomasi · B. H. Petkov · O. Drofa · M. Mazzola (✉)
Institute of Atmospheric Sciences and Climate (ISAC), Italian National Research Council (CNR), Bologna, Italy
e-mail: c.tomasi@isac.cnr.it; b.petkov@isac.cnr.it; o.drofa@isac.cnr.it; m.mazzola@isac.cnr.it

A. Kokhanovsky, C. Tomasi (eds.), *Physics and Chemistry of the Arctic Atmosphere*, Springer Polar Sciences, https://doi.org/10.1007/978-3-030-33566-3_2

the year, while the vertical profile of $p(z)$ was found to exhibit rather large seasonal variations at all the altitudes, since $p(z)$ was found to vary by more than 20% passing from winter to summer at stratospheric altitudes close to 30 km. The air temperature T_o at the surface-level was found to vary from 223 °K to 295 °K on average at the above-mentioned six Arctic sites, while the vertical profile of $T(z)$ presented large seasonal variations at both tropospheric and stratospheric altitudes, which have been estimated to be larger than 30 °K at the 5 km altitude and than 35 °K at the 25 km altitude. For instance, the annual thermal excursions recorded from January to July at Alert (82° 30′ N, 62° 21′ W, 65 m a.m.s.l.) were estimated to be wider than 40 °K at the surface level, and equal to: (i) 20.3 °K at the 5 km level, (ii) 10.5 °K at the 10 km level, (iii) 24.4 °K at the 20 km level, and (iv) 19.8 °K at the 30 km level (Tomasi et al. 2010).

The absorption processes produced by water vapour are known to strongly affect the incoming solar radiation and, at the same time, to cause important attenuation effects on both the infrared radiation emitted upward by the surface-atmosphere system and the infrared radiation emitted by the atmosphere toward the surface. Such absorption processes can vary largely as a function of altitude, since they depend closely on the air pressure, temperature and absolute humidity conditions, all varying with height. Similarly, the Rayleigh-scattering coefficient affecting the incoming solar radiation is mainly related to the air pressure and temperature conditions occurring at the various altitudes, since the Rayleigh-scattering cross section per molecule depends at each height on (i) the refractive index of air, (ii) the molecular number density, and (iii) the King factor for air depolarization, which all vary as a function of height (Tomasi et al. 2005). In particular, the refractive index of air varies as a function of wavelength as well as of altitude, being closely related to the total air pressure, air temperature and partial pressures of the main atmospheric constituents (i.e. of molecular nitrogen, oxygen, argon, water vapour, and carbon dioxide). In addition, the molecular number density $N(z)$ depends closely on total air pressure and the inverse of air temperature, as pointed out by Bodhaine et al. (1999), while the King factor for air depolarization varies as a function of pressure, temperature, water vapour partial pressure, and concentrations of the main anisotropic air molecules, and, hence, varies as a function of altitude.

Taking into account the above remarks, it is very important to know the pressure, thermal and moisture conditions of the Arctic atmosphere throughout the year and over the whole tropospheric and stratospheric altitude range. Knowing these data, it would be feasible to realistically evaluate the radiative effects affecting both the short-wave and long-wave radiation fluxes which regulate the energy balance of the Arctic surface-atmosphere system. An accurate study of the thermodynamic characteristics of the Arctic atmosphere is herein carried out by examining an overall set of 123029 RS measurements collected at 14 Arctic stations having: (i) latitudes ranging from about 67 °N (Sodankylä) to more than 82 °N (Alert), and (ii) longitudes varying over the western Arctic range from 156° 47′ W (Barrow) to 08° 32′ W (Jan Mayen) and over the eastern Arctic range from Ny-Ålesund (11° 53′ E) to Cherskij (161° 17′ E), in average steps of approximately 25° from one station to the other.

2.2 Analysis and Corrections of the Radiosounding Measurements

The RS measurements conducted from January 1, 2001, to December 31, 2015, were collected by us at the following 14 Arctic stations in order to study the thermodynamic characteristics of the Arctic atmosphere:

(1) Barrow (70026 PABR Barrow RS station) (71° 17′ N, 156° 47′ W, 19 m above mean sea level (a.m.s.l.)), the largest city of the North Slope Borough in Alaska, located on the Arctic Ocean, where the VIZ-B2 radiosondes were used from 2001 to 2012, and the Vaisala RS80-H radiosondes were subsequently employed from 2013 to 2015, carrying out the RS measurements of pressure *p(z)*, temperature *T(z)* and relative humidity *RH(z)* as a function of altitude. An overall number of 10721 RS measurements were collected at this site, of which 78% performed with VIZ-B2 radiosondes, and 22% with Vaisala RS80-H radiosondes.

(2) Inuvik (71957 YEV RS station) (68° 19′ N, 133° 32′ W, 103 m a.m.s.l.), placed on the west of the MacKenzie Delta, in the Northwest Territories of Canada, where Vaisala RS80-H, RS90 and RS92 radiosondes were routinely used from 2001 to 2015, to provide an overall set of 9105 RS measurements, of which 41% with RS80-H radiosondes, 13% with RS90 radiosondes, and 46% with RS92 radiosondes, respectively.

(3) Cambridge Bay (71925 YCB RS station) (69° 08′ N, 105° 04′ W, 25 m a.m.s.l.), located on Victoria Island in the Kitikmeot Region of Nunavut (Canada), where Vaisala RS80-H, RS90 and RS92 radiosondes were routinely used from 2001 to 2015, to provide an overall set of 10770 RS measurements, of which nearly 35% with RS80-H radiosondes, less than 3% with RS90 radiosondes, and nearly 63% with the RS92 radiosondes, respectively.

(4) Resolute (71924 YRB Resolute RS station) (74° 42′ N, 94° 58′ W, 46 m a.m.s.l.), located on Cornwallis Island in Nunavut (Canada), where RS80-H and RS92 Vaisala radiosondes were routinely employed from 2001 to 2015, to provide an overall set of 10685 RS measurements, of which 40% with RS80-H radiosondes, and 60% with RS92 radiosondes, respectively.

(5) Eureka (71917 WEU RS station) (79° 59′ N, 85° 56′ W, 10 m a.m.s.l.), a small research base located on the north side of Slidre Fiord, on Fosheim Peninsula, Ellesmere Island (Qikiqtaaluk Region, Nunavut, Canada), where Vaisala RS80-H and RS92 radiosondes were used during the whole period considered in the present study, to provide an overall set of 10758 RS measurements, of which 43% with RS80-H radiosondes, and 57% with RS92 radiosondes, respectively.

(6) Alert (71082 WLT RS station) (82° 30′ N, 62° 21′ W, 65 m a.m.s.l.), located in the Qikiqtaaluk Region of Nunavut, Canada (817 km from the North Pole), where Vaisala RS80-H radiosonde models and Vaisala RS92 radiosondes were regularly employed from January 2001 to December 2015, obtaining an overall set of 10646 RS measurements, of which 38% with RS80-H radiosondes, and 62% with RS92 radiosondes, respectively.

(7) Aasiaat (04220 BGEM RS station) (68° 42′ N, 52° 51′ W, 41 m a.m.s.l.), located in the Qeqertalik municipality in western Greenland, where Vaisala RS80-H, RS90 and RS92 radiosondes were routinely used from January 2001 to December 2009, collecting an overall set of 5955 RS measurements, of which 11% with RS80H radiosondes, 41% with RS90 radiosondes, and 48% with RS92 radiosondes, respectively.

(8) Danmarkshavn (Denmark's Harbour) (04320 BGDH RS station) (76° 46′ N, 18° 40′ W, 12 m a.m.s.l.), a small weather station located in Dove Bay, on the southern shore of the Germania Land Peninsula (Northeast Greenland National Park), where Vaisala RS80-H, RS90 and RS92 radiosonde models were regularly used from 2001 to 2011, to provide an overall set of 6067 RS measurements, of which 23% with RS80-H radiosondes, 34% with RS90 radiosondes, and 43% with RS92 radiosondes, respectively.

(9) Jan Mayen (01001 ENJA RS station) (70° 56′ N, 8° 40′ W, 9 m a.m.s.l.), a Norwegian volcanic island situated in the Arctic Ocean (located about 500 km east of Central Greenland and 1000 km west of the North Cape (Norway)), where Vaisala RS80-H and RS92 radiosonde models were regularly used from 2001 to 2015 to collect an overall set of 10718 RS measurements, of which 34% with RS80-H radiosondes, and 66% with RS92 radiosondes, respectively.

(10) Ny-Ålesund (01004 ENAS RS station) (78° 54′ N; 11° 53′ E; 11 m a.m.s.l.), a research town in Oscar II Land on the Spitsbergen Island (Svalbard, Norway), situated on the Brøgger peninsula and on the shore of the Kongsfjorden Bay (Kings Bay), where Vaisala RS90 and RS92 radiosondes were used from January 2005 to December 2015, obtaining an overall set of 3614 RS measurements, of which only 10% with RS90 radiosondes, and the residual 90% with RS92 radiosondes, respectively.

(11) Sodankylä (02836 EFSO RS station) (67° 22′ N, 26° 39′ E, 179 m a.m.s.l.), located in the region of Lapland (Finland), where Vaisala RS80-H, RS90 and RS92 radiosonde models were regularly used from January 2001 to December 2015, from which an overall set of 10624 RS measurements was collected, of which 7% with RS80-H radiosondes, nearly 26% with RS90 radiosondes, and more than 67% with RS92 radiosondes, respectively.

(12) Ostrov Dikson (20674 radiosounding station) (73° 30′ N, 80° 24′ E, 47 m a.m.s.l.), on the Dikson Island in Taymyrsky Dolgano-Nenetsky District (Krasnoyarsk Krai, Russia), situated in the Kara Sea near the mouth of the Yenisei River. The BAR, MRZ, Marz2-1 and Marz2-2 radiosonde models were mainly employed from 2005 to 2015 at this remote site of western Siberia, collecting an overall set of 7739 RS measurements, of which 51% with Marz2-2 radiosondes (including a few Marz2-1 radiosondes), 28% with BAR radiosondes, and 21% with MRZ radiosondes, respectively.

(13) Tiksi (21824 TK radiosounding station) (71° 35′ N, 128° 55′ E, 7 m a.m.s.l.), an urban locality in Northern Siberia, located on the shore of the Buor-Khaya Gulf of the Laptev Sea, in the Bulunsky District in the Sakha Republic (Russia). The BAR, MRZ, and Marz2-2 radiosonde models were most frequently used from 2001 to 2015 at this site, providing an overall set

of 7763 RS measurements, of which 8% with Marz2-2 radiosondes, 21% with BAR radiosondes, and 71% with MRZ radiosondes, respectively.

(14) Cherskij (25123 HA radiosounding station) (68° 45′ N, 161° 17′ E, 28 m a.m.s.l), an Arctic research station located in the Nizhnekolymsky District of the Sakha Republic (Russia), in North-eastern Siberia. The BAR, MRZ and Marz2-1 radiosonde models were employed at this site of eastern Siberia to collect an overall set of 6961 RS measurements from 2005 to 2015, of which 15% with Marz2-1 radiosondes, 30% with BAR radiosondes, and 55% with MRZ radiosondes, respectively.

The first six stations mentioned above (Barrow, Inuvik, Cambridge Bay, Resolute, Eureka, and Alert) are located in the North American sector, where the analysis of the RS data was made with the main purpose of providing an exhaustive picture of the thermodynamic conditions of the Arctic atmosphere observed over the whole year. Similarly, the data collected at the Aasiaat and Danmarkshavn stations were examined in order to represent the meteorological characteristics of the Arctic atmosphere occurring above Greenland, while the data collected at the Jan Mayen station were examined to represent the meteorological conditions occurring in the middle of the Arctic Ocean and those recorded at Ny-Ålesund were used to characterize the meteorological conditions of the Svalbard Archipelago area. The data collected at the Sodankylä station have been analyzed to define the meteorological features of the northern part of Scandinavia, and the data-sets recorded at the Ostrov Dikson, Tiksi and Cherskij RS stations have been considered for an accurate analysis aimed at describing exhaustively the meteorological characteristics of the Arctic atmosphere over the northern part of Siberia close to the Arctic Ocean, and the longitude range from 80.5 °E to 161.3 °E.

Figure 2.1 shows the geographical positions of the 14 above-selected RS stations. The number of RS measurements conducted at each station by using different radiosonde models in the various years are given in Table 2.1, together with the total number of RS measurements performed at each station. The most important instrumental characteristics of the Barocap, Thermocap, and Humicap sensors mounted on the Vaisala RS80-A, RS80-H, RS90, and RS92 radiosonde models are given in Table 2.2.

The RS measurements performed at the PABR Barrow RS station (Alaska) were taken over the period from 2001 to 2012 by using the VIZ-B2 radiosondes, while the Vaisala RS80-H radiosondes were regularly employed only during the last three years from 2013 to 2015. The VIZ-B2 radiosondes provided measurements of air pressure $p(z)$, temperature $T(z)$ and relative humidity $RH(z)$ every 1.2 s, and reported them every 6 s for archiving. A comparison between the performances of the VIZ-B2 and the Vaisala RS90 radiosondes was made by Mattioli et al. (2007) during the North Slope of Alaska Arctic Winter Radiometric Experiment (NSAAWRE) field campaign conducted at Barrow (Alaska, USA) from March 9 to April 9, 2004. The evaluations of the average day-time and night-time temperature differences $\Delta T(z)$ and the average differences $\Delta RH(z)$ between the performances of the VIZ-B2 and of the Vaisala RS90 radiosonde models are given in Table 2.3 together with their standard deviations, showing that: (1) The day-time difference $\Delta T(z)$ was evaluated

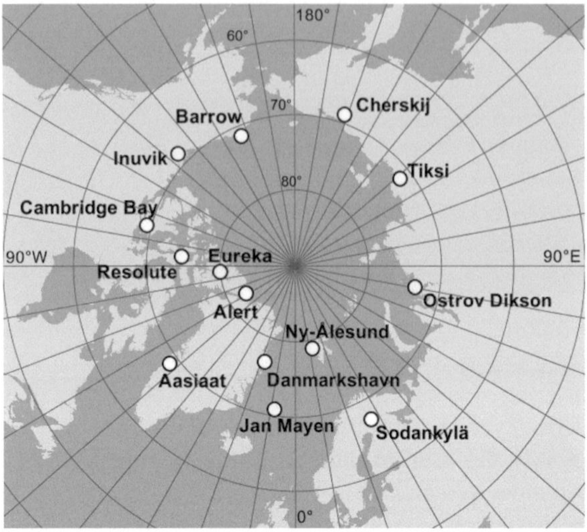

Fig. 2.1 Map of the Arctic radiosounding stations where the RS data were collected and used in the present analysis: (1) Barrow (71° 17′ N, Alaska, USA), (2) Inuvik (68° 19′ N, Northwest Territories of Canada), (3) Cambridge Bay (69° 08′ N, Nunavut, Canada), (4) Resolute (74° 42′ N, Nunavut, Canada), (5) Eureka (79° 59′ N, Nunavut, Canada), (6) Alert (82° 30′ N, Nunavut, Canada), (7) Aasiaat (68° 42′ N, Western Greenland), (8) Danmarkshavn (76° 46′ N, North-eastern Greenland), (9) Jan Mayen (70° 56′ N, Arctic Ocean between Greenland and Norway), (10) Ny-Ålesund (78° 54′ N, Spitsbergen, Svalbard), (11) Sodankylä (67° 22′ N, Northern Finland), (12) Ostrov Dikson (73° 30′ N, Northern Siberia, Russia), (13) Tiksi (71° 35′ N, Northern Siberia, Russia), and (14) Cherskij (68° 45′ N, Northern Siberia, Russia)

to range: (i) between +0.5 (± 1.4) °C at the surface-level and −0.1 (± 0.3) °C at an altitude close to 1.3 km (i.e. within the boundary layer of the Arctic atmosphere); (ii) between +0.0 (± 0.2) °C and + 0.0 (± 0.4) °C at the upper tropospheric levels, until to about 10 km height; and (iii) between +0.0 (± 0.5) °C and + 0.5 (± 1.1) °C at stratospheric altitudes from 10.5 to 26 km. (2) The night-time difference $\Delta T(z)$ was evaluated to range: (i) between −0.2 (± 0.2) °C at the 1.3 km level and +0.0 (± 1.3) °C at the surface-level; (ii) between −0.1 (± 0.3) °C and −0.5 (± 0.5) °C at the upper tropospheric levels, until to about 10 km height; and (iii) between −0.7 (± 0.6) °C and −3.5 (± 0.9) °C at stratospheric altitudes from 11 to 26 km. (3) The difference $\Delta RH(z)$ was evaluated to range: (i) between −0.6 (± 8.8) % at the surface-level and + 6.5 (± 9.0) % at the 0.7 km level; (ii) between +5.0 (± 7.7) % and + 6.3 (± 17.1) % at the upper tropospheric levels, until to about 10 km height; and (iii) between +14.5 (± 9.4) % and + 23.3 (± 10.2) % at the stratospheric altitudes ranging from 11 to 26 km. Considering that the accuracies of the RS90 radiosondes are of ±0.5 hPa for pressure, ± 0.5 °K for temperature, and ± 5% for relative humidity (RH), the comparison made in Table 2.3 indicates that the VIZ-B2 radiosondes are estimated to provide: (i) measurements of $T(z)$ with accuracies better than ±1.0 °K at tropospheric levels, and around ±2.0 °K on average at the low-stratospheric levels; and (ii) measurements of $RH(z)$ with accuracies of ±5% at the tropospheric levels.

Table 2.1 List of the 14 Arctic radiosounding stations considered in the present study, with their geographical coordinates and the total number of radiosoundings collected during various periods from 2001 to 2015 by using different radiosonde models at the various sites to collect the overall number of 123029 radiosounding measurements

Station	Altitude (above mean sea level)	Latitude	Longitude	Measurement period	Radiosonde models	Number of radiosoundings
Barrow	19	71° 17′ N	156° 47′ W	January 2001 – December 2015	VIZ-B2	8375
					RS80-H	2346
					With all radiosonde models	10721
Inuvik	103	68° 19′ N	133° 32′ W	January 2001 – December 2009 and January 2012 – December 2015	RS80-H	3727
					RS90	1216
					RS92	4162
					With all radiosonde models	9105
Cambridge Bay	25	69° 08′ N	105° 04′ W	January 2001 – December 2015	RS80-H	3718
					RS90	301
					RS92	6751
					With all radiosonde models	10770
Resolute	46	74° 42′ N	94° 58′ W	January 2001 – December 2015	RS80-H	4236
					RS92	6449
					With all radiosonde models	10685
Eureka	10	79° 59′ N	85° 56′ W	January 2001 – December 2015	RS80-H	4649
					RS90 and RS92	6109
					With all radiosonde models	10758
Alert	65	82° 30′ N	62° 21′ W	January 2001 – December 2015	RS80-H	4060
					RS92	6586
					With all radiosonde models	10646
Aasiaat	41	68° 42′ N	52° 51′ W	January 2001 – December 2009	RS80-H	651
					RS90	2459
					RS92	2845
					With all radiosonde models	5955

(continued)

Table 2.1 (continued)

Station	Altitude (above mean sea level)	Latitude	Longitude	Measurement period	Radiosonde models	Number of radiosoundings
Danmarkshavn	12	76° 46′ N	18° 40′ W	January 2001 – December 2009	RS80-H	1400
					RS90	2054
					RS92	2587
					With all radiosonde models	6041
Jan Mayen	9	70° 59′ N	08° 32′ W	January 2001 – December 2015	RS80-H	3646
					RS92	7072
					With all radiosonde models	10718
Ny-Ålesund	11	78° 54′ N	11° 53′ E	January 2005 – December 2015	RS90	373
					RS92	3241
					With all radiosonde models	3614
Sodankylä	179	67° 22′ N	26° 39′ E	January 2001 – December 2015	RS80-H	767
					RS90	2711
					RS92	7146
					With all radiosonde models	10624
Ostrov Dikson	47	73° 30′ N	80° 24′ E	January 2001 – December 2015	Marz 2-1	5
					Marz2-2	3970
					BAR	2157
					MRZ	1607
					With all radiosonde models	7739
Tiksi	7	71° 35′ N	128° 55′ E	January 2001 – December 2015	MRZ	6462
					BAR	1614
					Marz2-2	616
					With all radiosonde models	8692
Cherskij	28	68° 45′ N	161° 17′ E	January 2004 – December 2015	MRZ	3851
					BAR	2071
					Marz2-1	1039
					With all radiosonde models	6961

Table 2.2 Characteristics of the Barocap, Thermocap and Humicap sensors mounted on the four Vaisala radiosonde models, whose data have been considered in the present study

Characteristics	Vaisala radiosonde model			
	RS80-A	RS80-H	RS90	RS92
Barocap sensor (values measured in hPa)				
Measurement range	1060-3	1060-3	1080-3	1080-3
Declared resolution	0.1	0.1	0.1	0.1
Accuracy	± 0.5	± 0.5	± 0.5	± 0.5
Repeatability in calibration [range in hPa]	0.5	0.5	0.4 [1080-100] 0.3 [100-3]	0.4 [1080-100] 0.3 [100-3]
Reproducibility in sounding [range in hPa]	0.5	0.5	0.5 [1080-100] 0.3 [100-3]	0.5 [1080-100] 0.3 [100-3]
Thermocap sensor (values measured in °K)				
Measurement range	333-183	333-183	333–183	333–183
Declared resolution	0.1	0.1	0.1	0.1
Accuracy	± 0.2	± 0.5	± 0.5	± 0.5
Repeatability in calibration [range in K]	0.20	0.20	0.15	0.15
Reproducibility in sounding [range in K]	0.2 [1060-50] 0.3 [50-15] 0.4 [15-3]	0.2 [1060-50] 0.3 [50-15] 0.4 [15-3]	0.2 [1080-100] 0.3 [100-20] 0.5 [20-3]	0.2 [1080-100] 0.3 [100-20] 0.5 [20-3]
Humicap sensor (values measured in % RH)				
Measurement range	2–100	0–100	0–100	0–100
Declared resolution	1	1	1	1
Accuracy	<±3	± 5	± 5	±5
Repeatability in calibration [range in %]	2	2	2	2
Reproducibility in sounding [range in %]	<3	<3	<3	2

Table 2.3 Evaluations of the average day-time and night-time temperature differences $\Delta T(z)$ and the differences $\Delta RH(z)$ between the performances of the VIZ-B2 and the Vaisala RS90 radiosonde models (with their standard deviations), as determined by Mattioli et al. (2007) during the 2004 North Slope of Alaska Arctic Winter Radiometric Experiment (2004 NSAWRE) field campaign conducted at Barrow (Alaska, USA) from March 9 to April 9, 2004. The altitude values have been calculated according to the vertical profile of pressure $p(z)$ determined by Tomasi et al. (1998) in the "July-75 °N latitude" atmospheric model

Pressure level p (hPa)	Altitude z (km)	Day-time $\Delta T(z)$ (°C)	Night-time $\Delta T(z)$ (°C)	Difference $\Delta RH(z)$ (%)
1000	0.140	$+0.5 \pm 1.4$	$+0.0 \pm 1.3$	-0.6 ± 8.8
925	0.730	$+0.0 \pm 0.4$	-0.3 ± 0.4	$+6.5 \pm 9.0$
850	1.323	-0.1 ± 0.3	-0.2 ± 0.2	$+5.0 \pm 7.7$
700	2.957	$+0.0 \pm 0.2$	-0.1 ± 0.3	$+6.1 \pm 8.9$
500	5.516	$+0.1 \pm 0.2$	-0.1 ± 0.3	$+4.4 \pm 15.0$
400	7.136	$+0.2 \pm 0.2$	-0.1 ± 0.3	$+2.7 \pm 13.7$
300	9.117	$+0.1 \pm 0.2$	-0.3 ± 0.3	$+6.3 \pm 17.1$
250	10.344	$+0.0 \pm 0.4$	-0.5 ± 0.5	$+20.1 \pm 15.3$
200	11.828	$+0.0 \pm 0.5$	-0.7 ± 0.6	$+23.3 \pm 10.2$
150	13.769	$+0.2 \pm 0.4$	-0.8 ± 0.4	$+21.6 \pm 9.7$
100	16.520	$+0.2 \pm 0.6$	-1.2 ± 0.5	$+18.0 \pm 9.1$
70	18.944	$+0.0 \pm 0.6$	-1.6 ± 0.6	$+15.8 \pm 8.2$
50	21.759	$+0.1 \pm 0.9$	-1.8 ± 0.7	$+15.2 \pm 8.0$
30	25.984	$+0.5 \pm 1.1$	-3.5 ± 0.9	$+14.5 \pm 9.4$

At least 15 radiosonde models were employed at the Russian stations since 2001, each of them having a history usually complicated because of frequent alternations of up to seven instrument types employed at the same station. Major instruments used from 1973 to the late 1990s in the former Soviet Union were the A22, RKZ, MARS, and MRZ series (Zaitseva 1993). The radiosonde systems most frequently employed from 2001 to 2015 at both Ostrov Dikson and Tiksi stations were those named AVK-MRZ, AVK-BAR, MARL-A-MRZ, MARL-A-BAR, and Meteorit Marz2-1, while those most frequently employed at Cherskij were the Meteorit Marz2-2, AVK-MRZ, AVK-BAR, and Meteorit Marz2-1. The AVK, MARL-A and Meteorit acronyms refer to the radars or the radio-theodolites used to follow the radiosonde flights, and the MRZ, BAR, Marz2-1, and Marz2-2 acronyms refer to the sensor sets mounted onboard the radiosondes. The main technical characteristics of the MRZ and Marz2-2 radiosondes defined by the Central Aerological Observatory (2003) are the following: (a) the measurement range of temperature of the MRZ radiosondes is from +50 °C to −80 °C, with accuracy equal to ±0.6 °C; the measurement range of *RH* is from 15% to 100%, with accuracy equal to: (i) 8% over the *RH* range from 0% to 10%, (ii) 5% over the *RH* range from 10% to 90%, and (iii) 8% over the *RH* range varying from 90% to 100%; (b) the temperature measurement range of the Marz2-2 radiosondes is from +50 °C to −80 °C, with an accuracy of ±0.5 °C, and the measurement range of *RH* is from 15% to 98% over the temperature range from +50 °C to −40 °C, with accuracy of 15% for temperatures $T(z)$ lower than −40 °C. Comparing the various series of the BAR, MRZ, Marz2-1 and Marz2-2 radiosondes, it can be said that only slight differences exist between the various Russian radiosonde models, in practice mainly due to minor sensor differences and apparently effective corrections. For instance, the Marz2-1 and Marz2-2 radiosonde models have been evaluated by Schroeder (2009) to provide performances that are actually indistinguishable.

2.2.1 Correction of Raw Pressure Measurements Provided by the Various Radiosonde Models and Results

During its ascent, each Vaisala radiosonde has transmitted to the ground station a sequence of signals, each giving the measurements of p (in hPa), T (in °C), and RH (in %) at a certain measurement level. The Vaisala radiosondes were recorded at the ground station every 2 s and, hence, in altitude steps of 10–12 m, since the balloon ascent rate was regulated on average between 5 and 6 m s^{-1}. Therefore, each RS measurement performed with the various Vaisala radiosonde models consisted in general of: (1) more than 800 values of air pressure $p(z)$ and air temperature $T(z)$ measured at the standard and additional RS levels in the altitude range $0 \leq z \leq 10$ km, and further 900–1900 stratospheric levels up to a variable top level of 26–30 km usually reached by the radiosondes for favourable conditions for the launch, and (2) values of relative humidity $RH(z)$ recorded at the above-mentioned tropospheric levels and usually at only a relatively small number of stratospheric levels. In fact, because of the harder thermal conditions of the low Arctic stratosphere, the RS measurements were regularly recorded on the spring, autumn and winter days at no more than 800 levels from the surface to the tropopause level, and at further 200–400 levels only up to altitudes of 15–25 km, since the rubber balloon was subject to frequently break at the stratospheric levels because of the extremely cold air temperatures. When the radiosonde passes from the troposphere to the stratosphere, the values of $RH(z)$ decreases very sharply to values of only a few percent and, then, very often below the repeatability level of 2% established by the sensor manufacturers.

The multi-year data-sets examined in the present study were provided by the raw radiosonde measurements collected at the 14 Arctic stations listed in Table 2.1 and performed at 00:00 UTC and/or 12:00 UTC of each day characterized by favourable meteorological conditions for the RS launches. The following average numbers of RS measurements per month were carried out at the 14 above-selected stations, equal to: (i) only 27 at Ny-Ålesund (where the RS measurements were taken only once per day at 12:00 UTC, over the 11-year period from 2005 to 2015), (ii) 43 at Tiksi, (iii) 46 at Danmarkshavn, (iv) 50 at Inuvik, (v) 56 at Aasiaat, (vi) 58 at Cherskij, (vii) 59 at Resolute, Alert, and Sodankylä, (viii) 60 at Barrow, Cambridge

Bay, Eureka, and Jan Mayen, and (ix) 65 at Ostrov Dikson. The geometrical altitude z of each radiosonde level was calculated for each triplet of radiosonde signals p, T and RH, using the manufacturer's algorithm based on the assumption that the air is in hydrostatic equilibrium. Starting from the surface-level pressure p_o measured with good precision at each station, appropriate algorithms determined by the manufacturers allowed us to calculate the values of p, T and RH after each altitude step from the i-th to the $(i + 1)$-th level until to obtain the values of these meteorological parameters at all the levels. The geometrical altitude z was then calculated for each triplet of RS signals p, T and RH using the manufacturer algorithm based on the differential form of the hydrostatic equilibrium equation for air, in which the virtual temperature was properly calculated according to the criteria recommended by Dubin et al. (1966). The Vaisala Barocap measurements were assumed to have the characteristics of declared resolution, accuracy, repeatability in calibration and reproducibility in sounding given in Table 2.2. Therefore, no corrections were made "a priori" to the raw pressure data provided by the Vaisala Barocap sensors mounted on the RS-80A, RS-80H, RS90 and RS92 radiosondes and used at the above-chosen eleven Arctic RS stations. Some examples of vertical profiles of $p(z)$ determined with Vaisala radiosondes are shown in Fig. 2.2, as obtained at four Arctic stations with: (i) a Vaisala RS80-H radiosonde at Resolute (Nunavut, Canada), (ii) a Vaisala RS90 radiosonde at Danmarkshavn (North-eastern Greenland), and (iii) with Vaisala RS92 radiosondes at both Alert (Nunavut, Canada) and Ny-Ålesund (Spitsbergen, Svalbard). The comparison shows that these measurements of $p(z)$ covered in general the whole tropospheric and low-stratospheric height range up to altitudes of 27 km at Resolute, 39 km at Danmarkshavn, 36 km at Alert, and 31 km at Ny-Ålesund.

The VIZ-B2 radiosondes were employed at Barrow (Alaska) from 2001 to 2012 to carry out the measurements of parameters $p(z)$, $T(z)$, and RH(z) every 1.2 s, those taken every 6 s being collected for being archived (and, hence, in altitude steps of about 10–11 m, as indicated by Mattioli et al. (2005)). The VIZ-B2 meteorological data were first filtered and then fitted to a regular altitude grid through a linear interpolation procedure in height, to obtain a regular sequence of data in steps of 10 m over the altitude range from surface-level to 10 km altitude; and in steps of 100 m from level $z = 10$ km to the highest level reached by the radiosonde. An example of vertical profile of $p(z)$ determined with a VIZ-B2 radiosonde at Barrow is shown in Fig. 2.2 over the height range from surface up to the altitude of 33 km, giving an idea of the good results provided by the VIZ-B2 radiosondes at this coastal site of Alaska (USA).

Considering that the RS signals were transmitted every 2 seconds to the ground-station and that the ascent rate of the balloon was on average of 5–10 m s^{-1}, the Russian radiosonde measurements were estimated by us to provide values of parameters $p(z)$, $T(z)$, and RH(z) in altitude steps of about 10–20 m at the three Siberian stations located at Ostrov Dikson, Tiksi and Cherskij. The vertical profiles of $p(z)$ derived at stratospheric levels from the RS measurements were compared with the monthly mean vertical profiles of $p(z)$ obtained by Tomasi et al. (2011a)

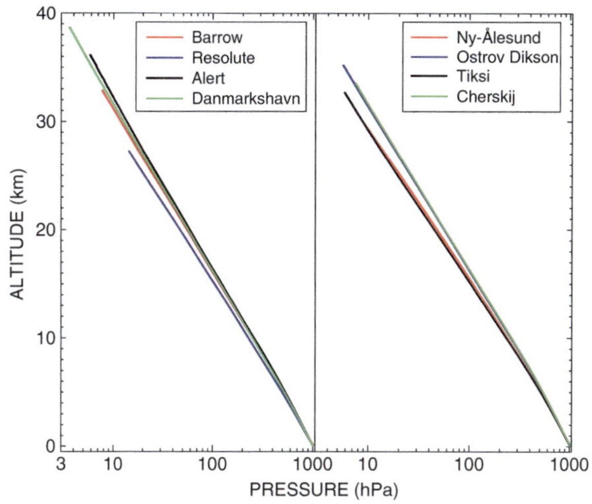

Fig. 2.2 Left part: Vertical profiles of air pressure *p(z)* obtained from the RS measurements conducted during different seasons and years at four of the 14 above-selected Arctic sites: (i) Barrow (Alaska, USA) using a VIZ-B2 radiosonde on April 21, 2007 (12:00 UTC); (ii) Resolute (Nunavut, Canada) using a Vaisala RS80-H radiosonde on January 1, 2001 (00:00 UTC); (iii) Alert (Nunavut, Canada) using a Vaisala RS92 radiosonde on July 22, 2014 (00:00 UTC); and (iv) Danmarkshavn (Greenland) using a Vaisala RS90 radiosonde on August 30, 2004 (00:00 UTC). Right part: As in the left part for the following four Arctic sites: (i) Ny-Ålesund (Spitsbergen, Svalbard) using a Vaisala RS92 radiosonde on October 30, 2008 (12:00 UTC); (ii) Ostrov Dikson (Siberia, Russia) using a MRZ radiosonde on September 8, 2005 (00:00 UTC); (iii) Tiksi (Siberia, Russia) using a BAR radiosonde on March 30, 2010 (00:00 UTC); and (iv) Cherskij (Siberia, Russia) using a Marz2-1 radiosonde on May 21, 2014 (00:00 UTC)

over the 12–30 km height range by averaging the monthly mean vertical profiles derived from MIPAS2D data-sets recorded by Dinelli et al. (2010) at the Arctic latitudes of 65 °N, 70 °N and 75 °N. The RS measurements giving pressure data differing from the MIPAS2D monthly averages by more than two standard deviations (i.e. by more than $\pm 48\%$) were rejected in the present analysis of pressure data, in order to avoid systematic instrumental errors in the pressure measurements conducted at the Ostrov Dikson, Tiksi and Cherskij stations. After the application of these selection criteria, the vertical profiles of *p(z)* recorded at the three above-mentioned Russian stations were examined to determine the monthly mean vertical profile of *p(z)* at the above-chosen Siberian sites. Three examples obtained by following such a procedure are shown in Fig. 2.2: (i) the first example, obtained at Ostrov Dikson (Siberia, Russia) using a MRZ radiosonde over the altitude range from surface to about 35 km, (ii) the second example, determined at Tiksi (Siberia, Russia) using a BAR radiosonde over the altitude range from surface to about 33 km, and (iii) the third example collected at Cherskij (Siberia, Russia) using a Marz2-1 radiosonde over the altitude range from surface to about 34 km.

2.2.2 Correction of Raw Temperature Measurements Provided by the Various Radiosonde Models

The raw temperature data provided by the above-described radiosonde models have been corrected by following different procedures for the various radiosonde models used in the present study:

(a) *Vaisala radiosondes*

A measurement range from 183 to 333 °K characterizes all the Thermocap sensors mounted on the four Vaisala radiosonde models considered in the present study, with the features of declared resolution, accuracy, repeatability in calibration and reproducibility in sounding defined in Table 2.2. The temperature data provided by the various Thermocap sensors are known to be affected by errors most commonly due to (i) heating by the incoming solar radiation and nocturnal infrared radiation, (ii) heat conduction from the other radiosonde components, and (iii) heat exchanges between the sensor and environment. The errors of this kind affecting the RS80-A and RS80-H Thermocap sensors were corrected following the procedure of Luers and Eskridge (1995), and those of the RS90 and RS92 F-Thermocap sensor were neglected, according to Luers (1997). The RS80 and RS90/92 Thermocap data are known to be also affected by lag errors depending mainly on air density and ventilation speed. The RS80-A and RS80-H lag errors of this kind were corrected over the whole altitude range, except within the temperature inversion layer sometimes formed near the ground, using the algorithm defined by Tomasi et al. (2004) for a radiosonde ascent rate of 6 m/s, which was calculated by taking into account the set of temperature measurements and Vaisala tests analyzed by Huovila and Tuominen (1991). This algorithm was found to yield time-lag estimates increasing on average from less than 3 s at sea-level to about 10 s at the 26 hPa level, which lead in general to lag errors smaller than 0.2 °K at all altitudes ranging from about 1 to 25 km. According to the characteristics given by the manufacturer, the time-constants of the RS90 and RS92 Thermocap sensors were estimated to increase from about 0.3 s at sea-level to 2.5 s at 10 hPa (~ 46 km altitude), considering that the RS90 Thermocap sensor has small sizes and its response time is sufficiently rapid to respond to the presence of temperature gradients without causing significant errors (Luers 1997). Taking into account these results and those of Rowe et al. (2008), the lag errors of the RS80 and RS90/92 Thermocap sensors due to the presence of temperature inversion features within the ground-layer were corrected by assuming that they are negligible for isothermal conditions of the ground-layer and increase linearly as the vertical temperature gradient $\gamma = dT/dz$ assumes increasing positive values, to reach an average RS80 value of 1.0 °K for $\gamma \geq 1$ °K/m, and an average RS90/92 value of 1.2 °K for $\gamma \geq 1.4$ °K/m.

Some examples of vertical profiles of *T(z)* derived from Vaisala radiosonde data-sets are shown in Fig. 2.3, pertaining to the RS measurements performed at: (i) Cambridge Bay on 30 October, 2002 (10:00 UTC), with a Vaisala RS80-H radiosonde that has reached the top-level of 34 km; (ii) Danmarkshavn on

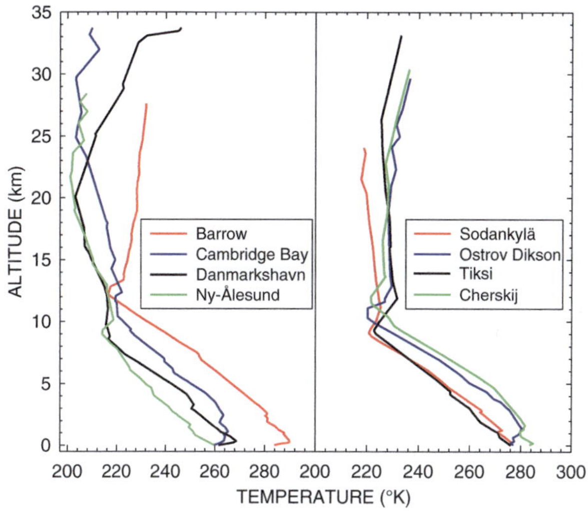

Fig. 2.3 Left part: Vertical profiles of air temperature *T(z)* obtained from the RS measurements conducted during different seasons and years at four of the 14 above-selected Arctic sites: (i) Barrow (Alaska, USA) using a VIZ-B2 radiosonde on July 31, 2002 (12:00 UTC) up to top-level $z_{top} = 28$ km; (ii) Cambridge Bay (Nunavut, Canada) using a Vaisala RS80-H radiosonde on October 30, 2002 (10:00 UTC) up to $z_{top} = 34$ km; (iii) Danmarkshavn (Greenland) using a Vaisala RS90 radiosonde on March 10, 2005 (12:00 UTC) up to $z_{top}z = 34$ km; and (iv) Ny-Ålesund (Spitsbergen, Svalbard) using a Vaisala RS92 radiosonde on December 4, 2013 (12:00 UTC) up to $z_{top} = 28$ km. Right part: As in the left part for the following four Arctic sites: (i) Sodankylä (Finland) using a Vaisala RS92 radiosonde on April 29, 2009 (00:00 UTC) up to $z_{top} = 24$ km; (ii) Ostrov Dikson (Siberia, Russia) using a Marz2–2 radiosonde on June 14, 2001 (12:00 UTC) up to $z_{top} = 30$ km; (iii) Tiksi (Siberia, Russia) using a BAR radiosonde on September 15, 2008 (00:00 UTC) up to $z_{top} = 33$ km; and (iv) Cherskij (Siberia, Russia) using a MRZ radiosonde on August 8, 2009 (12:00 UTC) up to $z_{top} = 30$ km

10 March, 2005 (12:00 UTC), with a Vaisala RS90 radiosonde up to the top-level of nearly 34 km; (iii) Ny-Ålesund on 4 December, 2013 (12:00 UTC), with a Vaisala RS92 radiosonde up to the top-level of 28 km; and (iv) Sodankylä on 29 April, 2009 (00:00 UTC) up to the top-level of 24 km. These profiles clearly indicate the presence of marked temperature inversions near the ground at Cambridge Bay and Danmarkshavn, and of a pronounced minimum at the tropopause levels ranging between 8 and 12 km.

(b) *VIZ-B2 radiosondes*

The temperature sensor used on the VIZ-B2 radiosondes employed at Barrow from 2001 to 2012 is a 4.5-cm-long white-coated rod (long rod) thermistor, whose white coating has a solar average absorptivity equal to 0.15 and an average emissivity equal to 0.86 (Luers and Eskridge 1998). Thus, the thermistor is a good reflector of solar radiation (85%) but also a strong absorber and emitter (86%) of infrared radiation. The thermistor is attached to a mount extending outward and above the body of the radiosonde by the thermistor lead wires (leading to the sensor), which are of small diameter and sufficiently long for allowing that only a small amount of heat can be conducted from the sensor to the mount. Overestimated positive values of $T(z)$ were estimated by Luers and Eskridge (1998) during the diurnal hours, no higher than +1 °C until to 35 km altitude, while no appreciable errors were found to occur during the night until to 20 km height, and a negative trend of temperature was observed at the upper levels, gradually increasing with altitude until reaching an error of −3.0 °C at the 40 km altitude.

The VIZ-B2 temperature data were in general fitted to a regular grid in altitude, by applying a linear interpolation procedure with: (i) a resolution of 10 m over the altitude range from surface-level to 10 km altitude; and (ii) a resolution of 100 m from $z = 10$ km to the highest level reached by the radiosonde (Mattioli et al. 2005). Comparing the vertical profiles of $T(z)$ obtained by using the National Weather Service (NWS) VIZ-B2 and Vaisala RS90 radiosondes at Barrow during the 2004 North Slope of Alaska Arctic Winter Radiometric Experiment conducted from March 9 to April 9, 2004, Mattioli et al. (2005) estimated that: (i) the standard deviation of the temperature difference $\Delta T(z)$ between the VIZ-B2 and the Vaisala RS90 increases with height at altitudes higher than 10 km, such a difference being on average equal to 0.42 °C for $z < 10$ km and to 1.7 °C for $z > 10$ km; and (ii) the vertical profiles of $T(z)$ derived from the VIZ-B2 and Vaisala RS90 data-sets basically agree during the day, but exhibit substantial differences during the night.

Comparing the temperature measurements conducted with the VIZ B2 radiosondes with those simultaneously carried out with the Vaisala RS90 radiosondes during the 2004 North Slope of Alaska Arctic Winter Radiometric Experiment (2004 NSAWRE) conducted at Barrow (Alaska) from March 9 to April 9, 2004, Mattioli et al. (2007) found that the night-time negative bias of the VIZ-B2 radiosondes with respect to the Vaisala RS90 measurements reached a maximum of −3.5 °C at the 30 hPa level (followed by a negative bias of −5 °C at the 20 hPa level), while no evident bias features were detected during the day over the pressure range from 915 to 50 hPa.

Examining the Mattioli et al. (2007) results, we determined an appropriate algorithm based on the day-time and night-time temperature differences $\Delta T(z)$ reported in Table 2.3 at the various tropospheric and low-stratospheric altitudes to: (i) correct the day-time and night-time measurements of temperature $T(z)$ recorded with the VIZ-B2 radiosondes, and (ii) render the temperature data provided by the VIZ-B2 radiosondes more homogeneous with those provided by the Vaisala RS90 radiosondes. For this purpose, we have used the estimates of the temperature difference $\Delta T(z)$ derived by Mattioli et al. (2007) from the comparison between the vertical profiles of temperature $T(z)$ measured at Barrow (Alaska) in spring 2004. The average night-time and day-time differences $\Delta T(z)$ obtained from overall sets consisting of 17 night-time comparison tests and 21 day-time comparison tests are given in Table 2.3 over the altitude range from the surface ($z_o = 0.14$ km) up to 25.98 km, showing that: (i) the day-time difference $\Delta T(z)$ was found to range on average between −0.1 and + 0.5 °K at tropospheric levels, and between +0.0 and + 0.5 °K at stratospheric levels, therefore without presenting pronounced bias features over the whole pressure range from 915 to 30 hPa; and (ii) the night-time difference $\Delta T(z)$ was estimated to vary between −0.3 and + 0.0 °K at tropospheric levels, and between −0.3 and −3.5 °K at stratospheric levels. Using the average correction values of Table 2.3, we converted the VIZ-B2 temperature measurements into equivalent data to those given by the Vaisala RS90 radiosondes and then analyzed these corrected values of $T(z)$ by following the procedure described above for analyzing the RS90/92 Thermocap data.

The differences between the temperature measurements performed using the VIZ-B2 and those made using the Vaisala RS80-H radiosondes were tested by Wang and Young (2005), who compared the measurements of $T(z)$ made with the VIZ-B2 radiosondes with those performed using the Vaisala RS80-H radiosondes, both the radiosonde models being operated by the Atmospheric Radiation Measurement (ARM) program during the 1996–2002 years at the two nearby RS stations of Norman (35° 13′ N, 97° 26′ W) and Purcell (35° 01′ N, 97° 22′W), both located in Oklahoma (USA) and about 25 km away one from the other. The VIZ-B2 data showed a good agreement with the RS80-H data near the surface, but become warmer aloft by about 0.5 °C above 11 km, suggesting that such a warm bias should be mainly due to solar radiation heating effects. Therefore, the corrections proposed by Mattioli et al. (2007) seem to be appropriate for taking into account the major part of the biases affecting the VIZ-B2 temperature measurements. On this matter it is also interesting to mention the analysis made by He et al. (2009), who compared the temperature profiles derived from the Global Positioning System (GPS) Radio Occultation (RO) data-sets recorded during the Constellation Observing System for Meteorology, Ionosphere, and Climate (COSMIC) mission with those provided by the MRZ, VIZ-B2 and Vaisala RS92 radiosonde systems over the altitude range from 12 to 25 km to assess the performance of the various radiosonde models in the upper troposphere and lower stratosphere. Despite the fact that COSMIC observations represent a weighted average over a cylindrical volume about 200 km long and a vertical scale of about 1 km, while the radiosondes carry out point measurements, the temperature profiles of each radiosonde type in general agree well with those of COSMIC, with: (1) a mean difference equal to −0.26 °K (determined for 4728 matches of the MRZ radiosonde model) and a standard deviation of 1.98 °K; (2) a mean difference equal to +0.23 °K (determined for 1474 matches of the VIZ-B2 radiosonde) and a standard deviation of 1.97 °K; and (3) a mean difference equal to −0.03 °K, as determined for 3786 matches of the Vaisala RS92 radiosonde model, and a standard deviation of 1.76 °K. These evaluations show that the temperature measurements performed with the Vaisala RS92 radiosonde systems agree well with those of COSMIC, while larger temperature biases were determined by examining the data collected with the MRZ and VIZ-B2 radiosonde systems relative to COSMIC, which are probably caused by diurnal radiative effects if one takes into account that these Russian radiosondes have an average solar absorptivity of 20% and a very low sensor emissivity equal to 0.04 (Luers and Eskridge 1998), while the VIZ-B2 radiosondes have a solar absorptivity of 0.15% and a higher sensor emissivity equal to 0.85, so that the VIZ-B2 sensors are likely subject to radiative heating during the day and radiative cooling during the night, which results in the opposite temperature biases relative to COSMIC. Luers and Eskridge (1998) identified also a cold bias in the VIZ-B2 radiosondes used during the night. The comparison with COSMIC data also shows that the VIZ-B2 radiosonde data are affected by a warm bias (about 0.37 °K) during the day, and a night-time cold bias ranging between 0.4 °K and 1.2 °K. Because of their long-wave radiation emissivity and solar radiation absorptivity characteristics, the performances of the VIZ-B2 temperature sensor are also markedly influenced by the heating of the sensor due to solar radiation absorption, causing an increase of the mean temperature difference between COSMIC and VIZ-B2 sensors ranging from a slight warm bias below the 20 km height to about 0.5 °K for solar elevation angles higher than 20°.

An example of vertical profile of *T(z)* determined using a VIZ-B2 radiosonde is shown in Fig. 2.3, relative to the RS measurements performed at Barrow (Alaska) on July 31, 2002 (12:00 UTC), which have reached the top-level of 28 km and showed marked features of a temperature inversion near the ground, and a tropopause minimum located at about 12.5 km.

(c) *Russian radiosondes*

The RS measurements of air temperature performed at the three Siberian stations of Ostrov Dikson, Tiksi and Cherskij were taken with the MRZ, BAR, Marz2-1 and Marz2-2 radiosonde models currently used in Russia. Luers and Eskridge (1998) pointed out that the Russian Marz2-2 radiosonde was completely automated first in 1971 and was largely used in the former USSR by 1984. The Marz2-2 radiosonde model uses the same temperature sensor employed in the RKZ radiosonde and designed to operate with the Meteorite radar in the Meteorite-RKZ-MARZ system series, which was introduced in the Soviet upper-air network in the 1958–1961 years (Zaitseva 1993). The Marz2-1 and Marz2-2 radiosondes have a bright surface on the support mount, typical of a tin or silver coating, which provides an average emittance equal to 0.04 and has absorptivity equal to 0.20. Therefore, the thermistor sensor becomes less sensitive to both infrared and solar radiation as the solar and infrared emission properties of the mount gradually decrease with time. Thanks to these improvements, the operational temperature measurement range of the MRZ, BAR, Marz2-1 and Marz2-2 radiosondes was estimated to be from −80 °C to +50 °C, with accuracies of ±0.6 °C for the MRZ and BAR radiosondes, and of ±0.5 °C for the Marz2-1 and Marz2-2 radiosondes. Luers and Eskridge (1998) evaluated that the temperature errors made with the Marz2-2 radiosondes are almost totally negligible during the night up to altitudes of 30–35 km, and do not exceed +1.0 °K as the solar elevation angle increases from 0° to 60° during the day.

The RS measurements of air temperature provided by the above-mentioned Russian radiosondes are expected to provide reliable measurements of *T(z)*, having an accuracy comparable with that of the VIZ-B2 radiosondes. With regard to this, Ho et al. (2017) estimated the global day-time temperature difference between mean RS data and RO data at upper tropospheric heights between 200 and 20 hPa levels to be equal to 0.71 °K for the VIZ-B2 radiosonde data, and 0.66 °K for the MRZ Russian radiosondes. Therefore, the temperature data provided by the above-mentioned Russian radiosondes were assumed to be right "a priori", and hence analyzed without corrections, since the measurement errors were assumed to be negligible at all the altitudes reached by the radiosondes.

The various Russian radiosonde models employed at the three above-selected Siberian sites provided RS measurements of *T(z)* covering the whole tropospheric height range and the stratosphere up to top-levels of 28–33 km, as can be seen in Fig. 2.3, which shows the three following vertical profiles of *T(z)*: (i) the vertical profile recorded at Ostrov Dikson on 14 June, 2001 (12:00 UTC) using a Marz 2-2 radiosonde, presenting a marked thermal inversion near the ground, a tropopause minimum at around 10–11 km altitude, and a top-level close to 30 km; (ii) the profile recorded at Tiksi on 15 September, 2008 (00:00 UTC) using a BAR radiosonde, without thermal inversion characteristics near the ground in such a period of the year, but with the presence of a marked tropopause minimum at the altitudes of around 9–10 km, and a top-level of more than 33 km; and (iii) the profile measured at Cherskij on 8 August, 2009 (12:00 UTC) with a MRZ radiosonde, presenting thermal inversion features near the ground and within the first 1.5 km of the atmosphere, and a pronounced temperature minimum at 11–12 km altitudes, followed by stratospheric temperature data up to the top-level of nearly 30 km reached by the radiosonde.

In order to check their reliability, the all the measurements of *T(z)* derived from the RS data taken at stratospheric levels above the Ostrov Dikson, Tiksi and Cherskij stations have been compared with the corresponding values determined by Tomasi et al. (2011a) to define the twelve monthly mean vertical profiles of *T(z)* over the 12–30 km altitude range by averaging the retrieved MIPAS2D data pertaining to the Arctic latitude of 70 °N. All the RS measurements giving values of *T(z)* differing by more than ± three standard deviations (i.e. by more than ±49.8%) from the MIPASS2D monthly averages were rejected in order to avoid systematic errors that may have affected the temperature sensors or exceptional events of stratospheric warming (or cooling). An example of this selection procedure is shown in Fig. 2.4, where the monthly mean vertical profiles of *T(z)* derived from the RS measurements performed at Tiksi in the months of January, April, July and October of the year 2009 are compared with those derived in the same months at stratospheric levels from the MIPAS2D satellite data collected by Tomasi et al. (2011a). It can be noted in these examples that only a few cases present strong stratospheric warming features in April and July, whereas numerous cases were observed in January and October 2009, the most part of which were found to vary largely within the limits of ± three standard deviations. Figure 2.4 also indicates that the performances of the thermal sensors mounted on the Russian radiosondes are fully reliable at stratospheric altitudes lower than 30 km, except a few cases observed in April and July 2009, in which only a few single vertical profiles of *T(z)* were found to provide air temperature values largely higher (by more than three standard deviations) than those predicted by the MIPAS2D satellite data in the low Arctic stratosphere.

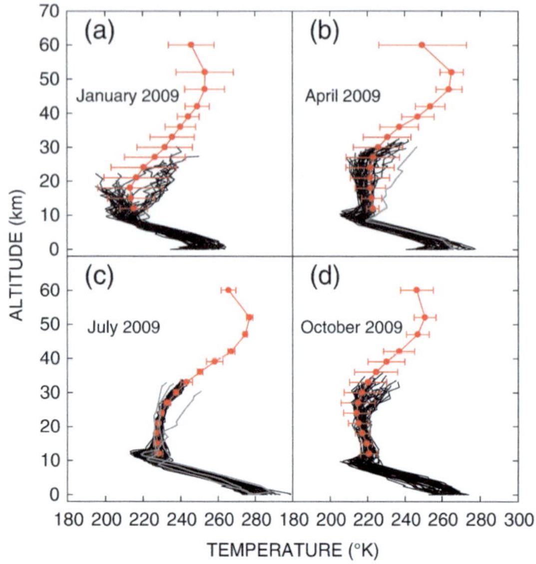

Fig. 2.4 Comparison between: (1) the monthly mean vertical profiles of air temperature *T(z)* (black curves) obtained at stratospheric altitudes from the RS measurements conducted in: (**a**) January 2009; (**b**) April 2009; (**c**) July 2009; and (**d**) October 2009 at Tiksi, and (2) the monthly mean vertical profiles of *T(z)* (red curves) derived by Tomasi et al. (2011a) over the 12–68 km altitude range from the MIPAS2D data-sets selected in the same months at the 70 °N latitude

2.2.3 Correction of Raw Relative Humidity Measurements Provided by the Various Radiosonde Models

Different procedures have been followed in the present analysis in order to correct the raw relative humidity measurements performed by the various radiosonde models:

(a) *Vaisala radiosondes*

The *RH* measurement range provided by all the Humicap sensors mounted on the Vaisala radiosondes extends from 0% to 100%, these sensors having the features of declared resolution, accuracy, repeatability in calibration and reproducibility in sounding reported in Table 2.2. The A-Humicap time-constant was estimated by Miloshevich et al. (2004) to increase in a nearly exponential fashion from 0.2 s to about 200 s in the RS80-A radiosondes, as air temperature *T(z)* decreases from +25 °C to −80 °C. Taking into account these evaluations, we adopted herein a correction algorithm for calculating such a time-constant as a function of *T*, and assumed that such a time-constant is subject to rapidly decrease as the radiosonde gradually ascends through the troposphere. Conversely, the time-constant of the H-Humicap sensor was estimated to assume appreciably smaller values than those of the A-Humicap sensor, due to the fact that a thinner polymer layer was used in this new sensor model, which assures a much faster response time at low temperatures (Miloshevich et al. 2006). Taking into account the characteristics of the Humicap time-constants, the *RH* data were corrected by following a complex procedure based on (i) the combined use of various algorithms and correction procedures (Wang et al. 2002; Miloshevich et al. 2004) and (ii) the criteria recommended by Turner et al.

(2003) and Miloshevich et al. (2006, 2009). To remove the most important errors made by the Vaisala radiosondes, we analyzed the original *RH* data through a careful procedure consisting of the following seven steps:

(1) A preliminary smoothing procedure, which was first applied to the raw *RH* data within all the height intervals containing sequences of constant values of *RH* with altitude, to construct a more schematic "skeleton" of the measured vertical profiles of *RH* (Miloshevich et al. 2004).
(2) The correction of the *RH* data for the so-called basic calibration model (BCM) errors, which was made by using the algorithms defined by Wang et al. (2002) for the RS80-A and RS80-H Humicap sensors, and those defined by Miloshevich et al. (2006, 2009) for the RS90 and RS92 Humicap sensors, which indicate that absolute errors of this kind in *RH* are usually assumed to increase from 0.6% to 2% in the Polar atmosphere, mainly presenting values of *RH* ranging between 0% and 80% (Tomasi et al. 2006). In particular, the *RH* measurements performed with the Vaisala RS92 radiosondes in the 2006–2007 biennium were corrected taking into account that they are in general affected by a systematic 1% increase of the mean calibration bias, while such an error was not found in the *RH* measurements performed with radiosondes manufactured before 2006 or after 2007 (Miloshevich et al. 2006, 2009).
(3) The correction of the chemical contamination (CC) dry biases affecting the *RH* data recorded with the RS80-A, RS80-H, RS90 and RS92 radiosonde models, which was performed by assuming that: (i) the CC biases affecting the RS80 data are very small and, hence, totally negligible in sensors having an age of a few years; (ii) those of the RS90 sensors are also fully negligible, since polymers with improved characteristics were employed in manufacturing such a new sensor model (Miloshevich et al. 2004) and (iii) those of the RS92 sensors were assumed to be very small, since such sensors were properly heated and regenerated before their launch.
(4) The correction of the temperature dependence (TD) dry biases, which was made by: (i) using the algorithms defined by Wang et al. (2002) to examine the RS80-A data, according to the calibration and temperature dependence features evaluated by Miloshevich et al. (2001); (ii) employing the calibration algorithm defined by Wang et al. (2002) to correct the RS80-H data; and (iii) totally neglecting the TD errors affecting the RS90 and RS92, according to the estimates made by Miloshevich et al. (2006) for Humicap sensors manufactured after June 2001.
(5) The correction of the *RH* data for the sensor aging (SA) dry biases, which was performed by: (i) using the algorithms defined by Wang et al. (2002) for the RS80-A Humicap sensor to remove the SA errors (estimated to vary on average between 0.8% for a 1-year sensor age and 1.3% for a 2-year sensor age) and for the RS80-H Humicap sensor to estimate the SA dry bias as a function of the radiosonde age (finding that such a dry bias did not exceed 2% in all cases; and (ii) neglecting the SA dry biases affecting the RS90 and RS92 Humicap data, according to Miloshevich et al. (2006).

(6) The correction of all the *RH* vertical profiles obtained at the previous step, which was made after the previous five steps to remove the *RH* lag errors by following the procedure of Miloshevich et al. (2004), consisting of (i) a first preliminary smoothing procedure, to minimize the tiny changes in slope at the skeleton points, (ii) a lag correction procedure properly adapted to the various RS80-A, RS80-H, RS90 and RS92 Humicap sensors utilizing the most suitable time-constant values, and (iii) a further smoothing procedure to reduce even more the discontinuities appearing in the *RH* vertical profiles.

(7) A final correction of all the dry biases affecting the RS92 data of *RH*, which was performed on the basis of the comparison tests conducted by Miloshevich et al. (2009) for the radiosonde data and the data provided by some reference sensors (cryogenic frost point hygrometer, microwave radiometer, and ARM Surface Temperature/Humidity Reference system) with the purpose of minimizing the day-time solar radiation error (SRE) and other empirical biases, such as the solar heating (SH) dry biases affecting the day-time RS80 data of *RH*, which were corrected by using the algorithm defined by Tomasi et al. (2012) for the highly transparent conditions usually observed in the Antarctic atmosphere. This algorithm was obtained by modifying the values of slope coefficient and effective optical depth of the atmosphere over the short-wave spectrum in the algorithm proposed by Cady-Pereira et al. (2008) to correct the RS80 radiosonde measurements performed at mid-latitudes (providing absolute corrections of about 5% in *RH* for solar zenith angle $SZA = 40°$ and lower than 0.4% in *RH* for $SZA = 86°$). The instrumental errors and empirical biases like those evidenced by Miloshevich et al. (2009) were indeed neglected in the analysis of the RS80 data, because it was taken into account that various instrumental errors were already corrected by using the algorithms of Wang et al. (2002) at the above steps (2)–(5). In particular, the SH dry biases affecting the day-time RS90 data were corrected by using the algorithm of Cady-Pereira et al. (2008) for the RS90 Humicap sensors, defined for the original value of the slope coefficient and effective atmospheric optical depth equal to 0.20, for which an average estimate of the SH dry bias was derived by us to be equal to about 10% in *RH* for $SZA = 40°$ and to about 0.5% in *RH* for $SZA = 86°$. The other instrumental errors affecting the RS90 Humicap sensor, such as the CC dry biases, the TD errors and the SA dry biases, were neglected according to the recommendations of Miloshevich et al. (2004, 2006). The empirical mean biases and SRE errors affecting the RS92 radiosonde data were corrected by using the pair of day-time and night-time algorithms proposed by Miloshevich et al. (2009) over the range $200 \le p \le 1030$ hPa.

To adapt this pair of algorithms to the pressure, temperature and moisture conditions of the polar atmosphere, we modified the algorithms proposed by Miloshevich et al. (2009) for the mid-latitude atmosphere and adapted them for being more properly used in the Arctic atmosphere, through an approach similar to that adopted by Tomasi et al. (2011b, 2012) for the Dome C atmosphere (in Antarctica) over the 250–640 hPa pressure range. For this purpose, the atmospheric model was extended to the pressure range from 1030 to 200 hPa and the *RH* range from less than 2% to more than 90%. The night-time (without solar radiation) and three day-time (with solar radiation) graphical representations of correction factor *G(p, RH)* are shown in Fig. 2.5, as determined for the RS92 Humicap sensor and different solar zenith angles *SZA*, according to the Miloshevich et al. (2009) assumptions. The night-time correction factor *G(p, RH)* calculated for *SZA* = 100° (night-time) can be appropriately used to correct the empirical mean biases, yielding values which are in general lower than unity until to $p = 400$ hPa and then slowly

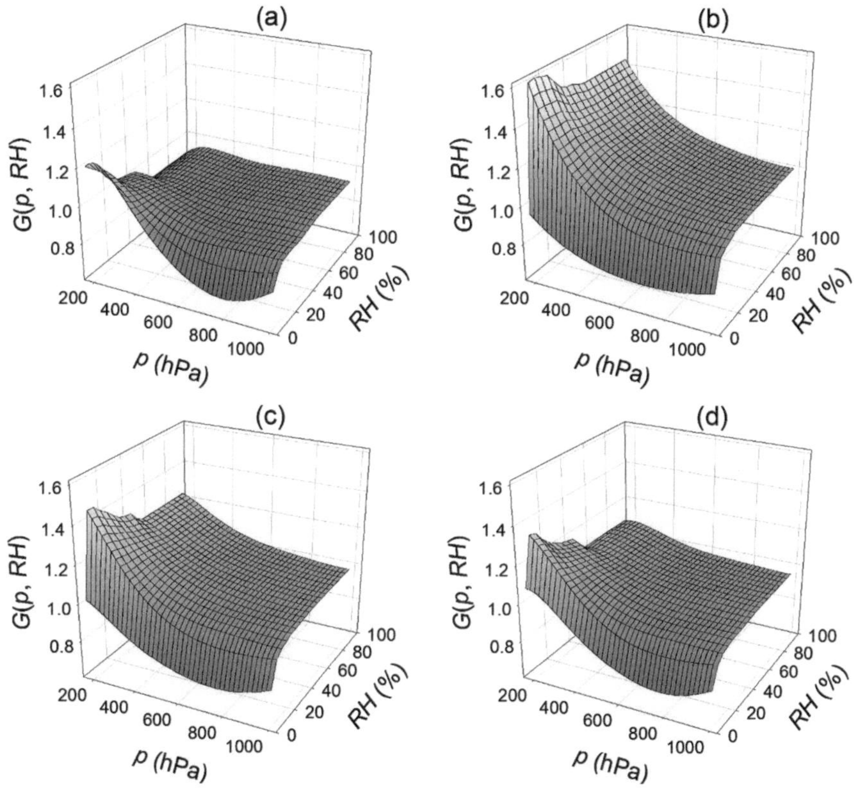

Fig. 2.5 Graphical representations of the correction factor *G(p, RH)* determined for the RS92 Humicap sensor as a function of air pressure *p* (hPa) and air relative humidity RH (%), for different values of solar zenith angles: (**a**) *SZA* = 100° (night-time), (**b**) *SZA* = 54° (day-time), (**c**) *SZA* = 74° (day-time), and (**d**) *SZA* = 86° (day-time). The four graphs have been calculated according to the criteria adopted by Miloshevich et al. (2009) in their procedure aimed at analyzing a large set of RS measurements performed over the 150–1030 hPa pressure range in mid-latitude atmospheres. The correction algorithms providing the values of *G(p, RH)* were estimated by Miloshevich et al. (2009) to be valid in the polar atmosphere over the same ranges of *p* and *RH* considered in the mid-latitude atmosphere and, hence, to be suitable for analyzing the RS data-sets collected at the Arctic coastal and inner low-altitude sites listed in Table 2.1 and examined in the present study

increase at upper altitudes until reaching a value close to 1.06 at the 200 hPa level. The three day-time algorithms shown in Fig. 2.5 and defining the correction factor *G(p, RH)* for *SZA* equal to 54°, 74° and 86°, respectively, include both the corrections made for the empirical mean biases and those accounting for the overall error due to solar heating (SH) by incoming solar radiation, which was assumed to slowly decrease as a function of *SZA* according to the Cady-Pereira et al. (2008) evaluations made over the 50° $\leq$ *SZA* $\leq$ 86° range. The day-time graphical representations of *G(p, RH)* presented in Fig. 2.5 can be reliably used to correct the RS data collected at the Arctic coastal sites located at different latitudes ranging from 67° N to 78° N, during all the summer months. The values of *G(p, RH)* obtained using the Miloshevich et al. (2009) algorithms described above are in general lower than unity for pressure values higher than about 600 hPa, and gradually increasing as a function of altitude until reaching values close to 1.2 at levels of around 200 hPa. Such algorithms of *G(p, RH)* defined according to Miloshevich et al. (2009) were used for each RS measurement, at all the pressure levels, to calculate the night-time and day-time corrections of the *RH* measurements.

Figure 2.6 shows some examples of vertical profiles of air relative humidity *RH(z)* obtained from the RS measurements conducted at: (i) Resolute (Nunavut, Canada) on January 7, 2001 (12:00 UTC), by using a Vaisala RS80-H radiosonde; (ii) Aasiaat (Western Greenland) on June 22, 2002 (12:00 UTC), by using a Vaisala RS90 radiosonde; and (iii) Jan Mayen (Arctic Ocean) on November 15, 2010 (00:00 UTC), by using a Vaisala RS92 radiosonde. It can be noted in these examples that large variations characterize the RH conditions of air at all the tropospheric levels from surface up to about 10 km altitude.

(b) *VIZ-B2 radiosondes*

The *RH* sensor mounted on the VIZ-B2 radiosondes is a carbon hygristor (CH), consisting of a small strip of plastic dipped in a mixture of carbon particles, celluloid resin, and other chemicals, and then dried. Thus, as electric current passes through the strip, the carbon particles allow it to act as a resistor, and the celluloid absorbs (diffuses) water vapour and expands (contracts) with changing *RH*, thus modifying the spacing between carbon particles and changing the resistance (Blackmore and Taubvurtzel 1999). The characteristics (measurement range, resolution, and accuracy) of the CH hygristor are comparable with those of the HUMICAP sensors employed by the Vaisala radiosondes. Actually, examining a set of *RH(z)* measurements performed with either Vaisala RS80-H and VIZ-B2 radiosondes launched at the Homestead (Oklahoma, USA) (36° 09′ N, 98° 24′ W) and Dodge City (37° 46′ N, 100° 01′ W) stations (about 210 km away one from the other) during the International H_2O Project (IHOP, 2002), Wang et al. (2003) found that: (i) the Vaisala RS-80-H radiosondes provided reliable values of *RH(z)* in the low and middle troposphere, and relatively small dry biases in the upper troposphere, and (ii) the VIZ-B2 radiosondes were affected by time-lag errors in measuring *RH(z)* at all the middle and upper troposphere levels, due to the slower response of the CH sensor caused by a less efficient ventilation. As pointed out by Mattioli et al. (2005), the VIZ-B2 data were in general filtered because of the presence of spikes in the *RH*

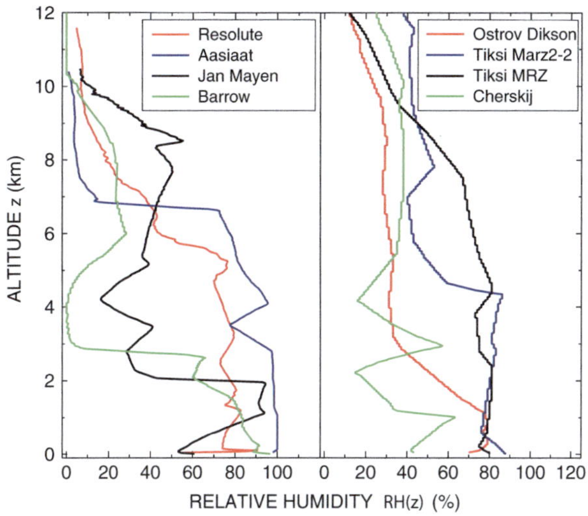

Fig. 2.6 Left part: Vertical profiles of air relative humidity *RH(z)* obtained from the RS measurements conducted in different seasons and years at the following Arctic sites: (i) Resolute (Nunavut, Canada) using a Vaisala RS80-H radiosonde on January 7, 2001 (12:00 UTC); (ii) Aasiaat (Western Greenland) using a Vaisala RS90 radiosonde on June 22, 2002 (12:00 UTC); (iii) Jan Mayen (Arctic Ocean) using a Vaisala RS92 radiosonde on November 15, 2010 (00:00 UTC); and (iv) Barrow (Alaska, USA) using a VIZ-B2 radiosonde on March 22, 2002 (12:00 UTC). Right part: As in the left part for the following RS measurements conducted at: (v) Ostrov Dikson (Siberia, Russia) using a Marz2-1 radiosonde on February 13, 2009 (12:00 UTC); (vi) Tiksi (Siberia, Russia) using a Marz2-2 radiosonde on September 12, 2015 (00:00 UTC); (vii) Tiksi (Siberia, Russia) using a MRZ radiosonde on September 15, 2001 (12:00 UTC); and (viii) Cherskij (Siberia, Russia) using a BAR radiosonde on May 25, 2008 (00:00 UTC)

measurements, and then fitted to a regular grid by applying a linear interpolation procedure in height, with: (i) a resolution of 10 m over the altitude range from surface-level to 10 km altitude; and (ii) a resolution of 100 m from $z = 10$ km to the highest level reached by the radiosonde. The statistical comparison of the RH measurements made with the VIZ-B2 and the Vaisala RS90 radiosondes showed that the RH profiles made with the VIZ radiosondes in the first 300 m above the surface are 5% lower on average than those provided by the Vaisala RS90 radiosondes. Above this level, the estimates of *RH(z)* provided by the VIZ sensor are larger on average than those of the Vaisala H-Humicap, this bias increasing gradually with height to reach a value of about 20% at around the 8 km level. The average RH difference was estimated to be smaller than 4% below 10 km, and to increase to 16% above 10 km (i.e. above the troposphere), the average root-mean-square difference being equal to 12% below the 10 km level and to 18% above the 10 km height. Some significant negative excursions were noted in some RS measurements: for instance, a difference of −70% was recorded between the 5 and 6 km heights in the soundings taken on March 21, at 23:00 UTC, when the RH from the VIZ sensor reached values as low as 1%; and a RH difference of −60% was measured at the altitude of 4 km on March 22 at 11:00 UTC. Thus, Mattioli et al. (2005) concluded that the RH measurements provided by the VIZ sensor can be subject to an apparent bias with respect to the Vaisala RS-80-H sensors above the tropopause level, with a difference of 16–20% on average. The CH hygristor measurements made with the VIZ-B2 radiosondes showed also to have a lower accuracy, presumably because of the different time lag and response time characteristics of the radiosonde models.

Mattioli et al. (2007) carried out a statistical comparison of the vertical profiles of *RH(z)* obtained from the VIZ-B2 and Vaisala RS90 data collected during the 2004 NSAWRE field campaign conducted at Barrow (Alaska) in spring 2004. At the 1000 hPa level, the VIZ profiles of *RH(z)* were found to be lower on average than those of the RS90 H-Humicap sensor by 2%, such a relative difference becoming larger at the upper levels until to 925 hPa. The difference $\Delta RH(z)$ was estimated to be close to 4% at the 250 hPa level (i.e. at ~10 km altitude), and to increase considerably at the upper troposphere levels, until reaching a value of about 23% and presenting an average value of 17% in the lower stratosphere. On the basis of an overall number of 38 comparison tests, Mattioli et al. (2007) estimated that the difference $\Delta RH(z)$ was equal to: (i) 5.1% over the range of $T(z) > -35$ °C; (ii) 10.3% over the range of *T(z)* from −35 °C to −50 °C; and (iii) 16.6% over the range of $T(z) < -50$ °C. The corresponding standard deviations were estimated to be equal to 12% over the upper troposphere height range and 9% over the lower stratosphere height range, respectively, such a bias being substantially due to the slow response time of the CH sensor at low temperatures (Blackmore and Taubvurtzel 1999), which cannot provide evaluations of *RH(z)* when the ambient temperature is lower than −60 °C.

Examining the Upper Tropospheric Humidity measurements retrieved from microwave satellite data, Moradi et al. (2013) estimated the performances of the VIZ-B2 radiosondes, finding that the VIZ-B2 data of RH show a large wet bias (of about 16% on average). Comparing the vertical profiles of *RH(z)* derived from the data-sets provided by an overall number of 490 pairs of VIZ-B2 and Vaisala RS80-H radiosondes launched within a half hour at the Norman and Purcell stations during the 1996–2002 ARM program, Wang and Young (2005) found that the VIZ-B2 measurements of *RH(z)* are higher than those provided by the Vaisala radiosondes in both the upper troposphere and lower stratosphere regions (i.e. over the altitude range from 8 to about 12 km), mainly because of a response lack of VIZ's carbon hygristor (CH), as previously pointed out by Wang et al. (2003).

The average correction terms calculated by Mattioli et al. (2007) through the comparison between the performances of the VIZ-B2 and the Vaisala RS90 radiosondes in measuring the RH conditions of air are given in Table 2.3. It can be seen that the difference $\Delta RH(z)$ is relatively small at the ground level (<1%) but assumes values ranging between 2.7% and more than 6% at tropospheric levels (with peaks at altitudes of 0.73 km, 2.96 km and 9.12 km), and increasing further at stratospheric altitudes, where it was found to range mainly between 14% and 23%. Using these average correction values, we have converted the VIZ-B2 measurements of *RH(z)* into equivalent evaluations of *RH(z)* that would be provided by the Vaisala RS90 radiosondes, which have been further corrected by following the procedure described above for analyzing the RS90/92 Thermocap data of *RH(z)* for the field errors made by the Vaisala RS90 radiosondes. The vertical profile of air relative humidity *RH(z)* obtained at Barrow (Alaska) on March 22, 2002 (12:00 UTC) by using a VIZ-B2 radiosonde is shown in Fig. 2.6 as an example of the performances provided by this radiosonde model up to levels higher than 12 km, yielding RH values of a few percents.

(c) *Russian radiosondes*

The *RH* measurement range of the MRZ and Marz2-2 radiosondes was estimated by the Central Aerological Observatory (2003) to be from 15% to 100% for temperatures varying between −40 °C and + 50 °C, while similar performances were evaluated to characterize the performances of the BAR and Marz2-1 radiosondes. The accuracy of the MRZ radiosondes was evaluated to be equal on average to: (i) 8% over the 0% −10% range of *RH*, (ii) 5% over the 10% −90% range of *RH*, and (iii) 8% over the 90–100% range of *RH*, while a higher accuracy of 15% was attributed by the Central Aerological Observatory (2003) to the *RH* measurements performed with Marz2-2 radiosondes over the 10–90% range of *RH* for temperatures lower than −40 °C. In order to give a measure of the efficiency of the various Russian radiosondes in determining the vertical profiles of air relative humidity *RH(z)*, the following four vertical profiles of *RH(z)* are shown in Fig. 2.6, as obtained at: (i) Ostrov Dikson, with a Marz2-1 radiosonde launched on February 13, 2009 (12:00 UTC); (ii) Tiksi, with a Marz2-2 radiosonde launched on September 12, 2015 (00:00 UTC); (iii) Tiksi, with a MRZ radiosonde launched on September 15, 2001 (12:00 UTC); and (iv) Cherskij, with a BAR radiosonde launched on May 25, 2008 (00:00 UTC). The examples shown in the left part of Fig. 2.6 clearly indicate that all the Russian radiosonde models can provide measurements of parameter *RH(z)* up to more than 12 km altitude with good accuracy, providing evidence of the variations occurring at the various tropospheric levels.

2.2.4 Calculation of the Vertical Profiles of Absolute Humidity for the RS Measurements Performed with the Various Radiosonde Models

Using the data-sets derived above to define the vertical profiles of air pressure *p(z)*, air temperature *T(z)*, and relative humidity *RH(z)*, the vertical profiles of absolute humidity *q(z)* were calculated for all the RS measurements listed in Table 2.1. Parameter *q(z)* was measured in g m^{-3} at each significant level *z* in terms of the well-known equation of state for water vapour,

$$q(z) = e(z)\,[R_w\,T(z)]^{-1} \tag{2.1}$$

where: (i) R_w is the gas constant for water vapour, equal to 0.4615 J g^{-1} $°K^{-1}$; (ii) the water vapour partial pressure $e(z)$ is measured in hPa and given at each level by the product $RH(z) \times E[T(z)]$, in which $E[T(z)]$ is the saturated water vapour pressure in the pure phase over a plane surface of pure liquid water; and (iii) air temperature $T(z)$ is measured in °K. Parameter $E[T(z)]$ can be reliably calculated by using the well-known formula of Murphy and Koop (2005), determined by taking into account some molar heat capacity measurements made over the $123 \leq T \leq 332$ °K temperature range.

Some examples of the vertical profiles of parameters $T(z)$, $RH(z)$, $e(z)$ and $q(z)$ are shown in Fig. 2.7, over the tropospheric altitude range from surface to 12 km, as obtained by applying the above-described correction procedures to the RS measurements recorded at the four Arctic sites of Danmarkshavn, Sodankylä, Barrow and Tiksi, chosen among those listed in Table 2.1. Since the Vaisala RS-80H,

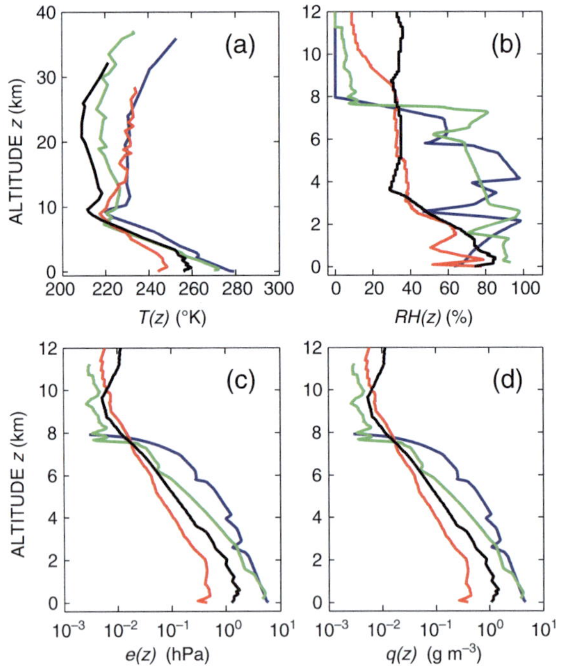

Fig. 2.7 Examples of the vertical profiles of: (**a**) air temperature $T(z)$, (**b**) relative humidity $RH(z)$, (**c**) water vapour partial pressure $e(z)$, and (**d**) absolute humidity $q(z)$, all obtained by applying the above-described correction procedures to the RS measurements recorded at the following four Arctic sites over the altitude range from the surface level to the tropopause level: (1) Danmarkshavn (Greenland) on June 19, 2002 (12:00 UTC) (blue curves) by using a Vaisala RS80-H radiosonde; (2) Sodankylä (Finland) on March 20, 2007 (00:00 (UTC) (green curves) by using a Vaisala RS92 radiosonde; (3) Barrow (Alaska, USA) on January 15, 2012 (00:00 UTC) (red curves) by using a VIZ-B2 radiosonde; and (4) Tiksi (Siberia, Russia) on October 24, 2015 (00:00 UTC) (black curves) by using a Marz2–2 radiosonde

Vaisala RS92, VIZ-B2 and Marz2-2 radiosondes were employed at these four sites, Fig. 2.7 provides a well-defined information on the thermodynamic characteristics of the Arctic atmosphere observed by means of the RS measurements carried out at various sites with different technical procedures. In addition, the comparison of the four vertical profiles of *T(z)* gives a measure of the large variability of the thermal conditions of the Arctic atmosphere, with variations recorded at the low tropospheric levels that result to be higher than 20 °C from one season to another. Even more dispersed features were found in the high troposphere and tropopause regions, where the levels of the temperature minima were estimated to vary from less than 9 km to more than 11 km, depending on both season and latitude. Largely scattered values of *RH* were recorded at the tropospheric levels lower than 8–9 km, clearly due to the strong changes in the meteorological characteristics of the polar atmosphere often occurring from one day to another and from one season to another. Nevertheless, the vertical profiles of *e(z)* were found to exhibit in general decreasing features with height, following an approximately exponential trend in all cases and presenting values differing by no more than one order of magnitude from one case to another at all the tropospheric levels. Correspondingly, the vertical profiles of *q(z)* were found to decrease as a function of height, presenting similar patterns substantially due to the fact that $\rho_w(z)$ is approximately proportional to *e(z)*, according to Eq. (2.1).

(a) *Vaisala radiosondes*

Some examples of the vertical profiles of air temperature *T(z)*, relative humidity *RH(z)*, water vapour partial pressure *e(z)*, and absolute humidity *q(z)*, obtained by applying the above-described correction procedures to the field measurements performed at (i) Danmarkshavn (Greenland) on June 19, 2002 (12:00 UTC) by using a Vaisala RS80-H radiosonde model, and (ii) Sodankylä (Finland) on March 20, 2007 (00:00 (UTC) by using a Vaisala RS92 radiosonde model are shown in Fig. 2.7, clearly indicating that the Vaisala radiosonde models are all suitable to provide reliable measurements of *T(z)* up to altitudes higher than 30 km, as well as corrected measurements of the moisture parameters up to the high-troposphere levels (i.e. until to about 8 km height in the example of Danmarkshavn and to about 11 km height in the example of Sodankylä).

(b) *VIZ-B2 radiosondes*

An accurate test on the reliability of the RH measurements carried out with the VIZ-B2 radiosondes at Barrow (Alaska) was made by Mattioli et al. (2007) during the above-mentioned 2004 NSAWRE experiment, finding that the RH comparison tests made between the VIZ-B2 and Vaisala RS-90 radiosondes attributed a set of biases made by the VIZ-B2 hygristor with respect to the RS-90 sensor assuming: (i) only a slightly negative value at the surface, (ii) values not exceeding +1% within the whole troposphere (i.e. until to an altitude close to 9 km), and (iii) values varying between +1% and + 2.5% at all the stratospheric levels up to 26 km, as can be seen in Table 2.3. On the basis of these results, the VIZ-B2 measurements of *RH(z)* were corrected in the present study by using the relative correction terms defined by Mattioli et al. (2007) at altitudes ranging between the surface-level and 26 km height. As reported in Table 2.3, such correction terms were found to vary between +20% (± 15%) at level $z = 10$ km, and + 15% (±9%) at level $z = 26$ km. Applying these average estimates of the correction factor determined as a function of altitude z, the $RH(z)$ values provided by the VIZ-B2 radiosondes were first corrected in order to obtain the equivalent values of $RH(z)$ to those given by the Humicap sensors mounted on the Vaisala RS90 radiosondes, and subsequently corrected for the errors affecting the RH data as a result of the wrong calibration of the RS90 Humicap sensors used for comparison. In carrying out these corrections, the corrections of the VIZ-B2 RH data were calculated at each radiosonde level by adding or subtracting the difference $\Delta RH(z)$ obtained through linear interpolation in altitude of the values of $\Delta RH(z)$ given in Table 2.3. An example of the vertical profiles of parameters $T(z)$, $RH(z)$, $e(z)$ and $q(z)$, derived from the VIZ-B2 radiosounding measurement carried out at Barrow (Alaska) on January 15, 2012 (00:00 UTC) is shown in Fig. 2.7, clearly indicating that the VIZ-B2 radiosonde models can provide accurate and reliable measurements of $T(z)$ up to altitudes higher than 28 km, and reliable measurements of the various moisture parameters up to altitudes higher than 12 km.

(c) *Russian radiosondes*

No corrections were made to the original RH measurements provided by the Russian radiosondes at the Ostrov Dikson, Tiksi and Cherskij RS stations, because we did not know with good accuracy the characteristics of the RH sensors mounted on the Russian radiosondes. However, a preliminary selection of such RH data was made at altitudes higher than 12 km, by (i) calculating the values of the stratospheric water vapour mixing ratio $Q(z)$ from the RS data recorded at the 12–30 km altitudes, and (ii) comparing systematically these evaluations of $Q(z)$ with those evaluated by Tomasi et al. (2011a) to determine the monthly mean vertical profiles of $Q(z)$ by averaging the MIPAS2D data collected at the 65 °N, 70 °N, 75 °N and 80 °N latitudes. All the radiosounding data giving values of $Q(z)$ differing from the MIPASS2D monthly average value at the same altitudes by more than ± two standard deviations were discarded from the present analysis, with the purpose of avoiding too high systematic errors made by the hygrometric sensors at the stratospheric altitudes. The RH measurements provided by the Russian radiosondes employed at the Ostrov Dikson, Tiksi and Cherskij stations have been found in some upper tropospheric layers to be considerably high during the crossing flight through ice particle clouds present at the higher tropospheric altitudes, presumably because of ice formation inside the dielectric matter of the humidity sensors. In order to remove such "wrong" data from the analysis of the RH data recorded at the three above-selected Russian stations of Ostrov Dikson, Tiksi and Cherskij, we have calculated the values of tropospheric precipitable water W_t within the layer from surface-level to 12 km height by: (i) determining the vertical profile of absolute humidity $q(z)$ in terms of Eq. (2.1) over the height range from surface to 12 km on the basis of the RS data recorded by the Russian radiosondes, and (ii) integrating each vertical profile of $q(z)$ to determine precipitable water. These evaluations of W_t were found to be very low and unrealistic in some cases, being W_t

smaller than 0.1 g cm^{-2} not only in winter but also on some spring and summer Arctic days. Therefore, in order to reject all the RS measurements affected by such underestimation errors and, at the same time, to remove the evaluations of W_t affected by high overestimation errors of *RH(z)* made at the various altitudes, we have adopted a rigorous selection procedure, consisting of the following two steps: (i) the relative frequency histograms of W_t were determined for each month of the year, from which the monthly average values of W_t were calculated with their standard deviations; and (ii) all the RS measurements providing values of W_t differing from the monthly average value by more than ± two standard deviations (i.e. by more than ±47.7%) were discarded, by assuming that such cases were due to the insensitiveness of the *RH* sensor to the real moisture conditions of the atmosphere or to the formation of ice crystals inside the dielectric matter of the humidity sensor (occurred when the radiosonde has crossed an ice particle cloud, as it occurs frequently at altitudes close to the tropopause level).

After this careful selection procedure, the residual RS measurements were analyzed by following the classical procedures described above. An example of the vertical profiles of *T(z)*, *RH(z)*, *e(z)* and *q(z)*, derived from the RS data collected at Tiksi (Siberia, Russia) on October 24, 2015 (00:00 UTC) with a Marz2-2 radiosonde is shown in Fig. 2.7, presenting a vertical profile of *T(z)* reaching an altitude higher than 34 km, and vertical profiles of the moisture parameters *RH(z)*, *e(z)* and *q(z)* up to top-levels higher than 12 km.

The daily vertical profiles of parameters *p(z)*, *T(z)* and *RH(z)* obtained in the previous section by correcting the raw data provided by the overall 123029 RS measurements performed at the above-selected 14 Arctic stations have been properly examined to give evidence of the annual cyclic variations of these vertical profiles from one year to another. The daily vertical profiles of absolute humidity *q*(*z*) were then calculated in terms of Eq. (2.1), from which the daily values of precipitable water *W* were determined by: (i) integrating *q(z)* over the whole altitude range covered by the radiosonde up to 12 km height (and, hence, determining the tropospheric water vapour content W_t), and (ii) adding the stratospheric water vapour content W_s derived from the MIPAS/ENVISAT limb-scanning measurements of water vapour mixing ratio carried out by Tomasi et al. (2011a). Accurate monthly mean calculations of W_s were performed by Tomasi et al. (2011a) in the polar atmosphere, over the 12–50 km altitude range, by analysing the set of MIPAS/ENVISAT limb-scanning measurements recorded from July 2002–April 2010 at latitudes from 65 °N to 90 °N taken in steps of 5°. It was found that stratospheric water vapour content W_s assumed values ranging: (i) between 4.7 10^{-4} g cm^{-2} and 7.0 10^{-4} g cm^{-2} during the year; (ii) between 4.7 10^{-4} and 5.3 10^{-4} g cm^{-2} in the winter months, and (iii) between 5.7 10^{-4} and 7.0 10^{-4} g cm^{-2} in the summer months. On the basis of these calculations, the daily values of total precipitable water *W* were determined by summing the values of W_t (obtained by integrating all the vertical profiles of *q*(*z*) derived from the 123029 RS measurements collected at the above-selected 14 Arctic stations, from the surface-level to the 12 km altitude) with those of W_s given in Table 2.4.

Table 2.4 Monthly mean values of stratospheric water vapour content W_s in the Arctic atmosphere (measured in mg cm^{-2}), as calculated with their standard deviations by Tomasi et al. (2011a) over the 12–50 km altitude range

	Latitude					
Month	65 °N	70 °N	75 °N	80 °N	85 °N	90 °N
January	0.51 ± 0.32	0.50 ± 0.27	0.49 ± 0.21	0.47 ± 0.22	0.48 ± 0.21	0.47 ± 0.23
February	0.51 ± 0.25	0.51 ± 0.26	0.50 ± 0.22	0.50 ± 0.20	0.51 ± 0.20	0.49 ± 0.19
March	0.49 ± 0.28	0.50 ± 0.23	0.51 ± 0.18	0.50 ± 0.19	0.51 ± 0.17	0.49 ± 0.19
April	0.55 ± 0.27	0.55 ± 0.25	0.54 ± 0.21	0.53 ± 0.18	0.53 ± 0.17	0.54 ± 0.26
May	0.53 ± 0.24	0.54 ± 0.20	0.54 ± 0.18	0.54 ± 0.18	0.54 ± 0.18	0.54 ± 0.18
June	0.57 ± 0.45	0.57 ± 0.27	0.59 ± 0.38	0.58 ± 0.19	0.57 ± 0.43	0.58 ± 0.83
July	0.70 ± 0.77	0.61 ± 0.61	0.60 ± 0.35	0.60 ± 0.42	0.58 ± 0.46	0.61 ± 0.98
August	0.70 ± 0.73	0.65 ± 0.48	0.60 ± 0.47	0.61 ± 0.29	0.60 ± 0.27	0.59 ± 0.32
September	0.65 ± 0.46	0.59 ± 0.26	0.61 ± 0.19	0.58 ± 0.15	0.56 ± 0.20	0.57 ± 0.20
October	0.62 ± 0.99	0.60 ± 0.20	0.59 ± 0.21	0.58 ± 0.18	0.58 ± 0.17	0.59 ± 0.18
November	0.57 ± 0.39	0.57 ± 0.30	0.55 ± 0.24	0.54 ± 0.20	0.53 ± 0.17	0.52 ± 0.18
December	0.53 ± 0.28	0.52 ± 0.24	0.50 ± 0.22	0.49 ± 0.20	0.49 ± 0.18	0.48 ± 0.18

2.3 Annual Variations in the Monthly Mean Vertical Profiles of Air Pressure and in the Monthly Mean Values of Surface-Level Pressure

The vertical profiles of air pressure *p(z)* shown in Fig. 2.2 clearly indicate that *p(z)* decreases as a function of altitude *z* in a nearly exponential fashion [according to the analytical function exp. $(-z/H_p)$], with best-fit daily values of scale height H_p ranging in general between 7.0 and 7.8 km throughout the year. The evaluations of H_p (made during the various seasonal periods and for different thermodynamic conditions of the atmosphere) clearly indicate that parameter H_p is subject to vary appreciably with season, being closely related to local effects (such as the wind speed and air density variations, caused by temperature and *RH* variations) and affected by the slight changes of air composition occurring along the vertical atmospheric path in the various periods of the year. The values of *p(z)* measured in summer are in general appreciably higher than those measured in winter at all the RS stations, with differences gradually increasing with height until becoming appreciably relevant at the tropopause level and particularly marked at the stratospheric altitudes higher than 30 km.

To better illustrate these aspects, the monthly mean vertical profiles of *p(z)* obtained in January and July by averaging the RS data collected during the above-selected years at the Barrow (Alaska), Alert (Nunavut, Canada), Ny-Ålesund (Spitsbergen, Svalbard), and Tiksi (Siberia, Russia) sites are shown in Fig. 2.8. Examining these data, it was found that the best-fit monthly values of scale height H_p are considerably different in January and July, being: (i) those measured in January equal to 7.175 km at Barrow, 7.100 km at Alert, 7.200 km at Ny-Ålesund

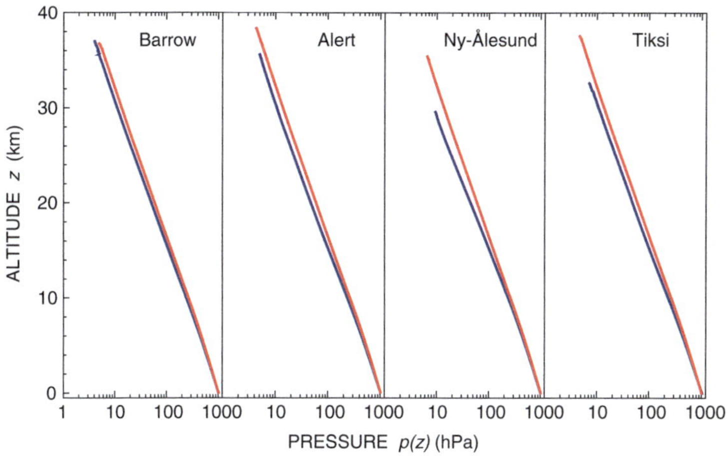

Fig. 2.8 Monthly mean vertical profiles of air pressure $p(z)$ obtained in January (blue curves) and July (red curves) from the RS measurements conducted over the 15-year period at the following four Arctic stations: (1) Barrow (Alaska, USA) (71° 17′ N, 156° 47′ W, 19 m above mean sea level (a.m.s.l.)); (2) Alert (Nunavut, Canada) (82° 30′ N, 62° 21′ W, 65 m a.m.s.l.); (3) Ny-Ålesund (Spitsbergen, Svalbard) (78° 54′ N, 11° 53′ E, 11 m a.m.s.l.); and (4) Tiksi (Siberia, Russia) (71° 35′ N, 128° 55′ E, 7 m a.m.s.l.)

and 7.000 km at Tiksi, and (ii) those measured in July equal to 7.725 km at Barrow, 7.675 km at Alert, 7.625 km at Ny-Ålesund and 7.725 km at Tiksi.

The time-patterns of the monthly mean values of air pressure p_o derived at the surface-level and of air pressure $p(z)$ measured at the three levels $z = 5$ km, $z = 12$ km and $z = 25$ km from the RS measurements conducted over the 2001–2015 period at the Resolute (Nunavut, Canada) and Ny-Ålesund (Spitsbergen, Svalbard) stations are shown in the left part of Fig. 2.9, while those obtained from the RS measurements carried out at Sodankylä (Finland) and Tiksi (Siberia, Russia) stations are shown in the right part. It can be noted that the surface-level data do not exhibit a periodical behaviour during the various years, indicating that the meteorological variations exert a marked influence on the local pressure conditions. The monthly mean values of p_o recorded at Resolute and Ny-Ålesund result to overlap ones with the others almost completely over the period from 2005 t0 2015, being the two stations located at levels close to the sea-level. Conversely appreciable differences between the time-patterns of p_o were found at the Sodankylä and Tiksi stations over the whole 2001–2015 period, being the monthly mean values of p_o recorded at Tiksi appreciably higher than those measured at Sodankylä, since Tiksi is located close to the coast, while Sodankylä is located in the inner of Lapland (Northern Finland), at the surface-level of 179 m a.m.s.l. The time-patterns of the monthly mean values of $p(z)$ recorded at the altitudes of 5, 12 and 25 km, result to be very similar, as evidenced in Fig. 2.9. The comparison between the data measured at Resolute and Ny-Ålesund (as shown in the left part of Fig. 2.9) and at Sodankylä and Tiksi (as shown in the right part of Fig. 2.9) gives evidence of the fact that the monthly mean

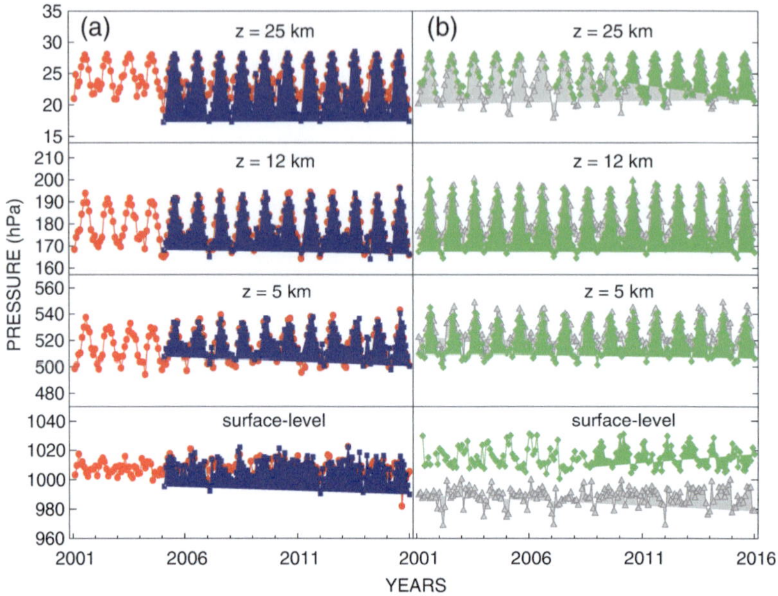

Fig. 2.9 Left part (**a**): Time-patterns of the monthly mean values of air pressure p_o derived at the surface-level and of air pressure $p(z)$ measured at the three levels $z = 5$ km, $z = 12$ km and $z = 25$ km from the RS measurements conducted over the 2001–2015 period at the Resolute (Nunavut, Canada) (red circles) and Ny-Ålesund (Spitsbergen, Svalbard) (blue squares) stations. Right part (**b**): As in the left part from the RS measurements performed at the Sodankylä (Finland) (grey triangles) and Tiksi (Siberia, Russia) (green diamonds) stations

values of air pressure measured at the heights of 5, 12 and 25 km exhibit very similar annual cycles during the 2001–2015 period, presenting in general marked minima in the winter months (from December to February) and pronounced maxima in the summer months (from June to August).

The daily values of surface-level pressure p_o derived from the RS measurements performed at the above-selected 14 Arctic sites over the 2001–2015 period were examined in order to determine the monthly mean values of this parameter at each site, over the whole 15-year period. Table 2.5 gives the 15-year average monthly mean values of p_o with their standard deviations. Examining the data of Table 2.5, it can be noted that the maximum and minimum values are recorded in the months of: (i) March and September at Barrow and Inuvik, (ii) May and August at Cambridge Bay and Resolute, (iii) April and August at Eureka, (iv) May and January at Alert, Aasiaat and Danmarkshavn, (v) June and January at Jan Mayen, (vi) May and November at Ny-Ålesund and Sodankylä, (vii) February and October at Ostrov Dikson, and (viii) January and July at Tiksi and Cherskij. These findings clearly show that similar variations characterize the surface-level pressure conditions within the various regional sectors of the Arctic. Analyzing the annual mean values of p_o determined at the 14 above-selected Arctic sites for the 15 years from 2001 to 2015,

Table 2.5 Monthly mean values of surface-level pressure p_o (hPa) determined with their standard deviations from the RS measurements performed at the 14 Arctic sites chosen in the present study

Station and RS measurement period	Month											
	Jan.	Feb.	March	Apr.	May	June	July	Aug.	Sept.	Oct.	Nov.	Dec.
Barrow (2001–2015)	1017.0 ± 13.0	1017.4 ± 11.0	1020.9 ± 12.4	1016.7 ± 9.4	1017.4 ± 7.7	1013.9 ± 6.9	1010.7 ± 7.4	1010.8 ± 7.2	1009.4 ± 8.8	1010.5 ± 9.3	1012.0 ± 11.8	1013.3 ± 11.8
Inuvik (2001–2009 and 2012–2015)	1005.1 ± 12.6	1006.4 ± 10.5	1009.1 ± 11.5	1005.1 ± 9.5	1005.7 ± 8.2	1001.2 ± 6.5	999.0 ± 6.5	999.1 ± 6.6	998.9 ± 8.8	1000.3 ± 9.8	1001.4 ± 11.3	1001.9 ± 12.6
Cambridge Bay (2001–2015)	1012.5 ± 10.4	1015.3 ± 9.7	1016.4 ± 10.3	1016.5 ± 10.2	1017.1 ± 7.5	1011.0 ± 7.2	1007.9 ± 7.1	1007.9 ± 6.3	1009.8 ± 8.1	1011.3 ± 9.2	1011.0 ± 10.5	1012.0 ± 11.0
Resolute (2001–2015)	1005.8 ± 9.6	1009.3 ± 10.0	1011.7 ± 9.9	1012.4 ± 10.0	1012.7 ± 7.7	1008.9 ± 7.8	1005.0 ± 6.6	1004.2 ± 8.0	1005.3 ± 7.5	1007.7 ± 9.5	1006.7 ± 9.0	1006.4 ± 10.8
Eureka (2001–2015)	1011.6 ± 10.2	1015.1 ± 10.4	1018.2 ± 10.4	1018.3 ± 10.6	1017.5 ± 8.0	1013.0 ± 8.1	1008.7 ± 7.3	1008.6 ± 7.7	1010.0 ± 7.8	1014.6 ± 9.8	1012.8 ± 9.3	1012.5 ± 11.3
Alert (2001–2015)	1003.1 ± 11.2	1006.3 ± 10.8	1010.2 ± 12.0	1010.2 ± 10.8	1011.2 ± 8.4	1006.7 ± 8.5	1003.5 ± 7.5	1003.4 ± 7.4	1003.1 ± 7.8	1007.2 ± 9.6	1004.4 ± 9.5	1004.2 ± 13.4
Aasiaat (2001–2009)	996.0 ± 10.9	1001.5 ± 12.1	1005.7 ± 11.6	1008.3 ± 10.5	1009.2 ± 8.6	1007.1 ± 8.5	1007.0 ± 6.8	1004.0 ± 7.4	1003.3 ± 7.1	1004.8 ± 10.5	1001.2 ± 10.4	996.2 ± 11.4
Danmarkshavn (2001–2009)	1006.4 ± 12.6	1012.7 ± 14.0	1016.0 ± 11.0	1017.2 ± 10.8	1018.5 ± 8.4	1014.3 ± 7.3	1011.3 ± 7.7	1012.3 ± 6.9	1010.6 ± 7.2	1014.0 ± 11.4	1010.6 ± 10.1	1007.8 ± 12.2
Jan Mayen (2001–2015)	1001.0 ± 14.4	1005.5 ± 14.4	1007.2 ± 14.6	1011.0 ± 11.8	1014.4 ± 8.4	1015.0 ± 6.8	1011.1 ± 7.3	1010.9 ± 7.7	1005.7 ± 10.3	1008.1 ± 13.2	1003.3 ± 12.7	1001.6 ± 14.9
Ny-Ålesund (2005–2015)	1003.2 ± 12.3	1007.9 ± 13.9	1007.5 ± 12.9	1010.1 ± 10.3	1015.2 ± 8.6	1011.2 ± 7.8	1010.8 ± 7.5	1008.6 ± 8.1	1006.8 ± 9.9	1007.2 ± 9.9	1002.6 ± 13.6	1002.6 ± 15.7
Sodankylä (2001–2015)	986.1 ± 15.4	987.9 ± 14.6	986.3 ± 12.1	990.1 ± 10.6	991.3 ± 8.5	989.4 ± 7.1	990.2 ± 7.4	990.5 ± 7.6	988.9 ± 10.8	987.8 ± 11.3	984.2 ± 13.9	984.6 ± 15.6
Ostrov Dikson (2001–2015)	1009.4 ± 14.9	1011.4 ± 14.5	1006.1 ± 13.5	1006.9 ± 13.7	1007.4 ± 9.4	1004.5 ± 8.0	1004.8 ± 7.6	1003.7 ± 9.1	1003.7 ± 10.6	1002.5 ± 11.6	1005.1 ± 13.2	1007.1 ± 14.6
Tiksi (2001–2015)	1022.8 ± 11.2	1021.9 ± 11.1	1019.5 ± 10.8	1016.5 ± 10.5	1012.9 ± 9.2	1009.1 ± 6.6	1008.8 ± 6.5	1009.1 ± 6.9	1010.8 ± 8.5	1012.9 ± 9.7	1016.6 ± 9.8	1022.4 ± 10.5
Cherskij (2004–2015)	1019.6 ± 16.6	1018.3 ± 15.1	1017.1 ± 10.3	1014.0 ± 11.3	1010.6 ± 8.8	1005.1 ± 6.3	1004.6 ± 7.3	1005.4 ± 11.0	1007.3 ± 11.7	1010.5 ± 11.4	1013.4 ± 13.9	1017.7 ± 13.1

it was found that parameter p_o was not subject to long-term trends to increase or decrease throughout the analyzed period of RS measurements. Actually, analyzing the time-series of the monthly mean values of p_o obtained at the herein-selected 14 Arctic sites from 2001 to 2015 in order to determine the slope coefficients β_p and the corresponding correlation coefficients r_p of the regression lines, it was found that: (a) in Northern America (Barrow, Inuvik, Cambridge Bay, Resolute and Eureka), coefficient β_p was estimated to vary between −0.73 hPa per decade at Eureka (with $r_p = -0.19$) and + 0.80 hPa per decade at Inuvik (with $r_p = +\ 0.29$); (b) in Greenland and Arctic Ocean (Alert, Aasiaat, Danmarkshavn, Jan Mayen and Ny-Ålesund), β_p was estimated to range from −1.06 hPa per decade at Jan Mayen (with $r_p = -0.18$) to +0.97 hPa per decade at Aasiaat (with $r_p = +0.17$); (c) in Northern Europe (Sodankylä) β_p was estimated to be equal to −0.03 hPa per decade (with $r_p = 1.5\ 10^{-3}$); and (d) in Northern Siberia (Ostrov Dikson, Tiksi and Cherskij) β_p was estimated to vary between −0.13 hPa per decade at Tiksi (with $r_p = -0.04$) and + 3.12 hPa per decade at Aasiaat (with $r_p = +\ 0.49$). Therefore, no evidence was inferred in the present analysis on the existence of appreciable long-period variations of the surface-level air pressure conditions taking place in the Arctic region.

2.4 Annual Variations in the Surface-Level Temperature Conditions of the Arctic Atmosphere

The vertical profiles of $T(z)$ and the values of surface-level temperature T_o recorded at the 14 selected Arctic stations were corrected for the lag and heating errors by applying the correction procedures described above in Sect. 2.2.2. Thereupon, the values of $T(z)$ were linearly interpolated in altitude at all the fixed levels by following the procedures described in Sect. 2.2.2 for the various radiosonde models, in such a way to define homogeneous vertical profiles of $T(z)$ for all the RS measurements collected at the 14 above-selected Arctic stations. Figure 2.10 shows eight vertical profiles of $T(z)$ recorded at eight of the above-selected Arctic stations and in different seasonal periods, which provide evidence that air temperature can vary largely by: (i) more than 30 °K at the surface, passing from winter to summer; (ii) 10–20 °K from winter to summer in the upper troposphere until to about 10 km altitude; and (iii) more than 20 °K throughout the year at stratospheric altitudes from 10 to 25 km.

The time-patterns of the monthly mean values of surface-level air temperature T_o and air temperature $T(z)$ defined at the 5, 12 and 25 km levels by examining the RS measurements conducted at Resolute and Ny-Ålesund over the period from 2001 to 2015 are shown in the left part of Fig. 2.11. It can be clearly seen that air temperature is subject to marked seasonal variations at the surface-level as well at the three upper levels selected above. Surface-level temperature T_o exhibits annual variations from about 240 °K in winter to nearly 280 °K in summer at Resolute, and more limited variations from about 260 °K to more than 280 °K at Ny-Ålesund. Comparable seasonal variations were obtained at both the stations at the altitude of

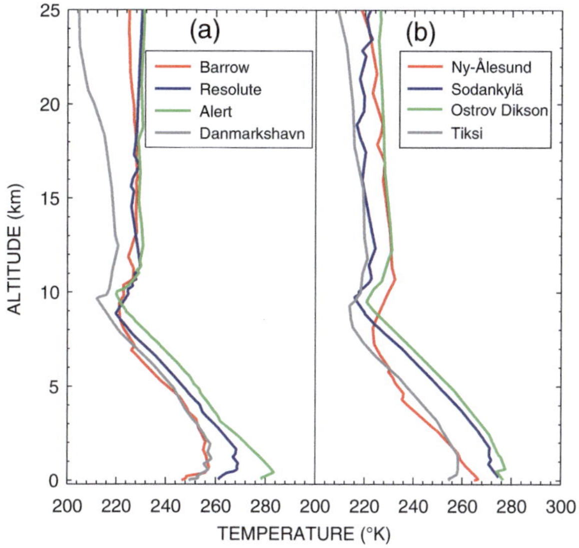

Fig. 2.10 Left part (**a**): Vertical profiles of air temperature *T(z)* obtained by applying the correction procedures described above to the RS measurements of *T(z)* recorded at the following four Arctic sites over the tropospheric altitude range from surface to z = 25 km: (i) Barrow (Alaska, USA) on January 31, 2003 (12:00 UTC) (red curve); (ii) Resolute (Nunavut, Canada) on May 21, 2004 (12:00 UTC) (blue curve); (iii) Alert (Nunavut, Canada), on July 31, 2007 (12:00 UTC) (green curve); and (iv) Danmarkshavn (Greenland) on October 30, 2005 (12:00 UTC) (grey curve). Right part (**b**): As in the left part for the vertical profiles of *T(z)* recorded at the following four Arctic sites: (i) Ny-Ålesund (Spitsbergen, Svalbard) on February 13, 2009 (12:00 UTC) (red curve), (ii) Sodankylä (Finland) on April 12, 2012 (12:00 UTC) (blue curve); (iii) Ostrov Dikson (Russia) on August 10, 2014 (12:00 UTC) (green curve); and (iv) Tiksi (Russia) on November 2, 2015 (00:00 UTC) (grey curve)

5 km (i.e. in the middle troposphere), from 230 °K in winter to no more than 260 °K in summer, while the seasonal variations recorded at the 12 km and 25 km levels were found to be wider on average at Ny-Ålesund than those measured at Resolute, being the temperature values measured at Ny-Ålesund varying from less than 205 °K to more than 230° K at the 12 km altitude and from less than 190 °K to about 230 °K at the 25 km level.

The comparison between the monthly mean values of surface-level air temperature T_o and air temperature *T(z)* at the 5, 12 and 25 km levels derived from the 2001–2015 RS measurements carried out at Sodankylä and Tiksi is shown in the right part of Fig. 2.11. The seasonal variations of surface-level temperature T_o measured at Sodankylä were found to range from minima of around 260 °K to maxima higher than 290 °K, while considerably larger seasonal variations were recorded at Tiksi, with minima of 230–240 °K in winter and maxima close to 290 °K in summer. Appreciably wider seasonal variations of the temperature *T(z)* recorded at level $z = 5$ km)were determined at Tiksi, ranging between 230 °K and 260 °K, while more limited variations were observed at Sodankylä, where minima close on

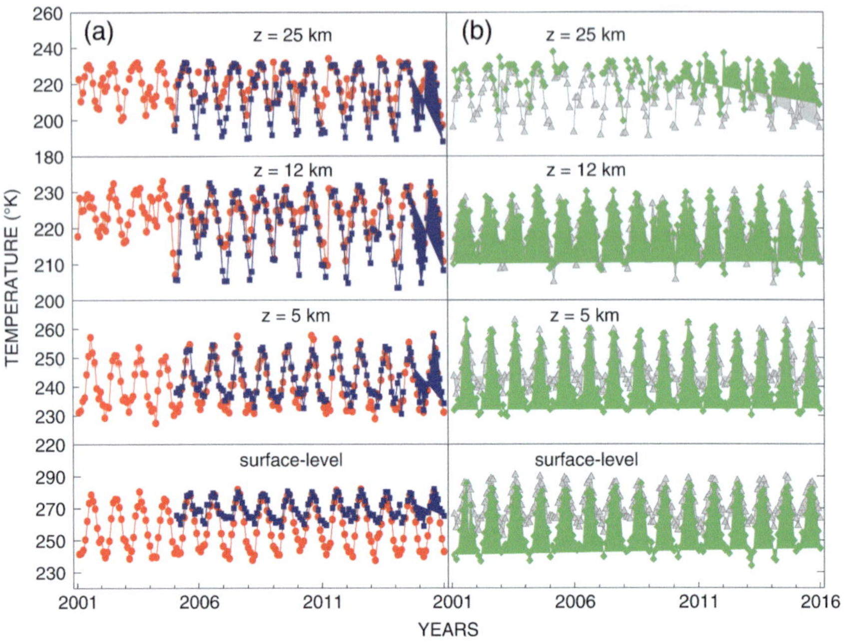

Fig. 2.11 Left part (a): Time-patterns of the monthly mean values of surface-level air temperature T_o and air temperature $T(z)$ derived at the four levels $z = 2$ km, $z = 5$ km, $z = 12$ km, and $z = 25$ km from the RS measurements conducted over the period from 2001 to 2015 at the Resolute (Nunavut, Canada) (red circles) and Ny-Ålesund (Spitsbergen, Svalbard) (blue squares) stations. Right part (b): As in the left part from the RS measurements performed at the Sodankylä (Finland) (grey triangles) and Tiksi (Russia) (green diamonds) stations

average to 240 °K were recorded in winter together with maxima slightly higher than 260 °K during summer. Comparable variations of $T(z)$ were recorded at the 12 km level at both Sodankylä and Tiksi, with winter minima ranging between less than 205 °K and 210 °K, and summer maxima most frequently higher than 230 °K at both sites. The seasonal variations of $T(z)$ measured at Sodankylä for $z = 25$ km were found to be wider in general than those observed at Tiksi, varying mainly between less than 190 °K in winter and more than 230 °K in summer, while those recorded at Tiksi ranged mainly between 200 °K and 235 °K.

Figures 2.10 and 2.11 provide a complete picture of the annual variations of air temperature occurring at the surface and some tropospheric and stratospheric levels in the various Arctic regions. Examining these results, it is worth to emphasize that surface-level temperature T_o is generally subject to seasonal variations that may considerably differ from 1 year to another. In order to give a measure of such year-to-year variations, the series of the monthly mean values of temperature T_o recorded over the period from 2001 to 2015 at the two Arctic stations of Eureka and Sodankylä are shown in Fig. 2.12, being represented by using the same graphical method adopted by Maturilli et al. (2013) in order to determine the time-series of

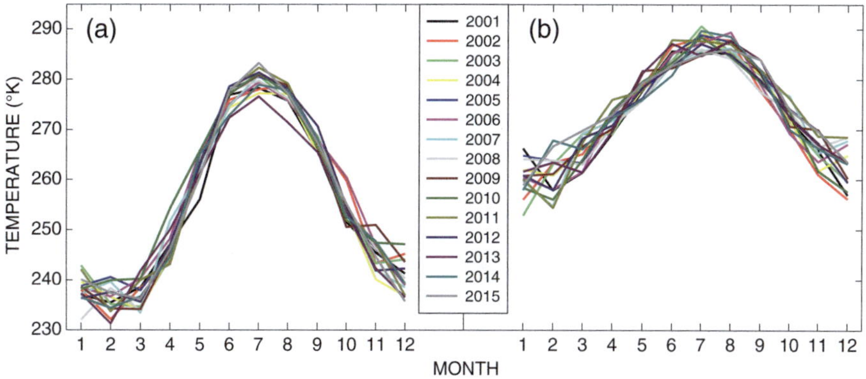

Fig. 2.12 Time-patterns of the monthly mean values of surface-level temperature T_o measured during the 15-year period from 2001 to 2015 at two Arctic sites: (**a**) Eureka (Nunavut, Canada) (10 m a.m.s.l., 79° 59′ N, 85° 56′ W); and (**b**) Sodankylä (Lapland, Finland) (179 m a.m.s.l., 67° 22′ N, 26° 39′ E)

T_o measured at the surface meteorological station managed by the Alfred Wegener Institute for Polar and Marine Research (AWI) located in the neighbourhood of Ny-Ålesund over the 17-year period from January 1, 1994, to December 31, 2010. It can be seen that the time-patterns of the monthly mean values of T_o obtained by examining the RS data recorded at Eureka (located at a latitude close to 80 °N) describe peaked curves in all the years, with the lowest values recorded in January, February and December, ranging mostly between 231 and 240 °K, and the highest values in July, varying between 275 and more than 280 °K. The monthly mean values of T_o result to describe rather similar cycles differing one from the other by no more than 5 °K on average, due to the diverse evolutionary features of the meteorological conditions occurring from 1 year to another in this remote region of the Canadian Arctic territory. The time-patterns of the monthly mean values of T_o obtained at Sodankylä exhibit more limited seasonal variations of T_o throughout the year, being this site located at a latitude not far from that of the Arctic polar cycle, and describe regular cyclic time-patterns during the year with minima mainly observed in January and February, and maxima recorded mainly in July and August. The lowest values of T_o were found to range between about 253 and 265 °K in the winter months, and the annual maxima of T_o between 285 and 291 °K in the summer months.

The examples shown in Fig. 2.12 clearly indicate that the surface-level temperature T_o (°K) is subject to seasonal cycles of variable intensity and amplitude during the various years. To give a more precise information on these variations, the monthly mean values of T_o obtained from the RS measurements examined in the present study at the 14 selected Arctic stations are given in Table 2.6, together with their standard deviations found to vary between less than 3.0 °K (most frequently in late spring and autumn) and more than 8.0 °K (mainly in winter and spring). It can be noted that the monthly mean values of T_o were estimated to range: (i) at Barrow

(Alaska), from 248.8 °K in both January and February to 277.5 °K in July; and (ii) at Inuvik (Northwest Territories of Canada), from 248.1 °K in January and 287.4 °K in July. At the four RS stations of Cambridge Bay, Resolute, Eureka and Alert, all located in Nunavut (Canada), the lowest monthly mean values of T_o were recorded in winter (most frequently in February), varying between 236.2 °K (at Eureka, in February) and 242.0 °K (at Alert, in March), and the highest values in July, ranging between 277.4 °K (at Alert) and 282.9 °K (at Cambridge Bay). In Greenland, the monthly mean values of T_o recorded at the stations of Aasiaat and Danmarkshavn were found to exhibit their minima in the winter-early spring period, equal to 260.0 °K at Aasiaat (in February) and 250.1 °K at Danmarkshavn (in March), while the maxima were observed in July at both stations, equal to 280.3 °K and 278.3 °K, respectively.

Examining the data-sets recorded at the stations of Jan Mayen, Ny-Ålesund, and Sodankylä, which are located in a very large Arctic region including Northern Europe and the Arctic Ocean region between Greenland and Svalbard Archipelago, the lowest monthly mean values of the year were observed in January at Sodankylä (equal to 260.2 °K), and in March at both Jan Mayen (268.6 °K) and Ny-Ålesund (262.3 °K), whereas the highest values were recorded in August at Jan Mayen (equal to 279.8 °K) and in July at both Ny-Ålesund (280.2 °K) and Sodankylä (287.5 °K).

Along the northern coastal regions of Siberia, at longitudes ranging from 80 °E to 161 °E, the monthly mean values of T_o were found to present the lowest values of the year in January at Cherskij (equal to 240.1 °K) and in February at both Ostrov Dikson (248.1 °K) and Tiksi (240.6 °K), while the highest values of the year were measured in July at all the three Russian stations, equal to 279.0 °K at Ostrov Dikson, 281.6 °K at Tiksi, and 286.9 °K at Cherskij, respectively.

Table 2.6 provides the overall set of monthly mean values of surface-level temperature T_o measured at the 14 above-selected Arctic stations during multi-year periods. These data allowed us to define the seasonal mean variations of T_o observed at the surface throughout the year in the various Arctic regions (North America, Greenland, Arctic Ocean (including the Spitsbergen Island), Scandinavia, and Siberia), as can be seen in Fig. 2.13, presenting an exhaustive picture of the seasonal variations of T_o observed over the last 15 years at the various stations, with their standard deviations. The comparison of the annual cycles of T_o measured at Barrow and Inuvik indicates that very close minimum values were obtained at these two sites during the winter months, equal to 248.8 °K at Barrow in both January and February, and to 248.1 °K at Inuvik in January, while the annual maximum was found in July for both these Arctic sites, equal to 277.5 °K at Barrow and 287.4 °K at Inuvik, the warmer months of the year being those from June to September at Barrow, and those from May to September at Inuvik. Very close annual minima of T_o were recorded at Cambridge Bay (equal to 240.9 °K) and Resolute (equal to 241.0 °K), both in February, while the highest summer values were both found in July, equal to 282.9 °K at Cambridge Bay and 278.5 °K at Resolute, the warmer months of the year being those from June to September at the first site and from June to August at Resolute. Comparing the annual cycles of T_o measured at Eureka and Alert, it can be noted that large seasonal variations of this temperature parameter

Table 2.6 Monthly mean values of surface-level temperature T_o (°K) determined with their standard deviations from the RS measurements performed at the 14 Arctic sites chosen in the present study

Station and RS measurement period	Month											
	Jan.	Feb.	March	Apr.	May	June	July	Aug.	Sept.	Oct.	Nov.	Dec.
Barrow (2001–2015)	248.8 ± 6.9	248.8 ± 7.3	249.1 ± 6.2	257.3 ± 6.3	266.9 ± 5.1	274.3 ± 3.4	277.5 ± 4.1	277.4 ± 3.9	274.3 ± 3.0	268.1 ± 4.0	258.0 ± 6.0	252.6 ± 6.1
Inuvik (2001–2009 and 2012–2015)	248.1 ± 7.1	249.2 ± 7.8	250.4 ± 7.1	261.3 ± 7.4	271.2 ± 7.6	284.8 ± 7.4	287.4 ± 6.1	284.7 ± 6.1	277.0 ± 5.3	267.4 ± 4.9	255.9 ± 5.6	251.3 ± 6.6
Cambridge Bay (2001–2015)	242.1 ± 6.3	240.9 ± 6.4	243.7 ± 7.0	253.0 ± 6.8	263.4 ± 6.2	274.8 ± 4.7	282.9 ± 4.3	280.7 ± 4.1	272.6 ± 4.0	263.8 ± 6.0	251.8 ± 6.8	245.3 ± 6.6
Resolute (2001–2015)	242.7 ± 6.3	241.0 ± 5.9	243.3 ± 7.1	252.2 ± 6.0	263.1 ± 5.4	272.8 ± 3.6	278.5 ± 3.8	275.8 ± 3.9	269.6 ± 3.2	262.0 ± 5.8	252.0 ± 6.3	246.3 ± 6.5
Eureka (2001–2015)	238.2 ± 6.5	236.2 ± 6.5	237.1 ± 7.5	246.9 ± 7.2	262.8 ± 5.9	275.9 ± 4.1	279.9 ± 3.4	276.7 ± 3.7	267.4 ± 4.7	254.4 ± 8.0	245.1 ± 7.2	240.4 ± 7.5
Alert (2001–2015)	244.0 ± 5.6	242.2 ± 6.1	242.0 ± 6.5	249.9 ± 6.2	261.8 ± 5.2	271.9 ± 3.8	277.4 ± 4.3	274.0 ± 4.5	265.3 ± 4.5	256.5 ± 6.0	249.6 ± 5.7	246.1 ± 6.4
Aasiaat (2001–2009)	262.1 ± 5.8	260.0 ± 7.2	260.2 ± 6.6	265.2 ± 4.4	272.1 ± 3.0	278.1 ± 2.8	280.3 ± 2.9	279.9 ± 2.1	276.0 ± 2.4	271.0 ± 2.8	268.0 ± 3.1	264.8 ± 5.0
Danmarkshavn (2001–2009)	251.7 ± 7.3	250.8 ± 8.0	250.1 ± 5.8	256.1 ± 6.5	266.4 ± 4.1	274.2 ± 3.3	278.3 ± 3.4	276.8 ± 3.2	269.7 ± 3.1	261.2 ± 5.2	255.6 ± 4.5	252.8 ± 8.1
Jan Mayen (2001–2015)	269.6 ± 4.3	268.9 ± 3.9	268.6 ± 4.6	270.0 ± 3.2	272.8 ± 2.5	276.1 ± 2.3	279.0 ± 1.8	279.8 ± 1.6	277.6 ± 2.4	273.4 ± 3.1	270.7 ± 3.7	269.8 ± 3.9
Ny-Ålesund (2005–2015)	265.0 ± 6.4	262.8 ± 5.9	262.3 ± 6.3	265.1 ± 5.7	271.3 ± 3.1	276.3 ± 2.4	280.2 ± 2.9	278.3 ± 2.1	274.0 ± 3.4	268.4 ± 4.1	265.9 ± 5.1	265.1 ± 5.2
Sodankylä (2001–2015)	260.2 ± 8.6	260.7 ± 8.8	265.3 ± 8.5	271.7 ± 6.4	278.0 ± 5.9	284.1 ± 5.7	287.5 ± 5.3	286.4 ± 5.6	280.8 ± 5.2	272.4 ± 5.5	265.8 ± 7.2	262.6 ± 8.9
Ostrov Dikson (2001–2015)	249.9 ± 7.8	248.1 ± 7.7	251.5 ± 7.7	258.1 ± 6.4	266.4 ± 4.1	274.6 ± 3.3	279.0 ± 4.2	278.4 ± 2.8	275.7 ± 2.7	267.3 ± 5.8	256.9 ± 6.5	252.6 ± 7.5
Tiksi (2001–2015)	243.8 ± 6.5	240.6 ± 7.4	246.7 ± 8.4	256.0 ± 8.0	268.7 ± 5.6	277.2 ± 4.7	281.6 ± 5.5	280.9 ± 9.5	275.2 ± 3.9	263.2 ± 6.4	252.0 ± 6.0	245.5 ± 6.8
Cherskij (2004–2015)	240.1 ± 7.2	240.3 ± 6.7	251.2 ± 8.9	260.0 ± 7.7	274.4 ± 6.9	284.4 ± 6.6	286.9 ± 6.2	283.8 ± 5.4	277.0 ± 4.1	265.2 ± 6.0	252.7 ± 7.9	243.8 ± 8.2

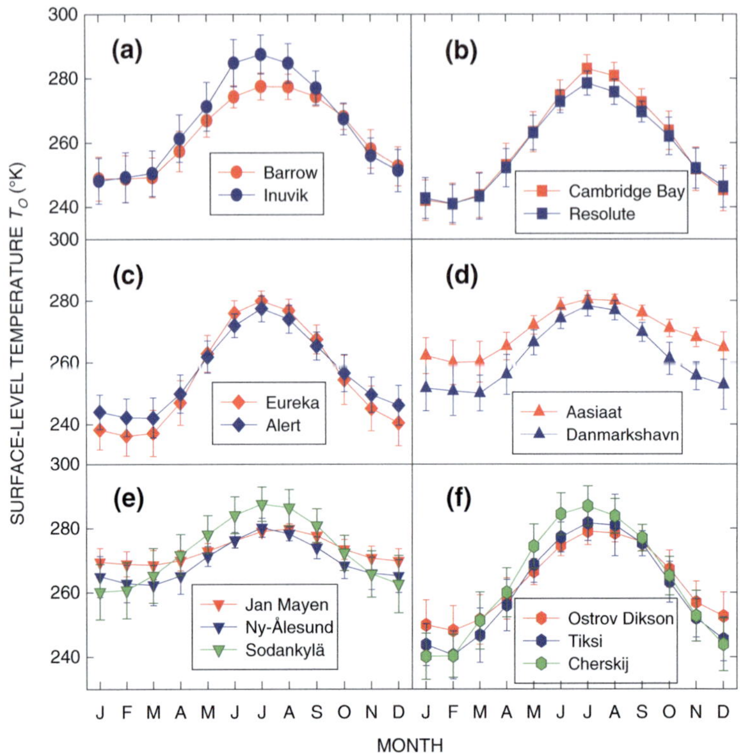

Fig. 2.13 Annual cycles of the monthly mean values of surface-level temperature T_o (°K) of air, derived from the multi-year sets of measurements recorded at the 14 Arctic stations selected in the present study, all represented with their standard deviations (vertical bars) derived from the monthly sets of RS data. Different colours are used in the various parts for the following stations: in part (**a**), Barrow (red circles) and Inuvik (blue circles); in part (**b**), Cambridge Bay (red squares) and Resolute (blue squares); in part (**c**), Eureka (red diamonds) and Alert (blue diamonds); in part (**d**), Aasiaat (red upward triangles) and Danmarkshavn (blue upward triangles); in part (e), Jan Mayen (red downward triangles), Ny-Ålesund (blue downward triangles), and Sodankylä (green downward triangles); and in part (f), Ostrov Dikson (red hexagons), Tiksi (blue hexagons), and Cherskij (green hexagons)

were observed at these two sites, with: (i) winter minima of 236.2 °K at Eureka in February and 242.0 °K at Alert in March, and (ii) summer maxima recorded in July at both sites, equal to 279.9 °K at Eureka and 277.4 °K at Alert, the warmer period of the year being from June to August at both sites. Examining the data relative to Greenland, it can be noted that higher thermal conditions were measured over the whole annual range at Aasiaat (located on the western coast of Greenland) than at Danmarkshavn (located on the north-eastern coast of Greenland, weakly affected by the effects of the North Atlantic Drift. Annual minima of 260.0 °K were recorded at Aasiaat in February and of 250.1 °K at Danmarkshavn in March, while the maxima were recorded in July at both sites, equal to 280.3 °K at Aasiaat and to 278.3 °K at Danmarkshavn, the warmer period of the year being from May to October at Aasiaat and from June to August at Danmarkshavn. Therefore, differences of about 10 °K are observed in winter between the two Greenlandic RS stations, and of about 2–4 °K in summer, these seasonal variations being less marked than those measured at the two higher-latitude sites of Eureka and Alert located in the northern part of Nunavut (Canada). Annual cycles of T_o similar to those recorded at the two Greenlandic sites were also observed at the Arctic Ocean and Northern Europe sites of Jan Mayen, Ny-Ålesund and Sodankylä. At Jan Mayen, the monthly minimum was recorded in March, equal to 268.6 °K, and the maximum of 279.8 °K in August, with the warmer period of the year from April to November, when monthly mean values of T_o higher than 270 °K were recorded. A monthly minimum of the year, equal to 262.3 °K, was measured at Ny-Ålesund in March, together with a maximum of 280.2 °K recorded in July, the warmer period of the year being observed from April to November, therefore with an average difference of about 6 °K in winter and of about 1 °K in summer. A considerably lower monthly minimum of T_o was measured at Sodankylä in winter, equal to 260.2 °K in January, while the monthly maximum of 287.5 °K was recorded in July, with the warmer period of the year from April to October, presenting monthly mean values ranging in summer between 284 and nearly 288 °K. Even more marked variations of the monthly mean values of T_o were observed at the three Siberian stations throughout the year, where appreciably larger seasonal cycles of T_o were determined than those observed at Jan Mayen and Ny-Ålesund, whose climatic conditions were in part mitigated by the warm effects of the North Atlantic Drift. In the coastal Siberian regions on the Arctic Ocean, the winter minima were estimated to be of 248.1 °K at Ostrov Dikson and 240.6 °K at Tiksi (both recorded in February) and of 240.1 °K at Cherskij in January. The monthly maxima were recorded in the summer month of July at all the three Siberian sites, equal to 279.0 °K at Ostrov Dikson, 281.6 °K at Tiksi, and 286.9 °K at Cherskij, the monthly summer average values of T_o higher than 270 °K being recorded from June to September at both Ostrov Dikson and Tiksi, and from May to September at Cherskij.

The monthly mean values of surface-level temperature T_o (°K) shown in Fig. 2.13 have been calculated over 15-year periods at ten of the 14 above-selected RS sites, except for the measurements collected at: (i) Aasiaat and Danmarkshavn over the 9 years from 2001 to 2009, (ii) Ny-Ålesund over the 11 years from 2005 to 2015, and (iii) Cherskij over the 12 years from 2004 to 2015. It seems evident that the surface-level temperature measured at the major part of these Arctic sites was subject to slowly increase with time during these multi-year periods, as shown by the analysis of the annual mean values of ground-level temperature measurements recorded at Ny-Ålesund by Maturilli et al. (2013) over the whole period from 1994 to 2010, which were found to yield an overall regression line of T_o having slope coefficient equal to +1.35 °K per decade. In order to verify the occurrence of this trend at other Arctic sites, the present multi-year data-sets of T_o (collected at 14 Arctic sites) have been analyzed through a regression procedure similar to that adopted by Maturilli et al. (2013), obtaining the results shown in Fig. 2.14, which presents the scatter plots of the annual mean values of T_o collected over

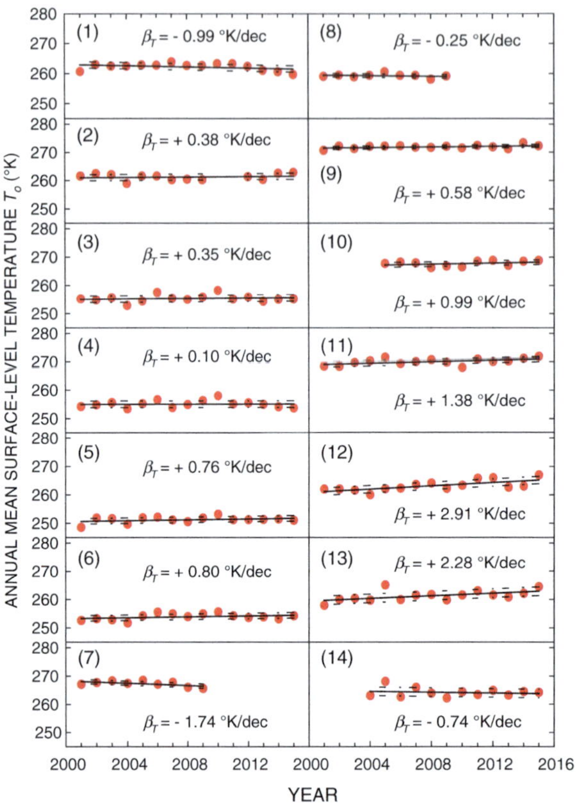

Fig. 2.14 Time-series of the annual mean values of surface-level air temperature T_o (red dots) measured at the 14 Arctic stations over their multi-year periods, with the regression lines (solid lines) and the corresponding parallel (dashed) lines drawn with intercepts equal to the best-fit intercept ± one standard deviation, all having the best-fit slope coefficient reported in each graph to give the average variation of T_o (measured in °K) per decade. The data derived from the RS measurements carried out at the stations of: (1) Barrow, (2) Inuvik, (3) Cambridge Bay, (4) Resolute, (5) Eureka, (6) Alert, and (7) Aasiaat] are shown in the left part, while those derived from the RS measurements collected at the stations of: (8) Danmarkshavn, (9) Jan Mayen, (10) Ny-Ålesund, (11) Sodankylä, (12) Ostrov Dikson, (13) Tiksi, and (14) Cherskij, are shown in the right part.

different sampling periods equal to: (i) 15 years at nine sites (Barrow, Cambridge Bay, Resolute, Eureka, Alert, Jan Mayen, Sodankylä, Ostrov Dikson and Tiksi), (ii) 13 years (from 2001 to 2009 and from 2012 to 2015) at Inuvik, (iii) 12 years (from 2004 to 2015) at Cherskij, (iv) 11 years (from 2005 to 2015) at Ny-Ålesund, and (v) 9 years at Aasiaat and Danmarkshavn. The values of the slope coefficients β_T (measured in °K per decade) and the corresponding correlation coefficients r_T (obtained by drawing the regression lines of the corresponding scatter plots of the annual mean values of surface-level temperature T_o (°K) given as a function

Table 2.7 Values of the slope coefficient β_T (measured in °K per decade) and corresponding correlation coefficient r_T, determined in Fig. 2.14 by drawing the regression lines through the corresponding scatter plots of the annual mean values of surface-level temperature T_o (°K) versus the measurement year, as determined from the RS measurements performed at the 14 Arctic sites chosen in the present study

	Regression line of T_o (°K) vs. year	
Station and RS measurement period	Slope coefficient β_T (°K/decade)	Correlation coefficient r_T
Barrow (2001–2015)	− 0.99	− 0.37
Inuvik (2001–2009 and 2012–2015)	+ 0.38	+ 0.13
Cambridge Bay (2001–2015)	+ 0.35	+ 0.13
Resolute (2001–2015)	+ 0.10	+ 0.05
Eureka (2001–2015)	+ 0.76	+ 0.32
Alert (2001–2015)	+ 0.80	+ 0.35
Aasiaat (2001–2009)	− 1.74	− 0.47
Danmarkshavn (2001–2009)	− 0.25	− 0.15
Jan Mayen (2001–2015)	+ 0.58	+ 0.39
Ny-Ålesund (2005–2015)	+ 0.99	+ 0.38
Sodankylä (2001–2015)	+ 1.38	+ 0.54
Ostrov Dikson (2001–2015)	+ 2.91	+ 0.73
Tiksi (2001–2015)	+ 2.28	+ 0.59
Cherskij (2004–2015)	−0.74	− 0.15

of the measurement year) are reported in Table 2.7, as determined from the RS measurements performed at the 14 Arctic sites chosen in the present study, over: (i) the 15-year period (from 2001 to 2015) at the above-mentioned nine RS stations, (ii) the 13-year period at Inuvik, (iii) the 12-year period at Cherskij, (iv) the 11-year period at Ny-Ålesund, and (v) the 9-year period at Aasiaat and Danmarkshavn.

It can be seen in Fig. 2.14 that the series of annual mean values of T_o, determined at Barrow was subject to appreciably decrease with time, yielding a negative value of slope coefficient $\beta_T = -0.99$ °K per decade, determined with a correlation coefficient $r_T = -0.37$, indicating that a good correlation exists in such a sequence of annual mean values of T_o. Considering that the Arctic Ocean maintain its ice-coverage in the surroundings of Barrow over the most part of the year, it seems realistic to assume that the ice-free surface of Arctic Ocean can exert only moderate mitigation effects on air warming throughout the year, mainly during the summer months. Conversely, a positive and relatively low value of β_T (equal to +0.38 °K per decade) was measured at Inuvik (Canada) with a low correlation coefficient ($r_T = +0.13$), associated with the significant dispersion of data. These results suggest that the trend of surface-level temperature T_o observed in Alaska over the 2001–2015 years and that estimated in the northern part of the Northwest Territories of Canada in the 2001–2009 and 2012–2015 years was not well-defined, being opposite in sign at the two sites. At Cambridge Bay and Resolute, located in the southern part of Nunavut (Canada) at latitudes ranging between 69 °N and 75 °N, rather low values of β_T were obtained, equal to +0.35 °K per decade and

+0.10 °K per decade at the second, respectively, as can be seen in Fig. 2.14, together with correlation coefficients r_T no higher than +0.13 in both cases. These findings indicate that the series of annual mean values of T_o are poorly aligned with time and provide relatively low values of the slope coefficients, very similar to those determined at Inuvik. Considerably higher values of β_T were determined at the stations of Eureka and Alert (located at latitudes close to 80 °N and 82 °N, respectively), equal to +0.76 °K per decade at Eureka, and +0.80 °K per decade at Alert, and found with correlation coefficients $r_T = +0.32$ and $r_T = +0.35$, respectively, which both indicate that general trend of T_o is clearly increasing from 2001 to 2015, as shown in Fig. 2.14.

Negative values of coefficient β_T were obtained by examining the multi-year scatter plots collected at the two sites of Greenland, equal to −1.74 °K per decade at Aasiaat (with $r_T = -0.47$) and −0.25 °K per decade at Danmarkshavn (with $r_T = -0.15$). These results provide evidence of the slow decrease of temperature T_o over the 2001–2009 period, at both these Greenlandic coastal sites, that can be reasonably explained by the fact that both the eastern and westerns coasts of Greenland are totally ice-free in the summer months, so that the ice-free oceanic surface can contribute to exert significant mitigation effects on air warming in these areas during an important part of the year. Conversely, positive values of β_T were determined at Jan Mayen, Ny-Ålesund and Sodankylä, located in the wide area of the Norwegian Sea and Arctic Ocean off the Greenland coasts and in Northern Europe, respectively. A best-fit value of $\beta_T = +\ 0.58$ °K per decade were determined at Jan Mayen over the 2001–2015 period, with correlation coefficient $r_T = +\ 0.39$, while a best-fit value of $\beta_T = +\ 0.99$ °K per decade was determined at Ny-Ålesund over the 2005–2015 period, with $r_T = +\ 0.38$, such a value of β_T being in good agreement with the value of $\beta_T = +\ 1.35$ °K per decade, which was determined by Maturilli et al. (2013) at the AWI surface-level meteorological station through the analysis of the 11-year data-set collected from 1994 to 2010. The present evaluations of β_T made at Jan Mayen and Ny-Ålesund indicate that a rather important increase of surface-level temperature T_o was observed over the last 15-year years in the Arctic Ocean to the east of Greenland, and in the Svalbard area. More marked surface-level warming effects were observed at Sodankylä over the 2001–2015 period, with a rather high value of β_T, equal to +1.38 °K per decade, and $r_T = +\ 0.54$, as can be seen in graph (11) of Fig. 2.14, these findings clearly indicating that the northern part of Scandinavian peninsula has been involved by the gradual increase of surface temperature conditions over the last 15 years.

Finally, examining the results obtained at the three Siberian stations of Ostrov Dikson, Tiksi, and Cherskij, the slope coefficient β_T was evaluated to be very high at both Ostrov Dikson (equal to +2.91 °K per decade over the 2001–2015 period, with a rather high value of the correlation coefficient $r_T = +\ 0.73$), and Tiksi (equal to +2.28 °K per decade over the same 15-year period, with $r_T = +\ 0.59$). Conversely, a negative value of β_T was determined at Cherskij, equal to −0.74 °K per decade, with $r_T = -0.15$ over the 2004–2015 period, this RS site being located at a latitude lower than 69 °N, and therefore lower than those of Ostrov Dikson, Tiksi and Barrow. The results obtained in the area including the Arctic Ocean coastal regions of western and central Siberia indicate that the trend of surface-level temperature T_o is markedly positive at the first two Siberian sites, where the Arctic Ocean is covered by ice over the major part of the year. A negative trend was indeed determined at Cherskij, where the features of parameter T_o were found to be very similar to those observed at Barrow, presumably due to the strong de-icing processes taking place in the Arctic Ocean during the local summer, when more efficient mitigative effects are induced by air warming and release large ice-free areas of the Arctic Ocean near the Eastern Siberian and Alaskan coasts.

On the basis of the above analysis, it can be pointed out that the multi-year trend of surface-level temperature T_o was found to be negative only in some limited areas of the Arctic region, such as the eastern coasts of Greenland (under the mitigative effects of the North Atlantic Drift) and the Bering Strait (across which a weak exchange of warm water masses takes place between Pacific Ocean and Arctic Ocean, causing appreciable effects on the climatic conditions of the nearby coastal Arctic regions). Conversely, the multi-year trend of T_o was estimated by us to be of opposite sign in the most part of Nunavut (Canada), Norwegian Sea and Arctic Ocean between Greenland and Svalbard, Northern Scandinavia in Europe, and Western and Central Siberia. These remarks confirm that atmospheric warming is not uniform over the Arctic region but exhibits some pronounced inequalities from one area to another, thus giving form to well-diversified trends of the time-variations of T_o in the various Arctic sectors.

2.4.1 Seasonal Variations in the Vertical Profiles of Air Temperature

The vertical profile of $T(z)$ measured at an Arctic site can often exhibit some particular features within the ground-layer of the atmosphere, due to the presence of a ground-based temperature inversion or of "elevated" layers above the surface, in which the gradient of $T(z)$ with height assumes negative or only moderately positive values, in general no higher than a few °K per km. These temperature inversion features characterizing the ground-layer of the atmosphere can be often observed at the Arctic sites. For instance, it can be noted in Fig. 2.3 that the vertical profiles of air temperature $T(z)$ obtained from the RS measurements exhibit evident thermal inversion features during all the seasons, as it can be clearly seen in the examples carried out at: (i) Barrow on July 31, 2002 (12:00 UTC); (ii) Cambridge Bay on October 30, 2002 (10:00 UTC); (iii) Danmarkshavn on March 10, 2005 (12:00 UTC); (iv) Ostrov Dikson on June 14, 2001 (12:00 UTC); and (v) Cherskij on August 8, 2009 (12:00 UTC). The occurrence of one or more temperature inversion sub-layers within the lower part of the atmosphere is due to the fact that such a lower part of the troposphere is strongly influenced by the infrared radiative exchange processes taking place at the surface and in the lower part of the troposphere, with up-ward and down-ward fluxes of infra-red radiation that can vary considerably as a function of altitude, due to the variable thermal and moisture characteristics of the low troposphere that are often observed at these high latitudes.

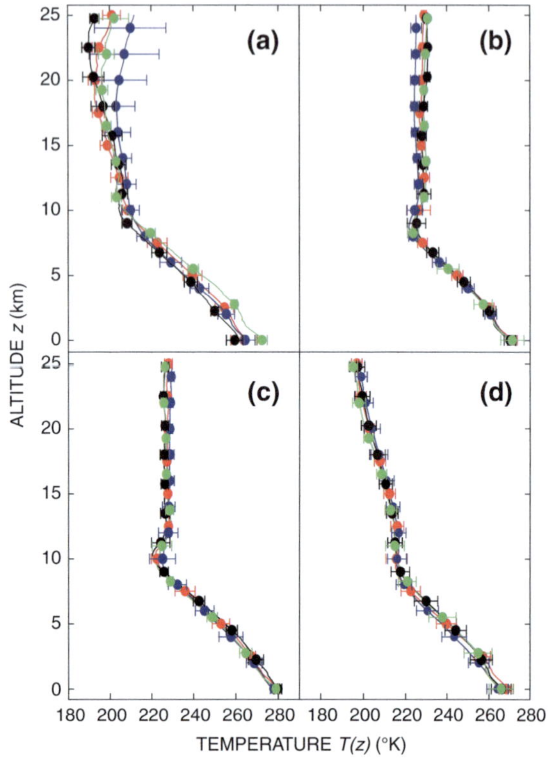

Fig. 2.15 Monthly mean vertical profiles of temperature $T(z)$ derived from the RS measurements conducted at Ny-Ålesund (Spitsbergen, Svalbard) over the altitude range from the surface to 25 km, as obtained from the monthly data-sets recorded in: (a) February, (b) May, (c) August, and (d) November of different years: 2005 (red curves and dots), 2008 (blue curves and dots), 2011 (black curves and dots) and 2014 (green curves and dots). The coloured dots and the horizontal bars provide the monthly mean values of $T(z)$ with their standard deviations at some selected altitudes

In order to analyze these aspects more deeply, the monthly mean vertical profiles of $T(z)$ derived from the RS measurements conducted at Ny-Ålesund (Spitsbergen, Svalbard) over the altitude range from the surface to 25 km are shown in Fig. 2.15, as obtained in February, May, August, and November of four different years from 2005 to 2014. Although obtained by averaging the measurements of $T(z)$ performed over 30-day periods and, hence, for different thermal conditions of the atmosphere from one day to another, it can be noted in Fig. 2.15 that the vertical profiles of $T(z)$ determined in February of four different years are characterized by rather stable temperature features near the surface observed in such 4 years (giving a 4-year mean vertical temperature gradient of -5.4 °K km^{-1} over the first kilometre above the surface, as determined in February by analyzing the overall data-set collected during the four selected years). This value results to be appreciably higher than those defined in Fig. 2.15 for the other months, which provide 4-year average temperature gradients over the first kilometre of the atmosphere estimated to be equal to -6.8 °K km^{-1} in May, -5.4 °K km^{-1} in August, and -6.0 °K km^{-1} in November.

The vertical profiles of $T(z)$ within the ground-layer of the Arctic atmosphere can be drastically altered by the radiative exchange processes occurring between the surface and the atmospheric ground-layer, which can generate in general marked ground-based temperature inversions sometimes associated with one or more “elevated” layers presenting weaker thermal inversion features with values of the vertical gradient of $T(z)$ assuming often only moderately positive values. These aspects will be analyzed more carefully in the Sect. 2.4.1.1, by studying a large set of RS measurements collected at various Arctic sites.

The conventional World Meteorological Organisation (WMO) thermal (lapse rate) definition of “tropopause” states that “the tropopause is the lowest level at which the lapse rate decreases below 2 °K km^{-1} and remains below this value for at least 2 km” (Holton et al. 1995). The vertical profiles of $T(z)$ shown in Fig. 2.15 provide evidence that the wide temperature minimum characterizing the tropopause region may exhibit important seasonal variations throughout the year, as pointed out by Highwood et al. (2007), with significant variations of the temperature inversion parameters T_{min} and z_{min} giving form to the tropopause minimum. On this matter, Fig. 2.15 clearly shows that the temperature minimum T_{min} observed in the tropopause region is subject to seasonal variations and varies considerably in its level z_{min} from one season to another of the same year as well as in the same season from one year to another, with features that may vary appreciably from one site to another, among those selected in the present study and listed in Table 2.1. The temperature minimum T_{min} characterizing the tropopause in the Arctic region is in general located at altitudes ranging throughout the year between around 8 km (in late spring) and more than 11 km (in late winter), giving values of z_{min} ranging (as indicated by the results shown in Fig. 2.15) between; (i) 9.8 and 10.9 km in February of the four selected years from 2005 to 2014, with a mean value of 10.3 km; (ii) 8.3 and more than 8.4 km in May of the four above-selected years, with a mean value of 8.4 km; (iii) 9.4 and 10.0 km in August of the four above-selected years, with a mean value of 9.6 km; and (iv) 8.9 and 9.7 km in November of the four above-selected years, with a mean value of 9.2 km. The seasonal variations of this pair of tropopause thermal parameters will be matter of a deeper analysis in the Sect. 2.4.1.2, based on the examination of a large set of RS measurements collected at the 14 Arctic sites listed in Table 2.1.

2.4.1.1 Seasonal Variations in the Structural Parameters of the Ground-Based Temperature Inversion

An interesting study on the structural parameters of the surface-based temperature inversion was carried out in the Arctic region by Bradley et al. (1992), who defined the annual cycle of the temperature inversions, as observed by means of routine RS measurements performed at 12:00 UTC of each day during the 20-year period from 1967 to 1986, at the nine following Arctic stations located in Alaska (USA) and in the Northwest Territories (Canada): Alert, Eureka, Mould Bay, Resolute, Sachs Harbour, Point Barrow, Barter Island, Inuvik and Kotzebue), five of such RS stations being also among those considered in the present analysis conducted on the basis of the RS measurements carried out during the 2001–

2015 multi-year period. Analyzing the 1967–1986 RS data, Bradley et al. (1992) showed that the surface-based temperature inversions observed in the winter months were primarily associated with strongly negative net radiation characteristics at the surface, whereas those observed in summer were most commonly due to near-surface cooling processes affecting the warm air masses. Consequently, the frequency of the inversion episodes was found to be appreciably higher in winter at the selected nine RS stations (generally, for more than 70% of the observed days), with the ground-based inversion depth Δz found to range mainly between 0.40 and 0.85 km. In particular, they found that: (i) both inversion depth Δz and strength ΔT are strongly related to surface temperature, and (ii) the surface-based inversions often involve temperature changes greater than 30 °K within the ground-layer observed at these Arctic sites, with values of Δz lower than 1 km in general, and values of vertical temperature gradient γ higher than 60 °K km^{-1} during the periods presenting features with advective transport of extremely warm air masses aloft. It was also stated by Bradley et al. (1992) that the mid-winter ground-based inversions persist very often for at least 2 to 4 days above the Arctic RS stations, and may remain sometimes undisturbed for several weeks.

An impressive study on the occurrence of the surface-based temperature inversion in the Arctic area was also carried out by Kahl et al. (1992), who analyzed the rawinsonde data-sets collected over periods of around 40 years (from 1948 to 1987 in most cases) at 13 Canadian stations located at latitudes ranging between 65.3 °N and 82.5 °N, among which those of Alert, Eureka, Resolute, Cambridge Bay and Inuvik (considered in the present study) were included. Analyzing these RS data, Kahl et al. (1992) pointed out that inversions at the northern sites (such as Alert and Eureka) were primarily located near the surface in winter, and were instead mainly "elevated" from mid-spring through summer, while those observed at the southern sites (such as Inuvik and Cambridge Bay) showed in general a bimodal pattern, with surface-based inversions occurring during late summer and winter, often due to interactions between short- and long-wave radiation exchanges occurring within the surface and low-troposphere system. The surface-based inversions observed in winter were found to exhibit the higher values of depth Δz and the most pronounced temperature differences ΔT mainly in February (for at least 98% of the soundings performed at all stations). Conversely, the lowest values of parameters Δz and ΔT were observed most frequently in the August and September months, with the presence of: (i) summer-time "elevated" inversions in numerous cases; (ii) irregular spatial patterns of the median height of the inversion base, often due to regional differences in terrain, snow cover and cloud cover; and (iii) little spatial variability of the median values of both Δz and ΔT. The occurrence of these inversion characteristics was found to closely depend often on the clear-sky percentages in all seasons, reflecting the controlling influence of cloud and clear-sky radiative forcing processes on the thermal characteristics of the inversion layer.

Figure 2.16 shows eight examples of vertical profiles of $T(z)$ determined over the altitude range from surface-level to 1.8 km, at different Arctic sites chosen among those listed in Table 2.1, and for different seasonal conditions of the Arctic troposphere. It can be noted that these profiles of $T(z)$ exhibit marked ground-

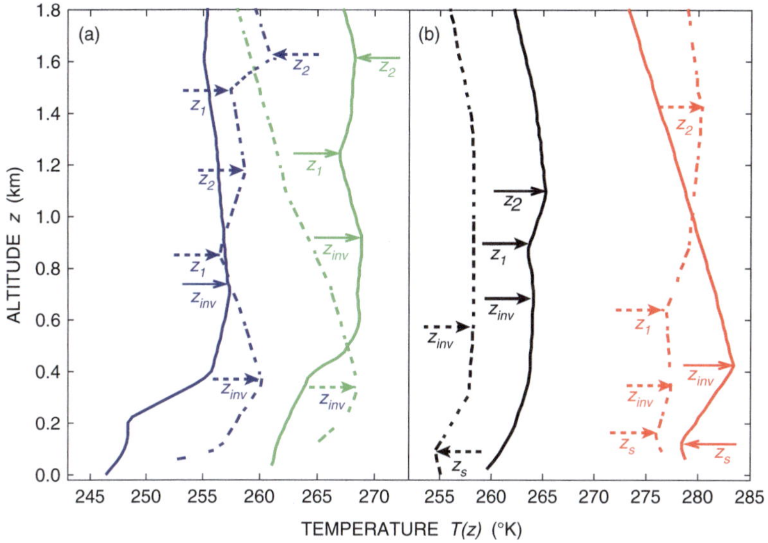

Fig. 2.16 Vertical profiles of temperature $T(z)$ obtained in the lower troposphere (from ground-level z_o to altitude z = 1.8 km) by examining the RS measurements performed at different Arctic sites and on days of the four seasonal periods. The left part (**a**) shows the vertical profiles of $T(z)$ recorded: (i) in winter at Barrow (Alaska, USA) on January 31, 2003 (12:00 UTC) (solid blue curve), and Eureka (Nunavut, Canada) on February 4, 2007 (00:00 UTC) (dashed blue curve), and (ii) in spring, at Resolute (Nunavut, Canada) on May 21, 2004 (12:00 UTC) (solid green curve), and Danmarkshavn (Greenland) on March 10, 2005 (12:00 UTC) (dashed green curve), with arrows of the same colour used to indicate the temperature inversion altitudes. The right part (**b**) shows the vertical profiles of $T(z)$ recorded: (i) in summer, at Alert (Nunavut, Canada) on July 31, 2007 (12:00 UTC) (solid red curve), and Ostrov Dikson (Siberia, Russia) on June 14, 2001 (12:00 UTC) (dashed red curve), and (ii) in autumn, at Cambridge Bay (Nunavut, Canada) on October 30, 2002 (10:00 UTC) (black solid curve), and Tiksi (Siberia, Russia) on November 2, 2015 (00:00 UTC) (dashed black curve), with arrows of the same colour used to indicate the altitudes of both ground-based and elevated temperature inversions

based temperature inversions, sometimes associated with one or more "elevated" thermal inversions generally located at upper tropospheric levels. As mentioned above, each vertical profile of $T(z)$ shown in Fig. 2.16 is substantially characterized in its lower part by three independent parameters defining the strength and the altitude-range of the thermal inversion formed near the ground: (i) the depth Δz of the temperature inversion, given by the difference between the top-level z_{inv} of the ground-based temperature inversion and surface-level z_o, or the difference between the upper and lower levels (called z_1 and z_2, respectively), which define the elevated thermal inversion layer; (ii) the difference $\Delta T = T_{inv} - T_o$, between temperature T_{inv} measured at the top-level of the ground-based thermal inversion and the surface-level temperature T_o, or in the cases of elevated thermal inversion layers between the temperatures measured at the two extreme levels z_1 and z_2; and (iii) the average vertical gradient of temperature γ, calculated in terms of ratio $\Delta T/\Delta z$ over the whole depth of the ground-based thermal inversion layer. In all the cases presenting also one or more elevated thermal inversions, the lower and upper levels defining the depth Δz of such an elevated inversion have been graphically indicated by horizontal arrows in Fig. 2.16.

Analyzing the vertical profiles of $T(z)$ shown in this graph, we have found that:

(1) The vertical profile of $T(z)$ measured at Barrow (Alaska) in winter, on January 31, 2003, at 12:00 UTC (and represented in Fig. 2.16 by a solid blue curve) exhibits a marked ground-based thermal inversion with: (i) $z_o = 19$ m a.m.s.l. and $z_{inv} = 725$ m, giving $\Delta z = 706$ m, and (ii) $T_o = 246.5$ °K and $T_{inv} = 257.4$ °K, yielding $\Delta T = 10.9$ °K and an average value of $\gamma = 15.1$ °K km^{-1}. The vertical profile of $T(z)$ shows that a weak "elevated" thermal inversion was also measured above Barrow on that day, with $z_1 = 1.625$ km and $z_2 = 2.075$ km, giving $\Delta z = 450$ m and $\Delta T = 0.8$ °K, which provide an average value of $\gamma = 1.8$ °K km^{-1}, clearly indicating that such an elevated thermal inversion was rather weak.

(2) The vertical profile of $T(z)$ measured at Eureka (Nunavut, Canada) in winter, on February 4, 2007, at 00:00 UTC (and represented by a dashed blue curve in Fig. 2.16) shows that a very strong ground-based thermal inversion was formed on that day, together with a pair of elevated thermal inversions located at altitudes from 0.86 to 1.63 km. The ground-based thermal inversion was found to have a top-level $z_{inv} = 340$ m against $z_o = 10$ m, an overall depth $\Delta z = 330$ m, and temperatures $T_o = 244.5$ °K and $T_{inv} = 260.1$ °K, yielding $\Delta T = 15.6$ °K and $\gamma = 47.3$ °K km^{-1}. The first elevated temperature inversion was estimated to have the levels $z_1 = 860$ m and $z_2 = 1190$ m, with $\Delta z = 330$ m, $\Delta T = 2.1$ °K and $\gamma = 6.4$ °K km^{-1}, while the second elevated temperature inversion was found to have levels $z_1 = 1490$ m and $z_2 = 1630$ m, with $\Delta z = 140$ m, $\Delta T = 3.4$ °K and $\gamma = 24.3$ °K km^{-1}.

(3) The vertical profile of $T(z)$ recorded at Resolute (Nunavut, Canada) on May 21, 2004 (12:00 UTC) (and represented by a solid green curve in Fig. 2.16) is characterized by a pronounced ground-based thermal inversion, and by a secondary elevated thermal inversion, which have: (i) top-level $z_{inv} = 925$ m (being $z_o = 46$ m), with $\Delta z = 879$ m, and temperatures $T_o = 261.1$ °K and $T_{inv} = 268.9$ °K, with $\Delta T = 7.8$ °K and $\gamma = 8.9$ °K km^{-1}; and (ii) levels $z_1 = 1250$ m and $z_2 = 1625$ m, with $\Delta z = 375$ m, $\Delta T = 1.3$ °K and $\gamma = 3.5$ °K km^{-1}.

(4) The vertical profile of $T(z)$ observed at Danmarkshavn (Greenland) on March 10, 2005 (12:00 UTC) (and represented by a dashed green curve in Fig. 2.16) exhibits a rather weak ground-based thermal inversion only, with $z_o = 12$ m, $z_{inv} = 340$ m, $\Delta z = 328$ m, $T_o = 261.5$ °K, $T_{inv} = 268.7$ °K, $\Delta T = 7.2$ °K and $\gamma = 22.0$ °K km^{-1}.

(5) The vertical profile of $T(z)$ measured at Alert (Nunavut, Canada) on July 31, 2007 (12:00 UTC) (and represented by a solid red curve in Fig. 2.16) shows only a marked thermal inversion suspended near the ground, with the bottom-level z_s of this thermal inversion equal to 125 m, being $z_o = 65$ m and with the

top-level z_{inv} = 425 m, Δz = 300 m, T_o = 278.5 °K, T_{inv} = 283.4 °K, and, hence, ΔT = 4.9 °K and γ = 16.3 °K km^{-1}.

(6) The vertical profile of $T(z)$ measured at Ostrov Dikson (Siberia, Russia) on June 14, 2001 (12:00 UTC) (and represented by a dashed red curve in Fig. 2.16) presents a weak thermal inversion suspended over the ground, together with an elevated thermal inversion over the altitude range from ~ 870 m to nearly 1100 m: (i) the suspended thermal inversion had levels z_{inv} = 370 m and z_s = 157 m (being z_o = 47 m), yielding Δz = 213 m, with T_o = 276.0 °K and T_{inv} = 277.4 °K, ΔT = 1.4 °K and γ = 6.6 °K km^{-1}; and (ii) the elevated temperature inversion was estimated to have levels z_1 = 652 m and z_2 = 1420 m, with Δz = 768 m, ΔT = 3.3 °K and γ = 4.3 °K km^{-1}.

(7) The vertical profile of $T(z)$ recorded at Cambridge Bay (Nunavut, Canada) on October 30, 2002 (10:00 UTC) (and represented by a solid black curve in Fig. 2.16) exhibits a pronounced ground-based thermal inversion, together with an elevated thermal inversion, of which: (i) the first with z_{inv} = 712 m (being z_o = 25 m), Δz = 687 m, T_o = 259.7 °K, T_{inv} = 264.1 °K, ΔT = 4.4 °K and γ = 6.4 °K km^{-1}; and (ii) the second with z_1 = 877 m, z_2 = 1096 m, Δz = 219 m, ΔT = 1.7 °K and γ = 7.8 °K km^{-1}.

(8) The vertical profile of $T(z)$ recorded at Tiksi (Siberia, Russia) on November 2, 2015 (00:00 UTC) (and represented by a dashed black curve in Fig. 2.16) is characterized by a well-marked suspended thermal inversion above the ground, with bottom-level z_s = 100 m (being z_o = 7 m) and top-level z_{inv} = 575 m, giving Δz = 475 m, with temperatures T_o = 254.7 °K, T_{inv} = 258.3 °K, ΔT = 3.6 °K and γ = 7.6 °K km^{-1}.

The eight examples shown in Fig. 2.16 suggest that the presence of a ground-based thermal inversion or alternatively of a thermal inversion structure having its bottom-level at a few tenths of meters above the ground has been often observed at all the Arctic sites during the colder months of the year (from September–October until to April–May), whereas the occurrence of these features was less frequently observed during the warmer period of the year. The examples shown in Fig. 2.16 also suggest that the formation of a ground-based thermal inversion may be accompanied in some cases by the presence of one or more "elevated" thermal inversions, usually located within the 0.6–1.6 km altitude range, resulting from the combined effects of radiative exchange processes and transport of moist air masses above the RS stations. Thus, considering that the presence of a ground-based thermal inversion near the ground can cause an important effect on the exchange of thermal radiation between the surface and the low atmosphere, therefore strongly influencing the radiative budget of the lower part of the Arctic troposphere, it could be useful to evaluate which seasonal impact may have the ground-based thermal inversions observed in the Arctic atmosphere on the climatic conditions of the Arctic RS stations. For this purpose, we have evaluated in Table 2.8 the seasonal percentages of the observed ground-based thermal inversions, including in these calculations also the thermal inversions that are suspended above the surface, like those shown in the

Table 2.8 Multi-year seasonal values of the following parameters measured at the 14 above-selected radiosounding stations listed in Table 2.1: (i) seasonal number N_{RS} of analyzed RS measurements; (ii) seasonal number N_{inv} of identified ground-based temperature inversions; (iii) seasonal percentage P_{inv} of recognized ground-based temperature inversions; (iv) mean seasonal value and seasonal quartiles of temperature difference $\Delta T = T_{inv} - T_o$; (v) mean seasonal value and seasonal quartiles of ground-based temperature inversion depth Δz (m); and (vi) mean seasonal value and seasonal quartiles of the vertical temperature gradient γ (°K km^{-1}) = $\Delta T/\Delta z$

Station and RS measurement period	Season	N_{RS}	N_{inv}	P_{inv} (%)	ΔT (°K)		Δz (m)		γ (°K km^{-1})	
					Seasonal mean	Seasonal quartiles	Seasonal mean	Seasonal quartiles	Seasonal mean	Seasonal quartiles
Barrow (2001–2015)	Winter (DJF)	2651	1314	50	6.8	1.5, 5.8, 10.7	321	83, 220, 522	24.4	10.7, 18.1, 31.1
	Spring (MAM)	2632	1018	39	6.6	2.0, 5.8, 10.0	328	110, 248, 495	22.9	10.6, 17.6, 30.1
	Summer (JJA)	2350	609	26	4.6	1.3, 4.3, 7.2	270	110, 248, 385	17.1	7.9, 14.0, 22.4
	Autumn (SON)	2524	694	28	3.8	0.4, 2.3, 6.0	166	55, 110, 247	22.8	7.2, 15.7, 30.9
Inuvik (2001–2009 and 2012–2015)	Winter (DJF)	2146	1239	58	6.9	2.4, 5.9, 10.6	317	110, 247, 467	24.2	14.3, 21.3, 29.9
	Spring (MAM)	1929	606	31	4.9	2.3, 4.3, 6.5	236	110, 193, 302	24.5	13.1, 19.6, 30.6
	Summer (JJA)	918	257	28	3.0	1.4, 3.0, 4.3	158	83, 138, 193	20.6	10.7, 16.7, 27.3
	Autumn (SON)	1682	629	37	4.2	1.1, 3.2, 5.9	202	55, 165, 274	21.1	10.9, 18.2, 28.1
Cambridge Bay (2001–2015)	Winter (DJF)	2645	1932	73	8.3	4.2, 7.9, 11.8	363	138, 302, 549	28.9	16.5, 24.0, 35.3
	Spring (MAM)	2574	1191	46	7.9	3.9, 7.8, 11.2	346	138, 274, 522	28.4	15.5, 23.0, 35.3
	Summer (JJA)	1087	180	17	3.9	1.8, 3.3, 6.0	179	83, 165, 247	22.5	12.8, 20.2, 29.8
	Autumn (SON)	2256	874	39	5.0	1.4, 4.0, 7.7	238	83, 165, 357	24.1	12.0, 19.9, 30.1

Resolute (2001–2015)	Winter (DJF)	2615	1876	72	5.6	1.6, 3.9, 9.2	252	55, 138, 384	28.7	14.3, 23.0, 35.8
	Spring (MAM)	2625	1318	50	4.9	1.5, 3.5, 7.8	233	55, 138, 357	26.6	12.1, 20.9, 32.8
	Summer (JJA)	1120	184	16	2.3	0.9, 1.9, 3.2	157	55, 138, 220	17.6	7.3, 13.8, 21.7
	Autumn (SON)	2451	1036	42	3.2	0.9, 2.2, 4.6	153	55, 83, 193	24.4	10.7, 19.1, 31.4
Eureka (2001–2015)	Winter (DJF)	2679	2163	81	10.8	5.0, 11.0, 16.0	434	193, 412, 631	28.6	18.2, 25.5, 34.6
	Spring (MAM)	2635	1704	65	9.2	3.2, 8.2, 14.4	397	165, 357, 604	25.2	14.7, 22.0, 32.4
	Summer (JJA)	1202	583	49	2.7	1.1, 2.1, 3.8	98	55, 83, 138	32.8	12.8, 23.2, 42.8
	Autumn (SON)	2575	1445	56	7.3	2.6, 6.4, 11.4	308	110, 247, 467	27.7	15.7, 23.6, 34.2
Alert (2001–2015)	Winter (DJF)	2640	2102	80	5.2	2.1, 3.9, 7.0	235	55, 138, 329	32.6	14.6, 24.3, 43.5
	Spring (MAM)	2624	1704	65	5.2	1.8, 3.9, 7.4	242	83, 138, 329	28.7	13.7, 22.2, 36.5
	Summer (JJA)	1255	199	16	2.6	0.7, 2.0, 4.2	158	55, 138, 219	18.4	7.7, 14.1, 23.2
	Autumn (SON)	2608	1710	66	3.9	1.4, 3.0, 5.4	181	55, 110, 219	28.7	12.8, 21.9, 36.5
Aasiaat (2001–2009)	Winter (DJF)	1376	475	35	1.5	0.6, 1.2, 2.1	88	28, 55, 110	21.4	7.3, 15.7, 28.6
	Spring (MAM)	1284	325	25	1.7	0.6, 1.2, 2.2	104	55, 83, 138	18.4	7.3, 14.3, 25.4
	Summer (JJA)	550	70	13	2.0	0.6, 1.4, 2.9	120	55, 138, 165	17.1	6.6, 13.3, 23.7
	Autumn (SON)	949	272	29	1.0	0.3, 0.8, 1.4	65	28, 55, 83	19.1	7.2, 14.3, 27.3
Danmarkshavn (2001–2009)	Winter (DJF)	1417	1107	78	4.5	1.3, 3.1, 6.1	184	55, 110, 247	29.1	13.7, 22.1, 36.4
	Spring (MAM)	1468	832	57	3.9	1.2, 2.9, 5.4	176	55, 138, 247	26.2	12.2, 20.2, 32.8
	Summer (JJA)	603	162	27	2.3	0.6, 1.4, 3.1	145	55, 110, 193	16.4	7.2, 12.2, 20.1
	Autumn (SON)	1272	771	61	2.1	0.6, 1.4, 2.9	121	28, 55, 138	22.9	9.6, 17.3, 28.6

(continued)

Table 2.8 (continued)

Station and RS measurement period	Season	N_{RS}	N_{inv}	P_{inv} (%)	ΔT (°K)		Δz (m)		γ (°K km^{-1})	
					Seasonal mean	Seasonal quartiles	Seasonal mean	Seasonal quartiles	Seasonal mean	Seasonal quartiles
Jan Mayen (2001–2015)	Winter (DJF)	1977	537	27	0.8	0.2, 0.4, .8	38	28, 28, 28	21.3	5.5, 10.7, 21.9
	Spring (MAM)	1856	278	15	0.7	0.2, 0.3, 0.9	42	28, 28, 28	16.6	5.5, 9.1, 21.5
	Summer (JJA)	923	141	15	0.9	0.2, 0.5, 1.0	55	28, 28, 55	18.2	7.2, 12.8, 21.5
	Autumn (SON)	1525	326	21	0.6	0.2, 0.4, 0.8	38	28, 28, 28	17.6	4.8, 10.7, 21.9
Ny- Ålesund (2005–2015)	Winter (DJF)	819	425	52	1.4	0.8, 1.2, 1.8	76	28, 55, 83	29.5	9.6, 20.0, 39.4
	Spring (MAM)	760	255	34	1.5	0.6, 1.2, 1.9	84	28, 55, 110	25.7	7.3, 16.4, 32.8
	Summer (JJA)	299	24	8	1.1	0.6, 1.0, 1.7	102	28, 55, 138	15.5	7.2, 12.8, 17.8
	Autumn (SON)	733	248	34	1.4	0.7, 1.2, 1.9	67	28, 55, 83	29.3	8.6, 20.0, 42.9
Sodankylä (2001–2015)	Winter (DJF)	2397	905	38	9.4	2.8, 8.4, 15.2	242	138, 247, 329	39.6	16.4, 35.2, 57.1
	Spring (MAM)	1758	527	30	6.5	2.1, 5.2, 10.4	167	83, 165, 219	40.9	16.7, 34.5, 58.0
	Summer (JJA)	1110	328	30	3.9	1.8, 3.4, 5.5	151	110, 138, 193	27.8	13.1, 22.5, 38.3
	Autumn (SON)	1545	409	27	5.5	1.6, 4.2, 8.0	198	110, 193, 274	28.1	11.9, 22.5, 39.2
Ostrov Dikson (2001–2015)	Winter (DJF)	1835	927	51	6.4	2.9, 6.2, 9.5	414	193, 358, 605	16.1	9.3, 14.5, 20.8
	Spring (MAM)	1899	803	42	4.8	1.2, 4.4, 7.9	322	110, 303, 468	15.1	7.7, 13.6, 20.5
	Summer (JJA)	1877	430	23	2.6	0.4, 1.6, 4.2	201	55, 138, 303	12.9	5.4, 10.7, 17.1
	Autumn (SON)	2013	567	28	3.2	0.6, 2.3, 5.0	244	55, 193, 358	13.7	6.5, 11.1, 17.9

Tiksi (2001–2015)	Winter (DJF)	1973	1189	60	8.8	2.8, 7.8, 14.2	568	193, 495, 825	16.8	9.1, 14.4, 21.1
	Spring (MAM)	2155	1071	50	6.7	1.4, 4.9, 11.1	391	110, 248, 633	18.5	9.5, 15.8, 23.9
	Summer (JJA)	2110	618	29	2.9	0.5, 1.4, 4.4	197	55, 138, 248	15.4	6.4, 10.9, 20.4
	Autumn (SON)	2237	908	41	4.2	0.7, 2.5, 6.2	302	83, 193, 413	14.1	6.8, 11.4, 18.5
Cherskij (2004–2015)	Winter (DJF)	1616	1162	72	11.1	5.1, 11.2, 16.6	587	303, 578, 798	19.6	11.8, 17.5, 24.7
	Spring (MAM)	1703	736	43	4.3	1.6, 3.0, 5.9	294	138, 193, 358	15.1	7.9, 12.7, 19.1
	Summer (JJA)	1810	516	29	2.4	1.2, 2.1, 3.3	163	110, 165, 220	15.7	7.3, 13.6, 19.3
	Autumn (SON)	1703	742	44	5.1	1.4, 3.6, 7.8	322	138, 248, 468	16.0	7.3, 13.2, 20.8

examples of Fig. 2.16 made for the RS carried out at the stations of Alert, Ostrov Dikson and Tiksi.

On the basis of the results obtained by examining all the RS data-sets collected at the 14 Arctic sites listed in Table 2.1, we have analyzed the evolutionary time-patterns of the three temperature-inversion parameters ΔT, Δz e γ, pertaining to all the ground-based thermal inversions and all the thermal inversions formed near the ground-level, as derived from the RS measurements carried out during the 5-year period from 2011 to 2015 at the four Arctic sites of Barrow (19 m a.m.s.l.), Alert (65 m a.m.s.l., Ny-Ålesund (11 m a.m.s.l.) and Tiksi (7 m a.m.s.l.). Figure 2.17 provides a set of useful data regarding the three inversion parameters Δz, ΔT and γ, and shows the evolutionary time-patterns of the daily and monthly mean values of the three above-mentioned inversion parameters defining the vertical characteristics of the surface-based temperature inversion observed at the above-selected four Arctic stations, on each day and in each month of the five years from

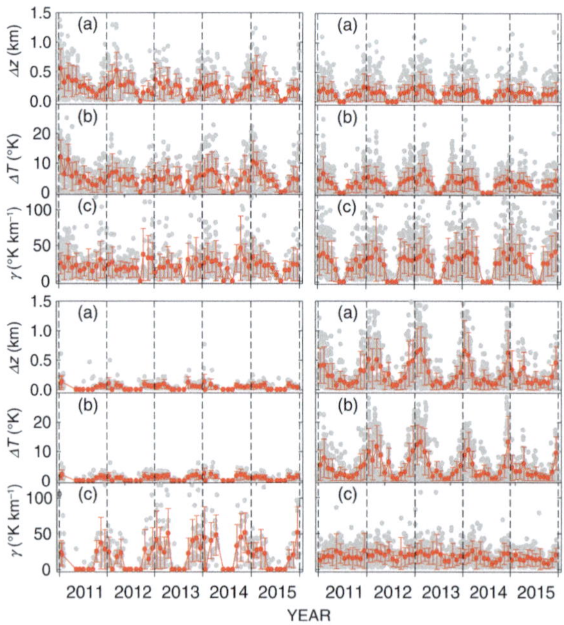

Fig. 2.17 Time-series of the daily mean values (grey dots) and monthly mean values (red dots) (the latter being represented with vertical red bars used to indicate the standard deviations) of the following parameters: (**a**) ground-based temperature inversion depth Δz (m); (**b**) difference ΔT between temperature T_{inv} (measured at the top-level of the ground-based thermal inversion) and surface-level temperature T_o, and (**c**) average vertical gradient $\gamma = \Delta T/\Delta z$ of temperature $T(z)$ calculated over the depth Δz, as obtained during the 5-year period from 2011 to 2015 at the following stations: (1) Barrow (Alaska, USA) (upper part, left side), (2) Alert (Nunavut, Canada) (upper part, right side), (3) Ny-Ålesund (Spitsbergen, Svalbard) (lower part, left side), and (4) Tiksi (Siberia, Russia) (lower part, right side). The vertical dashed lines have been drawn in each part to indicate the first and last day of each year, for the convenience of the reader

2011 to 2015. Before analyzing the results shown in Fig. 2.17, it is useful to mention that the seasonal mean values of surface-level temperature T_o was found to range at the four sites: (i) between 274 and 278 °K in summer, and between 249 and 252 °K in winter, at Barrow; (ii) between 274 and 277 °K in summer and between 242 and 244 °K in winter, at Alert; (iii) between 278 and 280 °K in summer and between 262 and 265 °K in winter, at Ny-Ålesund, thus denoting warmer air conditions than at the previous sites; and (iv) between 280 and 282 °K in summer and between 241 and 244 °K in winter, at Tiksi.

Examining the sequences of the daily and monthly mean values of ΔT, Δz and γ, shown in Fig. 2.17, it can be noticed that the ground-based temperature inversions formed above these Arctic sites exhibit in general its greatest strength during late autumn, the whole winter season and early spring, and shows weaker characteristics in the summer days, on which the ground-based thermal inversion is generally less strong (being limited to a very low surface-layer) or is totally vanished as a result of the strong ground heating by solar radiation and the relatively warm climate conditions. The time-patterns of the daily mean values of depth Δz result to be largely dispersed at Barrow, Alert and Tiksi, over ranges from a few tens of meters to more than 1.5 km, and appear to be more limited at Ny-Ålesund, where Δz results to be higher than 0.5 km only in a few cases. Thus, the monthly mean values of Δz were evaluated to describe well-defined annual cycles, with: (i) their maxima recorded usually in winter and early spring, and close to 0.2–0.3 km at Barrow, no higher than 0.3 km at Alert, equal to a few tens of meters at Ny-Ålesund, and often equal to 0.4–0.7 km at Tiksi, and (ii) their minima recorded in late summer and early autumn months, often denoting the absence of thermal inversion features near the ground.

Correspondingly, the difference ΔT was found to describe well-defined annual cycles, with: (a) daily values ranging between 0 °K and more than 25 °K in winter and spring at Barrow (giving monthly mean values of ΔT that varied most frequently between 5 and 12 °K), and appreciably lower values in the summer months, when ΔT was found to be lower than 1 °K in numerous cases; (b) daily values ranging between 0 °K and 25 °K in winter and spring at Alert, followed by very low values in summer, yielding monthly mean values ranging between 3 and 5 °K in the winter and spring months, and lower than a few °K in the summer months; (c) daily values no higher than 5 °K in winter and spring at Ny-Ålesund, and close to a few °K in summer and autumn, giving monthly mean values of 2–3 °K in winter and spring, and values close to 0 °K in the summer and autumn months; and (d) daily values ranging between 0 °K and more than 25 °K in winter and spring at Tiksi, followed by values close to a few tenths of °K in summer and early autumn, while the monthly mean values were estimated to exceed 10 °K often in the winter and spring months, and to be close to a few °K in the summer and autumn months.

As a result of these time-variations, the daily values of the mean vertical gradient γ of temperature were evaluated to be largely dispersed over the 5 years considered in Fig. 2.17, with: (a) daily values of gradient γ ranging between 0 °K km^{-1} and more than 100 °K km^{-1} in winter at Barrow, giving monthly mean values equal to around 20 °K km^{-1} in the winter and spring months, and often equal to a few °K km^{-1} in the summer months; (b) daily values of γ describing regular annual cycles at Alert, ranging between nearly 0 °K km^{-1} in summer and more than 100 °K km^{-1} in winter and early spring, and giving monthly mean values of around 40 °K km^{-1} in winter and spring, against values close to 0 °K km^{-1} in summer; (c) daily values of γ characterized by large dispersion features at Ny-Ålesund, with the highest values recorded in late autumn, winter and early spring, sometimes exceeding the value of 100 °K km^{-1}, which were found to yield monthly mean values no higher than 50 °K km^{-1} in the late autumn, winter and spring months, and rather low values (often close to 0 °K km^{-1}) in summer; and (d) daily values of γ mainly varying at Tiksi from nearly zero in summer to around 50 °K km^{-1} in winter, where a few cases ranging from 50 to more than 100 °K km^{-1} were also recorded), yielding monthly mean values ranging mainly between 10 and 30 °K km^{-1} in winter and early spring, and very low values observed in the summer months.

The comparison made in Fig. 2.17 among the time-sequences of the daily mean and monthly mean values of the three parameters Δz, ΔT and γ at the four above-chosen Arctic sites clearly indicate that the values of Δz and ΔT measured at Ny-Ålesund are considerably smaller than those measured at the other three sites, suggesting that the thermal inversions observed in winter and spring above Ny-Ålesund were considerably weaker than those monitored with RS measurements at the other sites (Barrow, Alert and Tiksi), presumably as a result of the mild conditions favoured by the effects of the North Atlantic Drift reaching the Svalbard Archipelago. A procedure similar to that adopted by us to carry out the results shown in Fig. 2.17 was also applied to the RS data recorded in the 2001–2010 years at the stations of Barrow, Alert, Ny-Ålesund and Tiksi, and to the 2001–2015 RS data sets recorded at the other ten stations (i. e., Inuvik, Cambridge Bay, Resolute, Eureka, Aasiaat, Danmarkshavn, Jan Mayen, Sodankylä, Ostrov Dikson, and Cherskij). Through the detailed analysis of such a multi-year set of RS data, we have determined the following results for each of the 14 above-mentioned RS stations: (1) the total number N_{RS} of RS measurements performed in winter (December–February), spring (March–May), summer (June–August) and autumn (September–November); (2) the seasonal number N_{inv} of ground-based temperature inversions observed in the four seasonal periods; (3) the seasonal percentages P_{inv} of ground-based temperature inversions identified in the present study; (4) the multi-year seasonal mean values of the difference ΔT calculated between temperature T_{inv} (measured at the top-level of the thermal inversion) and surface-level temperature T_o; (5) the multi-year seasonal quartiles (i.e. the 25th, 50th and 75th percentiles) of parameter ΔT; (6) the multi-year seasonal mean values of depth Δz; (7) the multi-year seasonal quartiles of Δz; (8) the multi-year seasonal mean values of the average vertical gradient $\gamma = \Delta T/\Delta z$; and (9) the multi-year seasonal quartiles of the average vertical gradient γ.

The seasonal evaluations of these nine parameters are reported in Table 2.8. It can be noted that the observed percentage of the days presenting clear ground-based temperature inversion features is particularly high in winter at the RS stations of Barrow (50%), Inuvik (58%), Cambridge Bay (73%), Resolute (72%), Eureka (81%), Alert (80%), Danmarkshavn (78%), Ny-Ålesund (52%), Ostrov Dikson (51%), Tiksi (60%) and Cherskij (72%), while such a percentage results to be lower than 50% at Aasiaat (35%), Jan Mayen (27%) and Sodankylä (38%). In particular, it is worth noting that Aasiaat and Jan Mayen are coastal stations subject to the significantly mitigative effects of warm oceanic currents during the winter period, where the ground-based thermal inversion features result to be rather weak, presenting values of ΔT generally lower than 0.5 °K. Lower percentages ranging between 25% and no more than 50% have been in fact found in spring and autumn at all the 14 Arctic stations examined in Table 2.8, together with percentages no higher than 30% in summer.

Very high seasonal mean values of ΔT were determined in winter at all the North-American stations, ranging between 5.2 and 10.8 °K, as well as at the North-European and Siberian stations, where ΔT was estimated to vary between 6.4 and 11.1 °K, while lower values were measured at Aasiaat (1.5 °K), Danmarkshavn (4.5 °K), Jan Mayen (0.8 °K) and Ny-Ålesund (1.4 °K). Appreciably lower values of ΔT were measured at all the 14 stations in spring, and even more in summer and autumn, as can be verified by examining the seasonal mean values of ΔT reported in Table 2.8. Correspondingly, the values of the ground-based thermal inversion depth Δz were estimated to vary on average from 200 m to 400 m in winter, and to largely exceed the threshold of 100 m in both summer and autumn, at all the North-American stations. Values of Δz lower than 100 m were indeed determined in winter at Aasiaat, Jan Mayen and Ny-Ålesund, together with similarly low values observed in the other seasons, which suggest that the thermal inversion structures monitored in this area are in general weaker throughout the whole year than in the other Arctic sectors. In fact, values of Δz greater than 400–500 m were determined in winter at the three Siberian stations, followed on average by values of 300–400 m in spring, and values of 200–300 m in both summer and autumn. The values of the 25th, 50th and 75th percentiles of Δz were determined with a certain accuracy at all the North-American and Siberian stations, whereas those evaluated at the Aasiaat, Danmarkshavn, Jan Mayen and Ny-Ålesund stations were determined with a lower accuracy in the various seasons because of: (i) the weakness of such ground-based thermal inversions during the whole year, and (ii) the narrowness of the RS measurements in describing the thermal features of these ground-based temperature inversions, since they consist of no more than 2–3 RS measurements recorded over the first 100 meters after the launch (in steps of 25–50 m), so that the resolution in height is low and not sufficiently accurate for identifying the true position in height of the thermal inversion layer. Consequently, the seasonal quartiles were not distinctly determined, as it is the case of Jan Mayen, where the first and second quartiles were estimated to be equal to 28 m above the sea-level in all the four seasonal periods.

As a result of these evaluations of seasonal mean and quartiles of ΔT and Δz reported in Table 2.8, the vertical gradient γ of temperature calculated within the ground-based temperature inversion layer (and herein measured in °K km^{-1}) was found to assume seasonal mean values ranging mainly: (i) between 18 and

33 °K km^{-1} at all the six North-American RS stations; (ii) between 16 and 30 °K km^{-1} at the stations of Aasiaat, Danmarkshavn, Jan Mayen and Ny-Ålesund; (iii) between 28 and 41 °K km^{-1} at Sodankylä (Finland), which is the unique European station considered in the present study; and (iv) between 13 and 20 °K km^{-1} at the three Siberian RS stations. Correspondingly, the quartiles of gradient γ were found to assume distinct values at all the 14 stations considered in the present analysis for all the four seasonal periods.

2.4.1.2 Seasonal Changes in the Thermal Characteristics of the Tropopause Region

The polar and sub-polar tropopause meteorological and structural characteristics of the Arctic atmosphere were deeply investigated by Zängl and Hoinka (2001) using the data-set of the European Centre for Medium-Range Weather Forecasts (ECMWF) Reanalysis (ERA) collected over the 1979–1993 period, together with the 1989–1993 radiosonde data-set. The results obtained through this analysis showed that the annual cycle of the Arctic tropopause can be classified into the two following different patterns: (i) a single wave with a tropopause pressure maximum in winter and a minimum in summer, which is typical for the sub-polar parts of Eastern Siberia and North America; and (ii) a double wave with pressure maxima in spring and autumn and minima in summer and winter, more frequently found above Northern Europe, Western Siberia, and more generally at all the high Arctic latitudes. A climatology of Arctic tropopause properties was also defined by Highwood et al. (2007) using the ECMWF – ERA data, which was evaluated to be consistent with that derived from the Historical Arctic Radiosonde Archive (HARA). The multi-decade radiosonde data-set, relative to the 1965–1990 years, showed some evidence of a long-term trend involving the Arctic tropopause properties: (i) the winter-time tropopause pressure has decreased by approximately 14 hPa per decade, on average, while that in other seasons has decreased on average by 5 hPa only per decade; and (ii) the tropopause temperature was subject to a significant trend only during the winter months, being decreased by 1.6 °K per decade since 1965. Highwood et al. (2007) discussed these changes also in relation to the changes in the atmospheric contents of ozone and greenhouse gases, and the strength of the stratospheric vortex, finding in particular that: (a) the winter tropopause properties appear to be strongly correlated with the Arctic Oscillation Index; (b) the area-weighted average tropopause pressure assumes its annual maximum (of about 300 hPa) in late March and the lowest value (close to 265 hPa) in early July, being more strongly influenced by changes in the tropospheric lapse rate due to dynamical mechanisms; and (c) the area-weighted average tropopause temperature usually reaches; (i) its annual minimum of about 211 °K in late December and/or in early January, and (ii) its annual maximum of around 224 °K in late June, this parameter being largely determined by the approximately isothermal conditions observed in the lower part of the stratosphere during this period of the year.

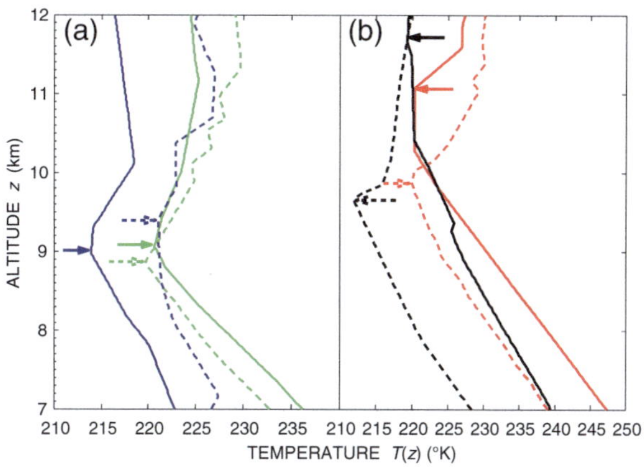

Fig. 2.18 Vertical profiles of temperature $T(z)$ derived in the altitude range from 7 to 12 km (including the upper troposphere, the tropopause region, and the lower stratosphere), as obtained from the RS measurements performed at eight different Arctic sites on days of different seasonal periods. The left part (**a**) shows the vertical profiles of $T(z)$ recorded: (i) in winter at Ny-Ålesund (Spitsbergen, Svalbard) on December 4, 2013 (12:00 UTC) (solid blue curve), and Barrow (Alaska, USA) on January 31, 2003 (12:00 UTC) (dashed blue curve), and (ii) in spring at Sodankylä (Finland) on April 29, 2009 (00:00 UTC) (solid green curve), and Resolute (Nunavut, Canada) on May 21, 2004 (12:00 UTC) (dashed green curve), with arrows of the same colour used to indicate the level z_{min}, at which the tropopause temperature reaches its minimum. The right part (**b**) shows the vertical profiles of $T(z)$ recorded: (i) in summer at Ostrov Dikson (Siberia, Russia) on June 14, 2011 (12:00 UTC) (solid red curve), and Alert (Nunavut, Canada), on July 31, 2007 (12:00 UTC) (dashed red curve), and (ii) in autumn at Cambridge Bay (Nunavut, Canada) on October 30, 2002 (10:00 UTC) (solid black curve), and Danmarkshavn (Greenland) on October 30, 2005 (12:00 UTC) (dashed black curve), with arrows of the same colour used to indicate the level z_{min} at which the tropopause temperature reaches its minimum

Taking into account the above remarks, we have shown in Fig. 2.18 a set of eight daily vertical profiles of temperature $T(z)$, recorded within the altitude range from 7 to 12 km for different seasonal conditions, to give a measure of the variations of the two tropopause parameters z_{min} and T_{min} occurring throughout the year. Evaluating in which altitude layer the temperature lapse rate is lower than 2 °K km^{-1}, the eight vertical profiles of $T(z)$ have been estimated to yield the following values of z_{min} and T_{min} in the four seasons: (1) in winter: $z_{min} = 9.05$ km and $T_{min} = 213.9$ °K at Ny-Ålesund on December 4, 2013 (12:00 UTC), and $z_{min} = 9.40$ km and $T_{min} = 221.1$ °K at Barrow on January 31, 2003 (12:00 UTC); (2) in spring: $z_{min} = 9.08$ km and $T_{min} = 220.7$ °K at Sodankylä on April 29, 2009 (00:00 UTC), and $z_{min} = 8.88$ km and $T_{min} = 219.8$ °K at Resolute on May 21, 2004 (12:00 UTC); (3) in summer: $z_{min} = 11.08$ km and $T_{min} = 220.3$ °K at Ostrov Dikson on June 14, 2011 (12:00 UTC), and $z_{min} = 9.88$ km, $T_{min} = 219.9$ °K at Alert on July 31, 2007 (12:00 UTC); and (4) in autumn, $z_{min} = 11.74$ km and $T_{min} = 219.3$ °K at Cambridge Bay on October 30, 2002 (10:00 UTC), and $z_{min} = 9.65$ km and $T_{min} = 212.0$ °K at Danmarkshavn on October 30, 2005 (12:00 UTC).

The above results clearly suggest that parameters T_{min} and z_{min} are subject to vary largely with season. In order to better investigate these variations, a more extended and accurate study of the seasonal variations of parameters T_{min} and z_{min} was conducted by analyzing the measurements performed at the stations of Resolute, Ny-Ålesund, Sodankylä and Cherskij over the period from 2011 to 2015. The results are shown in Fig. 2.19, clearly indicating that the time-patterns of the daily mean values of parameter T_{min} exhibit in general well-defined annual cycles at all the four stations, while the daily and monthly mean values of z_{min} are in general more widely dispersed and show more limited variations from one season to another.

The time-patterns of the daily mean values of parameter T_{min} determined in the tropopause region at these four Arctic RS stations during the 5 years from 2011 to 2015 were found to be largely dispersed. The daily values of T_{min} measured at Resolute were estimated to vary between about 200 °K and more than 240 °K, while the daily values of z_{min} varied from less than 5 km to about 10 km. The monthly mean values of T_{min} measured at Resolute ranged between about 210 °K and nearly 230 °K, being determined with average standard deviations of around ±5 °K. They described well-defined annual cycles, with maxima in June or July, and minima in the colder months of the year, from December to March. The annual maxima of

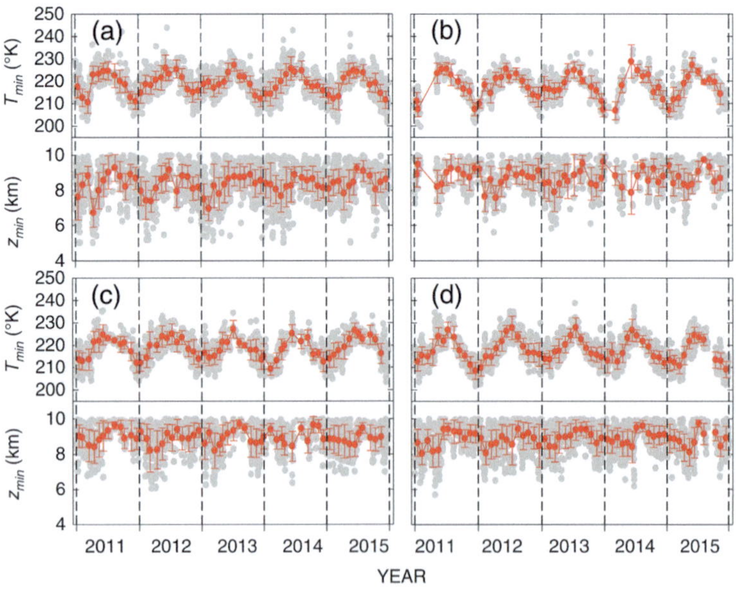

Fig. 2.19 Time-series of the daily mean values (grey dots) and monthly mean values (red dots) (the latter being represented with vertical red bars used to indicate the standard deviations) of temperature minimum T_{min} characterizing the tropopause region (upper part) and tropopause temperature minimum level z_{min} (lower part), as obtained during the 5-year period from 2011 to 2015 at the following four RS stations: (**a**) Resolute (Nunavut, Canada), (**b**) Ny-Ålesund (Spitsbergen, Svalbard), (**c**) Sodankylä (Finland), and (**d**) Cherskij (Siberia, Russia)

T_{min} recorded during the five years were found to vary between 225 °K and 230 °K, and the annual minima between 210 °K and 214 °K. Correspondingly, the monthly mean values of z_{min} were estimated to range between about 7 km and nearly 9 km, being determined with average standard deviations of ±1.2 km, and presenting their maxima in June–August and their minima in February–May.

The daily values of T_{min} measured at Ny-Ålesund were found to range between about 205 °K and 234 °K, while the daily values of z_{min} varied from about 5.5 km to more than 10 km. The monthly mean values of T_{min} ranged between about 205 °K and 235 °K, these values being determined with average standard deviations of ±4 °K. They were evaluated to describe annual cycles having the maxima in the summer months (June – July) and the minima in the winter months from December to February. The annual maxima of T_{min} were found to be quite regular over the five selected years, varying between 225 °K and 230 °K from one year to another, and the annual minima to be very close to 210 °K in all the years. The corresponding monthly mean values of z_{min} were found to range between about 7.8 and 9.5 km over the five years, being determined with average standard deviations of ±1.2 km. The annual maxima of z_{min} were observed usually in the early summer (June–July), and the minima in the winter months (December–February).

The daily values of T_{min} measured at Sodankylä were estimated to vary between 205 °K and 235 °K, and the daily values of z_{min} to range between about 6 km and 10 km. The monthly mean values of T_{min} varied between 210 °K and 228 °K, being found with average standard deviations of ±5 °K, and described regular annual cycles presenting their maxima most frequently in June–July, and the minima in December–January. The annual maxima of T_{min} were estimated to range between 224 °K and 226 °K, and the annual minima between 210 °K and 214 °K. The corresponding monthly mean values of z_{min} were found to vary between 8.1 and nearly 10 km, these estimates being obtained with average standard deviations of ±1.3 km. The annual maxima of z_{min} were observed in summer (June–August), and the minima in the months from March to May.

The daily values of T_{min} measured at Cherskij were estimated to vary from 202 °K to nearly 240 °K, and the daily values of z_{min} to range from less than 6 km to 10 km. The monthly mean values of T_{min} were estimated to: (i) vary between less than 210 °K and about 230 °K, having average standard deviations of ±5 °K, and (ii) describe annual cycles with maxima usually observed in June–July, and minima in the winter and early spring months, from December to March. The annual maxima of T_{min} were estimated to range between 226 °K and 229 °K, and the annual minima between 207 °K and 209 °K. The corresponding monthly mean values of z_{min} were found to vary between 8.0 and 9.8 km, these estimates being obtained with average standard deviations of ±1.0 km. The annual maxima of z_{min} were observed in the months from June to August, and the minima from February to May.

The results determined in Fig. 2.19 for four of the 14 RS stations have been fully confirmed by the analysis of the monthly mean values of T_{min} and z_{min} characterizing the tropopause region at the 14 RS stations considered in the present study (and listed in Table 2.1). The seasonal mean values and the quartiles of parameters T_{min} and z_{min} have been determined at all the 14 RS stations through a careful analysis of the daily vertical profiles of $T(z)$ derived from the RS data collected in the years from 2001 to 2015. The results are given in Table 2.9, showing that:

Table 2.9 Seasonal mean values and quartiles of the temperature minimum T_{min} (°K) and its altitude z_{min} (km) characterizing the tropopause region, as measured in the present analysis performed for the data-sets collected at the 14 above-selected RS stations

Station and RS measurement period	Season	T_{min} (°K) Mean value	Quartiles	z_{min} (km) Mean value	Quartiles
Barrow (2001–2015)	Winter (DJF)	216.3	213.3, 216.0, 219.3	8.4	7.8, 8.5, 9.1
	Spring (MAM)	218.9	216.3, 218.6, 221.3	8.5	7.9, 8.5, 9.1
	Summer (JJA)	223.1	220.4, 222.5, 225.3	9.2	8.9, 9.3, 9.7
	Autumn (SON)	217.2	214.6, 216.9, 219.4	8.8	8.4, 8.9, 9.4
Inuvik (2001–2009 and 2012–2015)	Winter (DJF)	216.5	213.6, 216.3, 219.3	8.5	8.0, 8.6, 9.1
	Spring (MAM)	218.4	215.8, 218.1, 220.6	8.5	8.0, 8.6, 9.1
	Summer (JJA)	221.7	219.4, 221.3, 223.3	9.3	9.1, 9.4, 9.7
	Autumn (SON)	216.9	214.3, 216.8, 219.0	8.9	8.5, 8.9, 9.4
Cambridge Bay (2001–2015)	Winter (DJF)	216.7	213.1, 216.4, 219.9	8.1	7.5, 8.2, 8.8
	Spring (MAM)	219.4	216.3, 219.3, 222.3	8.3	7.7, 8.3, 9.0
	Summer (JJA)	223.9	220.9, 223.3, 226.4	8.9	8.5, 9.1, 9.6
	Autumn (SON)	218.0	215.3, 217.9, 220.6	8.7	8.2, 8.8, 9.2
Resolute (2001–2015)	Winter (DJF)	215.9	212.1, 215.9, 219.6	8.0	7.3, 8.0, 8.7
	Spring (MAM)	220.0	217.0, 219.9, 223.3	7.9	7.3, 8.0, 8.7
	Summer (JJA)	224.2	221.1, 223.6, 226.6	8.7	8.2, 8.8, 9.4
	Autumn (SON)	217.9	215.3, 217.8, 220.3	8.5	8.0, 8.6, 9.1
Eureka (2001–2015)	Winter (DJF)	214.7	210.9, 214.6, 218.1	8.1	7.5, 8.2, 8.8
	Spring (MAM)	219.4	216.3, 219.5, 222.8	7.9	7.3, 8.0, 8.6
	Summer (JJA)	223.8	221.0, 223.3, 226.3	8.7	8.2, 8.9, 9.4
	Autumn (SON)	217.4	214.8, 217.3, 219.9	8.4	7.9, 8.5, 9.0
Alert (2001–2015)	Winter (DJF)	213.8	209.8, 213.5, 217.4	8.3	7.7, 8.4, 9.0
	Spring (MAM)	218.8	215.9, 219.1, 222.1	8.0	7.4, 8.1, 8.7
	Summer (JJA)	223.5	220.9, 223.1, 225.8	8.8	8.3, 8.9, 9.4
	Autumn (SON)	217.3	214.6, 217.3, 219.8	8.5	8.0, 8.5, 9.0
Aasiaat (2001–2009)	Winter (DJF)	215.2	211.5, 215.3, 218.8	8.2	7.6, 8.4, 9.0
	Spring (MAM)	219.7	216.6, 219.6, 222.6	8.3	7.6, 8.4, 9.0
	Summer (JJA)	222.7	220.3, 222.1, 224.3	9.1	8.7, 9.3, 9.6
	Autumn (SON)	218.3	215.6, 218.1, 220.9	8.6	8.1, 8.7, 9.2
Danmarkshavn (2001–2009)	Winter (DJF)	213.9	209.9, 213.4, 218.0	8.5	7.9, 8.7, 9.1
	Spring (MAM)	218.6	215.8, 218.8, 221.5	8.4	7.8, 8.4, 9.0
	Summer (JJA)	222.7	220.3, 222.1, 224.8	9.1	8.7, 9.2, 9.6
	Autumn (SON)	217.5	214.6, 217.5, 220.3	8.8	8.3, 8.8, 9.4

(continued)

Table 2.9 (continued)

Station and RS measurement period	Season	T_{min} (°K)		z_{min} (km)	
		Mean value	Quartiles	Mean value	Quartiles
Jan Mayen (2001–2015)	Winter (DJF)	214.7	210.8, 214.4, 218.0	8.6	8.1, 8.7, 9.3
	Spring (MAM)	219.2	215.9, 218.9, 222.3	8.4	7.9, 8.5, 9.1
	Summer (JJA)	223.6	221.1, 223.0, 225.3	9.0	8.6, 9.2, 9.5
	Autumn (SON)	218.5	215.8, 218.1, 221.3	8.8	8.3, 8.9, 9.4
Ny-Ålesund (2005–2015)	Winter (DJF)	212.5	208.6, 211.8, 215.9	8.7	8.3, 8.9, 9.3
	Spring (MAM)	218.2	215.0, 218.3, 221.1	8.2	7.6, 8.3, 8.9
	Summer (JJA)	224.0	221.3, 223.6, 225.9	8.8	8.3, 8.9, 9.4
	Autumn (SON)	217.2	214.0, 217.1, 219.9	8.6	8.2, 8.7, 9.2
Sodankylä (2001–2015)	Winter (DJF)	213.8	210.4, 213.3, 216.8	8.8	8.3, 8.9, 9.4
	Spring (MAM)	218.2	215.3, 218.3, 221.1	8.6	8.0, 8.7, 9.2
	Summer (JJA)	223.4	220.9, 222.8, 225.6	9.1	8.7, 9.2, 9.6
	Autumn (SON)	218.0	214.8, 217.6, 220.9	8.9	8.3, 8.9, 9.5
Ostrov Dikson (2001–2015)	Winter (DJF)	212.4	209.0, 212.3, 215.3	8.6	8.0, 8.7, 9.2
	Spring (MAM)	218.6	214.8, 218.8, 222.4	8.3	7.8, 8.4, 9.0
	Summer (JJA)	224.9	221.8, 224.3, 227.3	9.0	8.6, 9.1, 9.6
	Autumn (SON)	216.7	213.5, 216.6, 219.5	8.8	8.3, 8.9, 9.4
Tiksi (2001–2015)	Winter (DJF)	212.4	209.3, 212.3, 215.3	8.5	8.0, 8.7, 9.1
	Spring (MAM)	218.0	214.5, 217.9, 221.5	8.5	8.0, 8.6, 9.1
	Summer (JJA)	225.4	222.8, 225.0, 227.8	9.1	8.7, 9.2, 9.6
	Autumn (SON)	216.1	212.9, 215.9, 218.9	8.8	8.4, 8.9, 9.4
Cherskij (2004–2015)	Winter (DJF)	213.4	210.1, 213.3, 216.4	8.5	8.0, 8.6, 9.1
	Spring (MAM)	218.3	215.0, 218.0, 221.5	8.5	7.9, 8.5, 9.2
	Summer (JJA)	224.3	221.6, 223.8, 226.5	9.1	8.7, 9.3, 9.7
	Autumn (SON)	215.7	213.0, 215.6, 218.3	9.0	8.5, 9.0, 9.5

(1) The seasonal mean values of T_{min} determined at Barrow over the 2001–2015 period were estimated to be equal to 216.3 °K in winter, 218.9 °K in spring, 223.1 °K in summer and 217.2 °K in autumn, while those of z_{min} were found to be equal to 8.4 km in winter, 8.5 km in spring, 9.2 km in summer and 8.8 km in autumn.

(2) The seasonal mean values of T_{min} determined at Inuvik over two distinct periods from 2001 to 2009 and from 2012 to 2015 were estimated to be equal to 216.5 °K in winter, 218.4 °K in spring, 221.7 ° K in summer and 216.9 °K in autumn, these seasonal values being: (i) higher by 0.2 °K in winter, (ii) lower by 0.5 °K in spring, (iii) lower by 1.4 °K in summer, and (iv) lower by 0.3 °K in autumn than those determined at Barrow. Correspondingly, the seasonal mean values of z_{min} were estimated to be equal to 8.5 km in winter and spring, 9.3 km in summer and 8.9 km in autumn, these results differing only slightly from those measured at Barrow, by +0.1 km in all seasons except in spring,

when the difference between the Barrow and Inuvik seasonal mean values of z_{min} was found to be smaller than 0.1 km.

(3) The seasonal mean values of T_{min} determined at Cambridge Bay from 2001 to 2015 were estimated to be equal to 216.7 °K in winter, 219.4 °K in spring, 223.9 °K in summer and 218.0 °K in autumn, these seasonal evaluations being slightly higher on average than those measured at Inuvik, by 0.2 °K in winter, 1.0 °K in spring, 2.2 °K in summer, and 1.1 °K in autumn. The seasonal mean values of z_{min} were correspondingly found to be equal to 8.1 km in winter, 8.3 km in spring, 8.9 km in summer and 8.7 km in autumn, and, therefore, considerably lower on average than those measured at Inuvik, by 0.4 km in both winter and summer, and by 0.2 km in both spring and autumn.

(4) The seasonal mean values of T_{min} obtained at Resolute in the 15 years from 2001 to 2015 were equal to 215.9 °K in winter, 220.0 °K in spring, 224.2 °K in summer and 217.9 °K in autumn. Therefore, they result to be lower by 0.8 °K in winter and 0.1 °K in autumn than the values obtained at Cambridge Bay in the same seasons, being conversely higher by 0.6 °K in spring and 0.3 °K in autumn. The seasonal mean values of z_{min} were found to be equal to 8.0 km in winter, 7.9 km in spring, 8.7 km in summer and 8.5 km in autumn and, therefore, appreciably lower than those measured at Cambridge Bay by 0.1 km in winter, 0.4 km in spring, and 0.2 km in both summer and autumn.

(5) The seasonal mean values of T_{min} measured at Eureka from 2001 to 2015 were estimated to be equal to 214.7 °K in winter, 219.4 °K in spring, 223.8 °K in summer and 217.4 °K in autumn, these seasonal mean values being on average appreciably lower than those evaluated at Resolute, by 1.2 °K in winter, 0.6 °K in spring, 0.4 °K in summer, and 0.5 °K in autumn. In addition, the seasonal mean values of z_{min} were found to be equal to 8.1 km in winter, 7.9 km in spring, 8.7 km in summer and 8.4 km in autumn and, therefore, very close to those determined at Resolute, being the values obtained at Resolute and Eureka fully coincident in both spring and summer, and different by only −0.1 km in winter and autumn.

(6) The seasonal mean values of T_{min} determined at Alert from 2001 to 2015 were estimated to be equal to 213.8 °K in winter, 218.8 °K in spring, 223.5 °K in summer and 217.3 °K in autumn, being thus appreciably lower on average than those estimated at Eureka (by 0.9 °K in winter, 0.6 °K in spring, 0.3 °K in summer and 0.1 °K in autumn). The corresponding seasonal mean values of z_{min} were found to be equal to 8.3 km in winter, 8.0 km in spring, 8.8 km in summer and 8.5 km in autumn, resulting to be slightly higher than those measured at Eureka by 0.2 km in winter and by 0.1 km in the other three seasons.

(7) The seasonal mean values of T_{min} determined at Aasiaat from 2001 to 2009 were evaluated to be equal to 215.2 °K in winter, 219.7 °K in spring, 222.7 °K in summer and 218.3 °K in autumn, and, hence, appreciably higher than those observed at Eureka and Alert, both located at considerably higher latitudes, since the Aasiaat seasonal mean values of T_{min} were found to differ: (a) from those measured at Eureka by +0.5 °K in winter, by +0.3 °K in spring, by

−1.1 °K in summer, and by +2.3 °K in autumn, and (b) from those measured at Alert by +1.4 °K in winter, + 0.9 °K in spring, −0.8 °K in summer, and + 1.0 °K in autumn. The seasonal mean values of z_{min} were estimated at Aasiaat to be equal to 8.2 km in winter, 8.3 km in spring, 9.1 km in summer and 8.6 km in autumn, therefore being: (a) higher than those measured at Eureka in all the four seasons by 0.1 km in winter, 0.4 km in both spring and summer, and 0.2 km in autumn, and (b) slightly lower than those measured at Alert by 0.1 km in winter, 0.5 km in spring, and 0.3 km in summer, and conversely higher by 0.1 km in autumn.

(8) The seasonal mean values of T_{min} determined at Danmarkshavn from 2001 to 2009 were evaluated to be equal to 213.9 °K in winter, 218.6 °K in spring, 222.7 °K in summer and 217.5 °K in autumn, which were found to be appreciably lower on average than those measured at the Aasiaat station (by 1.3 °K in winter, 1.1 °K in spring, and 0.8 °K in autumn) and indeed coincident with the mean value measured at Aasiaat in summer. Correspondingly, the seasonal mean values of z_{min} were found to be equal to 8.5 km in winter, 8.4 km in spring, 9.1 km in summer and 8.8 km in autumn, which are: (i) appreciably higher than those evaluated at Aasiaat by 0.3 km in winter and 0.1 km spring, (ii) comparable with that determined at Aasiaat in summer, and (iii) higher than that recorded at Aasiaat by 0.2 km in autumn.

(9) The seasonal mean values of T_{min} determined at Jan Mayen from 2001 to 2015 were evaluated to be equal to 214.7 °K in winter, 219.2 °K in spring, 223.6 °K in summer and 218.5 °K in autumn, which were estimated to be lower by 0.5 °K than those measured at Aasiaat in both winter and spring, and higher by 0.9 °K in summer and by 0.2 °K in autumn. The seasonal mean values of z_{min} were found to be equal at Jan Mayen to 8.6 km in winter, 8.4 km in spring, 9.0 km in summer and 8.8 km in autumn, which result to be appreciably higher by 0.4 km than that measured at Aasiaat in winter and comparable with those recorded at this Greenlandic site in the other seasons, since the seasonal mean values determined at these two sites were found to differ by +0.1 km in spring, −0.1 km in summer and + 0.2 km in autumn. It is also worth noting that the seasonal mean values of T_{min} determined at Jan Mayen are appreciably higher than those determined at Danmarkshavn (which is located at a higher latitude by 6 °N and is therefore more weakly influenced by the temperate effects of the North Atlantic Drift) by 0.8 °K in winter, 0.6 °K in spring, 0.9 °K in summer and 1.0 °K in autumn. The seasonal mean values of z_{min} were evaluated at Jan Mayen to be equal to 8.6 km in winter, 8.4 km in spring, 9.0 km in summer and 8.8 km in autumn, which are very close to those measured at Danmarkshavn in the four seasons, which differ by +0.1 km in winter, −0.1 km in summer, and are in practice coincident in both spring and autumn. In addition, the seasonal mean values of T_{min} measured at Jan Mayen result to be higher by 1–2 °K on average than those determined at Ny-Ålesund (which is located at a 8 °N higher latitude than that of Jan Mayen), while the seasonal mean values of z_{min} are comparable within ±0.2 km with those observed at Ny-Ålesund in the same seasons.

(10) The seasonal mean values of T_{min} determined at Ny-Ålesund from 2005 to 2015 were evaluated to be equal to 212.5 °K in winter, 218.2 °K in spring, 224.0 °K in summer and 217.2 °K in autumn, which differ appreciably from those measured at Jan Mayen, by −2.2 °K in winter, −1.0 °K in spring, + 0.4 °K in summer and −1.3 °K in autumn. The corresponding seasonal mean values of z_{min} were estimated to be equal to 8.7 km in winter, 8.2 km in spring, 8.8 km in summer and 8.6 km in autumn, which are higher by 0.1 km only in winter than that measured at Jan Mayen, and lower by 0.2 km in the other three seasons.

(11) The seasonal mean values of T_{min} determined at Sodankylä (located in the northern part of the Scandinavian peninsula) from 2001 to 2015 were evaluated to be equal to 213.8 °K in winter, 218.2 °K in spring, 223.4 °K in summer and 218.0 °K in autumn, which are: (i) higher than that measured at Ny-Ålesund by +1.3 °K in winter, (ii) equal in spring, (iii) lower by 0.6 °K in summer, and (iv) higher by 0.8 °K in autumn. The corresponding seasonal mean values of z_{min} were estimated to be equal to 8.8 km in winter, 8.6 km in spring, 9.1 km in summer and 8.9 km in autumn, which are higher by 0.1 km in winter, by 0.4 km in spring, 0.3 km in summer, and 0.1 km in autumn than those measured at Ny-Ålesund in the same seasons, these seasonal increases being justified by the more than 11° N lower latitude of Sodankylä.

(12) The seasonal mean values of T_{min} determined at Ostrov Dikson from 2001 to 2015 were evaluated to be equal to 212.4 °K in winter, 218.6 °K in spring, 224.9 °K in summer and 216.7 °K in autumn, which appreciably differ from those measured at Sodankylä, being lower by 1.4 °K in winter, slightly higher by 0.4 °K in spring, considerably higher by 1.5 °K in summer, and markedly lower by 1.3 °K in autumn. The corresponding seasonal mean values of z_{min} were estimated to be equal to 8.6 km in winter, 8.3 km in spring, 9.0 km in summer and 8.8 km in autumn, which turn out to be lower than those measured at Sodankylä by 0.2 °K in winter, 0.3 °K in spring, and by 0.1 °K only in both summer and autumn.

(13) The seasonal mean values of T_{min} determined at Tiksi from 2001 to 2015 were evaluated to be equal to 212.4 °K in winter, 218.0 °K in spring, 225.4 °K in summer and 216.1 °K in autumn, and therefore comparable with that measured at Ostrov Dikson in winter, and differing in the other seasons by −0.6 °K in spring, + 0.5 °K in summer, and −0.6 °K in autumn. The seasonal mean values of z_{min} were found at Tiksi to be equal to 8.5 km in both winter and spring, 9.1 km in summer and 8.8 km in autumn, and therefore be very similar to those measured at Ostrov Dikson, with differences of −0.1 km in winter, + 0.2 km in spring, and + 0.1 km in summer, and of less than ±0.1 km in autumn.

(14) The seasonal mean values of T_{min} measured at Cherskij from 2001 to 2015 were evaluated to be equal to 213.4 °K in winter, 218.3 °K in spring, 224.3 °K in summer and 215.7 °K in autumn, differing from those measured at Tiksi by +1.0 °K in winter, + 0.3 °K in spring, − 1.1 °K in summer, and −0.4 °K

in autumn. The seasonal mean values of z_{min} were estimated at Charskij to be of 8.5 km in both winter and spring, 9.1 km in summer and 9.0 km in autumn, which result to be very similar to those measured at Tiksi, since they do not show appreciable differences in winter, spring and summer, and differ by +0.2 km only in autumn.

2.5 Annual Variations in the Vertical Profiles of Absolute Humidity

After the correction of the RS measurements for taking into account the numerous dry biases and lag errors affecting the measurements of *RH* made using the various radiosonde models (as described in Sect. 2.2.3), the vertical profiles of air relative humidity *RH(z)* obtained from the daily RS measurements carried out at the 14 above-selected Arctic sites have been performed for each RS observation, as shown in the examples of Fig. 2.6. The calculations of the vertical profiles of water vapour partial pressure *e(z)* and absolute humidity *q(z)* have been carried out for the known vertical profiles of air pressure *p(z)* and temperature *T(z)*, by following the procedures described in Sect. 2.2.4, which have been applied to the RS measurements performed with various radiosonde models over the altitude range from surface level to 12 km, as shown in the examples of Fig. 2.7 relative to a set of daily RS measurements performed at the stations of Danmarkshavn, Sodankylä, Barrow and Tiksi, the first two examples obtained by using Vaisala RS80-H and RS92 radiosondes, respectively, the third example with a VIZ-B2 radiosonde, and the fourth example with a Marz2-2 radiosonde. In general, the radiosonde measurements of *RH* were collected at more than 800 standard and additional levels in the altitude range from the surface to 10 km, and at a rather limited number of supplementary stratospheric levels from 10 to no more than 15 km. As mentioned in Sect. 2.2.4, the calculations of absolute humidity *q(z)* were made in terms of Eq. (2.1) at all the significant levels *z* of each RS measurement by using the Murphy and Koop (2005) formula to calculate the saturated water vapour pressure *E[T(z)]* in the pure phase over a plane surface of pure liquid water as a function of air temperature *T(z)* measured at the various atmospheric levels from surface to 12 km altitude. The monthly mean vertical profiles of absolute humidity *q(z)* were then calculated on the basis of the above daily measurements, calculated separately for each RS measurement year, as shown in Fig. 2.20. They were obtained by analyzing monthly sets of RS measurements carried out at Ny-Ålesund (Spitsbergen) over the altitude range from surface-level to 12 km, in the months of February, May, August, and November of the different years 2005, 2008, 2011, and 2014 (together with their standard deviations given at some fixed levels). Figure 2.20 clearly shows that the absolute humidity *q(z)* can vary largely with season and from one year to another within the lower part of the Arctic troposphere with altitude $z < 4$ km. In particular, the vertical profile of *q(z)* results to be characterized by: (i) values lower

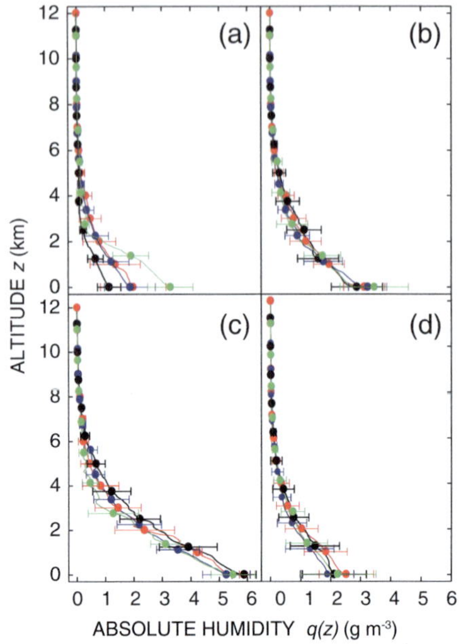

Fig. 2.20 Comparison among the monthly mean vertical profiles of absolute humidity *q(z)* derived from the RS measurements conducted at Ny-Ålesund (Spitsbergen, Svalbard) over the altitude range from the surface-level to 12 km, as obtained from the monthly data-sets recorded in (**a**) February, (**b**) May, (**c**) August, and (**d**) November of the years 2005 (red curves), 2008 (blue curves), 2011 (black curves) and 2014 (green curves). The coloured dots and the horizontal bars provide the monthly mean values of *T(z)* with their standard deviations at some selected levels

than 3.5 g m^{-3} in February of all the four selected years, at all the altitudes; (ii) surface-level values lower than 3.5 g m^{-3} in May of all the selected years, and values lower than 2 g m^{-3} at altitudes $z > 1$ km; (iii) considerably higher values in August, when *q(z)* was found to vary between 5 and 6 g m^{-3} at the surface-level and range between 3 and 4 g m^{-3} at levels of around 1.5 km, in all the selected years; and (iv) values lower than 3 g m^{-3} at all levels in November, when the vertical profile of *q(z)* was evaluated to exhibit features very similar to those measured in February.

The monthly mean vertical profiles of absolute humidity *q(z)* were subsequently calculated at all the 14 above-chosen RS Arctic stations over the whole multi-year periods in which the RS measurements have been collected at the various sites, over the altitude range from the surface-level to 12 km. An example of these multi-year monthly mean vertical profiles of *q(z)* are shown in Fig. 2.21, as obtained for the months of January, April, July and October from the RS measurements taken at the stations of Cambridge Bay, Eureka, Sodankylä and Tiksi over the whole period from 2001 to 2015. These vertical profiles of *q(z)* were found to vary largely with season in the lower part of the troposphere (for $z < 4$ km), presenting quite different characteristics of the atmospheric moisture conditions from one season to another. It can be clearly seen in Fig. 2.21 that: (i) the surface-level values q_o of absolute humidity, measured in July at the four RS stations, are considerably higher than those measured in the other months of January, April, and October at the same stations, since the multi-year monthly mean values of q_o calculated in July were estimated to be on average equal to 7.3 g m^{-3} at Cambridge Bay, 5.3 g m^{-3} at Eureka, 8.8 g m^{-3} at Sodankylä, and 6.9 g m^{-3} at Tiksi, and those determined in January, April and October to vary between 0.2 and 2.5 g m^{-3} at Cambridge Bay, 0.2

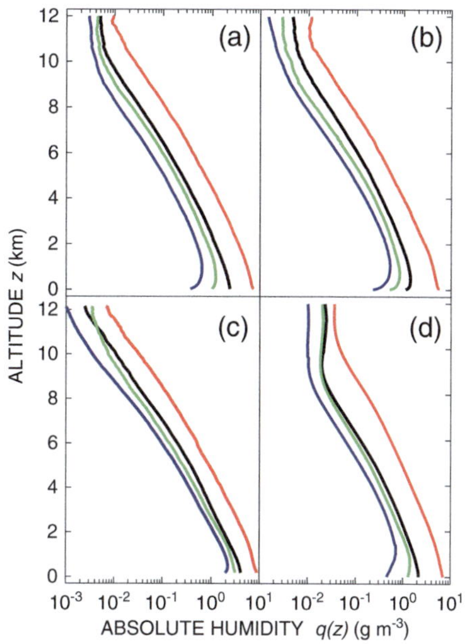

Fig. 2.21 Monthly mean vertical profiles of absolute humidity *q(z)* determined over the altitude range from the surface-level to 12 km, as derived from the RS measurements conducted over the period from 2001 to 2015 in the months of January (blue solid curve), April (green solid curve), July (red solid curve) and October (black solid curve) at the following four Arctic stations: (**a**) Cambridge Bay (Nunavut, Canada); (**b**) Eureka (Nunavut, Canada); (**c**) Sodankylä (Finland); and Tiksi (Siberia, Russia)

and 1.1 g m^{-3} at Eureka, 1.9 and 4.0 g m^{-3} at Sodankylä, and 0.4 and 2.1 g m^{-3} at Tiksi; and (ii) the monthly mean values of *q(z)* obtained in January, April and October result to be considerably lower than those measured in July over the whole atmosphere from surface-level until the altitude of 12 km.

Examining the monthly mean vertical profiles of absolute humidity *q(z)* calculated at the 14 RS Arctic stations by analyzing the multi-year sets of absolute humidity data, some examples of which are shown in Fig. 2.21, we have determined: (a) the monthly mean values of surface-level absolute humidity q_o in each year, and subsequently the multi-year monthly average values of q_o at each station; and (b) the best-fit values of the exponential coefficient k_w determined for each monthly mean vertical profile of *q(z)* by assuming that such a profile is best-fitted by the following analytical form:

$$q(z) = q_o \ \exp \ [-k_w \ (z - z_o)] \tag{2.2}$$

where q_o is the absolute humidity at the surface-level z_o of each RS station, and k_w is the best-fit coefficient (measured in km^{-1}), which in general defines the exponentially decreasing trend of *q(z)* as a function of altitude *z*. The results obtained at the two above points (a) and (b) are shown in Table 2.10 and Table 2.11, respectively.

Table 2.10 provides the monthly mean values of surface-level absolute humidity q_o (g m^{-3}), determined with their standard deviations by analysing the multi-year RS measurements performed at the 14 Arctic sites chosen in the present study, while Table 2.11 yields the monthly best-fit values of the exponential coefficient

Table 2.10 Monthly mean values of surface-level absolute humidity q_o (g m^{-3}) derived with their standard deviations from the RS measurements performed at the 14 Arctic sites chosen in the present study

Station and RS measurement period	Month											
	Jan.	Feb.	March	Apr.	May	June	July	Aug.	Sept.	Oct.	Nov.	Dec.
Barrow (2001–2015)	0.72 ± 0.53	0.74 ± 0.59	0.72 ± 0.48	1.47 ± 0.82	2.99 ± 1.11	4.80 ± 0.97	5.90 ± 1.36	6.02 ± 1.47	4.79 ± 1.16	3.08 ± 0.86	1.55 ± 0.80	0.99 ± 0.62
Inuvik (2001–2009 and 2012–2015)	0.66 ± 0.52	0.74 ± 0.74	0.71 ± 0.45	1.60 ± 0.89	3.21 ± 1.27	6.05 ± 1.75	8.12 ± 2.22	7.35 ± 1.77	4.98 ± 1.41	2.86 ± 0.99	1.22 ± 0.60	0.88 ± 0.94
Cambridge Bay (2001–2015)	0.37 ± 0.28	0.33 ± 0.27	0.45 ± 0.58	1.01 ± 0.63	2.35 ± 1.08	4.86 ± 1.22	7.32 ± 1.44	6.63 ± 1.47	4.10 ± 1.21	2.38 ± 1.05	0.93 ± 0.61	0.52 ± 0.59
Resolute (2001–2015)	0.42 ± 1.42	0.33 ± 0.61	0.44 ± 0.78	.88 ± 0.68	2.18 ± 0.97	4.22 ± 0.89	5.72 ± 0.91	5.11 ± 1.11	3.39 ± 0.78	2.02 ± 0.92	0.91 ± 1.08	0.56 ± 0.99
Eureka (2001–2015)	0.22 ± 0.18	0.19 ± 0.15	0.23 ± 0.25	0.51 ± 0.37	1.90 ± 0.89	4.26 ± 0.79	5.29 ± 0.66	4.81 ± 0.92	2.77 ± 0.86	1.12 ± 0.98	0.45 ± 0.55	0.30 ± 0.29
Alert (2001–2015)	0.39 ± 0.24	0.35 ± 0.26	0.35 ± 0.43	0.70 ± 0.42	1.89 ± 0.79	3.96 ± 0.88	5.23 ± 0.83	4.46 ± 0.93	2.55 ± 0.86	1.27 ± 0.82	0.69 ± 0.96	0.51 ± 0.62
Aasiaat (2001–2009)	1.78 ± 0.77	1.61 ± 0.99	1.64 ± 0.77	2.32 ± 0.83	3.67 ± 0.87	5.55 ± 0.74	6.48 ± 0.88	6.26 ± 0.86	4.71 ± 1.07	3.30 ± 0.86	2.64 ± 0.75	2.11 ± 0.73
Danmarkshavn (2001–2009)	0.81 ± 0.61	0.77 ± 0.65	0.64 ± 0.39	1.11 ± 0.73	2.46 ± 0.80	4.22 ± 0.59	5.06 ± 0.59	4.69 ± 0.86	2.91 ± 0.85	1.53 ± 0.98	0.97 ± 0.52	0.86 ± 0.63
Jan Mayen (2001–2015)	3.14 ± 1.11	2.95 ± 1.02	2.93 ± 0.99	3.18 ± 0.98	3.97 ± 0.99	5.18 ± 1.01	6.44 ± 0.94	6.75 ± 1.04	5.64 ± 1.33	3.99 ± 1.30	3.31 ± 1.21	3.12 ± 1.07
Ny-Ålesund (2005–2015)	2.10 ± 1.15	1.71 ± 1.00	1.66 ± 1.37	2.04 ± 1.05	3.20 ± 1.16	4.67 ± 0.77	6.00 ± 1.45	5.44 ± 0.90	4.01 ± 1.08	2.56 ± 0.99	2.14 ± 1.02	1.97 ± 0.92
Sodankylä (2001–2015)	1.94 ± 1.12	1.92 ± 1.11	2.26 ± 1.08	3.07 ± 1.14	4.42 ± 1.41	6.48 ± 2.04	8.82 ± 2.22	8.46 ± 2.25	6.48 ± 1.99	4.13 ± 1.51	2.83 ± 1.25	2.34 ± 1.27
Ostrov Dikson (2001–2015)	0.83 ± 0.63	0.71 ± 0.58	0.95 ± 0.69	1.51 ± 0.83	2.76 ± 0.87	4.91 ± 0.99	6.49 ± 1.48	6.11 ± 1.05	5.15 ± 1.12	3.02 ± 1.14	1.42 ± 0.78	1.04 ± 0.76
Tiksi (2001–2015)	0.45 ± 0.26	0.36 ± 0.26	0.63 ± 0.49	1.28 ± 0.76	3.15 ± 1.13	5.24 ± 1.10	6.95 ± 1.75	7.39 ± 1.56	4.79 ± 1.10	2.14 ± 0.90	0.88 ± 0.44	0.53 ± 0.34
Cherskij (2004–2015)	0.33 ± 0.28	0.33 ± 0.25	0.81 ± 0.53	1.49 ± 0.79	3.81 ± 1.37	6.90 ± 2.03	9.07 ± 2.59	7.90 ± 2.08	5.23 ± 1.43	2.46 ± 1.02	1.03 ± 0.68	0.51 ± 0.47

Table 2.11 Monthly mean values of the exponential best-fit coefficient k_W (measured in km^{-1}), derived with their standard deviations from the RS measurements performed at the 14 Arctic sites chosen in the present study and suitable for being used in Eq. (2.2) to define the best-fit vertical distribution curve of absolute humidity $q(z)$ as a function of altitude z (measured in km)

Station and RS measurement period	Month											
	Jan.	Feb.	March	Apr.	May	June	July	Aug.	Sept.	Oct.	Nov.	Dec.
Barrow (2001–2015)	0.12 ± 0.23	0.055 ± 0.140	0.061 ± 0.153	0.13 ± 0.24	0.31 ± 0.48	0.52 ± 0.78	0.67 ± 1.05	0.71 ± 1.17	0.60 ± 1.01	0.40 ± 0.66	0.16 ± 0.31	0.085 ± 0.188
Inuvik (2001–2009 and 2012–2015)	0.097 ± 0.206	0.048 ± 0.127	0.052 ± 0.133	0.12 ± 0.22	0.25 ± 0.35	0.47 ± 0.61	0.75 ± 1.15	0.75 ± 1.19	0.58 ± 0.83	0.37 ± 0.51	0.11 ± 0.20	0.065 ± 0.158
Cambridge Bay (2001–2015)	0.039 ± 0.119	0.034 ± 0.113	0.041 ± 0.119	0.085 ± 0.188	0.22 ± 0.36	0.35 ± 0.53	0.64 ± 0.95	0.68 ± 1.07	0.58 ± 0.93	0.30 ± 0.44	0.089 ± 0.208	0.048 ± 0.137
Resolute (2001–2015)	0.027 ± 0.093	0.024 ± 0.088	0.033 ± 0.103	0.070 ± 0.169	0.21 ± 0.34	0.30 ± 0.46	0.48 ± 0.65	0.55 ± 0.73	0.50 ± 0.75	0.24 ± 0.39	0.076 ± 0.191	0.046 ± 0.134
Eureka (2001–2015)	0.031 ± 0.111	0.029 ± 0.106	0.032 ± 0.110	0.049 ± 0.138	0.18 ± 0.30	0.30 ± 0.45	0.49 ± 0.66	0.49 ± 0.68	0.39 ± 0.53	0.11 ± 0.22	0.045 ± 0.134	0.039 ± 0.128
Alert (2001–2015)	0.030 ± 0.099	0.027 ± 0.093	0.030 ± 0.102	0.058 ± 0.152	0.17 ± 0.29	0.29 ± 0.45	0.42 ± 0.57	0.51 ± 0.70	0.31 ± 0.40	0.11 ± 0.21	0.047 ± 0.129	0.041 ± 0.121
Aasiaat (2001–2009)	0.16 ± 0.42	0.17 ± 0.37	0.20 ± 0.43	0.28 ± 0.53	0.37 ± 0.64	0.42 ± 0.62	0.66 ± 1.08	0.62 ± 1.00	0.64 ± 1.05	0.51 ± 0.85	0.39 ± 0.68	0.29 ± 0.59
Danmarkshavn (2001–2009)	0.061 ± 0.145	0.053 ± 0.132	0.045 ± 0.114	0.083 ± 0.179	0.24 ± 0.40	0.28 ± 0.45	0.43 ± 0.59	0.44 ± 0.61	0.36 ± 0.53	0.12 ± 0.21	0.067 ± 0.145	0.060 ± 0.144
Jan Mayen (2001–2015)	0.46 ± 0.74	0.34 ± 0.59	0.35 ± 0.61	0.34 ± 0.57	0.42 ± 0.67	0.38 ± 0.57	0.63 ± 0.93	0.72 ± 1.16	0.62 ± 0.96	0.53 ± 0.80	0.39 ± 0.62	0.39 ± 0.66
Ny-Ålesund (2005–2015)	0.15 ± 0.35	0.18 ± 0.40	0.19 ± 0.39	0.21 ± 0.39	0.28 ± 0.48	0.31 ± 0.48	0.59 ± 0.88	0.57 ± 0.82	0.54 ± 0.78	0.33 ± 0.51	0.25 ± 0.44	0.24 ± 0.45
Sodankylä (2001–2015)	0.31 ± 0.49	0.20 ± 0.35	0.28 ± 0.47	0.30 ± 0.47	0.44 ± 0.61	0.63 ± 0.88	0.89 ± 1.48	0.95 ± 1.62	0.77 ± 1.21	0.58 ± 0.85	0.36 ± 0.53	0.30 ± 0.46
Ostrov Dikson (2001–2015)	0.053 ± 0.151	0.055 ± 0.155	0.079 ± 0.193	0.14 ± 0.27	0.29 ± 0.43	0.49 ± 0.67	0.63 ± 0.92	0.65 ± 1.02	0.64 ± 1.00	0.35 ± 0.55	0.12 ± 0.25	0.084 ± 0.199
Tiksi (2001–2015)	0.016 ± 0.045	0.043 ± 0.120	0.054 ± 0.145	0.11 ± 0.23	0.29 ± 0.40	0.49 ± 0.66	0.66 ± 0.97	0.74 ± 1.25	0.54 ± 0.81	0.21 ± 0.34	0.063 ± 0.145	0.039 ± 0.111
Cherskij (2004–2015)	0.030 ± 0.100	0.031 ± 0.099	0.065 ± 0.152	0.14 ± 0.28	0.40 ± 0.58	0.73 ± 1.29	0.95 ± 1.91	0.87 ± 1.57	0.56 ± 0.89	0.25 ± 0.37	0.080 ± 0.178	0.046 ± 0.133

k_w defined in terms of Eq. (2.2), which can be appropriately used to give form to the monthly mean vertical distribution curves of $q(z)$ obtained throughout the year at the above-chosen 14 RS stations.

Examining the series of monthly mean values of q_o in Table 2.10, it can be seen that parameter q_o describes regular annual cycles at all the 14 RS stations, ranging between about 0.2 g m^{-3} (for the coldest and driest air conditions observed at Eureka in February) and more than 9 g m^{-3} (for the warmest and most wet air conditions observed at Cherskij in July). The monthly mean values of q_o were found to vary largely during the year at all the 14 RS stations. Actually: (1) at Barrow, from 0.72 g m^{-3} in both January and March to 6.02 g m^{-3} in August; (2) at Inuvik, from 0.66 g m^{-3} in January to 8.12 g m^{-3} in July; (3) at Cambridge Bay, from 0.33 g m^{-3} in February to 7.32 g m^{-3} in July; (4) at Resolute, from 0.33 g m^{-3} in February to 5.72 g m^{-3} in July; (5) at Eureka from 0.19 g m^{-3} in February to 5.29 g m^{-3} in July; (6) at Alert, from 0.35 g m^{-3} in January and February and March to 5.23 g m^{-3} in July; (7) at Aasiaat, 1.61 g m^{-3} in February to 6.48 g m^{-3} in July; (8) at Danmarkshavn, from 0.64 g m^{-3} in March to 5.06 g m^{-3} in July; (9) at Jan Mayen, from 2.93 g m^{-3} in March to 6.75 g m^{-3} in August; (10) at Ny-Ålesund, from 1.66 g m^{-3} in March to 6.00 g m^{-3} in July; (11) at Sodankylä, from 1.92 g m^{-3} in February to 8.82 g m^{-3} in July; (12) at Ostrov Dikson, from 0.71 g m^{-3} in February to 6.49 g m^{-3} in July; (13) at Tiksi, from 0.36 g m^{-3} in February to 7.39 g m^{-3} in August; and (14) at Cherskij, from 0.33 g m^{-3} recorded in both January and february to 9.07 g m^{-3} in July.

The results given in Table 2.10 provide an exhaustive picture of the average seasonal variations of surface-level absolute humidity q_o, obtained with their standard deviations throughout the year at the various Arctic stations located in North America, Greenland, Arctic Ocean, Scandinavia, and Siberia. The comparison between the annual cycles of q_o measured at Barrow and Inuvik indicates that very similar minima were measured at these two sites during the months from January to March, of around 0.72–0.74 g m^{-3} at Barrow, and ranging between 0.66 and 0.74 g m^{-3} at Inuvik, while the highest values of the year were observed at Barrow from June to September, varying between 4.8 and 6.0 g m^{-3}, and at Inuvik from June to September, ranging between 4.1 and 7.3 g m^{-3}. Very similar minima of q_o were recorded at Cambridge Bay and Resolute in the winter months from December to March, ranging between 0.33 and 0.52 g m^{-3} at Cambridge Bay, and between 0.33 and 0.56 g m^{-3} at Resolute, while values ranging between 4.86 and 7.32 g m^{-3} were found at Cambridge Bay in the summer months from June to August, and between 4.22 and 5.72 g m^{-3} at Resolute in the same period. The comparison between the annual cycles of q_o observed at Eureka and Alert shows that large seasonal variations in the moisture conditions at the surface occurred at both sites, with winter minima ranging between 0.19 and 0.30 g m^{-3} at Eureka in the months from December to March (giving an average winter value equal to 0.24 g m^{-3}), and varying between 0.35 and 0.51 g m^{-3} at Alert in the same winter months (giving an average winter value equal to 0.40 g m^{-3}), while monthly mean values ranging between 4.26 and 5.29 g m^{-3} were determined at Eureka in the summer period from June to August, and varying between 3.96 and 5.23 g m^{-3} at Alert in the same

3 months, clearly showing that higher summer maxima are most frequently recorded at Eureka than at Alert.

The surface-level moisture conditions measured over the whole year at Aasiaat and Danmarkshavn were found to be largely higher than those observed at the four Canadian stations located in Nunavut. In fact, over the period from December to March, the monthly mean values recorded at these two Greenlandic stations were found to range between 1.61 and 2.11 g m^{-3} at Aasiaat, and between 0.64 and 0.86 g m^{-3} at Danmarkshavn, thus giving winter mean values equal to 1.79 g m^{-3} at Aasiaat (which is more than seven times higher than that recorded at Eureka and more than four times higher than that measured at Alert) and equal to 0.57 g m^{-3} at Danmarkshavn (which is more than twice that measured at Eureka and higher by 42% than the Alert seasonal mean value). Rather high monthly mean values of q_o were also measured in the other seasons at Aasiaat and Danmarkshavn, with summer values ranging between 4.71 and 6.48 g m^{-3} at Aasiaat and between 4.22 and 5.06 g m^{-3} at Danmarkshavn, giving summer mean values of 6.10 g m^{-3} at Aasiaat and 4.66 g m^{-3} at Danmarkshavn. Therefore, the average value of q_o recorded at Aasiaat in summer is higher by 27% and 34%, respectively, than those measured at Eureka and Alert, while the summer average value of q_o recorded at Danmarkshavn turns out to be fully comparable with those recorded at the two Canadian sites in the same season.

Annual cycles presenting high values of q_o, comparable with those recorded at the two Greenlandic sites, have been also observed at Jan Mayen and Ny-Ålesund. The monthly mean values recorded at Jan Mayen in winter were found to vary between 2.95 g m^{-3} in February and 3.14 g m^{-3} in January, giving a mean winter value of 3.07 g m^{-3}, while those of Ny-Ålesund ranged between 1.71 g m^{-3} in February and 2.10 g m^{-3} in January, giving a mean winter value of 1.93 g m^{-3}, which is considerably lower than that measured at Jan Mayen. It is also worth noting that the winter mean value of q_o measured at Jan Mayen results to be largely higher than the winter mean value of 1.79 g m^{-3} measured at Aasiaat, and found to be in turn seven times higher than that recorded at Eureka. Very high monthly mean values of q_o were also measured in the other seasons at Jan Mayen, with summer values ranging between 5.18 g m^{-3} in June and 6.75 g m^{-3} in August, yielding a seasonal mean value of 6.12 g m^{-3}, which is very close to that recorded in summer at Aasiaat. Appreciably lower values of q_o were recorded at Ny-Ålesund in the summer months, varying between 4.67 g m^{-3} in June and 6.00 g m^{-3} in August, providing a summer mean value of 5.37 g m^{-3}, which results to be higher by 15% than that measured in this season at Danmarkshavn.

At Sodankylä, rather low values of q_o were measured in the winter months, ranging between 1.92 g m^{-3} in February and 2.34 g m^{-3} in December, for which a winter mean value equal to 2.07 g m^{-3} was determined in this continental low-latitude area. These winter values were considerably lower than those measured in the summer months, ranging between 6.48 g m^{-3} in June and 8.82 g m^{-3} in July, and yielding a summer mean value of 7.92 g m^{-3}. Even more marked variations of q_o were observed throughout the year at the three Siberian stations considered in the present analysis. The winter monthly mean values of q_o were found to vary: (i) between 0.71 g m^{-3} in February and 1.04 g m^{-3} in December at Ostrov Dikson, for which a winter mean value of q_o = 0.86 g m^{-3} was determined; (ii) between 0.36 g m^{-3} in February and 0.53 g m^{-3} in December at Tiksi, yielding a winter mean value of q_o = 0.45 g m^{-3} which is about half that recorded at Ostrov Dikson; and (iii) between the value of 0.33 g m^{-3} recorded in both January and February and the value of 0.51 g m^{-3} in December at Cherskij, for which a winter mean value of q_o = 0.39 g m^{-3} was determined, which is closer to that of Tiksi.

Table 2.11 provides the monthly best-fit values of exponential coefficient k_w determined in the troposphere, which can be most reliably used in Eq. (2.2) to define the analytical form of the twelve monthly mean vertical distribution curves of absolute humidity *q(z)* measured at the 14 RS stations chosen in the present study. It can be noted in Fig. 2.21 that: (i) the best-fit coefficient k_w determined in terms of Eq. (2.2) varies sharply as a function of altitude *z*, especially within the boundary layer the atmosphere. In fact, it assumes negative values during all the winter months and the most part of the spring and autumn months within the atmospheric ground-layer, which is often characterized in winter, early spring and late autumn by marked thermal inversions formed near the ground, so as to strongly inhibit the vertical transport of moist air masses from the surface toward the higher part of the temperature inversion layer, and subsequently changes its sign when it crosses the boundary layer until reaching the higher tropospheric levels. These features can be seen in Fig. 2.21, looking for instance at the vertical profiles of *q(z)* measured in January within the ground-based thermal inversion layer of the Arctic atmosphere. The best-fit coefficient k_w assumes usually positive values in summer within the whole boundary layer, decreasing gradually as altitude *z* reaches the higher levels of the low troposphere, and varies rapidly as *z* increases to reach levels higher than 4 km, as shown in Fig. 2.21 for different seasonal conditions occurring at various Arctic sites, and then changes more slowly as a function of *z* at the upper levels of the troposphere, and even more at the higher stratospheric levels of 10–12 km.

As a result of such a large variability (in both sign and magnitude) with altitude, the standard deviations of the monthly mean values of k_w determined at the 14 chosen Arctic sites were found to be largely greater than several hundreds per cent in most examined cases, as shown in Table 2.11. For instance, examining the vertical profile of *q(z)* shown in Fig. 2.21 for the case of January at Cambridge Bay it was estimated that coefficient k_w assumed an average negative value of about −0.22 km^{-1} over the range of *z* from surface-level to 1 km, and a series of positive values within the upper layers, estimated to be of: (i) + 0.14 km^{-1} for *z* ranging from 1 to 2 km, (ii) + 0.17 km for *z* ranging from 2 to 3 km, (iii) + 0.13 km for *z* ranging from 3 to 4 km, (iv) + 0.08 km for *z* ranging from 4 to 5 km, followed by gradually decreasing values of k_w within the upper layers of 1 km depth, until reaching a very small value of k_w (<5 10^{-3} km^{-1}) within the highest layer from 11 to 12 km height. This example gives a measure of the large variations of k_w that can be observed over the whole Arctic troposphere during the colder period of the year. Conversely, the vertical profiles of *q(z)* measured in the summer months provide usually positive values of k_w over the whole altitude range from surface-level to 12 km, since $q(z)$ decreases as a function of altitude with a rather stable rate from the surface-level to about 4 km altitude, yielding positive and gradually lower values of k_w over the whole altitude range from the surface-level to 12 km, as it can be deduced for instance by examining the vertical profile of $q(z)$ shown in Fig. 2.21 for Sodankylä (Finland) in July. For the above reasons, the monthly mean values of best-fit coefficient k_w have been determined at the 14 Arctic stations considered by us with standard deviations ranging between: (1) 20% in June and 157% in September at Barrow; (2) 24% in August and 191% in September at Inuvik; (3) 20% in July and 367% in September at Cambridge Bay; (4) 16% in July and 338% in January at Resolute; (5) 13% in July and 453% in September at Eureka; (6) 16% in July and 246% in November at Alert; (7) 13% in June and 66% in September at Aasiaat; (8) 12% in July and 110% in September at Danmarkshavn; (9) 15% in July and 45% in September at Jan Mayen; (10) 16% in June and 83% in March at Ny-Ålesund; (11) 25% in July and 104% in September at Sodankylä; (12) 17% in August and 158% in September at Ostrov Dikson; (13) 21% in June and 306% in September at Tiksi; and (14) 26% in August and 433% in September at Cherskij.

In spite of the large uncertainties affecting the monthly best-fit values of exponential coefficient k_w, the results given in Table 2.11 indicate that k_w varies during the year by assuming the highest value in summer and the lowest value in winter, varying at the various RS stations as follows: (1) at Barrow, from 0.055 km^{-1} in February to 0.71 km^{-1} in August; (2) at Inuvik, from 0.048 km^{-1} in February to 0.75 km^{-1} in both July and August; (3) at Cambridge Bay, from 0.034 km^{-1} in February to 0.68 km^{-1} in August; (4) at Resolute, from 0.024 km^{-1} in February to 0.55 km^{-1} in August; (5) at Eureka, from 0.029 km^{-1} in February to 0.49 km^{-1} in both July and August; (6) at Alert, from 0.027 km^{-1} in February to 0.51 km^{-1} in August; (7) at Aasiaat, from 0.16 km^{-1} in January to 0.66 km^{-1} in July; (8) at Danmarkshavn, from 0.045 km^{-1} in March to 0.44 km^{-1} in August; (9) at Jan Mayen, from 0.34 km^{-1} in February to 0.72 km^{-1} in August; (10) at Ny-Ålesund, from 0.15 km^{-1} in January to 0.59 km^{-1} in July; (11) at Sodankylä, from 0.20 km^{-1} in February to 0.95 km^{-1} in August; (12) at Ostrov Dikson, from 0.053 km^{-1} in January to 0.65 km^{-1} in August; (13) at Tiksi, from 0.016 km^{-1} in January to 0.74 km^{-1} in August; and (14) at Cherskij, from 0.030 km^{-1} in January to 0.95 km^{-1} in July.

In order to evaluate the multi-year trends of surface-level absolute humidity q_o at the 14 RS Arctic stations, the scatter plots of the annual mean values of q_o determined at these sites are shown in Fig. 2.22, as obtained over periods of: (i) 15 years at nine stations (Barrow, Cambridge Bay, Resolute, Eureka, Alert, Jan Mayen, Sodankylä, Ostrov Dikson and Tiksi), (ii) overall 13 years at Inuvik, (iii) 12 years at Cherskij, (iv) 11 years at Ny-Ålesund, and (v) 9 years at Aasiaat and Danmarkshavn. The multi-year values of the slope coefficient β_q (measured in g m^{-3} per decade at all the stations) and the corresponding correlation coefficient r_q (obtained for the regression lines of the corresponding scatter plots of the annual mean values of surface-level absolute humidity q_o drawn versus the measurement year) are reported in Table 2.12 for the 14 Arctic sites, as calculated over the above-

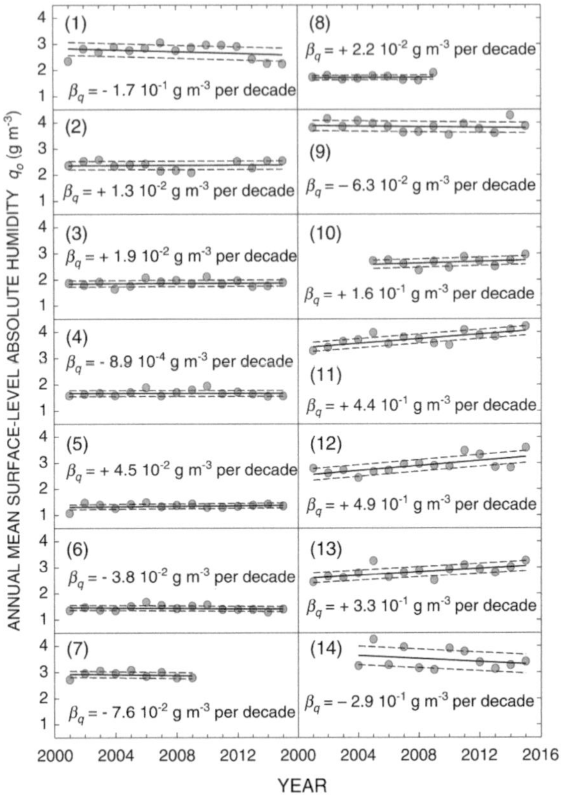

Fig. 2.22 Time-series of the annual mean values of surface-level absolute humidity q_o (green dots) measured in g m^{-3} at the 14 Arctic stations over their multi-year periods, with the regression lines (solid lines) and the corresponding parallel (dashed) lines drawn with intercepts equal to the best-fit intercept ± one standard deviation, all having the best-fit slope coefficient reported in each graph to give the average variation of q_o (measured in g m^{-3}) per decade. The data derived from the RS measurements carried out at the stations of (1) Barrow, (2) Inuvik, (3) Cambridge Bay, (4) Resolute, (5) Eureka, (6) Alert, and (7) Aasiaat are shown in the left part, while those derived from the RS measurements collected at the stations of (8) Danmarkshavn, (9) Jan Mayen, (10) Ny-Ålesund, (11) Sodankylä, (12) Ostrov Dikson, (13) Tiksi, and (14) Cherskij are shown in the right part.

mentioned multi-year periods. It can be seen in Fig. 2.22 that the series of annual mean values of q_o determined at Barrow was subject to an appreciable decrease as a function of time, yielding a negative value of slope coefficient $\beta_q = -\ 0.17$ g m^{-3} per decade (being the correlation coefficient $r_q = -\ 0.28$), which indicates that a satisfying correlation exists in this series of annual mean values of q_o. A positive and relatively low value of β_q (equal to $+1.3\ 10^{-2}$ g m^{-3} per decade) was obtained at Inuvik, associated with a very low correlation coefficient ($r_q = +\ 1.4\ 10^{-2}$) due to the significant dispersion features of the annual mean values of q_o. At Cambridge Bay, located at the latitude of 69 °N, a rather low and positive values of β_q was

Table 2.12 Values of the slope coefficients β_q (measured in g m^{-3} per decade) and corresponding correlation coefficients r_q, determined for the regression lines drawn for the corresponding scatter plots of the annual mean values of surface-level absolute humidity q_o (g m^{-3}) versus the measurement year, as determined from the RS measurements performed at the 14 Arctic sites chosen in the present study

	Regression line of q_o (g m^{-3}) vs. year	
Station and RS measurement period	Slope coefficient β_q (g m^{-3} per decade)	Correlation coefficient r_q
Barrow (2001–2015)	$-1.7\ 10^{-1}$	-0.28
Inuvik (2001–2009 and 2012–2015)	$+1.3\ 10^{-2}$	$+1.4\ 10^{-2}$
Cambridge Bay (2001–2015)	$+1.9\ 10^{-2}$	$+6.8\ 10^{-2}$
Resolute (2001–2015)	$-8.9\ 10^{-4}$	$-1.2\ 10^{-3}$
Eureka (2001–2015)	$+4.5\ 10^{-2}$	$+0.20$
Alert (2001–2015)	$-3.8\ 10^{-2}$	-0.15
Aasiaat (2001–2009)	$-7.6\ 10^{-2}$	-0.18
Danmarkshavn (2001–2009)	$+2.2\ 10^{-2}$	$+4.6\ 10^{-2}$
Jan Mayen (2001–2015)	$-6.3\ 10^{-2}$	-0.13
Ny-Ålesund (2005–2015)	$+1.6\ 10^{-1}$	$+0.30$
Sodankylä (2001–2015)	$+4.4\ 10^{-1}$	$+0.75$
Ostrov Dikson (2001–2015)	$+4.9\ 10^{-1}$	$+0.73$
Tiksi (2001–2015)	$+3.3\ 10^{-1}$	$+0.64$
Cherskij (2004–2015)	$-2.9\ 10^{-1}$	-0.27

obtained, equal to $+1.9\ 10^{-2}$ g m^{-3} per decade, with a correlation coefficient r_q equal to $+6.8\ 10^{-2}$, indicating that a weak increasing trend characterizes the moisture conditions of this site from one year to another, while the best-fit value of β_q determined at Resolute, located at the latitude of nearly 75 °N, was found to be equal to $-8.9\ 10^{-4}$ g m^{-3} per decade, with a correlation coefficient $r_q = -\ 6.8\ 10^{-2}$, which means that a slightly decreasing trend has been provided by the multi-year sequence of annual mean values of q_o over the 15-year period from 2001 to 2015. A pair of values of β_q with opposite signs were determined at the stations of Eureka and Alert, the first located a latitude of 80 °N, and the second at about 82 °N latitude: (i) the value of β_q determined at Eureka was equal to $+4.5\ 10^{-2}$ g m^{-3} per decade, with $r_q = +\ 0.20$ that clearly indicates that a positive multi-year trend of q_o was observed from 2001 to 2015, and (ii) the value of β_q determined at Alert was estimated to be equal to $-3.8\ 10^{-2}$ g m^{-3} per decade, with $r_q = -\ 0.15$, which suggests that an appreciably decreasing trend of q_o was recorded at this remote Canadian site from 2001 to 2015, as can be clearly seen in Fig. 2.22.

Values of coefficient β_q having opposite signs were also obtained by examining the multi-year scatter plots of surface-level absolute humidity q_o collected at the two Greenlandic sites of Aasiaat and Danmarkshavn. A best-fit value of $\beta_q = -\ 7.6\ 10^{-2}$ g m^{-3} per decade was determined at Aasiaat (with $r_q = -\ 0.18$), while a best-fit value of $\beta_q = +\ 2.2\ 10^{-2}$ g m^{-3} per decade was determined at Danmarkshavn (with a rather low value of $r_q = +\ 4.6\ 10^{-2}$). These opposite results were found in a region, where significant effects are caused on the thermal and moisture conditions of the atmospheric ground-layer by warm oceanic currents associated with the North Atlantic Drift and causing relevant effects on the surface-level moisture conditions. Similar features were also found at the stations of Jan Mayen and Ny-Ålesund, both located in the Arctic Ocean area between the western coasts of Greenland and the Svalbard Archipelago. A best-fit value of $\beta_q = -6.3\ 10^{-2}$ g m^{-3} per decade was evaluated at Jan Mayen, with correlation coefficient $r_q = -0.18$, which clearly indicates that the surface-level absolute humidity q_o was observed to very slowly decrease over the 2001–2015 period. Conversely, a positive best-fit value of $\beta_q = +0.16$ g m^{-3} per decade was determined at Ny-Ålesund over the 2005–2015 period, with $r_q = +0.30$, this value of β_q being of the same sign of the value of $\beta_q = +0.27$ g m^{-3} per decade, which has been derived by us for the average increase of water vapour mixing ratio equal to +0.22 g kg^{-1} per decade, which was determined by Maturilli et al. (2013) at Ny-Ålesund (Spitsbergen) by analyzing the 17-year data-set of this surface-level moisture parameter collected from 1994 to 2010 at the AWI surface-level meteorological station. The present evaluations of β_q made at Jan Mayen and Ny-Ålesund indicate that the trend of surface-level absolute humidity q_o was slowly decreasing at Jan Mayen over the last 15 years, and markedly increasing at Ny-Ålesund, while the multi-year trends of the surface-level temperature were observed to clearly increase at both the Arctic Ocean sites, by about +0.6 °K per decade at Jan Mayen and + 1.0 °K per decade at Ny-Ålesund, as shown in Fig. 2.14, presumably as a result of the combined effects due to global warming and heat transport by the North Atlantic Drift.

More marked variations in the moisture conditions of surface-level air were measured at Sodankylä over the 2001–2015 period, giving a rather high value of β_q (equal to +0.44 g m^{-3} per decade, with $r_q = +0.75$, such high values of both β_q and r_q clearly indicating that the northern part of Scandinavian peninsula has been involved during such a 15-year period by the gradual increase of surface-level moisture conditions, associated with the increase of surface-level temperature of around +1.4 °K per decade (as shown in Fig. 2.14). Examining the results obtained at the three Siberian stations of Ostrov Dikson, Tiksi, and Cherskij, the slope coefficient β_q was evaluated to be: (i) positive and rather high at Ostrov Dikson, being $\beta_q = +0.49$ g m^{-3} per decade, and found with a rather high correlation coefficient $r_q = +0.73$; (ii) high at Tiksi, being $\beta_q = +0.33$ g m^{-3} per decade, with $r_q = +0.64$, over the 2001–2015 period; and (iii) negative at Cherskij, being $\beta_q = -0.29$ g m^{-3} per decade, with $r_q = -0.27$ over the 2004–2015 period, this RS site being located at a latitude lower than 69 °N. The results obtained in the area including the Arctic Ocean coastal regions of western and central Siberia clearly indicate that the multi-year trend of surface-level absolute humidity q_o is: (i) clearly positive at both Ostrov Dikson and Tiksi, the values of best-fit slope coefficient β_q being comparable with that measured at Sodankylä in Northern Scandinavia, and (ii) markedly negative at Cherskij, where the surface-level temperature T_o has been evaluated to decrease from 2004 to 2015, with a best-fit slope coefficient equal to −0.74 °K per decade.

Summarizing the results given in Table 2.12, it can be noted in the three large longitudinal sectors of the Arctic that: (1) Among the evaluations of slope coefficient β_q and the corresponding correlation coefficients r_q, those determined at the North-American RS sites (Barrow, Inuvik, Cambridge Bay, Resolute, Eureka and Alert) exhibit the highest value of β_q at Eureka (equal to +0.045 g m^{-3} per decade) with a rather high value of $r_q = +\ 0.20$, and the lowest values of β_q at Barrow ($\beta_q = -$ 0.17 g m^{-3} per decade) with $r_q = -\ 0.28$. Considering only the Canadian stations, the lowest value of β_q was measured at Alert ($\beta_q = -\ 0.038$ g m^{-3} per decade) with $r_q = -\ 0.15$, to be compared with the above-mentioned maximum of +0.045 g m^{-3} per decade recorded at Eureka.

(2) In the Arctic sector including the Greenland area (with the stations of Aasiaat and Danmarkshavn) and the Arctic Ocean area between Greenland and Svalbard Islands (with the stations of Jan Mayen and Ny-Ålesund), the highest value of β_q was determined at Ny-Ålesund ($\beta_q = +\ 0.16$ g m^{-3} per decade) with coefficient $r_q = +\ 0.30$, and the lowest value of β_q was obtained at Aasiaat ($\beta_q = -0.076$ g m^{-3} per decade) with $r_q = -0.18$.

(3) In the Arctic sector including the Northern Europe area and the Northern Siberia, with the stations of Sodankylä, Ostrov Dikson, Tiksi and Cherskij, the highest value of $\beta_q = +\ 0.49$ g m^{-3} per decade was recorded at Ostrov Dikson, with a very high value of $r_q = +\ 0.73$, and the lowest value of β_q was measured at Cherskij, equal to -0.29 g m^{-3} per decade, with $r_q = -0.27$.

2.6 Annual Cycles of Precipitable Water

The value of tropospheric water vapour content W_t was calculated for each RS measurement by integrating the vertical distribution curve of absolute humidity $q(z)$ from the surface-level to 12 km altitude, such as those shown in Fig. 2.7, calculated for the RS measurements carried out for different seasonal conditions of the Arctic atmosphere at Danmarkshavn, Sodankylä, Barrow and Tiksi, by using different radiosonde models. Examining these daily data-sets, the various sets of daily values of precipitable water W have been then calculated in each year at the 14 above-chosen RS stations as the sum of: (i) the daily tropospheric water vapour content W_t, calculated by integrating the daily vertical profiles of $q(z)$ obtained from each single RS measurement over the altitude range from surface-level to 12 km, and (ii) the stratospheric water vapour content W_s, which was derived by Tomasi et al. (2011a) from the MIPAS/ENVISAT limb-scanning measurements of water vapour mixing ratio carried out over the 12–50 km altitude range, during the period from July 2002 to April 2010, at latitudes from 65 °N to 90 °N taken in steps of 5°. As reported in Table 2.4, the monthly mean values of W_s calculated at six fixed latitudes from 65 °N to 90 °N were estimated to range between 4.7 10^{-4} and 7.0 10^{-4} g cm^{-2}. Examining the yearly series of daily values of W calculated at the 14 chosen RS Arctic sites, the monthly mean values of precipitable water W were then determined for the various RS measurement years obtaining the monthly mean values of W from the overall multi-year data-sets collected at the 14 RS stations, which are given in Table 2.13 with their standard deviations.

A set of daily evaluations of W obtained at the four RS stations of Barrow, Eureka, Jan Mayen and Tiksi by analyzing the precipitable water data collected during the 2001–2015 years is shown in Fig. 2.23. In spite of the appreciably scattered data observed from one measurement day to another, the time-patterns of the daily mean values of W clearly exhibit a series of rather regular annual cycles, with: (i) winter minima of 0.065 g cm^{-2} at Barrow, 0.042 g cm^{-2}, 0.15 g cm^{-2} at Jan Mayen, and 0.013 g cm^{-2} at Tiksi, and (ii) summer maxima ranging from one year to another between 1.5 and 2.0 g cm^{-2} at Barrow, Jan Mayen and Tiksi, and between 1.1 and 1.5 g cm^{-2} at Eureka.

It can be also noticed that more pronounced dispersion features of the daily data of W were observed at Barrow and Jan Mayen than at Eureka and Tiksi, suggesting that such features may be probably caused by the more variable contributions given by the maritime air masses that are transported by winds moving from the Arctic Ocean toward the RS stations. According to these annual time-patterns, the monthly mean values of W observed at the four sites were also found to describe regular cycles during the observed five years from 2011 to 2015 with: (i) winter minima of about 0.18 g cm^{-2} at Barrow, around 0.11 g cm^{-2} at Eureka, around 0.35 g cm^{-2} at Jan Mayen, and about 0.18 g cm^{-2} at Tiksi, obtained with standard deviations varying mainly between ±0.1 and ± 0.2 g cm^{-2}; and (ii) summer maxima ranging between 1.0 and about 2.0 g cm^{-2} at Barrow, 0.8 and 1.6 g cm^{-2} at Eureka, 1.0 and 2.0 g cm^{-2} at both Jan Mayen and Tiksi, determined with standard deviations varying mainly between ±0.2 and ± 0.5 g cm^{-2}.

The monthly mean values of precipitable water W obtained from the RS measurements examined in the present study at the 14 selected Arctic stations are given in Table 2.13, together with their standard deviations, which have been found to vary: (i) between ±0.07 g cm^{-2} and ± 0.25 g cm^{-2} in the winter months, and (ii) between ±0.30 g cm^{-2} and ± 0.63 g cm^{-2} in the summer months. It can be noted in Table 2.13 that the monthly mean values of W varied: (i) at Barrow, from 0.29 g cm^{-2} in March to 1.40 g cm^{-2} in July, and (ii) at Inuvik, from 0.30 g cm^{-2} in March to 1.95 g cm^{-2} in July. At the other four RS stations located in Nunavut (Canada), we have found that the monthly mean values of W varied: (i) at Cambridge Bay, from 0.21 g cm^{-2} (measured in both January and February) to 1.71 g cm^{-2} in July; (ii) at Resolute, from 0.17 g cm^{-2} in February to 1.39 g cm^{-2} in July; (iii) at Eureka, from 0.17 g cm^{-2} in both January and February to 1.32 g cm^{-2} in July; and (iv) at Alert, from 0.18 g cm^{-2} in both February and March to 1.25 g cm^{-2} in July. Therefore, the lowest monthly mean values of W recorded in Nunavut (Canada) were measured in winter (most frequently in February), varying between 0.17 g cm^{-2} (measured at Resolute in February, and at Eureka in both January and February) and 0.21 g cm^{-2} (measured at Cambridge Bay, in both January and February), while the highest monthly mean values were recorded in July, varying between 1.25 g cm^{-2} (at Alert) and 1.71 g cm^{-2} (at Cambridge Bay).

In Greenland, the monthly mean values of W recorded at the stations of Aasiaat and Danmarkshavn were both found to have their minima in March, equal to

Table 2.13 Monthly mean values of precipitable water W (g cm^2) determined with their standard deviations from the RS measurements performed at the 14 Arctic sites chosen in the present study

Station and RS measurement period	Month											
	Jan.	Feb.	March	Apr.	May	June	July	Aug.	Sept.	Oct.	Nov.	Dec.
Barrow (2001–2015)	0.30 ± 0.20	0.31 ± 0.20	0.29 ± 0.17	0.44 ± 0.22	0.70 ± 0.30	1.14 ± 0.45	1.40 ± 0.50	1.32 ± 0.53	0.93 ± 0.39	0.61 ± 0.27	0.39 ± 0.22	0.32 ± 0.20
Inuvik (2001–2009 and 2012–2015)	0.32 ± 0.22	0.34 ± 0.19	0.30 ± 0.16	0.50 ± 0.23	0.82 ± 0.36	1.50 ± 0.52	1.95 ± 0.51	1.76 ± 0.56	1.14 ± 0.45	0.69 ± 0.29	0.42 ± 0.21	0.38 ± 0.21
Cambridge Bay (2001–2015)	0.21 ± 0.14	0.21 ± 0.13	0.25 ± 0.16	0.37 ± 0.21	0.59 ± 0.28	1.09 ± 0.45	1.71 ± 0.54	1.48 ± 0.51	0.84 ± 0.39	0.58 ± 0.28	0.31 ± 0.16	0.25 ± 0.16
Resolute (2001–2015)	0.18 ± 0.09	0.17 ± 0.07	0.20 ± 0.12	0.29 ± 0.14	0.54 ± 0.23	0.94 ± 0.30	1.39 ± 0.42	1.22 ± 0.44	0.72 ± 0.23	0.49 ± 0.25	0.27 ± 0.12	0.22 ± 0.12
Eureka (2001–2015)	0.17 ± 0.08	0.17 ± 0.08	0.18 ± 0.11	0.27 ± 0.13	0.51 ± 0.19	0.99 ± 0.30	1.32 ± 0.30	1.15 ± 0.35	0.69 ± 0.20	0.42 ± 0.21	0.25 ± 0.10	0.21 ± 0.11
Alert (2001–2015)	0.19 ± 0.09	0.18 ± 0.09	0.18 ± 0.08	0.26 ± 0.11	0.49 ± 0.18	0.87 ± 0.24	1.25 ± 0.28	1.06 ± 0.30	0.66 ± 0.19	0.40 ± 0.18	0.24 ± 0.11	0.21 ± 0.11
Aasiaat (2001–2009)	0.33 ± 0.18	0.33 ± 0.65	0.32 ± 0.22	0.44 ± 0.22	0.69 ± 0.28	1.19 ± 0.37	1.40 ± 0.43	1.34 ± 0.41	0.90 ± 0.33	0.62 ± 0.24	0.48 ± 0.23	0.37 ± 0.19
Danmarkshavn (2001–2009)	0.30 ± 0.20	0.29 ± 0.19	0.27 ± 0.15	0.35 ± 0.20	0.57 ± 0.20	0.94 ± 0.24	1.28 ± 0.30	1.14 ± 0.35	0.70 ± 0.28	0.45 ± 0.22	0.33 ± 0.19	0.30 ± 0.20
Jan Mayen (2001–2015)	0.56 ± 0.33	0.52 ± 0.27	0.51 ± 0.25	0.59 ± 0.28	0.76 ± 0.34	1.11 ± 0.41	1.53 ± 0.46	1.51 ± 0.46	1.19 ± 0.46	0.79 ± 0.39	0.61 ± 0.31	0.54 ± 0.27
Ny-Ålesund (2005–2015)	0.44 ± 0.24	0.35 ± 0.21	0.34 ± 0.19	0.42 ± 0.21	0.64 ± 0.22	1.02 ± 0.26	1.38 ± 0.36	1.24 ± 0.35	0.90 ± 0.32	0.56 ± 0.24	0.45 ± 0.23	0.40 ± 0.20
Sodankylä (2001–2015)	0.51 ± 0.24	0.49 ± 0.23	0.50 ± 0.24	0.68 ± 0.31	0.98 ± 0.40	1.47 ± 0.56	2.06 ± 0.61	1.91 ± 0.58	1.43 ± 0.54	0.88 ± 0.40	0.67 ± 0.29	0.58 ± 0.26
Ostrov Dikson (2001–2015)	0.31 ± 0.14	0.30 ± 0.14	0.34 ± 0.14	0.43 ± 0.18	0.66 ± 0.21	1.19 ± 0.37	1.63 ± 0.48	1.37 ± 0.34	1.09 ± 0.34	0.65 ± 0.24	0.43 ± 0.17	0.36 ± 0.15
Tiksi (2001–2015)	0.24 ± 0.11	0.22 ± 0.07	0.30 ± 0.11	0.41 ± 0.16	0.79 ± 0.29	1.32 ± 0.40	1.78 ± 0.58	1.60 ± 0.61	1.05 ± 0.32	0.55 ± 0.18	0.36 ± 0.12	0.27 ± 0.10
Cherskij (2004–2015)	0.25 ± 0.10	0.23 ± 0.09	0.31 ± 0.11	0.38 ± 0.13	0.84 ± 0.32	1.36 ± 0.44	1.91 ± 0.63	1.67 ± 0.52	1.09 ± 0.39	0.62 ± 0.21	0.41 ± 0.17	0.30 ± 0.13

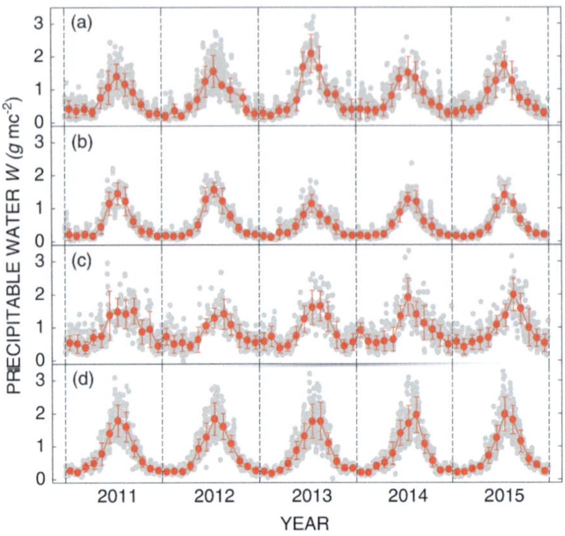

Fig. 2.23 Time-series of the daily mean values of precipitable water W (grey dots) and monthly mean values of W (red dots, represented with vertical red bars used to indicate the standard deviations), as derived from the RS measurements conducted during the 5-year period from 2011 to 2015 at the following four Arctic stations: (**a**) Barrow (Alaska, USA); (**b**) Eureka (Nunavut, Canada); (**c**) Jan Mayen (Arctic Ocean, Norway); and (**d**) Tiksi (Siberia, Russia). The values of W have been calculated by including the stratospheric water vapour content derived by Tomasi et al. (2011a) through the analysis of the MIPAS-ENVISAT limb-scanning measurements collected at the corresponding Arctic latitudes

0.32 g cm^{-2} at Aasiaat and 0.27 g cm^{-2} at Danmarkshavn, while their highest monthly mean values were both measured in July, equal to 1.40 g cm^{-2} at Aasiaat and to 1.28 g cm^{-2} at Danmarkshavn. Examining the data-sets recorded at the stations of Jan Mayen and Ny-Ålesund, located in the Arctic Ocean between the western coasts of Greenland and the Svalbard Archipelago, the lowest monthly mean values were observed in March at both the sites, equal to 0.51 g cm^{-2} at Aasiaat and to 0.34 g cm^{-2} at Ny-Ålesund, while the highest values were measured in July at both the sites, equal to 1.53 g cm^{-2} at Jan Mayen and to 1.38 g cm^{-2} at Ny-Ålesund.

On this matter, it is worth pointing out that regular measurements of precipitable water W were also carried out by Muscari et al. (2014) at the Thule Air Base (76° 30′ N, 68° 48′ W) located on the north-western coast of Greenland (1207 km north of the Arctic Circle and 1524 km from the North Pole), by using an example of Cimel CE-318 sun-photometer of the AEerosol RObotic NETwork (AERONET) over the 3-year period from March 2007 to late September 2009. Aerosol optical thickness (AOT) at 0.50 μm was found to vary during thus 3-year period between less than 0.02 (in the winter months) and more than 0.2 (in the May–August months). Correspondingly, precipitable water W was evaluated with a relative uncertainty varying between 5% and 12% during the March–September periods of such a 3-year period, and estimated to increase on average from less than 0.2 g cm^{-2} in March to more than 1 g cm^{-2} in June–August, and decrease in September, presenting a well-defined cycle of the monthly mean values from March to September, with values that are: (i) generally lower than 0.6 g cm^{-2} in spring and early autumn, and (ii) higher than 1.4 g cm^{-2} in summer, when maxima higher than 1.4 g cm^{-2} have been observed in 2007, equal to around 1.2 g cm^{-2} in 2008, and higher than 1.6 g cm^{-2} in 2009. Therefore, the measurements of W carried out by Muscari et al. (2014) at the Thule Air Base are in good agreement with those obtained by us at the stations of Aasiaat and Danmarkshavn and reported in Table 2.13.

At Sodankylä (Lapland, Finland), the lowest monthly mean values of the year were observed in February, equal to 0.49 g cm^{-2}, and the highest value was recorded in July, equal to 2.06 g cm^{-2}. At the three RS stations located in the northern coastal regions of Siberia, the monthly mean values of W were found to exhibit: (i) the lowest monthly mean values in February, equal to 0.30 g cm^{-2} at Ostrov Dikson, 0.22 g cm^{-2} at Tiksi, and 0.23 g cm^{-2} at Cherskij; and (ii) the highest values in July, equal to 1.63 g cm^{-2} at Ostrov Dikson, 1.78 g cm^{-2} at Tiksi, and 1.91 g cm^{-2} at Cherskij.

The results shown in Fig. 2.23 clearly suggest that analogous daily time-patterns and seasonal variations were also occurred at the other ten RS stations throughout each year, causing similar annual cycles to those shown in Fig. 2.23. In order to evaluate with which rates is varying the parameter W over the 15-year periods investigated in the present analysis of RS data, the procedure followed in Fig. 2.14 and in Fig. 2.22, with the aims of determining the average variations of surface-level temperature T_o and surface-level absolute humidity q_o, respectively, over the 15-year period from 2001 to 2015, was also applied to the time-series of the annual mean values of precipitable water W calculated at all the 14 RS stations over the corresponding multi-year periods, with the final purpose of evaluating the long-period trend of W and evaluate its average slope-coefficient β_W at each of the 14 RS stations.

The results obtained by following this regression procedure are shown in Fig. 2.24. They indicate that the series of annual mean values of W determined at Barrow has been subject to an appreciable decrease from 2001 to 2015, yielding a negative value of slope coefficient $\beta_W = -0.016$ g cm^{-2} per decade, found with correlation coefficient $r_W = -0.15$, which indicates that a satisfying correlation exists in this series of annual mean values of W. A negative and relatively low value of $\beta_W = -0.011$ g cm^{-2} per decade was obtained at Inuvik, associated with a correlation coefficient $r_W = -0.14$, arising from the significant dispersion features of the sequence of annual mean values of W. At Cambridge Bay, a negative value of β_W was obtained, equal to -0.022 g cm^{-2} per decade, with a correlation coefficient $r_W = -0.29$ clearly indicating a more pronounced decreasing trend then at thetwo previous sites. The best-fit value of β_W determined at Resolute was found to be even more negative than at Cambridge Bay, being $\beta_W = -0.030$ g cm^{-2} per decade, with $r_W = -0.46$, which indicates that such a decreasing trend is clearly followed by the sequence of annual mean values of W measured over the 15-year period

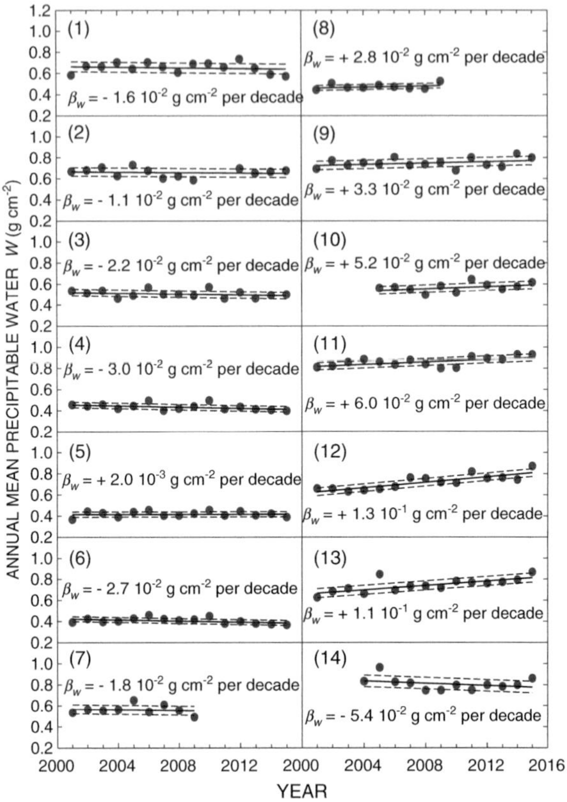

Fig. 2.24 Time-series of the annual mean values of precipitable water W (blue dots) measured in g cm^{-2} at the 14 Arctic stations over their multi-year periods, with the regression lines (solid lines) and the corresponding parallel (dashed) lines drawn with intercepts equal to the best-fit intercept $\pm$ one standard deviation, all having the best-fit slope coefficient β_W reported in each graph to give the average variation of W (measured in g cm^{-2}) per decade). The data derived from the RS measurements carried out at the stations of (1) Barrow, (2) Inuvik, (3) Cambridge Bay, (4) Resolute, (5) Eureka, (6) Alert, and (7) Aasiaat are shown in the left part, while those derived from the RS measurements collected at the stations of (8) Danmarkshavn, (9) Jan Mayen, (10) Ny-Ålesund, (11) Sodankylä, (12) Ostrov Dikson, (13) Tiksi, and (14) Cherskij are shown in the right part.

from 2001 to 2015. Values of β_W presenting opposite sign were determined at the stations of Eureka and Alert, such a result being similar to that found for the surface-level absolute humidity: (i) the value of β_W obtained at Eureka was estimated to be equal to $+2.0\ 10^{-3}$ g cm^{-2} per decade, having $r_W = +\ 3.4\ 10^{-3}$, which clearly indicates that such a positive multi-year trend of W observed from 2001 to 2015 was characterized by rather highly dispersed features; and (ii) the value of β_W determined at Alert was found to be equal to $-2.7\ 10^{-2}$ g cm^{-2} per decade, with a negative value of $r_W = -0.44$, as can be clearly seen in Fig. 2.24.

Values of coefficient β_W having opposite signs were also obtained by examining the multi-year scatter plots of precipitable water W measured at the two Greenlandic sites of Aasiaat and Danmarkshavn. The best-fit value of $\beta_W = -0.018$ g cm^{-2} per decade was determined at Aasiaat, with $r_W = -0.16$, while a best-fit value of $\beta_W = +0.028$ g cm^{-2} per decade was determined at Danmarkshavn, with a value of $r_q = +0.27$. Conversely, positive values of β_W were determined at both Jan Mayen and Ny-Ålesund, with: (i) a best-fit value of $\beta_W = +0.033$ g cm^{-2} per decade at Jan Mayen, together with $r_W = +0.35$, which clearly indicates that such a trend of W observed over the 2001–2015 period is very clear and without significant dispersion features, and (ii) an even more positive best-fit value of $\beta_W = +0.052$ g cm^{-2} per decade at Ny-Ålesund, found over the 2005–2015 period with $r_W = +0.44$.

Comparable results with those of Jan Mayen and Ny-Ålesund were obtained at Sodankylä (Finland) over the 2001–2015 period, obtaining a best-fit value of $\beta_W = +0.060$ g cm^{-2} per decade, with $r_W = +0.64$, clearly indicating that parameter W was subject to a slow increase over the 2001–2015 period. Examining the results obtained at the three Siberian stations of Ostrov Dikson, Tiksi, and Cherskij, the slope coefficient β_W was evaluated to be: (i) positive and equal to +0.13 g cm^{-2} per decade at Ostrov Dikson over the 2001–2015 period, with $r_W = +0.88$, which indicates a clearly increasing trend of W; (ii) positive and equal to +0.11 g cm^{-2} per decade at Tiksi over the 2001–2015 period, with $r_W = +0.78$, which indicates a clearly increasing trend of W characterized by features comparable with those of Ostrov Dikson; and (iii) negative and equal to −0.054 g cm^{-2} per decade at Cherskij over the 2004–2015 period, with $r_W = -0.26$, which indicates that the sequence of precipitable water data is rather dispersed.

The results given in Table 2.14 provide a general picture of the values of W and their trends, which characterize the Arctic atmosphere at the 14 above-chosen RS sites:

(1) At the North-American RS sites (Barrow, Inuvik, Cambridge Bay, Resolute, Eureka and Alert), the highest value of slope coefficient was determined at Eureka ($\beta_W = +0.002$ g cm^{-2} per decade) with a rather low correlation coefficient $r_W = +3.4\ 10^{-3}$, the values of β_W determined at the other North-American sites being all negative, and the most negative value of β_W being that of Resolute, equal to −0.030 g cm^{-2} per decade (with $r_W = -0.46$).

(2) In the Arctic sector including the Greenland area (with the stations of Aasiaat and Danmarkshavn) and the Arctic Ocean area between Greenland and Svalbard Islands (with the stations of Jan Mayen and Ny-Ålesund), the highest value of slope coefficient β_W was determined at Ny-Ålesund, equal to +0.052 g cm^{-2} per decade and found with $r_W = +0.44$, and the lowest value of β_W was obtained at Aasiaat ($\beta_W = -0.018$ g cm^{-2} per decade) with $r_W = -0.16$.

(3) In the Arctic sector including the Northern Europe and the Northern Siberia regions, with the stations of Sodankylä, Ostrov Dikson, Tiksi and Cherskij, it was found that the highest value of $\beta_W = +0.13$ g cm^{-2} per decade was recorded at Ostrov Dikson, with a very high value of $r_W = +0.88$, and the lowest value of β_W was found at Cherskij, equal to −0.054 g cm^{-2} per decade, with $r_W = -0.26$.

Table 2.14 Values of the slope coefficients β_W (measured in g cm^{-2} per decade) and corresponding correlation coefficients r_W, determined for the regression lines drawn for the corresponding scatter plots of the annual mean values of precipitable water W (g cm^{-2}) versus the measurement year, as determined from the RS measurements performed at the 14 Arctic sites chosen in the present study

	Regression line of W (g cm^{-2}) vs. year	
Station and RS measurement period	Slope coefficient β_W (g cm^{-2}/decade)	Correlation coefficient r_W
Barrow (2001–2015)	$-1.6\ 10^{-2}$	−0.15
Inuvik (2001–2009 and 2012–2015)	$-1.1\ 10^{-2}$	−0.14
Cambridge Bay (2001–2015)	$-2.2\ 10^{-2}$	−0.29
Resolute (2001–2015)	$-3.0\ 10^{-2}$	−0.46
Eureka (2001–2015)	$+2.0\ 10^{-3}$	$+3.4\ 10^{-3}$
Alert (2001–2015)	$-2.7\ 10^{-2}$	−0.44
Aasiaat (2001–2009)	$-1.8\ 10^{-2}$	−0.16
Danmarkshavn (2001–2009)	$+2.8\ 10^{-2}$	+0.27
Jan Mayen (2001–2015)	$+3.3\ 10^{-2}$	+0.35
Ny-Ålesund (2005–2015)	$+5.2\ 10^{-2}$	+0.44
Sodankylä (2001–2015)	$+6.0\ 10^{-2}$	+0.64
Ostrov Dikson (2001–2015)	$+1.3\ 10^{-1}$	+0.88
Tiksi (2001–2015)	$+1.1\ 10^{-1}$	+0.78
Cherskij (2004–2015)	$-5.4\ 10^{-2}$	−0.26

2.7 Conclusions: Trends of Surface-Level Temperature, Surface-Level Absolute Humidity and Precipitable Water over the 2001–2015 Years in the Arctic Region

The results shown in Fig. 2.14 (on the 2001–2015 trend of surface-level temperature T_o), Fig. 2.22 (on the 2001–2015 trend of surface-level absolute humidity q_o) and Fig. 2.24 (on the 2001–2015 trend of precipitable water W) clearly indicate that: (a) surface-level temperature T_o was found to increase over the considered 15-year period at ten RS stations (Inuvik, Cambridge Bay, Resolute, Eureka, Alert, Jan Mayen, Ny-Ålesund, Sodankylä, Ostrov Dikson, and Tiksi) by about +1.0 °K per decade on average, and decrease at only four RS stations (Barrow, Aasiaat, Danmarkshavn, and Cherskij) by about −0.9 °K per decade on average; (b) surface-level absolute humidity was estimated to increase from 2001 to 2015 at eight RS stations (Inuvik, Cambridge Bay, Eureka, Danmarkshavn, Ny-Ålesund, Sodankylä, Ostrov Dikson, and Tiksi) by about +0.2 g m^{-3} per decade on average, and decrease at six RS stations (Barrow, Resolute, Alert, Aasiaat, Jan Mayen, and Cherskij) by about −0.1 g m^{-3} per decade on average; and (c) precipitable water was evaluated to increase from 2001 to 2015 at seven RS stations (Eureka, Danmarkshavn, Jan Mayen, Ny-Ålesund, Sodankylä, Ostrov Dikson, and Tiksi) by about +0.06 g cm^{-2} per decade on average, and decrease at seven RS stations (Barrow, Inuvik, Cambridge Bay, Resolute, Alert, Aasiaat, and Cherskij) by nearly

-0.03 g cm^{-2} per decade on average. To better understand in which regions surface-level temperature and moisture conditions are subject to increase, we have divided the Arctic region into the following sub-regions: (i) the North-American area, including the RS stations of Inuvik, Cambridge Bay. Resolute, Eureka and Alert; (ii) the Greenland area and the Arctic Ocean sector between Greenland and Svalbard Archipelago, including the RS stations of Aasiaat, Danmarkshavn, Jan Mayen and Ny-Ålesund; (iii) the Scandinavian area (with Sodankylä) and the coastal regions of Siberia on the Arctic Ocean (including the RS stations of Ostrov Dikson and Tiksi); and (iv) the Arctic Ocean coastal area around the Bering Strait, including the RS stations of Cherskij (in Eastern Siberia) and Barrow (in Alaska).

Examining in Fig. 2.25 the values of best-fit slope coefficients β_T, β_q and β_W determined at the RS stations of Inuvik (IN), Cambridge Bay (CB), Resolute (RE), Eureka (EU), and Alert (AL) over the period from 2001 to 2015, it can be pointed out that in this Canadian sector of North America (presenting latitudes ranging from 68 °N to 82.5 °N): (a) the best-fit slope coefficient of surface-level temperature T_o observed over the 2001–2015 period was found to vary between $\beta_T = +0.10$ °K per decade (at Resolute) and $\beta_T = +0.80$ °K per decade at Alert, therefore assuming at all the five Canadian stations positive and rather low values within a few tenths of °K per decade; (b) the best-fit slope coefficient of surface-level absolute humidity q_o was estimated to vary between -0.038 g m^{-3} per decade at Alert and $+0.045$ g m^{-3} per decade at Eureka, therefore assuming mainly positive and rather low values at the five Canadian stations, within a few hundredths of g per cubic meter of air and per decade; and (c) the best-fit slope coefficient of precipitable water W was found to vary between -0.030 g cm^{-2} per decade at Resolute and $+0.002$ g cm^{-2} per decade at Eureka, therefore assuming mainly negative values at the five Canadian stations,

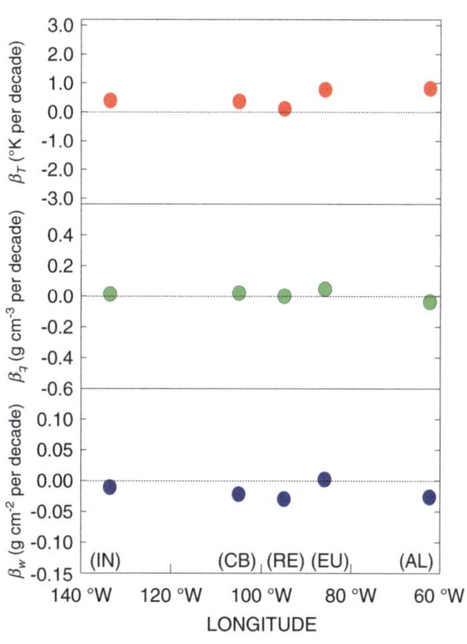

Fig. 2.25 Scatter plots of the multi-year average slope coefficients β_T (upper part with red dots), β_q (middle part with green dots) and β_W (lower part with blue dots) as a function of longitude over its range from 140 °W to 60 °W, including the five Canadian RS stations of Inuvik (IN), Cambridge Bay (CB), Resolute (RE), Eureka (EU), and Alert (AL)

which are within a few hundredths of g cm^{-2} per decade. Therefore, the results shown in Fig. 2.25 indicate that surface-level temperature T_o is clearly increasing in this area of North America, whereas surface-level absolute humidity q_o is rather stable, and precipitable water W has been found to exhibit slightly decreasing values from 2001 to 2015 in most cases, with rates within a few hundredths of g cm^{-2} per decade.

The values of the best-fit slope coefficients β_T, β_q and β_W determined at the RS stations of Aasiaat (AA), Danmarkshavn (DH), Jan Mayen (JM) and Ny-Ålesund (NÅ) are shown in Fig. 2.26, indicating that in this area including Greenland and the Arctic Ocean sector between the western Greenlandic coasts and the Svalbard Archipelago (for latitudes ranging between 68 °N and 79 °N): (a) the best-fit slope coefficient of parameter T_o observed over the period from 2001 to 2015 was found to vary between $\beta_T = -$ 1.74 °K per decade (at Aasiaat) and $\beta_T = +$ 0.99 °K per decade at Ny-Ålesund, being therefore largely variable within this region; (b) the best-fit slope coefficient of surface-level absolute humidity q_o was estimated to vary between −0.076 g m^{-3} per decade at Aasiaat and + 0.16 g m^{-3} per decade at Ny-Ålesund; and (c) the best-fit slope coefficient of precipitable water W was found to vary between −0.018 g cm^{-2} per decade at Aasiaat and + 0.052 g cm^{-2} per decade at Ny-Ålesund, presenting an overall positive trend in this region. Summarizing the results, it was found that: (i) the values of β_T measured at Aasiaat and Danmarkshavn are negative and, hence, of opposite signs with respect

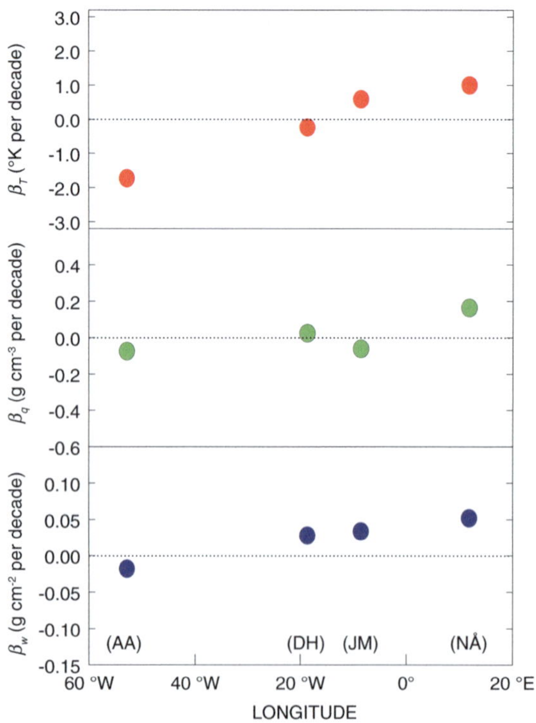

Fig. 2.26 Scatter plots of the multi-year average slope coefficients β_T (upper part with red dots), β_q (middle part with green dots) and β_W (lower part with blue dots) as a function of longitude over its range from 60 °W to 20 °E, including the Greenland and the Arctic Ocean area between Greenland and Svalbard Archipelago, with the stations of Asiaaat (AA), Danmarkshavn (DH), Jan Mayen (JM) and Ny-Ålesund (NÅ)

to those measured at Jan Mayen and Ny-Ålesund; (ii) the values of β_q measured at Aasiaat and Jan Mayen are negative and, hence, opposite to those evaluated at Danmarkshavn and Ny-Ålesund, which are both markedly positive and close to $+0.02$ g m^{-3} per decade; and (iii) only the value of β_W measured at Aasiaat from 2001 to 2009 was found to be negative, while those measured at Danmarkshavn (from 2001 to 2009), Jan Mayen (from 2001 to 2015) and Ny-Ålesund (from 2005 to 2015) were estimated to be positive and range between $+0.028$ g cm^{-2} per decade (at Danmarkshavn) and $+\ 0.052$ g cm^{-2} per decade (at Ny-Ålesund). These largely variable results indicate that no clear indications have been derived in this region, presenting: (i) negative trends of T_o at the two Greenlandic stations and positive trends at the two Arctic Ocean stations; (ii) negative trends of q_o at the low-latitude stations of Aasiaat and Jan Mayen, and positive trends at the stations of Danmarkshavn and Ny-Ålesund; and (iii) a negative trend of W (equal to -0.018 g cm^{-2} per decade) at Aasiaat, against positive trends of W measured at three of the four considered stations.

The results obtained in terms of the best-fit slope coefficients β_T, β_q and β_W evaluated at the RS stations of Sodankylä (SO), Ostrov Dikson (OD) and Tiksi (TI) are shown in Fig. 2.27. They indicate that the long-period trends of the three investigated parameters were found to be homogeneous at the three stations located in Northern Scandinavia and along the Arctic Ocean coasts of Siberia (at latitudes

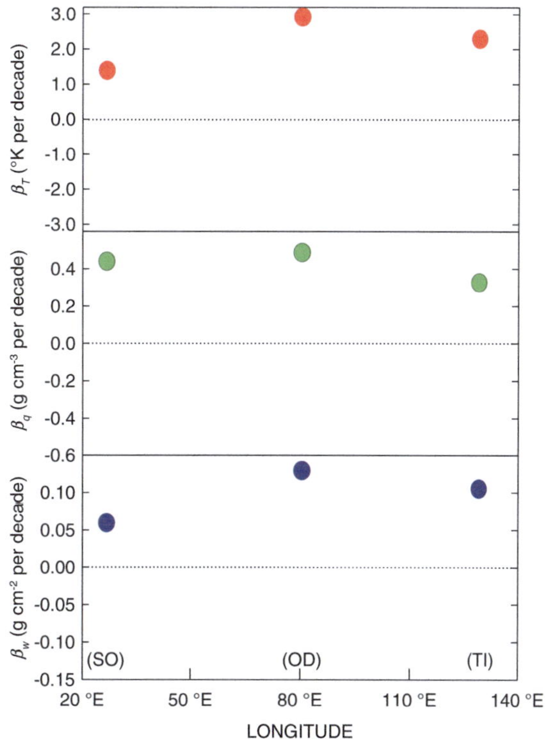

Fig. 2.27 Scatter plots of the multi-year average slope coefficients β_T (upper part with red dots), β_q (middle part with green dots) and β_W (lower part with blue dots) as a function of longitude over its range from 20 °E to 140 °E, including the Northern part of Scandinavian peninsula and the coastal regions of Western and Central Siberia on the Arctic Ocean, with the stations of Sodankylä (SO), Ostrov Dikson (OD) and Tiksi (TI)

varying between 67 °N and 73 °N): (a) the best-fit slope coefficient β_T of parameter T_o observed over the period from 2001 to 2015 was found to assume rather high values greater than +1.3 °K per decade at Sodankylä, + 2.9 °K per decade at Ostrov Dikson and + 2.2 °K per decade at Tiksi, therefore suggesting that pronounced warming effects are occurring in this region during the last years; (b) the best-fit slope coefficient β_q of surface-level absolute humidity q_o was estimated to assume positive and high values at all the three RS stations of this area, ranging between +0.33 g m^{-3} per decade at Tiksi and + 0.49 g m^{-3} per decade at Ostrov Dikson, with an intermediate value of $\beta_q = +\,0.44$ g m^{-3} per decade determined at Sodankylä; and (c) rather high values of β_W were found at the three stations, varying between +0.06 g cm^{-2} per decade at Sodankylä and + 0.13 g cm^{-2} per decade at Ostrov Dikson, the value of β_W determined at Tiksi being of +0.11 g cm ^{2} per decade, these findings suggesting that precipitable water W has been observed to appreciably increase over the period from 2001 to 2015.

Opposite trends to those observed at the Sodankylä, Ostrov Dikson and Tiksi stations have been found in the longitudinal sector of the Arctic region including Eastern Siberia (Cherskij, CE) located at nearly 69 °N latitude, and Alaska (Barrow, BA) located at a latitude only just higher than the 71 °N, both on the Arctic Ocean coasts that are sea-shores of the Bering Strait. The best-fit values of the slope coefficients β_T, β_q and β_W determined at these two stations are shown in Fig. 2.28,

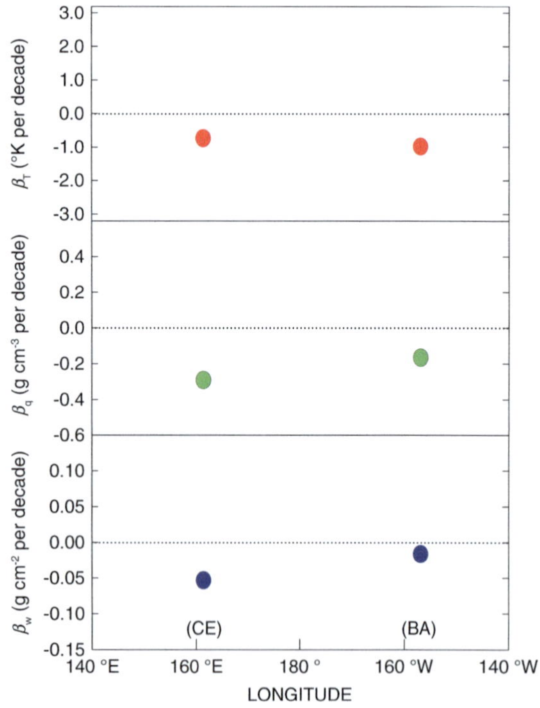

Fig. 2.28 Scatter plots of the multi-year average slope coefficients β_T (upper part with red dots), β_q (middle part with green dots) and β_W (lower part with blue dots) as a function of longitude over its range from 160 °E to 180 °E and from 180 °W to 140 °W, including the Eastern Siberia (Russia) and Alaska (USA) regions separated by the Bering Strait and, hence, the stations of Cherskij (CE) and Barrow (BR)

indicating that: (a) the best-fit slope coefficient β_T was estimated on the basis of the measurements of T_o collected from 2001 to 2015 at Barrow, and from 2004 to 2015 at Cherskij, this coefficient being found to assume rather low values, of around -1 °K per decade at Barrow and -0.7 °K per decade at Cherskij, which indicate that the annual mean temperature T_o is slowly decreasing over the last years; (b) the best-fit slope coefficient β_q was found to be negative at both these stations, equal to -0.17 g m^{-3} per decade at Barrow and -0.29 g m^{-3} per decade at Cherskij, suggesting that the annual mean ground-level moisture conditions were slowly decreased during the last 15 years; and (c) negative values of β_W were determined at both stations, equal to -0.016 g cm^{-2} per decade at Barrow, and to -0.054 g cm^{-2} per decade at Cherskij, both these values indicating that precipitable water W has been observed to slowly decrease during the multi-year period from 2001 to 2015.

The increasing trend of slope coefficient β_T shown in Figs. 2.25 and 2.27 clearly indicate that the surface temperature T_o was slowly increased during the last two decades in the major part of North America located at latitudes higher than 68 °N, in Northern Scandinavia and in the Arctic Ocean coastal regions of Northern Siberia. In addition, the results obtained herein for the various Arctic regions show that the slope coefficient β_q is in part appreciably decreasing (at some sites such as Alert, Aasiaat, Jan Mayen, Cherskij and Barrow), and slowly increasing at a number of RS stations (Inuvik, Cambridge Bay, Eureka, Danmarkshavn, Ny-Ålesund, Sodankylä, Ostrov Dikson and Tiksi). Different trends have been also determined for the slope coefficient bW evaluated to assume: (i) negative values at most of the North American stations (Barrow, Inuvik, Cambridge Bay, Resolute, Alert; see Fig. 2.25), as well at Aasiaat (Western Greenland; Fig. 2.26) and Cherskij (Eastern Siberia; Fig. 2.28), and (ii) positive and relatively high values of β_W at the stations of Danmarkshavn, Jan Mayen and Ny-Ålesund (all located in Greenland and in the Arctic Ocean between Greenland and Svalbard Archipelago; see Fig. 2.26) and at the Finnish station of Sodankylä and the Siberian stations of Ostrov Dikson and Tiksi (Fig. 2.27). Thus, the atmospheric content W of water vapour seems to slowly increase in some Arctic areas, such as the Arctic Ocean between Greenland and Svalbard Islands, the Northern Europe and the Arctic Ocean coasts of Western and Central Siberia.

On this matter it is worth point out that an appreciable increase in the atmospheric water vapour content measured in the Arctic region could lead to: (i) an increasing amount of incoming (short-wave) solar radiation absorbed on average by the Arctic atmosphere, thus causing a positive radiative forcing effect on radiation budget of the surface-atmosphere system; and (ii) increasing amounts of both infrared (long-wave) radiation emitted upward by the surface and absorbed by the atmosphere, and infrared (long-wave) radiation emitted downward by the Arctic atmosphere toward the surface. Should these changes in the radiation budget of the Arctic atmosphere take place during the future decades, an increase in the up-welling flux of the infrared radiation emitted by the surface could be induced together with an increase in the down-welling flux of infrared radiation emitted by the atmosphere toward the surface, which could both contribute to cause an increase in the green-house effect produced by the atmosphere inside the climate system. Actually, we have estimated in Chap. 10 of this book that: (1) the attenuation of short-wave solar radiation due to the water vapour absorption (when precipitable water W is equal to 0.8 g cm^{-2}, as measured on average by us at Ny-Ålesund during the spring season of the last decade) is equal to about 50 W m^{-2} during a cloudless day; (2) the long-wave radiative forcing produced by atmospheric water vapour (for $W = 0.8$ g cm^{-2}) on the upwelling flux emitted by the surface toward the atmosphere is of 11.9 W m^{-2} (absorbed by the atmosphere); and (3) the long-wave radiative forcing produced by atmospheric water vapour ($W = 0.80$ g cm^{-2}) on the down-welling flux emitted by the atmosphere toward the surface is of 74.5 W m^{-2} (entirely absorbed by the atmosphere). Therefore, on the basis of the present evaluations, a weak increase in W (such as that estimated above to be equal to 7 10^{-3} g cm^{-2} per decade, on average in the Arctic region) could cause an additional green-house effect of around 0.8 W m^{-2} in the Arctic region over the 2001–2015 period.

In addition, the increasing effects produced by the green-house gases could lead to a gradual increase of atmospheric warming in this area of our planet during the second half of the twenty-first century, when the permafrost methane feedback caused by the permafrost thawing is foreseen to produce a further strengthening of the greenhouse effect (Natali et al. 2015). Such a permafrost methane feedback caused by the permafrost thawing associated with the gradual warming of the Arctic atmosphere is positive, since the increase in the atmospheric concentration of methane produces a further increase of atmospheric temperature, such a warming being due to the greenhouse effect associated with the absorption of the long-wave terrestrial radiation by the methane molecules. Permafrost is a layer of permanently frozen soil that exists in cold Arctic regions, including Alaska, Northern Canada, and Siberia, whose top layer temporarily melts in summer, forming small shallow depressions filled with melt water and rain, known as "thaw ponds" (Jorgenson et al. 2006). Since wetlands and ponds constitute an important source of methane (Natali et al. 2015), and methane absorbs efficiently the long-wave radiation emitted by the surface-atmosphere system toward the outer space, with an intensity evaluated to be twenty-seven times stronger than that of carbon dioxide (the most important of the greenhouse gases), more methane will be released into the atmosphere as the Arctic temperature will gradually increase, and further warming effects will cause more thaw ponds, in a repeating cycle.

However, it is important to point out on this matter that the annual mean atmospheric concentration of methane measured at the Zeppelin Observatory (Spitsbergen, Svalbard) in 2017 was estimated to be equal to 1940.5 ppbv, therefore causing currently a direct radiative forcing in the upwelling overall long-wave radiance of 0.8 W m^{-2} only, which is therefore comparable with the additional weak absorption foreseen by us because of the small increase in precipitable water evaluated over the 2001–2015 period. The effects produced by the increase in the atmospheric concentration of methane due to permafrost thaw (which are estimated to be currently very small) are expected to become relevant for the overall radiation balance of our planet only during the last two decades of this century.

Acknowledgements The authors thank the Department of Atmospheric Science, University of Wyoming, for the downloading of the radio-sounding data sets collected from 2001 to 2015 at the 14 following radio-sounding stations located in the Arctic region: Barrow, Inuvik, Cambridge Bay, Resolute, Eureka, Alert, Aasiaat, Danmarkshavn, Jan Mayen, Ny-Ålesund, Sodankylä, Ostrov Dikson, Tiksi, and Cherskij.

References

Blackmore, W. H., & Taubvurtzel, B. (1999). Environmental chamber tests on NWS radiosonde humidity sensors. In: Preprints of the *11th symposium on meteorological observations and instrumentation.* American Meteorological Society, Dallas, Texas (USA), pp. 259–262.

Bodhaine, B. A., Wood, N. B., Dutton, E. G., & Slusser, J. R. (1999). On Rayleigh optical depth calculations. *Journal of Atmospheric and Oceanic Technology, 16*(11), 1854–1861. https://doi.org/10.1175/1520-0426(1999)016<1854:ORODC>2.0.CO;2.

Bradley, R. S., Keimig, F. T., & Diaz, H. F. (1992). Climatology of surface-based inversions in the North American Arctic. *Journal of Geophysical Research, Atmospheres, 97*(D14), 15699–15712. https://doi.org/10.1029/92JD01451.

Cady-Pereira, K. E., Shephard, M. W., Turner, D. D., Mlawer, E. J., Clough, S. A., & Wagner, T. J. (2008). Improved daytime column-integrated precipitable water vapor from Vaisala radiosonde humidity sensors. *Journal of Atmospheric and Oceanic Technology, 25*(6), 873–883. https://doi.org/10.1175/2007JTECHA1027.1.

Central Aerological Observatory (CAO). (2003). *Guide to hydrometeorological stations and posts, issue III, temperature – Wind Sounding of the Atmosphere*, St. Petersburg (Russia) (in Russian), 312 pp.

Dinelli, B. M., Arnone, E., Brizzi, G., Carlotti, M., Castelli, E., Magnani, L., Papandrea, E., Prevedelli, M., & Ridolfi, M. (2010). The MIPAS2D database of MIPAS/ENVISAT measurements retrieved with a multi-target 2-dimensional tomographic approach. *Atmospheric Measurement Techniques, 3*(2), 355–374. https://doi.org/10.5194/amt-3-355-2010.

Dubin, M., Sissenwine, N., & Teweles, S. (1966). *U. S. Standard Atmosphere Supplements, 1966.* Environmental Science Services Administration, National Aeronautics and Space Administration, United States Air Force, Washington, DC, 20402, 289 pp.

He, W., Ho, S., Chen, H., Zhou, X., Hunt, D., & Kuo, Y.-H. (2009). Assessment of radiosonde temperature measurements in the upper troposphere and lower stratosphere using COSMIC radio occultation data. *Geophysical Research Letters, 36*, L17807. https://doi.org/10.1029/2009GL038712 .

Highwood, E. J., Hoskins, B. J., & Berrisford, P. (2007). Properties of the Arctic tropopause. *Quarterly Journal of the Royal Meteorological Society, 126*(565). https://doi.org/10.1002/qj.49712656515.

Ho, S., Peng, L., & Vömel, H. (2017). Characterization of the long-term radiosonde temperature biases in the upper troposphere and lower stratosphere using COSMIC and Metop-A/GRAS data from 2006 to 2014. *Atmospheric Chemistry and Physics, 17*, 4493–4511. https://doi.org/10.5194/acp-17-4493-2017.

Holton, J. R., Haynes, P., McIntyre, M. E., Douglas, A., Rood, R., & Pfister, L. (1995). Stratosphere – Troposphere exchange. *Reviews of Geophysics, 33*(4), 403–439.

Huovila, S., & Tuominen, A. (1991, January 14–18). Influence of radiosonde lag errors on upper-air climatological data, paper presented at the Seventh Symposium on Meteorological Observations and Instrumentation, *Special Sessions on Laser Atmospheric Studies*, pp. 237–242, American Meteorological Society, New Orleans, LA.

Jorgenson, M. T., Shur, Y. L., & Pullman, E. R. (2006). Abrupt increase in permafrost degradation in Arctic Alaska. *Geophysical Research Letters, 33*, L02503. https://doi.org/10.1029/2005GL024960.

Kahl, J. D., Serreze, M. C., & Schnell, R. C. (1992). Tropospheric low-level temperature inversions in the Canadian Arctic. *Atmosphere-Ocean, 30*(4), 511–529. https://doi.org/10.1080/07055900.1992.9649453.

Luers, J. K. (1997). Temperature error of the Vaisala RS90 radiosonde. *Journal of Atmospheric and Oceanic Technology, 14*(6), 1520–1532. https://doi.org/10.1175/1520-0426(1997)014<1520:TEOTVR>2.0.CO;2.

Luers, J. K., & Eskridge, R. E. (1995). Temperature corrections for the VIZ and Vaisala radiosondes. *Journal of Applied Meteorology, 34*(6), 1241–1253. https://doi.org/10.1175/1520-0450(1995)034<1241:TCFTVA>2.0.CO;2

Luers, J. K., & Eskridge, R. E. (1998). Use of radiosonde temperature data in climate studies. *Journal of Climate, 11*(5), 1002–1019. https://doi.org/10.1175/1520-0442(1998)011<1002:UORTDI>2.0.CO;2.

Mattioli, V., Westwater, E. R., Cimini, D., Liljegren, J. S., Lesht, S, Gutman, S., &Schmidlin, F. (2005, March 14–18). Analysis of radiosonde and PWV data from the 2004 North slope of Alaska Arctic winter radiometric experiment. In *Proceedings of the Fifteenth ARM Science Team Meeting*, Daytona Beach, Florida, 29 pp.

Mattioli, V., Westwater, E. R., Cimini, D., Liljegren, J. C., Lesht, B. M., Gutman, S. I., & Schmidlin, F. J. (2007). Analysis of radiosonde and ground-based remotely sensed PWV data from the 2004 North slope of Alaska Arctic winter radiometric experiment. *Journal of Atmospheric and Oceanic Technology, 24*(3), 415–431. https://doi.org/10.1175/JTECH1982.1.

Maturilli, M., Herber, A., & König-Langlo, G. (2013). Climatology and time series of surface meteorology in Ny-Ålesund, Svalbard. *Earth System Science Data, 5*, 155–163. https://doi.org/10.5194/essd-5-155-2013.

Miloshevich, L. M., Vömel, H., Paukkunen, A., Heymsfield, A. J., & Oltmans, S. J. (2001). Characterization and correction of relative humidity measurements from Vaisala RS80-A radiosondes at cold temperatures. *Journal of Atmospheric and Oceanic Technology, 18*(2), 135–156. https://doi.org/10.1175/1520-0426(2001)018<0135:CACORH>2.0.CO;2.

Miloshevich, L. M., Paukkunen, A., Vömel, H., & Oltmans, S. J. (2004). Development and validation of a time-lag correction for Vaisala radiosonde humidity measurements. *Journal of Atmospheric and Oceanic Technology, 21*(9), 1305–1327. https://doi.org/10.1175/1520-0426(2004)021<1305:DAVOAT>2.0.CO;2.

Miloshevich, L. M., Vömel, H., Whiteman, D. N., Lesht, B. M., Schmidlin, F. J., & Russo, F. (2006). Absolute accuracy of water vapor measurements from six operational radiosonde types launched during AWEX-G and implications for AIRS validation, *Journal of Geophysical Research, Atmospheres*, *111*, D09S10, https://doi.org/10.1029/2005JD006083.

Miloshevich, L. M., Vömel, H., Whiteman, D. N., & Leblanc, T. (2009). Accuracy assessment and corrections of Vaisala RS92 radiosonde water vapor measurements. *Journal of Geophysical Research, Atmospheres, 114*, D11305. https://doi.org/10.1029/2008JD011565.

Moradi, I., Buehler, S. A., John, V. O., Reale, A., & Ferraro, R. R. (2013). Evaluating instrumental inhomogeneities in global radiosonde upper tropospheric humidity data using microwave satellite data. *IEEE Transactions on Geoscience and Remote Sensing, 51*(6), 3615–3624. https://doi.org/10.1109/TGRS.2012.2220551.

Murphy, D. M., & Koop, T. (2005). Review of the vapour pressures of ice and supercooled water for atmospheric applications. *Quarterly Journal of the Royal Meteorological Society, 131*(608), 1539–1565. https://doi.org/10.1256/qj.04.94.

Muscari, G., C. Di Biagio, A. di Sarra, M. Cacciani, S. E. Ascanius, P. P. Bertagnolio, C. Cesaroni1, R. L. de Zafra, P. Eriksen, G. Fiocco, I. Fiorucci, and D. Fuà (2014), Observations of surface radiation and stratospheric processes at Thule Air Base, Greenland, during the IPY, *Annals of Geophysics*, Volume 57, 3, SS0323, doi:https://doi.org/10.4401/ag-6382.

Natali, S. M., Schuur, E. A. G., Mauritz, M., Schade, J. D., Celis, G., Crummer, K. G., Johnston, C., Krapek, J., Pegoraro, E., Salmon, V. G., & Webb, E. E. (2015). Permafrost thaw and soil moisture driving CO_2 and CH_4 release from upland tundra. *Journal of Geophysical Research, Biogeosciences, 120*(3), 525–537. https://doi.org/10.1002/2014JG002872.

Rowe, P. M., Miloshevich, L. M., Turner, D. D., & Walden, V. P. (2008). Dry bias in Vaisala RS90 radiosonde humidity profiles over Antarctica. *Journal of Atmospheric and Oceanic Technology, 25*(9), 1529–1541. https://doi.org/10.1175/2008JTECHA1009.1.

Schroeder, S. R. (2009, January 11–15). Homogenizing the Russian Federation upper air climate record by adjusting radiosonde temperatures and dew points for instrument changes, Oral presentation at the *21st Conference on Climate Variability and Change, 89th American Meteorological Society Annual Meeting*, Phoenix (Arizona).

Tomasi, C., Vitale, V., & De Santis, L. V. (1998). Relative optical mass functions for air, water vapour, ozone and nitrogen dioxide in atmospheric models presenting different latitudinal and seasonal conditions, *Meteorology and Atmospheric Physics, 65*(1–2), 11–30. https://doi.org/10.1007/BF01030266.

Tomasi, C., Cacciari, A., Vitale, V., Lupi, A., Lanconelli, C., Pellegrini, A., & Grigioni, P. (2004). Mean vertical profiles of temperature and absolute humidity from a twelve-year radiosounding data-set at Terra Nova Bay (Antarctica). *Atmospheric Research, 71*(3), 139–169. https://doi.org/10.1016/j.atmosres.2004.03.009.

Tomasi, C., Vitale, V., Petkov, B., Lupi, A., & Cacciari, A. (2005). Improved algorithm for calculations of Rayleigh-scattering optical depth in standard atmospheres. *Applied Optics, 44*(16), 3320–3341. https://doi.org/10.1364/AO.44.003320.

Tomasi, C., Petkov, B., Benedetti, E., Vitale, V., Pellegrini, A., Dargaud, G., De Silvestri, L., Grigioni, P., Fossat, E., Roth, W. L., & Valenziano, L. (2006). Characterization of the atmospheric temperature and moisture conditions above Dome C (Antarctica) during austral summer and fall months. *Journal of Geophysical Research, Atmospheres, 111*, D20, D20305. https://doi.org/10.1029/2005JD006976.

Tomasi, C., Petkov, B., Stone, R. S., Benedetti, E., Vitale, V., Lupi, A., Mazzola, M., Lanconelli, C., Herber, A., & Von Hoyningen-Huene, W. (2010). Characterizing polar atmospheres and their effect on Rayleigh-scattering optical depth. *Journal of Geophysical Research, Atmospheres, 115*, D02205. https://doi.org/10.10129/2009JD012852.

Tomasi, C., Petkov, B., Dinelli, B. M., Castelli, E., Arnone, E., & Papandrea, E. (2011a). Monthly mean vertical profiles of pressure, temperature, and water vapour volume mixing ratio in the polar stratosphere and low mesosphere from a multi-year set of MIPAS-ENVISAT limb-scanning measurements. *Journal of Atmospheric and Solar-Terrestrial Physics, 73*(16), 2237–2271. https://doi.org/10.1016/j.jastp.2011.06.018.

Tomasi, C., Petkov, B., Benedetti, E., Valenziano, L., & Vitale, V. (2011b). Analysis of a 4 year radiosonde data set at Dome C for characterizing temperature and moisture conditions of the Antarctic atmosphere. *Journal of Geophysical Research, Atmospheres, 116*, D15, D15304. https://doi.org/10.1029/2011JD015803.

Tomasi, C., Petkov, B. H., & Benedetti, E. (2012). Annual cycles of pressure, temperature, absolute humidity and precipitable water from the radiosoundings performed at Dome C, Antarctica, over the 2005 – 2009 period. *Antarctic Science, 24*(6), 637–658. https://doi.org/10.1017/S0954102012000405.

Turner, D. D., Lesht, B. M., Clough, S. A., Liljegren, J. C., Revercomb, H. E., & Tobin, D. C. (2003). Dry bias and variability in Vaisala RS80-H radiosondes: The ARM experience. *Journal of Atmospheric and Oceanic Technology, 20*(1), 117–132. https://doi.org/10.1175/1520-0426(2003)020<0117:DBAVIV>2.0.CO;2.

Wang, J., & Young, K. (2005. June 19–23). *Comparisons of 7-year radiosonde data from two neighboring stations and estimation of random error variances for four types of radiosondes.* Presented at the *13th Symposium on Meteorological Observations and Instrumentation*, Savannah (Georgia, USA).

Wang, J., Cole, H. L., Carlson, D. J., Miller, E. R., Beierle, K., Paukkunen, A., & Laine, T. K. (2002). Corrections of humidity measurement errors from the Vaisala RS80 radiosonde – Application to TOGA COARE data. *Journal of Atmospheric and Oceanic Technology, Volume, 19*(7), 981–1002. https://doi.org/10.1175/1520-0426(2002)019<0981:COHMEF>2.0.CO;2.

Wang, J., Carlson, D. J., Parsons, D. B., Hock, T. F., Lauritsen, D., Cole, H. L., Beierle, K., & Chamberlain, E. (2003). Performance of operational radiosonde humidity sensors in direct comparison with a chilled mirror dew-point hygrometer and its climate implication. *Geophysical Research Letters, 30*(16), 1860. https://doi.org/10.1029/2003GL016985.

Zaitseva, N. A. (1993). Historical developments in radiosonde systems in the former Soviet Union. *Bullettin of the American Meteorological Society, 74*, 1893–1900. https://doi.org/10.1175/1520-0477(1993)074<1893:HDIRSI>2.0.CO;2.

Zängl, G., & Hoinka, K. P. (2001). The tropopause in the polar regions. *Journal of Climate, 14*(14), 3117–3139. https://doi.org/10.1175/1520-0442(2001)014<3117:TTITPR>2.0.CO;2.

Chapter 3
Trace Gases in the Arctic Atmosphere

Kimberly Strong, William R. Simpson, Kristof Bognar, Rodica Lindenmaier, and Sébastien Roche

Abstract The Arctic atmosphere is coupled to lower latitudes, both as a receptor for global pollution and as a driver for the global climate system. Arctic atmospheric composition is variable and changing, making measurements of trace gas concentrations essential for understanding atmospheric processes.

The atmospheric concentrations of carbon dioxide (CO_2) and methane in the Arctic are increasing in concert with global trends. Meanwhile, the Arctic represents up to 25% of the global land carbon sink and the Arctic Ocean accounts for 10–12% of the global ocean CO_2 sink. The Arctic thus has a strong influence on the global carbon cycle, while also responding more strongly to changes in climate than do mid-latitude regions. However, many processes that lead to carbon emissions and exchange in the Arctic are poorly understood and the region is sparsely sampled, resulting in large uncertainties in the quantification of carbon stocks, sources, and sinks.

The Arctic experiences poor air quality due to local sources and transport from diverse mid-latitude emission sources such as wildfires. The springtime Arctic troposphere frequently experiences ozone depletion episodes that are linked to surface-based production of reactive halogen species that then deplete ozone, particularly associated with bromine explosions. A major source of bromine in the Arctic is sea salt, but the importance of blowing snow and the mechanisms involved in the heterogeneous bromine release are the focus of ongoing research.

In the stratosphere, springtime ozone depletion continues in the Arctic, with significant interannual variability driven by atmospheric dynamics, transport, and temperature. Ozone recovery is anticipated due to reduction of chlorofluorocarbons

K. Strong (✉) · K. Bognar · R. Lindenmaier · S. Roche
Department of Physics, University of Toronto, Toronto, ON, Canada
e-mail: strong@atmosp.physics.utoronto.ca; kbognar@atmosp.physics.utoronto.ca; sebastien.roche@mail.utoronto.ca

W. R. Simpson
Department of Chemistry, Biochemistry, and Geophysical Institute, University of Alaska Fairbanks, Fairbanks, AK, USA
e-mail: wrsimpson@alaska.edu

A. Kokhanovsky, C. Tomasi (eds.), *Physics and Chemistry of the Arctic Atmosphere*, Springer Polar Sciences, https://doi.org/10.1007/978-3-030-33566-3_3

under the Montreal Protocol and its Amendments, but there are uncertainties due to coupling between stratospheric chemistry and climate.

This chapter provides a review of the trace gas composition of the Arctic atmosphere. It surveys our current knowledge and discusses outstanding questions, with a focus on tropospheric ozone and halogens, greenhouse gases, and the stratosphere.

Keywords Arctic atmospheric composition · Arctic air quality · Greenhouse gases · Carbon cycle · Tropospheric ozone · Halogen chemistry · Bromine explosion events · Stratospheric ozone depletion and recovery · Dynamical and chemical processes in the Arctic stratosphere

3.1 Introduction

The Arctic is a bellwether for global climate change, a receptor for global pollution, and a driver for the global climate system. It is a remote region with highly seasonal variations in temperature, radiation, and surface conditions. It has some local emission sources, but is also coupled to lower latitudes. High-quality trace gas measurements are essential for understanding what environmental changes are happening in the Arctic and why, as has been identified, for example, by the Arctic Monitoring and Assessment Programme: "*Long-term monitoring of atmospheric composition at existing stations needs to be continued and integrated into a Pan-Arctic observation network*" (AMAP 2015a).

Increases in the concentrations of the greenhouse gases CO_2 and CH_4 are well documented (IPCC 2013), and the Group on Earth Observations Carbon Strategy notes that "*Understanding the global carbon cycle, and predicting its evolution under future climate scenarios is one of the biggest challenges facing science today*" (Ciais et al. 2010). These challenges are due to lack of sufficient knowledge about carbon sources and sinks, feedbacks between climate change and carbon reservoirs, and anthropogenic emissions from fossil fuel burning and land use change. The issues are more acute in the Arctic where measurements are scarce and the magnitude and distribution of carbon sinks and sources are poorly known.

Meanwhile, Arctic chemistry, in both the troposphere and the stratosphere, remains a focus of attention. Research questions include the impact of rapid climate change on the Arctic, production of ozone from tropical and mid-latitude emissions of precursors such as methane, transport pathways of pollutants into the Arctic, new sources of local pollution from shipping and resource extraction, changes in local air quality, and deposition of harmful contaminants in snow. Arctic tropospheric ozone is greatly affected by severe ozone depletion events, first observed at Alert, Canada in the 1980s and since linked to extremely high concentrations of reactive bromine species in bromine explosion events. These are important because they increase the deposition of mercury to snow, causing harmful effects on ecosystems and humans, but the underlying processes are still not fully understood. Tropospheric ozone in the Arctic also acts a significant short-term climate forcer. In the stratosphere, ozone recovery is anticipated due to reduction of chlorofluorocarbons under the Montreal Protocol and its Amendments, but there are uncertainties due to coupling between stratospheric chemistry and climate. Springtime stratospheric ozone depletion continues in the Arctic, with significant interannual variability driven by atmospheric dynamics, transport, and temperature.

The Arctic experiences poor air quality due to local sources and transport from diverse mid-latitude emission sources. Wildfires are an episodic source of trace gases and particulates, affecting the carbon cycle, climate, air quality, and land ecology (e.g., Viatte et al. 2015; Lutsch et al. 2016). Fire frequency, intensity, and geographic distribution are strongly sensitive to climate change and are expected to increase (de Groot et al. 2013). Meanwhile, local emissions from oil and gas extraction, shipping, mining, and infrastructure development are already affecting the Arctic atmosphere (e.g., Aliabadi et al. 2015; Marelle et al. 2016; Schmale et al. 2018), while "*Changes in mid-latitude emissions, and climate-driven changes in transport patterns, coupled to increasing local emissions are expected to shift the balance among pollutant source contributions in the coming decades*" (Arnold et al. 2016). Multi-model studies exhibit large differences in their simulations of Arctic pollution (Emmons et al. 2015; Monks et al. 2015). To address knowledge gaps related to Arctic air pollution, the International Global Atmospheric Chemistry Project (IGAC) has established the air Pollution in the Arctic: Climate, Environment and Societies initiative (PACES, Arnold et al. 2016). One of its projects is ALaskan Pollution and Chemical Analysis (ALPACA), which aims to improve knowledge of atmospheric chemical mechanisms occurring under cold and dark conditions (Simpson et al. 2018).

In this section, we review the trace gas composition of the Arctic atmosphere, with a focus on tropospheric ozone and halogens, greenhouse gases, and the stratosphere. Air quality and transport of atmospheric pollutants to the Arctic are largely covered elsewhere in this book. Additional information about Arctic air quality can be found in results from field campaigns such as the Polar Study using Aircraft, Remote Sensing, Surface Measurements and Models, of Climate, Chemistry, Aerosols, and Transport (POLARCAT, Stohl et al. 2009; Law et al. 2014), Arctic Research of the Composition of the Troposphere from Aircraft and Satellites (ARCTAS, Jacob et al. 2010), and the Arctic Climate Change, Economy, and Society (ACCESS) aircraft campaign (Roiger et al. 2015). Arnold et al. (2016) provide an overview of our current understanding of Arctic air pollution, while various reports of the Arctic Monitoring and Assessment Programme address issues related to Arctic pollution mercury, persistent organic pollutants, and chemical of emerging concern (AMAP 2006, 2011, 2015b, 2017a).

3.2 Tropospheric Ozone and Halogens

3.2.1 Tropospheric Ozone

Interest in global oxidation chemistry led to measurements of ozone at remote sites around the world, with sustained Arctic observations beginning in the 1970s. It was soon discovered that upon the return of sunlight to Arctic locations, e.g., Alert, Nunavut, Canada and Utqiaġvik (formerly Barrow), Alaska, USA, in springtime, ozone mixing ratios decreased to near-zero levels (Bottenheim et al. 1986; Oltmans and Komhyr 1986). These events were called "ozone depletion episodes" or ODEs and Barrie et al. (1988) soon found that filterable bromine spiked as ozone was depleted, implicating reactive halogen species in this ozone depletion. Subsequent measurements at other Arctic locations have demonstrated that these ODEs are very common from March to May throughout the Arctic. A number of articles have reviewed ozone depletion events and the relationship to reactive halogens (Simpson et al. 2007, 2015; von Glasow and Crutzen 2007; Abbatt et al. 2012; Saiz-Lopez and von Glasow 2012). Measurements in Antarctica show that similar ozone depletion events occur during the Austral spring (Schnell et al. 1991; Oltmans 1993; Kalnajs and Avallone 2006; Jones et al. 2009, 2010). Schroeder et al. (1998) first showed that ozone depletion episodes were correlated with mercury oxidation and deposition, resulting in atmospheric mercury deposition events, which have also been the subject of reviews (Steffen et al. 2008; Cole and Steffen 2010; Berg et al. 2013). In this section, we focus on recent findings with respect to tropospheric ozone and halogens in the Arctic.

Trends in surface ozone have been analysed to determine if changing environmental conditions are altering ozone chemistry and levels. Global trends have been reviewed by Cooper et al. (2014), and an Arctic-specific trend has also been studied (Oltmans et al. 2012). Oltmans et al. (2012) presented a 38-year record of surface ozone mixing ratios from Utqiaġvik, and demonstrated that during March, but not April or May, there has been a significant increasing trend in ozone depletion event frequency, which the authors related the decline in multi-year sea ice in the Arctic Ocean. Using surface ozone data from Utqiaġvik, Alert, and Zeppelinfjellet (Norway), Koo et al. (2014) argued that ODE frequency in the Arctic might be affected by variability in teleconnection patterns in the Pacific. During the POLARCAT campaign, aircraft-based vertical profiles of ozone and CO were measured throughout the troposphere covering the Arctic Ocean, Greenland, Alaska and Canada during summertime (Ancellet et al. 2016). The WRF-Chem model generally agreed fairly well with these data.

The near-surface distribution of ozone during springtime typically shows smaller ozone concentrations at the surface than aloft. The origin of this distribution is considered to be surface-based production of reactive halogen species that then deplete ozone, as discussed below. Past (Bottenheim et al. 2002; Tarasick and Bottenheim 2002; Tackett et al. 2007) and recent work on ozone vertical profiles from balloon soundings show that ozone depletion is typically confined to the lowest 1000 m above ground level, with an average top of the ozone depleted layer of ~500 m (Helmig et al. 2012; Oltmans et al. 2012). Lidar observations also agree with this surface-based vertical domain of ozone depletion and showed that these ozone-depleted regions extend horizontally hundreds of kilometers over the sea ice (Seabrook et al. 2011, 2013). During the BRomine, Ozone, and Mercury EXperiment (BROMEX) campaign (Nghiem et al. 2013), Moore et al. (2014) showed that recovery of ozone following an ODE often occurred through vertical mixing caused by convection associated with open water in large cracks between sea ice floes, which are termed "open leads". This vertical delivery of ozone from aloft into the boundary layer was also accompanied by delivery of gaseous elemental mercury (Hg^0) to the halogen-rich boundary layer where it could be oxidized and then deposited to the snowpack.

Early observations of ozone depletion episodes were made primarily from coastal stations, but recent studies have used ships (Bottenheim et al. 2009; Gilman et al. 2010) and autonomous buoys (Knepp et al. 2010; Halfacre et al. 2014) to study ozone depletion directly over the frozen Arctic Ocean. It has generally been observed that ODEs are more extended and intense over the Arctic sea ice (Bottenheim et al. 2009) than at coastal sites, and that air from inland is seldom depleted in ozone. The horizontal dimensions of ozone-depleted airmasses were estimated to approach 1000 km for O_3 mixing ratios <15 nmol mol^{-1} and to be a third of that size for major ODEs (O_3 < 10 nmol mol^{-1}) (Halfacre et al. 2014). Antarctic measurements have confirmed that similarly large horizontal extents of ozone depletion events occur over sea ice in Antarctica (Jones et al. 2013).

At the near-snow surface, ozone can be deposited to the snowpack, or potentially affected by snowpack photochemical NO_x production (Jones et al. 2001; Honrath et al. 2002; Grannas et al. 2007). Helmig and co-workers have extensively studied ozone deposition, both in inland snowpacks at Greenland (Helmig et al. 2007a, b, 2009; Bocquet et al. 2011) and at the coast (Helmig et al. 2012). This work has shown that ozone deposition rates to snowpack are on the order of 0.01 cm/s, both in the absence and presence of atmospheric ODEs. The small deposition rate shows that direct deposition of ozone to snowpack is too slow to explain springtime ODEs. Summertime ozone deposition to Arctic tundra has also been studied and was found to be approximately an order of magnitude faster than deposition to snowpack (van Dam et al. 2016). While some of this deposition velocity was attributed to surface deposition to tundra, the authors hypothesize that surface emissions of biological volatile organic compounds could explain some of the ozone loss.

3.2.2 Halogen Chemistry

Springtime ODEs cannot be explained without the involvement of halogen species. Reactive halogens participate in ozone destruction cycles similar to those involved in stratospheric ozone loss. In the following, we focus on bromine (Br), as it is the main species responsible for ODEs (Thompson et al. 2015). The most important

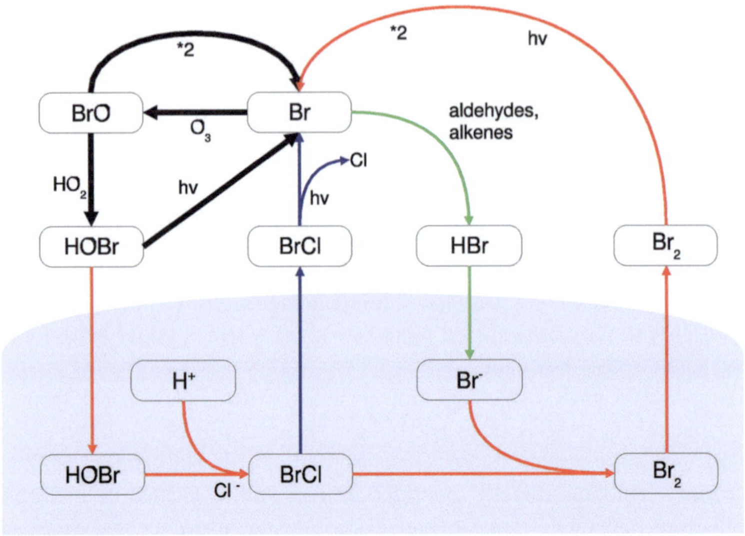

Fig. 3.1 A schematic diagram showing a simplified set of chemical reactions involved in bromine explosion events. The blue shading represents the condensed phase, either liquid brine or ice surface. (Reproduced from Figure 4, Simpson et al. 2007)

source of reactive bromine in the polar regions is a heterogeneous process that oxidizes bromide (Br^-) present in the condensed phase to molecular bromine that is released into the atmosphere, as illustrated in Fig. 3.1. Combined with the ozone destruction reactions, the result is an autocatalytic reaction cycle, meaning that the product acts as a catalyst to speed up the reaction. This process can lead to sudden buildup of reactive bromine in the atmosphere, and these events are often referred to as "bromine explosions" (Eigen and Kustin 1962; Fan and Jacob 1992; McElroy et al. 1999; Wennberg 1999).

Figure 3.2 shows how the boundary layer mixing ratios of ozone and the principal bromine species evolve during a bromine explosion event. Photolysis of Br_2 produces two Br radicals that often react with ozone, producing BrO, which is often used as the indicator for bromine explosions, since it is detectable using remote sensing. Other species, including Br_2, Br, HOBr (the catalyst in the reaction cycle), and HBr (important for bromine deposition) are more challenging to measure, although chemical ionization mass spectroscopy has recently detected more of these intermediates (Custard et al. 2017; Wang and Pratt 2017). The mechanisms involved in the bromine explosion reactions are the subject of several review papers (von Glasow and Crutzen 2007; Abbatt et al. 2012; Saiz-Lopez and von Glasow 2012; Simpson et al. 2015). The major source of bromine in the Arctic is sea salt, and the surfaces involved in the heterogeneous bromine release are the topic of ongoing research.

The proposed reaction cycle requires the condensed phase material to have a high surface area, high salinity, and low pH. Given the early realization that bromine

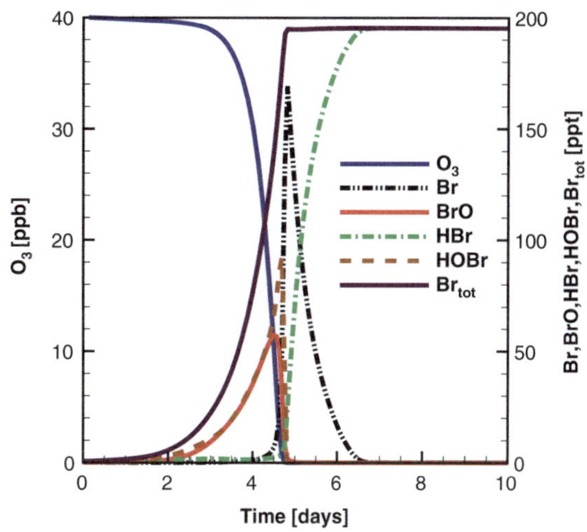

Fig. 3.2 The temporal evolution of ozone and the bromine species in a 200 m boundary layer during an ozone depletion event. (Reproduced from Figure 1, Cao et al. 2016)

enhancement is connected to first-year sea ice (Wagner et al. 2001), frost flowers (Style and Worster 2009; Barber et al. 2014) were initially thought to fulfil all the requirements to be a source of reactive bromine (Rankin et al. 2002; Kaleschke et al. 2004). Further research, however, showed that frost flowers have a surface area smaller than initially thought (Domine et al. 2005; Obbard et al. 2009), and their pH is likely too high (Kalnajs and Avallone 2006; Douglas et al. 2012). The most likely contribution of frost flowers to bromine activation is their indirect influence on the nearby or overlying snowpack (Hara et al. 2017).

In parallel to the frost flower theory, snow over sea ice and on land was also considered as a possible bromine source. Measurements by Foster et al. (2001) and Spicer et al. (2002) in Alert showed significant amounts of Br_2 and BrCl in and above the snowpack during ODEs. Outdoor snow chamber measurements by Pratt et al. (2013) in Utqiaġvik confirmed that snow exposed to sunlight produces bromine, and this process is more efficient in low-pH snow. Wren et al. (2013) obtained similar results in a laboratory setting. Bromine fluxes measured by Custard et al. (2017) in Utqiaġvik indicated that the snowpack could be a dominant source of reactive bromine to the atmosphere. Modelling studies, incorporating snowpack chemistry of varying complexity, corroborated the importance of the snowpack (Toyota et al. 2011, 2014; Thompson et al. 2015, 2017; Wang and Pratt 2017). While most studies focus on coastal sites, bromine production in snow has also been observed at Summit, Greenland, far from marine influences (Stutz et al. 2011; Thomas et al. 2011, 2012). Snowpack-related bromine explosions often occur in the presence of calm winds and a strong temperature inversion above the surface, conditions that are ubiquitous during the Arctic spring. The inversion confines reactants near the surface, while pumping by low winds enhances the exchange of trace gases between the snowpack interstitial air and the overlying airmass. In order to better understand the mechanisms involved in snowpack halogen production, and to be able to validate model simulations, vertically resolved measurements and improved detection limits for some species will be required.

Halogen activation, however, is not necessarily restricted to the surface. Airborne blowing snow and sea salt aerosol (SSA) are likely involved in bromine release as well. Yang et al. (2008) and Jones et al. (2009, 2010) argued that high wind speeds and blowing snow can create favorable conditions for bromine explosions. Correlation between bromine and high winds was observed by Frieß et al. (2011) and Liao et al. (2012) in Utqiaġvik, Alaska. Observations of long-range transport of large BrO plumes by Arctic cyclones (Begoin et al. 2010; Sihler et al. 2012; Blechschmidt et al. 2016; Zhao et al. 2016) lent further credibility to the view that bromine activation and recycling could proceed decoupled from the surface. In addition to blowing snow, 3-D modelling studies suggest that SSA-sourced bromine is sufficient to explain space-based BrO observations by the Global Ozone Monitoring Experiment (GOME), GOME-2, and the Ozone Monitoring Instrument (OMI) (e.g., Yang et al. 2010; Theys et al. 2011; Choi et al. 2018). An example is shown in Fig. 3.3, which is a case study from Choi et al. (2018) that illustrates the correlation between blowing snow-generated SSA and tropospheric BrO explosions. Peterson et al. (2017) reported aircraft measurements of an elevated BrO layer coincident with an enhancement in supermicron aerosol particles, and several studies in Antarctica provided evidence that SSA is depleted in bromide compared to sea water (Legrand et al. 2016; Giordano et al. 2018; Hara et al. 2018). In the absence of open water in the polar spring, the source of SSA is thought to be the sublimation of saline blowing snow (Yang et al. 2008; Huang and Jaeglé 2017; Zhao et al. 2017; Huang et al. 2018). More work is needed to determine whether blowing snow and SSA contribute to bromine release, or merely recycle halogens mixed up from the surface. It is also unclear if blowing snow has a direct role, or if lofted snow particles contribute mainly through SSA production.

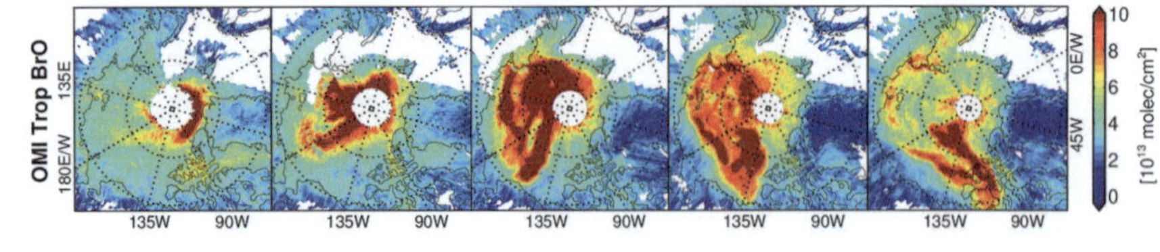

Fig. 3.3 Top panels: Column-integrated mass of blowing snow-generated SSAs simulated using the Goddard Earth Observing System (GEOS-5). Bottom panels: Tropospheric column BrO for 24–28 March 2007 measured by OMI. (Reproduced from Figure 7 of Choi et al. 2018)

The vertical distribution of bromine offers some clues about the release and recycling mechanisms. Measurements by Peterson et al. (2015) and Simpson et al. (2017) in Utqiaġvik, Alaska showed that BrO is often confined to the lowest 200–300 m of the atmosphere when a shallow, stable boundary layer is present, consistent with snowpack bromine production. This is seen in Fig. 3.4, which shows measurements from Simpson et al. (2017) of the lower-tropospheric (LT) vertical column density (VCD) of BrO, the aerosol optical depth (AOD), and surface ozone made in March 2012 at three sites: the Barrow Arctic Research Center (BARC) and two "IceLanders" that were deployed onto nearby sea ice (IL1, IL2). The vertical profile of BrO is variable, with the fraction of BrO LT-VCD in the lowest 200 m (f_{200}) varying from 0 to 0.8. Repartitioning of BrO_x due to reduced ozone levels results in less surface BrO, and aerosol extinction aloft was found to be necessary but not sufficient for BrO to be present aloft.

Peterson et al. (2015) also showed that well-distributed BrO events (up to 1–2 km) in good mixing conditions typically result in higher BrO columns than shallow events. Vertical mixing might be caused by convection over open leads (Moore et al. 2014), or high winds and storms. BrO retrievals from satellite data are not sensitive to shallow events, only to distributed BrO columns that are often

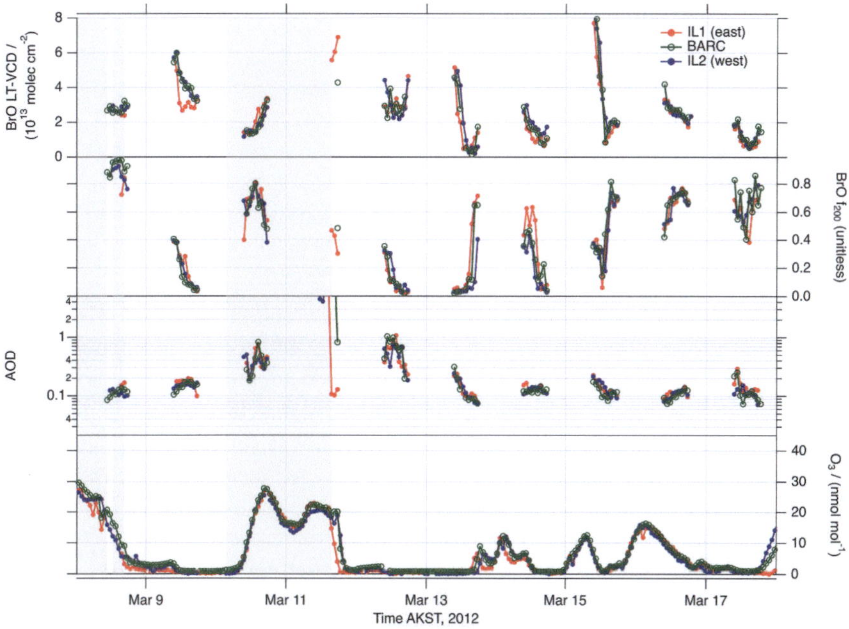

Fig. 3.4 Hourly BrO, aerosol optical depth, and ozone measured at three sites near Utqiaġvik, Alaska in March 2012 (see text). (**a**) BrO LT-VCD (from 0 to approximately 2000 m). (**b**) Fraction of BrO LT-VCD in the lowest 200 m; f_{200} = VCD (0–200 m) / LT-VCD. (**c**) AOD from 0 to 4000 m. (**d**) In situ surface ozone mixing ratio measured on the IL platforms and ~2 km northeast of the BARC building. (Reproduced from Figure 4 of Simpson et al. 2017)

accompanied by high winds and high aerosol load (Salawitch et al. 2010; Choi et al. 2012; Sihler et al. 2012). This might be the reason why satellite observations can be explained by SSA bromine release, while surface observations point to both the snowpack and aerosols as bromine sources (Simpson et al. 2017).

While bromine chemistry is the most prominent factor in ODEs, other species might also play important roles. Snowpack chlorine production was reported by Liao et al. (2014) and Custard et al. (2017) in Utqiaġvik. Chlorine does not efficiently destroy ozone in the troposphere, but could enhance bromine release via BrCl production (Thompson et al. 2015; Wang and Pratt 2017). Iodine has recently been observed in the Arctic springtime, as IO in Alert (Zielcke 2015), and in molecular form near Utqiaġvik (Raso et al. 2017). Although far less abundant than bromine, iodine might deplete ozone much more efficiently (Thompson et al. 2015). In addition, modelling studies suggest that NO_x (emitted naturally from the snowpack and sometimes present as fresh pollution from nearby combustion emissions) could enhance bromine production if sufficient surface area is present for heterogeneous reactions (Thomas et al. 2012; Cao et al. 2014; Wang and Pratt 2017). High levels of nitrogen oxides, however, might suppress BrO and reduce ozone loss (Custard et al. 2015), highlighting a potential impact of local pollution in the Arctic.

3.3 Greenhouse Gases

3.3.1 Introduction

The Arctic represents up to 25% of the global land carbon sink due to sequestration of CO_2 by the tundra and boreal forests (McGuire et al. 2009), while the Arctic Ocean accounts for 10–12% of the global ocean CO_2 sink even though it represents only 3% of the surface area of all oceans (MacGilchrist et al. 2014). The Arctic thus has a strong influence on the carbon cycle, while also responding more strongly to changes in climate than do mid-latitude regions. Mean temperatures have increased twice as fast in the Arctic as in the rest of the world, due to Arctic amplification processes (Cohen et al. 2014). As global surface temperatures rise due to anthropogenic emissions of greenhouse gases, sea ice and snow cover shrink, exposing darker ocean and land surfaces that absorb solar radiation and thereby enhance the warming (Pistone et al. 2014).

Higher air temperatures also increase the moisture content of the atmosphere, (a 1 °C temperature increase leads to approximately 7% water vapour increase), causing more warming because water vapour is the largest contributor to the natural greenhouse effect, by a factor of two to three times that of CO_2 (Myhre et al. 2013). The Arctic holds ∼1.5% of the Earth's atmospheric water content, in all its phases. The distribution of water vapour in the Arctic atmosphere is highly dependent on temperature, with the air being dryer in winter compared to summer, and dryer towards the pole (AMAP 2017b). Evaporation rates increase in response to sea ice loss, increasing the water vapour content of the atmosphere as well as cloud cover

(Serreze et al. 2012; Boisvert and Stroeve 2015). Transport from mid-latitudes also contributes to the moistening of the Arctic atmosphere (Zhang et al. 2013; Dufour et al. 2016), which can lead to increased precipitation (Bintanja and Selten 2014). Changes in moisture and temperature will also have an impact on terrestrial carbon sources (Parmentier et al. 2017).

3.3.2 The Arctic Carbon Cycle

Most of Earth's carbon is stored in rocks, but exchanges between the environment and this reservoir are slow, of the order of 0.01–0.1 Gt C per year (*note on conversions: 1 gigaton of carbon (1 Gt C) = 1 petagram of carbon (1 Pg C) = 3.667 Gt CO_2 = 1.338 Gt CH_4 = 1338 Tg CH_4*). The fast carbon cycle involves the exchanges of carbon between the atmosphere, land, oceans, and biosphere, dominated by atmospheric CO_2 (828 Gt C), but including methane (CH_4, 3.7 Gt C), carbon monoxide (CO, 0.2 Gt C) hydrocarbons, black carbon, and organic aerosols (Ciais et al. 2013). The influence of the biosphere on the carbon cycle is evident during the growing season, especially in the Northern Hemisphere, where the boreal forests remove CO_2 from the atmosphere via photosynthesis during the growing season and release it via respiration during the winter (Whitmarsh and Govindjee 1999). This cycle is illustrated in Fig. 3.5, which shows the impact of the growing season on the monthly means of the column-average dry-air mole fraction of CO_2 (XCO_2) over the Northern Hemisphere from the CarbonTracker model (Peters et al. 2007).

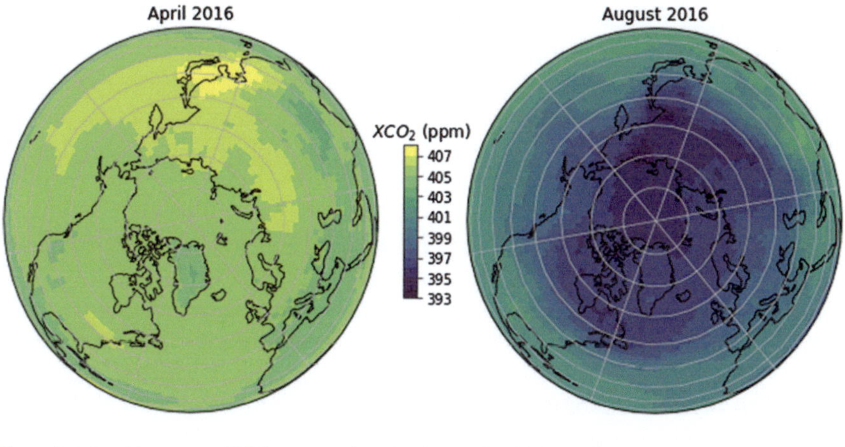

Fig. 3.5 Monthly mean XCO_2 over the Northern Hemisphere from CarbonTracker release CT2017, before (April, left panel) and at the end of (August, right panel) the 2016 growing season. (CarbonTracker data provided by the US National Oceanic and Atmospheric Administration (NOAA) Earth System Research Laboratory (ESRL), http://carbontracker.noaa.gov.)

The main components of the carbon cycle in the Arctic are the boreal forest, the tundra, freshwaters, and the Arctic Ocean. On land, the tundra is an overall sink of CO_2 estimated at 0.11 Gt C/y, however this estimate is associated with large uncertainties, from a 0.29 Gt C/y sink to a 0.08 Gt C/y source (McGuire et al. 2012). The Arctic Ocean is a significant carbon sink with 0.1–0.2 Gt C/y taken out of the atmosphere (Jeansson et al. 2011; Schuster et al. 2013; MacGilchrist et al. 2014), compared to 2.4 ± 0.5 Gt C/y for the global ocean sink (Le Quéré et al. 2018). A net flux of CO_2 from the atmosphere to the ocean is driven by a lower partial pressure of CO_2 in seawater than in air; this flux is more effective at high latitudes because CO_2 solubility is higher at colder temperatures (Wanninkhof 1992). The low partial pressure of CO_2 in surface waters is maintained by several "pump" processes that transport carbon deeper into the ocean (Parmentier et al. 2017). Dissolved CO_2 is photosynthesized by algae and consumed by shell-building organisms; when those die their bodies, and the carbon they contain, sinks to the bottom of the ocean. This is the biological pump (MacGilchrist et al. 2014). When sea ice forms, the water around it becomes highly concentrated in salt due to brine rejection (Worster and Rees Jones Worster and Rees Jones 2015) and thus denser than surrounding water. This cool, salty, dense surface water sinks and with it the dissolved carbon it contains (Rysgaard et al. 2007). Carbon is transported away from the Arctic Ocean by currents, with a net export of 0.225 ± 0.049 Gt C per year (MacGilchrist et al. 2014).

The Arctic is an overall CH_4 source. The principal natural source of CH_4 emissions is from the tundra and freshwater (10–30 Tg CH_4/y), with another 1–17 Tg CH_4/y emitted from the Arctic Ocean (AMAP 2015c). Unlike CO_2, the concentration of CH_4 in ocean surface waters is higher than in the air, even under aerobic conditions that should not favor methane production. This is because anaerobic conditions still exist in the cells of methane-producing archaea and bacteria (Damm et al. 2015). Temperature increases are expected to lead to greater CH_4 emissions from wetlands, however the magnitude of those emissions will also depend on changes in soil moisture and vegetation type (Klapstein et al. 2014). The eight Arctic nations (Canada, Denmark, Finland, Iceland, Norway, Russia, Sweden, USA) emit ~66 Tg CH_4/y, which is close to a fifth of global anthropogenic emissions (AMAP 2015c).

Large carbon pools are present in the Arctic in the form of permafrost, ground in which the temperature remain at or below 0 °C for at least two consecutive years. The size of this carbon pool represents 1400–1850 Gt C (Schuur et al. 2008, 2015; McGuire et al. 2009; Tarnocai et al. 2009; Hugelius et al. 2014), which is about half the global soil carbon content (Tarnocai et al. 2009; Hugelius et al. 2014). Recent emissions from permafrost amount to 0–1 Tg CH_4/y (USEPA 2010). The organic carbon content in the first meter of permafrost soils, most vulnerable to climate change, is shown in Fig. 3.6, which is based on data from the Northern Circumpolar Soil Database (Hugelius et al. 2013a, b). Measurements (Osterkamp 2007; Romanovsky et al. 2010; Brutsaert and Hiyama 2012; Elberling et al. 2013) and model projections (Schuur et al. 2015; Schaefer et al. 2014; Koven et al. 2015a, b; Burke et al. 2017; McGuire et al. 2018) indicate that

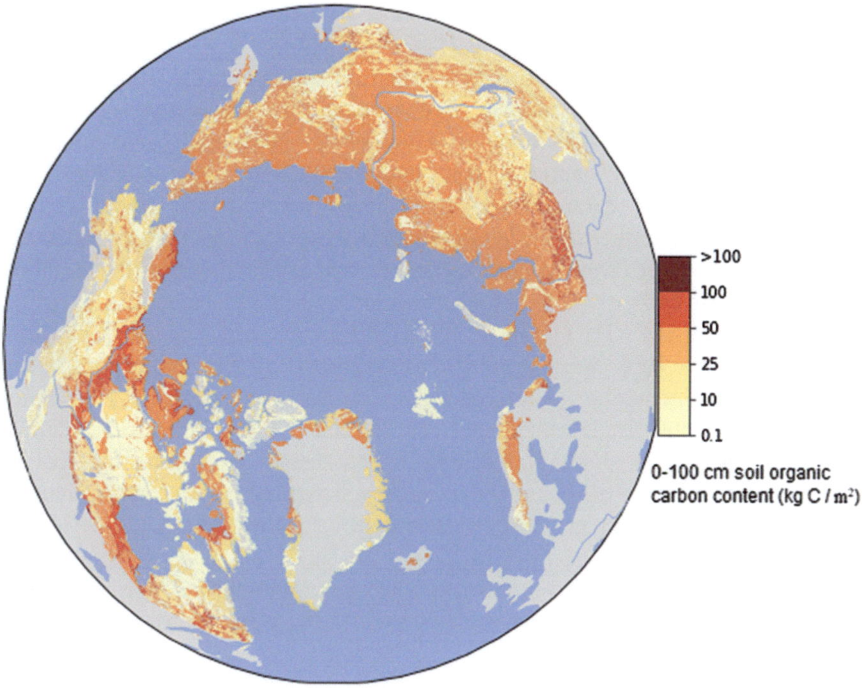

Fig. 3.6 Organic carbon content in the first meter of Arctic permafrost soils. Data from the Northern Circumpolar Soils Database (Hugelius et al. 2013a, b, 2014, https://bolin.su.se/data/ncscd/)

both CO_2 and CH_4 can be released from warming soil and thawing permafrost, with further enhancements possible, for example, due to abrupt thaw beneath thermokarst lakes (Walter Anthony et al. 2018). Decomposition of organic matter in dry aerobic soils leads to emissions of CO_2, while both CO_2 and CH_4 are emitted in wet anaerobic soils. In a permafrost carbon feedback, thawing permafrost would stimulate decomposition and ecosystem respiration, releasing greenhouse gases and contributing to further warming (e.g., Grosse et al. 2011; Belshe et al. 2013; Schuur et al. 2015; Schoolmeester et al. 2019). Based on incubation studies, the release of CO_2 from aerobic environments would be the main driver of that feedback (Schädel et al. 2016).

Increased GHG emissions from soils can be mitigated by enhanced primary production as high-latitude warming and CO_2 fertilization cause changes in plant physiology, availability of nutrients, a longer growing season, and the northward expansion of the boreal forest (Johnson and Kelley 1970; Shaver et al. 1992; Parazoo et al. 2018 and references therein), resulting in a negative feedback to climate change. Models vary widely in their estimates of the permafrost carbon-climate feedback due to uncertainties related to the role of changing soil moisture (Lawrence et al. 2015). Abbott et al. (2016) surveyed carbon emissions from permafrost and found that the main sources of uncertainty were changes in vegetation, water balance, and permafrost degradation. They estimated that the permafrost region would become an atmospheric carbon source by 2100, regardless of anthropogenic emission scenarios. Recent model-based estimates suggest that thawing of permafrost could result in emissions of carbon that could offset the Arctic land sink, but substantial carbon losses would not happen until after 2100 (McGuire et al. 2018). Parazoo et al. (2018) have investigated the impact of cold-season warming at northern high-latitudes on earlier spring thaw, longer non-frozen seasons, and increased depth and seasonal duration of soil thaw, finding that multiple processes govern carbon source-to-sink transitions at high latitudes.

Large amounts of CH_4 (1.146×10^6 Tg CH_4) are stored in ocean hydrates globally, with 116 Pg C north of 60° N (Kretschmer et al. 2015). Increases in sea water temperature could lead to emissions of CH_4 from those deposits. This would first occur in shallow Arctic waters, where most of the CH_4 would be oxidised in the water column (Ciais et al. 2013), resulting in more dissolved CO_2 and contributing to ocean acidification (Elliott et al. 2010). Although hydrate deposits and permafrost represent very large carbon pools, emissions from gas hydrates are likely to be small and spread over long time periods. 4.73 Tg CH_4/y may be emitted into the ocean from the seabed over the next 100 years (Kretschmer et al. 2015), but because of oxidation in the water column, most of those emissions will not reach the atmosphere.

3.3.3 Greenhouse Gas Measurements

The greenhouse gases CO_2 and CH_4 are monitored through national and international networks, and compiled, for example, by the World Data Centre for Greenhouse Gases (https://gaw.kishou.go.jp/). The NOAA/ESRL Global Greenhouse Gas Reference Network collects in situ measurements from 169 sites, 13 of which are in the Arctic (https://www.esrl.noaa.gov/gmd/ccgg/). The Total Carbon Column Observing Network (TCCON) measures column-average dry-air mole fractions of CO_2, CH_4, and other gases using infrared solar absorption spectroscopy, with three stations in the Arctic (https://tccondata.org/, Wunch et al. 2011, 2017).

The atmospheric concentration of CO_2 in the Arctic is increasing in concert with global trends, as seen in Fig. 3.7, which also shows the CO_2 growth rate. The underlying trend is due to anthropogenic emissions, while the sawtooth pattern is caused by the annual cycle of respiration and photosynthesis. The amplitude of this cycle is larger in the Northern Hemisphere than in the Southern Hemisphere due to the greater vegetation coverage, as can be seen in the surface record at Alert included in Fig. 3.7. This amplitude is increasing with time and increases more quickly at higher latitudes as plants respond to rising temperatures (Keeling et al. 1996; Forkel et al. 2016). Aircraft-based measurements of CO_2 indicate that the seasonal amplitude at altitudes of 3–6 km has increased by 50% for 45–90°N since 1958–1961, suggesting an increase of 30–60% in the seasonal exchange of CO_2 by northern extratropical land ecosystems (Graven et al. 2013).

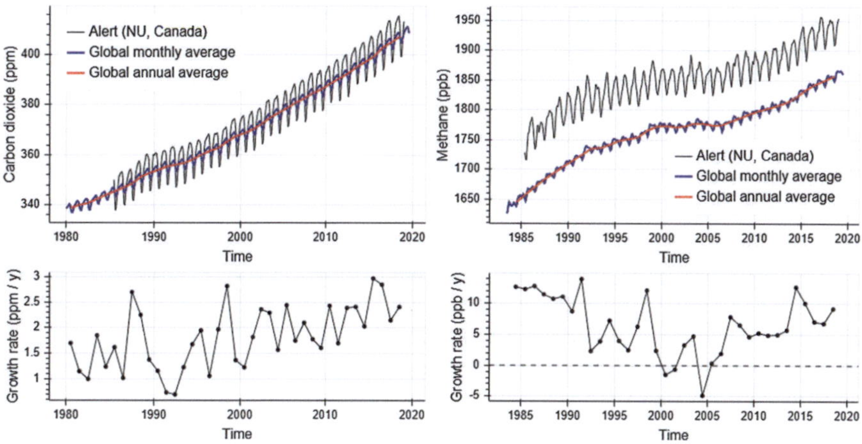

Fig. 3.7 Surface concentrations (top panels), and global atmospheric growth rates (bottom panels) of CO_2 (left) and CH_4 (right). The top panels show the monthly averaged surface flask measurements at Alert, Nunavut, Canada, 82.5°N (black) along with the global monthly average (blue) and global annual average (red) over the period 1980–2017 for CO_2 and 1983–2017 for CH_4. Data from NOAA's Cooperative Air Sampling Network (Dlugokencky 2018; Dlugokencky and Tans 2018; Dlugokencky et al. 2018a, b). Growth rate in ppm/y can be converted to Gt C/y with a factor of 2.12 Gt C/ppm (Prather et al. 2012)

Although CH_4 is about 200 times less abundant than CO_2 in the atmosphere, its contribution to global radiative forcing for the industrial era is $+0.97$ W/m^2, compared to $+1.82$ W/m^2 for CO_2 (Ciais et al. 2013; Myhre et al. 2013). It is also more chemically reactive than CO_2 and has a shorter removal lifetime of ~9 years in the troposphere (Prather et al. 2012), compared to CO_2, which is exchanged with the biosphere on ~5 year timescales, but only removed from the atmosphere/biosphere system on long timescales (>200 years). Figure 3.7 shows monthly averaged CH_4 surface concentration at Alert, along with the global monthly and annual averages. The smaller global concentrations indicate regional differences in the magnitude of sources and sinks. The seasonal variations are driven by concentrations of the OH radical, which lead to minimum CH_4 concentrations in July–August. The timing in emissions from wetlands and biomass burning, and atmospheric transport also affects the seasonal cycle (Dlugokencky et al. 2011).

Global surface concentrations of CH_4 more than doubled between 1750 and 2016, with a 257% increase from 771 to 1853 ppb (WMO 2018a). The growth rate slowed between 1999–2006, but averaged 6.7 ppb/y for 2007–2015 (WMO 2017). The reasons for these recent variations have been a topic of intense interest and numerous publications. Suggestions have included increased biogenic emissions from wetlands and agriculture in tropical regions (e.g., Nisbet et al. 2016), anthropogenic emissions from oil and gas production (e.g., Hausmann et al. 2016), and changes in OH concentration (e.g., Prather and Holmes 2017; Rigby et al. 2017; Turner et al. 2017).

Global and regional GHG emission budgets are built using two approaches. The bottom-up approach relies on extrapolating samples of measured emissions (natural sources and sinks) and inventory-based data (anthropogenic sources) to regional and global scales. The top-down inversion approach combines measurements of atmospheric concentrations with transport and chemistry models to infer sources and sinks. It is difficult to partition sources with the top-down approach, but it provides constraints on regional scales and can help determine if emission estimates from the bottom-up approach are over- or under-estimated (Kirschke et al. 2013). However, the Arctic is poorly sampled, and the lack of observational coverage introduces biases that make the estimates of carbon sources and sinks in this region uncertain (Ciais et al. 2014; Schimel et al. 2015).

Surface fluxes are derived from trace gas concentration and wind measurements on flux towers, using the eddy covariance method (Foken et al. 2012; Holl et al. 2019). FLUXNET is a major global network of such measurements with more than 500 sites, but with only 20 north of 60°N (FLUXNET 2015). Focussing on Alaska, Parazoo et al. (2016) demonstrated that current CO_2 observing capabilities are unlikely to detect CO_2 release from thawing permafrost thaw and changes in cold season respiration, and recommended year-round, vertically resolved, and spatially distributed sampling to quantify the changing carbon balance in the Arctic – Boreal Zone. Better spatial coverage of greenhouse gases is obtained from satellite observations, including the Scanning Imaging Absorption Spectrometer for Atmospheric Cartography (SCIAMACHY, Bovensmann et al. 1999; Schneising et al. 2011; Tan et al. 2016), the Atmospheric Infrared Sounder (AIRS, Aumann et al. 2003; Xiong et al. 2008; Boisvert and Stroeve 2015), the Tropospheric Emission Spectrometer (TES, Beer et al. 2001; Payne et al. 2009; Kulawik et al. 2010), Infrared Atmospheric Sounding Interferometer (IASI) (Xiong et al. 2013), the Greenhouse Gases Observing Satellite (GOSAT, Kuze et al. 2009; Belikov et al. 2014; Holl et al. 2016), the Orbiting Carbon Observatory (OCO-2, Crisp 2015; Eldering et al. 2017), and and the TROPOspheric Monitoring Instrument (TROPOMI, Hu et al. 2018). However, satellite retrievals over high albedo surfaces, like the snow and ice-covered regions of the Arctic, can be challenging (Merrelli et al. 2015).

The Carbon Arctic Reservoirs Vulnerability experiment (CARVE) used airborne measurements to estimate terrestrial fluxes over Alaska (Chang et al. 2014). Airborne measurements by CARVE were also combined with eddy covariance tower measurements to determine that cold season CH_4 emissions represent more than half of annual CH_4 emissions from the Alaskan tundra (Zona et al. 2016). Miller et al. (2016) used CARVE observations and a simple flux model to highlight that global-scale process models may underestimate CH_4 emissions from wetlands. Commane et al. (2017) also combined CARVE measurements with tower measurements and showed that Alaska was a net source of CO_2 in 2012–2014 because of early winter respiration. Similarly, the Arctic-Boreal Vulnerability Experiment (ABoVe) aims to combine satellite data with airborne and surface measurements over Alaska and northwestern Canada to enhance our understanding of the evolution of the Arctic and boreal ecosystems in a changing environment (Fisher et al. 2018).

3.3.4 Current Issues

Many processes that lead to carbon emissions and exchange in the Arctic are poorly understood and the region is sparsely sampled. This contributes to large uncertainties in the quantification of carbon stocks such as gas hydrates and subsea permafrost, and of current Arctic sources and sinks of carbon. Future emissions from thawing permafrost are difficult to estimate because they will depend on changes to the Arctic hydrological cycle and vegetation. Sea ice loss could be a large contribution to these changes as it will lead to warmer and wetter environments (Parmentier et al. 2017). There are few observations during the cold season, which may result in an underestimation of the sources of carbon in the Arctic (Zona et al. 2016; Pirk et al. 2016; Webb et al. 2016). Continued monitoring of the Arctic is essential to better understand the various processes controlling the Arctic carbon cycle and predict future changes. Models currently do not reproduce the observed sea ice decline (Li et al. 2017; Rosenblum and Eisenman 2017) and this may lead to an underestimation of the evolution of Arctic amplification. Satellite missions can sample regions otherwise inaccessible but must overcome systematic retrieval errors associated with measurements over the high albedo surfaces of the Arctic. Integration of ground-based data with airborne observations can lead to better regional estimates of sources and sinks of carbon and to improved understanding of the underlying processes that control them.

3.4 Stratospheric Ozone

3.4.1 Introduction

Measurements of total column ozone in the Arctic date back to the late 1920s, when early versions of the Dobson spectrophotometer were deployed at several locations (Dobson 1968). Measurements made at Spitzbergen by Götz in the summer of 1929 provided confirmation that ozone columns increased with latitude and decreased in the summer, and also led to the development of the Umkehr method to obtain the vertical distribution of ozone (Gotz Götz 1931; Gotz et al. Götz et al. 1934; Dobson 1968; Nicolet 1975). Over the following two decades, Arctic ozone measurements were made in Europe on a campaign basis (Brönnimann et al. 2003). However, with the creation of the International Ozone Commission in 1948, followed by International Geophysical Year in 1957–1958, an international total ozone monitoring network was gradually established, which is now overseen by the World Meteorological Organization/Global Atmosphere Watch, with data available from the World Ozone and Ultraviolet Radiation Data Centre (https://woudc.org/). The discovery of the Antarctic ozone hole in the early 1980s (Farman et al. 1985) led to intensive studies of stratospheric ozone chemistry over both polar regions, including multiple field campaigns and the establishment of the Arctic stations in the Network for the Detection of Stratospheric Composition (now the Network for the Detection of Atmospheric Composition Change, NDACC, www.ndacc.org, de Mazière et al. 2018). Current understanding of Arctic stratospheric ozone concentrations and chemistry is now based on measurements using multiple techniques from the ground, sondes, aircraft, and satellites (McElroy and Fogal 2008).

The first photochemical theory for formation and destruction of ozone was based on an oxygen-only chemical scheme and was proposed by Chapman (1930). During the following decades, it was discovered that stratospheric ozone was also chemically destroyed by catalytic reaction with hydrogen (Bates and Nicolet 1950; Hampson 1964), nitrogen oxides (Crutzen 1970, 1971; Johnston 1971), chlorine (Molina and Rowland 1974; Rowland and Molina 1975; Stolarski and Cicerone 1974), and bromine (Wofsy et al. 1975; Yung et al. 1980; McElroy et al. 1986). During these reactions, the initiating active species (e.g., ClO, BrO) are regenerated, and even small amounts of these gases can significantly deplete ozone. A very important step was the discovery that the anthropogenic chlorofluorocarbons (CFCs) are a major source of ozone-destroying stratospheric chlorine (Molina and Rowland 1974). A global ozone loss of 5–19% was estimated at that time for the following 50–100 years (Wuebbles 1983; Cicerone et al. 1983; Prather et al. 1984), but no one predicted that the Antarctic stratosphere was more likely than any other place on the globe to experience ozone depletion.

Figure 3.8 illustrates the distribution of monthly mean total column ozone in the Northern Hemisphere, using 2017 as an example. During the winter, air moves

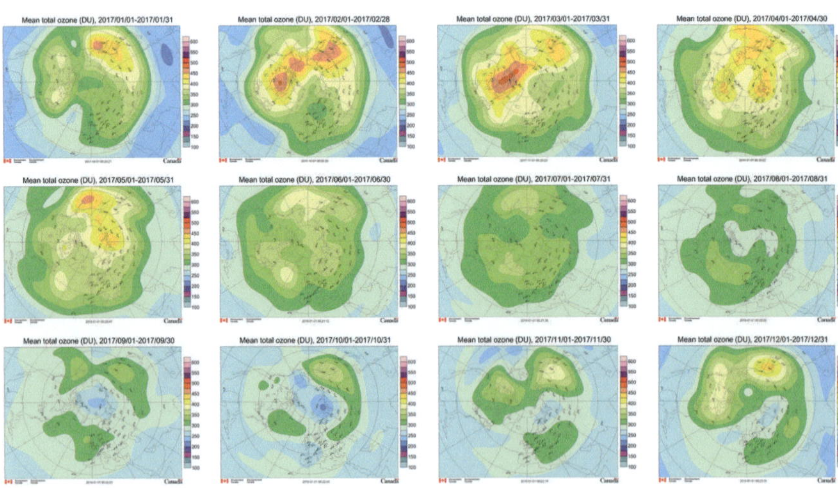

Fig. 3.8 Monthly mean total ozone columns for 2017 over the Northern Hemisphere. The total ozone maps are provided by Environment and Climate Change Canada (http://exp-studies.tor.ec.gc.ca/cgi-bin/selectMap) through the World Ozone and Ultraviolet Radiation Data Centre (https://woudc.org/) and are based on ground-based and satellite measurements. Total ozone values are given in Dobson units, with each number indicating observations taken from a ground station location corresponding to the bottom left corner of the number

toward the pole and descends, bringing ozone rich-air from the tropics. Ozone builds up, its concentration being the highest in February–March. However, under certain conditions, stratospheric ozone can be chemically depleted. In late spring, April–May, the polar air mixes with air from mid-latitudes, as the polar vortex (defined below) dissipates. Ozone total columns decrease through the summer, and reach minimum values by the fall. The Arctic stratospheric polar vortex in wintertime is more dynamically disturbed than that in Antarctica, resulting in higher temperatures and greater transport of ozone into the polar regions. This is why ozone losses observed in the Arctic are usually less severe than those observed in Antarctica. In this section, the dynamical and chemical processes controlling the Arctic stratospheric ozone budget are described and projections for ozone recovery are discussed.

3.4.2 Dynamical Processes in the Arctic Stratosphere

Atmospheric circulation plays an important role in the vertical and latitudinal distribution of trace gases characterized by relatively long atmospheric lifetimes, including the distribution of ozone in the Arctic stratosphere. The meridional circulation of the stratosphere, known as the Brewer-Dobson circulation, was identified from observations of stratospheric water vapour and ozone (Brewer 1949; Dobson 1956; Butchart 2014; Strahan 2015). In the lower stratosphere, the circulation consists of a two-structure cell characterized by a rising motion across the tropical tropopause, a slow stratospheric drift towards the middle and high latitudes, and descending motion through the tropopause at mid-latitudes and in the polar vortex (defined below). In the middle and upper stratosphere, the circulation consists of a single cell from the tropics into the winter hemisphere where planetary wave breaking occurs. A tropospheric return flow results from mass conservation. In the summer hemisphere, the circulation in the upper stratosphere is poleward and upward. The Brewer-Dobson circulation is driven by the breaking and dissipation of atmospheric waves (see Butchart 2014 and references therein). Figure 3.9 illustrates the Brewer-Dobson circulation and transport pathways in the middle atmosphere. There is a clear difference between the summer and winter hemispheres, with the wintertime stratosphere dominated by the large-amplitude, planetary-scale Rossby waves propagating upward from the troposphere. In the "surf zone" (McIntyre and Palmer 1983), these waves break intermittently, stirring the air across large distances of the winter stratosphere. Mixing of chemical constituents by the breaking Rossby waves is an efficient transport mechanism in the surf zone.

During polar night, when there is 24-hour darkness, the air above the pole becomes extremely cold. At the boundary between the cold polar air and the warmer air from lower latitudes, polar jets usually form. There is a strong latitudinal thermal gradient, which constrains the air to flow horizontally, and because of Earth's rotation, the air speeds up and transforms into a polar night jet stream (fast-flowing, narrow air current) or polar vortex. This comprises strong eastward winds encircling the pole (Andrews et al. 1987; Schoeberl and Newman 2015). Acting like a barrier,

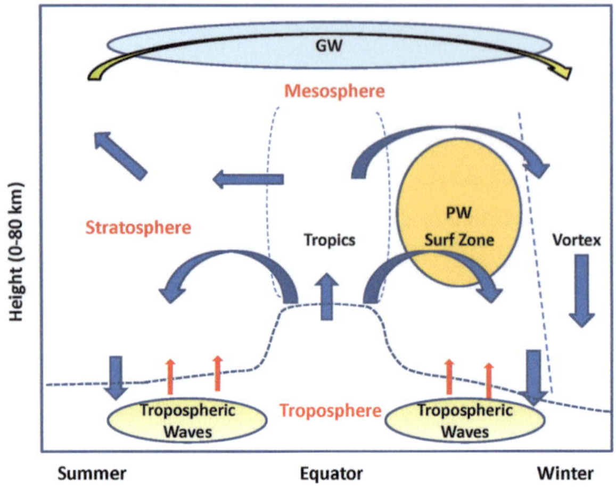

Fig. 3.9 Schematic diagram of the mean meridional circulation in the middle atmosphere. Transport by the Brewer-Dobson circulation is indicated by the blue arrows. The pole-to-pole transport in the mesosphere is shown by the yellow arrow, where gravity waves (GWs) are the primary source of wave energy. The year-round source of tropospheric wave energy propagating into the stratosphere is depicted by red arrows. Regions with restricted two-way transport ('mixing barriers') are indicated by vertical dotted lines. The dashed thick blue line above the troposphere is the tropopause. (Reproduced from Figure 1, Strahan 2015)

these very fast winds (~80 m/s at ~60 km altitude in the Northern Hemisphere) isolate the air inside the vortex from mid-latitude air masses (Schoeberl and Hartmann 1991; Wallace and Hobbs 2006). The barrier is impermeable to a first approximation, but not perfect, as shown by long-lived tracers whose occurrence inside the polar vortex was explained by mixing of surf zone air into the vortex (Michelsen et al. 1998; Kondo et al. 1999; Rex et al. 1999; Plumb et al. 2000). The temperature inside the vortex drops as the air cools radiatively, sometimes reaching values of 195 K or less. These temperatures promote the formation of polar stratospheric clouds described below.

The topographical features of the Northern Hemisphere polar region lead to stronger wave activity in the Northern Hemisphere compared to the Southern Hemisphere, where Antarctica is surrounded by water. As result, the Northern Hemisphere polar vortex is usually weaker and highly variable compared to the Southern Hemisphere. The vortex is sometimes interrupted by sudden stratospheric warmings, which are driven by the breaking of planetary waves propagating up from the troposphere and involve a rapid increase of stratospheric temperatures at high latitudes. In a major sudden stratospheric warming, the temperature enhancement leads to reversal of the mean zonal winds (as a result of the inversion of the mid-latitude-to-pole temperature gradient) from eastward to westward at heights near 30 km (*e.g.*, Chshyolkova et al. 2007; Kim and Flatau 2010; Butler et al. 2015 and references therein).

3.4.3 Chemical Processes in the Arctic Stratosphere

The chemical processes controlling the stratospheric ozone budget and polar ozone depletion are well documented in various reviews (e.g., Wardle et al. 1997; Solomon 1999; Staehelin et al. 2001; McConnell and Jin 2008; McElroy and Francis 2007; Solomon et al. 2007; Zerefos et al. 2009; Fergusson 2010; Godin-Beekmann et al. 2018; Wilmouth et al. 2018) and in the ozone assessment reports of the World Meteorological Organization (1985, 1988, 1989, 1991, 1994, 1998, 2003, 2007, 2011, 2014; 2018b).

After the discovery of significant ozone depletion in the Antarctic spring stratosphere, now known as the 'ozone hole' (Farman et al. 1985; Komhyr et al. 1986; Schoeberl et al. 1986; Solomon et al. 1986; Solomon et al. 1987a; b; Stolarski et al. 1986; de Zafra et al. 1987; Farmer et al. 1987; Krueger et al. 1987), attention turned to the Arctic. In the winter of 1987–1988, near-UV lunar measurements of OClO (Solomon et al. 1988) and scattered-sunlight and direct-moon observations of O_3 and NO_2 (Mount et al. 1988) at Thule, Greenland confirmed that the Arctic stratosphere was also chemically perturbed, indicating that ozone loss was possible via the mechanisms identified in Antarctica. The first major Arctic field campaign to focus on understanding stratospheric ozone destruction was the Airborne Arctic Stratospheric Experiment (AASE, Turco et al. 1990), which took place in 1989. It was followed by the European Arctic Stratospheric Ozone Experiment (EASOE) in the winter of 1991–1992 (Pyle et al. 1994); the Second European Stratospheric Arctic and Mid-Latitude Experiment (SESAME) in 1994–1995 (Farman et al. Farman et al. 1998); the Improved Limb Atmospheric Spectrometer (ILAS) validation balloon campaign in February–March 1997 (Kanzawa et al. 2003); the Photochemistry of Ozone Loss in the Arctic Region In Summer (POLARIS) campaign in 1997 (Newman et al. 1999); two major field experiments, the SAGE III Ozone Loss and Validation Experiment (SOLVE) and the Third European Stratospheric Experiment on Ozone (THESEO-2000), between 1999 and 2000 (Newman et al. 2002); and the second SAGE Ozone Loss and Validation (SOLVE II) and Validation of INTERnational Satellites and Study of Ozone Loss (VINTERSOL) campaign in 2002–2003 (Carslaw 2004). More recently, the "Reconciliation of essential process parameters for an enhanced predictability of Arctic stratospheric ozone loss and its climate interactions" (RECONCILE) project combined field and satellite measurements in 2009–2010 and 2010–2011, with laboratory and modelling studies, with a focus on improving the representation of polar ozone loss processes in global climate models (von Hobe et al. 2013).

These campaigns recorded a wealth of information about the Arctic stratosphere from instruments deployed on the ground, balloons, and aircraft. For example, vertical profile information for chlorine, bromine and nitrogen species was obtained from early balloon-borne in situ measurements (Arnold and Spreng 1994; Fabian et al. 1994; Kondo et al. 1994; Murcray et al. 1994; Pommereau and Piquard 1994a, b; Schmidt et al. 1994). Measurements made by aircraft instruments provided column amounts of O_3, HCl, ClO, and HNO_3 (e.g., Blom et al. 1994; Crewell et al. 1994).

Early ground-based measurements used millimeter-wave emission (de Zafra et al. 1994; Emmons et al. 1994), lidar (Beyerle et al. 1994; Neuber et al. 1994), UV-visible spectroscopy (Fish et al. 1994; Goutail et al. 1994), and Fourier transform infrared spectroscopy (Notholt 1994; Notholt et al. 1993, 1995, 1997).

During the AASE, polar stratospheric clouds (PSCs) were studied intensively (Hofmann 1990; Kruger 1990). PSCs occur between ~12 and 25 km and are associated with cold temperatures, below ~200 K, and they are critical to polar ozone depletion. It was shown that PSC type I (small, HNO_3-rich particles) and PSC type II (primarily large H_2O-ice particles and minor amounts of HNO_3 as hydrates) were both present in the Arctic stratosphere (Browell et al. 1990; McCormick et al. 1990). At that time, the composition of the PSCs was not well understood. Type I PSCs were classified into two forms: those containing solid particles (nitric acid trihydrate) and those containing liquid particles (supercooled ternary solutions of nitric acid, sulfuric acid and water) (Browell et al. 1990; Toon et al. 1990b). We now know that large nitric acid trihydrate particles ('NAT rocks') with diameters of tens of microns are a feature of the Arctic winter lower stratosphere (Fahey et al. 2001; Fueglistaler et al. 2002; Northway et al. 2002; Brooks et al. 2003; von Hobe et al. 2013).

Sporadic and limited denitrification and dehydration were also observed, in contrast with Antarctica (Fahey et al. 1990; Kawa et al. 1990; Khattatov et al. 1994; Kondo et al. 1994; Oelhaf et al. 1994; Rinsland et al. 1996; Arnold et al. 1998), and in some cases denitrification was seen without accompanying dehydration. This raised new questions regarding the mechanism of the denitrification process (Toon et al. 1990a; Gandrud et al. 1990; Salawitch et al. 1989; Koop et al. 1995). Another contrast to Antarctica was the rapid formation of $ClONO_2$ (consistent with the ClO decrease) by recombination of ClO with NO_2. NO_2 was photochemically released from HNO_3, after its release from type I PSCs upon their evaporation (Schoeberl et al. 1993; Webster et al. 1993). In Antarctica, the active chlorine was seen to convert into HCl through reaction with CH_4, since nitrogen was lost by denitrification (Toohey et al. 1993; Webster et al. 1993).

Building on these early investigations, subsequent measurements from campaigns, ground-based observatories, and satellite instruments (see summary in Appendix 3A, WMO 2014), combined with laboratory studies and atmospheric modelling, have furthered elucidated the processes controlling Arctic stratospheric ozone chemistry and chemical ozone depletion, summarized briefly as follows. These processes involve the formation of the winter polar vortex, which isolates cold dark air over the polar regions, allowing polar stratospheric clouds to form if temperatures are sufficiently cold, and dehydration and denitrification to remove water vapour and nitrogen oxides that would otherwise react with and neutralize chlorine. Chlorofluorocarbons (CFCs) transported to the polar stratosphere enhance the concentrations of chlorine and other halogen species. PSCs provide surfaces for the fast heterogeneous reactions that convert inactive chlorine (HCl and $ClONO_2$) to reactive Cl_2, and when sunlight returns in the spring, UV radiation breaks Cl_2 apart to form Cl, which is then available to destroy ozone through catalytic cycles in the lower stratosphere initiated by the ClO + ClO reaction. The release of bromocarbons, naturally occurring bromine-containing very short-lived substances (VSLSs), and the coupling of bromine and chlorine chemistry initiated by the BrO+ClO reaction also play an important role in the chemical depletion of ozone (Wofsy et al. 1975; Yung et al. 1980; Tung et al. 1986; McElroy et al. 1986; Daniel et al. 1999; Yang et al. 2014). These catalytic cycles continue until the Sun causes a dynamical breakdown of the winter vortex and PSCs evaporate.

Lower stratospheric temperatures in the Arctic exhibit high interannual variability due to strong wave activity in the Northern Hemisphere, sudden stratospheric warmings (Labitzke and van Loon 1999), and hence a weaker polar vortex. This variability in turn causes year-to-year variability in stratospheric ozone (Strahan et al. 2016), as column ozone loss can be related to the time-integrated volume of air within the vortex, between the 400 K and 700 K isentropic surfaces, where temperatures are cold enough to activate chlorine on PSCs (Rex et al. 2004; Tilmes et al. 2004). Figure 3.10 shows the mean springtime total column ozone averaged over 63°-90° latitude for the Arctic (March) and Antarctic (October) derived from satellite measurements dating back to 1970. The air poleward of 63°N can include both vortex and extravortex air, with the year-to-year differences in the Arctic due to both dynamical effects and chemical depletion (Petzoldt 1999; Tegtmeier et al. 2008; WMO 2014; Strahan et al. 2016). In comparison, Fig. 3.11 shows the

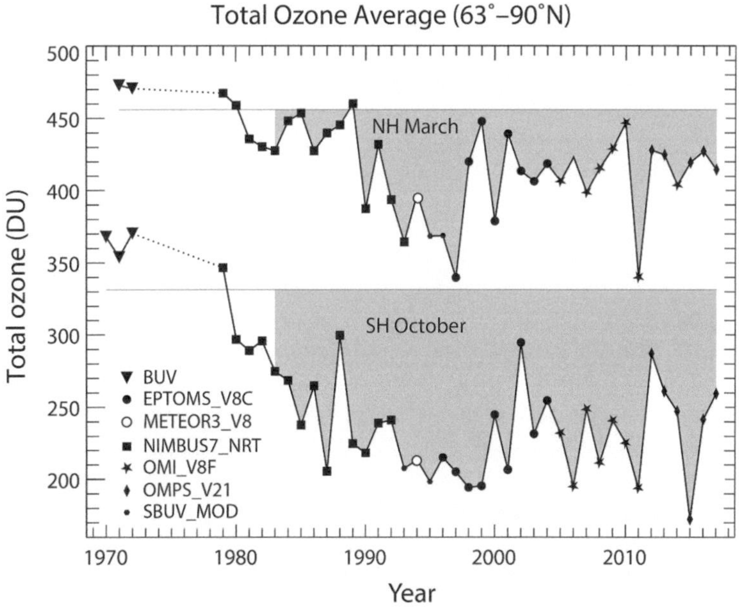

Fig. 3.10 Time series of total ozone column averaged over 63–90° latitude in March for the Northern Hemisphere (NH) and in October for the Southern Hemisphere (SH). Symbols indicate satellite datasets used in different years. The horizontal gray lines represent the average total ozone prior to 1983 in March for the NH and in October for the SH. (Reproduced from Fig. 4–4, WMO 2018b)

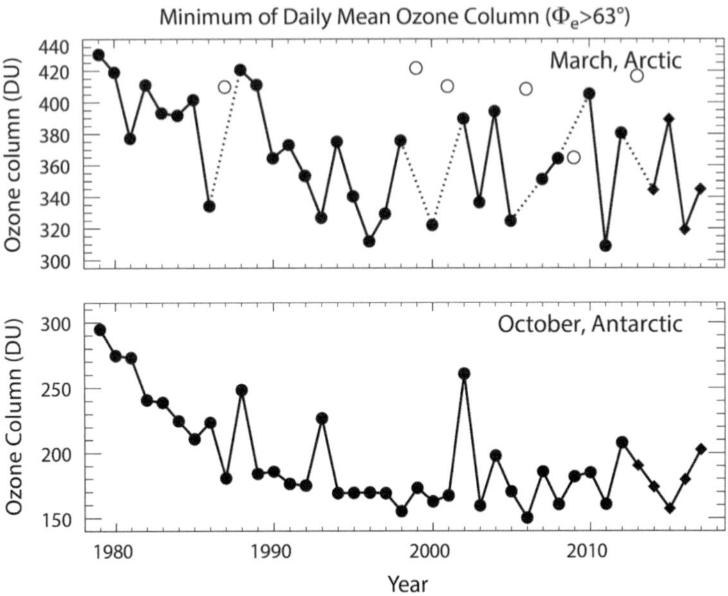

Fig. 3.11 Time series of the minimum of the daily average total ozone column within the 63° contour of equivalent latitude (Φe) in March for the Arctic and in October for Antarctica. Arctic winters in which the polar vortex broke up before March (1987, 1999, 2001, 2006, 2009, and 2013) are shown by open symbols; dotted lines connect surrounding years. Adapted from WMO (2014), updated using the Bodeker Scientific combined total column ozone data base (version 3.0; circles; Müller et al. 2008) through March 2013 and Aura OMI measurements thereafter (diamonds). (Reproduced from Fig. 4–5, WMO 2018b)

minimum daily average total ozone within the 63° contour of equivalent latitude; this provides a better indication of the position of the polar vortex, but both chemical ozone destruction and dynamical effects can contribute to the low column ozone values in this region.

Figures 3.10 and 3.11 show the downward trend in Arctic spring ozone columns in the late 1980s and early 1990s, as well as the year-to-year variability. Chemical ozone destruction typically accounts for one-third of the difference from the pre-1983 mean, with the remainder determined by the dynamical resupply of ozone (Strahan et al. 2016; WMO 2018b). The largest chemical ozone losses in the Arctic were observed in 1997, 2000, 2005, 2011, and 2016 (WMO 2007, 2011, 2014; Rex et al. 2006; Manney et al. 2011; Khosrawi et al. 2017; Wargan and Nielsen 2017). The 2010–2011 winter was particularly notable for its unusually weak tropospheric planetary wave activity and late final warming, which allowed the vortex to remain strong, stable, and cold for an extended period (Hurwitz et al. 2011; Isaksen et al. 2012; Strahan et al. 2013). This resulted in prolonged cold temperatures below the PSC threshold and unprecedented ozone destruction (Manney et al. 2011; Sinnhuber et al. Sinnhuber et al. 2011; Adams et al. 2012; Arnone et al. 2012; Lindenmaier et al. 2012; Pommereau et al. 2013). The minimum total ozone in spring 2011 was continuously below 250 DU for ~27 days, then dropped to ~220–230 DU for about a week in late March 2011 (Manney et al. 2011).

To illustrate the unusual conditions in 2011, Fig. 3.12 shows HNO_3, HCl, $ClONO_2$, and O_3, total columns normalized by the total columns of HF, as measured at Eureka, Canada (80°N, 86°W) from 1997 to 2018 using Fourier transform infrared (FTIR) solar absorption spectroscopy, updated from Lindenmaier et al. (2012). Normalizing with HF reduces the dynamical effects, thus highlighting interannual differences due to chemistry. Panel (e) shows the evolution of the scaled potential vorticity (sPV) on measurement days in 2011, derived at locations along the instrument line-of-sight for the 18, 20, 22, and 26 km altitude levels (Dunkerton and Delisi 1986; Manney et al. 2007 and references therein). For all these levels, the sampled air masses were well inside the vortex in 2011 from the start of the measurements until day 78 (March 19) and again on day 86 (March 27) and days 95 and 96 (April 5 and 6). The gray shading highlights the days in 2011 when the instrument sampled air masses on the edge or outside the vortex. The low values of normalized HNO_3, HCl, and $ClONO_2$ that occurred on days 68–71, followed by the minimum in normalized O_3 that occurred on day 77, are indicative of chlorine activation on cold aerosol particles followed by chemical ozone destruction. While several other years (1997, 2000, 2005, 2007, 2016) also show evidence of chlorine activation and chemical ozone loss, 2011 stands out as having the lowest normalized ozone values.

The 2015–2016 Northern Hemisphere winter stratosphere had the greatest potential yet seen for chemical ozone depletion, with temperatures in the lower stratosphere at record lows from December 2015 to early February 2016, exceptional denitrification and dehydration, and extensive chlorine activation throughout the polar vortex (Manney and Lawrence 2016). Ozone decreases began earlier and continued more rapidly than those in 2010–2011, but a major final sudden warming on 5–6 March terminated chlorine activation. Figure 3.13 shows maps of chlorine reservoir species, active chlorine, and ozone on the 480 K potential temperature surface (~ 20 km altitude) over the Northern Hemisphere on 11 January 2016, at the beginning of ozone depletion, and on 21 March, after the breakup of the polar vortex (Wargan and Nielsen 2017). In January, the low concentrations of $ClONO_2$ and HCl (reservoir species), combined with high amounts of ClO_x within the polar vortex, indicate chlorine activation, however ozone loss is not significant, as the depletion process starts when the vortex becomes illuminated by sunlight. In March, when there is only a remnant of the polar vortex left, $ClONO_2$ has elevated concentrations, ClO_x is almost gone, and ozone levels are low inside the vortex remnant.

3.4.4 Ozone Recovery

The Montreal Protocol on Substances that Deplete the Ozone Layer and its subsequent Amendments have resulted in reductions of ozone-depleting substances (ODSs) in the atmosphere over the last two decades, with an associated decline

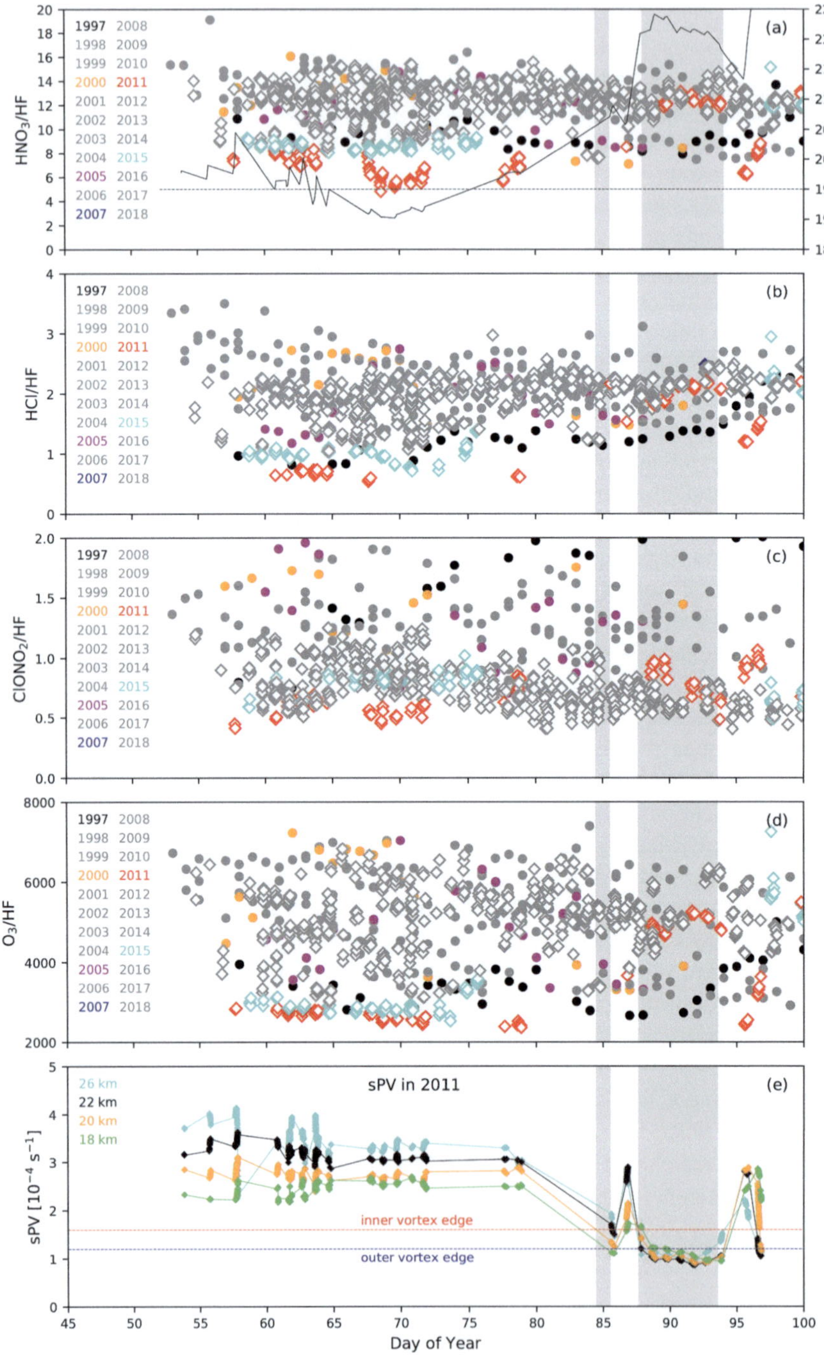

Fig. 3.12 Spring time series of daily averaged (**a**) HNO_3, (**b**) HCl, (**c**) $ClONO_2$, and (**d**) O_3, total columns, normalized by HF total columns, measured at Eureka, Canada from 1997 to 2018 using

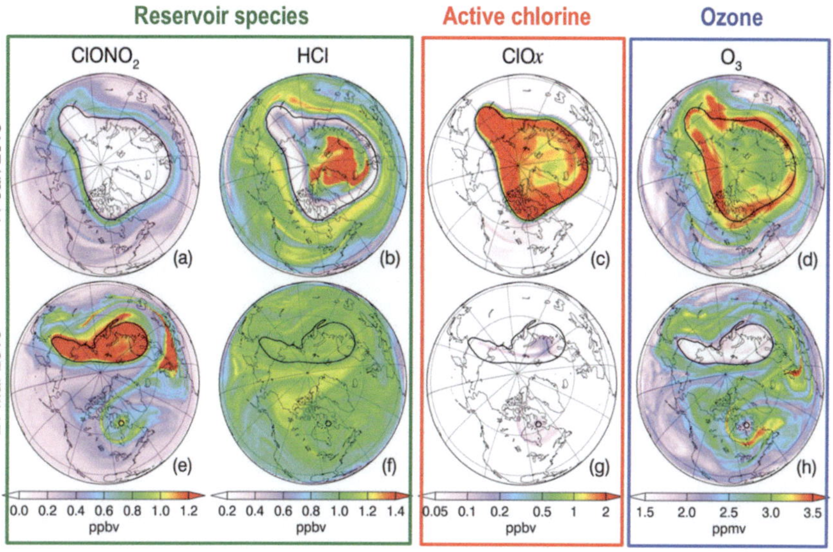

Fig. 3.13 Maps of $ClONO_2$ (**a** and **e**), HCl (b and f), ClOx (**c** and **g**), and O_3 (**d** and **h**) on the 480 K potential temperature surface on 11 January 2016 (**a–d**) and 21 March 2016 (**e–h**). The black contour is the edge of the polar vortex. The maps were generated using the Goddard Earth Observing System Data Assimilation System with a stratospheric chemistry module (STRATCHEM). The observations assimilated by the system consist of radiance data from satellite-borne sensors, conventional measurements and ozone observations. (Reproduced from Figure 4, Wargan and Nielsen 2017. Used by permission of NASA)

in Equivalent Effective Stratospheric Chlorine (EESC, the sum of chlorine and bromine derived from ODS tropospheric abundances weighted to reflect their expected depletion of stratospheric ozone). However, the long lifetimes of ODSs mean that stratospheric ozone depletion will continue for several decades. Meanwhile, ozone recovery is also linked to climate change, as increases in CO_2, CH_4, and N_2O cause changes in temperature, chemistry, and the circulation of the stratosphere, all of which affect ozone. Ravishankara et al. (2009) showed that

Fig. 3.12 (continued) FTIR spectroscopy. Filled/open circles represent ratios obtained using Bomem DA8/Bruker 125HR spectrometers. The colours correspond to different years as indicated in the legend. Panel (**a**) also shows the evolution of the temperature at 21 km along the line-of-site for the 2011 measurements at Eureka (black solid line) and the threshold temperature value for the formation of type I PSCs (black dotted line). Panel (**e**) shows the evolution of scaled potential vorticity in 2011 along the line-of-sight for the 18, 20, 22, and 26 km altitude levels, colour-coded as shown in the legend; sPV values are only plotted for days in 2011 when FTIR measurements were acquired. The horizontal lines show the inner (red) and the outer (blue) vortex edge in 2011, respectively. The gray shading highlights the days when the instrument sampled air masses on the edge or outside the vortex in 2011. (Figure courtesy of Erik Lutsch, updated from Lindenmaier et al. 2012)

N_2O, an important ozone-depleting and greenhouse gas, is the most important ODS emitted today. While increasing concentrations of CO_2 are cooling the stratosphere and thereby reducing global ozone loss rates, this cooling may generate more polar stratospheric clouds and enhance ozone depletion in the Arctic (e.g., Eyring et al. 2010).

Currently, it is not possible to discern a recovery in Arctic ozone in recent years above the dynamical variability (Knibbe et al. 2014; Solomon et al. 2016; Weber et al. 2018). However, chemistry-climate model runs from the Chemistry-Climate Model Initiative (CCMI, Eyring et al. 2013) predict that the Arctic total column ozone in March will recover to 1980 values between 2025 and 2043, compared with projected recovery dates (1σ) of 2055–2066 for Antarctica in October, 2020–2044 for the Northern Hemisphere, and 2043–2055 globally (Dhomse et al. 2018; WMO 2018b). The earlier recovery in the Northern Hemisphere is due to the greater influence of dynamics in this hemisphere.

CCMI model runs have been performed to quantify the relative effects of ODSs and GHGs on future Arctic ozone (e.g., Dhomse et al. 2018; Morgenstern et al. 2018). Figure 3.14 shows Arctic spring total column ozone simulations based on the work of Dhomse et al. (2018). The model simulations with increasing GHG and fixed ODS concentrations show a steady increase in ozone, which is attributed to the GHGs cooling the stratosphere and thus slowing down gas-phase ozone-destroying reactions while also increasing the poleward and downward transport of ozone by strengthening the Brewer-Dobson circulation (e.g., Oman et al. 2010; Oberländer et al. 2013). In contrast, the model simulations with fixed GHG and decreasing ODS concentrations show ozone diverging from the reference simulation around

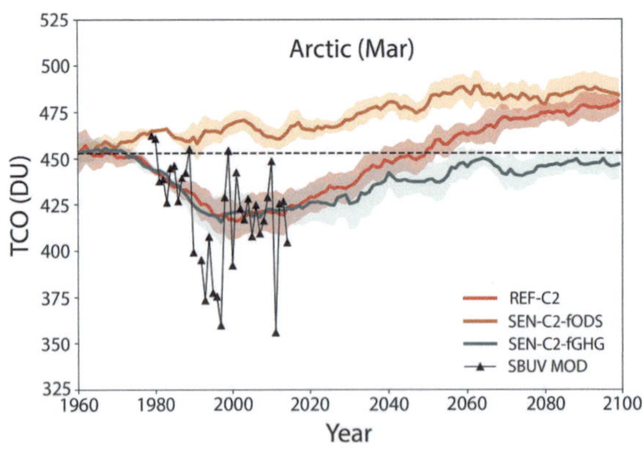

Fig. 3.14 Temporal evolution of multi-model means of total column ozone for Arctic spring (March). Results are shown for three CCMI scenarios. REF-C2 simulations use the most likely scenario of future ODS and GHG changes, SENC2-fODS simulations use fixed ODS levels from 1960, and SEN-C2-fGHG simulations use fixed GHG levels from 1960. Black triangles indicate observations derived from the Solar Backscatter Ultraviolet (SBUV) merged ozone dataset (MOD) (Frith et al. 2017). The black dashed line denotes the 1960 reference value. (Adapted from Figure 4–19, WMO 2018b)

2020, after which it returns to 1960 values. The model simulations that include the most likely scenario of future ODS and GHG changes have ozone increasing beyond 1960 values in the middle of the twenty-first century. These results indicate that previous Arctic ozone trends have primarily been controlled by ODSs but that GHGs will play an increasingly important role in the future. Other model studies have investigated the altitude dependence of Arctic ozone trends, showing that ODS changes primarily affect the lower stratosphere, while GHG changes primarily influence the upper stratosphere (Rieder et al. 2014; Douglass et al. 2014; Kirner et al. 2015).

Chemistry-climate models predict that higher GHG concentrations will cool the Arctic winter mid- and upper stratosphere (e.g., Oberländer et al. 2013; Rieder et al. 2014). There is less confidence in the projected temperature trends in the Arctic lower stratosphere (e.g., Langematz et al. 2014; Rieder et al. 2014; Bednarz et al. 2016) and hence in whether V_{PSC}, the volume of air inside the vortex at temperatures below the nitric acid trihydrate PSC formation threshold, is increasing as has been previously suggested (Rex et al. 2004, 2006). More recent studies indicate that multidecadal variability in V_{PSC} extremes can happen through internal variability (Pommereau et al. 2013; Rieder and Polvani 2013). Although the long-term recovery of Arctic stratospheric ozone is expected, the high dynamical variability may still lead to years with a cold and strong polar vortex, large V_{PSC}, and halogen-induced chemical ozone depletion events like that of 2010–2011 until ODS concentrations return to their 1960 values later in the twenty-first century (Langematz et al. 2014; Bednarz et al. 2016; Pommereau et al. 2018).

Acknowledgements Kimberly Strong and the co-authors of Chap. 3 from the University of Toronto would like to acknowledge the ongoing support from the Natural Sciences and Engineering Research Council of Canada (NSERC), the Canadian Space Agency (CSA), and Environment and Climate Change Canada (ECCC), as well as the many agencies who have supported the operation of the Polar Environment Atmospheric Research Laboratory (PEARL) by the Canadian Network for the Detection of Atmospheric Change (CANDAC). William Simpson (Chap. 3) acknowledges the support from the University of Alaska Fairbanks and ongoing support from NASA, NOAA, and NSF.

References

Abbatt, J. P. D., Thomas, J. L., Abrahamsson, K., Boxe, C., Granfors, A., Jones, A. E., King, M. D., Saiz-Lopez, A., Shepson, P. B., Sodeau, J., Toohey, D. W., Toubin, C., von Glasow, R., Wren, S. N., & Yang, X. (2012). Halogen activation via interactions with environmental ice and snow in the polar lower troposphere and other regions. *Atmospheric Chemistry and Physics, 12*(14), 6237–6271. https://doi.org/10.5194/acp-12-6237-2012.

Abbott, B. W., et al. (2016). Biomass offsets little or none of permafrost carbon release from soils, streams, and wildfire: An expert assessment. *Environmental Research Letters, 11*(3), 034014. https://doi.org/10.1088/1748-9326/11/3/034014.

Adams, C., Strong, K., Zhao, X., Bassford, M. R., Chipperfield, M. P., Daffer, W., Drummond, J. R., Farahani, E. E., Feng, W., Fraser, A., Goutail, F., Manney, G., McLinden, C. A., Pazmino, A., Rex, M., & Walker, K. A. (2012). Severe 2011 ozone depletion assessed with 11 years of ozone, NO_2, and OClO measurements at 80°N. *Geophysical Research Letters, 39*, L05806. https://doi.org/10.1029/2011GL050478.

Aliabadi, A. A., Staebler, R. M., & Sharma, S. (2015). Air quality monitoring in communities of the Canadian Arctic during the high shipping season with a focus on local and marine pollution. *Atmospheric Chemistry and Physics, 15*(5), 2651–2673. https://doi.org/10.5194/acp-15-2651-2015.

AMAP. (2006). *Arctic Pollution 2006: Acidification and Arctic Haze*. Arctic Monitoring and Assessment Programme, Oslo, Norway, vii + 28 pp.

AMAP. (2011). *AMAP Assessment 2011: Mercury in the Arctic*. Arctic Monitoring and Assessment Programme, Oslo, Norway, xiv + 193 pp.

AMAP. (2015a). *AMAP Assessment 2015: Black Carbon and Ozone as Arctic Climate Forcers*. Arctic Monitoring and Assessment Programme, Oslo, Norway, vii + 116 pp.

AMAP. (2015b). *AMAP Assessment 2015: Temporal Trends in Persistent Organic Pollutants in the Arctic*. Arctic Monitoring and Assessment Programme, Oslo, Norway, vi + 71 pp.

AMAP. (2015c). *AMAP Assessment 2015: Methane as an Arctic Climate Forcer*. Arctic Monitoring and Assessment Programme, Oslo, Norway, vii + 139 pp.

AMAP. (2017a). *AMAP Assessment 2016: Chemicals of Emerging Arctic Concern*. Arctic Monitoring and Assessment Programme, Oslo, Norway, xvi + 353pp.

AMAP (2017b), *Snow, Water, Ice and Permafrost in the Arctic (SWIPA) 2017*. Arctic Monitoring and Assessment Programme, Oslo, Norway, xiv + 269 pp.

Ancellet, G., Daskalakis, N., Raut, J. C., Tarasick, D., Hair, J., Quennehen, B., Ravetta, F., Schlager, H., Weinheimer, A. J., Thompson, A. M., Johnson, B., Thomas, J. L., & Law, K. S. (2016). Analysis of the latitudinal variability of tropospheric ozone in the Arctic using the large number of aircraft and ozonesonde observations in early summer 2008. *Atmospheric Chemistry and Physics, 16*(20), 13341–13358. https://doi.org/10.5194/acp-16-13341-2016.

Andrews, D. G., Holton, J. R., & Leovy, C. B. (1987). *Middle atmosphere dynamics*. San Diego: Academic.

Arnold, F., & Spreng, S. (1994). Balloon-borne mass spectrometer measurements of HCl and HF in the winter Arctic stratosphere. *Geophysical Research Letters, 21*(13), 1255–1258. https://doi.org/10.1029/93GL03229.

Arnold, F., Burger, V., Gollinger, K., Roncossek, M., Schneider, J., & Spreng, S. (1998). Observations of nitric acid perturbations in the winter Arctic stratosphere: Evidence for PSC-sedimentation. *Journal of Atmospheric Chemistry, 30*, 49–59. https://doi.org/10.1023/A:1006014003391.

Arnold, S. R., Law, K. S., Brock, C. A., Thomas, J. L., Starkweather, S. M., von Salzen, K., Stohl, A., Sharma, S., Lund, M. T., Flanner, M. G., Petäjä, T., Tanimoto, H., Gamble, J., Dibb, J. E., Melamed, M., Johnson, N., Fidel, M., Tynkkynen, V.-P., Baklanov, A., Eckhardt, S., Monks, S. A., Browse, J., & Bozem, H. (2016). Arctic air pollution: Challenges and opportunities for the next decade. *Elementa: Science of the Anthropocene, 4*, p. 000104. https://doi.org/10.12952/journal.elementa.000104.

Arnone, E., Castelli, E., Papandrea, E., Carlotti, M., & Dinelli, B. M. (2012). Extreme ozone depletion in the 2010-2011 Arctic winter stratosphere as observed by MIPAS/ENVISAT using a 2-D tomographic approach. *Atmospheric Chemistry and Physics, 12*(19), 9149–9165. https://doi.org/10.5194/acp-12-9149-2012.

Aumann, H. H., et al. (2003). AIRS/AMSU/HSB on the aqua mission: Design, science objectives, data products, and processing systems. *IEEE Transactions on Geoscience and Remote Sensing, 41*(2), 253–264. https://doi.org/10.1109/TGRS.2002.808356.

Barber, D. G., Ehn, J. K., Pućko, M., Rysgaard, S., Deming, J. W., Bowman, J. S., Papakyriakou, T., Galley, R. J., & Søgaard, D. H. (2014). Frost flowers on young Arctic Sea ice: The climatic, chemical, and microbial significance of an emerging ice type. *Journal of Geophysical Research: Atmospheres, 119*(20), 11593–11612. https://doi.org/10.1002/2014JD021736.

Barrie, L. A., Bottenheim, J. W., Schnell, R. C., Crutzen, P. J., & Rasmussen, R. A. (1988). Ozone destruction and photochemical reactions at polar sunrise in the lower Arctic atmosphere. *Nature, 334*(6178), 138–141. https://doi.org/10.1038/334138a0.

Bates, D. R., & Nicolet, M. (1950). Atmospheric hydrogen. *Publications of the Astronomical Society of the Pacific, 62*(365), 106–110. https://doi.org/10.1086/126244.

Bednarz, E. M., Maycock, A. C., Abraham, N. L., Braesicke, P., Dessens, O., & Pyle, J. A. (2016). Future Arctic ozone recovery: The importance of chemistry and dynamics. *Atmospheric Chemistry and Physics, 16*(18), 12159–12176. https://doi.org/10.5194/acp-16-12159-2016.

Beer, R., Glavich, T. A., & Rider, D. M. (2001). Tropospheric emission spectrometer for the earth observing System's Aura satellite. *Applied Optics, 40*, 2356–2367. https://doi.org/10.1364/AO.40.002356.

Begoin, M., Richter, A., Weber, M., Kaleschke, L., Tian-Kunze, X., Stohl, A., Theys, N., & Burrows, J. P. (2010). Satellite observations of long range transport of a large BrO plume in the Arctic. *Atmospheric Chemistry and Physics, 10*(14), 6515–6526. https://doi.org/10.5194/acp-10-6515-2010.

Belikov, D. A., Bril, A., Maksyutov, S., Oshchepkov, S., Saeki, T., Takagi, H., Yoshida, Y., Ganshin, A., Zhuravlev, R., Aoki, S., & Yokota, T. (2014). Column-averaged CO_2 concentrations in the subarctic from GOSAT retrievals and NIES transport model simulations. *Polar Science, 8*, 129–145. https://doi.org/10.1016/j.polar.2014.02.002.

Belshe, E. F., Schuur, E. A. G., & Bolker, B. M. (2013). Tundra ecosystems observed to be CO_2 sources due to differential amplification of the carbon cycle. *Ecology Letters, 16*, 1316–1324. https://doi.org/10.1111/ele.12164.

Berg, T., Pfaffhuber, K. A., Cole, A. S., Engelsen, O., & Steffen, A. (2013). Ten-year trends in atmospheric mercury concentrations, meteorological effects and climate variables at Zeppelin, Ny-Ålesund. *Atmospheric Chemistry and Physics, 13*(13), 6575–6586. https://doi.org/10.5194/acp-13-6575-2013.

Beyerle, G., Neuber, R., Schrems, O., Wittrock, F., & Knudsen, B. (1994). Multiwavelength lidar measurements of stratospheric aerosols above Spitsbergen during winter 1992/93. *Geophysical Research Letters, 21*(1), 57–60. https://doi.org/10.1029/93GL02846.

Bintanja, R., & Selten, F. M. (2014). Future increases in Arctic precipitation linked to local evaporation and sea-ice retreat. *Nature, 509*, 479–482. https://doi.org/10.1038/nature13259.

Blechschmidt, A.-M., Richter, A., Burrows, J. P., Kaleschke, L., Strong, K., Theys, N., Weber, M., Zhao, X., & Zien, A. (2016). An exemplary case of a bromine explosion event linked to cyclone development in the Arctic. *Atmospheric Chemistry and Physics, 16*(3), 1773–1788. https://doi.org/10.5194/acp-16-1773-2016.

Blom, C., Fischer, H., Glatthor, N., Gulde, T., & Hopfner, M. (1994). Airborne measurements during the European Arctic stratospheric ozone experiment: Column amounts of HNO_3 and O_3 derived from FTIR emission sounding. *Geophysical Research Letters, 21*(13), 1351–1354. https://doi.org/10.1029/93GL02000.

Bocquet, F., Helmig, D., van Dam, B. A., & Fairall, C. W. (2011). Evaluation of the flux gradient technique for measurement of ozone surface fluxes over snowpack at summit, Greenland. *Atmospheric Measurement Techniques, 4*(10), 2305–2321. https://doi.org/10.5194/amt-4-2305-2011.

Boisvert, L. N., & Stroeve, J. C. (2015). The Arctic is becoming warmer and wetter as revealed by the atmospheric infrared sounder. *Geophysical Research Letters, 42*(11), 4439–4446. https://doi.org/10.1002/2015GL063775.

Bottenheim, J. W., Gallant, A. G., & Brice, K. A. (1986). Measurements of NO_Y species and O_3 at 82° N latitude. *Geophysical Research Letters, 13*(2), 113–116. https://doi.org/10.1029/GL013i002p00113.

Bottenheim, J. W., Fuentes, J. D., Tarasick, D. W., & Anlauf, K. G. (2002). Ozone in the Arctic lower troposphere during winter and spring 2000 (ALERT2000). *Atmospheric Environment, 36*(15–16), 2535–2544. https://doi.org/10.1016/S1352-2310(02)00121-8.

Bottenheim, J. W., Netcheva, S., Morin, S., & Nghiem, S. V. (2009). Ozone in the boundary layer air over the Arctic Ocean: Measurements during the TARA transpolar drift 2006–2008. *Atmospheric Chemistry and Physics, 9*(14), 4545–4557. https://doi.org/10.5194/acp-9-4545-2009.

Bovensmann, H., Burrows, J. P., Buchwitz, M., Frerick, J., Noël, S., Rozanov, V. V., Chance, K. V., & Goede, A. P. (1999). SCIAMACHY: Mission objectives and measurement modes. *Journal of Atmospheric Science, 56*(2), 127–150. https://doi.org/10.1175/1520-0469(1999)056<0127:SMOAMM>2.0.CO;2.

Brewer, A. W. (1949). Evidence for a world circulation provided by the measurements of helium and water vapour distribution in the stratosphere. *Quarterly Journal of the Royal Meteorological Society, 75*, 351–363. https://doi.org/10.1002/qj.49707532603.

Brönnimann, S., Staehelin, J., Farmer, S. F. G., Cain, J. C., Svendby, T., & Svenøe, T. (2003). Total ozone observations prior to the IGY. I: A history. *Quarterly Journal of the Royal Meteorological Society, 129*(593), 2797–2817. https://doi.org/10.1256/qj.02.118.

Brooks, S. D., Baumgardner, D., Gandrud, B., Dye, J. E., Northway, M. J., W, D., Fahey, T. P., Bui, O. B. T., & Tolbert, M. A. (2003). Measurements of large stratospheric particles in the Arctic polar vortex. *Journal of Geophysical Research: Atmospheres, 108*(D20), 4652. https://doi.org/10.1029/2002JD003278.

Browell, E. V., Butler, C. F., Ismail, S., Robinette, P. A., Carter, A. F., Higdon, N. S., Toon, O. B., Schoeberl, M. R., & Tuck, A. F. (1990). Airborne lidar observations in the wintertime Arctic stratosphere: Polar stratospheric clouds. *Geophysical Research Letters, 17*(4), 385–388. https://doi.org/10.1029/GL017i004p00385.

Brutsaert, W., & Hiyama, T. (2012). The determination of permafrost thawing trends from long-term streamflow measurements with an application in eastern Siberia. *Journal of Geophysical Research: Atmospheres, 117*(D22110), 1984–2012. https://doi.org/10.1029/2012JD018344.

Burke, E. J., Ekici, A., Huang, Y., Chadburn, S. E., Huntingford, C., Ciais, P., Friedlingstein, P., Peng, S., & Krinner, G. (2017). Quantifying uncertainties of permafrost carbon–climate feedbacks. *Biogeosciences, 14*(12), 3051–3066. https://doi.org/10.5194/bg-14-3051-2017.

Butchart, N. (2014). The Brewer-Dobson circulation. *Reviews of Geophysics, 52*(2), 157–184. https://doi.org/10.1002/2013RG000448.

Butler, A. H., Seidel, D. J., Hardiman, S. C., Butchart, N., Birner, T., & Match, A. (2015). Defining sudden stratospheric warmings. *Bulletin of the American Meteorological Society, 96*, 1913–1928. https://doi.org/10.1175/BAMS-D-13-00173.1.

Cao, L., Sihler, H., Platt, U., & Gutheil, E. (2014). Numerical analysis of the chemical kinetic mechanisms of ozone depletion and halogen release in the polar troposphere. *Atmospheric Chemistry and Physics, 14*(7), 3771–3787. https://doi.org/10.5194/acp-14-3771-2014.

Cao, L., He, M., Jiang, H., Grosshans, H., & Cao, N. (2016). Sensitivity of the reaction mechanism of the ozone depletion events during the Arctic spring on the initial atmospheric composition of the troposphere. *Atmosphere, 7*(10), 124. https://doi.org/10.3390/atmos7100124.

Carslaw, K. (Ed.). (2004). Special issue on SAGE III Ozone loss and validation experiment II and the validation of international satellites and study of ozone loss (SOLVE-II/VINTERSOL). *Atmospheric Chemistry and Physics*. https://www.atmos-chem-phys.net/special_issue12.html

Chang, R. Y.-W., Miller, C. E., Dinardo, S. J., Karion, A., Sweeney, C., Daube, B. C., Henderson, J. M., Mountain, M. E., Eluszkiewicz, J., Miller, J. B., Bruhwiler, L. M. P., & Wofsy, S. C. (2014). Methane emissions from Alaska in 2012 from CARVE airborne observations. *Proceedings of the National Academy of Sciences, 111*, 12953–16699. https://doi.org/10.1073/pnas.1412953111.

Chapman, S. (1930). On ozone and atomic oxygen in the upper atmosphere. *Philosophical Magazine, 10*(64), 369–383. https://doi.org/10.1080/14786443009461588.

Choi, S., Wang, Y., Salawitch, R. J., Canty, T., Joiner, J., Zeng, T., Kurosu, T. P., Chance, K., Richter, A., Huey, L. G., Liao, J., Neuman, J. A., Nowak, J. B., Dibb, J. E., Weinheimer, A. J., Diskin, G., Ryerson, T. B., da Silva, A., Curry, J., Kinnison, D., Tilmes, S., & Levelt, P. F. (2012). Analysis of satellite-derived Arctic tropospheric BrO columns in conjunction with aircraft measurements during ARCTAS and ARCPAC. *Atmospheric Chemistry and Physics, 12*(3), 1255–1285. https://doi.org/10.5194/acp-12-1255-2012.

Choi, S., Theys, N., Salawitch, R. J., Wales, P. A., Joiner, J., Canty, T. P., Chance, K., Suleiman, R. M., Palm, S. P., Cullather, R. I., Darmenov, A. S., da Silva, A., Kurosu, T. P., Hendrick, F., & van Roozendael, M. (2018). Link between Arctic tropospheric BrO explosion observed from space and sea-salt aerosols from blowing snow investigated using ozone monitoring instrument BrO data and GEOS-5 data assimilation system. *Journal of Geophysical Research: Atmospheres, 123*(13), 6954–6983. https://doi.org/10.1029/2017JD026889.

Chshyolkova, T., Manson, A. H., Meek, C. E., Aso, T., Avery, S. K., Hall, C. M., Hocking, W., Igarashi, K., Jacobi, C., Makarov, N., Mitchell, N., Murayama, Y., Singer, W., Thorsen, D., & Tsutsumi, M. (2007). Polar vortex evolution during Northern Hemispheric winter 2004/05. *Annales Geophysicae, 25*(6), 1279–1298. https://doi.org/10.5194/angeo-25-1279-2007.

Ciais, P., Dolman, A. J., Dargaville, R., Barrie, L., Bombelli, A., Butler, J., Canadell, P., & Moriyama, T. (2010). *GEO Carbon Strategy*. Rome: GEO Secretariat Geneva, FAO. 48 pp.

Ciais, P., et al. (2013). Carbon and other biogeochemical cycles. In T. F. Stocker, D. Qin, G.-K. Plattner, M. Tignor, S. K. Allen, J. Boschung, A. Nauels, Y. Xia, V. Bex, & P. M. Midgley (Eds.), *Climate Change 2013: The Physical Science Basis. Contribution of Working Group I to the Fifth Assessment Report of the Intergovernmental Panel on Climate Change*. Cambridge/New York: Cambridge University Press.

Ciais, P., et al. (2014). Current systematic carbon-cycle observations and the need for implementing a policy-relevant carbon observing system. *Biogeosciences, 11*(13), 3547–3602. https://doi.org/10.5194/bg-11-3547-2014.

Cicerone, R. J., Walters, S., & Liu, S. C. (1983). Nonlinear response of stratospheric ozone column to chlorine injections. *Journal of Geophysical Research: Oceans, 88*(C6), 3647–3661. https://doi.org/10.1029/JC088iC06p03647.

Cohen, J., Screen, J. A., Furtado, J. C., Barlow, M., Whittleston, D., Coumou, D., Francis, J., Dethloff, K., Entekhabi, D., Overland, J., & Jones, J. (2014). Recent Arctic amplification and and extreme mid-latitude weather. *Nature Geoscience, 7*, 627–637. https://doi.org/10.1038/ngeo2234.

Cole, A. S., & Steffen, A. (2010). Trends in long-term gaseous mercury observations in the Arctic and effects of temperature and other atmospheric conditions. *Atmospheric Chemistry and Physics, 10*(10), 4661–4672. https://doi.org/10.5194/acp-10-4661-2010.

Commane, R., Lindaas, J., Benmergui, J., Luus, K. A., Chang, R. Y.-W., Daube, B. C., Euskirchen, E. S., Henderson, J. M., Karion, A., Miller, J. B., Miller, S. M., Parazoo, N. C., Randerson, J. T., Sweeney, C., Tans, P., Thoning, K., Veraverbeke, S., Miller, C. E., & Wofsy, S. C. (2017). Carbon dioxide sources from Alaska driven by increasing early winter respiration from Arctic tundra. *Proceedings of the National Academy of Sciences, 114*(21), 5361–5366. https://doi.org/10.1073/pnas.1618567114.

Cooper, O. R., Parrish, D. D., Ziemke, J., Balashov, N. V., Cupeiro, M., Galbally, I. E., Gilge, S., Horowitz, L., Jensen, N. R., Lamarque, J.-F., Naik, V., Oltmans, S. J., Schwab, J., Shindell, D. T., Thompson, A. M., Thouret, V., Wang, Y., & Zbinden, R. M. (2014). Global distribution and trends of tropospheric ozone: An observation-based review. *Elementa: Science of the Anthropocene, 2*, 000029. https://doi.org/10.12952/journal.elementa.000029.

Crewell, S., Künzi, K., Nett, H., Wehr, T., & Hartogh, P. (1994). Aircraft measurements of ClO and HCl during EASOE 1991/92. *Geophysical Research Letters, 21*(13), 1267–1270. https://doi.org/10.1029/93GL02499.

Crisp, D. (2015). Measuring atmospheric carbon dioxide from space with the Orbiting Carbon Observatory-2 (OCO-2). *Proceeding of the Society of Photo-Optical Instrumentation Engineers (SPIE), 9607*, Earth Observing Systems XX, 960702. https://doi.org/10.1117/12.2187291.

Crutzen, P. J. (1970). The influence of nitrogen oxide on the atmospheric ozone content. *Quarterly Journal of the Royal Meteorological Society, 96*(408), 320–327. https://doi.org/10.1002/qj.49709640815.

Crutzen, P. J. (1971). Ozone production rates in an oxygen-hydrogen-nitrogen oxide atmosphere. *Journal of Geophysical Research: Oceans and Atmospheres, 76*(30), 7311–7327. https://doi.org/10.1029/JC076i030p07311.

Custard, K. D., Thompson, C. R., Pratt, K. A., Shepson, P. B., Liao, J., Huey, L. G., Orlando, J. J., Weinheimer, A. J., Apel, E., Hall, S. R., Flocke, F., Mauldin, L., Hornbrook, R. S., Pöhler, D., General, S., Zielcke, J., Simpson, W. R., Platt, U., Fried, A., Weibring, P., Sive, B. C., Ullmann, K., Cantrell, C., Knapp, D. J., & Montzka, D. D. (2015). The NO_X dependence

of bromine chemistry in the Arctic atmospheric boundary layer. *Atmospheric Chemistry and Physics, 15*(18), 10799–10809. https://doi.org/10.5194/acp-15-10799-2015.
Custard, K. D., Raso, A. R. W., Shepson, P. B., Staebler, R. M., & Pratt, K. A. (2017). Production and release of molecular bromine and chlorine from the Arctic coastal snowpack. *ACS Earth and Space Chemistry, 1*(3), 142–151. https://doi.org/10.1021/acsearthspacechem.7b00014.
Damm, E., Thoms, S., Beszczynska-Möller, A., Nöthig, E. M., & Kattner, G. (2015). Methane excess production in oxygen-rich polar water and a model of cellular conditions for this paradox. *Polar Science, 9*(3), 327–334. https://doi.org/10.1016/j.polar.2015.05.001.
Daniel, J. S., Solomon, S., Portmann, R. W., & Garcia, R. R. (1999). Stratospheric ozone destruction: The importance of bromine relative to chlorine. *Journal of Geophysical Research: Atmospheres, 104*(D19), 23871–23880. https://doi.org/10.1029/1999JD900381.
de Groot, W. J., Flannigan, M. D., & Cantin, A. S. (2013). Climate change impacts on future boreal fire regimes. *Forest Ecology and Management, 294*, 35–44. https://doi.org/10.1016/j.foreco.2012.09.027.
de Mazière, M., Thompson, A. M., Kurylo, M. J., Wild, J. D., Bernhard, G., Blumenstock, T., Braathen, G. O., Hannigan, J. W., Lambert, J.-C., Leblanc, T., McGee, T. J., Nedoluha, G., Petropavlovskikh, I., Seckmeyer, G., Simon, P. C., Steinbrecht, W., & Strahan, S. E. (2018). The network for the detection of atmospheric composition change (NDACC): History, status and perspectives. *Atmospheric Chemistry and Physics, 18*(7), 4935–4964. https://doi.org/10.5194/acp-18-4935-2018.
de Zafra, R. L., Jaramillo, M., Parrish, A., Solomon, P., Connor, B., & Barnett, J. (1987). High concentrations of chlorine monoxide at low altitudes in the Antarctic spring stratosphere: Diurnal variation. *Nature, 328*, 408–411. https://doi.org/10.1038/328408a0.
de Zafra, R. L., Emmons, L. K., Reeves, J. M., & Shindell, D. T. (1994). An overview of millimeter-wave spectroscopic measurements of chlorine monoxide at Thule, Greenland, February–March, 1992: Vertical profiles, diurnal variation, and longer-term trends. *Geophysical Research Letters, 21*(13), 1271–1274. https://doi.org/10.1029/93GL01677.
Dhomse, S. S., Kinnison, D., Chipperfield, M. P., Salawitch, R. J., Cionni, I., Hegglin, M. I., Abraham, N. L., Akiyoshi, H., Archibald, A. T., Bednarz, E. M., Bekki, S., Braesicke, P., Butchart, N., Dameris, M., Deushi, M., Frith, S., Hardiman, S. C., Hassler, B., Horowitz, L. W., Hu, R.-M., Jöckel, P., Josse, B., Kirner, O., Kremser, S., Langematz, U., Lewis, J., Marchand, M., Lin, M., Mancini, E., Marécal, V., Michou, M., Morgenstern, O., O'Connor, F. M., Oman, L., Pitari, G., Plummer, D. A., Pyle, J. A., Revell, L. E., Rozanov, E., Schofield, R., Stenke, A., Stone, K., Sudo, K., Tilmes, S., Visioni, D., Yamashita, Y., & Zeng, G. (2018). Estimates of ozone return dates from chemistry-climate model initiative simulations. *Atmospheric Chemistry and Physics, 18*(11), 8409–8438. https://doi.org/10.5194/acp-18-8409-2018.
Dlugokencky, E. J. (2018). *Trends in atmospheric methane*. National Oceanic and Atmospheric Administration, Earth System Research Laboratory (NOAA/ESRL). Available at https://www.esrl.noaa.gov/gmd/ccgg/trends_ch4/
Dlugokencky, E. J., & Tans, P. (2018). *Trends in atmospheric carbon dioxide*. National Oceanic and Atmospheric Administration, Earth System Research Laboratory (NOAA/ESRL). Available at http://www.esrl.noaa.gov/gmd/ccgg/trends/global.html
Dlugokencky, E. J., Nisbet, E. G., Fisher, R., & Lowry, D. (2011). Global atmospheric methane: Budget, changes and dangers. *Philosophical Transactions of the Royal Society A, 369*, 2058–2072. https://doi.org/10.1098/rsta.2010.0341.
Dlugokencky, E. J., Lang, P. M., Mund, J. W., Crotwell, A. M., Crotwell, M. J., & Thoning, K. W. (2018a). Atmospheric carbon dioxide dry air mole fractions from the NOAA ESRL carbon cycle cooperative global air sampling network, 1968–2017, Version: 2018-07-31. Available at ftp://aftp.cmdl.noaa.gov/data/trace_gases/co2/flask/surface/
Dlugokencky, E. J., Lang, P. M., Crotwell, A. M., Mund, J. W., Crotwell, M. J., & Thoning, K. W. (2018b). Atmospheric methane dry air mole fractions from the NOAA ESRL carbon cycle cooperative global air sampling network, 1983–2017, Version: 2018-08-01. Available at ftp://aftp.cmdl.noaa.gov/data/trace_gases/ch4/flask/surface/

Dobson, G. M. B. (1956). Origin and distribution of the polyatomic molecules in the atmosphere. *Proceedings of the Royal Society of London Series A, 236*(1205), 187–193. https://doi.org/10.1098/rspa.1956.0127.

Dobson, G. M. B. (1968). Forty years' research on atmospheric ozone at Oxford: A history. *Applied Optics, 7*(3), 387–405. https://doi.org/10.1364/AO.7.000387.

Domine, F., Taillandier, A. S., Simpson, W. R., & Severin, K. (2005). Specific surface area, density and microstructure of frost flowers. *Geophysical Research Letters, 32*(13), L13502. https://doi.org/10.1029/2005GL023245.

Douglas, T. A., Domine, F., Barret, M., Anastasio, C., Beine, H. J., Bottenheim, J., Grannas, A., Houdier, S., Netcheva, G., Rowland, R. S., & Steffen, A. (2012). Frost flowers growing in the Arctic ocean-atmosphere – Sea ice – Snow interface: 1. Chemical composition. *Journal of Geophysical Research: Atmospheres, 117*(D14), D00R09. https://doi.org/10.1029/2011JD016460.

Douglass, A. R., Strahan, S. E., Oman, L. D., & Stolarski, R. S. (2014). Understanding differences in chemistry climate model projections of stratospheric ozone. *Journal of Geophysical Research: Atmospheres, 119*(8), 4922–4939. https://doi.org/10.1002/2013JD021159.

Dufour, A., Zolina, O., & Gulev, S. K. (2016). Atmospheric moisture transport to the Arctic: Assessment of Reanalyses and analysis of transport components. *Journal of Climate, 29*, 5061–5081. https://doi.org/10.1175/JCLI-D-15-0559.1.

Dunkerton, T. J., & Delisi, D. P. (1986). Evolution of potential vorticity in the winter stratosphere of January-February 1979. *Journal of Geophysical Research: Atmospheres, 91*(D1), 1199–1208. https://doi.org/10.1029/JD091iD01p01199.

Eigen, M., & Kustin, K. (1962). The kinetics of halogen hydrolysis. *Journal of the American Chemical Society, 84*(8), 1355–1361. https://doi.org/10.1021/ja00867a005.

Elberling, B., Michelsen, A., Schädel, C., Schuur, E. A. G., Christiansen, H. H., Berg, L., Tamstorf, M. P., & Sigsgaard, C. (2013). Long-term CO_2 production following permafrost thaw. *Nature Climate Change, 3*, 890–894. https://doi.org/10.1038/nclimate1955.

Eldering, A., O'Dell, C. W., Wennberg, P. O., Crisp, D., Gunson, M. R., Viatte, C., Avis, C., Braverman, A., Castano, R., Chang, A., Chapsky, L., Cheng, C., Connor, B., Dang, L., Doran, G., Fisher, B., Frankenberg, C., Fu, D., Granat, R., Hobbs, J., Lee, R. A. M., Mandrake, L., McDuffie, J., Miller, C. E., Myers, V., Natraj, V., O'Brien, D., Osterman, G. B., Oyafuso, F., Payne, V. H., Pollock, H. R., Polonsky, I., Roehl, C. M., Rosenberg, R., Schwandner, F., Smyth, M., Tang, V., Taylor, T. E., To, C., Wunch, D., & Yoshimizu, J. (2017). The orbiting carbon Observatory-2: First 18 months of science data products. *Atmospheric Measurement Techniques, 10*(2), 549–563. https://doi.org/10.5194/amt-10-549-2017.

Elliott, S., Reagan, M., Moridis, G., & Smith, P. C. (2010). Geochemistry of clathrate derived methane in Arctic Ocean waters. *Geophysical Research Letters, 37*, L12607. https://doi.org/10.1029/2010GL043369.

Emmons, L., Reeves, J., Shindell, D., & deZafra, R. (1994). N_2O as an indicator of Arctic vortex dynamics: Correlations with O_3 over Thule, Greenland in February and March, 1992. *Geophysical Research Letters, 21*(13), 1275–1278. https://doi.org/10.1029/93GL01054.

Emmons, L. K., Arnold, S. R., Monks, S. A., Huijnen, V., Tilmes, S., Law, K. S., Thomas, J. L., Raut, J.-C., Bouarar, I., Turquety, S., Long, Y., Duncan, B., Steenrod, S., Strode, S., Flemming, J., Mao, J., Langner, J., Thompson, A. M., Tarasick, D., Apel, E. C., Blake, D. R., Cohen, R. C., Dibb, J., Diskin, G. S., Fried, A., Hall, S. R., Huey, L. G., Weinheimer, A. J., Wisthaler, A., Mikoviny, T., Nowak, J., Peischl, J., Roberts, J. M., Ryerson, T., Warneke, C., & Helmig, D. (2015). The POLARCAT model Intercomparison project (POLMIP): Overview and evaluation with observations. *Atmospheric Chemistry and Physics, 15*(12), 6721–6744. https://doi.org/10.5194/acp-15-6721-2015.

Eyring, V., Cionni, I., Bodeker, G. E., Charlton-Perez, A. J., Kinnison, D. E., Scinocca, J. F., Waugh, D. W., Akiyoshi, H., Bekki, S., Chipperfield, M. P., Dameris, M., Dhomse, S., Frith, S. M., Garny, H., Gettelman, A., Kubin, A., Langematz, U., Mancini, E., Marchand, M., Nakamura, T., Oman, L. D., Pawson, S., Pitari, G., Plummer, D. A., Rozanov, E., Shepherd, T. G., Shibata, K., Tian, W., Braesicke, P., Hardiman, S. C., Lamarque, J. F., Morgenstern, O., Pyle, J. A., Smale, D., & Yamashita, Y. (2010). Multi- model assessment of stratospheric ozone return dates and ozone recovery in CCMVal-2 models. *Atmospheric Chemistry and Physics, 10*(19), 9451–9472. https://doi.org/10.5194/acp-10-9451-2010.

Eyring, V., Lamarque, J.-F., Hess, P., Arfeuille, F., Bowman, K., Chipperfield, M. P., Duncan, B., Fiore, A., Gettelman, A., Giorgetta, M. A., Granier, C., Hegglin, M., Kinnison, D., Kunze, M., Langematz, U., Luo, B., Martin, R., Matthes, K., Newman, P. A., Peter, T., Robock, A., Ryerson, T., Saiz-Lopez, A., Salawitch, R., Schultz, M., Shepherd, T. G., Shindell, D., Staehelin, J., Tegtmeier, S., Thomason, L., Tilmes, S., Vernier, J.-P., Waugh, D. W., & Young, P. J. (2013). Overview of IGAC/SPARC Chemistry-Climate Model Initiative (CCMI) community simulations in support of upcoming ozone and climate assessments. *SPARC Newsletter, 40*, 48–66.

Fabian, P., Borchers, R., & Kourtidis, K. (1994). Bromine-containing source gases during EASOE. *Geophysical Research Letters, 21*(13), 1219–1222. https://doi.org/10.1029/93GL03044.

Fahey, D. W., Kelly, K. K., Kawa, S. R., Tuck, A. F., Loewenstein, M., Chan, K. R., & Heidt, L. E. (1990). Observations of denitrification and dehydration in the winter polar stratospheres. *Nature, 344*, 321–324. https://doi.org/10.1038/344321a0.

Fahey, D. W., Gao, R. S., Carslaw, K. S., Kettleborough, J., Popp, P. J., Northway, M. J., Holecek, J. C., Ciciora, S. C., McLaughlin, R. J., Thompson, T. L., Winkler, R. H., Baumgardner, D. G., Gandrud, B., Wennberg, P. O., Dhaniyala, S., McKinney, K., Peter, T., Salawitch, R. J., Bui, T. P., Elkins, J. W., Webster, C. R., Atlas, E. L., Jost, H., Wilson, J. C., Herman, R. L., Kleinböhl, A., & von König, M. (2001). The detection of large HNO_3-containing particles in the winter Arctic stratosphere. *Science, 291*(5506), 1026–1031. https://doi.org/10.1126/science.1057265.

Fan, S. M., & Jacob, D. J. (1992). Surface ozone depletion in Arctic spring sustained by bromine reactions on aerosols. *Nature, 359*(6395), 522–524. https://doi.org/10.1038/359522a0.

Farman, J. C., Gardiner, B. G., & Shanklin, J. D. (1985). Large losses of total ozone in Antarctica reveal seasonal ClO_x/NO_x interaction. *Nature, 315*, 207–210. https://doi.org/10.1038/315207a0.

Farman, J. C., Harris, N. R. P., Isaksen, I., Megie, G., Peter, T., Pyle, J. A., & Simon, P. (1998). Second European Stratospheric Arctic and Mid-latitude Experiment (SESAME): Part I, 1994–1995: Preface. *Journal of Atmospheric Chemistry, 30*(1), 1–2. https://doi.org/10.1023/A:1017148226356.

Farmer, C., Toon, O., Schaper, P., Blavier, J.-F., & Lowes, L. (1987). Stratospheric trace gases in the spring 1986 Antarctic atmosphere. *Nature, 329*, 126–130. https://doi.org/10.1038/329126a0.

Fergusson, A. (2010). *The Arctic ozone layer: how the Arctic ozone layer is responding to ozone-depleting chemicals and climate change*. Environment Canada, Catalogue No. En164-18/2010E. ISBN: 978-1-100-10787-5.

Fish, D., Jones, R., Freshwater, R., Roscoe, H., Oldham, D., & Harries, J. (1994). Total ozone measured during EASOE by a UV-visible spectrometer which observes stars. *Geophysical Research Letters, 21*(13), 1387–1390. https://doi.org/10.1029/93GL01869.

Fisher, J. B., et al. (2018). Missing pieces to modeling the Arctic-boreal puzzle. *Environmental Research Letters, 13*, 020202. https://doi.org/10.1088/1748-9326/aa9d9a.

FLUXNET. (2015). Oak Ridge National Laboratory Distributed Active Archive Center (ORNL DAAC). *FLUXNET Web Page*. Available online [http://fluxnet.ornl.gov] from ORNL DAAC, Oak Ridge, Tennessee, U.S.A.

Foken, T., Aubinet, M., & Leuning, R. (2012). The Eddy covariance method. In M. Aubinet, T. Vesala, & D. Papale (Eds.), *Eddy Covariance: A practical guide to measurement and data analysis*. Springer Atmospheric Sciences. Dordrecht: Springer. https://link.springer.com/book/10.1007/978-94-007-2351-1.

Forkel, M., Carvalhais, N., Rödenbeck, C., Keeling, R., Heimann, M., Thonicke, K., Zaehle, S., & Reichstein, M. (2016). Enchanced seasonal CO_2 exchange caused by amplified plant productivity in northern ecosystems. *Science, 351*(6274), 696–699. https://doi.org/10.1126/science.aac4971.

Foster, K. L., Plastridge, R. A., Bottenheim, J. W., Shepson, P. B., Finlayson-Pitts, B. J., & Spicer, C. W. (2001). The role of Br_2 and BrCl in surface ozone destruction at polar sunrise. *Science, 291*(5503), 471–474. https://doi.org/10.1126/science.291.5503.471.

Frieß, U., Sihler, H., Sander, R., Pöhler, D., Yilmaz, S., & Platt, U. (2011). The vertical distribution of BrO and aerosols in the Arctic: Measurements by active and passive differential optical absorption spectroscopy. *Journal of Geophysical Research: Atmospheres, 116*, D14, D00R04. https://doi.org/10.1029/2011JD015938.

Frith, S. M., S, R., Stolarski, N. A. K., & McPeters, R. D. (2017). Estimating uncertainties in the SBUV version 8.6 merged profile ozone data set. *Atmospheric Chemistry and Physics, 17*(23), 14695–14707. https://doi.org/10.5194/acp-17-14695-2017.

Fueglistaler, S., Luo, B. P., Voigt, C., Carslaw, K. S., & Peter, T. (2002). NAT-rock formation by mother clouds: A microphysical model study. *Atmospheric Chemistry and Physics, 2*(2), 93–98. https://doi.org/10.5194/acp-2-93-2002.

Gandrud, B. W., Dye, J. E., Baumgardner, D., Ferry, G. V., Loewenstein, M., Chan, K. R., Sanford, L., Gary, B., & Kelly, K. (1990). The January 30, 1989, Arctic polar stratospheric clouds (PSC) event: Evidence for a mechanism of dehydration. *Geophysical Research Letters, 17*(4), 457–460. https://doi.org/10.1029/GL017i004p00457.

Gilman, J. B., Burkhart, J. F., Lerner, B. M., Williams, E. J., Kuster, W. C., Goldan, P. D., Murphy, P. C., Warneke, C., Fowler, C., Montzka, S. A., Miller, B. R., Miller, L., Oltmans, S. J., Ryerson, T. B., Cooper, O. R., Stohl, A., & de Gouw, J. A. (2010). Ozone variability and halogen oxidation within the Arctic and sub-Arctic springtime boundary layer. *Atmospheric Chemistry and Physics, 10*(21), 10223–10236. https://doi.org/10.5194/acp-10-10223-2010.

Giordano, M. R., Kalnajs, L. E., Goetz, J. D., Avery, A. M., Katz, E., May, N. W., Leemon, A., Mattson, C., Pratt, K. A., & DeCarlo, P. F. (2018). The importance of blowing snow to halogen-containing aerosol in coastal Antarctica: Influence of source region versus wind speed. *Atmospheric Chemistry and Physics, 18*(22), 16689–16711. https://doi.org/10.5194/acp-18-16689-2018.

Godin-Beekmann, S., Newman, P. A., & Petropavlovskikh, I. (Eds.). (2018). 30th anniversary of the Montreal Protocol: From the safeguard of the ozone layer to the protection of the Earth Climate. *Comptes Rendus Geoscience* special issue, *350*(7), 331–448.

Götz, F. W. P. (1931). Zum Strahlungsklima des Spitzbergensommers. Strahlungs- und Ozonmessungen in der Königsbucht 1929. *Gerlands Beitrage zur Geophysik, 31*, 119–154.

Götz, F. W. P., Meetham, A. R., & Dobson, G. M. B. (1934). The vertical distribution of ozone in the atmosphere. *Proceedings of the Royal Society of London Series A, 145*, 416–446.

Goutail, F., Pommereau, J., Sarkissian, A., Kyro, E., & Dorokhov, V. (1994). Total nitrogen dioxide at the Arctic polar circle since 1990. *Geophysical Research Letters, 21*(13), 1371–1374. https://doi.org/10.1029/93GL01783.

Grannas, A. M., Jones, A. E., Dibb, J., Ammann, M., Anastasio, C., Beine, H. J., Bergin, M., Bottenheim, J., Boxe, C. S., Carver, G., Chen, G., Crawford, J. H., Dominé, F., Frey, M. M., Guzmán, M. I., Heard, D. E., Helmig, D., Hoffmann, M. R., Honrath, R. E., Huey, L. G., Hutterli, M., Jacobi, H. W., Klán, P., Lefer, B., McConnell, J., Plane, J., Sander, R., Savarino, J., Shepson, P. B., Simpson, W. R., Sodeau, J. R., von Glasow, R., Weller, R., Wolff, E. W., & Zhu, T. (2007). An overview of snow photochemistry: Evidence, mechanisms and impacts. *Atmospheric Chemistry and Physics, 7*(16), 4329–4373. https://doi.org/10.5194/acp-7-4329-2007.

Graven, H. D., Keeling, R. F., Piper, S. C., Patra, P. K., Stephens, B. B., Wofsy, S. C., Welp, L. R., Sweeney, C., Tans, P. P., Kelley, J. J., Daube, B. C., Kort, E. A., Santoni, G. W., & Bent, J. D. (2013). The amplitude of the seasonal cycle of carbon dioxide in high northern latitudes has increased by 50% since 1960. *Science, 341*(6150), 1085–1089. https://doi.org/10.1126/science.1239207.

Grosse, G., Harden, J., Turetsky, M., McGuire, A. D., Camill, P., Tarnocai, C., Frolking, S., Schuur, E. A. G., Jorgenson, T., Marchenko, S., Romanovsky, V., Wickland, K. P., French, N., Waldrop, M., Bourgeau-Chavez, L., & Striegl, R. G. (2011). Vulnerability of high-latitude soil organic carbon in North America to disturbance. *Journal of Geophysical Research: Biogeosciences, 116*, G00K06. https://doi.org/10.1029/2010JG001507.

Halfacre, J. W., Knepp, T. N., Shepson, P. B., Thompson, C. R., Pratt, K. A., Li, B., Peterson, P. K., Walsh, S. J., Simpson, W. R., Matrai, P. A., Bottenheim, J. W., Netcheva, S., Perovich, D. K., & Richter, A. (2014). Temporal and spatial characteristics of ozone depletion events from measurements in the Arctic. *Atmospheric Chemistry and Physics, 14*(10), 4875–4894. https://doi.org/10.5194/acp-14-4875-2014.

Hampson, J. (1964). Photochemical behavior of the ozone layer, *Tech. Note TN 1627/64*. Canadian Armament Research and Development Establishment, Valcartier, QC, Canada.

Hara, K., Matoba, S., Hirabayashi, M., & Yamasaki, T. (2017). Frost flowers and sea-salt aerosols over seasonal sea-ice areas in northwestern Greenland during winter – Spring. *Atmospheric Chemistry and Physics, 17*(13), 8577–8598. https://doi.org/10.5194/acp-17-8577-2017.

Hara, K., Osada, K., Yabuki, M., Takashima, H., Theys, N., & Yamanouchi, T. (2018). Important contributions of sea-salt aerosols to atmospheric bromine cycle in the Antarctic coasts. *Scientific Reports, 8*(1), 13852. https://doi.org/10.1038/s41598-018-32287-4.

Hausmann, P., Sussmann, R., & Smale, D. (2016). Contribution of oil and natural gas production to renewed increase in atmospheric methane (2007-2014): Top-down estimate from ethane and methane column observations. *Atmospheric Chemistry and Physics, 16*(5), 3227–3244. https://doi.org/10.5194/acp-16-3227-2016.

Helmig, D., Oltmans, S. J., Morse, T. O., & Dibb, J. E. (2007a). What is causing high ozone at summit, Greenland? *Atmospheric Environment, 41*(24), 5031–5043. https://doi.org/10.1016/j.atmosenv.2006.05.084.

Helmig, D., Bocquet, F., Cohen, L., & Oltmans, S. J. (2007b). Ozone uptake to the polar snowpack at summit, Greenland. *Atmospheric Environment, 41*(24), 5061–5076. https://doi.org/10.1016/j.atmosenv.2006.06.064.

Helmig, D., Cohen, L. D., Bocquet, F., Oltmans, S., Grachev, A., & Neff, W. (2009). Spring and summertime diurnal surface ozone fluxes over the polar snow at summit, Greenland. *Geophysical Research Letters, 36*(8), L08809. https://doi.org/10.1029/2008GL036549.

Helmig, D., Boylan, P., Johnson, B., Oltmans, S., Fairall, C., Staebler, R., Weinheimer, A., Orlando, J., Knapp, D. J., Montzka, D. D., Flocke, F., Frieß, U., Sihler, H., & Shepson, P. B. (2012). Ozone dynamics and snow-atmosphere exchanges during ozone depletion events at Barrow, Alaska. *Journal of Geophysical Research: Atmospheres, 117*, D20303. https://doi.org/10.1029/2012JD017531.

Hofmann, D. (1990). Stratospheric cloud microlayers and small-scale temperature variations in the Arctic in 1989. *Geophysical Research Letters, 17*(4), 369–372. https://doi.org/10.1029/GL017i004p00369.

Holl, G., Walker, K. A., Conway, S., Saitoh, N., Boone, C. D., Strong, K., & Drummond, J. R. (2016). Methane cross-validation between three Fourier transform spectrometers: SCISAT ACE-FTS, GOSAT TANSO-FTS, and ground-based FTS measurements in the Canadian high Arctic. *Atmospheric Measurement Techniques, 9*(5), 1961–1980. https://doi.org/10.5194/amt-9-1961-2016.

Holl, D., Wille, C., Sachs, T., Schreiber, P., Runkle, B. R. K., Beckebanze, L., Langer, M., Boike, J., Pfeiffer, E.-M., Fedorova, I., Bolshianov, D. Y., Grigoriev, M. N., & Kutzbach, L. (2019). A long-term (2002 to 2017) record of closed-path and open-path eddy covariance CO_2 net ecosystem exchange fluxes from the Siberian Arctic. *Earth System Science Data, 11*(1), 221–240. https://doi.org/10.5194/essd-11-221-2019.

Honrath, R. E., Lu, Y., Peterson, M. C., Dibb, J. E., Arsenault, M. A., Cullen, N. J., & Steffen, K. (2002). Vertical fluxes of NO_X, HONO, and HNO_3 above the snowpack at summit, Greenland. *Atmospheric Environment, 36*(15–16), 2629–2640. https://doi.org/10.1016/S1352-2310(02)00132-2.

Hu, H., Landgraf, J., Detmers, R., Borsdorff, T., de Brugh, J. A., Aben, I., Butz, A., & Hasekamp, O. (2018). Toward global mapping of methane with TROPOMI: First results and intersatellite comparison to GOSAT. *Geophysical Research Letters, 45*(8), 3682–3689. https://doi.org/10.1002/2018GL077259.

Huang, J., & Jaeglé, L. (2017). Wintertime enhancements of sea salt aerosol in polar regions consistent with a sea ice source from blowing snow. *Atmospheric Chemistry and Physics, 17*(5), 3699–3712. https://doi.org/10.5194/acp-17-3699-2017.

Huang, J., Jaeglé, L., & Shah, V. (2018). Using CALIOP to constrain blowing snow emissions of sea salt aerosols over Arctic and Antarctic Sea ice. *Atmospheric Chemistry and Physics, 18*(22), 16253–16269. https://doi.org/10.5194/acp-18-16253-2018.

Hugelius, G., Tarnocai, C., Broll, G., Canadell, J. G., Kuhry, P., & Swanson, D. K. (2013a). The northern circumpolar soil carbon database: Spatially distributed datasets of soil coverage and soil carbon storage in the northern permafrost regions. *Earth System Science Data, 5*(1), 3–13. https://doi.org/10.5194/essd-5-3-2013.

Hugelius, G., Bockheim, J. G., Camill, P., Elberling, B., Grosse, G., Harden, J. W., Johnson, K., Jorgenson, T., Koven, C. D., Kuhry, P., Michaelson, G., Mishra, U., Palmtag, J., Ping, C.-L., O'Donnell, J., Schirrmeister, L., Schuur, E. A. G., Sheng, Y., Smith, L. C., Strauss, J., & Yu, Z. (2013b). A new data set for estimating organic carbon storage to 3 m depth in soils of the northern circumpolar permafrost region. *Earth System Science Data, 5*(2), 393–402. https://doi.org/10.5194/essd-5-393-2013.

Hugelius, G., Strauss, J., Zubrzycki, S., Harden, J. W., Schuur, E. A. G., Ping, C. L., Schirrmeister, L., Grosse, G., Michaelson, G. J., Koven, C. D., O'Donnel, J. A., Elberling, B., Mishra, U., Camill, P., Yu, Z., Palmtag, J., & Kuhry, P. (2014). Estimated stocks of circumpolar permafrost carbon with quantified uncertainty ranges and identified data gaps. *Biogeosciences, 11*(23), 6573–6593. https://doi.org/10.5194/bg-11-6573-2014.

Hurwitz, M. M., Newman, P. A., & Garfinkel, C. I. (2011). The Arctic vortex in March 2011: A dynamical perspective. *Atmospheric Chemistry and Physics, 11*(22), 11447–11453. https://doi.org/10.5194/acp-11-11447-2011.

IPCC. (2013). Climate Change 2013: The Physical Science Basis. Contribution of Working Group I to the Fifth Assessment Report of the Intergovernmental Panel on Climate Change. In T. F. Stocker, D. Qin, G.-K. Plattner, M. Tignor, S. K. Allen, J. Boschung, A. Nauels, Y. Xia, V. Bex, & P. M. Midgley (Eds.) (1535 pp). Cambridge/New York: Cambridge University Press.

Isaksen, I. S. A., Zerefos, C., Wang, W.-C., Balis, D., Eleftheratos, K., Rognerud, B., Stordal, F., Berntsen, T. K., LaCasce, J. H., Søvde, O. A., Olivié, D., Orsolini, Y. J., Zyrichidou, I., Prather, M., & Tuinder, O. N. E. (2012). Attribution of the Arctic ozone column deficit in March 2011. *Geophysical Research Letters, 39*, L24810. https://doi.org/10.1029/2012GL053876.

Jacob, D., Crawford, J., Maring, H., Clarke, A., Dibb, J., Emmons, L., Ferrare, R., Hostetler, C., Russell, P. B., Singh, H., Thompson, A. M., Shaw, G., McCauley, E., Pederson, J. R., & Fisher, J. A. (2010). The Arctic research of the composition of the troposphere from Aircraft and Satellites (ARCTAS) mission: Design, execution, and first results. *Atmospheric Chemistry and Physics, 10*(11), 5191–5212. https://doi.org/10.5194/acp-10-5191-2010.

Jeansson, E., Olsen, A., Eldevik, T., Skjelvan, I., Omar, A. M., Lauvset, S. K., Nilsen, J. E. Ø., Bellerby, R. G. J., Johannessen, T., & Falck, E. (2011). The Nordic seas carbon budget: Sources, sinks, and uncertainties. *Global Biogeochemical Cycles, 25*, GB4010. https://doi.org/10.1029/2010GB003961.

Johnson, P. L., & Kelley, J. J. (1970). Dynamics of carbon dioxide and productivity in an Arctic biosphere. *Ecology, 51*(1), 73–80. https://doi.org/10.2307/1933600.

Johnston, H. S. (1971). Reduction of stratospheric ozone by nitrogen oxide catalysts from supersonic transport exhaust. *Science, 173*(3996), 517–522. https://doi.org/10.1126/science.173.3996.517.

Jones, A. E., Weller, R., Anderson, P. S., Jacobi, H.-W., Wolff, E. W., Schrems, O., & Miller, H. (2001). Measurements of NO_X emissions from the Antarctic snowpack. *Geophysical Research Letters, 28*(8), 1499–1502. https://doi.org/10.1029/2000GL011956.

Jones, A. E., Anderson, P. S., Begoin, M., Brough, N., Hutterli, M. A., Marshall, G. J., Richter, A., Roscoe, H. K., & Wolff, E. W. (2009). BrO, blizzards, and drivers of polar tropospheric ozone depletion events. *Atmospheric Chemistry and Physics, 9*(14), 4639–4652. https://doi.org/10.5194/acp-9-4639-2009.

Jones, A. E., Anderson, P. S., Wolff, E. W., Roscoe, H. K., Marshall, G. J., Richter, A., Brough, N., & Colwell, S. R. (2010). Vertical structure of Antarctic tropospheric ozone depletion events: Characteristics and broader implications. *Atmospheric Chemistry and Physics, 10*(16), 7775–7794. https://doi.org/10.5194/acp-10-7775-2010.

Jones, A. E., Wolff, E. W., Brough, N., Bauguitte, S. J.-B., Weller, R., Yela, M., Navarro-Comas, M., Ochoa, H. A., & Theys, N. (2013). The spatial scale of ozone depletion events derived from an autonomous surface ozone network in coastal Antarctica. *Atmospheric Chemistry and Physics, 13*(3), 1457–1467. https://doi.org/10.5194/acp-13-1457-2013.

Kaleschke, L., Richter, A., Burrows, J., Afe, O., Heygster, G., Notholt, A. M., Rankin, H. K., Roscoe, J., Hollwedel, T. W., & Jacobi, H.-W. (2004). Frost flowers on sea ice as a source of sea salt and their influence on tropospheric halogen chemistry. *Geophysical Research Letters, 31*(16), L16114. https://doi.org/10.1029/2004GL020655.

Kalnajs, L. E., & Avallone, L. M. (2006). Frost flower influence on springtime boundary-layer ozone depletion events and atmospheric bromine levels. *Geophysical Research Letters, 33*(10), L10810. https://doi.org/10.1029/2006GL025809.

Kanzawa, H., et al. (2003). Validation and data characteristics of nitrous oxide and methane profiles observed by the Improved Limb Atmospheric Spectrometer (ILAS) and processed with the Version 5.20 algorithm. *Journal of Geophysical Research: Atmospheres, 108*(D16), 8003. https://doi.org/10.1029/2002JD002458.

Kawa, S. R., Fahey, D. W., Anderson, L. C., Loewenstein, M., & Chan, K. R. (1990). Measurements of total reactive nitrogen during the airborne Arctic stratospheric expedition. *Geophysical Research Letters, 17*(4), 485–488. https://doi.org/10.1029/GL017i004p00485.

Keeling, C. D., Chin, J. F. S., & Whorf, T. P. (1996). Increased activity of northern vegetation inferred from atmospheric CO_2 measurements. *Nature, 382*, 146–149. https://doi.org/10.1038/382146a0.

Khattatov, V., Yushkov, V., Khaplanov, M., Zaitzev, I., Rosen, J., & Kjome, N. (1994). Some results of water vapor, ozone and aerosol balloon-borne measurements during EASOE. *Geophysical Research Letters, 21*(13), 1299–1302. https://doi.org/10.1029/93GL03559.

Khosrawi, F., Kirner, O., Sinnhuber, B.-M., Johansson, S., Höpfner, M., Santee, M. L., Froidevaux, L., Ungermann, J., Ruhnke, R., Woiwode, W., Oelhaf, H., & Braesicke, P. (2017). Denitrification, dehydration and ozone loss during the 2015/2016 Arctic winter. *Atmospheric Chemistry and Physics, 17*(21), 12893–12910. https://doi.org/10.5194/acp-17-12893-2017.

Kim, Y.-J., & Flatau, M. (2010). Hindcasting the January 2009 Arctic sudden stratospheric warming and its influence on the Arctic Oscillation with unified parametrization of orographic drag in NOGAPS. Part I: extended-range stand-alone forecast. *Weather and Forecasting, 25*(6), 1628–1644. https://doi.org/10.1175/2010WAF2222421.1.

Kirner, O., Ruhnke, R., & Sinnhuber, B.-M. (2015). Chemistry-climate interactions of stratospheric and mesospheric ozone in EMAC long-term simulations with different boundary conditions for CO_2, CH_4, N_2O, and ODS. *Atmosphere-Ocean, 53*(1), 140–152. https://doi.org/10.1080/07055900.2014.980718.

Kirschke, S., Bousquet, P., Ciais, P., Saunois, M., Canadell, J. G., Dlugokencky, E. J., Bergamaschi, P., Bergmann, D., Blake, D. R., Bruhwiler, L., Cameron-Smith, P., Castaldi, S., Chevallier, F., Feng, L., Fraser, A., Heimann, M., Hodson, E. L., Houweling, S., Josse, B., Fraser, P. J., Krummel, P. B., Lamarque, J.-F., Langenfelds, R. L., Le Quéré, C., Naik, V., O'Doherty, S., Palmer, P. I., Pison, I., Plummer, D., Poulter, B., Prinn, R. G., Rigby, M., Ringeval, B., Santini, M., Schmidt, M., Shindell, D. T., Simpson, I. J., Spahni, R., Steele, L. P., Strode, S. A., Sudo, K., Szopa, S., van der Werf, G. R., Voulgarakis, A., van Weele, M., Weiss, R. F., Williams, J. E., & Zeng, G. (2013). Three decades of global methane sources and sinks. *Nature Geoscience, 6*, 813–823. https://doi.org/10.1038/ngeo1955.

Klapstein, S. J., Turetsky, M. R., McGuire, A. D., Harden, J. W., Czimczik, C. I., Xu, X., Chanton, J. P., & Waddington, J. M. (2014). Controls on methane released through ebullition in peatlands affected by permafrost degradation. *Journal of Geophysical Research: Biogeosciences, 119*(3), 418–431. https://doi.org/10.1002/2013JG002441.

Knepp, T. N., Bottenheim, J., Carlsen, M., Carlson, D., Donohoue, D., Friederich, G., Matrai, P. A., Netcheva, S., Perovich, D. K., Santini, R., Shepson, P. B., Simpson, W., Valentic, T., Williams, C., & Wyss, P. J. (2010). Development of an autonomous sea ice tethered buoy for the study of ocean-atmosphere-sea ice-snow pack interactions: The O-buoy. *Atmospheric Measurement Techniques, 3*(1), 249–261. https://doi.org/10.5194/amt-3-249-2010.

Knibbe, J. S., van der A, R. J., & de Laat, A. T. J. (2014). Spatial regression analysis on 32 years of total column ozone data. *Atmospheric Chemistry and Physics, 14*(16), 8461–8482. https://doi.org/10.5194/acp-14-8461-2014.

Komhyr, W. D., Grass, R. D., & Leonard, R. K. (1986). Total ozone decrease at South pole, Antarctica, 1964–1985. *Geophysical Research Letters, 13*(12), 1248–1251. https://doi.org/10.1029/GL013i012p01248.

Kondo, Y., Schmidt, U., Sugita, T., Aimedieu, P., Koike, M., Ziereis, H., & Iwasaka, Y. (1994). Total reactive nitrogen, N_2O, and ozone in the winter Arctic stratosphere. *Geophysical Research Letters, 21*(13), 1247–1250. https://doi.org/10.1029/93GL03071.

Kondo, Y., Koike, M., Engel, A., Schmidt, U., Mueller, M., Sugita, T., Kanzawa, H., Nakazawa, T., Aoki, S., Irie, H., Toriyama, N., Suzuki, T., & Sasano, Y. (1999). NOy-N_2O correlation observed inside the Arctic vortex in February 1997: Dynamical and chemical effects. *Journal of Geophysical Research: Atmospheres, 104*(D7), 8215–8224. https://doi.org/10.1029/1999JD900020.

Koo, J.-H., Wang, Y., Jiang, T., Deng, Y., Oltmans, S. J., & Solberg, S. (2014). Influence of climate variability on near-surface ozone depletion events in the Arctic spring. *Geophysical Research Letters, 41*(7), 2582–2589. https://doi.org/10.1002/2014GL059275.

Koop, T., Biermann, U. M., Raber, W., Luo, B. P., Crutzen, P. J., & Peter, T. (1995). Do stratospheric aerosols freeze above the ice frost point? *Geophysical Research Letters, 22*(8), 917–920. https://doi.org/10.1029/95GL00814.

Koven, C. D., Lawrence, D. M., & Riley, W. J. (2015a). Permafrost carbon-climate feedback is sensitive to deep soil carbon decomposability but not deep soil nitrogen dynamics. *Proceedings of the National Academy of Sciences, 112*(12), 3752–3757. https://doi.org/10.1073/pnas.1415123112.

Koven, C. D., Schuur, E. A. G., Schädel, C., Bohn, T. J., Burke, E. J., Chen, G., Chen, X., Ciais, P., Grosse, G., Harden, J. W., Hayes, D. J., Hugelius, G., Jafarov, E. E., Krinner, G., Kuhry, P., Lawrence, D. M., MacDougall, A. H., Marchenko, S. S., McGuire, A. D., Natali, S. M., Nicolsky, D. J., Olefeldt, D., Peng, S., Romanovsky, V. E., Schaefer, K. M., Strauss, J., Treat, C. C., & Turetsky, M. (2015b). A simplified, data-constrained approach to estimate the permafrost carbon – Climate feedback. *Philosophical Transactions of the Royal Society A, 373*, 20140423. https://doi.org/10.1098/rsta.2014.0423.

Kretschmer, K., Biastoch, A., Rüpke, L., & Burwicz, E. (2015). Modeling the fate of methane hydrates under global warming. *Global Biogeochemical Cycles, 29*, 610–625. https://doi.org/10.1002/2014GB005011.

Krueger, A. J., Schoeberl, M. R., & Stolarski, R. S. (1987). TOMS observations of total ozone in the 1986 Antarctic spring. *Geophysical Research Letters, 14*(5), 527–530. https://doi.org/10.1029/GL014i005p00527.

Kruger, B. (1990). Observations of polar stratospheric clouds in the Arctic winter 1989 at 79°N. *Geophysical Research Letters, 17*(4), 365–268. https://doi.org/10.1029/GL017i004p00365.

Kulawik, S. S., Jones, D. B. A., Nassar, R., Irion, F. W., Worden, J. R., Bowman, K. W., Machida, T., Matsueda, H., Sawa, Y., Biraud, S. C., Fischer, M. L., & Jacobson, A. R. (2010). Characterization of Tropospheric Emission Spectrometer (TES) CO_2 for carbon cycle science. *Atmospheric Chemistry and Physics, 10*(12), 5601–5623. https://doi.org/10.5194/acp-10-5601-2010.

Kuze, A., Suto, H., Nakajima, M., & Hamazaki, T. (2009). Thermal and near infrared sensor for carbon observation Fourier-transform spectrometer on the greenhouse gases observing satellite for greenhouse gases monitoring. *Applied Optics, 48*(35), 6716–6733. https://doi.org/10.1364/AO.48.006716.

Labitzke, K., & van Loon, H. (1999). *The stratosphere: Phenomena, history, and relevance* (p. 179). Berlin: Springer.

Langematz, U., Meul, S., Grunow, K., Romanowsky, E., Oberländer, S., Abalichin, J., & Kubin, A. (2014). Future Arctic temperature and ozone: The role of stratospheric composition changes. *Journal of Geophysical Research: Atmospheres, 119*(5), 2092–2112. https://doi.org/10.1002/2013JD021100.

Law, K. S., Stohl, A., Quinn, P. K., Brock, C. A., Burkhart, J. F., Paris, J.-D., Ancellet, G., Singh, H. B., Roiger, A., Schlager, H., Dibb, J., Jacob, D. J., Arnold, S. R., Pelon, J., & Thomas, J. L. (2014). Arctic air pollution: New insights from POLARCAT-IPY. *Bulletin of the American Meteorological Society, 95*(12), 1873–1895. https://doi.org/10.1175/BAMS-D-13-00017.1.

Lawrence, D. M., Koven, C. D., Swenson, S. C., Riley, W. J., & Slater, A. G. (2015). Permafrost thaw and resulting soil moisture changes regulate projected high-latitude CO_2 and CH_4 emissions. *Environmental Research Letters, 10*(9), 094011. https://doi.org/10.1088/1748-9326/10/9/094011.

Le Quéré, C., et al. (2018). Global Carbon Budget 2017. *Earth System Science Data, 10*(1), 405–448. https://doi.org/10.5194/essd-10-405-2018.

Legrand, M., Yang, X., Preunkert, S., & Theys, N. (2016). Year-round records of sea salt, gaseous, and particulate inorganic bromine in the atmospheric boundary layer at coastal (Dumont d'Urville) and central (Concordia) East Antarctic sites. *Journal of Geophysical Research: Atmospheres, 121*(2), 997–1023. https://doi.org/10.1002/2015JD024066.

Li, D., Zhang, R., & Knutson, T. R. (2017). On the discrepancy between observed and CMIP5 multi-model simulated Barents Sea winter sea ice decline. *Nature Communications, 8*, 14991. https://doi.org/10.1038/ncomms14991.

Liao, J., Huey, L. G., Tanner, D. J., Flocke, F. M., Orlando, J. J., Neuman, J. A., Nowak, J. B., Weinheimer, A. J., Hall, S. R., Smith, J. N., Fried, A., Staebler, R. M., Wang, Y., Koo, J.-H., Cantrell, C. A., Weibring, P., Walega, J., Knapp, D. J., Shepson, P. B., & Stephens, C. R. (2012). Observations of inorganic bromine (HOBr, BrO, and Br_2) speciation at Barrow, Alaska, in spring 2009. *Journal of Geophysical Research: Atmospheres, 117*(D14), D00R16. https://doi.org/10.1029/2011JD016641.

Liao, J., Huey, L. G., Liu, Z., Tanner, D. J., Cantrell, C. A., Orlando, J. J., Flocke, F. M., Shepson, P. B., Weinheimer, A. J., Hall, S. R., Ullmann, K., Beine, H. J., Wang, Y., Ingall, E. D., Stephens, C. R., Hornbrook, R. S., Apel, E. C., Riemer, D., Fried, A., Mauldin, R. L., III, Smith, J. N., Staebler, R. M., Neuman, J. A., & Nowak, J. B. (2014). High levels of molecular chlorine in the Arctic atmosphere. *Nature Geoscience, 7*(2), 91–94. https://doi.org/10.1038/NGEO2046.

Lindenmaier, R., Strong, K., Batchelor, R. L., Chipperfield, M. P., Daffer, W. H., Drummond, J. R., Duck, T. J., Fast, H., Feng, W., Fogal, P. F., Kolonjari, F., Manney, G. L., Manson, A., Meek, C., Mittermeier, R. L., Nott, G. J., Perro, C., & Walker, K. A. (2012). Unusually low ozone, HCl, and HNO_3 column measurements at Eureka, Canada during winter/spring 2011. *Atmospheric Chemistry and Physics, 12*(8), 3821–3835. https://doi.org/10.5194/acp-12-3821-2012.

Lutsch, E., Dammers, E., Conway, S., & Strong, K. (2016). Long-range transport of NH_3, CO, HCN, and C_2H_6 from the 2014 Canadian wildfires. *Geophysical Research Letters, 43*(15), 8286–8297. https://doi.org/10.1002/2016GL070114.

MacGilchrist, G. A., Naveira Garabato, A. C., Tsubouchi, T., Bacon, S., Torres-Valdés, S., & Azetsu-Scott, K. (2014). The Arctic Ocean carbon sink. *Deep-Sea Research, 1*(86), 39–55. https://doi.org/10.1016/j.dsr.2014.01.002.

Manney, G. L., & Lawrence, Z. D. (2016). The major stratospheric final warming in 2016: Dispersal of vortex air and termination of Arctic chemical ozone loss. *Atmospheric Chemistry and Physics, 16*(23), 15371–15396. https://doi.org/10.5194/acp-16-15371-2016.

Manney, G. L., Daffer, W. H., Zawodny, J. M., Bernath, P. F., Hoppel, K. W., Walker, K. A., Knosp, B. W., Boone, C., Remsberg, E. E., Santee, M. L., Harvey, V. L., Pawson, S., Jackson, D. R., Deaver, L., McElroy, C. T., McLinden, C. A., Drummond, J. R., Pumphrey, H. C., Lambert, A., Schwartz, M. J., Froidevaux, L., McLeod, S., Takacs, L. L., Suarez, M. J., Trepte, C. R., Cuddy, D. C., Livesey, N. J., Harwood, R. S., & Waters, J. W. (2007). Solar occultation satellite data and derived meteorological products: Sampling issues and comparisons with Aura microwave limb sounder. *Journal of Geophysical Research: Atmospheres, 112, D24S50.* https://doi.org/10.1029/2007JD008709.

Manney, G. L., Santee, M. L., Rex, M., Livesey, N. J., Pitts, M. C., Veefkind, P., Nash, E. R., Wohltmann, I., Lehmann, R., Froidevaux, L., Poole, L. R., Schoeberl, M. R., Haffner, D. P., Davies, J., Dorokhov, V., Gernandt, H., Johnson, B., Kivi, R., Kyrö, E., Larsen, N., Levelt, P. F., Makshtas, A., McElroy, C. T., Nakajima, H., Parrondo, M. C., Tarasick, D. W., von der

Gathen, P., Walker, K. A., & Zinoviev, N. S. (2011). Unprecedented Arctic ozone loss in 2011. *Nature, 478*, 469–475. https://doi.org/10.1038/nature10556.

Marelle, L., Thomas, J. L., Raut, J.-C., Law, K. S., Jalkanen, J.-P., Johansson, L., Roiger, A., Schlager, H., Kim, J., Reiter, A., & Weinzierl, B. (2016). Air quality and radiative impacts of Arctic shipping emissions in the summertime in northern Norway: From the local to the regional scale. *Atmospheric Chemistry and Physics, 16*(4), 2359–2379. https://doi.org/10.5194/acp-16-2359-2016.

McConnell, J. C., & Jin, J. J. (2008). Stratospheric ozone chemistry. *Atmosphere-Ocean, 46*(1), 69–92. https://doi.org/10.3137/ao.460104.

McCormick, M., Kent, G., Hunt, W., Osborn, M., Poole, L., & Pitts, M. (1990). Arctic polar stratospheric cloud observations by airborne lidar. *Geophysical Research Letters, 17*(4), 381–383. https://doi.org/10.1029/GL017i004p00381.

McElroy, C. T., & Fogal, P. F. (2008). Understanding ozone depletion: Measurements and models. *Atmosphere-Ocean, 46*(1), 15–37. https://doi.org/10.3137/ao.4601002.

McElroy, C. T., & Francis, D. R. (Eds.). (2007). Ozone science 2007: A Canadian contribution to the twentieth anniversary of the Montreal Protocol. Environment Canada, Catalogue No. En56-218/2007E, ISBN: 978-0-662-46843-1

McElroy, M. B., Salawitch, R. J., Wofsy, S. C., & Logan, J. A. (1986). Reductions of Antarctic ozone due to synergistic interactions of chlorine and bromine. *Nature, 321*, 759–762. https://doi.org/10.1038/321759a0.

McElroy, C. T., McLinden, C. A., & McConnell, J. C. (1999). Evidence for bromine monoxide in the free troposphere during the Arctic polar sunrise. *Nature, 397*, 338–341. https://doi.org/10.1038/16904.

McGuire, A. D., Anderson, L. G., Christensen, T. R., Dallimore, S., Guo, L., Hayes, D. J., Heimann, M., Lorenson, T. D., Macdonald, R. W., & Roulet, N. (2009). Sensitivity of the carbon cycle in the Arctic to climate change. *Ecological Monographs, 79*(4), 523–555. https://doi.org/10.1890/08-2025.1.

McGuire, A. D., Christensen, T. R., Hayes, D., Heroult, A., Euskirchen, E., Kimball, J. S., Koven, C., Lafleur, P., Miller, P. A., Oechel, W., Peylin, P., Williams, M., & Yi, Y. (2012). An assessment of the carbon balance of Arctic tundra: Comparisons among observations, process models, and atmospheric inversions. *Biogeosciences, 9*(8), 3185–3204. https://doi.org/10.5194/bg-9-3185-2012.

McGuire, D. A., Lawrence, D. M., Koven, C., Clein, J. S., Burke, E., Chen, G., Jafarov, E., MacDougall, A. H., Marchenko, S., Nicolsky, D., Peng, S., Rinke, A., Ciais, P., Gouttevin, I., Hayes, D. J., Ji, D., Krinner, G., Moore, J. C., Romanovsky, V., Schädel, C., Schaefer, K., Schuur, E. A. G., & Zhuang, Q. (2018). Dependence of the evolution of carbon dynamics in the northern permafrost region on the trajectory of climate change. *Proceedings of the National Academy of Sciences, 115*(15), 3882–3887. https://doi.org/10.1073/pnas.1719903115.

McIntyre, M. E., & Palmer, T. N. (1983). Breaking planetary waves in the stratosphere. *Nature, 305*, 593–594. https://doi.org/10.1038/305593a0.

Merrelli, A., Bennartz, R., O'Dell, C. W., & Taylor, T. E. (2015). Estimating bias in the OCO-2 retrieval algorithm caused by 3-D radiation scattering from unresolved boundary layer clouds. *Atmospheric Measurement Techniques, 8*(4), 1641–1656. https://doi.org/10.5194/amt-8-1641-2015.

Michelsen, H. A., Manney, G. L., Gunson, M. R., & Zander, R. (1998). Correlations of stratospheric abundances of NO_y, O_3, N_2O, and CH_4 derived from ATMOS measurements. *Journal of Geophysical Research: Atmospheres, 103*(D21), 28347 – 28359. https://doi.org/10.1029/98JD02850.

Miller, S. M., Miller, C. E., Commane, R., Chang, R. Y.-W., Dinardo, S. J., Henderson, J. M., Karion, A., Lindaas, J., Melton, J. R., Miller, J. B., Sweeney, C., Wofsy, S. C., & Michalak, A. M. (2016). A multiyear estimate of methane fluxes in Alaska from CARVE atmospheric observations. *Global Biogeochemical Cycles, 30*(10), 1441–1453. https://doi.org/10.1002/2016GB005419.

Molina, M. J., & Rowland, F. S. (1974). Stratospheric sink for chlorofluoromethanes: Chlorine atom catalyzed destruction of ozone. *Nature, 249*, 820–812. https://doi.org/10.1038/249810a0.

Monks, S. A., Arnold, S. R., Emmons, L. K., Law, K. S., Turquety, S., Duncan, B. N., Flemming, J., Huijnen, V., Tilmes, S., Langner, J., Mao, J., Long, Y., Thomas, J. L., Steenrod, S. D., Raut, J. C., Wilson, C., Chipperfield, M. P., Diskin, G. S., Weinheimer, A., Schlager, H., & Ancellet, G. (2015). Multi-model study of chemical and physical controls on transport of anthropogenic and biomass burning pollution to the Arctic. *Atmospheric Chemistry and Physics, 15*(6), 3575–3603. https://doi.org/10.5194/acp-15-3575-2015.

Moore, C. W., Obrist, D., Steffen, A., Staebler, R. M., Douglas, T. A., Richter, A., & Nghiem, S. V. (2014). Convective forcing of mercury and ozone in the Arctic boundary layer induced by leads in sea ice. *Nature, 506*(7486), 81–84. https://doi.org/10.1038/nature12924.

Morgenstern, O., Stone, K. A., Schofield, R., Akiyoshi, H., Yamashita, Y., Kinnison, D. E., Garcia, R. R., Sudo, K., Plummer, D. A., Scinocca, J., Oman, L. D., Manyin, M. E., Zeng, G., Rozanov, E., Stenke, A., Revell, L. E., Pitari, G., Mancini, E., Di Genova, G., Visioni, D., Dhomse, S. S., & Chipperfield, M. P. (2018). Ozone sensitivity to varying greenhouse gases and ozone-depleting substances in CCMI-1 simulations. *Atmospheric Chemistry and Physics, 18*(2), 1091–1114. https://doi.org/10.5194/acp-18-1091-2018.

Mount, G., Solomon, S., Jakoubek, R., & Schmeltekopf, A. (1988). Observations of stratospheric NO_2 and O_3 at Thule, Greenland. *Science, 242*(4878), 555–558. https://doi.org/10.1126/science.242.4878.555.

Müller, R., Grooß, J.-U., Lemmen, C., Heinze, D., Dameris, M., & Bodeker, G. (2008). Simple measures of ozone depletion in the polar stratosphere. *Atmospheric Chemistry and Physics, 8*(2), 251–264. https://doi.org/10.5194/acp-8-251-2008.

Murcray, F., Starkey, J., Williams, W., Matthews, W., Schmidt, U., Aimedieu, P., & Camy-Peyret, C. (1994). HNO_3 profiles obtained during the EASOE campaign. *Geophysical Research Letters, 21*(13), 1223–1226. https://doi.org/10.1029/93GL01056.

Myhre, G., et al. (2013). Anthropogenic and natural radiative forcing. In T. F. Stocker, D. Qin, G.-K. Plattner, M. Tignor, S. K. Allen, J. Boschung, A. Nauels, Y. Xia, V. Bex, & P. M. Midgley (Eds.), *Climate change 2013: The physical science basis. Contribution of working group I to the fifth assessment report of the Intergovernmental Panel on Climate Change*. Cambridge/New York: Cambridge University Press.

Neuber, R., Beyerle, G., Fiocco, G., di Sarra, A., Fricke, K.-H., Knudsen, B., David, C., Godin, S., Stefanutti, L., & Vaughanet, G. (1994). Latitudinal distribution of stratospheric aerosols during the EASOE winter 1991/92. *Geophysical Research Letters, 21*(13), 1283–1286. https://doi.org/10.1029/93GL02890.

Newman, P. A., Fahey, D. W., Brune, W. H., Kurylo, M. J., & Kawa, S. R. (1999). Preface [to special section on Photochemistry of Ozone Loss in the Arctic Region in Summer (POLARIS)]. *Journal of Geophysical Research: Atmospheres, 104*(D21), 26481–26495. https://doi.org/10.1029/1999JD900832.

Newman, P. A., Harris, N. R. P., Adriani, A., Amanatidis, G. T., Anderson, J. G., Braathen, G. O., Brune, W. H., Carslaw, K. S., Craig, M. S., DeCola, P. L., Guirlet, M., Hipskind, R. S., Kurylo, M. J., Küllmann, H., Larsen, N., Mégie, G. J., Pommereau, J.-P., Poole, L. R., Schoeberl, M. R., Stroh, F., Toon, O. B., Trepte, C. R., & van Roozendael, M. (2002). An overview of the SOLVE/THESEO 2000 campaign. *Journal of Geophysical Research: Atmospheres, 107*(D20), 8259. https://doi.org/10.1029/2001JD001303.

Nghiem, S. V., Clemente-Colón, P., Douglas, T., Moore, C., Obrist, D., Perovich, D. K., Pratt, K. A., Rigor, I. G., Simpson, W., Shepson, P. B., Steffen, A., & Woods, J. (2013). Studying bromine, ozone, and mercury chemistry in the Arctic. *Eos, Transactions American Geophysical Union, 94*(33), 289–291. https://doi.org/10.1002/2013EO330002.

Nicolet, M. (1975). Stratospheric ozone: An introduction to its study. *Reviews of Geophysics, 13*(5), 593–636. https://doi.org/10.1029/RG013i005p00593.

Nisbet, E. G., Dlugokencky, E. J., Manning, M. R., Lowry, D., Fisher, R. E., France, J. L., Michel, S. E., Miller, J. B., White, J. W. C., Vaughn, B., Bousquet, P., Pyle, J. A., Warwick, N. J., Cain, M., Brownlow, R., Zazzeri, G., Lanoisellé, M., Manning, A. C., Gloor, E., Worthy, D. E. J.,

Brunke, E.-G., Labuschagne, C., Wolff, E. W., & Ganesan, A. L. (2016). Rising atmospheric methane : 2007–2014 growth and isotopic shift. *Global Biogeochemical Cycles, 30*(9), 1356–1370. https://doi.org/10.1002/2016GB005406.

Northway, R. S., Gao, P. J., Popp, J. C., Holecek, D. W., Fahey, K. S., Carslaw, M. A., Tolbert, L. R., Lait, S., Dhaniyala, R. C., Flagan, P. O., Wennberg, M. J., Mahoney, R. L., Herman, G. C. T., & Bui, T. P. (2002). An analysis of large HNO_3-containing particles sampled in the Arctic stratosphere during the winter of 1999/2000. *Journal of Geophysical Research: Atmospheres, 107*(D20), 8298–8319. https://doi.org/10.1029/2001JD001079.

Notholt, J. (1994). The moon as a light source for FTIR measurements of stratospheric trace gases during the polar night: Application for HNO_3 in the Arctic. *Journal of Geophysical Research: Atmospheres, 99*(D2), 3607–3614. https://doi.org/10.1029/93JD03040.

Notholt, J., Neuber, R., Schrems, O., & Clarmann, T. V. (1993). Stratospheric trace gas concentrations in the Arctic polar night as derived by FTIR-spectroscopy with the moon as IR light source. *Geophysical Research Letters, 20*(19), 2059–2062. https://doi.org/10.1029/93GL01971.

Notholt, J., von der Gathen, P., & Peil, S. (1995). Heterogeneous conversion of HCl and $ClONO_2$ during the Arctic winter 1992/1993 initiating ozone depletion. *Journal of Geophysical Research: Atmospheres, 100*(D6), 11269–11274. https://doi.org/10.1029/95JD00850.

Notholt, J., Toon, G., Stordal, F., Solberg, S., Schmidbauer, N., Becker, E., Meier, A., & Sen, B. (1997). Seasonal variations of atmospheric trace gases in the high Arctic at 79°N. *Journal of Geophysical Research: Atmospheres, 102*(D11), 12855–12861. https://doi.org/10.1029/97JD00337.

Obbard, R. W., Roscoe, H. K., Wolff, E. W., & Atkinson, H. M. (2009). Frost flower surface area and chemistry as a function of salinity and temperature. *Journal of Geophysical Research: Atmospheres, 114*, D20305. https://doi.org/10.1029/2009JD012481.

Oberländer, S., Langematz, U., & Meul, S. (2013). Unraveling impact factors for future changes in the Brewer-Dobson circulation. *Journal of Geophysical Research: Atmospheres, 118*(18), 10,296–10,312. https://doi.org/10.1002/jgrd.50775.

Oelhaf, H., von Clarmann, T., Fischer, H., Friedl-Vallon, F., Fritzsche, C., Linden, A., Piesch, C., Seefeldner, M., & Völker, W. (1994). Stratospheric $ClONO_2$ and HNO_3 profiles inside the Arctic vortex from MIPAS-B limb emission spectra obtained during EASOE. *Geophysical Research Letters, 21*(13), 1263–1266. https://doi.org/10.1029/93GL01303.

Oltmans, S. J. (1993). Climatology of Arctic and Antarctic tropospheric ozone. In *The tropospheric chemistry of ozone in the polar regions* (pp. 25–40). Berlin/Heidelberg: Springer Berlin Heidelberg.

Oltmans, S. J., & Komhyr, W. (1986). Surface ozone distributions and variations from 1973–1984 measurements at the NOAA Geophysical Monitoring for Climate Change Baseline observatories. *Journal of Geophysical Research: Atmospheres, 91*(D4), 5229–5236. https://doi.org/10.1029/JD091iD04p05229.

Oltmans, S. J., Johnson, B. J., & Harris, J. M. (2012). Springtime boundary layer ozone depletion at Barrow, Alaska: Meteorological influence, year-to-year variation, and long-term change. *Journal of Geophysical Research: Atmospheres, 117*(D14), D00R18. https://doi.org/10.1029/2011JD016889.

Oman, L. D., Plummer, D. A., Waugh, D. W., Austin, J., Scinocca, J. F., Douglass, A. R., Salawitch, R. J., Canty, T., Akiyoshi, H., Bekki, S., Braesicke, P., Butchart, N., Chipperfield, M. P., Cugnet, D., Dhomse, S., Eyring, V., Frith, S., Hardiman, S. C., Kinnison, D. E., Lamarque, J.-F., Mancini, E., Marchand, M., Michou, M., Morgenstern, O., Nakamura, T., Nielsen, J. E., Olivié, D., Pitari, G., Pyle, J., Rozanov, E., Shepherd, T. G., Shibata, K., Stolarski, R. S., Teyssèdre, H., Tian, W., Yamashita, Y., & Ziemke, J. R. (2010). Multimodel assessment of the factors driving stratospheric ozone evolution over the 21st century. *Journal of Geophysical Research: Atmospheres, 115*, D24306. https://doi.org/10.1029/2010JD014362.

Osterkamp, T. E. (2007). Characteristics of the recent warming of permafrost in Alaska. *Journal of Geophysical Research: Earth Surface, 112*, F02S02. https://doi.org/10.1029/2006JF000578.

Parazoo, N. C., Commane, R., Wofsy, S. C., Koven, C. D., Sweeney, C., Lawrence, D. M., Lindaas, J., Chang, R. Y. W., & Miller, C. E. (2016). Detecting regional patterns of changing CO_2 flux in Alaska. *Proceedings of the National Academy of Science, 113*(28), 7733–7738. https://doi.org/10.1073/pnas.1601085113.

Parazoo, N. C., Koven, C. D., Lawrence, D. M., Romanovsky, V., & Miller, C. E. (2018). Detecting the permafrost carbon feedback: Talik formation and increased cold-season respiration as precursors to sink-to-source transitions. *The Cryosphere, 12*(1), 123–144. https://doi.org/10.5194/tc-12-123-2018.

Parmentier, F.-J. W., Christensen, T. R., Rysgaard, S., Bendtsen, J., Glud, R. N., Else, B., van Huissteden, J., Sachs, T., Vonk, J. E., & Sejr, M. K. (2017). A synthesis of the arctic terrestrial and marine carbon cycles under pressure from a dwindling cryosphere. *Ambio, 46*(Suppl 1), 53–69. https://doi.org/10.1007/s13280-016-0872-8.

Payne, V. H., Clough, S. A., Shephard, M. W., Nassar, R., & Logan, J. A. (2009). Information-centered representation of retrievals with limited degrees of freedom for signal: Application to methane from the tropospheric emission spectrometer. *Journal of Geophysical Research: Atmospheres, 114*, D10307. https://doi.org/10.1029/2008JD010155.

Peters, W., Jacobson, A. R., Sweeney, C., Andrews, A. E., Conway, T. J., Masarie, K., Miller, J. B., Bruhwiler, L. M. P., Petron, G., Hirsch, A. I., Worthy, D. E. J., van der Werf, G. R., Randerson, J. T., Wennberg, P. O., Krol, M. C., & Tans, P. P. (2007). An atmospheric perspective on North American carbon dioxide exchange: CarbonTracker. *Proceedings of the National Academy of Sciences, 104*, 18925–18930. https://doi.org/10.1073/pnas.0708986104.

Peterson, P. K., Simpson, W. R., Pratt, K. A., Shepson, P. B., Frieß, U., Zielcke, J., Platt, U., Walsh, S. J., & Nghiem, S. V. (2015). Dependence of the vertical distribution of bromine monoxide in the lower troposphere on meteorological factors such as wind speed and stability. *Atmospheric Chemistry and Physics, 15*(4), 2119–2137. https://doi.org/10.5194/acp-15-2119-2015.

Peterson, P. K., Pöhler, D., Sihler, H., Zielcke, J., General, S., Frieß, U., Platt, U., Simpson, W. R., Nghiem, S. V., Shepson, P. B., Stirm, B. H., Dhaniyala, S., Wagner, T., Caulton, D. R., Fuentes, J. D., & Pratt, K. A. (2017). Observations of bromine monoxide transport in the Arctic sustained on aerosol particles. *Atmospheric Chemistry and Physics, 17*(12), 7567–7579. https://doi.org/10.5194/acp-17-7567-2017.

Petzoldt, K. (1999). The role of dynamics in total ozone deviations from their long-term mean over the Northern hemisphere. *Annales Geophysicae, 17*(2), 231–241. https://doi.org/10.1007/s00585-999-0231-1.

Pirk, N., Tamstorf, M. P., Lund, M., Mastepanov, M., Pedersen, S. H., Mylius, M. R., Parmentier, F.-J. W., Christiansen, H. H., & Christensen, T. R. (2016). Snowpack fluxes of methane and carbon dioxide from high Arctic tundra. *Journal of Geophysical Research: Biogeosciences, 121*(11), 2886–2900. https://doi.org/10.1002/2016JG003486.

Pistone, K., Eisenman, I., & Ramanathan, V. (2014). Observational determination of albedo decrease caused by vanishing Arctic Sea ice. *Proceedings of the National Academy of Sciences, 111*(9), 3322–3326. https://doi.org/10.1073/pnas.1318201111.

Plumb, R. A., Waugh, D. W., & Chipperfield, M. P. (2000). The effects of mixing on tracer relationships in the polar vortices. *Journal of Geophysical Research: Atmospheres, 105*(D8), 10047–10062. https://doi.org/10.1029/1999JD901023.

Pommereau, J.-P., & Piquard, J. (1994a). Observations of the vertical distribution of stratospheric OClO. *Geophysical Research Letters, 21*(13), 1231–1234. https://doi.org/10.1029/94GL00390.

Pommereau, J., & Piquard, J. (1994b). Ozone and nitrogen dioxide vertical distributions by UV-visible solar occultation from balloons. *Geophysical Research Letters, 21*(13), 1227–1230. https://doi.org/10.1029/94GL00389.

Pommereau, J.-P., Goutail, F., Lefèvre, F., Pazmino, A., Adams, C., Dorokhov, V., Eriksen, P., Kivi, R., Stebel, K., Zhao, X., & van Roozendael, M. (2013). Why unprecedented ozone loss in the Arctic in 2011? Is it related to climate change? *Atmospheric Chemistry and Physics, 13*(10), 5299–5308. https://doi.org/10.5194/acp-13-5299-2013.

Pommereau, J.-P., Goutail, F., Pazmino, A., Lefèvre, F., Chipperfield, M. P., Feng, W., von Roozendael, M., Jipsen, N., Hansen, G., Kivi, R., Bognar, K., Strong, K., Walker, K.,

Kuzmichev, A., Khattatov, S., & Sitnikova, V. (2018). Recent Arctic ozone depletion: Is there an impact of climate change? *Comptes Rendus Geoscience, 350*(7), 347–353. https://doi.org/10.1016/j.crte.2018.07.009.

Prather, M. J., & Holmes, C. D. (2017). Overexplaining or underexplaining methane's role in climate change. *Proceedings of the National Academy of Sciences, 114*(21), 5324–5326. https://doi.org/10.1073/pnas.1704884114.

Prather, M. J., McElroy, M. B., & Wofsy, S. C. (1984). Reductions in ozone at high concentrations of stratospheric halogens. *Nature, 312*, 227–231. https://doi.org/10.1038/312227a0.

Prather, M. J., Holmes, C. D., & Hsu, J. (2012). Reactive greenhouse gas scenarios: Systematic exploration of uncertainties and the role of atmospheric chemistry. *Geophysical Research Letters, 39*, L09803. https://doi.org/10.1029/2012gl051440.

Pratt, K. A., Custard, K. D., Shepson, P. B., Douglas, T. A., Pöhler, D., General, S., Zielcke, J., Simpson, W. R., Platt, U., Tanner, D. J., Huey, L. G., Carlsen, M., & Stirm, B. H. (2013). Photochemical production of molecular bromine in Arctic surface snowpacks. *Nature Geoscience, 6*(5), 351–356. https://doi.org/10.1038/NGEO1779.

Pyle, J. A., Harris, N. R. P., Farman, J. C., Arnold, F., Braathen, G., Cox, R. A., Faucon, P., Jones, R. L., Megie, G., O'Neill, A., Platt, U., Pommereau, J.-P., Schmidt, U., & Stordal, F. (1994). An overview of the EASOE campaign. *Geophysical Research Letters, 21*(13), 1191–1194. https://doi.org/10.1029/94GL00004.

Rankin, A. M., Wolff, E. W., & Martin, S. (2002). Frost flowers: Implications for tropospheric chemistry and ice core interpretation. *Journal of Geophysical Research: Atmospheres, 107*(D23), AAC-4. https://doi.org/10.1029/2002JD002492.

Raso, A. R. W., Custard, K. D., May, N. W., Tanner, D., Newburn, M. K., Walker, L., Moore, R. J., Huey, L. G., Alexander, L., Shepson, P. B., & Pratt, K. A. (2017). Active molecular iodine photochemistry in the Arctic. *Proceedings of the National Academy of Sciences, 114*(38), 10053–10058. https://doi.org/10.1073/pnas.1702803114.

Ravishankara, A. R., Daniel, J. S., & Portmann, R. W. (2009). Nitrous oxide (N_2O): The dominant ozone-depleting substance emitted in the 21st century. *Science, 326*(5949), 123–125. https://doi.org/10.1126/science.1176985.

Rex, M., Salawitch, R. J., Toon, G. C., Sen, B., Margitan, J. J., Osterman, G. B., Blavier, J.-F., Gao, R. S., Donnelly, S., Keim, E., Neuman, J., Fahey, D. W., Webster, C. R., Scott, D. C., Herman, R. L., May, R. D., Moyer, E. J., Gunson, M. R., Irion, F. W., Chang, A. Y., Rinsland, C. P., & Bui, T. P. (1999). Subsidence, mixing, and denitrification of Arctic polar vortex air measured during POLARIS. *Journal of Geophysical Research: Atmospheres, 104*(D21), 26611–26623. https://doi.org/10.1029/1999JD900463.

Rex, M., Salawitch, R. J., von der Gathen, P., Harris, N. R. P., Chipperfield, M. P., & Naujokat, B. (2004). Arctic ozone loss and climate change. *Geophysical Research Letters, 31*, L04116. https://doi.org/10.1029/2003GL018844.

Rex, M., Salawitch, R. J., Deckelmann, H., von der Gathen, P., Harris, N. R. P., Chipperfield, M. P., Naujokat, B., Reimer, E., Allaart, M., Andersen, S. B., Bevilacqua, R., Braathen, G. O., Claude, H., Davies, J., de Backer, H., Dier, H., Dorokhov, V., Fast, H., Gerding, M., Godin-Beekmann, S., Hoppel, K., Johnson, B., Kyrö, E., Litynska, Z., Moore, D., Nakane, H., Parrondo, M. C., Risley, A. D., Jr., Skrivankova, P., Stübi, R., Viatte, P., Yushkov, V., & Zerefos, C. (2006). Arctic winter 2005: Implications for stratospheric ozone loss and climate change. *Geophysical Research Letters, 33*, L23808. https://doi.org/10.1029/2006GL026731.

Rieder, H. E., & Polvani, L. M. (2013). Are recent Arctic ozone losses caused by increasing greenhouse gases? *Geophysical Research Letters, 40*(16), 4437–4441. https://doi.org/10.1002/grl.50835.

Rieder, H. E., Polvani, L. M., & Solomon, S. (2014). Distinguishing the impacts of ozone-depleting substances and well-mixed greenhouse gases on Arctic stratospheric ozone and temperature trends. *Geophysical Research Letters, 41*(7), 2652–2660. https://doi.org/10.1002/2014GL059367.

Rigby, M., Montzka, S. A., Prinn, R. G., White, J. W. C., Young, D., O'Doherty, S., Lunt, M. F., Ganesan, A. L., Manning, A. J., Simmonds, P. G., Salameh, P. K., Harth, C. M., Mühle, J.,

Weiss, R. F., Fraser, P. J., Steele, L. P., Krummel, P. B., McCulloch, A., & Park, S. (2017). Role of atmospheric oxidation in recent methane growth. *Proceedings of the National Academy of Sciences, 114*(21), 5373–5377. https://doi.org/10.1073/pnas.1616426114.

Rinsland, C. P., Gunson, M. R., Salawitch, R. J., Newchurch, M. J., Zander, R., Abbas, M. M., Abrams, M. C., Manney, G. L., Michelsen, H. A., Chang, A. Y., & Goldman, A. (1996). ATMOS measurements of $H_2O + 2CH_4$ and total reactive nitrogen in the November 1994 Antarctic stratosphere: Dehydration and denitrification in the vortex. *Geophysical Research Letters, 23*(17), 2397–2400. https://doi.org/10.1029/96GL00048.

Roiger, A., Thomas, J., Schlager, H., Law, K. S., Kim, J., Schäfler, A., Weinzierl, B., Dahlkötter, F., Krisch, I., Marelle, L., Minikin, A., Raut, J., Reiter, A., Rose, M., Scheibe, M., Stock, P., Baumann, R., Bouarar, I., Clerbaux, C., George, M., Onishi, T., & Flemming, J. (2015). Quantifying emerging local anthropogenic emissions in the Arctic region: The ACCESS aircraft campaign experiment. *Bulletin of the American Meteorological Society, 96*, 441–460. https://doi.org/10.1175/BAMS-D-13-00169.1.

Romanovsky, V., Smith, S., & Christiansen, H. (2010). Permafrost thermal state in the polar Northern Hemisphere during the International Polar Year 2007–2009: a synthesis. *Permafrost and Periglacial Processes, 21*(2), 106–116. https://doi.org/10.1002/ppp.689.

Rosenblum, E., & Eisenman, I. (2017). Sea ice trends in climate models only accurate in runs with biased global warming. *Journal of Climate, 30*(16), 6265–6278. https://doi.org/10.1175/JCLI-D-16-0455.1.

Rowland, F. S., & Molina, M. J. (1975). Chlorofluoromethanes in the environment. *Reviews of Geophysics, 13*(1), 1–35. https://doi.org/10.1029/RG013i001p00001.

Rysgaard, S., Glud, R. N., Sejr, M. K., Bendtsen, J., & Christensen, P. B. (2007). Inorganic carbon transport during sea ice growth and decay: A carbon pump in polar seas. *Journal of Geophysical Research: Oceans, 112*, C03016. https://doi.org/10.1029/2006JC003572.

Saiz-Lopez, A. and R. von Glasow (2012), Reactive halogen chemistry in the troposphere, *Chemical Society Reviews*, Volume 41, 19, 6448–6472, doi:https://doi.org/10.1039/c2cs35208g.

Salawitch, R. J., Gobbi, G. P., Wofsy, S. C., & McElroy, M. B. (1989). Denitrification in the Antarctic stratosphere. *Nature, 339*, 525–527. https://doi.org/10.1038/339525a0.

Salawitch, R. J., Canty, T., Kurosu, T., Chance, K., Liang, Q., da Silva, A., Pawson, S., Nielsen, J. E., Rodriguez, J. M., Bhartia, P. K., Liu, X., Huey, L. G., Liao, J., Stickel, R. E., Tanner, D. J., Dibb, J. E., Simpson, W. R., Donohoue, D., Weinheimer, A., Flocke, F., Knapp, D., Montzka, D., Neuman, J. A., Nowak, J. B., Ryerson, T. B., Oltmans, S., Blake, D. R., Atlas, E. L., Kinnison, D. E., Tilmes, S., Pan, L. L., Hendrick, F., van Roozendael, M., Kreher, K., Johnston, P. V., Gao, R. S., Johnson, B., Bui, T. P., Chen, G., Pierce, R. B., Crawford, J. H., & Jacob, D. J. (2010). A new interpretation of total column BrO during Arctic spring. *Geophysical Research Letters, 37*(21), L21805. https://doi.org/10.1029/2010GL043798.

Schädel, C., Bader, M. K. F., Schuur, E. A. G., Biasi, C., Bracho, R., Čapek, P., de Baets, S., Diáková, K., Ernakovich, J., Estop-Aragones, C., Graham, D. E., Hartley, I. P., Iversen, C. M., Kane, E., Knoblauch, C., Lupascu, M., Martikainen, P. J., Natali, S. M., Norby, R. J., A, J., O'Donnell, T., Roy Chowdhury, H., Šantrůčková, G., Shaver, V. L., Sloan, C. C., Treat, M. R., Turetsky, M. P. W., & Wickland, K. P. (2016). Potential carbon emissions dominated by carbon dioxide from thawed permafrost soils. *Nature Climate Change, 6*, 950–953. https://doi.org/10.1038/nclimate3054.

Schaefer, K., Lantuit, H., Romanovsky, V. E., Schuur, E. A. G., & Witt, R. (2014). The impact of the permafrost carbon feedback on global climate. *Environmental Research Letters, 9*, 085003. https://doi.org/10.1088/1748-9326/9/8/085003.

Schimel, D., Pavlick, R., Fisher, J. B., Asner, G. P., Saatchi, S., Townsend, P., Miller, C., Frankenberg, C., Hibbard, K., & Cox, P. (2015). Observing terrestrial ecosystems and the carbon cycle from space. *Global Change Biology, 21*(5), 1762–1776. https://doi.org/10.1111/gcb.12822.

Schmale, J., Arnold, S. R., Law, K. S., Thorp, T., Anenberg, S., Simpson, W. R., Mao, J., & Pratt, K. A. (2018). Local Arctic air pollution: A neglected but serious problem. *Earth's Future, 6*(10), 1385–1412. https://doi.org/10.1029/2018EF000952.

Schmidt, U., Bauer, R., Engel, A., Borchers, R., & Lee, J. (1994). The variation of available chlorine, Cly, in the Arctic polar vortex during EASOE. *Geophysical Research Letters, 21*(13), 1215–1218. https://doi.org/10.1029/93GL01055.

Schneising, O., Buchwitz, M., Reuter, M., Heymann, J., Bovensmann, H., & Burrows, J. P. (2011). Long-term analysis of carbon dioxide and methane column-averaged mole fractions retrieved from SCIAMACHY. *Atmospheric Chemistry and Physics, 11*(6), 2863–2880. https://doi.org/10.5194/acp-11-2863-2011.

Schnell, R. C., Liu, S. C., Oltmans, S. J., Stone, R. S., Hofmann, D. J., Dutton, E. G., Deshler, T., Sturges, W. T., Harder, J. W., Sewell, S. D., Trainer, M., & Harris, J. M. (1991). Decrease of summer tropospheric ozone concentrations in Antarctica. *Nature, 351*(6329), 726–729. https://doi.org/10.1038/351726a0.

Schoeberl, M. R., & Hartmann, D. L. (1991). The dynamics of the stratospheric polar vortex and its relation to springtime ozone depletions. *Science, 251*(4989), 46–52. https://doi.org/10.1126/science.251.4989.46.

Schoeberl, M. R., & Newman, P. A. (2015). Middle atmosphere | polar vortex, Encyclopedia of atmospheric sciences (second edition), reference module in earth systems and environmental sciences, pp. 12–17. https://doi.org/10.1016/B978-0-12-382225-3.00228-0.

Schoeberl, M. R., Krueger, A. J., & Newman, P. A. (1986). The morphology of Antarctic total ozone as seen by TOMS. *Geophysical Research Letters, 13*(12), 1217–1220. https://doi.org/10.1029/GL013i012p01217.

Schoeberl, M. R., Douglass, A. R., Stolarski, R. S., Toohey, D., Avallone, L., Anderson, J. G., Brune, W., Fahey, D. W., & Kelly, K. (1993). The evolution of ClO and NO along air parcel trajectories. *Geophysical Research Letters, 20*(22), 2511–2514. https://doi.org/10.1029/93GL01690.

Schoolmeester, T., Gjerdi, H. L., Crump, J., Alfthan, B., Fabres, J., Johnsen, K., Puikkonen, L., Kurvits, T., & Baker, E. (2019). *Global linkages – A graphic look at the changing Arctic*. Nairobi and Arendal: UN Environment and GRID-Arendal. http://www.grida.no/publications/431.

Schroeder, W. H., Anlauf, K. G., Barrie, L. A., Lu, J. Y., Steffen, A., Schneeberger, D. R., & Berg, T. (1998). Arctic springtime depletion of mercury. *Nature, 394*(6691), 331–332. https://doi.org/10.1038/28530.

Schuster, U., G.A. McKinley, N. Bates, F. Chevallier, S.C. Doney, A.R. Fay, M. González-Dávila, N. Gruber, Jones, J. Krijnen, P. Landschützer, N. Lefèvre, M. Manizza, J. Mathis, N. Metzl, A. Olsen, A. F. Rios, C. Rödenbeck, J. M. Santana-Casiano, T. Takahashi, R. Wanninkhof, and A. J. Watson (2013), An assessment of the Atlantic and Arctic sea–air CO_2 fluxes, 1990–2009, *Biogeosciences*, Volume 10, 1, 607–627. https://doi.org/10.5194/bg-10-607-2013.

Schuur, E. A. G., Bockheim, J., Canadell, J. G., Euskirchen, E., Field, C. B., Goryachkin, S. V., Hagemann, S., Kuhry, P., Lafleur, P. M., Lee, H., Mazhitova, G., Nelson, F. E., Rinke, A., Romanovsky, V. E., Shiklomanov, N., Tarnocai, C., Venevsky, S., Vogel, J. G., & Zimov, S. A. (2008). Vulnerability of permafrost carbon to climate change: Implications for the global carbon cycle. *Bioscience, 58*(8), 701–714. https://doi.org/10.1641/B580807.

Schuur, E. A. G., McGuire, A. D., Schädel, C., Grosse, G., Harden, J. W., Hayes, D. J., Hugelius, G., Koven, C. D., Kuhry, P., Lawrence, D. M., Natali, S. M., Olefeldt, D., Romanovsky, V. E., Schaefer, K., Turetsky, M. R., Treat, C. C., & Vonk, J. E. (2015). Climate change and the permafrost carbon feedback. *Nature, 520*, 171–179. https://doi.org/10.1038/nature14338.

Seabrook, J. A., Whiteway, J., Staebler, R. M., Bottenheim, J. W., Komguem, L., Gray, L. H., Barber, D., & Asplin, M. (2011). LIDAR measurements of Arctic boundary layer ozone depletion events over the frozen Arctic Ocean. *Journal of Geophysical Research: Atmospheres, 116*(D17), D00S02. https://doi.org/10.1029/2011JD016335.

Seabrook, J. A., Whiteway, J. A., Gray, L. H., Staebler, R., & Herber, A. (2013). Airborne lidar measurements of surface ozone depletion over Arctic Sea ice. *Atmospheric Chemistry and Physics, 13*(12), 6023–6029. https://doi.org/10.5194/acp-13-6023-2013.

Serreze, M. C., Barrett, A. P., & Stroeve, J. (2012). Recent changes in tropospheric water vapor over the Arctic as assessed from radiosondes and atmospheric reanalyses. *Journal of Geophysical Research: Atmospheres, 117*, D10104. https://doi.org/10.1029/2011JD017421.

Shaver, G. R., Billings, W. D., Chapin, F. S., III, Giblin, A. E., Nadelhoffer, K. J., Oechel, W. C., & Rastetter, E. B. (1992). Global change and the carbon balance of Arctic ecosystems. *Bioscience, 42*(6), 433–441. https://doi.org/10.2307/1311862.

Sihler, H., Platt, U., Beirle, S., Marbach, T., Kühl, S., Dörner, S., Verschaeve, J., Frieß, U., Pöhler, D., Vogel, L., Sander, R., & Wagner, T. (2012). Tropospheric BrO column densities in the Arctic derived from satellite: Retrieval and comparison to ground-based measurements. *Atmospheric Measurement Techniques, 5*(11), 2779–2807. https://doi.org/10.5194/amt-5-2779-2012.

Simpson, W. R., von Glasow, R., Riedel, K., Anderson, P., Ariya, P., Bottenheim, J., Burrows, J., Carpenter, L. J., Frieß, U., Goodsite, M. E., Heard, D., Hutterli, M., Jacobi, H.-W., Kaleschke, L., Neff, B., Plane, J., Platt, U., Richter, A., Roscoe, H., Sander, R., Shepson, P., Sodeau, J., Steffen, A., Wagner, T., & Wolff, E. (2007). Halogens and their role in polar boundary-layer ozone depletion. *Atmospheric Chemistry and Physics, 7*(16), 4375–4418. https://doi.org/10.5194/acp-7-4375-2007.

Simpson, W. R., Brown, S. S., Saiz-Lopez, A., Thornton, J. A., & von Glasow, R. (2015). Tropospheric halogen chemistry: Sources, cycling, and impacts. *Chemical Reviews, 115*(10), 4035–4062. https://doi.org/10.1021/cr5006638.

Simpson, W. R., Peterson, P. K., Frieß, U., Sihler, H., Lampel, J., Platt, U., Moore, C., Pratt, K., Shepson, P., Halfacre, J., & Nghiem, a. S. V. (2017). Horizontal and vertical structure of reactive bromine events probed by bromine monoxide MAX-DOAS. *Atmospheric Chemistry and Physics, 17*(15), 9291–9309. https://doi.org/10.5194/acp-17-9291-2017.

Simpson, W., Law, K., Schmalem, J., Pratt, K., Arnold, S., & Mao, J. (2018). Alaskan Layered Pollution And Chemical Analysis (ALPACA) White Paper. https://alpaca.community.uaf.edu/

Sinnhuber, B.-M., Stiller, G., Ruhnke, R., von Clarmann, T., Kellmann, S., & Aschmann, J. (2011). Arctic winter 2010/2011 at the brink of an ozone hole. *Geophysical Research Letters, 38*(24), L24814. https://doi.org/10.1029/2011GL049784.

Solomon, S. (1999). Stratospheric ozone depletion: A review of concepts and history. *Reviews of Geophysics, 37*(3), 275–316. https://doi.org/10.1029/1999RG900008.

Solomon, S., Garcia, R. R., Rowland, F. S., & Wuebbles, D. J. (1986). On the depletion of Antarctic ozone. *Nature, 321*, 755–758. https://doi.org/10.1038/321755a0.

Solomon, P. M., Connor, B., deZafra, R., Parrish, A., Barrett, J., & Jaramillo, M. (1987a). High concentrations of chlorine monoxide at low altitudes in the Antarctic spring stratosphere: Secular variation. *Nature, 328*, 411–413. https://doi.org/10.1038/328411a0.

Solomon, S., Mount, G., Sanders, R., & Schmeltekopf, A. (1987b). Visible spectroscopy at McMurdo Station, Antarctica 2. *Observations of OClO, Journal of Geophysical Research: Atmospheres, Volume, 92*(D7), 8329–8338. https://doi.org/10.1029/JD092iD07p08329.

Solomon, S., Mount, G. H., Sanders, R. W., Jakoubek, R. O., & Schmeltekopf, A. L. (1988). Observations of the nighttime abundance of OClO in the winter stratosphere above Thule, Greenland. *Science, 242*(4878), 550–555. https://doi.org/10.1126/science.242.4878.550.

Solomon, S., Portmann, R. W., & Thompson, D. W. J. (2007). Contrasts between Antarctic and Arctic ozone depletion. *Proceedings of the National Academy of Sciences, 104*(2), 445–449. https://doi.org/10.1073/pnas.0604895104.

Solomon, S., Ivy, D. J., Kinnison, D., Mills, M. J., Neely, R. R., III, & Schmidt, A. (2016). Emergence of healing in the Antarctic ozone layer. *Science, 353*(6296), 269–274. https://doi.org/10.1126/science.aae0061.

Spicer, C. W., Plastridge, R. A., Foster, K. L., Finlayson-Pitts, B. J., Bottenheim, J. W., Grannas, A. M., & Shepson, P. B. (2002). Molecular halogens before and during ozone depletion events in the Arctic at polar sunrise: Concentrations and sources. *Atmospheric Environment, 36*(15–16), 2721–2731. https://doi.org/10.1016/S1352-2310(02)00125-5.

Staehelin, J., Harris, N. R. P., Appenzeller, C., & Eberhard, J. (2001). Ozone trends: A review. *Reviews of Geophysics, 39*(2), 231–290. https://doi.org/10.1029/1999RG000059.

Steffen, A., Douglas, T., Amyot, M., Ariya, P., Aspmo, K., Berg, T., Bottenheim, J., Brooks, S., Cobbett, F., Dastoor, A., Dommergue, A., Ebinghaus, R., Ferrari, C., Gardfeldt, K., Goodsite, M. E., Lean, D., Poulain, A. J., Scherz, C., Skov, H., Sommar, J., & Temme, C. (2008). A synthesis of atmospheric mercury depletion event chemistry in the atmosphere and snow. *Atmospheric Chemistry and Physics, 8*(6), 1445–1482. https://doi.org/10.5194/acp-8-1445-2008.

Stohl, A., Law, K., Bottenheim, J. W., Monks, P., & Quinn, P. (Eds.). (2009). POLARCAT (polar study using aircraft, remote sensing, surface measurements and models, of climate, chemistry, aerosols, and transport). *Atmospheric Chemistry and Physics* special issue. https://www.atmos-chem-phys.net/special_issue182.html

Stolarski, R. S., & Cicerone, R. J. (1974). Stratospheric chlorine: A possible sink for ozone. *Canadian Journal of Chemistry, 52*, 1610–1615. https://doi.org/10.1139/v74-233.

Stolarski, R. S., Krueger, A. J., Schoeberl, M. R., McPeters, R. D., Newman, P. A., & Alpert, J. C. (1986). Nimbus 7 satellite measurements of the springtime Antarctic ozone decrease. *Nature, 322*, 808–811. https://doi.org/10.1038/322808a0.

Strahan, S. E. (2015). Middle atmosphere | transport circulation, Encyclopedia of atmospheric sciences (second edition), reference module in earth systems and environmental sciences, pp. 41–49. https://doi.org/10.1016/B978-0-12-382225-3.00231-0.

Strahan, S. E., Douglass, A. R., & Newman, P. A. (2013). The contributions of chemistry and transport to low arctic ozone in March 2011 derived from Aura MLS observations. *Journal of Geophysical Research: Atmospheres, 118*(3), 1563–1576. https://doi.org/10.1002/jgrd.50181.

Strahan, S. E., Douglass, A. R., & Steenrod, S. D. (2016). Chemical and dynamical impacts of stratospheric sudden warmings on Arctic ozone variability. *Journal of Geophysical Research: Atmospheres, 121*(19), 11836–11851. https://doi.org/10.1002/2016JD025128.

Stutz, J., Thomas, J. L., Hurlock, S. C., Schneider, M., von Glasow, R., Piot, M., Gorham, K., Burkhart, J. G., Ziemba, L., Dibb, J. E., & Lefer, B. L. (2011). Longpath DOAS observations of surface BrO at summit, Greenland. *Atmospheric Chemistry and Physics, 11*(18), 9899–9910. https://doi.org/10.5194/acp-11-9899-2011.

Style, R. W., & Worster, M. G. (2009). Frost flower formation on sea ice and lake ice. *Geophysical Research Letters, 36*, L11501. https://doi.org/10.1029/2009GL037304.

Tackett, P. J., Cavender, A. E., Keil, A. D., Shepson, P. B., Bottenheim, J. W., Morin, S., Deary, J., Steffen, A., & Doerge, C. (2007). A study of the vertical scale of halogen chemistry in the Arctic troposphere during Polar Sunrise at Barrow, Alaska. *Journal of Geophysical Research: Atmospheres, 112*(D7), D07306. https://doi.org/10.1029/2006JD007785.

Tan, Z., Zhuang, Q., Henze, D. K., Frankenberg, C., Dlugokencky, E., Sweeney, C., Turner, A. J., Sasakawa, M., & Machida, T. (2016). Inverse modeling of pan-Arctic methane emissions at high spatial resolution: What can we learn from assimilating satellite retrievals and using different process-based wetland and lake biogeochemical models? *Atmospheric Chemistry and Physics, 16*(19), 12649–12666. https://doi.org/10.5194/acp-16-12649-2016.

Tarasick, D. W., & Bottenheim, J. W. (2002). Surface ozone depletion episodes in the Arctic and Antarctic from historical ozonesonde records. *Atmospheric Chemistry and Physics, 2*(3), 197–205. https://doi.org/10.5194/acp-2-197-2002.

Tarnocai, C., Canadell, J. G., Schuur, E. A. G., Kuhry, P., Mazhitova, G., & Zimov, S. (2009). Soil organic carbon pools in the northern circumpolar permafrost region. *Global Biogeochemical Cycles, 23*, GB2023. https://doi.org/10.1029/2008GB003327.

Tegtmeier, S., Rex, M., Wohltmann, I., & Krüger, K. (2008). Relative importance of dynamical and chemical contributions to Arctic wintertime ozone. *Geophysical Research Letters, 35*, L17801. https://doi.org/10.1029/2008GL034250.

Theys, N., van Roozendael, M., Hendrick, F., Yang, X., de Smedt, I., Richter, A., Begoin, M., Errera, Q., Johnston, P. V., Kreher, K., & de Mazière, M. (2011). Global observations of tropospheric BrO columns using GOME-2 satellite data. *Atmospheric Chemistry and Physics, 11*(4), 1791–1811. https://doi.org/10.5194/acp-11-1791-2011.

Thomas, J. L., Stutz, J., Lefer, B., Huey, L. G., Toyota, K., Dibb, J. E., & von Glasow, R. (2011). Modeling chemistry in and above snow at summit, Greenland – Part 1:

Model description and results. *Atmospheric Chemistry and Physics, 11*(10), 4899–4914. https://doi.org/10.5194/acp-11-4899-2011.

Thomas, J. L., Dibb, J. E., Huey, L. G., Liao, J., Tanner, D., Lefer, B., von Glasow, R., & Stutz, J. (2012). Modeling chemistry in and above snow at summit, Greenland – Part 2: Impact of snowpack chemistry on the oxidation capacity of the boundary layer. *Atmospheric Chemistry and Physics, 12*(14), 6537–6554. https://doi.org/10.5194/acp-12-6537-2012.

Thompson, C. R., Shepson, P. B., Liao, J., Huey, L. G., Apel, E. C., Cantrell, C. A., Flocke, F., Orlando, J., Fried, A., Hall, S. R., Hornbrook, R. S., Knapp, D. J., Maudlin, R. L., III, Montzka, D. D., Sive, B. C., Ullmann, K., Weibring, P., & Weinheimer, A. (2015). Interactions of bromine, chlorine, and iodine photochemistry during ozone depletions in Barrow, Alaska. *Atmospheric Chemistry and Physics, 15*(16), 9651–9679. https://doi.org/10.5194/acp-15-9651-2015.

Thompson, C. R., Shepson, P. B., Liao, J., Huey, L. G., Cantrell, C., Flocke, F., & Orlando, J. (2017). Bromine atom production and chain propagation during springtime Arctic ozone depletion events in Barrow, Alaska. *Atmospheric Chemistry and Physics, 17*(5), 3401–3421. https://doi.org/10.5194/acp-17-3401-2017.

Tilmes, S., Müller, R., Grooß, J.-U., & Russell III, J. M. (2004). Ozone loss and chlorine activation in the Arctic winters 1991-2003 derived with the tracer-tracer correlations. *Atmospheric Chemistry and Physics, 4*(8), 2181–2213. https://doi.org/10.5194/acp-4-2181-2004.

Toohey, D. W., Avallone, L. M., Lait, L. R., Newman, P. A., Schoeberl, M. R., Fahey, D. W., Woodbridge, E. L., & Anderson, J. G. (1993). The seasonal evolution of reactive chlorine in the Northern Hemisphere stratosphere. *Science, 261*(5125), 1134–1136. https://doi.org/10.1126/science.261.5125.1134.

Toon, O., Turco, R., & Hamill, P. (1990a). Denitrification mechanisms in the polar stratospheres. *Geophysical Research Letters, 17*(4), 445–448. https://doi.org/10.1029/GL017i004p00445.

Toon, O. B., Browell, E. V., Kinne, S., & Jordan, J. (1990b). An analysis of lidar observations of polar stratospheric clouds. *Geophysical Research Letters, 17*(4), 393–396. https://doi.org/10.1029/GL017i004p00393.

Toyota, K., McConnell, J. C., Lupu, A., Neary, L., McLinden, C. A., Richter, A., Kwok, R., Semeniuk, K., Kaminski, J. W., Gong, S.-L., Jarosz, J., Cjipperfield, M. P., & Sioris, C. E. (2011). Analysis of reactive bromine production and ozone depletion in the Arctic boundary layer using 3-D simulations with GEM-AQ: Inference from synoptic-scale patterns. *Atmospheric Chemistry and Physics, 11*(8), 3949–3979. https://doi.org/10.5194/acp-11-3949-2011.

Toyota, K., McConnell, J. C., Staebler, R. M., & Dastoor, A. P. (2014). Air – Snowpack exchange of bromine, ozone and mercury in the springtime Arctic simulated by the 1-D model PHANTAS – Part 1: In-snow bromine activation and its impact on ozone. *Atmospheric Chemistry and Physics, 14*(8), 4101–4133. https://doi.org/10.5194/acp-14-4101-2014.

Tung, K. K., Ko, M. K. W., Rodriguez, J. M., & Sze, N. D. (1986). Are Antarctic ozone variations a manifestation of dynamics or chemistry? *Nature, 333*, 811–814. https://doi.org/10.1038/322811a0.

Turco, R., Plumb, A., & Condon, E. (1990). The airborne Arctic stratospheric expedition: Prologue. *Geophysical Research Letters, 17*(4), 313–316. https://doi.org/10.1029/GL017i004p00313.

Turner, A. J., Frankenberg, C., Wennberg, P. O., & Jacob, D. J. (2017). Ambiguity in the causes for decadal trends in atmospheric methane and hydroxyl. *Proceedings of the National Academy of Sciences, 114*(21), 5367–5372. https://doi.org/10.1073/pnas.1616020114.

USEPA. (2010). *Office of Atmospheric Programs (6207J), Methane and Nitrous Oxide Emissions From Natural Sources*, U.S. Environmental Protection Agency, EPA 430-R-10-001, Washington, DC, 20460. Available at http://nepis.epa.gov/

van Dam, B., Helmig, D., Doskey, P. V., & Oltmans, S. J. (2016). Summertime surface O_3 behavior and deposition to tundra in the Alaskan Arctic. *Journal of Geophysical Research: Atmospheres, 121*(13), 8055–8066. https://doi.org/10.1002/2015JD023914.

Viatte, C., Strong, K., Hannigan, J., Nussbaumer, E., Emmons, L. K., Conway, S., Paton-Walsh, C., Hartley, J., Benmergui, J., & Lin, J. (2015). Identifying fire plumes in the Arctic with tropospheric FTIR measurements and transport models. *Atmospheric Chemistry and Physics, 15*(5), 2227–2246. https://doi.org/10.5194/acp-15-2227-2015.

von Glasow, R., & Crutzen, P. J. (2007). Tropospheric halogen chemistry. In *The atmosphere* (R. F. Keeling, Ed., Vol. 4) *Treatise on geochemistry* (H. D. Holland and K. K. Turekian, Eds.) (pp. 21–64). Elsevier-Pergamon.

von Hobe, M., et al. (2013). Reconciliation of essential process parameters for an enhanced predictability of Arctic stratospheric ozone loss and its climate interactions (RECONCILE): Activities and results. *Atmospheric Chemistry and Physics, 13*(18), 9233–9268. https://doi.org/10.5194/acp-13-9233-2013.

Wagner, T., Leue, C., Wenig, M., Pfeilsticker, K., & Platt, U. (2001). Spatial and temporal distribution of enhanced boundary layer BrO concentrations measured by the GOME instrument aboard ERS-2. *Journal of Geophysical Research: Atmospheres, 106*(D20), 24225–24235. https://doi.org/10.1029/2000JD000201.

Wallace, J. M., & Hobbs, P. V. (2006). *Atmospheric science: an introductory survey*, 2nd edition, Elsevier Inc. In *pp 483*.

Walter Anthony, K., Schneider von Deimling, T., Nitze, I., Frolking, S., Emond, A., Daanen, R., Anthony, P., Lindgren, P., Jones, B., & Grosse, G. (2018). 21st-century modeled permafrost carbon emissions accelerated by abrupt thaw beneath lakes. *Nature Communications, 9*(1). https://doi.org/10.1038/s41467-018-05738-9.

Wang, S., & Pratt, K. A. (2017). Molecular halogens above the Arctic snowpack: Emissions, diurnal variations, and recycling mechanisms. *Journal of Geophysical Research: Atmospheres, Volume, 122*(21), 11991–12007. https://doi.org/10.1002/2017JD027175.

Wanninkhof, R. (1992). Relationship between wind speed and gas exchange over the ocean. *Journal of Geophysical Research: Oceans, Volume, 97*(C5), 7373–7382. https://doi.org/10.1029/92JC00188.

Wardle, D. I., Kerr, J., McElroy, C. T., & Francis, D. R. (Eds.). (1997). *Ozone Science*. Environment Canada: A Canadian Perspective on the Changing Ozone Layer.

Wargan, K., & Nielsen, J. E. (2017). Quantifying chemical ozone loss in the Arctic stratosphere with GEOS-STRATCHEM data assimilation system. *GMAO Research Brief, 2017-2*.

Webb, E. E., Schuur, E. A. G., Natali, S. M., Oken, K. L., Bracho, R., Krapek, J. P., Risk, D., & Nickerson, N. R. (2016). Increased wintertime CO_2 loss as a result of sustained tundra warming. *Journal of Geophysical Research: Biogeosciences, 121*(2), 249–265. https://doi.org/10.1002/2014JG002795.

Weber, M., Coldewey-Egbers, M., Fioletov, V. E., Frith, S., Wild, J. D., Burrows, J. P., Long, C. S., & Loyola, D. (2018). Total ozone trends from 1979 to 2016 derived from five merged observational datasets – The emergence into ozone recovery. *Atmospheric Chemistry and Physics, 18*(3), 2097–2117. https://doi.org/10.5194/acp-18-2097-2018.

Webster, C. R., May, R. D., Toohey, D. W., Avallone, L. M., Anderson, J. G., Newman, P., Lait, L., Schoeberl, M. R., Elkins, J. W., & Chan, K. R. (1993). Chlorine chemistry on polar stratospheric cloud particles in the Arctic winter. *Science, 261*(5125), 1130–1133. https://doi.org/10.1126/science.261.5125.1130.

Wennberg, P. (1999). Bromine explosion. *Nature, 397*, 299–301. https://doi.org/10.1038/16805.

Whitmarsh, J., & Govindjee (1999). The photosynthetic process. In: G. S. Singhal, G. Renger, S. K. Sopory, K. D. Irrgang, Govindjee (Eds.), *Concepts in photobiology: Photosynthesis and photomorphogenesis*. Boston: Kluwer Academic Publishers.

Wilmouth, D. M., Salawitch, R. J., & Canty, T. P. (2018). Stratospheric ozone depletion and recovery. In B. Torok, & T. Dransfield (Eds.), *Green chemistry: An inclusive approach* (pp. 177–209). Oxford: Elsevier Publishing.

WMO. (1985). *Atmospheric Ozone 1985* (WMO Global Ozone Research and Monitoring Project – Report no. 16). Geneva: World Meteorological Organization.

WMO. (1988). *Report of the International Ozone Trends Panel 1988* (WMO Global Ozone Research and Monitoring Project – Report no. 18). Geneva: World Meteorological Organization.

WMO. (1989). *Scientific Assessment of Stratospheric Ozone: 1989* (Global Ozone Research and Monitoring Project – Report no. 20). Geneva: World Meteorological Organization.

WMO. (1991). *Scientific Assessment of Ozone Depletion: 1991* (Global Ozone Research and Monitoring Project – Report no. 25). Geneva: World Meteorological Organization.

WMO. (1994). *Scientific Assessment of Ozone Depletion: 1994* (Global Ozone Research and Monitoring Project – Report no. 37). Geneva: World Meteorological Organization.

WMO. (1998). *Scientific Assessment of Ozone Depletion: 1998* (Global Ozone Research and Monitoring Project – Report no. 44). Geneva: World Meteorological Organization.

WMO. (2003). *Scientific Assessment of Ozone Depletion: 2002* (Global Ozone Research and Monitoring Project – Report no. 47). Geneva: World Meteorological Organization.

WMO. (2007). *Scientific Assessment of Ozone Depletion: 2006* (Global Ozone Research and Monitoring Project – Report no. 50). Geneva: World Meteorological Organization.

WMO. (2011). *Scientific Assessment of Ozone Depletion: 2010* (Global Ozone Research and Monitoring Project – Report no. 52). Geneva: World Meteorological Organization.

WMO. (2014). *Scientific Assessment of Ozone Depletion: 2014* (Global Ozone Research and Monitoring Project – Report no. 55). Geneva: World Meteorological Organization.

WMO. (2017). WMO WDCGG Data Summary (WDCGG no. 41). GAW data, Volume IV-Greenhouse Gases and Other Atmospheric Gases, Japan Meteorological Agency in co-operation with World Meteorological Organization. Available at https://gaw.kishou.go.jp/static/publications/summary/sum41/sum41.pdf

WMO. (2018a). *WMO Statement on the Status of the Global Climate in 2017* (WMO-no. 1212). Geneva: World Meteorological Organization. ISBN: 978-92-63-11212-5.

WMO. (2018b). *Scientific Assessment of Ozone Depletion: 2018* (Global Ozone Research and Monitoring Project – Report no. 58). Geneva: World Meteorological Organization.

Wofsy, S. C., McElroy, M. B., & Yung, Y. L. (1975). The chemistry of atmospheric bromine. *Geophysical Research Letters, 2*(6), 215–218. https://doi.org/10.1029/GL002i006p00215.

Worster, M. G., & Rees Jones, D. W. (2015). Sea-ice thermodynamics and brine drainage. *Philosophical Transactions of the Royal Society A, 373*, 20140166. https://doi.org/10.1098/rsta.2014.0166.

Wren, S. N., Donaldson, D. J., & Abbatt, J. P. D. (2013). Photochemical chlorine and bromine activation from artificial saline snow. *Atmospheric Chemistry and Physics, 13*(19), 9789–9800. https://doi.org/10.5194/acp-13-9789-2013.

Wuebbles, D. J. (1983). Chlorocarbon emission scenarios: Potential impact on stratospheric ozone. *Journal of Geophysical Research: Oceans, 88*(C2), 1433–1443. https://doi.org/10.1029/JC088iC02p01433.

Wunch, D., Toon, G. C., Blavier, J.-F. L., Washenfelder, R. A., Notholt, J., Connor, B. J., Griffith, D. W. T., Sherlock, V., & Wennberg, P. O. (2011). The Total Carbon Column Observing Network. *Philosophical Transactions of the Royal Society A, 369*, 2087–2112. https://doi.org/10.1098/rsta.2010.0240.

Wunch, D., P. O. Wennberg, G. Osterman, B. Fisher, B. Naylor, C. M. Roehl, C. O'Dell, L. Mandrake, C. Viatte, M. Kiel, D.W.T. Griffith, N. M. Deutscher, V. A. Velazco, J. Notholt, T. Warneke, T, C. Petri, M. de Maziere, M. K. Sha, R. Sussmann, M. Rettinger, D. Pollard, J. Robinson, I. Morino, O. Uchino, F. Hase, T. Blumenstock, D. G. Feist, S. G. Arnold, K. Strong, J. Mendonca, R. Kivi, P. Heikkinen, L. Iraci, J. Podolske, P. Hillyard, S. Kawakami, M. Dubey, H. A. Parker, E. Sepulveda, O.E. Garcia, Y. Te, P. Jeseck, M. R. Gunson, D. Crisp, and A. Eldering (2017). Comparisons of the Orbiting Carbon Observatory-2 (OCO-2) XCO_2 measurements with TCCON, *Atmospheric Measurement Techniques*, 10, 6, 2209–2238, doi:https://doi.org/10.5194/amt-10-2209-2017.

Xiong, X., Barnet, C., Maddy, E., Sweeney, C., Liu, X., Zhou, L., & Goldberg, M. (2008). Characterization and validation of methane products from the Atmospheric Infrared Sounder (AIRS). *Journal of Geophysical Research: Biogeosciences, 113*, G00A01. https://doi.org/10.1029/2007JG000500.

Xiong, X., Barnet, C., Maddy, E. S., Gambacorta, A., King, T. S., & Wofsy, S. C. (2013). Mid-upper tropospheric methane retrieval from IASI and its validation. *Atmospheric Measurement Techniques, 6*(9), 2255–2265. https://doi.org/10.5194/amt-6-2255-2013.

Yang, X., Pyle, J. A., & Cox, R. A. (2008). Sea salt aerosol production and bromine release: Role of snow on sea ice. *Geophysical Research Letters, 35*(16), L16815. https://doi.org/10.1029/2008GL034536.

Yang, X., Pyle, J. A., Cox, R. A., Theys, N., & van Roozendael, M. (2010). Snow-sourced bromine and its implications for polar tropospheric ozone. *Atmospheric Chemistry and Physics, 10*(16), 7763–7773. https://doi.org/10.5194/acp-10-7763-2010.

Yang, X., Abraham, N. L., Archibald, A. T., Braesicke, P., Keeble, J., Telford, P. J., Warwick, N. J., & Pyle, J. A. (2014). How sensitive is the recovery of stratospheric ozone to changes in concentrations of very shortlived bromocarbons? *Atmospheric Chemistry and Physics, 14*(19), 10,431–10,438. https://doi.org/10.5194/acp-14-10431-2014.

Yung, Y. L., Pinto, J. P., Watson, R. T., & Sander, S. P. (1980). Atmospheric bromine and ozone perturbations in the lower stratosphere. *Journal of Atmospheric Science, 37*, 339–353. https://doi.org/10.1175/1520-0469(1980)037<0339:ABAOPI>2.0.CO;2.

Zerefos, C., Contopoulos, G., & Skalkeas, G. (Eds.). (2009). Twenty years of ozone decline. In *Proceedings of the symposium for the 20th anniversary of the montreal protocol*. Springer, XII, 470 pp. https://doi.org/10.1007/978-90-481-2469-5

Zhang, X., He, J., Zhang, J., Polyakov, I., Gerdes, R., Inoue, J., & Wu, P. (2013). Enhanced poleward moisture transport and amplified northern high-latitude wetting trend. *Nature Climate Change, 3*, 47–51. https://doi.org/10.1038/nclimate1631.

Zhao, X., Strong, K., Adams, C., Schofield, R., Yang, X., Richter, A., Friess, U., Blechschmidt, A.-M., & Koo, J.-H. (2016). A case study of a transported bromine explosion event in the Canadian high arctic. *Journal of Geophysical Research: Atmospheres, 121*(1), 457–477. https://doi.org/10.1002/2015JD023711.

Zhao, X., Weaver, D., Bognar, K., Manney, G., Millán, L., Yang, X., Eloranta, E., Schneider, M., & Strong, K. (2017). Cyclone-induced surface ozone and HDO depletion in the Arctic. *Atmospheric Chemistry and Physics, 17*(24), 14955–14974. https://doi.org/10.5194/acp-17-14955-2017.

Zielcke, J. (2015). *Observations of reactive bromine, iodine and chlorine species in the Arctic and Antarctic with differential optical absorption spectroscopy*. PhD thesis. Ruperto-Carola University of Heidelberg.

Zona, D., Gioli, B., Commane, R., Lindaas, J., Wofsy, S. C., Miller, C. E., Dinardo, S. J., Dengel, S., Sweeney, C., Karion, A., Chang, R. Y.-W., Henderson, J. M., Murphy, P. C., Goodrich, J. P., Moreaux, V., Liljedahl, A., Watts, J. D., Kimball, J. S., Lipson, D. A., & Oechel, W. C. (2016). Cold season emissions dominate the Arctic tundra methane budget. *Proceedings of the National Academy of Sciences, 113*(1), 40–45. https://doi.org/10.1073/pnas.1516017113.

Chapter 4
Arctic Aerosols

Roberto Udisti, Rita Traversi, Silvia Becagli, Claudio Tomasi, Mauro Mazzola, Angelo Lupi, and Patricia K. Quinn

Abstract Aerosols play an important role in the climatic system through their direct and indirect effects on radiation. Beside this, they are also part of the complex chain of chemical reactions taking place in the atmosphere. Indirect effects involve aerosols acting as cloud and ice condensation nuclei, brightening of clouds, modification of precipitation capabilities etc. After deposition, they also change reflectivity properties of bright surfaces, particularly important in polar regions.

In the Artic few natural aerosol sources exist, except oceans for sea-salt and soil for dust, both of them increasing in magnitude due to global warming. Beside this, anthropogenic aerosols are easily transported to the Arctic by atmospheric transport from middle latitudes, in particular during winter and early spring, forming the so-called Artic Haze.

In this contribution the processes causing the development of Arctic Haze and its characteristics are introduced. Following, a review of the physical and optical properties as well as chemical composition of Arctic aerosols are reviewed using data obtained from numerous monitoring stations in the Arctic.

R. Udisti
Department of Chemistry "Ugo Schiff", University of Florence, Florence, Italy

Institute of Atmospheric Sciences and Climate (ISAC), Italian National Research Council (CNR), Bologna, Italy
e-mail: roberto.udisti@unifi.it

R. Traversi · S. Becagli
Department of Chemistry "Ugo Schiff", University of Florence, Florence, Italy
e-mail: rita.traversi@unifi.it; silvia.becagli@unifi.it

C. Tomasi · M. Mazzola (✉) · A. Lupi
Institute of Atmospheric Sciences and Climate (ISAC), Italian National Research Council (CNR), Bologna, Italy
e-mail: c.tomasi@isac.cnr.it; m.mazzola@isac.cnr.it; a.lupi@isac.cnr.it

P. K. Quinn
NOAA PMEL, Seattle, WA, USA
e-mail: Patricia.k.quinn@noaa.gov

A. Kokhanovsky, C. Tomasi (eds.), *Physics and Chemistry of the Arctic Atmosphere*, Springer Polar Sciences, https://doi.org/10.1007/978-3-030-33566-3_4

Keywords Aerosol · Arctic haze · Atmospheric transport · Chemistry · Absorption · Scattering · Size distribution

4.1 Arctic Haze

4.1.1 Introduction to Arctic Haze

Observations of a strange haze of unknown origin in the Arctic atmosphere were reported as far back as the late 1800s by British, American, and Swedish explorers and scientists (Garrett and Verzella 2008). Observations continued into the late 1940s and 1950s by pilots flying reconnaissance weather missions in the Canadian and Alaskan Arctic who reported "the high incidence of haze at flight altitudes" (Mitchell 1956). Shaw (1975) first measured the vertical structure of the haze in an Alaskan "bush" airplane near Barrow, Alaska with a hand-held sunphotometer and attributed it to ice crystals seeded by open leads or blowing dust from riverbeds. Later, "chemical fingerprinting" of the haze revealed that it was anthropogenic in origin, containing high concentrations of vanadium, a tracer of heavy oil combustion, along with titanium, chromium, manganese, iron and nickel (Rahn et al. 1977; Rahn and McCaffrey 1980). By the late 1970s, the anthropogenic source of the haze was documented but surprising since the ability of aerosol to be transported thousands of kilometers from its source region was not yet fully recognized. An informal measurement network to investigate properties and sources of the haze was designed during the first Arctic Chemistry Symposium held in Lillestrom, Norway in 1978. Since then, a combination of intensive field programs and long-term measurements has revealed the large spatial extent of the anthropogenic haze and its origination primarily from emissions from northern Eurasia that are transported to and trapped in the Arctic during winter and early spring.

The haze is composed of a mixture of non-sea salt (nss) sulfate and particulate organic matter (POM) and, to a lesser extent, ammonium, nitrate, dust, and black carbon (BC) (Li and Barrie 1993; Quinn et al. 2002). It also contains distinct heavy metals, which has allowed for the identification of particular industrial sources as per "chemical fingerprinting" (Shaw 1983; Polissar et al. 2001). Particles within the haze are well aged with a mass median diameter of about 0.2 microns or less (Heintzenberg 1980; Clarke 1989; Leaitch et al. 1989) and, hence, are very efficient at scattering visible solar radiation since the peak in the particle surface area size distribution is near the maximum efficiency for Mie scattering (Shaw 1987; Waggoner and Weiss 1980). The haze is also weakly absorbing due to the presence of black carbon (Hansen and Rosen 1984; Noone and Clarke 1988). The result of scattering and absorption of solar radiation by the haze is a noticeable reduction in visibility to a few kilometers or less.

In addition to changes in visibility, the haze can impact the Arctic's radiation balance. Particulate nss sulfate, BC, and POM are known as short-lived climate pollutants (SLCPs) as they are short-lived compared to greenhouse gases with lifetimes from days to weeks and can perturb the radiation balance of the earth-atmosphere system in a number of ways. The reflective components of the haze, primarily nss sulfate, prevent solar radiation from reaching the Earth's surface and, hence, lead to a cooling of the surface as well as the surface-atmosphere column. The impact of BC, an absorbing component of the haze, on the Arctic's radiation balance depends on the altitude of the haze (Flanner 2013). High altitude BC (above about 430 hPa) cools the Arctic surface by absorbing solar radiation that would otherwise reach the surface (Shaw and Stamnes 1980). In contrast, low altitude BC deposits energy near the surface leading to warming (Sand et al. 2013). In addition, low altitude BC-containing aerosol may be deposited to snow and ice surfaces resulting in a darkening of the surface, an increase in the absorption of solar radiation, and a surface warming (Flanner et al. 2007). Since the radiative effects of aerosols depend strongly on the altitude of the aerosol layer (Samset and Myhre 2011), knowledge of not only the total column loading but also the vertical distribution of the aerosols is crucial for accurate estimates of the impacts on Arctic climate.

4.1.2 Atmospheric Transport of Aerosol Pollutants to the Arctic

Several seasonally dependent mechanisms contribute to the formation of Arctic haze, which typically builds up during winter and early spring (roughly December through April). Surfaces of constant potential temperature form shells over the Arctic, with minimum values in the Arctic boundary layer, which results in the so-called polar dome (Klonecki et al. 2003; Stohl 2006) (Fig. 4.1).

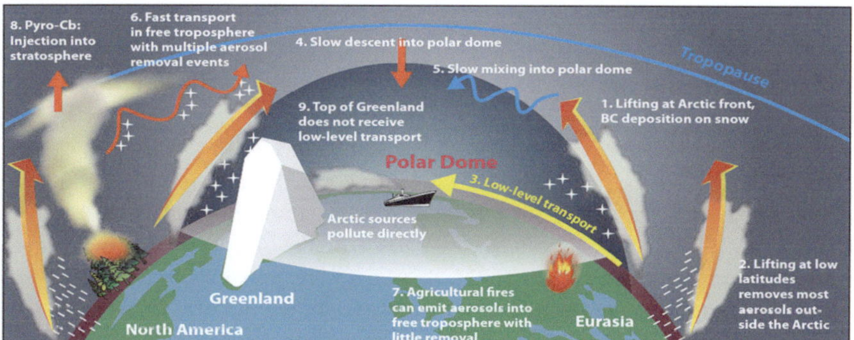

Fig. 4.1 Schematic illustration of processes relevant for the transport of trace pollutants into the Arctic based on the study of Stohl (2006), from AMAP (2011) (AMAP 2011). The polar dome is asymmetric and its extent is temporally highly variable. Its southernmost extent is greatest over Eurasia. The placement of the polar dome as depicted in the figure is more typical of the winter/spring situation. In the summer the dome is more confined to northerly latitudes. The dome is not homogeneous but is highly stratified with strong vertical gradients

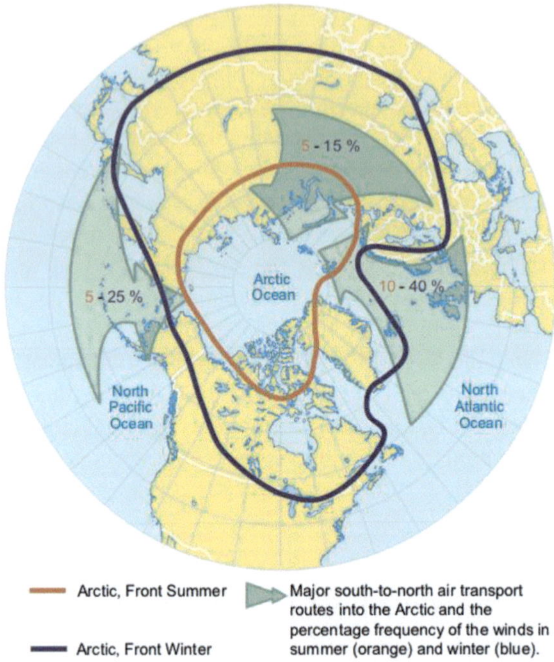

Fig. 4.2 The mean position of the Arctic Front in winter (January) and summer (July) superimposed on the frequency of major poleward transport routes (Iversen 1996)

Strong surface-based temperature inversions inside the dome limit vertical mixing between the boundary layer and the free troposphere and lead to slow dry deposition of pollutants to the surface and, hence, inhibit the deposition of BC to snow and ice surfaces (Strunin et al. 1997). In addition, the Arctic lower troposphere is isolated from lower latitudes by a transport barrier referred to as the "Arctic Front" (Barrie 1986), which shifts seasonally as shown in Fig. 4.2.

In the winter, the Arctic Front can be located as far south as 40°N (Barrie 1986) due to strong diabatic cooling of air at snow-covered surfaces which helps to establish the polar dome and also allows polluted air from northern Eurasia to enter the Arctic at low altitudes as per pathway 3 in Fig. 4.1 (Stohl 2006). During the summer, as lower latitudes warm, the Arctic front approximately follows the northern coastline of Eurasia which inhibits the transport route from northern Eurasia into the Arctic. The northward retreat of the Arctic front is one of the reasons why aerosol concentrations in the Arctic are much lower in the summer than in the winter. In addition, increased efficiency of wet scavenging in summer is also important for driving the seasonal cycle of aerosols in the Arctic (Garrett et al. 2010).

Pollution emitted into relatively warm air masses south of the Arctic front, during both winter and summer, can only reach the Arctic by following sloping surfaces of constant potential temperature (isentropic surfaces) upwards into the Arctic middle or upper troposphere (see pathways 1 and 2 in Fig. 4.1) (Carlson 1981; Iversen 1984; Barrie 1986). The lifting is typically associated with cloud formation and precipitation which can lead to the scavenging of soluble pollutants from the atmosphere. These are the primary pathways for emissions from the densely populated areas of eastern North America and southeastern Asia into the Arctic.

Air can also penetrate the polar dome by slow descent from above (see pathway 4 in Fig. 4.1) and by incorporation of extra-Arctic air via mixing into the polar dome (pathway 5). However, descent from the upper troposphere to the surface is a slow process, taking several weeks. Air descending from the upper troposphere is composed of a mixture of air that has been cleaned during uplift (pathway 6), stratospheric air, as well as agricultural (pathway 7) and biomass burning plumes (pathway 8) from various mid-latitude source regions (Brock et al. 2011; Warneke et al. 2009; Warneke et al. 2010).

Due to the position of the Arctic front and resulting transport pathways, source regions to the Arctic vary with altitude. Low altitude transport into the Arctic during winter and spring (pathway 3 in Fig. 4.1) brings emissions from northern Eurasia into the boundary layer. During Arctic haze season (winter/spring) as well as summer, emissions from lower latitudes can be transported into the Arctic at higher altitudes. The result is an Arctic atmosphere that is highly heterogeneous and composed of superimposed aerosol layers of different composition, origin, and transport history (Brock et al. 2011).

Several seasonally dependent mechanisms contribute to the formation of Arctic haze. As described above, strong surface-based temperature inversions form in the polar night causing the atmosphere to stabilize. This cold and stable atmosphere inhibits turbulent transfer between atmospheric layers as well as the formation of cloud systems and precipitation, the major removal pathway for particulates from the atmosphere (Barrie et al. 1981; Shaw 1995). In addition, meridional transport from the mid-latitudes to the Arctic intensifies during the winter and spring (Iversen and Joranger 1985). The combination of these factors results in the transport of precursor gases and particulates to the Arctic and the trapping of the pollutant haze for up to 15–30 d (Shaw 1995). The seasonality of nss sulfate observed at Barrow, Alaska is shown in Fig. 4.3.

Transport pathways not only vary seasonally but also on interannual time scales. A primary driver of this variability is the North Atlantic Oscillation (NAO), which results in a redistribution of atmospheric mass between the Arctic and the subtropical Atlantic (Hurrell 1995). During the positive phase of the NAO, when below normal atmospheric pressure occurs across the high latitudes of the North Atlantic and above normal pressure occurs over the central North Atlantic, transport from all three northern hemisphere continents (Europe, North America, and Asia) into the Arctic is enhanced (Duncan and Bey 2004; Eckhardt et al. 2004).

The historical definition of Arctic haze includes the enhancement of lower and upper atmosphere pollutants observed during winter and spring. In a departure from this traditional view, Brock et al. (2011) have suggested that Arctic haze is best defined as the low variance enhancement in aerosol concentrations measured at Arctic surface monitoring sites during winter and spring. Chemical transport models (Koch and Hansen 2005; Fisher et al. 2010) and chemical "fingerprinting" analyses (Rahn and McCaffrey 1980) identify the combustion of fossil fuel, biomass, and

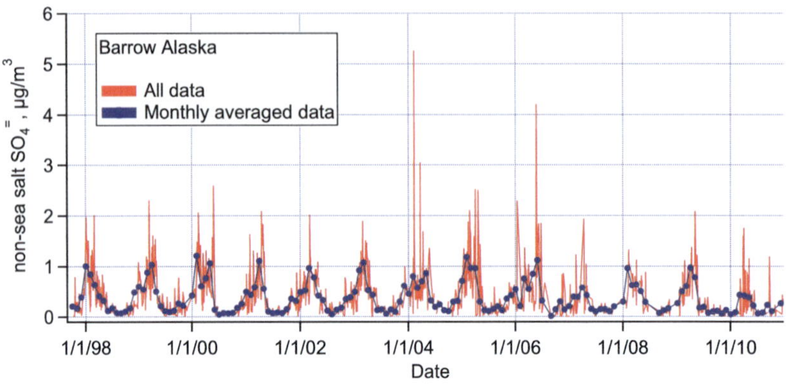

Fig. 4.3 Seasonal cycle of non-sea salt sulfate measured at Barrow, Alaska for the years 1997 to 2014

biofuel from northeastern Europe and Siberia as the principle sources of this chronic pollution that is observed every winter and spring at ground-based monitoring sites across the Arctic. Brock et al. (2011) argue that episodic events involving transport of anthropogenic and biomass burning emissions from a variety of mid- and northern latitude sources are distinct from seasonally varying Arctic haze. Episodic transport events, resulting in dense aerosol layers at higher altitudes, can occur during any season (Brock et al. 1989; Stohl et al. 2006; Paris et al. 2009) and, as such, are not strictly "Arctic haze". In fact, these episodic transport events are similar to the long-range, intercontinental transport of aerosol layers in the free troposphere that occur throughout the globe (Stohl et al. 2002). Although transported to the Arctic in the upper atmosphere, some of the dense layers aloft may ultimately contribute to chronic Arctic haze through atmospheric mixing. Sources of the layers associated with episodic transport events appear to be diverse and include biomass burning in southeastern Siberia and agricultural fires in southern Russia (Warneke et al. 2010) as well as fossil fuel combustion (Jacob et al. 2010).

4.1.3 Long Term Observations of Arctic Haze

Seasonality and Sources of Arctic Haze Long term observations of Arctic haze are available from surface sites at several locations in the Arctic, with three of the longest, most continuous, and publically available data records from Alert (82.46°N) (ALT) in the Canadian Arctic; Mt. Zeppelin (78.90°N) near Ny Alesund Svalbard, Norway (NYA); and Barrow (71.3 °N) (BRW) in Alaskan Arctic (Fig. 4.4).

The seasonality of Arctic haze has been observed at each of these sites as well as several others as illustrated by the time series of nss $SO_4^{=}$ data shown in Fig. 4.5. Each site has observations of a winter-early spring increase in nss $SO_4^{=}$ with

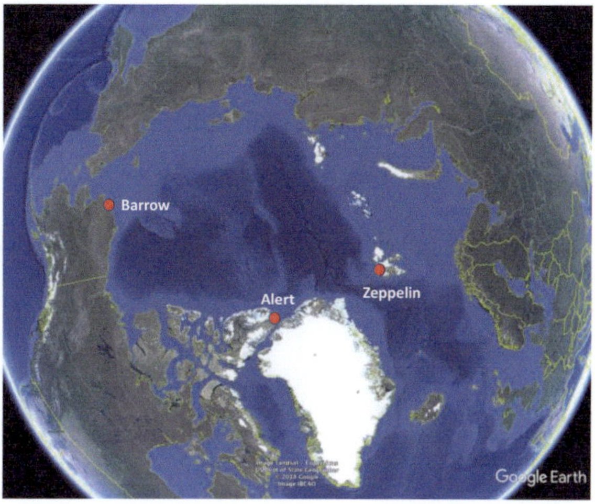

Fig. 4.4 Locations of three long-term monitoring stations in the Arctic

maximum concentrations reaching up to about 1 μg S m^{-3}. Summertime monthly average concentrations are less than 0.03 μg S m^{-3}. Nss $SO_4{}^{=}$ makes up about 30% of the submicrometer aerosol mass during the haze season (Barrie et al. 1981; Quinn et al. 2002). Figure 4.6 shows the time series of submicrometer aerosol nitrate at Alert and Barrow, two sites which have a clear seasonal cycle for this chemical species.

Maximum concentrations are near 0.04 μg N m^{-3}. Other species also indicative of continental sources (e.g., nss K^+ for biomass burning and Mg^{+2} and Ca^{+2} for dust) have maximum concentrations in winter and spring indicating long range transport to the Arctic (Quinn et al. 2002).

Sources of Arctic haze aerosol have been identified through back trajectory analyses which employ measured or modeled meteorological fields to estimate the most likely transport pathway for an air mass from a source region to a receptor site. Two examples of frequently used trajectory models include FLEXPART (Flexible Particle Dispersion Model), which is a Lagrangian transport and dispersion model (Stohl 2006) and HYSPLIT (Hybrid Single Particle Lagrangian Integrated Trajectory Model) (Frossard et al. 2011). In addition to transport modeling, chemical tracer species and isotopic analysis have been used to identify source regions and/or source types that impact the Arctic atmosphere. For example, using daily PM10 (particulate matter with aerodynamic equivalent diameter <10 μm) aerosol samples collected at Ny Alesund during the spring and summer of 2014 and tracer species for sources of nss sulfate, Udisti et al. (2016) assessed the relative source contributions of nss sulfate from dust, sea salt, and biogenic and anthropogenic sources. Sea salt sodium was used as a sea spray marker, nss calcium as a crustal marker, and methane sulfonic acid (MSA) as a biogenic marker. They found that crustal, sea salt, biogenic and anthropogenic sources accounted for 3.3, 12, 11, and 75% of the sulfate measured during spring, 2014.

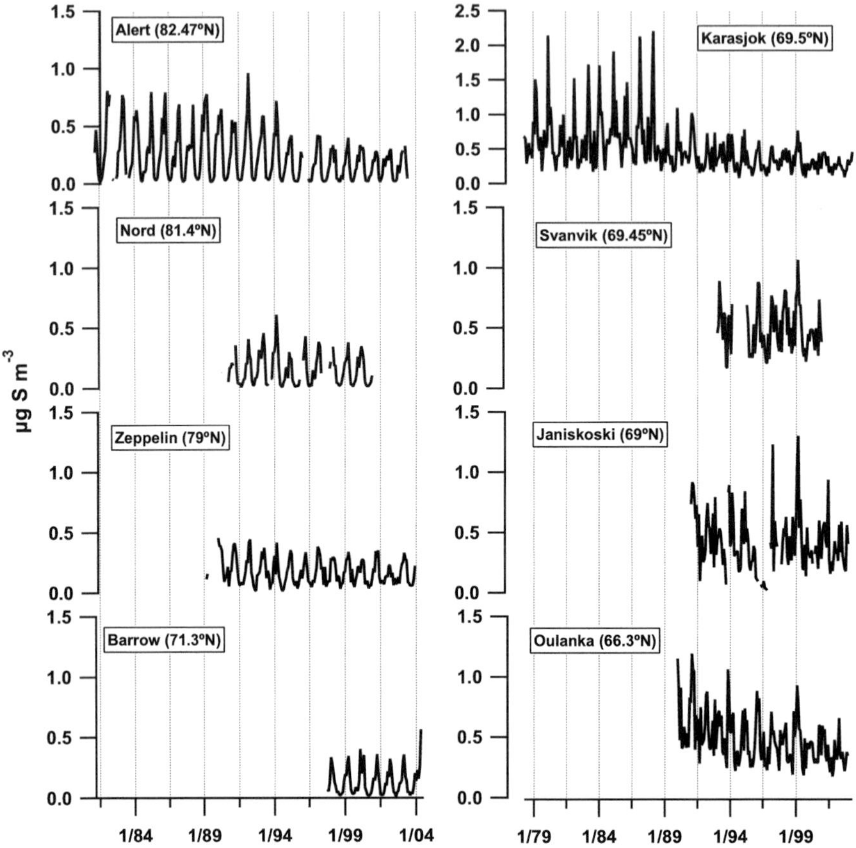

Fig. 4.5 Time series of monthly averaged nss $SO_4^{=}$ concentrations in $\mu g\ S\ m^{-3}$ for eight Arctic monitoring sites. Data made available for Alert by the Canadian National Atmospheric Chemistry (NAtChem) Database and Analysis System, for Barrow by NOAA PMEL (http://saga.pmel.noaa.gov/data/), and for the other stations by EMEP (http://www.emep.int/)

Winiger et al. (2015) used the first dual-carbon isotope based source apportionment of elemental carbon (EC), the mass-based correspondent to optically defined eBC, to identify sources of EC during high pollution events at NYA in winter and spring of 2009. They found that during these events, 52 ± 15% of the EC was a result of biomass burning with varying components from fossil fuel combustion sources, likely including liquid fossil and gas flaring. These types of "chemical fingerprinting" analyses to identify source of SLCPs to the Arctic help to guide policy actions to mitigate the climate impacts of emissions (Winiger et al. 2015)

Trends in Arctic Haze An advantage of long-term observations is the ability to use the data to construct trends to assess changes in emissions, transport pathways, and removal and to validate and improve transport models. The first trend analysis for Arctic aerosol focused on the longest term aerosol data available at that point –

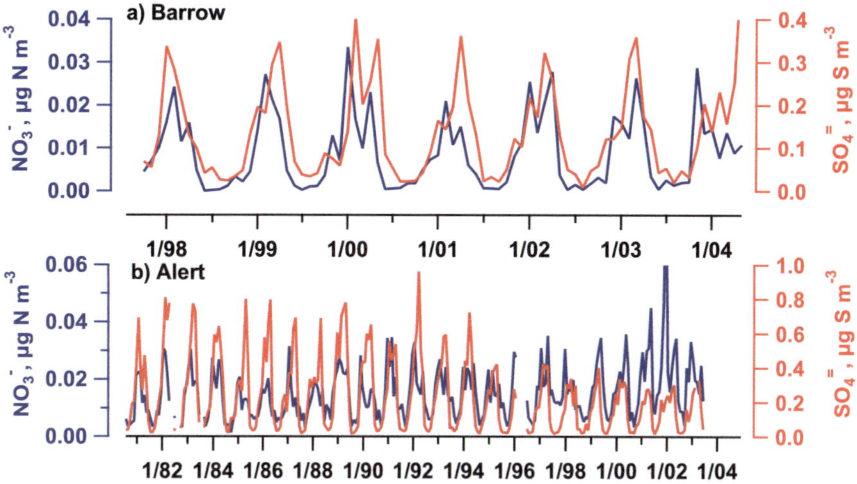

Fig. 4.6 Time-series of monthly averaged particulate nss sulfate and nitrate concentrations in µg S m^3 and µg N m^3, respectively, for (**a**) Barrow, Alaska and (**b**) Alert, Canada. Data made available for Alert by the Canadian National Atmospheric Chemistry (NAtChem) Database and Analysis System and for Barrow by NOAA PMEL (http://saga.pmel.noaa.gov/data/)

light scattering by aerosol particles and aerosol optical depth (AOD) measured at Barrow since 1977 (Bodhaine and Dutton 1993) and Zeppelin since 1991 (Stone et al. 2014). AOD is a measure of the fraction of solar radiation removed through the scattering and absorption of light by aerosol over the atmospheric path length of the measurement. It is not specific to any one chemical component but it reveals information about trends in aerosol concentrations over the entire atmospheric column.

Between 1977 and 1993, both AOD and light scattering measured during the springtime Arctic haze period at BRW peaked in 1982 followed by a factor of two decrease between 1982 and 1993 (Bodhaine and Dutton 1993). The observed decrease was likely due to reductions in emissions from the former Soviet Union resulting from economic factors and from Europe due to stricter pollution controls. Stone et al. (2014) assessed the trend in AOD at BRW, ALT, and NYA and found an increase between 2001 and 2012, primarily due to high aerosol loadings during the springs of 2008 and 2009. The high 2008 loadings were a result of biomass burning smoke from Asia that was prevalent throughout the atmosphere from the surface up to about 6.5 km (Warneke et al. 2009). Atmospheric circulation patterns during spring 2009 resulted in the transport of haze and dust into the central Arctic (Stone et al. 2010). Tomasi et al. (2012) attributed the increase at NYA to enhanced aerosol loading in the stratosphere due to minor volcanic eruptions. Their analysis for BRW indicates additional contributions from Eurasian emissions and/or coal burning in China.

Long-term continuous records of black carbon, based on measurements of aerosol light absorption, also are available at several locations in the Arctic. The longest records are from BRW and ALT (1988 and 1989, respectively, to present). Measurements at NYA began in the late 1990s. Black carbon at each of these sites is derived from measurements of aerosol light absorption using an assumed mass absorption efficiency in units of $m^2\ g^{-1}$ to convert light absorption to BC mass concentration (Sharma et al. 2013). The resulting black carbon concentration is referred to as equivalent black carbon or eBC (Petzold et al. 2013).

All trend analyses of eBC to date, report a decrease during the Arctic haze season over the past three decades. Quinn et al. (2007) found a 60% decrease in average March monthly aerosol light absorption between 1988 and 2006 at BRW. Sharma et al. (2006) found a 49% and 33% decrease in wintertime eBC concentrations between 1989 and 2003 at ALT and BRW, respectively. Sharma et al. (2013) and Stone et al. (2014) updated the trend analyses for BRW, ALT, and NYA and found an overall decline of 40% in eBC during winter/spring (January to April) between 1990 and 2012 with most of the change occurring during the early 1990s. Coupling a model analysis with the observations indicated that the maximum contribution to surface BC concentrations at ALT, BRW, and NYA was from the combined former Soviet Union and European regions and that trends in wintertime BC are primarily due to changes in emissions in the former Soviet Union (Sharma et al. 2013). Recent linear trend analyses published in 2015 are shown in Fig. 4.7 and indicate a 2.6–3.0% decrease in eBC per year at NYA, ALT, and BRW (AMAP 2015).

Long-term records of nss sulfate also are available for ALT, BRW, and NYA. For all three stations, concentrations have decreased during the haze season at rates between 2 and 3.5% per year (AMAP 2015) (Fig. 4.8).

There are no long-term records of organic aerosol concentrations for the Arctic. Sporadic measurements have been made but are too infrequent to reveal anything about trends.

4.1.4 Vertical Distributions of Arctic Pollution

Aircraft campaigns have occurred sporadically in the Arctic since the early 1980s. These campaigns have primarily targeted the spring time Arctic haze. In the spring of 1983, six aircraft participated in the Arctic Gas and Aerosol Sampling Program (AGASP) I (Schnell 1984). AGASP II was conducted in the spring of 1986 (Parungo et al. 1990). These early flights revealed large inter-annual variability in the vertical structure of eBC. Other aircraft campaigns aimed at the study of spring time Arctic pollution include TOPSE (Tropospheric Ozone Production about the Spring Equinox) (Scheuer et al. 2003) in February through May of 2000, ASTAR (Arctic Study of Tropospheric Aerosol and Radiation) conducted in May and June 2004 (Engvall et al. 2008), ARCTAS (Arctic Research of the Composition of the Troposphere from Aircraft and Satellites in April 2008 (Jacob et al. 2010), ARCPAC (Aerosol, Radiation, and Cloud Processes affecting Arctic Climate) in April 2008

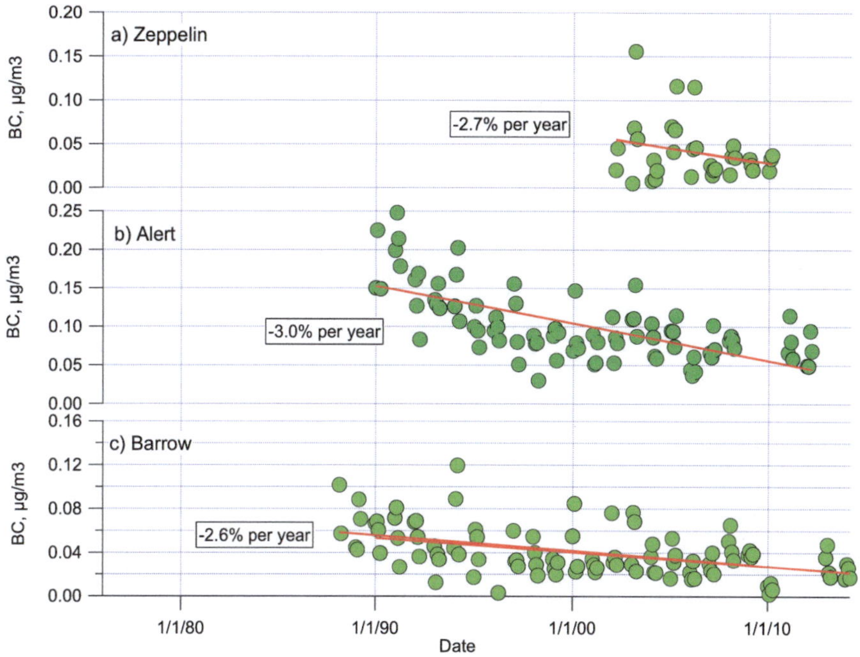

Fig. 4.7 Linear trends in eBC averaged over the Arctic haze months (January to April) for (**a**) Zeppelin, (**b**) Alert, and (**c**) Barrow (AMAP 2015)

(Brock et al. 2011), and PAMARCMiP (Polar Airborne Measurements and Arctic Regional Climate Model Simulation Project) conducted in 2009, 2011, 2013, and 2015 (Stone et al. 2014). In addition, 2 years (2011–2012) of tethered balloon flights (up to 1200 m a.g.l.) were conducted by (Ferrero et al. 2016).

With the development of instrumentation able to measure sulfate concentrations on a relatively fast time scale, the first detailed vertical profiles of nss sulfate came out of TOPSE (Scheuer et al. 2003). It was found that in early February there were significant enhancements in sulfate that were confined to near the surface. As the season progressed, polluted parcels of air aloft were observed and were attributed to transport into the Arctic along vertically higher isentropes tracing back to warmer source regions in northern Eurasia. During April, below 3000 m, sulfate concentrations began to decrease as solar surface heating began to mix the lowest levels of the atmosphere. In addition, removal processes dominated by wet deposition also likely became increasingly important (Barrie 1986). Measurements during ASTAR confirmed the TOPSE results indicating that the Arctic free troposphere is composed of aged, polluted air masses.

ARCTAS and ARCPAC were both conducted during April of 2008 with ARCTAS covering a wider geographic region of the Arctic and ARCPAC focusing on the Alaskan Arctic. Flights during ARCTAS observed a pervasive background of Asian pollution throughout the depth of the Arctic tropospheric column. Superimposed

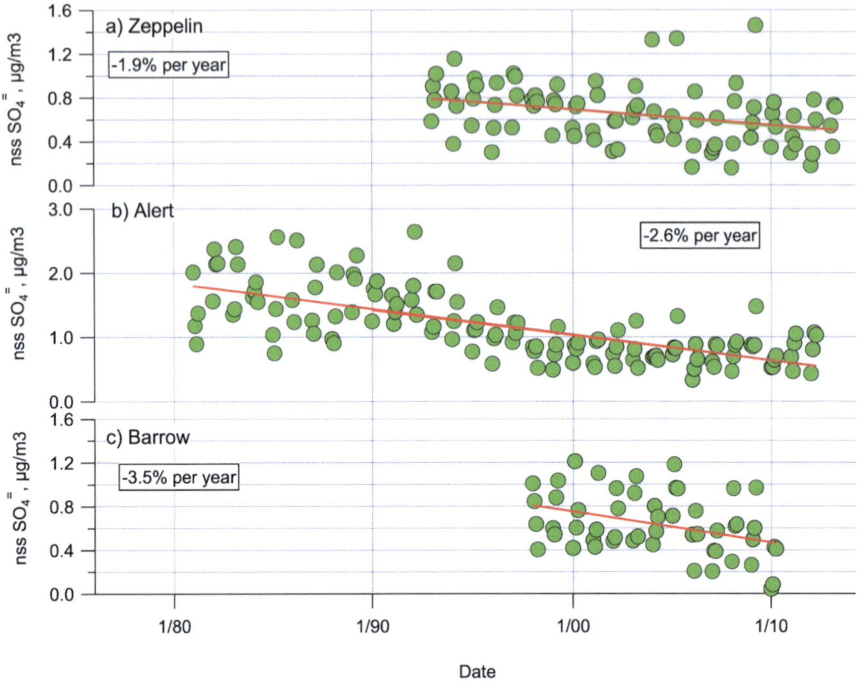

Fig. 4.8 Linear trends in nss sulfate averaged over the Arctic haze months (January to April) at (**a**) Zeppelin, (**b**) Alert, and (**c**) Barrow

upon this background were distinct industrial pollution plumes from Europe, Asia, and North America as well as Siberian biomass burning plumes. Aerosols below about 2 km in altitude were anthropogenic in origin (fossil fuel combustion) whereas those at 2 to 4 km were mainly from biomass burning (Jacob et al. 2010). The striking feature dominating measurements during ARCPAC was the frequent occurrence of layers of biomass burning aerosol as identified by enhancements in CO and acetonitrile (Warneke et al. 2009). Two primary sources of smoke were identified based on FLEXPART and RAQMS (Realtime Air Quality Modeling System) simulations – a broad region located along the Russian/Kazakhstan border which involved agricultural fires that developed into forest fires and southeastern Siberia.

Figure 4.9 displays a chronology of equivalent black carbon (eBC) and refractory black carbon (rBC) ensemble vertical profiles obtained during many of these Arctic aircraft campaigns.

Refractory black carbon (rBC) is based on an incandescence measurement. The incandescence emission is measured and calibrated using black carbon proxies such as fullerene. Unlike long-term surface-based observations, there are no monitoring programs in the Arctic that provide regular vertical profile information of aerosol concentrations. The aircraft campaigns that have occurred have been conducted in

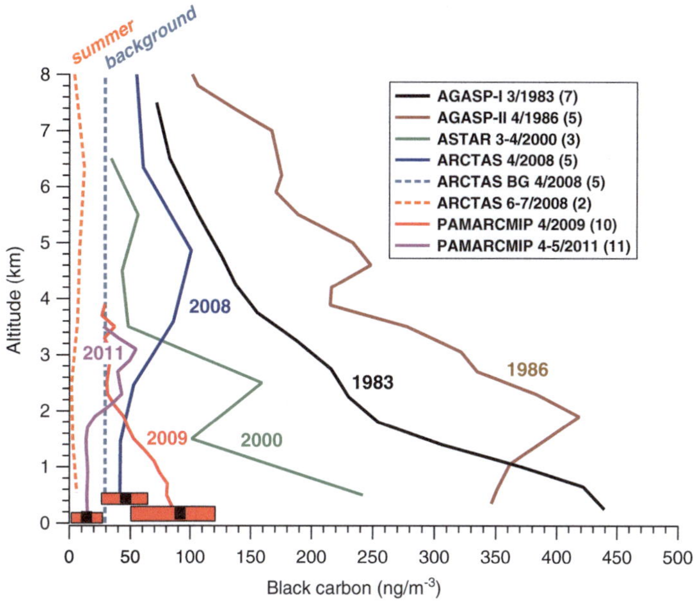

Fig. 4.9 Averaged profiles of eBC (AGASP and ASTAR) and rBC (ARCTAS, PAMARCMIP) conducted between 1983 and 2011. The number of flights included in each average profile is given in parentheses. The grey dashed line represents clean background values measured in the Alaskan Arctic during April 2008. The yellow dashed line is for the summer of 2008. Climatological means (± 1 s.d.) of equivalent black carbon (eBC) measured during April at Barrow (Alaska), Alert (Canada), and Zeppelin (Svalbard) (shown by altitude of site; Barrow is lowest at 11 m above sea level, Alert is at 210 m, and Zeppelin is at 474 m) are indicated by black symbols and red bars. (Source: Stone et al. 2014)

different parts of the Arctic and have employed different measurements techniques. As a result, direct comparison of observations is problematic. That said, the vertical profiles displayed in Fig. 4.9 provide evidence that concentrations of black carbon, and presumably other aerosol species, have decreased throughout the atmospheric column between 1983 and 2011 (Stone et al. 2014). The large eBC concentrations measured during AGASP I and II are consistent with higher AOD values also observed during the experiments (Bodhaine 1989; Dutton et al. 1989). The decrease in concentrations throughout subsequent years is consistent with the downward trends in eBC measured at BRW, ALT, and NYA surface sites. Overall, for regions where repeat measurements have been made, there appears to be a downward trend in atmospheric aerosol concentrations at the surface and aloft in the Arctic atmosphere.

4.1.5 Potential Impacts of a Changing Climate on Arctic Haze

The European sector of the Arctic saw unprecedented warmth during the first few months of 2006. Temperatures measured at Ny Alesund were the highest ever recorded to date during January, April, and May. Due to this abnormal warmth, the sea surrounding the Svalbard archipelago was almost completely free of closed ice at the end of April for the first time. Concurrently, a large number of fires occurred in the Baltic countries including western Russia, Belarus, and the Ukraine (Stohl et al. 2006). The fires originated as agricultural fires designed to burn fields before the start of the next growing season. A synoptic situation developed very similar to the conditions when Arctic haze is observed at Svalbard in winter and early. In this case, however, instead of the pollution source region being extremely cold as normally happens during the transport of pollution to the Arctic, the Arctic receptor region became unusually warm. Should the warming of the Arctic continue to proceed more quickly than that of the mid-latitudes, such transport conditions are likely to become more frequent in the future.

4.2 Background Arctic Aerosol

4.2.1 Introduction

Aerosol particles suspended in the Arctic atmosphere originate from both primary and secondary processes including natural (e.g., marine phytoplanktonic activity, sea salt, continental dust) and anthropogenic (e.g., traffic, domestic heating, electrical power production, industrial activities) sources. Therefore, their chemical composition can vary considerably over the spatial and temporal (both seasonal and inter-annual) scales, depending on the timing of local and remote sources as well as on the direction and intensity of the long-range atmospheric transport processes. In all these cases, the multimodal aerosol size distribution curve is characterized by three modes: (i) the nucleation mode, covering the size range from 5 10^{-3} to 10^{-1} μm, which has in general a mean geometric diameter equal to about 2 10^{-2} μm and a geometric standard deviation equal to about 1.8, mainly consisting in general of small sulfuric acid droplets or containing combustion matter; (ii) the accumulation mode, covering the size-range from 0.1 to around 1 μm, which has in general mean diameter equal to 0.15 μm, and geometric standard deviation equal to 1.4; and (iii) the coarse particle mode, consisting of particles having diameters greater than 1 μm, which shows in general its concentration maximum at diameters ranging from 5 μm to no more than 8 μm, mainly containing crustal material (hereinafter often referred to as "mineral dust") and sea salt.

Both real and imaginary parts of the refractive index of the Arctic aerosol particles belonging to the accumulation and coarse particle modes may vary largely from one case to another, due to the different origins and, hence, to their different chemical composition features. In fact, the radiative effects of Arctic aerosol on the solar radiation budget at both surface-level and top-level of the atmosphere are closely related to the absorption properties of: (i) anthropogenic substances due to pollution from industrial regions, or (ii) combustion substances generated by forest fires. With regard to these processes, Shaw et al. (1993) estimated that a dense aerosol layer of anthropogenic aerosol particles can cause a local warming rate of: (a) about 1.7 K/day for surface albedo equal to 0.65 and close to the 35° solar elevation angle; and (b) about 1.0 K/day for a solar elevation angle of 15°. For the same surface albedo conditions, the presence of the above particle layer was evaluated to change the net flux at the tropopause level by about 5 W m^{-2}. These evaluations indicate that tropospheric aerosols present in the Arctic region can perturb the radiation balance of the surface-atmosphere system through direct forcing effects induced by the particle absorption and scattering processes affecting the solar (short-wave) radiation. Quinn et al. (2008) estimated that the tropospheric aerosol direct effects occurring at the surface are particularly marked in spring and summer at latitudes ranging between 60 °N and 90 °N, presenting seasonally averaged values of direct aerosol-induced radiative forcing nearly equal to −1 W m^{-2} in summer. In addition, Albrecht (1989) pointed out that particularly marked indirect effects may be produced by tropospheric aerosol particles in summer, since aerosol-cloud interactions cause an increase of the cloud cover fraction, as a result of the increased cloud droplet number concentrations. This leads to a decrease in cloud drop size, which in turn causes a decrease in precipitation and an increase in cloud life-time.

Therefore, in order to achieve realistic estimates of the radiative forcing effects produced by the aerosol particles suspended over the various Arctic regions in the different seasonal periods, it is of basic importance to know the chemical composition of these aerosols. On the basis of these data, the aerosol optical properties can be more accurately defined by taking into account the specific characteristics of the fine and coarse particle modes. This classification can be made by examining the chemical composition features of particulate matter samples collected during several years at the following Arctic sites or regions:

(1) Ny-Ålesund (78° 56′ N, 11° 55′ E), in the Oscar II Land on the island of Spitsbergen (Svalbard, Norway), located on the shore of the Bay of Kongsfjorden in the Brøgger peninsula, where particulate matter chemical composition measurements and optical measurements were performed during March–September campaigns from 2010 to 2015;
(2) Thule Air Base (76° 32′ N, 68° 42′ W), the United States Air Force's northernmost base, situated on the northwest side of Greenland, 1207 km north of the Arctic Circle and 1524 km from the North Pole, where all-year round measurements of PM 10 chemical composition were performed in 2012, 2013 and 2014;
(3) Central Arctic Ocean, in a large off-shore area located at latitudes ranging between 77.9 °N and 87.5 °N, and longitudes varying between 11.1 °W and 9.6 °E, far from Northern Greenland and Spitsbergen Island, where Chang et al. (2011) performed measurements of submicron aerosol chemical composition

from 5 August to 8 September 2008 (Arctic Summer Cloud Ocean Study, or ASCOS, campaign), using an aerosol mass spectrometer;

(4) Barrow (71° 19′ N, 156° 37′ W), now called with the new first nation name Utqiagvik, the largest city of the North Slope Borough in Alaska (USA), located above the Arctic Circle, on the Arctic Ocean coast, where an important and accurate analysis of the chemical composition of aerosol particles was performed by Quinn et al. (2002) over the period from October 1997 to December 2000; and.

(5) Alert (82° 30′ N, 62° 20′ W), located in the Qikiqtaaluk region (Nunavut, Canada) at 817 km only from the North Pole, using the results provided by Sirois and Barrie (1999) from 1980 to 1995 and Quinn et al. (2007, 2008) from the early 1990s to 2003 on the particulate matter chemical composition, integrated with equivalent BC and OC *in-situ* measurements carried out by Sharma et al. (2002, 2004, 2006) over the period from 1989 to 2003.

4.2.2 Measurements of Aerosol Chemical Composition at Ny-Ålesund

The Italian National Research Council (CNR) is coordinating and supporting the Italian scientific activities carried out since 1997 at the "Dirigibile Italia" station, located in Ny-Ålesund, Svalbard Islands. In particular, since 2010, spring-summer campaigns for aerosol chemical and physical characterization were continuously carried out in the new "Gruvebadet" (GVB) Observatory, located at about 50 m a.m.s.l., about 800 m south-west from the Ny-Ålesund village (78°55′N, 11°56′E), which is located in the Kongsfiorden fiord that develops in the north-west to south-east direction, so that the dominant winds are channelled in the same direction. In this way, the geographical position of the GVB Observatory guarantees for the lowest impact of local anthropic emissions (Udisti et al. 2016). Besides, the research infrastructure "Amundsen-Nobile Climate Change Tower" was built on 2009 at the Kolhaugen site, 30 m a.m.s.l. and about 1 km far from Ny-Ålesund village: it is an instrumented tower, 34 m high, that is part of the ESFRI (European Strategy Forum on Research Infrastructures) project "Svalbard Integrated Observing System" (SIOS). This atmospheric monitoring system hosts a series of instrumentations, installed at different heights from the soil level to monitor various atmospheric physical parameters, greenhouse gases and micro-meteorological parameters (Mazzola et al. 2016).

The aerosol sampling devices used at the GVB Observatory included: (i) a Scanning Mobility Particle Sizer (SMPS, 54 size classes in the 10–500 nm range); (ii) an Aerodynamic Particle Sizer (APS, 52 size classes from 0.5 to 20 μm); (iii) a pair of low- and medium-volume PM10 (particulate matter smaller than 10 μm aerodynamic equivalent diameter) samplers; and (iv) a set of multi-stage impactors (4 to 12 stages). The PM10 samplers were continuously used to collect aerosol particles on Teflon (for ions and metal composition measurements) and quartz (EC and OC fraction determination) filters with temporal resolution of 1–4 days. The Dekati 4-stage (DK-4; <1, 1–2.5, 2.5–10, >10 μm)) and 12-stage (DK-12; from 0.045 to >8.5 μm) impactors were specifically devoted to study the size distribution characteristics of ions (DK-4, by Ion Chromatography) and elements (DK-12, by Particle Induced X-ray Emission – PIXE – measurements).

4.2.2.1 PM10 Sampling and Calculation of Chemical Components

PM10 aerosol samples were continuously collected on Teflon filters (Pall R2PJ047, 47 mm diameter, 2 μm nominal porosity) on daily basis (00:01–23:59, Universal Time Coordinated – UTC) from March to September (with some exceptions) of the years from 2010 to 2015. The samplings were carried out in "actual conditions", the pressure and temperature being continuously monitored to maintain a constant flow rate of 38.3 L/min (EN12341 European rules), corresponding to a 24-h air volume of about 55 m^3.

PM10 mass was measured by weighing the filters with a microbalance (0.01 mg sensitivity) before and after each exposure. The filters were conditioned at low humidity (silica gel) and controlled temperature (25 $\pm$ 1 °C) for at least 24 h before the weighing. At the analysis time, one half of each PM10 Teflon filter was extracted in about 10 ml Milli-Q water (accurately evaluated by weighing) by ultrasonic bath for 20 min, for ionic content determination. Every filter manipulation was carried out under a class-100 laminar-flow hood, in order to minimize contamination risks. Inorganic anions and cations, as well as selected organic anions, were simultaneously measured by using a three Dionex ion-chromatography system, equipped with electrochemical-suppressed conductivity detectors. The sample handling during the ion chromatography (IC) injection was minimized by using a specifically-designed Flow-Injection Analysis (IC-FIA) device (Morganti et al. 2007). Cations (Na^+, NH_4^+, K^+, Mg^{2+} and Ca^{2+}) were determined by using a Dionex CS12A-4 mm analytical column with 20 mM H_2SO_4 eluent. Inorganic anions (Cl^-, NO_3^-, SO_4^{2-} and $C_2O_4^{2-}$) were measured by a Dionex AS4A-4 mm analytical column with a 1.8 mM Na_2CO_3 / 1.7 mM $NaHCO_3$ eluent. F^- and some organic anions (acetate, glycolate, formate and methanesulfonate) were determined by a Dionex AS11 separation column by a gradient elution (0.075 mM to 2.5 mM $Na_2B_4O_7$ eluent). A six-standard calibration curve was daily used for quantification. Further details are reported in Udisti et al. (2004) and Becagli et al. (2011).

The Ny-Ålesund aerosol particles contain not-negligible contributions of sea salt and crustal components, so that Na^+ and Ca^{2+}, which originate from both these sources, cannot be used as univocal sea spray and crustal markers, respectively. In order to quantify the sea-salt (ss) and non-sea-salt (nss) contributions of these elements in each sample, we used a simple equation system (Becagli et al. 2012; Udisti et al. 2012) consisting of the following equations (ss stands for sea-salt while nss for non sea-salt): (i) tot-Na^+ = ss-Na^+ + nss-Na^+, (ii) tot-Ca^{2+} = ss-Ca^{2+} + nss-Ca^{2+}, (iii) ss-Na^+ = tot-Na^+ − 0.95 nss-Ca^{2+}, and (iv) nss-Ca^{2+} = tot-

Ca^{2+} – 0.038 ss-Na^{+}, where 0.95 represents the Na^{+}/Ca^{2+} weight-to-weight (w/w) ratio in the Earth upper crust (Rudnick and Gao 2003), and 0.038 is the Ca^{2+}/Na^{+} w/w ratio in sea-water (Nozaki 1997).

In this section, we report the mean values of six particle categories: sea spray, mineral dust, non-sea-salt sulfate, nitrate, ammonium, biogenic emissions.

(a) Sea spray was evaluated by multiplying the ss-Na concentration by a numerical factor equal to 3.248 (Kerminen et al. 2000).
(b) Mineral dust was evaluated by multiplying the nss-Ca concentration by a factor equal to 33.78, evaluated by ion chromatography (IC-nss-Ca). This value is close to the mineral dust/Ca ratio calculated in the Earth upper crust (Ca = 2.56%, corresponding to a mineral dust/Ca = 39.06), and was experimentally obtained as the mean value of the mineral dust/IC-nss-Ca ratios measured on the PM10 samples collected at the GVB Observatory during the 2015 campaign. In this campaign, 217 PM10 samples were collected and filters were analyzed for ion chromatography (water-soluble fraction) and PIXE (total content). Total mineral dust was calculated as the sum of the contributions of all the main crustal element oxides measured by PIXE: [crustal content] = 2.14[Si] + 1.89[Al] + 1.43[Fe] + 1.40[PIXE-nss-Ca] + 1.35[PIXE-nss-Na] + 1.66[Mg] + 1.21[K] + 1.67[Ti]. The IC-nss-Ca and PIXE-nss-Ca values were calculated by the previous equation series. Since only ion chromatographic measurements were carried out on the 2010–2014 PM10 samples, we used the 33.78 (w/w) mineral dust/IC-nss-Ca mean ratio obtained on the 2015 samples to calculate the mineral dust contribution by the ion chromatographic nss-Ca measurements.
(c) The non-sea-salt fraction of sulfate (nss-SO_4^{2-}) was calculated by subtracting the sea spray contribution ($ssSO_4^{2-} = ssNa^{+} \times 0.253$, indicating the SO_4^{2-}/Na^{+} w/w ratio in sea-water (Bowen 1979).
(d) Nitrate and ammonium were directly measured by ion chromatography.
(e) The biogenic contribution was evaluated by summing the methanesulfonic acid (MSA) and the biologic sulfate (bio-nss-SO_4^{2-}) concentrations, the last term being evaluated by multiplying the MSA concentration by 3.0 (Udisti et al. 2016).

4.2.2.2 Chemical Composition of PM10 Collected at Ny-Ålesund

The chemical characterization of the aerosol particles sampled in the Arctic stations of Alert (Canadian Arctic), Station Nord (Northern Greenland), Zeppelin (Svalbard Islands), Barrow (Alaska), Karasjok and Svanvic (Northern Norway), Oulanka (Northern Finland) and Janiskoski (Northern Russia) revealed a typical seasonal cycle in the time-patterns of the monthly mean mass concentrations of the Arctic aerosol particles. This cycle is characterized by high concentrations of aged anthropogenic particles, mainly belonging to the accumulation mode, such aerosol components being transported in the late winter and early spring from lower latitudes. These high concentrations are followed by a sharp drop to low concentrations of smaller particles in the summer, locally generated mainly through nucleation processes (Browse et al. 2012).

The late winter – early spring transport of contaminants from polluted continental areas is called "Arctic Haze". The chemical composition of Arctic Haze particles has been well characterized in the last years (Quinn et al. 2007), showing that these particles are mainly composed by a mixture of sulfate, organic particulate matter, ammonium, nitrate, dust, black carbon and heavy metals (Li and Barrie 1993; Quinn et al. 2002, 2007). The highest observed concentrations of aerosol mass were observed in transport episodes from Russia and central Europe (Tunved et al. 2013). During the summer months (June–August), a much larger fraction of air mass transport takes place over the Atlantic Ocean, and the aerosol is characterized by a higher fraction of smaller-sized particles, which is suggested to have been formed locally. Tunved et al. (2013) suggested that wet removal largely controls the properties of the aerosol size distribution observed in Svalbard. Although precipitations initially remove aerosol mass and number, the wet removal taking place in summer seems to facilitate the conditions favouring new particle formation and growth. This is shown as a gradual increase in the concentration of Aitken nuclei as the amount of precipitation increased further during the summer months. This evidence is consistent with the fact that photochemistry plays a central role in formation of these small particles that only takes place when the balance between generation and removal of nucleating gases is favourable (i.e. in the presence of sources of precursor gases, photochemical activity and low condensation sinks).

The ionic composition of PM10 collected at Ny-Ålesund during the 2010–2015 spring – summer period shows that the ionic balances were dominated by two main components: (i) sea spray (marked by Na^+, Cl^-, Mg^{2+} and, partially, $SO_4{}^{2-}$ (sea-salt sulfate: ss-$SO_4{}^{2-}$), and (ii) secondary aerosol (marked by non-sea-salt sulfate (nss-$SO_4{}^{2-}$), ammonium, nitrate and nitrite). For instance, Fig. 4.10 shows the PM10 ionic balances, calculated on all the samples collected during the 2014 campaign (total) and on seasonally segregated data-sets (in spring and summer).

The overall-period plot shows that Na^+ and $NH_4{}^+$ were the major cations, accounting for 51% and 25% of the cation content, respectively. Similarly, chloride and sulfate dominate the anion budget (45% and 38%, respectively). However, such components were found to be not equally distributed in the spring and summer

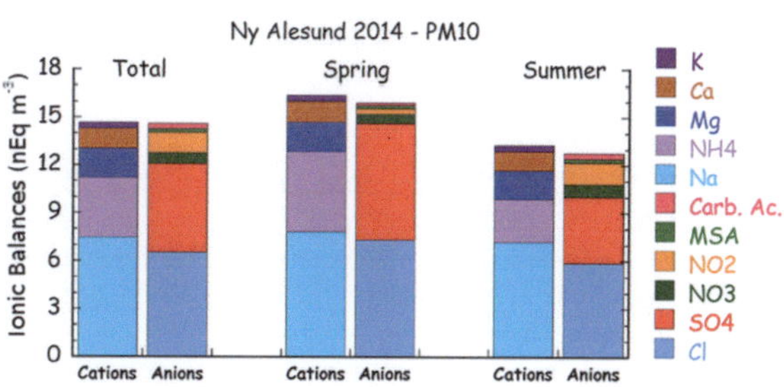

Fig. 4.10 Ionic balances (measured in nEq m^{-3}) of the PM10 samples collected at Ny-Alesund (Spisbergen, Svalbard) during the 2014 campaign (Udisti et al. 2016) over the total, and seasonal (spring (MAM) and summer (JJAS)) periods

samples, showing sharp seasonal differences. Conversely, sulfate and ammonium seem to provide the highest overall contribution in spring (both in absolute quantity and in ion percentage). In this season, NH_4^+ accounts for 30% of the cation budget, while SO_4^{2-} reaches a contribution as high as 46% of the anion content. In particular, sulfate accounts for 19% (5.6 nEq m^{-3}),[1] 23% (7.3 nEq m^{-3}) and 16% (4.1 nEq m^{-3}) of the total ion content (anions plus cations) in the whole period, and for the spring and summer samples, respectively. All-period and seasonal ionic balances show that anions and cations are almost balanced, indicating a quite completely neutralized aerosol (Udisti et al. 2016).

On a multi-annual scale, a large variability in atmospheric load and chemical composition (including the acid-base balance) was found in the spring-summer samples collected in the 2010–2015 period. Figures 4.11a and 4.11b show the seasonal and inter-annual distributions (box plots) of the main aerosol components (PM10, sea salt, nss-sulfate, ammonium, MSA, OC, nss-K^+, dust and nitrate) along the 6-year sampling campaigns. In these plots, each box contains the 50% of data, with the median value displayed as a horizontal line. The bottom and the top of the box mark the limits of 25% of the variable population (i.e. the 25th and 75th percentiles, respectively). The vertical lines extending from the bottom and top of each box respectively mark the minimum and maximum value that fall within an acceptable range (equal to 1.5 times the box width). The outliers are shown as small circles.

Details on the chemical composition of PM10 collected at Ny-Ålesund (Svalbard) are given in the following sub-paragraphs.

(A) **PM10 mass concentrations measured at Ny-Ålesund**

Figure 4.11a shows the distribution (box plots) of the atmospheric mass concentrations of the PM10 particulate fraction. Along the sampling period (2010–2015), a seasonal behaviour is well evident, with spring concentration values significantly higher than the summer ones. The spring median values range from 2.83 $\mu g\ m^{-3}$ (measured in 2015) to 4.39 $\mu g\ m^{-3}$ (in 2013), while the summer median values varied from 1.46 $\mu g\ m^{-3}$ (in 2011) to 2.28 $\mu g\ m^{-3}$ (in 2012). Also the data variability was higher in spring, associated with the higher frequency of the Arctic Haze transport events. Spring 2013 recorded the highest mean and standard deviation.

As concerning the long period (6 years) inter-annual variability, both spring and summer concentrations do not show any clear temporal trend or peculiar spikes in particular years. Mean values were estimated to range between 3 and 5 $\mu g\ m^{-3}$,

[1] "Eq" means "chemically equivalent" and represents the measurement unit of the substance amount that reacts or can react with an other substance: it implies that 1 Eq of a certain substance reacts in all cases with 1 Eq of another substance. Eq is equal to the number of moles (as in the cases when the ion has only one charge, as it is, for instance, in ions Na^+ and K^+) or to the number of moles multiplied by n, where n is the charge of the ion, as in the case of 1 mole of SO_4^{2-}, which contains 2 Eq of SO_4^{2-}). Actually, quantity Eq is needed to define the ionic balance and evaluate the acidity or the alcalinity conditions of a chemical compound.

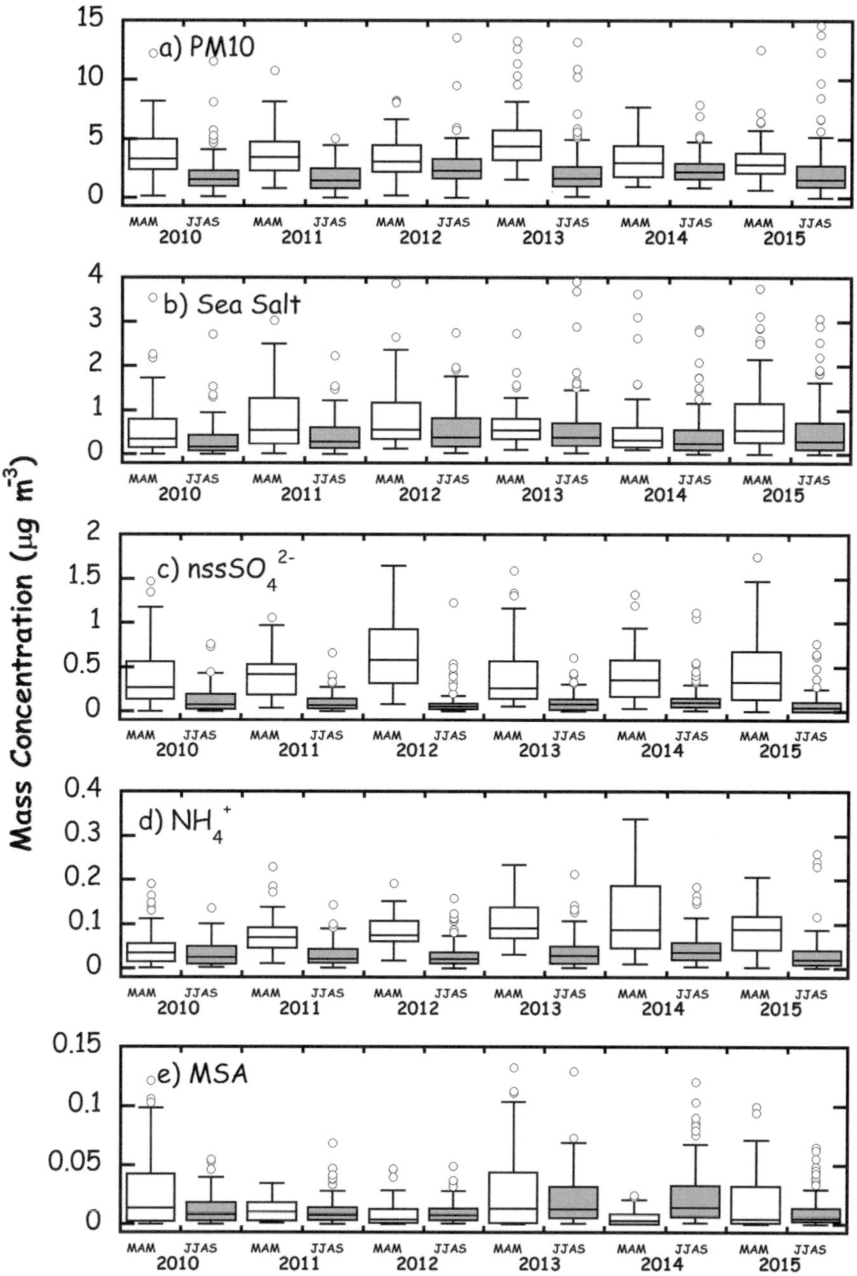

Fig. 4.11a Mass concentrations of PM10 and selected chemical components of the PM10 samples collected at Ny Alesund during the spring (MAM) and summer (JJAS) periods of the 2010–2015 years for: (**a**) PM10, (**b**) sea salt, (**c**) nss SO_4^{2-}, (**d**) NH_4^+, and (**e**) MSA. In all the graphs, each box contains the 50% of data, with the median value given by a horizontal line. The bottom and the top of the box indicate the values of the 25th and 75th percentiles, respectively, while the vertical lines extending from the bottom to the top of each box indicate the minimum and maximum values that fall within an acceptable range (equal to 1.5 times the box width). The outliers are shown as small circles

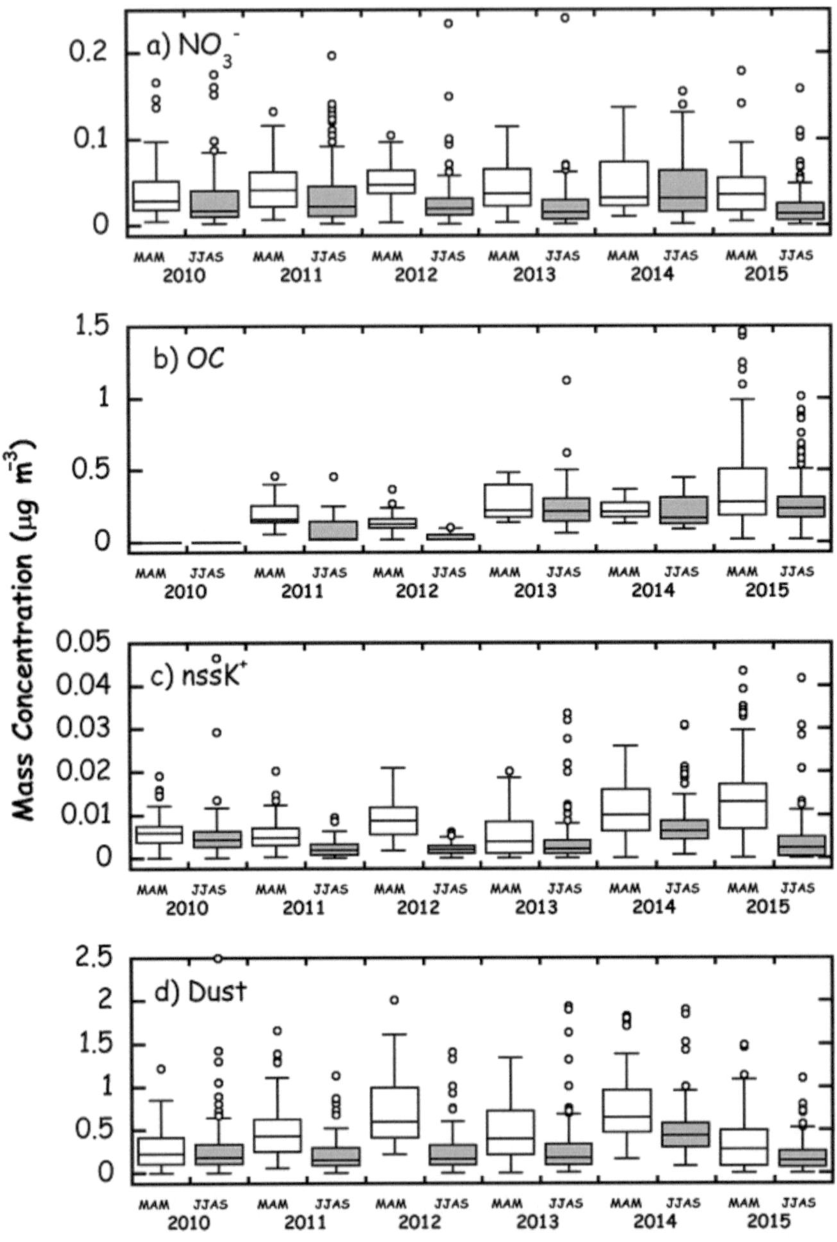

Fig. 4.11b Mass concentrations of PM10 and selected chemical components of the PM10 samples collected at Ny Alesund during the spring (MAM) and summer (JJAS) periods of the 2010–2015 years) for: (**a**) NO_3^-, (**b**) organic carbon OC, (**c**) nss K^{+-}, and (**d**) dust. In all the graphs, each box contains the 50% of data, with the median value given by a horizontal line. The bottom and the top of the box indicate the values of the 25th and 75th percentiles, respectively, while the vertical lines extending from the bottom to the top of each box indicate the minimum and maximum values that fall within an acceptable range (equal to 1.5 times the box width). The outliers are shown as small circles

with a standard deviation of around 2–3 μg m^{-3}. These values were found to be about ten times lower than the European air-quality standard limit of 40 μg m^{-3} (annual mean of 24-h PM10 values).

(B) **Sea salt in PM10 collected at Ny-Ålesund**

Sea spray is a relevant component of the primary aerosol, especially in the marine boundary layer (MBL) and affects the Earth climate through its interaction with solar radiation. The scatter effects produced by aerosols on the incoming solar radiation cause an important direct radiative forcing effect. In the Antarctic MBL, sea salt particles contribute significantly to determine the optical properties of aerosols in the summer months (Quinn et al. 1996; Murphy et al. 1998). Since sea salt particles are hygroscopic, they constitute a major source of cloud condensation nuclei (CCN) in remote marine areas. In this way, sea spray particles exert important effects on the cloud coverage, and consequently on both atmospheric albedo and hydrological cycle. It is important to study the sources and transport processes of particulate matter, as well as the seasonality of the size distribution characteristics of sea-salt aerosol particles, in order to better understand sign and intensity of its impact on the Earth climate (IPCC 2007). This is particularly relevant for the Polar regions, since they are characterized by high surface albedo conditions, where the radiative forcing exerted by the aerosols can produce opposite effects with respect to those commonly observed in not glaciated regions.

Sea-salt aerosol originates from the bubble bursting during whitecap formation by surface winds over open ocean waters (Monahan et al. 1986). Once the sea-water droplets from bubble bursting are ejected into the surface air layer, they evaporate forming saline droplets or minute crystals of airborne sea salt with radii of 0.8–4 μm and salt contents of 4–50 pg per particle. Recently, the formation of sea salt aerosol from wind mobilization of saline snow or from "frost flower" structures on the sea-ice surface around Antarctica (Rankin et al. 2000, 2002; Wolff et al. 2003; Huang et al. 2017) addressed towards ss-Na^{+} as a proxy-marker of past changes in sea-ice extension and persistence (e.g., Wolff et al. 2006). However, the relative contribution of these sources is still poorly constrained over polar regions, affecting our ability to predict their impact on cloud formation and climate, as well as to achieve a reliable interpretation of chemical records from the ice cores sampled in the polar regions.

In the Svalbard Archipelago area, the main source of sea salt seems to be the open ocean (Huang and Jaeglé 2017). However, the local orography was observed to control the wind regimes in many sampling sites, (Tomasi et al. 2007; Mazzola et al. 2012), making difficult the identification of larger-scale sea-spray advection processes. Indeed, the wind circulation is channelled along the fjords especially in the Svalbard region (e.g., Argentini et al. 2003). Recent meteorological studies (e.g., Maturilli et al. 2013) confirmed that the prevalent winds in the Ny-Ålesund area blow along the axis of the fjord (Kongsfjorden). However, also katabatic winds and thermal land-sea breeze circulation can influence local wind direction (Esau and Repina 2012).

Teinilä et al. (2003) showed that sodium and chloride were the most abundant ions after sulfate in samples of size-segregated aerosol collected at Ny-Ålesund both during "polar night" and "polar day". The sodium, chloride, and magnesium elements are the main components of the sea spray source, whereas minor contributions are furnished by potassium, calcium, and sulfate. Since all these components have other sources than sea-water, their sea-spray (ss-) fractions were calculated using $ssNa^+$ as specific marker and knowing the X/ss-Na^+ weight/weight (w/w) ratios in sea-water of each X-type component (Kulshrestha et al. 1996; Das et al. 2005), which were assumed to be equal to 0.036 for K^+, 0.129 for Mg^{2+}, 0.038 for Ca^{2+}, 1.81 for Cl^-, and 0.253 for $SO_4{}^{2-}$ (Henderson and Henderson 2009) More than 96% of the total Na^+, Cl^-, and Mg^{2+} mass in PM10 collected in the 2013 campaign was found to be originated by sea spray (Giardi et al. 2016). Conversely, Ca^{2+}, K^+, and $SO_4{}^{2-}$ have relevant contributions from other sources, especially crustal aerosol (Ca^{2+}) and industrial/anthropogenic emissions ($SO_4{}^{2-}$, K^+) by long-range transport from continental regions (Fisher et al. 2011). The sea salt aerosol shown in Fig. 4.11b was calculated by multiplying the ss-Na concentration by a numerical factor equal to 3.248 (Kerminen et al. 2000).

Sea salt aerosol monitored in the PM10 samples shows weak seasonal dependence time-patterns, with spring concentrations that are slightly higher than summer values. In particular years (especially in spring 2011, 2012 and 2015), a large variability was observed since: (i) spring median values were found to range between 0.32 $\mu g\ m^{-3}$ (in 2014) and 0.56 $\mu g\ m^{-3}$ (in 2012); and (ii) summer median values were estimated to be equal to 0.1–0.2 $\mu g\ m^{-3}$ along the entire data set. The highest mean concentrations and variabilities were measured in spring samples collected on 2011 (0.93 $\pm$ 0.93 $\mu g\ m^{-3}$), 2012 (0.83 $\pm$ 0.69 $\mu g\ m^{-3}$) and 2015 (0.95 $\pm$ 1.29 $\mu g\ m^{-3}$). These seasonal variations in the sea-spray concentration can be plausibly attributed to the different atmospheric pathways supplying sea salt to Ny-Ålesund. A 72-h back-trajectory analysis carried out on the 2013 PM10 samples showed that Ny-Ålesund aerosol was enriched in sea-salt components (especially in the super-micrometric fraction) when air masses travelled at least 2 days at low altitude over open sea areas before reaching Ny-Ålesund (Giardi et al. 2016). Conversely, low sea-salt concentrations (with a size distribution shifted toward sub-micrometric modes) were observed when air masses travelled over open sea-water at altitude higher than 1000 m a.m.s.l. or at low altitudes over sea-ice areas.

(C) Sulfate in PM10 collected at Ny-Ålesund

Anthropic sulfate is a major component of the Arctic Haze. In order to distinguish between sulfate originated by sea spray and anthropic or biogenic sulfate, hereinafter we refer to the sulfate species not coming from sea water, as non-sea-salt sulfate (nss-$SO_4{}^{2-}$). A comprehensive knowledge of the atmospheric concentration of nss-sulfate is useful in evaluating the relevance of the impact of anthropogenic aerosol on the critical Arctic ecosystem and in improving climate models based on aerosol-solar radiation feedback processes. Several studies were carried out to understand the inter- and intra-annual trends of sulfur-cycle compounds in the Arctic aerosol (e.g., Hara et al. 2003; Scheuer et al. 2003; Quinn et al. 2007; VanCuren et al. 2012; Nguyen et al. 2013) and in ice cores drilled in the Northern hemisphere (e.g., Isaksson et al. 2005; Goto-Azuma and Koerner 2001). In particular, extensive data sets of aerosol chemical composition are available for the Zeppelin Station (Ny-Ålesund, 78° 54′ N, 11° 53′ E, 474 m a.m.s.l.) at the web site ebas.nilu.no (e.g., Aas et al. 2015).

Part (c) of Fig. 4.11a shows the statistical distribution of nss-sulfate along the 2010–2015 spring-summer sampling campaigns. A strong seasonal feature is well evident, associated with a higher atmospheric concentration in spring, subject to vary largely from one period to another, and lower and quite constant values recorded in summer. Spring median concentrations were found to range from 0.26 $\mu g\ m^{-3}$ (2013) to 0.58 $\mu g\ m^{-3}$ (2012). The highest mean concentration and variability was measured in 2012 (0.65 $\pm$ 0.40 $\mu g\ m^{-3}$). Summer concentrations were relatively stable around 0.1–0.2 $\mu g\ m^{-3}$, while the summer mean concentration was found to vary between 0.08 $\pm$ 0.01 $\mu g\ m^{-3}$ in 2015 and 0.15 $\pm$ 0.18 $\mu g\ m^{-3}$ in 2014.

PM10 total sulfate concentrations measured at the GVB Observatory in the 2014 spring and early summer were compared by Udisti et al. (2016) with the aerosol measurements of Total Suspended Particulate (TSP) carried out at the Zeppelin station up to mid July (Aas et al. 2015). It is worth noting that the Zeppelin data are more representative of free tropospheric circulation (since this station is located at the 474 m a.m.s.l. altitude), while the measurements carried out at the GVB Observatory (located at the altitude of 50 m a.m.s.l.) are strongly related to the Planetary Boundary Layer (PBL) dynamics, so that the aerosol concentrations are affected by the depth of the mixing layer near the surface. The sulfate profiles matched quite well in the late spring-early summer (from May to mid-July). Conversely, the spring profiles were quite different, likely due to the different aerosol atmospheric loads below and above the PBL in such a period, when the vertical atmospheric structure is well layered.

In order to evaluate the different sulfate contributions, a sulfate source apportionment was performed on the 2014 data set (Udisti et al. 2016). Figure 4.12 shows the absolute contributions of the different sulfate sources (measured in ng m^{-3}) (Udisti et al. 2013). The sea-salt fraction of sulfate (ss-SO_4^{2-}) was calculated by multiplying the concentration of ss-Na^+ (used as sea spray marker) by factor 0.253, which accounts for the SO_4^{2-}/Na^+ w/w ratio in sea-water (Bowen 1979). The crustal fraction of sulfate (cr-SO_4^{2-}) was estimated by multiplying the content of nss-Ca^{2+} (used as crustal marker) by 0.59 (which is the SO_4^{2-}/Ca^{2+} w/w ratio in the uppermost Earth crust (Wagenbach et al. 1996)). The sulfate contribution from marine phytoplanktonic activity, via atmospheric oxidation of dimethyl sulfide (DMS) emitted from micro-algal population, was estimated by multiplying the methanesulfonic acid (MSA) concentration (being MSA considered to be a specific marker of the marine biogenic emissions) by 3.0 (bio-sulfate) (Udisti et al. 2016). The anthropic sulfate (anthr-SO_4^{2-}) was calculated by subtracting ss-sulfate, crustal-sulfate and bio-sulfate from the total sulfate content in PM10 samples.

In the overall-period samples, the ss-, crustal, biologic and anthropic sulfate fractions account for 15.2, 4.2, 21.2 and 61.2% of the total sulfate budget,

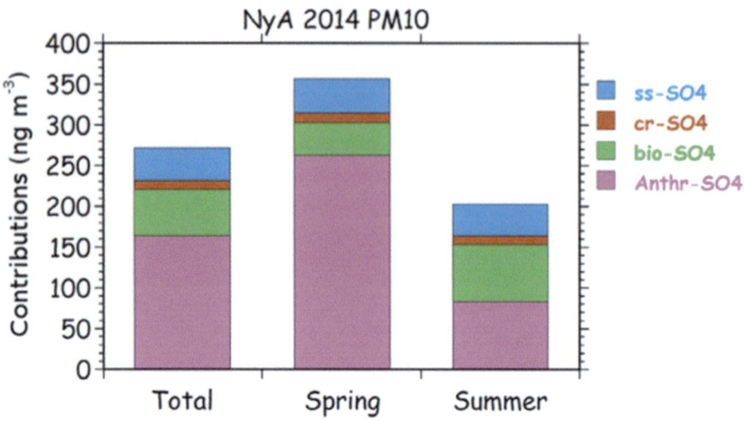

Fig. 4.12 Overall-period (total) and seasonal contributions of sea salt (ss-SO_4), crustal (cr-SO_4), biogenic (bio-SO_4) and anthropogenic (anthr-SO_4) fractions to the sulfate budget of aerosol mass concentration, as evaluated by analyzing the PM10 samples collected in 2014 at Ny-Ålesund (Spitsbergen, Svalbard)

respectively. It is evident that the anthropogenic emissions play a dominant role in the sulfate atmospheric concentration measured at GVB Observatory, with a minor (but significant) contribution from sea spray and biogenic sources. Conversely, the crustal contribution is quite low. More interesting information can be derived by analyzing the seasonal time-patterns of the anthropic sulfate fractions. In the spring samples, the anthropogenic contribution is very high (covering the 74.8% of the sulfate budget), so demonstrating the effect of the so-called "Arctic Haze" transport events. Indeed, these pollution episodes are more probable and intense in spring (Quinn et al. 2007), due to the inefficient pollutants dispersal, slow removal rates and isentropic transport into the Arctic as low-pressure systems run-up against the quasi-stationary Siberian anticyclonic area frequently standing over North Siberia (Barrie 1986). In summer, the anthropogenic source impact is lower (42%), providing a contribution comparable to that coming from biogenic emissions (35%), which reaches absolute and percentage values about two times higher than those measured in spring. The ss-contribution is quite constant (as absolute concentration) in the two seasons, but the percentage contribution is significantly higher in summer (19.8%) than in spring (12.0%). The crustal source plays a minority role. The all-period contribution is 4.2% of the sulfate budget (Udisti et al. 2016).

(D) **Ammonium in PM10 collected at Ny-Ålesund**

Ammonium salts are the reaction products of gas-phase ammonia (NH_3) with acidic species, other than the product of the dissolution of NH_3 in water (as NH_4OH in atmospheric moisture, cloud droplets, rainwater drops, superficial waters, etc.). NH_3 (and its species in solution, NH_4OH) is one of the most efficient neutralizing agent for atmospheric acidic species, such as H_2SO_4 and HNO_3, leading to the formation of ammonium salts, and is an important pH-regulator of the MBL aerosol

(Bouwman et al. 1997). NH_3 is emitted into the atmosphere by anthropic and natural sources. Anthropic sources include agricultural practices (especially fertilizer production and application), excreta from factory farming, biomass burning and combustion of fossil fuels (mainly traffic, ship emissions, domestic and industrial heating, production of thermo-electric power, and industrial emissions). Natural sources include excreta from wild animals and emissions from soils, vegetation and ocean (Quinn et al. 2002).

In clean marine areas, NH_3 mainly originates from biological decomposition of organic matter, but there is a highly variable contribution by zooplankton as well (see Johnson et al. 2007, and references therein). Reduced nitrogen plays also an important role as nutrient for many phytoplankton species. The biogenic emissions of sulfur and nitrogen species could be partly linked with each other. The processes within the ocean surface layer and the interaction with the MBL are very complex, depending on parameters such as temperature, sea water pH and wind stress. Johnson and Bell (2008) developed a co-emission concept, in which an interaction mechanism was suggested between H_2SO_4 from biogenic emissions (via DMS atmospheric oxidation) and NH_3 in the MBL. This system is mainly driven by the acidity of the dimethylsulfoniopropionate (DMSP) oxidation products (mainly sulfuric acid) that controls the bidirectional NH_3 flux from the ocean surface towards neutralization. The equilibrium of this process depends on the sea surface temperature and, hence, on latitude.

Part (d) of Fig. 4.11a shows the multi-annual distribution of NH_4^+ in the PM10 collected at Ny-Ålesund in the 2010–2015 campaigns. Together with sulfate, ammonium is one of the major components of the Arctic Haze. Therefore, its distribution shows a sharp seasonal pattern, with higher spring atmospheric concentrations. Spring median concentrations range from 0.035 $\mu g\ m^{-3}$ (in 2010) to 0.091 $\mu g\ m^{-3}$ (in 2013). Summer median values are quite constant from 0.020 $\mu g\ m^{-3}$ (in 2015) to 0.036 $\mu g\ m^{-3}$ (in 2014). Spring atmospheric concentrations show also the largest variability, with the highest values equal to 0.108 $\pm$ 0.056 $\mu g\ m^{-3}$ in 2013 and to 0.120 $\pm$ 0.091 $\mu g\ m^{-3}$ in 2014. Means and standard deviations in summer were found to be close to 0.03 $\pm$ 0.03 $\mu g\ m^{-3}$ in both cases. The reaction of NH_3 with sulfuric acid follows two steps, giving the still acidic species NH_4HSO_4 or the completely neutralized $(NH_4)_2SO_4$, depending on the relative abundance of ammonia and sulfuric acid equivalents. To evaluate the neutralizing effect of NH_3 on H_2SO_4, it is necessary to observe the relationship between NH_4^+ and sulfate species mainly originated in acidic form. Sea-salt sulfate originates mainly as Na_2SO_4, and crustal sulfate as $CaSO_4$. Therefore, their contribution has to be subtracted from the total sulfate content. By using ss-Na^+ as sea spray marker and nss-Ca^{2+} as crustal indicator, the ss-SO_4^{2-} and cr-SO_4^{2-} fractions were calculated as follows: (i) ss-SO_4^{2-} was calculated by multiplying the ss-Na^+ concentration (used as sea spray marker) by the SO_4^{2-}/Na^+ w/w ratio in sea-water, assumed to be equal to 0.253 (Bowen 1979); and (ii) the crustal fraction of sulfate (cr-SO_4^{2-}) was estimated by multiplying the content of nss-Ca^{2+} (used as crustal marker) by the SO_4^{2-}/Ca^{2+} w/w ratio in the uppermost Earth crust, assumed to be equal to 0.59 (Wagenbach et al. 1996).

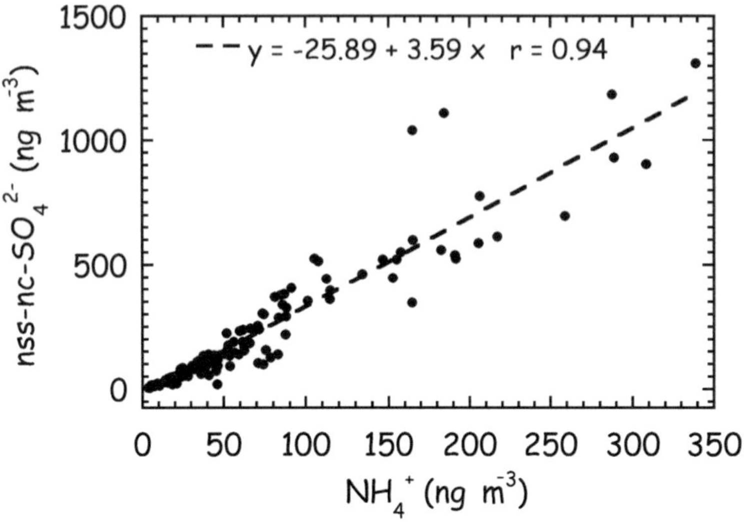

Fig. 4.13 Scatter plot of nss-nc-SO_4^{2-} mass concentration versus the NH_4^+ mass concentration (both measured in ng m^{-3}), obtained for a set of 136 PM10 daily samples collected at Ny-Ålesund during the spring – summer campaign of 2014 (Udisti et al. 2016). The correlation coefficient *r* of the regression (dashed) line was found to be equal to +0.94

The values of ss-SO_4^{2-} and cr-SO_4^{2-} fractions were removed from the tot-SO_4^{2-} concentration, then obtaining the non-sea-salt, non crustal sulfate (nss-nc-SO_4^{2-}) fraction. By plotting the values of nss-nc-SO_4^{2-} concentration collected at Ny-Ålesund during the spring – summer campaign of 2014 versus the NH_4^+ concentration, it was possible to evaluate the neutralization degree of H_2SO_4 by NH_3. Figure 4.13 shows the scatter plot evaluated for a set of 136 daily PM10 measurements performed on the 2014 samples at the GVB Observatory (Udisti et al. 2016). Nss-nc-SO_4^{2-} and NH_4^+ concentrations result to be significantly correlated, since their regression line was determined with a correlation coefficient *r* equal to +0.94, as shown in the scatter plot of Fig. 4.13, in which the slope coefficient is equal to +3.59, which is an intermediate value between the SO_4^{2-}/NH_4^+ ratio of 2.66 (corresponding to the w/w ratio in $(NH_4)_2SO_4$ salt) and the SO_4^{2-}/NH_4^+ w/w ratio of 5.33 in NH_4HSO_4. Therefore, the sulfate was found to be present in both these species when examining the aerosol particles collected at Ny-Ålesund during the 2014 spring – summer period.

On a multi-annual and seasonal scale, the nss-nc-SO_4^{2-}/NH_4^+ ratio was found to change in a significant way. Figure 4.14 shows the time-patterns of this ratio, obtained from the analysis of the daily samples collected over the whole six-year sampling measurement period from 2010 to 2015. A sharp seasonal trend is evident, yielding values higher than 5.33 in March–May (MAM), indicating the presence of free H_2SO_4, and values equal or lower than 2.66 in summer (June to Septermber, JJAS), showing that the samples were completely neutralized in this seasonal period.

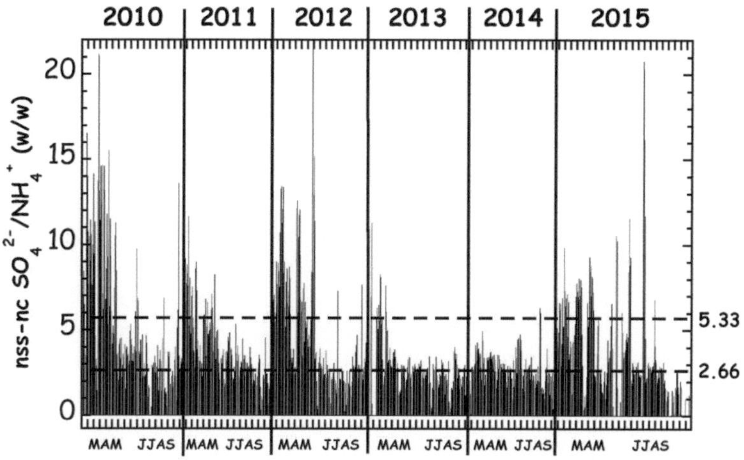

Fig. 4.14 Multi-annual trend of the nss-nc-SO_4^{2-}/NH_4^+ ratio in the PM10 daily aerosol samples collected at the GVB Observatory (Ny-Ålesund) during the spring (March to May, MAM) and summer (June to September, JJAS) periods of the 2010–2015 years

The acidic aerosol monitored in early spring can be mainly attributed to Arctic Haze transport events.

(E) MSA in PM10 collected at Ny-Ålesund

Methanesulfonic acid (MSA) arises from the atmospheric oxidation of DMS, a volatile compound, which is ubiquitous in the global surface ocean, being produced by phytoplankton. Emission of oceanic DMS into the atmosphere contributes to both MSA and non-sea-salt sulfate (nss-SO_4^{2-}) production (Bates et al. 1992; Kettle and Andreae 2000), which efficiently affects the Earth's radiation budget by exerting both direct and indirect effects (Kaufman et al. 2002). Recently, MSA concentration in Arctic aerosol was found to be closely related with the occurrence of new particle formation (NPF) processes, typically during the summer months (dall'Osto et al. 2017). Indeed, solar flux can exert some photochemical control mechanisms on the marine biological processes by generating particle precursors and/or favouring the occurrence of atmospheric photochemical processes, which produce condensable vapours from precursor gases (Kulmala et al. 2005).

Several studies have been performed to deeply understand the relation between MSA and the parameters that regulate the biogenic production (such as clorophyll-a concentration and sea-ice extent) on the basis of recent observations and ice core data collected in both hemispheres (Sharma et al. 2012; Isaksson et al. 2005; O'Dwyer et al. 2000; Curran et al. 2003; Becagli et al. 2009; Abram et al. 2013). However, cause and effect associations among MSA concentration in the atmosphere, changes in sea-ice extent, and phytoplankton productivity are yet to be further investigated because several other processes (in addition to the source strength) play an important role in determining the concentration of MSA at the sampling sites (Gondwe et al. 2003, 2004; Sharma et al. 2012; Park et al. 2013).

In order to better understand these complex interactions between ocean and atmosphere, strongly influenced by the biogenic activity, the long term studies of aerosol MSA concentration in the Arctic region are particularly relevant. Part (e) of Fig. 4.11a shows the seasonal distribution of MSA in PM10 along the six sampling years. The features of the MSA distribution during the two seasonal periods present a large year-to-year variability: while the MSA concentration recorded in the spring months of 2010, 2013 and 2015 shows higher concentrations than those measured in summer, in the other years (2011, 2012 and 2014) the summer concentrations are higher than the spring ones. On the multi-annual scale, the highest mean values were observed in spring 2010 and 2013 (equal to 0.013 $\mu g\ m^{-3}$ in both cases), these values being similar to those measured in summer 2013 and 2014 (equal to 0.013 $\mu g\ m^{-3}$ and 0.014 $\mu g\ m^{-3}$, respectively). A previous study carried out at Ny-Ålesund shows that MSA presents seasonal time-patterns with a concentration increase in March/April and a decrease to background concentration in September (Becagli et al. 2016).

The spring-summer high concentrations of MSA depend on the behaviour of the DMS source (Lana et al. 2011), on the efficiency of the photochemical processes and gas-to-particle conversion producing MSA (in the aerosol) from gaseous DMS (e.g. Karl et al. 2007; Gondwe et al. 2004), and on the transport processes occurring in the atmosphere. The large year-by-year variability of MSA concentration observed in part (e) of Fig. 4.11a can be attributed to: (i) the different timing of sea ice melting, (ii) the distinct features of the micro-algal assemblages at the DMS source areas, which in turn are linked to the sea-ice extent, and (iii) the different efficiency of transport processes from the source area to the sampling site.

A previous study (Becagli et al. 2016) reports a significant correlation between MSA and primary production (PP), calculated by a bio-optical model in the Barents Sea, during the spring and summer periods. Conversely, the linear regression between MSA and PP in the Greenland Sea is significant only during the summer period. A close correlation has been also obtained between MSA and P^B (chlorophyll specific production index or assimilation number, a parameter defining phytoplanktonic stress) in the summer period, underlining the concurrent presence of the so-called "bloom-dominated regime" and "stress-forced regimes" in both the Barents and Greenland Seas. The source intensity expressed by PP is able to explain more than 30% of MSA variability. The other factors explaining such a variability are the taxonomic differences in the phytoplanktonic assemblages and the transport processes from the DMS source areas to the sampling sites.

A significant correlation and consistent dependence features were found at Ny-Ålesund between MSA and sea ice conditions in the Barents Sea (sea ice melting and ice free marginal ice zone extension). This suggests that the source intensity (i.e., primary production) is related to both sea ice melting and extension of marginal sea ice areas, and that these factors are the main drivers for MSA concentration at this site (Becagli et al. 2016). These authors found that the mean MSA concentration is 2.5 times lower at the GVB Observatory than that measured at the higher Zeppelin site. This could be explained by the fact that MSA in the PM10 is affected by the boundary layer dynamics, while conversely the gaseous precursor of MSA (DMS) is mixed into the free troposphere during frontal passages and most of the DMS oxidation to MSA occurs in the free troposphere, where the particle surface area is lower and the actinic flux is higher (Bates et al. 1992; Le Quéré and Saltzman 2009). The vertical stratification reduces MSA transport from the free troposphere to the Arctic boundary layer. Therefore, it is important to take into account that the Zeppelin site is generally located above the boundary layer in spring and summer, so that the MSA concentration results to be higher than that expected at the lower altitudes.

(F) **Nitrate in PM10 collected at Ny-Ålesund**

Nitrate is a minor component of the Arctic Haze (Quinn et al. 2007). Nevertheless, other sources contribute to the nitrate load of the polar atmosphere. Nitrate sources are a complex mixture of natural and anthropogenic inputs, including marine emissions, photochemical transformation of gaseous precursors, biomass burning and combustion processes in continental regions. The relative contribution of the different sources active during the year, and the role of solar radiation in the chemical reactions producing and recycling nitrate in the atmosphere are factors far to be fully understood (Traversi et al. 2017). As a result of the industrial revolution and the increasing human population, the concentration of the so called atmospheric "reactive nitrogen" has increased appreciably, as documented by nitrate concentrations measured in ice cores collected around the Northern Hemisphere (Goto-Azuma and Koerner 2001; Isaksson et al. 2003; Hastings et al. 2009). The reactive nitrogen – enriched air, produced as a result of energy and food production (Galloway et al. 2003), can be subject to long-range atmospheric transport reaching the nutrient-limited Arctic regions, where local pollutions sources are very few (Dickerson 1985). Oxidized nitrogen, mainly originating as nitric oxide (NO) and nitrogen dioxide (NO_2), is transported during winter in the reservoirs of peroxyacyl nitrates and is deposited in the Arctic, after conversion to gaseous nitric acid (HNO_3) or as particulate bound nitrate (p-NO_3) through wet or dry deposition (e.g. Bergin et al. 1995; Seinfeld and Pandis 2006). The relative importance of these two processes is poorly quantified for Arctic regions, where dry deposition of nitrate compounds is particularly hard to be evaluated (Björkman et al. 2013, 2014).

Constraining nitrate sources is quite complex. This is also confirmed by the fact that nitrate is the only parameter appearing in a source factor yielded by Positive Matrix Factorization (PMF) analysis on 2010 and 2015 sampling campaigns (unpublished data). No significant correlation with other chemical markers was found in the data presented herein, if not for limited time periods, when a dominating source can be identified. For instance, despite being a component of Arctic Haze, nitrate is not clearly linked to nss-$SO_4{}^{2-}$, as reported by Teinilä et al. (2003) from size-segregated aerosol samples collected at Ny-Ålesund. Conversely, it was found to be closely correlated with nss-Ca^{2+} during a sampling campaign carried out in the period from 23 April to 17 May 2001, such a result being probably due to a common local source for $NO_3{}^-$ and nss-Ca^{2+} in this period (Teinilä et al. 2003). In the same work, during the "dark" campaign (from 21 February to 7 March 2001), most of the nitrate was likely bound to sea-salt particles. The strong acidity of the sub-micrometric sulfate aerosol prevents the preservation of nitrate in the sub-micrometric size fraction. As a consequence, nitrate is mainly distributed in the coarse aerosol size classes, with an opposite atmospheric behaviour with respect to sub-micrometric H_2SO_4 particles, which could explain the lack of correlation between nitrate and nss-$SO_4{}^{2-}$ in the Arctic Haze.

Examining the nitrate seasonal time-patterns determined at the GVB Observatory over the 2010–2015 period (see Fig. 4.11b, plot (a)), it can be seen that nitrate concentration in PM10 shows always higher average and median values in spring than in summer. The mean concentrations range between 0.039 $\mu g\ m^{-3}$ and 0.049 $\mu g\ m^{-3}$ in spring. A similar range can be observed in the same season for the median values, showing their lowest levels in 2014 (0.031 $\mu g\ m^{-3}$) and the highest ones in 2012 (0.047 $\mu g\ m^{-3}$). As mentioned above, summer levels are always lower, with averages ranging between 0.022 $\mu g\ m^{-3}$ (in 2015) and 0.043 $\mu g\ m^{-3}$ (in 2014). Similar multi-annual patterns are shown by the median values, equal to 0.013 $\mu g\ m^{-3}$ in 2015 and 0.031 $\mu g\ m^{-3}$ in 2014. It was also found that the concentration variability (expressed in terms of its standard deviation) was usually wider in spring than in summer, except for 2012 and 2014. In these years, it has however to be noticed that the concentration time-patterns of nitrate ions are quite similar during the two periods, as shown in part (a) of Fig. 4.11b by the box containing the 50% of the values.

(G) **Organic Carbon in PM10 collected at Ny-Ålesund**

Organic Carbon (OC) and Black Carbon (BC) are relevant components of Arctic Haze, together with sulfate, ammonium, nitrate and dust (Quinn et al. 2002). In particular, OC exerts a negative forcing on the radiation balance of the surface-atmosphere system by; (i) favouring the solar light scattering with respect to the absorption processes, and (ii) acting on the cloud formation. Through such direct and indirect effects, OC constitutes (together with BC and sulfate) a key parameter in modelling the impact of atmospheric aerosols on Arctic and global climate (e.g., Yang et al. 2014). However, despite the large efforts in reducing uncertainties associated with atmospheric organic aerosols, their roles in global aerosol radiative forcing, atmospheric chemistry and climate changes remain still poorly understood (Pandis et al. 2013). These uncertainties are in part due to our limited knowledge of the atmospheric secondary organic aerosols, which constitute a significant fraction (ranging from 10% to 40%) of organic aerosols (Volkamer et al. 2006), since they are more hygroscopic and oxygenated than those of primary origin (Saxena and Hildemann 1996; Zhang et al. 2007).

OC originates from both anthropic and natural sources. Anthropic emissions include vehicular exhausts, biomass burning, cooking, agricultural practices and a large variety of industrial and manufactory processes involving the emission of Volatile Organic Carbon compounds (VOCs) into the atmosphere (see Kawamura and Bikkina (2016) and references cited herein). Natural sources include biogenic emissions (Rinaldi et al. 2011) in the marine atmospheric boundary layer, along with sea-spray (Steinberg and Bada 1984; Tedetti et al. 2006), forest emissions, biological metabolic processes and natural forest fires. However, they are mainly produced in the atmosphere by the photochemical oxidation of various organic precursors, most likely in aqueous aerosol phase. More recently, biogenic VOCs, such as isoprene and monoterpenes, are considered to be the sources of low molecular weight di-carboxylic acids and related compounds (see Kawamura and Bikkina (2016) and references cited herein).

Part (b) of Fig. 4.11b shows the statistical distribution of OC along the 2011–2015 spring – summer campaigns (OC concentration was not measured in the 2010 samples). The OC fraction was determined on pre-fired quartz filters collected by a PM10 sampler, operating in parallel with the PM10 and DK-4 sampling devices for chemical composition. The aerosol samples were collected with a resolution of 4 days (2011–2014) or 24 h (2015). OC analyses were carried out on a punch (1.5 × 1 cm) of the quartz filter by a thermo – optical sunset analyzer, by using the NIOSH (National Institute for Occupational Safety and Health) protocol in order to separate the Elemental Carbon (EC) contribute. Although OC is a major component of Arctic Haze, its seasonal time-patterns are not so evident as those observed for nss-SO_4^{2-} and NH_4^+. The highest seasonal differences were measured in the samples collected in 2011 (with spring and summer median values of 0.16 and 0.024 μg m^{-3}, respectively) and in 2012 (with spring and summer median values of 0.12 and 0.020 μg m^{-3}, respectively). As an example, Fig. 4.15 shows the clearly decreasing trend of the Arctic Haze contribution to OC concentration observed in the 2011 spring samples. Conversely, spring and summer concentrations measured in the 2013–2015 samples result to show less different features, with median values ranging from 0.17 μg m^{-3} (in summer 2014) to 0.28 μg m^{-3} (in spring 2015). Mean values are quite different from the median ones, in this way demonstrating that only a few events with large concentrations dominate the seasonal OC atmospheric load (see, for instance, the summer profile in Fig. 4.15). The highest seasonal mean

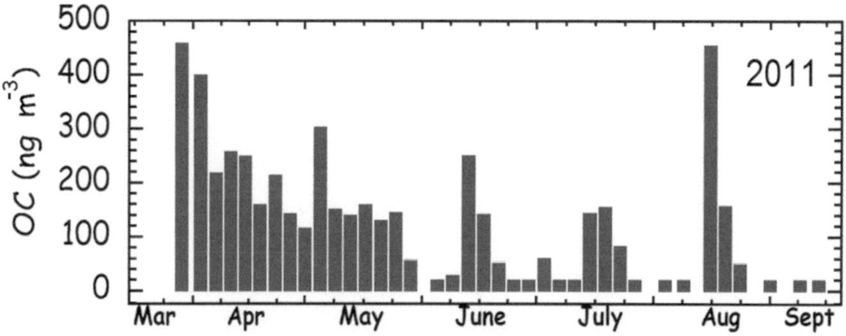

Fig. 4.15 Temporal trend of OC concentration (measured in ng m^{-3}), as obtained by analyzing the PM10 samples collected at Ny Ålesund (Spitsbergen, Svalbard) during the 2011 spring (March–May) and summer (June–September) measurement periods

values were recorded in the spring samples collected in 2013 (0.28 ± 0.12 $\mu g\ m^{-3}$) and in 2015 (0.52 ± 0.67 $\mu g\ m^{-3}$).

(H) **nss-K^+ in PM10 collected at Ny-Ålesund**

Biomass burning is a significant source of atmospheric gases and particulate matter (including EC, OC and inorganic salts) on both regional and global scales (Andreae and Merlet 2001; Moroni et al. 2015, 2017). Particles produced by biomass burning can influence the global climate by absorbing and scattering solar radiation (Chand et al. 2009) and acting as CCN, which modify the cloud microphysical properties (Mochida and Kawamura 2004; Petters et al. 2009). Recent studies on the mineral chemistry and geochemical signature of aerosols, sampled at ground level and along the vertical atmospheric path at Ny-Ålesund, have shown a significant contribution of biomass burning forest fires in the Svalbard area, especially in summertime (Giardi et al. 2016; Moroni et al. 2016), when the prevailing northern circulation is capable of transporting wild-land fire emissions far from their sources located in North America and Eurasia tundra regions (Lavoué et al. 2000; Sharma et al. 2013).

Among the inorganic ions collected in aerosol particles enriched or mainly formed during wildfires, non-sea-salt potassium (nss-K^+) is one of the main components (Zhai et al. 2017). Potassium chloride crystals were observed by using the Transmission Electron Microscopy (*TEM*) technique to examine fresh biomass burning particles (Li et al. 2015). Even though freshly emitted, biomass burning particles can be markedly coated by secondary species, such as ammonium nitrate and ammonium sulfate, within a very short period (Leskinen et al. 2007).

A devoted SEM-EDS (Scanning Electron Microscopy – Energy Dispersive Spectroscopy) analysis of aerosol samples collected during an intense biomass burning event, recorded in the PM10 collected at Ny-Ålesund in July 2015, showed the presence of potassium chloride particles associated with ammonium sulfate particles (Moroni et al. 2017). KCl is a typical component of fresh smoke, and its content tends to decrease during transport due to the reaction with the secondary sulfate (nss-sulfate) emitted from natural and/or anthropogenic sources to form K_2SO_4 (Li et al. 1993a). Ammonium sulfate is a minor component of the smoke and its content in the aerosol tends to increase due to the mixing with secondary haze (Pósfai et al. 2003). The fact that KCl was found instead of K_2SO_4 suggests a lack of reaction with sulfate. This can be explained considering that the reaction between sulfuric acid (resulting from the slow oxidation of sulfur dioxide) and ammonia in a wildfire plume is much faster than the exchange reaction between sulfuric acid and potassium chloride (Moroni et al. 2017).

The nss-K^+ concentrations in PM10 collected at Ny-Ålesund from 2010 to 2015 (as can be seen in Part (*c*) of Fig. 4.11b) are always higher in spring than in summer, in terms of both average and median values. Such a recurring trend can be likely ascribed to the seasonal evolutionary patterns of the transport processes to the Arctic. Indeed, during the Arctic Haze occurrence periods, aerosol loads produced by both natural and anthropic sources, including wildfires (especially from

Alaska and Greenland) and biomass burning related to human activities (e.g., from Northern Russia) can more likely reach the High Arctic latitudes, thus contributing to increase the nss-K^+ concentration in the particulate matter. The average spring nss-K^+ concentration was estimated to range between 5.4 ng m^{-3} (measured in both 2011 and 2013) and 13.6 ng m^{-3} (in 2015), while similar levels were observed in spring of 2013 and 2015, with median values of 3.8 ng m^{-3} (in 2013 and 13.0 ng m^{-3} (in 2015). In summer, such mean values were found to decrease by a factor 2 with respect to the spring ones, with the lowest mean values observed again in 2011 (2.1 ng m^{-3}) and the highest summer values in 2014 (8.8 ng m^{-3}). Data variability between the two seasons (expressed in terms of standard deviations) is usually comparable, showing larger dispersion data in the spring of 2010, 2011, 2012, and 2014, and lower dispersion in 2013 and 2015.

(l) **Mineral dust in PM10 collected at Ny-Ålesund**

Dust entrainment in deserts and semiarid regions and its mobilization by strong winds is the most important source of mineral dust in the global atmosphere, followed by emissions from dry agricultural lands (wind-driven erosion and soil particle suspension during harvesting and tilling (Ginoux et al. 2010)). In the populated regions, anthropogenic activities may cause significant loads of local air pollution particles by traffic (exhaust, road dust, tire wear, and break abrasion), power generation (coal fly ash), smelters and other metallurgical industries (metal oxides), and construction activities. Continuous emissions from these pollution sources have been found to be more important than natural sources in Europe, Eastern Asia, and North America, except on days when transport of desert dust is significant (Almeida et al. 2006; Viana et al. 2008; Evan et al. 2009; Moreno et al. 2011). The quantification of dust load and the identification of its temporal trends in the Arctic atmosphere can improve the understanding of: (i) the seasonal and multi-year evolutionary transport processes from continental mid-latitude region to the Arctic (Groot Zwaaftink et al. 2016); (ii) the features of dust ice nucleating capability (Hoose et al. 2010; Fan 2013); and (iii) the relevance of nutrient supply on the low nutrient high productivity oceanic area, such as that of the Arctic Ocean (Moore et al. 2004).

As reported in Sect. 4.2.2.1, mineral dust load was evaluated by multiplying the nss-Ca concentration by 33.78. This value was experimentally obtained as the mean value of the mineral dust/nss-Ca ratio measured by Ion Chromatography (IC) techniques in the PM10 samples collected at the GVB Observatory during the 2015 campaign. The value is close to the mineral dust/Ca ratio calculated in the Earth upper crust and estimated to be equal to 39.06 (Henderson and Henderson 2009). Part (*d*) of Fig. 4.11b shows the statistical distribution of dust along the spring (MAM) and summer (JJAS) periods of the 2010–2015 years. As observed for Arctic Haze markers, the dust load shows strong seasonal time-patterns characterized by higher concentrations in spring than in summer. Previous studies revealed that dust aerosols observed in the Arctic from December to March arises from both pollution (dust tracers mixed with anthropogenic pollution tracers) and natural (not mixed with pollution) sources, with a contribution of about 50% from each of them (Maenhaut et al. 1989; Li and Winchester 1990; Nguyen et al. 2013). The source apportionment of particulate matter was based on statistical analysis of chemical/elemental compositions of aerosols measured in the Arctic. The pollution sources of dust aerosol become relatively important in winter, as the natural sources become gradually weaker. Indeed, dust entrainment in the deserts of Central and Eastern Asia exhibits the lowest values of the annual cycle during winter (Luo et al. 2003). Transport of Saharan dust toward the high-latitude areas of the Northern Hemisphere also occurs with the lowest frequency in winter (Meloni et al. 2007), contributing to provide on the average less than 10% of the particulate matter (PM10) mass monitored in Northern Europe at latitudes higher than 45 °N, (Pey et al. 2013).

The global and parcel models considered together suggest that these dust-anthropic mixed aerosols transported toward the Arctic (due to their low concentrations and the presence of acid coated-dust particles) are responsible for the occurrence of mixed-phase clouds, and can cause significant perturbations on both Arctic ice nuclei and clouds in winter (Fan 2013). It is interesting to notice in part (*d*) of Fig. 4.11b that a high number of outlier values is present in summer (in spite of the higher variability observed in spring than in summer), which are sometimes of the same order of magnitude of the maximum values measured in spring. The cause of the summer (and sometimes autumn) spikes in dust aerosol is not completely understood, but it may be due to local particle sources (Barrie and Barrie 1990). A recent study, based on the Rare Earth Elements determination on high temporal resolution PM10 samples collected at Ny-Ålesund, confirmed that crustal aerosol in summer is mainly related to local source (Giardi et al. 2018). Spring median concentrations were found to range from 0.22 $\mu g\ m^{-3}$ (in 2010) to 0.64 $\mu g\ m^{-3}$ (in 2014). Summer mean values are quite stable, with the exception of those obtained by analyzing the 2014 samples. The 2010–2013 and 2015 summer averages were found to be close to 0.16 $\mu g\ m^{-3}$, while the 2014 mean value was about three times higher (0.43 $\mu g\ m^{-3}$) in 2014. The summer mean value of 0.21 $\mu g\ m^{-3}$ was obtained over the whole six-year period.

The spring and summer average mass concentrations of PM10 and its main components (as obtained through the analysis of the PM10 samples collected at the GVB Observatory during the three campaigns carried out in 2012, 2013 and 2014) are given in Table 4.1, yielding values of PM10 equal to 4.02 $\mu g\ m^{-3}$ in spring, and 2.42 $\mu g\ m^{-3}$ in summer, with relative standard deviations of 63% and 73%, respectively. As can be seen in Table 4.1, mineral dust is the second most important mass fraction among the components of PM10 measured in the spring and summer months (equal to 0.652 $\mu g\ m^{-3}$ in the MAM months, and 0.387 $\mu g\ m^{-3}$ in the JJAS months, respectively) after the sea spray particle component (equal to 0.805 $\mu g\ m^{-3}$ in the MAM months and 0.593 $\mu g\ m^{-3}$ in the JJAS months), followed by nss-sulfate ions (0.533 $\mu g\ m^{-3}$ in spring and 0.119 $\mu g\ m^{-3}$ in summer) and gradually lower mass concentrations of OC, ammonium ions, nitrate ions, MSA^- ions, and nss-K^+ ions in both spring and summer periods.

Table 4.1 Values of the three-year average mass concentrations of PM10 aerosol particles and their components (measured in μg m^{-3}), as determined with their standard deviations by Udisti et al. (2016) during the spring (March–May) and summer (June–September) seasons at the GVB Observatory (Ny-Ålesund, Spitsbergen, Svalbard) during the three campaigns conducted in 2012, 2013 and 2014

Seasonal period	PM10	Sea spray	nss-SO_4^{2-}	NH_4^+	MSA	NO_3^-	OC	nss-K^+	Mineral dust
Spring (MAM)	4.02± 2.55	0.805± 0.901	0.533± 0.418	0.101± 0.061	0.0160± 0.0274	0.0464± 0.0260	0.215± 0.111	0.0082± 0.0056	0.652± 0.413
Summer (JJAS)	2.42± 1.77	0.593± 0.817	0.119± 0.152	0.0382± 0.0345	0.0185± 0.0209	0.0317± 0.0352	0.174± 0.171	0.0050± 0.0080	0.387± 0.595

4.2.2.3 Multi-stage Sampling and Analysis of the Chemical Components at Ny-Ålesund

During the spring and summer months of the 2010–2014 years, size-segregated aerosol samplings were performed at the GVB Observatory by using a 4-stage Dekati PM10 device (described above in Sect. 4.2.2). This sampler was used to collect aerosol particles on stages with four cut-off values: greater than 10 μm, 2.5 μm, 1 μm, and smaller than 1 μm. The first three impactor stages (collecting the three size fractions having >10 μm, 10–2.5 μm and 2.5–1 μm ranges) were constituted by polycarbonate membranes (25 mm diameter), while the last stage (backup filter, collecting the fraction with sizes <1 μm) was a 47 mm Teflon filter (Pall R2PJ047, 2 μm nominal porosity). Samplings were carried out in "actual conditions" of pressure and temperature, which were continuously monitored to maintain a constant flow rate of 29.0 L min^{-1}. Each sampling was performed over a period of about 4 days, in order to collect a sufficient amount of particulate matter for the chemical analysis planned on each sampling substrate. Each sampling started at 00:01 UTC of the first measurement day (contemporaneously to the start of the PM10 daily sampling) and was finished in the evening of the fourth day. The time of the sampling stop was chosen between the 18:00 and 22:00 UTC, depending on the solar light conditions (and usually avoiding to work during the dark hours in order to reduce the risk of meeting a polar bear). After the filter replacement, the next sampling was planned to start at 00:01 UTC of the subsequent day.

In order to compare the 4-stage impactor with the PM10 chemical composition, the fraction of particulate matter with sizes >10 μm was not included in the data discussion. The super-micrometric fraction was calculated as the sum of the particle mass concentrations measured over the 1–2.5 and 2.5–10 μm size ranges (i.e. for the overall 1–10 μm size fraction), and the sub-micrometric fraction was determined over the size-range < 1 μm (47 mm backup Teflon filter).

The chemical analysis (ionic composition by ion chromatography) was carried out on one-half of each polycarbonate membrane and on a quarter of each Teflon filter, by following the same procedure described above for the PM10 samples (see Sect. 4.2.2.1). The gravimetric determination of PM10 was not carried out, due to the very low particulate mass deposited on the membranes in all the four stages. The calculations of the ss- and nss-fractions and those of the sea spray and mineral dust (crustal) contributions were made by following the procedure described in Sect. 4.2.2.1.

Chemical Composition of Sub-Micrometric and Super-Micrometric Aerosol Collected at Ny-Ålesund (Svalbard) from 2010 to 2014

Figure 4.16 (consisting of four parts (a) – (d)) and Fig. 4.17 (consisting of three parts (a) – (c)) show the box plots of the seven chemical components of the sub-micrometric particle fraction, while Fig. 4.18 (consisting of four parts (a) – (d)) and Fig. 4.19 (consisting of four parts (a) – (d)) show the box plots of the eight chemical components of the super-micrometric particle fraction.

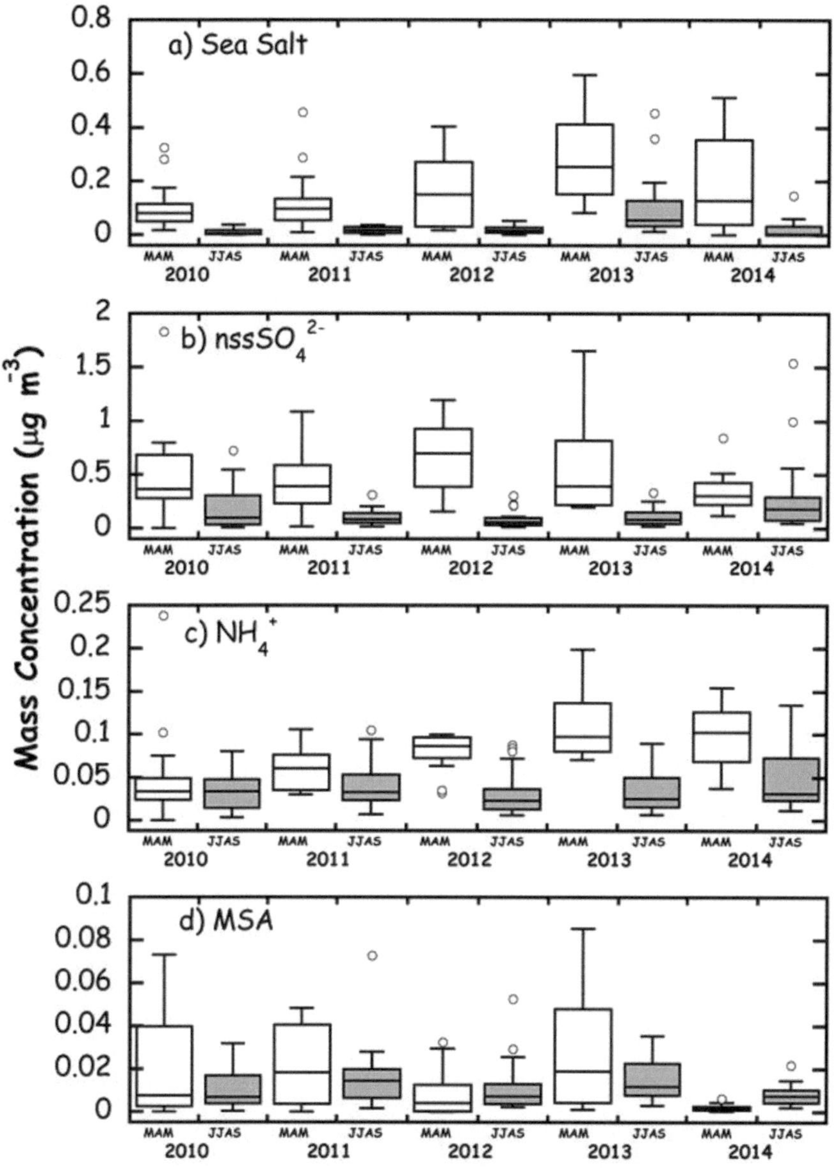

Fig. 4.16 Box-plot distribution of the ionic components of the sub-micrometric fraction of the aerosol collected at Ny-Ålesund (Spitsbergen, Svalbard) in the spring (white) and summer (grey) months of the 2010–2014 campaigns. In all the graphs, each box contains the 50% of data, with the median value given by a horizontal line. The bottom and the top of the box indicate the values of the 25th and 75th percentiles, respectively, while the vertical lines extending from the bottom to the top of each box indicate the minimum and maximum values that fall within an acceptable range (equal to 1.5 times the box width). The outliers are shown as small circles

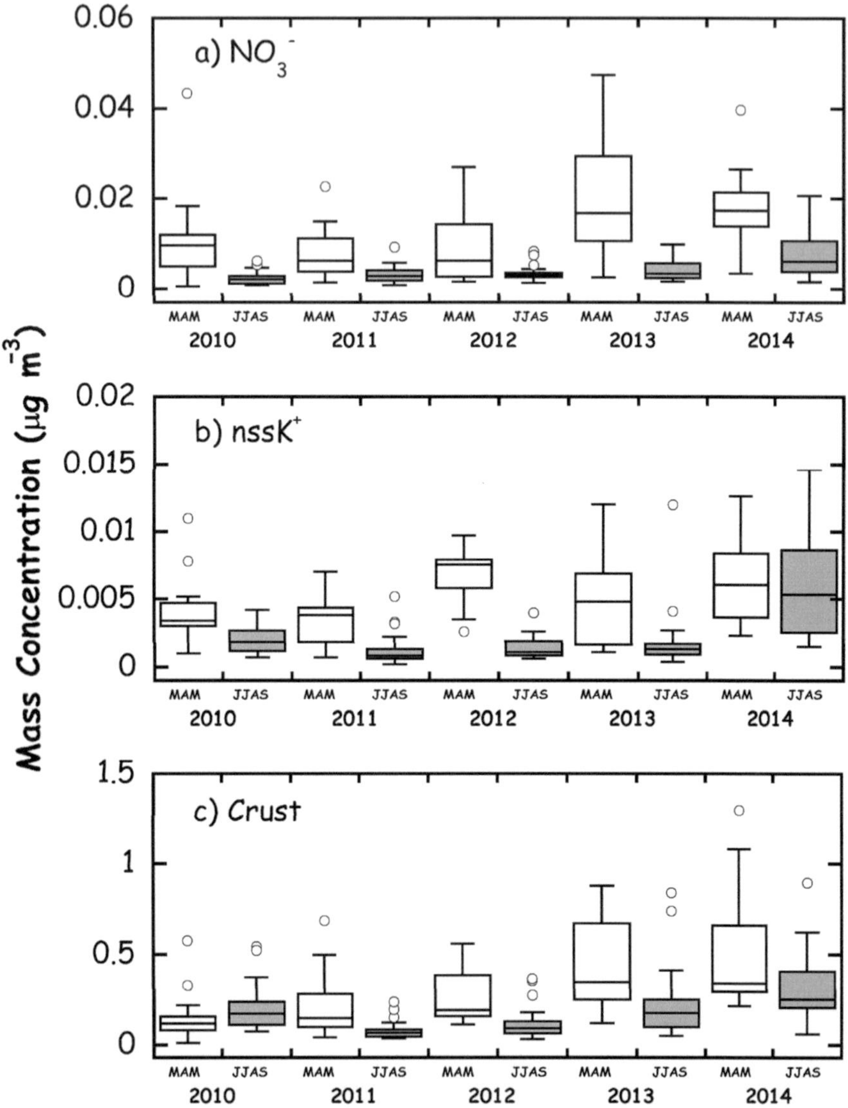

Fig. 4.17 Box-plot distribution of the ionic components of the sub-micrometric fraction of the aerosol collected at Ny-Ålesund (Spitsbergen, Svalbard) in the spring (white) and summer (grey) months of the 2010–2014 campaigns. In all the graphs, each box contains the 50% of data, with the median value given by a horizontal line. The bottom and the top of the box indicate the values of the 25th and 75th percentiles, respectively, while the vertical lines extending from the bottom to the top of each box indicate the minimum and maximum values that fall within an acceptable range (equal to 1.5 times the box width). The outliers are shown as small circles

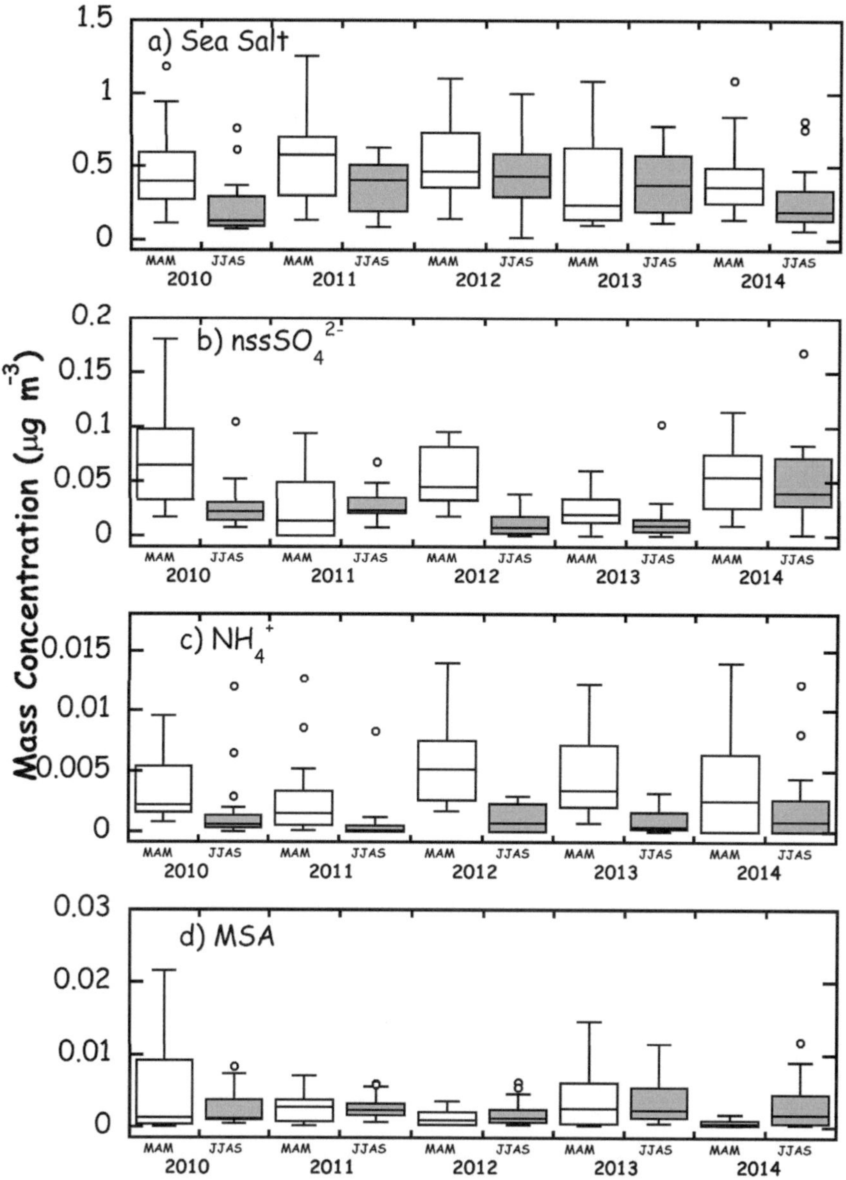

Fig. 4.18 Box-plot distribution of the ionic components of the super-micrometric fraction of the aerosol collected at Ny-Ålesund (Spitsbergen, Svalbard) in the spring (white) and summer (grey) months of the 2010–2014 campaigns. In all the graphs, each box contains the 50% of data, with the median value given by a horizontal line. The bottom and the top of the box indicate the values of the 25th and 75th percentiles, respectively, while the vertical lines extending from the bottom to the top of each box indicate the minimum and maximum values that fall within an acceptable range (equal to 1.5 times the box width). The outliers are shown as small circles

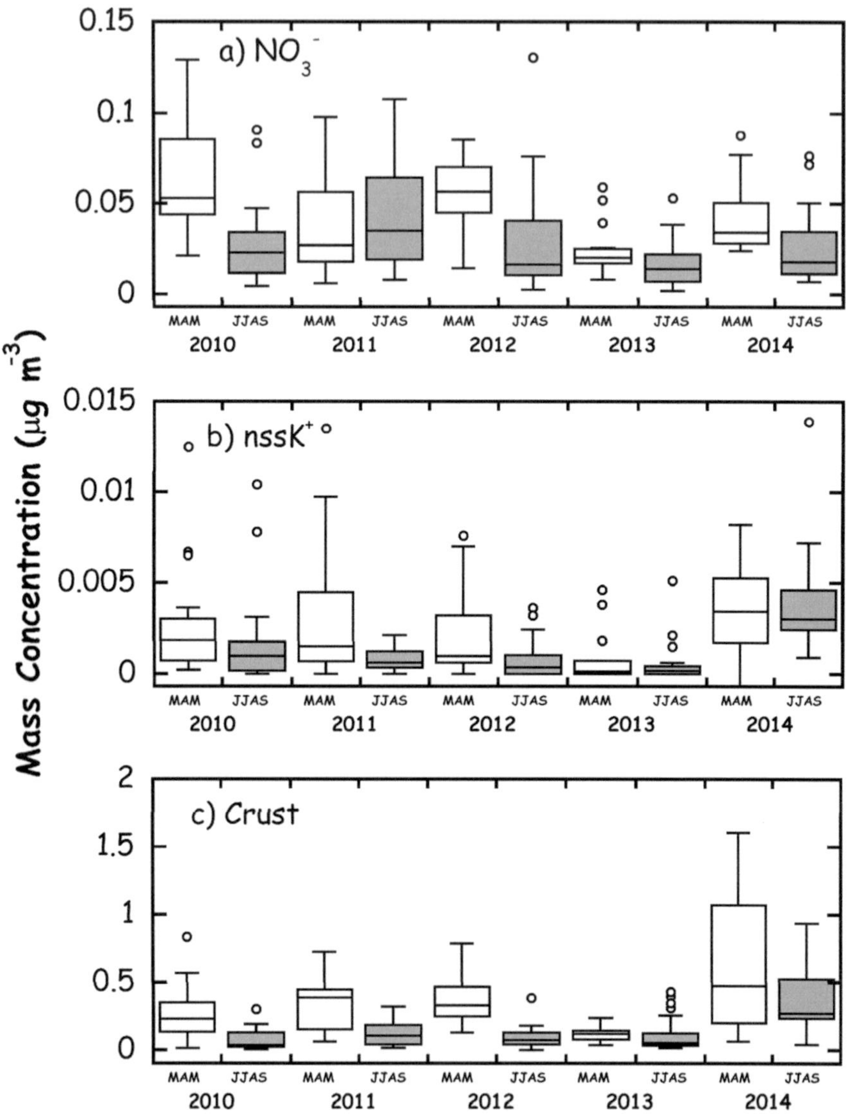

Fig. 4.19 Box-plot distribution of the ionic components of the super-micrometric fraction of the aerosol collected at Ny-Ålesund (Spitsbergen, Svalbard) in the spring (white) and summer (grey) months of the 2010–2014 campaigns. In all the graphs, each box contains the 50% of data, with the median value given by a horizontal line. The bottom and the top of the box indicate the values of the 25th and 75th percentiles, respectively, while the vertical lines extending from the bottom to the top of each box indicate the minimum and maximum values that fall within an acceptable range (equal to 1.5 times the box width). The outliers are shown as small circles

The results shown in Fig. 4.16 and in Fig. 4.17 indicate that the sub-micrometric fraction is enriched in the chemical components originated by secondary processes, such as nss-sulfate, ammonium, MSA and (although to a minor extent) nss-potassium. Conversely, the primary aerosol coming from sea spray and continental dust was found to be the predominant component of the super-micrometric particle fraction. Nitrate is mainly originated from secondary processes and constitutes an exception, since it presents higher mean concentrations in the super-micrometric fraction, probably as a result of the growth of hygroscopic NH_4NO_3 particles, formed into the atmosphere by reaction of ammonia with nitric acid.

The seasonal and inter-annual trends of the various chemical components are shortly described in the following sub-paragraphs.

(A) **Sea salt**

The sea spray aerosol is mainly present in the super-micrometric fraction, with mean mass concentration values around 0.2–0.5 $\mu g\ m^{-3}$ that turn out to be about 2 times higher than the sub-micrometric mean values, usually found to be around 0.1–0.3 $\mu g\ m^{-3}$. In the sub-micrometric fraction, the sea salt contribution was estimated to vary between $0.010 \pm 0.012\ \mu g\ m^{-3}$ in summer 2010 and $0.29 \pm 0.17\ \mu g\ m^{-3}$ in spring 2013. The seasonal cycle is well evident, with higher concentrations and more pronounced variability features in spring. Other than 2013, the highest spring mean concentrations were recorded in the samples collected in 2012 ($0.16 \pm 0.13\ \mu g\ m^{-3}$) and in 2014 ($0.19 \pm 0.18\ \mu g\ m^{-3}$). Conversely, the summer mean concentration was found to be always lower than 0.1 $\mu g\ m^{-3}$, with a maximum of $0.098 \pm 0.106\ \mu g\ m^{-3}$ recorded in 2013. The multi-annual time-patterns show an increasing trend of both spring and summer mean concentrations over the last 2–3 years, with the highest values found in the 2013 samples collected in both spring and summer periods.

In the super-micrometric mode, the seasonal and inter-annual time-variations were both found to be less marked, with spring and summer mean concentrations ranging between 0.2 and 0.5 $\mu g\ m^{-3}$ without evidence of variations with season and sampling year. The lowest mean concentration was measured in summer 2010 ($0.22 \pm 0.19\ \mu g\ m^{-3}$), while the highest mean value was recorded in spring 2011 ($0.59 \pm 0.38\ \mu g\ m^{-3}$). The variability patterns were found to be relatively high in all cases, with a maximum in spring 2013, characterized by some concentration spikes (as demonstrated by the median value of 0.24 $\mu g\ m^{-3}$ that is significantly lower than the mean value of $0.49 \pm 0.55\ \mu g\ m^{-3}$).

(B) **Non-sea-salt sulfate (nss-SO_4^{2-})**

Since nss-SO_4^{2-} originates from secondary processes leading to the oxidation of atmospheric SO_2, the mean concentration of this component is higher in the sub-micrometric size range than in the super-micrometric one. In fact, the values of the sub-micrometric mean concentration were found to range between 0.2 and 0.6 $\mu g\ m^{-3}$, showing mass concentration values that are for all the samples about 10 times higher than those determined in the super-micrometric range (usually ranging between 0.02 and 0.07 $\mu g\ m^{-3}$). Due to their dominant anthropic origin

(Arctic Haze), nss-SO_4^{2-} shows clear seasonal variations, with spring average concentrations of both sub-micrometric and super-micrometric particle components that result to be significantly higher than those observed in summer.

In the sub-micrometric particle mode, the spring mean concentrations of nss-sulfate were found to range between 0.35 ± 0.18 μg m^{-3} (in 2014) and 0.66 ± 0.31 μg m^{-3} (in 2012). In summer, when the anthropic emissions play a less important role than that due to the biogenic particle sources, the mean concentration was estimated to range between 0.085 ± 0.077 μg m^{-3} (in 2012) and 0.28 ± 0.34 μg m^{-3} (in 2014). No particular trends were observed in the inter-annual time-patterns, the highest mean concentrations and largest data dispersion features being observed in the 2010, 2012 and 2013 samples.

In the super-micrometric particle fraction, a more limited difference was observed between the spring and summer mean concentrations of nss-sulfate. In some years (specifically in 2011 and 2014), the summer mean concentration values of this component were found to be equal to 0.028 ± 0.013 μg m^{-3} in 2011 and 0.056 ± 0.046 μg m^{-3} in 2014, and therefore estimated to be comparable or slightly higher than the spring mean values of 0.027 ± 0.032 μg m^{-3} (in 2011) and 0.055 ± 0.032 μg m^{-3} (in 2014). More generally, the super-micrometric seasonal mean concentration of nss-sulfate was found to range between 0.014 ± 0.020 μg m^{-3} (in summer 2013) and 0.070 ± 0.045 μg m^{-3} (in spring 2010). Similarly to the sub-micrometric fraction, the super-micrometric nss-SO_4^{2-} concentration did not present clear inter-annual trend features during the multi-year sampling period.

(C) **Ammonium**

The seasonal mean concentration of ammonium follows the trend of nss-sulfate, especially during the spring months. Actually, the spring mean values are significantly higher than the summer ones, with a partial exception in 2010, when the sub-micrometric spring and summer mean concentrations were found to be comparable, being equal to 0.049 ± 0.054 μg m^{-3} and 0.035 ± 0.022 μg m^{-3}, respectively. In opposition to nss-sulfate, the NH_4^+ mean concentrations were found to be always higher in spring than in summer, also when considering the super-micrometric fraction.

The NH_4^+ mean concentration of sub-micrometric particles was found to range between 0.031 ± 0.025 μg m^{-3} in summer 2012, and 0.110 ± 0.042 μg m^{-3} in spring 2013. The inter-annual time-patterns of the summer mean values of this parameter did not show a temporal trend, being the concentration data all included within the narrow range from 0.03 to 0.04 μg m^{-3}. Conversely, the spring time-patterns of the mean mass concentration of ammonium ions showed a gradual increase from 2010 to 2014, with a relative maximum in 2013.

The spring mean concentration of super-micrometric particles was found to vary largely, ranging between 2.7 ± 3.4 ng m^{-3} (in 2011) and 5.6 ± 3.4 ng m^{-3} (in 2012). Conversely, the summer mean concentrations were found to be strictly confined within a quite narrow interval from 0.7 ± 1.8 ng m^{-3} (in 2011) to

2.5 ± 4.5 ng m^{-3} (in 2014). A clear inter-annual trend of the seasonal super-micrometric mode concentrations was not evident neither in spring nor in summer.

(D) **MSA**

As previously observed for the other components belonging to secondary aerosol, MSA is mainly present in the sub-micrometric aerosol particle fraction, as it can be also noticed by looking at the different ordinate scales used in Fig. 4.16 and in Fig. 4.18, having the full scales equal to 0.1 μg m^{-3} and 0.03 μg m^{-3} for the sub-micrometric and super-micrometric particle classes, respectively. The average mass concentration was found to be generally higher in spring than in summer, with the exception of 2012, when the summer mean concentration was significantly higher than the spring mean value for both the size fractions. This was mainly due to the large occurrence of high MSA spikes, especially during May, which have strongly driven the mean trend of the concentration values. In fact, the spring average values recorded in 2012 were found to be: (i) equal to about 77% of the summer mean value recorded for the sub-micrometric mode, since the spring seasonal mean value was 8.7 ± 11.0 ng m^{-3} and the summer mean value was 11.0 ± 12.0 ng m^{-3}, and (ii) equal to about 68% of the summer mean value determined for the super-micrometric mode, since the spring mean value was 1.3 ± 1.1 ng m^{-3} and the summer mean value was 1.9 ± 1.7 ng m^{-3}.

In the sub-micrometric fraction, the MSA seasonal mean values were estimated to range between 8.7 ± 11.0 ng m^{-3} in spring 2012 (as mentioned above) and 34.0 ± 48.0 ng m^{-3} in summer 2010. No evident inter-annual variations were observed neither for the seasonal mean values and the related data dispersion in this size class, nor for the median values. A particularly large variability was observed in the spring periods of 2010, 2011 and 2013.

In the super-micrometric mode, the spring concentration values and their related variability were found to be considerably lower than those determined for the sub-micrometric particle class, with averages ranging between 0.66 ± 0.05 ng m^{-3} in spring 2014 and 7.3 ± 9.6 ng m^{-3} in spring 2010. The background MSA concentration was found to reach a minimum in 2014 (equal to 0.5 ng m^{-3}), and to exhibit the highest value in 2011 (2.8 ng m^{-3}). Also in this size class, no clear inter-annual trend features were observed for both spring and summer mean concentrations.

(E) **Nitrate**

The nitrate mean mass concentration of sub-micrometric particles was found to be always lower than that of super-micrometric particles, in all the observed years and seasons. A further general remark can be formulated on the basis of the data recorded over the whole multi-year period (with an exception in 2011 only) and for both the size ranges, the spring nitrate mean concentration being higher than that measured in summer in all the observation years. In particular, nitrate showed mean concentration values of the sub-micrometric particles ranging between 2.4

and 21.0 ng m^{-3} over the whole sampling period, whereas about 5–10 times higher values of the super-micrometric particle mode concentration were recorded, varying between 16 and 66 ng m^{-3}.

As mentioned above, the sub-micrometric nitrate seasonal mean concentrations were always higher in spring than in summer, probably because of the dominant role of the Arctic Haze transport events, frequently occurring during the spring months. In fact, mean values of nitrate mass concentration were measured in spring, ranging between 6.2 and 17.0 ng m^{-3}, while the summer mean concentration was found to vary between 2.4 and 7.5 ng m^{-3}, presenting a more limited data dispersion in summer than in spring. The inter-annual variability shows similar features over the 2010–2012 period, while those observed in 2013 and 2014 exhibited appreciably higher mean concentration values and more marked variations in spring.

The nitrate mean concentration of super-micrometric particles was found to range between 25.0 and 66.0 ng m^{-3}, considering the whole data-set. Also for this particle mode, the mean and background spring values were found to be both higher than those measured in summer, although with less marked differences: while the summer mean values span the 16.0–43.0 ng m^{-3} range, the spring values were found to vary between 20.0 and 66.0 ng m^{-3}, yielding a roughly twice spring-to-summer ratio. The only exception was recorded in 2011, when comparable nitrate levels were observed in both spring and summer equal to 38.0 and 43.0 ng m^{-3}, respectively. Unlike the sub-micrometric mode, the data dispersion observed in spring and summer for this size fraction was found to be often comparable, probably due to the different mixtures of sources generating the super-micrometric particles from one transport episode to another. Examining the inter-annual trend of these measurements, quite specular time-patterns were observed with respect to the sub-micrometric fraction, with lower spring mean concentrations than those recorded in summer in both years 2013 and 2014 (especially in the former one).

(F) **Non-sea-salt potassium (nss-K^+)**

As it can be easily inferred by examining part (*b*) of Fig. 4.17 (pertaining to the sub-micrometric particle mode) and part (*b*) of Fig. 4.19 (super-micrometric particle mode), both mean and median values of the nss-K^+ mean concentration were evaluated to be in general higher in spring than in summer. These features are probably associated with the seasonality of the transport processes occurring from the mid-latitude regions to the Arctic, which deliver important loads of anthropogenic aerosol emitted in the mid-low latitude regions preferably in spring (Arctic Haze). In addition, being nss-K^+ a marker of biomass burning emissions, it is also related to the larger occurrence of these transport phenomena in late spring.

In the sub-micrometric fraction, the spring average values were found to be significantly higher than the summer ones in all the above-considered years, except in 2014, when the mean values were estimated to be comparable in the two seasons, being equal to 6.4 ± 3.1 ng m^{-3} in spring and 7.6 ± 9.6 ng m^{-3} in summer.

The spring average concentrations measured in the five years were found to range between 5.0 ± 3.7 ng m^{-3} and 6.8 ± 1.9 ng m^{-3}, whereas the summer mean concentrations were found to vary between 1.2 ± 1.3 ng m^{-3} and 7.6 ± 9.6 ng m^{-3}. Background levels showed similar time-patterns, with higher values in all the spring periods of the 2010–2013 years and almost comparable values in 2014. The inter-annual variability of the sub-micrometric particle concentration was slightly different over the spring and summer periods, since the spring means showed variable time-patterns with higher values in 2012 and 2014, and the summer means exhibited a particularly high value only in 2014.

The seasonal nss-K^+ mean concentration of the super-micrometric particle fraction was found to be lower than the corresponding values pertinent to the finer particle fraction in all the observation years, due to the dominant origin of this component (mainly generated by biomass combustion processes producing particles mainly smaller than 1 μm). In spring, the mean concentration values of the super-micrometric particle mode varied between 0.83 ± 1.50 ng m^{-3} (in 2013) and 3.5 ± 2.3 ng m^{-3} (in 2014), which corresponds to 30–60% of the values measured for the sub-micrometric particle mode. In summer, the average concentration of the aerosols belonging to such a larger particle fraction varied between 0.48 ± 1.1 ng m^{-3} (in 2013) and 3.9 ± 2.8 ng m^{-3} (in 2014), covering a larger percentage (30–90%) of the corresponding values determined for the sub-micrometric particle mode. The spring average concentration values determined for the super-micrometric particle class were found to be higher than the summer ones in all the years except the 2014 year, when the spring and the summer values were found to be comparable, being equal to 3.5 ± 2.3 ng m^{-3} and 3.9 ± 2.8 ng m^{-3}, respectively.

No clear inter-annual variability features were observed for the spring and summer mean values, since 2013 was the year with the lowest seasonal mean values and data variability, as well as that with the lowest background values. Different inter-annual variations were observed in spring and summer, with larger data dispersion affecting the spring data of 2011 and 2014.

(G) **Mineral dust**

The box-plot distribution of the crustal contribution is shown in parts (c) of Fig. 4.17 for the sub-micrometric mode, and of Fig. 4.19 for the super-micrometric particle fraction. Since the mineral dust is originated by primary processes of aerosol formation, these particles should mainly belong to the super-micrometric mode. However, it is important to take into account that the major part of the continental dust is transported toward Ny-Ålesund and the Svalbard Archipelago from remote continental areas at lower latitudes, and that the large particles are preferentially settled, so that the mineral dust content becomes enriched in smaller particles. Actually, we have observed similar dust contributions in both the super-micrometric and the sub-micrometric particle modes, with overall mean concentrations ranging from 0.1 to 0.4 μg m^{-3} in summer and from 0.2 to 0.6 μg m^{-3} in spring.

The mean concentration of the sub-micrometric crustal particle fraction was found to show sharp seasonal variations throughout the year, with: (i) relatively constant and lower summer values varying between 0.086 ± 0.059 μg m^{-3} (in 2011) and 0.36 ± 0.35 μg m^{-3} (in 2014), and (ii) largely variable and higher spring values ranging between 0.15 ± 0.13 μg m^{-3} (in 2010) and 0.50 ± 0.32 μg m^{-3} (in 2014). The samples collected in 2010 represented an exception, yielding a summer mean concentration about twofold the spring one. The inter-annual time-patterns showed an increasing trend in both spring and summer, with more marked increasing trends in 2013 and 2014.

The summer mean concentration of mineral dust in the super-micrometric mode was found to be always significantly lower than that measured in spring, assuming relatively constant values of around 0.1 μg m^{-3} (with the exception of the 2014 value, which was found to be equal to 0.37 ± 0.26 μg m^{-3}). Conversely, the spring samples were found to provide higher and more variable mean mass concentration values, ranging between 0.12 ± 0.06 μg m^{-3} in 2013 and 0.64 ± 0.52 μg m^{-3} in 2014. Examining the time-patterns of the mass concentration of mineral dust shown in part (c) of Fig. 4.19, no particular trends were observed to characterize the inter-annual time-patterns of the crust component, although significantly higher mean concentrations were measured in the 2014 spring and summer samples.

The spring and summer average concentrations of the sub-micrometric particle mode determined by analyzing the particulate matter sampled collected during the three campaigns conducted at the GVB Observatory in 2012, 2013 and 2014 are presented in Table 4.2. The higher atmospheric concentrations partial mass fractions measured in spring were found to be those of nss-sulfate (0.537 μg m^{-3}), mineral dust (0.387 μg m^{-3}) and sea spray (0.209 μg m^{-3}) components, together with lower concentrations of ammonium, nitrate, MSA, and nss-K^+ ions. In summer, the predominant fractions were found to be those of mineral dust (0.233 μg m^{-3}) and nss-sulfate (0.157 μg m^{-3}), with lower and decreasing concentrations of sea spray, ammonium, MSA, nitrate and nss-K^+.

The spring and summer average concentrations of the chemical components sampled in the super-micrometric particles during the three campaigns carried out in 2012, 2013 and 2014 at the GVB Observatory are given in Table 4.3. The higher concentrations measured in spring were found to be those of sea spray (0.487 μg m^{-3}) and mineral dust (0.380 μg m^{-3}), followed by lower concentrations of nss-sulfate (0.044 μg m^{-3}), nitrate (0.042 μg m^{-3}), ammonium, MSA, and nss-K^+, the last three components with concentrations ranging between 2 and 5 ng m^{-3}. Similarly, rather high concentrations were measured in summer for sea spray (0.432 μg m^{-3}) and mineral dust (0.187 μg m^{-3}), together with lower concentrations of nss-sulfate (27 ng m^{-3}), nitrate (24 ng m^{-3}), MSA, ammonium, and nss-K^+, the last three chemical components having summer mean concentrations ranging between 1.7 and 2.9 ng m^{-3}.

Table 4.2 Values of the three-year average mass concentrations of the DK4 sub-micrometric aerosol particles and their components (measured in $\mu g\ m^{-3}$), as determined with their standard deviations by Udisti et al. (2016) during the spring (March–May) and summer (June–September) seasons at the GVB Observatory (Ny-Ålesund, Spitsbergen, Svalbard) during the three campaigns conducted in 2012, 2013 and 2014

Seasonal period	Sea spray	nss-SO_4^{2-}	NH_4^+	MSA	NO_3^-	nss-K^+	Mineral dust
Spring (MAM)	0.209 ± 0.167	0.537 ± 0.358	0.0965 ± 0.0358	0.0130 ± 0.0201	0.0156 ± 0.0111	0.0061 ± 0.0030	0.387 ± 0.262
Summer (JJAS)	0.0469 ± 0.0756	0.157 ± 0.218	0.0377 ± 0.0282	0.0117 ± 0.0098	0.0050 ± 0.0036	0.0036 ± 0.0063	0.233 ± 0.251

Table 4.3 Values of the three-year average mass concentrations of the DK4 super-micrometric aerosol particles and their components (measured in $\mu g\ m^{-3}$), as determined with their standard deviations by Udisti et al. (2016) during the spring (March–May) and summer (June–September) seasons at the GVB Observatory (Ny-Ålesund, Spitsbergen, Svalbard) during the three campaigns conducted in 2012, 2013 and 2014

Seasonal period	Sea spray	nss-SO_4^{2-}	NH_4^+	MSA^-	NO_3^-	nss-K^+	Mineral dust
Spring (MAM)	0.487 ± 0.374	0.0440 ± 0.0284	0.0047 ± 0.0038	0.0019 ± 0.0027	0.0418 ± 0.0218	0.0022 ± 0.0023	0.380 ± 0.374
Summer (JJAS)	0.432 ± 0.410	0.0267 ± 0.0357	0.0017 ± 0.0035	0.0029 ± 0.0029	0.0236 ± 0.0227	0.0017 ± 0.0024	0.187 ± 0.214

4.2.2.4 *In-situ* Measurements of the Aerosol Optical Parameters at Ny-Ålesund

Accurate ground-based measurements of aerosol optical thickness *AOT(λ)* at various visible and near-infrared wavelengths have been performed at Ny-Ålesund (10 m a.m.s.l.) in the Spitsbergen Island (Svalbard) by the Alfred Wegener Institute for Polar and Marine Research (AWI) group based in Bremerhaven and Potsdam, Germany, over the period from 1991 to 2013 by employing the SP1H and SP2H sun-photometer models and the STAR01 star-photometer (Tomasi et al. 2012). The above sun-photometer measurements were found to provide daily mean values of $AOT(0.50\ \mu m)$ that vary most frequently between 0.04 and 0.12 in the spring months and between 0.03 and 0.08 in the summer months, while correspondingly the Ångström exponent $\alpha(0.40\text{–}0.87\ \mu m)$ ranged mainly between 1.00 and 1.70 in spring and between 1.00 and 1.50 in summer (Tomasi et al. 2015). These results clearly indicate that the incoming direct solar radiation is predominantly extinguished at the visible and near-infrared wavelengths by the fine particle mode (during both spring and summer periods) in all the cases characterized by values of $\alpha(0.40\text{–}0.87\ \mu m)$ higher than 1.20, and by the coarse particle mode in the less numerous cases yielding values of $\alpha(0.40\text{–}0.87\ \mu m)$ lower than 1.20. In order to evaluate the so-called "scale height" of the vertical profile of aerosol extinction coefficient (Tomasi 1982), the above-mentioned AWI sun-photometric measurements of $AOT(0.50\ \mu m)$ can be usefully compared with the surface-level measurements of the volume aerosol extinction coefficient derived from the ground-based measurements of the volume scattering and absorption coefficients conducted by the CNR group based in Bologna, Italy, at the GVB Observatory during the spring and summer months (i.e. from late March to the end of September or to early October).

In order to better characterize the optical properties of the aerosol particles at Ny-Ålesund, regular ground-based measurements of the volume aerosol scattering coefficient $\beta_{sca}(0.532\ \mu m)$ were regularly conducted at the GVB Observatory during the spring and summer months of 2011 and 2015. A portable low-power M903 Radiance Research nephelometer was employed for this purpose, equipped with: (i) a monochromatic light source, to provide the measurements of the forward scattering and back-scattering coefficients of aerosol particles having mass concentration varying from about 1 to $10^3\ \mu g\ m^{-3}$, and (ii) an internal data logger to store the scattering coefficient averages. The time-patterns of the daily mean values of volume scattering coefficient $\beta_{sca}(0.532\ \mu m)$ are shown in parts (a) and (b) of Fig. 4.20 for 2011 and 2015, respectively, giving evidence of: (i) the gradual decrease of this optical parameter during spring 2011, from values higher than 10 Mm^{-1} in late March and April to values ranging between nearly null and 5 Mm^{-1} in the summer months; and (ii) the quite stable features of $\beta_{sca}(0.532\ \mu m)$ observed during summer 2015, found to be in general appreciably lower than those recorded in summer 2011. On this matter, it is worth mentioning that the sharp and pronounced peak of $\beta_{sca}(0.532\ \mu m)$ recorded in July 2015 was caused by an intense transport episode of biomass burning particles generated by forest fires in northern Canada and then transported over the Svalbard Archipelago area (Markowicz et al. 2016; Moroni et al. 2017). In particular, examining the sun-photometer measurements of $AOT(0.55\ \mu m)$ performed at the GVB Observatory, Markowicz et al. (2016) estimated that the daily mean values of $AOT(0.55\ \mu m)$ were lower than 0.10 during the first decade of July 2015, and varied between 0.20 and 0.80 during the 10–16 July period, in which the smoke particles were evaluated to produce attenuation effects varying between 20% and 50% of the overall particulate extinction, whereas more limited attenuation effects were produced by the anthropogenic and biogenic fine particles, together with weaker extinction effects by dust and sea salt particles. On July 10 and 11, 2015, the volume scattering coefficient $\beta_{sca}(0.532\ \mu m)$ was found by Markowicz et al. (2016) to exhibit a marked peak higher than 130 Mm^{-1}, as shown in Fig. 4.20.

Regular ground-level measurements of the volume aerosol absorption coefficient $\beta_{abs}(0.530\ \mu m)$ were simultaneously performed at the GVB Observatory from 2011 to 2015, by using the three-wavelength Particle Soot Absorption Photometer (hereinafter referred to as 3λ-PSAP) manufactured by the Radiance Research to investigate the light absorption features produced by aerosol particles, and equipped with a triplet of wavelengths equal to 0.467, 0.530 and 0.660 μm. This instrument was routinely employed at the GVB Observatory to measure in near real time the light absorption coefficient of aerosol particles at the three above-mentioned wavelengths, and investigate the light absorption properties of aerosols and their wavelength-dependence features. The calibration and correction procedures defined by Virkkula et al. (2005) were followed in analysing the GVB field data, through the use of an integrating plate technique, in which the change in the optical transmission of the internal filter (on which particles have been deposited) is related to the volume absorption coefficient $\beta_{abs}(\lambda)$ of the incident light, through a relationship based on the Beer extinction law separately considered for each of the three above-mentioned wavelengths. Following this procedure, daily mean triplets of the ground-level volume aerosol absorption coefficient $\beta_{abs}(\lambda)$ were obtained from the 3λ-PSAP measurements. The measurements of $\beta_{abs}(0.530\ \mu m)$ derived from the 3λ-PSAP measurements conducted in 2011 and 2015 are shown in Fig. 4.20, to provide evidence of: (i) the gradual decrease of $\beta_{abs}(0.530\ \mu m)$ during spring 2011, until showing a prevailing range lower than 0.6 Mm^{-1} in summer; and (ii) the gradually decreasing time-patterns of $\beta_{abs}(0.530\ \mu m)$ in 2015, describing a marked peak in late March and a further more pronounced peak in early July 2015, both due to the transport of relevant loads of biomass burning particles from Canada. Markowicz et al. (2016) examined the time-patterns of the absorption coefficients measured at the 0.467, 0.530 and 0.660 μm wavelengths using the 3λ-PSAP photometer at the GVB Observatory, and found that a triplet of very pronounced peaks was measured at the three wavelengths, equal to 10–15 Mm^{-1} on July 10 and 11, 2015, against values largely lower than 2 Mm^{-1} on the previous days.

The nephelometer measurements of coefficient $\beta_{sca}(0.532\ \mu m)$ and the 3λ-PSAP measurements of coefficient $\beta_{abs}(0.530\ \mu m)$ performed during the years 2011 and 2015 were herein analyzed in order to determine the corresponding values of volume extinction coefficient $\beta_{ext}(0.531\ \mu m)$ (obtained by summing the

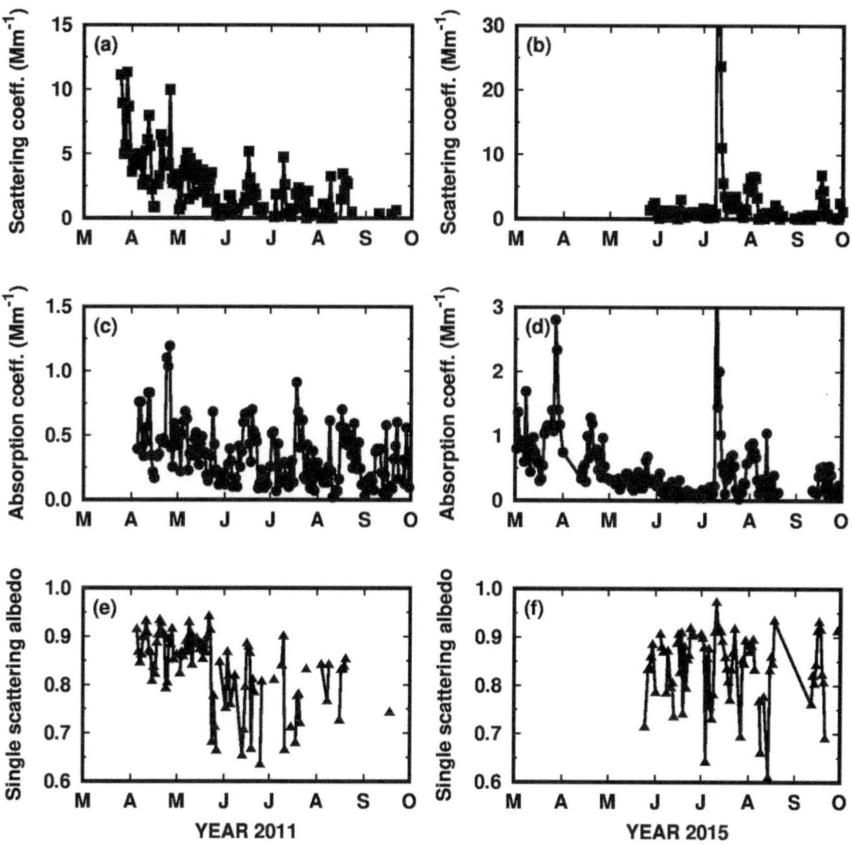

Fig. 4.20 Time-patterns of the daily mean values of (i) volume aerosol scattering coefficient $\beta_{sca}(0.532\ \mu m)$ measured in Mm^{-1} at the GVB Observatory (Ny-Ålesund) with the portable M903 Radiance Research nephelometer in the spring and summer months of the years 2011 (part (**a**)) and 2015 (part (**b**)); (ii) volume aerosol absorption coefficient $\beta_{abs}(0.530\ \mu m)$ measured in Mm^{-1} at the same site with the 3λ-PSAP Radiance Research photometer during the years 2011 (part (**c**)) and 2015 (part (**d**)); and (iii) single scattering albedo $\omega(0.531\ \mu m)$ obtained as ratio $\beta_{sca}(0.532\ \mu m)/\beta_{ext}(0.531\ \mu m)$ between the daily mean values of $\beta_{sca}(0.532\ \mu m)$ shown above and the daily mean values of $\beta_{ext}(0.531\ \mu m)$, obtained by summing the daily mean values of $\beta_{sca}(0.532\ \mu m)$ and $\beta_{abs}(0.530\ \mu m)$ shown above in parts (**a–d**) for the years 2011 (part (**e**)) and 2015 (part (**f**))

above coefficients $\beta_{sca}(0.532\ \mu m)$ and $\beta_{abs}(0.530\ \mu m)$). Using the daily mean values of coefficients $\beta_{sca}(0.532\ \mu m)$ and $\beta_{ext}(0.531\ \mu m)$ determined above and shown in parts (a), (b), (c) and (d) of Fig. 4.20, we have also determined the corresponding daily mean values of single scattering albedo $\omega(0.531\ \mu m)$ in terms of the $\beta_{sca}(0.532\ \mu m)/\beta_{ext}(0.531\ \mu m)$ ratio. The time-patterns of $\omega(0.531\ \mu m)$ are shown in Fig. 4.20, separately for the years 2011 (part (e)) and 2015 (part (f)) to give evidence of the fact that $\omega(0.531\ \mu m)$ ranged mainly between 0.80 and

0.95 during spring 2011, and was then subject to vary more largely in the summer months of 2011, between 0.62 and about 0.85. These results clearly indicate that the solar radiation absorption by anthropogenic and natural aerosols can produce relevant effects at this remote Arctic site on the particulate matter optical properties, especially during the particularly intense transport episodes of combustion particles. Part (f) of Fig. 4.20 fully confirms these results, since $\omega(0.531\ \mu m)$ was estimated to vary even more largely in summer 2015, between 0.60 and more than 0.90 throughout the whole season, with sharp day-to-day variations due to the abrupt changes in the origins of particulate matter transported over this remote site. In particular, it is worth noting that the lowest values of $\omega(0.531\ \mu m)$ were recorded in half-July and during the first two weeks of August 2015, as estimated by Markowicz et al. (2016), who determined values of this optical parameter largely lower than 0.50 on both July 13 and 14 and on July 16, 2015.

4.2.2.5 Measurements of the BC, OC, EC and TC Mass Concentrations at Ny-Ålesund (Spitsbergen)

Regular measurements of equivalent BC mass concentration were carried out by Eleftheriadis et al. (2009) at the Zeppelin Observatory (78° 54′ N, 11° 53′ E, 474 m a.m.s.l.), located a few km far from Ny-Ålesund on the top of Zeppelinfjellet (a mountain dominating the Kongsfjorden), by analyzing the measurements performed with: (i) the Magee Scientific AE-9 aethalometer from August 1998 to July 1999 (with peak wavelength $\lambda_p = 0.880\ \mu m$ and site-specific mass absorption coefficient $\alpha_{AE} = 15.2\ m^2\ g^{-1}$) and (ii) the Magee Scientific AE-31 aethalometer from May 2001 to spring 2007 (with $\lambda_p = 0.880\ \mu m$ and $\alpha_{AE} = 15.9\ m^2\ g^{-1}$). The time-patterns of the monthly mean values of BC mass concentration obtained from the overall set of the 1998–2007 measurements are shown in Fig. 4.21, presenting: (i) rather high winter – spring values, ranging between 42 ng m^{-3} in May and 69 ng m^{-3} in March; (ii) considerably lower values in the summer months, varying mainly between 7 and 10 ng m^{-3}; and (iii) higher values in the autumn and early winter months, gradually increasing from 14 ng m^{-3} in October to more than 40 ng m^{-3} in December. Examining the seasonal variations of BC recorded over this 10-year period, with respect to the potential source regions estimated by means of a Potential Source Contribution Function (PSCF) model based on the calculated air mass trajectories, Eleftheriadis et al. (2009) estimated that the BC concentration values obtained at the Zeppelin Observatory were mainly influenced by various source regions mostly located in northern and central Russia.

Appreciably lower values were more recently determined by Sinha et al. (2017) at Ny-Ålesund, through the analysis of a set of measurements conducted by applying the COntinuous Soot MOnitoring System (COSMOS) procedure to the PM10 data-sets collected during the 4-year period from 2012 to 2015, and assuming an average value of mass absorption cross-section $\sigma_{abs} = 8.73\ m^2\ g^{-1}$. The analysis of these data provided monthly mean values of BC mass concentration that were estimated to: (i) to vary between 10 ng m^{-3} (in May) and 33 ng m^{-3} (in February) during the

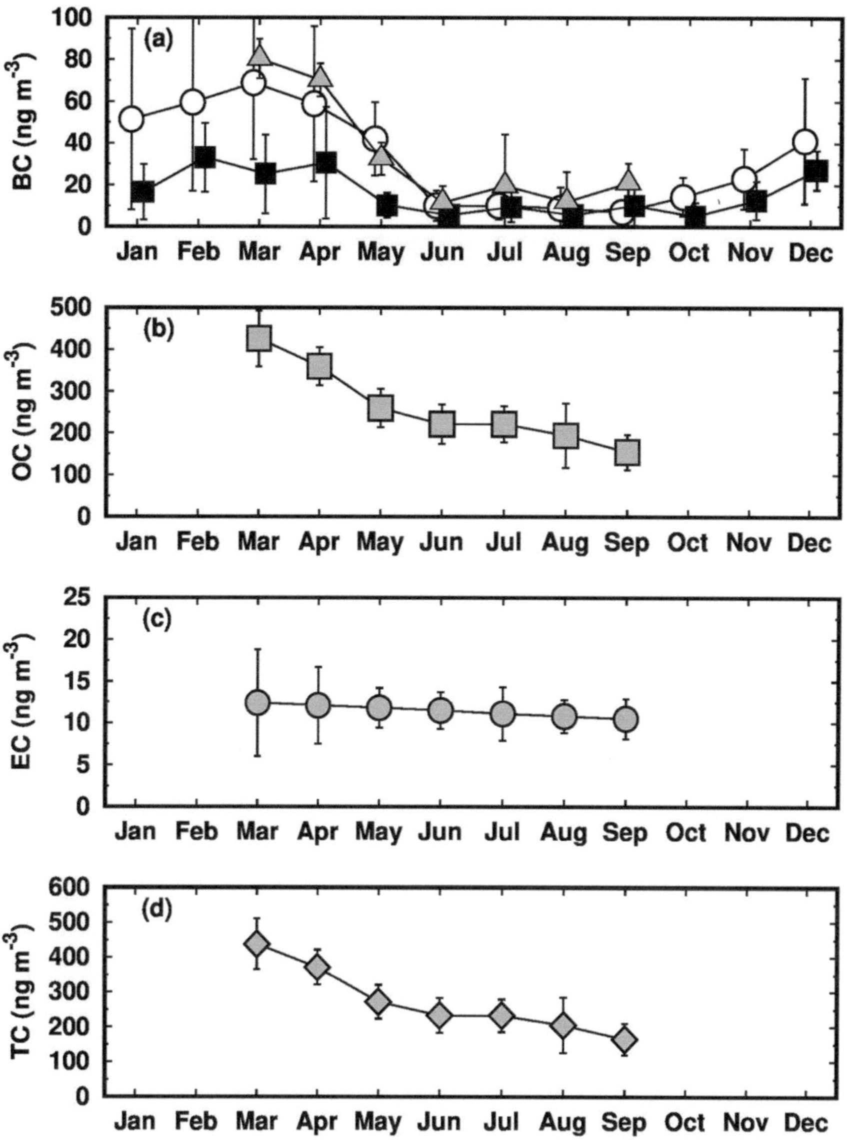

Fig. 4.21 Part (**a**): Time-patterns of the monthly mean values of the Black Carbon (BC) mass concentration derived from the measurements performed by: (i) Eleftheriadis et al. (2009) at the Zeppelin Observatory near Ny-Ålesund) over the 1998–2008 period (open circles); (ii) Sinha et al. (2017) at Ny-Ålesund over the 2012–2015 period (solid squares); and (iii) Traversi et al. (2017) at the GVB Observatory (Ny-Ålesund) during the March – September months of the 2011–2015 period (grey triangles). Part (**b**): Organic Carbon (OC) mass concentration derived from the measurements performed by Traversi et al. (2017) by analyzing the *in-situ* samples of PM10. Part (**c**): As in part (**b**) but for the Elemental Carbon (EC) mass concentration. Part (**d**): as in part (**b**) but for the Total Carbon (TC) mass concentration, given by the sum of OC and EC. The vertical bars represent the standard deviations of the monthly mean values of BC, OC, EC and TC, respectively

first half of the year; (ii) range between 5.5 ng m^{-3} (in June) and 10.1 ng m^{-3} (in September) during summer; and (iii) slowly increase from 5.0 ng m^{-3} (in October) to 27 ng m^{-3} (in December) during the last part of the year. These results are compared with those of Eleftheriadis et al. (2009) in Fig. 4.21, providing evidence of the particularly marked decrease of the BC concentration during the winter and early spring, as shown by the measurements performed during the last two decades from 1998 to 2015.

Monthly mean values of equivalent BC mass concentration were also derived by Traversi et al. (2017) at the GVB Observatory during the 2011–2015 years from the measurements of coefficient $\beta_{abs}(0.670\ \mu m)$ performed with the 3λ-PSAP. The monthly mean values of BC were found to: (i) vary during the spring months between 23 ng m^{-3} (in May) and 70 ng m^{-3} (in April); (ii) range in the summer months between 24 ng m^{-3} (in June) and 45.0 ng m^{-3} (in September) (and, therefore; presenting considerably higher than those measured by Sinha et al. (2017)); and (iii) slowly increase from 5.0 ng m^{-3} (in October) to 27 ng m^{-3} (in December). The analysis of these measurements provided a spring average value of BC mass concentration equal to 50.1 $\pm$ 11.0 ng m^{-3}, a summer average value of 34.2 $\pm$ 8.3 ng m^{-3}, and a spring-summer average value of 41.0 $\pm$ 8.7 ng m^{-3} (over the whole 7-month period from March to September). As mentioned above, the results obtained by Traversi et al. (2017) and shown in Fig. 4.21 are appreciably higher than those measured by Sinha et al. (2017) in all the spring and summer months of the years from 2011 to 2015.

Monthly mean values of OC (Organic Carbon), EC (Elemental Carbon) and TC (Total Carbon, given by the sum of OC and EC) mass concentrations were determined by Traversi et al. (2017) at the GVB Observatory by Thermo-Optical analysis of the PM10 samples collected on quartz filters during the March–September campaigns from 2011 to 2015. As can be seen in Fig. 4.21, the monthly mean values of OC concentration were estimated to gradually decrease from 426 ng m^{-3} (in March) to 154 ng m^{-3} (in September), yielding a spring average value of 348 $\pm$ 145 ng m^{-3} and a summer average value of 198 $\pm$ 143 ng m^{-3}. These measurements of OC concentration agree very closely with the seasonal average concentrations of OC reported in Table 4.1.

The monthly mean values of EC were correspondingly found to: (i) vary between 12.4 ng m^{-3} (in March) and 11.8 ng m^{-3} (in May), yielding a spring mean value equal to 12.1 ng m^{-3}; and (ii) slowly decrease from 11.5 ng m^{-3} (in June) to 10.5 ng m^{-3} (in September), giving a summer mean value of 11.0 ng m^{-3}. Thus, the monthly mean values of TC mass concentration were evaluated to decrease: (i) from 438 ng m^{-3} in March to 272 ng m^{-3} in May, giving a spring average value of 360 ng m^{-3}; and (ii) from 233 ng m^{-3} in June to 165 ng m^{-3} in September, yielding a summer average value of 209 ng m^{-3}.

The seasonal average estimates of the mass concentrations of BC, OC, EC and TC shown in Fig. 4.21 can be used to evaluate the optical properties of particulate matter at the GVB Observatory, which can be subject to vary appreciably from one year to the other, as evidenced by Traversi et al. (2017) on the basis of the field-measurements conducted at this site with a nephelometer and a 3λ-PSAP photometer. The time-patterns of the daily mean concentrations of BC, OC, EC and TC derived from the *in-situ* PM10 measurements performed at the GVB Observatory were determined by Traversi et al. (2017) during the 2011–2015 spring (March–May) and summer (June–September) periods, finding that these parameters varied largely during all the five observation years.

The time-patterns of the seasonal and overall mean mass concentrations of BC, OC, EC and TC obtained from the *in-situ* PM10 measurements carried out from 2011 to 2015 are shown in Fig. 4.22. The mean concentration values were determined with relatively high standard deviations: (i) the relative standard deviations of BC concentration measured in the spring months varied between 9% and 37% during the five years, while those measured in summer varied between 5% and 37%, and those obtained over the whole 7-month period between 7% and 36%; (ii) the relative standard deviations of OC concentration measured in the spring months were found to vary between 6% and 26% during the five years, while those measured in the summer months ranged between 6% and 46%, and those obtained over the whole 7-month period between 6% and 34%; (iii) the relative standard deviations of EC concentration measured in spring were found to vary between 6% and 67% during the five years, while those measured in the summer months varied between 6% and 55%, and those pertaining to the whole 7-month period between 6% and 59%; and (iv) the relative standard deviations of TC concentration measured in spring were found to vary between 6% and 40% during the five above-selected years, while those measured in summer varied between 6% and 46%, and those obtained over the whole 7-month period between 7% and 35%. These results provide a clear measure of the strong day-by-day variations affecting the BC, OC, EC and TC concentrations at Ny-Ålesund, as a result of the complex transport processes of natural and anthropogenic particulate matter from the mid-latitude regions toward the Arctic Ocean.

4.2.2.6 A General Picture of the Chemical Composition Characteristics of Sub-Micrometric, Super-Micrometric and Overall Aerosol Particles at Ny-Ålesund (Spitsbergen, Svalbard)

The chemical composition of PM10 sampled at the GVB Observatory was investigated to determine the percentage mass fractions of the main components of PM10 in spring (from March to May) and summer (from June to September). A graphical representation of the chemical composition results obtained from these measurements is shown in Fig. 4.23 for PM10, sub-micrometric particle mode and super-micrometric particle mode. The total average concentration of PM10 was found to be: (i) 2.82 μg m^{-3} in spring, with mass percentages of 42.3% for sea salt, 27.6% for mineral dust, 12.5% for sulfate and nitrate, 12.5% for OC, 3.6% for MSA, and 1.6% for BC; and (ii) 1.40 μg m^{-3} in summer, with mass percentages of 36.1% for sulfate and nitrate, 28.5% for sea salt, 23.1% for mineral dust, 7.1% for OC, 2.7% for MSA, and 2.4% for BC. The total average mass concentration of

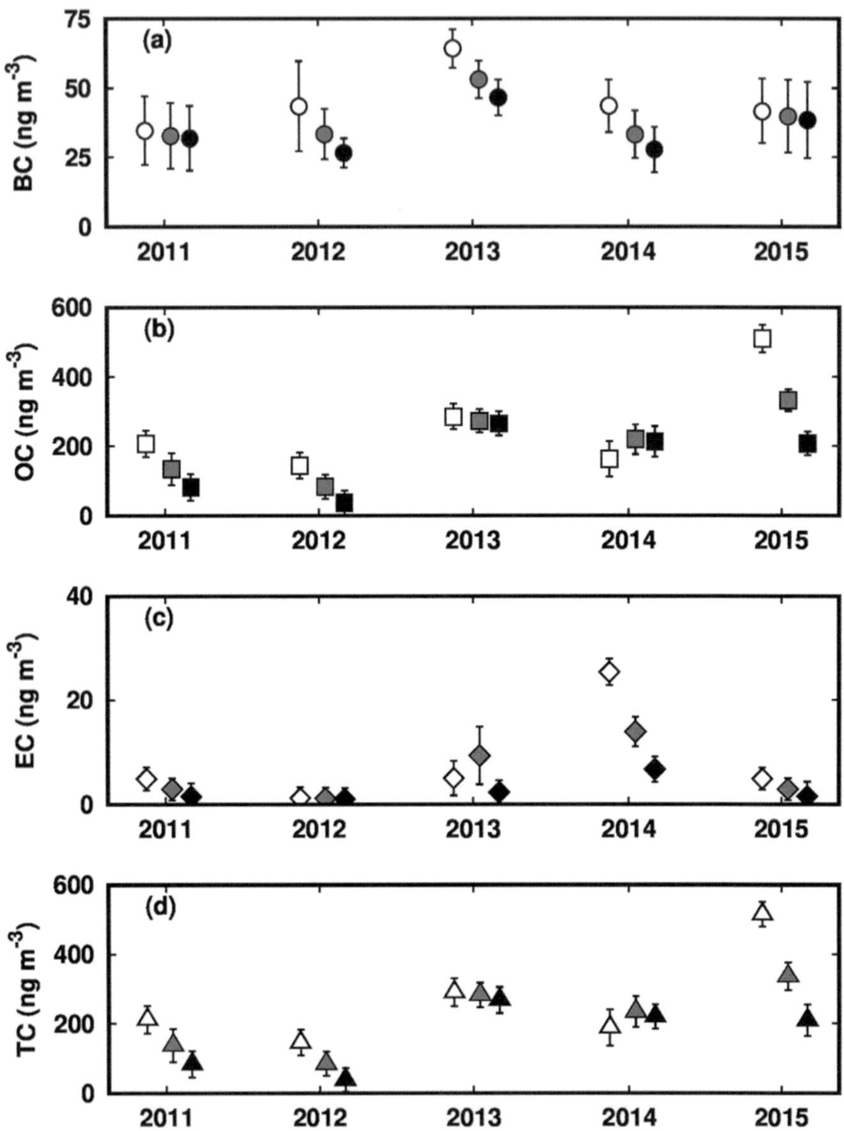

Fig. 4.22 Seasonal average and annual mean mass concentrations of Black Carbon (BC), Organic Carbon (OC), Elemental Carbon (EC) and Total Carbon (TC, equal to OC + EC), derived by Traversi et al. (2017) from the *in-situ* PM10 measurements conducted at the GVB Observatory (Ny-Ålesund, Spitsbergen, Svalbard) during the 2011–2015 period in spring (March–May) (open symbols), summer (June–September) (solid symbols) and over the whole period from March to September (grey symbols). Part (**a**) shows the spring average values (open circles), summer average values (solid circles) and spring – summer average values (grey circles) of the Black Carbon (BC) mass concentration. Part (**b**) shows the spring average values (open squares), summer average values (solid squares) and spring – summer average values (grey squares) of the Organic

sub-micrometric particles was found to be: (i) 1.66 $\mu g\ m^{-3}$ in spring, with mass percentages of 42.9% for sulfate and nitrate, 23.4% for mineral dust, 12.6% for sea salt, 12.5% for OC, 5.8% for MSA, and 2.8% for BC; and (ii) 0.70 $\mu g\ m^{-3}$ in summer, with mass percentages of 33.1% for mineral dust, 25.2% for sulfate and nitrate, 24.8% for OC, 6.7% for sea salt, 5.4% for MSA, and 4.8% for BC.

The total average concentration of super-micrometric particles was found to be: (i) 0.96 $\mu g\ m^{-3}$ in spring, with mass percentages of 50.6% for sea salt, 39.5% for mineral dust, 9.4% for sulfate and nitrate, and 0.5% for MSA; and (ii) 0.89 $\mu g\ m^{-3}$ in summer, with mass percentages of 48.7% for sea salt, 42.8% for mineral dust, 8.3% for sulfate and nitrate, and 0.2% for MSA.

4.2.3 Chemical Composition of PM10 Collected at Thule (Greenland) from 2010 to 2014

Thule Air Base (76° 32′ N, 68° 42′ W, 77 m a.m.s.l.) is the United States Air Force's northernmost base, located 1207 km north of the Arctic Circle and 1524 km from the North Pole, in the north-west sector of Greenland. The Thule High Arctic Atmospheric Observatory (THAAO) is located on the South Mountain, at a 20-minute drive from main base, and is super-intended by the Danish Meteorological Institute (DMI). Measurements of the optical properties of aerosol particles and atmospheric gases (O_3, N_2O, CO, HNO_3 and columnar water vapour) are regularly performed at THAAO, together with LIDAR (Light Detection and Ranging) observations of the upper tropospheric and lower stratospheric aerosols, and measurements of the middle atmospheric temperature profiles. These ground-based instruments are part of the international Network for Detection of Atmospheric Composition Change (NDACC, http://www.ndsc.ncep.noaa.gov). The Italian effort at Thule started in 1990, based on a collaboration between personnel of University of Rome "La Sapienza", the Italian National Agency for New Technologies, Energy and Economic Sustainable Development (ENEA) and the Italian National Institute of Geophysics and Volcanology (INGV). These Italian institutions installed a large number of ground-based instruments devoted to the observation of the Arctic environment, and continuously maintain and operate them.

PM10 sampling measurements at THAAO started in January 2010 and was performed by the University of Florence research group. The PM10 sampler (low-

Fig. 4.22 (continued) Carbon (OC) mass concentration. Part (**c**) shows the spring average values (open diamonds), summer average values (solid diamonds) and spring – summer average values (grey diamonds) of the EC mass concentration. Part (**d**) shows the spring average values (open up-ward triangles), summer average values (solid up-ward triangles) and spring – summer average values (grey up-ward triangles) of the TC mass concentration. The vertical bars represent the standard deviations of the monthly mean values of BC, OC, EC and TC, represented in parts (**a**), (**b**), (**c**) and (**d**), respectively

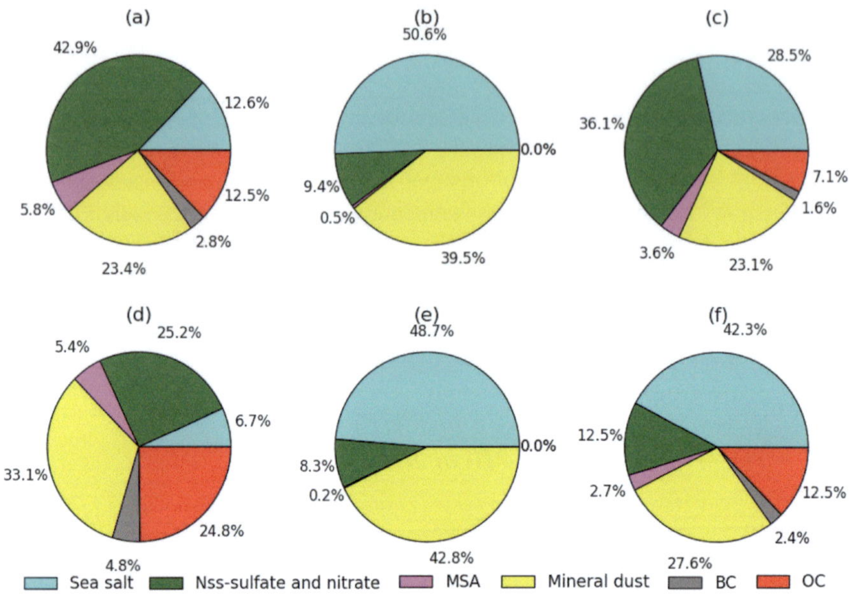

Fig. 4.23 Average composition diagrams of the ground-level particulate matter sampled at the GVB Observatory (Ny-Ålesund, Spitsbergen, Svalbard) during different seasonal periods and pertaining to: (**a**) PM10 sampled during the spring months (March–May); (**b**) PM10 sampled during the summer months (June–September); (**c**) sub-micrometric particles sampled during the spring months (March–May); (**d**) sub-micrometric particles sampled during the summer months (June–September); (**e**) super-micrometric particles sampled during the spring months (March–May); and (**f**) super-micrometric particles sampled during the summer months (June–September). Different colours are used to indicate the main particulate matter constituents (sea salt, nss-sulfate and nitrate, methanesulfonic acid (MSA), mineral dust, black carbon (BC), and organic carbon (OC)

volume Tecora Sky Post sequential sampler with a PM10 sampling head) was located in the building 1971 (see Fig. 4.24), a few km up-wind the airport and the main base. The main wind directions indicate that this sampling site is not affected by air masses coming from local anthropogenic activity (Muscari et al. 2014). PM10 sampling was all-year-round performed (since January 2010 to the present) with a 48-hour resolution (i.e. with continuous sampling over 2-day periods), in collaboration with the Danish Meteorological Institute. The sequential PM10 sampler contains 15 filters (47 mm Teflon filter Pall R2PJ047, 2 μm nominal porosity) and allows 1-month autonomy. Samplings were carried out in "actual conditions": pressure and temperature were continuously monitored to maintain a constant flow rate of 38.3 L/min (EN 12341 European rules), corresponding to a 48-h air volume of about 110 m^3. During the sampling campaigns, some interruptions occurred, due to sampler technical problems and maintenance operations. PM10 mass was measured by weighing the filters with a microbalance having a sensitivity of 0.01 mg before and after each exposure. The filters were conditioned at low

Fig. 4.24 Position of the sampling site (Building 1971) in the Thule area

humidity (silica gel) and controlled temperature (25 ± 1 °C) for at least 24 h before the weighing. Chemical analysis (ionic composition by ion chromatography) was carried out on 25% of each Teflon filter, by following the same procedure described above (in Sect. 4.2.2.1) for the PM10 samples collected at Ny-Ålesund. The calculations of ss- and nss-fractions and the sea spray and mineral dust (crustal) fractions were made by following the same procedure described in Sect. 4.2.2.1.

The box plots of the chemical components measured in the PM10 samples collected at Thule during the 2010–2014 annual campaigns are shown in parts (a) – (d) of Fig. 4.25 and in parts (a) – (d) of Fig. 4.26. Each box contains the 50% of data, with the median value displayed as a horizontal line. The bottom and the top of each box mark the limits of 25% and 75% of the variable population (indicating the 25th and 75th percentiles, respectively). The vertical lines extending from the bottom and top of each box respectively mark the minimum and maximum value that fall within an acceptable range (1.5 times the box width). The outliers are shown as small circles.

Although all-year-round samplings were carried out at Thule, here we report only the March to May (MAM) and June to September (JJAS) average values, in order to obtain a more reliable comparison between the Ny Alesund and Thule data sets. The seasonal and inter-annual trends of the chemical components are shortly described in the following sub-paragraphs.

(A) **PM10 mass concentration**

Part (a) of Fig. 4.25 shows the statistical distribution of PM10 spring and summer mean concentrations measured on the samples collected at Thule during the 2010–2014 annual campaigns. The seasonal variations are less evident than those observed at Ny Alesund, showing that the impact of the Arctic Haze was less effective in Northern Greenland, with respect to the Svalbard Islands area. However, the spring mean concentrations were estimated to be generally higher than the summer ones, with the exception of the 2011 samples, which provided a summer mean value

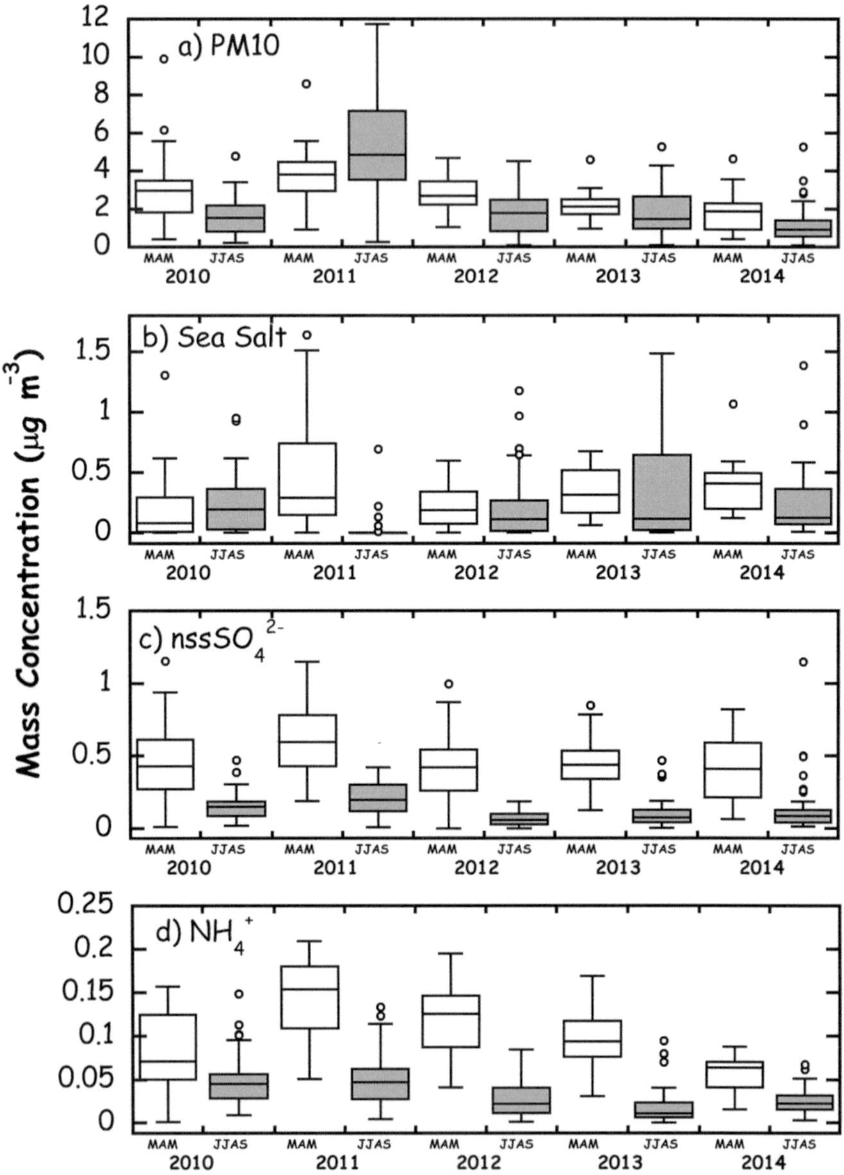

Fig. 4.25 Box-plot distribution of the ionic components of the spring (white) and summer (grey) PM10 aerosol samples collected during the 2010–2014 campaigns conducted at Thule (Greenland)

of 5.31 ± 2.77 μg m^{-3}, which was appreciably higher than the spring average value of 3.65 ± 1.46 μg m^{-3}. The PM10 spring mean values varied between 1.94 ± 1.16 μg m^{-3} in 2014 and 3.65 ± 1.46 μg m^{-3} in 2011, whereas the summer averages ranged from 1.16 ± 1.00 μg m^{-3} in 2014 to 5.31 ± 2.77 μg m^{-3} in 2011. With the exception of the 2011 summer concentration, these values resulted to be slightly lower than those measured at Ny-Ålesund, where the spring concentration values were found to range between 2.83 and 4.39 μg m^{-3}, and the summer values between 1.46 and 2.28 μg m^{-3}. In summer, we have observed a higher variability in 2011, 2012 and 2013. The data dispersion reached its maxima in 2011 for the summer samples and in 2010 for the spring samples.

The inter-annual temporal trend was similar in both seasons: the mean concentration was found to increase from 2010 to 2011 and then to continuously and progressively decrease from 2011 to 2014. As already noted, the seasonal mean concentrations of PM10 measured at Thule were found to be significantly lower than those measured at Ny-Ålesund, where the PM10 multi-annual values ranged between 2.8 and 4.3 μg m^{-3} in spring and between 1.5 and 2.3 μg m^{-3} in summer.

(B) **Sea Salt**

Part (b) of Fig. 4.25 shows the box-plot distribution of the spring and summer mean concentrations of sea salt measured on the samples collected at Thule in the 2010–2014 annual campaigns. The multi-annual time-patterns of this concentration parameter did not show significant seasonal variations in both spring and summer periods, during which very high data dispersion characteristics were observed, with large differences between the mean and median values, and high standard deviation values. Such data varied appreciably year-by-year, with maxima in spring 2011 and summer 2013. Spring mean concentrations were found to be usually higher than the summer values, with the exception of the samples collected in 2010 and 2013, which showed summer mean concentrations slightly higher than in the other years, together with higher data dispersions.

Sea salt spring mean concentration was found to vary between 0.24 ± 0.43 μg m^{-3} in 2010 and 0.54 ± 0.60 μg m^{-3} in 2011, while the summer mean values ranged from 0.04 ± 0.14 μg m^{-3} in 2011 to 0.35 ± 0.46 μg m^{-3} in 2013. These values are significantly lower than those measured at Ny-Ålesund, which were found to range from 0.65 ± 0.82 μg m^{-3} in spring 2014 to 0.95 ± 1.29 μg m^{-3} in spring 2015, and from 0.44 ± 0.54 μg m^{-3} in summer 2014 to 0.61 ± 0.97 μg m^{-3} in summer 2015.

(C) **Non-sea-salt sulfate (nss-SO_4^{2-})**

Part (c) of Fig. 4.25 shows the statistical distribution of the spring and summer mean concentrations of non-sea-salt sulfate measured on the samples collected at Thule in the 2010–2014 annual campaigns. For this PM10 component, a strong seasonal variation in the time-patterns is clearly evident, since the spring mean values are 2–3 times higher than summer mean concentrations. Therefore, the contribution given by the Arctic Haze transport events significantly affect the atmospheric concentration of nss-SO_4^{2-} at Thule in spring (inversely to the small overall contribution given by Arctic Haze to the PM10 mass concentration in such seasonal period, being the spring values only slightly higher than those measured in summer). The multi-annual time-variations of nss-SO_4^{2-} did not show a particular trend. In spring, the mean concentration of nss-sulfate was found to vary around 0.4–0.5 μg m^{-3}, with the exception of 2011 samples, which showed significantly higher mean values (0.61 $\pm$ 0.26 μg m^{-3}). In summer, the mean concentration was found to be of around 0.1 μg m^{-3}, showing the highest value of 0.21 $\pm$ 0.12 μg m^{-3} in 2011. The spring nss-SO_4^{2-} mean concentration was found to vary between 0.41 $\pm$ 0.22 μg m^{-3} in 2012, and 0.61 $\pm$ 0.26 μg m^{-3} in 2011, while the summer mean ranged from 0.07 $\pm$ 0.05 μg m^{-3} in 2012 to 0.21 $\pm$ 0.12 μg m^{-3} in 2011. These values are very similar to those measured at Ny-Ålesund, found to range in spring from 0.41 $\pm$ 0.31 μg m^{-3} in 2014 to 0.65 $\pm$ 0.40 μg m^{-3} in 2012, and in summer from 0.08 $\pm$ 0.01 μg m^{-3} in 2015 to 0.15 $\pm$ 0.18 μg m^{-3} in 2014.

(D) **Ammonium**

Part (d) of Fig. 4.25 shows the statistical distribution of spring and summer mean concentrations of ammonium measured on the samples collected at Thule in the 2010–2014 annual campaigns. The multi-annual time-patterns of ammonium seasonal concentration show similar features in spring and summer. The average concentration was found to reach a maximum in 2011 and then to continuously and progressively decrease up to reach rather low values in 2013 (summer samples) and in 2014 (spring samples). This seasonal variation is more evident in the spring samples, for which the highest data dispersion was observed in the 2010–2012 years. The seasonal variations are well evident, since the spring mean values were always higher than the summer average concentrations. Similarly to nss-SO_4^{2-}, the spring samples showed mean concentrations 2–3 times higher than those measured in summer. In particular, the spring mean concentration measured in 2013 (0.097 $\pm$ 0.036 μg m^{-3}) was found to be 5.2 times higher than the summer mean concentration measured in 2014 (0.018 $\pm$ 0.021 μg m^{-3}).

The spring mean concentration was found to range between 0.057 $\pm$ 0.022 μg m^{-3} in 2014 and 0.14 $\pm$ 0.05 μg m^{-3} in 2011, while the summer mean values varied between 0.018 $\pm$ 0.021 μg m^{-3} in 2014 and 0.051 $\pm$ 0.033 μg m^{-3} in 2010. These values are similar to those measured at Ny-Ålesund, found to vary in spring from 0.046 $\pm$ 0.042 μg m^{-3} in 2010 to 0.12 $\pm$ 0.09 μg m^{-3} in 2014, and in summer from 0.030 $\pm$ 0.029 μg m^{-3} in 2012 to 0.045 $\pm$ 0.036 μg m^{-3} in 2014.

(E) **MSA**

Part (a) of Fig. 4.26 shows the statistical distribution of spring and summer mean concentrations of MSA measured analyzing the samples collected at Thule in the 2010–2014 annual campaigns. The MSA mean concentration of PM10 aerosol particles at Thule was found to exhibit evident seasonal time-variations, with higher values in summer than in spring, except 2012, when the spring and summer mean values were found to be quite similar, being equal to 10.0 $\pm$ 12.0 ng m^{-3} in spring and 9.5 $\pm$ 8.0 ng m^{-3} in summer. In the other years, the mean spring values were found to account for 25–60% of the summer means.

The spring mean concentration of MSA was evaluated to range between 10.0 ± 12.0 ng m^{-3} in 2012 and 1.9 ± 1.0 ng m^{-3} in 2014. The median values, which can be taken as representative of back-ground levels, showed different time-patterns, with very low values observed in 2013 (1.6 ng m^{-3}) and in 2014 (1.8 ng m^{-3}), and a maximum in 2010 (6.2 ng m^{-3}). Considering both the mean and median values, a gradual lowering trend of the MSA concentration was detected from 2010 to 2014. The summer mean values were also observed to reach a minimum in 2014 (7.8 ± 5.8 ng m^{-3}), similarly to the spring mean concentrations described above, and show a maximum in 2010, equal to 17.0 ± 14.0 ng m^{-3}. The related data dispersion was found to exhibit the same trend. The median values showed a maximum in 2010 (11.0 ng m^{-3}) and a minimum in 2014 (6.0 ng m^{-3}), even though their values are quite similar along the overall multi-year sampling period. Similarly to the spring averages and relative standard deviations, the summer averages appear to be also characterized by a decreasing trend from 2010 to 2014. The PM10 spring and summer mean concentrations of MSA detected at Thule are significantly lower than those measured at Ny-Ålesund, which were found to vary in spring from 5.7 ± 7.0 ng m^{-3} in 2014 to 31.0 ± 40.0 ng m^{-3} in 2013, and in summer from 10.0 ± 13.0 ng m^{-3} in 2015 to 24.0 ± 26.0 ng m^{-3} in 2014.

(F) **Nitrate**

Part (b) of Fig. 4.26 shows the statistical distribution of spring and summer mean concentrations of nitrate measured on the samples collected at Thule in the 2010–2014 annual campaigns. Nitrate concentration shows markedly higher mean values in spring than in summer over the whole 2010–2014 period, with spring averages 2 to 4 times higher than the summer ones. The median values were also found to exhibit the same behaviour, with a similar spring-to-summer ratio. In spring, the mean concentration values ranged from 0.057 ± 0.045 μg m^{-3} in 2014 to 0.105 ± 0.138 μg m^{-3} in 2010. The median values showed slightly lower values than the average ones, with the lowest value (0.037 μg m^{-3}) measured in 2014 and the highest value (0.088 μg m^{-3}) found in 2013.

As mentioned above, the summer mean concentrations were lower than the spring ones, showing averages ranging from 0.017 ± 0.015 μg m^{-3} in 2012 to 0.046 ± 0.027 μg m^{-3} in 2011. Median values showed similar intra-annual patterns, reaching very low values in both 2012 and 2014 (0.014 and 0.013 μg m^{-3}, respectively) and the maximum of 0.042 μg m^{-3} in 2011. Concentration variability (expressed in terms of standard deviations) was also found to be wider in spring than in summer, with the exception of 2013, when data dispersion was found to be comparable in the two seasons. The inter-annual time-patterns of the nitrate seasonal concentrations did not show any clear temporal trend over the 2010–2014 period, neither in spring nor in summer. The nitrate spring mean concentration values obtained at Thule were about 2 times higher than those measured at Ny-Ålesund in the same season, found to vary from 0.039 ± 0.029 μg m^{-3} in 2015 to 0.049 ± 0.022 μg m^{-3} in 2012. Similarly, the summer mean concentration values were also found to be significantly lower than those measured at Ny-Ålesund in the

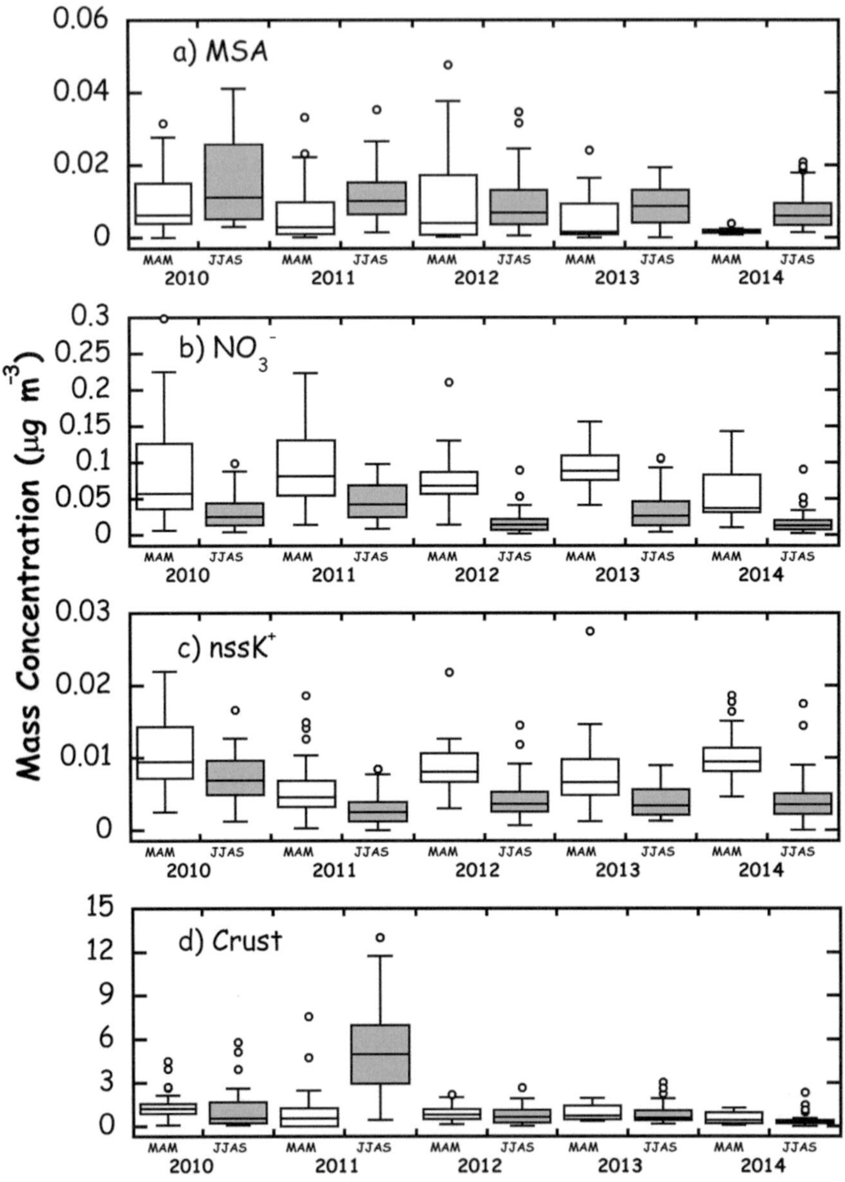

Fig. 4.26 Box-plot distribution of the ionic components of the spring (white) and summer (grey) PM10 aerosol samples collected during the 2010–2014 campaigns conducted at Thule (Greenland)

same season, which were found to range from 0.022 ± 0.026 μg m^{-3} in 2015 to 0.043 ± 0.034 μg m^{-3} in 2014.

(G) **Non-sea-salt potassium (nss K^+)**

Part (c) of Fig. 4.26 shows the statistical distribution of spring and summer mean concentrations of non-sea-salt potassium measured on the samples collected at Thule in the 2010–2014 annual campaigns. The nss-K^+ concentrations in PM10 collected at Thule show always higher values in spring than in summer, both as mean and median values, with similar trends to those observed at Ny-Ålesund. This behaviour can be associated to the higher intensity of the transport processes occurring during the spring months, causing more marked concentrations of both natural and anthropogenic nss-K^+ in the Arctic atmosphere. The average spring concentration of nss-K^+ ranges between 6.0 ± 4.4 ng m^{-3} measured in 2012 and 11.0 ± 6.6 ng m^{-3} observed in 2011. The annual median values were found to span a similar range from 4.6 ng m^{-3} in 2012 to 9.5 ng m^{-3} in 2013. The mean summer concentration values of nss-K^+ decrease by a factor of about 2 with respect to spring (similarly to the observations made at Ny-Ålesund), with the lowest average value measured in 2012 (2.7 ± 1.9 ng m^{-3}) and the highest value measured in 2011 (7.2 ± 3.5 ng m^{-3}). Data dispersion is always larger in spring than in summer, with spring standard deviations ranging from 3.5 ng m^{-3} (in 2013) to 6.6 ng m^{-3} (in 2011), while the summer standard deviations vary from 1.9 ng m^{-3} (in 2012) to 3.5 ng m^{-3} (in 2011). No clear time-patterns of the inter-annual variability were observed in the 2010–2014 period, since both spring and summer values usually reached a minimum in 2012 and a maximum in 2013. The Thule PM10 nss-K^+ spring and summer mean concentrations are very similar to those measured at Ny-Ålesund, where mean values were measured ranging in spring from 5.4 ± 5.3 ng m^{-3} (in 2013) to 14.0 ± 10.0 ng m^{-3} (in 2015), and in summer from 2.1 ± 1.4 ng m^{-3} (in 2012) to 8.8 ± 11.0 ng m^{-3} (in 2014).

(H) **Mineral dust**

Part (d) of Fig. 4.26 shows the statistical distribution of the spring and summer mean concentrations of the mineral dust measured by analyzing the samples collected at Thule in the 2010–2014 annual campaigns. The multi-annual time-patterns of mineral dust concentration yield spring and summer mean concentrations that maintain quite constant values (around 1–2 μg m^{-3}), with very limited variations of these data in all the summer samples, except those of summer 2011. In this year, the highest value of the average concentrations was measured, associated with the highest seasonal mean value, equal to 5.40 ± 3.53 μg m^{-3} and with a maximum concentration of about 13 μg m^{-3}. Relatively high data dispersions were also measured in summer 2010 (when a seasonal mean value of 1.19 ± 1.48 μg m^{-3} was measured), and in spring 2011 (with a mean value of 0.93 ± 1.40 μg m^{-3}). No clear seasonal variations were observed over the 5-year period, except in summer 2011, the spring and summer mean concentrations being found to be very similar, with only a light predominance of the spring concentration values.

In fact, the spring mean concentrations of mineral dust in PM10 were found to range between 0.57 ± 0.42 $\mu g\ m^{-3}$ in 2014 and 1.37 ± 0.96 $\mu g\ m^{-3}$ in 2010. Summer average values, excluding 2011 samples, were found to lie within the range from 0.39 ± 0.43 $\mu g\ m^{-3}$ (measured in 2014) to 1.19 ± 1.48 $\mu g\ m^{-3}$ (in 2010). Therefore, the spring values observed at Thule were found to be about twice the mean concentrations measured at Ny-Ålesund during the same period, which were found to vary from 0.28 ± 0.23 $\mu g\ m^{-3}$ (in 2010) to 0.77 ± 0.44 $\mu g\ m^{-3}$ (in 2014), whereas the summer values ranged from 0.21 ± 0.20 $\mu g\ m^{-3}$ (in 2011) to 0.58 ± 0.83 $\mu g\ m^{-3}$ (in 2014).

The spring and summer average mass concentrations of PM10 and of its main chemical components measured during the 2010–2014 annual campaigns carried out at Thule are given in Table 4.4. PM10 mean concentration was equal to 2.37 $\mu g\ m^{-3}$ in spring (with a relative standard deviation of 41%) and 1.60 $\mu g\ m^{-3}$ in summer (with a relative standard deviation of 70%). The three prevailing PM10 components in both seasonal periods were estimated to be: (i) mineral dust (0.839 $\mu g\ m^{-3}$ in spring and 0.679 $\mu g\ m^{-3}$ in summer); (ii) sea spray (0.322 $\mu g\ m^{-3}$ in spring and 0.278 $\mu g\ m^{-3}$ in summer); and (iii) nss-sulfate (0.430 $\mu g\ m^{-3}$ in spring and 0.110 $\mu g\ m^{-3}$ in summer). Gradually lower concentrations were measured in spring for ammonium (0.097 $\mu g\ m^{-3}$), nitrate (0.077 $\mu g\ m^{-3}$), nss-K^+ (8.2 ng m^{-3}), and MSA (7.1 ng m^{-3}), and in summer for ammonium (0.024 $\mu g\ m^{-3}$), nitrate ions (0.023 $\mu g\ m^{-3}$), MSA (8.8 ng m^{-3}) and nss-K^+ (3.7 ng m^{-3}).

4.2.3.1 Evaluation of the Optical Characteristics and Chemical Composition Features of Aerosol Particles at Thule

A Cimel sun-photometer of the AEerosol RObotic NETwork (AERONET) was regularly employed since 2007 at the Thule station, located in north-west Greenland, to collect measurements of direct sun irradiance at various visible and near-infrared wavelengths and simultaneous sky radiance measurements, from which the scattering and absorption properties of columnar aerosol were evaluated by Muscari et al. (2014). Examining the multi-year data-sets of these sun-photometric measurements, Tomasi et al. (2015) found that the monthly mean values of $AOT(0.50\ \mu m)$ measured at this remote site decrease slowly in general during the spring and summer seasons, from about 0.10 in April–May to around 0.05 in June–September (with standard deviations close to ±0.03). Correspondingly, the Ångström wavelength exponent $\alpha(0.40–0.87\ \mu m)$ was found to exhibit rather stable monthly mean values, ranging from 1.40 to 1.50 from March to September (with standard deviations close to ±0.3). The winter – spring relative frequency histogram of $AOT(0.50\ \mu m)$ was estimated to have a mean value of 0.093 and an asymmetric shape whose long-tailed right-hand wing is influenced by the rather frequent occurrences of Arctic haze episodes. The summer–autumn relative frequency histogram of $AOT(0.50\ \mu m)$ was found to exhibit more symmetric features, with mean and median values equal to 0.058 and 0.049, respectively. Very similar shapes of both seasonal relative frequency histograms of the Ångström exponent $\alpha(0.40–0.87\ \mu m)$ were obtained,

Table 4.4 Values of the three-year average mass concentrations of PM10 aerosol particles and their components (measured in μg m^{-3}), as determined with their standard deviations by Udisti et al. (2016) during the spring (March–May) and summer (June–September) seasons at Thule (Greenland) during the three campaigns conducted in 2012, 2013 and 2014

Seasonal period	PM10	Sea spray	nss-SO_4^{2-}	NH_4^+	MSA	NO_3^-	nss-K^+	Mineral dust
Spring (MAM)	2.371 = 0.971	0.322 ± 0.309	0.430 ± 0.209	0.0971 ± 0.0418	0.0071 ± 0.0098	0.0768 ± 0.0392	0.0082 ± 0.0045	0.839 ± 0.530
Summer (JJAS)	1.600 = 1.115	0.278 ± 0.377	0.110 ± 0.176	0.0236 ± 0.0198	0.0088 ± 0.0065	0.0233 ± 0.0220	0.0037 ± 0.0027	0.679 ± 0.586

with mean values equal to 1.38 in winter-spring and 1.41 in summer-autumn, indicating that no relevant seasonal changes in the aerosol size-distribution curves occurred on average from one period to the other of the year. Optical measurements of the absorption properties of aerosol particles were not directly performed at Thule. In order to try some general indications on the aerosol absorption properties at Thule, it is useful to consider some remarks of Moroni et al. (2017), who noted that important loads of forest fire particles are often transported over whole Arctic Ocean area from Alaska and Northern America to reach the Greenland region and subsequently the Svalbard Islands, as occurred in July 2015.

The Thule Air Base is located at 76° 32′ N latitude and 68° 42′ W longitude, with a distance of around 570 km from Alert (82° 30 N, 68° 20′ W). Therefore, the Alert and Thule sites were probably both involved during such events by common transport episodes of combustion particles from the continental areas of Nunavut and Alaska in Northern America. On the basis of these remarks, it seems reasonable to assume that the monthly mean and seasonal average values of the BC and OC concentrations measured at Alert (over the period from 1981 to 2007) could be representative of the seasonal time-patterns of these mass concentration parameters occurred at Thule, although reduced by a certain percentage (of ~ 20%) with respect to the Alert data, because of the dispersion effects on the pollutant concentrations during such transport episodes. At Alert, we have considered the following two sets of measurements: (a) those performed by Sharma et al. (2002, 2006) during the periods from November 1998 to May 1999, and from June to August in 1999; and (b) those carried out at Alert by Gong et al. (2010) over the multi-year period from 1981 to 2007. The time-patterns of the monthly mean mass concentrations of equivalent BC and OC mass concentrations derived from the Sharma et al. (2002, 2006) measurements provided values of BC mass concentration of around 90 ng m^{-3} in March and April, a lower value in May (close to 55–60 ng m^{-3}), and values ranging between 15 and 8 ng m^{-3} in the summer months. The monthly mean values of equivalent BC concentration derived from the Gong et al. (2010) measurements were found to be: (i) higher than 100 ng m^{-3} during the first three months of the year; (ii) rapidly decreasing in March and April; (iii) equal to around 30 ng m^{-3} in May; (iv) ranging between 10 and 20 ng m^{-3} in summer, from June to October; and (v) increasing to nearly 80 ng m^{-3} in December. Taking into account that the decrease effects observed by Sinha et al. (2017) over the last 10 years have probably involved the whole area including the Alert and Thule stations, it seems reasonable to assume that the BC mass concentration measured at Thule: (i) varied between 72 and 94 ng m^{-3} in January and February; (ii) decreased from 95 to 56 ng m^{-3} in the spring months; (iii) ranged between 15 ng m^{-3} in June and 6 ng m^{-3} in September; and (iv) gradually increased in October and November, until reaching a value of nearly 40 ng m^{-3} in the last month of the year. Similarly, on the basis of analogous assumptions, the OC mass concentration at Thule was evaluated to: (i) increase from around 70 ng m^{-3} in January to about 170 ng m^{-3} in March; (ii) assume an average value close to 250 ng m^{-3} in April; (iii) gradually decrease from May to October, until reaching a value close to 20 ng m^{-3} in October; and (iv) increase again in November and December, when a monthly mean value of around 45 ng m^{-3} was reached.

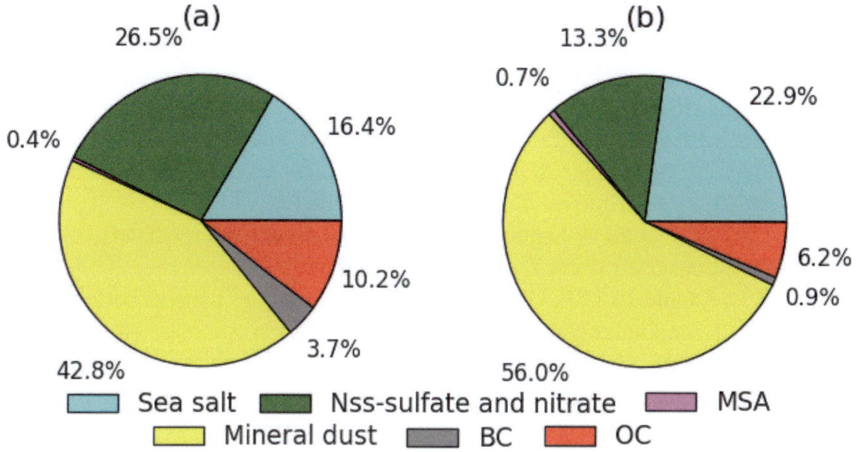

Fig. 4.27 Average composition diagrams of the ground-level particulate matter sampled at Thule (North-western Greenland) during different seasons and pertaining to: (**a**) PM10 sampled during the spring months (March–May); and (**b**) PM10 sampled during the summer months (June–September). Different colours are used to indicate the main particulate matter constituents (sea salt, nss-sulfate and nitrate, methanesulfonic acid (MSA), mineral dust, black carbon (BC), and organic carbon (OC)

Analyzing the chemical composition characteristics of PM10 particles sampled at Thule during the five annual campaigns from 2010 to 2014, we have determined the percentage mass fractions of the main components in spring (from March to May) and summer (from June to September), finding that the total average mass concentration of PM10 was equal to: (i) 1.96 $\mu g\ m^{-3}$ in spring, the mass percentages being of 42.8% for mineral dust, 26.5% for sulfate and nitrate, 16.4% for sea salt, 10.2% for OC, 3.7% for BC, and 0.4% for MSA; and (ii) 1.21 $\mu g\ m^{-3}$ in summer, with mass percentages of 56.0% for mineral dust, 22.9% for sea salt, 13.3% for sulfate and nitrate, 6.2% for OC, 0.9% for BC, and 0.7% for MSA. The graphical representation of these chemical composition characteristics is given in Fig. 4.27 for the PM10 particles sampled at Thule during the spring and summer periods.

4.2.4 Measurements of Sub-Micrometric Aerosol Chemical Composition in the Central Arctic Ocean

Measurements of sub-micrometric aerosol chemical composition were conducted by Chang et al. (2011) over the central Arctic Ocean from 5 August to 8 September 2008, as a part of ASCOS. This expedition took place on the Swedish icebreaker *Oden* from 2 August to 9 September 2008, starting from Longyearbyen (Spitsbergen, Svalbard) with an open water station (OW1) on 3 August 2008 (from 0:00 to

12:00 UTC), followed by a 24-hour station in the marginal ice zone (MIZ1) starting on 4 August 2008 (12:00 UTC), located at 79.9 °N latitude and 6.1 °E longitude. The ship headed north through the Arctic Ocean drifting pack ice as far as 87.4 °N, 1.5 °W on 12 August 2008, when an ice camp was established on an ice floe (IF). The *Oden* drifted with the ice for 21 days at the so-called IF station by sampling aerosol, and departed on 2 September 2008 to return southwards. A second marginal ice zone station (MIZ2) was performed on 6 September 2008 (09:00 UTC) to 7 September 2008 (04:00 UTC) at 80.7 °N latitude and 8.9 °E longitude, immediately followed by a final 12-hour open water station (OW2), ending on 7 September 2008 (16:00 UTC) at 80.4 °N and 10.1 °E. Overall, the cruise ranged over the area between 11.1 °W and 9.6 °E longitudes and between 77.9 °N to 87.5 °N latitudes. Therefore, the ASCOS measurements were performed for the larger part at the IF station, which moved in the meantime from 12 August to 2 September 2008 at longitudes varying between 11 °W and 0° and at latitudes from 87.1 to 87.5 °N.

A compact time-of-flight aerosol mass spectrometer (Aerodyne Research Inc.) was employed during the whole cruise to measure the sub-micrometric aerosol non-refractory chemical composition, obtaining the following estimates of the particle mass concentration (Table 4.5):

(1) The organic matter (OM) concentration was estimated to be equal to 0.264 μg m^{-3} during the whole ASCOS cruise measurement period, and with local average values of 0.624 μg m^{-3} at the MIZ1 site, 0.206 μg m^{-3} at the IF station, 0.361 μg m^{-3} at the MIZ2 station, and 0.206 μg m^{-3} at the OW2 site.
(2) The average value of nss-sulfate concentration was estimated to be 0.235 μg m^{-3} during the whole ASCOS measurement period, with local average values of 0.386 μg m^{-3} at the MIZ1 site, 0.202 μg m^{-3} at the IF station, 0.055 μg m^{-3} at the MIZ2 station, and 0.276 μg m^{-3} at the OW2 site.
(3) The average value of methanesulfonic acid (MSA) concentration was evaluated to be 0.040 μg m^{-3} during the whole ASCOS cruise, with local average values

Table 4.5 Sub-micrometric average mass concentrations (measured in μg m^{-3}) determined by Chang et al. (2011) (and assumed in the present study) during the whole ASCOS experimental campaign and at the MIZ1, IF, MIZ2 and OW2 stations

	Over the whole ASCOS campaign	MIZ1 station (79.9 °N, 6.1 °E)	IF station (87.1–87.5 °N, 11 °W - 0°)	MIZ2 station (80.7 °N, 8.9 °E)	OW2 station (80.4 °N, 10.1 °E)
Components	Concentration	Concentration	Concentration	Concentration	Concentration
Organic matter	0.2640	0.6240	0.2064	0.100	0.2064
Nss-sulfate	0.2346	0.3864	0.2024	0.0552	0.2760
MSA	0.0400	0.0950	0.0350	0.0900	0.0500
Nitrate	0.0206	0.0240	0.0192	0.0432	0.0288
Black carbon	0.0875	0.0900	0.0875	0.0860	0.0875

of 0.095 μg m^{-3} at the MIZ1 site, 0.035 μg m^{-3} at the IF station, 0.090 μg m^{-3} at the MIZ2 site, and 0.050 μg m^{-3} at the OW2 station.

(4) The average value of nitrate concentration was estimated to be 0.021 μg m^{-3} over the whole ASCOS measurement period, with local average values of 0.024 μg m^{-3} at the MIZ1 site, 0.019 μg m^{-3} at the IF station, 0.043 μg m^{-3} at the MIZ2 site, and 0.029 μg m^{-3} at the OW2 station.

(5) An average value of mineral dust concentration of 0.233 μg m^{-3} was assumed at all the measurement sites of the ASCOS cruise, according to the three-year average evaluations made by us at the GVB Observatory in August and early September days of the 2012–2014 years.

(6) An average value of sea salt concentration of 0.047 μg m^{-3} was assumed over the whole ASCOS measurement period on the basis of the measurements performed by us at the GVB Observatory during the August and early September period of the three years from 2012 to 2014.

(7) An average value of Black Carbon (BC) concentration equal to 0.034 μg m^{-3} was assumed over the whole ASCOS measurement period, by considering the measurements performed at the GVB Observatory in August and early September of the 2012–2014 years, from which local average values of BC were derived, equal to 0.035 μg m^{-3} at the MIZ1 site, 0.034 μg m^{-3} at the IF station, 0.033 μg m^{-3} at the MIZ2 site, and 0.034 μg m^{-3} at the OW2 station.

On the basis of the above ASCOS measurements (Chang et al. 2011) and our assumptions on the average mass concentration of mineral dust, sea salt and BC, the average composition diagrams of the sub-micrometric aerosol particles during the whole ASCOS experiment and at the four local ASCOS stations were obtained, as shown in Fig. 4.28 for the overall set of measurements collected during the whole campaign onboard the icebreaker *Oden*, as well as at the MIZ1 (79.9 °N, 6.1 °E), IF (from 11 °W to 0° longitude and from 87.1 to 87.5 °N latitude), MIZ2 (80.7 °N, 8.9 °E) and OW2 (80.4 °N, 10.1 °E) stations. The obtained data-sets allows us to point out the following remarks:

(a) Considering the data-set collected during the whole ASCOS measurement period, including both the remote Arctic Ocean area (where the IF measurements were performed) and the other areas close to the MIZ1, MIZ2 and OW2 sites (which are all not far from the Svalbard Archipelago), the mass percentages of organic matter (OM) and of nss-sulfate plus nitrate, as observed on average during the ASCOS measurement period, were found to be similar (30.2% and 29.2%, respectively). Similar features were also observed at the IF station, with relative mass percentages of OM and nss-sulfate (plus nitrate) equal to 26.5% and 28.5%, respectively. Conversely, different mass percentages were measured at the MIZ1, MIZ2 and OW2 sites closer to the Spitsbergen Island, ranging the mass concentration of OM from 23.6% (at the OW2 station) to 43.2% (at the MIZ2 site) and the mass concentration of nss-sulfate (plus nitrate) from 28.4% (at the MIZ1 site) to 34.8% (at the OW2 station).

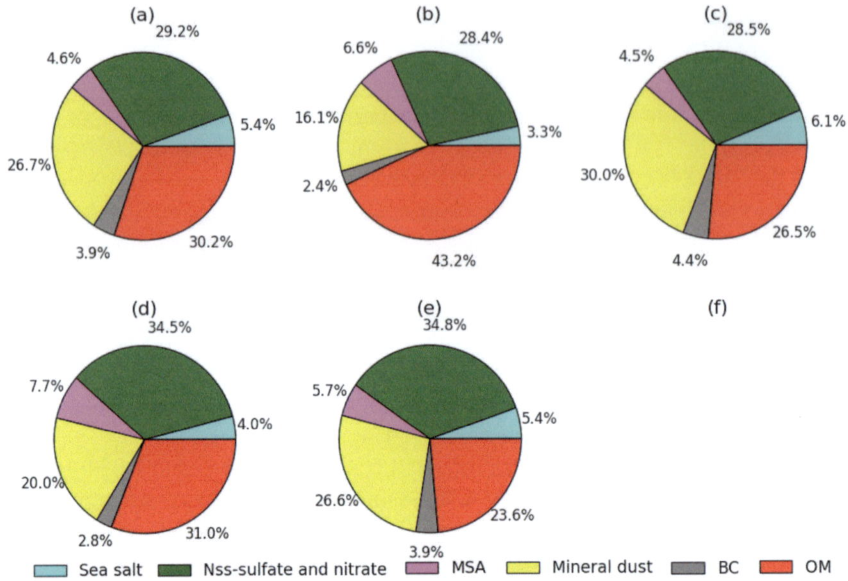

Fig. 4.28 Average composition diagrams of the sub-micrometric particles derived from the particle samples collected during the ASCOS experiment, conducted in the Central Arctic Ocean from 5 August to 8 September, 2008: (**a**) from the overall set of particle samples collected during the whole ASCOS experiment; (**b**) from the set of particle samples collected at the MIZ1 station; (**c**) from the set of particle samples collected at the IF station; (**d**) from the set of particle samples collected at the MIZ2 station; and (**e**) from the set of particle samples collected at the OW2 station. Different colours are used to indicate the main constituents of the sampled sub-micrometric aerosol particles (sea salt, nss-sulfate and nitrate, methanesulfonic acid (MSA), mineral dust, black carbon (BC), and organic matter (OM)

(b) The MSA mass percentage was estimated to be equal to 4.6% during the whole ASCOS measurement period, therefore very close to the average percentage of 4.5% estimated at the IF remote station, and appreciably lower than the percentages evaluated at the MIZ1 station (6.6%), MIZ2 station (7.7%) and OW2 station (5.7%). On the basis of the above results, the ASCOS chemical composition diagrams of sub-micrometric aerosol particles were derived, which are shown in Fig. 4.28, indicating that nss-sulfate (plus nitrate) and OM components contribute by about 29% and 30% of the aerosol mass detected by the aerosol mass spectrometer. The mass concentration range of nss-sulfate (from 0.202 to 0.386 $\mu g\ m^{-3}$) and MSA (from 0.035 to 0.095 $\mu g\ m^{-3}$) determined during the ASCOS experiment were estimated to agree with the measurements performed at other polar sites by: (i) Leck and Persson (1996) during the International Arctic Ocean Expedition (IAOE) carried out from August to mid-October 1991, yielding a concentration range of nss-sulfate from 3 10^{-3} to 0.66 $\mu g\ m^{-3}$ and a MSA concentration range from 2 10^{-4} to 0.13 $\mu g\ m^{-3}$; (ii) Kerminen and Leck (2001), during the AOE-96 Expedition carried out onboard the Swedish icebreaker *Oden* from July 20 to August 24, 1996, when a nss-sulfate concentration range from 3.6 10^{-3} to 1.185 $\mu g\ m^{-3}$ and a MSA concentration range from 2 10^{-3} to 0.104 $\mu g\ m^{-3}$ were estimated; (iii) Norman et al. (1999) at Alert (Nunavut, Canada), who found a nss-sulfate concentration range from 3 10^{-2} to 0.22 $\mu g\ m^{-3}$; (iv) Li and Barrie (1993) at Alert (Nunavut, Canada), who found a MSA range from 3 10^{-2} to 5 10^{-2} $\mu g\ m^{-3}$; and (v) Heintzenberg and Leck (1994), who analyzed a set of fine aerosol particle samples collected at the Zeppelin Observatory (near Ny-Ålesund) during the 26-month period from March 1990 to April 1992, determining average values of nss-sulfate and MSA concentrations of 0.39 $\mu g\ m^{-3}$ and 0.018 $\mu g\ m^{-3}$, respectively.

In general, higher aerosol loadings were measured during the two MIZ measurement periods for all the constituents than in the OW2 period, in accordance with the past Arctic Ocean expeditions. The relatively high MSA concentrations measured at the MIZ2 station in early September suggested that marine biology was still active on those days. No evidence was given by the present analyses to the hypothesis that other components would be present in the fine particulate matter besides organic matter, nss-sulfate, MSA, nitrate and ammonium.

The ratio of methanesulfonic acid to nss-sulfate (MSA/nss-$SO_4{}^{2-}$) calculated from the Marine Biogenic factor was estimated by Chang et al. (2011) to be equal to 0.25 ± 0.02, which is similar to other summertime values measured in polar regions, such as that calculated from sub-micrometric filter measurements during the IAOE-91 experiment (Leck and Persson 1996), excluding the influence of fog (which has an average ratio between MSA and nss-$SO_4{}^{2-}$ concentrations equal to 0.22). In addition, Chang et al. (2011) noted that the MSA/nss-$SO_4{}^{2-}$ ratio is within the lower bounds of uncertainty found by Norman et al. (1999) at Alert (Canada) in July and August of 1993 and 1994 (giving a value of such a ratio equal to 0.39 ± 0.21). Approximately 50% of the total nss-sulfate mass was apportioned to the marine biogenic factor, while the remaining 50% was attributed to the Continental factor, without percents ascribed to the Organic factor and the ship emission factor. The marine biogenic factor is dominated by the sulfate peaks. Since MSA is only significantly formed via the photo-oxidation of dimethyl sulfide (DMS), which mainly originates from biological activities, Chang et al. (2011) also estimated that the major part of the MSA mass (about 71%) is related to Organic factor, the inorganic component being attributed to secondary marine biogenic sources. The ship emission factor was found to present time-series generally characterised by low signals for the most part of the ASCOS campaign, with intermittent periods of high loading which coincide with times when the air was contaminated by ship exhaust.

The chemical composition determined by Chang et al. (2011) for sub-micrometric aerosol particles in the remote Arctic Ocean can be reliably completed with the average BC and OC mass concentrations measured by us at the GVB Observatory on the same days of the ASCOS campaign, from early August to half September of the 2012–2014 period. On the basis of these measurements, the BC and OC mass concentrations were assumed to be 4.5 10^{-3} and 9 10^{-3} ng m^{-3}, respectively, giving mass percentages of 3.9% and 7.8%. Lower percentages of BC (2.4%) and OC (3.5%) were estimated at the MIZ1 station, while: (1) relative mass concentrations of BC equal to 4.4% and of OC equal to 8.0% were attributed to the particle composition measurements conducted at the IF station; (2) percentages of BC equal to 2.8% and of OC equal to 5.9% were estimated at the MIZ2 station; and (3) percentages of BC equal to 3.9% and of OC equal to 6.8% were determined at the OW2 station. These evaluations of the BC mass concentration agree very well with the estimates obtained by Sinha et al. (2017) at Ny-Ålesund, showing that the monthly mean concentration of BC was estimated to be close to 5 ng m^{-3} in August 2008.

On the basis of the evaluations made by Chang et al. (2011), the total mass concentration of sub-micrometric particles, evaluated over the whole ASCOS measurement period, was estimated to be 0.87 μg m^{-3}, with mass percentages of 30.2% for Organic Matter (OM), 29.2% for sulfate (plus nitrate), 26.7% for mineral dust, 5.4% for sea salt, 4.6% for MSA, and 3.9% for BC, as shown in Fig. 4.28. The mass percentages of sea salt, sulfate (plus nitrate), mineral dust, MSA, OM and BC determined at the MIZ1, IF, MIZ2 and OW2 stations are given in Fig. 4.28.

4.2.5 Measurements of Aerosol Chemical Composition at Barrow (Alaska, USA)

Simultaneous measurements of aerosol chemical composition and volume scattering and absorption coefficients at the visible wavelength $\lambda = 0.55$ μm were conducted at Barrow (Alaska, USA) by Quinn et al. (2002) over the period from October 1997 to December 2000, for rather low relative humidity (RH) conditions of air during selected sampling periods, in which the samples of air were properly heated to achieve values of *RH* ranging between 20 and 40% (with seasonal average air temperatures of 31 ± 8 °C in spring, 14 ± 3 °C in summer, 25 ± 8 °C in fall, and 38 ± 7 °C in winter). The chemical composition of particulate matter has been analyzed, and the physical and optical measurements of the aerosol particles were all performed at these *RH* reference conditions. Thereupon, using the above measurements made at a low reference *RH* and appropriate growth factors for the various particulate components, Quinn et al. (2002) adjusted the measured parameters to a desired value of *RH*. In order to determine the aerosol chemical composition and the mass concentration, the air was sampled by using a Berner-type multi-jet cascade impactor (Berner et al. 1979) with aerodynamic D50 cut-off diameters of 10 and 1 μm. Using the techniques carefully described by Quinn et al. (2002), the samples were analyzed at the NOAA's Pacific Marine Environmental Laboratory (PMEL) to determine the concentrations of major cations (Na^+, $NH_4{}^+$, K^+, Mg^{2+}, Ca^{2+}) and anions (MSA^-, Cl^-, $NO_3{}^-$, $SO_4{}^{2-}$) with the ion chromatography technique (Quinn et al. 1998). In this analysis, the concentration of $NO_3{}^-$ was not measured during the various seasons, since it was taken into account that large uncertainties were often associated with heating of the aerosol particles to maintain the sample *RH* below 40%, and that such a heating may have caused the substantial volatilization of ammonium nitrate from the substrate, thus causing artificially low nitrate concentrations. Examining the main components of the particulate matter, Quinn et al. (2002) found the following experimental features:

(1) The non-sea-salt sulfate (nss-SO_4^{2-}) concentration was initially calculated from Na^+ concentrations by assuming that the mass ratio of sulfate to sodium in sea-water is 0.252. Negative nss-SO_4^{2-} concentrations resulting from this approach have been attributed to depletion of sea-salt SO_4^{2-} through fractionation processes, as also found by Norman et al. (1999) for an Antarctic winter aerosol case and for some sea-salt SO_4^{2-} depletion cases observed in the Canadian Arctic aerosol particles, on the basis of sulfur isotope analysis applications. In this case, the calculation of nss-SO_4^{2-} within the sub-micrometric size range using the SO_4^{2-}/Na^+ sea-water mass ratio of 0.252 did not yield negative values during observations. Following the method of Wagenbach et al. (1998), negative values of nss-SO_4^{2-} concentration were not obtained for both sub-micrometric particles and super-micrometric particles. The nss-SO_4^{2-} concentration of super-micrometric particles was calculated by plotting a linear regression of nss-SO_4^{2-} versus Na^+ for the winter, spring, and summer seasons, using the 0.252 mass ratio in all cases. The obtained negative slope was then added to the conventional ratio of 0.252, obtaining super-micrometric SO_4^{2-}/Na^+ sea-water mass ratios equal to: (i) 0.13 for the winter cases, (ii) 0.082 for the spring cases, and (iii) 0.23 for the summer cases.
(2) The sea-salt aerosol concentrations were evaluated by Quinn et al. (2002) by assuming that the sea salt concentration is given by the sum of the concentrations of Cl^- and Na^+, the latter concentration being multiplied by 1.47, which is the sea-water ratio between the total concentration given by the sum of Na^+, K^+, Mg^{2+}, Ca^{2+}, SO_4^{2-}, and HCO_3^- and that of Na^+ (Holland 1978). The concentration of Cl^- in excess of the Cl^-/Na^+ sea-water ratio of 1.8 was not added to the sea salt mass. This approach prevented from including the K^+, Mg^{2+} Ca^{2+} SO_4^{2-}, and HCO_3^- ions in the evaluation of the sea salt mass, and allowed for the loss of Cl^- mass through Cl^- depletion processes, also assuming that all Na^+ ions were derived from sea-water.

4.2.5.1 Determination of the Chemical Composition of Sub-Micrometric Particles at Barrow

The sub-micrometric aerosol mass concentrations of the various components were determined by Quinn et al. (2002) by weighing the Millipore filters at PMEL, before and after each sample collection, using a Mettler UMT2 microbalance. The super-micrometric aerosol mass concentrations were instead determined by weighing the Tedlar films with a Cahn Model 29 microbalance. The measurements of sub-micrometric aerosol mass concentrations included the water mass associated with the aerosol on the filter at the glove box RH. The monthly average values of the mass concentrations of the most important components of sub-micrometric particles are given in Table 4.6, all normalized to 0 °C air temperature and 1013 hPa air pressure, for the RH conditions of the various air samples. The ±95% uncertainties for the low and high mass concentrations estimated by Quinn et al. (2002) were respectively equal for the above-assumed normal temperature and pressure conditions to: (i) 0.6 ± 5.5% and 3.0 ± 5.2% μg m^{-3} for the total mass; (ii) 0.2 ± 9.6% and 1.0 ± 6.1% μg m^{-3} for sea salt; (iii) 0.05 ± 7.0% and 1.0 ± 6.0% μg m^{-3} for nss-SO_4^{2-}; (iv) 0.01 ± 39% and 0.2 ± 7.8% μg m^{-3} for NH_4^+; (v) 0.0005 ± 6.0% and 0.01 ± 6.0% μg m^{-3} for MSA; (vi) 0.001 ± 14% and 0.03 ± 6.1% μg m^{-3} for nss-K^+; (vii) 0.0005 ± 16% and 0.09 ± 6.2% μg m^{-3} for nss-Mg^{2+};.(viii) 0.004 ± 69% and 0.03 ± 48% μg m^{-3} for nss-Ca^{2+}; (ix) 0.3 ± 51% and 1.0 ± 21% μg m^{-3} for the residual mass (including errors due to the concentrations of the ionic species, the aerosol mass, and water); and (x) 0.11 ± 33% and 3.0 ± 5.3% μg m^{-3} for water (at $RH = 33\%$). These uncertainties in aerosol mass include the errors due to weighing, blank levels, storage and transport, and those made in estimating the volume of air sampled during the measurements. The residual mass concentrations are given by the overall mass of the chemically unanalyzed species, as calculated from the gravimetrically determined aerosol mass less the mass of the ionic species and water.

Since the chemical analysis of aerosol particle load was made by Quinn et al. (2002) analyzing only the ionic species, the measurements of the concentrations of these species and the total aerosol mass were used to determine the fraction of the aerosol mass. Thus, the mass fractions of the sub-micrometric ionic components were calculated by dividing the mass concentration of each component by the gravimetric mass, together with the percentiles for the sub-micrometric and super-micrometric mass fractions of the ionic mass (including associated water at $RH = 33\%$), sea salt, nss-SO_4^{2-}, H_2O associated with the ionic species at $RH = 33\%$, and the residual mass. The last quantity was estimated as the difference between the gravimetric mass and the overall mass of the ionic components and associated water. Sea salt, nss-SO_4^{2-}, and the residual component were found to dominate the aerosol mass, while: (i) the remaining ionic species (MSA, nss-K^+, nss-Mg^{2+}, and nss-Ca^{2+}) are useful tracers of the source of the aerosol and contribute less than 10% to the aerosol sub-micrometric mass; (ii) NH_4^+ yields relative contributions to the sub-micrometric mass that are on average equal to 6.5% (± 4.4%) during the winter (October to January), 11.0% (± 4.3%) during the spring (March to June), and 4.2% (± 2.9%) during the summer (July to September).

The ionic mass and associated water make up 80% to 100% of the sub-micrometric aerosol mass during the months of November to May, as shown in Fig. 4.29, in which sea salt and nss-SO_4^{2-} are the dominant species of the ionic mass. It can be seen that: (i) the sea salt mass fraction is larger than that of nss-SO_4^{2-} during the winter months (October to January); (ii) sea salt and nss-SO_4^{2-} mass fractions are comparable in February and March; and (iii) the nss-SO_4^{2-} mass fraction is larger than that of sea salt from April to June. Sea salt aerosol particles originate from wind-driven formation of sea spray droplets and their subsequent evaporation, so that sub-micrometric sea salt concentration increases

Table 4.6 Values of the mass concentrations of sub-micrometric aerosol particles (measured in $\mu g\ m^{-3}$), as obtained by Quinn et al. (2002) through the analysis of the particulate matter samples collected at Barrow (Alaska) from October 1997 to December 2007. The concentration data are reported with their arithmetic mean and standard deviation values determined for air temperature equal to 0 °C and air pressure equal to 1013 hPa

Month	Mass	Sea salt	Nss-SO_4^{2-}	NH_4^+	MSA	Nss-K^+	Nss-Mg^{2+}	nss-Ca^{2+}	Residual
January	3.80 ± 2.20	1.10 ± 1.10	0.78 ± 0.43	0.23 ± 0.09	$<10^{-4}$	0.03 ± 0.03	0.09 ± 0.11	0.03 ± 0.03	0.62 ± 1.00
February	3.50 ± 1.70	0.93 ± 1.10	0.91 ± 0.39	0.26 ± 0.09	$<10^{-4}$	0.03 ± 0.02	0.06 ± 0.07	0.02 ± 0.03	0.56 ± 0.57
March	2.60 ± 1.30	0.68 ± 0.73	0.71 ± 0.44	0.21 ± 0.12	$<10^{-4}$	0.02 ± 0.01	0.03 ± 0.04	0.01 ± 0.01	0.41 ± 0.69
April	2.10 ± 1.10	0.43 ± 0.61	0.72 ± 0.36	0.19 ± 0.07	$2\ 10^{-3} \pm 2\ 10^{-3}$	0.01 ± 0.01	$8\ 10^{-3} \pm 10^{-2}$	$3\ 10^{-3} \pm 5\ 10^{-3}$	0.32 ± 0.40
May	1.20 ± 0.60	0.15 ± 0.15	0.60 ± 0.50	0.14 ± 0.06	0.01 ± 0.01	$5\ 10^{-3} \pm 4\ 10^{-3}$	$2\ 10^{-3} \pm 3\ 10^{-3}$	$6\ 10^{-3} \pm 9\ 10^{-3}$	0.08 ± 0.10
June	0.89 ± 1.10	0.05 ± 0.10	0.19 ± 0.21	0.06 ± 0.05	0.01 ± 0.02	$2\ 10^{-3} \pm 2\ 10^{-3}$	$3\ 10^{-4} \pm 6\ 10^{-4}$	$4\ 10^{-3} \pm 8\ 10^{-3}$	0.45 ± 1.10
July	0.60 ± 0.49	0.08 ± 0.09	0.10 ± 0.10	0.03 ± 0.03	0.01 ± 0.01	$10^{-3} \pm 10^{-3}$	$<10^{-4}$	$5\ 10^{-3} \pm 10^{-2}$	0.27 ± 0.25
August	0.81 ± 0.82	0.17 ± 0.22	0.08 ± 0.06	0.02 ± 0.01	0.01 ± 0.01	$10^{-3} \pm 10^{-3}$	$<10^{-4}$	$10^{-3} \pm 2\ 10^{-3}$	0.40 ± 0.33
September	0.61 ± 0.44	0.17 ± 0.11	0.09 ± 0.12	0.02 ± 0.04	$7\ 10^{-3} \pm 4\ 10^{-3}$	$10^{-3} \pm 10^{-3}$	$<10^{-4}$	$2\ 10^{-3} \pm 5\ 10^{-3}$	0.26 ± 0.37
October	2.00 ± 1.60	0.74 ± 0.98	0.13 ± 0.16	0.05 ± 0.06	$9\ 10^{-4} \pm 10^{-3}$	$2\ 10^{-3} \pm 7\ 10^{-3}$	$3\ 10^{-4} \pm 10^{-3}$	0.02 ± 0.07	0.93 ± 1.20
November	1.40 ± 1.20	0.62 ± 0.86	0.13 ± 0.09	0.05 ± 0.04	$2\ 10^{-4} \pm 5\ 10^{-4}$	$3\ 10^{-3} \pm 6\ 10^{-3}$	$10^{-3} \pm 3\ 10^{-3}$	$9\ 10^{-3} \pm 2\ 10^{-2}$	0.30 ± 0.45
December	3.10 ± 3.00	1.10 ± 1.20	0.32 ± 0.23	0.09 ± 0.05	$10^{-4} \pm 3\ 10^{-4}$	0.01 ± 0.01	0.02 ± 0.02	$9\ 10^{-3} \pm 10^{-2}$	0.55 ± 0.44

in October, peaks in December and January, begins a steady decrease in February, and reaches its lowest concentration in the summer months from May to September (as can be seen in Fig. 4.29), the sub-micrometric particle concentration being equal to 1.1 $\mu g\ m^{-3}$ on average. The winter maximum characterizing the Arctic sub-micrometric sea salt concentration has been attributed to seasonally high winds in the high-latitude source regions of the Pacific and Atlantic Oceans and long-range transport to the Arctic (Sturges and Barrie 1988; Quinn et al. 2000). On the basis of the results shown in Fig. 4.29 for the sub-micrometric particles, the average mass concentration of ionic mass (including sea salt, nss-SO_4^{2-}, nss-K^+, nss-Mg^{2+}, nss-Ca^{2+}, NH_4^+, MSA, Cl^- not associated with sea salt, NO_3^-, and water associated with the ionic components at $RH = 33\%$) was found to be 0.79 $\mu g\ m^{-3}$ during the winter season (from October to May) and 0.49 $\mu g\ m^{-3}$ during the summer season (from June to September), while that of sea salt was found to be 0.32 $\mu g\ m^{-3}$ in the 8-month winter season and 0.18 $\mu g\ m^{-3}$ in the 4-month summer season. Calculated on the basis of the results shown in Fig. 4.29, the average seasonal concentration of sub-micrometric nss-SO_4^{2-} ions was estimated to be 0.30 $\mu g\ m^{-3}$ in the long winter season and 0.18 $\mu g\ m^{-3}$ in summer, while the water mass content was evaluated to be 0.14 $\mu g\ m^{-3}$ in winter and 0.15 $\mu g\ m^{-3}$ in summer, and the mass concentration of residual components was estimated to be 0.19 $\mu g\ m^{-3}$ in winter and 0.45 $\mu g\ m^{-3}$ in summer.

Non-sea-salt (nss) SO_4^{2-} concentration measured in the Arctic region is originated by several sources, since: (i) marine biogenic nss-SO_4^{2-} is due to the oxidation of atmospheric dimethyl sulfide (DMS) which, in turn, results from oceanic phytoplankton processes; and (ii) nss-SO_4^{2-} ions also originate from anthropogenic sources, such as the burning of fossil fuels and smelting of sulfide ores in Eurasia. As can be seen in Fig. 4.30, the concentration of sub-micrometric nss-SO_4^{2-} ions observed at Barrow begins to increase in December, and describes a broad peak from January to May, followed by a marked drop in June and subsequent low values from July to November. The broad peak observed from January to May is due to several factors, such as the long-range transport of anthropogenic primary nss-SO_4^{2-}, the long-range transport of anthropogenic SO_2 that is then photo-oxidized to nss-SO_4^{2-} as the light levels increase, and the local production of biogenic nss-SO_4^{2-} from the oxidation of DMS. Examining the time-patterns shown in Fig. 4.30, the following evaluations of the average seasonal concentrations were estimated for the various components: (i) total mass of sub-micrometric particles was equal to 2.36 $\mu g\ m^{-3}$ in the winter season and 0.67 $\mu g\ m^{-3}$ in summer; (ii) the average sub-micrometric sea salt concentration was 0.77 $\mu g\ m^{-3}$ in winter and 0.22 $\mu g\ m^{-3}$ in summer; and (iii) the average concentration of sub-micrometric nss-SO_4^{2-} was 0.51 $\mu g\ m^{-3}$ in winter and 0.17 $\mu g\ m^{-3}$ in summer. Studying these processes, Ferek et al. (1995) measured the presence of DMS in Arctic waters under the ice as early as April, suggesting that DMS may be converted to nss-SO_4^{2-} as the ice recedes, so contributing to late spring concentrations. High nss-SO_4^{2-} concentrations in May coincide with an increase in MSA concentration, this species being of pure biogenic origin and contributing in this way to yield the high nss-SO_4^{2-} concentration levels observed in May.

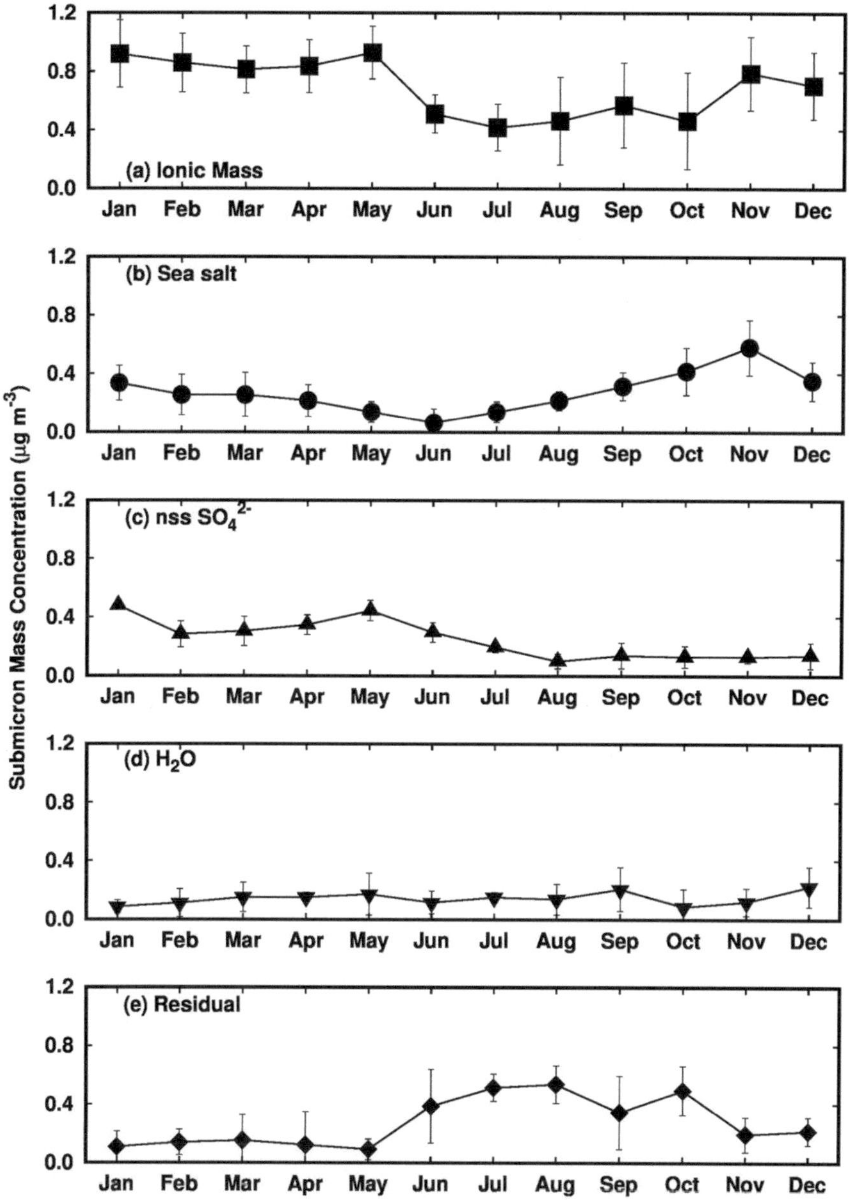

Fig. 4.29 Time-patterns of the monthly mean values of the mass concentration components of sub-micrometric particles measured by Quinn et al. (2002) at Barrow (Alaska): (**a**) ionic mass (including sea salt, nss $SO_4{}^{2-}$ ions, nss K^+ ions, nss Mg^{2+} ions, nss Ca^{2+} ions, $NH_4{}^+$ ions, MSA^- ions, Cl^- ions not associated with sea salt, $NO_3{}^-$ ions, and water associated with the ionic components at 33% RH); (**b**) sea salt; (**c**) nss $SO_4{}^{2-}$ ions; (**d**) water associated with the ionic components at 33% RH; and (**e**) residual mass (calculated as the difference between the gravimetric mass and the mass of ionic species and associated water). The vertical bars denote the standard deviations of the various mass concentration parameters

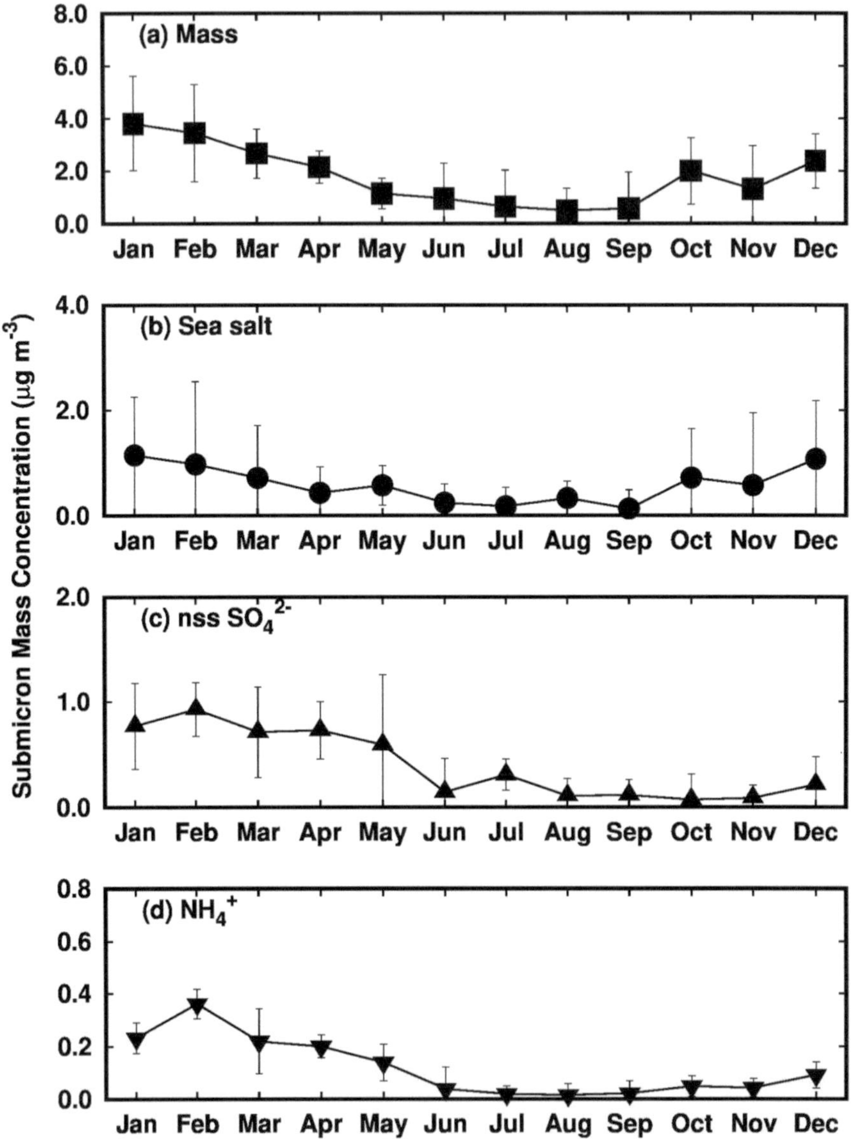

Fig. 4.30 Time-patterns of the monthly mean values of the mass concentration components of sub-micrometric particles measured by Quinn et al. (2002) at Barrow (Alaska): (**a**) gravimetrically determined aerosol mass, (**b**) sea salt, (**c**) nss SO_4^{2-} ions, and (**d**) NH_4^+ ions. The vertical bars denote the standard deviations of the various mass concentration parameters

The sub-micrometric concentration of NH_4^+ was found to exhibit the highest values of the year from January to May, as can be noted in Fig. 4.30, and subsequently to describe a minimum in summer, from June to September, providing an average concentration of NH_4^+ of 0.17 $\mu g\ m^{-3}$ in winter and 0.023 $\mu g\ m^{-3}$ only in summer. In addition, Fig. 4.31 shows that the sub-micrometric concentration of MSA exhibits the highest values (> 10^{-2} $\mu g\ m^{-3}$) in late spring and summer (from May to August). On this matter, it is worth to notice that the seasonal cycle of MSA concentration was observed at Barrow to start a month later than at Alert, and to be followed by peaked monthly mean values in June and August, so that the ratio between MSA and nss-SO_4^{2-} concentrations measured at Barrow resulted to be equal to 0.08 ± 0.09 during the first MSA peak in June, and to 0.18 ± 0.08 during the second peak in August. As a result of these variations, the average seasonal mass concentration of sub-micrometric MSA was estimated to be 3.78 ng m^{-3} in winter and 16.7 ng m^{-3} in summer.

Non-sea-salt K^+ in sub-micrometric particles is a useful tracer of aerosol derived from biomass burning processes. At Barrow, the sub-micrometric nss-K^+ concentration was observed to start to increase in December, present a peak in January, and drop off from February to April, showing the lowest values over the period from May to November. Therefore, the evolutionary time-patterns of nss-K^+ are typical of other species transported over long distances during the winter/spring haze season. Quinn et al. (2002) found that the linear regression of sub-micrometric nss-K^+ versus nss-SO_4^{2-} yields a poor correlation, substantially confirming that these ionic species are derived from different sources, since nss-SO_4^{2-} is primarily produced by fossil fuel combustion and smelting of sulfide ores during this time of year, while nss-K^+ results from biomass burning. The average seasonal concentration of sub-micrometric nss-K^+ was evaluated to be equal to 13.2 ng m^{-3} in winter and to 1.2 ng m^{-3} only in summer.

As shown in Fig. 4.31, the monthly mean concentrations of sub-micrometric nss Mg^{2+} and Ca^{2+} describe seasonal cycles that exhibit the highest values of both ions during the period from December to March, suggesting that these species are both transported from Eurasia over long distances. In fact, soil dust from Asia has been observed in haze layers over Alaska (Rahn et al. 1977) and the North Pacific (Uematsu et al. 1985) during the spring months. Elevated aerosol Ca^{2+} concentrations measured in the Norwegian Arctic during an episode of long-range transport of pollutants occurred in March 1983 from the former Soviet Union were attributed to metal emissions from coal combustion (Pacyna and Ottar 1989). Measurements of other trace elements found in soil dust (Al, Si, Ti, and Fe) would be useful in order to determine the winter/spring maximum sources of nss-Mg^{2+} and Ca^{2+} measured at Barrow, where these measurements provide rather low value during the summer months for both the above-cited sub-micrometric ions, indicating that the local sources of nss-Mg^{2+} and Ca^{2+} are very weak. The following average seasonal mass concentrations of sub-micrometric ions were obtained from the data-set shown in Fig. 4.31: (i) the average seasonal concentration of nss- Mg^{2+} is 31.4 ng m^{-3} in winter, and 5.6 ng m^{-3} in summer; (ii) the average seasonal concentration of nss-Ca^{2+} is 13.2 ng m^{-3} in winter and 0.22 $\mu g\ m^{-3}$ in summer; and (iii) the average concentration of nss-SO_4^{2-} is 0.51 $\mu g\ m^{-3}$ in winter and 3.8 ng m^{-3} in summer.

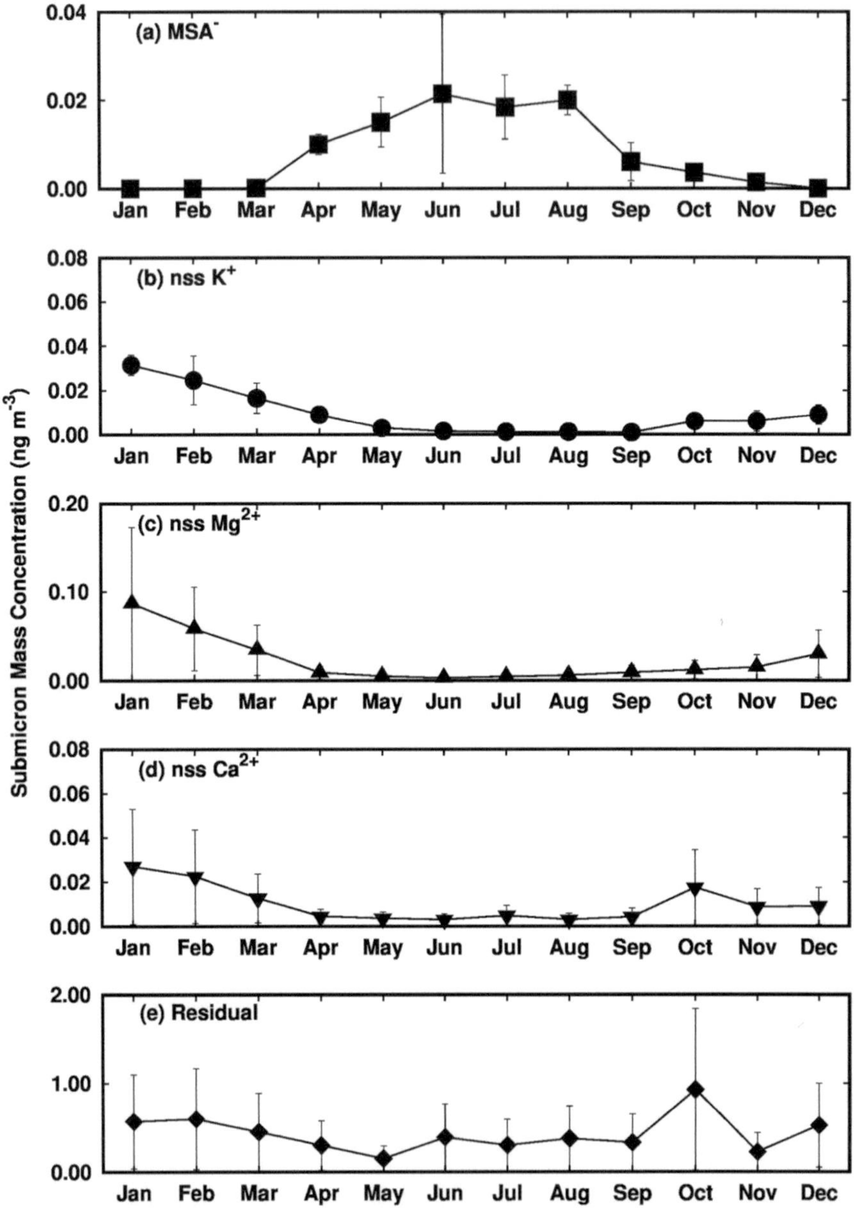

Fig. 4.31 Time-patterns of the monthly mean values of the mass concentration components of sub-micrometric particles measured by Quinn et al. (2002) at Barrow (Alaska): (**a**) MSA^-, (**b**) nss K^+, (**c**) nss Mg^{2+}, (**d**) nss Ca^{2+}, and (**e**) residual particulate matter. The vertical bars denote the standard deviations of the various mass concentration parameters

The residual mass fraction is lower than 20% during the months from November to May, and becomes comparable to the ionic mass fraction during the local summer from June to October. Particulate organic matter is most likely a large component of the residual mass, with comparable contributions given to the aerosol mass by organic acid anions (formate, acetate, propionate, and pyruvate) and inorganic anions (Li and Winchester 1967). The residual mass fractions measured during the summer months most likely are of biogenic origin, since strong sources of anthropogenic aerosol are not active during this season and, hence, the residual mass results to be rather low during the warmest period of the year. The results presented in Fig. 4.31 provided average seasonal concentrations of residual mass of 468.8 ng m^{-3} in winter and 348.8 ng m^{-3} in summer.

4.2.5.2 Determination of the Chemical Composition of Super-Micrometric Particles at Barrow

As shown above for the sub-micrometric chemical components, the mass fractions of the super-micrometric ionic components were calculated by dividing the mass concentration of each component by the gravimetric mass. On the basis of the monthly averages, the ionic mass and associated water were estimated by Quinn et al. (2002) to constitute about the 60–80% of the super-micrometric aerosol mass throughout the year, as shown in Fig. 4.32. Examining the monthly mean values of ionic mass (including sea salt, nss-$SO_4{}^{2-}$, nss-K^+, nss-Mg^{2+}, nss-Ca^{2+}, $NH_4{}^+$, MSA, Cl^- not associated with sea salt, $NO_3{}^-$, and water associated with the ionic components at $RH = 33\%$), which do not show marked variations throughout the year, the average seasonal ion mass concentration was found to be 0.71 μg m^{-3} in winter and 0.70 μg m^{-3} in summer.

Considering the ionic mass, sea salt results to be the dominant species, which contributes by 60% to 98% on a monthly basis. The largest contribution of sea salt to the ionic mass (> 90%) occurs between July and December. Super-micrometric sea salt concentration exhibits a very different seasonal cycle from that of sub-micrometric aerosol particles, with concentrations increasing in July, peaking in August to October, and decreasing in November and December until showing the lowest concentrations in January through June. During the summer peak months, the arithmetic mean of the super-micrometric sea salt concentration ranges from 1.4 to 2.1 μg m^{-3}, as obtained by summing the contributions of the various components given in Table 4.7. Maximum super-micrometric sea salt concentrations was found to occur at Barrow during the summer, when: (i) the ice pack extent is at its annual minimum, (ii) the long-range south-to-north transport is weak, and (iii) aerosol removal by wet deposition is stronger, so that the summer maximum in super-micrometric sea salt appears to result from local open leads and oceanic waters. As a result of these variations, the average seasonal super-micrometric sea salt particle

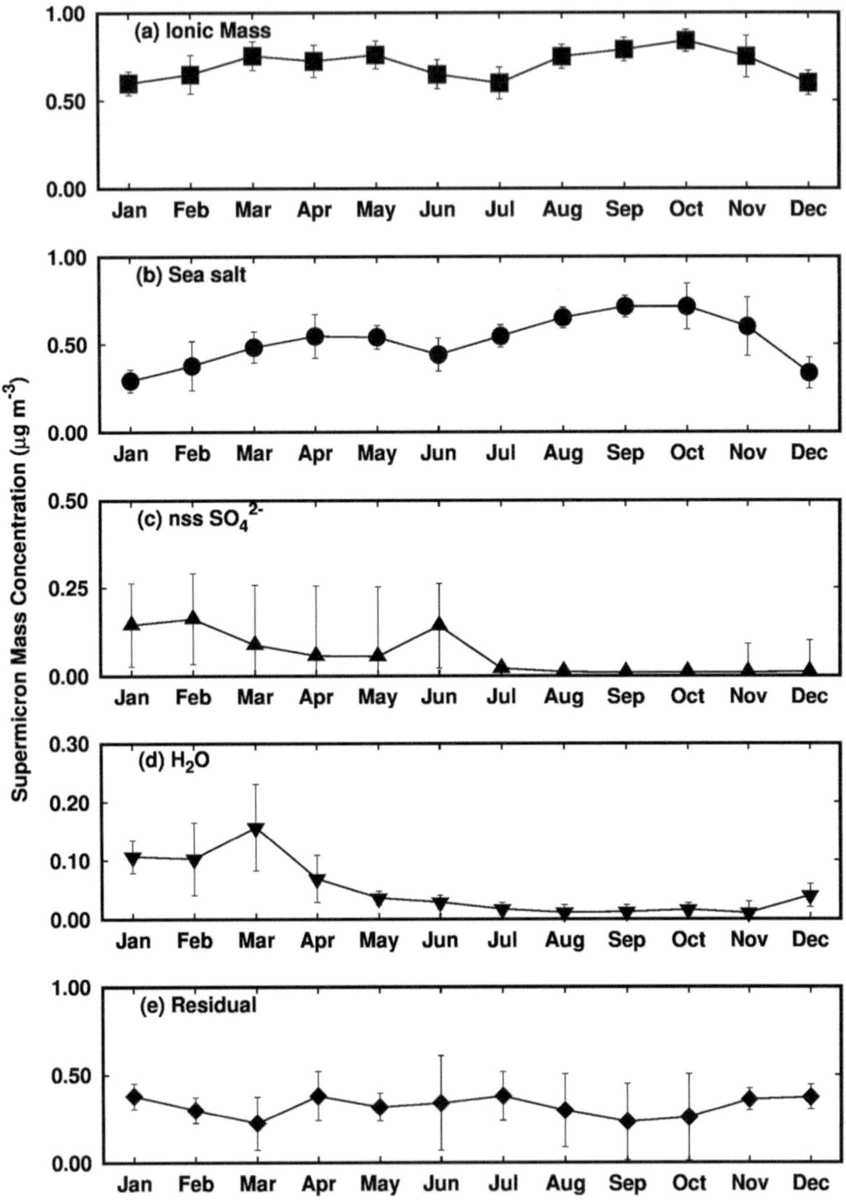

Fig. 4.32 Time-patterns of the monthly mean values of the mass concentration components of super-micrometric particles measured by Quinn et al. (2002) at Barrow (Alaska): (**a**) ionic mass (including sea salt, nss $SO_4{}^{2-}$ ions, nss K^+ ions, nss Mg^{2+} ions, nss Ca^{2+} ions, $NH_4{}^+$ ions, MSA^- ions, Cl^- ions not associated with sea salt, $NO_3{}^-$ ions, and water associated with the ionic components at 33% RH); (**b**) sea salt; (**c**) nss $SO_4{}^{2-}$; (**d**) water associated with the ionic components at 33% RH; and (**e**) residual particulate matter (calculated as the difference between the gravimetric mass and the mass of ionic species and associated water). The vertical bars denote the standard deviations of the various mass concentration parameters

Table 4.7 Values of the mass concentrations of super-micrometric aerosol particles (measured in μg m^{-3}), as obtained by Quinn et al. (2002) through the analysis of the particulate matter samples collected at Barrow (Alaska) from October 1997 to December 2007. The concentration data are reported with their arithmetic mean and standard deviation values determined for air temperature equal to 0 °C and air pressure equal to 1013 hPa

Month	Mass	Sea salt	nss-SO_4^{2-}	NH_4^+	MSA	nss-Mg^{2+}	nss-Ca^{2+}	Residual
January	0.87 ± 0.55	0.32 ± 0.28	0.08 ± 0.08	$6\ 10^{-3} \pm 3\ 10^{-3}$	$<10^{-4}$	$9\ 10^{-3} \pm 7\ 10^{-3}$	$4\ 10^{-3} \pm 6\ 10^{-3}$	0.32 ± 0.19
February	0.63 ± 0.67	0.28 ± 0.39	0.06 ± 0.03	$7\ 10^{-3} \pm 4\ 10^{-3}$	$<10^{-4}$	0.01 ± 0.01	$3\ 10^{-3} \pm 4\ 10^{-3}$	0.18 ± 0.20
March	0.72 ± 0.60	0.33 ± 0.44	0.05 ± 0.04	$6\ 10^{-3} \pm 5\ 10^{-3}$	$<10^{-4}$	$7\ 10^{-3} \pm 6\ 10^{-3}$	$4\ 10^{-3} \pm 4\ 10^{-3}$	0.13 ± 0.10
April	0.64 ± 0.31	0.29 ± 0.23	0.04 ± 0.03	$4\ 10^{-3} \pm 4\ 10^{-3}$	$10^{-4} \pm 2\ 10^{-4}$	$6\ 10^{-3} \pm 7\ 10^{-3}$	$4\ 10^{-3} \pm 5\ 10^{-3}$	0.24 ± 0.18
May	0.51 ± 0.58	0.34 ± 0.35	0.04 ± 0.02	$3\ 10^{-3} \pm 5\ 10^{-3}$	$4\ 10^{-4} \pm 5\ 10^{-4}$	$5\ 10^{-3} \pm 5\ 10^{-3}$	$5\ 10^{-3} \pm 3\ 10^{-3}$	0.21 ± 0.18
June	0.43 ± 0.39	0.26 ± 0.35	0.04 ± 0.01	$3\ 10^{-3} \pm 4\ 10^{-3}$	$7\ 10^{-4} \pm 7\ 10^{-4}$	$10^{-3} \pm 4\ 10^{-4}$	$2\ 10^{-3} \pm 2\ 10^{-3}$	0.11 ± 0.07
July	0.82 ± 0.78	0.55 ± 0.62	0.01 ± 0.01	$4\ 10^{-4} \pm 5\ 10^{-4}$	$5\ 10^{-4} \pm 5\ 10^{-4}$	$5\ 10^{-4} \pm 10^{-3}$	$10^{-3} \pm 2\ 10^{-3}$	0.23 ± 0.14
August	3.10 ± 0.49	2.00 ± 0.33	0.01 ± 0.01	$2\ 10^{-3} \pm 2\ 10^{-3}$	$10^{-3} \pm 10^{-3}$	$2\ 10^{-3} \pm 3\ 10^{-3}$	$10^{-3} \pm 10^{-3}$	0.87 ± 0.17
September	1.90 ± 0.92	1.40 ± 0.69	0.01 ± 0.01	$5\ 10^{-4} \pm 5\ 10^{-4}$	$8\ 10^{-4} \pm 8\ 10^{-4}$	$3\ 10^{-3} \pm 3\ 10^{-3}$	$< 10^{-4}$	0.43 ± 0.24
October	2.90 ± 1.90	2.10 ± 1.40	0.04 ± 0.04	$6\ 10^{-4} \pm 10^{-3}$	$<10^{-4}$	$6\ 10^{-3} \pm 7\ 10^{-3}$	$2\ 10^{-3} \pm 4\ 10^{-3}$	0.82 ± 0.41
November	2.00 ± 0.78	1.00 ± 0.67	0.02 ± 0.03	$6\ 10^{-4} \pm 10^{-3}$	$<10^{-4}$	$6\ 10^{-3} \pm 4\ 10^{-3}$	$2\ 10^{-3} \pm 2\ 10^{-3}$	0.82 ± 0.50
December	1.50 ± 0.89	0.69 ± 0.41	0.01 ± 0.01	$2\ 10^{-3} \pm 3\ 10^{-3}$	$<10^{-4}$	$2\ 10^{-3} \pm 2\ 10^{-3}$	$10^{-3} \pm 10^{-3}$	0.57 ± 0.33

mass concentration was estimated to be 0.49 μg m^{-3} in winter and 0.59 μg m^{-3} in summer.

Non-sea-salt SO_4^{2-} yields a relatively small contribution to the super-micrometric aerosol mass, varying from less than 1% to 16% on a monthly basis, the more marked inputs being provided in the late winter and early spring months, as can be seen in Fig. 4.32. The monthly mean super-micrometric nss-SO_4^{2-} concentrations result to be about an order of magnitude lower than the respective monthly mean sub-micrometric concentrations, showing a seasonal trend with the highest mean concentrations in January and February, rather low values from March to June, and the lowest concentrations from July to December (with the exception of October). Examining these data, the average seasonal concentration of super-micrometric nss-SO_4^{2-} were estimated to be 67 ng m^{-3} ‾in winter and 46 ng m^{-3} in summer. Similarly, the monthly mean water mass concentration in super-micrometric particles was found to vary throughout the year, presenting the higher values in winter and the lower values in summer, giving average seasonal concentrations of 67 ng m^{-3} in winter and 17 ng m^{-3} in summer. The mass fraction of residual mass was found to be fairly constant throughout the year, presenting monthly averages ranging from 20% to 39%, which lead to obtain average seasonal mass concentrations of 0.325 μg m^{-3} in winter and 0.313 μg m^{-3} in summer. Organic species most likely comprise a large portion of this residual mass (Li and Winchester 1990).

Using the SO_4^{2-} to Na^+ mass ratio for sea-water of 0.252, in order to calculate the monthly mean values of super-micrometric nss-SO_4^{2-} concentration, one finds that 31% of the samples turn out to have negative nss-SO_4^{2-} concentrations. In fact, negative nss-SO_4^{2-} concentrations have been attributed to sea salt sulfate depletion, since it was hypothesized by Wagenbach et al. (1998) that sea salt sulfate depletion is driven by the crystallization of mirabilite ($Na_2SO_4 \cdot 10\ H_2O$) on the surface of the sea ice when the air temperature is lower than −8.2 °C. The remaining brine becomes depleted in SO_4^{2-}. However, it is not known how the brine becomes preferentially airborne to form sea salt aerosol. The negative nss-SO_4^{2-} concentrations derived from using a sea-water SO_4^{2-} to Na^+ ratio of 0.252 indicate that sea salt sulfate may also be depleted at Barrow in the super-micrometric size range, the degree of depletion being greatest in the winter and spring months, and the lack of a strong depletion in summer being consistent with the brine/sea ice hypothesis.

The time-patterns of the monthly mean values of gravimetrically determined particulate mass shown in Fig. 4.33 appear to be subject to marked variations from one month to another, providing an average seasonal value of 1.58 μg m^{-3} in winter and 1.96 μg m^{-3} in summer. Correspondingly, the average seasonal mass concentrations of sea salt super-micrometric particles were found to be 0.77 μg m^{-3} in winter and 0.43 μg m^{-3} in summer, while the average seasonal dry-air mass concentrations of super-micrometric nss-SO_4^{2-} was estimated to be 42 ng m^{-3} in winter and 23 ng m^{-3} in summer, thus resulting to be lower than the mass concentrations determined for moist air conditions by about 40% and 50%, respectively. In

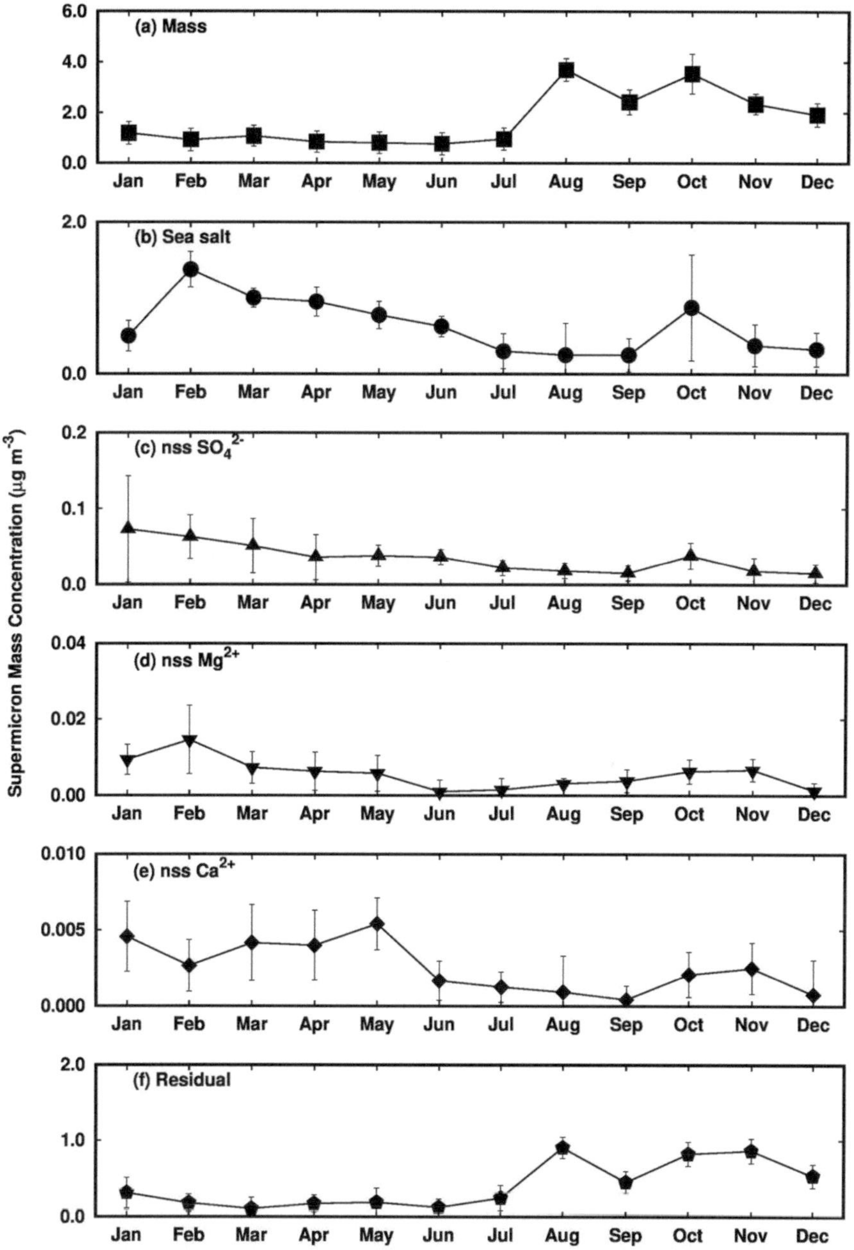

Fig. 4.33 Time-patterns of the monthly mean values of the mass concentration components of super-micrometric particles measured by Quinn et al. (2002) at Barrow (Alaska): (**a**) gravimetrically determined aerosol mass; (**b**) sea salt; (**c**) nss SO_4^{2-}; (**d**) nss Mg^{2+}; (**e**) nss Ca^{2+}; and (**f**) residual particulate matter. The vertical bars denote the standard deviations of the various mass concentration parameters

addition, analyzing the results shown in Fig. 4.33, it can be noted that the remaining ionic species (MSA, nss-K^+, nss-Mg^{2+}, and nss-Ca^{2+}) are useful tracers of the various aerosol sources, which contribute by less than 4% to the aerosol super-micrometric mass, while NH_4^+ contributes on average by less than 1%. Monthly mean super-micrometric nss-K^+ concentrations were estimated to range between 3 and 6 ng m^{-3}, making up less than 0.3% of the super-micrometric mass. Monthly mean super-micrometric nss-Mg^{2+} concentrations recorded at Barrow were found to describe a wide maximum in January–May and a further peak in September–November, giving seasonal average concentrations of 7.3 ng m^{-3} in winter and 2.4 ng m^{-3} in summer. Monthly mean super-micrometric concentrations of nss-Ca^{2+} ions were found to exhibit a similar double-peaked annual cycle, yielding average seasonal concentrations of 3.3 ng m^{-3} in winter and 1.1 ng m^{-3} in summer. These peaked patterns suggest the presence of a distant source in winter – spring and of a local source in summer. Barrie and Barrie (1990) reported a similar two-peaked annual cycle in soil dust at Alert. Since the September soil dust peak does not coincide with elevated dust concentrations from northern hemisphere dust storms, they attributed it to the onset of snow on the ground and the resuspension of soil promoted by snow drifting over relatively bare soil, while the presence of Mg^{2+} and Ca^{2+} in the aerosol was also attributed to wind-blown erosion of the soil. Finally, considering the residual component, it can be pointed out that the results shown in Fig. 4.33 give evidence to the fact that the seasonal average mass concentration of such residuals was equal to 0.40 μg m^{-3} in winter and 0.43 μg m^{-3} in summer, without large variations from August–September to the last 3 months of the year.

4.2.5.3 Measurements of the Optical Characteristics of Aerosol Particles at Barrow

Measurements of the volume scattering coefficients produced by aerosol particles were carried out at the 0.450, 0.550, and 0.700 μm wavelengths by using a TSI 3563 nephelometer (for RH ranging between 20 and 40%) (Delene and Ogren 2002). Two Berner-type impactors were also operated upstream of the nephelometer. The sample airflow was switched every 5 min between the two impactors so that scattering by either sub-micrometric or sub-10 μm particles was measured. The overall uncertainty of the volume aerosol scattering coefficient $\beta_{sca}(0.55\ \mu m)$ was estimated to be equal to 7.8% for a 24-hour averaging time, mainly due to noise and calibration errors. The time-patterns of the monthly mean values of $\beta_{sca}(0.55\ \mu m)$ are shown in Fig. 4.34, highlighting that this optical parameter (quantifying the scattering properties of aerosol particles) exhibited values higher than 10 Mm^{-1} in winter (December–February) and values ranging between 1.4 and 3.1 Mm^{-1} in summer months (June–September).

The volume absorption coefficients of aerosol particles were measured by monitoring the change in transmission through a filter with a Particle Soot Absorption Photometer (PSAP) of the Radiance Research, which was located downstream of the impactors to determine both the sub-micrometric absorption coefficient

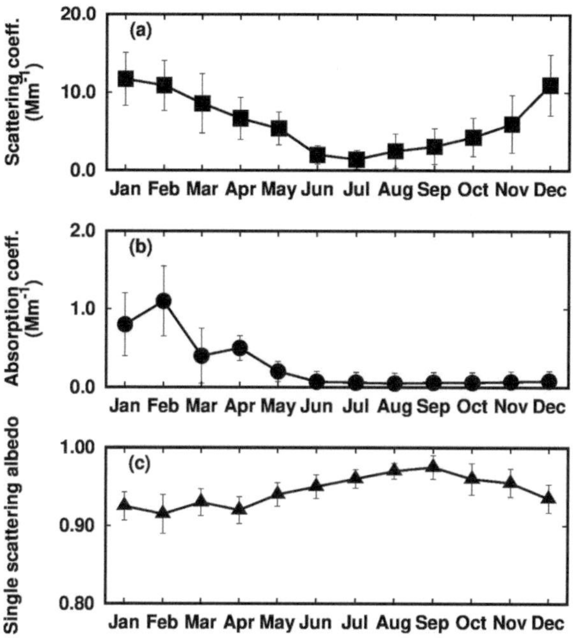

Fig. 4.34 Time-patterns of the monthly average values (dashed curve) and their standard deviations (denoted by the vertical bars), as calculated for the following particulate matter optical parameters measured at Barrow (Alaska, USA) by Quinn et al. (2002) during the period from October 1997 to December 2000: (**a**) volume scattering coefficient $\beta_{sca}(0.55\ \mu m)$, measured in Mm^{-1} with a TSI 3563 nephelometer; (**b**) volume absorption coefficient $\beta_{abs}(0.55\ \mu m)$, measured in Mm^{-1} with a Particle Soot Absorption Photometer (PSAP, Radiance Research); and (**c**) single scattering albedo $\omega(0.55\ \mu m)$, calculated as the ratio between the monthly average values of $\beta_{sca}(0.55\ \mu m)$ and the volume extinction coefficient given by the sum $(\beta_{sca}(0.55\ \mu m) + \beta_{abs}(0.55\ \mu m))$

$\beta_{abs\text{-}1}(0.55\ \mu m)$ and the sub-10 μm absorption coefficient $\beta_{abs\text{-}10}(0.55\ \mu m)$. The uncertainty associated with these absorption measurements included errors due to calibration, noise, and adjustment to STP conditions and wavelength $\lambda = 0.55$ μm. The overall uncertainty of the overall absorption coefficient $\beta_{abs}(0.55\ \mu m)$ was estimated to be 22% for a 24-hour averaging time. The time-patterns of the monthly mean values of overall volume coefficient $\beta_{abs}(0.55\ \mu m)$ are also shown in Fig. 4.34, together with those of single scattering albedo $\omega(0.55\ \mu m)$, calculated as the ratio between volume scattering coefficient $\beta_{sca}(0.55\ \mu m)$ and volume extinction coefficient $\beta_{ext}(0.55\ \mu m)$, which was calculated as the sum of volume coefficients $\beta_{sca}(0.55\ \mu m)$ and $\beta_{sca}(0.55\ \mu m)$. It can be seen that the aerosol absorption coefficient $\beta_{abs}(0.55\ \mu m)$ presents: (i) the higher monthly mean values in winter, being equal to 0.8 and 1.1 Mm^{-1} in January and February, respectively; and (ii) the lowest monthly mean values from June to November, ranging between 5 10^{-2} and 7 10^{-2} Mm^{-1}. Correspondingly, the monthly mean values of single scattering albedo $\omega(0.55\ \mu m)$ were estimated to range between 0.950 and 0.975 in summer, and slowly decrease from 0.960 to 0.915 during the October–February period, subsequently increasing in March–May to reach values varying between 0.920 and 0.940 during the late spring.

4.2.5.4 Determination of the Mass Fractions of the Chemical Components of Sub-Micrometric, Super-Micrometric and Overall Aerosol Particles at Barrow and Their Optical Properties

Figure 4.34 shows that the volume scattering coefficient of sub-micrometric aerosol particles at a visible wavelength is relatively low in summer, clearly indicating that the increase in number concentration is mainly due to small particles, which do not cause by themselves intense scattering effects. For this reason, the wide maximum in particle concentration observed from June through September, when the light volume scattering coefficient of sub-micrometric aerosol particles is relatively low, is due to the particularly intense formation of biogenic sulfur particles taking place in summer, as suggested by Bodhaine (1989) and Polissar et al. (1999, 2001).

The magnitude of the volume scattering coefficient of a chemical component depends on its size-dependent mass (or volume) concentration. Particles suspended in a unit volume and having a refractive index at wavelength $\lambda = 0.55\ \mu m$ with real part $n(0.55\ \mu m) = 1.5$ and imaginary part $\chi(0.55\ \mu m) = 10^{-7}$ (which is near that of $(NH_4)_2SO_4$ and sea salt for low *RH* conditions) are calculated to provide a volume scattering efficiency, which is log-normally distributed with the most efficient size range for scattering produced by particles having diameters ranging principally between 0.2 and 1.0 μm (Quinn et al. 1996). The seasonal cycle of the volume scattering coefficient $\beta_{sca}(0.55\ \mu m)$ produced by sub-micrometric aerosol particles and measured at Barrow since October 1997 is similar to that reported by Bodhaine (1989) and Bodhaine and Dutton (1993) for the bulk aerosol. The scattering effects produced by sub-micrometric aerosol particles were found to increase in October, describing a wide maximum over the period from December to April, followed by a pronounced decrease in May, and a subsequent drop to lowest levels in summer (June–September), as can be seen in Fig. 4.34. This trend was found to follow the combined time-patterns of sub-micrometric ss-SO_4^{2-} and nss-SO_4^{2-}, which are the two predominant ionic components of the aerosol monitored at Barrow. Drawing the regression line of $\beta_{sca}0.55\ \mu m)$ versus the sea-salt concentration measurements made during the October–January period (presenting the higher sea salt concentrations of the year, together with relatively low nss-SO_4^{2-} concentrations), a square correlation coefficient $r^2 = +\ 0.64$ of the regression line was determined, which indicates that sea salt turns out to explain about 60% of the variance in scattering that is due to the sub-micrometric particles during this period of the year. The October through January regression line drawn by plotting coefficient $\beta_{sca}(0.55\ \mu m)$ versus nss-SO_4^{2-} provided a low value of $r^2 = +\ 0.08$, which suggests that only a weak correlation exists between the scattering properties of the sub-micrometric particles and nss-SO_4^{2-} concentration, therefore confirming the hypothesis of Bodhaine (1989), based on the correlation found between measured $\beta_{sca}(0.55\ \mu m)$ and sea salt concentration, and showing that sea salt strongly influences light scattering during the October–November period.

The ground-based sun-photometer measurements conducted at Barrow with both GMD/NOAA (Global Monitoring Division of the National Oceanic and Atmospheric Administration, USA) sun-photometer and AERONET sun-photometer from March 2000 to September 2012 (Tomasi et al. 2015) clearly indicated that the aerosol optical thickness $AOT(0.55\ \mu m)$ varied mainly between 0.05 and 0.18 in the late winter and spring (February–May), the higher values of $AOT(0.55\ \mu m)$ being due to the frequent Arctic haze episodes occurred at this coastal site of northern Alaska on the Arctic Ocean, while lower values ranging mainly between 0.04 and 0.08 were measured in summer (June–September). Correspondingly, Ångström's (1964) exponent $\alpha(0.40–0.87\ \mu m)$ was estimated to vary mainly between 1.10 and 1.80 in both spring and summer periods. The most frequent cases presenting values of $\alpha(0.40–0.87\ \mu m)$ higher than 1.40–1.50 are characterized by predominant extinction features due to fine particles, while the cases giving values of $\alpha(0.40–0.87\ \mu m)$ lower than 1.30 clearly indicate that solar radiation extinction is predominantly produced by the coarse particle mode. The knowledge of the optical visible and near-infrared characteristics of the aerosol particles suspended in the vertical atmospheric column allow one to determine the aerosol-induced radiative forcing effects, provided that the size-distribution features of both fine and coarse particles modes are well known, together with the spectral features of the complex refractive index in the visible and near-infrared wavelength range. These optical properties of airborne aerosol particles can be reliably defined only if the particulate matter chemical composition is known with good accuracy. The evaluations of the particulate mass fractions of the main chemical components, made by Quinn et al. (2002) examining a 3-year set of simultaneous measurements, were taken into account in the present study to determine the monthly mean concentrations of several aerosol components, such as sea salt (ss-) and non-sea salt (nss-) sulfate, methanesulfonic acid (MSA), ammonium, and nss- K^+, Mg^{2+} and Ca^{2+}, over both the sub-micrometric and super-micrometric particle size ranges. The Quinn et al. (2002) data presented in Figs. 4.29, 4.30, 4.31, 4.32, and 4.33 were herein integrated by taking into account that: (i) the mass concentrations of Black Carbon (BC) and Organic Carbon (OC) were measured by Sharma et al. (2006) at Barrow during the various periods of the year; and (ii) the conversion from the above nss-Mg^{2+} and Ca^{2+} concentrations to the Al – Si components can be reliably made by using the soil factors proposed by Polissar et al. (1998) for the North-west Alaska. In this wide region: (i) the sea-salt component was found at the various sites; (ii) high concentrations of BC, H^+, and K^+ were identified as markers of forest fires and local wood smokes; (iii) appreciable concentrations of anthropogenic aerosol transported over long distances were measured during the most intense Arctic haze episodes; and (iv) local soil dust was often measured, formed through the mobilization of soil particles by strong winds. For these particulate matter compositional data, the percentage mass fractions were determined by Tomasi et al. (2012) for nss-sulfate, nitrate, sea salt, mineral dust, water-soluble organic matter (WSOM) and BC components, over both the summer period from June to

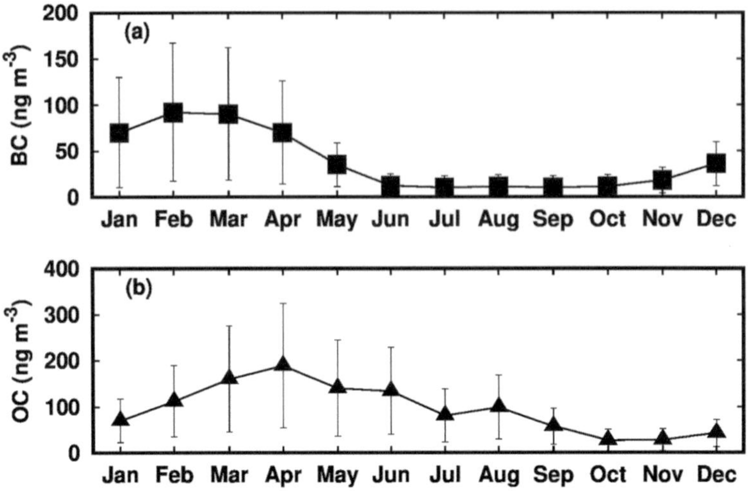

Fig. 4.35 Time-patterns of the monthly mean values of the mass concentrations of particulate matter: (**a**) black carbon (BC) mass concentration; and (**b**) organic carbon (OC) mass concentration, both measured by Sharma et al. (2002, 2006) at Point Barrow (Alaska, USA)

September, and the rest of the year, and separately for the sub-micrometric and super-micrometric size-ranges, as well for the overall aerosol size-distribution. It is worth mentioning on this matter that appreciable percentage mass concentrations of ammonium nitrate were determined by Behrenfeldt et al. (2008) as a fraction of the components belonging to the "undefined organic" class, containing sub-micrometric spherical 'organic like' particles of Eurasian origin (presumably originated from forest fires), together with predominant fractions of mineral dust and sea salt (oceanic) particles.

Regular measurements of equivalent black carbon were derived by Sharma et al. (2006) at Point Barrow (Alaska, USA) during the 15-year period from 1989 to 2003 by examining a large set of hourly light absorption data carried out by using an AE-8 Magee Scientific aethalometer. The seasonal cycle of BC was determined from these data, showing higher concentrations in winter than in summer, as can be seen in Fig. 4.35. The BC concentration was estimated to be close to 70 ng m^{-3} in January and exhibit an annual maximum of about 90 ng m^{-3} in February, then starting to decrease in March to reach a value of around 70 ng m^{-3} in April. This parameter was found to subsequently reach the values of 35 ng m^{-3} in May and 12 ng m^{-3} in June, followed by stable values close to 10 ng m^{-3} in both August and September, and then slowly increasing values in autumn, until reaching a value of around 35 ng m^{-3} in December.

A recent study conducted by Sinha et al. (2017) showed that the equivalent BC concentration measurements performed at Barrow from 1998 to 2015 provided winter mean values of this component, which were subject to slowly decrease over such a 18-year period with an average slope coefficient equal to -0.56 ± 0.45 ng m^{-3} per year, while the measurements of summer mean concentration of BC were estimated to be decreased with a slope coefficient of-0.53 ± 0.14 ng m^{-3} per year. Taking into account these results, which clearly indicate that a general decrease in the intensity of BC sources was occurred during the last two decades, we have updated the monthly mean estimates of BC concentration determined by Sharma et al. (2006) by correcting them according to the Sinha et al. (2017) evaluations, in order to take realistically into account the decreasing rate of BC mass concentration observed from 1998 to 2015. These evaluations of the monthly mean concentrations of BC are shown in Fig. 4.35, indicating that the monthly mean values of BC were estimated to range between 6 and 8 ng m^{-3} in summer, and then to increase gradually from 10 to 64 ng m^{-3} in the October–January period, and subsequently reach a maximum of 86 ng m^{-3} in February, followed by a gradual decrease in April and May. The time-patterns of the monthly mean values of the organic carbon (OC) mass concentration at surface-level are also shown in Fig. 4.35, as derived according to the seasonal evaluations made by Sharma et al. (2006), and properly corrected to take into account the decreasing trend estimated by Sinha et al. (2017) over the last 20 years. On the basis of these evaluations, the OC concentration was estimated to range from 35 to 80 ng m^{-3} in summer, from 40 to 55 ng m^{-3} in autumn, and from 60 to more than 180 ng m^{-3} in the winter – spring months.

It is worth mentioning that Sharma et al. (2006) evidenced (on the basis of a set of aethalometer measurements regularly conducted at Barrow) that the BC mass concentration was 40% higher during the positive phase of the North Atlantic Oscillation (observed from 1990 to 1994) than the average measured during the negative phase observed since 1994. In addition, using backward trajectory techniques, Sharma et al. (2006) found that the BC concentration measurements performed at Barrow were appreciably influenced by frequent long-range transport episodes occurring in the Arctic sector during the entire year, for the major part moving from the Eastern- and Western-Russian sectors toward the Arctic coast of Alaska in winter and spring.

Analyzing the features of particulate matter composition described above, we have determined the percentage mass fractions of sulfate, nitrate, sea salt, mineral dust, WSOM (including the OC component) and black carbon (BC) components, in the sub-micrometric, super-micrometric and overall size-distributions of aerosol particles, separately considered during the summer period from June to September and the winter period from October to May. The results are presented in Fig. 4.36. They indicate that:

(1) The total average mass concentration of sub-micrometric particles was estimated to be 0.89 μg m^{-3} in summer, with mass percentages of 34.4% for mineral dust, 24.4% for sea salt, 22.0% for sulfate and nitrate, 18.0% for WSOM and OC, and only 1.2% for BC.

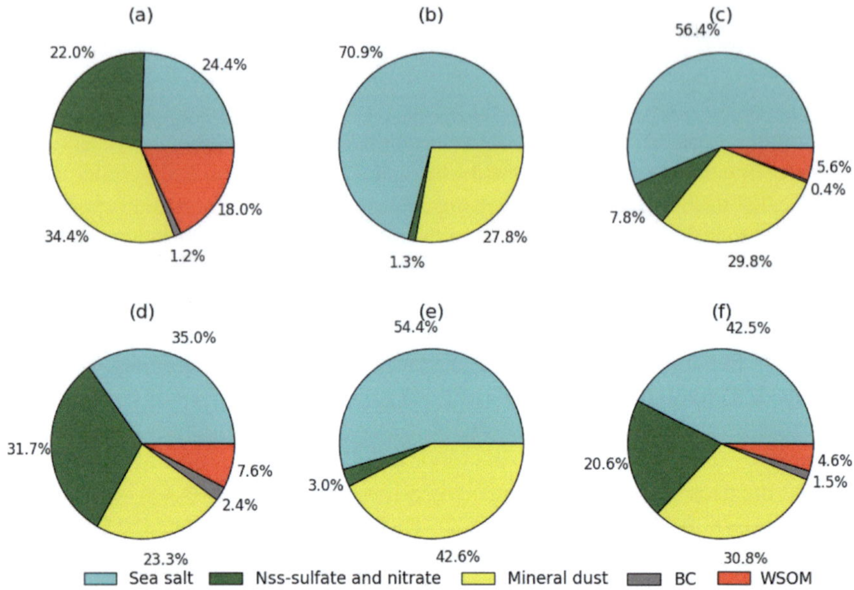

Fig. 4.36 Average composition diagrams of the ground-level particulate matter sampled at Barrow during different seasonal periods, pertaining to: (**a**) sub-micrometric aerosol particles sampled during the summer months (from June to September); (**b**) super-micrometric aerosol particles sampled during the summer months (June–September); (**c**) overall aerosol particles sampled during the summer months (June–September); (**d**) sub-micrometric aerosol particles sampled during the "winter" months (from October to May); (**e**) super-micrometric aerosol particles sampled during the "winter" months (October–May); and (**f**) overall aerosol particles sampled during the "winter" months (October–May). Different colours are used to indicate the main particulate matter constituents (sea salt, nss-sulfate and nitrate, mineral dust, black carbon (BC), and water-soluble organic matter (WSOM)

(2) The total average mass concentration of super-micrometric particles was estimated to be 1.96 $\mu g\ m^{-3}$ in summer months, with mass percentages of 70.9% for sea salt, 27.8% for mineral dust, and low percentages of sulfate and nitrate, WSOM and OC, and BC.

(3) The total average mass concentration of the overall particle size-distribution was estimated to be 2.86 $\mu g\ m^{-3}$ in summer, with mass percentages of 56.4% for sea salt, 29.8% for mineral dust, 7.8% for sulfate and nitrate components, and lower percentages for WSOM and OC, and BC.

(4) The total average mass concentration of sub-micrometric particles was estimated to be 2.21 $\mu g\ m^{-3}$ in winter (October–May), with mass percentages of 35.0% for sea salt, 31.7% for sulfate and nitrate, 23.3% for mineral dust, 7.6% for WSOM (including OC), and 2.4% for BC.

(5) The total average mass concentration of super-micrometric particles was estimated to be 1.42 $\mu g\ m^{-3}$ in winter (October–May), with mass percentages of

54.4% for sea salt, 42.6% for mineral dust, 3.0% for sulfate and nitrate, and very low percentages of WSOM and OC, and BC.

(6) The total average mass concentration of the overall particle size-distribution was estimated to be 3.63 $\mu g\ m^{-3}$ during the 8-month winter period, with mass percentages of 42.5% for sea salt, 30.8% for mineral dust, 20.6% for sulfate and nitrate, 4.6% for WSOM (including OC), and 1.5% for BC.

4.2.6 Measurements of Aerosol Chemical Composition at Alert (Nunavut, Canada)

Alert is located in the Canadian Arctic on the northern tip of Ellesmere Island (82.5 °N, 62.3 °W), and is a station where a long-term set of aerosol chemistry data have been recorded from 1980 to 1995 by Sirois and Barrie (1999). Sub-micrometric and super-micrometric particle concentrations were also measured and analyzed by Quinn et al. (2002) to carry out a more complete picture of the Alert data on the aerosol chemical composition, which were derived using a high-volume sampler with no size segregation. Carrying out a comparison of the annual cycles of sea salt, nss-SO_4^{2-}, and MSA concentrations measured at both Barrow and Alert sites, Quinn et al. (2002) found that (i) a winter maximum in sea salt was recorded at Alert, with the highest monthly mean concentrations measured in the November–February period, and (ii) a decrease in the sea salt concentrations was observed in March and April, followed by a wide minimum from late May to the end of September. The arithmetic mean concentrations obtained at Alert during the winter maximum were found to be lower than those measured at Barrow by a factor of 2 to 3, most likely due to the longer distance of Alert from the oceanic source regions of sea salt particles.

4.2.6.1 Evaluation of Sub-Micrometric and Super-Micrometric Particle Chemical Composition at Alert

The monthly mean concentrations of sea salt measured at Alert were found to be in general lower than those measured at Barrow by a factor of 10 to 18 during the winter and spring months. The time-patterns of the sea salt concentration are shown in Fig. 4.37, indicating that it was relatively low from June to September, when it was estimated to assume the lowest values of the year. In fact, the mass concentration of overall sea salt particles was found to: (i) be close to around 0.50 $\mu g\ m^{-3}$ in January and February; (ii) gradually decrease from 0.40 to 0.16 $\mu g\ m^{-3}$ in March–May; (iii) range between 0.10 and 0.12 $\mu g\ m^{-3}$ in June–September; and (iv) increase gradually from 0.20 to 0.46 $\mu g\ m^{-3}$ during the last 3 months of the year. Therefore, the average seasonal concentration of overall sea

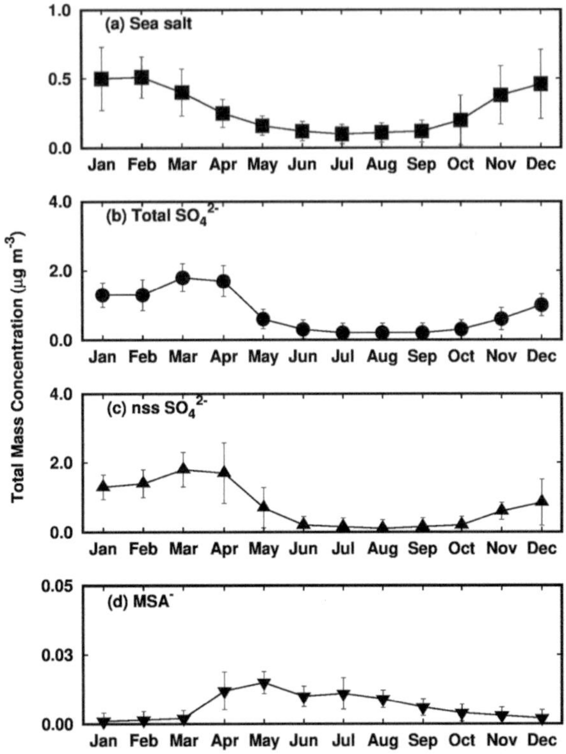

Fig. 4.37 Time-patterns of the monthly average values (solid symbols) calculated for the following mass concentrations measured at Alert (Nunavut, Canada) by Quinn et al. (2002) during the period from July 1980 to May 1995: (**a**) sea salt; (**b**) total SO_4^{2-}; (**c**) nss SO_4^{2-}, and (**d**) MSA^-. The vertical bars denote the standard deviations of the various mass concentrations

salt particles was estimated to be 0.36 µg m^{-3} in winter and 0.11 µg m^{-3} in summer.

The seasonal cycles of total sulfate (tot-SO_4^{2-}) and nss-SO_4^{2-} obtained at Alert over the sub-micrometric and super-micrometric particle size range were found to be similar to those observed at Barrow. In fact, tot-SO_4^{2-} and nss-SO_4^{2-} concentrations were found to be rather low at Alert, from June to September, and to increase steadily from October to February. In March and April, both these concentrations measured at Alert continued to increase, while those measured at Barrow were observed to drop rapidly after February. Similar mean concentrations of tot-SO_4^{2-} and nss-SO_4^{2-} were also measured in May at both these sites. Such a divergence observed between the seasonal cycles recorded in March and April may be due to the different transport patterns of anthropogenic sulfate involving the Barrow and Alert sites, as well as to the differences in the conversion processes of SO_2 into SO_4^{2-} taking place at the two monitoring stations. A long-term trend in sulfate concentration may be also in part responsible for this behaviour, since the Alert measurements were taken during the period from 1980 to 1995 and the Barrow measurements from 1997 to 2000, while Sirois and Barrie (1999) reported that a decrease in sulfate occurred during the winter/spring months at Alert between 1991 and 1995. Should this decrease be an Arctic-wide phenomenon that has persisted through 2000, it could reasonably explain some of the differences observed in March and April between the sulfate concentration measurements performed at Alert and Barrow.

The total concentration of SO_4^{2-} was estimated to assume: (i) monthly mean values of around 0.2 μg m^{-3} in summer, as shown in Fig. 4.37; (ii) increasing values in the subsequent months, from 0.3 μg m^{-3} in October to 1.0 μg m^{-3} in December; (iii) values ranging between 1.3 and 1.8 μg m^{-3} in January–March; and (iv) values gradually decreasing from 1.7 to 0.6 μg m^{-3} in April–May. For these estimates, the average seasonal mass concentration of tot-SO_4^{2-} was found to be 1.08 μg m^{-3} in winter, and 0.23 μg m^{-3} in summer.

Similar mass concentration features were observed for nss-SO_4^{2-}, with monthly mean concentrations: (i) ranging between 0.10 and 0.20 μg m^{-3} from June to September (showing that this species constitutes the predominant part of tot-SO_4^{2-} in summer); (ii) increasing during the winter period (and varying between 0.20 and 0.85 μg m^{-3} in October–December); (iii) varying between 1.3 and 1.4 μg m^{-3} in January and February; (iv) showing a maximum of 1.8 μg m^{-3} in March; and (v) decreasing in April to reach a value of 0.7 μg m^{-3} in May. It can be clearly seen in Fig. 4.37 that the tot-SO_4^{2-} concentration consisted mostly of nss-SO_4^{2-}. On the basis of these monthly mean data showing large variations from one season to another, the average seasonal concentration of nss-SO_4^{2-} was estimated to be 1.07 μg m^{-3} in winter and 0.15 μg m^{-3} in summer.

Examining a set of six-year observations of aerosol particle composition performed at Alert from 1980 to 1986, Barrie and Barrie (1990) found that the concentration of Na^+ in the Canadian high Arctic is mainly associated with the presence of sea salt particles that were either unmodified or changed by anthropogenic activities. The concentration of NH_4^+ results from the reaction of gas phase NH_3 (originated from both natural and anthropogenic sources) with sulfuric acid or other acidic species (such as nitric acid) (Prospero et al. 1996). In general, natural sources of NH_3 include excreta from wild animals and emissions from soils, vegetation, and oceanic surface, while anthropogenic sources include fertilizer production and other agricultural applications, such as biomass burning and excreta from domestic animals. The sub-micrometric NH_4^+ seasonal cycle observed at Barrow follows the time-patterns described by the cycle of nss-SO_4^{2-} ions, showing a marked concentration increase in December and a broad peak in January–May, followed by a sharp decrease in June and very low values in summer and autumn until to November. The similarity in the behaviour of NH_4^+ and nss-SO_4^{2-} is most likely due to the fast reaction of NH_3 with sulfuric acid near source regions located outside of the Arctic. The monthly mean values of the molar ratio between the NH_4^+ and nss-SO_4^{2-} concentrations fall within a relatively narrow range of 1.50 to 1.75 during the winter and spring seasons, indicating a molecular composition between ammonium bisulfate and ammonium sulfate. Seasonal mean values of the (NH_4^+/nss-SO_4^{2-}) ratio ranging mainly between 1.1 and 1.4 were found in summer, as a result of a more limited availability of NH_4^+ able to neutralize sulfuric acid. The super-micrometric concentration of NH_4^+ was more than one order of magnitude lower than its sub-micrometric concentration, so that it contributes to cause a relative increase of the overall super-micrometric ionic mass not exceeding 1% of its sub-micrometric value.

In the Arctic atmosphere, the methanesulfonic acid (MSA) is derived solely from the oxidation of biogenically produced dimethyl sulfide (DMS). The seasonal cycle of sub-micrometric MSA was found at Alert to be out of phase relative to the other chemical species, since their sub-micrometric concentrations were observed to start to increase in April and subsequently describe a wide peak in May through September, until dropping sharply in October. The super-micrometric MSA concentration was estimated to show the same seasonal behaviour, but making up only by 4% to 10% of the overall sub-micrometric and super-micrometric concentrations during the months from May to October, when it is clearly detectable. The high MSA concentrations observed in late spring may be the result of long-range transport processes from the oceanic source regions of the North Pacific (Li and Barrie 1993; Li et al. 1993b). By late June, as the ice recedes in the Arctic and Bering Sea, phytoplankton productivity in surface waters begins and becomes gradually more intense, causing an increasing DMS emission into the atmosphere. Figure 4.37 shows the time-patterns of the monthly mean concentration of MSA, found to: (i) be equal to 1 ng m^{-3} in January and increase appreciably in February and March; (ii) reach values of 12–15 ng m^{-3} in April–May; (iii) range between 6 and 11 ng m^{-3} in summer; and (iv) decrease appreciably from 4 to 2 ng m^{-3} in October–December. As a result of these monthly variations, the average seasonal concentration of MSA was estimated to be 5 ng m^{-3} in winter and 9 ng m^{-3} in summer, showing pronounced peaks in both spring and summer, since the Arctic Ocean can become in late spring and summer a substantial source of DMS through open leads and open ocean waters (Ferek et al. 1995), as the ice melt continues and extends because of the relatively high air temperature conditions near the surface. In addition, DMS released by ice algae during the ice breakup may contribute to the late spring and early summer peak in DMS (Levasseur et al. 1994). The comparison made by Quinn et al. (2002) between the Barrow and Alert seasonal cycles of the MSA concentration provides evidence that they show similar features with a maximum in spring and summer, and some slight differences since the monthly mean concentration of MSA at Alert increases in March, peaks in May, and shows a second peak in July, while that measured at Barrow starts to increase a month later (in April), subsequently describing a pair of peaks in June and August. The monthly mean concentrations measured at Barrow and Alert from May to September are both within ±60%. Using the ratio between the MSA and nss-$SO_4{}^{2-}$ concentrations to determine the fraction of nss-$SO_4{}^{2-}$ that is biogenically produced (Li et al. 1993a, b), the monthly mean values of this ratio were found to be 0.02 at Alert, during the first MSA peak, and 0.17 during the second peak, which occurred in summer, when anthropogenic nss-$SO_4{}^{2-}$ concentrations were appreciably lower.

The time-patterns of the monthly mean values of equivalent BC and OC concentrations recorded at Alert were evaluated using the median values of the ratio between BC and OC mass concentrations determined by Sharma et al. (2002), on the basis of local measurements performed with aethalometers and PSAP photometers in the 1998/99 winter and summer months. Taking into account that the winter and

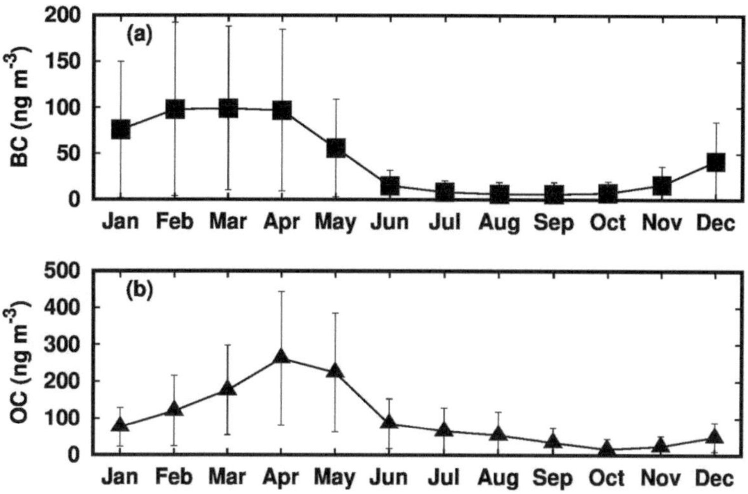

Fig. 4.38 Time-patterns of the monthly average values of the mass concentrations of: (**a**) black carbon (BC), and (**b**) organic carbon (OC), both derived from the measurements conducted by Sharma et al. (2006) at Alert (Nunavut, Canada) and updated by taking into account the decrease investigated by Sinha et al. (2017), which has presumably influenced the BC and OC mass concentrations over the last 10 years

summer mean concentrations of BC measured at Barrow were found by Sinha et al. (2017) to exhibit a slow decrease over the 1998–2015 period, with negative slope coefficients of -0.56 ± 0.45 ng m^{-3} per year in winter and -0.53 ± 0.14 ng m^{-3} per year in summer, we have assumed that a decrease rate in BC concentration should be occurred at Alert as well, although with a weaker long-term variation than that observed at Barrow, because of the greater distance of this remote site in Nunavut (Canada) from the most important sources of BC in northern America. Therefore, a negative trend of the BC concentration was assumed at Alert, of around 3 ng m^{-3} per year in the winter months and 2 ng m^{-3} per year in summer. Maintaining the same monthly mean values of ratio BC/OC determined by Sharma et al. (2006) at Alert (Canada) throughout the year, we have calculated the corresponding monthly mean values of the BC and OC concentrations shown in Fig. 4.38, which shows that the monthly mean values of BC are rather low in summer, decreasing from 20 ng m^{-3} in June to 13 ng m^{-3} in July and then ranging between 11 and 12 ng m^{-3} from August to October, subsequently increasing to 22 ng m^{-3} in November and reaching a value close to 50 ng m^{-3} in December. Thus, the seasonal average values of BC were estimated to be of 78 ng m^{-3} in winter (DJF), 90 ng m^{-3} in spring (MAM), and 15 ng m^{-3} in summer (JJA) and autumn (SON).

On the basis of the monthly mean values of BC/OC estimated by Sharma et al. (2006), the OC sub-micrometric concentration was evaluated to: (i) increase from 76 ng m^{-3} in January to 278 ng m^{-3} in April; (ii) decrease from 248 ng m^{-3} in May to 105 ng m^{-3} in July; (iii) continue to decrease from 99 ng m^{-3} in August to

62 ng m^{-3} in September; and (iv) increase from 28 ng m^{-3} in October to 48 ng m^{-3} in December. For these evaluations, the OC concentration was estimated to have an average value of 90 ng m^{-3} in winter (DJF), 238 ng m^{-3} in spring (MAM), 106 ng m^{-3} in summer (JJA), and 41 ng m^{-3} in autumn (SON).

4.2.6.2 Determination of the Chemical Composition of Sub-Micrometric, Super-Micrometric and Overall Aerosol Particles at Alert

In order to estimate the mass percentages of the chemical components of overall aerosol particles monitored in the Alert atmosphere, it is useful to take into account not only the seasonal mass concentration estimates of sea salt, tot-SO_4^{2-} (or nss-SO_4^{2-}), NH_4^+, BC and OC, but also the following considerations:

(1) Sharma et al. (2002) provided evidence that some of the aerosol constituents may be light absorbing, such as soil dust. In fact, dust may be in part mobilized in the Central Asia desert regions and transported toward the Arctic region over long distances, and in part over the continental areas of North-America, so that it seems realistic to assume that a considerable fraction of particulate matter monitored at Alert consists of mineral dust, especially during the spring and summer periods.
(2) The seasonal variations of the average concentration of nss-SO_4^{2-} were measured at Alert by Gong et al. (2010) over the multi-year period from 1981 to 2007, in order to investigate the influences of anthropogenic emissions and long-range transport processes over the chemical composition of atmospheric aerosols. These measurements have provided monthly mean concentrations found to be: (i) higher than 1 μg m^{-3} during the first 5 months of the year, (ii) lower than 20 ng m^{-3} during the summer months, and (iii) gradually increasing from October to December, until reaching a value of around 750 ng m^{-3} in the last month of the year. These variations were found to be in close agreement with those estimated by Quinn et al. (2002), which have been presented in the previous sub-section.
(3) The BC concentrations measured at Alert by Quinn et al. (2002) using an aethalometer were found to be appreciably lower than those measured at Barrow. This difference may be reasonably attributed in part to the larger distance of this high-latitude site from the North American, European and Asian source regions of anthropogenic and industrialized aerosols.
(4) The concentrations of equivalent BC measured at Alert by Sharma et al. (2002) were estimated to range mainly from 10 ng m^{-3} in summer to 400 ng m^{-3} in winter, the highest concentration values being usually measured during the winter because of the frequent transport events of polluted matter from major Russian industrial sites and the most populated areas of North America. The evaluations of BC obtained by Sharma et al. (2002) (and corrected for the Sinha et al. (2017) estimates of the reduction effects observed over the last two decades) agree very well with those obtained by Gong et al. (2010),

who examined a large data-set collected from 1981 to 2007, finding that BC concentration was: (i) often higher than 100 ng m^{-3} during the first 3 months of the year; (ii) gradually decreasing in spring until reaching a value of around 30 ng m^{-3} in May; (iii) ranging between 10 and 20 ng m^{-3} from June to October; and (iv) subsequently increasing in November until reaching a value close to 80 ng m^{-3} in December.

(5) Sharma et al. (2004) pointed out that the Arctic aerosols are influenced in the Alert area by regional sources providing appreciable concentrations of biogenic sulfur, often derived from the DMS-rich waters along the fringes of the pack ice area.

On the basis of the above remarks, the evaluations of the sulfate plus nitrate, sea salt, mineral dust, WSOM including OC, and BC were made for the overall mass concentration of aerosol particles sampled at Alert, as shown in Fig. 4.39. As mentioned above, these results were obtained by taking into account: (i) the evaluations made by Quinn et al. (2002) analyzing a data-set collected at Alert over the period from July 1980 to May 1995, and yielding the mass concentrations of sea salt, total SO_4^{2-}, nss-SO_4^{2-}, and MSA; and (ii) the mass concentrations of BC and OC measured by Sharma et al. (2002, 2004, 2006) at Alert over various multi-year periods and properly reduced to take into account the effects of the gradually decreasing trend observed by Sinha et al. (2017) over the last 20 years. The results derived from these studies indicate that: (a) the overall aerosol particle mass concentration was estimated to be 2.05 μg m^{-3} in the winter period (from October to May), with mass percentages of 55.7% for nss-sulfate plus nitrate, 18.6% for mineral dust, 17.8% for sea salt, 5.4% for WSOM and OC (considered all together), and 2.5% for BC; and (b) the overall aerosol particle mass concentration was estimated to be 0.80 μg m^{-3} during the summer (from June to September), with mass percentages of 36.5% for mineral dust, 30.7% for nss-sulfate plus nitrate, 17.0% for WSOM and OC, 14.6% for sea salt, and 1.2% for BC (Table 4.8).

4.2.7 Concluding Remarks

In order to obtain reliable estimates of the optical properties characterizing the overall sub-micrometric and super-micrometric Arctic aerosol particles monitored at the Ny-Ålesund, Thule, Barrow and Alert stations and the sub-micrometric aerosol particles analyzed during the ASCOS cruise in the Central Arctic Ocean, we have decided to take into account the mass percentages of the overall particle size-distributions reported in the chemical composition features shown in: (i) Fig. 4.23 for the total particulate matter monitored at Ny-Ålesund in spring and summer; (ii) Fig. 4.27 for the chemical composition characteristics observed at Thule in spring and summer; (iii) Fig. 4.36 for those determined at Barrow in winter and summer, (iv) Fig. 4.39 for those defined at Alert in winter and summer; and (v) Fig. 4.28 defined over the Central Arctic Ocean from early August to early September

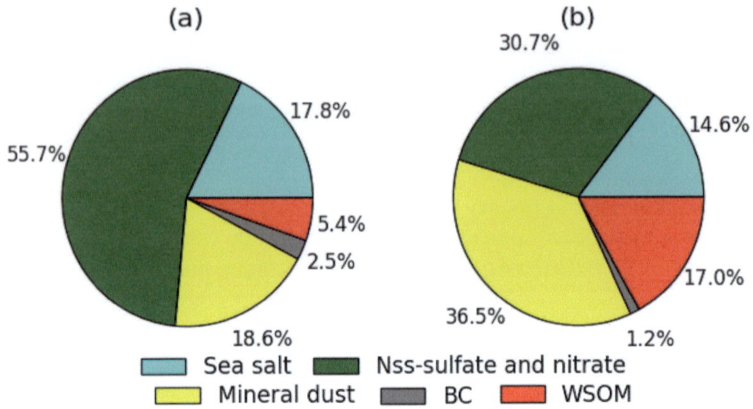

Fig. 4.39 Left part (**a**): average composition diagram of the ground-level particulate matter sampled at Alert in the winter months (from October to May). Right part (**b**): average composition diagram of the summer ground-level particulate matter sampled at Alert in the summer months (from June to September). Different colours are used to indicate the main particulate matter constituents (sea salt, nss-sulfate and nitrate, mineral dust, black carbon (BC), and water-soluble organic matter (WSOM)

Table 4.8 Values of the monthly mean values of the average mass concentrations of the sea salt, nss-sulfate and MSA components of overall (fine + coarse) aerosol particles derived from the measurements performed at Alert (Nunavut, Canada) by Quinn et al. (2002) from July 1980 to May 1995, and of the average mass concentrations of Black Carbon (BC) and Organic Carbon (OC) .determined by Sharma et al. (2006) in the same period and corrected according to the decrease investigated by Sinha et al. (2017)

Month	Sea salt (μg m^{-3})	nss-SO_4^{2-} (μg m^{-3})	MSA (ng m^{-3})	BC (ng m^{-3})	OC (ng m^{-3})
January	0.50 ± 0.23	1.3 ± 0.35	1.0 ± 1.3	82 ± 74	76 ± 53
February	0.51 ± 0.15	1.4 ± 0.40	1.5 ± 1.6	104 ± 94	137 ± 96
March	0.40 ± 0.17	1.8 ± 0.50	2.0 ± 2.9	105 ± 89	187 ± 122
April	0.25 ± 0.10	1.7 ± 0.87	12.0 ± 6.8	103 ± 88	278 ± 181
May	0.16 ± 0.03	0.7 ± 0.58	15.0 ± 4.0	62 ± 53	248 ± 161
June	0.12 ± 0.03	0.2 ± 0.13	10.0 ± 3.7	20 ± 17	113 ± 68
July	0.10 ± 0.03	0.15 ± 0.08	11.0 ± 5.7	13 ± 12	105 ± 63
August	0.11 ± 0.05	0.10 ± 0.07	9.0 ± 3.2	11 ± 10	99 ± 64
September	0.12 ± 0.08	0.15 ± 0.08	6.0 ± 3.0	11 ± 10	62 ± 40
October	0.20 ± 0.18	0.20 ± 0.14	4.0 ± 1.9	12 ± 11	28 ± 20
November	0.38 ± 0.21	0.60 ± 0.20	3.0 ± 1.3	22 ± 18	33 ± 23
December	0.46 ± 0.25	0.85 ± 0.67	2.0 ± 0.9	48 ± 43	56 ± 40

of 2008, for the chemical composition features of the sub-micrometric aerosol particles samples collected during the ASCOS cruise. The mass percentages of the chemical components were used to determine the average seasonal values of the dry-air mass particulate mass concentration of each component (separately for

Table 4.9 Spectral values of real part $n(\lambda)$ and imaginary part $\chi(\lambda)$ of particulate matter refractive index defined by Vermote et al. (1997) at eleven selected wavelengths from 0.30 to 3.75 μm for dry-air conditions of the four 6S basic aerosol components DL (dust-like) with dry-air density ρ of particulate matter equal to 2.36 g cm^{-3}, WS (water-soluble), with $\rho = 1.86$ g cm^{-3}, OCEAN (oceanic), with $\rho = 2.25$ g cm^{-3}, and SO (soot), with $\rho = 1.62$ g cm^{-3}

	Dust-like (DL)		Water soluble (WS)		Oceanic (OCEAN)		Soot (SO)	
Wavelength λ (μm)	$n(\lambda)$	$\chi(\lambda)$	$n(\lambda)$	$\chi(\lambda)$	$n(\lambda)$	$\chi(\lambda)$	$n(\lambda)$	$\chi(\lambda)$
0.300	1.530	8.0 10^{-3}	1.530	5.0 10^{-3}	1.388	1.00 10^{-8}	1.750	4.7 10^{-1}
0.400	1.530	8.0 10^{-3}	1.530	5.0 10^{-3}	1.385	9.90 10^{-9}	1.750	4.6 10^{-1}
0.488	1.530	8.0 10^{-3}	1.530	5.0 10^{-3}	1.382	6.41 10^{-9}	1.750	4.5 10^{-1}
0.515	1.530	8.0 10^{-3}	1.530	5.0 10^{-3}	1.381	3.70 10^{-9}	1.750	4.5 10^{-1}
0.550	1.530	8.0 10^{-3}	1.530	6.0 10^{-3}	1.381	4.26 10^{-9}	1.750	4.4 10^{-1}
0.633	1.530	8.0 10^{-3}	1.530	6.0 10^{-3}	1.377	1.62 10^{-8}	1.750	4.3 10^{-1}
0.694	1.530	8.0 10^{-3}	1.530	7.0 10^{-3}	1.376	5.04 10^{-8}	1.750	4.3 10^{-1}
0.860	1.520	8.0 10^{-3}	1.520	1.2 10^{-2}	1.372	1.09 10^{-6}	1.750	4.3 10^{-1}
1.536	1.400	8.0 10^{-3}	1.510	2.3 10^{-2}	1.359	2.43 10^{-4}	1.770	4.6 10^{-1}
2.250	1.220	9.0 10^{-3}	1.420	1.0 10^{-2}	1.334	8.50 10^{-4}	1.810	5.0 10^{-1}
3.750	1.270	1.1 10^{-2}	1.452	4.0 10^{-3}	1.398	2.90 10^{-3}	1.900	5.7 10^{-1}

each seasonal period) and calculate the weighted average values of the following important radiative parameters of dry particulate matter: (i) real part $n(0.55\ \mu m)$ of refractive index; (ii) imaginary part $\chi(0.55\ \mu m)$ of refractive index; (iii) single scattering albedo $\omega(0.55\ \mu m)$; and (iv) asymmetry factor $g(0.55\ \mu m)$, which have been defined in the Second Simulation of the Satellite Signal in the Solar Spectrum (hereinafter 6S) radiative transfer code by Vermote et al. (1997) for the four basic aerosol components: (a) dust-like (DL) particulate matter, assumed to have dry-air density $\rho = 2.36$ g cm^{-3}; (b) water-soluble (WS) particulate matter, with $\rho = 1.86$ g cm^{-3}; (c) oceanic (OCEAN) particulate matter, with $\rho = 2.25$ g cm^{-3}; and (d) soot (SO) particulate matter, with $\rho = 1.62$ g cm^{-3}. The spectral values of real part $n(\lambda)$ and imaginary part $\chi(\lambda)$ of the four Vermote et al. (1997) particulate matter refractive index are given in Table 4.9 for eleven selected wavelengths from 0.30 to 3.75 μm, while the spectral values of aerosol single scattering albedo $\omega(\lambda)$ and asymmetry factor $g(\lambda)$ are given in Table 4.10 at the same wavelengths. The procedure was applied to the five particle data-sets described above as follows:

(A) *Calculations of PM10 optical properties at Ny-Ålesund in spring and summer.*

On the basis of the mass percentages given in Fig. 4.23 for the chemical components of the PM10 particles sampled at Ny-Ålesund in spring and summer, we calculated the mass percentages of the 6S components, finding the following percentages: (i) in spring, 28.5% for the OCEAN (sea salt) component, 46.8% for the WS component, 23.1% for the DL component, and 1.6% only for the SO component; and (ii) in summer, 42.3% for OCEAN, 27.7% for WS, 27.6% for DL and 2.4% for SO. For these percentages, we have evaluated the following average seasonal values of: (1) dry-air mass density equal to 2.08 g cm^{-3} in spring and

Table 4.10 Spectral values of aerosol single scattering albedo $\omega(\lambda)$ and asymmetry factor $g(\lambda)$ defined by Vermote et al. (1997) at eleven selected wavelengths from 0.30 to 3.75 μm for dry-air conditions of the four 6S basic aerosol components DL (dust-like), WS (water-soluble), OCEAN (oceanic) and SO (soot)

	Dust-Like (DL)		Water Soluble (WS)		Oceanic (OCEAN)		Soot (SO)	
Wavelength λ (μm)	$\omega(\lambda)$	$g(\lambda)$	$\omega(\lambda)$	$g(\lambda)$	$\omega(\lambda)$	$g(\lambda)$	$\omega(\lambda)$	$g(\lambda)$
0.300	0.616	0.861	0.970	0.653	1.000	0.793	0.261	0.417
0.400	0.630	0.854	0.963	0.642	1.000	0.790	0.267	0.395
0.488	0.646	0.850	0.964	0.633	1.000	0.785	0.230	0.357
0.515	0.650	0.848	0.964	0.631	1.000	0.785	0.219	0.346
0.550	0.656	0.846	0.957	0.628	1.000	0.783	0.208	0.334
0.633	0.669	0.841	0.957	0.621	1.000	0.781	0.181	0.308
0.694	0.678	0.837	0.950	0.616	1.000	0.780	0.162	0.289
0.860	0.700	0.828	0.913	0.610	1.000	0.778	0.120	0.244
1.536	0.766	0.830	0.805	0.571	0.996	0.782	0.0372	0.119
2.250	0.797	0.904	0.839	0.549	0.988	0.797	0.0145	0.0576
3.750	0.820	0.871	0.881	0.432	0.972	0.749	0.0029	0.0205

2.16 g cm^{-3} in summer; (2) real part *n(0.55 μm)* of particulate matter refractive index equal to 1.491 in spring and 1.472 in summer; (3) imaginary part $\chi(0.55\ \mu m)$ of particulate matter refractive index equal to 1.16 10^{-2} in spring and 1.45 10^{-2} in summer; (4) single scattering albedo $\omega(0.55\ \mu m)$ equal to 0.888 in spring and 0.874 in summer; and (5) asymmetry factor *g(0.55 μm)* equal to 0.718 in spring and 0.747 in summer.

(B) *Calculations of PM10 optical properties at Thule in spring and summer.*

On the basis of the mass percentages given in Fig. 4.27 for the chemical components of the PM10 particles sampled at Thule, the mass percentages of the 6S components were estimated to be: (i) in spring, 16.4% for the OCEAN (sea salt), 37.1% for WS, 42.8% for DL, and 3.7% for SO; and (ii) in summer, 22.9% for OCEAN, 20.2% for WS, 56.0% for DL, and 0.9% only for SO. For these percentages, we have evaluated the following average seasonal values of: (1) dry-air mass density equal to 2.13 g cm^{-3} in spring and 2.24 g cm^{-3} in summer; (2) real part *n(0.55 μm)* of particulate matter refractive index equal to 1.514 in spring and 1.498 in summer; (3) imaginary part $\chi(0.55\ \mu m)$ of particulate matter refractive index equal to 2.2 10^{-2} in spring and 9.7 10^{-3} in summer; (4) single scattering albedo $\omega(0.55\ \mu m)$ equal to 0.808 in spring and 0.792 in summer; and (5) asymmetry factor *g(0.55 μm)* equal to 0.736 in spring and 0.783 in summer.

(C) *Calculations of the optical properties of overall sub-micrometric and super-micrometric particles at Barrow in winter and summer.*

On the basis of the mass percentages given in Fig. 4.36 for the chemical components of the overall (sub-micrometric + super-micrometric) particles monitored at Barrow by Quinn et al. (2002), the mass percentages of the 6S components were estimated to be: (i) in winter (from October to May), 42.5% for OCEAN (sea salt), 25.2% for WS, 30.8% for DL, and 1.5% for SO; and (ii) in summer (from June to September), 56.4% for OCEAN, 13.4% for WS, 29.8% for DL, and 0.4% only for SO. For these percentages, we have evaluated the following average seasonal values of: (1) dry-air mass density equal to 2.18 g cm^{-3} in winter and 2.23 g cm^{-3} in summer; (2) real part $n(0.55\ \mu m)$ of particulate matter refractive index equal to 1.470 in winter and 1.447 in summer; (3) imaginary part $\chi(0.55\ \mu m)$ of particulate matter refractive index equal to 1.1 10^{-2} in winter and 5.0 10^{-3} in summer; (4) single scattering albedo $\omega(0.55\ \mu m)$ equal to 0.871 in winter and 0.889 in summer; and (5) asymmetry factor $g(0.55\ \mu m)$ equal to 0.737 in spring and 0.779 in summer.

(D) *Calculations of the optical properties of overall sub-micrometric and super-micrometric particles at Alert in winter and summer.*

Taking into account the mass concentration percentages of the five 6S components considered in Fig. 4.39 for the overall aerosol particle load monitored at Alert in winter and in summer, the dry-air mass density of particulate matter and the four radiative particle parameters $n(0.55\ \mu m)$, $\chi(0.55\ \mu m)$, $\omega(0.55\ \mu m)$ and $g(0.55\ \mu m)$ have been calculated by assuming that the aerosol components investigated at Alert and defined in Fig. 4.39 are characterized by the dry-air mass density and optical parameters given in Table 4.8 for the four 6S basic aerosol components DL, WS, OC and SO defined by Vermote et al. (1997). Besides these assumptions, we used as weights the following mass percentages determined in Fig. 4.39 over the 8-month winter and 4-month summer periods: (a) 17.8% for OCEAN (sea salt), 59.6% for WS, 18.6% for DL, and 2.5% for SO, during the winter period; and (b) 14.6% for OCEAN, 47.7% for WS, 36.5% for DL, and 1.2% for SO, during the summer period. On the basis of these assumptions, the following results were obtained: (1) the dry-air mass density was evaluated to be 1.99 g cm^{-3} in winter and 2.07 g cm^{-3} in summer; (2) the real part $n(0.55\ \mu m)$ of particulate matter refractive index equal to 1.486 in winter and 1.511 in summer; (3) the imaginary part $\chi(0.55\ \mu m)$ of overall particles equal to 1.6 10^{-2} in winter and 1.1 10^{-2} in summer; (4) the single scattering albedo $\omega(0.55\ \mu m)$ of particulate matter equal to 0.876 in winter and 0.844 in summer; and (5) the asymmetry factor $g(0.55\ \mu m)$ of particulate matter equal to 0.680 in winter and 0.727 in summer.

(E) *Calculations of the optical properties of sub-micrometric aerosol particles in the Central Arctic Ocean obtained during the ASCOS cruise in summer 2008.*

On the basis of the mass percentages given in Fig. 4.28 for the chemical components of the sub-micrometric particles sampled over the Central Arctic Ocean during the ASCOS cruise in summer 2008, the mass percentages of the 6S components were estimated to be 5.4% for OCEAN (sea salt), 64.0% for WS, 26.7% for DL, and 3.9% for SO. Using these percentages, we have evaluated the following average seasonal values: (1) dry-air mass density equal to 2.01 g cm^{-3}; (2) real part $n(0.55\ \mu m)$ of particulate matter refractive index equal to 1.531; (3) imaginary part $\chi(0.55\ \mu m)$ of particulate matter refractive index equal to 2.3 10^{-2}; (4) single scattering albedo $\omega(0.55\ \mu m)$ equal to 0.850; and (5) asymmetry factor $g(0.55\ \mu m)$ equal to 0.683.

A complete picture of the above-considered dry-air density and optical parameters $n(0.55\ \mu m)$, $\chi(0.55\ \mu m)$, $\omega(0.55\ \mu m)$ and $g(0.55\ \mu m)$ of the Arctic particles observed at the five stations can be obtained by applying the above-described calculation procedure to the data-sets provided in Table 4.9 for: (i) the particulate matter dry-air density; (ii) the spectral values of the real and imaginary parts of particle refractive index of the four 6S components (Vermote et al. 1997), as given in Table 4.10 for dry-air conditions at eleven visible and near-infrared wavelengths; and (iii) the spectral values of aerosol single scattering albedo $\omega(\lambda)$ and asymmetry factor $g(\lambda)$ of the four 6S components defined by Vermote et al. (1997) for dry-air conditions at the eleven above-selected wavelengths.

Acknowledgements This study was partially funded by the Italian Ministry of University and Research (MIUR) within the framework of the PRIN 2007 (2007L8Y4NB – "Dirigibile Italia") and PRIN 2009 (20092C7KRC – "Arctica") projects. The logistic assistance of the Polar Support Unit of the Department of Earth System Sciences and Environmental Technologies of the National Research Council is gratefully acknowledged.

References

Aas, W., Platt, S., Solberg, S., & Yttri, K. E. (2015). Monitoring of long-range transported air pollutants in Norway, annual report 2014, environmental monitoring M-562, Kjeller, NILU (Miljødirektoratet rapport, M-367/2015) (NILU OR, 20/2015).

Abram, N. J., Mulvaney, R., Wolff, E. W., Triest, J., Kipfstuhl, S., Trusel, L. D., Vimeux, F., Fleet, L., & Arrowsmith, C. (2013). Acceleration of snow melt in an Antarctic Peninsula ice core during the twentieth century. *Nature Geoscience, 6*, 404–411. https://doi.org/10.1038/ngeo1787.

Albrecht, B. A. (1989). Aerosols, cloud microphysics, and fractional cloudiness. *Science, 245*(4923), 1227–1230. https://doi.org/10.1126/science.245.4923.1227.

Almeida, S. M., Pio, C. A., Freitas, M. C., Reis, M. A., & Trancoso, M. A. (2006). Source apportionment of atmospheric urban aerosol based on weekdays/weekend variability: Evaluation of road re-suspended dust contribution. *Atmospheric Environment, 40*(11), 2058–2067. https://doi.org/10.1016/j.atmosenv.2005.11.046.

AMAP. (2011). *The impact of black carbon on Arctic climate*. Oslo: Arctic Monitoring and Assessment Programme.

AMAP. (2015). *AMAP assessment 2015: Black carbon and ozone as Arctic climate forcers*. Oslo: AMAP.

Andreae, M. O., & Merlet, P. (2001). Emission of trace gases and aerosols from biomass burning. *Global Biogeochemical Cycles, 15*(4), 955–966. https://doi.org/10.1029/2000GB001382.

Ångström, A. (1964). The parameters of atmospheric turbidity. *Tellus, 16*(1), 64–75. https://doi.org/10.1111/j.2153-3490.1964.tb00144.x.

Argentini, S., Viola, A. P., Mastrantonio, G., Maurizi, A., Georgiadis, T., & Nardino, M. (2003). Characteristics of the boundary layer at Ny-Ålesund in the Arctic during the ARTIST field experiment. *Annals of Geophysics, 46*(2), 185–196. https://doi.org/10.1007/s10546-013-9861-7.

Barrie, L. A. (1986). Arctic air pollution: An overview of current knowledge. *Atmospheric Environment, 20*(4), 643–663. https://doi.org/10.1016/0004-6981(86)90180-0.

Barrie, L. A., & Barrie, M. J. (1990). Chemical components of lower tropospheric aerosols in the high Arctic: Six years of observations. *Journal of Atmospheric Chemistry, 11*(3), 211–226. https://doi.org/10.1007/BF00118349.

Barrie, L. A., Hoff, R. M., & Daggupaty, S. M. (1981). The influence of mid-latitudinal pollution sources on haze in the Canadian Arctic. *Atmospheric Environment, 15*(8), 1407–1418.

Bates, T. S., Calhoun, J. A., & Quinn, P. K. (1992). Variations in the methanesulfonate to sulfate molar ratio in submicrometer marine aerosol particles over the South Pacific Ocean. *Journal of Geophysical Research, Atmospheres, 97*(D9), 9859–9865. https://doi.org/10.1029/92JD00411.

Becagli, S., Castellano, E., Cerri, O., Curran, M., Frezzotti, M., Marino, F., Morganti, A., Proposito, M., Severi, M., Traversi, R., & Udisti, R. (2009). Methanesulphonic acid (MSA) stratigraphy from a Talos Dome ice core as a tool in depicting sea ice changes and southern atmospheric circulation over the previous 140 years. *Atmospheric Environment, 43*(5), 1051–1058. https://doi.org/10.1016/j.atmosenv.2008.11.015.

Becagli, S., Ghedini, C., Peeters, S., Rottiers, A., Traversi, R., Udisti, R., Chiari, M., Jalba, A., Despiau, S., Dayan, U., & Temara, A. (2011). MBAS (methylene blue active substances) and LAS (linear Alkylbenzene Sulphonates) in Mediterranean coastal aerosols: Sources and transport processes. *Atmospheric Environment, 45*(37), 6788–6801. https://doi.org/10.1016/j.atmosenv.2011.04.041.

Becagli, S., Scarchilli, C., Traversi, R., Dayan, U., Severi, M., Frosini, D., Vitale, V., Mazzola, M., Lupi, A., Nava, S., & Udisti, R. (2012). Study of present-day sources and transport processes affecting oxidised sulphur compounds in atmospheric aerosols at Dome C (Antarctica) from year-round sampling campaigns. *Atmospheric Environment, 52*(C), 98–108. https://doi.org/10.1016/j.atmosenv.2011.07.053.

Becagli, S., Lazzara, L., Marchese, C., Dayan, U., Ascanius, S. E., Cacciani, M., Caiazzo, L., Di Biagio, C., Di Iorio, T., di Sarra, A., Eriksen, P., Fani, F., Giardi, F., Meloni, D., Muscari, G., Pace, G., Severi, M., Traversi, R., & Udisti, R. (2016). Relationships linking primary production, sea ice melting, and biogenic aerosol in the Arctic. *Atmospheric Environment, 136*, 1–15. https://doi.org/10.1016/j.atmosenv.2016.04.002.

Behrenfeldt, U., Krejci, R., Ström, J., & Stohl, A. (2008). Chemical properties of Arctic aerosol particles collected at the Zeppelin station during the aerosol transition period in May and June of 2004. *Tellus Series B, 60*(3), 405–415. https://doi.org/10.1111/j.1600-0889.2008.00349.x.

Bergin, M. H., Davidson, C. I., Dibb, J. E., Jaffrezo, J. L., Kuhns, H. D., & Pandis, S. N. (1995). A simple model to estimate atmospheric concentrations of aerosol chemical species based on snow core chemistry at summit, Greenland. *Geophysical Research Letters, 22*(24), 3517–3520. https://doi.org/10.1029/95GL03196.

Berner, A., Lürzer, C., Pohl, F., Preining, O., & Wagner, P. (1979). The size distribution of the urban aerosol in Vienna. *The Science of Total Environment, 13*(3), 245–261. https://doi.org/10.1016/0048-9697(79)90105-0.

Björkman, M. P., Kühnel, R., Partridge, D. G., Roberts, T. J., Aas, W., Mazzola, M., Viola, A., Hodson, A., Ström, J., & Isaksson, E. (2013). Nitrate dry deposition in Svalbard. *Tellus, Series B, 65*(1). https://doi.org/10.3402/tellusb.v65i0.19071.

Björkman, M. P., Vega, C. P., Kühnel, R., Spataro, F., Ianniello, A., Esposito, G., Kaiser, J., Marca, A., Hodson, A., Isaksson, E., & Roberts, T. J. (2014). Nitrate postdeposition processes in Svalbard surface snow. *Journal of Geophysical Research, Atmospheres, 119*(22), 12953–12976. https://doi.org/10.1002/2013JD021234.

Bodhaine, B. A. (1989). Barrow surface aerosol: 1976– 1986. *Atmospheric Environment, 23*(11), 2357–2369. https://doi.org/10.1016/0004-6981(89)90249-7.

Bodhaine, B. A., & Dutton, E. G. (1993). A long-term decrease in Arctic haze at Barrow, Alaska. *Geophysical Research Letters, 20*(10), 947–950. https://doi.org/10.1029/93GL01146.

Bouwman, A. F., Lee, D. S., Asman, W. A. H., Dentener, F. J., Van Der Hoek, K. W., & Olivier, J. G. J. (1997). A global high-resolution emission inventory for ammonia. *Global Biogeochemical Cycles, 11*(4), 561–597. https://doi.org/10.1029/97GB02266.

Bowen, H. J. M. (1979). *Environmental chemistry of the elements*. London/New York: Academic. 333 pp.

Brock, C. A., Radke, L. F., Lyons, J. H., & Hobbs, P. V. (1989). Arctic hazes in summer over Greenland and the North American Arctic. I. Incidence and origins. *Journal of Atmospheric Chemistry, 9*, 129–148.

Brock, C. A., Cozic, J., Bahreini, R., Froyd, K. D., Middlebrook, A. M., McComiskey, A., Brioude, J., Cooper, O. R., Stohl, A., Aikin, K. C., de Gouw, J. A., Fahey, D. W., Ferrare, R. A., Gao, R.-S., Gore, W., Holloway, J. S., Hübler, G., Jefferson, A., Lack, D. A., Lance, S., Moore, R. H., Murphy, D. M., Nenes, A., Novelli, P. C., Nowak, J. B., Ogren, J. A., Peischl, J., Pierce, R. B., Pilewskie, P., Quinn, P. K., Ryerson, T. B., Schmidt, K. S., Schwarz, J. P., Sodemann, H., Spackman, J. R., Stark, H., Thomson, D. S., Thornberry, T., Veres, P., Watts, L. A., Warneke, C., & Wollny, A. G. (2011). Characteristics, sources, and transport of aerosols measured in spring 2008 during the aerosol, radiation, and cloud processes affecting Arctic Climate (ARCPAC) Project. *Atmospheric Chemistry and Physics, 11*, 2423–2453. https://doi.org/10.5194/acp-11-2423-2011.

Browse, J., Carslaw, K. S., Arnold, S. R., Pringle, K., & Boucher, O. (2012). The scavenging processes controlling the seasonal cycle in Arctic sulphate and black carbon aerosol. *Atmospheric Chemistry and Physics, 12*(15), 6775–6798. https://doi.org/10.5194/acp-12-6775-2012.

Carlson, T. N. (1981). Speculations on the movement of polluted air to the Arctic. *Atmospheric Environment, 15*, 1473–1477.

Chand, D., Wood, R., Anderson, T. L., Satheesh, S. K., & Charlson, R. J. (2009). Satellite-derived direct radiative effect of aerosols dependent on cloud cover. *Nature Geoscience, 2*, 181–184. https://doi.org/10.1038/ngeo437.

Chang, R. Y.-W., Leck, C., Graus, M., Müller, M., Paatero, J., Burkhart, J. F., Stohl, A., Orr, L. H., Hayden, K., Li, S.-M., Hansel, A., Tjernström, M., Leaitch, W. R., & Abbatt, J. P. D. (2011). Aerosol composition and sources in the Central Arctic Ocean during ASCOS. *Atmospheric Chemistry and Physics, 11*(20), 10619–10636. https://doi.org/10.5194/acp-11-10619-2011.

Clarke, A. D. (1989). In-stiu measurements of the aerosol size distributions, physicochemistry, and light absorption properties of Arctic haze. *Journal of Atmospheric Chemistry, 9*, 255–267.

Curran, M. A. J., Van Ommen, T. D., Morgan, V. I., Phillips, K. L., & Palmer, A. S. (2003). Ice core evidence for Antarctic sea ice decline since 1950s. *Science, 302*(5648), 1203–1206. https://doi.org/10.1126/science.1087888.

Dall'Osto, M., Beddows, D. C. S., Tunved, P., Krejci, R., Ström, J., Hansson, H.-C., Yoon, Y. J., Park, K.-T., Becagli, S., Udisti, R., Onasch, T., D, C., O'Dowd, R. S., & Harrison, R. M. (2017). Arctic sea ice melt leads to atmospheric new particle formation. *Scientific Reports, 7*, 3318. https://doi.org/10.1038/s41598-017-03328-1.

Das, B., Vinebrooke, R. D., Sanchez-Azofeifa, A., Rivard, B., & Wolfe, A. P. (2005). Inferring sedimentary chlorophyll concentrations with reflectance spectroscopy: A novel approach to reconstructing historical changes in the trophic status of mountain lakes. *Canadian Journal of Fisheries and Aquatic Sciences, 62*(5), 1067–1078. https://doi.org/10.1139/f05-016.

Delene, D. J., & Ogren, J. A. (2002). Variability of aerosol optical properties at four North American surface monitoring sites. *Journal of the Atmospheric Sciences, 59*(6), 1135–1150. https://doi.org/10.1175/1520-0469(2002)059<1135:VOAOPA>2.0.CO;2.

Dickerson, R. R. (1985). Reactive nitrogen-compounds in the Arctic. *Journal of Geophysical Research, Atmospheres, 90*(D6), 10739–10743. https://doi.org/10.1029/JD090iD06p10739.

Duncan, B. N., & Bey, I. (2004). A modeling study of the export pathways of pollution from Europe: Seasonal and interannual variations (1987–1997). *Journal of Geophysical Research D: Atmospheres, 109*. https://doi.org/10.1029/2003JD004079.

Dutton, E. G., Deluisi, J. J., & Herbert, G. (1989). Shortwave aerosol optical depth of Arctic haze measured on board the NOAA WP-3D during AGASP-II, April 1986. *Journal of Atmospheric Chemistry, 9*, 71–79.

Eckhardt, S., Stohl, A., Wernli, H., James, P., Forster, C., & Spichtinger, N. (2004). A 15-year climatology of warm conveyor belts. *Journal of Climate, 17*, 218–237.

Eleftheriadis, K., Vratolis, S., & Nyeki, S. (2009). Aerosol black carbon in the European Arctic: Measurements at Zeppelin station, Ny-Ålesund, Svalbard from 1998–2007. *Geophysical Research Letters, 36*, L02809. https://doi.org/10.1029/2008GL035741.

Engvall, A. C., Krejci, R., Strom, J., Minikin, A., Treffeisen, R., Stohl, A., & Herber, A. (2008). In-situ airborne observations of the microphysical properties of Arctic tropospheric aerosol during late spring and summer. *Tellus B, 60*, 392–404.

Esau, I., & Repina, I. (2012). Wind climate in Kongsfjorden, Svalbard, and attribution of leading wind driving mechanisms through turbulence-resolving simulations. *Advances in Meteorology, 2012*. Article ID 568454. https://doi.org/10.1155/2012/568454.

Evan, A. T., Vimont, D. J., Heidinger, A. K., Kossin, J. P., & Bennartz, R. (2009). The role of aerosols in the evolution of tropical North Atlantic Ocean temperature anomalies. *Science, 324*, 778–781. https://doi.org/10.1126/science.1167404.

Fan, S.-M. (2013). Modeling of observed mineral dust aerosols in the Arctic and the impact on winter season low-level clouds. *Journal of Geophysical Research, Atmospheres, 118*(19), 11161–11174. https://doi.org/10.1002/jgrd.50842.

Ferek, R. J., Hobbs, P. V., Radke, L. F., Herring, J. A., Sturges, W. T., & Cota, G. F. (1995). Dimethyl sulfide in the Arctic atmosphere. *Journal of Geophysical Research, Atmospheres, 100*(D12), 26093–26104. https://doi.org/10.1029/95JD02374.

Ferrero, L., Cappelletti, D., Busetto, M., et al. (2016). Vertical profiles of aerosol and black carbon in the Arctic: A seasonal phenomenology along 2 years (2011–2012) of field campaigns. *Atmospheric Chemistry and Physics, 16*, 12601–12629.

Fisher, J. A., Jacob, D. J., Purdy, M. T., Kopacz, M., Le Sager, P., Carouge, C. C., Holmes, C. D., Yantosca, R. M., Batchelor, R. L., Strong, K., Diskin, G., Fuelberg, H., Holloway, J., Hyer, E. J., McMillan, W. W., Warner, J., Streets, D., Zhang, Q., Wang, Y., & Wu, S. (2010). Source attribution and interannual variability of Arctic pollution in spring constrained by aircraft (ARCTAS, ARCPAC) and satellite (AIRS) observations of carbon monoxide. *Atmospheric Chemistry and Physics, 10*, 977–996.

Fisher, J. A., Jacob, D. J., Wang, Q., Bahreini, R., Carouge, C. C., Cubison, M. J., Dibb, J. E., Diehl, T., Jimenez, J. L., Leibensperger, E. M., Lu, Z., Meinders, M. B. J., Pye, H. O. T., Quinn, P. K., Sharma, S., Streets, D. G., Donkelaar, A., & Yantosca, R. M. (2011). Sources, distribution, and acidity of sulfate - ammonium aerosol in the Arctic in winter – Spring. *Atmospheric Environment, 45*(39), 7301–7318. https://doi.org/10.1016/j.atmosenv.2011.08.030.

Flanner, M. G. (2013). Arctic climate sensitivity to local black carbon. *Journal of Geophysical Research D: Atmospheres, 118*(4), 1840–1851.

Flanner, M. G., Zender, C. S., Randerson, J. T., & Rasch, P. J. (2007). Present day climate forcing and response from black carbon in snow. *Journal of Geophysical Research D: Atmospheres, 112*. https://doi.org/10.1029/2006JD008003.

Frossard, A. A., Shaw, P. M., Russell, L. M., Kroll, J. H., Canagaratna, M. R., Worsnop, D. R., Quinn, P. K., & Bates, T. S. (2011). Springtime Arctic haze contributions of submicron organic particles from European and Asian combustion sources. *Journal of Geophysical Research D: Atmospheres, 116*. https://doi.org/10.1029/2010JD015178.

Galloway, J. N., Aber, J. D., Erisman, J. W., Seitzinger, S. P., Howarth, R. W., Cowling, E. B., & Cosby, B. J. (2003). The nitrogen cascade. *Bioscience, 53*(4), 341–356. https://doi.org/10.1641/0006-3568(2003)053[0341:TNC]2.0.CO;2.

Garrett T.J., & Verzella, L. L. (2008). An evolving history of Arctic aerosols. *Bulletin of the American Meteorological Society*, 299–302. https://doi.org/10.1175/BAMS-89-3-299.

Garrett, T. J., Zhao, C., & Novelli, P. C. (2010). Assessing the relative contributions of transport efficiency and scavenging to seasonal variability in Arctic aerosol. *Tellus, 62B*, 190–196.

Giardi, F., Becagli, S., Traversi, R., Frosini, D., Severi, M., Caiazzo, L., Ancillotti, C., Cappelletti, D., Moroni, B., Grotti, M., Bazzano, A., Lupi, A., Mazzola, M., Vitale, V., Abollino, O., Ferrero, L., Bolzacchini, E., Viola, A., & Udisti, R. (2016). Size distribution and ion composition of aerosol collected at Ny-Ålesund in the spring–summer field campaign 2013. *Rendiconti Lincei, Scienze Fisiche e Naturali, 27*(Suppl 1), 47–58. https://doi.org/10.1007/s12210-016-0529-3.

Giardi, F., Traversi, R., Becagli, S., Severi, M., Caiazzo, L., Ancillotti, C., & Udisti, R. (2018). Determination of rare Earth elements in multi-year high-resolution Arctic aerosol record by double focusing inductively coupled plasma mass spectrometry with desol-

vation nebulizer inlet system. *Science of the Total Environment, 613–614*, 1284–1294. https://doi.org/10.1016/j.scitotenv.2017.09.247.
Ginoux, P., Garbuzov, D., & Hsu, H. C. (2010). Identification of anthropogenic and natural dust sources using Moderate Resolution Imaging Spectroradiometer (MODIS) deep blue level 2 data. *Journal of Geophysical Research, Atmospheres, 115*, D05204. https://doi.org/10.1029/2009JD012398.
Gondwe, M., Krol, M., Gieskes, W., Klaassen, W., & de Baar, H. (2003). The contribution of ocean-leaving DMS to the global atmospheric burdens of DMS, MSA, SO_2, and nssSO_4^{2-}, *Global Biogeochemical Cycles, 17*, 2, 1056, 25-1–25-18. https://doi.org/10.1029/2002GB001937.
Gondwe, M., Krol, M., Klaassen, W., Gieskes, W., & de Baar, H. (2004). Comparison of modeled versus measured MSA: nss SO_4^{2-} ratios: A global analysis. *Global Biogeochemical Cycles, 18*(2), GB2006. https://doi.org/10.1029/2003GB002144.
Gong, S. L., Zhao, T. L., Sharma, S., Toom-Sauntry, D., Lavoué, D., Zhang, X. B., Leaitch, W. R., & Barrie, L. A. (2010). Identification of trends and interannual variability of sulfate and black carbon in the Canadian high Arctic: 1981–2007. *Journal of Geophysical Research, Atmospheres, 115*, D07305. https://doi.org/10.1029/2009JD012943.
Goto-Azuma, K., & Koerner, R. M. (2001). Ice core studies of anthropogenic sulfate and nitrate trends in the Arctic. *Journal of Geophysical Research, Atmospheres, 106*(D5), 4959–4969. https://doi.org/10.1029/2000JD900635.
Groot Zwaaftink, C. D., Grythe, H., Skov, H., & Stohl, A. (2016). Substantial contribution of northern high-latitude sources to mineral dust in the Arctic. *Journal of Geophysical Research, Atmospheres, 121*(22), 13678–13697. https://doi.org/10.1002/2016JD025482.
Hansen, A. D. A., & Rosen, H. (1984). Vertical distribution of particulate carbon, Sulphur and bromine in the Arctic haze and comparison with ground-level measurements at Barrow, Alaska. *Geophysical Research Letters, 11*, 381–384.
Hara, K., Yamagata, S., Yamanouchi, T., Sato, K., Herber, A., Iwasaka, Y., Nagatani, M., & Nakata, H. (2003). Mixing states of individual aerosol particles in spring Arctic troposphere during ASTAR 2000 campaign. *Journal of Geophysical Research, Atmospheres, 108*, D7, 4209, AAC 1–AAC 12. https://doi.org/10.1029/2002JD002513
Hastings, M. G., Jarvis, J. C., & Steig, E. J. (2009). Anthropogenic impacts on nitrogen isotopes of ice-core nitrate. *Science, 324*(5932), 1288. https://doi.org/10.1126/science.1170510.
Heintzenberg, J. (1980). Particle size distribution and optical properties of Arctic haze. *Tellus, 32B*, 251–260.
Heintzenberg, J., & Leck, C. (1994). Seasonal variation of the atmospheric aerosol near the top of the marine boundary layer over Spitsbergen related to the Arctic sulphur cycle. *Tellus B, 46*(1), 52–67. https://doi.org/10.3402/tellusb.v46i1.15751.
Henderson, P., & Henderson, G. M. (2009). *The Cambridge handbook of earth science data*. Cambridge: Cambridge University Press. 277 pp.
Holland, H. D. (1978). *The chemistry of the atmosphere and oceans*. New York: Wiley-Interscience. 369 pp.
Hoose, C., Kristjansson, J. E., Chen, J.-P., & Hazra, A. (2010). A classical-theory-based parameterization of heterogeneous ice nucleation by mineral dust, soot and biological particles in a global climate model. *Journal of the Atmospheric Sciences, 67*(8), 2483–2503. https://doi.org/10.1175/2010JAS3425.1.
Huang, J., & Jaeglé, L. (2017). Wintertime enhancements of sea salt aerosol in polar regions consistent with a sea ice source from blowing snow. *Atmospheric Chemistry and Physics, 17*(5), 3699–3712. https://doi.org/10.5194/acp-17-3699-2017.
Huang, G., Brook, R., Crippa, M., Janssens-Maenhout, G., Schieberle, C., Dore, C., Guizzardi, D., Muntean, M., Schaaf, E., & Friedrich, R. (2017). Speciation of anthropogenic emissions of non-methane volatile organic compounds: A global gridded data set for 1970–2012. *Atmospheric Chemistry and Physics, 17*, 7683–7701. https://doi.org/10.5194/acp-17-7683-2017.
Hurrell, J. W. (1995). Decadal trends in the North Atlantic oscillation: Regional temperatures and precipitation. *Science, 269*, 676–679.

IPCC. (2007). *Climate change 2007: Synthesis report, contribution of Working Groups I, II and III to the fourth assessment report of the Intergovernmental Panel on Climate Change*. Core Writing Team (R. K. Pachauri, and A. Reisinger, Eds.) (pp 104). Geneva: IPCC.

Isaksson, E., Hermanson, M., Hicks, S., Igarashi, M., Kamiyama, K., Kokichi, J., Moore, J., Motoyama, H., Muir, D., Pohjola, V., Vaikmäe, R., van de Wal, R. S. W., & Watanabe, O. (2003). Ice cores from Svalbard – Useful archives of past climate and pollution history. *Physics and Chemistry of the Earth, 28*(28–32), 1217–1228. https://doi.org/10.1016/j.pce.2003.08.053.

Isaksson, E., Kekonen, T., Moore, J., & Mulvaney, R. (2005). The methanesulfonic acid (MSA) record in a Svalbard ice core. *Annals of Glaciology, 42*, 345–351. https://doi.org/10.3189/172756405781812637.

Iversen, T. (1984). On the atmospheric transport of pollution to the Arctic. *Geophysical Research Letters, 11*, 457–460.

Iversen, T. (1996). Atmospheric transport routes for the Arctic. In: E. Wolff, R. C. Bales (Eds.), *Chemical exchange between the atmosphere and polar snow* (Global environmental chnage, Vol. 43. NATO ASi series 1 (pp. 71–92). Berlin: Springer.

Iversen, T., & Joranger, E. (1985). Arctic air pollution and large scale atmospheric flows. *Atmospheric Environment, 19*, 2099–2108.

Jacob, D. J., Crawford, J. H., Maring, H., Clarke, A. D., Dibb, J., Emmons, L. K., Ferrare, R. A., Hostetler, C. A., Russell, P. B., Singh, H. B., Thompson, A. M., Shaw, G. E., McCauley, E., Pederson, J. R., & Fisher, J. A. (2010). The Arctic research of the composition of the troposphere from aircraft and satellites (ARCTAS) mission: Design, execution, and first results. *Atmospheric Chemistry and Physics, 10*, 5191–5212.

Johnson, M. T., & Bell, T. G. (2008). Coupling between dimethylsulfide emissions and the ocean–atmosphere exchange of ammonia. *Environmental Chemistry, 5*(4), 259–267. https://doi.org/10.1071/EN08030.

Johnson, M., Sanders, R., Avgoustidi, V., Lucas, M., Brown, L., Hansell, D., Moore, M., Gibb, S., Liss, P., & Jickells, T. (2007). Ammonium accumulation during a silicate-limited diatom bloom indicates the potential for ammonia emission events. *Marine Chemistry, 106* (1–2 special issue), 63–75. https://doi.org/10.1016/J.MARCHEM.2006.09.006.

Karl, M., Gross, A., Leck, C., & Pirjola, L. (2007). Intercomparison of dimethylsulfide oxidation mechanisms for the marine boundary layer: Gaseous and particulate sulfur constituents. *Journal of Geophysical Research, Atmospheres, 112*, D15304. https://doi.org/10.1029/2006JD007914.

Kaufman, Y. J., Tanré, D., & Boucher, O. (2002). A satellite view of aerosols in the climate system. *Nature, 419*(6903), 215–223. https://doi.org/10.1038/nature01091.

Kawamura, K., & Bikkina, S. (2016). A review of dicarboxylic acids and related compounds in atmospheric aerosols: Molecular distributions, sources and transformation. *Atmospheric Research, 170*(15), 140–160. https://doi.org/10.1016/j.atmosres.2015.11.018.

Kerminen, V.-M., & Leck, C. (2001). Sulfur chemistry over the Central Arctic Ocean during the summer: Gas-to-particle transformation. *Journal of Geophysical Research, Atmospheres, 106*(D23), 32087–32099. https://doi.org/10.1029/2000JD900604.

Kerminen, V.-M., Teinilä, K., & Hillamo, R. (2000). Chemistry of sea-salt particles in the summer Antarctic atmosphere. *Atmospheric Environment, 34*(17), 2817–2825. https://doi.org/10.1016/S1352-2310(00)00089-3.

Kettle, A. J., & Andreae, M. O. (2000). Flux of dimethylsulfide from the oceans: A comparison of updated data sets and flux models. *Journal of Geophysical Research, Atmospheres, 105*(D22), 26793–26808. https://doi.org/10.1029/2000JD900252.

Klonecki, A., Hess, P., Emmons, L. K., Smith, L., Orlando, J. J., & Blake, D. (2003). Seasonal changes in the transport of pollutants into the Arctic troposphere-model study. *Journal of Geophysical Research D: Atmospheres, 108*(D4). https://doi.org/10.1029/2002JD002199.

Koch, D., & Hansen, J. (2005). Distant origins of Arctic black carbon: A Goddard Institute for space studies model eexperiment. *Journal of Geophysical Research, 110*(D4.):14 pp). https://doi.org/10.1029/2004jd005296.

Kulmala, M., Petäjä, T., Mönkkönen, P., Koponen, I. K., Dal Maso, M., Aalto, P. P., Lehtinen, K. E. J., & Kerminen, V.-M. (2005). On the growth of nucleation mode particles: Source rates of condensable vapor in polluted and clean environments. *Atmospheric Chemistry and Physics, 5*, 409–416. https://doi.org/10.5194/acp-5-409-2005.

Kulshrestha, U. C., Sarkar, A. K., Srivastava, S. S., & Parashar, D. C. (1996). Investigation into atmospheric deposition through precipitation studies at New Delhi (India). *Atmospheric Environment, 30*(24), 4149–4154. https://doi.org/10.1016/1352-2310(96)00034-9.

Lana, A., Bell, T. G., Simo, R., Vallina, S. M., Ballabrera-Poy, J., Kettle, A. J., Dachs, J., Bopp, L., Saltzman, E. S., Stefels, J., Johnson, J. E., & Liss, P. S. (2011). An updated climatology of surface dimethlysulfide concentrations and emission fluxes in the global ocean. *Global Biogeochemical Cycles, 25*(1), GB1004., 17 pp. https://doi.org/10.1029/2010GB003850.

Lavoué, D., Liousse, C., Cachier, H., Stocks, B. J., & Goldammer, J. G. (2000). Modeling of carbonaceous particles emitted by boreal and temperate wildfires at northern latitudes. *Journal of Geophysical Research, Atmospheres, 105*(D22), 26871–26890. https://doi.org/10.1029/2000JD900180.

Le Quéré, C., & Saltzman, E. S. (2009). Introduction to surface ocean – Lower atmosphere processes. In C. Le Quere & E. S. Saltzman (Eds.), *Surface Ocean – Lower atmosphere processes*. Washington, DC: American Geophysical Union. https://doi.org/10.1029/2009GM000915.

Leaitch, W. R., Hoff, R. M., & MacPherson, J. L. (1989). Airborne and Lidar measurements of aerosol and cloud particles in the troposphere over alert Canada in April 1986. *Journal of Atmospheric Chemistry, 9*, 187–212.

Leck, C., & Persson, C. (1996). Seasonal and short-term variability in dimethyl sulfide, sulfur dioxide and biogenic sulfur and sea salt aerosol particles in the arctic marine boundary layer during summer and autumn. *Tellus B, 48*(2), 272–299. https://doi.org/10.1034/j.1600-0889.48.issue2.1.x.

Leskinen, A. P., Jokiniemi, J. K., & Lehtinen, K. E. J. (2007). Characterization of aging wood chip combustion aerosol in an environmental chamber. *Atmospheric Environment, 41*(17), 3713–3721. https://doi.org/10.1016/j.atmosenv.2006.12.016.

Levasseur, M., Gosselin, M., & Michaud, S. (1994). A new source of dimethylsulfide (DMS) for the arctic atmosphere. *Marine Biology, 121*(2), 381–387. https://doi.org/10.1007/BF00346748.

Li, S.-M., & Barrie, L. A. (1993). Biogenic sulfur aerosol in the Arctic troposphere: 1. Contributions to total sulfate. *Journal of Geophysical Research, Atmospheres, 98*(D11), 20613–20622. https://doi.org/10.1029/93JD02234.

Li, S. M., & Winchester, J. W. (1967). Geochemistry of organic and inorganic ions of late winter Arctic aerosols. *Atmospheric Environment, 23*(11), 2401–2415. https://doi.org/10.1016/0004-6981(89)90252-7.

Li, S.-M., & Winchester, J. W. (1990). Haze and other aerosol components in late winter Arctic Alaska, 1986. *Journal of Geophysical Research, Atmospheres, 95*(D2), 1797–1810. https://doi.org/10.1029/JD095iD02p01797.

Li, S. M., Barrie, L. A., & Sirois, A. (1993a). Biogenic sulfur aerosol in the Arctic troposphere: 2. Trends and seasonal variations. *Journal of Geophysical Research, Atmospheres, 98*(D11), 20623–20631. https://doi.org/10.1029/93JD02233.

Li, S.-M., Barrie, L. A., Talbot, R. W., Harriss, R. C., Davidson, C. I., & Jaffrezo, J.-L. (1993b). Seasonal and geographic variations of methanesulfonic acid in the Arctic troposphere. *Atmospheric Environment, Part A, General Topics, 27*(17), 3011–3024. https://doi.org/10.1016/0960-1686(93)90333-T.

Li, F., Wang, H., & Gao, Y. (2015). Extratropical Ocean warming and winter Arctic Sea ice cover since the 1990s. *Journal of Climate, 28*(14), 5510–5522. https://doi.org/10.1175/JCLI-D-14-00629.1.

Luo, J.-J., Masson, S., Behera, S., Delecluse, P., Gualdi, S., Navarra, A., & Yamagata, T. (2003). South Pacific origin of the decadal ENSO-like variation as simulated by a coupled GCM. *Geophysical Research Letters, 30*(24). https://doi.org/10.1029/2003GL018649.

Maenhaut, W., Cornille, P., Pacyna, J. M., & Vitols, V. (1989). Trace element composition and origin of the atmospheric aerosol in the Norwegian Arctic. *Atmospheric Environment, 23*(11), 2551–2569. https://doi.org/10.1016/0004-6981(89)90266-7.

Markowicz, K. M., Pakszys, P., Ritter, C., Zielinski, T., Udisti, R., Cappelletti, D., Mazzola, M., Shiobara, M., Xian, P., Zawadzka, O., Lisok, J., Petelski, T., Makuch, P., & Karasiński, G. (2016). Impact of North American intense fires on aerosol optical properties measured over the European Arctic in July 2015. *Journal of Geophysical Research, Atmospheres, 121*(8211), 14487–14512. https://doi.org/10.1002/2016JD025310.

Maturilli, M., Herber, A., & König-Langlo, G. (2013). Climatology and time series of surface meteorology in Ny-Ålesund, Svalbard. *Earth System Science Data, 5*, 155–163. https://doi.org/10.5194/essd-5-155-2013.

Mazzola, M., Stone, R. S., Herber, A., Tomasi, C., Lupi, A., Vitale, V., Toledano, C., Cachorro, V. E., Torres, B., Berjon, A., Ortiz, J. P., Neill, N. T. O.'., Masataka, S., Stebel, K., Aaltonen, V., Zielinski, T., Petelski, T., Goloub, P., Blarel, L., Li, Z., Abboud, I., Cuevas, E., Stock, M., Schulz, K.-H., & Virkkula, A. (2012). Evaluation of Sun-photometer capabilities for the retrievals of aerosol optical depth at high latitudes: The POLAR-AOD intercomparison campaigns. *Atmospheric Environment, 52C*, 4–17. https://doi.org/10.1016/j.atmosenv.2011.07.042.

Mazzola, M., Viola, A. P., Lanconelli, C., & Vitale, V. (2016). Atmospheric observations at the Amundsen-Nobile climate change tower in Ny-Ålesund, Svalbard. *Rendiconti Lincei, Scienze Fisiche e Naturali, 27*(Suppl 1), 7–18. https://doi.org/10.1007/s12210-016-0540-8.

Meloni, D., di Sarra, A., Biavati, G., DeLuisi, J. J., Monteleone, F., Pace, G., Piacentino, S., & Sferlazzo, D. M. (2007). Seasonal behavior of Saharan dust events at the Mediterranean island of Lampedusa in the period 1999–2005. *Atmospheric Environment, 41*(14), 3041–3056. https://doi.org/10.1016/j.atmosenv.2006.12.001.

Mitchell, M. (1956). Visual range in the polar regions with particular reference to the Alaskan Arctic. *Journal of Atmospheric and Terrestrial Physics (Special Supplement)*, 195–211.

Mochida, M., & Kawamura, K. (2004). Hygroscopic properties of levoglucosan and related organic compounds characteristic to biomass burning aerosol particles. *Journal of Geophysical Research, Atmospheres, 109*(D21). https://doi.org/10.1029/2004JD004962.

Monahan, E. C., Spiel, D. E., & Davidson, K. L. (1986). A model of marine aerosol generation via whitecaps and wave disruption. In E. C. Monahan & G. Mac Niocaill (Eds.), *Oceanic whitecaps* (pp. 167–174). Norwell: D. Reidel.

Moore, J. K., Doney, S. C., & Lindsay, K. (2004). Upper Ocean ecosystem dynamics and iron cycling in a global three-dimensional model. *Global Biogeochemical Cycles, 18*(4). https://doi.org/10.1029/2004GB002220.

Moreno, T., Querol, X., Alastuey, A., Reche, C., Cusack, M., Amato, F., Pandolfi, M., Pey, J., Richard, A., Prévôt, A. S. H., Furger, M., & Gibbons, W. (2011). Variations in time and space of trace metal aerosol concentrations in urban areas and their surroundings. *Atmospheric Chemistry and Physics, 11*, 9415–9430. https://doi.org/10.5194/acp-11-9415-2011.

Morganti, A., Becagli, S., Castellano, E., Severi, M., Traversi, R., & Udisti, R. (2007). An improved flow analysis-ion chromatography method for determination of cationic and anionic species at trace levels in Antarctic ice cores. *Analytica Chimica Acta, 603*(2), 190–198. https://doi.org/10.1016/j.aca.2007.09.050.

Moroni, B., Becagli, S., Bolzacchini, E., Busetto, M., Cappelletti, D., Crocchianti, S., Ferrero, L., Frosini, D., Lanconelli, C., Lupi, A., Maturilli, M., Mazzola, M., Perrone, M. G., Sangiorgi, G., Traversi, R., Udisti, R., Viola, A., & Vitale, V. (2015). Vertical profiles and chemical properties of aerosol particles upon Ny-Ålesund (Svalbard Island). *Advances in Meteorology, 2015*. Article ID 292081. https://doi.org/10.1155/2015/292081.

Moroni, B., Cappelletti, D., Ferrero, L., Crocchianti, S., Busetto, M., Mazzola, M., Becagli, S., Traversi, R., & Udisti, R. (2016). Local vs. long-range sources of aerosol particles upon Ny-Ålesund (Svalbard Islands): Mineral chemistry and geochemical records. *Rendiconti Lincei, Scienze Fisiche e Naturali, 27*(Suppl 1), 115–127. https://doi.org/10.1007/s12210-016-0533-7.

Moroni, B., Cappelletti, D., Crocchianti, S., Becagli, S., Caiazzo, L., Traversi, R., Udisti, R., Mazzola, M., Markowicz, K., Ritter, C., & Zielinski, T. (2017). Morphochemical characteristics and mixing state of long range transported wildfire particles at Ny-Ålesund (Svalbard Islands). *Atmospheric Environment, 156*(5), 135–145. https://doi.org/10.1016/j.atmosenv.2017.02.037.

Murphy, D. M., Thomson, D. S., & Mahoney, M. J. (1998). In situ measurements of organics, meteoritic material, mercury, and other elements in aerosols at 5 to 19 kilometers. *Science, 282*(5394), 1664–1669. https://doi.org/10.1126/science.282.5394.1664.

Muscari, G., Di Biagio, C., di Sarra, A., Cacciani, M., Ascanius, S. E., Bertagnolio, P. P., Cesaroni, C., de Zafra, R. L., Eriksen, P., Fiocco, G., Fiorucci, I., & Fuà, D. (2014). Observations of surface radiation and stratospheric processes at Thule air base, Greenland, during the IPY. *Annals of Geophysics, 57*(3), SS0323. https://doi.org/10.4401/ag-6382.

Nguyen, Q. T., Kristensen, T. B., Hansen, A. M. K., Skov, H., Bossi, R., Massling, A., Sørensen, L. L., Bilde, M., Glasius, M., & Nøjgaard, J. K. (2013). Characterization of humic-like substances in Arctic aerosols. *Journal of Geophysical Research, Atmospheres, 119*(8), 5011–5027. https://doi.org/10.1002/2013JD020144.

Noone, K. J., & Clarke, A. D. (1988). Soot scavenging measurements in Arctic snowfall. *Atmospheric Environment, 22*, 2773–2779.

Norman, A. L., Barrie, L. A., Toom-Sauntry, D., Sirois, A., Krouse, H. R., Li, S. M., & Sharma, S. (1999). Sources of aerosol sulfate at alert: Apportionment using stable isotopes. *Journal of Geophysical Research, Atmospheres, 104*(D9), 11619–11631. https://doi.org/10.1029/1999JD900078.

Nozaki, Y. (1997). A fresh look at element distribution in the North Pacific Ocean. *Eos, Transactions of the American Geophysical Union, 78*(21), 221–223. https://doi.org/10.1029/97EO00148.

O'Dwyer, J., Isaksson, E., Vinje, T., Jauhiainen, T., Moore, J., Pohjola, V., Vaikmie, R., & van de Wal, R. S. W. (2000). Methanesulfonic acid in a Svalbard ice core as an indicator of ocean climate. *Geophysical Research Letters, 27*(8), 1159–1162. https://doi.org/10.1029/1999GL011106.

Pacyna, J. M., & Ottar, B. (1989). Origin of natural constituents in the Arctic aerosol. *Atmospheric Environment, 23*(4), 809–811. https://doi.org/10.1016/0004-6981(89)90485-X.

Pandis, S. N., Donahue, N. M., Murphy, B. N., Riipinen, I., Fountoukis, C., Karnezi, E., Patoulias, D., & Skyllakou, K. (2013). Introductory lecture: Atmospheric organic aerosols: Insights from the combination of measurements and chemical transport models. *Faraday Discussions, 165*, 9–24. https://doi.org/10.1039/C3FD00108C.

Paris, J. D., Stohl, A., Nédélec, P., Arshinov, M. Y., Panchenko, M. V., Shmargunov, V. P., Law, K. S., Belan, B. D., & Ciais, P. (2009). Wildfire smoke in the Siberian Arctic in summer: Source characterization and plume evolution from airborne measurements. *Atmospheric Chemistry and Physics, 9*(23), 9315–9327. https://doi.org/10.5194/acp-9-9315-2009.

Park, K.-T., Lee, K., Yoon, Y.-J., Lee, H.-W., Kim, H.-C., Lee, B.-Y., Hermansen, O., Kim, T.-W., & Holmen, K. (2013). Linking atmospheric dimethyl sulfide and the Arctic Ocean spring bloom. *Geophysical Research Letters, 40*(1), 155–160. https://doi.org/10.1029/2012GL054560.

Parungo, F. P., Nagamoto, C. T., Sheridan, P. J., & Schnell, R. C. (1990). Aerosol characteristics of Arctic haze sampled during AGASP-II. *Atmospheric Environment, 24*, 937–949.

Petters, M. D., Parsons, M. T., Prenni, A. J., DeMott, P. J., Kreidenweis, S. M., Carrico, C. M., Sullivan, A. P., McMeeking, G. R., Levin, E., Wold, C. E., Collett, J. L., Jr., & Moosmüller, H. (2009). Ice nuclei emissions from biomass burning. *Journal of Geophysical Research, Atmospheres, 114*, D07209. https://doi.org/10.1029/2008JD011532.

Petzold, A., Ogren, J. A., Fiebig, M., Laj, P., Li, S. M., Baltensperger, U., Holzer-Popp, T., Kinne, S., Pappalardo, G., Sugimoto, N., Wehrli, C., Wiedensohler, A., & Zhang, X. Y. (2013). Recommendations for reporting "black carbon" measurements. *Atmospheric Chemistry and Physics, 13*(16), 8365–8379. https://doi.org/10.5194/acp-13-8365-2013.

Pey, J., Querol, X., Alastuey, A., Forastiere, F., & Stafoggia, M. (2013). African dust outbreaks over the Mediterranean Basin during 2001–2011: PM10 concentrations, phenomenology and trends, and its relation with synoptic and mesoscale meteorology. *Atmospheric Chemistry and Physics, 13*(3), 1395–1410. https://doi.org/10.5194/acp-13-1395-2013.

Polissar, A. V., Hopke, P. K., Paatero, P., Malm, W. C., & Sisler, J. F. (1998). Atmospheric aerosol over Alaska. 2. Elemental composition and sources. *Journal of Geophysical Research, Atmospheres, 103*(D15), 19045–19057. https://doi.org/10.1029/98JD01212.

Polissar, A. V., Hopke, P. K., Paatero, P., Kaufmann, Y. J., Hall, D. K., Bodhaine, B. A., Dutton, E. G., & Harris, J. M. (1999). The aerosol at Barrow, Alaska: Long-term trends and source locations. *Atmospheric Environment, 33*(16), 2441–2458. https://doi.org/10.1016/S1352-2310(98)00423-3.

Polissar, A. V., Hopke, P. K., & Harris, J. M. (2001). Source regions for atmospheric aerosol measured at Barrow, Alaska. *Environmental Science & Technology, 35*(21), 4214–4226. https://doi.org/10.1021/es0107529.

Pósfai, M., Simonics, R., Li, J., Hobbs, P. V., & Buseck, P. R. (2003). Individual aerosol particles from biomass burning in southern Africa: 1. Compositions and size distributions of carbonaceous particles. *Journal of Geophysical Research, 108*(D13), 8483. https://doi.org/10.1029/2002JD002291.

Prospero, J. M., Barrett, K., Church, T., Dentener, F., Duce, R. A., Galloway, J. N., Levy, H., Moody, J., & Quinn, P. (1996). Atmospheric deposition of nutrients to the North Atlantic Basin. *Biogeochemistry, 35*(1) (Nitrogen cycling in the North Atlantic Ocean and its watersheds, October 1996), 27–73.

Quinn, P. K., Kapustin, V. N., Bates, T. S., & Covert, D. S. (1996). Chemical and optical properties of marine boundary layer aerosol particles of the mid-Pacific in relation to sources and meteorological transport. *Journal of Geophysical Research, Atmospheres, 101*(D3), 6931–6951. https://doi.org/10.1029/95JD03444.

Quinn, P. K., Coffman, D. J., Kapustin, V. N., Bates, T. S., & Covert, D. S. (1998). Aerosol optical properties in the marine boundary layer during ACE 1 and the underlying chemical and physical aerosol properties. *Journal of Geophysical Research, Atmospheres, 103*(D13), 16547–16563. https://doi.org/10.1029/97JD02345.

Quinn, P. K., Bates, T. S., Miller, T. L., Coffman, D. J., Johnson, J. E., Harris, J. M., Ogren, J. A., Forbes, G., Anderson, T. L., Covert, D. S., & Rood, M. J. (2000). Surface submicron aerosol chemical composition: What fraction is not sulfate? *Journal of Geophysical Research, Atmospheres, 105*(D5), 6785–6805. https://doi.org/10.1029/1999JD901034.

Quinn, P. K., Miller, T. L., Bates, T. S., Ogren, J. A., Andrews, E., & Shaw, G. E. (2002). A 3-year record of simultaneously measured aerosol chemical and optical properties at Barrow, Alaska. *Journal of Geophysical Research, Atmospheres, 107*(D11), AAC 8-1–AAC 8-15. https://doi.org/10.1029/2001JD001248.

Quinn, P. K., Shaw, G., Andrews, E., Dutton, E. G., Ruoho-Airola, T., & Gong, S. L. (2007). Arctic haze: Current trends and knowledge gaps. *Tellus Series B, 59*(1), 99–114. https://doi.org/10.1111/j.1600-0889.2006.00238.x.

Quinn, P. K., Bates, T. S., Baum, E., Doubleday, N., Fiore, A. M., Flanner, M., Fridlind, A., Garrett, T. J., Koch, D., Menon, S., Shindell, D., Stohl, A., & Warren, S. G. (2008). Short-lived pollutants in the Arctic: Their climate impact and possible mitigation strategies. *Atmospheric Chemistry and Physics, 8*(6), 1723–1735. https://doi.org/10.5194/acp-8-1723-2008.

Rahn, K. A., & McCaffrey, R. J. (1980). On the origin and transport of the winter Arctic Aerosol. *Annals of the New York Academy of Sciences, 338*, 486–503.

Rahn, K. A., Borys, R. D., & Shaw, G. E. (1977). The Asian source of Arctic haze bands. *Nature, 268*, 713–715. https://doi.org/10.1038/268713a0.

Rankin, A. M., Auld, V., & Wolff, E. W. (2000). Frost flowers as a source of fractionated sea salt aerosol in the polar regions. *Geophysical Research Letters, 27*(21), 3469–3472. https://doi.org/10.1029/2000GL011771.

Rankin, A. M., Wolff, E. W., & Martin, S. (2002). Frost flowers: Implications for tropospheric chemistry and ice core interpretation. *Journal of Geophysical Research, Atmospheres, 107*(D23), 4683. https://doi.org/10.1029/2002JD002492.

Rinaldi, M., Decesari, S., Carbone, C., Finessi, E., Fuzzi, S., Ceburnis, D., D, C., O'Dowd, J., Sciare, J. P., Burrows, M., Vrekoussis, B., Ervens, K. T., & Facchini, M. C. (2011). Evidence of a natural marine source of oxalic acid and a possible link to glyoxal. *Journal of Geophysical Research, 116*, D16204. https://doi.org/10.1029/2011JD015659.

Rudnick, R. L., & Gao, S. (2003). Composition of the continental crust. *Treatise on Geochemistry, 3*, 12, 1–64, https://doi.org/10.1016/B0-08-043751-6/03016-4.

Samset, B. H., & Myhre, G. (2011). Vertical dependence of black carbon, sulfate and biomass burning aerosol radiative forcing. *Geophysical Research Letters*. https://doi.org/10.1029/2011gl049697.

Sand, M., Berntsen, T., Seland, Ø., & Kristjánsson, J. E. (2013). Arctic surface temperature change to emissions of black carbon within Arctic or midlatitudes. *Journal of Geophysical Research D: Atmospheres, 118*(14), 7788–7798.

Saxena, P., & Hildemann, L. (1996). Water-soluble organics in atmospheric particles: A critical review of the literature and application of thermodynamics to identify candidate compounds. *Journal of Atmospheric Chemistry, 24*(Suppl 1), 57–109. https://doi.org/10.1007/BF00053823.

Scheuer, E., Talbot, R. W., Dibb, J. E., Seid, G. K., DeBell, L., & Lefer, B. (2003). Seasonal distributions of fine aerosol sulfate in the North American Arctic basin during TOPSE. *Journal of Geophysical Research, Atmospheres, 108*(D4), 8370. https://doi.org/10.1029/2001JD001364.

Schnell, R. C. (1984). Arctic haze and the Arctic gas and aerosol sampling program (AGASP). *Geophysical Research Letters, 11*(5), 361–364.

Seinfeld, J. H., & Pandis, S. N. (2006). *Atmospheric chemistry and physics: From air pollution to climate change* (2nd ed.). New York: Wiley.

Sharma, S., Brook, J. R., Cachier, H., Chow, J., Gaudenzi, A., & Lu, G. (2002). Light absorption and thermal measurements of black carbon in different regions of Canada. *Journal of Geophysical Research, Atmospheres, 107*(D24), 4771. https://doi.org/10.1029/2002JD002496.

Sharma, S., Lavoué, D., Cachier, H., Barrie, L. A., & Gong, S. L. (2004). Long-term trends of the black carbon concentrations in the Canadian Arctic. *Journal of Geophysical Research, Atmospheres, 109*, D15203. https://doi.org/10.1029/2003JD004331.

Sharma, S., Andrews, E., Barrie, L. A., Ogren, J. A., & Lavoué, D. (2006). Variations and sources of the equivalent black carbon in the high Arctic revealed by long-term observations at Alert and Barrow: 1989–2003. *Journal of Geophysical Research, Atmospheres, 111*, D14208. https://doi.org/10.1029/2005JD006581.

Sharma, S., Chan, E., Ishizawa, M., Toom-Sauntry, D., Gong, S. L., Li, S. M. D., Tarasick, W., Leaitch, W. R., Norman, A., Quinn, P. K., Bates, T. S., Levasseur, M., Barrie, L. A., & Maenhaut, W. (2012). Influence of transport and ocean ice extent on biogenic aerosol sulfur in the Arctic atmosphere. *Journal of Geophysical Research, Atmospheres, 117*, D12. https://doi.org/10.1029/2011JD017074.

Sharma, S., Ishizawa, M., Chan, D., Lavoué, D., Andrews, E., Eleftheriadis, K., & Maksyutov, S. (2013). 16-year simulation of Arctic black carbon: Transport, source contribution, and sensitivity analysis on deposition. *Journal of Geophysical Research, Atmospheres, 118*(2), 943–964. https://doi.org/10.1029/2012JD017774.

Shaw, G. E. (1975). The vertical distribution of atmospheric aerosols at Barrow, Alaska. *Tellus, 27*, 39–50.

Shaw, G. E. (1983). Evidence for a central Eurasian source area of Arctic haze in Alaska. *Nature, 299*, 815–818.

Shaw, G. E. (1987). Aerosols as climate regulators: A climate biosphere linkage? *Atmospheric Environment, 21*, 985–1086.

Shaw, G. E. (1995). The Arctic haze phenomenon. *Bulletin of the American Meteorological Society, 76*, 2403–2412.

Shaw, G. E., Stamnes, K. (1980). Arctic haze: perturbation of the polar radiation budget. *Annals of the New York Academy of Sciences, 338*, 533–539.

Shaw, G. E., Stamnes, K., & Hu, Y. X. (1993). Arctic haze: Perturbation to the radiation field. *Meteorology and Atmospheric Physics, 51*(3), 227–235. https://doi.org/10.1007/BF01030496.

Sinha, P. R., Kondo, Y., Koike, M., Ogren, J. A., Jefferson, A., Barrett, T. E., Sheesley, R. J., Ohata, S., Moteki, N., Coe, H., Liu, D., Irwin, M., Tunved, P., Quinn, P. K., & Zhao, Y. (2017). Evaluation of ground-based black carbon measurements by filter-based photometers at two Arctic sites. *Journal of Geophysical Research, Atmospheres, 122*(6), 3544–3572. https://doi.org/10.1002/2016JD025843.

Sirois, A., & Barrie, L. A. (1999). Arctic lower tropospheric aerosol trends and composition at alert, Canada: 1980 – 1995. *Journal of Geophysical Research, Atmospheres, 104*(D9), 11599–11618. https://doi.org/10.1029/1999JD900077.

Steinberg, S. M., & Bada, J. L. (1984). Oxalic, glyoxalic and pyruvic acids in Eastern Pacific-Ocean waters. *Journal of Marine Research, 42*(3), 697–708. https://doi.org/10.1357/002224084788506068.

Stohl, A. (2006). Characteristics of atmospheric transport into the Arctic troposphere. *Journal of Geophysical Research D: Atmospheres, 111*. https://doi.org/10.1029/2005JD006888.

Stohl, A., Eckhardt, S., Forster, C., James, P., & Spichtinger, N. (2002). On the pathways and timescales of intercontinental air pollution transport. *Journal of Geophysical Research D: Atmospheres, 107*. https://doi.org/10.1029/2002JD002862.

Stohl, A., Andrews, E., Burkhart, J. F., Forster, C., Herber, A., Hoch, S. W., Kowal, D., Lunder, C., Mefford, T., Ogren, J. A., Sharma, S., Spichtinger, N., Stebel, K., Stone, R. S., Storm, J., Torseth, K., Wehrli, C., & Yttri, K. E. (2006). Pan-Arctic enhancements of light absorbing aerosol concentrations due to North American boreal forest fires during summer 2004. *Journal of Geophysical Research D: Atmospheres, 111*. https://doi.org/10.1029/2006JD007216.

Stone, R. S., Herber, A., Vitale, V., Mazzola, M., Lupi, A., Schnell, R. C., Dutton, E. G., Liu, P. S. K., Li, S. M., Dethloff, K., Lampert, A., Ritter, C., Stock, M., Neuber, R., & Maturilli, M. (2010). A three-dimensional characterization of Arctic aerosols from airborne Sun photometer observations: PAM-ARCMIP, April 2009. *Journal of Geophysical Research: Atmospheres, 115*(D13), D13203. https://doi.org/10.1029/2009JD013605.

Stone, R. S., Sharma, S., Herber, A., Eleftheriadis, K. S., & Nelson, D. W. (2014). A characterization of Arctic aerosols on the basis of aerosol optical depth and black carbon measurements. *Elementa, 2*. https://doi.org/10.12952/journal.elementa.000027.

Strunin, M. A., Postnov, A. A., & Mezrin, M. Y. (1997). Meteorological potential for contamination of arctic troposphere: Boundary layer structure and turbulent diffusion characteristics. *Atmospheric Research, 44*, 37–51.

Sturges, W. T., & Barrie, L. A. (1988). Chlorine, bromine, and iodine in Arctic aerosols. *Atmospheric Environment, 22*(6), 1179–1194. https://doi.org/10.1016/0004-6981(88)90349-6.

Tedetti, M., Kawamura, K., Charrière, B., Chevalier, N., & Sempéré, R. (2006). Determination of low molecular weight dicarboxylic and ketocarboxylic acids in seawater samples. *Analytical Chemistry, 78*(17), 6012–6018. https://doi.org/10.1021/ac052226w.

Teinilä, K., Hillamo, R., Kerminen, V.-M., & Beine, H. J. (2003). Aerosol chemistry during the NICE dark and light campaigns. *Atmospheric Environment, 37*, 563–575. https://doi.org/10.1016/S1352-2310(02)00826-9.

Tomasi, C. (1982). Features of the scale height for particulate extinction in hazy atmospheres. *Journal of Applied Meteorology, 21*(7), 931–944. https://doi.org/10.1175/1520-0450(1982)0210931:FOTSHF>2.0.CO;2.

Tomasi, C., Vitale, V., Lupi, A., Di Carmine, C., Campanelli, M., Herber, A., Treffeisen, R., Stone, R. S., Andrews, E., Sharma, S., Radionov, V., von Hoyningen-Huene, W., Stebel, K., Hansen, G. H., Myhre, C. L., Wehrli, C., Aaltonen, V., Lihavainen, H., Virkkula, A., Hillamo, R., Ström, J., Toledano, C., Cachorro, V., Ortiz, P., de Frutos, A., Blindheim, S., Frioud, M., Gausa, M., Zielinski, T., Petelski, T., & Yamanouchi, T. (2007). Aerosols in polar regions: A historical overview based on optical depth and in situ observations. *Journal of Geophysical Research, 112*, D16205. https://doi.org/10.1029/2007JD008432.

Tomasi, C., Lupi, A., Mazzola, M., Stone, R. S., Dutton, E. G., Herber, A., Radionov, V. F., Holben, B. N., Sorokin, M. G., Sakerin, S. M., Terpugova, S. A., Sobolewski, P. S., Lanconelli, C., Petkov, B. H., Busetto, M., & Vitale, V. (2012). An update on polar aerosol optical properties using POLAR-AOD and other measurements performed during the International Polar Year. *Atmospheric Environment, 52*(C), 29–47. https://doi.org/10.1016/j.atmosenv.2012.02.055.

Tomasi, C., Kokhanovsky, A. A., Lupi, A., Ritter, C., Smirnov, A., O'Neill, N. T., Stone, R. S., Holben, B. N., Nyeki, S., Wehrli, C., Stohl, A., Mazzola, M., Lanconelli, C., Vitale, V., Stebel, K., Aaltonen, V., de Leeuw, G., Rodriguez, E., Herber, A. B., Radionov, V. F., Zielinski, T., Petelski, T., Sakerin, S. M., Kabanov, D. M., Xue, Y., Mei, L., Istomina, L., Wagener, R.,

McArthur, B., Sobolewski, P. S., Kivi, R., Courcoux, Y., Larouche, P., Broccardo, S., & Piketh, S. J. (2015). Aerosol remote sensing in polar regions. *Earth-Science Reviews, 140*, 108–157. https://doi.org/10.1016/j.earscirev.2014.11.001.

Traversi, R., Becagli, S., Caiazzo, L., Calzolai, G., Chiari, M., Giardi, F., Lucarelli, F., Lupi, A., Mazzola, M., Petkov, B., Severi, M., Vitale, V., & Udisti, R. (2017, April 4–7). *Aerosol chemical composition at Ny-Ålesund (Svalbard Islands): A focus on elemental and organic carbon*. Oral presentation held at the Arctic Science Summit Week (ASSW) 2017, Prague.

Tunved, P., Ström, J., & Krejci, R. (2013). Arctic aerosol life cycle: Linking aerosol size distributions observed between 2000 and 2010 with air mass transport and precipitation at Zeppelin station, Ny-Ålesund, Svalbard. *Atmospheric Chemistry and Physics, 13*, 3643–3660. https://doi.org/10.5194/acp-13-3643-2013.

Udisti, R., Becagli, S., Benassai, S., Castellano, E., Fattori, I., Innocenti, M., Migliori, A., & Traversi, R. (2004). Atmosphere-snow interaction by a comparison between aerosol and uppermost snow-layers composition at Dome C, East Antarctica, Seventh International Symposium on Antarctic Glaciology. *Annals of Glaciology, 39*(6), 53–61. https://doi.org/10.3189/172756404781814474.

Udisti, R., Dayan, U., Becagli, S., Busetto, M., Frosini, D., Legrand, M., Lucarelli, F., Preunkert, S., Severi, M., Traversi, R., & Vitale, V. (2012). Sea spray aerosol in central Antarctica. Present atmospheric behaviour and implications for paleoclimatic reconstructions. *Atmospheric Environment, 52*(C), 109–120. https://doi.org/10.1016/j.atmosenv.2011.10.018.

Udisti, R., Becagli, S., Frosini, D., Ghedini, C., Rugi, F., Severi, M., Traversi, R., Zanini, R., Calzolai, G., Chiari, M., Lucarelli, F., Nava, S., Ardini, F., Grotti, M., Vione, D., Malandrino, M., Bolzacchini, E., Ferrero, L., Perrone, M. G., Sangiorgi, G., Francesconi, S., Giannarelli, S., Cappelletti, D., Moroni, B., Ceccato, D., Mittner, P., & Sartori, P. (2013) Activity and preliminary results from the 2011 and 2012 field seasons at Ny Ålesund. In: Research Activity in Ny Ålesund 2011–12. CNR Editions DTA/14-2013. ISSN 2239-5172, pp. 53–68.

Udisti, R., Bazzano, A., Becagli, S., Bolzacchini, E., Caiazzo, L., Cappelletti, D., Ferrero, L., Frosini, D., Giardi, F., Grotti, M., Lupi, A., Malandrino, M., Mazzola, M., Moroni, B., Severi, M., Traversi, R., Viola, A., & Vitale, V. (2016). Sulfate source apportionment in the Ny-Ålesund (Svalbard Islands) Arctic aerosol. *Rendiconti Lincei, Scienze Fisiche e Naturali, 27*(Suppl 1), 85–94. https://doi.org/10.1007/s12210-016-0517-7.

Uematsu, M., Duce, R. A., & Prospero, J. M. (1985). Deposition of atmospheric mineral particles in the North Pacific Ocean. *Journal of Atmospheric Chemistry, 3*(1), 123–138. https://doi.org/10.1007/BF00049372.

VanCuren, R. A., Cahill, T., Burkhart, J., Barnes, D., Zhao, Y., Perry, K., Cliff, S., & McConnel, J. (2012). Aerosols and their sources at summit Greenland – First results of continuous size- and time-resolved sampling. *Atmospheric Environment, 52*(C), 82–97. https://doi.org/10.1016/j.atmosenv.2011.10.047.

Vermote, E. F., Tanré, D., Deuzé, J. L., Herman, M., & Morcrette, J.-J. (1997). Second simulation of the satellite signal in the solar spectrum (6S): An overview. *IEEE Transactions on Geoscience and Remote Sensing, 35*(3), 675–686. https://doi.org/10.1109/36.581987.

Viana, M., Kuhlbusch, T. A. J., Querol, X., Alastuey, A., Harrison, R. M., Hopke, P. K., Winiwarter, W., Vallius, M., Szidat, S., Prévôt, A. S. H., Hueglin, C., Bloemen, H., Wåhlin, P., Vecchi, R., Miranda, A. I., Kasper-Giebl, A., Maenhaut, W., & Hitzenberger, R. (2008). Source apportionment of particulate matter in Europe: A review of methods and results. *Journal of Aerosol Science, 39*(10), 827–849. https://doi.org/10.1016/j.jaerosci.2008.05.007.

Virkkula, A., Ahlquist, N. C., Covert, D. S., Arnott, W. P., Sheridan, P. J., Quinn, P. K., & Coffman, D. J. (2005). Modification, calibration and a field test of an instrument for measuring light absorption by particles. *Aerosol Science and Technology, 39*(1), 68–83. https://doi.org/10.1080/027868290901963.

Volkamer, R., Jimenez, J. L., San Martini, F., Dzepina, K., Zhang, Q., Salcedo, D., Molina, L. T., Worsnop, D. R., & Molina, M. J. (2006). Secondary organic aerosol formation from anthropogenic air pollution: Rapid and higher than expected. *Geophysical Research Letters, 33*, L17811. https://doi.org/10.1029/2006GL026899.

Wagenbach, D., Preunkert, S., Schäfer, J., Jung, W., & Tomadin, L. (1996). Northward transport of Saharan dust recorded in a deep alpine ice core. In: S. Guerzoni & R. Chester (Eds.), *The impact of desert dust across the Mediterranean* (Environmental science and technology library, Vol. 11, pp. 291–300), Dordrecht: Springer, .

Wagenbach, D., Ducroz, F., Mulvaney, R., Keck, L., Minikin, A., Legrand, J., Hall, J. S., & Wolff, E. W. (1998). Sea-salt aerosol in coastal Antarctic regions. *Journal of Geophysical Research, Atmospheres, 103*(D9), 10961–10974. https://doi.org/10.1029/97JD01804.

Waggoner, A. P., & Weiss, R. E. (1980). Comparison of the fine particle mass concentration and light scattering extinction in ambient aerosol. *Atmospheric Environment, 14*, 623–626.

Warneke, C., Bahreini, R., Brioude, J., Brock, C. A., de Gouw, J., Fahey, D. W., Froyd, K. D., Holloway, J., Middlebrook, A., Miller, L., Montzka, S., Murphy, D. M., Peischl, J., Ryerson, T. B., Schwarz, J. P., Spackman, J. R., & Veres, P. (2009). Biomass burning in Siberia and Kazakhstan as an important source for haze over the Alaskan Arctic in April 2008. *Geophysical Research Letters, 36*. https://doi.org/10.1029/2008GL036194.

Warneke, C., Froyd, K. D., Brioude, J., Bahreini, R., Brock, C. A., Cozic, J., de Gouw, J., Fahey, D. W., Ferrare, R. A., Holloway, J., Middlebrook, A., Miller, L., Montzka, S., Schwarz, J. P., Sodemann, H., Spackman, J. R., & Stohl, A. (2010). An important contribution to springtime Arctic aerosol from biomass burning in Russia. *Geophysical Research Letters, 37*(1). https://doi.org/10.1029/2009GL041816.

Winiger, P., Andersson, A., Yttri, K. E., Tunved, P., & Gustafsson, O. (2015). Isotope-based source apportionment of EC aerosol particles during winter high-pollution events at the zeppelin observatory, Svalbard. *Environmental Science and Technology, 49*, 11959–11966.

Wolff, E. W., Rankin, A. M., & Röthlisberger, R. (2003). An ice core indicator of Antarctic Sea ice production? *Geophysical Research Letters, 30*(22). https://doi.org/10.1029/2003GL018454.

Wolff, E. W., Fischer, H., Fundel, F., Ruth, U., Twarloh, B., Littot, G. C., Mulvaney, R., Röthlisberger, R., de Angelis, M., Boutron, C. F., Hansson, M., Jonsell, U., Hutterli, M. A., Lambert, F., Kaufmann, P., Stauffer, B., Stocker, T. F., Steffensen, J. P., Bigler, M., Siggaard-Andersen, M. L., Udisti, R., Becagli, S., Castellano, E., Severi, M., Wagenbach, D., Barbante, C., Gabrielli, P., & Gaspari, V. (2006). Southern Ocean sea-ice extent, productivity and iron flux over the past eight glacial cycles. *Nature, 440*, 491–496. https://doi.org/10.1038/nature04614.

Yang, M., Beale, R., Liss, P., Johnson, M., Blomquist, B., & Nightingale, P. (2014). Air–sea fluxes of oxygenated volatile organic compounds across the Atlantic Ocean. *Atmospheric Chemistry and Physics, 14*, 7499–7517. https://doi.org/10.5194/acp-14-7499-2014.

Zhai, J., Lu, X., Li, L., Zhang, Q., Zhang, C., Chen, H., Yang, X., & Chen, J. (2017). Size-resolved chemical composition, effective density, and optical properties of biomass burning particles. *Atmospheric Chemistry and Physics, 17*, 7481–7493. https://doi.org/10.5194/acp-17-7481-2017.

Zhang, Q., Jimenez, J. L., Canagaratna, M. R., Allan, J. D., Coe, H., Ulbrich, I., Alfarra, M., Takami, A., Middlebrook, A. M., Sun, Y., Dzepina, K., Dunlea, E., Docherty, K., DeCarlo, P. F., Salcedo, D., Onasch, T., Jayne, J. T., Miyoshi, T., Shimono, A., Hatakeyama, S., Takegawa, N., Kondo, Y., Schneider, J., Drewnick, F., Borrmann, S., Weimer, S., Demerjian, K., Williams, P., Bower, K., Bahreini, R., Cottrell, L., Griffin, R. J., Rautiainen, J., Sun, J. Y., Zhang, Y. M., & Worsnop, D. R. (2007). Ubiquity and dominance of oxygenated species in organic aerosols in anthropogenically-influenced Northern Hemisphere midlatitudes. *Geophysical Research Letters, 34*, L13801. https://doi.org/10.1029/2007GL029979.

Chapter 5
A Climatological Overview of Arctic Clouds

Abhay Devasthale, Joseph Sedlar, Michael Tjernström, and Alexander Kokhanovsky

Abstract The Arctic climate system is complex and clouds are one of its least understood components. Since cloud processes occur from micrometer to synoptic scales, their couplings with the other components of the Arctic climate system and their overall role in modulating the energy budget at different spatio-temporal scales is challenging to quantify. The in-situ measurements, as limited in space and time as they are, still reveal the complex nature of cloud microphysical and thermodynamical processes in the Arctic. However, the synoptic scale variability of cloud systems can only be obtained from the satellite observations. A considerable progress has been made in the last decade in understanding cloud processes in the Arctic due to the availability of valuable data from the multiple campaigns in the Central Arctic and due to the advances in the satellite remote sensing. This chapter provides an overview of this progress.

First an overview of the lessons learned from the recent in-situ measurement campaigns in the Arctic is provided. In particular, the importance of supercooled liquid water clouds, their role in the radiation budget and their interaction with the vertical thermodynamical structure is discussed. In the second part of the chapter, a

A. Devasthale
Atmospheric remote sensing unit, Research and development department, Swedish Meteorological and Hydrological Institute (SMHI), Norrköping, Sweden
e-mail: abhay.devasthale@smhi.se

J. Sedlar
Cooperative Institute for Research in Environmental Sciences (CIRES), University of Colorado Boulder, Boulder, CO, USA
e-mail: joseph.sedlar@Colorado.edu

M. Tjernström
Department of Meteorology, University of Stockholm (MISU), Stockholm, Sweden

Bert Bolin Center for Climate Research, University of Stockholm, Stockholm, Sweden
e-mail: michaelt@misu.su.se

A. Kokhanovsky (✉)
Vitrociset Belgium SPRL, Darmstadt, Germany
e-mail: a.kokhanovsky@vitrocisetbelgium.com

A. Kokhanovsky, C. Tomasi (eds.), *Physics and Chemistry of the Arctic Atmosphere*, Springer Polar Sciences, https://doi.org/10.1007/978-3-030-33566-3_5

climatological overview of cloud properties using the state-of-the-art satellite based cloud climate datasets is provided. The agreements and disagreements in these datasets are highlighted. The third and the fourth parts of the chapter highlight two most important processes that are currently being researched, namely cloud response to the rapidly changing sea-ice extent and the role of moisture transport in to the Arctic in governing cloud variability. Both of these processes have implications for the cloud feedback in the Arctic.

Keywords Arctic clouds · Cloud properties · Supercooled liquid clouds · Surface radiation budget · Satellite remote sensing · Moisture transport · Thermodynamic structure · Cloud variability · Arctic sea-ice · Climate data records

We understand little about the enigmatic character of clouds and their role in the Arctic climate system. The in-situ measurements reveal, to a certain extent, the complex nature of cloud and thermodynamical processes in the Arctic. However, assessing the extent of the observed cloud processes and their couplings with the surface in the entire Arctic remain a challenge. Strong and heterogeneous seasonal changes in the Arctic ocean surface and atmospheric variability complicate such assessment. It mainly comes down to the fact that the current observational systems are still not good enough to provide research and climate quality information on cloud properties in the polar regions. At the same time, considerable progress has also been made in the last decade in understanding cloud processes in the Arctic. The aim of this chapter is to provide an overview of the climatological properties of Arctic clouds, based on the state-of-the-art datasets. This chapter is divided into three parts. First the lessons learned on the Arctic clouds from the campaign measurements in the central Arctic are discussed. Then the climatological perspective of clouds is presented based on space based satellite observations. And lastly the two key processes affecting Arctic clouds in relation to rapid decline in the Arctic sea ice are discussed. The climatological microphysical properties of Arctic clouds, including liquid and ice water paths, from the space based passive sensors are not discussed here, as large uncertainties still exist in their retrievals due to underlying bright surfaces typical for the Arctic, thus preventing any meaningful quantitative analysis.

5.1 Lessons Learned from the Campaign Measurements

5.1.1 Frequent Presence of Cloud Liquid

During the summer of 1980, a series of flight campaigns across the Beaufort Sea measured in situ cloud droplet size distributions and concentrations through lower tropospheric clouds (Herman and Curry 1984). Most of these flight profiles occurred across warm clouds, clouds with ambient temperatures above freezing. However three profiles occurred across layers where the temperatures were significantly below freezing, −9° to −15 °C. The FSSP instrument identified supercooled liquid droplets in these clouds, with integrated liquid water paths (LWPs) ranging from 5 to 15 g m^{-2} (Curry and Herman 1985). A subsequent surface-based field campaign based on a drifting ice floe north of Alaska in April 1992 (Ruffieux et al. 1992) combined radiative fluxes, laser ceilometer and radiosounding measurements to characterize the cloud phase and its impact on the surface energy budget. Analysis of these data indicated the common presence of liquid droplets, often coincidentally with the presence of ice crystals, known as mixed-phase clouds (Pinto et al. 1997). Cloud liquid occurred at in-cloud temperatures ranging from −25° to −13 °C (Pinto et al. 1997). Supercooled liquid-only and mixed-phase clouds over the Beaufort Sea were later observed from in situ flight campaigns during September and October in 1994 at temperatures as cold as −18 °C (BASE; Curry et al. 1997). While these campaigns were restricted to the Pacific Arctic north of Alaska, they were significant in that they observed the presence of cloud liquid during spring, summer and autumn seasons.

The first indication that liquid water is present within Arctic clouds over the entire year came from the unprecedented year-long ice drift campaign the Surface Heat Budget of the Arctic during 1997–1998 (SHEBA, Uttal et al. 2002). Profiles of backscatter and depolarization ratio from near-continuous operation of the Depolarization and Backscatter Unattended Lidar instrument were combined with independent thermodynamic profiles from radiosoundings and column-integrated LWP retrievals from the microwave radiometer. The synergy of instruments provided a unique capacity to classify the occurrence of cloudiness and cloud phase. Together with cloud radar measurements, the observations revealed that cloud liquid, either in liquid-only or a mixed-phase composition, was present during all months of the year in the Beaufort Sea, least frequent during December 1997 and most frequent during July 1998 (Intrieri et al. 2002a). At SHEBA, liquid was observed at temperatures as cold as −34 °C (Intrieri et al. 2002a). Comparing the SHEBA annual observations of cloud liquid with multiyear surface-based remote sensing stations surrounding the Arctic, Shupe (2011) corroborated the finding of frequent mixed-phase and liquid-only clouds during all seasons across the lower troposphere below 2 km.

Profiles of the vertical distribution of liquid water content (LWC) in Arctic clouds have been available from aircraft profiling and from surface-based cloud radar retrievals. In general, LWC is observed to increase with height across the cloud layer, reaching a maximum slightly below the observed cloud top (e.g., Herman and Curry 1984; Curry and Herman 1985; Curry 1986; Shupe et al. 2001). Increasing LWC from cloud base to top mimics the profile shape of LWC from adiabatic ascent, however considerable dispersion and inhomogeneity often causes a departure in the observed or retrieved LWC from the adiabatic estimate at a particular level (Curry 1986; Shupe et al. 2005). When these liquid cloud layers support the formation of ice crystals, generally the ice water content increases downward, as crystals grow and fall from the cloud layer towards the surface, often accumulating mass from collisions with cloud droplets (Shupe et al. 2005).

5.1.2 Surface Cloud Radiative Forcing at High Latitudes

The surface energy budget over the Arctic Ocean is largely controlled by the relatively large fluxes of radiation. Surface radiation energy balance of the high-latitude polar region is dominated by infrared, or longwave, radiation during much of the year because (1) the sun is below the horizon from late autumn to late spring; and (2) the surface albedo is high due to highly reflective snow and sea ice cover blanketing the Arctic Ocean (e.g., Curry et al. 1996). The presence of clouds, and their microphysical composition, represents an important medium interacting with visible and infrared electromagnetic radiation. Liquid and ice hydrometeors scatter and absorb visible, near-infrared, and infrared radiation differently, impacting the effective optical depth of the cloud layer for each of these portions of the radiative spectrum. For example, constraining radiative transfer model simulations with observed broadband surface radiative fluxes from Arctic clouds revealed that longwave optical depth and downwelling fluxes reach an asymptotic value; here the clouds mimic blackbody radiation emitters, and the longwave flux becomes solely a function of the emission temperature of the cloud layer (Stephens 1978; Pinto et al. 1997). For shortwave radiation simulations, larger optical depths were required for liquid-only clouds relative to ice-only clouds in order to arrive at the observed surface radiative measurements (Pinto et al. 1997). The influence of shortwave optical depth on the net surface flux is also critically dependent on the surface albedo, which can vary drastically with the myriad of surface characteristics over the Arctic Ocean.

Cloud radiative forcing (CRF) combines the competing effects of the cloud greenhouse surface warming in the longwave with the surface cooling in the shortwave due to reduced solar transmission and reflection back to space. The presence of a cloud drastically increases the downwelling longwave radiative flux, and this increase was easily distinguishable in a time series relative to the observed downwelling longwave flux signature when clouds were absent (clear skies) in autumn of 1994 (e.g., Curry et al. 1997). The difference between the all-sky and clear-sky longwave radiative flux is a measure of the longwave CRF. In terms of shortwave radiation, Herman and Curry (1984) highlighted the important similar relationship between shortwave radiation reflectivity and transmissivity as a function of cloud optical depth by combining radiative transfer modeling and in situ observations performed in summer of 1980.

The SHEBA campaign provided the first observationally-based evidence detailing how CRF varies over the Arctic sea ice for a full annual cycle. Monthly averages of the longwave (LW) CRF typically varied between 20 and 60 W m^{-2}, resulting in a longwave surface warming, while shortwave (SW) CRF resulted in a surface cooling and ranged from 0 W m^{-2} during the dark months and greater than -50 W m^{-2} in summer (Intrieri et al. 2002b; Shupe and Intrieri 2004). Total CRF, the net of LW and SW CRF, was positive for the majority of the year, except a brief period in early July when a lower surface albedo increased the net SW radiation at the surface. Thus the presence of clouds over the Arctic sea ice typically led to a local surplus of radiative energy at the surface. This total CRF magnitude was dependent upon a number of factors. The amount of fractional cloudiness, the cloud phase and cloud particle size, the amount of integrated LWP, the emission temperature of the cloud layer, the solar zenith angle and the surface albedo were all shown to be critically important in determining the net effect of the CRF at the surface (Shupe and Intrieri 2004; Sedlar et al. 2011; Persson et al. 2017).

Outside of peak summer months, Arctic clouds typically warm the surface. This forcing becomes especially important during the transition from the summer sea ice melt season to the initiation of the freeze up in autumn. During August to early September of 2008, measurements from a central Arctic ice-drift campaign, the Arctic Summer Cloud Ocean Study (ASCOS, Tjernström et al. 2014), managed to observe the transition from melt to freeze and quantify the role of synoptic variability and associated cloud variations on the CRF and surface energy budget. During ASCOS, the CRF was always positive, due to a combination of supercooled liquid within low-level mixed-phase clouds and increasing solar zenith angles (Sedlar et al. 2011). A frontal passage associated with a large scale synoptic shift in atmospheric flow led to a dissipation of the low level mixed-phase cloud for a period of 2 days; an upper tropospheric ice cloud advected over the ice-camp and subsequently the LW CRF dropped from ~ 75 W m^{-2} down to ~ 10 W m^{-2} (Sedlar et al. 2011). The ice cloud had a lower LW emissivity due to the absence of liquid droplets. Subsequently, the near-surface air temperature dropped from near 0 °C down to -6 °C. Melt ponds at the surface initially began to freeze and become covered with hoar frost. This process, combined with a synoptic snowfall event at the very end of these 2 days, led to an increase in the surface albedo. This prevented additional surface absorption of shortwave radiation. For 1 week after this event, lower level mixed-phase clouds returned and the total CRF responded back to the 50–70 W m^{-2} range; this week long period was pivotal in mitigating the full transition from melt to freeze. When the mixed-phase cloud dissipated once again at the end of the week, the combination of increased surface albedo, increasing solar zenith angles, and lack of supercooled liquid had preconditioned the surface for the onset of the seasonal freeze (Sedlar et al. 2011).

Sotiropoulou et al. (2016) concluded that a similar series of events, a large-scale synoptic shift in atmospheric thermodynamics and brief periods of cloud variability, ultimately caused the transition from melt to freeze season during the Arctic Clouds during Summer Experiment (ACSE) in 2014. These observational studies stress the critical forcing that clouds, particularly those containing supercooled liquid, have on the Arctic surface energy budget and subsequent seasonal transitions.

5.1.3 Cloud Interaction with Thermodynamic Structure

From aircraft flight data during the BASE campaign during June 1980, Curry (1986) identified an intimate connection between the low-level cloud layers and temperature inversion structures near cloud top. Observations of LWC occasionally penetrated into the stable temperature inversion layer, coinciding with sharp but often geometrically thin specific humidity mixing ratios that increased with height. Cloud condensate penetrating above the temperature inversion base is in contrast to the fundamental understanding of marine stratiform clouds in the subtropics and midlatitudes. These southerly clouds are limited in their vertical extent by dramatic decreases in specific humidity at and within the inversion layer which is the linkage between the boundary layer and free troposphere (e.g., Paluch and Lenschow 1991). Direct in situ measurements from the early flight campaigns indicated that a strong decrease in specific humidity across the temperature inversion structure over the Arctic may not be as common a phenomenon as at lower latitudes.

Frequent specific humidity inversions simultaneously located with low-level temperature inversions over the central Arctic sea ice were again identified from the Arctic Ocean Expedition in summer 2001 (Tjernström et al. 2004). Thermodynamic profiles from 2 consecutive mid-morning radiosoundings over the sea ice during ASCOS in 2008 (Fig. 5.1) highlight the relationship between temperature and moisture inversions. Both profiles are characterized by a strong temperature inversion, with an inversion base height that has risen from approximately 700–1100 m over the course of 6 hours (Fig. 5.1a, b); the temperature inversion structure also becomes more well-defined and geometrically thinner with time. Both profiles also capture a specific humidity inversion that is collocated with the temperature inversion, albeit with different physical properties. The specific humidity inversion during the early morning sounding (Fig. 5.1c, gray) is relatively weak in magnitude, about 0.3 g kg^{-1} increase in specific humidity, compared to over 1 g kg^{-1} observed at mid day (Fig. 5.1c, black). Additionally, they are not as deep as the temperature inversions with which they are associated (Fig. 5.1a, c), although a secondary, stronger specific humidity inversion located higher within the temperature inversion was also observed (Fig. 5.1c, gray).

Profiles of relative humidity (Fig. 5.1e, f) and cloud boundaries derived from vertically-pointing remote sensing instruments (dashed horizontal lines) indicate a unique connection between the cloud layers and the inversion layers. The cloud top during both radiosounding profiles is not limited vertically to the temperature inversion base level, but instead is roughly 100–200 m above inversion base. This cloud top inversion penetration has been examined using a number of in situ and remotely-sensed observations across the Arctic. Results show that clouds over sea ice were twice as likely to penetrate and persist within the lower tropospheric temperature inversion than to be capped at the inversion base level like their canonical lower latitude cloud counterparts (Sedlar and Tjernström 2009; Sedlar et al. 2012). And when the clouds penetrate above the inversion base, they tend to do so by 75–200 m. A specific humidity inversion was present within the temperature

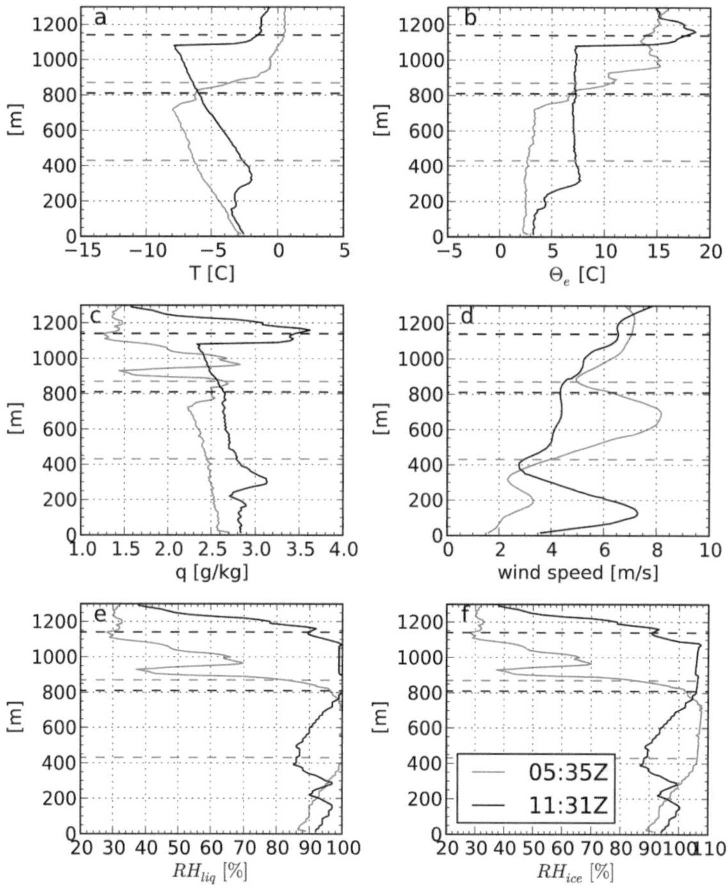

Fig. 5.1 Radiosounding profiles from the ASCOS field campaign on 28 August 2008 at 05:35 UTC (gray) and 11:31 UTC (black). The soundings were launched from within the sea ice at 87.3°N, 9.4°W. (**a**) Temperature [°C]; (**b**) equivalent potential temperature [°C]; (**c**) specific humidity [g kg^{-1}]; (**d**) wind speed [m s^{-1}]; (**e**) relative humidity with respect to liquid [%]; (**f**) relative humidity with respect to ice [%]. Dashed horizontal lines indicate the cloud base (lower) and cloud top (upper) levels at the local sounding time. All profiles a function of height [m above surface]

inversion for over 90% of the observations when clouds tops were within the inversion layer. Even 60% clouds that were capped at the temperature inversion base contained collocated specific humidity and temperature inversion structures (Sedlar et al. 2012). Specific humidity inversions were stronger and slightly deeper when clouds were observed to penetrate the inversion, rather than when clouds were capped by the inversion (Sedlar et al. 2012). Thus, the presence of a specific humidity inversion appears to be a necessary but not sufficient physical feature permitting cloud top penetration within the temperature inversion.

Tjernström et al. (2004) and Sedlar et al. (2012) hypothesized that increasing specific humidity could potentially serve as an important source of moisture to the boundary layer; however the transport of moisture across the stable boundary layer – free troposphere interface would require the production of turbulent motions, like those generated from cloud entrainment and detrainment processes. The observational results of Sedlar and Tjernström (2009) and Sedlar et al. (2012) indicate that clouds typically interact with the lower tropospheric inversion structures in a manner that potentially could provide the link between moisture inversions due to advection and the transport of moisture down into the turbulent boundary layer. A large eddy simulation modeling study by Solomon et al. (2011) managed to simulate a mixed-phase cloud layer with a cloud top slightly penetrating and surviving within the inversion layer. The authors found that weak radiative cooling forced by equally weak turbulent motions within this statically stable layer were enough the sustain liquid condensate formation. Budgets of the total water content indicated that moisture could be entrained due to cloud-generated turbulent mixing and transported into the cloud layer. Together with weak sedimentation of droplets from above the inversion, these processes could successfully replenish cloud water lost from ice crystal precipitation, and potentially serve as a fundamental mechanism for Arctic cloud persistence (Solomon et al. 2011).

5.1.4 Surface to Cloud Decoupling Structure

While strong static stability is common for the wintertime Arctic boundary layer (e.g., Kahl 1990), during summer the situation is more complex and dependent upon a number of physical atmospheric features. Tjernström and Graversen (2009) examined all soundings from SHEBA and found stably stratified conditions about half the time in winter, but hardly at all in summer. Tjernström et al. (2004) showed that the 2001 Arctic summer boundary layer was most frequently well mixed and often only a few hundred meters in depth; surface-based inversions were virtually never observed. Analysis from more recent in situ summer and autumn field campaigns within the high latitude Arctic sea ice have further improved the understanding of the physical composition of the cloudy boundary layer structure.

These studies have revealed how low-level stratiform clouds are an important feature contributing to the stability structure of the lower troposphere. Through radiative divergence and cloud-top cooling, cloud layers are able to generate turbulent eddies and the turbulent kinetic energy causes a mixing of air parcels between the cloud and sub-cloud layers. When turbulent mixing extends from the cloud layer down to the surface, the cloud is statically coupled with the surface. The 05:35 UTC sounding in Fig. 5.1 (gray) shows an example of a coupled surface-to-cloud layer. A decoupled surface-cloud layer indicates a stability structure where a stable interface separates the cloud driven mixed layer above from a turbulent surface layer below. An example of a decoupled cloud layer is shown in the 11:31 UTC sounding (Fig. 5.1, black). Detailed analysis of cloud and thermodynamic observations from the

ASCOS 2008 and ACSE 2014 summer-autumn field campaigns revealed that the surface-cloud decoupled structure is the most frequent stratification regime, nearly at a ratio of 3:1 to the coupled structure (Shupe et al. 2013; Sedlar and Shupe 2014; Sotiropoulou et al. 2014; Brooks et al. 2017). These studies found that the decoupling height, the height where the cloud-driven mixed layer and the surface-mixed layer become thermodynamically separated, often ranges from 100–300 m above the surface (Brooks et al. 2017). However, clouds that had base heights that were more elevated above the surface often had a higher decoupling height (Sotiropoulou et al. 2014).

When decoupling between the cloud and surface exist, the cloud cannot directly interact with the surface. The source of heat and moisture (and potentially cloud condensation nuclei) for the cloud layer that the surface represents, via vertical transport through turbulent eddies, is essentially absent when decoupling occurs (Brooks et al. 2017). Sedlar and Shupe (2014) performed a spectral analysis of in-cloud vertical velocities derived from 35 GHz cloud radar retrievals to determine how the coupling state of the surface-cloud layer impacts cloud-driven buoyancy overturning. They concluded that the dominant overturning circulation timescales ranged from 2–6 min when the system was coupled, and 5–9 min when decoupling occurred. Relatively similar peak horizontal wavelengths associated with these modest overturning timescale differences suggested that the dominant variability within the in-cloud vertical motions was driven by the cloud layer, and not necessarily dependent upon the surface-cloud coupling state; the cloud layer was resilient to the state of static coupling between cloud layer and surface (Sedlar and Shupe 2014). Sedlar and Shupe (2014) did find that the absolute magnitude of vertical velocity variance was larger when the cloud and surface layer were coupled, compared to decoupled. And with this increased variance, the amount of cloud liquid also increased (Shupe et al. 2008; Shupe et al. 2013; Sedlar and Shupe 2014), suggesting a feedback loop where increased vertical motion drives cloud condensate and cloud lifecycle.

However, the decoupled state has been concluded as the dominant surface-cloud coupling state, and yet these clouds still persist for long durations (Sedlar et al. 2011; Shupe et al. 2011). Here the unique thermodynamic structure where the cloud top penetrates the stable temperature inversion, often within a collocated specific humidity inversion layer, may be critical to sustaining the cloud (e.g., Morrison et al. 2012; Sedlar et al. 2012). South of the sea ice edge, heat and moisture may be transported horizontally in over the sea ice towards the central Arctic. Because sea ice is melting during the summer months, the temperature is constrained, causing the near surface air mass to transform to the sea ice boundary condition (e.g., Tjernström et al. 2015). Eventually, radiative divergence occurs and cloud layers form (Herman and Goody 1976). The condensation of water vapor into cloud may further lead to thin, interstitial vertical layers of specific humidity advected in over the sea ice. These layers are likely to form the specific humidity inversions so readily present across the Arctic, and may also be critical in determining the coupling nature between the surface and cloud layer.

5.2 Climatological View of Arctic Clouds from the Satellite Climate Data Records

The harsh environmental conditions in the Arctic Ocean make it difficult to perform permanent in-situ measurements. The spaceborne observing systems have been helping to remotely observe clouds in the Arctic since the late seventies. Especially, the data from the Advanced Very High Resolution Radiometers (AVHRRs), a five-channel imager flown onboard a series of NOAA and also recently onboard EUMETSAT's MetOp satellites, have been useful in observing cloud variability and relevant processes in the Arctic (Key et al. 2016; Karlsson et al. 2017). However, retrieving information on cloud amount and cloud properties over the ice and snow covered Arctic Ocean under extreme illumination conditions has always proven challenging. Separating visible and thermal cloud signatures from similar surface signatures using just five spectrally broad channels of AVHRR instruments is difficult. However, a considerable progress has been made in the recent decade to make use of the surface blind observations from the active sensors (mainly CALIOP-CALIPSO and CloudSat) to train, improve and validate cloud property retrieval algorithms that use AVHRR like imager data (Heidinger et al. 2014; Key et al. 2016; Karlsson et al. 2017; Karlsson and Devasthale 2018). Nearly 40 years of observations from AVHRRs provide the longest space based observations of clouds over the Arctic and, therefore, are valuable to study multidecadal variability and to understand climatological features of clouds.

Currently, there exist three long-term climate data records (CDR) that are exclusively based on AVHRR data. The first CDR, called CM SAF Cloud, Albedo And Surface Radiation dataset from AVHRR data – second edition (CLARA-A2), uses hierarchical decision tree thresholding approach that is anchored in physical principles to retrieve cloud properties (Karlsson et al. 2017). The second CDR, NOAA's Pathfinder Atmospheres – Extended program (PATMOS-x), uses naïve Bayesian approach (Heidinger et al. 2014). The third CDR, generated in the framework of ESA Climate Change Initiative (CCI), Cloud CCI (Version 3), uses neural network and optimal estimation techniques to provide cloud property retrievals (Stengel et al. 2017). It is important to note that the current versions of these three datasets use the exactly the same (inter-)calibration coefficients to relate radiances to reflectances and brightness temperatures.

Figure 5.2 shows the climatological total cloud fraction from these three CDRs over the Arctic averaged over the 26-year period from 1984 to 2009. As expected, the cloudiness is highest in the Atlantic and Pacific sectors of the Arctic. For comparison, a similar climatology from the most recent International Satellite Cloud Climatology Project dataset (ISCCP-HGM) is shown (Young et al. 2018). Both in the pan-Arctic and central Arctic regions, PATMOS-x shows highest cloud fraction followed by ISCCP, ESA CCI and CLARA-A2 respectively. There is nearly 10% difference in the estimates of total climatological cloudiness between PATMOS-x and CLARA-A2. CALIOP-CALIPSO provides the most reliable description of cloudiness over the Arctic as it does not rely upon the surface conditions. Therefore,

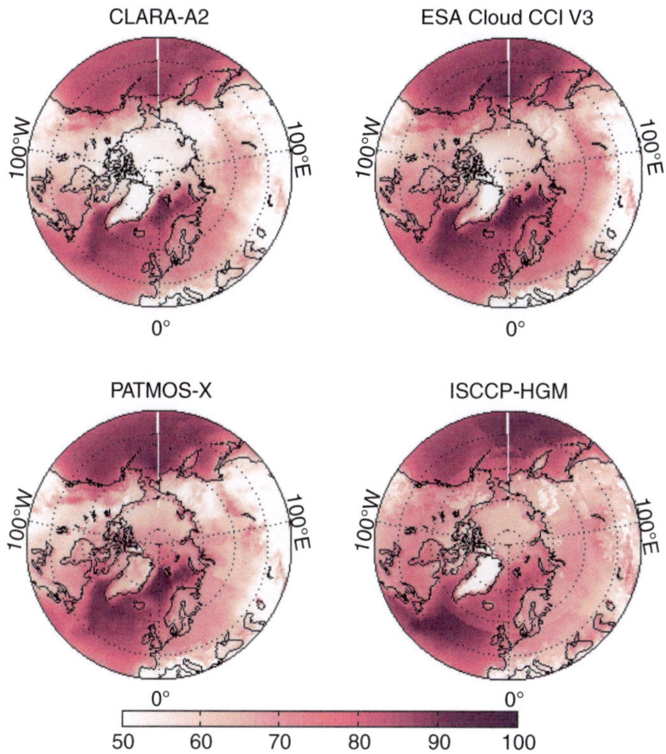

Fig. 5.2 Climatological spatial distribution of the total cloud fraction (in %) based on four, 26-year long climate data records (1984–2009). In the case of CLARA-A2, ESA Cloud CCI and PATMOS-x, the monthly mean data records based on AVHRR data from prime afternoon satellites (NOAA-7, −9, −11, −14, −16 and − 18) are used for the analysis

cloud fraction derived from CALIOP measurements can be considered as a reference in evaluating the CDRs that are based on passive imagers. Figure 5.3 shows the difference in total cloud fraction derived from four CDRs mentioned above with the 10-year climatology (2007–2016) from CALIOP. It can be seen that all CDRs underestimate cloudiness compared to CALIOP. The underestimation is largest in CLARA-A2 and smallest in PATMOS-x over the Arctic Ocean. This underestimation is mainly due to the fact that the subvisual cirrus cannot be seen with passive imagers like AVHRR and the lack of contrast in the spectral signatures makes it difficult to distinguish clouds from the surface in the polar winter.

In comparison to winter, the retrievals of cloud properties are more reliable in the polar summer. As the Arctic surface temperature experience amplified warming, investigating the role of air-sea interactions and clouds in governing the intraseasonal and interannual sea ice variability is becoming even more relevant. The low level clouds are an important piece of puzzle in the Arctic as they directly influence the surface radiation budget. The reliable information on the variability

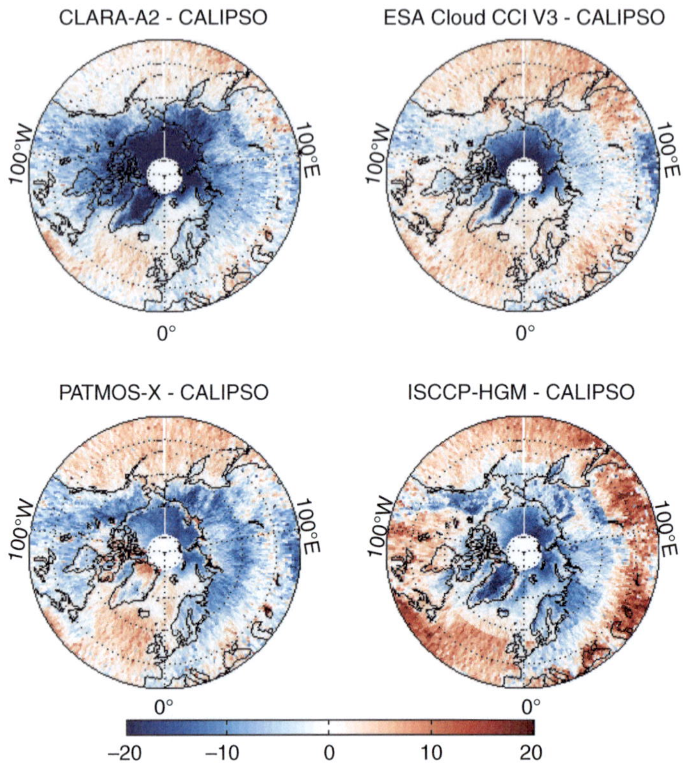

Fig. 5.3 Differences in total cloud fraction (in %) between four CDRs (26-year average) and CALIOP-CALIPSO (10-year average)

of these clouds is needed. How well do the longest, AVHRR based CDRs agree with one other in the description of low clouds in the polar summer? Figure 5.4 shows the comparison of low cloud fractions from the three CDRs. When compared to CLARA-A2, ESA Cloud CCI (both versions 2 and 3) and PATMOS-x overestimate low level cloudiness, especially PATMOS-x by up to 30% in early autumn. The differences between CLARA-A2 and ESA CCI remain within 5% over the central Arctic Ocean. Over the oceanic areas where low level cloudiness is usually persistent, such as Norwegian Sea, Greenland Sea, Barents and Kara Seas, the differences are typically below 15%.

Figure 5.5 shows the 27-year time-series' of monthly mean total cloudiness from 1984–2009 based on the four CDRs over the Arctic (60 N–90 N, 180 W-180E). It shows the seasonal and interannual variability as well as trends in the total cloudiness. The seasonal amplitude of total cloudiness is largest in CLARA-A2 (about 40%) and lowest in ISCCP (<20%). The former is due to the fact that CLARA-A2 underestimates cloud fraction in winter, while the latter is most likely due to misclassification in the ISCCP. Except ESA Cloud CCI, all remaining three

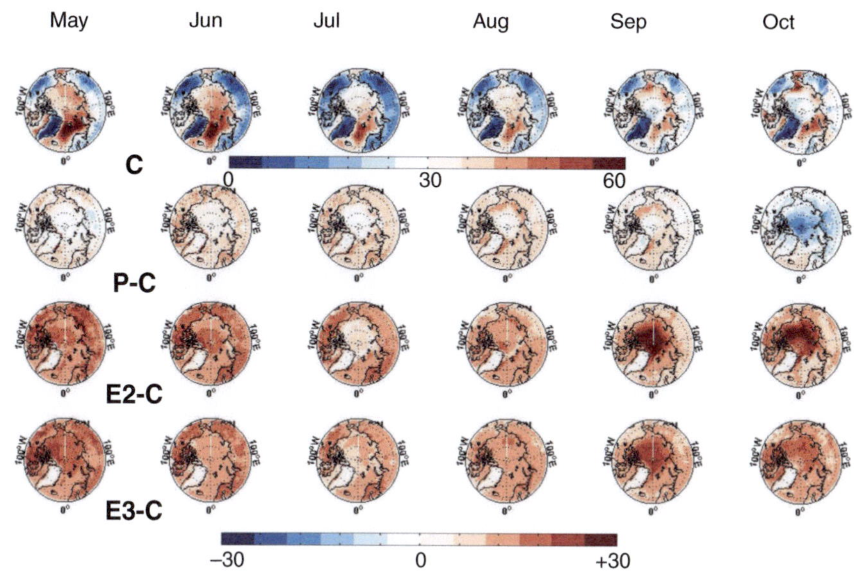

Fig. 5.4 Climatological mean low level cloud fraction (in %) from CLARA-A2 for individual summer months (top row). The remaining rows show differences of low level cloud fraction from PATMOS-x, ESA Cloud CCI V2 and V3 with CLARA-A2, denoted as P-C, E2-C and E3-C respectively

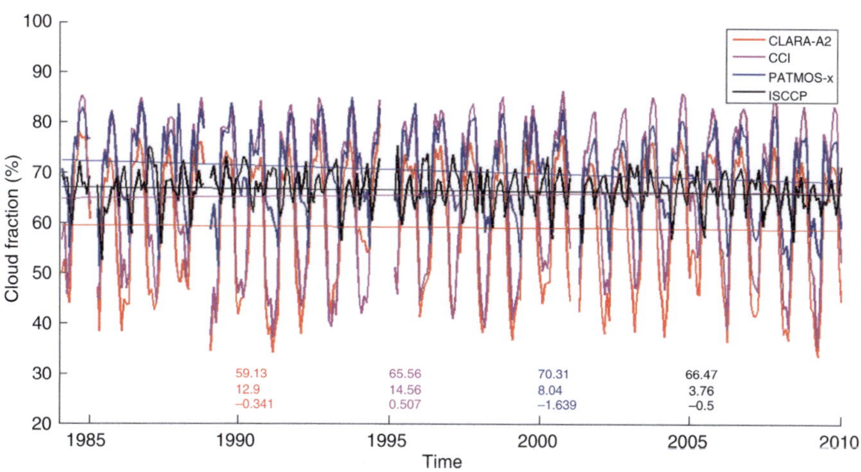

Fig. 5.5 Time series of monthly mean total cloud fraction from the four CDRs over the Arctic (60 N–90 N). For each CDR, the three numbers indicate mean, standard deviation and trend per decade respectively

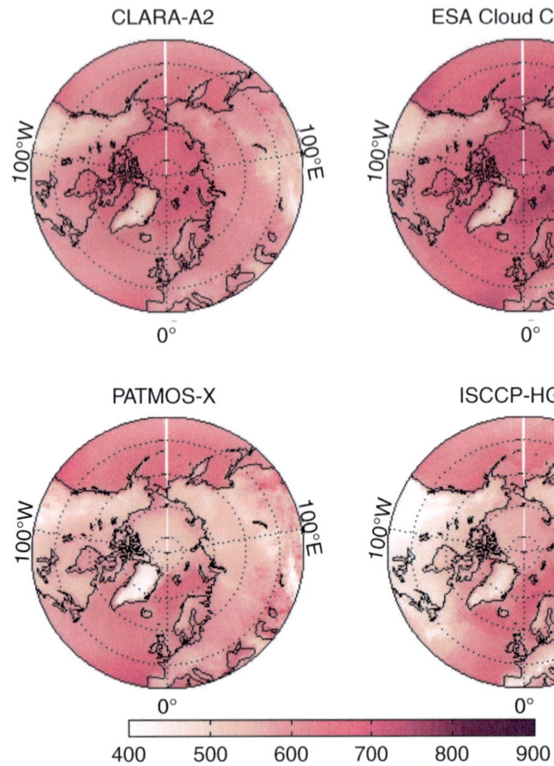

Fig. 5.6 Climatological spatial distribution of cloud top pressures (hPa) based on four, 26-year long climate data records (1984–2009)

CDRs show a declining trend in the total cloudiness. However, based on Mann-Kendall test, none of the trends are statistically significant at 95% confidence interval.

Figure 5.6 shows the climatological spatial distribution of cloud top pressures (CTPs) and their intercomparison. The CTPs in the Atlantic and Pacific sectors of the Arctic are higher due to the dominance of low level clouds in these regions. Figure 5.7 further shows the time-series' of CTPs for the same 26-year period. The seasonal amplitude is CTPs in PATMOS-x is twice as large as in CLARA-A2. While all four CDRs show increasing trends in CTPs, none of the trends are statistically significant (Mann-Kendall test). The CDRs analysed here also provide retrievals of cloud microphysical properties (during polar summer), but the differences among them are extremely high, preventing any meaningful comparison over the permanent, sea ice covered areas in the Arctic.

To fully grasp the role of clouds in the Arctic climate system and in the net radiation budget, a detailed understanding of the cloud radiative effects (CRE) is needed. Given the uniqueness of its environment, the CREs play a special role in the

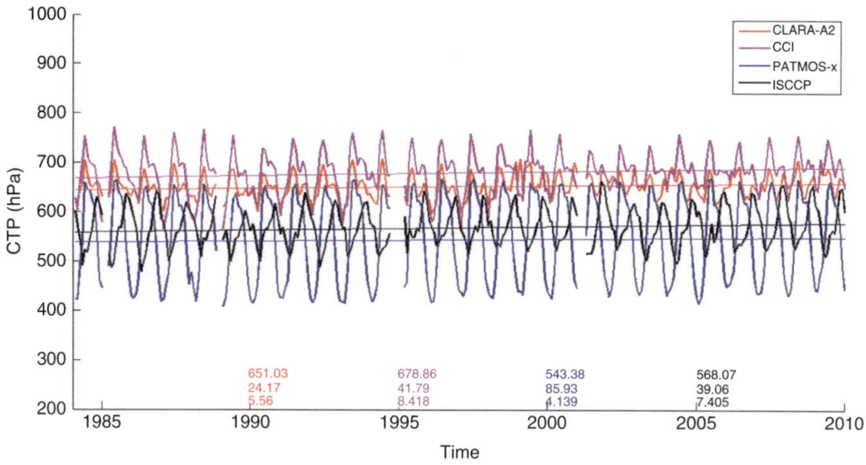

Fig. 5.7 Time series of monthly mean cloud top pressures from the four CDRs over the Arctic (60 N–90 N). For each CDR, the three numbers indicate mean, standard deviation and trend per decade respectively

local radiation budget in the Arctic. For example, the lack of sufficient insolation means that the longwave CREs dominate the net CREs during a large part of the year. Since the low level clouds are most prevalent in the Arctic, the longwave CREs at surface influence the seasonal and interannual sea ice variability. Processes, such as heat and moisture transport in to the Arctic, can also heavily influence the CREs by modulating cloud distribution and properties. The understanding of aerosol indirect effects over the Arctic still remains very low and inconclusive. This is mainly due to the fact that most of modern aerosol and cloud satellite retrieval techniques give large biases or no results at all for areas with underlying snow and ice surfaces. One possibility to solve this problem is to use acrive remote sensing techniques based on spaceborne lidar/radar measurements.

A number of previous studies have investigated the CREs in the Arctic (Curry et al. 1996; Walsh and Chapman 1998; Shupe and Intrieri 2004; Wang and Key 2005; Kay et al. 2016; to name a few). Kay and L'Ecuyer (2013) presented the estimates of CREs constrained by the state-of-the-art observational datasets, especially including cloud descriptions from CALIOP and CPR/CloudSat sensors. The surface blind observations of clouds from these sensors provide an important constraint to obtain more quantitatively accurate estimates of CREs. They concluded that the clouds over the Arctic Ocean warm the surface by 10 W/m^2 in annual average and cool the top of the atmosphere (TOA) by -12 W/m^2. The combined observations from CALIOP and CloudSat however lack the sufficient spatial coverage to investigate CREs at monthly scales. The Arctic experiences rapid, dynamic changes in its surface properties at a monthly scale. The atmosphere, and thus clouds, also responds to these changes.

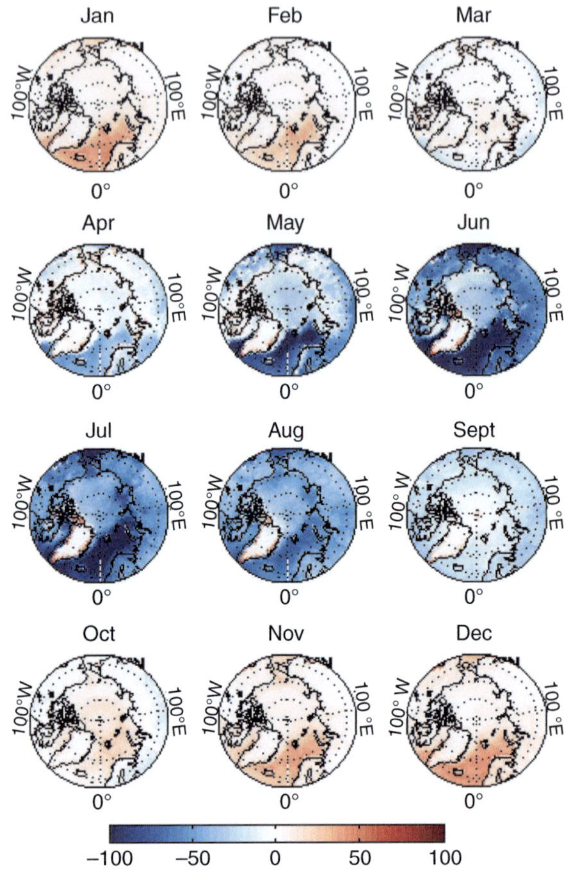

Fig. 5.8 Monthly climatological top of the atmosphere net cloud radiative effects (W/m^2) in the Arctic based on the 15-year EBAF CERES data (2002–2016)

Nearly two decades of observations from NASA's Clouds and Earth's Radiant Energy System (CERES) sensor flying in the A-Train era provides enough observations to examine climatological CREs covering the entire Arctic (Loeb et al. 2009). In contrast to the top of the atmosphere, the estimates of CREs at surface derived from CERES data are based on a number of retrievals assumptions. Nonetheless, they provide first order insights into the intraannual variability in the CREs. Figures 5.8 and 5.10 provide 15-year (2002–2016) average of monthly CREs based on CERES Energy Balanced and Filled (EBAF) data at the top of the atmosphere and surface respectively. The Fig. 5.9 shows the climatological annual cycle of CREs both at the surface and top of the atmosphere averaged over 60 N–90 N using the same dataset.

The cloud radiative effects lead to cooling at the TOA during a large part of the year, except few months in winter. The seasonal amplitude it strongest over the Atlantic sector of the Arctic, the region characterized by the persistent low level stratocumulus clouds. The intraannual amplitude of CREs over the permanently sea ice covered part of the Arctic Ocean and over the Greenland is relatively much lower.

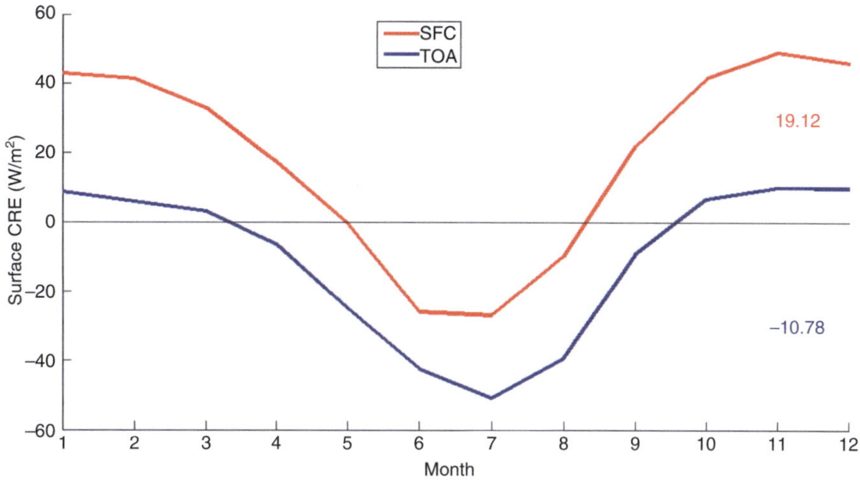

Fig. 5.9 Climatological seasonal cycle of the top of the atmosphere and surface net cloud radiative effects (W/m^2) in the Arctic (60 N–90 N) based on the 15-year EBAF CERES data (2002–2016). The numbers indicate annual mean CREs

In contrast to TOA CREs, the CREs predominantly produce warming at the surface except in few months in summer, with annual average of 19.11 W/m^2 over 70 N–90 N. However, a strong spatial variability is seen in the distribution and intraannual cycle of CREs, indicating the importance of investigating processes affecting CREs at monthly scales rather than averaging over the seasons.

5.3 Cloud Response to Decreasing Sea Ice Extent in the Arctic

As the Arctic sea ice extent declines at an unprecedented rate in the last few decades (Comiso et al. 2008; Cavalieri and Parkinson 2012; Parkinson and DiGirolamo 2016), quantifying contributions of various components of the Arctic climate system to the observed amplification in the Arctic surface temperatures becomes ever more important (Screen and Simmonds 2010; Serreze and Barry 2011; Döscher et al. 2014; Vihma et al. 2016). In this context, the exact role of clouds in the Arctic climate system and future cloud feedbacks in a warming world remain an enigma to the research community (Vavrus 2004; Kay and Gettelman 2009; Vavrus et al. 2011). Understanding the response of clouds to decreasing sea ice extent will help to eventually quantify the relevance of this response to the surface energy budget. Apart from the long-term changes in cloudiness, short-term variations in cloud properties at intra-seasonal scales are also becoming increasingly important in the Arctic. As the Arctic sea ice becomes thinner with time (Kwok and Rothrock

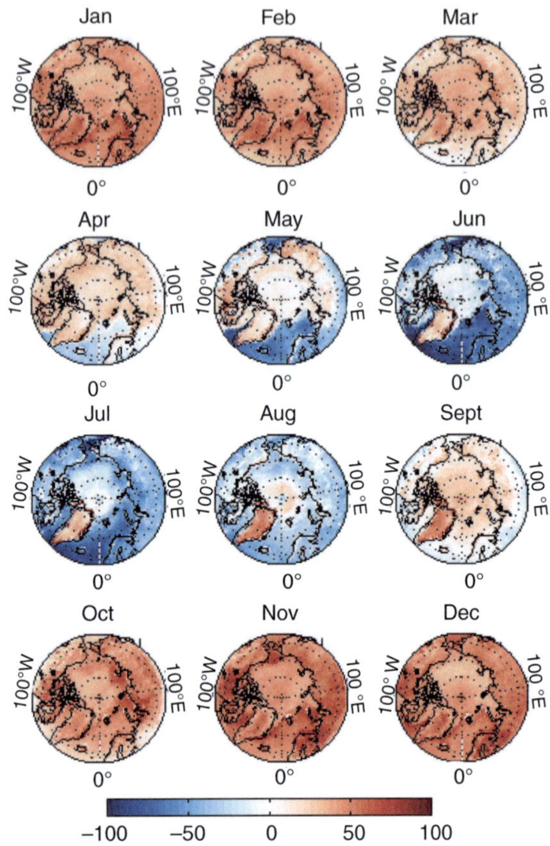

Fig. 5.10 Monthly climatological surface net cloud radiative effects (W/m^2) in the Arctic based on the 15-year EBAF CERES data (2002–2016)

2009), it also becomes increasingly sensitive and vulnerable to the short-term fluctuations in the atmosphere, i.e. so-called preconditioning (Graversen et al. 2011; Devasthale et al. 2013; Kapsch et al. 2013; Tjernström et al. 2015). The effect of such preconditioning was evident in the case of the last three record minimum sea ice extent events in the Arctic in 2007, 2012 and 2016 (Graversen et al. 2011; Devasthale et al. 2013).

In-situ measurements in the Arctic have shown the complex nature of the interactions between low-level clouds and local thermodynamics (Sedlar et al. 2012; Cox et al. 2015; Qiu et al. 2015). These interactions are governed by variations in sea ice among other factors, since sea ice influences atmospheric thermodynamics and energy fluxes. Therefore, few studies have previously investigated the response of clouds to sea ice (Schweiger et al. 2008; Kay and Gettelman 2009; Eastman and Warren 2010; Vavrus et al. 2011; Liu et al. 2012; Wu and Lee 2012; Taylor et al. 2015; Devasthale et al. 2016). They have found that cloudiness during the late summer and early autumn increases in response to the decreasing sea ice.

However, these studies are based either on short-term observational data, use reanalysis models or do not cover the entire Arctic Ocean. Recent improvements in the preprocessing, calibration, training, and retrieval algorithms for the space based heritage sensors such as Advanced Very High Resolution Radiometers (AVHRR) flown onboard of meteorological satellites from NOAA since 1982 enable us to investigate long-term, climate quality data records to quantify cloud response to sea ice (Karlsson et al. 2017; Heidinger et al. 2014). This section uses two long-term datasets derived from these multi-decadal AVHRR data to address the aforementioned question.

The rapidly changing sea ice concentration (SIC) in the Arctic Ocean during the summer half year (May–October) in the last three decades, specifically trends (per decade), are shown in Fig. 5.11. It is evident that the SIC has decreased consistently from early summer to until the end of the melt season. The strongest decrease occurs in August, September and October months, when sea ice is shrinking at rates larger than 15% per decade over the Beaufort and East Siberian Seas. The observed decreasing trend is not limited to these areas however. The marginal sea ice zones in the Eurasian and Atlantic sectors of the Arctic also show strong, decreasing trends in SIC. Statistically significant trends in low-level cloudiness (cloud top pressures greater than 680 hPa) are also shown in Fig. 5.11. Low-level cloudiness is increasing in September and October months. The trends in cloudiness are strongest in October over East Siberian, Chukchi and Beaufort Seas exceeding 5% per decade, after the

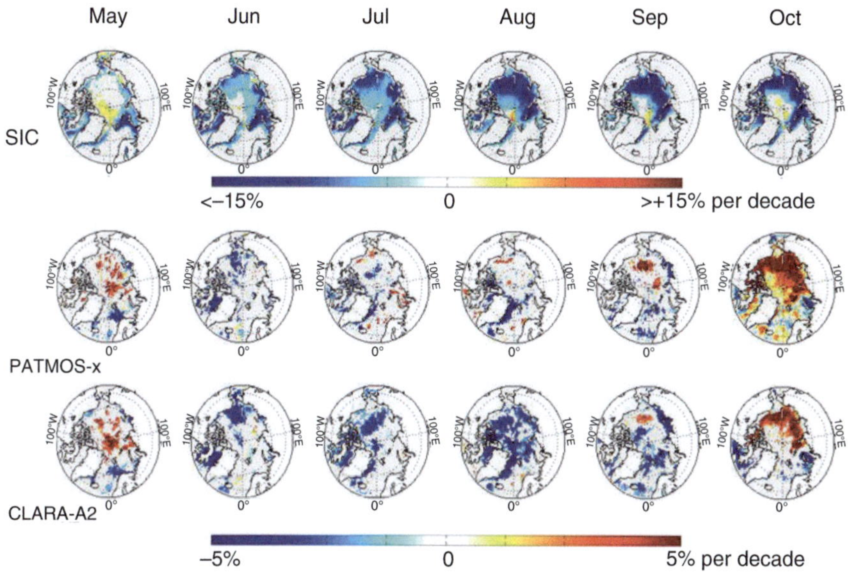

Fig. 5.11 Trends per decade in sea ice concentration (SIC) and low-level cloudiness for the 33-year period from 1982 to 2014. For cloud fraction data, only statistically significant trends exceeding 90% confidence (Student t test) are shown

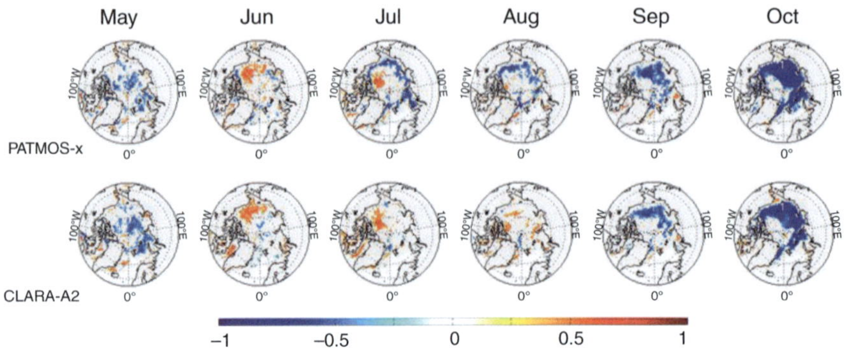

Fig. 5.12 Correlations of low cloud fraction (cloud top pressure greater than 680 hPa) with sea ice concentration based on PATMOS-x and CLARA-A2 datasets. Only those correlations that are significant at 90% confidence level are shown

annual minimum in SIC is reached. In JJA, cloudiness has decreased over areas with multiyear sea ice. These decreasing trends, although present in both CLARA-A2 and PATMOS-x datasets, are not equally robust. In the beginning of the summer half year (May), both datasets further agree on increasing trends in low-level cloudiness over the central Arctic.

The observed cloud response to changing SIC is investigated in Fig. 5.12. It is evident that the low-level cloudiness is strongly (negatively) correlated with SIC at the end of melt season (in August, September and October when the rate of decrease of SIC is also strongest). Statistically significant negative correlations, consistent in both CLARA-A2 and PATMOS-x datasets, are observed over marginal sea ice areas in the Beaufort and East Siberian Seas as well as over northern Barents and Kara Seas. No significant correlations are observed over the Central Arctic Ocean, the area dominated with multiyear sea ice. In June, the correlations over Beaufort Sea are positive in contrast to other months. The summer months (JJA) in general have weaker correlations compared to autumn months (September and October) when sea ice extent reaches its annual minimum and freeze up begins. This feature is also consistent in both CLARA-A2 and PATMOS-x datasets. The analysis of long-term observations presented here further suggests that the similar co-variability observed in the last decade by previous studies is indeed robust (Schweiger et al. 2008; Kay and Gettelman 2009; Eastman and Warren 2010; Vavrus et al. 2011; Liu et al. 2012; Wu and Lee 2012; Taylor et al. 2015; Devasthale et al. 2016).

This co-variability between cloudiness and sea ice can be explained by the changes in the thermodynamical characteristics in the boundary layer. In the absence of sea ice cover, the stratification and stability of the lowermost troposphere is weakened over the newly ice-free water, allowing vertical transport of heat and moisture, creating conditions favourable for cloud formation. This scenario is witnessed in the Arctic when the warm (relatively speaking) ice-free Arctic Ocean is exposed to the cold and dry Arctic atmosphere in the spring and autumn. *What are the implications of the observed co-variability of SIC and low-level cloudiness?* It is known that the surface cloud forcing remains largely positive in the Arctic throughout the year, except few months in summer (Walsh and Chapman 1998; Shupe and Intrieri 2004; Kay and L'Ecuyer 2013; Boeke and Taylor 2016). The lack of incoming solar radiation during the remainder of the year means that the outgoing long-wave radiation trapped in the lowermost troposphere as well as re-radiated by clouds back to the surface contributes most to the surface forcing of clouds, leading to net warming effect of clouds in the Arctic. If cloudiness increases during and after the sea ice melt season, it could trigger a positive feedback loop wherein increased long-wave forcing at the surface could hinder sea ice recovery after the melt season and, as a result, accelerate melting in the following season. This would in turn lead to increased cloudiness over marginal sea ice areas, thus closing the positive feedback loop.

5.4 How Do Moisture Intrusions Affect Clouds in the Arctic?

As the Arctic sea ice declines rapidly, it is becoming increasingly vulnerable to the vagaries of the weather. It is also getting moister (Boisvert and Stroeve 2015). Certain anomalous and persistent weather conditions can have a lasting impact not only on the seasonal sea ice growth and melting, but also on the state of the multiannual sea ice. Ever since the record sea ice minima of September 2007 and 2012 (Devasthale et al. 2013), a considerable attention is drawn on to the role of moisture intrusions into the Arctic (Woods et al. 2013; Johansson et al. 2017; Liu et al. 2018) and how they would precondition the Arctic atmosphere, mainly via their longwave forcing at the surface, thus facilitating a rapid sea ice melt (Graversen et al. 2011). Apart from this direct radiative impact of moisture intrusions, they also affect clouds, not only by increasing their amount, but also by changing their macro- and microphysical properties. Recent observations based on the decade long data from the advanced A-Train sensors characterize to what extent clouds respond to moisture intrusions into the Arctic (Johansson et al. 2017). If the cloud cover increases as a result of intrusions, as these studies show, it has implications for the seasonal sea ice variability. For example, if the intrusions occur in autumn or winter, they might hinder the recovery of sea ice after the melt season both through their direct radiative impact and also via changing cloud properties, making it further vulnerable to melt in the following year. The increased cloudiness can exert additional longwave forcing at the surface, leading to additional surface warming. If the intrusions occur in spring or summer, they could accelerate the seasonal sea ice melt. Therefore, precise characterization of intrusions and their direct and indirect radiative impacts is needed not only to study the processes involved, but also to evaluate how well the climate models take into account these impacts of intrusions.

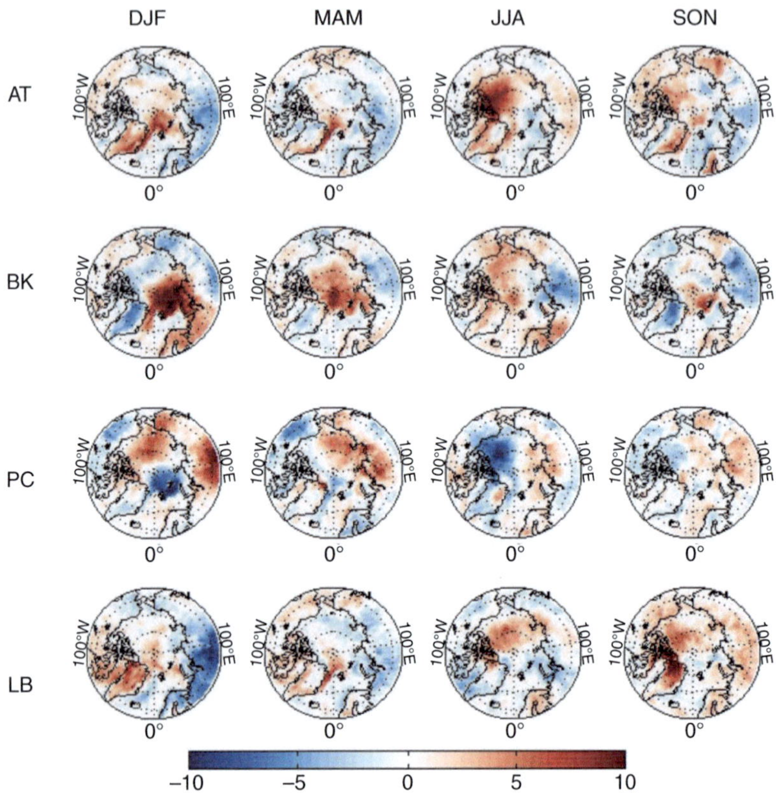

Fig. 5.13 Average anomalies in total cloud fraction (in %) observed during the intrusion events compared to climatology, based on 34-year (1982–2015) cloud property dataset CLARA-A2

The recent observational studies have mainly used limited data from newer satellite sensors to characterize the response of clouds to intrusions. Then the question arises as to how robust is this observed cloud response from the climate perspective and can we corroborate this response in the multidecadal observations? The current section investigates this question in detail. Previous studies have identified four important pathways of the moisture intrusions into the Arctic, namely, Atlantic sector of the Arctic (AT), Barents and Kara Seas (BK), Pacific sector (PC) and via Labrador Sea (LB). Figure 5.13 shows the anomalies of total cloud fraction observed when the moisture intrusions occur from these sectors during different seasons. When the northward moisture flux into the Arctic across 70 N exceeds its 90%ile value (as a function of season and sector) and lasts for at least 24 h (based on ERA-Interim Reanalysis Daily data), an intrusion event is said to have occurred. The cloud property data analysed here are from the CLARA-A2 climate data record stretching from 1982 to 2015 (34-year).

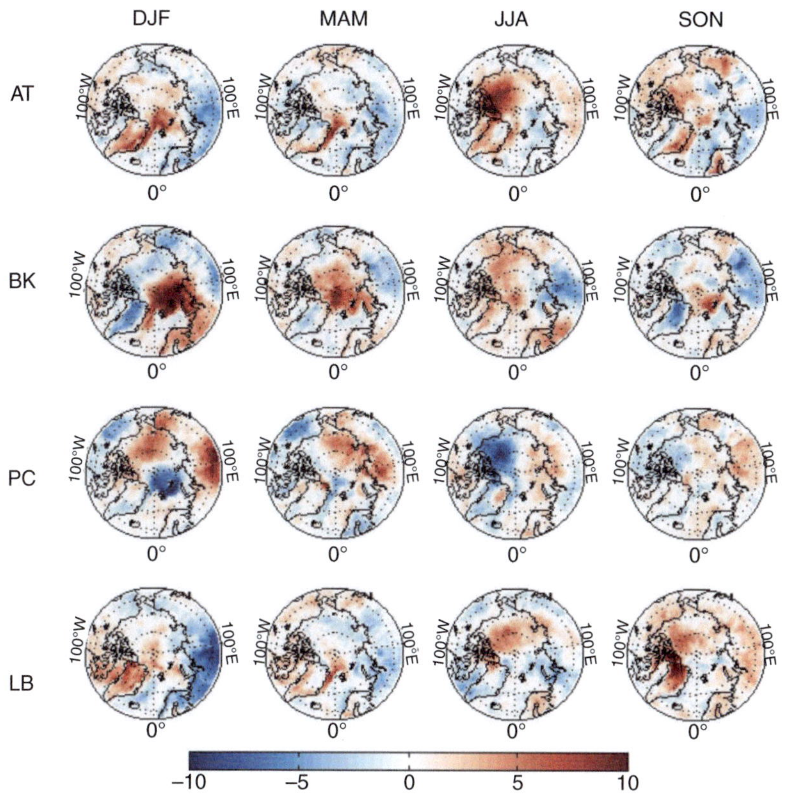

Fig. 5.14 As shown in Fig. 5.13, but for the cloud top pressures (in hPa)

As evident in Fig. 5.13, among all the sectors and seasons, the largest change in cloudiness is observed when the moisture intrusions occur from the BK sector during the DJF months. A significant increase in total cloudiness, far exceeding 10% locally when averaged over all events, is observed over the Central Arctic and the marginal sea ice zones in the northeast Atlantic. When the intrusions occur from the AT sector, cloudiness is increased in the Fram Strait. This is noteworthy, since the Fram Strait is one of the main pathways of sea ice export out of the Arctic. An increase in cloudiness during winter and spring months could accelerate this export by destabilizing the sea ice via longwave forcing. In general, the cloud response is stronger during winter and spring compared to other two seasons. Figure 5.14 shows the corresponding changes in cloud top pressures (CTP). The areas where an increase in cloudiness is observed, associated in intrusions, the CTPs are also increased. This suggests the increase in low level cloudiness, which is consistent with the previous study by Johansson et al. (2017). Figure 5.15 further shows the change in downward longwave radiation (DLR) at the surface. Consistent with the changes in cloudiness, increases in low level cloud amount result in increases in

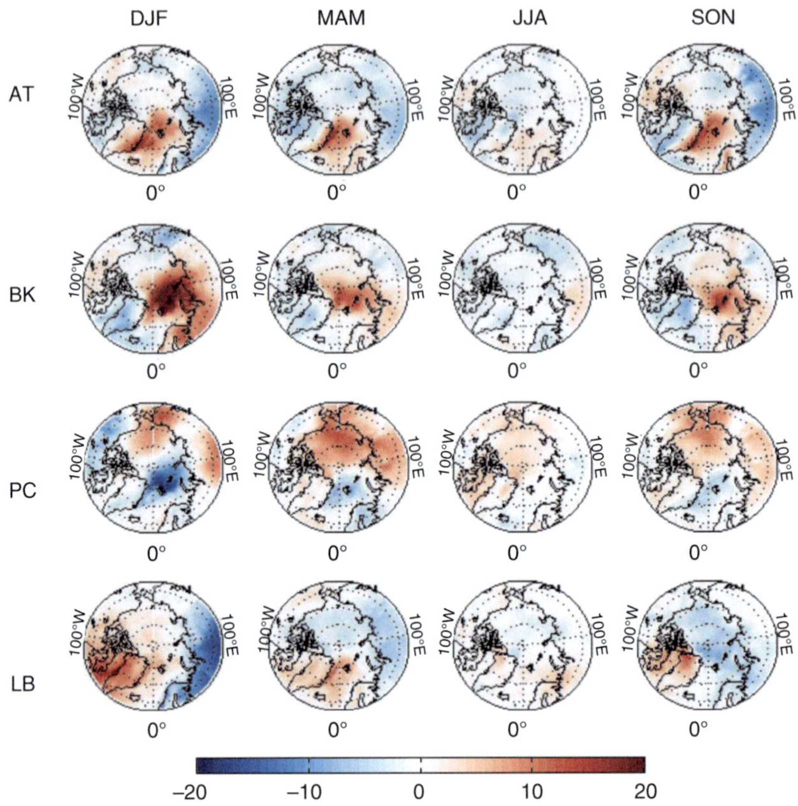

Fig. 5.15 As shown in Fig. 5.13, but for the downward longwave radiation at the surface (in W/m^2)

longwave radiation reaching at the surface. The increase is largest in the case of intrusions from the BK sector during winter, wherein the central parts of the Arctic are warmed significantly, with local anomalies in downward longwave radiation exceeding 30 W/m^2. The increases in the DLR are also seen in the Fram Strait, one of the major pathways of sea ice export out of the Arctic. The intrusions from the PC sector however lead to the most widespread increases in the DLR. Interestingly the cloud response is weakest in the summer months (JJA) as the surface temperatures in the Arctic warm rapidly. Figure 5.16 shows the corresponding changes in sea ice concentrations, averaged over a week after an intrusion event is occurred. It is clearly seen that the sea ice concentrations are reduced in the case of each sector, when the intrusions occur from that sector, providing potential evidence for the impact of intrusions on sea ice variability.

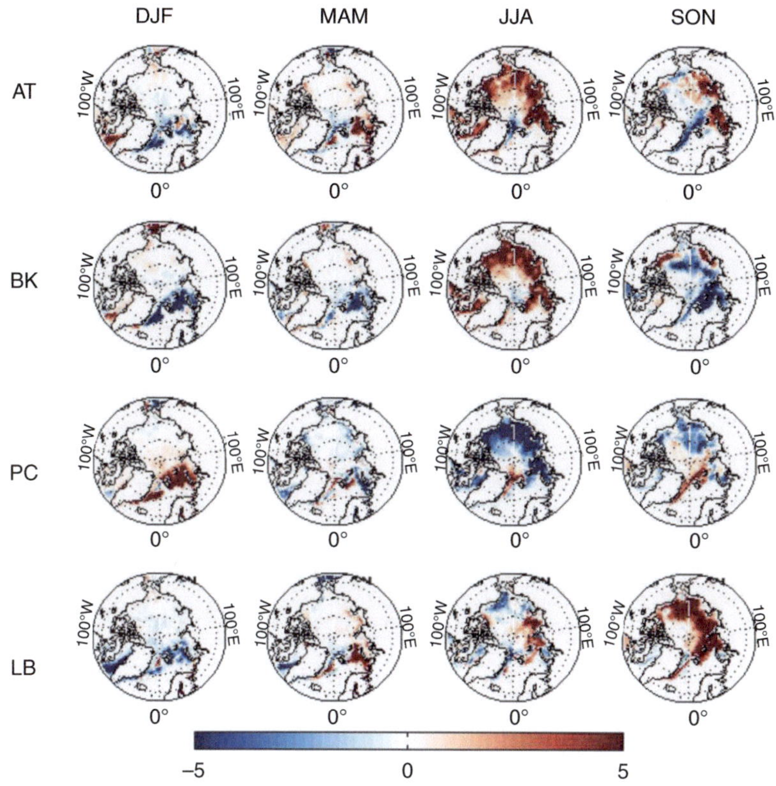

Fig. 5.16 Same as in Fig. 5.13, but for the sea ice concentrations (%)

The sectorial analysis approach presented above focuses on large-scale moisture intrusions into the Arctic. It does not take into account the local variability in moisture and those intrusions that may have large local impact, but do not qualify as intrusion events when averaged over bigger sectors. Therefore, in order to take into account this local variability, the moisture intrusions are further defined based on 90%ile thresholds computed at each $1^0 x 1^0$ grid instead of big sectors as before. Figure 5.17 shows the results of this analysis. The cloud response over three areas stands out, namely over, Barents and Kara Seas, Pacific sector of the Arctic, and Labrador Sea. Strong local increases in cloudiness are seen over these sectors during winter and spring seasons. The anomalies in cloud top pressures, downward longwave radiation reaching at the surface and sea ice concentrations are consistent with one another over these regions and also consistent with the findings from the previous sectorial analysis. In general, the magnitude of all anomalies is stronger when the local moisture variability is taken into account, indicating the sectorial analysis may underestimate the cloud response.

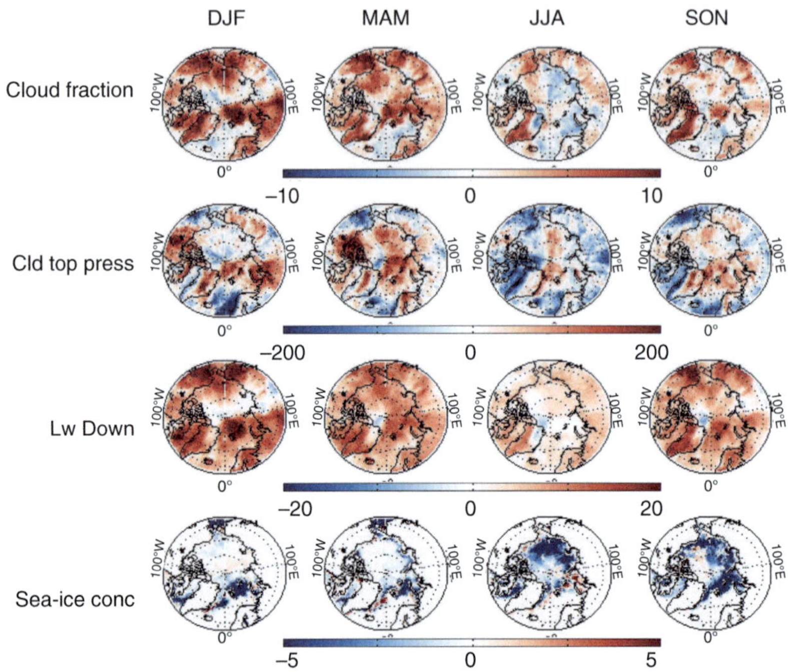

Fig. 5.17 The changes in cloud fraction (%) and top pressure (hPa), downward longwave radiation at the surface (W/m^2), and sea ice concentrations (%) observed during the moisture intrusion events, when the local variability in moisture is taken into account to define the intrusion events

Acknowledgements Abhay Devasthale would like to acknowledge the support from SMHI, EU Horizon 2020 project INTAROS and Swedish Research Council (Vetenskapsrådet). The availability of data records from EUMETSAT/CM-SAF, EUMETSAT/OSI-SAF, NOAA/NCDC/PATMOS-X, NOAA/NCDC/ISCCP, ESA Cloud CCI, and NASA/AIRS is gratefully acknowledged. The authors also thank the Science Teams behind these climate data records.

References

Boeke, R. C., & Taylor, P. C. (2016). Evaluation of the Arctic surface radiation budget in CMIP5 models. *Journal of Geophysical Research – Atmospheres, 121*, 8525–8548. https://doi.org/10.1002/2016JD025099.

Boisvert, L. N., & Stroeve, J. C. (2015). The Arctic is becoming warmer and wetter as revealed by the atmospheric infrared sounder. *Geophysical Research Letters, 42*, 4439–4446. https://doi.org/10.1002/2015GL063775.

Brooks, I. M., Tjernström, M., Persson, P. O. G., Shupe, M. D., Atkinson, R. A., Canut, G., Birch, C. E., Mauritsen, T., Sedlar, J., & Brooks, B. J. (2017). The turbulent structure of the Arctic summer boundary layer during the Arctic summer Cloud-Ocean study. *Journal of Geophysical Research, [Atmospheres], 122*, 9685–9704. https://doi.org/10.1002/2017JD027234.

Cavalieri, D. J., & Parkinson, C. L. (2012). Arctic Sea ice variability and trends, 1979–2010. *The Cryosphere, 6*, 881–889. https://doi.org/10.5194/tc-6-881-2012.

Comiso, J. C., Parkinson, C. L., Gersten, R., & Stock, L. (2008). Accelerated decline in the Arctic Sea ice cover. *Geophysical Research Letters, 35*, L01703. https://doi.org/10.1029/2007GL031972.

Cox, C. J., et al. (2015). Humidity trends imply increased sensitivity to clouds in a warming Arctic. *Nature Communications, 6*, 10117. https://doi.org/10.1038/ncomms10117.

Curry, J. A. (1986). Interactions among turbulence, radiation and microphysics in Arctic stratus clouds. *Journal of the Atmospheric Sciences, 43*(1), 90–106.

Curry, J. A., & Herman, G. F. (1985). Infrared radiative properties of summertime Arctic stratus clouds. *Journal of Climate and Applied Meteorology, 24*, 525–538.

Curry, J. A., Rossow, W. B., Randall, D., & Schramm, J. L. (1996). Overview of Arctic cloud and radiation characteristics. *Journal of Climate, 9*, 1731–1764.

Curry, J. A., Pinto, J. O., Benner, T., & Tschudi, M. (1997). Evolution of the cloudy boundary layer during the autumnal freezing of the Beaufort Sea. *Journal of Geophysical Research, 102*(D12), 13851–13860.

Devasthale, A., Sedlar, J., Koenigk, T., & Fetzer, E. J. (2013). The thermodynamic state of the Arctic atmosphere observed by AIRS: Comparisons during the record minimum sea ice extents of 2007 and 2012. *Atmospheric Chemistry and Physics, 13*, 7441–7450. https://doi.org/10.5194/acp-13-7441-2013.

Devasthale, A., Sedlar, J., Kahn, B. H., Tjernström, M., Fetzer, E. J., Tian, B., Teixeira, J., & Pagano, T. S. (2016). A decade of space borne observations of the Arctic atmosphere: Novel insights from NASA's atmospheric infrared sounder (AIRS) instrument. *Bulletin of the American Meteorological Society, 97*, 2163–2176. https://doi.org/10.1175/BAMS-D-14-00202.1.

Döscher, R., Vihma, T., & Maksimovich, E. (2014). Recent advances in understanding the Arctic climate system state and change from a sea ice perspective: A review. *Atmospheric Chemistry and Physics, 14*, 13571–13600. https://doi.org/10.5194/acp-14-13571-2014.

Eastman, R., & Warren, S. G. (2010). Interannual variations of Arctic cloud types in relation to sea ice. *Journal of Climate, 23*, 4216–4232.

Graversen, R. G., Mauritsen, T., Drijfhout, S., Tjernström, M., & Mårtensson, S. (2011). Warm winds from the Pacific caused extensive Arctic Sea ice melt in summer 2007. *Climate Dynamics, 36*, 2103–2112. https://doi.org/10.1007/s00382-010-0809-z.

Heidinger, A. K., Foster, M. J., Walther, A., & Zhao, X. (2014). The pathfinder atmospheres–extended AVHRR climate dataset. *Bulletin of the American Meteorological Society, 95*, 909–922. https://doi.org/10.1175/BAMS-D-12-00246.1.

Herman, G. F., & Curry, J. A. (1984). Observational and theoretical studies of solar radiation in Arctic stratus clouds. *Journal of Climate and Applied Meteorology, 23*, 5–24.

Herman, G., & Goody, R. (1976). Formation and persistence of summertime Arctic stratus clouds. *Journal of the Atmospheric Sciences, 33*, 1537–1553.

Intrieri, J. M., Shupe, M. D., Uttal, T., & McCarty, B. J. (2002a). An annual cycle of Arctic cloud characteristics observed by radar and lidar at SHEBA. *Journal of Geophysical Research, 107*(C10), 1–15.

Intrieri, J. M., Fairall, C. W., Shupe, M. D., Persson, P. O. G., Andreas, E. L., Guest, P. S., & Moritz, R. E. (2002b). An annual cycle of Arctic surface cloud forcing at SHEBA. *Journal of Geophysical Research, 107*(C10), 1–14.

Johansson, E., Devasthale, A., Tjernström, M., Ekman, A. M. L., & L'Ecuyer, T. (2017). Response of the lower troposphere to moisture intrusions into the Arctic. *Geophysical Research Letters, 44*, 2527–2536. https://doi.org/10.1002/2017GL072687.

Kahl, J. D. (1990). Characteristics of the low-level temperature inversion along the Alaskan Arctic coast. *International Journal of Climatology, 10*, 537–548.

Kapsch, M.-L., Graversen, R. G., & Tjernström, M. (2013). Springtime atmospheric energy transport and the control of Arctic summer sea ice extent. *Nature Climate Change, 3*, 744–748. https://doi.org/10.1038/nclimate1884.

Karlsson, K.-G., & Devasthale, A. (2018). Inter-comparison and evaluation of the four longest satellite-derived cloud climate data records: CLARA-A2, ESA cloud CCI V3, ISCCP-HGM, and PATMOS-x. *Remote Sensing, 10*, 1567.

Karlsson, K.-G., Anttila, K., Trentmann, J., Stengel, M., Fokke Meirink, J., Devasthale, A., Hanschmann, T., Kothe, S., Jääskeläinen, E., Sedlar, J., Benas, N., van Zadelhoff, G.-J., Schlundt, C., Stein, D., Finkensieper, S., Håkansson, N., & Hollmann, R. (2017). CLARA-A2: the second edition of the CM SAF cloud and radiation data record from 34 years of global AVHRR data. *Atmospheric Chemistry and Physics, 17*, 5809–5828. https://doi.org/10.5194/acp-17-5809-2017.

Kay, J. E., & Gettelman, A. (2009). Cloud influence on and response to seasonal Arctic Sea ice loss. *Journal of Geophysical Research, 114*, D18204. https://doi.org/10.1029/2009JD011773.

Kay, J. E., & L'Ecuyer, T. (2013). Observational constraints on Arctic Ocean clouds and radiative fluxes during the early 21st century. *Journal of Geophysical Research – Atmospheres, 118*, 7219–7236. https://doi.org/10.1002/jgrd.50489.

Kay, J. E., L'Ecuyer, T., Chepfer, H., Loeb, N., Morrison, A., & Cesana, G. (2016). Recent advances in Arctic cloud and climate research. *Current Climate Change Reports, 2*, 159. https://doi.org/10.1007/s40641-016-0051-9.

Key, J., Wang, X., Liu, Y., Dworak, R., & Letterly, A. (2016). The AVHRR polar pathfinder climate data records. *Remote Sensing, 8*(3), 167. https://doi.org/10.3390/rs8030167.

Kwok, R., & Rothrock, D. A. (2009). Decline in Arctic Sea ice thickness from submarine and ICESat records: 1958–2008. *Geophysical Research Letters, 36*, L15501. https://doi.org/10.1029/2009GL039035.

Liu, Y., Key, J. R., Liu, Z., Wang, X., & Vavrus, S. J. (2012). A cloudier Arctic expected with diminishing sea ice. *Geophysical Research Letters, 39*, L05705. https://doi.org/10.1029/2012GL051251.

Liu, Y., Key, J., Vavrus, S., & Woods, C. (2018). Time evolution of cloud response to moisture intrusions into the Arctic during winter. *Journal of Climate, 31*(22), 9389–9405. https://doi.org/10.1175/JCLI-D-17-0896.1.

Loeb, N. G., Wielicki, B. A., Doelling, D. R., Smith, G. L., Keyes, D. F., Kato, S., Manalo-Smith, N., & Wong, T. (2009). Toward optimal closure of the Earth's top-of-atmosphere radiation budget. *Journal of Climate, 22*, 748–766. https://doi.org/10.1175/2008JCLI2637.1.

Morrison, H., de Boer, G., Feingold, G., Harrington, J., Shupe, M. D., & Sulia, K. (2012). Resilience of persistent Arctic mixed-phase clouds. *Nature Geoscience, 5*, 11–17. https://doi.org/10.1038/NGE01332.

Paluch, I. R., & Lenschow, D. H. (1991). Stratiform cloud formation in the marine boundary layer. *Journal of the Atmospheric Sciences, 48*(19), 2141–2158.

Parkinson, C. L., & DiGirolamo, N. E. (2016). New visualizations highlight new information on the contrasting Arctic and Antarctic Sea ice trends since the late 1970's. *Remote Sensing of Environment, 183*, 198–204. https://doi.org/10.1016/j.res.2016.05.020.

Persson, P. O. G., Shupe, M. D., Perovich, D., & Solomon, A. (2017). Linking atmospheric synoptic transport, cloud phase, surface energy fluxes, and sea ice-growth: Observations of midwinter SHEBA conditions. *Climate Dynamics, 49*, 1341–1364. https://doi.org/10.1007/s00382-016-3383-1.

Pinto, J. O., Curry, J. A., & Fairall, C. W. (1997). Radiative characteristics of the Arctic atmosphere during spring as inferred from ground-based measurements. *Journal of Geophysical Research, 102*(D6), 6941–6952.

Qiu, S., Dong, X., Xi, B., & Li, J.-L. F. (2015). Characterizing Arctic mixed-phase cloud structure and its relationship with humidity and temperature inversion using ARM NSA observations. *Journal of Geophysical Research – Atmospheres, 120*, 7737–7746. https://doi.org/10.1002/2014JD023022.

Ruffieux, D., Persson, P. O. G., Fairall, C. W., & Wolfe, D. E. (1992). Ice pack and lead surface energy budgets during LEADEX 1992. *Journal of Geophysical Research, 100*, 4593–4612.

Schweiger, A. J., Lindsay, R. W., Vavrus, S., & Francis, J. A. (2008). Relationships between Arctic Sea ice and clouds during autumn. *Journal of Climate, 21*, 4799–4810.

Screen, J. A., & Simmonds, I. (2010). The central role of diminishing sea ice in recent arctic temperature amplification. *Nature, 464*, 1334–1337.

Sedlar, J., & Shupe, M. D. (2014). Characteristic nature of vertical motions observed in Arctic mixed-phase stratocumulus. *Atmospheric Chemistry and Physics, 14*, 3461–3478. https://doi.org/10.5195/acp-14-3461-2014.

Sedlar, J., & Tjernström, M. (2009). Stratiform cloud – Inversion characterization during the Arctic melt season. *Boundary-Layer Meteorology, 132*, 455–474. https://doi.org/10.1007/s10546-009-9407-1.

Sedlar, J., Tjernström, M., Mauritsen, T., Shupe, M. D., Brooks, I. M., Persson, P. O. G., Birch, C. E., Leck, C., Sirevaag, A., & Nicolaus, M. (2011). A transitioning Arctic surface energy budget: The impacts of solar zenith angle, surface albedo and cloud radiative forcing. *Climate Dynamics, 37*, 1643–1660. https://doi.org/10.1007/s00382-010-0937-6.

Sedlar, J., Shupe, M. D., & Tjernström, M. (2012). On the relationship between thermodynamic structure and cloud top, and its climate significance in the Arctic. *Journal of Climate, 25*, 2374–2393. https://doi.org/10.1175/JCLI-D-11-00186.1.

Serreze, M. C., & Barry, R. G. (2011). Processes and impacts of Arctic amplification: A research synthesis. *Global and Planetary Change, 77*, 85–96. https://doi.org/10.1016/j.gloplacha.2011.03.004.

Shupe, M. D. (2011). Clouds at Arctic atmospheric observatories. Part II: Thermodynamic phase characteristics. *Journal of Applied Meteorology and Climatology, 50*, 645–661. https://doi.org/10.1175/2010JAMC2468.1.

Shupe, M. D., & Intrieri, J. M. (2004). Cloud radiative forcing at the Arctic surface: The influence of cloud properties, surface albedo, and solar zenith angle. *Journal of Climate, 17*, 616–628.

Shupe, M. D., Uttal, T., Matrosov, S. Y., & Frisch, A. S. (2001). Cloud water contents and hydrometeor sizes during the FIRE Arctic clouds experiment. *Journal of Geophysical Research, 106*(D14), 15015–15028.

Shupe, M. D., Uttal, T., & Matrosov, S. Y. (2005). Arctic cloud microphysics retrievals from surface-based remote sensors at SHEBA. *Journal of Applied Meteorology, 44*, 1544–1562.

Shupe, M. D., Kollias, P., Persson, P. O. G., & McFarquhar, G. M. (2008). Vertical motions in Arctic mixed-phase Stratiform clouds. *Journal of the Atmospheric Sciences, 65*, 1304–1322. https://doi.org/10.1175/2007JAS2479.1.

Shupe, M. D., Walden, V. P., Eloranta, E., Uttal, T., Campbell, J. R., Starkweather, S. M., & Shiobara, M. (2011). Clouds at Arctic atmospheric observatories. Part I: Occurrence and macrophysical properties. *Journal of Applied Meteorology and Climatology, 50*, 626–644. https://doi.org/10.1175/2010JAMC2467.1.

Shupe, M. D., Persson, P. O. G., Brooks, I. M., Tjernström, M., Sedlar, J., Mauritsen, T., Sjogren, S., & Leck, C. (2013). Cloud and boundary layer interactions over the Arctic Sea ice in late summer. *Atmospheric Chemistry and Physics, 13*, 9379–9400. https://doi.org/10.5194/acp-13-9379-2013.

Solomon, A., Shupe, M. D., Persson, P. O. G., & Morrison, H. (2011). Moisture and dynamical interactions maintaining decoupled Arctic mixed-phase stratocumulus in the presence of a humidity inversion. *Atmospheric Chemistry and Physics, 11*, 10127–10148. https://doi.org/10.5194/acp-11-10127-2011.

Sotiropoulou, G., & coauthors. (2016). Atmospheric Conditions during the Arctic Clouds in Summer Experiment (ACSE): Contrasting Open Water and Sea Ice Surfaces during Melt and Freeze-Up Seasons. *Journal of Climate, 29*, 8721–8744. https://doi.org/10.1175/JCLI-D-16-0211.1.

Sotiropoulou, G., Sedlar, J., Tjernström, M., Shupe, M. D., Brooks, I. M., & Persson, P. O. G. (2014). The thermodynamic structure of summer Arctic stratocumulus and the dynamic coupling to the surface. *Atmospheric Chemistry and Physics, 14*, 12573–12592. https://doi.org/10.5194/acp-14-12573-2014.

Stengel, M., Stapelberg, S., Sus, O., Schlundt, C., Poulsen, C., Thomas, G., Christensen, M., Carbajal Henken, C., Preusker, R., Fischer, J., et al. (2017). Cloud property datasets retrieved

from AVHRR, MODIS, AATSR and MERIS in the framework of the Cloud_cci project. *Earth System Science Data, 9*, 881–904.
Stephens, G. L. (1978). Radiation profiles in extended water clouds. II: Parameterization schemes. *Journal of the Atmospheric Sciences, 35*, 2123–2132.
Taylor, P. C., Kato, S., Xu, K.-M., & Cai, M. (2015). Covariance between Arctic Sea ice and clouds within atmospheric state regimes at the satellite footprint level. *Journal of Geophysical Research – Atmospheres, 120*, 12656–12678. https://doi.org/10.1002/2015JD023520.
Tjernström, M., & coauthors. (2014). The Arctic Summer Cloud Ocean Study (ASCOS): overview and experimental design. *Atmospheric Chemistry and Physics, 14*, 2823–2869. https://doi.org/10.5194/acp-14-2823-2014.
Tjernström, M., & coauthors. (2015). Warm-air advection, air mass transformation and fog causes rapid ice melt. *Geophysical Research Letters, 42*, 1–9. https://doi.org/10.1002/2015GL064373.
Tjernström, M., & Graversen, R. G. (2009). The vertical structure of the lower Arctic troposphere analysed from observations and ERA-40 reanalysis. *Quarterly Journal of the Royal Meteorological Society, 135*, 431–433. https://doi.org/10.1002/qj.380.
Tjernström, M., Leck, C., Persson, P. O. G., Jensen, M. L., Oncley, S. P., & Targino, A. (2004). The summertime Arctic atmosphere. Meteorological measurements during the Arctic Ocean experiment 2001. *Bulletin of the American Meteorological Society, 85*(9), 305–321.
Uttal, T., & coauthors. (2002). Surface heat budget of the Arctic Ocean. *Bulletin of the American Meteorological Society, 83*(2), 255–276.
Vavrus, S. (2004). The impact of cloud feedbacks on Arctic climate under greenhouse forcing. *Journal of Climate, 17*, 603–615.
Vavrus, S., Holland, M. M., & Bailey, D. A. (2011). Changes in Arctic clouds during intervals of rapid sea ice loss. *Climate Dynamics, 36*, 1475. https://doi.org/10.1007/s00382-010-0816-0.
Vihma, et al. (2016). The atmospheric role in the Arctic water cycle: A review on processes, past and future changes, and their impacts. *Journal of Geophysical Research – Biogeosciences, 121*, 586–620. https://doi.org/10.1002/2015JG003132.
Walsh, J. E., & Chapman, W. L. (1998). Arctic cloud-radiation-temperature associations in observational data and atmospheric reanalyses. *Journal of Climate, 11*, 3030–3045.
Wang, X., & Key, J. (2005). Arctic surface, cloud, and radiation properties based on the AVHRR polar pathfinder data set. Part I: Spatial and temporal characteristics. *Journal of Climate, 18*(14), 2558–2574.
Woods, C., Caballero, R., & Svensson, G. (2013). Large-scale circulation associated with moisture intrusions into the Arctic during winter. *Geophysical Research Letters, 40*, 4717–4721. https://doi.org/10.1002/grl.50912.
Wu, D. L., & Lee, J. N. (2012). Arctic low cloud changes as observed by MISR and CALIOP: Implication for the enhanced autumnal warming and sea ice loss. *Journal of Geophysical Research, 117*, D07107. https://doi.org/10.1029/2011JD017050.
Young, A. H., Knapp, K. R., Inamdar, A., Hankins, W., & Rossow, W. B. (2018). The international satellite cloud climatology project H-series climate data record product. *Earth System Science Data, 10*, 1–11.

Chapter 6
Arctic Ice Fog: Its Microphysics and Prediction

Ismail Gultepe, Andrew J. Heymsfield, and Martin Gallagher

Abstract Ice fog consists of suspended small ice crystals with maximum sizes less than about 200 μm, having similar fall velocities as fog droplets, and that often reduces visibility to less than 1 km. Its formation is strongly dependent on high number concentrations of available heterogeneous ice nuclei (IN) at temperatures (T) > −40 °C, homogeneous nucleation below −40 °C, and available moisture in the air. Radiative cooling, advective cooling, and cold air subsidence, particularly over the Polar region or high elevation mountainous geographical regions, play an important role in its formation and development. Ice fog crystals form at cold T when the relative humidity with respect to ice (RH_i) is ≥100%. Favorable ice nucleation conditions typically occur at T < −15 °C and its microphysical characteristics and their evolution needs to be better understood for a physically based representation in numerical forecast models. This is likely to be of growing societal importance due to the known sensitivity of the Arctic environment to climate change. Accidents related to low visibility over the northern latitudes may increase tenfold over the Arctic regions because of increasing population and traffic. This suggests that ice fog conditions can have major impacts on aviation and ground/water-based transportation, as well as on climate change and ecosystem. These open issues, as well as challenges related to ice fog measurements and predictions, are discussed in detail, and its importance for evaluating weather and climate conditions over cold environments are emphasized.

I. Gultepe (✉)
Environment and Climate Change Canada, Meteorological Research Division, Toronto, ON, Canada

Faculty of Engineering and Applied Science, Ontario Technical University, Oshawa, ON, Canada
e-mail: ismail.gultepe@canada.ca

A. J. Heymsfield
NCAR, Boulder, CO, USA

M. Gallagher
Faculty of Natural Sciences, Department of Earth & Environmental Sciences, University of Manchester, Manchester, UK

A. Kokhanovsky, C. Tomasi (eds.), *Physics and Chemistry of the Arctic Atmosphere*, Springer Polar Sciences, https://doi.org/10.1007/978-3-030-33566-3_6

Keywords Ice fog · Visibility · Ice nucleation · Ice fog prediction · Cold temperature · Arctic · Microphysics · Homogeneous nucleation · Heterogeneous nucleation · Small ice crystals · Ice nucleation

6.1 Introduction

Cold fog consists of suspended small ice crystals in the air that often reduces visibility to less than 1 km and occurs as ice fog at temperatures (T) usually less ≤ -10 °C and as freezing fog at T ≤ 0 °C at the saturation levels with respect to ice and water, respectively. Its formation is strongly dependent on a high number concentration of available heterogeneous ice nuclei (IN) at T > -40 °C and homogeneous nucleation at T below -40 °C, and available moisture in the air (Gultepe et al. 2015). Its formation is usually related to strong radiative cooling particularly over the Polar region or high elevation mountainous geographical regions (Wexler 1936, 1949; Ohtake 1967; Wendler 1969; Gultepe et al. 2014, 2015, 2016a), but it may also occur because of cold air advection, cold and moist air subsidence, frontal passages, and broken ice surfaces.

Formation of ice fog generally occurs following strong radiative cooling of the surface layer air trapped below a strong temperature inversion (Fig. 6.1a) as described by Wexler (1936, 1949) and Gultepe et al. (2014). Radiative cooling is initiated close to the surface. This results in a layer of increasing relative humidity (Fig. 6.1b) which deepens as air temperature decreases in the early morning. During this development, a strong surface inversion forms, but its strength gradually decreases with time as increasing solar radiation warms the surface resulting in heating that reduces the relative humidity and weakens the inversion layer. At higher levels an isothermal layer forms whose thickness increases with time. The inversion height can decrease as a result of subsidence because of radiative cooling at higher elevations just above the fog top, resulting in a weak upper inversion after mixing of cold and warm air towards the end of the fog event. Ice fog can also form due to advection of a cold air dome e.g. from Siberia over Alaska, from a slow moving anti-cyclonic circulation (Bowling et al. 1968). The subsidence of resulting cold air leads to the colder surface layer air, allowing ice fog to form. However, ice fog formation depends on the availability of heterogeneous IN. These studies suggest that probability of ice fog formation can be related to both regional and larger scale dynamical and microphysical processes.

There have been relatively few studies on ice fog characteristics, such as Thuman and Robinson (1954), Robinson et al. (1957), Benson (1965), Benson and Rogers (1965), and Ohtake (1967). For this reason, some information available for warm fog and ice cloud research are also used for supporting ice fog observation and their prediction. Wendler (1969) compared ice fog conditions to a low level cirrus cloud, but with high optical thickness. Two extreme ice fog events were studied by Gotaas and Benson (1965) during the winter of 1961–1962 with air T below -40 °C, and they suggested that cooling near the surface is not completely attributable to

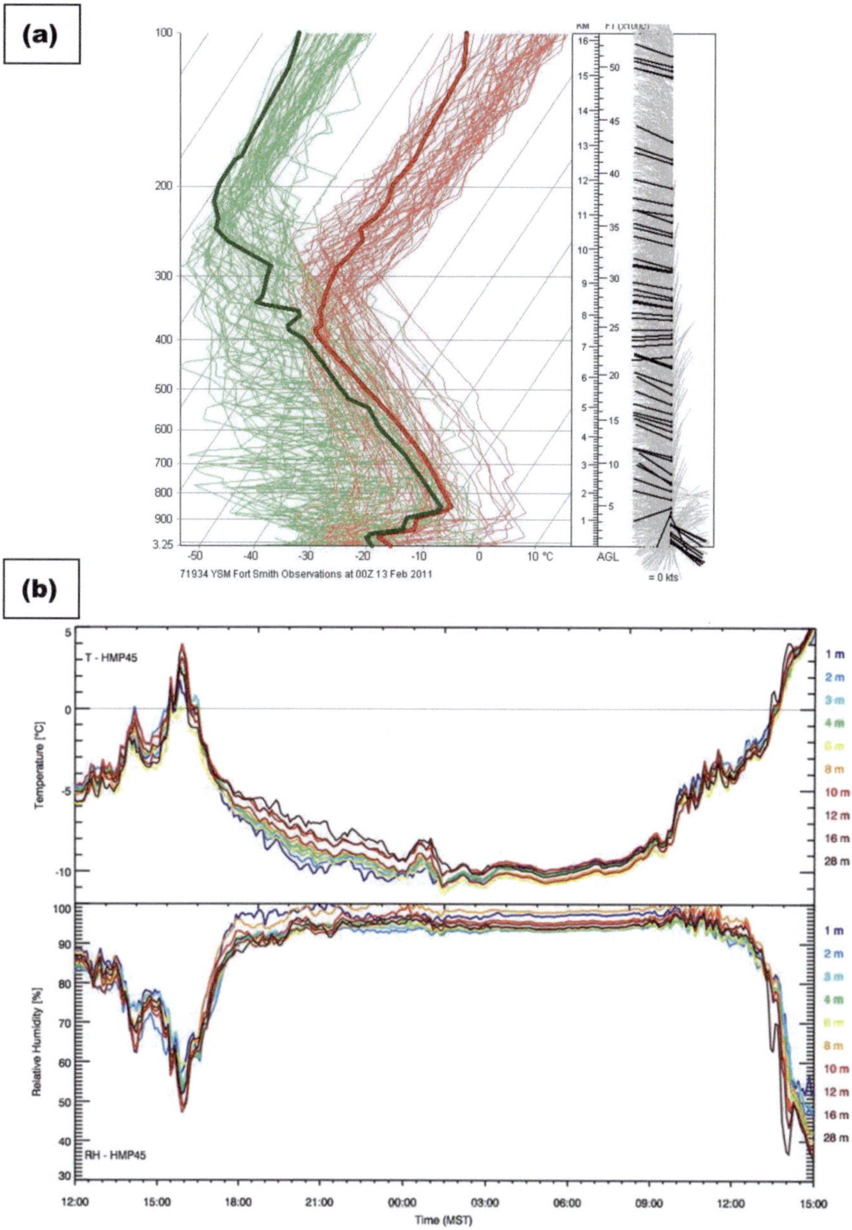

Fig. 6.1 A typical cold region boundary layer profile shown in box a: T (red) and Td (green) profiles from radiosonde measurements for entire January 2011 (top); thick lines are for January 13, 2011 at 0000 UTC. The T and RH_w time series given at 9 levels (box b) for a mountain ice fog event occurred during the MATERHORN project on January 16, 2015

cold air advection or heat losses from the air and snow surface. They hypothesized that radiative cooling due to the ice crystals contributed to formation of a strong temperature gradient close to the initial ice crystal zone formation leading to supersaturation with respect to ice and subsequent ice fog formation.

The sizes of ice fog particles are usually less than 200 μm (Gultepe et al. 2015) with fall speeds similar to 10 μm droplets observed in warm fog/freezing fog (Koenig 1970). Assuming that these ice crystals grow by vapor deposition, their growth rate and habit will depend on the temperature and supersaturation of the environment. These may be calculated using the method described by Koenig (1970), assuming supercooled droplets do not play a role in ice fog formation, and vapor deposition is the major source for their formation and growth (Hobbs 1965; Hobbs et al. 1974). Koenig (1970) also showed that diffusional growth of small ice crystals was much larger than this due to ventilation effects, and is a strong function of crystal habit. Although IN composition and concentration play an important role for ice initiation (Zelenyuk and Imre 2005), the IN concentration cannot be easily predicted or measured (Gultepe and Isaac 2002; Niemand et al. 2012; Connolly et al. 2009; DeMott et al. 2010; Murray et al. 2012; Gultepe et al. 2014). Szyrmer and Zawadzki (1997) and recently both Murray et al. (2012) and Emersic et al. (2015) suggested that ice forming nuclei issues remain an area of debate because of the large variability in IN concentrations and different measurement approaches; therefore, it is the key uncertainty of ice fog prediction (Gultepe et al. 2015). Hoose and Möhler (2012) reviewed six decades of laboratory ice nucleation studies on many different atmospheric ice nuclei and various freezing mechanisms (Heymsfield et al. 2017; Kanji et al. 2017; Gultepe et al. 2017a, b). Possible ice nuclei includes natural and anthropogenic mineral dusts (Ansmann et al. 2008), soot particles, biological, organic and ammonium sulfate particles. Interestingly, they note that the majority of studies show a reduction in nucleation ability due to the presence of various coatings on IN. They showed that typical temperature-supersaturation regions can be identified for ice initiation for these different particle types with dust particles typically active below −10 °C, although some biological particles, such as ice positive bacteria can nucleate ice at much warmer temperatures.

Several global modeling studies have been conducted to assess the importance of biological particles on cloud ice and precipitation formation (Sesartic et al. 2012). Phillips et al. (2009), using a mesoscale model, studied ensembles of continental-clouds over the US state of Oklahoma to better understand their formation. These studies generally conclude that biogenic IN inclusion leads to no significant changes in cloud formation or precipitation on local and global scales due to their low number concentrations. However, these studies do not consider biogenic IN impact on surface cloud or fog at cold temperatures. Recent aircraft measurements, using novel real-time biogenic particle spectrometers for the first time, have detected higher than expected concentrations of biogenic particles, up to 100 L^{-1} in the boundary layer and at altitudes greater than 2 km (Twohy et al. 2016). These concentrations, significantly under-predicted by models, were attributed to either missing emission sources or inadequate capture of their vertical transport. Regional modeling supported by real-time measurements of biogenic aerosols in Europe, Hummel et al. (2015), also suggest that biogenic emissions may be underestimated and that their emissions parameterizations need to be improved, particularly in response to surface wetness and meteorological drivers. Gosselin et al. (2016) describe measurements of biogenic emissions from surfaces using the first quantitative comparison between various biogenic aerosol measurement techniques at a forest location. They pointed to several persistent fog events where increased biogenic particle concentrations were observed suggestive of a possible source of these particles not influenced by more common rain initiated mechanisms. Clearly, more work is required to investigate their potential contribution to surface fogs.

Recently, Kim et al. (2014) and Gultepe et al. (2015) used a version of the WRF model to study ice fog formation and its microphysical characteristics, and they concluded that improvements for ice fog microphysical schemes e.g. for nucleation parameterizations are needed for better prediction of ice fog conditions, such as visibility.

Ice nucleation parameterizations can be done using IN measurements or assumed that small ice crystals represent IN. But, many previous cloud spectrometers, particularly those used on aircraft had difficulty with accurate discrimination and quantification of small ice particles (less than 100 μm) because of issues due to optical resolution and shattering of ice particles (Lawson 2011; Lawson et al. 2006a, b; Field et al. 2006). Lawson et al. (2010, 2006a, b) and Baker and Lawson (2006) studied small ice crystals at cold T on both aircraft and surfaced based in-situ observations. These studies emphasize that, using various in-situ instruments including a cloud imaging probe (CPI) and polar nephelometer (PN) at the South Pole Station, N_i was up to 500 L^{-1} for sizes <200 μm. The largest N_i was usually at about 50 μm maximum size, suggesting that N_i at small sizes were abundant. It was concluded that the collection efficiency of the sensors due to wind effects and aspirator issues were not applicable for the full range of ice fog conditions.

There are a few studies on ice fog microphysics. Most of these works (Kumai and O'Brien 1964; Hobbs 1965) state that the small ice crystals were spherical particles. Similar to their work, Korolev and Isaac (2003), using cloud particle imager probe measurements, also stated that ice particles were spherical when sizes were less than 100 μm. Gultepe et al. (2014) and Kim et al. (2014), on the other hand, based on ice crystal measurements at T close to −30 °C suggested that the spherical particle assumption may not be correct because of various pristine/single ice crystals that likely grow at those temperatures. Recent technological improvements in cloud particle spectrometers are now beginning to probe the small ice particle populations in clouds. Lloyd et al. (2015) found a very good agreement for small ice crystal size distributions measured by a number of different spectral instruments, including a 2D stereo probe (2DS) and a holographic spectrometer down to 50 μm (Henneberger et al. 2013). Below this size range, Vochezer et al. (2016) used the Small Ice Detector (SID) and Particle Phase Discriminator (PPD) probes to measure the full 2D scattering pattern of ice crystals. This collected data is then used to estimate crystal size and shape for ice particle size between 4 and 50 μm. Recently, Gultepe et al. (2017b) studied ice fog microphysical properties based on a review paper.

They stated that the current knowledge of how ice fog forms, evolves, dissipates, and impacts the environment is complicated and its scientific issues are in a way similar to these of ice clouds. These issues are related to ice nuclei, crystal growth, surface and boundary layer processes, and ice crystal optical properties.

Recent studies on cold fog (Velde et al. 2010; Gultepe et al. 2007a, b) highlighted that accurate fog prediction critically depends on model horizontal and vertical resolutions, initial conditions, and the microphysical parameterizations used. The ice nucleation process, and their composition, PBL turbulence, radiation, land-surface interactions, and air quality also play an important role on fog formation and development (Bergot et al. 2007; Bergot and Guedalia 1994). Recent field and modeling studies of ice fog microphysical properties reported by Gultepe et al. (2015) and Kim et al. (2014) have focused on ice fog formation due to both deposition nucleation (over Yellowknife, NWT, Canada) and homogeneous ice nucleation (over Fairbanks Alaska, U.S.) processes, respectively. Both studies provided detailed observations of ice fog events and their numerical modeling, and emphasized the importance of high ice crystal number concentration on visibility and measurement issues.

Based on our current knowledge, ice fog microphysical characteristics and their evolution still need to be better understood, while a physically-based representation needs to be developed for inclusion in numerical forecast models and their applications. This is likely to be of growing societal importance due to the known sensitivity of the Arctic environment to climate change. Perturbations of IN due to anthropogenic pollution, coupled with significant changes in moisture and temperature regimes, may lead to increased frequency of ice fog in these regions. A study by Gultepe et al. (2012, 2019b) showed that accidents related to weather and low visibility over the northern latitudes, as a result of increased traffic and population, will likely result in a tenfold increase over the Arctic regions. This suggests that ice fog conditions can have major impacts on aviation and ground/water-based transportation, as well as on climate change at low temperatures, where its impacts will be largest.

6.2 Ice Fog Physical Properties and Formation Mechanisms

Ice fog composed of suspended ice crystals in the air forms at cold temperatures when the relative humidity with respect to ice (RH_i) is $\geq$100% with sufficient IN. Favorable ice fog nucleation conditions typically occur at temperatures < -15 °C, as described by Gultepe et al. (2014, 2015). Measurements of ice crystal sizes in ice fogs suggest that ice crystal fall velocities (V_f) are similar to those of warm fog droplets, but their sizes are typically smaller than 200 μm (Gultepe et al. 2015). However, this does not preclude ice formation at temperatures > -15 °C. Standard fog definitions (e.g. as used in aviation forecasting) are based on visibility metrics with values <1 km, referred to simply as fog and whose intensity is a simple function of the fog particle size distribution and the extinction efficiency (Q_{ext}) for visible light (at a wavelength of 0.790 μm). Q_{ext} is however a strong function of particle size, crystal shape, and light wavelength when ice crystal sizes less than 10 μm.

The ice fog term usually is used for deposition ice fog formation by the forecasters at T below −15 °C in northern latitudes (Gultepe et al. 2015). Its formation is due to either spontaneously freezing of droplets at T < −40 °C by homogeneous IN (Heymsfield and Sabin 1989; Heymsfield et al. 2017; Vali 1994) or heterogeneous IN (Gultepe et al. 2017a, b). Ice fog is a class of cold fogs that may also include freezing fog (also called supercooled fog; droplets freeze after touching cold surfaces when surface and air T < 0 °C) and frozen fog (droplets freeze in air when air T < 0 °C) (Heymsfield et al. 2017; Gultepe et al. 2014, 2017a, b). Formation mechanisms of freezing fog (freezing on contact with a cold surface) or frozen fog (spherical frozen droplets in air) and their effect on surface conditions compared to ice fog are different. In earlier work, ice fog was defined as a man-made phenomenon and referred to as anthropogenic fog. However, this generic term (ice fog) can also refer to fog formed by secondary ice processes, e.g. anthropogenic nucleation processes in urban environments.

These fog types can be separated based on their physical and thermodynamic properties, which are related to temperature, relative humidity, IN number concentration, and IN chemical composition. For practical applications, RH_i and T are normally only used for defining the occurrence of such events. IN concentration is the limiting factor for ice fog formation, and it does not generally occur even if other physical conditions are favorable for heterogeneous ice crystal formation.

Ice fog can be important for significant reduction in local visibility which can affect aviation, transportation, and sporting activities. It also affects the local surface radiation budget and albedo, acting like a blanket reducing outgoing long-wave radiation to space which can be important for local ecosystems. Ice fog formation mechanisms are related to cooling processes within the BL (leading to saturation with respect to ice) IN and T. The cooling of boundary layer occurs usually by means of cold air mass advection, radiative cooling, inversion and mixing processes, and nocturnal subsidence. Anthropogenic pollution can also be a factor for increasing the frequency of ice fog events if these particles have hygroscopic characteristics. Here, we consider ice fog formation and development, if IN exist, by definition when $RH_w < 95\%$ but $RH_i > 100\%$ and $T < -7$ °C.

Here, freezing, immersion, and contact ice nucleation processes (Young 1974; Meyers et al. 1992) are not considered to be part of ice fog occurrence because ice fog crystals are usually single pristine particles formed by deposition nucleation at T usually colder than −15 °C (Gultepe et al. 2015). In their work, they suggested that ice crystals formed at the $RH_w < 100\%$ when IN exist. Clearly, heterogeneous nucleation can happen at cold temperatures when IN exist with $RH_i \geq 100\%$. As the ice fog evolves and then enters the dissipation stage, both mixing and radiative heating/cooling can lead to larger ice crystals, which may then start to aggregate and precipitate. However, it is possible that with internally mixed aerosols, which include both soluble CCN and insoluble IN (DeMott et al. 2015), subsequent ice crystals may form by both freezing and deposition nucleation processes with condition that their fall speeds are less than about a few cm s^{-1} and particles sizes less than about 200 μm. It is also possible that secondary ice nucleation processes can play an important role for ice fog but here this is not verified and further observations are needed.

6.3 Measurements of Ice Fog and Visibility

6.3.1 Ground Based In-Situ Measurements

Ice fog crystal measurements at surface locations have generally been performed using cloud sensors developed mainly for aircraft platforms (Gultepe et al. 2014; Lawson et al. 2006a, b; Lloyd et al. 2015). As part of the Fog Remote Sensing and Modeling (FRAM) project (Gultepe et al. 2015), ice fog properties were measured in detail using various sensors that provided ice crystal size spectra, visibility, snowfall rate and accumulation, solar and infrared broadband fluxes, vertical profiles of T, vapor mixing ratio (q_v) and liquid water content (LWC) together with 3D turbulence intensities. Details of the instruments and analysis techniques are described in Gultepe et al. (2014) and will only briefly be described here. Ice crystal number concentration size distributions were measured by the optical sensors shown in Fig. 6.2a, b, c. Figure 6.2d shows an ice fog event that formed at the Yellow Knife International Airport. These sensors include an Ultra High Sensitivity Aerosol Spectrometer (DMT UHSAS), a fog measuring device (FMD), a ground cloud imaging probe (GCIP), a laser precipitation monitor (LPM), and a precipitation disdrometer (OTT ParSiVel). Some images of ice crystals recorded by a microscope and a camera with macro capability are shown in Fig. 6.3a. Figure.6.3b shows aerosol size distribution measurements over 8 channels between 0.3 and 10 μm obtained from a Climatronic Aerosol Profiler (CAP, MetOne Inc.), and details can be found in Gultepe et al. (2012). Between 1200 UTC and 1800 UTC, aerosol concentrations decreased, and that is an indication of the ice nucleation process that occurred during FRAM project at −20 °C with RH_i of about 100%, while visibility decreased to less than a few 100 s of meters. Small ice crystal size distributions and shapes of fog and snow particles were recorded at 1 s intervals using the GCIP and FMD sensors. The GCIP resolution was 15 μm, and min and max sizes were about 7.5 and 930 μm, respectively. An example of ice crystals images representing an ice fog event is shown in Fig. 6.4 where ice crystal sizes between 095445 UTC and 095459 UTC were usually less than 200 μm.

Another detailed field project on ice fog was performed by Kim et al. (2014). In their work, they used a Video Ice Particle Sampler (VIPS) probe. VIPS works by impact of ice crystals via an inlet onto a transparent moving belt, subsequently imaged and recorded by a video microscope (Miloshevich and Heymsfield 1997). Ice crystal images are digitized from the video and analyzed with image analysis software as described by McFarquhar and Heymsfield (1997). Ice crystal sizes from 10 to 350 μm can be measured normally. Operation and calibration of this version

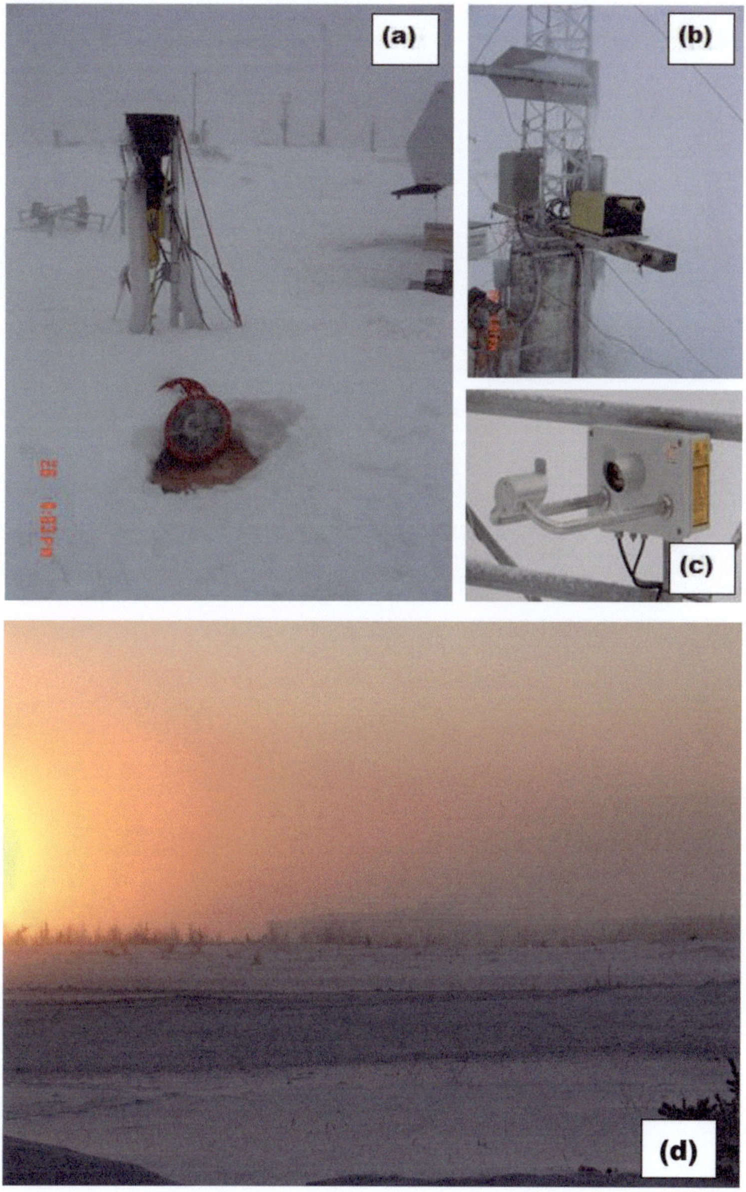

Fig. 6.2 Microphysical instruments used for ice fog crystal spectral measurements: GCIP (**a**), FMD (FM-100) (**b**), and LPM sensors (**c**). An ice fog event occurred at the Yellow Knife International Airport during FRAM ice fog project is shown on January 16, 2011 (**d**)

of the instrument is described by (Kim et al. 2014), who suggested a minimum resolvable particle shape of 5 μm. Their results were similar to Gultepe et al. (2014), but they were mostly related to freezing/homogeneous nucleation processes.

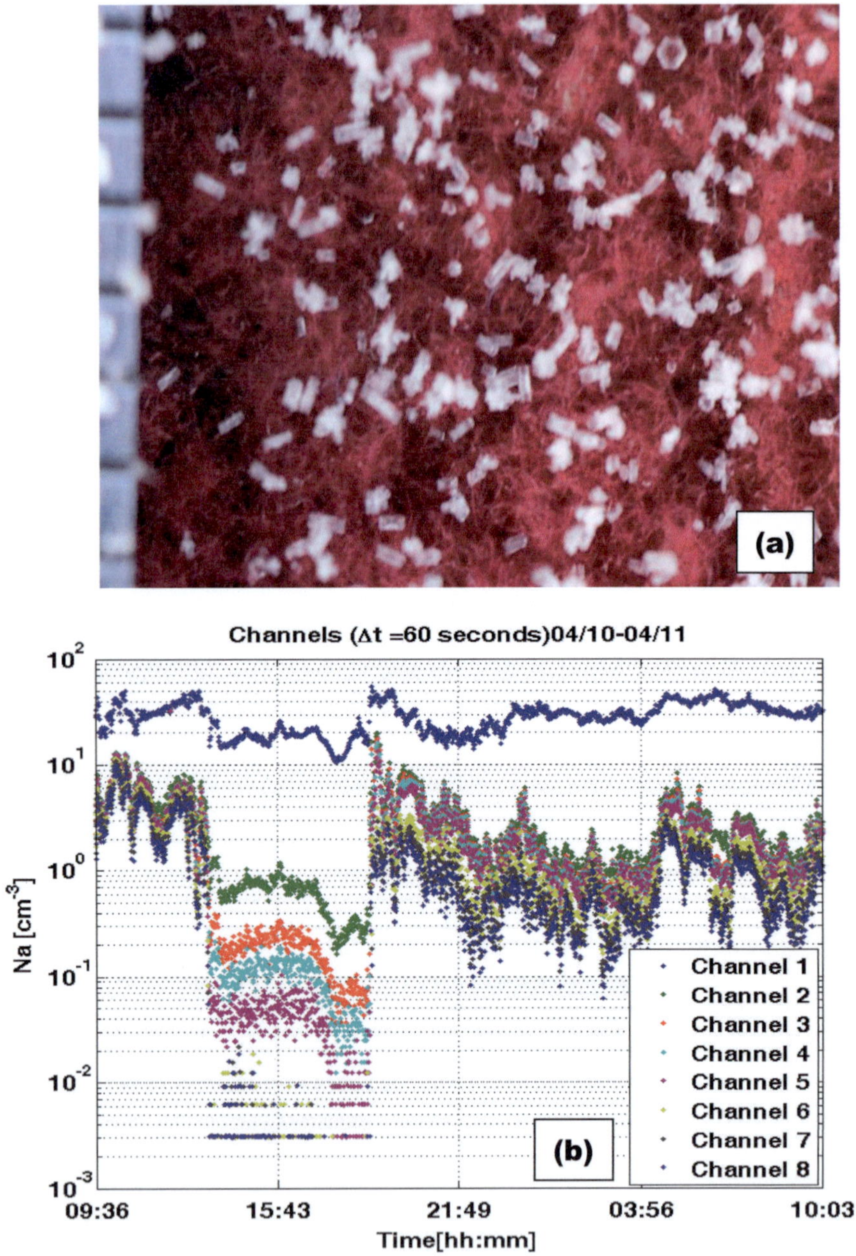

Fig. 6.3 Ice crystal images taken by a microscope () on April 102,011 . The maximum crystal size is about 100 μm. Time series of aerosol number concentration (Na) from the CAP aerosol measurements obtained over 8 channels between 0.3 and 10 μm is shown in (**b**). Between 1200 UTC and 1800 UTC, decreasing of aerosols is an indication of ice nucleation process

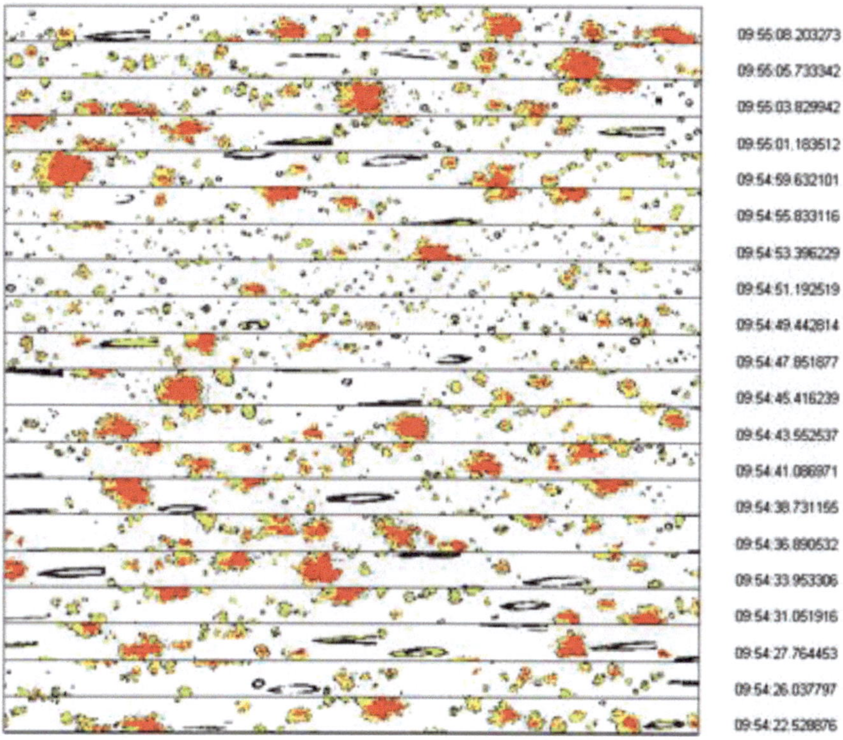

Fig. 6.4 The GCIP sensor probe measurements of ice fog and light snow precipitation. The small ice crystals size distributions of fog and snow particles over 62 channels were obtained at 1 s intervals on Jan 17, 2011. The resolution of the spectra is 15 μm, and min and max sizes are about 7.5 and 930 μm, respectively. The smallest image obtained represents particles less than about 10 μm, but shape is not recognizable. Time is shown on the right hand side of the boxes

Ice fog particle spectra can be provided based on various ice fog types, but this info is still in a progressive stage, and needs to be studied more carefully because of measurement difficulties at cold temperatures and small size ranges (<200 μm). Figure 6.5 shows ice fog spectra together with some light snow particles that occurred on the 17th of January 2011 over Yellowknife, Canada during FRAM ice fog project.

6.3.2 Remote Sensing Based Observations

6.3.2.1 Satellite Based Monitoring

Passive low earth orbit and geostationary satellite measurements have been used for monitoring ice fog in the absence of higher level cloud layers and the probability

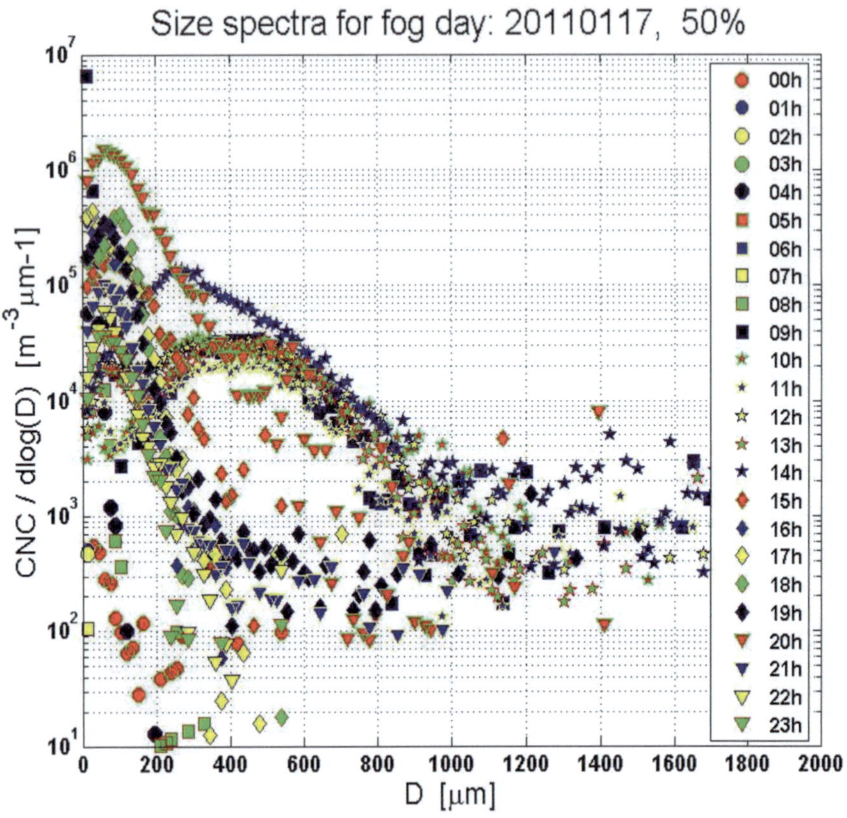

Fig. 6.5 Ice fog particle spectra per hr. averaged over 30 s time periods on January 17, 2011 based on GCIP measurements with 50 shadowing probabilities. The particles with maximum diameter less than 200 μm represent ice fog crystals. Particles larger than 200 μm represent transition from ice fog particles to lightly snowing ice crystals

of clouds with hazardous low ceilings (Pavolonis 2010a, b, Calvert and Pavolonis 2011; Gultepe et al. 2015; Schmit et al. 2005, 2008). Further discussion of methods for satellite detection and characterization of fog and low clouds can be found in Gultepe et al. (2007, 2014), Wetzel et al. (1996), Lee et al. (1997), Cermak and Bendix (2007, 2008, 2011), and Bendix et al. (2005). During the ice fog project (Gultepe et al. 2014, 2015) which took place over Barrow AL, both NOAA AVHRR and Modis satellite images were available through NASA LARC (Minnis et al. 1998, 2011). All satellite based techniques for fog use a difference between spectral irradiance values from the 3.9 μm (SWIR) and an IR channels for warm fog or cold fog research. The shortwave radiation (SW) contribution at the 3.9 μm channel for ice fog detection during daytime can complicate warm versus cold fog detection and retrieval of the ice fog microphysical parameters. At the 3.9 μm channel, the single scattering albedo for ice crystal sizes >10 μm can be about 20–30% less than

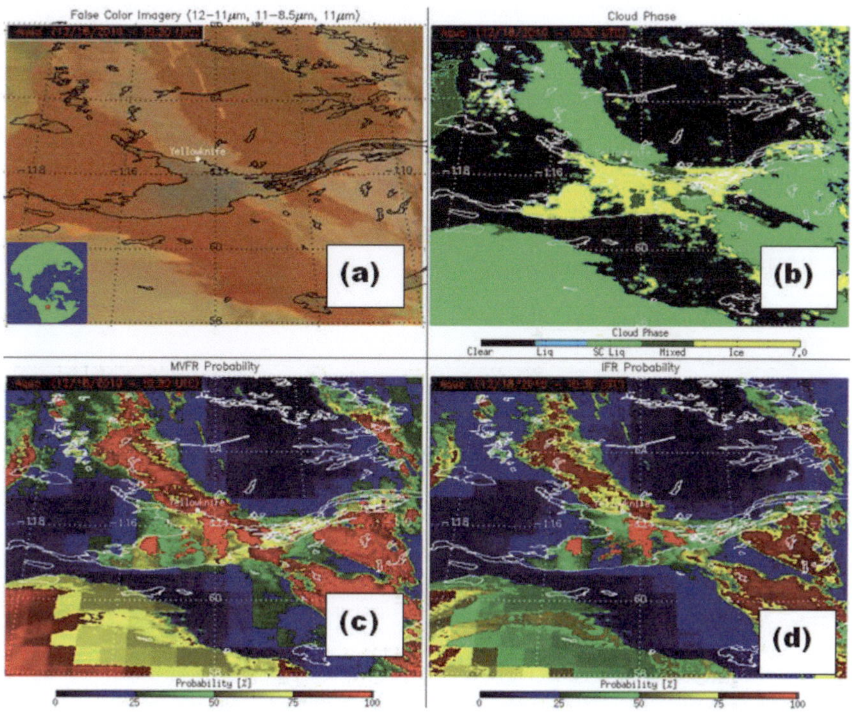

Fig. 6.6 Aqua MODIS passed over on Dec 182,010; 3.9 and 11 micron temperature difference over Yellow Knife International Airport occurred during FRAM project (**a**) and cloud phase, MFR and IFR images in (**b**), (**c**), and (**d**), respectively (Pavolonis 2010b)

this for fog droplets. But scattering of ice crystals in the visible light range, which depends on the particle shape, can be much larger than this for droplets and that reduces ice fog visibility strongly. However, operational satellite coverage over the Arctic regions is currently limited, and geostationary satellite observations suffer from reduced resolution in these areas (Gultepe et al. 2015). Figure 6.6a shows a false color image of severe ice fog event on a nighttime overpass of Aqua MODIS on the tenth of April 10 of 2008. For the same ice fog event, Fig. 6.6b, c, d show cloud particle phase, MFR, IFRV, respectively.

6.3.2.2 Microwave Radiometer (MWR) Based Monitoring

The Radiometrics profiling microwave radiometer (PMWR) measurements are commonly used for atmospheric profiling in which liquid regions can be discriminated (i.e. Gultepe et al. 2014; Ware et al. 2003, 2013; Cadeddu et al. 2009; Cimini et al. 2010; Bianco et al. 2005; Solheim et al. 1998). The magnitude of the LWC values in boundary layer clouds should be used cautiously, but as an indicator it can be very

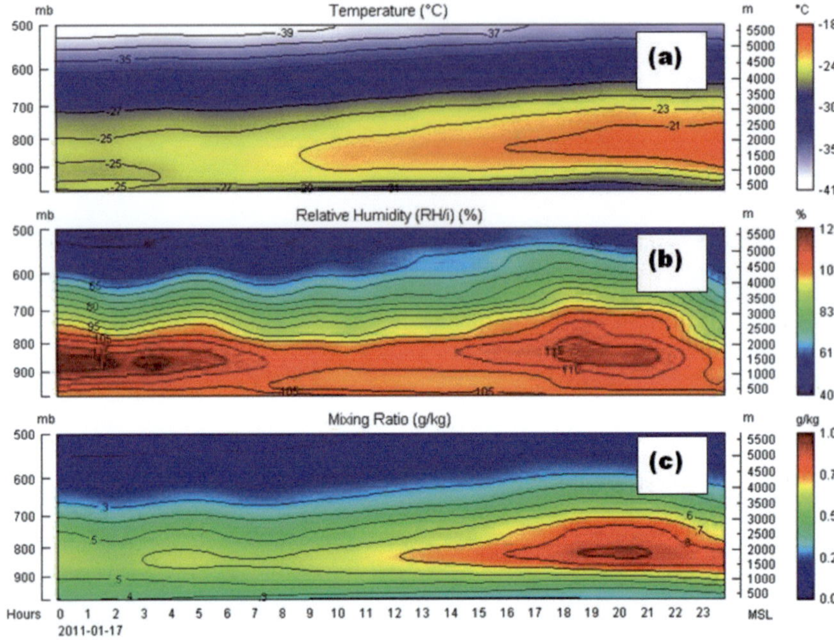

Fig. 6.7 An ice fog event occurred on Jan 172,011 that was observed by Radiometric during FRAM ice fog project in Yellow Knife, NWT, Canada; temperature, RH with respect water, and vapor mixing ratio (qv) are shown in (**a**), (**b**), and (**c**), respectively

useful for ice fog occurrence process and detecting inversion layers. Performing this can provide information underlying the reason for ice fog occurrences. An example of PMWR measurements of from which T and RH_w retrieved are shown in Fig. 6.7a, b, respectively. Below the 1 km level, T was at about −30 °C (Fig. 6.7a). Above 1 km, there was strong inversion layer with average T ~ −15 °C where RH_w values (>70%) were observed in the lowest layer of boundary layer (Fig. 6.7b). Ice fog formation in that layer (Fig. 6.7c) was observed with vapor mixing ratio (q_v) at about 0.8 g kg^{-1}. The visibility was approximately between 0.5 and 1 km during event (observed by FD12p present weather sensor).

6.3.2.3 LiDAR, Radar, and Ceilometer Based Monitoring

Information on ice fog conditions can also be obtained using LiDAR (Light Detection and Ranging), cloud radar, or a ceilometer. Backscattering from ice fog crystal particles is related to LiDAR transmitted wavelength, ice crystal surface area, and number of ice crystals. Sassen (1991) and Sassen and Zhu (2009) used LiDAR explicitly for retrievals of ice crystal microphysical properties representing cirrus clouds. Using polarization LiDAR data they provided an extensive review on LiDAR capabilities to discriminate ice particle shape. The backscatter coefficient (β) of LiDAR transmitted radiation reflects the volumetric properties of ice crystal number concentration spectra and their mass density. However, attenuation by liquid clouds limits the ability of the LiDAR and ceilometer to detect high level ice clouds, and a similar result is considered to be valid for ice fog conditions. Except for precipitating ice clouds and polluted environments, ice fog and boundary layer ice clouds can be considered having similar physical conditions if they are not associated with frontal systems. Figure 6.8 shows time-height cross sections of backscatter from and linear depolarization ratio from the DOE (Department Of Energy)lidar in (a) and (b),

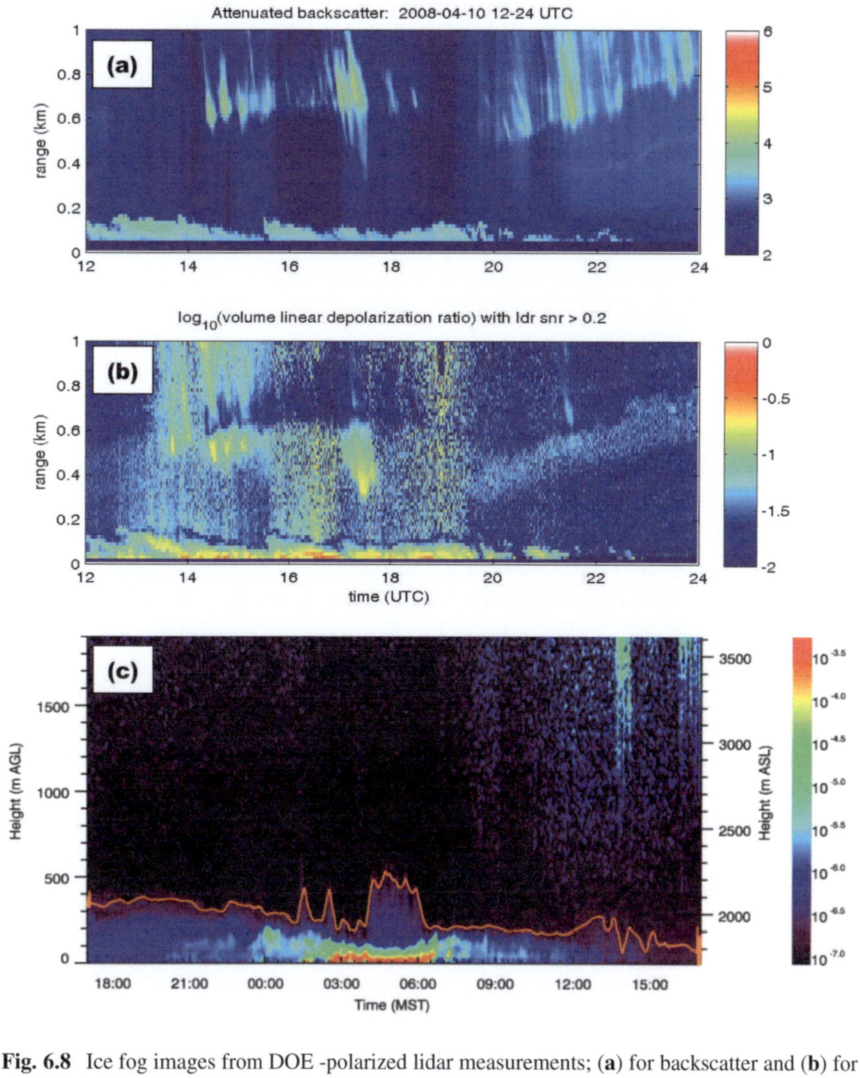

Fig. 6.8 Ice fog images from DOE -polarized lidar measurements; (**a**) for backscatter and (**b**) for depolarization ratio (during an ice fog event occurred at the DOE NSA site, Barrow, Alaska on April 10 of 2008. A CL31 ceilometer backscatter coefficient image is shown on (**c**). Ice fog event was observed between 1000 and 2000 UTC

respectively, during an ice fog event which occurred at the DOE NSA (North Slope Alaska) site in Barrow Alaska on April 10 of 2008 (Department Of Energy (DOE) Indirect and Semi-Direct Aerosol Campaign (ISDAC) project; Flynn et al. 2007; Lindeman et al. 2011). A CL31 ceilometer backscatter coefficient image for an ice fog event occurred during MATERHORN project (Gultepe et al. 2016b) is shown on (c). Ice fog event was observed between 1000 and 2000 UTC. All these images indicated that ice fog layers can be clearly monitored by the polarized lidar systems and ceilometers.

Millimeter-wavelength Cloud Radar (35 GHz) reflectivity can be used for ice cloud detection and monitoring (Matrosov 2010; Gultepe et al. 2015). Radar reflectivity (Z) is related to the integral of the product of particle number concentration and particle size to the sixth power over the size spectrum, and high Z values are usually indicative of larger particle sizes occurring in the clouds. Previous studies using mm cloud radar (Haeffelin et al. 2010; Uematsu et al. 2005; Hamazu et al. 2003; Mead et al. 1989; Boers et al. 2013) have investigated warm fog distribution and its microphysical characteristics for nowcasting applications. These suggest that mm cloud radars are more readily adaptable for detection of ice fogs due to the presence of larger particle sizes compared to small droplets found in warm fogs, but the measured radar reflectivity factor needs to be modified for ice dielectric constants. Other than cloud radars, lidars can also be used for ice fog detection because of less optical thickness compared to warm fog conditions. Sassen and Zhu (2009) provided a global survey of CALIPSO (Cloud-Aerosol Lidar and Infrared Pathfinder Satellite Observations) satellite linear depolarization ratio (Δ) for ice clouds that included all types of ice clouds but mainly for cirrus and optically thin low level clouds. It collected observations in both the nadir and off-nadir viewing directions, at a vertical resolution of 60 m in the upper troposphere and with a ground footprint of 333 m (Winker et al. 2007; Hunt et al. 2009). They found out that Δ decreased with increasing latitude and it was about 0.3–0.4 near the surface during 2007–2008 winter. The criteria for identifying cirrus and other ice clouds are that they must be transparent to lidar probing (i.e., cloud optical depth < _3.0–4.0). This criteria also fits well with ice clouds but vertical resolution of CALIPSO lidar close to earth is very large (60 m) for fog detection and that may limit the its capability for ice fog detection (having ice fog thickness < 100 m) and this needs to be further researched (Grenier et al. 2009).

Ceilometers have been used in many fog studies (Nowak et al. 2008) and typically use wavelengths centered at 0.905 μm with low power to measure fog macro-physical characteristics and evolution (Gultepe et al. 2009, 2014; Haeffelin et al. 2010; Gultepe et al. 2016a, b). The ceilometer provides a backscatter profile between 0 and 10 km with a resolution of approximately 10 m, using 110 ns laser pulses with a repetition rate of 6.5 KHz with an average laser power of 19.5 mW and a beam divergence of ~0.02° (Nowak et al. 2008). Its operation is similar to a LiDAR; however, compared to radar, it uses shorter wavelengths (ultraviolet in contrast to infrared), which allows detection of smaller particles. If clouds are optically thick; the transmitted power is however quickly attenuated by limiting the measurement range.

Integration of satellite based retrievals of ice fog microphysical parameters with surface based sensors including a LiDAR, cloud radar, ceilometer, and microwave radiometer, as well as surface in-situ fog sensors e.g. fog and standard visibility sensors can be used to improve the prediction of ice fog micro and macro properties. These types of integration systems were developed by the Gultepe et al. (2014, 2019a) for use in ice fog observations during Arctic field projects.

6.4 Ice Fog Prediction

Visibility (Vis) parameterizations, based on relationships between Vis and extinction (β_{ext}) can be applied to model based simulations for ice fog prediction. The applicable day- and nighttime Vis parameterizations can also be obtained as a function of a brightness contrast threshold coefficient and particle size spectra described below.

6.4.1 *Visibility Definition*

The definition of Meteorological Observation Range (MOR) by the World Meteorological Organization (WMO) is based on Koschmieder's law. Assuming a brightness contrast threshold (ε) of 0.05, the daytime visibility (Vis_d) is given by

$$Vis_{d5} = \ln\left(\frac{1}{\varepsilon}\right) / \beta_{ext} \tag{6.1}$$

where $\ln(1/\varepsilon) = 2.996$. β_{ext} is the extinction coefficient given as

$$\beta_{ext} = \int_{i=1}^{m} \pi Q_{ext}(r) n(r) r^2 dr \tag{6.2}$$

where m signifies number of particle spectral bins and Q_{ext} is the extinction efficiency, which is ~2 for spherical water droplets at visible wavelengths based on Mie theory. In Eq. 6.2, $n(r)$ and r^2 represents the particle number concentration and spherical particle cross sectional area, respectively. For ice crystals, Q_{ext} depends on particle shape and cross sectional area, which may be represented by the maximum dimension (L_i) and width (w_i) of the particle, particle spectra, and optical scattering properties of the ice crystal at visible wavelengths.

6.4.2 Parameterization for Ice Fog Visibility

The prediction of ice fog may be represented by conditions where $RH_i > 100\%$ and the presence of ice crystals inferred when the total IWC > 0.001 g m^{-3} and $N_i > 0.01$ L^{-1}. Ice fog intensity can be evaluated by *Vis,* which is function of both N_i and IWC predicted by a forecast model. If ice fog parameters are obtained from a numerical model simulation, then Vis for ice fog, assuming that N_i, IWC, and mean equivalent mass diameter (d) are known, can be obtained as described by Ohtake and Huffman (1969). Their result was developed for ice fog occurrence in Fairbanks, Alaska, representing likely homogeneous nucleation processes. In their work, N_i and average particle diameter (D_i), were observed to be 200 cm^{-3} and 7.2 μm respectively for high IWC conditions, (i.e. IWC > 0.1 g m^{-3}), and 80 cm^{-3} and 4.5 μm for low IWC's (0.01 < IWC < 0.1 g m^{-3}). If ice crystals are assumed to form due to deposition of vapor directly onto ice nuclei at cold T, then N_i can be parameterized as a function of RH_i. A relationship between N_i and $RH_{i,}$ or T, or IN versus RH_i (or S_i) have been commonly used (Fletcher 1962, Meyers et al. 1992, Cooper 1986, Cooper and Vali 1981). DeMott et al. (1997, 1999, 2015) studied aerosols mixed with soluble and insoluble components to determine the influence on the activity of ice nuclei, showing the relationships between thermodynamic and physical aerosol properties. Sometimes, N_i predicted values at cold T were found to be high and so that N_i was artificially modified (Koenig and Murray 1976; Morrison et al. 2003, 2005).

Gultepe et al. (2014), using low level aircraft observations and surface in-situ observations, obtained a parameterization for the extinction coefficient (β_{ext}) as a function of N_i. Using the cross-sectional surface area (A_c) of an ice crystal and a mass-area relationship, e.g. $m_i = aA_c{}^b$ (Lawson 2011), $\beta_{ext,}$ is calculated using measurements obtained in low level Arctic cloud measurements over the North Slope Alaska (NSA) site during the FRAM project (Gultepe et al. 2014). Measurements were made by a 2DS optical imaging probe capable of imaging particles from 10 to 1280 μm with a resolution of 10 μm. Using this information β_{ext} is written as a function of both IWC and total N_i (similar to Gultepe et al. 2007a), by using the relationship between Vis and β_{ext} above, a Vis parameterization for ice crystals (between 10 and 500 μm in this case) was derived, referred to here as Vis_1,

$$Vis_{i1} = 1.19(IWC\ N_i)^{-0.5066}. \tag{6.3}$$

The subscript i1 in Eq. 6.3 represents the method 1 based on 2DS ice fog measurements. Using a different instrument, the FMD (DMT Inc.), for optical scattering measurements for particles with sizes less than 50 μm, (collected during the FRAM-IF project in Barrow, Alaska, on April 102,008), a further Vis parameterization (subscript i2 representing method 2 based on FMD ice fog measurements) was derived as

$$Vis_{i2} = 0.24(IWC\ N_i)^{-0.5147}. \tag{6.4}$$

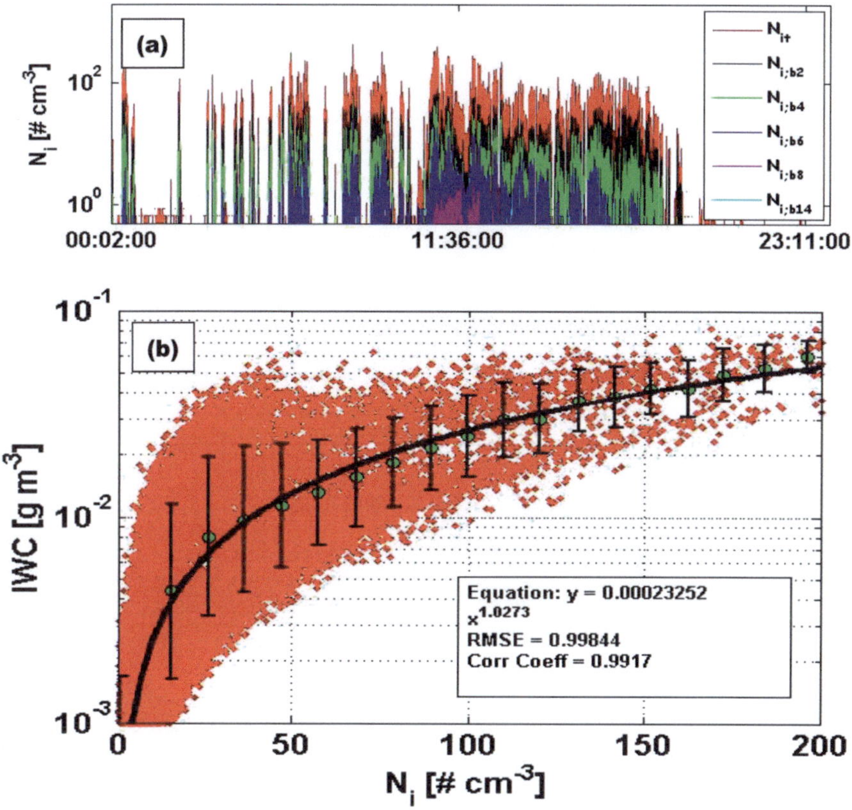

Fig. 6.9 Time series of N_i obtained by FMD (FM-100) fog instrument for total ice crystal number concentration (red line) and specific bins (other lines) for an ice fog event (**a**) that occurred on January 16, 2015 during the MATERHORN project and scatter plot for 1 s observations of IWC versus N_i (**b**). The fit to observations is also shown on the plot (**b**)

If the temporal evolution of IWC and N_i are available based on forecasting model outputs, then Vis (or β_{ext}) can be obtained from either Eq. 6.3 or Eq. 6.4. Gultepe et al. (2014) used measurements from an FMD probe to derive N_i for deposition nucleation events and reported concentrations as high as 1000 L^{-1} or more. The time series for N_i for a mountain ice fog event that occurred on the 16th of January 2015 during the Mountain Terrain Atmospheric Modeling and Observations (MATERHORN; Fernando et al. 2015) project Utah, US, is shown in Fig. 6.9a. This data set was used to derive a relationship between IWC and N_i (Fig. 6.9b) to demonstrate the importance of N_i in Vis parameterizations. To compare Vis values representing ice fog and warm fog conditions, Fig. 6.10 is provided. This figure shows the difference between ice fog Vis (Vis_i) and warm fog Vis (Vis_w) as a function of condensed water content based on assumed number

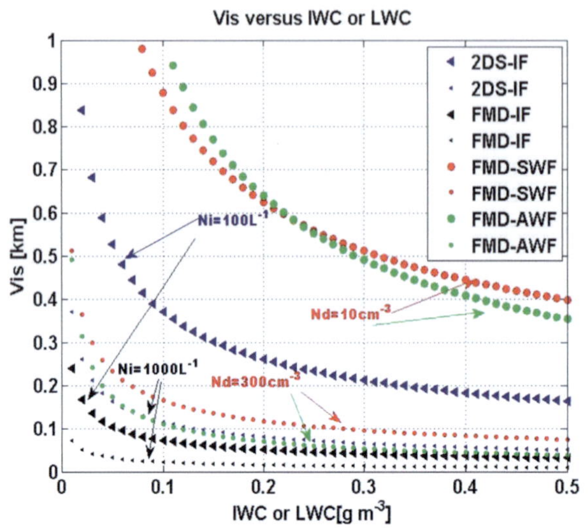

Fig. 6.10 Visibility (Vis) versus IWC and LWC for ice fog and warm fog conditions. The lines for ice fog are indicated by filled triangles based on FMD Eq. 6.4 and 2DS Eq. 6.3) observations representing low flying aircraft data, and for warm fog by filled circles where SWF represents surface warm fog observations and AWF represent low level aircraft warm fog observations. For warm fog based on a marine fog case study from FMD measurements: $Vis = 0.88(LWC \cdot N_d)^{-0.49}$ and from an aircraft droplet measurements: $Vis = 1.002(LWC \cdot N_d)^{-0.65}$

concentrations (N). It is seen that both Vis_i and Vis_w decrease with increasing N_i and N_d (number concentration of I ice crystals and droplets), respectively. Note that these relationships may change based on various ice microphysical characteristics e.g. particle shape and size. Assuming N_i changes from 1 to 10^4 L^{-1} while IWC changes from 0.005 to 0.03 g m^{-3}, Gultepe et al. (2015) suggested that based on FMD and 2DS measurements, Vis can be significantly different, representing measurement limitations. Vis measured by different present weather sensors may show significant differences for the same fog events at cold temperatures (Fig. 6.11). This figure shows that uncertainty in Vis can be as high as 50% based on different Vis sensors for ice fog events. Note that Eq. 6.3 represents relatively large particles compared to Eq. 6.4, which shows much higher contributions from small particles (<50 μm). When $N_i > 1000$ L^{-1}, both equations result in similar values of Vis of about 100 m. This result should be considered cautiously because of limited measurements conducted during ice fog conditions.

The N_i predictions for an ice fog event from a forecast model can be directly related to the number concentration of IN. Therefore, the N_i prediction can be obtained by either homogeneous nucleation or heterogeneous nucleation parameterizations, or both, and these parameterizations need to be studied and researched in detail. Some IN parameterizations differentiate aerosol species that can also be important for generating ice fog crystals and its intensity. In the following, the IN

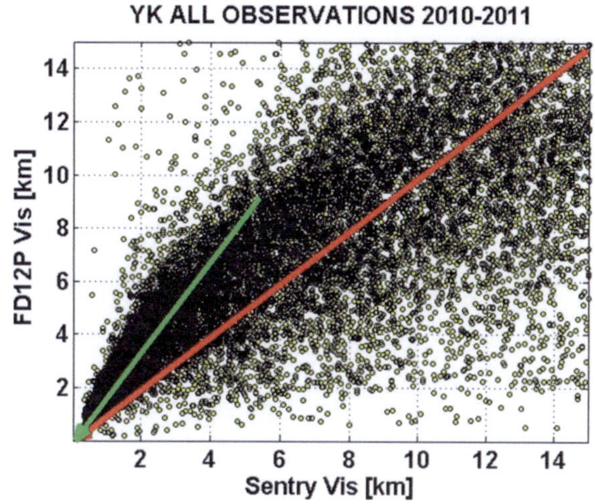

Fig. 6.11 The FD12p visibility and Sentry visibility measurements used to generate a scatter plot for the entire ice fog events occurred during the FRAM project. The red line is for a 1-1 line and green line is fit to observations, indicating FD12P Vis is overestimated compared to Sentry Vis

parameterizations provided are very important for describing ice fog occurrences at T > −40 °C (heterogeneous nucleation) and at T < −40 °C (homogeneous nucleation).

6.4.3 Heterogeneous Ice Nucleation

Deposition nucleation, a source of ice crystals and thus ice fog formation, generally requires an insoluble IN, which may be either natural or anthropogenic in origin. When T is above −40 °C, heterogeneous nucleation becomes the dominant nucleation mechanism including deposition, immersion and contact freezing. Ice nucleation parameterizations are usually based on laboratory measurements (Atkinson et al. 2013; Connolly et al. 2009; Lohmann and Diehl 2006; Murray et al. 2011; Niedermeier et al. 2010; Niemand et al. 2012, and Welti et al. 2012), atmospheric field projects (DeMott et al. 2003, 2010, 2015, Huffman and Vali 1973, Meyers et al. 1992, Phillips et al. 2008), combinations of both, or theoretical approaches (Hoose et al. 2010; Khvorostyanov and Curry 2004, 2005). Laboratory measurements in controlled environments are limited to the experimental conditions. Therefore, the results may not represent real atmospheric conditions and they should be used cautiously over large temperature and saturation ranges (e.g. for Arctic weather conditions). When results are applied to the model predictions of ice fog conditions, this issue should be recognized. Unfortunately understanding the phase transition of water droplets to ice crystals is not trivial, and theoretical approaches are therefore not fully constrained (Ickes et al. 2015). In fact, most parameterizations are developed for either condensation or immersion freezing because these processes are dominant in mixed-phase clouds; but, explicit formulation for ice fog deposition nucleation doesn't exist presently. Therefore, the parameterizations developed for heterogeneous nucleation of cloud studies should be used with caution. Some of the frequently used parameterization schemes developed for ice clouds and ice fog at $RH_i \geq 100\%$ are described below.

6.4.3.1 Fletcher (1962) Parameterization

One of the original IN nucleation parameterizations was provided by Fletcher (1962), which was based on field and laboratory measurements. This parameterization predicts the ice crystal number concentration using an exponential temperature function as

$$N_i(T) = N_s e^{-\beta_s[T - T_0]} \tag{6.5}$$

where T is the temperature in °C, T_0 as 273.15 °C, N_s the ice nuclei number concentration assumed to be 10^{-5} L^{-1}, and β_s =0.6 °C^{-1}. The Fletcher parameterization typically under-predicts N_i at warmer temperatures, but over-predicts at lower temperatures <245 K. Meyers et al. (1992) modified Eq. 6.5 according to the Huffman and Vali (1973) formulation to include the supersaturation dependence of deposition ice nucleation (e.g. between −10 and −20 °C such that

$$N_i(T) = N_s\left[(S_i - 1)(S_0 - 1)^{-1}\right]^b e^{-\beta_s[T - T_0]} \tag{6.6}$$

where $b = 4.5$ and $S_i - 1$ is the fractional ice supersaturation and S_o is the fractional ice supersaturation at water saturation.

6.4.3.2 Koenig and Murray (1976) Parameterization

The origin of IN can either be natural or anthropogenic. Here, discussions are based on the natural sources of IN. Based on Koenig and Murray (1976), the maximum number of new particles (N_i) without a liquid phase occurrence can be obtained as a function of T as

$$N_i = Aexp\left\{\frac{\ln(10)}{B}(T - T_o)\right\}, \tag{6.7}$$

where A and B are constants given as 1 and 4, respectively. T_o is 0 °C, and T is the temperature in °C. For a given T = −30 °C, IN is estimated to be about 10 cm^{-3}. This equation was derived based on classical nucleation theory for heterogeneous nucleation.

6.4.3.3 Meyers et al. (1992) Parameterization

A subsequent improvement to N_i parameterization was provided by Meyers et al. (1992) which (Eq. 6.8) is still widely used and was based on a limited number of aircraft observations.

$$N_i = \exp\{a + b[100(S_i - 1)]\} \tag{6.8}$$

where a = −0.639 and b = 0.1296, and S_i is the super saturation with respect to ice. The original parameterization was valid only over the temperature range − 7 °C to −20 °C, ice supersaturations from +2 to +25%, water supersaturations between −5 and +45% although it has been extrapolated outside these limits. The maximum values of N_{id} predicted by Eq. 6.7 can reach up to 100 L^{-1} at T = −20 °C, which are significantly less than N_i observed in typical ice fog, e.g. during the Fog Remote Sensing and Modeling-Ice Fog (FRAM-IF) project over Yellowknife International Airport during winter of 2010–2011. For the WRF simulations discussed later, N_i was subsequently multiplied by a factor of 10 to better represent the observed N_i values in these cases.

6.4.3.4 Cooper (1986) Parameterization

Thompson et al. (2004) suggested use of the Cooper parameterization (1986) given below for N_i prediction in order to reduce the IN estimated by the Fletcher parameterization as

$$N_i = 0.005\exp\{0.304(T_o - T)\}. \tag{6.9}$$

The resulting Cooper curve lies between that N_i predicted by the Meyers et al. and Fletcher parameterizations when T > −25 °C. Note that Meyers et al. parameterization used for modeling studies was obtained for air saturated with respect to water using limited in-situ measurements.

6.4.3.5 Latest IN Parameterizations

Using the work of Meyers et al. (1992) related to the IN activation spectrum based on various observational data sets, a new parameterization for the number of ice nuclei is provided by Phillips et al. (2007) as

$$N_{IN} = 1000\big[\exp(12.96(S_i - 1.1))\big]^{0.3} \tag{6.10}$$

where N_{IN} is the ice nuclei concentration in m^{-3}. The Meyers et al. (1992) parameterization is usually valid for T > −30 °C, and it is modified for T between −30 and −80 °C, utilizing DeMott et al. (2003) observations based on continuous Flow Diffusion Chamber (CFDC) measurements collected on aircraft. These measurements represent IN typically averaged over 10–30 min flight periods.

Phillips et al. (2008) modified Eq. 6.9 for various temperature and ice supersaturation ranges in order to account for three main types of freezing aerosols (e.g. dust and metallic compounds, black carbon, and insoluble organics). Using appropriate scaling to estimate IN concentrations to be matched that of observations from DeMott et al. (2003) at T = −30 °C, Phillips et al. (2008) provided N_i parameterization as

$$N_{IN} = 1000 \exp\left[2.96\,(S_i - 1) - 0.639\right]\alpha, \tag{6.11}$$

where α is an adjustment factor (0.06). Equations 6.10 and 6.11 should be used with caution for ice fog conditions because of low IN number concentrations. Demott et al. (2010, 2015) provided further improved parameterizations for global ice nuclei concentrations for mineral dust types (Mamouri and Ansmann 2015). Later work assumes that mineral dust can coexist with natural and soluble aerosols for immersion freezing conditions; however, they cautioned that use of their IN parameterization may result in underestimations of N_i at low temperatures ($T < -30$ °C) by up to a factor of 100. The Demott et al. (2015) parameterization is a modification of the proposed parameterization by Tobo et al. (2013) that is used to predict IN concentrations (assuming mainly dust particles) based on aerosol number concentrations (N_a) with sizes greater than 0.5 μm, so that

$$N_{IN} = {N_a}^{[\alpha(273.16 - T_c) + \beta]} \exp\left[\gamma\,(273.16 - T_c) + \delta\right] \tag{6.12}$$

where T_c is the cloud temperature, $\alpha = 0.074$, $\beta = 3.8$, $\gamma = 0.414$, and $\delta = -9.671$ (Tobo et al. 2012, 2013; Mamouri and Ansmann 2015). For mixed phase cloud conditions using data from multiple locations based on $RH_w > 100\%$ for aerosol sizes >0.5 μm, IN parameterization is studied by DeMott et al. (2010), and similar to his work, Lloyd et al. (2015) based on $RH_w < 80\%$ and aerosol sizes between 0.5 and 1.6 μm provided IN parameterization as

$$N_{IN}\left[L^{-1}\right] = a(273.16 - T_c\,[K])^b\, N_a^{(c(273.16 - T_c) + d)}\left[cm^{-3}\right], \tag{6.13}$$

where a = 0.0000594, b = 3.33, c = 0.0264, and d = 0.0033. Lloyd et al. (2015) stated that high values of IN based on Eq. 6.13 and measurements up to 1000 L^{-1} was not easy to explain for their mountain cloud study project. Tobo et al. (2013) stated that global IN concentrations as a function of T in which IN reaches 10^5 L^{-1} at $T < -35$ °C for soil dust conditions. This IN value for ice clouds found by Lloyd et al. (2015) resembles ice fog N_i values that reduced Vis down to 100 m, which occurred during the FRAM ice fog project.

Niemand et al. (2012) demonstrated a new particle surface area based parameterization by defining an aerosol active site density, $n_s(T)$, obtained from cloud freezing experiments at the AIDA (Aerosol Interactions and Dynamics in the Atmosphere) chamber. The $n_s(T)$ was first formulated by Connolly et al. (2009). The results of Niemand et al. (2012) were compared to field measurements of IN concentrations conducted at the Taunus Observatory, Kleiner Feldberg, Germany. In that study, aerosols were collected in the size range 10 nm–20 μm by electrostatic precipitation onto silicon wafers and their freezing activity was automatically determined using a CCD camera based instrument in the Frankfurt Ice Nuclei Deposition Freezing Chamber (FRIDGE, Klein et al. 2010a, b; Bundke et al. 2008). They used the parameter S_{ae} defined as the surface area of the aerosols in question, such that the ice active fraction, in a particular particle size bin (f_{ij}) can be parameterized using

$$f_{i,j}=\frac{N_{i,j}}{N_{tot,j}}=1-\exp\left[-S_{ae,j}n_s(T)\right] \tag{6.14}$$

which can be used to parameterize immersion freezing of dust particles in the temperature range − 12 to −36 °C in cloud models. Niemand et al. (2012) provided the following equation (Fig. 6.12a) for n_s as a function of T

$$n_s(T)\left[m^{-2}\right]=\exp\left[-0.517\,(T-273.15)+8.934\right]. \tag{6.15}$$

Comparisons based on various studies (Fig. 6.12b) also indicate the variability of n_s as a function of T (Murray et al. 2012) such that

$$N_i(T)=N_a\Big(1-\exp\left(-n_s(T)S\right). \tag{6.16}$$

For large particles (>3 μm) and temperature < −30 °C, Eq. 6.16 approaches 1 and needs to be calculated explicitly for each size bin.

The above equations usually represent IN number concentrations for environmental conditions saturated with respect to water and mixed phase conditions; therefore, their applications to ice fog predictions are limited and because of that they should be used cautiously.

6.4.4 *N_i Estimation by Homogeneous Freezing Nucleation*

Ice crystal formation at T < −40 °C can be due to instantaneous freezing of haze particles or droplets, referred to as homogeneous nucleation (Heymsfield and Sabin 1989). In this section, we will focus on homogeneous nucleation of ice crystals that play an important role in ice fog formation at T < −40 °C.

The most common theoretical framework to describe homogeneous freezing based on thermodynamics and kinetics of supercooled water is the Classical Nucleation Theory (CNT). However, many thermodynamic and kinetic processes within supercooled water are not very well understood and cannot be estimated experimentally (Ickes et al. 2015). Nonetheless, it is still the only theoretical

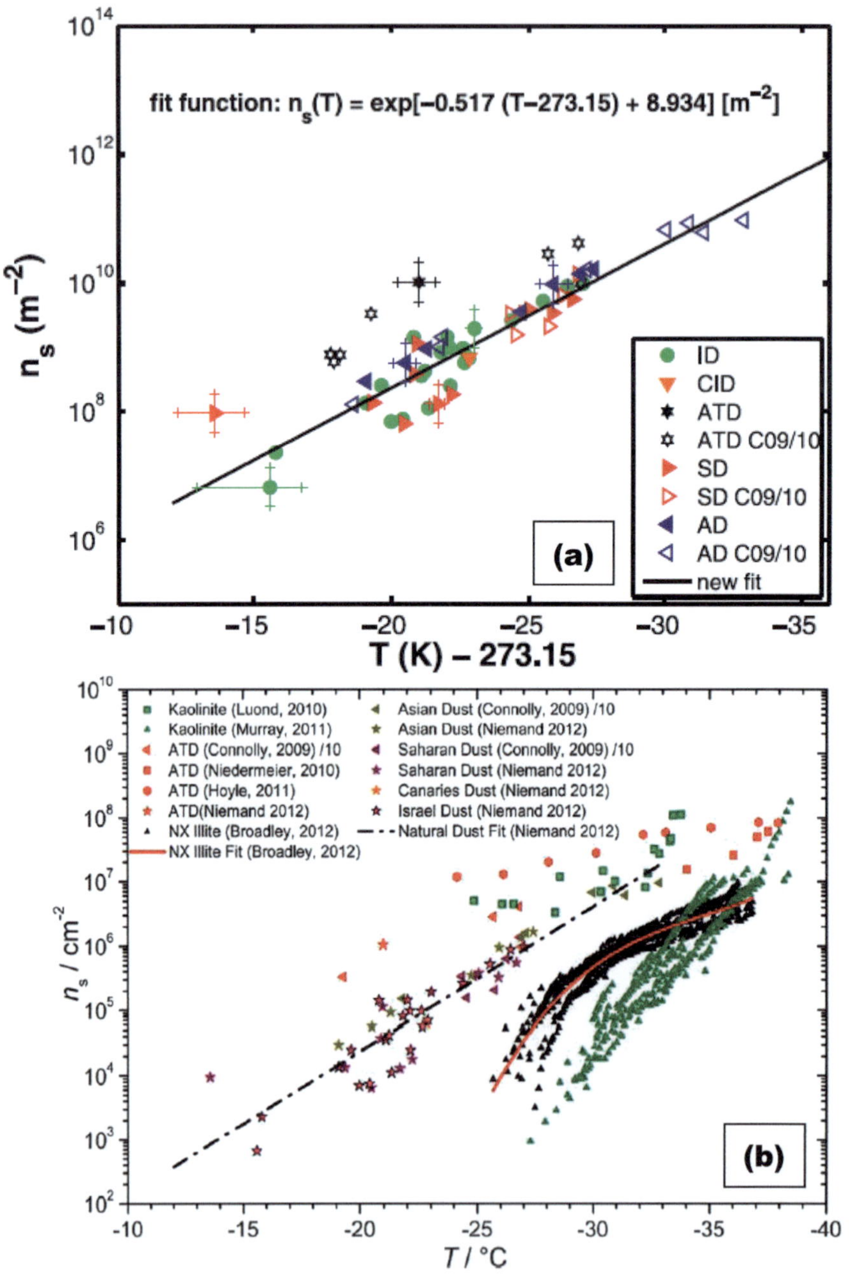

Fig. 6.12 A particle surface area based parameterization by defining an aerosol active site density ($n_s(T)$) versus T obtained from the cloud freezing experiments is shown in box (**a**) (Niemand et al. 2012, © American Meteorological Society. Used with permission). The black solid line is a T-dependent fit to all different natural dust types. The plot also includes the data from Connolly et al. (2009) reduced by a factor of 10 (**a**). The summary of the n_s parameter as a function of T based

framework, which can be used in the context of prediction of ice crystals in a weather model. To reduce the uncertainty of CNT estimations, the physics of supercooled water have to be better understood, so that thermodynamic and kinetic processes, and thus the theoretical framework can be better constrained. An overview of parameterization schemes for homogeneous nucleation is given below.

The classical nucleation theory of homogeneous nucleation rate of liquid drops from the vapor (J_{VL}) is described by Dunning (1969) and Anderson et al. (1980). The corresponding homogeneous nucleation rate of ice crystals from water vapor is given by (Hobbs et al. 1974) as

$$J_{vi} = \frac{P 4\pi r_{vi}^{*2} \beta n_v}{(2\pi mkT)^{1/2}} \exp\left\{ \frac{-16\pi \sigma_{vi}^3 \varepsilon}{3\left[n_s kTln\left(\frac{P}{P_s}\right)\right]^2} \right\}, \tag{6.17}$$

where $4\pi r_{vi}^{*2}\beta$ is the surface area of the critical embryo, n_v the number of single water molecules per unit volume of the vapor phase, P the water vapor pressure, P_s the saturated vapor pressure over a plane surface, ϵ the shape factor for the ice embryo, σ_{vi}^3 the interface free energy between ice and vapor, n_s is the number of molecules per unit volume of ice and m the mass of a water molecule, k and T are the Boltzmann constant and temperature, respectively.

Anderson et al. (1980) stated that the total nucleation rate for ice particle formation (J_{IF}) can then be represented by the sum of the rate of droplet nucleation from vapor (J_{vw}) and their subsequent freezing rate, which can be used to determine the total number of ice crystals formed directly as:

$$J_{IF} = J_{vw} + J_{vi}. \tag{6.18}$$

The total number of ice crystals nucleated should contain both freezing droplets (due to the existence of IN inside the droplet) and ice crystals formed directly by deposition, but these freezing droplets may not result in perfect spheres after the freezing process. The probability of a drop with volume V that will freeze at time t is given by Hobbs et al. (1974) as

$$P = 1 - \exp(-VJ_{wi}t) \tag{6.19}$$

where J_{wi} is the liquid to solid homogeneous/freezing nucleation rate.

Fig. 6.12 (continued) on studies of Murray et al. (2011), Connolly et al. (2009), Niemand et al. (2012), Murray et al. (2011), *Lüönd* et al. (2010), Hoyle et al. (2011), Niedermeier et al. (2010), and Broadley et al. (2012), is shown in (**b**). (Adapted from Murray et al. 2012, reproduced by permission of The Royal Society of Chemistry). Niemand et al. (2012) corrected the values from Connolly et al. (2009) by adjusting down by a factor of 10 in order to correct an earlier error in Connolly et al. (2009). (Details on these figures are found in Niemand et al. (2012))

For simplification, to calculate the homogeneous nucleation rate (T < −40 °C), parameterizations are used extensively. Ice crystal number concentration calculations based on the classical theory of homogeneous nucleation suggests that its value is usually less than observed values and unreliable; and therefore, its parameterization is generally used for N_i prediction that over the ocean can reach up to 10^4 particles per cm^{-3} for aerosol sizes less than 0.1 μm (haze particles). A homogeneous nucleation rate, based on liquid water and sulfuric acid, can be estimated using the equation given by Jaecker-Voirol and Mirabel (1989) as

$$J_{\mathrm{hom};h} = 10^x, \tag{6.20}$$

where

$$x = 7.0 - \left(64.24 + 4.7RH_f\right) + \left(6.13 + 1.95RH_f\right)\log_{10} N_{H_2SO_4}, \tag{6.21}$$

J_{hom} is the new production of sulfuric acid-water particles (# cm^{-3} s^{-1}) and RH_f is the fractional RH with respect to water. $N_{H_2SO_4}$ is the number concentration of gas-phase sulfuric acid molecules (molecules cm^{-3}). This equation does not depend on temperature directly, but on the RH fraction and is valid between 230 K and 300 K. Koop et al. (2000) also provided a formulation for a homogeneous nucleation rate as a function of water activity for haze particles, which was used in the study by Kim et al. (2014). The later stated that high concentrations of ice fog particles were present when homogeneous nucleation events occurred during cold Alaska winter weather conditions.

An equation similar to Eq. 6.20 for homogeneous freezing nucleation rate of the droplets as a function of temperature is given by DeMott et al. (1994) as follows

$$J_{\mathrm{hom};f} = 10^y, \tag{6.22}$$

where

$$y = -606.3952 - (52.6611T) - \left(1.7439T^2\right) - \left(0.0265T^3\right) - \left(1.536 * 10^{-4}T^4\right) \tag{6.23}$$

The above equations can be applied to ice fog visibility prediction that may occur due to homogeneous nucleation events associated with power generation units, Arctic leads or polynyas (Gultepe et al. 2003, 2008), or low level inversion layers near the surface. When the nucleation rate is known, the number concentration of freezing droplets can be obtained using equations given in Heymsfield and Miloshevich (1993), DeMott et al. (1994) and Kim et al. (2014) as

$$N_{i;\mathrm{hom}} = \sum\nolimits_{i=1}^{k} ((1 - \exp - J_{hom} V \Delta t)\, n(D) \Delta D), \tag{6.24}$$

where V is the droplet volume, t time, and n(D) the droplet size spectra. Using either Eq. 6.3 or Eq. 6.4 and knowing both N_i and IWC from an ice fog model or a forecast model, ice fog Vis can be predicted.

6.5 Autoconversion Processes and Ice Fog

Autoconversion proces, defined as a process that transfers cloud or fog ice to the precipitation, is directly related to IN (N_i) spectra and both T and S_i parameters. To better understand and estimate atmospheric vapor mixing ratio profiles and ice fog/ice cloud effect on visibility, fog ice water content needs to be estimated accurately. In NWPs this depends on microphysical schemes (Gultepe et al. 2017a, b). Autoconversion of fog/cloud particles here is defined as removal or transferring of ice crystal/snow particles in a cloud layer by microphysical processes such as aggregation, gravitational settling (after growing of ice crystals by deposition), or riming. Ice fog dissipation is strongly affected by the autoconversion of ice crystals to snow above the BL or within the BL. Autoconversion of droplets/ice crystals to precipitation are important for sizes >20–40 μm that can be influenced by collision and coalescence. Mineral dust IN coated with soluble materials (Wex et al. 2014; Mamouri and Ansmann 2015; Niedermeier et al. 2010) can affect ice crystal formation. This kind of nucleation can occur between −9 and −35 °C where ice fog can form due to freezing droplets (Fukuta and Schaller 1982), and deposition nucleation processes.

Autoconversion of droplets to precipitation generally uses a parameterization similar to Lin et al. (1983) parameterization, but in some studies the threshold of q_w (water mixing ratio) is replaced by the product of predicted N_d and a droplet critical size (r_c) of about 20 μm (Phillips et al. 2007). This critical size can be taken as less than 10 μm for warm fog conditions (Gultepe et al. 2009). Autoconversion of ice crystals to snow is the main contributor for defining the intensity of ice fog visibility during its dissipation. Similar to warm fog conditions, autoconversion of the ice fog crystals to precipitation in the frontal systems can be performed when the ice crystal size exceeds a threshold value which is about 200–300 μm size range (Gultepe et al. 2014). Based on Phillips et al. (2007) and Khairoutdinov and Kogan (2000) studies, ice crystal autoconversion to snow can be written as

$$\left(\frac{\partial q_{sn}}{\partial t}\right)_{auto} = K {q_i}^{7/3} N_i^{-1/3} \tag{6.25}$$

where K can be taken as frozen spherical ice crystals, q_i the ice water mixing ratio and q_{sn} the snow mixing ratio. Eq. 6.1.25 should be used judicially for ice crystal conversion to snow because of the various shapes of small ice crystals. Autoconversion of ice crystals to snow, when they grow to larger size, can also be obtained based on a critical threshold of ice crystal size (Ferrier 1994; Phillips et al. 2007). When ice crystals with dimensions, L_i > 200 μm, autoconversion of ice fog crystals by aggregation and diffusion processes to snow occurs. L_i can be obtained as a function of N_i, q_i, q_{sn} and ρ_i. For solid ice crystals, ρ_i can be assumed to be 0.90 g cm^{-3}, and final values of N_{sn} as a function of time can be obtained as

$$\left(\frac{\partial N_{sn}}{\partial t}\right)_{auto} = \frac{\left(\frac{\partial q_{sn}}{\partial t}\right)_{auto}}{\frac{\pi \rho_i w_i^2 L_i}{4\rho_a}}. \tag{6.26}$$

In this equation, w and L are the width and maximum size (length), respectively, of an ice crystal. Increasing size results in increased aggregation and snow precipitation, but decreasing N_i and IWC leads to ice fog dissipation. The product of w_i^2 and L_i can be approximated as L_i^2 for columnar ice crystals and $0.10\ w_i^2$ for plate like ice crystals. The critical maximum size of an ice fog particle can be assumed to be 200 μm from (Fig. 6.13), which compares ice crystal fall velocity versus fog droplet fall velocity, assuming various relationships for particle fall velocity versus size. The denominator in Eq. 6.26 should be used appropriately when ice crystals have a distinct habit.

The ice fog dissipation stage usually proceeds when solar radiation starts to influence the ice fog layer through heating processes. This may lead to decreasing RH and increasing competition of vapor among the remaining ice particles resulting in diffusional growth and light precipitation. Therefore, ice fog formation and dissipation processes should include detailed radiative transfer models with improved optical properties of ice crystals.

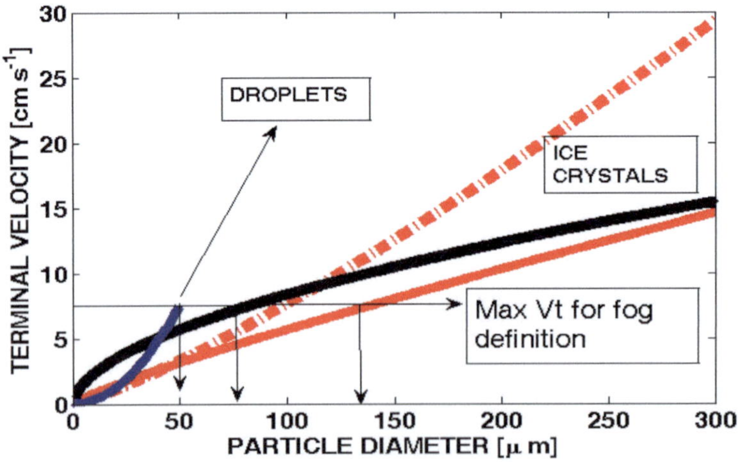

Fig. 6.13 Terminal velocities of ice crystals and droplets, representing ice fog and liquid fog, respectively. Warm fog particles have usually sizes less than about 50 μm that correspond to an ice crystal size of about 150–200 μm that is the upper threshold of ice fog crystal sizes

6.6 Challenges Related to Ice Fog Modeling for Weather and Climate

The prediction of ice fog using a forecast model requires accurate simulation of fog microphysical parameters based on the various microphysical schemes described above. Here, the application of models for ice fog prediction, including the Canadian High Resolution Deterministic Prediction System (HRDPS), US North American Model (NAM), Weather Reasearch and Forecasting (WRF), and Large Eddy Simulation (LES) models are briefly discussed. Then, a summary of ice fog studies by the climate models is emphasized. It should be noted that ice fog effects related to climate change and LES simulations are very limited or non-exist; therefore, results of these sections should be used carefully.

6.6.1 HRDPS Model

The HRDPS model (Kehler et al. 2016) is based on the Canadian Global Environmental Multi-scale (GEM) (Côté et al. 1998) forecast model. The HRDPS model uses various microphysical algorithms with ice nuclei parameterization of Meyers et al. (1992). Therefore, HRDPS simulations usually result in an underestimate of N_i. In fact, Gultepe et al. (2001) suggested that N_i versus T relationships for large ice crystals at sizes >100 micron (as used in Meyers et al. 1992) do not show statistically significant correlations. On the other hand, Heymsfield and Platt (1984) provided a parameterization for ice crystal spectra as a function of T and IWC that can represent N_i in the numerical models better than simple use of T. The Vis prediction is based on the MY2 bulk microphysical scheme (the 2-moment version of the Milbrandt and Yau (2005a, b), hereafter "MY2"). Accurate prediction of N_i from HRDPS for ice fog conditions can only be performed if microphysical algorithms are improved. In the MY2 scheme, the clouds and precipitation are predicted by the grid-scale condensation scheme. Ice crystals in the MY2 scheme are represented by two categories: (1) "ice" representing pristine crystals and (2) "snow" representing larger crystals (D > 250 μm) and/or aggregates. Particle size distributions (PSDs) are represented by functions, in this case as complete gamma functions with the two prognostic variables (for each category type x), total N_i, N_{ix}, and the mass mixing ratio (q_x). Some simulations done with the MY2 scheme result in low values of N_i and underestimate Vis. This suggests that ice fog predictions with this model need to be further researched.

6.6.2 North American Mesoscale (NAM) Model

For operational applications, the National Centers for Environmental Prediction's (NCEP) 12-km NAM model (Rogers et al. 2009; Ferrier et al. 2002) has been used for regular weather guidance over the continental US (i.e. the CONUS), including

Alaska. The NAM runs are for 4 times per day (00, 06, 12 and 18 UTC), and provide forecast products but not for ice fog. The North American Mesoscale Model post processor calculates Vis based on the work of Stoelinga and Warner (1999) that ignores the N_i. Some test simulations done during the FRAM ice fog project resulted in large uncertainties in Vis simulations (Gultepe et al. 2009). The model usually has a high bias in RH and fog IWC. Therefore, two diagnostic methods for identifying ice fog conditions have been proposed. Method 1 uses boundary layer parameters from the NAM simulations (Zhou and Du 2010) and then predicts the occurrence of ice fog, but doesn't provide microphysical parameters such as IWC. Method 2 utilizes physically-based post-processing analysis and provides IWC at a height z. Based on the work of Zhou and Ferrier (2008) and Zhou (2011), IWC during the FRAM-IF project was predicted and Vis was improved. It is concluded that the boundary layer moisture advection and turbulence schemes need to be improved.

6.6.3 WRF Model

The WRF (Weather Research and Forecasting) model has been used for ice fog detection over mountains (Pu et al. 2016) and Arctic regions with version 3.4.1 (Nygaard et al. 2011a, b). Gultepe et al. (2015) and Kim et al. (2014) also used the WRF model for ice fog predictions. In Gultepe et al. (2015), two nested domains were used with a horizontal grid of 100 × 78 (30 km resolution) and 88 × 64 (10 km resolution), respectively. The 35 vertical layers were utilized between the surface and the 100 hPa level. Both Initial and boundary conditions were from NARR (North American Regional Reanalysis) data. The WRF model for ice fog prediction was run using three different microphysical algorithms (Gultepe et al. 2015) for the simulations of q_v, IWC, N_i, and Vis. These algorithms were; (1) Milbrandt and Yau scheme (Milbrandt and Yau 2005a, b), (2) the Morrison scheme (Morrison et al. 2005; Morrison and Milbrandt 2011), and (3) the Thompson scheme (Thompson et al. 2004, 2008). N_i prediction based on the deposition ice nucleation mode underestimated N_i for ice fog conditions. Therefore, N_i values were increased by a factor of 10 based on observed N_i by Gultepe et al. (2014). Figures 6.14a–f are given for the MY2 scheme, and Figs. 6.14g–l are for the Morrison scheme. The Thompson scheme uses an *a priori* fixed N_d value; therefore, it is not shown here. All three of these schemes utilize a 2-moment treatment for the ice crystal microphysical categories, and predict q_v and N_i. The nucleation and growth processes in these schemes are not the same. Overall, simulated q_v and Vis (Fig. 6.14) based on the above schemes are found to be significantly different.

6.6.4 Ice Fog and LES Models

Ice fog models are usually 1-D models with prognostic equations for fog microphysical parameters such as the IWC and N_i. Presently, there is not a 1-D ice fog model for Vis predictions as there is for warm fog applications. The 1-D warm fog models

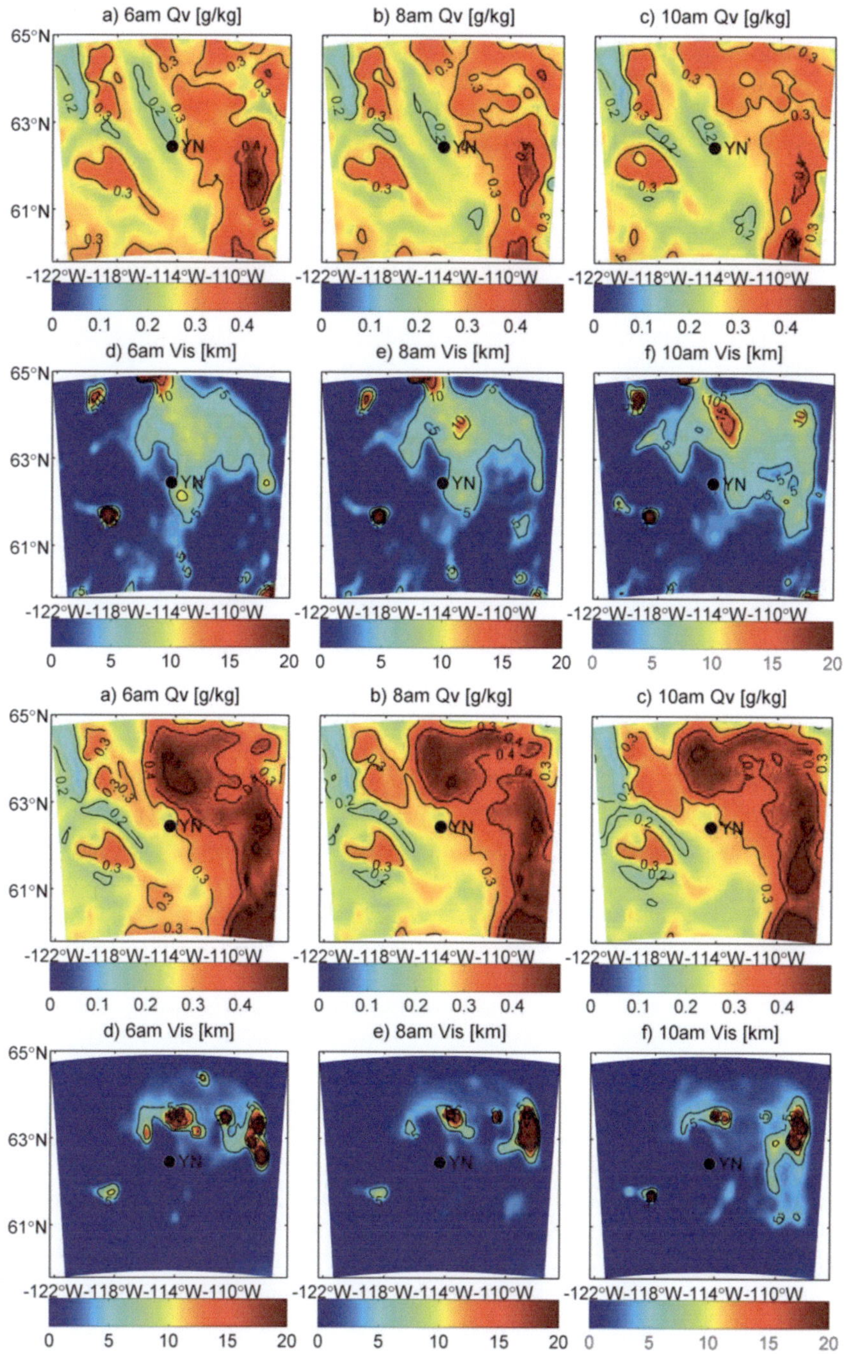

Fig. 6.14 The WRF model simulations of an ice fog event occurred during January 17, 2011: the top two rows are for vapor mixing ratio (Qv) and Vis from the microphysical algorithm of Milbrandt and Yau (2005a, b) and the bottom two rows are for the same parameters, but for Morrison et al. (2005). The YN indicates location for the project site

are initialized with output of mesoscale or regional climate models, and assume horizontal homogeneity in external forcing and at surface BL conditions. The Code de Brouillard a'l'Echelle Locale (COBEL) column model (Bergot and Guedalia 1994; Bergot et al. 2005) and the microphysical fog (MIFOG) and parameterized fog (PAFOG) (Bott et al. 1990; Bott and Trautmann 2002) models with detailed radiative transfer method for fog microphysics, and surface-vegetation scheme have been used for warm fog predictions. The surface turbulent heat and moisture fluxes can depend strongly on surface properties; therefore, the variability of these fluxes in heterogeneous terrain is difficult to predict by single-column models (Bergot et al. 2007). However, mesoscale forecast models do account for the horizontal heterogeneities at the scales pre-described and limited by coarser vertical resolutions (Pagowski et al. 2004). It is possible that the use of LES models for BL cloud studies (Chung and Matheou 2014) can further help to develop LES modeling for ice fog predictions. Inoue et al. (2014) also used an LES model for studying cloud occurrences over the ocean. These high resolution LES models can be used for further developing the capability of ice fog predictions at cold temperatures, but presently no work exist on this subject.

6.6.5 Climate Models and Ice Fog Radiative Impact

Ice fog plays an important role in the atmospheric heat budget. A high ice fog crystal number concentration can reduce the further cooling due to infrared radiation being lost to space, and their precipitation rate can be less than 0.01 mm h^{-1}. A detailed analysis of ice fog conditions affecting radiative budget have been studied by Girard and Blanchet (2001). Using a regional aerosol climate model, they stated that ice fog occurrence can be up to 30–40% of the time over the Arctic regions (Curry et al. 1990; Girard and Blanchet 2001). Because of reducing solar radiative fluxes, these studies emphasized that ice fog effects on local climate can be up to 60 W m^{-2}. An increase of the surface warming up to 30 W m^{-2} was also possible (Gultepe et al. 2009; Girard and Blanchet 2001). Ice fog crystal growth can also result in light snow formation (Gultepe et al. 2017a; Gultepe 2015). This can be responsible for upper atmospheric cooling that needs to be researched using regional or global climate models with detailed ice nucleation processes, such as one used by Girard and Blanchet (2001). In addition to arctic fog formation and its effect on climate, Gultepe et al. (2017a) studied mountain ice fog and noted that ice fog can affect the surface energy budget significantly; decreases IR cooling ($\sim$30 W m^{-2}) and increases heating by SW radiation leading to dissipation of ice fog (up to 100 W m^{-2}).

Overall, ice fog studies are needed to improve its prediction but neither fog/cloud models, or NWP and climate models, are ready for this purpose. A better understanding of ice fog conditions can be tested using large eddy simulations (LES) models, which resolve smaller scales compared to NWPs and climate models. The, knowledge gained can be used for developing new schemes for NWP and climate studies.

6.7 Discussion

Presently, ice fog measurements and modeling studies are very limited because of issues related to accurate measurement of IN, N_i, and Vis at cold temperatures, and a lack of understanding of ice microphysical processes (Gultepe et al. 2017a, b). Although laboratory measurements of these properties are improving, their applicability is currently limited due to dynamic and radiation feedback processes relevant to real atmospheric conditions. Therefore, ice fog predictions using standard forecasts, fog, and climate models remains limited. Most of the modeling work described is based on cloud microphysical processes that may not hold for particular ice fog and freezing fog conditions. In the following subsections, the current understanding of ice fog measurements, its predictions, and its effects due to climate change are summarized.

6.7.1 Measurement Issues and Instruments

Recently, ice fog crystal measurements for small crystals (sizes <200 μm), have been performed using improved optical probes. Other than the general difficulty in making such measurements, primarily due to the need to discriminate between small ice particles and super-cooled droplets, issues may arise due to artifacts caused by shattering of ice crystals under high wind and turbulence conditions on instrument surfaces (McFarquhar et al. 2017). This has been of particular concern for high speed aircraft measurements. For surface based measurements this effect may be lessened, however contribution to ice particle number concentration by blowing snow and other mechanisms need to be considered. Choularton et al. (2008) showed that strongly glaciated cloud periods at the high altitude Sphinx Laboratory, on the Jungfraujoch mountain in Switzerland, were linked to more polluted air masses impacting the site, characterized by high loadings of sulfate aerosol. Clouds were found to contain variable N_i concentrations, ranging from 1 L^{-1} up to as high as 1000 L^{-1} as measured by different instruments. Targino et al. (2009), using observations from the same location, showed that the lifetime of these glaciated clouds correlated with higher concentrations of organic, inorganic and black carbon aerosols. Once again, N_i values could be in excess of 1000 L^{-1}. Previously, similar high concentrations of ice crystals were found in mountain cap clouds by Rogers and Vali (1987), who conducted ground-based and airborne measurements at the snow covered Elk Mountain Observatory, Wyoming. They discounted blowing snow from the surface as a contributor to N_i as the observed ice crystal sizes and shapes were not consistent with this mechanism. Lloyd et al. (2015) presented surface cloud microphysical measurements from over 400 individual cloud events recorded at the Jungfraujoch mountain top site in 2013 and 2014. Ice crystal sizes were found to be generally <300 μm, but number size distributions could exhibit a distinct bimodality with peaks at 80 μm and 200 μm, which was confirmed by different instruments. Again high N_i concentrations up to 1000 L^{-1} were observed at temperatures < -15 °C. They analyzed potential contributions to these high concentrations from wind blown snow, pre-activated aerosols, surface based rime splintering and breakup of vapor grown ice, including surface frost at the surface. They were able to discount blowing snow contributions as in the Rogers and Vali (1987) study, showing these events were infrequent at their location (13% of cloud events in 2013 and 27% in 2014) and when these did occur they only accounted for concentrations up to 100 L^{-1}. They also discounted contributions due to the Hallet-Mossop secondary ice production mechanism (Hallett and Mossop 1974) due to the low temperatures.

Gultepe et al. (2014, 2015) measured small ice crystals concentrations in ice fogs over the Arctic regions for $T < -15$ °C using a Fog Measuring Device (DMT FMD) and a ground based Cloud Imaging Probe (DMT G-CIP), which are adapted from the instruments originally developed for aircraft observations. The FMD is a Mie-scattering particle spectrometer that measures particles in the size range from 1 to 50 μm. Particle size is inferred from a measured scatter cross section. The GCIP is a 1-D imaging spectrometer that is able to record particles with sizes 15 to 960 μm with a resolution of 15 μm. In addition, this instrument provides a 3-level grey scale intensity measurement per pixel allowing size corrections to be applied for out of focus particles in the sample volume. Precipitation was measured using a Laser Precipitation Monitor (Thies Inc. LPM), which measures the particle size spectrum and fall speed for precipitation particles about 200 μm to 1 cm size range (Gultepe et al. 2015). It discriminates particle shape based on the particle fall speeds versus size relationships (Bloemink and Lanzinger 2005; Lanzinger et al. 2006).

Aerosol size distribution measurements were obtained using a laser scattering optical particle counter (MetOne) covering a size range of 0.3–10 μm. Concentrations from this sensor were used to infer IN concentrations. Although these sensors contribute significantly to IF measurements and predictions, routine measurements of IN concentrations and their characteristics remain a major challenge to improving the understanding of IF and its prediction. Presently, IN parameterizations for aerosol concentrations at sizes >0.3 μm (DeMott et al. 2010) underestimate observed N_i and that shows a very large variability depending on T and RH_i, and IN composition as well as its crystalline structure (Fitzner et al. 2015).

Another issue is to describe deposition nucleation and freezing nucleation processes when both CCN and IN may contribute to the nucleation processes at the same time. Understanding of physical processes between freezing droplets and deposition ice crystal nucleation needs to be studied in detail. Freezing of supercooled droplets as spherical particles or a splintering mechanism can result in a very large number of ice crystals (Hallett and Mossop 1974), however this process occurs over a relatively narrow temperature range, typically –6 to –8 °C, but it can play an important role for freezing fog conditions. Other secondary

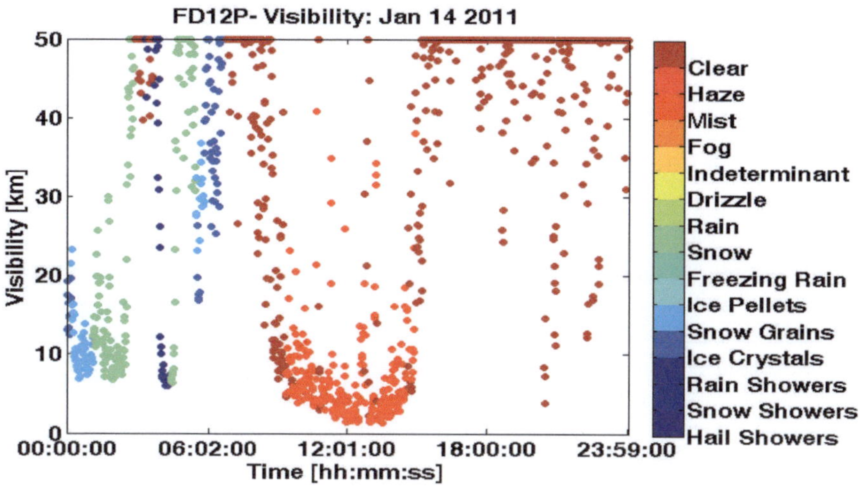

Fig. 6.15 The time series of FD12P Vaisala present weather sensor visibility as a function of precipitation type measurements. The dots with red color at about 1200 UTC represent ice fog crystals. Other particle shapes are seen along the color bar

processes such as freezing nucleation and contact nucleation can also play an important role for ice crystal nucleation but not verified in this stage. To evaluate freezing nucleation and deposition nucleation processes, better measurements of IN and Ni physical characteristics are needed from both laboratory and environmental conditions (Gultepe et al. 2019a). Figure 6.15 shows recorded values of Vis as a function of particle types, including fog and snow events that occurred during the FRAM project that took place over Yellow Knife International Airport with temperatures about −30 °C. This shows that snow and ice crystal shapes as well as Ni are important for Vis calculations.

Aerosols are closely linked to fog and cloud formation since some of them can act as condensation nuclei (CN) or ice nuclei (IN). Borys (1989) has shown that the aerosol composition plays a major role in the ice deposition nuclei efficiency. Blanchet and Girard (1994) have hypothesized that the anthropogenic sulfuric aerosols in large concentrations during the arctic winter, alter the ice particle formation and reduce the ice crystal number concentration, increase their size, and enhance the dehydration rate of the lower troposphere leading to a surface cooling. Garrett and Verzella (2008) in a review of Arctic aerosols summarized the history of observations that have been made of the various types of particles. Various studies have shown that aerosol particles in Arctic haze and also deposited on the surface that cannot be attributed to local sources (Shantz et al. 2012; Kanji et al. 2017), for example, dust with varying compositions of metals can only come from combustion, and soot made up of varying fractions of organic and elemental carbon, and bio-aerosols and that are related to local blooms of algae. Other bio-aerosols are likely transported from distant sources, along with the pollutants associated with industrial processes.

6.7.2 Modeling Issues and Future Challenges

Compared to clouds aloft, ice fog or warm fog conditions are related to surface BL processes, including radiation processes, vegetation, anthropogenic sources, elevation of ground, and turbulence and mixing especially during dissipation stages (Gultepe et al. 2007a, b). Therefore, the above conditions in conjunction with parameterized equations for Vis (indicating fog intensity), need to be addressed properly for modeling studies. Another factor that needs to be considered is the large scale advection and subsidence of cold air systems over the targeted areas that face ice fog conditions. Arctic regions are especially more susceptible to cold fog conditions because of large ice melting processes if climate warming occurs.

Ice fog intensity can be described based on visibility, and its prediction is strongly related to both IWC and N_i (Ohtake and Huffman 1969; Gultepe et al. 2008, 2014). Both IWC and N_i values, in the end, describe particle size, which is usually is less than 200 μm for characterizing an ice fog (Gultepe et al. 2015). Heymsfield (1972) and Gultepe et al. (2015) stated that V_t is less than about 10 cm s^{-1} (Fig. 6.13) for ice fog crystals. Present ice nucleation parameterizations do not include physical conditions occurring within the ice fog, but mostly cloudy conditions. IN concentrations, especially in the deposition nucleation mode, are much lower than that of ice crystal number concentrations and this needs to be better understood (Steinke et al. 2015). Deposition nuclei and its effect on radiative processes within the ice fog in both the clear Arctic boundary layer conditions and the existence of a weak or strong inversion have to be studied and new microphysical parameterizations need to be developed that can help accurately estimate ice fog visibility prediction and related ice microphysical parameters. Occurrence of ice fog over short time and space scales requires a study of its occurrence within a 3D volume and that can complicate the accuracy of its prediction because of microphysical algorithms (Mueller et al. 2010).

Ice fog is an important meteorological event occurring at low temperatures and quite often specifically at high surface elevations and northern high latitudes. Ice fog occurs very often when temperatures go below −15 °C. In fact, IN can be found at T as high as −8 °C (Gultepe et al. 2008). IN parameterizations as a function of T are obtained by many studies (Murray et al. 2012), however large discrepancies between these exist as discussed by Emersic et al. (2015) (Fig. 6.16). Figure 6.16a, b, c shows IN versus T based on DeMott et al. (2010), Hoose et al. (2010), Phillips et al. (2008), and Niemand et al. (2012) parameterizations. Differences of approximately 2 orders of magnitude exist between e.g. the DeMott et al. (2010) parameterization and others. Similarly, a summary by Murray et al. (2012) shows that large variability exists among various parameterizations some of which may be due to different experimental methods used in laboratory studies (Emersic et al. 2015) (Fig. 6.16d).

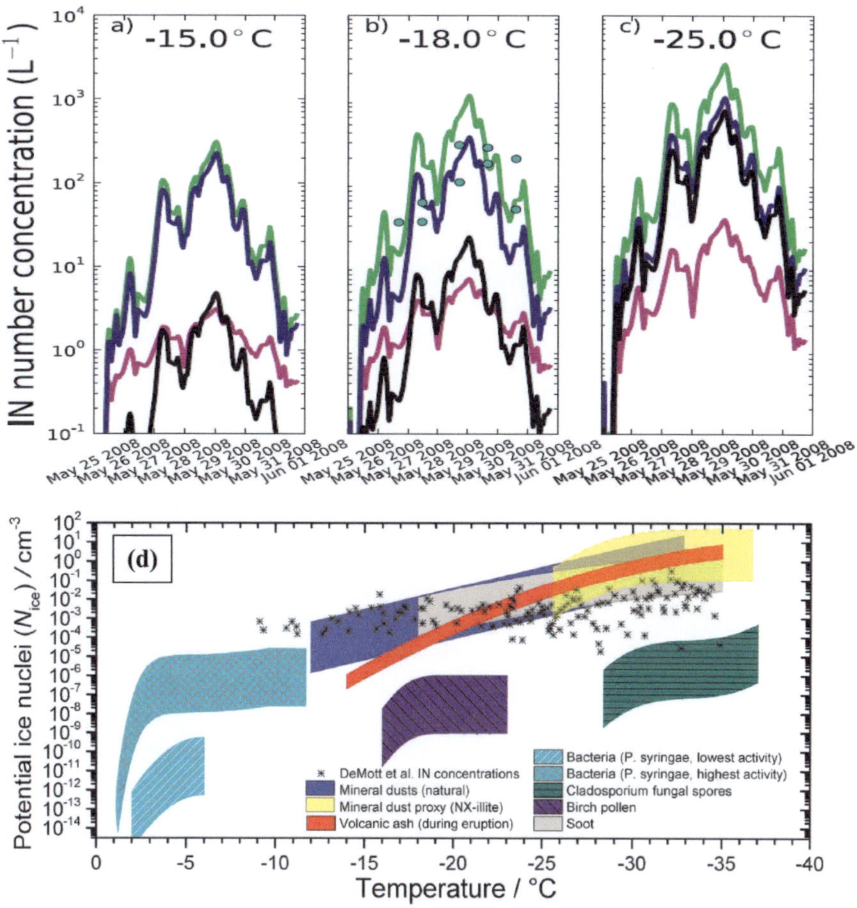

Fig. 6.16 Comparison of different ice nucleation parameterizations obtained at the temperatures of −15 °C (**a**), −18 °C (**b**), and − 25 °C (**c**) where DeMott et al. (2010), Hoose et al. (2010), Phillips et al. (2008), and Niemand et al. (2012) studies are represented by purple, green, blue, and black lines, respectively. The turquoise dots in box b represent the measured IN after removal of the background aerosols (Niemand et al. 2012,© American Meteorological Society. Used with permission). Potential immersion mode ice nuclei concentrations as a function of T for a range of atmospheric aerosol species are shown in (**d**). (Adapted from Murray et al. 2012, reproduced by permission of The Royal Society of Chemistry). Also provided are ice crystal number concentrations from DeMott et al. (2010) for comparison, which were taken using a continuous flow diffusion chamber at water saturation. Note that these were obtained for the ice crystal numbers produced using global averages of IN numbers, except for volcanic ash cases. For bacterial IN, it is assumed that 1% of the total number of airborne bacteria is IN active in line with Phillips et al. (2009). (Details can be found in Murray et al. (2012))

In some cases more than 3–4 orders of magnitude of difference can exist between the various parameterizations. These results suggest that currently IN values using such parameterizations may not be sufficiently accurate for predictive use in ice fog modeling studies.

There has been much discussion recently concerning the impact of aerosols on global climate through a variety of aerosol radiative and aerosol cloud interaction feedbacks. Bio-aerosols are just one component of these feedback cycles, and whilst they represent only a minor fraction of the total aerosol burden in the atmosphere an increasing number of studies are showing aerosols potentially contribute to large scale cloud and precipitation processes (Despres et al. 2012; Sesartic et al. 2012). This is mainly due to the long established observation that some biological aerosols appear to be the most efficient natural ice nuclei, with some strains of bacteria and fungi able to facilitate ice formation at high temperatures. Common examples are Pseudomonas syringae and Erwinia herbicola. Whilst the role of these ice nuclei is still under debate, it is likely that their impact will be more important at regional emissions scales. This suggests that in low temperature surface clouds, such as ice fogs, and in regions generally deficient in more common ice nuclei, their relative contributions, particularly close to surfaces, may be more important. Recent model studies have shown that inclusion of common bacterial and fungi ice nucleators does not lead to a significant global impact on cloud evolution and precipitation, although there is a small but significant regional impact, particularly at Northern hemisphere high latitudes, e.g. Sesartic et al. (2012). This study does not look at surface clouds, and neither does Phillips et al. (2009). The concentration gradient of bioaerosols can be large and, in remote locations, they can sometime dominate the local coarse aerosol loadings (Huffman et al. 2013; Gabey et al. 2010). Whilst their concentration gradients can be large locally and in some regions can dominate local coarse aerosol loadings (Huffman et al. 2013, Gabey et al. 2010), their potential contribution to, for example, surface cloud glaciation processes has not yet been quantified.

6.7.3 Remote Sensing Issues for Ice Fog Detection and Microphysical Parameters

Remote sensing issues for detecting ice fog and to retrieve its microphysical properties are related to the measurement platforms and retrieval algorithms. These are performed in a way similar to measure or retrieve ice cloud properties. The most important issue for ice fog remote sensing is ice fog particle spectral properties such as their particle size distribution and phase. Another issue is a very shallow ice fog occurrence in the boundary layer. During these conditions, detection of ice fog by the active remote sensing platforms is not feasible. Usually it is believed that the ice crystal number concentrations in the forecast models is less than 100 L^{-1}, and that doesn't usually play an important role for reducing ice fog visibility to a few hundreds of meters.

Ice fog remote sensing from a ground based platform can be performed using a ceilometer, LiDAR, and microwave radiometer, or their combinations. Extinction of remote sensing platforms' transmitted beam depends on ice crystal size parameter, extinction efficiency, ice particle spectra, and total number concentration of ice crystals in a given air volume. Without knowing accurate ice fog particle spectra and its formation mechanism e.g. homogeneous or heterogeneous nucleation process, remote sensing based algorithms would not provide accurate fog detection and retrieve microphysical parameters.

Ice fog detection and microphysical analysis from satellites can also include large uncertainties because of assumed particle size distributions, inversion layers very close to surface, snow cover, and a limited coverage of satellite passes over the cold climates above 60°N, as well as high level clouds obscuring low clouds and fog. It is possible to reduce these uncertainties when multi-channel satellite observations, similar to GOES-R (Pavolonis 2010b) are adapted for Arctic regions. Therefore, presently, Arctic low level clouds or fog can only be observed with polar orbiting satellites, but they can also have issues with time resolution, and therefore, their use for ice fog prediction can be limited. Therefore, special orbiting satellites are needed for Arctic operations (Kidder and Vonder Haar 1990).

6.7.4 Overall Research Needs for Ice Fog Analysis

Ice fog effects on various components of cold climatological regions, including transportation, sporting activities, aviation, marine environments, and climate; therefore, improvements for ice fog measurements and predictions are needed. For these reasons, ice fog particles spectra, ice nuclei, atmospheric moisture and temperature profiles, cloud conditions, and surface and inversion layer thermodynamic and radiation characteristics are needed to be observed, and then they can be used for developing ice microphysical parameterizations and visibility predictions. Both in-situ and remote sensing platforms such as Raman LiDAR, profiling microwave radiometers, and cloud radar can also play a significant role for observations of ice fog, and their integrations with NWPs and detailed 3D ice fog models, including LES models, can be used to better evaluate ice fog physical and thermodynamic characteristics. In addition, to improve ice fog predictions in the northern latitudes, new operational satellites, with new spectral channels response to small ice crystals, based on special overpass orbits are also needed.

The INP (Ice nucleating particles) can be related to various origins of aerosols in the high latitudes. Wex et al. (2019) studied annual variability of ice-nucleating particle concentration at different Arctic locations using filter techniques. Their work suggests that ice crystal number concentrations are about 0.1 L^{-1} at about −25 °C and less than 0.0001 L^{-1} at −10 °C. They also indicated that biogenic source of INPs can increase due to Arctic changes due to opening of ice-covered ocean surfaces. In addition, they observed the INPs at T ~−5 °C. Hartmann et al. (2019) also studied INPs over the European Arctic during previous centuries (~500 years) and found out that biogenic INPs played an important role in the Arctic INP distributions but both volcanic and anthropogenic origin INPs did not play a role in contributing to INPs. In their analysis, they related INP found in the precipitation samples to INPs (<10 L^{-1} at −25 °C) in the air, assuming that LWC in clouds is ~0.4 g m^{-3}. This may not be hold as a constant because the LWC ranges from 0.01 g m^{-3} up to 1 g m^{-3} depends on vertical air motions (Heymsfield 1972; Gultepe et al. 1995, 2004; Khvorostyanov et al. 2003). Zhao et al. (2019) further stated that IN by aerosols from anthropogenic pollution at cold temperatures can play an important role at T > −37 °C, affecting negatively homogeneous IN occurrence at colder temperatures. They used 11 years of A-Train satellite data and cloud resolving simulations in their analysis. Their results suggest that heterogeneous nucleation and subsequent ice crystal growth can efficiently compete with and even prevent homogeneous freezing. These studies suggest that INPs are highly variable in ice fog research studies; therefore, INPs physical and chemical properties over the cold temperatures e.g. Arctic environment needs to be better understood to improve prediction of ice fog visibility and coverage.

Ice fog impact on aviation and local weather (flight delays and severe icing hazard) and climate can be very significant (Girard and Blanchet 2001; Gultepe et al. 2014, 2019b). Ice fog may lead to frost formation, but frost formation may not always be related to ice fog occurrence. In addition, the combination of ice fog and frost conditions may compromise power lines, leading to disruptions in electrical energy distribution. When T is very low, RH_i can easily be saturated with respect to ice with less moisture compared to warmer T conditions. Ice fog is strongly related to frost formation, and it might be more dangerous to aircraft than snow because of the stronger surface adhesion compared to snow particles at cold temperatures.

Overall ice fog impact on environment and social activities suggests that better observations and modeling skills in the northern climates are needed to improve its prediction.

6.8 Conclusions

Ice fog microphysics and its impact on weather and climate are important for improving our understanding of atmospheric systems, including clouds and fog-climate relationships. To summarize, the key issues and challenges identified related to ice fog observations and modeling are,

1. Ice fog can frequently occur at colder temperatures such as T < −10 °C if ice nuclei exist and boundary layer cooling processes occur.
2. Ice fog microphysics can be very different than those of elevated clouds because of its low visibilities and anthropogenic effects such as man-made air pollution generated by factories, cars, aircraft, and heating systems.
3. Ice fog crystal number concentrations (~1000 L^{-1}) can be significantly higher than those of clouds because of the large amount of available moisture and ice nuclei exist in the boundary layer.

4. Present ice crystal microphysical parameterizations are developed for elevated clouds; therefore, their use in numerical models can lead to significant errors in fog formation, development, and dissipation.
5. Radiative effects are not considered in the ice or droplet growth equations in NWPs; therefore, ice fog forms earlier and dissipates later because of underestimates of IR radiative cooling and overestimates SW warming, respectively.
6. Visibility parameterizations in NWP models also need to use ice crystal number concentration and their sizes, not only ice water content. If not included, Vis prediction can have more than 50% uncertainty.
7. Accurate prediction of ice fog Vis and IWC from the NWP models can be used advantageously to help improve visibility that can significantly affect transportation and aviation applications.
8. Remote sensing methods for observing ice fog conditions are needed over the Arctic regions because of a lack of operational observing systems in that region. Currently, operational satellites such as GOES do not cover the necessary short time and space scales over the Arctic regions.
9. Vertical distribution of ice fog conditions suggested (Fig. 6.8) that ice fog microphysical properties were not homogeneous, indicated by variable backscatter ratio and depolarization coefficients. This might be due to changing T and RHi conditions at the inversion levels (Fig. 6.7).
10. Ice fog models in 3D, compared to 1D fog column models, should be developed for better ice fog forecasting because ice fog formation and decay change quickly over 3D space as a function of time, indicating impact of boundary layer physical and dynamical processes.
11. Possible climate warming over the Arctic regions can supply more moisture into the boundary layer system, and relatively colder temperatures and increasing IN can result in an increase in ice and freezing fog occurrences. Therefore, ice fog-climate interactions should be better understood over cold climatic regions.
12. Reduced visibility and increased icing can be a major factor in aviation accidents over the Arctic regions (Gultepe et al. 2019b). The per capita rate of aviation related fatalities in northern regions can be up to 18 times higher compared with mid-latitudes.

Overall, this work suggests that ice fog observations, microphysical methods, and its relationship to the climate change need to be researched and its impact on aviation, transportations, climate, and weather should be mitigated. The new research in this area is needed for a better understanding of ice fog measurement issues and to improve prediction methods, as well as to determine fog-climate interactions.

Acknowledgements Funding for this work was provided by the Canadian National Search and Rescue Secretariat (SAR) under the project # SN 201221/EC S2001-T104-H029-MR029-165 and Environment Canada (EC). The authors also thank EC technicians Rob Reed, Mike Harwood, and S. Bacic for their help for instrument development, installations, and calibration of the sensors used in FRAM Ice Fog project, which took place nearby Yellow Knife International Airport, NWT,

Canada. Support to NCAR for Andrew J. Heymsfield from the National Science Foundation is appreciated.

This work is supported by the Environment and Climate Change Canada, Toronto, ON, Canada and Faculty of Engineering and Applied Science of UOIT (Ontario Technical University), Oshawa, ON, Canada.

References

Anderson, R. J., Miller, R. C., Kassner, J. L., & Hagen, D. E. (1980). A study of homogeneous condensation-freezing nucleation of small water droplets in an expansion cloud chamber. *Journal of the Atmospheric Sciences, 37*, 2508–2520.

Ansmann, A., Tesche, M., Althausen, M., Müller, D., Freudenthaler, V., Heese, B., Wiegner, M., Pisani, G., Knippertz, P., & Dubovik, O. (2008). Influence of Saharan dust on cloud glaciation in southern Morocco during SAMUM. *Journal of Geophysical Research, 113*(D04210), 25757–25761.

Atkinson, J. D., Murray, B. J., Woodhouse, M. T., Whale, T. F., Baustian, K. J., Carslaw, K. S., Dobbie, S., O'Sullivan, D., & Malkin, T. L. (2013). The importance of feldspar for ice nucleation by mineral dust in mixed-phase clouds. *Nature, 498*(7454), 355–358.

Baker, B., & Lawson, R. P. (2006). Improvement in determination of ice water content from two-dimensional particle imagery. Part I: Image-to-mass relationships. *Journal of Applied Meteorology and Climatology, 45*, 1282–1290.

Bendix, J., Thies, B., Cermak, J., & Nauß, T. (2005). Ground fog detection from space based on MODIS daytime data—A feasibility study. *Weather and Forecasting, 20*, 989–1005.

Benson, C. S. (1965). *Ice fog: Low temperature air pollution defined with Fairbanks, Alaska*, Geophysical Research Institute, UAG R-173. CRREL Research Report , 121 pp.

Benson, C. S., & Rogers, G. W. (1965). *Alaskan air pollution-The nature of ice fog and its development and settlement implications*. In Proceedings of the 16th AAAS Alaskan Science Conference, Juneau, AL, pp. 233–245.

Bergot, T., & Guedalia, D. (1994). Numerical forecasting of radiation fog. Part I: Numerical model and sensitivity tests. *Monthly Weather Review, 122*, 1218–1230.

Bergot, T., Carrer, D., Noilhan, J., & Bougeault, P. (2005). Improved site specific numerical prediction of fog and low clouds: A feasibility study. *Weather and Forecasting, 20*, 627–646.

Bergot, T., Terradellas, E., Cuxart, J., Mira, A., Liechti, O., Mueller, M., & Nielsen, N. W. (2007). Intercomparison of single-column numerical models for the prediction of radiation fog. *Journal of Applied Meteorology and Climatology, 46*, 504–521.

Bianco, L., Cimini, D., Marzano, F. S., & Ware, R. (2005). Combining microwave radiometer and wind profiler radar measurements for high-resolution atmospheric humidity profiling. *Journal of Atmospheric and Oceanic Technology, 22*, 949–965.

Blanchet, J.-P., & Girard, E. (1994). Arctic greenhouse cooling. *Nature, 371*, 383.

Bloemink, H. I., & Lanzinger, E. (2005). *Precipitation type from the Thies dis- drometer*. TECO-2005 Meeting, Bucharest, Romania, 4–7 May 2005, IOM 82(TD 1265).

Boers, R., Baltink, H. K., Hemink, H. J., Bosveld, F. C., & Moerman, M. (2013). Ground-based observations and modeling of the visibility and radar reflectivity in a radiation fog layer. *Journal of Atmospheric and Oceanic Technology, 30*, 288–300.

Borys, R. D. (1989). Studies of ice nucleation by arctic aerosols on AGASP-II. *Journal of Atmospheric Chemistry, 9*, 169–185.

Bott, A., & Trautmann, T. (2002). PAFOG—A new efficient forecast model of radiation fog and low-level stratiform clouds. *Atmospheric Research, 64*, 191–203.

Bott, A., Sievers, U., & Zdunkowski, W. (1990). A radiation fog model with a detailed treatment of the interaction between radiative transfer and fog microphysics. *Journal of the Atmospheric Sciences, 47*, 2153–2166.

Bowling, S. A., Ohtake, T., & Benson, C. S. (1968). Winter pressure systems and ice fog in Fairbanks, Alaska. *Journal of Applied Meteorology, 7*, 961–968.

Broadley, S. L., Murray, B. J., Herbert, R. J., Atkinson, J. D., Dobbie, S., Malkin, T. L., Condliffe, E., & Neve, L. (2012). Immersion mode heterogeneous ice nucleation by al illite rich powder representative of atmospheric mineral dust. *Atmospheric Chemistry and Physics, 12*, 287–307.

Bundke, U., Nillius, B., Jaenicke, R., Wetter, T., Klein, H., & Bingemer, H. (2008). The fast ice nucleus chamber FINCH. *Atmospheric Research, 90*, 180–186.

Cadeddu, M., Turner, D., & Liljegren, J. (2009). A neural network for real-time retrievals of PWV and LWP from Arctic millimeter-wave ground-based observations. *IEEE Transactions on Geoscience and Remote Sensing, 47*, 1887–1900. https://doi.org/10.1109/TGRS.2009.2013205.

Calvert, C., & Pavolonis, M. J. (2011). GOES-R Advanced Baseline Imager (ABI) algorithm theoretical basis document for fog and low cloud detection, Version 2.0., 67 pp.

Cermak, J., & Bendix, J. (2007). Dynamical nighttime fog/low stratus detection based on meteosat SEVIRI data – A feasibility study. *Pure and Applied Geophysics, 164*, 1179–1192.

Cermak, J., & Bendix, J. (2008). A novel approach to fog/low stratus detection using meteosat 8 data. *Atmospheric Research, 87*: 3–4, 279–292.

Cermak, J., & Bendix, J. (2011). Detecting ground fog from space-A microphysics-based approach. *International Journal of Remote Sensing, 32*(12), 3345–3371.

Choularton, T. W., Bower, K. N., Weingartner, E., Crawford, I., Coe, H., Gallagher, M. W., Flynn, M., Crosier, J., Connolly, P., Targino, A., Alfarra, M. R., Baltensperger, U., Sjogren, S., Verheggen, B., Cozic, J., & Gysel, M. (2008). The influence of small aerosol particles on the properties of water and ice clouds. *Faraday Discussions, 137*, 205–222.

Chung, D., & Matheou, G. (2014). Large-eddy simulation of stratified turbulence. Part I: A vortex-based subgrid-scale model. *Journal of the Atmospheric Sciences, 71*, 1863–1879. https://doi.org/10.1175/JAS-D-13-0126.1.

Cimini, D., Campos, E., Ware, R., Albers, S., Giuliani, G., Oreamuno, J., Joe, P., Koch, S. E., Cober, S., & Westwater, E. (2010). Thermodynamic atmospheric profiling during the 2010 winter Olympics using ground-based microwave radiometry. *IEEE Transactions on Geoscience and Remote Sensing, 49*(12), 4959–4969.

Connolly, P. J., Möhler, O., Field, P. R., Saathoff, H., Burgess, R., Choularton, T., & Gallagher, M. (2009). Studies of the heterogenous freezing by three different desert dust samples. *Atmospheric Chemistry and Physics, 9*, 2805–2824.

Cooper, W. A. (1986). *Ice initiation in natural clouds. Precipitation enhancement—A scientific challenge* (Meteor. Monogr., no. 43, Amer. Meteor. Soc. pp. 29–32).

Cooper, W. A., & Vali, G. (1981). The origin of ice in mountain cap clouds. *Journal of the Atmospheric Sciences, 38*, 1244–1259.

Côté, J., Gravel, S., Méthot, A., Patoine, A., Roch, M., & Staniforth, A. (1998). The operational CMC-MRB global environmental multiscale (GEM) model. Part I: Design considerations and formulation. *Monthly Weather Review, 126*, 1373–1395.

Curry, J. A., Meyers, F. G., Radke, L. F., Brock, C. A., & Ebert, E. E. (1990). Occurrence and characteristics of lower tropospheric ice crystals in the Arctic. *International Journal of Climatology, 10*, 749–764.

DeMott, P. J., Meyers, M. P., & Cotton, W. R. (1994). Parameterization and impact of ice initiation processes relevant to numerical model simulations of cirrus clouds. *Journal of the Atmospheric Sciences, 51*, 77–90.

DeMott, P. J., Rogers, D. C., & Kreidenweiss, S. M. (1997). The susceptibility of ice formation in upper tropospheric clouds to insoluble aerosol components. *Journal of Geophysical Research, 102*, 19575–19584.

DeMott, P. J., Chen, Y., Kreidenweis, S. M., Rogers, D. C., & Sherman, D. E. (1999). Ice formation by black carbon particles. *Geophysical Research Letters, 26*, 2429–2432.

DeMott, P. J., Sassen, K., Poellot, M., Baumgardner, D., Rogers, D. C., Brooks, S., Prenni, A. J., & Kreidenweis, S. M. (2003). African dust aerosols as atmospheric ice nuclei. *Geophysical Research Letters, 30*, 1732.

DeMott, P. J., Prenni, A. J., Liu, X., Kreidenweis, S. M., Petters, M. D., Twohy, C. H., Richardson, M. S., Eidhammer, T., & Rogers, D. C. (2010). Predicting global atmospheric ice nuclei distributions and their impacts on climate. *Proceedings of the National Academy of Sciences of the United States of America, 107*, 11217–11222.

DeMott, P. J., Hill, T. C. J., McCluskey, C. S., & Franc, G. D. (2015). Sea spray aerosol as a unique source of ice nucleating particles. *Proceedings of the National Academy of Sciences of the United States of America*. https://doi.org/10.1073/pnas.1514034112, 7 pp.

Despres, V. R., Huffman, J. A., Burrows, S. M., Hoose, C., Safatov, A. S., Buryak, G., Frohlich-Nowoisky, J., Elbert, W., Andreae, M. O., Pöschl, U., & Jaenicke, R. (2012). Primary biological aerosol particles in the atmosphere: A review. *Tellus Series B: Chemical and Physical Meteorology, 64*, 1–58.

Dunning, W. J. (1969). *Nucleation and growth processes in the dehydration of salt hydrate crystals*. In Proceedings of an International Symposium on Special Topics in Ceramics, held June 18–23, 1967, at Alfred University, Alfred, New York, pp. 132–135. https://doi.org/10.1007/978-1-4899-6461-8_7.

Emersic, C., Connolly, P. J., Boult, S., Campana, M., & Li, Z. (2015). Investigating the discrepancy between wet-suspension- and dry-dispersion-derived ice nucleation efficiency of mineral particles. *Atmospheric Chemistry and Physics, 15*, 11311–11326.

Fernando, H. J. S., & Co-authors. (2015). The MATERHORN: Unraveling the Intricacies of Mountain Weather. *Bulletin of the American Meteorological Society*. https://doi.org/10.1175/BAMS-D-13-00131.1.

Ferrier, B. S. (1994). A double-moment multiple-phase fourclass bulk ice scheme. Part I: Description. *Journal of the Atmospheric Sciences, 51*, 249–280. https://doi.org/10.1175/1520-0469(1994)051,0249:ADMMPF.2.0.CO;2.

Ferrier, B. S., Jin, Y., Lin, Y., Black, T., Rogers, E., & DiMego, G. (2002). *Implementation of a new grid-scale cloud and precipitation scheme in the NCEP eta model*. Preprints, 15th Conference on numerical weather prediction, San Antonio, TX, American Meteor Society, 10.1.

Field, P. R., Heymsfield, A. J., & Bansemer, A. (2006). Shattering and particle interarrival times measured by optical array probes in ice clouds. *Journal of Atmospheric and Oceanic Technology, 23*, 1357–1371.

Fitzner, M., Sosso, G. C., Cox, S. J., & Michaelides, A. (2015). The many faces of heterogeneous ice nucleation: Interplay between surface morphology and hydrophobicity. *Journal of the American Chemical Society, 137*, 13658–13669.

Fletcher, N. H. (1962). *The physics of rainclouds*. Cambridge: Cambridge University Press, 386 pp.

Flynn, C. J., Mendoza, A., Zheng, Y., & Mathur, S. (2007). Novel polarization-sensitive micropulse LiDAR measurement technique. *Optics Express, 15*, 2785–2790.

Fukuta, N., & Schaller, R. C. (1982). Ice nucleation by aerosol particles: Theory of condensation-freezing nucleation. *Journal of the Atmospheric Sciences, 39*, 648–655.

Gabey, A. M., Gallagher, M. W., Whitehead, J., Dorsey, J. R., Kaye, P. H., & Stanley, W. R. (2010). Measurements and comparison of primary biological aerosol above and below a tropical forest canopy using a dual channel fluorescence spectrometer. *Atmospheric Chemistry and Physics, 10*, 4453–4466.

Garrett, T. J., & Verzella, L. L. (2008). Looking back: An evolving history of Arctic aerosols. *Bulletin of the American Meteorological Society, 89*, 299–302. https://doi.org/10.1175/BAMS-89-3-299.

Girard, E., & Blanchet, J. P. (2001). Microphysical parameterization of Arctic diamond dust, ice fog, and thin stratus for climate models. *Journal of the Atmospheric Sciences, 58*, 1181–1198.

Gosselin, M. I., Rathnayake, C. M., Crawford, I., Pöhlker, C., Fröhlich-Nowoisky, J., Schmer, B., Després, V. R., Engling, G., Gallagher, M., Stone, E., Pöschl, U., & Huffman, J. A. (2016). Fluorescent bioaerosol particle, molecular tracer, and fungal spore concentrations during dry and rainy periods in a semi-arid forest. *Atmospheric Chemistry and Physics, 16*, 15165–15184. https://doi.org/10.5194/acp-16-15165-2016.

Gotaas, Y., & Benson, C. S. (1965). The effect of suspended ice crystals on radiative cooling. *Journal of Applied Meteorology, 4*, 446–453.

Grenier, P., Blanchet, J. P., & Alpizar, R. M. (2009). Study of polar thin ice clouds and aerosols seen by CloudSat and CALIPSO during midwinter 2007. *Journal of Geophysical Research, 114*, D09201. https://doi.org/10.1029/2008JD010927.

Gultepe, I. (2015). Mountain weather: Observations and Modeling. *Advances in Geophysics, 56*, 229–312.

Gultepe, I., & Coauthors. (2007a). Fog research: A review of past achievements and future perspectives. *Pure and Applied Geophysics, 164*, 1121–1159.

Gultepe, I., & Isaac, G. A. (2002). The effects of air-mass origin on Arctic cloud microphysical parameters during FIRE.ACE. *Journal of Geophysical Research, Oceans, 107*(C10)., SHE 4-1 to 4-12.

Gultepe, I., O'C Starr, D., Heymsfield, A. J., Uttal, T., Ackerman, T. P., & Westphal, D. L. (1995). Dynamical characteristics of cirrus clouds from aircraft and radar observations in micro and meso-gamma scales. *Journal of the Atmospheric Sciences, 52*, 4060–4078.

Gultepe, I., Isaac, G. A., & Cober, S. G. (2001). Ice crystal number concentration versus temperature for climate studies. *International Journal of Climatology, 21*, 1281–1302.

Gultepe, I., Isaac, G., Williams, A., Marcotte, D., & Strawbridge, K. (2003). Turbulent heat fluxes over leads and polynyas and their effect on Arctic clouds during FIRE-ACE: Aircraft observations for April 1998. *Atmosphere-Ocean, 41*, 15–34.

Gultepe, I., Isaac, G. A., Key, J., Intrieri, J., O'C Starr, D., & Strawbridge, K. B. (2004). Dynamical and microphysical characteristics of the Arctic clouds using integrated observations collected over SHEBA during the April 1998 FIRE.ACE flights of the Canadian Convair. *Meteorology and Atmospheric Physics, 85*, 235–263.

Gultepe, I., Pagowski, M., & Reid, J. (2007b). Using surface data to validate a satellite based fog detection scheme. *Weather and Forecasting, 22*, 444–456.

Gultepe, I., Minnis, P., Milbrandt, J., Cober, S. G., Nguyen, L., Flynn, C., & Hansen, B. (2008). The Fog Remote Sensing and Modeling (FRAM) field project: visibility analysis and remote sensing of fog. In W. F. Feltz & J. J. Murray (Eds.), *Remote sensing applications for aviation weather hazard detection and decision support*. ISBN: 9780819473080, Proceedings of SPIE Vol. 7088-2, 12 pp.

Gultepe, I., Pearson, G., Milbrandt, J. A., Hansen, B., Platnick, S., Taylor, P., Gordon, M., Oakley, J. P., & Cober, S. G. (2009). The fog remote sensing and modeling (FRAM) field project. *Bulletin of the American Meteorological Society, 90*, 341–359.

Gultepe, I., Kuhn, T., Pavolonis, M., Calver, C., Gurka, J., Isaac, G. A., Heymsfield, A. J., Liu, P., Zhou, B., Ware, R., Sloan, J., Milbrandt, J., & Bernstein, B. (2012). *Ice fog (pogonip) and frost in Arctic during FRAM-Ice Fog project: Aviation and nowcasting applications*. The 16th International Conference on Clouds and Precipitation (ICCP), 30 July-3 August 2012, Leipzig, Germany. Extended abstract in CD.

Gultepe, I., Kuhn, T., Pavolonis, M., Calvert, C., Gurka, J., Heymsfield, A. J., Liu, P. S. K., Zhou, B., Ware, R., Ferrier, B., Milbrandt, J., & Bernstein, B. (2014). Ice fog in Artic during FRAM-ICE for project: Aviation and nowcasting applications. *Bulletin of the American Meteorological Society, 95*, 211–226.

Gultepe, I., Zhou, B., Milbrandt, J., Bott, A., Li, Y., Heymsfield, A. J., Ferrier, B., Ware, R., Pavolonis, M., Kuhn, T., Gurka, J., Liu, P., & Cermak, J. (2015). A review on ice fog measurements and Modeling. *Atmospheric Research, 151*, 2–19.

Gultepe, I., Rabin, R., Ware, R., & Pavolonis, M. (2016a). Light Snow Precipitation and Effects on Weather and Climate. *Advances in Geophysics, 57*, 147–210.

Gultepe, I., Fernando, H. J. S., Pardyjak, E. R., Hoch, S. W., Silver, Z., Creegan, E., Leo, L. S., Pu, Z., De Wekker, S. F. J., & Hang, C. (2016b). An overview of the MATERHORN fog project: Observations and predictability. *Pure and Applied Geophysics, 173*, 2983–3010.

Gultepe, I., Heymsfield, A. J., Field, P. R., & Axisa, D. (2017a). Ice-phase precipitation. *Meteorological Monographs, 58*, 6.1–6.36. https://doi.org/10.1175/AMSMONOGRAPHS-D-16-0013.1.

Gultepe, I., Heymsfield, A. J., Gallagher, M., Ickes, L., & Baumgardner, D. (2017b). Ice fog: The current state of knowledge and future challenges. *Meteorological Monographs, 58*, 4.1–4.24. https://doi.org/10.1175/AMSMONOGRAPHS-D-17-0002.1.

Gultepe, I., Agelin-Chaab, M., Komar, J., Elfstrom, G., Boudala, F., & Zhou, B. (2019a). A meteorological supersite for aviation and cold weather applications. *Pure and Applied Geophysics, 176*(5), 1977–2017. https://doi.org/10.1007/s00024-018-1880-3.

Gultepe, I., Sharman, R., Williams, P. D., Zhou, B., Ellrod, G., Minnis, P., Trier, S., Griffin, S., Yum, S. S., Gharabaghi, B., Feltz, W., Temini, M., Pu, Z., Storer, L. N., Kneringer, P., Weston, M. J., Chuang, H. Y., Thobois, L., Dimri, A. P., Dietz, S. J., Gutenberg, M. V. A., & Neto, F. L. A. (2019b). A review of high impact weather for aviation meteorology. *Pure and Applied Geophysics, 176*(5), 1869–1923.

Haeffelin, M., & Coauthors. (2010). PARISFOG: Shedding new light on fog physical processes. *Bulletin of the American Meteorological Society, 91*, 767–783.

Hallett, J., & Mossop, S. C. (1974). Production of secondary ice crystals during the riming process. *Nature, 249*, 26–28.

Hamazu, K., Hashiguchi, H., Wakayama, T., Matsuda, T., Doviak, R. J., & Fukao, S. (2003). A 35-GHz scanning doppler radar for fog observations. *Journal of Atmospheric and Oceanic Technology, 20*, 972–986.

Hartmann, M. T., Blunier, S. O., Brügger, J., Schmale, M. S., Vogel, A., Wex, H., & Stratmann, F. (2019). Variation of Ice Nucleating Particles in the European Arctic Over the Last Centuries. *GRL*, 10. https://doi.org/10.1029/2019GL082311.

Henneberger, J., Fugal, J. P., Stetzer, O., & Lohman, U. (2013). HOLIMO II, 2013: A digital holographic instrument for ground-based in situ observations of microphysical properties of mixed-phase clouds. *Atmospheric Measurement Techniques, 6*, 2975–2987.

Heymsfield, A. (1972). Ice crystal terminal velocities. *Journal of the Atmospheric Sciences, 29*, 1348–1357.

Heymsfield, A. J., & Miloshevich, L. M. (1993). Homogenous ice nucleation and supercooled liquid water in orographic wave clouds. *Journal of the Atmospheric Sciences, 50*, 2335–2353.

Heymsfield, A. J., & Platt, C. M. R. (1984). A parameterization of the particle size spectrum of ice clouds in terms of the ambient temperature and the ice water content. *Journal of the Atmospheric Sciences, 41*, 846–855.

Heymsfield, A. J., & Sabin, R. M. (1989). Cirrus crystal nucleation by homogeneous freezing of solution droplets. *Journal of the Atmospheric Sciences, 46*, 2252–2264.

Heymsfield, A. J., Krämer, M., Luebke, A., Brown, P., Cziczo, D. J., Franklin, C., Lawson, P., Lohmann, U., McFarquhar, G., Ulanowski, Z., & Van Tricht, K. (2017). Cirrus clouds. *Meteorological Monographs, 58*, 2.1–2.26. https://doi.org/10.1175/AMSMONOGRAPHS-D-16-0010.1.

Hobbs, P. V. (1965). The aggregation of ice particles in clouds and fogs at low temperatures. *Journal of the Atmospheric Sciences, 22*, 296–300.

Hobbs, P. V., Chang, S., & Locatelli, J. D. (1974). The dimension and aggregation of ice crystals in natural clouds. *Journal of Geophysical Research, 79*, 2199–2206.

Hoose, C., & Möhler, O. (2012). Heterogeneous ice nucleation on atmospheric aerosols: A review of results from laboratory experiments. *Atmospheric Chemistry and Physics, 12*, 9817–9854.

Hoose, C., Kristjansson, J. E., Chen, J. P., & Hazra, A. (2010). A classical theory-based parameterization of heterogeneous ice nucleation by mineral dust, soot, and biological particles in a global climate model. *Journal of the Atmospheric Sciences, 67*, 2483–2503.

Hoyle, C. R., Pinti, V., Welti, A., Zobrist, B., Marcolli, C., Luo, B., Höskuldsson, Á., Mattsson, H. B., Stetzer, O., Thorsteinsson, T., Larsen, G., & Peter, T. (2011). Ice nucleation properties of volcanic ash from Eyjafjallajokull. *Atmospheric Chemistry and Physics, 11*, 9911–9926.

Huffman, P. J., & Vali, G. (1973). The effects of vapor depletion on ice nucleus measurements with membrane filters. *Journal of Applied Meteorology, 12*, 1018–1024.

Huffman, J. A., et al. (2013). High concentrations of biological aerosol particles and ice nuclei during and after rain. *Atmospheric Chemistry and Physics Discussions, 13*, 6151–6164.

Hummel, M., Hoose, C., Gallagher, M., Healy, D. A., Huffman, J. A., O'Connor, D., Pöschl, U., Pöhlker, C., Robinson, N. H., Schnaiter, M., Sodeau, J. R., Stengel, M., Toprak, E., & Vogel, H. (2015). Regional-scale simulations of fungal spore aerosols using an emission parameterization adapted to local measurements of fluorescent biological aerosol particles. *Atmospheric Chemistry and Physics, 15*, 6127–6146. https://doi.org/10.5194/acp-15-6127-2015.

Hunt, W. H., Winker, D. M., Vaughan, M. A., Powell, K. A., Lucker, P. L., & Weimer, C. (2009). CALIPSO lidar description and performance assessment. *Journal of Atmospheric and Oceanic Technology, 26*, 1214–1228.

Ickes, L., Welti, A., Hoose, C., & Lohmann, U. (2015). Classical nucleation theory of homogeneous freezing of water: Thermodynamic and kinetic parameters. *Physical Chemistry Chemical Physics, 17*, 5514–5537.

Inoue, M., Matheou, G., & Teixeira, J. (2014). LES of a spatially developing atmospheric boundary layer: Application of a fringe method for the stratocumulus to shallow cumulus cloud transition. *Monthly Weather Review, 142*, 3418–3423.

Jaecker-Voirol, A., & Mirabel, P. (1989). Heteromolecular nucleation in the sulfuric acid-water system. *Atmospheric Environment, 23*, 2053–2057.

Kanji, Z. A., Ladino, L. A., Wex, H., Boose, Y., Burkert-Kohn, M., Cziczo, D. J., & Krämer, M. (2017). Overview of ice nucleating particles. *Meteorological Monographs, 58*, 1.1–1.33. https://doi.org/10.1175/AMSMONOGRAPHS-D-16-0006.1.

Kehler, S., Hanesiak, J., Curry, M., Sills, D., & Taylor, N. (2016). High Resolution Deterministic Prediction System (HRDPS) Simulations of Manitoba Lake Breezes. *Atmosphere-Ocean*. https://doi.org/10.1080/07055900.2015.1137857.

Khairoutdinov, M., & Kogan, Y. (2000). A new cloud physics parameterization in a large-eddy simulation model of marine stratocumulus. *Monthly Weather Review, 128*, 229–243.

Khvorostyanov, V. I., Curry, J. A., Gultepe, I., & Strawbridge, K. (2003). A springtime cloud over the Beaufort Sea polynya: 3D simulation with explicit microphysics and comparison with observations. *Journal of Geophysical Ressearch-Atmospheres, 108*(D9), 4296. https://doi.org/10.1029/2001JD001489.

Khvorostyanov, V. I., & Curry, J. A. (2004). The theory of ice nucleation by heterogeneous freezing of deliquescent mixed CCN. Part I: Critical radius, energy, and nucleation rate. *Journal of the Atmospheric Sciences, 61*, 2676–2691.

Khvorostyanov, V. I., & Curry, J. A. (2005). The theory of ice nucleation by heterogeneous freezing of deliquescent mixed CCN. Part II: Parcel model simulation. *Journal of the Atmospheric Sciences, 62*, 261–285.

Kidder, S. Q., & Vonder Haar, T. H. (1990). On the use of satellites in Molniya orbits for meteorological observation of middle and high latitudes. *Journal of Atmospheric and Oceanic Technology, 7*, 517–522.

Kim, C. K., Stuefer, M., Schmitt, C. G., Heymsfield, A. J., & Thompson, G. (2014). A numerical modeling of ice fog in interior Alaska using weather research and forecasting model. *Pure and Applied Geophysics, 171*, 1963–1982.

Klein, H., Haunold, W., Bundke, U., Nillius, B., Wetter, T., Schallenberg, S., & Bingemer, H. (2010a). A new method for sampling of atmospheric ice nuclei with subsequent analysis in a static diffusion chamber. *Atmospheric Research, 96*, 218–224.

Klein, H., Nickovic, S., Haunold, W., Bundke, U., Nillius, B., Eber, M., Weinbruch, S., Schuetz, L., Levin, Z., Barrie, L. A., & Bingemer, H. (2010b). Saharan dust and ice nuclei over Central Europe. *Atmospheric Chemistry and Physics, 10*, 10211–10221.

Koenig, L. R. (1970). Numerical modeling of ice deposition. *Journal of the Atmospheric Sciences, 28*, 226–237.

Koenig, L. R., & Murray, F. W. (1976). Ice bearing cumulus cloud evolution: Numerical simulation and general comparison against observations. *Journal of Applied Meteorology, 15*, 747–761.

Koop, T., Luo, B., Tsias, A., & Peter, T. (2000). Water activity as the determinant for homogeneous ice nucleation in aqueous solutions. *Letters to Nature, 404*, 611–614.

Korolev, A. V., & Isaac, G. A. (2003). Roundness and aspect ratio of particles in ice clouds. *Journal of the Atmospheric Sciences, 60*, 1795–1808.

Kumai, M, & O'Brien, H. W. (1964). A study of ice fog and ice-fog splintering nuclei at Fairbanks, Alaska. U.S. Army Cold Regions Research and Engineering Lab. (USA CRREL) research report, 150 pp.

Lanzinger, E., Theel, M., & Windolph, H. (2006). *Rainfall amount and intensity measured by the Thies laser precipitation monitor*. TECO-2006 Meeting, Geneva, Switzer- land, 4–6 December 2006, IOM94 (TD1354).

Lawson, R. P. (2011). Effects of ice particles shattering on optical cloud particle probes. *Atmospheric Measurement Techniques Discussions, 4*, 939–968.

Lawson, R. P., Baker, B. A., Zmarzly, P., O'Connor, D., Mo, Q., Gayet, J. F., & Schherbakov, V. (2006a). Microphysical and optical properties of atmospheric ice crystals at south pole station. *Journal of Applied Meteorology and Climatology, 45*, 1505–1524.

Lawson, R. P., O'Connor, D., Zmarzly, P., Weaver, K., Kaker, B., Mo, Q., & Jonsson, H. (2006b). The 2D-S (stereo) probe: Design and preliminary tests of a new airborne, high-speed, high-resolution particle imaging probe. *Journal of Atmospheric and Oceanic Technology, 23*, 1462–1477.

Lawson, R. P., Jensen, E., Mitchell, D., Baker, B., Mo, Q., & Pilson, B. (2010). Microphysical and radiative properties of tropical clouds investigated in TC4 and NAMM. *Journal of Geophysical Research, 115*, D00J08. https://doi.org/10.1029/2009JD013017, 16 pp.

Lee, T. F., Turk, F. J., & Richardson, K. (1997). Stratus and Fog Products Using GOES-8–9 3.9-μm Data. *Weather and Forecasting, 12*, 664–677.

Lin, Y. L., Farley, R. D., & Orville, H. D. (1983). Bulk Parameterization of the Snow Field in a Cloud Model. *Journal of Applied Meteorology and Climatology, 22*, 1065–1092.

Lindeman, J., Boybeyi, Z., & Gultepe, I. (2011). The aerosol semi-direct effect during a polluted case of the ISDAC field campaign. The aerosol semi-direct effect during a polluted case of the ISDAC field campaign. *Journal of Geophysical Research -Atmosphere, 116*, D00T10. https://doi.org/10.1029/2011JD015649.

Lloyd, G., Choularton, T. W., Bower, K. N., Gallagher, M. W., Connolly, P. J., Flynn, M., Farrington, R., Crosier, J., Schlenczek, O., Fugal, J., & Henneberger, J. (2015). The origins of ice crystals measured in mixed-phase clouds at the high-alpine site Jungfraujoch. *Atmospheric Chemistry and Physics, 15*, 12,953–12,969.

Lohmann, U., & Diehl, K. (2006). Sensitivity studies of the importance of dust ice nuclei for the indirect aerosol effect on stratiform mixed-phase clouds. *Journal of the Atmospheric Sciences, 63*, 968–982.

Lüönd, F., Stetzer, O., Welti, A., & Lohmann, U. (2010). Experimental study on the ice nucleation ability of size-selected kaolinite particles in the immersion mode. *Journal of Geophysical Research, 115*, D14201. https://doi.org/10.1029/2009JD012959.

Mamouri, R. E., & Ansmann, A. (2015). Estimated desert-dust ice nuclei profiles from polarization LiDAR: methodology and case studies. *Atmospheric Chemistry and Physics, 15*, 3463–3477.

Matrosov, S. Y. (2010). Synergetic use of millimeter- and centimeter-wavelength radars for retrievals of cloud and rainfall parameters. *Atmospheric Chemistry and Physics, 10*, 3321–3331.

McFarquhar, G., & Heymsfield, A. J. (1997). Parameterization of tropical cirrus ice crystal size distributions and implications for radiative transfer: results from CEPEX. *Journal of the Atmospheric Sciences, 54*, 2187–2200.

McFarquhar, G. M., Baumgardner, D., Bansemer, A., Abel, S. J., Crosier, J., French, J., Rosenberg, P., Korolev, A., Schwarzoenboeck, A., Leroy, D., Um, J., Wu, W., Heymsfield, A. J., Twohy, C., Detwiler, A., Field, P., Neumann, A., Cotton, R., Axisa, D., & Dong, J. (2017). Processing of ice cloud in situ data collected by bulk water, scattering, and imaging probes: fundamentals, uncertainties, and efforts toward consistency. *Meteorological Monographs, 58*, 11.1–11.33. https://doi.org/10.1175/AMSMONOGRAPHS-D-16-0007.1.

Mead, J. B., McIntosh, R. E., Vandemark, D., & Swift, C. T. (1989). Remote Sensing of Clouds and Fog with a 1.4-mm Radar. *Journal of Atmospheric and Oceanic Technology, 6*, 1090–1097.

Meyers, M. P., DeMott, P. J., & Cotton, W. R. (1992). New primary ice-nucleation parameterizations in an explicit cloud model. *Journal of Applied Meteorology, 31*, 708–721.

Milbrandt, J. A., & Yau, M. K. (2005a). A multi-moment bulk microphysics parameterization. Part I: Analysis of the role of the spectral shape parameter. *Journal of the Atmospheric Sciences, 62*, 3051–3064.

Milbrandt, J. A., & Yau, M. K. (2005b). A multimoment bulk microphysics parameterization. Part II: A proposed three-moment closure and scheme description. *Journal of the Atmospheric Sciences, 62*, 3065–3081.

Miloshevich, L. M., & Heymsfield, A. J. (1997). A balloon-borne continuous cloud particle replicator for measuring vertical profiles of cloud microphysical properties: Instrument design, performance, and collection efficiency analysis. *Journal of Atmospheric and Oceanic Technology, 14*, 753–768.

Minnis, P., Garber, D. P., Young, D. F., Arduini, R. F., & Takano, Y. (1998). Parameterization of reflectance and effective emittance for satellite remote sensing of cloud properties. *Journal of the Atmospheric Sciences, 55*, 3313–3339.

Minnis, P., Sun-Mack, S., Young, D. F., Heck, P. W., Garber, D. P., Chen, Y., Spangenberg, D. A., Arduini, R. F., Trepte, Q. Z., Smith, W. L., Jr., Ayers, J. K., Gibson, S. C., Miller, W. F., Chakrapani, V., Takano, Y., Liou, K.-N., Xie, Y., & Yang, P. (2011). CERES Edition-2 cloud property retrievals using TRMM VIRS and Terra and Aqua MODIS data, Part I: Algorithms. *IEEE Transactions on Geoscience and Remote Sensing, 49*(11), 4374–4400. https://doi.org/10.1109/TGRS.2011.2144601.

Morrison, H., & Milbrandt, J. (2011). Comparison of two-moment bulk microphysics schemes in idealized supercell thunderstorm simulations. *Monthly Weather Review, 139*, 1103–1130.

Morrison, H., Shupe, M. D., & Curry, J. A. (2003). Modeling clouds observed at SHEBA using a bulk parameterization implemented into a single-column model. *Journal of Geophysical Research, 108*, 4255. https://doi.org/10.1029/2002JD002229.

Morrison, H., Curry, J. A., & Khvorostyanov, V. I. (2005). A new double-moment microphysics parameterization for application in cloud and climate models. Part I: Description. *Journal of the Atmospheric Sciences, 62*, 1665–1677.

Mueller, M. D., Masbou, M., & Bott, A. (2010). Three-dimensional fog forecasting in complex terrain. *Quarterly Journal of the Royal Meteorological Society, 136*, 2189–2202.

Murray, S. L., Broadley, T. W., Wilson, J. D. A., & Wills, R. H. (2011). Heterogeneous freezing of water droplets containing kaolinite particles. *Atmospheric Chemistry and Physics, 11*, 4191–4207.

Murray, B. J., O'Sullivan, D., Atkinson, J. D., & Webb, M. E. (2012). Ice nucleation by particles immersed in supercooled cloud droplets. *Chemical Society Reviews, 41*, 6519–6554.

Niedermeier, D., Hartmann, S., Shaw, R. A., Covert, D., Mentel, T. F., Schneider, J., Poulain, L., Reitz, P., Spindler, C., Clauss, T., Kiselev, A., Hallbauer, E., Wex, H. K., Mildenberger, K., & Stratmann, F. (2010). Heterogeneous freezing of droplets with immersed mineral dust particles – measurements and parameterization. *Atmospheric Chemistry and Physics, 10*, 3601–3614.

Niemand, M., Möhler, O., Vogel, B., Vogel, H., Hoose, C., Connolly, P., Klein Bingemer, H., DeMott, P., Skrotzki, J., & Leisner, T. (2012). Parameterization of immersion freezing on mineral dust particles: an application in a regional scale model. *Journal of the Atmospheric Sciences, 69*, 3077–3092.

Nowak, D., Ruffieux, D., Agnew, J. L., & Vuilleumier, L. (2008). Detection of fog and low cloud boundaries with ground-based remote sensing systems. *Journal of Atmospheric and Oceanic Technology, 25*, 1357–1368.

Nygaard, K., Egil, B., Kristjánsson, J. E., & Makkonen, L. (2011a). Prediction of in-cloud icing conditions at ground level using the WRF model. *Journal of Applied Meteorology and Climatology, 50*, 2445–2459.

Nygaard, K., Kristjánsson, J. E., & Makkonen, L. (2011b). Relationships between road slipperiness, traffic accident risk and winter road activity. *Climate Research, 15*, 185–193.

Ohtake, T. (1967). *Alaskan ice fog*. In Proceedings of International Conference on Physics of Snow and Ice, Part I, Hokkaido University, Sapporo, Japan, pp. 105–118.

Ohtake, T., & Huffman, P. J. (1969). Visual range in ice fog. *Journal of Applied Meteorology, 8*, 499–501.

Pagowski, M., Gultepe, I., & King, P. (2004). Analysis and modeling of an extremely dense fog event in southern Ontario. *Journal of Applied Meteorology, 43*, 3–16.

Pavolonis, M. J. (2010a). Advances in extracting cloud composition information from spaceborne infrared radiances: A robust alternative to brightness temperatures Part I: Theory. *Journal of Applied Meteorology and Climatology, 49*, 1992–2012.

Pavolonis, M. J. (2010b) *GOES-R Advanced Baseline Imager (ABI) algorithm theoretical basis document for cloud type and cloud phase, version 2.0*. Available from NOAA, Madison, Wisconsin, 86 pp.

Phillips, V. T. J., & Coauthors. (2009). Potential impacts from biological aerosols on ensembles of continental clouds simulated numerically. *Biogeosciences, 6*, 987–1014.

Phillips, V. T., Donner, L. J., & Garner, S. (2007). Nucleation processes in deep convection simulated by a cloud-system resolving model with double-moment bulk microphysics. *Journal of the Atmospheric Sciences, 64*, 738–761.

Phillips, V. T. J., DeMott, P. J., & Andronache, C. (2008). An empirical parameterization of heterogeneous ice nucleation for multiple chemical species of aerosol. *Journal of the Atmospheric Sciences, 65*, 2757–2783.

Pu, Z., Chachere, C., Pardyjak, E., Hoch, S., & Gultepe, I. (2016). Prediction of ice fog with WRF model: performance and early evaluation during MATERHORN-Fog. *Pure and Applied Geophysics, 173*, 3165–3186.

Robinson, E., Thuman, W. C., & Wiggins, E. J. (1957). Ice fog as a problem of air pollution in the Arctic. *Arctic, 10*, 88–104.

Rogers, D. C., & Vali, G. (1987). Ice crystal production by mountain surfaces. *Journal of Applied Meteorology and Climatology, 26*, 1152–1168.

Rogers E., DiMego, G., Black, T., Ek, M., Ferrier, B., Gayno, G., Janjic, Z., Lin, Y., Pyle, M., Wong, V., Wu, W.-S., & Carley, J. (2009). *The NCEP North American Mesoscale Modeling System: Recent changes and future plans*. In Proceedings of 23rd conference on weather analysis and forecasting/19th conference on numerical weather prediction, American Meteorological Society, Omaha, NE. Preprints 2A.4.

Sassen, K. (1991). The polarization LiDAR technique for cloud research: a review and current assessment. *Bulletin of the American Meteorological Society, 72*, 1848–1866.

Sassen, K., & Zhu, J. (2009). A global survey of CALIPSO linear depolarization ratios in ice clouds: Initial findings. *Journal of Geophysical Research, 114*, D00H07. https://doi.org/10.1029/2009JD012279, 12 pp.

Schmit, T. J., Gunshor, M. M., Menzel, W. P., Li, J., Bachmeier, S., & Gurka, J. J. (2005). Introducing the next-generation Advanced Baseline Imager (ABI) on GOES-R. *Bulletin of the American Meteorological Society, 86*, 1079–1096.

Schmit, T. J., Li, J., Gurka, J. J., Goldberg, M. D., Schrab, K. J., Li, J., & Feltz, W. F. (2008). The GOES-R Advanced Baseline Imager and the continuation of current sounder products. *Journal of Applied Meteorology, 47*, 2696–2711.

Sesartic, A., Lohmann, U., & Storelvmo, T. (2012). Bacteria in the ECHAM5-HAM Global climate model. *Atmospheric Chemistry and Physics, 12*, 8645–8661.

Shantz, N., Gultepe, I., Liu, P., Earle, M., & Zelenyuk, A. (2012). Spatial and temporal variability of aerosol particles in Arctic spring. *Quarterly Journal of the Royal Meteorological Society, 138*(669), 2229–2240. https://doi.org/10.1002/qj.1940.

Solheim, F., Godwin, J., Westwater, E., Han, Y., Keihm, S., Marsh, K., & Ware, R. (1998). Radiometric profiling of temperature, water vapor, and liquid water using various inversion methods. *Radio Science, 33*, 393–404.

Steinke, I., Hoose, C., Möhler, O., Connolly, P., & Leisner, T. (2015). A new temperature- and humidity-dependent surface site density approach for deposition ice nucleation. *Atmospheric Chemistry and Physics, 15*, 3703–3717.

Stoelinga, T. G., & Warner, T. T. (1999). Non-hydrostatic, mesobeta-scale model simulations of cloud ceiling and visibility for an east coast winter precipitation event. *Journal of Applied Meteorology, 38*, 385–404.

Szyrmer, W., & Zawadzki, I. (1997). Biogenic and anthropogenic sources of ice forming-ice nuclei: A review. *Bulletin of the American Meteorological Society, 78*, 209–227.

Targino, A. C., Coe, H., Cozic, J., Crosier, J., Crawford, I., Bower, K., Flynn, M., Gallagher, M., Allan, J., Verheggen, B., Weingartner, E., Baltensperger, U., & Choularton, T. (2009). Influence of particle chemical composition on the phase of cold clouds at a high-alpine site in Switzerland. *Journal of Geophysical Research, 114*, 1–20.

Thompson, G., Rasmussen, R. M., & Manning, K. (2004). Explicit forecasts of winter precipitation using an improved bulk microphysics scheme. Part I: Description and sensitivity analysis. *Monthly Weather Review, 132*, 519–542.

Thompson, G., Field, P. R., Rasmussen, R. M., & Hall, W. D. (2008). Explicit forecasts of winter precipitation using an improved bulk microphysics scheme. Part II: Implementation of a new snow parameterization. *Monthly Weather Review, 136*, 5095–5115.

Thuman, W. C., & Robinson, E. (1954). Studies of Alaskan ice-fog particles. *Journal of Meteorology, 11*, 151–156.

Tobo, Y., DeMott, P. J., Raddatz, M., Niedermeier, D., Hartmann, S., Kreidenweis, S. M., Stratmann, F., & Wex, H. (2012). Impacts of chemical reactivity on ice nucleation of kaolinite particles: a case study of levoglucosan and sulfuric acid. *Geophysical Research Letters, 39*, L19803. https://doi.org/10.1029/2012GL053007.

Tobo, Y., Prenni, A. J., DeMott, P. J., Huffman, J. A., McCluskey, C. S., Tian, G., Pöhlker, C., Pöschl, U., & Kreidenweis, S. M. (2013). Biological aerosol particles as a key determinant of 5 ice nuclei populations in a forest ecosystem. *Journal of Geophysical Research, 118*, 10,100–10,110.

Twohy, C. H., McMeeking, G. R., DeMott, P. J., McCluskey, C. S., Hill, T. C. J., Burrows, S. M., Kulkarni, G. R., Tanarhte, M., Kafle, D. N., & Toohey, D. W. (2016). Abundance of fluorescent biological aerosol particles at temperatures conducive to the formation of mixed-phase and cirrus clouds. *Atmospheric Chemistry and Physics, 16*, 8205–8225. https://doi.org/10.5194/acp-16-8205-2016.

Uematsu, A., Hashiguchi, H., Teshiba, M., Tanaka, H., Hirashima, K., & Fukao, S. (2005). Moving cellular structure of fog echoes obtained with a millimeter-wave scanning doppler radar at Kushiro, Japan. *Journal of Applied Meteorology, 44*, 1260–1273.

Vali, G. (1994). Freezing rate due to heterogeneous nucleation. *Journal of the Atmospheric Sciences, 51*, 1843–1856.

Velde, D. V. I. R., Steeneveld, G. J., Schreur, B. G. J. W., & Holtslag, A. M. (2010). Modeling and Forecasting the onset and duration of severe radiation fog under frost conditions. *Monthly Weather Review, 138*, 4237–4253.

Vochezer, P., Järvinen, E., Wagner, R., Kupiszewski, P., Leisner, T., & Schnaiter, M. (2016). In situ characterization of mixed phase clouds using the small ice detector and the particle phase discriminator. *Atmospheric Measurement Techniques, 9*, 159–177.

Ware, R., Carpenter, R., Guldner, J., Liljegren, J., Nehrkorn, T., Solheim, F., & Vandenberghe, F. (2003). A multichannel radiometric profiler of temperature, humidity, and cloud liquid. *Radio Science, 38*(4), 8079. https://doi.org/10.1029/2002RS002856.

Ware, R., Cimini, D., Campos, E., Giuliani, G., Albers, S., Nelson, M., Koch, S. E., Joe, P., & Cober, S. (2013). Thermodynamic and liquid profiling during the 2010 Winter Olympics. *Atmospheric Research*, 278–290.

Welti, A., Lüönd, F., Kanji, Z. A., Stetzer, O., & Lohmann, U. (2012). Time dependence of immersion freezing. *Atmospheric Chemistry and Physics, 12*, 9893–9907.

Wendler, G. (1969). Heat balance studies an ice fog period in Fairbanks, Alaska. *Monthly Weather Review, 7*, 512–520.

Wetzel, M. A., Borys, R. D., & Xu, L. E. (1996). Satellite microphysical retrievals for land-based fog with validation by balloon profiling. *Journal of Applied Meteorology, 35*, 810–829.

Wex, H., DeMott, P. J., Tobo, Y., Hartmann, S., Rösch, M., Clauss, T., Tomsche, L., Niedermeier, D., & Stratmann, F. (2014). Kaolinite particles as ice nuclei: Learning from the use of different kaolinite samples and different coatings. *Atmospheric Chemistry and Physics, 14*, 5529–5546.

Wex, H., Huang, L., Zhang, W., Hung, H., Traversi, R., Becagli, S., et al. (2019). Annual variability of ice nucleating particle concentrations at different Arctic locations. *Atmospheric Chemistry and Physics Discussions*, 5293–5311. https://doi.org/10.5194/acp-19-5293-2019.

Wexler, H. (1936). Cooling in the lower atmosphere and structure of polar continental air. *Monthly Weather Review, 64*, 122–136.

Wexler, H. (1949). Observations of nocturnal radiation at Fairbanks, Alaska, and Fargo, North Dakota. *Monthly Weather Review, 6*, 368–369.

Winker, D. M., Hunt, W. H., & McGill, M. J. (2007). Initial performance assessment of CALIOP. *Geophysical Research Letters, 34*, L19803. https://doi.org/10.1029/2007GL030135.

Young, C. K. (1974). The role of contact nucleation in ice phase initiation in clouds. *Journal of the Atmospheric Sciences, 31*, 768–776.

Zelenyuk, A., & Imre, D. G. (2005). Single particle laser ablation time-of-flight mass spectrometer: An introduction to SPLAT. *Aerosol Science and Technology, 39*, 554–568.

Zhao, B., Wang, Y., Gu, Y., Liou, K. N., Jonathan, H., Jiang, J. H., Fan, J., Liu, X., Huang, L., & Yung, Y. L. (2019). Ice nucleation by aerosols from anthropogenic pollution. *Nature Geoscience, 12*, 602–607.

Zhou, B. (2011). *Introduction to a new fog diagnostic scheme* (NCEP Office Technical Note 466) (p. 32). U.S. Department of Commerce, NOAA, National Weather Service, NCEP.

Zhou, B., & Du, J. (2010). Fog prediction from a multi-model mesoscale ensemble prediction system. *Weather and Forecasting, 25*, 303–322.

Zhou, B., & Ferrier, B. S. (2008). Asymptotic analysis of equilibrium in radiation fog. *Journal of Applied Meteorology and Climatology, 47*, 1704–1722.

Chapter 7
Polar Stratospheric Clouds in the Arctic

Francesco Cairo and Tiziana Colavitto

Abstract Polar ozone depletion is a major environmental issue, for the alterations induced on the chemical-physical equilibrium of the stratosphere and their impact on planetary climate and ecosystem.

Early studies recognized that Polar Stratospheric Clouds (PSCs) play a crucial role in this process, for being the primary surfaces where heterogeneous chemical reactions promote the formation of species responsible for ozone removal. In addition, PSC particle formation scavenges nitric acid and water vapor from the gas phase, causing denitrification and dehydration. As denitrification enhances chlorine radical lifetime, PSCs become a significant element for the chemical balance of the polar stratosphere. A brief survey of the chemistry of the stratospheric ozone and of stratospheric particles is here given, together with an outline of the dynamic of the stratosphere with emphasis on the polar regions. It is then presented the effect of the CFCs in the perturbation of the ozone chemistry in the winter polar stratosphere, and the twofold role of Polar Stratospheric Clouds, promoting the release of active chlorine from its reservoir species and denitrifying the stratosphere upon condensation of nitric acid and further removal by particle gravitational settling, is discussed. A survey of the main scientific activities carried out in the arctic to study the processes in which PSC are involved is then provided. An illustration on PSC formation microphysics theories, and related open issues, ends the chapter.

Keywords Polar Stratospheric Clouds · ClO dimer · Lidar · Balloon campaigns · Aircraft campaigns · Microphysics · Ozone · Nitric acid · Stratospheric aerosol · Polar vortex · Sulphuric acid · CFCs · Nitric acid Trihydrate · Type Ia · Type Ib · Type II · Sulphuric acid Tetrahydrate ·

F. Cairo (✉)
Institute of Atmospheric Sciences and Climate, National Research Council of Italy, Rome, Italy
e-mail: F.cairo@isac.cnr.it

T. Colavitto
State Scientific High School "Giuseppe Peano", Rome, Italy
e-mail: t.colavitto@libero.it

A. Kokhanovsky, C. Tomasi (eds.), *Physics and Chemistry of the Arctic Atmosphere*, Springer Polar Sciences, https://doi.org/10.1007/978-3-030-33566-3_7

Supercooled ternary solution · Stratospheric sulphate aerosol · Chlorine · Bromine

7.1 Historical Overview

The history of the ozone hole is with no doubt one of the most fascinating examples of a scientific journey that has seen the development of an entire research activity carried out in remote and unique environments such as the Polar Regions. The issue of the stratospheric ozone depletion raised a great interest in the scientific community since the middle 70s and even more after the discovery of the ozone hole over South Pole in 1984. The whole scientific community immediately recognized the primary importance of this phenomenon, due to the well-known role of the ozone molecule in protecting the living systems through its interactions with the solar UV radiation. Several international programs were set up to focus on this issue and in several years of intense research activity the pieces of the puzzle have been progressively composed into a settled picture. Laboratory investigations, atmospheric observations and theoretical and modeling studies have provided over the years a deeper understanding of role of the stratosphere in the climate system. However, if the understanding of the main mechanisms for the ozone depletion resulted to be quite simple under a general point of view, the correlations between the many factors involved in such phenomenon turned out to be extremely complex. In particular, the ongoing changes of the chemical – physical properties of the polar stratospheres still challenge the attempt of estimating the "ozone-hole" recovery. The tight connections between troposphere and stratosphere suggest that future ozone trends will be also linked to changes occurring in the chemical composition and dynamic of the troposphere, and consequently to global climate changes. As such, it cannot be said that the ozone "hole" is a completely solved scientific subject, being still one of the most important environmental problems of our time. Despite this, there has been a natural slowing in the rate of discoveries and in the development of new theories, because of both the progresses made in the understanding and the difficulty of proving the theories through specific observations and accurate modeling. This chapter aims at a review of the main scientific works made on these topics, with special emphasis on the role Polar Stratospheric Clouds (PSCs) play in the polar stratospheric chemistry, and on the research carried out in the Arctic. The leading theme is the advancing of the characterization and understanding of the fundamental processes related to the ozone depletion, with the aim of establishing the state of the art of the knowledge on the Arctic stratosphere and issues related to the ozone hole. We tried to resume all the main recognized topics linked to the stratospheric ozone depletion, such as measurements and characterization of the stratospheric aerosols, PSCs, chemistry (i.e., heterogeneous chemistry occurring on the surface of PSCs, the catalytic cycles leading to the ozone depletion and the impact of both PSC formation and ozone loss to the stratospheric chemical-physical equilibrium), the dynamics (polar vortex and waves of different nature and origin). Thus, this work can be thought as a journey through the research activities related to the polar stratosphere. We followed a chronological approach in giving the main achievements within each topic addressed, so to give a broad description of how stratospheric observations developed and changed their targets as soon as new achievements were got. Going along the most important steps, we will attempt to rebuild decades of observation activity in the Arctic, so to provide an overview of the investigations that represented the basis of the theoretical advancing that led to the actual knowledge.

7.2 The Stratospheric Ozone

The existence and the main properties of the stratosphere were known since the early investigations of the atmosphere thermal structure, as a result of studies focused on the characterization of its radiative properties. The existence of an "ozone layer" and the awareness of the benefits that ozone provides both for the living systems by filtering the ultraviolet radiation and maintaining the radiative budget in the stratosphere, through the absorption of the infrared radiation emitted from the Earth (Andrews et al., 1987) were a matter of fact, and since the late 50s ozone measurements were systematically carried out.

7.2.1 Opening Remarks

The stratospheric ozone issues induced serious scientific and public concerns soon after it was found that significant stratospheric ozone loss had been occurring over Antarctica since the late 70s, as revealed by observations covering several decades (Farman et al. 1985). Yet, it was understood that such progressive decreasing trends of springtime ozone could not be explained neither by changes in stratospheric meteorology nor in solar radiation. The role of chlorine and halocarbons gases was recognized, since there was already concern on some effects that chlorine released by sprays and air conditioning systems could have on the ozone molecule. However scientists had predicted those effects to be visible only in a minor amount and over long times. In late August 1986, the Antarctic expedition in McMurdo station headed by Dr. Susan Solomon proved that certain chlorine compounds were more abundant in the stratosphere than at the ground (Solomon et al. 1986). That represented the beginning of a new conception of the stratospheric chemistry. Consequently, a frenetic research activity started, leading to new theories continuously arising from measurement campaigns organized all over the Antarctic continent and, later on, when it was realized that the phenomenon was also affecting the counterpart to the north, over the Arctic. While laboratory studies were keeping on investigating and improving the knowledge of the reaction between ozone and chlorine, direct experiments carried out in 1987 by aircraft equipped with different instruments for measuring a large number of chemical species were flown over Antarctica from the Southern tip of South America. The relationship existing between chlorine oxides and springtime ozone depletion was confirmed by the anti-correlated amounts of these two species as measured during this and other airplanes campaigns.

Although ozone depletion regularly characterizes both Polar Regions, the most dramatic changes have been historically registered over Antarctica, due to the particular meteorological and dynamical properties of the Southern polar winter stratosphere.

7.2.2 The Chapman Cycle and the Catalytic Cycles

Stratospheric ozone mostly resides in a sharp region of the atmosphere centered at approximately 25 km, called ozone layer. The molecule was already known in late 1800s, but the first successful attempt of understanding its chemistry has to be attributed to Chapman (1930). According to a very simple "natural" cycle of creation and destruction, ozone forms in the stratosphere from the photolysis of molecular oxygen.

$$\begin{aligned} &O_2 + h\nu \rightarrow O + O \quad (\nu < 240nm) \\ &O + O_2 + M \rightarrow O_3 + M \end{aligned} \tag{7.1}$$

and it is quickly destroyed by the absorption of the UV radiation in the range 200–320 nm.

The recombination of odd oxygen atoms reproduces O_2.

$$\begin{aligned} &O_3 + h\nu \rightarrow O_2 + O\left(^1D\right) \\ &O\left(^1D\right) + M \rightarrow O + M \\ &O + O + M \rightarrow O_2 + M \\ \\ &net\ result\ \ O + O_3 \rightarrow 2O_2 \end{aligned} \tag{7.2}$$

This process, known as the Chapman scheme, represents the basis of the stratospheric ozone chemistry.

Adding to these oxygen-only reactions, various catalytic cycles concur to shape the distribution of ozone in the atmosphere. Such cycles are of the form:

$$\begin{aligned} &X + O_3 \rightarrow XO + O_2 \\ &XO + O \rightarrow X + O_2 \\ &net\ result\ \ O + O_3 \rightarrow 2O_2 \end{aligned} \tag{7.3}$$

where the species X of greatest importance in the stratosphere are H, OH, NO, Cl, Br and, possibly, I. The relative importance of such cycles vary with altitude, and they are moderated by the creation of stable species involving X, such as for the chlorine, $ClONO_2$, HCl or HOCl. Furthermore, these catalytic cycles are particularly effective where atomic oxygen is abundant, i.e. higher than where ozone effectively maximizes.

As anticipated, the layered structure of the ozone distribution owes its amount to peak at 25 km, because the radiation in the lower atmosphere is too weak for effectively dissociating the O_2 molecules, and upper stratospheric O_2 is not abundant enough for generating significant amounts of O atoms.

It is known that the O_3 column is higher at high latitudes in the winter, although most of the ozone production occurs in the more irradiated equatorial regions. This depends on the circulation of the coupled troposphere-stratosphere system, described in the studies of Brewer (1949) and Dobson (1956), and reviewed by Holton et al. (1995).

According to this scheme known as Brewer-Dobson circulation, the air masses rise into the stratosphere through the tropical tropopause and descend back to the troposphere at high latitudes.

Between these two vertical motions, the air is moved poleward by a quasi-horizontal flux. Such transport is stronger in wintertime, so the polar stratospheres are the places where ozone is more abundantly found during winters. Further, the poleward transport of other man-made gases creates the conditions favorable for the ozone-hole formation. One important requirement for the formation of the ozone hole is the confinement of stratospheric air above the poles. This is achieved with the establishment of a polar vortex, the dynamical feature that characterizes the wintertime stratospheres.

7.2.3 Dynamics of the Stratosphere

The air motion and the general dynamical features of the polar stratosphere are well defined since the early 50s. The stratospheric circulation can be described by two main terms, i.e., a mean overturning and a quasi-horizontal meridional transport. The first term, the already mentioned Brewer-Dobson circulation, implies upwelling of tropospheric air from the tropical regions, transport poleward and down welling back in the extra tropics. Such a circulation is stronger in the Northern hemisphere wintertime than in the Southern one. A schematic of such circulation is reproduced in Fig. 7.1. The second term, called Eddy Transport, arises from zonally asymmetric processes generally represented by the breaking of planetary waves propagating from the troposphere. These eddies cause horizontal mixing of constituents and are generally fast, except at approximately 20° and 65° in both hemispheres. The first region determines the isolation of the ascending tropical air from the mid-latitudes (Mote et al. 1996; Volk et al. 1996), a mechanism known as “tropical pipe” (Plumb 1996). The barriers at 65°, existing only during the winter, keep

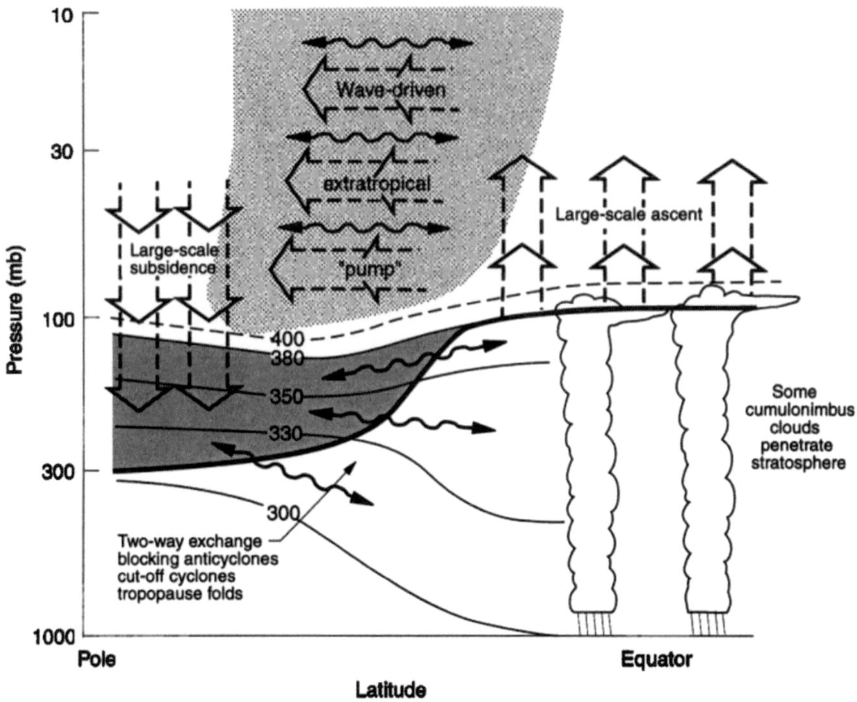

Fig. 7.1 Schematic representation of the tropospheric-stratospheric transport known as Brewer-Dobson circulation (doubled-headed arrows). The black thick line is the tropopause. (Source: Holton et al. 1995; copyright 1995 by the American Geophysical Union, reproduced with permission of John Wiley & Sons)

away the mid-latitude air from the polar region, thus maintaining the isolation of the air above the poles. This polar vortex forms because, as the winter approaches, the thermal contrast between polar and mid latitude regions originates a pressure gradient that, through the Coriolis force, induces a circular airflow around the poles. The poleward side of this high-speed wind region -known as "polar night jet"- marks its boundaries. This cyclone, extending from the upper troposphere into the mesosphere, spins up and gets stronger as dark polar night progresses. Usually, the polar vortex is defined by values of potential vorticity (PV), a suitable parameter for capturing its isolation (Nash et al. 1996).

The polar vortex, even with a certain year-to-year variability, presents two constant properties: low temperatures due to the lack of solar irradiance and absence of horizontal mixing with midlatitudes. The vortex edge, associated to strong gradients of PV, keeps away the mid latitudes air, thus maintaining extremely low temperatures and unchanged chemical composition during the winter (Schoeberl and Hertmann 1991; Bowman 1993; Manney et al. 1994a, b).

The vortex is not static: with the ongoing of the season, the air enclosed inside its boundaries descends due to subsidence induced by radiative cooling (Rosenfield et al. 1994).

When at the end of the winter the sunlight returns in the Polar Regions, the vortex is weakened by the reduced speed of the polar jet, and the propagation of planetary waves became more favorable, generating disturbances or warmings. The major warming marks the end of the vortex, by changing the direction of the polar jet towards the summer regime (Hess 1991) and allowing significant mixing between mid-latitude and the polar stratospheric air.

Dramatic differences exist between the two Hemispheres in the strength of the polar vortexes. Because the Northern Hemisphere is characterized by stronger upward-propagating planetary wave activity, the Arctic vortex is weaker and more variable.

Fundamental reviews (e.g. Holton 1992; Andrews et al. 1987; WMO 1986; Salby 2012; Dessler 2000) accurately cover both general dynamics and specific winter features of the stratosphere as well as the inter-hemispheric differences.

7.2.3.1 Waves in the Stratosphere

Stratospheric dynamics is mechanically induced by the propagation and absorption of waves generated in the troposphere. These waves deposit momentum when they are absorbed in the stratosphere and induce a poleward shift of air. Several forms of wave motion can influence the stratosphere. During winter, large scale Rossby (planetary) waves (Rossby 1939) can propagate up the stratosphere, transferring momentum and disturbing the stratosphere itself, since their vertical amplitude increases with the propagation. The dissipation of planetary waves in the stratosphere has a critical role for the polar vortex stability and it is recognized as the main factor driving the springtime vortex break-up by displacement out of its radiative equilibrium. Gravity (or buoyancy) waves are another form of air motion that can vertically propagate into the stratosphere, but with shorter space and timescales. Locally induced waves (orographic waves or leewaves) are instead small scale phenomena originating from the interaction of the main stratospheric circulations with orographic features, such as mountain chains. The winds, directed perpendicularly respect to the orography, generates usually rapid temperature fluctuations of several degrees of amplitude that have a significant impact on the local thermal fields of the downwind region. Such fluctuations, if superimposed on already cold synoptic conditions, can trigger the condensation of stratospheric water vapors, or other trace gases, which in the polar stratosphere may form PSCs on local scales. Rapid temperature changes due to lee wave events can be often observed in the temperature profiles taken by meteorological radio soundings and the effect of their propagation in the upper stratosphere can be also seen in lidar temperature profiles (Duck et al. 1998). Finally, significant differences exist between the main dynamics of the two hemispheres in terms of stability of the polar vortex and planetary wave activity. The Northern stratosphere is visibly more disturbed both because of a much stronger Brewer-Dobson wintertime circulation, and because of the presence of larger wave activity caused by the more irregular distribution of the lands, mountains and ocean around the Arctic Circle.

The detection and parameterization of waves in the stratosphere has been addressed for understanding their role in the vortex stability or in driving other phenomena (such as the formation of PSCs). Similarly, to the early measurements of stratospheric temperatures and ozone levels, the refereed scientific literature reports work on the stratospheric dynamics since the early 60s (Hines 1960; Charney and Drazin 1961; Gossard 1962). By the end of the 70s, work was also been done on the description and modeling of the planetary waves during the winter stratosphere, outlining the future work focused on their role for the polar vortex stability.

By the beginning of the 80s, this topic has been addressed as the attention to stratospheric processes was becoming major. Lidar measurements of upper stratospheric and mesospheric temperatures turned useful also for the detection of wave-like structures due to gravity wave perturbations (e.g. Chanin and Hauchecorne 1981; Jenkins et al. 1987). The detection of such structures is mentioned in most of the work on temperature measurements by both ground-based and airborne instruments (Gary 1989) or from gradients of potential vorticity or chemical species (Hartmann et al. 1989), and comparative studies between models and observations have been also performed (Bacmeister et al. 1990). Inter hemispheric comparisons for the middle atmosphere dynamics started to be performed yet in late 80s, in relation to the discovery of the ozone depletion (Andrews 1989 and references therein). Such studies highlighted the impact of stationary and transient waves of different types on the zonal mean climatological state of the winter stratosphere, influencing stratospheric warmings and polar ozone minima. The dissipation of planetary waves in the stratosphere, known as the Brewer-Dobson circulation and "wave-driven pump" (Holton et al. 1995) has been largely studied and put in relation with the Brewer-Dobson circulation and ozone distribution and depletion. The planetary wave activity in the upper stratosphere and lower mesosphere has been probed also using gaseous species as tracers of diabatic air motion. Allen et al. (1999, 2000) used CO measurements from the Improved Stratospheric and Mesospheric Sounder (ISAMS) boarded on the UARS satellite mission for tracking the horizontal mixing in the polar regions caused by large-scale wave breaking events. Studies have shown considerable variability in the width of the Antarctic vortex edge when it is perturbed, which results in transport and irreversible mixing of vortex air into mid-latitudes.

Gravity waves also play a crucial role in controlling the large scale circulation of the middle atmosphere; as the planetary waves, they break, deposit their energy and momentum and alter the flow. Particularly the short scale gravity waves have been largely investigated and their modeling attempted, but with difficulties because of the lack of knowledge on their sources and physical mechanism of their propagation. Information on sources has been inferred from horizontal winds and temperature variations in the lower stratosphere and upper troposphere, but the interpretation resulted complex. In addition, it has been found difficult to parametrize them, when they are not generated by the orography but rather by convection or shear

(McLandress 1998). Despite already since the middle 1980s radio soundings and rocket measurements have provided long-term averages of background winds and static stability of the stratosphere, (e.g. Eckermann 1995; Allen and Vincent 1995), a climatology and inter-hemispheric comparisons of gravity waves in the stratosphere have been possible only after the advent of long-term, high resolution temperature data acquired by satellite instruments with global coverage. (McLandress et al. 2000). These results allowed significant advancements in our ability to model lower stratospheric gravity waves and to resolve such waves in global meteorological analyses (Dörnbrack et al. 2001).

7.2.3.2 Polar Vortex

The polar vortex is the main dynamical feature of the polar winter stratospheres. Such structure and the movement of the air enclosed within its boundaries have been characterized through several methods, thus providing different definitions of its boundaries. Early works were based on the use of limited in situ observations and some of the first data available from the polar orbiting satellites. After the ozone-hole discovery all the research activities experienced a remarkable growth. In 1987, the first aircraft campaigns in Antarctica provided in situ gradients of chemical species that allowed a first characterization of the Antarctic polar vortex chemical boundary (Podolske et al. 1989; Proffitt et al. 1989b). The horizontal distribution of active species such as ClO, BrO, H_2O and NO_y (Fahey et al. 1989; Schoeberl et al. 1989) and the individuation of a coinciding dynamical boundary as defined by physical parameters like temperature, wind speed, potential vorticity and mixing, was achieved (Proffitt et al. 1989a; Fahey et al. 1989a). In particular, the fast-response instruments demonstrated a rapid rise of the active forms of chlorine (ClO) while the aircraft was proceeding towards the inner vortex, and narrow transitions were observed also for the other tracked species, indicating a chemically perturbed region coinciding with high-latitude areas of large ozone decreases (Jones et al. 1989; Anderson et al. 1989; Austin et al. 1989). A very complete review of this early knowledge can be found in the work of Tuck (1989) and references therein. The previously mentioned works of Schoeberl (Schoeberl and Hartmann 1991 and references therein; Schoeberl et al. 1992 and references therein) are considered a fundamental basis for what concern the characterization of a polar vortex. In the first one, the general chemical, dynamical, physical properties of the polar vortex and phenomena occurring there in relation with the springtime ozone depletion are summarized; moreover, evidences for year-to-year variability of the Antarctic vortex and a first inter-hemispheric comparison of the northern and southern vortices are presented. In the second one, more emphasis is given to the inter-hemispheric polar vortices comparison. This has been accomplished through a reconstruction of an integrated picture of the polar vortices (meteorological conditions, constituents, dynamics) from the observations made during the AAOE (Antarctic, 1987) and the AASE (Arctic, 1989) aircraft campaigns, supported with Lagrangian modelling. It was found that during the Northern winter, indeed, the circulation can be highly disturbed by a marked amplification of the planetary wave activity that weakens the Arctic polar vortex by slowing down the speed of the polar night jet and allows for intrusion of mid-latitude air into the Polar Regions. These phenomena are called Sudden Stratospheric Warmings; they can be termed as "major" or "minor" depending on whether they lead to the unrecoverable disruption of the polar vortex, or not; in the first case, a change in the main wind direction (from westerlies to easterlies) occurs and the sudden warming becomes definitive, as the typical springtime phenomenon that leads to the vortex break up.

McIntyre and Palmer (1983) were among the first to describe the erosion of the vortex in the middle stratosphere and coined the term "Rossby wave-breaking" to describe the disruption of the vortex during the stratospheric warming. The quantification of such erosion was first addressed by Butchart and Remsberg (1986). After the ozone-hole discovery, the attention was directed more towards the lower stratosphere dynamics and the polar vortex development in the lower winter stratosphere became subject of considerable debate, with the aim of characterizing both the amount and patterns of air descent within the vortex and the degree of horizontal mixing of polar air during different times of the winter. Important studies have been published by using analyzed fields of geopotential height and temperature produced daily by the National Meteorological Center (NMC) (Manney and Zurek 1993) or by the U.K. Meteorological Office (UKMO) (O'Neill et al. 1994) and dynamical fields based on temperature measurements from the Improved Stratospheric and Mesospheric Sounder on the UARS (Rosier et al. 1994). A large amount of work has been done concerning the diabatic descent of air inside the vortex, since there was clear evidence of polar air subsidence during the wintertime in both Arctic and Antarctic vortices by in situ (e.g. Tuck 1989; Proffitt et al. 1989b; Heidt et al. 1989; Schmidt et al. 1991) and remote observations (Parrish et al. 1988; Toon et al. 1989a, b, 1992). Estimates of air descent speed and calculations of cooling rates have been performed by using radiative transfer models (Kiehl and Solomon 1986; Rosenfield et al. 1987), satellite instruments such as the HALogen Occultation Experiment (HALOE) on UARS (Russell et al. 1993; Schoeberl et al. 1995), radiative transfer models along with temperature observations (Rosenfield et al. 1994), trajectory models (Schoeberl and Sparling 1994). More studies confirming the polar air descent and observations related with this downward motion have been published by using measurements from either satellite or meteorological analyses (e.g. Rood et al. 1997; Allen et al. 2000; Rosenfield and Schoeberl 2001). In addition to the subsidence, the motion of air through the polar vortex has been also addressed, as part of the research activity concerning the dynamics of vortex break-up and mixing of polar and middle latitude air after the final warming. Many studies confirmed the isolation of the Antarctic vortex during the winter time prior to its breakdown and the vortex erosion after the main warming by means of models (Bowman 1993 and references therein for previous work; Fisher et al. 1993; Bowman and Chen 1994), analyzed winds (Bowman 1993; Chen et al. 1994; Chen 1994), and similar analysis addressed the Arctic vortex (Pierce and Fairlie 1993; Manney et al. 1994a; Waugh et al. 1994; Plumb et al. 1994). Some of these studies (Waugh and Plumb 1994; Plumb et al. 1994) highlighted the existence of thin "filaments" of vortex air exiting from the edge of the lower stratospheric vortex and directed towards mid-latitudes at the end of the winter, associated to the vortex erosion due to "wave-breaking" events. They have been detected by in-situ aircrafts observations (Newman et al. 1996), sondes and lidar (Teitelbaum et al. 2001) and satellites (Manney et al. 1998), related to Rossby wave breaking (Waugh and Dritschel 1999) and transport variations within the vortex (Manney et al. 2001b). Appenzeller and Holton (1997) attempted to diagnose the production of such tracer laminae using satellite data and meteorological analyses, trying to determine their contribution to horizontal transport. Lee et al. (2001) presented evidences that the vortex is separated into two regions (a mixed vortex core and a weakly mixed ring of air extending to the vortex boundary), providing a more complete picture of the evolution of the vortex and its boundaries.

By the mid-90s, the studies for the meteorological characterization of the Northern polar vortex intensified, showing that even with a great inter-annual variability, the Arctic vortex could be as cold as the Antarctic in some year, when also the ozone loss could be significant (e.g. Proffitt et al. 1993; Rosier et al. 1994; Manney et al. 1994a, b, 1996, 1997; Coy et al. 1997; Newman et al. 1997, Pawson and Naujokat 1999; Manney and Sabutis 2000). The comparison of the 3D motion of air inside the vortices started in the 1990s, performed by Manney et al. (1994a, b), by both horizontal winds derived from observations using the UKMO data assimilation system and vertical velocities from a middle atmosphere radiation code. A similar study was been performed later by using 19 years (1979–1998) of NCEP/NCAR reanalysis data and potential vorticity (PV) area diagnostics (Zhou et al. 2000), trying to focus on the comparison of the vortices persistence. Inter-hemispheric comparisons of the stratospheric polar vortices taking also account of the interannual variability of the two hemispheres followed (Kuroda and Kodera 1998; Scaife et al. 2000), and the role of wave-forcing (Newman et al. 2001) and the contribution of photochemical and dynamical processes (Chippereld and Jones 1999; Hadjinicolaou et al. 2002) were modeled.

7.3 CFCs and Heterogeneous Chemistry in the Polar Stratosphere

As already pointed out, early laboratory studies conducted by Molina and Rowland (1974) and Stolarski and Cicerone (1974) had revealed that halocarbons could have a significant impact on ozone chemistry. Not by chance, chlorofluorocarbons, methyl chloroform and halons were monitored since 1978, even if their effects were not expected to be so strong and to occur in such remote locations. However, when the ozone depletion was first assessed, different possible explanations were proposed for the phenomenon, trying to link it to solar-cycles or dynamical effects - a complete review is given by Solomon (Solomon et al. 1986). Finally, the chlorofluorocarbons theory was widely accepted. On the other hand, reactions involving radicals (Bates and Nicolet 1950; Crutzen 1970; Stolarski and Cicerone 1974) had already been proposed as candidates for possible catalytic ozone destruction, as in the following cycle.

$$
\begin{aligned}
&CF_2Cl_2 + h\nu \rightarrow CF_2Cl + Cl\\
&Cl + O_3 \rightarrow ClO + O_2\\
&ClO + O \rightarrow Cl + O_2 \qquad .
\end{aligned}
\tag{7.4}
$$

$$net\ result\quad O + O_3 \rightarrow 2O_2$$

Initially, these catalytic reactions were thought to have maximum effect around 40 km, where studies and direct observations of chlorine partitioning were showing high ClO levels (Anderson et al. 1977 and 1989). However, since most of the ozone resides below 30 km, this reaction could only lead to a 5% of column ozone loss, and not to the 50% as reported by Farman (Farman et al. 1985) in Antarctica. Most of the total ozone was indeed depleted between 15 and 25 km, thus below the altitude predicted by gas phase photochemistry. This indicated the existence of a more complex set of circumstances and chemical reactions in which several elements were playing a role. A collection of field observations, laboratory measurements and modeling studies has in fact provided evidence that suspended particles are very important in ozone depletion in the lower stratosphere as they play a key role in halogen activation and nitrogen deactivation. Stratospheric aerosol and PSC resulted to be those particles.

7.3.1 Stratospheric Sulphate Aerosol

The Stratospheric Sulphate Aerosol (SSA) is formed of liquid droplets composed of sulphuric acid H_2SO_4 and water. Observations of the aerosol layer (Rosen 1964; Fiocco and Grams 1964; McCormick et al. 1979) allowed researchers to determine the global characteristics of this aerosol. It was found that these particles are concentrated in a layer extending several km above the tropopause, that they are present at all latitudes, that the aerosol undergoes seasonal variations. After the major volcanic events of El Chichon in 1982 and Mt. Pinatubo in 1991, it was observed that the aerosol was highly enhanced in the following months as a result of injections of water and sulphur dioxide into the stratosphere (McCormick et al. 1984). These droplets have diameters on the order of 0.02–0.5 μm, increasing by factors of 2–4 after large volcanic eruptions.

The source of the sulphur for the aerosol is: i. through direct injection from sulphur-rich volcanic eruptions that are capable of injecting huge amounts of sulphur, usually in the form of SO_2 or H_2S directly in the stratosphere. The SO_2 is subsequently oxidized to H_2SO_4 (McKeen et al. 1984) on a time-scale of about 1 month (Bluth et al. 1992; Hoff 1992); ii. from tropical injection of tropospheric air containing OCS, SO_2, and sulfate particles produced at the surface from natural as well as anthropogenic activities. These species are transported into the stratosphere as part of the overall general circulation and are oxidized there to H_2SO_4 (Turco et al. 1979; McKeen et al. 1984.) We refer to these nonvolcanic sources as background sources of aerosols.

The majority of the aerosol particles in the stratosphere are in its lower part, in the so called "Junge layer" discovered by Christian Junge in 1960, extending from the tropopause to about 30 km altitude. A recent and extensive overview of the current knowledge on stratospheric aerosol is provided in Kremser et al. (2016).

The SSA is very important for the ozone depletion; observations of enhanced polar and midlatitude ozone depletion following the eruption of Pinatubo confirm the impact of liquid aerosol surfaces on chlorine and nitrogen partitioning chemistry. Precisely, these droplets can perturb the chemical composition of the stratosphere in two ways. First, they provide surfaces on which chemical heterogeneous reactions can take place. Second, they intercept and backscatter solar radiation and absorb solar near-infrared and terrestrial infrared radiation, which changes the heating of the stratosphere and the rates of photolytic reactions (e.g. Stenchikov et al. 1998; Rozanov et al. 2002; Timmreck et al. 2003). Moreover, in the cold polar stratosphere (Hamill et al. 1988) they might form solid hydrates as the sulphuric acid tetrahydrate (SAT) - although there is still no experimental confirmation of its existence in the stratosphere - that could act as an efficient freezing nucleus and possibly promoting the deposition of other gases.

In the midlatitudes SSA interact with both Nitrogen compounds and chlorine reservoirs, thus indirectly with ozone. In the polar regions they may act as condensation sites for PSCs which in turn provide the surfaces necessary to convert inactive to active chlorine leading to polar ozone loss.

The most important heterogeneous reaction in SSA in the stratosphere is:

$$N_2O_5 + H_2O \overset{aerosol}{\rightarrow} 2HNO_3 \tag{7.5}$$

If the water is in the liquid phase this reaction is clearly an hydrolysis.

During the night, N_2O_5 is formed in the following reaction:

$$NO_2 + NO_3 \rightarrow N_2O_5 \tag{7.6}$$

This reaction consumes two molecules of the "NO_x family" ((NO_x) = (NO) + (NO_2)) for each N_2O_5 created. The net effect is to shift the partition inside the NO_y family ((NO_y) = (NO_x) + their oxidized products) from NO_x toward nitric acid through the reaction (7.5).

HNO_3 is then destroyed by photolysis and oxidation by *OH*:

$$HNO_3 + h\nu \rightarrow OH + NO_2 \tag{7.7}$$

$$HNO_3 + OH \rightarrow NO_3 + H_2O \tag{7.8}$$

In the upper stratosphere the creation of nitric acid causes a depletion in nitrogen oxides, a process called denoxification. The reduced availability of nitrogen oxide will decrease slightly the abundance of chlorine nitrate (or bromine), following this reaction.

$$ClO + NO_2 \rightarrow ClONO_2 \tag{7.9}$$

$$(BrO + NO_2 \rightarrow BrONO_2)$$

and the decreasing of the nitrogen oxide mixing ratio will lead to increase the active halogens that, at these stratospheric chlorine abundances, may promote more ozone destruction via the catalytic Cl cycle.

The net result is a decline in column O_3 over most of the globe. However, after a few years the effect of volcanic particles on ozone is diminished by their gradual removal from the stratosphere by natural air circulation. Because of particle removal, the two large volcanic eruptions on the last two decades cannot account for the long-term decreases observed in ozone over the same period.

The effect of a future major volcanic eruption will depend on chlorine levels. If the chlorine level is low, heterogeneous chemistry can lead to ozone increases in the stratosphere because the suppression of ozone destruction via the NO catalytic cycle would be prevalent, whereas if the chlorine level is high, as observed in recent years, volcanic aerosols lead to additional ozone depletion via the enhancement of the Cl catalytic cycle (Tie and Brasseur 1995). Both changes in aerosols and chlorine led to changes in ozone levels and it is difficult to separate their impact on ozone.

Besides, the sulphate aerosol injected into stratosphere can cause tropospheric cooling (e.g. Hansen et al. 1992; Santer et al. 2001; Yokohata et al. 2005) that would be expected to change the tropospheric circulation and the interaction between the stratosphere and the troposphere (Stenchikov et al. 2002). When volcanoes are quiescent, SSA have a typical radius of 0.1 μm, mass mixing ratio of 10^{-9} and number density of 10 cm^{-3} (Turco et al. 1982; Hofmann 1990).

The production of SSA droplets begins with the oxidation of compounds such as sulphure dioxide (SO_2) or carbonyl sulphide (OCS) and the end product is the gaseous H_2SO_4. The exact mechanism of gas-to –particle conversion has not been fully elucidated. SSA could be formed in situ either by homogeneous nucleation or heterogeneous nucleation on solid materials present in the stratosphere (Turco et al. 1982; Zolensky et al. 1989; Blake and Kato 1995; Pueschel 1996).

Assuming that the background (non volcanic) stratospheric aerosols are formed by homogeneous nucleation, Hamill et al. (1997) was the first to describe the life cycle of these particles.

The authors assume that the aerosol particles are formed by homogeneous nucleation in rising tropical air and are carried aloft into stratosphere as suggested by some observational evidence showing that the stratospheric aerosol resulting from the June 1991 eruptions of Mt. Pinatubo formed through homogeneous nucleation processes, as nearly all analysed sulphate particles showed no evidence of a solid or dissolved nucleus particle (Sheridan et al. 1992). However, conditions favourable for homogeneous nucleation occur infrequently in the stratosphere. The homogeneous nucleation of sulphate particles in the upper troposphere at midlatitudes is negligible (Hamill et al. 1982) and even in the equatorial region the nucleation rate is small except at the coldest temperature, where the nucleation rate is so large that every sulphuric acid molecule is transformed into a tiny but stable particle containing about 10 water molecules for every sulphuric acid molecules (Hamill et al. 1997). The newly nucleated particles coalesce to form larger particles. As the particles rises with the ascending air, their composition changes due to exchanges with ambient water vapour and eventually they move to mid and high latitudes where they are finally removed from the stratosphere by a variety of processes (e.g. gravitational sedimentation and stratospheric – tropospheric exchange). These removal processes need to be further quantified, although it appears reasonable to assume that isentropic mixing across the midlatitude tropopause is the major removal process. (Hamill et al. 1997).

Stratospheric aerosol particles play an important role for the stratospheric sulphur cycle and for transport meteoric material. The stratospheric background aerosol is mainly composed of sulphuric acid and water with some inclusions of meteoric material. Measurements on the number concentrations of total and non-volatile aerosol with size diameters >0.01 μm as well as particle size distributions, were conducted onboard the Russian research aircraft Geophysica using the low-pressure Condensation Nucleus Counter COPAS (Condensation Particle Counter System). It was found that at altitudes around 19 km the total number of particles with size diameter > 0.01 μm increases inside the polar vortex which indicates a source of particles in the above polar stratosphere or mesosphere. Inside the polar vortex a much higher fraction of particles contained non-volatile residues than outside the vortex. This is most likely due to a strongly increased fraction of meteoric material in the particles which is transported downward from the mesosphere inside the polar vortex. Therefore, there the nucleation might be rather a heterogeneous than a homogeneous process most likely condensation of H_2SO_4 and H_2O on cores of meteoric material (Curtius et al. 2005).

We remind that large fraction of the meteoroids entering the Earth's atmosphere ablate in the mesosphere at altitudes above 75 km. The ablates materials composed of elements as magnesium, iron and condenses to form particles termed meteoric smoke (e.g., Cziczo et al. 2001; Plane 2003).

Although solid hydrates would be the thermodynamically stable form of SSA at low stratospheric temperatures, many observations confirm that SSA are supercooled liquid droplets rather than solid crystals. Laboratory experiments show that liquid H_2SO_4/H_2O solutions do not freeze readily, even when highly supercooled (Molina et al. 1993; Anthony et al. 1995; Koop et al. 1997) or when certain solid nuclei are present (Koop et al. 1995; Biermann et al. 1996). Theoretical studies reach a similar conclusion (Luo et al. 1994; MacKenzie et al. 1995).

Although there is a general consistency of SSA observations with microphysical and dynamical modelling results, one outstanding issue to be resolved is the quantification of the non-volcanic background SSA level. This is required for

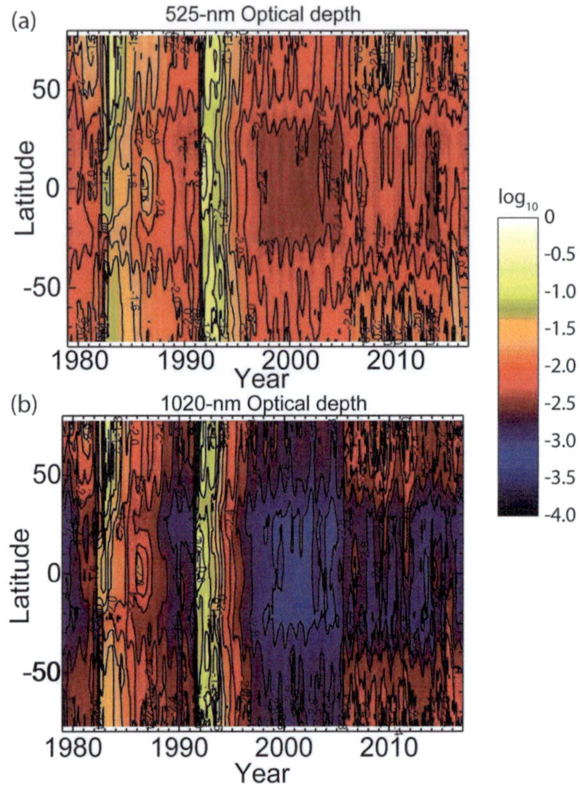

Fig. 7.2 A space-based evaluation of total stratospheric aerosol optical depth at 525 (upper panel) and 1020 nm (lower panel). Contours are shown at 0.0006, 0.0008, 0.001, 0.002, 0.004, 0.006, 0.008, 0.01, 0.02, 0.04, 0.06, 0.08, 0.1, 0.15, 0.2, and 0.4. (Source: Thomason et al. 2018)

many reasons: to assess the validity of models of SSA production, to determine the portioning of ozone destruction amongst the chlorine, hydrogen and nitrogen radical families (Wennberg et al. 1994) and to allow detection of possible future anthropogenic trends.

Unfortunately, there are very limited periods without volcanic eruptions since systematic measurements began. However, there are indications that there is no long trend – term in the background aerosol level (Deshler et al. 2006, 2008; Kovilakam and Deshler 2015), as depicted in Fig. 7.2 that reports the stratospheric aerosol optical depth at visible and NIR, based on aerosol climatology extending from 1979 to 2016 from spaceborne observations (Thomason et al. 2018).

A thoughtful assessment of the SSA system can be found in the SPARC *Assessment of Stratospheric Aerosol Properties* (ASAP) (SPARC 2006).

7.3.2 Polar Stratospheric Clouds

> The light was especially good today: the sun was directly reflected by a single twisted iridescent cloud in the North, a brilliant and most beautiful object. (R. F. Scott, August 1st, 1911)

Fig. 7.3 Polar Stratospheric Clouds (PSCs) high above Arctic low clouds over central Norway on 14 January 2003 at ~8Z from the cockpit of the NASA DC-8 during the SOLVE-II campaign. A strong tropospheric jet from the west-northwest was flowing onshore into southern Scandinavia, right under a portion of the stratospheric polar vortex. This resulted in very strong gravity wave activity in the stratosphere. (Photo courtesy of Paul A. Newman (NASA))

These words that Robert F. Scott wrote during one of his Antarctic expedition are probably the most famous example documenting the early knowledge of "strange-looking" clouds existing in the polar stratosphere, although their observations had been reported for more than a century (Stanford and Davis 1974), of iridescent and pearly appearance as is clearly shown in Fig. 7.3.

The PSC delivers the other major class of stratospheric particles. The presence of PSCs has been recognized for many years from ground based observations. But the extent of PSCs in both polar regions was not known fully until PSCs were observed by a satellite instrument in the late 1970s.

The term "Polar Stratospheric Cloud" was coined by McCormick et al. (1982), who first presented satellite observations of high-altitude clouds in the Antarctic and Arctic stratospheres. PSCs were observed in the Arctic stratosphere at altitudes between about 17 and 25 km during January 1979 and in the Antarctic stratosphere from June to October 1979 at altitudes from the tropopause up to about 23 km. The PSCs were observed when stratospheric temperatures were very low with a high probability of observation when temperatures were colder than 190 K and a low probability when temperatures were warmer than 198 K.

Several years after the first satellite observations, it became clear that PSCs played an important role in the ozone depletion. The particles of which they are formed support chemical reactions leading to active chlorine formation, which can catalytically destroy ozone and, since nitric acid can condense into their particles, this removal of HNO_3 from the gas phase increases ozone loss by perturbing the reactive chlorine and nitrogen chemical cycles in late winter and early spring.

PSCs have, then, a twofold role in the stratospheric chemistry, making available the surfaces where heterogeneous chemistry allows inactive chlorine compounds to be converted into forms directly responsible for the springtime ozone loss, and scavenging nitric acid from the gas phase (denoxification). When such a removal becomes irreversible (the so-called denitrification) upon sedimentation of the particles, the chemical balance of the stratosphere is changed and the lifetime of active chlorine is prolonged, causing delay in the recombination of halogen compounds into inactive forms (WMO 1998).

Solomon et al. (1986) proposed that PSCs were involved in ozone depletion because extensive cloud formation was observed in the region when the highest ozone loss was registered (12–25 km). In addition, Molina and Molina (1987) argued that $ClONO_2$ rapidly reacts with HCl adsorbed on ice surfaces, and at the same time, ice was recognized as one of the main components of PSCs (Toon et al. 1986; Crutzen and Arnold 1986). It seemed thus reliable that chlorine reservoir species (HCl and $ClONO_2$) arising from CFC photo dissociation, could *heterogeneously* react on PSC surfaces through mechanisms either involving chlorine.

$$
\begin{aligned}
&HCl_{(s)} + ClONO_{2(g)} \rightarrow Cl_{2(g)} + HNO_{3(s)} \\
&H_2O_{(s)} + ClONO_{2(g)} \rightarrow HOCl_{(g)} + HNO_{3(s)} \\
&HCl_{(s)} + N_2O_{5(g)} \rightarrow ClNO_{2(g)} + HNO_{3(s)} \\
&HCl_{(s)} + HOCl_{(g)} \rightarrow Cl_{2(g)} + H_2O_{(s)}
\end{aligned} \tag{7.10}
$$

or bromine compounds (McElroy et al. 1986).

$$
\begin{aligned}
&H_2O_{(s)} + BrONO_{2(g)} \rightarrow HOBr_{(g)} + HNO_{3(s)} \\
&HCl_{(s)} + BrONO_{2(g)} \rightarrow BrCl_{(g)} + HNO_{3(s)} \\
&HCl_{(s)} + HOBr_{(g)} \rightarrow BrCl_{(g)} + H_2O_{(s)}
\end{aligned} \tag{7.11}
$$

So, heterogeneous processes on the PSC surface transform inert chlorine forms (hydrogen chloride and chlorine nitrate) in to active chlorine (molecular chlorine and hypochlorous acid).

Satellite, ground-based and aircraft measurements confirmed the suggested anti-correlation between stratospheric ozone and chlorine abundances. After having risen from the natural level of 0.6 ppbv (early 1980s) to about 2.5 ppbv in 1990 and 3.2 ppbv in 2000, stratospheric measurements indicate that total chlorine is today at or near the peak, while bromine abundances are probably still increasing (WMO 2018).

Early ground-based and airborne lidar measurements indicated two distinct growth stages of PSCs (Iwasaka et al. 1985; Poole and McCormick 1988) giving rise to subdivide them into two classes: Type I PSCs that are thought to be relatively small particles containing nitric acid (HNO_3) as a major component (either in the form of liquid ternary solutions with water and sulfuric acid or as solid hydrates of nitric acid, as we shall see) form at temperatures above the ice frost point temperature T_{ICE}, and Type II PSCs that are larger and thicker, are primarily formed of H_2O ice particles. Lidar depolarization measurements in Arctic showed that Type I PSCs occur in at least two forms (Browell et al. 1990; Toon et al. 1990): Type Ia solid particles (with low backscatter and high depolarization) and Type Ib liquid particles (with high backscatter and low depolarization). Type II PSC have both high backscatter and high depolarization, as expected from large ice particles ($r \geq 1$ μm). For the Type I PSC several studies suggested that particles may remain liquid, out of thermodynamic equilibrium, well below the temperature at which thermodynamically favored solid hydrates of nitric acid might have formed. The most probable candidate of such hydrates was the nitric acid trihydrate (NAT) which, given the relative abundances of water vapor and nitric acid in the stratosphere, could form at temperatures below T_{NAT}, a few degrees above T_{ICE}. Thermodynamic models predict a steep increase in HNO_3 solubility into liquid H_2SO_4/H_2O solutions a few degrees below T_{NAT}, leading to the formation of non-equilibrium supercooled ternary solution (STS) droplets (Beyer et al. 1994; Carslaw et al. 1994; Tabazadeh et al. 1994; Luo et al. 1995). Above 260 K, the liquid fully evaporate (which explains the evaporation of droplets at the upper boundary of Junge particle layer), while upon cooling, H_2O and possibly acid hydrates may precipitate.

STS droplets would not coexist for long along with NAT particles, as the latter would be expected to grow slowly at the expense of the former, due to lower HNO_3 saturation vapor pressures over NAT.

For the thermodynamics of Type II PSC, it was fairly well established that they were primarily water ice particles. Laboratory experiments suggested that ice nucleates inside liquid Type Ib PSC particles or on the surface of solid Type Ia PSC particles once the temperature falls to 2–4 K below T_{ICE} (Koop et al. 1995). After ice formation in liquid STS droplets, the fate of the acidic components (HNO_3, H_2SO_4) is not well known: it is not clear whether soluble species could be trapped within the ice matrix or might lead to a solid coating of the ice particle, influencing its microphysical properties.

In order to derive PSC characteristics from remote sensing data, the refractive index of PSC particles must be known. The real part of the refractive index (m) for PSC particles have been estimated using the size distributions measured by single-wavelength lidar and optical particle counters (Adriani et al. 1995; Scarchilli et al. 2005; Deshler et al. 2000). At a wavelength of 532 nm, estimates values are $m = 1.42 \pm 0.04$ for NAT, $m = 1.39 \pm 0.03$ for STS and $m = 1.32 \pm 0.01$ for Type II PSCs.

PSC particle size distributions have been measured in situ by several instruments (Ferry et al. 1989; Dye et al. 1992, 1996; Goodman et al. 1989, 1997; Hofmann and

Deshler 1991; Hofmann et al. 1989; Adriani et al. 1995). Unimodal Type Ib PSC size distributions were generally observed during fast synoptic cooling between 4 and 10 K day^{-1} (Deshler et al. 1991; Beyerle et al. 1994; Adriani et al. 1995; Larsen et al. 2004; Voigt et al. 2003). Bimodal Type Ia PSC distributions are observed during slow synoptic cooling (1–3 K day^{-1}; Beyerle et al. 1994; Adriani et al. 1995; Deshler et al. 2003). Type II particles captured in situ have size distribution reminding those of thin tropospheric high cirrus clouds, with particles with sizes on the order of 5–50 μm (Goodman et al. 1989; Larsen et al. 2002).

The first in situ PSC measurements were performed in Antarctica in mid-to late winter 1987 and indicated that some clouds could be composed of NAT (Fahey et al. 1989). Observed size distributions were often bimodal (Type Ia) (Hofmann and Deshler 1991; Deshler et al. 1991). Later observations from the Arctic indicated that the composition of the particles (Rosen et al. 1989; Schlager et al. 1990) was far from the NAT stoichiometry, with unimodal size distributions. Ground-based lidar, satellite, balloon borne backscatter sonde, airborne and balloon borne chemical in situ characterization gave the evidence that such Type Ib PSCs were to be associated with an STS composition.

For the spatial variability of PSCs, many observations suggest that they can exist on synoptic scales, providing a large area through which air parcels can be processed. In particular, PSC observations by backscatter sondes in the Arctic (Larsen et al. 1996, 1997) also suggested that most liquid Type Ib PSC particles are observed during fast synoptic cooling, shortly (<1 day) after entering a cold region. Solid Type Ia PSCs were observed when temperatures had been below T_{NAT} for a longer time during both synoptic cooling and heating cycles.

Gravity mountain waves can cause mesoscale temperature fluctuations promoting the development of PSC with different characteristics on synoptic and regional spatial scales (e.g., Volkert and Intes 1992). Such strong temperature perturbations influence the cloud formation, as observed by airborne lidars (e.g., Godin et al. 1994).

The mechanisms that were suggested to account for the formation of nitric acid particles were both homogeneous and heterogeneous freezing (Tabazadeh et al. 2001; Tolbert and Toon 2001; Drdla et al. 2002). Although it was clear that SSA could take up water upon cooling and grow to STS droplets, it was unclear whether NAT could nucleate above T_{ICE}. The synoptic-scale temperature field can provide favorable cold conditions for solid PSCs to form in Antarctica, but synoptic-scale formation below the frost point T_{ICE} is much too infrequent to account for the widespread formation of solid nitric acid particles in the Arctic. Although some Arctic PSC observations could not be explained by mesoscale processes (Pagan et al. 2004; Larsen et al. 2004; Voigt et al. 2005), mesoscale temperatures below T_{ICE} may surely sometime provide a pathway for NAT formation, through nucleation of nitric acid hydrates on mountain wave-induced ice-cloud (e.g. Carslaw et al. 1999; Hu et al. 2002; Luo et al. 2003). However, studies have also demonstrated that heterogeneous freezing at temperatures above the ice frost point appears to be the most likely mechanism for producing nitric acid hydrate particles at synoptic scales in the Arctic (Drdla and Browell 2004) and the details of such process remain poorly defined, as well as the nuclei involved and the factors that control the freezing rate.

The presence of solid PSC particles can be detected by lidar depolarization measurements; many lidar studies suggest that are the large solid particles that cause denitrification, which are widely observed in both hemispheres.

Once formed, these PSC particles move downward because of gravitational settling. Because most PSCs contain nitric acid, their downward motion removes nitric acid from regions of the ozone layer: that process is called *denitrification*. Over Antarctica, temperatures are lower than T_{NAT} for several months and often fall below T_{ICE}. The lower temperatures and longer lifetime of individual PSC particles favor the formation of large NAT particles, commonly called "NAT-rocks". They have a low number densities ($<10^{-2}$ cm^{-3}) and so they have a long particle/gas phase equilibration times. This allows the particles to grow to the observed sizes which enable them to sediment rapidly and to lead to more efficient denitrification, as compared to Arctic conditions. However, very low number concentrations of very large particles have also occasionally been measured in the Arctic stratosphere (Fahey et al. 2001; Northway et al. 2002). Populations of very few large PSC particles have been observed previously (e.g., Deshler et al. 1991, 1994; Deshler and Oltmans 1998; Dye et al. 1992), but only with extensive field campaign in the Arctic in the early 2000 we have learned that such large particles are enriched in nitric acid only thanks to these measurements (Fahey et al. 2001). Moreover, several studies have concluded that such particles are widespread, both from analyses of remote measurements (Poole et al. 2003) and from in situ observations during the 2002/2003 Arctic winter (Larsen et al. 2004; Voigt et al. 2005).

Fueglistaler et al. (2002a, b) performed model studies showing that PSCs of type Ia containing high number density NAT particles ($n = 0.01$–1.0 cm^{-3}) can act as mother clouds for these NAT-rocks particles, which efficiently denitrify the polar vortex. The model indicates that severe denitrification occurs only if the NAT particles sediment in ambient air with temperatures below the NAT existence temperature for several days. Even in a very cold arctic winter like the 1999/2000 (Manney et al. 2001a), this requirement reduces the number of type Ia clouds which are suited for the development of NAT-rocks.

Fahey et al. (2001) have demonstrated that these few large nitric acid particles are capable of efficiently denitrifying the polar stratosphere. Many observations of denitrifications in both hemispheres have been documented in the last years. For the NH all the observations show that some Arctic airmasses were severely denitrified (>50%) near 20 km in several cold winters, but the short persistence of PSC events in the NH limits the extent of severe denitrification to only a few kilometers, as observed in many past cod Arctic winters (e.g. Sugita et al. 1998; Popp et al. 2001). Conversely, Tabazadeh et al. (2001) have shown that denitrification over a broad altitude range (>10 km in depth) occurs rapidly in the Antarctic when the duration of an average PSC event is about 2 weeks. Antarctic denitrification rapidly sets in during the mid-to late-June time period.

7.3.3 The ClOOCl Dimer and His Role in the Ozone Loss

What makes the polar stratosphere so peculiar with respect to the chlorine chemistry are the air confinement, the cold temperatures and the sunlight at the end of the winter. These are the ingredients of the "Ozone hole".

We have seen that chlorine (or bromine) atoms may be very efficient catalysts for ozone destruction according to the process (7.3). This cycle needs atomic oxygen to close. Moreover the efficiency of these cycles is limited by formation of reservoir species, as:

$$\mathrm{CH_4} \;\; + \mathrm{Cl} \rightarrow \mathrm{CH_3} + \mathrm{HCl} \tag{7.12}$$

$$\mathrm{ClO} + \mathrm{NO_2} \;\; + \mathrm{M} \rightarrow \mathrm{ClONO_2} + \mathrm{M} \tag{7.13}$$

and we have seen how heterogeneous reaction on PSC particles can release chlorine from these reservoir species. However the concentration of atomic oxygen is far too small, where PSCs dwell, to cause significant ozone depletion. So despite the heterogeneous chemistry on PSC surfaces is well represented, the catalytic cycle (7.3) could not account for any strong ozone removal, due to the relative scarcity of stratospheric atomic oxygen; other catalytic radical mechanisms have to be envisaged.

What was known was that the effective depletion starts when chlorine and similar compounds accumulated during the winter undergo photo dissociation with the return of the sunlight in early spring. The sunlight also dissociates the remaining few gaseous HNO_3, thus eventually facilitating some recombination of ClO and NO_2 into the inactive form $ClONO_2$.

Molina and Molina (1987) argued that discrepancies between estimated and observed ozone losses in the cold polar stratosphere could be resolved by introducing reactions involving the ClO dimer (CL_2O_2 also known as ClOOCl) more efficient in destroying ozone through the following chlorine catalytic cycle:

$$\begin{array}{l} ClO + ClO + M \underset{k_{diss}}{\overset{k_{rec}}{\rightleftarrows}} Cl_2O_2 + M \\ Cl_2O_2 + h\nu \overset{J}{\rightarrow} Cl + ClO_2 \; photolysis \\ ClO_2 + M \rightarrow Cl + O_2 + M \\ 2 \times (Cl + O_3 \rightarrow ClO + O_2) \\ net\,result \;\; 2O_3 + h\nu \rightarrow 3O_2 \end{array} \tag{7.14}$$

A similar cycle involving bromine and chlorine oxides was also suggested (McElroy et al. 1986).

$$\begin{aligned} &BrO + ClO + h\nu \rightarrow Br + Cl + O_2 \\ &Cl + O_3 \rightarrow O_2 + ClO \\ &Br + O_3 \rightarrow BrO + O_2 \\ net\ result\ &2O_3 + h\nu \rightarrow 3O_2 \end{aligned} \tag{7.15}$$

Note how in the first reaction of (14), the thermal decomposition of the ClO dimer leads to a null cycle that brings no changes in ozone. It is both the high concentration of chlorine monoxide and the very low temperature and that allows the formation of the dimer and its existence in the cold, confined air of the winter polar stratosphere. During the spring sunrise, Cl_2 photo dissociates in Cl which reacts rapidly with O_3 to form ClO, starting the cycle (14) based on the dimer ClOOCl. As temperature rises in late spring, the vortex is disrupted and mid-latitude air delivers new NO_y to reform chlorine reservoir species.

7.4 PSC Observations in the Arctic

For the critical role PSCs play in the chemical loss of stratospheric polar ozone, PSC have attracted a wealth of theoretical studies and experimental work. Here we will focus on observational activities carried out in the Arctic.

7.4.1 Ground Based Measurements

Many lidars have operated in the Arctic from the eighties although only few stations are continuing PSC measurements up to date. The longest PSC data record available is from the German Station Koldewey in Ny-Ålesund (79 N,12E), where measurements have been taken since late eighties; another long data record came from Thule (76 N, 68 W), where a lidar of the Italian ENEA was installed in 1990 and has operated since then, and from Esrange, Kiruna (67 N, 20E) where a lidar from the Bonn university is still operating.

As stated earlier, the Arctic polar vortex tends to easily break its circular symmetry and move frequently towards mid-latitudes, this leading to a great variability of the temperature and consequently a strong interannual variability of the characteristics of the stratosphere above the measurement stations. Moreover, being on the average warmer than its Antarctic counterpart, Type Ib and Type II PSC are strongly linked to mesoscale features rather than synoptic temperature fields.

Thule, as instance, is on the average close to the polar vortex boundary; there, as reported in Di Sarra et al. (2002). type II PSC observations are virtually absent and type Ib clouds are seldom observed; the more frequent clouds, whose frequency is highly variable through the winters, are mixed type Ia clouds, although it is not rare to not observe any PSC at all during the winter.

Indeed, the influence of gravity waves is much weaker at Thule than at Esrange, located close to the Scandinavian mountain ridge, which is a major source of orographically induced gravity waves that can rapidly cool the atmosphere below Type Ib and Type II formation temperatures. There, PSC presence is dominated by orographically induced gravity waves, and ice and STS clouds are more likely to occur (Blum et al. 2005; Khosrawi et al. 2011).

Conversely, unperturbed synoptic temperature conditions are usually prevalent in Ny-Ålesund, generally located at the center of the Arctic vortex; there, Type Ib account for the majority of all PSCs observed, Type Ia clouds are much less frequent (Beyerle et al. 2001) and very faint, and ice PSC events are extremely rare (Massoli et al. 2006; Biele et al. 2001). Due to the large geographical and interannual variability in the Arctic stratospheric conditions, it is difficult to draw general conclusions from measurements from single stations and one should rather rely on the global picture delivered by satellite observations.

Müller et al. (2001) compared lidar measurements from Sodankylä (67 N, 26 E), and Ny-Ålesund, representative of, respectively, vortex edge and vortex center conditions, finding that PSC occurrence is favoured at the edge of the Arctic polar vortex, probably because of an uneven distribution of water vapor, more present at the vortex edge due to an uneven descent of air inside the vortex, faster at its edges, that could in turn induce a non-uniform threshold formation temperature and different PSC occurrences within the vortex. A common feature of PSC observations from the various stations during winters season is the downward shifting in the vertical distribution of their occurrence during the season, following the stratospheric temperature minimum trend.

The international Network for the Detection of Atmospheric Composition Change (NDACC) gathers many of those datasets, stores them in common formats and make them available. The network is composed of more than 70 globally distributed, ground-based, remote-sensing research stations with more than 160 currently active instruments, providing high quality, consistent, standardized, long-term measurements of atmospheric temperatures and trace gases, particles, spectral UV radiation reaching the Earth's surface.

7.4.2 Airborne Campaigns

In 1989, the first Airborne Arctic Stratospheric Expedition (AASE-I), deployed in Stavanger, Norway (59 N 5E), the NASA ER 2 and the NASA DC 8, the same aircraft employed for AAOE Antarctic campaign (Turco et al. 1990). One of the main finding of the research activities was that while NO_x and to some degree NO_y were perturbed within the arctic vortex, there was little evidence for dehydration, due to the rarity of type II PSC in the Arctic, in contrast to the Antarctic. However, ClO was still found to be dramatically enhanced such that a large fraction of the available chlorine resided in the form of ClO and its dimer ClOOCl.

The second Airborne Arctic Stratospheric Expedition (AASE II) (Anderson and Toon 1993) deployed again the NASA DC 8 and ER 2 in the winter 91–92, first from Fairbanks, Alaska (65 N, 147 W) then from Bangor, Maine (45 N, 69 W) where they would have a high probability of encountering air in the arctic vortex, but could still have acceptable weather for aircraft operations. The eruption of Mt. Pinatubo (15 N, 120E) in June 1991, which injected a dense volcanic cloud into the stratosphere, provided the opportunity to investigate stratospheric transport processes, volcanic cloud properties, as well as heterogeneous chemistry on sulfuric acid aerosol during the campaign (Rodriguez et al. 1994, Kawa et al. 1997).

The analysis of the campaign observations showed again how also in the Arctic high ClO concentrations could be found, notably in air parcels whose back trajectories experienced temperatures within a few degrees of those needed to form nitric acid containing PSCs (Schoeberl et al. 1993). It was also found how the species reservoir of Cl were reformed following the decline of ClO with exposure to sunlight along the trajectories in a manner that is consistent with the photolysis of HNO_3 to produce NO_x followed by reaction with ClO to form $ClONO_2$. Little evidence of denitrification was found this time (Loewenstein et al. 1993) related to the relatively warm temperature in the arctic vortex in 1992 (Newman et al. 1993) that prevented extensive PSC formation.

The APE-POLECAT mission (Stefanutti et al. 1999) took place between December 1996 and January 1997 from Rovaniemi, Finland (66.5 N, 25.7E). That was the inaugural mission of the high-altitude research aircraft, M-55 Geophysica, operating together with the DLR Falcon-20 which was fitted with an aerosol lidar. The two aircraft were equipped with a payload specifically aimed to probe the chemistry and microphysics PSCs at the aircraft altitude. Unfortunately, due to a rather weak vortex, offset toward 90 E and with synoptic temperatures everywhere too high for PSC existence close to the M-55 flight altitudes, it was decided a shift in the priorities of the campaign, which concentrated on studies of transport and chemistry around the polar vortex, and on remote sensing of very high, mountain-wave-induced, PSC. The cloud remotely probed showed intermediate optical characteristics which prompted for a revision of the current PSC classification, indicating how type I formation may produce a continuum of cloud optical properties, rather than the binary division into type Ia and type Ib currently assumed (Tsias et al. 1999). The measurements of stratospheric aerosol loading, at middle latitudes and in the Arctic, showed values lower than those measured before the eruption of Mr. Pinatubo, suggesting how the increases stratospheric aerosols following the volcanic event had been completely removed.

Between November 1999 and April 2000, two major field experiments, the NASA sponsored Stratospheric Aerosol and Gas Experiment (SAGE) III Ozone Loss and Validation Experiment (SOLVE) and the EU Third European Stratospheric Experiment on Ozone (THESEO 2000), were conducted in the Arctic (Newman et al. 2002). The campaign included satellites, research balloons, six aircraft, ground stations, ozonesondes. The NASA deployed the stratospheric aircraft ER 2 and the DC 8, while the THESEO 2000 campaign provided the French ARAT, the DLR Falçon, the Swiss Learjet, and the French Mystère 20 aircraft. Extensive PSCs formed across the Arctic, and very large particles containing nitric acid trihydrate, the since then so-called "NAT rocks", were observed for the first time, showing that denitrification can occur without the formation of ice particles (Fahey et al. 2001). These particles, with sizes exceeding 10 μm, were found over extensive portions of the polar vortex, at temperatures above the frost point, large enough to considerably denitrify the stratosphere upon sedimentation.

The EUPLEX project took place between 2002 and 2004 as part of the pan-european project Validation if International Satellites and Study of Ozone Loss (VINTERSOL), with the deployment of the high altitude research aircraft M55-Geophysica in January/February 2003 from Kiruna in cooperation with the ongoing activities of the NASA-led second SOLVEIII (SAGEIII Validation and Ozone Loss Experiment) project deploying the NASA DC 8, both aircraft shown in Fig. 7.4, and with other project devoted to the validation of the ENVISAT satellite, then just launched. EUPLEX specifically aimed at studying the processes leading to PSC formation in lee-waves and their properties, compared to those formed on synoptic scales, and the induced halogen activation and chemical ozone loss. A set of meteorological parameters, dynamical tracers, reactive species involved in the ozone depletion, and chemical and microphysical characteristics of aerosols and PSC were gathered in the course of 14 flights. One of the main finding was that small NAT particles could form at very small number densities without pre-existence of ice particles (Voigt et al. 2005). These small NAT particles, which may subsequently act as giant NAT rocks embryos, were likely to form by heterogeneous nucleation and the most likely candidate for starting such process was meteoritic dust particles.

Fig. 7.4 The M55-Geophysica and NASA DC8 under take-off preparation during VINTERSOL-EUPLEX field campaign at Kiruna airport in Late January 2003 around noon time. (Photo courtesy of P. Mazzinghi (CNR))

The formation of NAT containing particles within lee-wave events was shown to potentially have a significant role for the formation of synoptic NAT PSCs, leading to extensive denitrification in the Arctic region.

Such observations pointed to the importance of presence of non-equilibrium liquid particles, and quasi-lagrangian flights sampling of PSCs in lee-wave regions allowed to test detailed non-equilibrium PSC models; in fact, measured ClO measurements suggested ongoing Cl activation on liquid aerosol particles, and a new ClO/ClO dimer equilibrium constant derived from the measurements, together with higher Br_y mixing ratios (Lowe et al. 2006) allowed CTM models to reproduce the integrated ozone loss in the observed magnitude.

The analysis of tracers allowed to depict the vortex edge's dynamical features, and the mixing in of extra vortex air masses was modelled and even predicted employing Lagrangian models (Werner et al. 2010).

The last great activity of coordinated extensive field measurements was carried out in the framework of the EU funded project RECONCILE (von Hobe and The RECONCILE Team 2013), that conducted one airborne campaign in winter 2010/11 and two balloon campaigns in the same winter and in 2011/12. The project aimed at studying transport and mixing within and at the edge of the polar vortex, PSC microphysics, chlorine chemistry and ozone loss. The early winter 2010/11 was characterized by a sequence of tropospheric disturbances that developed a great planetary wave activity influencing the formation and evolution of the polar vortex. Luckily, exceptionally cold temperatures established in December and January, leading to extensive PSC formation. At the end of January 2010 an event of sudden stratospheric warming disrupted and concluded the anomalously cold vortex (Dornbrack et al. 2012). In spring 2011 ozone levels over the Arctic dropped to values lower than ever before (Manney et al. 2011), because of a combination of inhibited dynamic supply of ozone from lower latitudes and substantial chemical ozone loss, fostered by very cold and stable polar vortex conditions.

The in situ optical particle spectrometers onboard the Geophysica FSSP (Forward Scattering Spectrometer Probe) and CIP (Cloud Imaging Probe) observed PSCs during the first five RECONCILE flights. Measurements with these instruments show the existence of so-called NAT-rocks up to sizes of roughly 25 μm in diameter (Molleker et al. 2014), confirming earlier observations (Fahey et al. 2001; Northway et al. 2002). The probably first images of NAT-rocks recorded by the greyscale CIP (optical array probe) provide an additional proof of their existence with an optical detection technique other than forward scattering. The presence of nitric acid in these large particles was confirmed. The micro cameras on board the Geophysica recorded images of particles of roughly 15 μm to 30 μm in diameter, with number concentration of about 10^{-3} cm^{-3}. These observations suggested how the occurrence of large NAT particles may be a regular feature of synoptic-scale PSCs in the Arctic.

During the project activities, it was also demonstrated how significant chlorine activation can occur not only on polar stratospheric clouds (PSCs) but also on cold binary aerosol (Wegner et al. 2012), as suggested in previous studies (Drdla 2005; Drdla 2006).

It was also found how PSCs can form more rapidly and at higher temperatures than previously thought by heterogeneous nucleation models (Hoyle et al. 2013; Engel et al. 2013). Background aerosol were collected and heterogeneous condensation nuclei were subsequently characterized by chemical analysis of the aerosol non-volatile fraction. Interestingly, while the presence of meteoritic particles was also detected in previous field activities (Curtius et al. 2005) and therefore largely expected, other components like crustal material and particles of anthropogenic origin were also found (Ebert et al. 2016) raising concerns about the current understanding of the origin and transport mechanisms of aerosol in the stratosphere.

7.4.3 Balloon Campaigns

Many balloon research activities aimed at measuring in-situ the optical, microphysical and chemical properties of stratospheric particles and PSC and thus assessing the pathways to their formation, combining these field measurements with laboratory simulations and microphysical modelling, have been carried out. The first balloon-borne gondola carrying a mass spectrometer system for particle analysis, a backscatter sonde, and pressure and temperature sensors was launched from Kiruna, Sweden, on 25 January 1998. Upon traversing a PSC, the data showed an H_2O/HNO_3 molar ratios consistently above 10 at atmospheric temperatures between 189 and 192 K, indicating ternary solution particles of H_2O, HNO_3, and H_2SO_4 rather than solid nitric acid hydrates (Schreiner et al. 1999).

During the activities of the SOLVE/Theseo 2000 campaign, a comprehensive balloon borne payload was launched from Esrange, Kiruna on 19 January and 25 January. The payload consisted of four instruments, an optical particle counter (Deshler et al. 2000), an aerosol composition mass spectrometer (ACMS) (Schreiner et al. 1999; Voigt et al. 2000), backscattersondes (Adriani et al. 1999, Rosen and Kjome 1991) and a frost point hygrometer (Ovarlez 1991). Voigt et al. (2000) and Schreiner et al. (2002) reported observations of the composition of PSCs, showing presence of STS particles near the frost point and solid NAT particles near and slightly above the NAT equilibrium temperature. A few observations of large size NAT that were in thin cloud layers below the other cloud layers, probably settling, were also reported.

These observations were reproduced by numerical simulation (Larsen et al. 2002), using a detailed microphysical model to represent the simultaneously measured chemical compositions and particle sizes.

The following EU funded project CIPA (Comprehensive investigations of polar stratospheric aerosols) brought to fly the same comprehensive scientific payload on a large balloon, this time including three optical particle counters (OPCs) (Deshler et al. 2000), able to deliver particle number density, volume and size distribution of optically detectable particles ($0.15 < r < 10$ μm) and the number density of condensation nuclei (CN, $r > 0.01$ μm). Particle composition and size was again detected with an aerosol composition mass spectrometer (ACMS). Particle optical properties were measured with two backscatter sondes. Water vapor was detected with a frost point hygrometer, together with accurate temperature and pressure data. Two flights were conducted in the winter 2002/2003 over Scandinavia (Weisser et al. 2006)., providing a unique set of data for mountain-wave induced PSC events, and NAT particles stoichiometry was measured, confirming their presence. Synoptic scale PSCs were observed in early December 2002 from Kiruna using the same set of balloon-borne instruments. On traversing a PSC, solid particles were observed whenever the temperature was below the equilibrium temperature for nitric acid trihydrate and liquid particles appeared when the temperature fell below an even lower threshold about 3 K above the ice frost point, with solid particles still present to form an externally mixed cloud, with a high correlation of liquid super-cooled ternary solution aerosols with local temperatures, confirming the predictions made by current STS models (Carslaw et al. 1994; Tabazadeh et al. 1994). The liquid particles contained HCl as well (below 1% in weight), which was then measured for the first time in STS.

Balloon borne ECC ozonesondes have been launched routinely since mid-eighties from many stations in the Arctic (Alert, Canada (82.50° N 62.33°), Heiss Island, Russian Federation (80.6° N 58.1°E), Eureka, Canada (80.05° N, 86.42° W), Summit, Greenland (72.34° N, 38.29° W) Scoresbysund, Greenland (70.48° N, 21.97°W), Andøya, Norway (69.3°N, 16.0°E) both routinely and on campaign basis. An extremely fruitful approach to the balloon research on ozone loss has been the Match technique: the idea of the Match method consists in probing an individual air parcel twice in order to obtain information on the temporal evolution of the concentration of chemical species (Rex et al. 1999) so the air masses probed by a first balloon are tracked with the aid of meteorological observations, and a second balloon, eventually from a different launching station is released when there is a chance for it to sample the same air masses. During such activity, 300–600 ozonesondes are typically launched at about 30 stations in the northern hemisphere. A similar approach was adopted to test the current microphysical models for PSC evolution, during the RECONCILE campaign under Arctic winter conditions (Engel et al. 2013; Khaykin et al. 2013).

During the RECONCILE campaign in January 2010, balloon borne backscatter-sondes were launched when PSC observations were carried out by the lidar station in Ny-Ålesund, reaching altitudes higher than the M55-Geophysica ceiling altitude. Matches with the Geophysica aircraft and the Cloud-Aerosol Lidar (CALIPSO) satellite were accomplished allowing for a comparison of different particle backscatter measurements and for the observation of cloud changes. Concomitant with the RECONCILE activities, balloon-borne soundings has been carried out in the framework of LAPBIAT-II (Lapland Atmosphere-Biosphere Facility) campaign held at Sodankylä, Finland (360 km east of Kiruna, Sweden) under the premises of Finnish Meteorological Institute's Arctic Research Center (FMI-ARC). 172 individual balloon instruments including radio and water vapour sondes, aerosol and ozone sensors were flown during the LAPBIAT-II campaign, in coordination with the RECONCILE aircraft flights, in two phases, with 18 payloads flown between 17 January and 06 February 2010 and 15 payloads flown between 10 and 24 March

2010. These activities registered an unprecedented evidence of water redistribution in the Arctic stratosphere as a consequence of ice PSC formation and concurrent irreversible dehydration by up to 1.5 ppm within 20–24 km altitude range, followed by a rehydration in a 2 km thick layer below. The source region and the spatiotemporal evolution of the dehydrated air masses were well established using AURA MLS observations (Khaykin et al. 2013). These observations provided a unique high-resolution snapshot of repartitioning of water vapour into ice particles, bearing important implications for the PSCs formation thresholds, and are analyzed in detail by Engel et al. (2013, 2014) using microphysical box modelling.

7.4.4 The Satellite Picture

Over the last three decades several satellite instruments have taken aerosol extinction vertical profiles, giving impulse to PSC studies in polar stratosphere.

The Stratospheric Aerosol Measurements (SAM) II instrument on the Nimbus 7 spacecraft was launched in October 1978. SAM II was a single channel solar photometer measuring stratospheric aerosol extinction profiles at a wavelength of 1.0 micron at latitudes from 64 to 80 degrees in both hemispheres. Observations from the Arctic for winter 1978/9 showed some extinction profiles with values unexpectedly large in the lower stratosphere, at altitudes between about 15 and 25 km. These were attributed to PSCs which were observed only when the local stratospheric temperature was very low (185–200 K) (McCormick et al. 1982).

Similar instruments providing long term dataset on Arctic PSC climatology are the POAMs and SAGEs; The Stratospheric Aerosol and Gas Experiment (SAGE) I was launched in February 1979, SAGE II in October 1984 and SAGE III in 2011. They measured sunlight through the limb of the Earth's atmosphere in the visible and near infrared, to convert it into vertical profiles of ozone, water vapor, nitrogen dioxide and aerosol concentrations.

The Polar Ozone and Aerosol Measurement (POAM) II and III instruments were both 9-channel photometers measuring atmospheric extinction in the visible and near infrared, to retrieve density profiles of ozone, nitrogen dioxide, and water vapor, as well as wavelength-dependent aerosol extinction. The measurements were confined to the latitude range of 55 N–71 N in the northern hemisphere, and 63S–88S in the southern hemisphere (Fromm et al. 1999; Bevilacqua et al. 2002).

The Improved Limb Atmospheric Spectrometer (ILAS) on board the Advanced Earth Observing Satellite (ADEOS) observed more than 60 polar stratospheric cloud (PSC) profiles during the winter 1996/1997 in the Northern Hemisphere, when the polar vortex was particularly long-lasting and significant ozone loss was observed (Saitoh et al. 2002; Irie et al. 2004). The correlation between the aerosol extinction coefficient and temperature showed that the extinction data increase only when temperatures decreased well below NAT temperature, suggesting the formation of particles uptaking HNO_3, and correlation with nitric acid profiles showed in some cases a marked decrease in the mixing ratio, suggesting the uptake of nitric acid into particles. The highest probability of sighting PSCs was obtained in mid-January at an altitude of approximately 23 km, and subsequent occurrences of PSCs were found intermittently at lower altitudes until mid-March.

The availability of a large number of observations on synoptic scale has allowed a satellite based Match technique to infer quantitative chemical ozone loss rates. Terao et al. (2012) reports an evaluation of ozone loss rates at at the 475 K isentropic surface inside the Arctic polar vortex, from six winters (January through March) using POAM II for 1994–1996, the ILAS for 1997, and the POAM III for 1999–2000. They found a high correlation between ozone loss and the PSC probability.

A 22 year, high latitude, stratospheric aerosol and cloud database combining SAM II, SAGE II, POAM II, and POAM III 1 μm aerosol extinction profiles is presented in Fromm et al. (2003). The database contains over 36,000 polar vortex region profiles in the Arctic with over 2000 PSC observations. Antarctic observations are present as well. The climatology of background stratospheric aerosol shows the dominance of volcanic effects in the evolution of background extinctions out of the vortex, although perturbations apparently not related to volcanoes are also reported. Interannual variations of background extinction inside the boreal vortex show a considerable volcanic influence, contrary to what seems to happen in Antarctica. An analysis of long term PSC sighting is presented by computing PSC probability for midwinter PSC at 20 km, in a fixed temperature range. They show that this PSC probability is nearly identical between hemispheres. However, while in the Antarctic the inter annual PSC probability pattern is markedly cyclic, although influenced by aperiodic volcanic perturbations in background aerosol, in the Arctic the PSC probability has much less temporal coherence and it is more impacted by background aerosol variability due to volcanic perturbations. PSC observations are seen to be strongly associated with elevated tropopause heights, suggesting that tropospheric, synoptic-scale flow perturbations are the primary forcing mechanism for Arctic PSC formation, finding later confirmed by Achtert et al. (2012).

Satellite instruments observing limb infrared emission measurements have operated on different missions with the advantage of allowing day- and nighttime measurements. Moreover, in some cases their spectral resolution allows to discriminate between specific PSC types.

The Improved Stratospheric and Mesospheric Sounder (ISAMS) (Taylor et al. 1994) on board the Upper Atmosphere Research Satellite (UARS), observed northern PSCs during the 1991/92 season, comparing them with analog Antarctic observations, and correlating their appearance with UARS/Microwave Limb Sounder observations of enhanced chlorine monoxide.

The synergy of ISAMS with the other instruments on the same UARS satellite, the Microwave Limb Sounder (MLS) and the Cryogenic Limb Array Etalon Spectrometer (CLAES), allowed to perform simultaneous observations of PSC and HNO_3 mixing ratios over Scandinavia. (Massie et al. 1997) PSC observations were correlated with temperature, and importantly, with HNO_3 mixing ratio showing PSC volume increases and simultaneous loss of HNO_3 as temperatures decreased, indicative of PSC growth processes which transform sulfate droplets into ternary STS or NAT particles.

In recent times, a suite of satellite missions is providing a complete picture of PSC morphology and composition on polar vortex-wide scales with unprecedented horizontal, vertical, temporal resolution and coverage. The ILAS spectrometer on ADEOS (Sasano et al. 2005), the Michelson Interferometer for Passive Atmospheric Sounding (MIPAS) (Spang et al. 2005a, b) on Envisat (2002–2012) and the Cloud-Aerosol Lidar with Orthogonal Polarization (CALIOP) (Pitts et al. 2013) on the Cloud-Aerosol-Lidar and Infrared Pathfinder Satellite Observations (CALIPSO) satellite (2006–present) are able to retrieve PSC types, and their datasets have already motivated numerous studies that extend our present knowledge of PSC processes (e.g. Peter and Grooß 2012, and references therein). Spang et al. (2018) presented an extensive climatology of Antarctic and Arctic PSC observations from MIPAS. Although placing caveats on the representativeness of a climatology that extends for only 10 years, affected by very large interannual variability due to the great variability for NH winter conditions, with warm seasons without any PSC sightings and other winters showing synoptic-scaled widespread PSC coverage, that study showed how in the NH the STS is always the dominating PSC type for all observed winter periods; Ice PSCs are rare but have been observed sporadically with the earliest events in the first half of December and the latest by the end of February, depending on the individual winter; and the denitrification in the middle stratosphere and consequent re-nitrification in the lower stratosphere due to the gravitational settling and evaporation of NAT particles is effective, but less pronounced than in the SH.

Pitts et al. (2018) presented an extensive climatology over both poles based on the 12 year CALIOP observations. Again they found a significant year to year variability in the Arctic, much larger than in the Antarctic due to a more disturbed polar vortex. As shown in Fig. 7.5, the most likely occurrence of Arctic PSC is between longitudes 60° W and 90° E, coincident with the preferential location of the Arctic vortex. Arctic PSC are prevalently seen as STS and STS-NAT mixtures, while ice PSC are seldom observed. In fact, a comparison with the Antarctic climatology shows that the largest differences in PSC composition are in ice, which comprised nearly 25% of Antarctic PSCs compared to less than 5% in the Arctic (a result of the much colder southern vortex) and to a minor extent in NAT mixtures, which comprised nearly 60% of Arctic PSCs, but only about 40% of Antarctic PSCs.

7.5 PSC Composition and Modelling

As we have seen, PSCs may be formed by different particles containing water, sulphuric acid and nitric acid, either in solid or liquid phases. While it is clear that only below the ice frost point T_{ICE} – which at those altitudes and given the relative scarcity of water vapor of roughly 5 ppmv is around 188 K – ice can exist, solid nitric acid trihydrate particles are thermodynamically stable at higher temperatures up to a T_{NAT} that – taking into account the stratospheric abundances of water and nitric acid – is around 188 K. The existence of other similar hydrates has

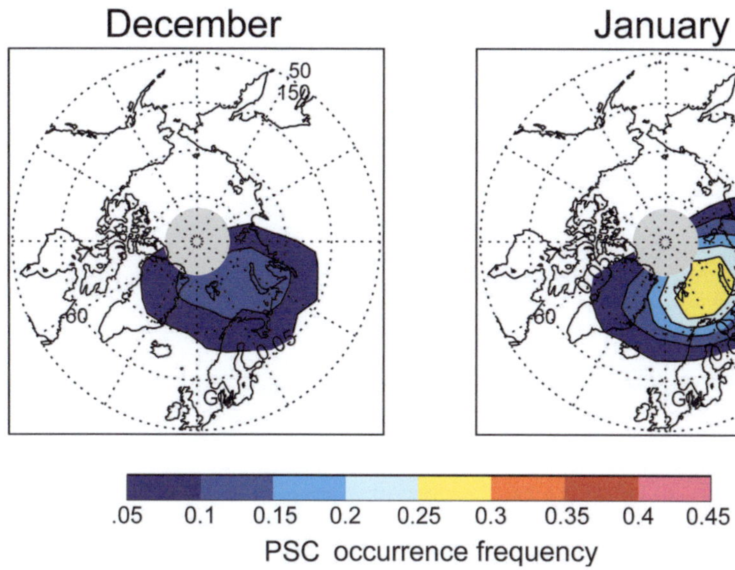

Fig. 7.5 Eleven-year average, monthly mean polar maps of Arctic PSC occurrence frequency at $\theta = 500$ K (~20 km). Light gray regions indicate latitudes not sampled by CALIOP. (Source: Pitts et al. 2018)

been also conceived, such as Nitric Acid Dihydrate or NAD (HNO_3: H_2O = 1:2), which is stable at temperatures lower than T_{NAT} ($T_{NAD} = T_{NAT}$ - 2.5 K) although their existence in the atmosphere has not been confirmed experimentally. While the nucleation of NAT and ice is a threshold process, the ubiquitous liquid SSA can continuously take up nitric acid from the gas phase and grow upon cooling, forming droplets of supercooled solution of sulphuric and nitric acid and water (STS), with volumes and relative concentrations varying with temperatures and dimensions up to 0.3 um in radius, as shown in Fig. 7.6.

As we have seen, either the existence of NAT and STS have been confirmed experimentally by in-situ measurements of PSC particle composition, so that above T_{ICE} PSC may be composed both of NAT particles (Type Ia clouds), or STS droplets (Type Ib) with greater number density, or by an external mixture of the two particle kind (mixed phase clouds).

The typical size distribution of type Ia PSCs is bimodal, with the larger mode radius (2 μm) representing the 1% of the available condensation nuclei. These particles are typically observed during slow synoptic cooling (1–3 K/day) and very often as thick layers (WMO 1998), and are rarely in thermodynamic equilibrium with the nitric acid gas phase due to the relatively low number density. Size distribution measurements revealed that type Ib PSCs have typical unimodal size distribution, with particles smaller than 1 μm representing a large fraction of condensation nuclei (WMO 1998).

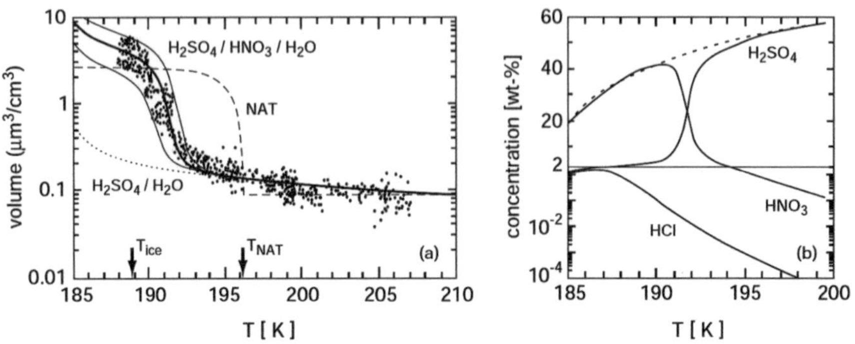

Fig. 7.6 The left panel of figure report volumes and relative concentrations of STS function of temperatures. The thick solid lines report model calculations assuming 5 ppmv of water and 10 ppbv of nitric acid, the thin solid lines assumed 5 and 15 ppbv of nitric acid. A temperature around 192 K can be considered as the one below which the STS PSC may form. The dotted line reports the growth of sulphate aerosol as it would be in absence of HNO3. Right panel reports the relative concentration of sulphuric, chloridric and nitric acid, in STS droplets, the remaining up to 100% being water. (Source: Peter, T., & Grooß, J.-U. in "Stratospheric Ozone Depletion and Climatic Change,", R. Muller ed., RCS publishing, 2012, ISBN 978–1–84,973-002-0. Reproduced by permission of The Royal Society of Chemistry)

All the nitric acid-containing particles remove HNO_3 from the gas phase, but above T_{ICE} only the NAT particles can grow big enough to sediment down to sufficiently lower altitudes and thus irreversibly deplete the polar stratosphere of nitric acid. In fact, type Ib PSCs do not remove HNO_3 in significant amount, due to their non-selective nucleation mechanism that limits the dimension reached by each liquid particle (radius ≤ 1 μm). Thus, sedimentation of type Ib PSCs does not occur to a sufficient extent and the HNO_3 is released back to the gas phase once these short-lifetime PSC particles have evaporated. It has to be mentioned that Type II PSCs are the responsible for the dehydration of the winter stratosphere regularly observed in Antarctica. In the Arctic, dehydration never occurs on large scale due to synoptic temperatures in the vortex rarely reaching T_{ICE}. Some hypotheses suggested that ice clouds could significantly incorporate NO_y as well. Because of the effective sedimentation of ice crystals, type II PSCs have been thought to cause the highest HNO_3 removal. However, evidences of denitrification without dehydration in the Arctic have demonstrated that type Ia PSCs are effective and the major responsible for scavenging HNO_3 from the gas phase. Therefore, understanding the exact mechanism of formation and growth of NAT is vitally important to properly simulate Arctic denitrification in the models.

According to earlier microphysical schemes, as the temperature decreases, the ubiquitous liquid aerosol sulfate gradually takes up HNO_3 and H_2O from the gas phase, turning into type Ib at approximately 190 K which represent a sort of "dew point" T_{STS} for the nitric acid. An additional temperature drop may freeze the water fraction of liquid particles into ice at temperatures around 182 K, 3–4 K below T_{ICE} and the nitric acid fraction would condensate during this phase as well. In an ascending branch of this thermal cycle, ice remains stable up to T_{ICE}, whereas NAT is supposed to be released after ice evaporation and to be solid up to approximately 195 K. Higher temperatures would lead to NAT evaporation, leaving behind the sulfate fraction of the original STS as ordinary stratospheric sulphate aerosol.

The main implication of such scenario was that Ia PSCs would require ice particles to form. Airborne lidar measurements in a Scandinavian mountain wave event (Carslaw et al. 1998; Wirth et al. 1999) supported this pathway to NAT formation, revealing tails of type Ia PSCs emerging downwind of the location where ice previously formed. Ice-mediated formation of NAT particles is indeed suggested to occur frequently in the Arctic (Carslaw et al. 1999).

However this pathway, although it is certainly at work and effective, cannot explain the totality of type Ia PSC observations. As instance a vortex-wide Arctic PSC perspective for the winter 2009–2010 from the satellite borne lidar CALIOP showed the presence of NAT clouds in the early winter, many days before any ice PSC was ever detected, suggesting that these early winter NAT PSC had been formed at temperature above T_{ICE} (Pitts et al. 2011). So, the way type Ia PSCs would form is a question still not completely clarified; according to the present view, two nucleation mechanisms should coexists, and one should allow NAT nucleation above T_{ICE}.

Such nucleation should proceed heterogeneously on preexisting solid inclusions into the SSA/STS droplet, which may be either meteoric dust, volcanic ash or the like or possibly an H_2SO_4 hydrate (SAT). Upon cooling, NAT would form, most likely by deposition, and subsequently ice would nucleate heterogeneously at T_{ICE}. Conversely, if such solid inclusions are absent, SSA would take up nitric acid, became STS and remain liquid until ice nucleates homogeneously. This latter pathway would allow the existence of liquid supercooled droplets at temperatures few degrees below T_{ICE}. A schematic of the different PSC particle formation pathways with respect to temperature is reported in Fig. 7.7.

However, it has to be stressed that only the NAT heterogeneous nucleation on ice is supported by observational evidences, both from laboratory and field experiments, while NAT heterogeneous nucleation on solid inclusions, which had to be invoked to explain field measurements, lacks of firm experimental grounds, and further field and laboratory studies are needed to specifically address such nucleation pathway. The capability of NAT nucleation on ice is well known and documented in laboratory experiments (Hanson and Mauersberger 1988; Iannarelli and Rossi 2015; Weiss et al. 2016) and in mountain wave ice PSCs (Carslaw et al. 2002; Fueglistaler et al. 2002a, b; Luo et al. 2003; Voigt et al. 2003). Interesting, in view of the possibility of NAT particle formation above T_{ICE}, is the possibility of such NAT particles acting as an ice nuclei, as suggested in early PSC studies (e.g. Peter 1997) and more recently revived (Voigt et al. 2018). Ice nucleation on NAT may be important also in other regions of the atmosphere, as in the tropical tropopause region, where NAT particles have been detected in situ (Popp et al. 2007; Voigt et al. 2008), and may promote cirrus formation in a region crucial for fixing the amount of waer vapour entering the stratosphere.

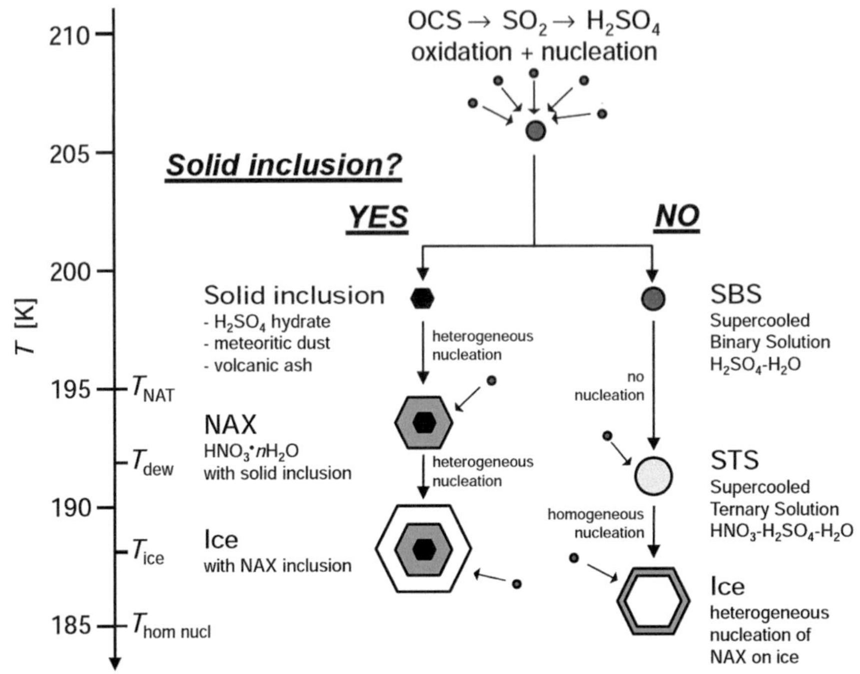

Fig. 7.7 PSC formation pathways. Left branch: nitric acid hydrate, NAX (e.g. NAT or NAD), nucleates heterogeneously at or slightly below T_{NAT} on pre-existing solid inclusion (e.g., a dust or ash particle or an H_2SO_4 hydrate particle); subsequently ice nucleates heterogeneously on NAX at or slightly below T_{ICE}. Right branch: SSA droplets form STS droplets at T_{dew} (T_{STS} in our text); subsequently ice nucleates homogeneously in STS below T_{ICE} – 3 K; finally NAT may nucleate heterogeneously on ice. (Source: Peter, T., & Grooß, J.-U. in "Stratospheric Ozone Depletion and Climatic Change,", R. Muller ed., RCS publishing, 2012, ISBN 978-1-84973-002-0. Reproduced by permission of The Royal Society of Chemistry)

Extensive reviews of the current status of PSC microphysics theories have been presented in Zondlo et al. (2000), Lowe and MacKenzie (2008) and, more recently in Peter and Grooß (2012).

7.6 Conclusions

More than 30 years of research have clarified the crucial role that PSCs play in the depletion of stratospheric ozone, namely how heterogeneous reactions in and on their particles free Cl from its stable reservoirs of HCl and $ClONO_2$ and release chlorine radicals that destroy ozone catalytically and how they are able to removes HNO_3 from the polar stratosphere (denitrification) via sedimentation, thus enhancing ozone depletion by delaying the reformation of chlorine reservoirs.

Both the rates of these heterogeneous reactions (Shi et al. 2001), and the effectiveness of sedimentation (Woiwode et al. 2014) depends on particle composition, which includes liquid binary H_2SO_4/H_2O droplets, liquid ternary $HNO_3/H_2SO_4/H_2O$ droplets, solid nitric acid trihydrate particles, and H_2O ice particles.

In spite of intense laboratory experiments, field activities and theoretical works, the role of cold background binary stratospheric aerosol on chlorine activation is still uncertain, and most important, the details of how NAT particles nucleate, grow, and sediment, are still not settled. These open issues somehow limit our capacity to properly simulate PSC in global models (Kirner et al. 2011; Smith et al. 2014) and affect our prognostic abilities concerning future ozone recovery in an evolving atmosphere. This is of specific concern in the Arctic, where winter temperatures float close to the PSC formation thresholds and where the expected future stratospheric cooling and changes in the amount of nitric acid, water vapour and sulphates in response to a changing climate, could lead to variation in PSC extension and lifetimes, impacting ozone processes (Butchart et al. 2010; Langematz et al. 2014; Schmidt et al. 2013; Rieder and Polvani 2013, Pommereau et al. 2013).

References

Achtert, P., Karlsson Andersson, M., Khosrawi, F., & Gumbel, J. (2012). On the linkage between tropospheric and polar stratospheric clouds in the Arctic as observed by space–borne lidar. *Atmospheric Chemistry and Physics, 12*, 3791–3798. https://doi.org/10.5194/acp-12-3791-2012.

Adriani, A., Deshler, T., Di Donfrancesco, G., & Gobbi, G. P. (1995). Polar stratospheric clouds and volcanic aerosol during spring 1992 over McMurdo Station, Antartica: Lidar and particle counter comparisons. *Journal of Geophysical Research, 100*, 25877–25897.

Adriani, A., Cairo, F., Pulvirenti, L., Cardillo, F., Viterbini, M., Di Donfrancesco, G., & Pommereau, J. P. (1999). Stratospheric background aerosol and polar cloud observations by laser backscattersonde within the framework of the European project "Stratospheric Regular Sounding". *Annales Geophysicae, 17*(10), 1352–1360.

Allen, S. J., & Vincent, R. A. (1995). Gravity-wave activity in the lower atmosphere: Seasonal and latitudinal variations. *Journal of Geophysical Research, 100*, 1327–1350.

Allen, D. R., Standford, J. L., Lope'z-Valverde, M. A., Nakamura, N., Lary, D. J., Douglass, A. R., Cerniglia, M. C., Remedios, J. J., & Taylor, F. W. (1999). Observations of middle atmosphere CO from the UARS ISAMS during the early northern winter 1991/1992. *Journal of the Atmospheric Sciences, 56*, 563–583.

Allen, D. R., Stanford, J. L., Nakamura, N., Lo'pez-Valverde, M. A., Lo'pez-Puertas, M., Taylor, F. W., & Remedios, J. J. (2000). Antarctic polar descent and planetary wave activity observed in ISAMS CO from April to July 1992. *Geophysical Research Letters, 27*, 665–668.

Anderson, J. G., & Toon, O. B. (1993). Airborne Arctic stratospheric expedition II: An overview. *Geophysical Research Letters, 20*(22), 2499–2502.

Anderson, J. G., Margitan, J. J., & Stedman, D. H. (1977). Atomic chlorine and the chlorine monoxide radical in the stratosphere: Three in-situ measurements. *Science, 198*, 501–503.

Anderson, J. G., Brune, W. H., & Proffitt, M. H. (1989). Ozone destruction by chlorine radicals within the Antarctic vortex: The spatial and temporal evolution of ClO-O3 anticorrelation based on in situ ER-2 data. *Journal of Geophysical Research, 94*, 11,465–11,479.

Andrews, D. G. (1989). Some comparisons between the middle atmosphere dynamics of the southern and northern hemispheres, Southern and Northern Hemispheres. *Journal Pure and Applied Geophysics, 130*, 213–232.

Andrews, D. G., Holton, J. R., & Leovy, C. B. (1987). *Middle atmosphere dynamics*. San Diego: Academic.

Anthony, S. E., Tisdale, R. T., Disselkamp, R. S., Tolbert, M. A., & Wilson, J. C. (1995). FTIR studies of low temperature sulphuric-acid aerosols. *Geophysical Research Letters, 22*, 1105–1108.

Appenzeller, C., & Holton, J. R. (1997). Tracer filamentation in the stratosphere – a global climatology. *Journal of Geophysical Research, 102*, 13555–13569.

Austin, J., Jones, R. L., Mckenna, D. S., Buckland, A. T., Anderson, J., Fahey, D. W., Tuck, A., Farmer, B. C., Heidt, L. E., Proffitt, M., & Vedder, F. J. (1989). Lagrangian photochemical modeling studies of the 1987 Antarctic spring vortex. 2. Seasonal trends in ozone. *Journal of Geophysical Research, 94*, 16717–16735. https://doi.org/10.1029/JD094iD14p16717.

Bacmeister, J. T., Schoeberl, M. R., Lait, L. R., Newman, P. A., & Gary, B. (1990). Small-scale waves encountered during AASE. *Geophysical Research Letters, 17*, 349–352.

Bates, D. R., & Nicolet, M. (1950). Atmospheric nitrogen. *Publications of the Astronomical Society of the Pacific, 62*, 106–110.

Bevilacqua, R. M., Fromm, M. D., Alfred, J. M., Hornstein, J. S., Nedoluha, G. E., Hoppel, K. W., Lumpe, D., Randall, C. E., Shettle, E. P., Browell, E. V., Butler, C., Dornbrack, A., & Strawa, A. W. (2002). Observations and analysis of polar stratospheric clouds detected by POAM III during the 1999/2000 Northern hemisphere winter. *Journal of Geophysical Research: Atmospheres, 107*(D20).

Beyer, K. D., Seago, S. W., Chang, H. Y., & Molina, M. J. (1994). Composition and freezing of aqueous H2SO4/HNO3 solutions under polar stratospheric conditions. *Geophysical Research Letters, 21*, 871–874.

Beyerle, G., Neuber, R., Schrems, O., Wittrock, F., & Knudsen, B. (1994). Multiwavelenght lidar measurements of stratospheric aerosols above Spitsbergen during winter 1992/93. *Geophysical Research Letters, 21*, 57–60.

Beyerle, G., Deckelmann, H., Neuber, R., Rosen, J. M., Reimer, E., & Schoeberl, M. R. (2001). Occurrence of solid particles in the winter polar stratosphere above the nitric acid trihydrate coexistence temperature inferred from ground-based polarization lidar observations at Ny Ålesund, Spitsbergen. *Journal of Geophysical Research: Atmospheres, 106*(D3), 2979–2992.

Biele, J., Tsias, A., Luo, B. P., Carslaw, K. S., Neuber, R., Beyerle, G., & Peter, T. (2001). Nonequilibrium coexistence of solid and liquid particles in Arctic stratospheric clouds. *Journal of Geophysical Research: Atmospheres, 106*(D19), 22991–23007.

Biermann, U. M., Presper, T., Koop, T., Möβinger, J., Crutzen, P. J., & Peter, T. (1996). The unusuitability of meteoritic and other nuclei for polar stratospheric cloud freezing. *Geophysical Research Letters, 23*, 1693–1696.

Blake, D. F., & Kato, K. (1995). Latitudinal distribution of black carbon soot in the upper troposphere and lower stratosphere. *Journal of Geophysical Research, 100*, 7195–7202.

Blum, U., Fricke, K. H., Müller, K. P., Siebert, J., & Baumgarten, G. (2005). Long-term lidar observations of polar stratospheric clouds at Esrange in northern Sweden. *Tellus B: Chemical and Physical Meteorology, 57*(5), 412–422.

Bluth, G. J., Doiron, S. D., Schnetzler, C. C., Krueger, A. J., & Walter, L. S. (1992). Global tracking of the SO2 clouds from the June, 1991 Mount Pinatubo eruptions. *Geophysical Research Letters, 19*(2), 151–154.

Bowman, K. P. (1993). Barotropic simulation of large-scale mixing in the Antarctic polar vortex. *Journal of the Atmospheric Sciences, 50*, 2901–2914.

Bowman, K. P., & Chen, P. (1994). Mixing by barotropic instability in a nonlinear model. *Journal of the Atmospheric Sciences, 51*, 3692–3705.

Brewer, A. W. (1949). Evidence for a world circulation provided by measurements of the helium and water vapor distribution in the stratosphere. *Journal of the Atmospheric Sciences, 51*, 3006–3018.

Browell, E. V., Butler, C. F., Ismail, S., Robinette, P. A., Carter, A. F., Higdon, N. S., Toon, O. B., Schoeberl, M. R., & Tuck, A. F. (1990). Airborne lidar observations in the wintertime Arctic stratosphere: Polar stratospheric clouds. *Geophysical Research Letters, 17*, 385–388.

Butchart, N., & Remsberg, E. E. (1986). The area of the stratospheric polar vortex as a diagnostic for tracer transport on an isentropic surface. *Journal of the Atmospheric Sciences, 43*, 1319–1339.

Butchart, N., Cionni, I., Eyring, V., Waugh, D. W., Akiyoshi, H., Austin, J., Brühl, C., Chipperfield, M. P., Cordero, E., Dameris, M., Deckert, R., Frith, S. M., Garcia, R. R., Gettelman, A., Giorgetta, M. A., Kinnison, D. E., Li, F., Mancini, E., Manzini, E., McLandress, C., Pawson, S., Pitari, G., Rozanov, E., Sassi, F., Shepherd, T. G., Shibata, K., & Tian, W. (2010). Chemistry-climate model simulations of 21st century stratospheric climate and circulations changes. *Journal of Climate, 23*, 5349–5374. https://doi.org/10.1175/2010JCLI3404.1.

Carslaw, K. S., Luo, B. P., Clegg, S. L., Peter, T., Brimblecombe, P., & Critzen, P. J. (1994). Stratospheric aerosol growth and HNO3 gas phase depletion from coupled HNO3 and water uptake by liquid particles. *Geophysical Research Letters, 21*, 2479–2482.

Carslaw, K. S., Wirth, M., Tsias, A., Luo, B. P., Dörnbrack, A., Leutbecher, H., Volkert, W. R., Bacmeister, J. T., & Peter, T. (1998). Particle microphysics and chemistry in remotely observed mountain polar stratospheric clouds. *Journal of Geophysical Research: Atmospheres, 103*(D5), 5785–5796.

Carslaw, K. S., Peter, T., Bacmeister, J. T., & Eckermann, S. D. (1999). Widespread solid particle formation by mountain waves in the Arctic stratosphere. *Journal of Geophysical Research, 104*(D1), 1827–1836.

Carslaw, K. S., Kettleborough, J. A., Northway, M. J., Davies, S., Gao, R., Fahey, D. W., Baumgardner, D. G., Chipperfield, M. B., & Kleinbohl, A. (2002). A vortex-scale simulation of the growth and sedimentation of large nitric acid hydrate particles. *Journal of Geophysical Research, 107*, D20, 8300. https://doi.org/10.1029/2001JD000467.

Chanin, M. L., & Hauchecorne, A. (1981). Lidar observation of gravity and tidal waves in the stratosphere and mesosphere. *Journal of Geophysical Research, 86*, 9715–9721.

Chapman, S. (1930). A theory of upper atmospheric ozone. *Memoirs of the Royal Meteorological Society, 3*, 103–125.

Charney, J. G., & Drazin, P. D. (1961). Propagation of planetary disturbances from the lower into the upper atmosphere. *Journal of Geophysical Research, 66*, 83–109.

Chen, P. (1994). The permeability of the Antarctic vortex edge. *Journal of Geophysical Research, 99*(D10), 20563–20572.

Chen, P., Holton, J. R., O'Neill, A., & Swinbank, R. (1994). Quasi-horizontal transport and mixing in the Antarctic stratosphere. *Journal of Geophysical Research, 99*(D8), 16851–16866.

Chippereld, M. P., & Jones, R. L. (1999). Relative influence of atmospheric chemistry and transport on Arctic ozone trends. *Nature, 400*, 551–554.

Coy, L., Nash, E., & Newman, P. (1997). Meteorology of the polar vortex: Spring 1997. *Geophysical Research Letters, 24*(22), 2693–2696.

Crutzen, P. J. (1970). The influence of nitrogen oxyde on the atmospheric ozone content. *Quarterly Journal of the Royal Meteorological Society, 96*, 320–325.

Crutzen, P. J., & Arnold, F. (1986). Nitric acid cloud formation in the cold Antarc-tic stratosphere: A major cause for the springtime "ozone hole". *Nature, 324*, 651–655.

Curtius, J., Weigel, R., Vössing, H.-J., Wernli, H., Werner, A., Volk, C.-M., Konopka, P., Krebsbach, M., Schiller, C., Roiger, A., Schlager, H., Dreiling, V., & Borrmann, S. (2005). Observations of meteoritic material and implications for aerosol nucleation in the winter Arctic lower strato15 sphere derived from in situ particle measurements. *Atmospheric Chemistry and Physics, 5*, 3053–3069.

Cziczo, D. J., Thomson, D. S., & Murphy, D. M. (2001). Ablation, flux, and atmospheric implications of meteors inferred from stratospheric aerosol. *Science, 291*, 1772–1775.

Deshler, T. (2008). A review of global stratospheric aerosol: Measurements, importance, life cycle, and local stratospheric aerosol. *Atmospheric Research, 90*, 223–232.
Deshler, T., & Oltmans, S. J. (1998). Vertical profiles of volcanic aerosol and polar stratospheric clouds above Kiruna, Sweden: Winters 1993 and 1995. *Journal of Atmospheric Chemistry, 30*, 11–23.
Deshler, T., Adriani, A., Hofmann, D. J., & Gobbi, G. P. (1991). Evidence for denitrification in the 1990 Antarctic spring stratosphere: II. Lidar and aerosol measurements. *Geophysical Research Letters, 18*, 1999–2002.
Deshler, T., Johnson, B. J., & Rozier, W. R. (1994). Changes in the character of the polar stratospheric clouds over the Antarctica in 1992 due to the Pinatubo volcanic aerosol. *Geophysical Research Letters, 21*, 273–276.
Deshler, T., Nardi, B., Adriani, A., Cairo, F., Hansen, G., Fierli, F., & Pulvirenti, A. H. L. (2000). Determining the index of refraction of polar stratospheric clouds above Andoya (69° N) by combining size-resolved concentration and optical scattering measurements. *Journal of Geophysical Research: Atmospheres, 105*(D3), 3943–3953.
Deshler, T., Larsen, N., Weissner, C., Schreiner, J., Mauersberger, K., Cairo, F. A. A., Donfrancesco, G. D., Ovarlez, J., Ovarlez, H., Blum, U., Fricke, K. H., & Dörnbrack, A. (2003). Large nitric acid particles at the top of an Arctic stratospheric cloud. *Journal of Geophysical Research: Atmospheres, 108*(D16).
Deshler, T., Anderson-Sprecher, R., Jäger, H., Barnes, J., Hofmann, D. J., Clemesha, B., Simonich, D., Osborn, M., Grainger, R. G., & Godin-Beekmann, S. (2006). Trends in the nonvolcanic component of stratospheric aerosol over the period 1971-2004. *Journal of Geophysical Research, 111*, D01201. https://doi.org/10.1029/2005JD006089.
Dessler, A. (2000). *The chemistry and physics of stratospheric ozone*. Academic, 214pp.
Di Sarra, A., Cacciani, M., Fiocco, G., Fua, D., & Jørgensen, T. S. (2002). Lidar observations of polar stratospheric clouds over northern Greenland in the period 1990–1997. *Journal of Geophysical Research: Atmospheres, 107*(D12).
Dobson, G. M. B. (1956). Origin and distribution of the polyatomic molecules in the atmosphere. *Proceedings of the Royal Society of London. Series A, Mathematical and Physical Sciences, A236*, 187–193.
Dörnbrack, A., Leutbecher, M., Reichardt, J., Behrendt, A., Muller, K.-P., & Baumgarten, G. (2001). Relevance of mountain wave cooling for the formation of polar stratospheric clouds over 25 Scandinavia: Mesoscale dynamics and observations for January 1997. *Journal of Geophysical Research, 106*, 1569–1581.
Dornbrack, A., Pitts, M. C., Poole, L. R., Orsolini, Y. J., Nishii, K., & Nakamura, H. (2012). The 2009–2010 Arctic stratospheric winter – General evolution, mountain waves and predictability of an operational weather forecast model. *Atmospheric Chemistry and Physics, 12*, 3659–3675.
Drdla, K. (2005). Temperature thresholds for polar stratospheric ozone, Eos Trans. AGU, 86, Fall Meet. Suppl., Abstract A31D-03.
Drdla, K. (2006). Re-examining polar ozone loss: Do polar stratospheric clouds really control chlorine activation? *Geophysical Research*. Abstracts, 8, 09979, sRef-ID: 1607–7962/gra/EGU06-A-09979.
Drdla, K., & Browell, E. V. (2004). Microphysical modeling of the 1999-2000 Arctic winter: 3. Impact of homogeneous freezing on polar stratospheric clouds. *Journal of Geophysical Research, 109*, D10201. https://doi.org/10.1029/2003-JD004352.
Drdla, K., Schoeberl, M. R., & Browell, E. V. (2002). Microphysical modeling of the 1999–2000 Arctic winter: 1. Polar stratospheric clouds, denitrification, and dehydration. *Journal of Geophysical Research: Atmospheres, 107*(D5).
Duck, T. J., Whiteway, J. I., & Carswell, A. (1998). Lidar observations of gravity wave activity and Arctic stratospheric vortex core warming. *Geophysical Research Letters, 25*, 2813–2816. https://doi.org/10.1029/98GL02113.
Dye, J. E., Baumgardner, D., Gandrud, B. W., Kawa, S. R., Kelly, K. K., Loewenstein, M., Ferry, G. V., Chan, K. R., & Gary, B. L. (1992). Particle size distribution in Arctic polar stratospheric

clouds, growth and freezing of sulphuric acid droplets, and implications for cloud formation. *Journal of Geophysical Research, 97*, 8015–8034.

Dye, J. E., Baumgardner, D., Gandrud, B. W., Drdla, K., Barr, K., Fahey, D. W., Del Negro, L. A., Tabazadeh, A., Jonsson, H. H., Wilson, J. C., Loewenstein, M., Podolske, J. R., & Chan, K. R. (1996). In-situ observations of an Antarctic polar stratospheric cloud: Similarities with Arctic observations. *Geophysical Research Letters, 23*, 1913–1916.

Ebert, M., Weigel, R., Kandler, K., Günther, G., Molleker, S., Grooß, J.-U., Vogel, B., Weinbruch, S., & Borrmann, S. (2016). Chemical analysis of refractory stratospheric aerosol particles collected within the arctic vortex and inside polar stratospheric clouds. *Atmospheric Chemistry and Physics, 16*, 8405–8421. https://doi.org/10.5194/acp-16-8405-2016.

Eckermann, S. D. (1995). Effect of background winds on vertical wavenumber spectra of atmospheric gravity waves. *Journal of Geophysical Research, 100*, 14,097–14,112.

Engel, I., Luo, B. P., Pitts, M. C., Poole, L. R., Hoyle, C. R., Grooß, J.-U., Dörnbrack, A., & Peter, T. (2013). Heterogeneous formation of polar stratospheric clouds – Part 2: Nucleation of ice on synoptic scales. *Atmospheric Chemistry and Physics, 13*. https://doi.org/10.5194/acp-13-10769-2013.

Engel, I., Luo, B. P., Khaykin, S. M., Wienhold, F. G., Vömel, H., Kivi, R., Hoyle, C. R., Grooß, J.-U., Pitts, M. C., & Peter, T. (2014). Arctic stratospheric dehydration – Part 2: Microphysical modeling. *Atmospheric Chemistry and Physics, 14*, 3231–3246. https://doi.org/10.5194/acp-14-3231-2014.

Fahey, D. W. (1989). Measurements of nitric oxide and total reactive nitrogen in the Antarctic stratosphere: Observations and chemical implications. *Journal of Geophysical Research, 94*, 16665–16681.

Fahey, D. W., Kelly, K. K., Ferry, V. G., Loewenstein, M., Chan, R. K., Poole, L., & Wilson, J. C. (1989a). In situ measurements of total reactive nitrogen, total water and aerosol in a polar stratospheric cloud in the Arctic. *Journal of Geophysical Research, 94*, 11299–11315.

Fahey, D. W., Kelly, K. K., Ferry, G. V., Poole, L. R., Wilson, J. C., Murphy, D. M., Loewenstein, M., & Chan, K. R. (1989b). In situ measurements of total reactive nitrogen, total water and aerosol in a polar stratospheric cloud in the Antarctic. *Journal of Geophysical Research, 94*, 11,299–11,315.

Fahey, D. W., Gao, R. S., Carslaw, K. S., Kettleborough, J., Popp, P. J., Northway, M. J., Holecek, J. C., Ciciora, S. C., McLaughlin, R. J., Thompson, T. L., Winkler, R. H., Baumgardner, D. G., Gandrud, B., Wennberg, P. O., Dhaniyala, S., McKinney, K., Peter, T., Salawitch, R. J., Bui, T. P., Elkins, J. W., Webster, C. R., Atlas, E. L., Jost, H., Wilson, J. C., Herman, R. L., Kleinböhl, A., & von König, M. (2001). The detection of large HNO3-containing particles in the winter Arctic stratosphere. *Science, 291*(5506), 1026–1031.

Farman, J. C., Gardiner, B. G., & Shanklin, J. D. (1985). Large losses of ozone in Antarctica reveal seasonal ClOX /NOX interaction. *Nature, 315*, 207–210.

Ferry, G. V., Neish, E., Schultz, M., & Pueschel, R. F. (1989). Concentrations and size distributions of Antarctic stratospheric aerosols. *Journal of Geophysical Research, 94*, 16459–16474.

Fiocco, G., & Grams, G. W. (1964). Observations of the aerosol layer at 20 km by optical radar. *Journal of the Atmospheric Sciences, 21*, 323–324.

Fisher, M., O'Neill, A., & Sutton, R. (1993). Rapid descent of mesospheric air in the stratospheric polar vortex. *Geophysical Research Letters, 20*, 1267–1270.

Fromm, M. D., Bevilacqua, R. M., Hornstein, J., Shettle, E., Hoppel, K., & Lumpe, J. D. (1999). An analysis of polar ozone and aerosol measurement (POAM) II Arctic polar stratospheric cloud observations, 1993–1996. *Journal of Geophysical Research: Atmospheres, 104*(D20), 24341–24357.

Fromm, M., Alfred, J., & Pitts, M. (2003). A unified, long-term, high-latitude stratospheric aerosol and cloud database using SAMII, SAGE II, and POAM II/III data: Algorithm description, database definition, and climatology. *Journal of Geophysical Research, 108*(D12), 4366. https://doi.org/10.1029/2002JD002772.

Fueglistaler, S., Luo, B. P., Buss, S., Wernli, H., Voigt, C., Müller, M., Neuber, R., Hostetler, C. A., Poole, L. R., Flentje, H., Fahey, D. W., Northway, M. J., & Peter, T. (2002a). Large

NAT particle formation by mother clouds: Analysis of SOLVE/THESEO-2000 observations. *Geophysical Research Letters, 29*(12), 52–51.

Fueglistaler, S., Luo, B. P., Voigt, C., Carslaw, K. S., & Peter, T. (2002b). NAT-rock formation by mother clouds: A microphysical model study. *Atmospheric Chemistry and Physics, 2*(2), 93–98.

Gary, B. L. (1989). Observational results using the microwave temperature profiler during the airborne Antarctic ozone experiment. *Journal of Geophysical Research, 94*, 11223–11231.

Godin, S., Mégie, G., David, C., Haner, D., Flésia, C., & Emery, Y. (1994). Airborne lidar observation of mountain-wave-induced polar stratospheric clouds during EASOE. *Geophysical Research Letters, 21*(13), 1335–1338.

Goodman, J., Toon, O. B., Pueschel, R. F., & Snetsinger, K. G. (1989). Antarctic stratospheric ice crystals. *Journal of Geophysical Research, 94*, 16449–16457.

Goodman, J., Verma, S., Pueschel, R. F., Hamill, P., Ferry, G. V., & Webster, D. (1997). New evidence of size and composition of polar stratospheric cloud particles. *Geophysical Research Letters, 24.5*(615–18), 1997.

Gossard, E. E. (1962). Vertical flux of energy into the lower ionosphere from internal gravity waves generated in the troposphere. *Journal of Geophysical Research, 67*, 745.

Hadjinicolaou, P., Jrrar, A., Pyle, J. A., & Bishop, L. (2002). The dynamically driven long-term trend in stratospheric ozone over northern middle latitudes. *Quarterly Journal of the Royal Meteorological Society, 128*(583), 1393–1412.

Hamill, P., Turco, R. P., Kiang, C. S., Toon, O. B., & Whitten, R. C. (1982). An analysis of various nucleation mechanism for sulfate particles in the stratosphere. *Journal of Aerosol Science, 13*, 561–585.

Hamill, P., Turco, R. P., & Toon, O. B. (1988). On the growth of nitric and Sulfuric acid aerosol particles under stratospheric conditions. *Journal of Atmospheric Chemistry, 7*, 287–315.

Hamill, P., Jensen, E. J., Russell, P. B., & Mauman, J. J. (1997). The life cycle of stratospheric aerosol properties. *Bulletin of the American Meteorological Society, 78*, 1–6.

Hanson, D., & Mauersberger, K. (1988). Laboratory studies of the nitric acid trihydrate: Implications for the south polar stratosphere. *Geophysical Research Letters, 15*(8), 855–858. https://doi.org/10.1029/GL015i008p00855.

Hansen, J., Lacis, A., Ruedy, R., & Mki, S. (1992). Potential climate impact of Mount Pinatubo eruption. *Geophysical Research Letters, 19*, 215–218. https://doi.org/10.1029/91GL02788.

Hartmann, D. L., Heidt, L. E., Loewenstein, M., Podolske, J. R., Vedder, J., Starr, W. L., & Strahan, S. E. (1989). Transport into the south polar vortex in early spring. *Journal of Geophysical Research, 94*, 16,779–16,795.

Heidt, L. E., Vedder, J. F., Pollock, W. H., Lueb, R. A., & Henry, B. E. (1989). Trace gases in the Antarctic atmosphere. *Journal of Geophysical Research, 94*(D9), 11,599–11,611.

Hess, P. G. (1991). Mixing processes following the final stratospheric warming. *Journal of the Atmospheric Sciences, 48*. https://doi.org/10.1175/1520-0469(1991)048<1625:MPFTFS>2.0.CO;2.

Hines, C. O. (1960). Internal atmospheric gravity waves at ionospheric heights. *Canadian Journal of Physics, 38*, 1441.

Hoff, R. M. (1992). Differential SO2 column measurements of the Mt. Pinatubo volcanic plume. *Geophysical Research Letters, 19*(2), 175–178.

Hofmann, D. J. (1990). Increase in the stratospheric background sulphuric acid aerosol mass in the past 10 years. *Science, 248*, 996–1000.

Hofmann, D. J., & Deshler, T. (1991). Stratospheric cloud observation during formation of the Anctarctic ozone hole in 1989. *Journal of Geophysical Research, 96*(D), 2897–2912.

Hofmann, D. J., Rosen, J. M., Harder, J. W., & Hereford, J. V. (1989). Balloon-borne measurements of aerosol, condensation nuclei, and cloud particles in the stratosphere at McMurdo Station, Antarctica, during the spring of 1987. *Journal of Geophysical Research, 94*(D9), 11253–11269. https://doi.org/10.1029/JD094iD09p11253.

Holton, J. R. (1992). *An introduction to dynamic meteorology* (4th ed., p. 535). San Diego: Academic Press.

Holton, J. R., Haynes, P. H., McIntyre, M. E., Douglass, A. R., Rood, R. B., & Pfister, L. (1995). Stratosphere-troposphere exchange. *Reviews of Geophysics, 33*, 403–439.

Hoyle, C. R., Engel, I., Luo, B. P., Pitts, M. C., Poole, L. R., Grooß, J.-U., & Peter, T. (2013). Heterogeneous formation of polar stratospheric clouds – Part 1: Nucleation of nitric acid trihydrate (NAT). *Atmospheric Chemistry and Physics, 13*, 10769–10785. https://doi.org/10.5194/acp-13-10769-2013.

Hu, R.-M., Carslaw, K. S., Hostetler, C., Poole, L. R., Luo, B., Peter, T., Fueglistaler, S., McGee, T. J., & Burris, J. F. (2002). Microphysical properties of wave polar stratospheric clouds retrieved from lidar measurements during SOLVE/THESEO 2000. *Journal of Geophysical Research, 107*(D20), 8294. https://doi.org/10.1029/2001JD001125.

Iannarelli, R., & Rossi, M. J. (2015). The mid-IR Absorption Cross Sections of α- and β-NAT ($HNO_3 \cdot 3H_2O$) in the range 170 to 185 K and of metastable NAD ($HNO_3 \cdot 2H_2O$) in the range 172 to 182 K. *Journal of Geophysical Research: Atmospheres, 120*, 11707–11727. https://doi.org/10.1002/2015JD023903.

Irie, H., Pagan, K. L., Tabazadeh, A., Legg, M. J., & Sugita, T. (2004). Investigation of polar stratospheric cloud solid particle formation mechanisms using ILAS and AVHRR observations in the Arctic. *Geophysical Research Letters, 31*(15).

Iwasaka, Y., Hirasawa, T., & Fukunishi, H. (1985). Lidar measurements on the Antarctic stratospheric aerosol layer: I, winter enhancement. *Journal of Geomagnetism and Geoelectricity, 37*, 1087–1095.

Jenkins, D. B., Wareing, D. P., Thomas, L., & Vaughan, G. (1987). Upper stratospheric and mesospheric temperatures derived from lidar observations at Aberystwyth. *Journal of Atmospheric and Terrestrial Physics, 49*, 287–298.

Jones, R. L., Austin, J., McKenna, D. S., Anderson, J. G., Fahey, D. W., Farmer, C. B., Heidt, L. E., Kelly, K. K., Murphy, D. M., Proffitt, M. H., Tuck, A. F., & Vedder, J. F. (1989). Lagrangian photochemical modeling studies of the 1987 Antarctic spring vortex, 1, comparison with AAOE observations. *Journal of Geophysical Research, 94*, 11,529–11,558.

Kawa, S. R., Newman, P. A., Lait, L. R., Schoeberl, M. R., Stimpfle, R. M., Kohn, D. W., Webster, C. R., May, R. D., Baumgardner, D., Dye, J. E., Wilson, J. C., Chan, K. R., & Loewenstein, M. (1997). Activation of chlorine in sulfate aerosol as inferred from aircraft observations. *Journal of Geophysical Research: Atmospheres, 102*(D3), 3921–3933.

Khaykin, S. M., Engel, I., Vömel, H., Formanyuk, I. M., Kivi, R., Korshunov, L. I., Krämer, M., Lykov, A. D., Meier, S., Naebert, T., Pitts, M. C., Santee, M. L., Spelten, N., Wienhold, F. G., Yushkov, V. A., & Peter, T. (2013). Arctic stratospheric dehydration – Part 1: Unprecedented observation of vertical redistribution of water. *Atmospheric Chemistry and Physics, 13*, 11503–11517. https://doi.org/10.5194/acp-13-11503-2013.

Khosrawi, F., Urban, J., Pitts, M. C., Voelger, P., Achtert, P., Kaphlanov, M., Santee, M. L., Manney, G. L., Murtagh, D., & Fricke, K.-H. (2011). Denitrification and polar stratospheric cloud formation during the Arctic winter 2009/2010. *Atmospheric Chemistry and Physics, 11*, 8471–8487. https://doi.org/10.5194/acp-11-8471-2011.

Kiehl, J., & Solomon, S. (1986). On the radiative balance of the stratosphere. *Journal of the Atmospheric Sciences, 43*, 1525–1534.

Kirner, O., Ruhnke, R., Buchholz-Dietsch, J., Jöckel, P., Brühl, C., & Steil, B. (2011). Simulation of polar stratospheric clouds in the chemistry-climate-model EMAC via the submodel PSC. *Geoscientific Model Development, 4*, 169–182. https://doi.org/10.5194/gmd-4-169-2011.

Koop, T., Biermann, U. M., Raber, W., Luo, B. P., Crutzen, P. J., & Peter, T. (1995). Do stratospheric aerosol droplets freeze above the ice frost point ? *Geophysical Research Letters, 22*, 917–920.

Koop, T., Luo, B. P., Biermann, U., Crutzen, P. J., & Peter, T. (1997). Freezing of HNO3/H2SO4/H2O solutions at stratospheric temperatures: Nucleation statistics and experiments. *The Journal of Physical Chemistry. A, 101*, 1117–1133.

Kovilakam, M., & Deshler, T. (2015). On the accuracy of stratospheric aerosol extinction derived from in situ size distribution measurements and surface area density derived from remote SAGE

II and HALOE extinction measurements. *Journal of Geophysical Research – Atmospheres, 120.* https://doi.org/10.1002/2015JD023303.

Kremser, S., Thomason, W. L., Von Hobe, M., Hermann, M., Deshler, T., Timmreck, C., Toohey, M., Stenke, A., Schwarz, J, Weigel, R., Fueglistaler, S., Prata, F., Vernier, J, Schlager, H. , Barnes, J., Antuña-Marrero, J. C., Fairlie, T., Palm, M., Mahieu, E., & Meland, B. (2016). Stratospheric aerosol – Observations, processes, and impact on climate. *Reviews of Geophysics, 54*: n/a–n/a. https://doi.org/10.1002/2015RG000511.

Kuroda, Y., & Kodera, K. (1998). Interannual variability in the troposphere and stratosphere of the southern hemisphere winter. *Journal of Geophysical Research, 103*(D12), 13787–13799.

Langematz, U., Meul, S., Grunow, K., Romanowsky, E., Oberländer, S., Abalichin, J., & Kubin, A. (2014). Future Arctic temperature and ozone: The role of stratospheric composition changes. *Journal of Geophysical Research: Atmospheres, 119*(5), 2092–2112.

Larsen, N., Knudsen, B., Rosen, J. M., Kjome, N. T., & Kyrö, E. (1996). Balloonborne backscatter observations of type 1 PSC formation: Inference about physical state from trajectory analysis. *Geophysical Research Letters, 23*, 1091–1994.

Larsen, N., Knudsen, B. M., Rosen, J. M., Kjome, N. T., Neuber, R., & Kyrö, E. (1997). Temperature histories in liquid and solid polar stratospheric cloud formation. *Journal of Geophysical Research, 102*, 23505–23517.

Larsen, N., Svendsen, S. H., Knudsen, B. M., Voigt, C., Weisser, C., Kohlmann, A., Schreiner, J., Mauersberger, K., Deshler, T., Kroger, C., Rosen, J. M., Kjome, N. T., Adriani, A., Cairo, F., Di Donfrancesco, G., Ovarlez, J., Ovarlez, H., Dornbrack, A., & Birner, T. (2002). Microphysical mesoscale simulations of polar stratospheric cloud formation constrained by in situ measurements of chemical and optical cloud properties. *Journal of Geophysical Research: Atmospheres, 107*(D20). https://doi.org/10.1029/2001JD000999.

Larsen, N., Knudsen, B. M., Svendsen, S. H., Deshler, T., Rosen, J. M., Kivi, R., Weisser, C., Schreiner, J., Mauerberger, K., Cairo, F., Ovarlez, J., Oelhaf, H., & Spang, R. (2004). Formation of solid particles in synoptic-scale Arctic PSCs in early winter 2002/2003. *Atmospheric Chemistry and Physics, 4*(7), 2001–2013.

Lee, A. M., Roscoe, H. K., Jones, A. E., Haynes, P. H., Shuckburgh, E. F., Morrey, M. W., & Pumphrey, H. C. (2001). The impact of the mixing properties within the Antarctic stratospheric vortex on ozone loss in spring. *Journal of Geophysical Research, 106*, 3202–3211.

Loewenstein, M., Podolske, J. R., Fahey, D. W., Woodbridge, E. L., Tin, P., Weaver, A., Newman, P. A., Strahan, S. E., Kawa, S. R., Schoeberl, M. R., & Lait, L. R. (1993). New observations of the NOy/N2O correlation in the lower stratosphere. *Geophysical Research Letters, 20*(22), 2531–2534.

Lowe, D., & Mackenzie, A. (2008). Polar stratospheric cloud microphysics and chemistry. *Journal of Atmospheric and Solar - Terrestrial Physics, 70*, 13–40. https://doi.org/10.1016/j.jastp.2007.09.011.

Lowe, D., MacKenzie, A. R., Schlager, H., Voigt, C., Dörnbrack, A., Mahoney, M. J., & Cairo, F. (2006). Liquid particle composition and heterogeneous reactions in a mountain wave polar stratospheric cloud. *Atmospheric Chemistry and Physics, 6*(11), 3611–3623.

Luo, B. P., Peter, T., & Crutzen, P. J. (1994). Freezing of stratospheric aerosol droplets. *Geophysical Research Letters, 21*, 1447–1450.

Luo, B. P., Carslaw, K. S., Peter, T., & Clegg, S. L. (1995). Vapor-pressures of H2SO4/HNO3/HCl/HBr/H2O solutions to low stratospheric temperatures. *Geophysical Research Letters, 22*, 247–250.

Luo, B. P., Voigt, C., Fueglistaler, S., & Peter, T. (2003). Extreme NAT supersaturations in mountain wave ice PSCs: A clue to NAT formation. *Journal of Geophysical Research, 108*(D15), 4441. https://doi.org/10.1029/2002JD003104.

MacKenzie, A. R., Kulmala, M., Laaksonen, A., & Vesala, T. (1995). On the theories of type 1 polar stratospheric cloud formation. *Journal of Geophysical Research, 100*, 11275–11288.

Manney, G. L., & Sabutis, J. L. (2000). Development of the polar vortex in the 1999–2000 Arctic winter stratosphere. *Geophysical Research Letters, 27*(17), 2589–2592.

Manney, G. L., & Zurek, R. W. (1993). Interhemispheric comparison of the development of the stratospheric polar vortex during fall: A 3-din~ensional perspective for 1991-1992. *Geophysical Research Letters, 20*, 1275–1278.

Manney, G. L., Zurek, R. W., O'Neill, A., Swinbank, R., Kumer, J. B., Mergenthaler, J. L., & Roche, A. E. (1994a). Stratospheric Warmings during February and March 1993. *Geophysical Research Letters, 21*, 813–816, 1994.

Manney, G. L., Farrara, J. D., & Mechoso, C. R. (1994b). Simulations of the February 1979 stratospheric sudden warming: Model comparisons and three-dimensional evolution. *Monthly Weather Review, 122*, 1115–1140.

Manney, G. L., Santee, M. L., Froidevaux, L., Waters, J. W., & Zurek, R. W. (1996). Polar vortex conditions during the 1995- 96 Arctic winter: Meteorology and MLS ozone. *Geophysical Research Letters, 23*, 3203–3206.

Manney, G. L., Froidevaux, L., Santee, M. L., Zurek, R. W., & Waters, J. W. (1997). MLS observations of Arctic ozone loss in 1996-97. *Geophysical Research Letters, 24*, 2697–2700.

Manney, G. L., Bird, J. C., Donovan, D. P., Duck, T. J., Whiteway, J. A., Pal, S. R., & Carswell, A. I. (1998). Modeling ozone laminae in ground-based Arctic wintertime observations using trajectory calculations and satellite data. *Journal of Geophysical Research, 103*(d5), 5797–5814.

Manney, G. L., Sabutis, J. L., & Swinbank, R. (2001a). A unique stratospheric warming event in November 2000. *Geophysical Research Letters, 28*(13). https://doi.org/10.1029/2001GL012973.

Manney, G. L., Michelsen, H. A., Bevilacqua, R. M., Gunson, M. R., Irion, F. W., Livesey, N. J., Oberheide, J., Riese, M., Russell, J. M., Toon, G. C., & Zawodny, J. (2001b). Comparison of satellite ozone observations in coincident air masses in early November 1994. *Journal of Geophysical Research, 106*(D9), 9923–9943.

Manney, G. L., Santee, M. L., Rex, M., Livesey, N. J., Pitts, M. C., Veefkind, P., Nash, E. R., Wohltmann, I., Lehmann, R., Froidevaux, L., Poole, L. R., Schoeberl, M. R., Haffner, D. P., Davies, J., Dorokhov, V., Gernandt, H., Johnson, B., Kivi, R., Kyro, E., Larsen, N., Levelt, P. F., Makshtas, A., McElroy, C. T., Nakajima, H., Concepcion Parrondo, M., Tarasick, D. W., von derGathen, P., Walker, K. A., & Zinoviev, N. S. (2011). Unprecedented Arctic ozone loss in 2011. *Nature, 478*, 469–475. https://doi.org/10.1038/nature10556.

Massie, S. T., Dye, J. E., Baumgardner, D., Randel, W. J., Wu, F., Tie, X., Pan, L., Figarol, F., Brasseur, G. P., Santee, M. L., Read, W. G., Grainger, R. G., Lambert, A., Mergenthaler, J. L., & Tabazadeh, A. (1997). Simultaneous observations of polar stratospheric clouds and HNO3 over Scandinavia in January, 1992. *Geophysical Research Letters, 24*(5), 595–598.

Massoli, P., Maturilli, M., & Neuber, R. (2006). Climatology of Arctic polar stratospheric clouds as measured by lidar in Ny Ålesund, Spitsbergen (79° N, 12° E). *Journal of Geophysical Research: Atmospheres, 111*(D9).

McCormick, M. P., Chu, W. P., McMaster, L. R., Hamill, P., Swissler, T. J., & Pepin, T. J. (1979). Satellite studies of the stratospheric aerosol. *Bulletin of the American Meteorological Society, 60*, 1038–1046.

McCormick, M. P., Steele, H. M., Hamill, P., Chu, W. P., & Swissler, T. J. (1982). Polar stratospheric cloud sightings by SAM II. *Journal of the Atmospheric Sciences, 39*(6), 1387–1397.

McCormick, M. P., Swissler, T. J., Fuller, W. H., Hunt, W. H., & Osborne, M. T. (1984). Airborne and ground-based lidar measurements of the El Chichón stratospheric aersol from 90°N to 56°S. *Geofisica International, 23*, 187–221.

McElroy, M. B., Salawitch, R. J., Wofsy, S. C., & Logan, J. A. (1986). Reductions of Antarctic ozone due to synergistic interactions of chlorine and bromine. *Nature, 321*, 759–762.

McIntyre, M. E., & Palmer, T. N. (1983). Breaking planetary waves int the stratosphere. *Nature, 305*, 593–6001983.

McKeen, S. A., Liu, S. C., & Kiang, C. S. (1984). On the chemistry of stratospheric SO2 from volcanic eruptions. *Journal of Geophysical Research: Atmospheres, 89*(3), 4873–4881.

McLandress, C. (1998). On the importance of gravity waves in the middle atmosphere and their parameterization in general circulation models. *Journal of Atmospheric and Solar-Terrestrial Physics, 60*(14), 1357–1383.

McLandress, C. D., Alexander, M. J., & Wu, L. (2000). Microwave limb sounder observations of gravity waves in the stratosphere: A climatology and interpretation. *Journal of Geophysical Research, 105*, 11947–11196.

Molina, L. T., & Molina, M. J. (1987). Production of Cl2O2 from the self-reaction of the ClO radical. *Journal of Physical Chemistry, 91*(2), 433–436.

Molina, M. J., & Rowland, F. S. (1974). Stratospheric sink for chlorofluoromethanes: Chlorine atom-catalysed destruction of ozone. *Nature, 249*, 810–812.

Molina, M. J., Zhang, R., Wooldridge, P. J., McMahon, J. R., Kim, J. E., Chang, H. Y., & Beyer, K. (1993). Physical chemistry of the H2SO4/HNO3/H2O system: Implications for polar stratospheric clouds. *Science, 251*, 1418–1423.

Molleker, S., Borrmann, S., Schlager, H., Luo, B., Frey, W., Klingebiel, M., Weigel, R., Ebert, M., Mitev, V., Matthey, R., Woiwode, W., Oelhaf, H., Dörnbrack, A., Stratmann, G., Grooß, J. U., Günther, G., Vogel, B., Müller, R., Krämer, M., Meyer, J., & Cairo, F. (2014). Microphysical properties of synoptic-scale polar stratospheric clouds: In situ measurements of unexpectedly large HNO3 containing particles in the arctic vortex. *Atmospheric Chemistry and Physics, 14*(19), 10785–10801.

Mote, P. W., Rosenlof, K. H., McIntyre, M. E., Carr, E. S., Gille, J. C., Holton, J. R., Kinnersley, J. S., Pumphrey, H. C., Russell, J. M., III, & Waters, J. W. (1996). An atmospheric tape recorder: The imprint of tropical tropopause temperatures on stratospheric water vapor. *Journal of Geophysical Research, 101*(D2), 3989–4006.

Müller, M., Neuber, R., Beyerle, G., Kyrö, E., Kivi, R., & Wöste, L. (2001). Non-uniform PSC occurrence within the Arctic polar vortex. *Geophysical Research Letters, 28*(22), 4175–4178.

Nash, E. R., Newman, P. A., Rosenfield, J. E., & Schoeberl, M. R. (1996). An objective determination of the polar vortex using Ertel's potential vorticity. *Journal of Geophysical Research, 101*, 9471–9478.

Newman, P., Lait, L. R., Schoeberl, M., Nash, E. R., Kelly, K., Fahey, D. W., Nagatani, R., Toohey, D., Avallone, L., & Anderson, J. (1993). Stratospheric meteorological conditions in the Arctic polar vortex, 1991 to 1992. *Science, 261*(5125), 1143–1146.

Newman, P. A., Lait, L. R., Schoeberl, M. R., Seablom, M., Coy, L., Rood, R., Swinbank, R., Proffitt, M., Loewenstien, M., Podolske, J. R., Elkins, J. W., Webster, C. R., May, R. D., Fahey, D. W., Dutton, T. M. G. S., & Chan, K. R. (1996). Measurements of polar vortex air in the midlatitudes. *Journal of Geophysical Research, 101*(D8), 12879–12891.

Newman, P. A., Gleason, J. F., & Stolarski, R. S. (1997). Anomalously low ozone over the Arctic. *Geophysical Research Letters, 24*(22), 2689–2692.

Newman, P. A., Nash, E. R., & Rosenfield, J. E. (2001). What controls the temperature of the Arctic stratosphere during the spring? *Journal of Geophysical Research, 106*(D17), 19999. (2000JD000061).

Newman, A., Harris, N. R. P., Adriani, A., Amanatidis, G. T., Anderson, J. G., Braathen, G. O., Brune, W. H., Carslaw, K. S., Craig, M. S., Decola, P. L., Guirlet, M., Hipskind, R. S., Kurylo, M. J., Küllmann, H., Larsen, N., Mégie, G. J., Pommereau, J.-P., Poole, L. R., Schoeberl, M. R., Stroh, F., Toon, O. B., Trepte, C. R., & van Roozendael, M. (2002). An overview of the SOLVE/THESEO 2000 campaign. *Journal of Geophysical Research: Atmospheres, 107*(D20).

Northway, M. J., Gao, R., Popp, P., Holecek, J. C., Fahey, D., Carslaw, K. S., Tolbert, M., Lait, L. R., Dhaniyala, S., Flagan, R. C., Wennberg, P., Mahoney, M. J., Herman, R. L., Toon, G., & Bui, T. P. (2002). An analysis of large HNO3-containing particles sampled in the Arctic stratosphere during the winter of 1999/2000. *Journal of Geophysical Research, 107*, 8298. https://doi.org/10.1029/2001JD001079.

O'Neill, A., O'Neill, A., Grose, W. L., Pope, V. D., Maclean, H., & Swinbank, R. (1994). Evolution of the stratosphere during northern winter 1991 /92 as diagnosed from U.K. meteorological office analyses. *Journal of the Atmospheric Sciences, 51*, 2800–2817.

Ovarlez, J. (1991). Stratospheric water vapor measurements during CHEOPS-3. *Geophysical Research Letters, 18*(4), 771–774.

Pagan, K. L., Tabazadeh, A., Drdla, K., Hervig, M. E., Eckermann, S. D., Browell, E. V., Legg, M. J., & Foschi, P. G. (2004). Observational evidence against mountain-wave generation of ice nuclei as a prerequisite for the formation of three solid nitric acid polar stratospheric clouds observed in the Arctic in early December 1999. *Journal of Geophysical Research, 109*, D04312. https://doi.org/10.1029/2003JD003846.

Parrish, A., deZafra, R. L., Solomon, P. M., & Barrett, J. W. (1988). A ground-based technique for millimeter wave spectroscopic observations of stratospheric trace constituents. *Radio Science, 23*, 106–118.

Pawson, S., & Naujokat, B. (1999). The cold winters of the middle 1990s in the northern lower stratosphere. *Journal of Geophysical Research, 104*, 14209–14222.

Peter, T. (1997). Microphysics and heterogeneous chemistry of polar stratospheric clouds. *Annual Review of Physical Chemistry, 48*(1), 785–822.

Peter, T., & Grooß, J.-U. (2012). Polar stratospheric clouds and sulfate aerosol particles: Microphysics, denitrification and heterogeneous chemistry. In R. Muller (Ed.), *Stratospheric ozone depletion and climate change* (pp. 108–144). RCS Publishing. ISBN 978-1-84973-002-0.

Pierce, R. B., & Fairlie, T. D. A. (1993). Chaotic advection in the stratosphere: Implications for the dispersal of chemically perturbed air from the polar vortex. *Journal of Geophysical Research, 98*(D10), 18589–18595.

Pitts, M. C., Poole, L. R., Dörnbrack, A., & Thomason, L. W. (2011). The 2009-2010 Arctic polar stratospheric cloud season: A CALIPSO perspective. *Atmospheric Chemistry and Physics, 11*(5), 2161.

Pitts, M. C., Poole, L. R., Lambert, A., & Thomason, L. W. (2013). An assessment of CALIOP polar stratospheric cloud composition classification. *Atmospheric Chemistry and Physics, 13*, 2975–2988. https://doi.org/10.5194/acp-13-2975-2013.

Pitts, M. C., Poole, L. R., & Gonzalez, R. (2018). Polar stratospheric cloud climatology based on CALIPSO spaceborne lidar measurements from 2006 to 2017. *Atmospheric Chemistry and Physics, 18*, 10881–10913. https://doi.org/10.5194/acp-18-10881-2018.

Plane, J. M. C. (2003). Atmospheric chemistry of meteoric metals. *Chemical Reviews, 103*, 4963–4984.

Plumb, R. A. (1996). A "tropical pipe" model of stratospheric transport. *Journal of Geophysical Research, 101*, 3957–3972.

Plumb, R. A., Waugh, D. W., Atkinson, R. J., Newman, P. A., Lait, L. R., Schoeberl, M. R., Browell, E. V., Simmons, A. J., & Loewenstein, M. (1994). Intrusions into the lower stratospheric Arctic vortex during the winter of 1991/92. *Journal of Geophysical Research, 99*, 1089–1105.

Podolske, J. P., Loewenstein, M., Strahan, S. E., & Chan, K. R. (1989). Stratospheric nitrous oxide distribution in the southern hemisphere. *Journal of Geophysical Research, 94*, 16,767–16,772.

Pommereau, J.-P., Goutail, F., Lefèvre, F., Pazmino, A., Adams, C., Dorokhov, V., Eriksen, P., Kivi, R., Stebel, K., Zhao, X., & van Roozendael, M. (2013). Why unprecedented ozone loss in the Arctic in 2011? Is it related to climate change? *Atmospheric Chemistry and Physics, 13*, 5299–5308. https://doi.org/10.5194/acp-13-5299-2013.

Poole, L. R., & McCormick, M. P. (1988). Airborne lidar observations of Arctic polar stratospheric clouds: Indications of two distinct growth stages. *Geophysical Research Letters, 15*, 21–23.

Poole, L. R., Trepte, C. R., Harvey, V. L., Toon, G. C., & VanValkenburg, R. L. (2003). SAGE III observations of Arctic polar stratospheric clouds – December 2002. *Geophysical Research Letters, 30*(23), 2216–2220.

Popp, P. J., Northway, M. J., Holecek, J. C., Gao, R. S., Fahey, D. W., Elkins, J. W., Hurst, D. F., Romashkin, P. A., Toon, G. C., Sen, B., Schauffler, S. M., Salawitch, R. J., Webster, C. R., Herman, R. L., Jost, H., Bui, T. P., Newman, P. A., & Lait, L. R. (2001). Severe and extensive denitrification in the 1999/2000 Arctic winter stratosphere. *Journal of Geophysical Research, 28*, 2875–2878.

Popp, P. J., Marcy, T. P., Watts, L. A., Gao, R. S., Fahey, D. W., Weinstock, E. M., Smith, J. B., Herman, R. L., Troy, R. F., Webster, C., Christensen, L. E., Baumgardner, D. G., Voigt, C., Karcher, B., Wilson, J. C., Mahoney, M. J., Jensen, E. J., & Bui, T. P. (2007). Condensed-phase nitric acid in a tropical subvisible cirrus clouds. *Geophysical Research Letters, 34*, L24812. https://doi.org/10.1029/2007GL031832.

Proffitt, M. H., Steinkamp, M. J., Powell, J. A., McLaughlin, R. J., Mills, O. A., Schmeltekopf, A. L., Thompson, T. L., Tuck, A. F., Tyler, T., Winkler, R. H., & Chan, K. R. (1989a). In situ ozone measurements within the 1987 Antarctic ozone hole from a high-altitude ER-2 aircraft. *Journal of Geophysical Research, 94*(D14), 16547–16555. https://doi.org/10.1029/89JD00905.

Proffitt, M. H., Powell, J. A., Tuck, A. F., Fahey, D. W., Kelly, K. K., Krueger, A. J., Schoeberl, M. R., Gary, B. L., Margitan, J. J., & Chan, K. R. (1989b). A chemical definition of the boundary of the Antarctic ozone hole. *Journal of Geophysical Research, 94*(D9), 11437–11448.

Proffitt, M., Aikin, K., Margitan, J., Loewenstein, M. R., Podolske, J., Weaver, A. R., Chan, K., Fast, H., & Elkins, J. (1993). Ozone loss inside the northern polar vortex during the 1991-1992 winter. *Science, 261*, 1150–1154.

Pueschel, R. F. (1996). Stratospheric aerosols: Formation, properties, effects. *Journal of Aerosol Science, 27*, 383–402.

Rex, M., Von Der Gathen, P., Braathen, G. O., Harris, N. R. P., Reimer, E., Beck, A., Alfier, R., Krüger-carstensen, R., Chipperfield, M., De Backer, H., Balis, D., O'Connor, F., Dier, H., Dorokhov, V., Fast, H. A., Gamma, M. G., Kyrö, E., Litynska, Z., Mikkelsen, I. S., Molyneux, M., Murphy, G., Reid, S. J., Rummukainen, M., & Zerefos, C. (1999). Chemical ozone loss in the Arctic winter 1994/95 as determined by the Match technique. *Journal of Atmospheric Chemistry, 2*, 35–59.

Rieder, H. E., & Polvani, L. M. (2013). Are recent Arctic ozone losses caused by increasing greenhouse gases? *Geophysical Research Letters, 40*(16), 4437–4441.

Rodriguez, J. M., Ko, M. K. W., Sze, N. D., Heisey, C. W., Yue, G. K., & McCormick, M. P. (1994). Ozone response to enhanced heterogeneous processing after the eruption of Mt. Pinatubo. *Geophysical Research Letters, 21*(3), 209–212.

Rood, R. B., Douglass, A. R., Cerniglia, M. C., & Read, W. G. (1997). Synoptic scale mass exchange from the troposphere to the stratosphere. *Journal of Geophysical Research, 102*, 23,467–23,485.

Rosen, J. M. (1964). The vertical distribution of dust to 30 kilometers. *Journal of Geophysical Research, 69*(21), 4673–4676.

Rosen, J. M., & Kjome, N. T. (1991). Backscattersonde: A new instrument for atmospheric aerosol research. *Applied Optics, 30*(12), 1552–1561.

Rosen, J. M., Oltman, S. J., & Evans, W. F. (1989). Balloon borne observations of PSCs, frost point, ozone, and nitric acid in the north polar vortex. *Geophysical Research Letters, 16*, 791–794.

Rosenfield, J. E., & Schoeberl, M. R. (2001). On the origin of polar vortex air. *Journal of Geophysical Research, 106*, 33485–33497.

Rosenfield, J. E., Schoeberl, M. R., & Geller, M. A. (1987). A computation of the stratospheric diabatic circulation using an accurate radiative transfer model. *Journal of the Atmospheric Sciences, 44*, 859–876.

Rosenfield, J. E., Newman, P. A., & Schoeberl, M. R. (1994). Computations of diabatic descent in the stratospheric polar vortex. *Journal of Geophysical Research, 99*, 16,677–16,689.

Rosier, S. M., Lawrence, B. N., Andrews, D. G., & Taylor, F. W. (1994). Dynamical evolution of the northern stratosphere in the early winter 1991/92, as observed by the improved stratospheric and mesospheric sounder. *Journal of the Atmospheric Sciences, 51*, 2783–2799.

Rossby, C.-G. (1939). Relation between variations in the intensity of the zonal circulation of the atmosphere and the displacement of the semipermanent centres of action. *Journal of Marine Research, 2*, 38–55.

Rozanov, E., Egorova, T., Fröhlich, C., Haberreiter, M., Peter, T., & Schmutz, W. (2002). Estimation of the ozone and temperature sensitivity to the variation of spectral solar flux. *European Space Agency SP, 508*, 181–184.

Russell, J. M., Gordley, L. L., Park, J. H., Drayson, S. R., Tuck, A. F., Harries, J. E., Cicerone, R. J., Crutzen, P. J., & Frederick, J. E. (1993). The halogen occultation experiment. *Journal of Geophysical Research, 89*, 5115–5124.

Saitoh, N., Hayashida, S., Sasano, Y., & Pan, L. L. (2002). Characteristics of Arctic polar stratospheric clouds in the winter of 1996/1997 inferred from ILAS measurements. *Journal of Geophysical Research: Atmospheres, 107*(D24).

Salby, M. (2012). *Physics of the atmosphere and climate*. Cambridge: Cambridge University Press. https://doi.org/10.1017/CBO9781139005265.

Santer, B. D., Wigley, T. M. L., Doutriaux, C., Boyle, J. S., Hansen, J. E., Jones, P. D., Meehl, G. A., Roeckner, E., Sengupta, S. S., & Taylor, K. E. (2001). Accounting for the effects of volcanoes and ENSO in comparisons of modeled and observed temperature trends. *Journal of Geophysical Research, 106*, 28033–28060.

Sasano, Y., Oshchepkov, S., Yokota, T., & Nakajima, H. (2005). Characterization of stratospheric liquid ternary solution aerosol from broadband infrared extinction measurements. *Journal of Geophysical Research, 110*(D18). https://doi.org/10.1029/2004JD005709.

Scaife, A., Butchart, N., Warner, C., Stainforth, D., Norton, W., & Austin, J. (2000). Realistic quasi-biennial oscillations in a simulation of the global climate. *Geophysical Research Letters, 27*, 3481–3484.

Scarchilli, C., Adriani, A., Cairo, F., Di Donfrancesco, G., Buontempo, C., Snels, M., Moriconi, M. L., Deshler, T., Larsen, N., Luo, B., Mauersberger, K., Ovarlez, J., Rosen, J., & Schreiner, J. (2005). Determination of polar stratospheric cloud particle refractive indices by use of in situ optical measurements and T-matrix calculations. *Applied Optics, 44*(16), 3302–3311.

Schlager, H., Arnold, F., Hofmann, D. J., & Deshler, T. (1990). Balloon observations of nitric acid aerosol formation in the arctic stratosphere: I gaseous nitric acid. *Geophysical Research Letters, 17*, 1275–1278.

Schmidt, U., Bauer, R., Khedim, A., Klein, E., Kulessa, G., & Schiller, C. (1991). Profile observations of long-lived trace gases in the Arctic vortex. *Geophysical Research Letters, 18*, 767–770.

Schmidt, H., Rast, S., Bunzel, F., Esch, M., Giorgetta, M., Kinne, S., Krismer, T., Stenchikov, G., Timmreck, C., Tomassini, L., & Walz, M. (2013). Response of the middle atmosphere to anthropogenic and natural forcings in the CMIP5 simulations with the Max Planck institute earth system model. *Journal of Advances in Modeling Earth Systems, 5*(1), 98–116.

Schoeberl, M. R., & Hartmann, D. L. (1991). The dynamics of the stratospheric polar vortex and its relation to springtime ozone depletions. *Science, 251*, 46–52.

Schoeberl, M. R., & Sparling, L. C. (1994). Trajectory modeling: Diagnostic tools in Atmospheric physics. In G. Fiocco & C. Visconti (Eds.), *S. I. F. Course CXVI*. Amsterdam: North-Holland.

Schoeberl, M. R., Lait, L. R., Newman, P. A., Martin, R. L., Proffitt, M. H., Hartmann, D. L., Loewenstein, M., Podolske, J., Strahan, S. E., & Gary, B. (1989). Reconstruction of the constituent distribution and trends in the Antarctic polar vortex from ER-2 flight observations. *Journal of Geophysical Research, 94*, 16815–16845.

Schoeberl, M. R., Lait, L. R., Newman, P. A., & Rosenfield, J. E. (1992). The structure of the polar vortex. *Journal of Geophysical Research, 97*, 7859–7882.

Schoeberl, M. R., Stolarski, R. S., Douglass, A. R., Newman, P. A., Lait, L. R., Waters, J. W., Froidevaux, L., & Ready, W. G. (1993). MLS ClO observations and Arctic polar vortex temperatures. *Geophysical Research Letters, 20*(24), 2861–2864.

Schoeberl, M. R., Luo, M., & Rosenfield, J. E. (1995). An analysis of the Antarctic HALOE trace gas observation. *Journal of Geophysical Research, 100*. https://doi.org/10.1029/94JD02749.

Schreiner, J., Voigt, C., Kohlmann, A., Arnold, F., Mauersberger, K., & Larsen, N. (1999). Chemical analysis of polar stratospheric cloud particles. *Science, 283*(5404), 968–970.

Schreiner, J., Voigt, C., Weisser, C., Kohlmann, A., Mauersberger, K., Deshler, T., Kroger, C., Rosen, J., Kjome, N., Larsen, N., Adriani, A., Cairo, F., Di Donfrancesco, G., Ovarlez, J., Ovarlez, H., & Dornbrack, A. (2002). Chemical, microphysical, and optical properties of polar stratospheric clouds. *Journal of Geophysical Research, 107*. https://doi.org/10.1029/2001JD000825.

Sheridan, P. J., Schnell, R. C., Hofmann, D. J., & Deshler, T. (1992). Electron microscope studies of Mt. Pinatubo aerosol layers over Laramie, Wyoming during summer 1991. *Geophys.Res.Lett., 19*, 203–206.

Shi, Q., Jayne, J. T., Kolb, C. E., Worsnop, D. R., & Davidovits, P. (2001). Kinetic model for reaction of ClONO2 with H2O and HCl and HOCl with HCl in sulfuric acid solutions. *Journal of Geophysical Research: Atmospheres, 106*(D20), 24259–24274.

Smith, K. L., Neely, R. R., Marsh, D. R., & Polvani, L. M. (2014). The specified chemistry whole atmosphere community climate model (SC-WACCM). *Journal of Advances in Modeling Earth Systems, 6*(3), 883–901.

Solomon, S., Garcia, R. R., Roland, F. S., & Wuebbles, D. J. (1986). On the depletion of Antarctic ozone. *Nature, 321*, 755–758.

Spang, R., Remedios, J. J., Kramer, L. J., Poole, L. R., Fromm, M. D., Müller, M., Baumgarten, G., & Konopka, P. (2005a). Polar stratospheric cloud observations by MIPAS on ENVISAT: Detection method, validation and analysis of the northern hemisphere winter 2002/2003. *Atmospheric Chemistry and Physics, 5*, 679–692. https://doi.org/10.5194/acp-5-679-2005.

Spang, R., Remedios, J. J., Tilmes, S., & Riese, M. (2005b). MIPAS observation of polar stratospheric clouds in the Arctic 2002/2003 and Antarctic 2003 winters. *Advances in Space Research, 36*, 868–878.

Spang, R., Hoffmann, L., Müller, R., Grooß, J.-U., Tritscher, I., Höpfner, M., Pitts, M., Orr, A., & Riese, M. (2018). A climatology of polar stratospheric cloud composition between 2002 and 2012 based on MIPAS/Envisat observations. *Atmospheric Chemistry and Physics, 18*, 5089–5113. https://doi.org/10.5194/acp-18-5089-2018.

SPARC (Stratospheric Process And their Role in Climate). (2006). *SPARC assessment of stratospheric aerosol properties* (Thomason, L., & Peter, Th., eds., World Climate Research Program report 124, SPARC report 4, 346 pp.). Verrieres le Buisoon, France. 2006.

Stanford, J. L., & Davis, J. S. (1974). A century of stratospheric cloud reports: 1870–1972. *Bulletin of the American Meteorological Society, 55*, 213–219.

Stefanutti, L., MacKenzie, A. R., Balestri, S., Khattatov, V., Fiocco, G., Kyro, E., & Peter, T. (1999). Airborne Polar Experiment-Polar Ozone, Leewaves, Chemistry, and Transport (APE-POLECAT): Rationale, road map and summary of measurements. *Journal of Geophysical Research, 104*(D19), 23941–23959.

Stenchikov, G. L., Kirchner, I., Robock, A., Graf, H.-F., Antuña, J. C., Grainger, R. G., Lambert, A., & Thomason, L. (1998). Radiative forcing from the 1991 Mount Pinatubo volcanic eruption. *Journal of Geophysical Research, 103*, 13837–13857.

Stenchikov, G., Robock, A., Ramaswamy, V., Schwarzkopf, M. D., Hamilton, K., & Ramachandran, S. (2002). Arctic oscillation response to the 1991 Mount Pinatubo eruption: Effects of volcanic aerosols and ozone depletion. *Journal of Geophysical Research, 107*, 4803–4818.

Stolarski, R. S., & Cicerone, R. J. (1974). Stratospheric chlorine: A possible sink for ozone. *Canadian Journal of Chemistry, 52*(8), 1610–1615. https://doi.org/10.1139/v74-233.

Sugita, T., Kondo, Y., Nakajima, H., Schmidt, U., Engel, A., Oelhaf, H., Wetzel, G., Koike, M., & Newman, P. A. (1998). Denitrification observed inside the Arctic vortex in February 1995. *Journal of Geophysical Research, 103*(D13), 16,221–16,234.

Tabazadeh, A., Turco, R. P., & Jacobson, M. Z. (1994). A model for studying the composition and chemical effects of stratospheric aerosols. *Journal of Geophysical Research, 99*, 12897–12914.

Tabazadeh, A., Jensen, E. J., Toon, O. B., Drdla, K., & Schoeberl, M. R. (2001). Role of the stratospheric polar freezing belt in denitrification. *Science, 291*, 2591–2594.

Taylor, F. W., Lambert, A., Grainger, R. G., Rodgers, C. D., & Remedios, J. J. (1994). Properties of northern hemisphere polar stratospheric clouds and volcanic aerosol in 1991/2 from UARS/ISAMS satellite measurements. *Journal of the Atmospheric Sciences, 51*, 3019–3026.

Teitelbaum, H., Moustaoui, M., & Fromm, M. (2001). Exploring polar stratospheric cloud and ozone minihole formation: The primary importance of synoptic-scale flow perturbations. *Journal of Geophysical Research, 106*, 28173–28188.

Terao, Y., Sugita, T., & Sasano, Y. (2012). Ozone loss rates in the Arctic winter stratosphere during 1994–2000 derived from POAM II/III and ILAS observations: Implications for relationships

among ozone loss, PSC occurrence, and temperature. *Journal of Geophysical Research, 117*, D05311. https://doi.org/10.1029/2011JD016789.
Thomason, L. W., Ernest, N., Luis, M. L., Rieger, L., Bourassa, A., Vernier, J.-P., Manney, G., Luo, B., Arfeuille, F., & Peter, T. (2018). A global space-based stratospheric aerosol climatology: 1979–2016. *Earth System Science Data, 10*, 469–492. https://doi.org/10.5194/essd-10-469-2018.
Tie, X., & Brasseur, G. (1995). The response of stratospheric ozone to volcanic eruptions: Sensitivity to atmospheric chlorine loading. *Geophysical Research Letters, 22*(22), 3035–3038.
Timmreck, C., Graf, H.-F., & Steil, B. (2003). Aerosol chemistry interactions after the Mt. Pinatubo eruption. In A. Robock & C. Oppenheimer (Eds.), *Volcanism and the Earth's atmosphere* (AGU monograph) (Vol. 139, pp. 213–225).
Tolbert, M. A., & Toon, O. B. (2001). Solving the PSC mystery. *Science, 292*, 61.
Toon, O. B., Hamill, P., Turco, R. P., & Pinto, J. (1986). Condensation of HNO3 and HCl in the winter polar stratosphere. *Geophysical Research Letters, 13*, 1284–1287.
Toon, G. C., Farmer, C. B., Lowes, L. L., Schaper, P. W., Blavier, J.-F., & Norton, R. H. (1989a). Infrared aircraft measurements of stratospheric composition over Antarctica during September 1987. *Journal of Geophysical Research, 94*(D14), 16571.
Toon, G. C., Farmer, C. B., Schaper, P. W., Blavier, J.-F., & Lowes, L. L. (1989b). Ground-based infrared measurements of tropospheric source gases over Antarctica during the 1986 austral spring. *Journal of Geophysical Research, 94*, 11,613–11,624.
Toon, O. B., Browell, E. V., Kinne, S., & Jordan, J. (1990). An analysis of lidar observations of polar stratospheric clouds. *Geophysical Research Letters, 17*, 393–396.
Toon, G. C., Farmer, C. B., Schaper, P. W., Lowes, L. L., & Norton, R. H. (1992). Composition measurements of the 1989 arctic winter stratosphere by airborne infrared solar absorption spectroscopy. *Journal of Geophysical Research, 97*, 7939.
Tsias, A., Wirth, M., Carslaw, K. S., Biele, J., Mehrtens, H., Reichardt, J., Wedekind, C., WeiB, V., Renger, W., Neuber, R., von Zahn, U., Stein, B., Santacesaria, V., Stefanutti, L., Fierli, F., Bacmeister, J., & Peter, T. (1999). Aircraft lidar observations of an enhanced type Ia polar stratospheric clouds during APE-POLECAT. *Journal of Geophysical Research, 104*(D19), 23961–23969.
Tuck, A. F. (1989). Synoptic and chemical evolution of the Antarctic vortex in late winter and early spring, 1987. *Journal of Geophysical Research, 94*, 11,687–11,737.
Turco, R. P., Hamill, P., Toon, O. B., Whitten, R. C., & Kiang, C. S. (1979). A one-dimensional model describing aerosol formation and evolution in the stratosphere: I. physical processes and mathematical analogs. *Journal of the Atmospheric Sciences, 36*(4), 699–717.
Turco, R. P., Whitten, R. C., & Toon, O. B. (1982). Stratospheric aerosols: Observations and theory. *Reviews of Geophysics and Space Physics, 20*, 233–279.
Turco, R., Plumb, A., & Condon, E. (1990). The airborne Arctic stratospheric expedition: Prologue. *Geophysical Research Letters, 17*(4), 313–316.
Voigt, C., Schreiner, J., Kohlmann, A., Zink, P., Mauersberger, K., Larsen, N., Deshler, T., Kröger, C., Rosen, J., Adriani, A., Cairo, F., DonfrancescoG, D., Viterbini, M., Ovarlez, J., Ovarlez, H., David, C., & Dörnbrack, A. (2000). Nitric acid trihydrate (NAT) in polar stratospheric clouds. *Science, 290*(5497), 1756–1758.
Voigt, C., Larsen, N., Deshler, T., Kröger, C., Schreiner, J., Mauersberger, K., Luo, B., Adriani, A., Cairo, F., Donfrancesco, G. D., Ovarlez, J., Ovarlez, H., Dornbrack, A., Knudsen, B., & Rosen, J. (2003). In situ mountain wave polar stratospheric cloud measurements: Implications for nitric acid trihydrate formation. *Journal of Geophysical Research: Atmospheres, 108*(D5).
Voigt, C., Schlager, H., Luo, B. P., Dörnbrack, A., Roiger, A., Stock, P., Curtius, J., Vössing, H., Borrmann, S., Davies, S., Konopka, P., Schiller, C., Shur, G., & Peter, T. (2005). Nitric Acid Trihydrate (NAT) formation at low NAT supersaturation in polar stratospheric clouds (PSCs). *Atmospheric Chemistry and Physics, 5*, 1371–1380. https://doi.org/10.5194/acp-5-1371-2005.
Voigt, C., Schlager, H., Roiger, A., Stenke, A., de Reus, M., Borrmann, S., Jensen, E., Schiller, C., Konopka, P., & Sitnikov, N. (2008). Detection of reactive nitrogen containing particles in

the tropopause region – evidence for a tropical nitric acid trihydrate (NAT) belt. *Atmospheric Chemistry and Physics, 8*, 7421–7430. https://doi.org/10.5194/acp-8-7421-2008.

Voigt, C., Dörnbrack, A., Wirth, M., Groß, S. M., Pitts, M. C., Poole, L. R., Baumann, R., Ehard, B., Sinnhuber, B.-M., Woiwode, W., & Oelhaf, H. (2018). Widespread polar stratospheric ice clouds in the 2015–2016 Arctic winter – implications for ice nucleation. *Atmospheric Chemistry and Physics, 18*, 15623–15641. https://doi.org/10.5194/acp-18-15623-2018.

Volk, C. M., Elkins, J. W., Fahey, D. W., Salawitch, R. J., Dutton, G. S., Gilligan, J. M., Proffitt, M. H., Loewenstein, M., Podolske, J. R., Minschwaner, K., Margitan, J. J., & Chan, K. R. (1996). Quantifying transport between the tropical and midlatitude lower stratosphere. *Science, 272*, 1763–1768.

Volkert, H., & Intes, D. (1992). Orographically forced stratospheric waves over northern Scandinavia. *Geophysical Research Letters, 19*, 1205–1208.

von Hobe, M., & The RECONCILE Team. (2013). Reconciliation of essential process parameters for an enhanced predictability of Arctic stratospheric ozone loss and its climate interactions (RECONCILE): activities and results. *Atmospheric Chemistry and Physics, 13*(18), 9233–9268.

Waugh, D. W., & Dritschel, D. G. (1999). The dependence of Rossby wave breaking on the vertical structure of the polar vortex. *Journal of the Atmospheric Sciences, 56*, 2359–2375.

Waugh, D. W., & Plumb, R. A. (1994). Contour advection with surgery: A technique for investigating fine scale structure in tracer transport. *Journal of the Atmospheric Sciences, 51*(4), 530–540.

Waugh, D. W., Plumb, R. A., Atkinson, R. J., Schoeberl, M. R., Lait, L. R., Newman, P. A., Loewenstein, M., Toohey, D. W., Avallone, L. M., Webster, C. R., & May, R. D. (1994). Transport of material out of the stratospheric Arctic vortex by Rossby wave breaking. *Journal of Geophysical Research, 99*(D1), 1071–1088.

Wegner, T., Grooß, J.-U., Von Hobe, M., Stroh, F., Sumińska-Ebersoldt, O., Volk, C., Hösen, E., Mitev, V., Shur, G., & Müller, R. (2012). Chlorine activation on stratospheric aerosols: Uncertainties in parameterizations and surface area. *Atmospheric Chemistry & Physics Discussions, 12*, 20561–20591. https://doi.org/10.5194/acpd-12-20561-2012.

F. Weiss, F. Kubel, Ó. Gálvez, M. Hoelzel, S. F. Parker, P. Baloh, R. Iannarelli, M. J. Rossi, H. Grothe, (2016), Angew. Chem. Int. Ed., 55, 3276.

Weisser, C., Mauersberger, K., Schreiner, J., Larsen, N., Cairo, F., Adriani, A., Ovarlez, J., & Deshler, T. (2006). Composition analysis of liquid particles in the Arctic stratosphere under synoptic conditions. *Atmospheric Chemistry and Physics, 6*, 689–696. https://doi.org/10.5194/acp-6-689-2006.

Wennberg, P. O., Anderson, J. G., & Weisenstein, D. K. (1994). Kinetics of reactions of ground state nitrogen atoms (4S3/2) with NO and NO2. *Journal of Geophysical Research, 99*(D9), 18839–18846. https://doi.org/10.1029/94JD01823.

Werner, A., Volk, C. M., Ivanova, E. V., Wetter, T., Schiller, C., Schlager, H., & Konopka, P. (2010). Quantifying transport into the Arctic lowermost stratosphere. *Atmospheric Chemistry and Physics, 10*(23), 11623–11639.

Wirth, M., Tsias, A., Dörnbrack, A., Weiß, V., Carslaw, K. S., Leutbecher, M., Renger, W., Volkert, H., & Peter, T. (1999). Model guided Lagrangian observation and simulation of mountain polar stratospheric clouds. *Journal of Geophysical Research: Atmospheres, 104*(D19), 23971–23981.

WMO (World Meteorological Organization), Atmospheric ozone. (1986). *Assessment of our understanding of the processes controlling its present distribution and change*. Geneva: Global Ozone Research and Monitoring Project, 1986.

WMO (World Meteorological Organization) , Scientific assessment of ozone depletion. (1998). *Global Ozone Research and Monitoring Project report 44* (p. 1999). Geneva.

WMO (World Meteorological Organization), Scientific Assessment of Ozone Depletion. (2018). Global ozone research and monitoring project–report no. 58, 588 pp., Geneva, 2018.

Woiwode, W., Grooß, J.-U., Oelhaf, H., Molleker, S., Borrmann, S., Ebersoldt, A., Frey, W., Gulde, T., Khaykin, S., Maucher, G., Piesch, C., & Orphal, J. (2014). Denitrification by large NAT particles: The impact of reduced settling velocities and hints on particle characteristics. *Atmospheric Chemistry and Physics, 14*, 11525–11544. https://doi.org/10.5194/acp-14-11525-2014.

Yokohata, T., Emori, S., Nozawa, N., Tsushima, Y., Ogura, T., & Kimoto, M. (2005). Climate response to volcanic forcing: Validation of climate sensitivity of a coupled atmosphere–ocean general circulation model. *Geophysical Research Letters, 32*, L21710.

Zhou, S. T., Gelman, M. E., Miller, A. J., & McCormack, J. P. (2000). An inter-hemisphere comparison of the persistent stratospheric polar vortex. *Geophysical Research Letters, 27*, 1123–1126.

Zolensky, M. E., McKay, D. S., & Kaczor, L. A. (1989). A tenfold increase in the abundance of large solid particles in the stratosphere, as measured over the period 1976-1984. *Journal of Geophysical Research, 94*, 1047–1056.

Zondlo, M. A., Hudson, P. K., Prenni, A. J., & Tolbert, M. A. (2000). Chemistry and microphysics of polar stratospheric clouds and cirrus clouds. *Annual Review of Physical Chemistry, 51*(1), 473–499.

Chapter 8
Noctilucent Clouds: General Properties and Remote Sensing

Christian von Savigny, Gerd Baumgarten, and Franz-Josef Lübken

Abstract Noctilucent clouds (NLC) – also known as polar mesospheric clouds (PMC) – occur at mid and high latitudes during the summer months in each hemisphere and are with altitudes of about 83 km the highest clouds in the terrestrial atmosphere. NLC are an optically thin phenomenon. They are known to consist of H_2O ice particles with radii of less than about 100 nm. The first reported sightings of NLC occurred in 1885, two years after the eruption of the volcano Krakatoa in 1883. They exhibit spatial and temporal variability over a large range of scales and react very sensitively to variations in ambient conditions. This high sensitivity makes them highly relevant observables for the investigation of a wide variety of different atmospheric processes including dynamical effects, solar-terrestrial interactions and long-term changes of the Earth's middle atmosphere. The role of NLC as indicators of long-term changes in the mesopause region in particular has been a topic of intensive debate.

This chapter on NLC serves two purposes. First, it provides an overview of the current understanding of the basic characteristics of NLC. Secondly, it introduces the observational techniques employed by the science community to remotely sense NLC, including both passive and active methods.

Keywords Noctilucent clouds · NLC · Polar mesospheric clouds · PMC · Polar summer mesopause · Solar impact · Planetary waves · Gravity waves · Tides · Remote sensing of NLC · Middle atmosphere climate change

---

C. von Savigny (✉)
Institute of Physics, University of Greifswald, Greifswald, Germany
e-mail: csavigny@physik.uni-greifswald.de

G. Baumgarten · F.-J. Lübken
Leibniz-Institute of Atmospheric Physics (IAP), Kühlungsborn, Germany
e-mail: baumgarten@iap-kborn.de; luebken@iap-kborn.de

A. Kokhanovsky, C. Tomasi (eds.), *Physics and Chemistry of the Arctic Atmosphere*, Springer Polar Sciences, https://doi.org/10.1007/978-3-030-33566-3_8

8.1 Introduction

Noctilucent clouds (NLC) are with an altitude of about 83 km the highest clouds in the terrestrial atmosphere. They occur at mid and high latitudes during the summer months in each hemisphere, slightly below the polar summer mesopause. NLC are optically thin and are known consist of H_2O ice particles with radii of less than about 100 nm. NLC exhibit variability – both spatially and temporally – over a large range of scales and react very sensitively to variations in ambient conditions, particularly to temperature variations. This high sensitivity makes them highly relevant observables for the investigation of a wide variety of different atmospheric processes such as gravity and planetary waves, vertical coupling of the atmosphere or extra-terrestrial impacts on the middle atmosphere. The role of NLC as indicators of long-term changes in the mesopause region has also been a topic of intensive debate. The current chapter on NLC aims at serving two purposes. First, it provides an overview of the current understanding of the basic characteristics of NLC, and secondly, it introduces the observational techniques employed by the science community to remotely sense NLC, including both passive and active methods.

Before discussing the basic characteristics and the remote sensing of NLC, a few sentences on nomenclature are appropriate. The term "noctilucent clouds" or "nightshining" clouds originates from the fact that the clouds are not visible to the bare eye during the day, but only when the sun is between about 5° and 15° below the horizon. If this condition is fulfilled, the lower atmosphere is in darkness and the clouds are still illuminated. During the day, scattered radiation in the lower atmosphere dominates strongly over the radiation scattered by NLC, making them invisible to the bare eye. If the sun is too far below the horizon, the clouds are not illuminated any more and will not be visible.

It should also be mentioned that different authors use different terms, which refer to the same phenomenon. Ground-based observers usually use the term *noctilucent cloud* (NLC). In combination with space-borne observations the term *polar mesospheric clouds* (PMC) is frequently used (e.g. Donahue et al. 1972; DeLand et al. 2007). There exist, however, numerous exceptions. Several satellite groups use the term NLC (e.g. von Savigny et al. 2005), and some ground-based lidar observers use the term PMC (e.g. Gardner et al. 2001; Chu et al. 2003), arguing that it is not dark, when ground-based lidar measurements in the polar summer regions are carried out. Hence, the word *noctilucent* is misleading. The terms NLC and PMC can be used synonymously and in this chapter we only use NLC. NLC pioneer Georg Witt attempted to introduce the term *Noctilucent Polar Mesospheric Clouds* (NLPMC), which was, however, so far not accepted by the scientific community.

In the following we will distinguish between *measurements* and *observations* in the following sense: Measurements are considered to correspond to the data directly provided by a scientific instrument, e.g. spectra measured with a spectrometer, whereas observations correspond to derived data products, e.g. NLC particle size derived from spectral NLC measurements.

The chapter is structured as follows. Section 8.2 provides a brief review of the history of NLC research. The following Sect. 8.3 discusses the fundamental characteristics of NLC and their atmospheric environment, their microphysical properties and global morphology. The drivers for variability in NLC as well as their role as indicators of climate change in the polar summer mesopause region are discussed in Sect. 8.4. Section 8.5 introduces the most important measurement techniques to remotely sense NLC, including passive (Sect. 8.5.1) and active (Sect. 8.5.2) methods. A brief summary is provided in the concluding Sect. 8.6.

8.2 Brief History of NLC Research

The first reported sightings of NLC presumably occurred in 1885 (Backhouse 1885; Jesse 1885; Leslie 1885), about two years after the eruption of Krakatoa (Sunda Strait, Indonesia) in August 1883, which caused various atmospheric optical phenomena (Symons et al. 1888). Robert C. Leslie observed NLC on the evening of June 6, 1885 from Southampton and wrote in his short tellingly letter to Nature "*But I have seen nothing in the way of twilight effect so strange as that of Monday evening, the 6th, when about 10 p.m. a sea of luminous silvery white cloud lay above a belt of ordinary clear twilight sky, which was rather low in tone and colour. These clouds were wave-like in form, and evidently at a great elevation* ..." (Leslie 1885). As a visual illustration of the described characteristics, Fig. 8.1 shows a modern photograph of an NLC.

The fact that these first reports occurred less than 2 years after the eruption of Krakatoa led to ongoing speculations on the role of this volcanic eruption for the formation and sightings of NLC. The Krakatoa eruption was with a Volcanic Explosivity Index (VEI) of 6 one of the three biggest volcanic eruptions of the nineteenth century. The hypotheses proposed to describe a causal relationship between the Krakatoa eruption and the (enhanced) occurrence of NLC, range from injection of small volcanic ash particles acting as nucleation nuclei (Schroeder 1999), injection of large amounts of H_2O into the middle atmosphere (Thomas et al. 1989; Lübken et al. 2018) to indirect dynamical effects caused by the thermal effects in the stratosphere associated with the eruption. It is, however, fair to state that the potential effects of the eruption on NLC and the mesosphere in general are only poorly understood.

The first observation of the altitude of NLC was carried out a few years later by Otto Jesse applying a triangulation technique to photographs taken from several different locations in the Berlin area (Germany) (Jesse 1896). These NLC height observations can be considered highly accurate and are with a mean value of 82.08 $\pm$ 0.009 km in remarkable agreement with current observations based on much more sophisticated methods (see also Sect. 8.4.5). In the late 1950s Georg Witt from the Meteorological Institute of Stockholm University started his pioneering work on ground-based imaging of NLC (e.g. Witt 1957).

Fig. 8.1 Sample NLC photograph taken from Kühlungsborn on July 14, 2009 (© Gerd Baumgarten)

The first satellite observations of NLC were performed in 1969 with two limb-scanning photometers (measuring at 589.0 and 557.7 nm) on the OGO-6 (Orbital Geophysical Observatory) and reported in 1972 by Donahue et al. (1972). The first lidar observation of NLC was performed in August 1989 at Andenes (Norway) and reported by Hansen et al. (1989). A great deal of the current understanding of day-to-day, seasonal and long-term variations of NLC is based on a long-term satellite data set retrieved from measurements with the nadir-viewing SBUV(/2) (Solar Backscatter Ultraviolet) instruments on various U.S. satellites (e.g. Thomas et al. 1991; DeLand et al. 2007; DeLand and Thomas 2015) (see also Sect. 8.5.1.2). It is also worth mentioning that NASA's AIM (Aeronomy of Ice in the Mesosphere) (Russell et al. 2009) spacecraft, carrying the CIPS (Cloud Imaging and Particle Size) and the SOFIE (Solar Occultation for Ice Experiment) instruments was the first dedicated satellite mission to study NLC.

8.3 Basic Characteristics of NLC

This section starts with a general description of the polar summer mesopause region, which hosts NLC. The following sections deal with different relevant features of NLC.

8.3.1 The Polar Summer Mesopause Region

NLC occur at altitudes of about 83 km at mid and high latitudes in the summer hemisphere near the polar summer mesopause. Figure 8.2 shows the latitude and altitude dependence of the atmospheric temperature from the surface up to 100 km altitude for the month of July. The dashed lines indicate the tropopause, stratopause and mesopause. As the Figure demonstrates, for altitudes up to the lower mesosphere the temperature above the summer pole is larger than the temperature at the winter pole, which is consistent with the expectation based solely on radiative equilibrium, because the summer pole is continuously sunlit and the winter pole is in darkness. In the upper mesosphere, however, the summer pole is colder than the winter pole, which is unexpected at first glance and inconsistent with the view that the atmosphere is in radiative equilibrium. The reason for this apparent deviation from radiative equilibrium lies in a wave-driven residual upper mesospheric circulation, which causes upwelling – and subsequent adiabatic cooling – over the summer pole and downwelling – and subsequent adiabatic heating – over

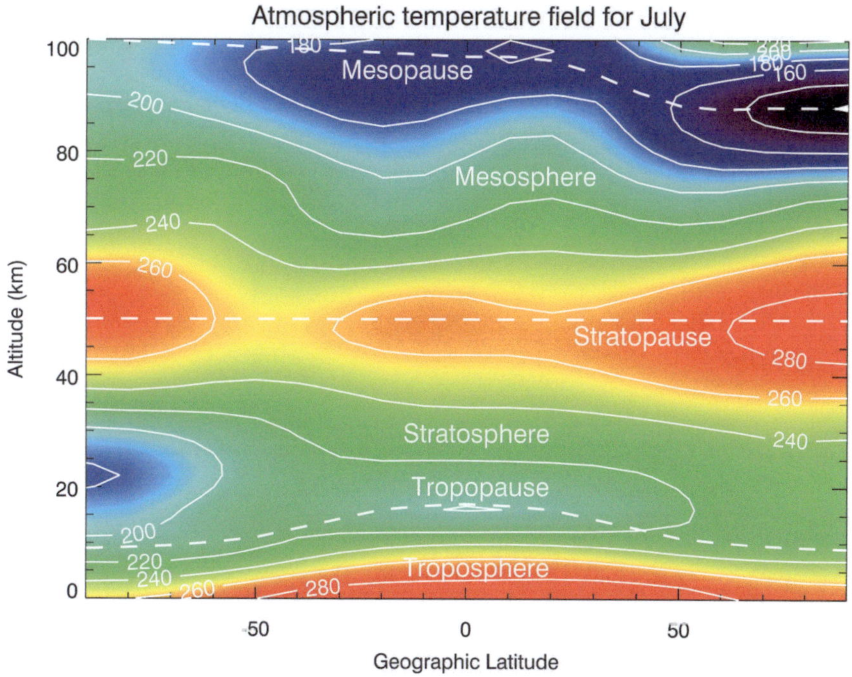

Fig. 8.2 Atmospheric temperature (in K) as a function of latitude and altitude based on the NRL-MSIS00 Model for July 1, 2000. From the surface up to the lower mesosphere the temperatures are higher at the summer pole than at the winter pole. In the middle and upper mesosphere, this behaviour is reversed, as described in the text. NLC form below the cold polar summer mesopause region

the summer pole. The temperature differences between mesopause temperature at the summer and winter pole can exceed 100 K and temperatures below 100 K have been observed at the polar summer mesopause (e.g. Lübken et al. 2014). The low temperature at the polar summer mesopause is one of the key conditions for the formation of mesospheric ice clouds.

8.3.2 Global Morphology of NLC

Information on the global morphology of NLC can only be provided by satellite observations. Measurements with numerous satellite instruments (see also Sect. 8.5.1.2) were used in the past to study NLC. It is interesting that with very few exceptions (e.g. CIPS (McClintock et al. 2009) and SOFIE on AIM) these satellite instruments were not specifically designed to observe NLC, i.e. in most cases, NLC data products are serendipitous data products.

Figure 8.3 shows as an example the temporal and latitude dependence of NLC occurrence rate for the northern (left panel) and the southern (right panel) hemisphere NLC seasons, averaged over all seasons from 2002 to 2012 and based on limb-scatter measurements with the SCIAMACHY instrument on Envisat. Occurrence rate is defined as the ratio of the number of SCIAMACHY limb measurements with detected NLC and the total number of limb measurements on a given day and in a certain latitude bin (here 3° or 5° wide). The abscissa corresponds to day relative to solstice, which is assumed to occur on June 21st (December 21) in the northern (southern) hemisphere in all years considered, irrespective of the actual

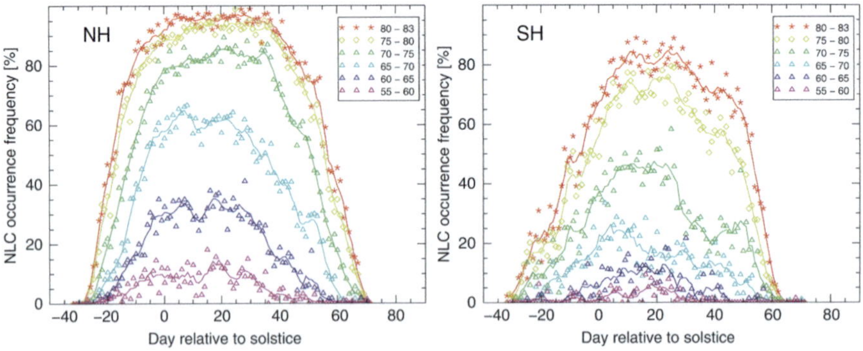

Fig. 8.3 Temporal variation of NLC occurrence rate (in %) for the northern (left panel) and the southern hemisphere (right panel), for different latitude ranges and based on limb-scatter measurements with the SCIAMACHY instrument on the Envisat satellite. The shown values correspond to multi-annual mean variations for the years 2002 to 2012. Note that the lower NLC occurrence rates in the southern hemisphere are in part caused by different viewing geometries of the SCIAMACHY limb-scatter measurements in the two hemispheres. (Figure adapted from Köhnke et al. 2018)

solstice dates. The plots illustrate that the NLC season in the northern hemisphere starts in late May and lasts until about the end of August. In the southern hemisphere the NLC season commences in late November and lasts until about mid-February. Note that the apparent season start and end dates depend on the sensitivity of the employed instrumentation and may differ between different data sets. Figure 8.3 also shows that NLC tend to occur first at the highest latitudes at the beginning of the season and disappear latest at the highest latitudes at the end of the season. Note that the systematic difference in NLC occurrence rate between the two hemispheres is in part related to the SCIAMACHY observation geometry, associated with small scattering angles ($\Theta < 60°$) in the northern hemisphere and large scattering angles ($\Theta > 120°$) in the southern hemisphere during the corresponding NLC seasons. The NLC occurrence rates inferred from the SCIAMACHY limb-scatter observations exceed 90% at the highest northern latitudes during the core period of essentially all NLC seasons. The multi-annual mean occurrence rate variations shown in Fig. 8.3 is still characterized by significant variability. This variability is driven by different processes, which will be discussed in more detail in Sect. 8.4.

8.3.3 Microphysical Properties of NLC

Scientists have assumed for a long time that NLC are mainly composed of solid H_2O (Wegener 1912), but experimental proof was only provided relatively recently based satellite occultation measurements with the HALOE (Hervig et al. 2001) and ACE-FTS (Eremenko et al. 2005) instruments. Additional confirmation was provided by measurements with the MIPAS instrument (Lopez-Puertas et al. 2010). The generally accepted NLC formation mechanism is heterogeneous nucleation on nanometer-sized meteoric smoke particles (e.g. Rapp and Thomas 2006), which are believed to be formed by recondensation of meteoric material evaporated when entering the Earth's atmosphere.

In many studies on NLC Mie-theory is used to simulate their scattering properties, i.e. spherical particles are implicitly assumed. This assumption is not made, because strong evidence for the sphericity of NLC particles exists, but rather, because of a lack of accurate knowledge on particle shape combined with the availability and simple applicability of Mie scattering codes. Several studies presented experimental evidence a non-spherical shape of NLC particles. Baumgarten et al. (2002b) reported on measurements of the depolarization of laser radiation by scattering on NLC particles, which were inconsistent with spherical particles. Assuming cylinder shaped particles, the observed depolarization is consistent with elongated particles with length-over-diameter ratios exceeding a value of 2.5. Eremenko et al. (2005) reported satellite observations of NLC transmission spectra in the OH stretch region near 3.2 μm that were inconsistent with spherical particles. In addition, Rapp et al. (2007) reported on the analysis of different multi-spectral data sets of radiation scattered by NLC particles. Assuming a Gaussian particle size distribution, the results are consistent with aspherical particles with axial ratios of about 0.2 (needles) or 5.0 (plates). The multi-spectral NLC IR extinction measurements carried out with the SOFIE instrument on AIM also allow estimating the shape of NLC particles. The measurements are consistent with axial ratios of about 2 (Hervig and Gordley 2010), i.e. oblate particles.

It is now well established that the size of NLC particles has an upper limit of about 100 nm (e.g von Cossart et al. 1999; von Savigny et al. 2005; Baumgarten and Fiedler 2008; Hervig et al. 2009a; Rusch et al. 2017). Observations of the NLC particle size are typically based on optical measurements of the cloud extinction or scattering at multiple wavelengths or on measurements of cloud scattering at multiple scattering angles (see Sect. 8.5.3).

An intriguing feature of NLC particles is that their size is a function of altitude. The mean particle size increases with decreasing altitude, reaching several tens of nm near the bottom of the ice layer at around 82 km. The ice particle layer extends over a vertical range of about 10 km, but only the lower part of the ice layer – where the largest particles occur – is visible to the eye or to optical instrumentation. This is due to the fact that the particle scattering cross section scales with a power of the particle radius (6th power in the Rayleigh-limit and a somewhat lower power for particles with sizes of tens of nm). The characteristic vertical variation of NLC particle sizes is seen in different satellite, rocket or ground-based data sets (Gumbel and Witt 1998; von Savigny et al. 2005; Baumgarten and Fiedler 2008) and is also reproduced by model simulations.

The size distribution of NLC particles is still not fully established. Many studies on the retrievals of NLC size information assumed a log-normal distribution, despite the fact that this assumption is not well justified. Several studies on microphysical model simulations of NLC formation and growth produced size distributions that were better approximated by Gaussian distributions than by log-normal distributions (e.g. Rapp and Thomas 2006; Berger and von Zahn 2002). Some studies even assumed monodisperse NLC, i.e. a Dirac δ size distribution (Gumbel and Witt 2001). In addition, exponential size distributions were also employed based on vertically integrated model simulations with the WACCM model (Scott Baily, pers. comm.).

NLC are optically thin with slant optical depths smaller than 5×10^{-3} at a wavelength of at 550 nm in occultation geometry (Debrestian et al. 1997), which corresponds to a vertical optical depth of less than 5×10^{-5}. This implies a negligible effect of NLC on the overall atmospheric transmission and hence on surface climate. NLC are also known to interact with the chemical composition of the ambient air masses (Summers et al. 2001; Plane et al. 2004; Murray and Plane 2005). In particular, the sedimentation of the ice particles leads to a vertical redistribution of water vapour, which in turn affects the abundance of HO_x species and influences the vertical distribution of O_3 (e.g. Siskind et al. 2018), because of the catalytic destruction of O_3 by HO_x.

8.4 Variability of NLC on Different Spatial and Temporal Scales

NLC are a highly variable phenomenon both in space and time. Many studies deal with the characterization of the variability of NLC on spatial scales from about 10 m to thousands of km and on temporal scales from seconds to a century. A major motivation for the continued investigation of NLC is their role as sensitive indicators of changes in the physical and chemical conditions in the polar summer mesopause region (see Sect. 8.4.5). In order to fully understand these potential long-term changes all other sources of variability in NLC have to be understood as well. They include extraterrestrial effects like solar variability (see Sect. 8.4.1), as well as dynamical effects associated with planetary waves (see Sect. 8.4.2), a phenomenon generally know as *interhemispheric coupling* (see section 8.4.3) and lunar gravitational impacts (see Sect. 8.4.4).

8.4.1 Solar Influence

Solar influence on NLC and the polar summer mesopause region was investigated with respect to both the solar 11-year cycle and the solar 27-day rotational cycle. An 11-year solar cycle signature has been discussed in the literature in the context of the ground-based visual NLC observations. A quasi decadal variation in visual NLC occurrence has been identified in the North-Western European data base (Gadsden 1998) and also in the data base of observations near Moscow (Romejko et al. 2003). Interestingly, there is a 2–3 year time lag between solar maximum and the minimum in NLC occurrence, which is inconsistent with the hypothesis that enhanced H_2O photolysis during periods of high solar activity is responsible for the signature identified in the visual NLC data bases. A pronounced quasi-decadal signature is also observed in the satellite NLC record provided by the SBUV(/2) instrument series (Thomas et al. 1991; DeLand et al. 2007), at least for the period from 1979 until about 2000. A novel explanation for the strong apparent decadal variation in NLC parameters during solar cycles 21–23 has been discussed in recent years (e.g. Hervig et al. 2016): the atmospheric perturbations associated with the volcanic eruptions of El Chichon (1982) and Mount Pinatubo (1991) may have affected NLC. Hervig et al. (2016) conclude that the solar impact on NLC may have been overestimated in previous studies, because potential volcanic influence was not considered. The exact physico-chemical mechanisms linking stratospheric volcanic injections to the polar summer mesopause are, however, not well understood.

NLC are also affected by the solar rotational cycle that has an average period of about 27 days. Robert et al. (2010) presented first evidence for a solar-driven 27-day signature in NLC parameters observed with the SCIAMACHY and the SBUV satellite instruments. The study also identified a 27-day signature in temperature observations in the polar summer mesopause region with MLS (Microwave Limb

Sounder) on NASA's Aura spacecraft. The authors concluded that the 27-day signature in polar mesopause temperature is the immediate cause of the corresponding signature in NLC. Several more recent studies also dealt with solar 27-day signatures in NLC and/or background conditions at the polar summer mesopause. Thomas et al. (2015) extracted 27-day signatures in temperature and H_2O profiles observed with SOFIE/AIM. A main finding of this study is that the phase relationship between the 27-day signature in temperature and H_2O is inconsistent with the simple view that the atmospheric signatures are driven by variable H_2O photolysis and variable solar diabatic heating. The phase relationship is, however, consistent with a dynamically driven mechanism, i.e. a 27-day modulation of the upwelling in the polar upper mesosphere during summer. It is also worth mentioning that a solar 27-day signature was also identified in NLC altitude (Thurairajah et al. 2016; Köhnke et al. 2018). The 27-day signature has an amplitude of about 100 m. The studies by Thurairajah et al. (2016) and Köhnke et al. (2018) are partly inconsistent, because in the former work the 27-day signature in NLC altitude is more pronounced in the southern hemisphere, whereas it is stronger in the northern hemisphere in Köhnke et al. (2018).

A potential effect on NLC associated with solar proton events (SPEs) has also been discussed in the literature. von Savigny et al. (2007b) reported on a severe reduction of the NLC occurrence rate during Antarctic summer 2004/2005, which coincided with the January 2005 solar proton event. The NLC reduction coincided with a significant increase in polar summer mesopause temperature of more than 10 K. Becker and von Savigny (2010) further investigated this hypothesis by dedicated model simulations with the Kühlungsborn Mechanistic Circulation Model (KMCM). In this study an anomalous mesospheric cooling was introduced in order to mimic the reduced solar diabatic heating associated with the SPE-induced catalytic O_3 destruction in the polar summer mesosphere. The affected meridional temperature gradients caused changes in the zonal wind, subsequently in the filtering and propagation of gravity waves, which in turn affected the residual circulation in the upper mesosphere. The main effect was a significant heating at the polar summer mesopause, which was qualitatively consistent with the observed temperature increase and the reduction in NLC occurrence during the January 2005 SPE (von Savigny et al. 2007b). Bardeen et al. (2016) studied potential effects of the January 2012 SPE on NLC, but were not able to fully confirm the results presented in the earlier studies.

8.4.2 Planetary Wave Signatures in NLC

The summer middle atmosphere is generally not expected to be affected by significant planetary wave activity, because of the Charney-Drazin-criterion (Charney and Drazin 1961) according to which the vertical propagation of planetary waves is suppressed for an easterly background flow. However, the summer hemisphere is partly affected by fairly strong quasi-5-day-wave and quasi-2-day-wave activity.

These waves are thought to be generated in-situ in the summer mesosphere as a consequence of baroclinic instabilities of the summertime jet (Garcia et al. 2005). Corresponding wave signatures have been reported in temperature measurements in various studies.

Quasi 5-day wave signatures in NLC were identified in visual NLC observations from the ground, e.g. in the works by Gadsden (1985) and Kirkwood and Stebel (2003), highlighting the usefulness of the visual NLC observations. Quasi 5-day-waves were also reported in several satellite-based studies on NLC, e.g. by Merkel et al. (2003, 2008), who employed SNOE (Student Nitric Oxide Experiment) NLC observations, or by von Savigny et al. (2007a), who used the SCIAMACHY-limb NLC data set in combination with Aura/MLS temperature observations. In addition, Nielsen et al. (2010) employed satellite observations with the CIPS and SOFIE instruments on the AIM satellite to investigate 5-day-wave signatures in NLC. The available studies suggest that the quasi-5-day-wave signatures in NLC are primarily driven by variations in the upper mesospheric temperature field. The studies by von Savigny et al. (2007a), Merkel et al. (2008) and Nielsen et al. (2010) show that the quasi-5-day-wave signatures in NLC are highly inversely correlated with the corresponding signatures in temperature. Nielsen et al. (2010) highlight the importance of 5-day-wave signatures to extend the duration of the NLC season near its end in mid-August.

Figure 8.4 shows as an example an analysis of SCIAMACHY NLC observations during the 2005 NLC season in the northern hemisphere. Panel (a) displays the seasonal variation of the zonally and daily averaged NLC occurrence rate. Panel (b) shows a Hovmöller-plot, i.e. NLC occurrence rate as a function of longitude and time. Panel (c) shows the wavelet power spectrum of the NLC occurrence rate, which clearly shows the presence of periodic variations with a period of about 5 days.

In summary, the available studies on planetary wave signatures in NLC demonstrate the importance of NLC observations for the investigation of dynamical processes in the polar summer mesopause region.

8.4.3 Interhemispheric Coupling

Interhemispheric coupling is a fascinating atmospheric phenomenon and corresponds to a gravity-wave mediated teleconnection between the stratosphere in one hemisphere and the polar mesopause region in the other hemisphere. One of the first studies on interhemispheric coupling was the one by Becker and Fritts (2006), providing an explanation for the anomalously low PMSE (Polar Mesospheric Summer Echo) occurrence rates during the northern NLC/PMSE season 2002. The link between the dynamics of the polar winter stratosphere and the polar summer mesopause region was further investigated in the work by Karlsson et al. (2007), who carried out a combined analysis of NLC radii retrieved from Odin/OSIRIS (Llewellyn et al. 2004) limb scatter observations and ECWMF temperature in the

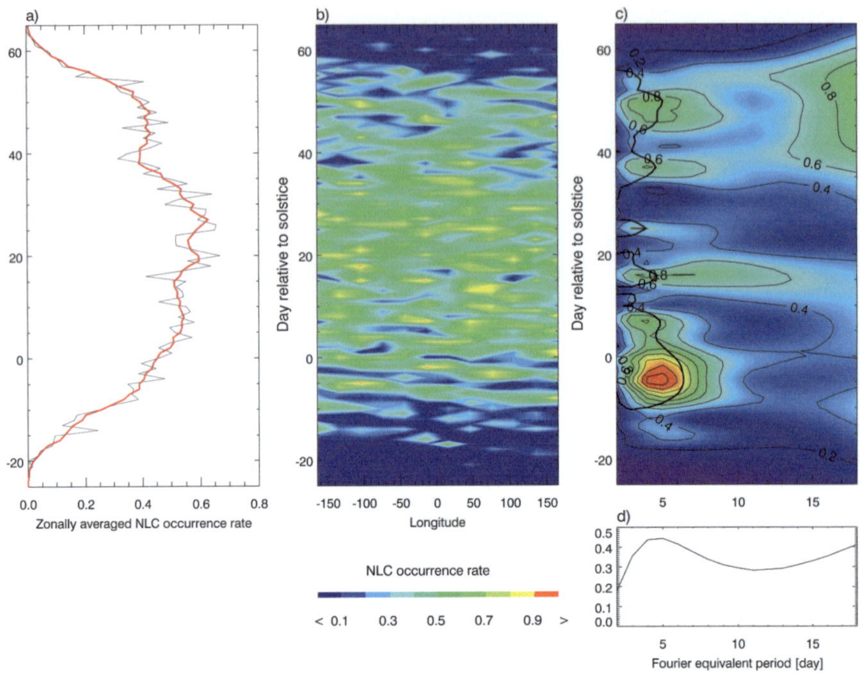

Fig. 8.4 Panel (**a**) Temporal evolution of daily and zonally averaged NLC occurrence rate for the northern NLC season 2005. Panel (**b**) Hovmöller-plot of the longitude and time dependence of NLC occurrence rate. Panel (**c**) Normalized wavelet power spectrum of the NLC occurrence rates (the 95% confidence level is indicated by the thick black line). Panel (**d**) shows the wavelet power averaged between days −20 and 60 relative to solstice. (Figure adapted from von Savigny et al. 2007a)

polar winter stratosphere. An inverse correlation between the temperature in the polar winter stratosphere and NLC sizes was found, indicating a positive correlation between polar winter stratosphere temperature and polar summer mesopause temperature. For an easy-to-follow introduction to the basic mechanism behind interhemispheric coupling we refer to Körnich and Becker (2010).

8.4.4 Lunar Influence

Lunar gravitational effects not only affect the Earth's oceans, but are also detectable in the atmosphere. The amplitudes of these lunar tidal signatures generally increase with increasing altitude in the atmosphere, implying that their detection is easier in the middle and upper atmosphere than close to the surface. The number of studies on potential lunar gravitational effects on NLC is quite limited. Several studies attempted to identify lunar tidal signatures in ground-based visual NLC observations

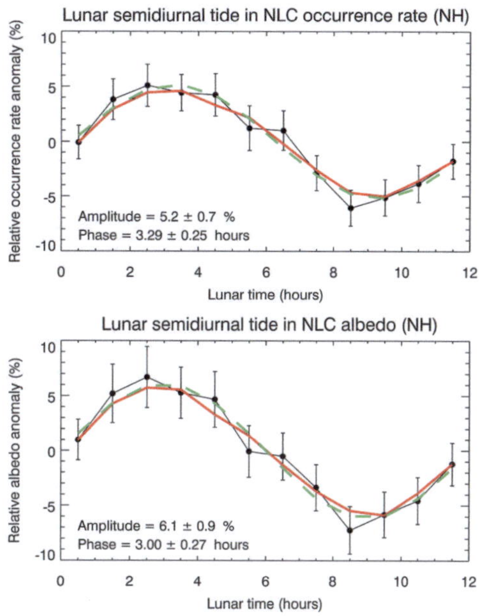

Fig. 8.5 Lunar semidiurnal tidal signatures (in %) in NLC occurrence rate (**a**) and NLC albedo (**b**), based on SBUV(/2) observations and for the latitude range from 55° N to 75° N. The black circles show the signature extracted with the superposed epoch analysis technique. The red line is a smoothed version (running mean over 3 lunar hours) and the green dashed line displays a sinusoidal fit to the black data points. (Figure adapted from von Savigny et al. 2017)

(Kropotkina and Shefov 1975; Gadsden 1985; Dalin et al. 2006; Pertsev et al. 2015). The results obtained in these studies are partly contradictory and some studies identify unrealistically large lunar tidal variations in NLC occurrence. Nevertheless, lunar tidal signatures should be detectable in ground-based visual NLC observations and some of the reported findings have to be considered real.

Recently, the lunar semidiurnal tide was identified in satellite observations of NLC with the SBUV(/2) instrument series, employing a data set covering a period of more than three decades (von Savigny et al. 2017). The amplitudes of the relative lunar tidal variations in NLC occurrence rate, NLC albedo and ice water content are about 5% in the northern hemisphere and slightly smaller in the southern hemisphere. Figure 8.5 shows as an example the lunar semidiurnal tide in NLC occurrence rate and NLC albedo extracted from the SBUV(/2) data set provided by DeLand and Thomas (2015). Lunar time quantifies the azimuthal angular difference between the observer and the position of the moon, similar to the definition of local solar time. The black points correspond to the mean occurrence rate (or albedo) anomaly averaged in a given lunar time bin. The error bars are the corresponding error of the mean. The red line shows the original data smoothed with a 3-h running mean filter and the green dashed line is a sinusoidal fit to the data points. The lunar tidal signatures were extracted using the superposed epoch analysis (or composite analysis) as described in more detail in von Savigny et al. (2017). The main findings of this study were recently confirmed by analyses of satellite NLC observations with SOFIE/AIM by Hoffmann et al. (2018) and by ground-based Lidar observations in the work of Fieder and Baumgarten (2018).

Note that lunar tidal signatures have been identified in previous studies in several other mesospheric and thermospheric parameters, e.g. in temperature, winds, and airglow emissions (e.g, Forbes et al. 2013; Stening et al. 2003; von Savigny et al. 2015).

8.4.5 *Noctilucent Clouds as Potential Indicators for Climate Change*

The long term evolution of noctilucent cloud (NLC) frequency and intensity is often considered to be a potential indicator of long term trends. This includes a discussion on the first reported appearance of NLC in 1885 and the first measurement of NLC altitudes, $z_{NLC} = 82.08$ km, by Jesse (1896). The general conception is that an increase of greenhouse gases (here mainly CO_2) leads to a cooling of the middle atmosphere due to an enhancement of the escape of infrared photons to space. If the atmosphere gets colder one would naively expect NLC to appear more frequently and perhaps also more intense since they consist of ice particles.

NLC exist at approximately 83 km and only at polar and middle latitudes during the summer season, where temperatures are typically lower than 150 K. Unfortunately, observations of temperatures and other basic atmospheric parameters in the mesosphere and lower thermosphere (MLT) for periods extending back to the preindustrial era are not available. NLC basically constitute the only long term observation in this part of the atmosphere. Measurements of reflection heights of radio waves in the ionosphere are available since the late 1950s. Indeed, they show a systematic decrease with time which is explained by a cooling of the atmosphere below the reflection height (Peters et al. 2017). We note that the decrease of a given pressure level (p_{ref}) in the MLT (closely related to the radio wave reflection height) summarizes all temperature changes below p_{ref} but does not include changes at(!) p_{ref}. Direct observations in the MLT are rather challenging and, on a regular basis, exist only since satellites are in orbit, i.e. since 1979. Measurements in the MLT during summer at high latitudes are even more complicated and sporadic due to permanent sunshine conditions.

The role of NLC as potential indicators for climate change is complicated by the fact that several processes on time scales of decades interfere with long term trends. This concerns, for example, the impact of solar cycle, ozone decline/recovery, volcanic eruptions, and secular modulation of water vapor concentrations in the stratosphere. Model studies are often used to study the impact of these processes on the morphology of NLC. Here we apply the LIMA and MIMAS models. LIMA (Leibniz Institute Model of the middle Atmosphere) reflects the main features of the middle atmosphere, in particular the cold summer mesopause. LIMA is nudged to the real atmosphere in the tropo- and lower stratosphere. In MIMAS (Mesospheric Ice Microphysics And tranSport model) several million ice particles are exposed to these background conditions and, for example, grow in size or sublimate. This allows to derive mean NLC parameters, e.g. the maximum backscatter signal, β_{max}, as observed by lidars. In this chapter we give all β_{max}-values in units of $10^{-10}/(m\ sr)$.

Model simulations covering the period 1961–2013 have shown that trends of mesospheric clouds derived from satellite observations are at the edge of being significant (DeLand and Thomas 2015; Berger and Lübken 2015). In this section we concentrate on trends on centennial timescales, i.e. for the time period 1871–2008. Details of the LIMA and MIMAS models and the setup for these runs are described in Lübken et al. (2018). In brief, concentrations of carbon dioxide (CO_2) and methane (CH_4) increase in the model according to well established observations. In the MLT nearly all methane is converted to water vapor which promotes the generation of ice particles in the summer mesopause region. We have performed several model runs, e.g. raising CO_2 and keeping CH_4 constant (see Lübken et al. 2018 for more details). In this section we consider only runs were both CO_2 and CH_4 increase with time. Note that some of the decadal variations mentioned above are ignored in these model runs. Furthermore, we do not consider any potential trend of the mean circulation since very little is known about such long term trends.

In Fig. 8.6 the distribution of NLC altitudes at 69 ± 3° N is shown for two years, namely for 1880 (preindustrial era), and for 2007 (modern times). The latter was chosen to be in the middle of a solar cycle. More precisely, the altitudes of maximum backscatter are shown in Fig. 8.6 where we have considered NLC of all brightness classes, i.e. including faint clouds. As can be seen from this figure the mean NLC altitude in 1880 was approximately 1.5 km higher compared to modern times. A typical variability (standard deviation) within one year is ± 0.5 km. In Fig. 8.6 we also show the annual mean of all lidar measurements at the ALOMAR

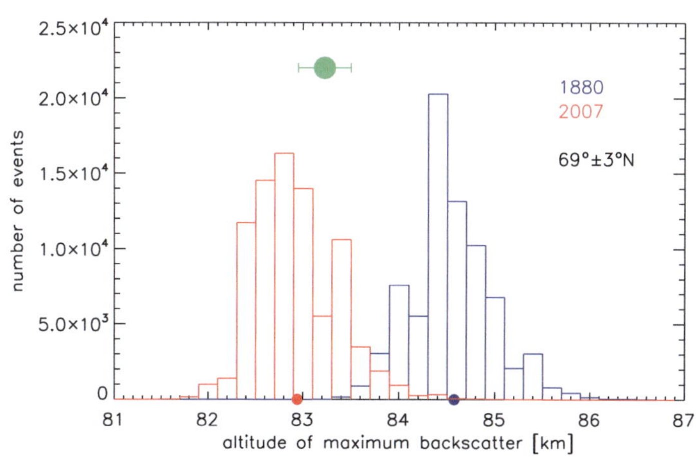

Fig. 8.6 Histogram of NLC altitudes from LIMA/MIMAS at 69° ± 3° N from the years 1880 (blue) and 2007 (red), respectively. The red and blue dots mark the means in that years. The large green dot shows the mean NLC altitude observed by the ALOMAR lidar at 69° N. The horizontal bar indicates the year-to-year variability

observatory (69° N) from the year 2007. We also show the standard deviation of all annual mean values available so far (1997–2018). Again, we consider all NLC classes, e.g. including faint NLC. NLC altitudes from LIMA/MIMAS nicely agree with lidar observations. We note that NLC altitudes observed by this lidar show only small trends of some decameters per decade (Fiedler et al. 2017). Generally speaking, NLC altitudes vary comparatively little over centennial time periods. This implies that NLC altitudes are rather poor indicators of long term changes, at least compared to other NLC parameters (see below). On the other hand, NLC altitudes were actually measured at the end of the nineteenth century and are presumably the only quantitative information available from the MLT in that period. Note that comparing visual NLC observations with LIMA/MIMAS is non-trivial for various reasons (see discussion in Lübken et al. 2018).

The temporal development of the annual mean ice water content (IWC) at 69° ± 3° N is shown in Fig. 8.7. IWC gives the amount of water vapor bound in ice particles in a column of unit area. Compared to lower latitudes there is significantly more ice (per unit area) available at polar latitudes (not shown). This is because it is generally colder in the MLT at polar latitudes. As can be seen in Fig. 8.7, IWC increases by ∼ 14 g/km2 or approximately 40% from preindustrial to modern times. In Fig. 8.7 we also show the increase of tropospheric methane leading to a corresponding increase of water vapor in the MLT. The strong correlation between IWC and CH_4 (or H_2O) comes from the fact that IWC is mainly driven by the amount of water vapor available for ice formation. We note that CO_2 also increases strongly in the period shown in Fig. 8.7. However, when we allow CO_2 to increase in LIMA/MIMAS but keep CH_4 (H_2O) constant, IWC changes only little in time.

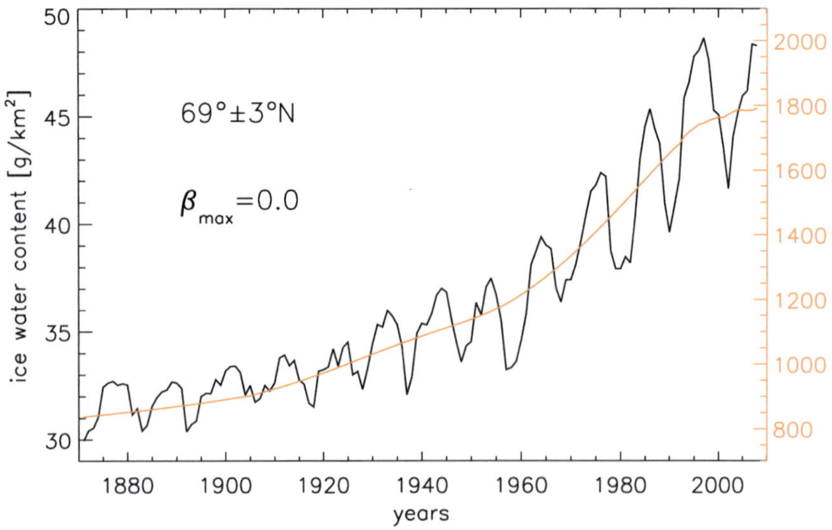

Fig. 8.7 Black and left axis: Time series of ice water content at 69° ± 3° N from LIMA/MIMAS, considering NLC of all β_{max}-classes. Brown and right axis: Time series of tropospheric methane

In summary, IWC is a sensitive indicator for anthropogenic impact on NLC on centennial time scales, more precisely on the impact of water vapor in the MLT caused by methane increase.

In summary, some features of noctilucent clouds (e.g. IWC) are sensitive indicators of long term changes caused by anthropogenic increase of greenhouse gases, namely CO_2 and CH_4. This applies only to centennial timescales because the morphology of NLC may be impacted by other processes on shorter periods, e. g. by ozone recovery or solar cycle. In the future we intend to study the potential impact of future scenarios of greenhouse gas emissions on NLC parameters.

8.5 Remote Sensing of NLC

Different remote sensing techniques have been employed to study NLC since the first reported sighting in the late nineteenth century. Section 8.5.1 provides an overview of the passive remote sensing techniques applied to investigate NLC, followed by a discussion of active techniques in Sect. 8.5.2. Finally, the techniques to infer information on NLC particle size are summarized in Sect. 8.5.3.

8.5.1 Passive Remote Sensing Methods

8.5.1.1 Ground-Based Visual Observations

Noctilucent clouds become visible to the human eye just before sunrise and after sunset. They appear silver-blue and show rich structures (Fig. 8.8). Ever since

Fig. 8.8 Noctilucent clouds seen from Trondheim (about 63N) in late July when the sun is about 8 degrees below the horizon

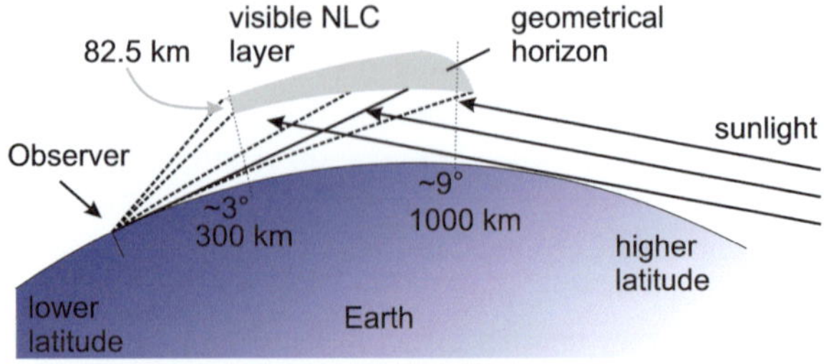

Fig. 8.9 Idealized NLC observation geometry: The observer looks towards the sun, which is below the horizon at the location of the observer (Baumgarten et al. 2009a)

Fig. 8.10 Small scale structures (<50 km) as seen in noctilucent clouds on July 11, 2016. The structures are caused by gravity waves and their transition to turbulence

the first observations of noctilucent clouds from northern Europe their annual re-occurrence has attracted researchers to speculate about their composition and effect on the atmosphere (Backhouse 1885; Leslie 1885; Jesse 1885). It was quickly realized that their extreme height of about 83 km makes them an extraordinary indicator for the thermal state of the upper atmosphere (Foerster and Jesse 1892; Lübken et al. 2008). However their visual ground based observations are limited to favorable illumination conditions with solar elevations from about −5° to −15° and latitudes from about 50° to 60° with a few examples at latitudes down to about 40° and up to about 65° (e.g. Gadsden and Schroeder 1989; Baumgarten et al. 2009a). Observations from even lower and higher latitudes are sparse and limited to a few examples, usually observed at shallow angles which means that the actual clouds are located more than 500 km poleward of the observer (Fig. 8.9).

Rich structures in NLC (Fig. 8.10) and their temporal variations are used to study dynamical processes, giving the first fingerprint of the importance of small scale structures for the state of the middle atmosphere (Witt 1962; Baumgarten and Fritts 2014). The fast appearance and variations of these structures that change within minutes allowed researchers to speculate that they are made of ice crystals which was confirmed about 100 years later by satellite observations (Kohlrausch 1887; Hervig et al. 2001).

The rich structures are used to triangulate the altitude of the clouds which is essential to understand the surrounding atmosphere and the scale of structures (Jesse 1896; Størmer 1935). The altitude measurements are today performed with a much higher precision by active remote sensing methods allowing a more detailed view on the morphology of the clouds and properties of particles forming the clouds.

8.5.1.2 Satellite Observations of NLC

Limb-Scatter Observations

The first satellite-based observations of NLC were based on limb-scatter measurements and were carried out with photometers on the OGO-6 satellite in 1969 (Donahue et al. 1972), as already mentioned above. During the past decades, limb-scatter measurements by a variety of satellite instruments were employed to detect NLC, to estimate the NLC particle size and to study different characteristics of NLC. It is important to mention that none of these instruments was specifically designed to investigate NLC. The list of instruments includes the UV spectrometer on the Solar Mesosphere Explorer (SME) (e.g. Rusch et al. 1991), the WIND Imaging Interferometer (WINDII) (e.g. Evans et al. 1995; Wiens et al. 1995), the Student Nitric Oxide Experiment (SNOE) (e.g. Merkel et al. 2003; Bailey et al. 2005, 2007), the Ultraviolet and Visible Imaging and Spectrographic Imaging (UVISI) instrument on MSX (Midcourse Space Experiment) (e.g. Carbary et al. 2003, 2004), the Global Ozone Monitoring by Occultation of Stars (GOMOS) instrument (e.g. Pérot et al. 2010), the Optical Spectrograph and InfraRed Imager System (OSIRIS) (e.g. von Savigny et al. 2005; Petelina et al. 2006; Karlsson et al. 2007), the SCanning Imaging Absorption spectroMeter for Atmospheric CHartographY (SCIAMACHY) (e.g. von Savigny et al. 2004; Robert et al. 2009) and the Spatial Heterodyne Imager for Mesospheric Radicals (SHIMMER) (e.g. Stevens et al. 2009).

The NLC detection principle for limb-scatter observations is quite simple. NLC in the field of view of the instrument lead to enhanced limb-scattering and characteristic peaks in the limb-radiance profiles. Figure 8.11 shows as an example SCIAMACHY limb-radiance measurements at several UV wavelengths with NLC (left panel) and without NLC (right panel) in the instruments's field of view. The overall shape of the limb-radiance profile without NLC can be explained as follows: starting at the highest tangent height, the limb-radiance will increase with decreasing tangent height, because the Rayleigh-scattered radiance is to first order proportional to the neutral density at the tangent point. This increase in limb-radiance is roughly exponential, as long as the line of sight is optically thin. At a sufficiently low tangent height this condition is not fulfilled any more and the limb-radiance does not increase any more with decreasing tangent height.

If an NLC is present in the instruments's field of view, the limb-radiance at the tangent height corresponding to the NLC altitude (if the cloud occurs at or near the tangent point) is enhanced. If the cloud does not occur at the tangent point, but in the near or far field, the cloud signature will appear at a lower tangent height. Algorithms for detecting NLC in limb-scatter observations are typically based on threshold criteria being applied to the limb-radiance near 83 km itself, and/or to the

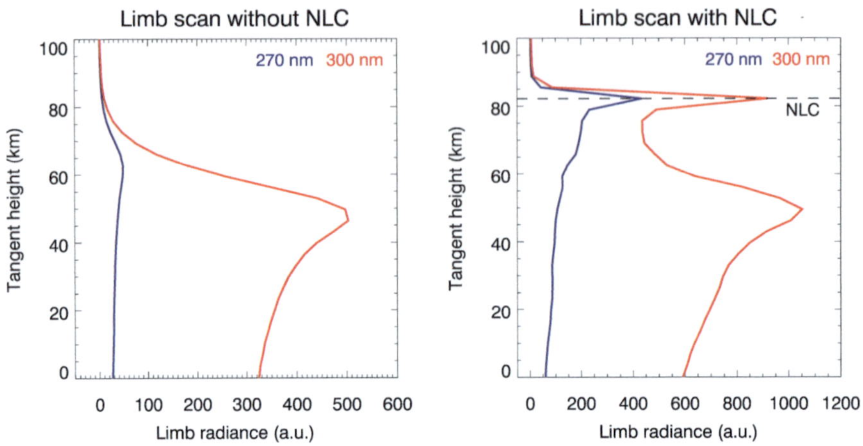

Fig. 8.11 Limb-radiance profiles at wavelengths of 270 nm (blue) and 300 nm (red) measured by SCIAMACHY/Envisat for scenes without (left panel) and with (right panel) NLC in the instrument's field of view. The SCIAMACHY limb-scatter measurements were carried out on July 3, 2002 at latitudes of 61°N (left panel) and 79°N (right panel). The limb-radiance enhancement caused by NLC is clearly visible in the right panel

limb-radiance gradients below or above the NLC peak. For more information we refer to, e.g. Robert et al. (2009).

Note that the contribution of the NLC scatter signal to the limb radiance near cloud altitude can be significantly larger than the Rayleigh-scattering contribution. In contrast, the NLC contribution in Nadir viewing geometry is only a small fraction of the total nadir backscatter radiance (see below).

Occultation Measurements

Satellite solar occultation measurements were also used to observe NLC and to retrieve physical cloud properties. Among the satellite instruments used to observe NLC are POAM II (Polar Ozone and Aerosol Measurement) (Debrestian et al. 1997), HALOE (Halogen Occultation Experiment) (Hervig and Siskind 2006) and SOFIE (Solar Occultation For Ice Experiment) (Gordley et al. 2009). SOFIE is the only instrument specifically built to study NLC and their atmospheric environment and allowed for major advances in the understanding of NLC and their microphysical properties (e.g. Hervig et al. 2009a,b; Hervig and Gordley 2010).

Limb-Emission Measurements

To our best knowledge only a single satellite instrument was used to study NLC by exploiting their thermal emission. Lopez-Puertas et al. (2010) and Garcia-Comas et al. (2016) employed limb-emission measurements in the infrared (10–13 μm) spectral range with MIPAS (Michelson Interferometer for Passive Atmospheric Sounding) on Envisat and found – amongst other results – that the spectral distribution of the measured IR radiances can be well described by modelled emissions of H_2O ice particles, again confirming that ice is the main component of NLC particles.

Nadir Measurements

The basic principle behind detecting NLC in nadir UV backscatter measurements from space is to exploit the small enhancement in nadir radiances over the Rayleigh-scattering background, which is caused by the NLC (see also Fig. 8.12 and below). In contrast to limb-scatter observations, where the Rayleigh-background originates from the upper mesosphere – and is consequently quite weak – nadir backscatter measurements are associated with a significantly stronger background originating from lower altitudes. In addition, the long slant path through the atmosphere associated with limb measurements further enhances the relative contribution of NLC-scattered radiation to the overall radiance for limb geometry. Note that nadir detections of NLC would essentially be impossible for wavelengths larger than about 300 nm. This is related to the absorption of solar radiation by O_3 below 300 nm, because of which all detected photons detected in nadir geometry were

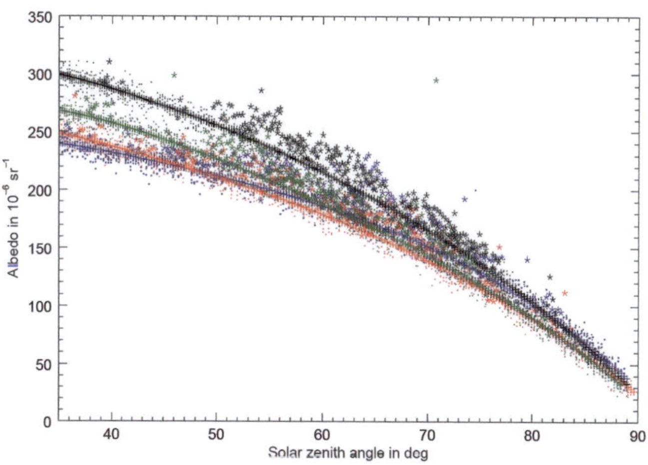

Fig. 8.12 Illustration of the detection of NLC in satellite nadir UV backscatter measurements. The small dots correspond to spectrally averaged (from 258 to 278 nm) albedo as a function of the SZA (solar zenith angle) for SCIAMACHY Nadir measurements on June 22, 2007 for the four different nadir measurement geometries (indicated in different colours). The plus symbols show a 4th order polynomial fit after 50 iterations and the identified NLC are marked as asterisks. (Adapted from Langowski et al. 2018)

(mostly singly) scattered in the atmosphere and the contribution by surface reflection is essentially zero. Nadir UV backscatter measurements with the SBUV (Solar Backscatter Ultraviolet) and SBUV/2 instruments on various NASA and NOAA satellites were very successfully employed to investigate NLC, their morphology and time evolution (e.g. DeLand et al. 2007). The NLC detection technique was initially introduced by Thomas et al. (1991). The SBUV NLC time series started in 1979 and now covers 40 years, which is truly unique. The latest data version includes NLC ice water content (IWC) as a data product (DeLand and Thomas 2015). The SBUV NLC detection algorithm is based on nadir backscatter radiances at the five shortest SBUV wavelengths in the spectral range between 252 and 292.3 nm and carries out a number of tests (as described in detail in DeLand et al. (2003)) to identify NLC.

Benze et al. (2009) modified the original cloud detection scheme applied to SBUV/2 nadir measurements by employing a single wavelength only. This was done in order to allow direct comparisons between SBUV/2 observations with observations carried out with the Cloud Imaging and Particle Size (CIPS) experiment (McClintock et al. 2009), which does not have multi-spectral capability. However, due to CIPS' imaging capability, a given atmospheric volume can be observed sequentially from different satellite positions, i.e. at different scattering angles. This allows estimating the NLC scattering phase function and subsequently particle size information can be retrieved.

In a recent study, nadir measurements with the SCIAMACHY instrument were also used to detect NLC (Langowski et al. 2018). Because of the poorer signal-to-noise (S/N) ratio of the SCIAMACHY nadir measurements compared to the SBUV measurements, the original SBUV algorithm could not be implemented directly. Instead, a single wavelength approach – similar to the one described in Benze et al. (2009) – was employed and the nadir UV radiances were averaged over the 258 to 278 nm spectral range in order to enhance the S/N ratio. The disadvantage of this approach is that the consistency tests applied to the multi-spectral SBUV(/2) measurements are not possible. Figure 8.12 illustrates the NLC detection approach applied to nadir UV backscatter measurements. The nadir radiances are plotted against the solar zenith angle (SZA). The *background* albedo is determined by fitting a 4th-order polynomial to the SZA-dependence of the nadir albedo values. Potential cloudy pixels are identified using a 2-σ threshold criterion and are excluded from the background fit in subsequent iterations of the retrieval scheme. More information can be found in Langowski et al. (2018).

8.5.2 Active Remote Sensing Methods

8.5.2.1 Lidar

Active lidar remote sensing of Noctilucent clouds was proposed already in 1887 by illuminating the clouds from the ground (Jesse 1887). Although there were several groups working carefully on the detection of NLC by lidar since the 1960s it took until 1989 to develop the required sensitivity (Fiocco and Grams 1966; Hansen et al. 1989). Since the initial observations by a lidar in northern Norway the lidar technique was significantly improved and the clouds were detected by several instruments from near the north pole to the south pole (Nussbaumer et al. 1996; Stebel et al. 2000; Gardner et al. 2001; Chu et al. 2001; Wickwar 2002; Blum et al. 2005). Sophisticated lidars are optimized to separate photons generated by the laser and scattered back by the clouds from solar photons scattered in the atmosphere. This allows for observations during daytime and at high latitudes (von Zahn et al. 1998; Höffner et al. 2003). The most efficient way of separating photons generated by the laser from solar photons is to minimize the field of view of the lidar system to just the area illuminated by the laser beam with diameter of about 5 m at 100 km distance. Using narrow band spectral filters of a few picometers spectral bandwidth allows to separate laser photons from background photons according to their color (von Zahn et al. 2000). This enables detecting the diurnal variations of NLC parameters, e.g. the centroid altitude or the frequency of occurrence (Fiedler et al. 2005; Gerding et al. 2013a,b).

The ALOMAR RMR-lidar in northern Norway was built to measure noctilucent clouds with different colors of the outgoing laser beam, and also to measure the polarization of the backscattered light (von Cossart et al. 1999; Baumgarten et al. 2002b). Figure 8.13 shows a typical time evolution of the cloud as seen by the ALOMAR RMR-Lidar at a wavelength of 532 nm.

Three widely separated colors ($\lambda = 355$, 532 and 1064 nm) and the depolarization of the light at $\lambda = 532$ nm due to scattering on NLC is used to calculate the size of the particles to be about 50 nm and shape to be non-spherical prolate or oblate with an axis ratio of about 2 (Baumgarten et al. 2007, 2008; Kiliani et al.

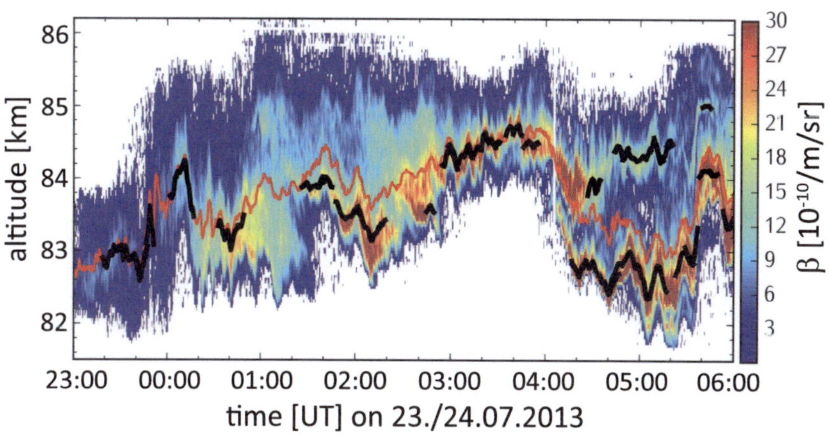

Fig. 8.13 Time evolution of vertical structure of NLC measured by lidar (color contours). The red solid line indicates the centroid altitude of the layer z_c. Black lines indicate the altitude of wave like structures in NLC. (After Ridder et al. 2017)

2015). Lidar instruments were used to link satellite and ground based observations as simultaneous and common volume measurements are difficult due to local time restrictions of the different satellite instruments.

As lidars measure a small spot of a few meters in diameter in the NLC layer no information about the horizontal structure of the cloud layer is available from single lidar observations (Kaifler et al. 2013a,b). Baumgarten et al. (2002a) used a twin lidar to infer horizontal structures of NLC with a separation of the sounding volumes of about 40 km. During an extensive rocket campaign first simultaneous and common volume observations of camera and lidar were performed by Baumgarten et al. (2009b) and allowed estimating the sensitivity of ground based cameras showing that even weak NLC are detected by ground based cameras. Simultaneous lidar and satellite based observations using the CIPS instrument on the AIM satellite were used to investigate the decay time of structures in noctilucent clouds. It was found that structures in the clouds are primarily advected without significant microphysical modifications on timescales shorter than about an hour similar to results from a microphysical model (Baumgarten et al. 2012; Kiliani et al. 2013).

Combining the time evolution of wavelike features on timescales shorter than their decay time with wind measurements by Meteor radar, the horizontal structures were inferred from single lidar measurements and compared to those from ground based camera observations (Ridder et al. 2017). The results allowed to quantify systematic differences of noctilucent cloud altitudes as seen by different instruments. For example, ground based triangulation of NLC altitudes yields values that are about 300 m lower than the centroid altitude of the cloud layer. The reason for this systematic difference is that the triangulation of NLC altitudes requires the presence structures in the layer and NLC with structures are found at lower altitudes as can be seen in Fig. 8.14.

Continuous lidar measurements over two decades have recently been analyzed to quantify diurnal variations as well as lunar tides. It was found that the solar diurnal variation of NLC altitudes is about two times larger than the lunar tides (Fiedler et al. 2017; Fieder and Baumgarten 2018). Furthermore year to year variations of NLC parameters are usually larger than decadal scale variations as can be seen for example in Fig. 8.15. This is in accordance to model simulations that were used to study cloud variations on centennial timescales (see also Sect. 8.4.5 and Lübken et al. 2018).

8.5.3 *Particle Size Observations*

An essential microphysical parameter of NLC is the size of the NLC particles and a significant fraction of NLC studies deals with retrieving information on NLC particle size. In the optical spectral range there are at least three different approaches that can be employed to remotely sense size information of NLC particles. First, the spectral dependence of the particle scattering cross section depends on particle size, if the particle size is large enough to be outside the Rayleigh-regime. Second,

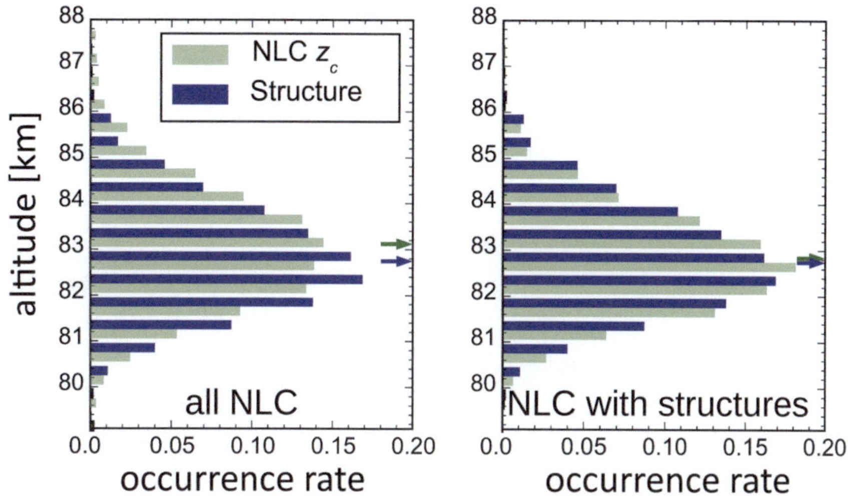

Fig. 8.14 Histogram of NLC altitudes identified by the centroid of the layer z_c and the altitude of structures found in NLC. Arrows indicate the mean of the distributions. The left panel shows all NLC and all structures found, the right panel shows the same structures as the left panel but the NLC were limited to those where structures were found. (After Ridder et al. 2017)

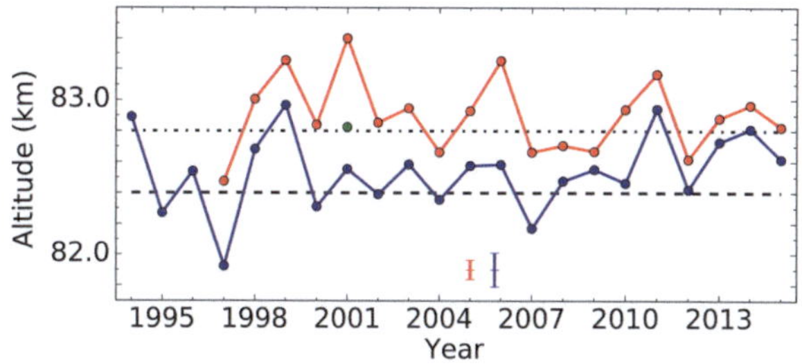

Fig. 8.15 Decadal observation of NLC centroid altitude by lidar for different brightness classes. The red line corresponds to weak and strong clouds ($\beta_{\text{max}} > 4$) while the blue line corresponds to strong clouds only ($\beta_{\text{max}} > 13$) restricted to measurement times of 16 to 8 UT. (After Fiedler et al. 2017)

particle size can also be estimated from measurements of the scattering phase function. And third, the degree of polarization of electromagnetic radiation scattered by NLC particles also contains information on the particle size.

Particle size observations based on the first approach, i.e. the spectral dependence approach, can be based on ground-based multi-colour Lidar measurements (e.g. von Cossart et al. (1999); Baumgarten et al. (2007)), satellite based limb-scatter measurements (e.g. Rusch et al. (1991); Carbary et al. (2004); von Savigny et al. (2004); Carbary et al. (2003); Karlsson et al. (2007); Robert et al. (2009)) or solar occultation measurements (e.g. Debrestian et al. 1997). In the following paragraphs

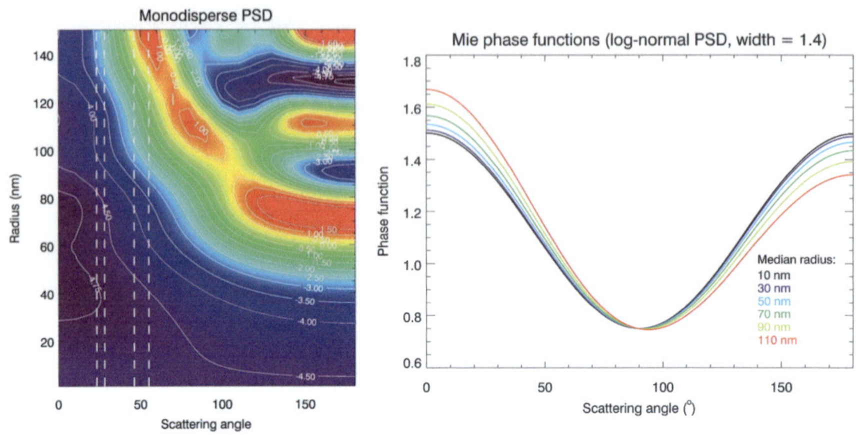

Fig. 8.16 Left panel: Ångström exponents for satellite limb-scatter measurements in the 265–300 nm spectral range as a function of scattering angle and particle radius for ice particles with a monodisperse particle size distribution (PSD). The white dashed lines correspond to the range of scattering angles of SCIAMACHY limb-scatter measurements in the northern hemisphere for the ascending (around 25°) and descending (around 50°) parts of the orbit and for a latitude of 60° N. Right panel: Mie Scattering phase functions for a log-normal particle size distribution with a width parameter of $\sigma = 1.4$ and different values of the median radius (Calculations performed with the Oxford Mie-code)

the basic approaches to infer NLC particle size information from spectral and phase function measurements are briefly explained.

All NLC particle size retrievals are based on different pieces of a priori information. Assumptions have to be made in terms of the NLC particle size distribution (see also Sect. 8.3.3). Typically, a normal or log-normal particle size distribution is assumed and the width parameter is fixed. The median (or mode) radius is then fitted until the modelled spectral dependence (or phase function) matches the measured one within measurement uncertainties. An exception to this approach are the ground-based Lidar measurements with the Alomar Lidar. In this case, measurements are carried out at three wavelengths, i.e. 1064, 532 and 355 nm and both median radius and width of an assumed log-normal distribution are retrieved (von Cossart et al. 1999; Baumgarten et al. 2007).

As an illustration of the first particle size retrieval approach, the left panel of Fig. 8.16 shows as an example the Ångström-exponent of spherical ice particles with a monodisperse particle size distribution in the 265–300 nm spectral range for the satellite limb-scatter geometry and as a function of scattering angle and particle radius.

Here, the Ångström-exponent α describes the spectral dependence of the differential scattering cross section $\sigma(\lambda, \Theta)$ according to the following equation:

$$\sigma(\lambda, \Theta) \propto \lambda^{\alpha} \tag{8.1}$$

with λ corresponding to wavelength and Θ to the scattering angle of the limb-scatter observations. The white dashed lines in Fig. 8.16 illustrate the range of scattering angles associated with SCIAMACHY limb-scatter observations in the northern hemisphere for the ascending (scattering angles around 25°) and descending (scattering angles around 50°) parts of the Envisat orbit at a tangent point latitude of 60° N. For small particles, the Ångström-exponent is apparently close to the Rayleigh-value, as expected. The absolute value of the Ångström-exponent decreases for increasing particle radii. For scattering angles lower than about 90° there exists a unique relationship between particle size and Ångström-exponent in the expected particle size range (up to about 100 nm). However, for larger scattering angles, a unique identification of particle size is not possible any more, as a consequence of the Mie-resonances. This behaviour complicates the retrieval of NLC particle size information from satellite limb-scatter measurements. It also illustrates that the possibility of NLC particle size retrievals from limb-scatter observations in a given hemisphere depends on the scattering angle and therefore directly on the orbit parameters of the spacecraft and the viewing direction of the instrument relative to spacecraft flight direction.

In order to illustrate the phase function approach, the right panel of Fig. 8.16 shows simulated scattering phase functions for spherical ice particles with a log-normal particle size distributions (width $\sigma = 1.4$) and for different median radii. For small (<80°) and large (>120°) scattering angles, there are significant differences between the phase functions for the particle populations with different median radii. This effect is exploited, when retrieving NLC particle size information from multi-angle scattering measurements. The phase function technique was, e.g. applied to rocket photometer measurements of the NLC scattering signal (Gumbel et al. 2001). As the spinning rocket ascends or descends through the NLC layer, the radiation scattered by NLC particles is measured at a range of different scattering angles. The rocket measurements described by Gumbel and Witt (2001) – carried out during the MIDAS-DROPSS campaign in Norway in 1999 – are remarkable, because they allowed the determination of NLC particles size information based on both the phase function approach and the spectral approach. The particle sizes retrieved from the two approaches were found to be in good overall agreement. In addition, the phase function technique is also applied to satellite measurements of the NLC nadir backscatter made with the CIPS (Cloud Imaging and Particle Size Experiment) (McClintock et al. 2009) on the AIM (Aeronomy of Ice in the Mesosphere) satellite (Russell et al. 2009) at wavelengths around 265 nm.

8.6 Conclusions

This chapter provided an overview of the current understanding of the main characteristics of noctilucent clouds (NLC), also known as polar mesospheric clouds (PMCs). NLC are optically thin H_2O ice clouds that occur a few kilometers below the mesopause in the summer hemisphere at mid and high latitudes. They are highly transient phenomena exhibiting variability over large spatial and temporal scales. NLC are used as sensitive indicators for the investigation of different dynamical processes as well as long-term variations in the polar summer mesopause region. A second focus of the present chapter was on remote sensing methods employed by the scientific community from the ground and from satellite platforms to investigate NLC, their global morphology and variability, as well as their microphysical properties. NLC will without doubt remain an interesting and relevant topic for future atmospheric research.

Acknowledgements This work was supported by the University of Greifswald and the Leibniz-Institute of Atmospheric Physics (IAP) in Kühlungsborn.

References

Backhouse, T. W. (1885). The luminous cirrus clouds of June and July. *Meteorological Magazine, 20*, 133.

Bardeen, C. G., Marsh, D. R., Jackman, C. H., Hervig, M. E., & Randall, C. E. (2016). Impact of the January 2012 solar proton event on polar mesospheric clouds. *Journal of Geophysical Research-Atmospheres, 121*, 9165–9173. https://doi.org/10.1002/2016JD0248202.

Baumgarten, G., Lübken, F.-J., & Fricke, K. H. (2002a). First observation of one noctilucent cloud by a twin lidar in two different directions. *Annales Geophysicae, 20*(11), 1863–1868.

Baumgarten, G., Fricke, K. H., & von Cossart, G. (2002b). Investigation of the shape of noctilucent cloud particles by polarization lidar technique. *Geophysical Research Letters, 29*, 13. https://doi.org/10.1029/2001gl013877.

Baumgarten, G., Fiedler, J., & von Cossart, G. (2007). The size of noctilucent cloud particles above ALOMAR (69N, 16E): optical modeling and method description. *Advances in Space Research, 40*(6), 772–784. https://doi.org/10.1016/j.asr.2007.01.018.

Baumgarten, G., & Fiedler, J. (2008). Vertical structure of particle properties and water content in noctilucent clouds. *Geophysical Research Letters, 35*, L10811. https://doi.org/10.1029/2007GL033084

Baumgarten, G., Fiedler, J., Lübken, F.-J., & von Cossart, G. (2008). Particle properties and water content of noctilucent clouds and their interannual variation. *Journal of Geophysical Research-Atmospheres, 113*, D6. https://doi.org/10.1029/2007jd008884.

Baumgarten, G., Gerding, M., Kaifler, B., & Müller, N. (2009a). A trans-European network of cameras for observation of noctilucent clouds from 37N to 69N. In *Proceedings of the 19th ESA Symposium on European Rocket and Balloon Programmes and Related Research*, Bad Reichenhall, Germany, 7–11

Baumgarten, G., Fiedler, J., Fricke, K. H., Gerding, M., Hervig, M., Hoffmann, P., Müller, N., Pautet, P.-D., Rapp, M., Robert, C., Rusch, D., von Savigny, C., & Singer, W. (2009b). The noctilucent cloud (NLC) display during the ECOMA/MASS sounding rocket flights on 3 August 2007: morphology on global to local scales. *Annales Geophysicae, 27*, 953–965. https://doi.org/10.5194/angeo-27-953-2009.

Baumgarten, G., Chandran, A., Fiedler, J., Hoffmann, P., Kaifler, N., Lumpe, J., Merkel, A., Randall, C. E., Rusch, D., & Thomas, G. (2012). On the horizontal and temporal structure of noctilucent clouds as observed by satellite and lidar at ALOMAR (69N). *Geophysical Research Letters, 39*, 1. https://doi.org/10.1029/2011gl049935.

Baumgarten, G., & Fritts, D. C. (2014). Quantifying Kelvin-Helmholtz instability dynamics observed in noctilucent clouds: 1. Methods and observations. *Journal of Geophysical Research-Atmospheres, 119*(15), 9324–9337. https://doi.org/10.1002/2014jd021832.

Bailey, S. M., Merkel, A. W., Thomas, G. E., & Carstens, J. N. (2005). Observations of polar mesospheric clouds by the student nitric oxide explorer. *Journal of Geophysical Research, 110*, D13203. https://doi.org/10.1029/2004JD005422.

Bailey, S. M., Merkel, A. W., Thomas, G. E., & Rusch, D. W. (2007). Hemispheric differences in polar mesospheric cloud morphology observed by the Student Nitric Oxide Explorer. *Journal of Atmospheric and Solar – Terrestrial Physics, 69*, 1407–1418. https://doi.org/10.1016/j.jastp.2007.02.008

Becker, E., & Fritts, D. C. (2006). Enhanced gravity-wave activity and interhemispheric coupling during the MaCWAVE/MIDAS northern summer program 2002. *Annales Geophysicae, 24*, 1175–1188. https://doi.org/10.5194/angeo-24-1175-2006.

Becker, E., & von Savigny, C. (2010). Dynamical heating of the polar summer mesopause induced by solar proton events. *Journal of Geophysical Research, 115*, D00I18. https://doi.org/10.1029/2009JD012561, [printed 116(D1), 2011].

Benze, S., Randall, C. E., DeLand, M. T., Thomas, G. E., Rusch, D. W., Bailey, S. M., Russell, J. M., McClintock, W., Merkel, A. W., & Jeppesen, C. (2009). Comparison of polar mesospheric cloud measurements from the Cloud Imaging and Particle Size experiment and the solar backscatter ultraviolet instrument in 2007. *Journal of Atmospheric and Solar – Terrestrial Physics, 71*(3–4), 365–372. https://doi.org/10.1016/j.jastp.2008.07.014.

Berger, U., & von Zahn, U. (2002). Icy particles in the summer mesopause region: Three-dimensional modeling of their environment and two-dimensional modeling of their transport. *Journal of Geophysical Research, 107*(A11), 1366. https://doi.org/10.1029/2001JA000316.

Berger, U., & Lübken, F.-J. (2015). Trends in mesospheric ice layers in the Northern Hemisphere during 1961–2013. *Journal of Geophysical Research, 120*, 11277–11298. https://doi.org/10.1002/2015JD023355.

Blum, U., & Fricke, K. H. (2005). The Bonn University lidar at the Esrange: technical description and capabilities for atmospheric research. *Annales Geophysicae, 23*(5), 1645–1658. https://doi.org/10.5194/angeo-23-1645-2005

Carbary, J. F., Morrison, D., & Romick, G. J. (2003). Ultraviolet imaging and spectrographic imaging of polar mesospheric clouds. *Advances in Space Research, 31*(9), 2091.

Carbary, J. F., Morrison, D., & Romick, G. J. (2004). Evidence for bimodal particle distribution from the spectra of polar mesospheric clouds. *Geophysical Research Letters, 31*, L13108. https://doi.org/10.1029/2004GL020101.

Charney, J. G., & Drazin, P. G. (1961). Propagation of planetary-scale disturbances from the lower into the upper atmosphere. *Journal of Geophysical Research, 66*(1), 83–109.

Chu, X., Gardner, C. S., & Papen, G. (2001). Lidar observations of polar mesospheric clouds at South Pole: Seasonal variations. *Geophysical Research Letters, 28*(7), 1203–1206. https://doi.org/10.1029/2000gl012524.

Chu, X., Gardner, C. S., & Roble, R. G. (2003). Lidar studies of interannual, seasonal and diurnal variations of polar mesospheric clouds at the South Pole. *Journal of Geophysical Research, 108*(D8), 8447. https://doi.org/10.1029/2002JD002524

Dalin, P. A., Pertsev, N. N., & Romejko, V. A. (2006). Significance of lunar impact on noctilucent clouds. *Journal of Atmospheric and Solar – Terrestrial Physics, 68*, 1653–1663.

Debrestian, D. J., Lumpe, J. D., Shettle, E. P., Bevilacqua, R. M., Olivero, J. J., Hornstein, J. S., Glaccum, W., Rusch, D. W., Randall, C. E., & Fromm, M. D. (1997). An analysis of POAM II solar occultation observations of polar mesospheric clouds in the southern hemisphere. *Journal of Geophysical Research, 102*, 1971–1981.

DeLand, M. T., Shettle, E. P., Thomas, G. E., & Olivero, J. J. (2003). Solar backscattered ultraviolet (SBUV) observations of polar mesospheric clouds (PMCs) over two solar cycles. *Journal of Geophysical Research, 108*(D8), 8445. https://doi.org/10.1029/2002JD002398

DeLand, M. T., Shettle, E. P., Thomas, G. E., & Olivero, J. J. (2007). Latitude-dependent long-term variations in polar mesospheric clouds from SBUV version 3 PMC data. *Journal of Geophysical Research, 112*, D10315. https://doi.org/10.1029/2006JD007857.

DeLand, M. T., & Thomas, G. E. (2015). Updated PMC trends derived from SBUV data. *Journal of Geophysical Research-Atmospheres, 120*. https://doi.org/10.1002/2014JD022253.

Donahue, T. M., Guenther, B., & Blamont, J. E. (1972). Noctilucent clouds in daytime: circumpolar particulate layers near the summer mesopause. *Journal of the Atmospheric Sciences, 29*, 1205–1209.
Eremenko, M. N., Petelina, S. V., Zasetsky, A. Y., Karlsson, B., Rinsland, C. P., Llewellyn, E. J., & Sloan, J. J. (2005). Shape and composition of PMC particles derived from satellite remote sensing measurements. *Geophysical Research Letters, 32*, L16S06. https://doi.org/10.1029/2005GL023013.
Evans, W. F. J., Laframboise, L. R., Sine, K. R., Wiens, R. H., & Shepherd, G. G. (1995). Observation of polar mesospheric clouds in summer 1993 by the WINDII instrument on UARS. *Geophysical Research Letters, 22*, 2793–2796. https://doi.org/10.1029/95GL03018.
Fiedler, J., Baumgarten, G., & von Cossart, G. (2005). Mean diurnal variation of noctilucent clouds during 7 years of lidar observations at ALOMAR. *Annales Geophysicae, 23*, 1175–1181.
Fiedler, J., Baumgarten, G., Berger, U., & Lübken, F.-J. (2017). Long-term variations of noctilucent clouds at ALOMAR. *Journal of Atmospheric and Solar – Terrestrial Physics, 162*, 79–89. https://doi.org/10.1016/j.jastp.2016.08.006.
Fiedler, J., & Baumgarten, G. (2018). Solar and lunar tides in noctilucent clouds as determined by ground-based lidar. *Atmospheric Chemistry and Physics, 18*, 16051-16061. https://doi.org/10.5194/acp-18-16051-2018.
Fiocco, G., & Grams, G. (1966). Observations of the upper atmosphere by optical radar in Alaska and Sweden during the summer 1964. *Tellus, 18*(1), 34–48. https://doi.org/10.1111/j.2153-3490.1966.tb01441.x.
Foerster, W., & Jesse, O. (1892). Invitation to observe the luminous night clouds. *Nature, 46*, 1199, 589–590. https://doi.org/10.1038/046589a0.
Forbes, J. M., Zhang, X., Bruinsma, S., & Oberheide, J. (2013). Lunar semidiurnal tide in the thermosphere under solar minimum conditions. *Journal of Geophysical Research, Space Physics, 118*, 1788–1801.
Gadsden, M. (1985). Observations of noctilucent clouds from North-West Europe. *Annales Geophysicae, 3*, 119–126.
Gadsden, M., & Schroeder, W. (1989). *Noctilucent Clouds* (p. 165). New York: Springer.
Gadsden, M. (1998). The North-West Europe data on noctilucent clouds: a survey. *Journal of Atmospheric and Solar – Terrestrial Physics, 60*(12), 1163–1174. https://doi.org/10.1016/S1364-6826(98)00072-8.
Garcia, R. R., Liebermann, R., Russell III, J. M., & Mlynczak, M. G. (2005). Large-scale waves in the mesosphere and lower thermosphere observed by SABER. *Journal of the Atmospheric Sciences, 62*, 4384–4399.
García-Comas, M., López-Puertas, M., Funke, B., Jurado-Navarro, Á. A., Gardini, A., Stiller, G. P., von Clarmann, T., & Höpfner, M. (2016). Measurements of global distributions of polar mesospheric clouds during 2005–2012 by MIPAS/Envisat. *Atmospheric Chemistry and Physics, 16*, 6701–6719. https://doi.org/10.5194/acp-16-6701-2016.
Gardner, C. S., Papen, G. C., Chu, X., & Pan, W. (2001). First lidar observations of middle atmosphere temperatures, Fe densities, and polar mesospheric clouds over the North and South Poles. *Geophysical Research Letters, 28*, 1199–1202.
Gerding, M., Kopp, M., Hoffmann, P., Höffner, J., & Lübken, F.-J. (2013a). Diurnal variations of midlatitude NLC parameters observed by daylight-capable lidar and their relation to ambient parameters. *Geophysical Research Letters, 40*(24), 6390–6394. https://doi.org/10.1002/2013gl057955.
Gerding, M., Höffner, J., Hoffmann, P., Kopp, M., & Lübken, F.-J. (2013). Noctilucent cloud variability and mean parameters from 15 years of lidar observations at a mid-latitude site (54N, 12E). *Journal of Geophysical Research-Atmospheres, 118*(2), 317–328. https://doi.org/10.1029/2012jd018319.
Gordley, L. L., et al. (2009). The solar occultation for ice experiment. *Journal of Atmospheric and Solar – Terrestrial Physics, 71*, 300–315.
Gumbel, J., & Witt, G. (1998). In situ measurements of the vertical structure of a noctilucent cloud. *Geophysical Research Letters, 25*(4), 493–496.

Gumbel, J., & Witt, G. (2001). Rocket-borne photometry of NLC particle populations. *Advances in Space Research, 28*(7), 1053–1058.

Gumbel, J., Stegman, J., Murtagh, D. P., & Witt, G. (2001). Scattering phase functions and particle sizes in noctilucent clouds. *Geophysical Research Letters, 28*(8), 1415–1418. https://doi.org/10.1029/2000GL012414.

Hansen, G., Serwazi, M., & von Zahn, U. (1989). First detection of a noctilucent cloud by lidar. *Geophysical Research Letters, 16*, 12, 1445–1448. https://doi.org/10.1029/gl016i012p01445.

Hervig, M., Thompson, R. E., McHugh, M., Gordley, L. L., Russell III, J. M., & Summers, M. E. (2001). First confirmation that water ice is the primary component of polar mesospheric clouds. *Geophysical Research Letters, 28*, 971–974.

Hervig, M., & Siskind, D. (2006). Decadal and inter-hemispheric variability in polar mesospheric clouds, water vapor, and temperature. *Journal of Atmospheric and Solar – Terrestrial Physics, 68*(1), 30–41. https://doi.org//10.1016/j.jastp.2005.08.010.

Hervig, M. E., Gordley, L. L., Russell III, J. M., & Bailey, S. M. (2009a). SOFIE PMC observations during the northern summer of 2007. *Journal of Atmospheric and Solar – Terrestrial Physics, 71*, 331–339. https://doi.org/10.1016/j.jastp.2008.08.010.

Hervig, M. E., Gordley, L. L., Deaver, L. E., Siskind, D. E., Stevens, M. H., Russell III, J. M., Bailey, S. M., Megner, L., & Bardeen, C. G. (2009b). First Satellite Observations of Meteoric Smoke in the Middle Atmosphere. *Geophysical Research Letters, 36*, L18805. https://doi.org/10.1029/2009GL039737.

Hervig, M. E., & Gordley, L. L. (2010). Temperature, shape, and phase of mesospheric ice from Solar Occultation for Ice Experiment observations. *Journal of Geophysical Research, 115*, D15208. https://doi.org/10.1029/2010JD013918.

Hervig, M. E., Berger, U., & Siskind, D. E. (2016). Decadal variability in PMCs and implications for changing temperature and water vapor in the upper mesosphere. *Journal of Geophysical Research-Atmospheres, 121*, 2383–2392. https://doi.org/10.1002/2015JD024439.

Höffner, J., Fricke-Begemann, C., & Lübken, F.-J. (2003). First observations of noctilucent clouds by lidar at Svalbard, 78N. *Atmospheric Chemistry and Physics, 3*(4), 1101–1111. https://doi.org/10.5194/acp-3-1101-2003.

Hoffmann, C., von Savigny, C., Hervig, M., & Oberbremer, E. (2018). The lunar semidiurnal tide at the polar summer mesopause observed by SOFIE. *Journal of Atmospheric and Solar – Terrestrial Physics, 167*, 134–145.

Jesse, O. (1885). Auffallende Abenderscheinungen am Himmel. *Meteorologische Zeitschrift, 2*, 311.

Jesse, O. (1887). Die Beobachtung der leuchtenden Wolken. *Meteorologische Zeitschrift, 4*, 179–181.

Jesse, O. (1896). Die Höhe der leuchtenden Nachtwolken. *Astronomische Nachrichten, 140*, 161–168.

Kaifler, N., Baumgarten, G., Fiedler, J., & Lübken, F.-J. (2013a). Quantification of waves in lidar observations of noctilucent clouds at scales from seconds to minutes. *Atmospheric Chemistry and Physics, 13*(23), 11757–11768. https://doi.org/10.5194/acpd-13-7397-2013.

Kaifler, N., Baumgarten, G., Klekociuk, A. R., Alexander, S. P., Fiedler, J., & Lübken, F.-J. (2013b). Small scale structures of NLC observed by lidar at 69N/69S and their possible relation to gravity waves. *Journal of Atmospheric and Solar – Terrestrial Physics, 104*, 244–252. https://doi.org/10.1016/j.jastp.2013.01.004.

Karlsson, B., Körnich, H., & Gumbel, J. (2007). Evidence for interhemispheric stratosphere-mesosphere coupling derived from noctilucent cloud properties. *Geophysical Research Letters, 34*, L16806. https://doi.org/10.1029/2007GL030282.

Kiliani, J., Baumgarten, G., Lübken, F.-J., Berger, U., & Hoffmann, P. (2013). Temporal and spatial characteristics of the formation of strong noctilucent clouds. *Journal of Atmospheric and Solar – Terrestrial Physics, 104*, 151–166. https://doi.org/10.1016/j.jastp.2013.01.005.

Kiliani, J., Baumgarten, G., Lübken, F.-J., & Berger, U. (2015). Impact of particle shape on the morphology of noctilucent clouds. *Atmospheric Chemistry and Physics, 15*(22), 12897–12907. https://doi.org/10.5194/acp-15-12897-2015.

Kirkwood, S., & Stebel, K. (2003). Influence of planetary waves on noctilucent cloud occurrence over NW Europe. *Journal of Geophysical Research, 108*(D8), 8440. https://doi.org/10.1029/2002JD002356.

Köhnke, M., von Savigny, C., & Robert, C. (2018). Observation of a 27-day solar signature in noctilucent cloud altitude. *Advances in Space Research, 61*(10), 2531–2539.

Körnich, H., & Becker, E. (2010). A simple model for the interhemispheric coupling of the middle atmosphere circulation. *Advances in Space Research, 45*(5), 661–668. https://doi.org/10.1016/j.asr.2009.11.001.

Kohlrausch, E. (1887). Zur Höhe der Wolken. *Annalen der Physik und Chemie, 267*(8), 1047–1048. https://doi.org/10.1002/andp.18872670835.

Kropotkina, Y. P., & Shefov, N. N. (1975). Influence of lunar tides on the probability of appearance of noctilucent clouds. *Russian Academy of Sciences, Atmospheric and Oceanic Physics, 11*, 1184–1186.

Langowski, M., von Savigny, C., Bachmann, T., & DeLand, M. (2018). First results on the retrieval of noctilucent cloud albedo and occurrence rate from SCIAMACHY nadir measurements. *Journal of Atmospheric and Solar – Terrestrial Physics, 175*, 31–39.

Leslie, R. C. (1885). *Sky glows. Nature, 32*, 245–245. https://doi.org/10.1038/032245a0.

López-Puertas, M., García-Comas, M., Funke, B., Bermejo-Pantaleón, D., Höpfner, M., Grabowski, U., Stiller, G. P., von Clarmann, T., & von Savigny, C. (2010). Measurements of polar mesospheric clouds in infrared emission by MIPAS/ENVISAT. *Journal of Geophysical Research, 114*, D00I07. https://doi.org/10.1029/2009JD012548, [printed 116(D1), 2011].

Llewellyn, E. J., Lloyd, N. D., Degenstein, D. A., Gattinger, R. L., Petelina, S. V., Bourassa, A. E., Wiensz, J. T., Ivanov, E. V., McDade, I. C., Solheim, B. H., McConnell, J. C., Haley, C. S., von Savigny, C., Sioris, C. E., McLinden, C. A., Griffioen, E., Kaminski, J., Evans, W. F. J., Puckrin, E., Strong, K., Wehrle, V., Hum, R. H., Kendall, D. J. W., Matsushita, J., Murtagh, D. P., Brohede, S., Stegman, J., Witt, G., Barnes, G., Payne, W. F., Piche, L., Smith, K., Warshaw, G., Deslauniers, D.-L., Marchand, P., Richardson, E. H., King, R. A., Wevers, I., McCreath, W., Kyrölä, E., Oikarinen, L., Leppelmeier, G. W., Auvinen, H., Megie, G., Hauchecorne, A., Lefevre, F., de La Nöe, J., Ricaud, P., Frisk, U., Sjoberg, F., von Scheele, F., & Nordh, L. (2004). The OSIRIS instrument on the Odin satellite. *Canadian Journal of Physics, 82*, 411–422. https://doi.org/10.1139/P04-005.

Lübken, F.-J., Baumgarten, G., Fiedler, J., Gerding, M., Höffner, J., & Berger, U. (2008). Seasonal and latitudinal variation of noctilucent cloud altitudes. *Geophysical Research Letters, 35*, 6. https://doi.org/10.1029/2007gl032281.

Lübken, F.-J., Höffner, J., Viehl, T. P., Kaifler, B., & Morris, R. J. (2014). Winter/summer mesopause temperature transition at Davis (69°S) in 2011/2012. *Geophysical Research Letters, 41*, 5233–5238. https://doi.org/10.1002/2014GL060777.

Lübken, F.-J., Berger, U., & Baumgarten, G. (2018). On the anthropogenic impact on long-term evolution of noctilucent clouds. *Geophysical Research Letters, 45*. https://doi.org/10210.1029/2018GL077719.

McClintock, W. E., Rusch, D. W., Thomas, G. E., Merkel, A. W., Lankton, M. R., Drake, V. A., Bailey, S. M., & Russell, J. M. (2009). The cloud imaging and particle size experiment on the Aeronomy of Ice in the mesosphere mission: Instrument concept, design, calibration, and on-orbit performance. *Journal of Atmospheric and Solar – Terrestrial Physics, 71*, 3–4, 340–355. https://doi.org/10.1016/j.jastp.2008.10.011.

Merkel, A. W., Thomas, G. E., Palo, S. E., & Bailey, S. M. (2003). Observations of the 5-day planetary wave in PMC measurements from the Student Nitric Oxide Explorer Satellite. *Geophysical Research Letters, 30*(4), 1194. https://doi.org/10.1029/2002GL016524.

Merkel, A. W., Garcia, R. R., Bailey, S. M., & Russell III, J. M. (2008). Observational studies of planetary waves in PMCs and mesospheric temperature measured by SNOE and SABER. *Journal of Geophysical Research, 113*, D14202. https://doi.org/10.1029/2007JD009396.

Murray, B. J., & Plane, J. M. C. (2005). Modelling the impact of noctilucent cloud formation on atomic oxygen and other minor constituents of the summer mesosphere. *Atmospheric Chemistry and Physics, 5*, 1027–1038.

Nielsen, K., Siskind, D. E., Eckermann, S. D., Hoppel, K. W., Coy, L., McCormack, J. P., Benze, S., Randall, C. E., & Hervig, M. E. (2010). Seasonal variation of the quasi 5 day planetary wave: Causes and consequences for polar mesospheric cloud variability in 2007. *Journal of Geophysical Research, 115*, D18111. https://doi.org/10.1029/2009JD012676.

Nussbaumer, V., Fricke, K. H., Langer, M., Singer, W., & von Zahn, U. (1996). First simultaneous and common volume observations of noctilucent clouds and polar mesosphere summer echoes by lidar and radar. *Journal of Geophysical Research-Atmospheres, 101*(D14), 19161–19167. https://doi.org/10.1029/96jd01213.

Pérot, K., Hauchecorne, A., Montmessin, F., Bertaux, J.-L., Blanot, L., Dalaudier, F., Fussen, D., & Kyrölä, E. (2010). First climatology of polar mesospheric clouds from GOMOS/ENVISAT stellar occultation instrument. *Atmospheric Chemistry and Physics, 10*, 2723–2735. https://doi.org/10.5194/acp-10-2723-2010.

Pertsev, N. N., Dalin, P. A., & Perminov, V. I. (2015). Influence of semidiurnal and semimonthly lunar tides on the mesopause as observed in hydroxyl layer and noctilucent clouds characteristics. *Geomagnetism and Aeronomy, 55*, 811. https://doi.org/10.1134/S0016793215060109.

Petelina, S. V., Llewellyn, E. J., Degenstein, D. A., & Lloyd, N. D. (2006). Odin/OSIRIS limb observations of polar mesospheric clouds in 2001–2003. *Journal of Atmospheric and Solar – Terrestrial Physics, 68*(1), 42–55. https://doi.org/10.1016/j.jastp.2005.08.004.

Peters, D. H. W., Entzian, G., & Keckhut, P. (2017). Mesospheric temperature trends derived from standard phase-height measurements. *Journal of Atmospheric and Solar – Terrestrial Physics, 163*, 23–30. https://doi.org/10.1016/j.jastp.2017.04.007.

Plane, J. M. C., Murray, B. J., Chu, X. Z., & Gardner, C. S. (2004). Removal of meteoric iron on polar mesosperic clouds. *Science, 304*, 426–428.

Rapp, M., & Thomas, G. E. (2006). Modeling the microphysics of mesospheric ice particles: Assessment of current capabilities and basic sensitivities. *Journal of Atmospheric and Solar – Terrestrial Physics, 68*(7), 715–744. https://doi.org/10.1016/j.jastp.2005.10.015.

Rapp, M., Thomas, G. E., & Baumgarten, G. (2007). Spectral properties of mesospheric ice clouds: Evidence for nonspherical particles. *Journal of Geophysical Research, 112*, D03211. https://doi.org/10.1029/2006JD007322.

Ridder, C., Baumgarten, G., Fiedler, J., & Lübken, F.-J., & Stober, G. (2017). Analysis of small-scale structures in lidar observations of noctilucent clouds using a pattern recognition method. *Journal of Atmospheric and Solar – Terrestrial Physics, 162*, 48–56. https://doi.org/10.1016/j.jastp.2017.04.005.

Robert, C. E., von Savigny, C., Burrows, J. P., & Baumgarten, G. (2009). Climatology of noctilucent cloud radii and occurrence frequency using SCIAMACHY. *Journal of Atmospheric and Solar – Terrestrial Physics, 71*(3–4), 408–423.

Robert, C. E., von Savigny, C., Rahpoe, N., Bovensmann, H., Burrows, J. P., DeLand, M. T., & Schwartz, M. J. (2010). First evidence of a 27 day solar signature in noctilucent cloud occurrence frequency. *Journal of Geophysical Research, 115*, D00I12. https://doi.org/10.1029/2009JD012359, [printed 116(D1), 2011].

Romejko, V. A., Dalin, P. A., & Pertsev, N. N. (2003). Forty years of noctilucent cloud observations near Moscow: Database and simple statistics. *Journal of Geophysical Research, 108*(D8), 8443. https://doi.org/10.1029/2002JD002364.

Rusch, D. W., Thomas, G. E., & Jensen, E. J. (1991). Particle size distributions in polar mesospheric clouds derived from Solar Mesosphere Explorer measurements. *Journal of Geophysical Research, 96*, 12933–12939.

Rusch, D., Thomas, G., Merkel, A., Olivero, J., Chandran, A., Lumpe, J., Carstans, J., Randall, C., Bailey, S., & Russell III, J. (2017). Large ice particles associated with small ice water content observed by AIM CIPS imagery of polar mesospheric clouds: Evidence for microphysical coupling with small-scale dynamics. *Journal of Atmospheric and Solar – Terrestrial Physics, 162*, 97–105. https://doi.org/10.1016/j.jastp.2016.04.018.

Russell, J. M., Bailey, S. M., Gordley, L. L., Rusch, D. W., Horányi, M., Hervig, M. E., Thomas, G. E., Randall, C. E., Siskind, D. E., Stevens, M. H., Summers, M. E., Taylor, M. J., Englert, C. R., Espy, P. J., McClintock, W. E., & Merkel, A. W. (2009). The Aeronomy of Ice in the

Mesosphere (AIM) mission: Overview and early science results. *Journal of Atmospheric and Solar – Terrestrial Physics, 71*, 3–4, 289–299. https://doi.org/10.1016/j.jastp.2008.08.011.

Schroeder, W. (1999). Were Noctilucent Clouds Caused by the Krakatoa Eruption? A Case Study of the Research Problems before 1885. *Bulletin of the American Meteorological Society, 80*, 2081–2018.

Siskind, D. E., Merkel, A. W., Marsh, D. R., Randall, C. E., Hervig, M. E., Mlynczak, M. G., & Russell III, J. M. (2018). Understanding the effects of polar mesospheric clouds on the environment of the upper mesosphere and lower thermosphere. *Journal of Geophysical Research-Atmospheres, 123*. https://doi.org/10.1029/2018JD028830.

Stebel, K., Barabash, V., Kirkwood, S., Siebert, J., & Fricke, K. H. (2000). Polar mesosphere summer echoes and noctilucent clouds: Simultaneous and common-volume observations by radar, lidar and CCD camera. *Geophysical Research Letters, 27*, 5, 661–664. https://doi.org/10.1029/1999gl010844.

Stening, R. J., Tsuda, T., & Nakamura, T. (2003). Lunar tidal winds in the upper atmosphere over Jakarta. *Journal of Geophysical Research, 108*, 1192.

Stevens, M. H., Englert, C. R., Hervig, M., Petelina, S. V., Singer, W., & Nielsen, K. (2009). The diurnal variation of polar mesospheric cloud frequency near 55°N observed by SHIMMER. *Journal of Atmospheric and Solar – Terrestrial Physics, 71*, 401–407.

Størmer, C. (1935). Measurements of Luminous Night Clouds in Norway 1933 and 1934. *Astrophysica Norvegica, 1*, 87.

Summers, M. E., Conway, R. R., Englert, C. R., Siskind, D. E., Stevens, M. H., Russell, J. M., Gordley, L. L., & McHugh, M. J. (2001). Discovery of a water vapor layer in the Arctic summer mesosphere: Implications for polar mesospheric clouds. *Geophysical Research Letters, 28*, 3601–3604.

Symons, G., Judd, J., Strachey, R., Wharton, W., Evans, F., Russell, F., Archibald, D., & Whipple, G. (1888). The Eruption of Krakatoa: And Subsequent Phenomena, Krakatoa Committee of the Royal Society (Great Britain).

Thomas, G. E., Olivero, J. J., Jensen, E. J., Schröder, W., & Toon, O. B. (1989). Relation between increasing methane and the presence of ice clouds at the mesopause. *Nature, 338*, 490–492.

Thomas, G. E., McPeters, R. D., & Jensen, E. J. (1991). Satellite observations of polar mesospheric clouds by the solar backscattered ultraviolet spectral radiometer: Evidence of a solar cycle dependence. *Journal of Geophysical Research, 96*(D1), 927–939. https://doi.org/10.1029/90JD02312.

Thomas, G. E., Thurairajah, B., Hervig, M. E., von Savigny, C., & Snow, M. (2015). Solar-Induced 27-day Variations of Mesospheric Temperature and Water Vapor from the AIM SOFIE experiment: Drivers of Polar Mesospheric Cloud Variability. *Journal of Atmospheric and Solar – Terrestrial Physics, 134*, 56–68.

Thurairajah, B., Thomas, G. E., von Savigny, C., Snow, M., Hervig, M. E., Bailey, S. M., & Randall, C. E. (2016). Solar-induced 27-day variations of polar mesospheric clouds from AIM SOFIE and CIPS Experiments. *Journal of Atmospheric and Solar – Terrestrial Physics, 162*, 122–135.

von Cossart, G., Fiedler, J., & von Zahn, U. (1999). Size distributions of NLC particles as determined from 3-color observations of NLC by ground-based lidar. *Geophysical Research Letters, 26*, 11, 1513–1516. https://doi.org/10.1029/1999gl900226.

von Savigny, C., Kokhanovsky, A., Bovensmann, H., Eichmann, K.-U., Kaiser, J. W., Noël, S., Rozanov, A. V., Skupin, J., & Burrows, J. P. (2004). NLC Detection and Particle Size Determination: First Results from SCIAMACHY on ENVISAT. *Advances in Space Research, 34*(4), 851–856.

von Savigny, C., Petelina, S. V., Karlsson, B., Llewellyn, E. J., Degenstein, D. A., Lloyd, N. D., & Burrows, J. P. (2005). Vertical variation of NLC particle sizes retrieved from Odin/OSIRIS limb scattering observations. *Geophysical Research Letters, 32*, L07806. https://doi.org/10.1029/2004GL021982.

von Savigny, C., Robert, C., Bovensmann, H., Burrows, J. P., & Schwartz, M. (2007a). Satellite observations of the quasi 5-day wave in noctilucent clouds and mesopause temperatures. *Geophysical Research Letters, 34*, L24808. https://doi.org/10.1029/2007GL030987.

von Savigny, C., Sinnhuber, M., Bovensmann, H., Burrows, J. P., Kallenrode, M.-B., & Schwartz, M. (2007b). On the disappearance of noctilucent clouds during the January 2005 solar proton events. *Geophysical Research Letters, 34*, L02805. https://doi.org/10.1029/2006GL028106.

von Savigny, C., Lednyts'kyy, O., Forbes, J.M., & Zhang, X. (2015). Lunar semidiurnal tide in the terrestrial airglow. *Geophysical Research Letters, 42*, 3553–3559.

von Savigny, C., DeLand, M. T., & Schwartz, M. J. (2017). First identification of lunar tides in satellite observations of noctilucent clouds. *Journal of Atmospheric and Solar – Terrestrial Physics, 162*, 116–121.

von Zahn, U., von Cossart, G., Fiedler, J., & Rees, D. (1998). Tidal variations of noctilucent clouds measured at 69N latitude by groundbased lidar. *Geophysical Research Letters, 25*(9), 1289–1292. https://doi.org/10.1029/98gl00546.

von Zahn, U., von Cossart, G., Fiedler, J., Fricke, K. H., Nelke, G., Baumgarten, G., Rees, D., Hauchecorne, A., & Adolfsen, K. (2000). The ALOMAR Rayleigh/Mie/Raman lidar: objectives, configuration, and performance. *Annales Geophysicae, 18*(7), 815–833. https://doi.org/10.1007/s00585-000-0815-2.

Wegener, A. (1912). Die Erforschung der obersten Atmosphärenschichten. *Gerlands Beitraege zur Geophysik, 11*, 104–124.

Wickwar, V. B. (2002). Visual and lidar observations of noctilucent clouds above Logan, Utah, at 41.7N. *Journal of Geophysical Research, 107*, D7. https://doi.org/10.1029/2001jd001180.

Wiens, R. H., Evans, W. F. J., Zalcik, M. S., Manson, A. H., & Shepherd, G. G. (1995). WINDII observation of a PMC breakup event during ANLC-93. *Geophysical Research Letters, 22*(20), 2797–2800. https://doi.org/10.1029/95GL03023.

Witt, G. (1957). Noctilucent cloud observations. *Tellus, IX*(3), 365–371.

Witt, G. (1962). Height, structure and displacements of noctilucent clouds. *Tellus, 14*(1), 1–18. https://doi.org/10.3402/tellusa.v14i1.9524.

Chapter 9
Remote Sensing of Arctic Atmospheric Aerosols

Alexander Kokhanovsky, Claudio Tomasi, Alexander Smirnov, Andreas Herber, Roland Neuber, André Ehrlich, Angelo Lupi, Boyan H. Petkov, Mauro Mazzola, Christoph Ritter, Carlos Toledano, Thomas Carlund, Vito Vitale, Brent Holben, Tymon Zielinski, Simon Bélanger, Pierre Larouche, Stefan Kinne, Vladimir Radionov, Manfred Wendisch, Jason L. Tackett, and David M. Winker

Abstract In this chapter remote sensing techniques as applied to studies of Arctic aerosol are surveyed. They include the analysis of ground and shipborne observations of atmospheric aerosol using sunphotometers and also airborne/satellite observations using optical instrumentation (lidars, imagers, radiometers).

Keywords Aerosol · Aerosol optical thickness · Radiative transfer · Lidar · Sunphotometer · AERONET · Satellite remote sensing · Airborne remote sensing · Ship-borne remote sensing · CALIOP · CALIPSO

A. Kokhanovsky
Vitrociset Belgium SPRL, Darmstadt, Germany
e-mail: a.kokhanovsky@vitrocisetbelgium.com

C. Tomasi · A. Lupi · B. H. Petkov · M. Mazzola (✉) · V. Vitale
Institute of Atmospheric Sciences and Climate (ISAC), Italian National Research Council (CNR), Bologna, Italy
e-mail: c.tomasi@isac.cnr.it; a.lupi@isac.cnr.it; b.petkov@isac.cnr.it; m.mazzola@isac.cnr.it; v.vitale@isac.cnr.it

A. Smirnov
Science Systems and Applications, Inc., Lanham, MA, USA

Biospheric Sciences Laboratory, NASA Goddard Space Flight Center, Greenbelt, MD, USA
e-mail: Alexander.Smirnov-1@nasa.gov

A. Herber
Alfred Wegener Institute Helmholtz Centre for Polar and Marine Research, Bremerhaven, Germany
e-mail: Andreas.Herber@awi.de

R. Neuber · C. Ritter
Alfred Wegener Institute Helmholtz Centre for Polar and Marine Research, Potsdam, Germany
e-mail: Roland.Neuber@awi.de; Christoph.Ritter@awi.de

A. Kokhanovsky, C. Tomasi (eds.), *Physics and Chemistry of the Arctic Atmosphere*, Springer Polar Sciences, https://doi.org/10.1007/978-3-030-33566-3_9

9.1 Remote Sensing of Arctic Aerosols Using Ground-Based Optical Measurements

9.1.1 Introduction

The ground-based remote sensing measurements of aerosol optical parameters in the Arctic atmosphere are for the most part conducted with multi-wavelength sun-photometers. A sun-photometer is oriented towards the Sun to detect the incoming direct solar irradiance at the surface-level after its attenuation occurred during the passage of the atmosphere. The atmospheric aerosol load causes a decrease in the

A. Ehrlich · M. Wendisch
Leipzig Institute for Meteorology (LIM), University of Leipzig, Leipzig, Germany
e-mail: a.ehrlich@uni-leipzig.de; m.wendisch@uni-leipzig.de

C. Toledano
Faculty of Science, University of Valladolid, Valladolid, Spain
e-mail: toledano@goa.uva.es

T. Carlund
Swedish Meteorological and Hydrological Institute (SMHI), Norrköping, Sweden
e-mail: Thomas.Carlund@smhi.se

B. Holben
Biospheric Sciences Laboratory, NASA Goddard Space Flight Center, Greenbelt, MD, USA
e-mail: Brent.N.Holben@nasa.gov

T. Zielinski
Institute of Oceanology, Polish Academy of Sciences, Sopot, Poland
e-mail: tymon@iopan.pl

S. Bélanger
Département de biologie, chimie et géographie, Université du Québec à Rimouski, Rimouski, QC, Canada
e-mail: simon_belanger@uqar.ca

P. Larouche
Institut Maurice-Lamontagne, Pêches et Océans Canada, Mont-Joli, QC, Canada
e-mail: Pierre.Larouche@dfo-mpo.gc.ca

S. Kinne
Max-Planck Institute for Meteorology, Hamburg, Germany
e-mail: stefan.kinne@mpimet.mpg.de

V. Radionov
Arctic and Antarctic Research Institute, Saint Petersburg, Russia
e-mail: vradion@aari.ru

J. L. Tackett
Science Systems and Applications, Inc., Hampton, VA, USA
e-mail: jason.l.tackett@nasa.gov

D. M. Winker
NASA Langley Research Center, Hampton, VA, USA
e-mail: david.m.winker@nasa.gov

incoming direct solar irradiance at the top-of-atmosphere (TOA) level, which is expressed in terms of the aerosol optical thickness *AOT(λ)*, given by the integral of the volume aerosol extinction coefficient along the vertical path of the atmosphere. The spectral dependence of *AOT(λ)* in the visible and near-infrared is described with good approximation by the so-called Ångström (1964) formula, which establishes that *AOT(λ)* is proportional to the wavelength power term $\lambda^{-\alpha}$ (with λ measured in μm), where α is the Ångström's exponent. The Ångström formula implies that *AOT(λ)* substantially varies as a function of the optical weights of the size-distribution modes consisting of fine particles (giving by themselves relatively high values of α) and of accumulation and coarse particles (giving rather low values of exponent α, generally close to null in the cases presenting predominant extinction features due to coarse particles).

The largest networks of sun-photometers operated in the Arctic are those of AERONET (Holben et al. 1998), GAW (Wehrli 2000), and POLAR-AOD (Tomasi et al. 2007). Spectral measurements of *AOT(λ)* are performed at various Arctic sites with the Cimel CE-318 sun-photometers of the AERONET network, which are equipped with: (i) an entrance quartz window, to prevent dust entering the instrument; (ii) an optical head with a pair of collimators; (iii) an angular field of view (hereinafter referred to as FOV) having a diameter of 1.2°, for both direct solar irradiance and sky-diffuse radiation measurements; (iv) a step-by-step rotating filter wheel, which allow one to place a series of seven narrow-band interference filters peaked at wavelengths ranging from 0.34 to 1.60 μm in order to measure the values of *AOT(λ)* on the optical axis of the instrument and record the overall spectral sequence of monochromatic direct solar irradiance measurements; and (v) an UV enhanced silicon Hamamatsu detector (model 51,336) used as sensor for the direct solar irradiance measurements, its operating temperature varying usually from –30 °C to +60 °C. To avoid thermal errors of the sun-photometer measurements, the corrections of the output voltages *V(λ)* measured by this instrument have been made by taking into account the measurements of the detector internal temperature recorded with a thermistor, in such a way to compensate the output signal for any temperature dependence effect produced on the silicon detector. In the current AERONET algorithm (version 3), each spectral channel has an individual temperature correction properly measured in a thermal chamber.

The peak-wavelengths of the above-mentioned narrow-band interference filters are equal to 340, 380, 440, 500, 675, 870, and 1020 nm (and 1600 nm in some more advanced examples equipped with a supplemental filter), as shown in Fig. 9.1. An additional narrow-band interference filter peaked at the 936 nm wavelength is also usually inserted inside the filter wheel to measure the vertical atmospheric content of water vapour (hereinafter referred to as precipitable water (PW) or as atmospheric water vapour content). The full-width-half-maximum (FWHM) values of the first two filters centred at the 340 and 380 nm wavelengths and mounted within the filter wheel of the Cimel CE-318 sun-photometers are equal to 3 nm, and those of the other interference filters centred at wavelengths ranging from 440 to 1020 (or 1640) nm (as well as the water vapour filter centred at the 936 nm wavelength) are equal to about 10 nm.

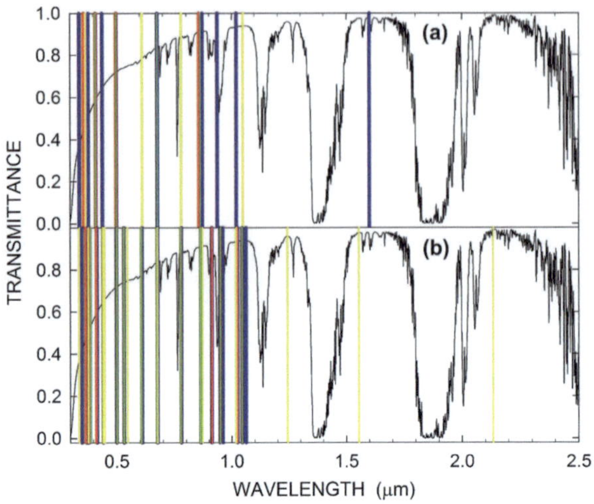

Fig. 9.1 Part (**a**): Atmospheric transmittance curve of the 75° N atmospheric model of Tomasi et al. (1998) having: (i) an atmospheric water vapour content of 0.8 g cm^{-2}, (ii) an aerosol vertical load giving optical thickness *AOT(0.50 μm)* = 0.05 and wavelength exponent $\alpha = 1.40$, and (iii) ground-level visual range $V_o = 50$ km. The peak wavelengths of the narrow-band interference filters mounted on the various sun-photometer models are indicated with different colours for: (i) the Cimel CE-318 sun-photometers of the AERONET network (blue colour), (ii) the PFR sun-photometers of the Global Atmosphere Watch (GAW) network (red colour), and (iii) the SP02 (green colour) and SP022 Carter Scott (yellow colour) sun-photometers used by the GMD/NOAA team at Barrow (Alaska). Part (**b**): As in the part (**a**) for (i) the AWI SP1A sun-photometer (blue colour), the AWI SP2H sun-photometer (red colour) and the AWI STAR01 star-photometer (green colour); and (ii) the new Portable Sun-photometer (SPM) manufactured by the IAO-SB-RAS Institute (Tomsk, Russia) (yellow colour)

The calculations of the Ångström exponent α are made using the values of *AOT(λ)* obtained at the 440, 500, 675, and 870 nm wavelengths, with an accuracy usually ranging between 0.01 and 0.02 at the visible and near-infrared wavelengths. The field measurements performed with the Cimel CE-318 sun-photometers employed in the AERONET and AEROCAN networks are regularly conducted at the following nine Arctic sites: (i) Barrow (Alaska, USA) by NASA/GSFC (USA), AERONET; (ii) Resolute Bay (Nunavut, Canada) by AEROCAN (Canada); (iii) Eureka – PEARL (Nunavut, Canada) by AEROCAN (Canada); (iv) Thule (North-western Greenland) by NASA/GSFC (USA.), AERONET; (v) Ittoqqortoomiit (Eastern Greenland) by NASA/GSFC (USA), AERONET; (vi) Hornsund (Spisbergen, Svalbard) by NASA/GSFC (USA.), AERONET, in cooperation with the Warsaw University (PAS, Poland); (vii) Andenes/Andøya Rocket Range (Northern Norway) by the GOA (Grupo de Óptica Atmosférica), Universidad de Valladolid (Valladolid, Spain), in cooperation with AERONET (NASA/GSFC, USA), to conduct routine sun-photometer measurements at the Arctic Lidar Observatory for Middle Atmosphere Research (ALOMAR), located at Andenes (Andøya

Table 9.1 List of the Arctic stations where the Cimel CE-318 sun-photometers of the AERONET and AEROCAN networks have been regularly operated over the field measurement periods of the last decade

Sun-photometer station	Geographical coordinates and altitude	Managing institution	Measurement period
Barrow (Alaska, USA)	71° 19′ N, 156° 40′ W, 2 m a.m.s.l.	AERONET (NASA/GSFC, U.S.A.)	March 2002–September 2013
Resolute Bay (Nunavut, Canada)	74° 25′ N, 94° 34′ W, 65 m a.m.s.l.	AEROCAN, (Canada)	April 2008–September 2011
Eureka – PEARL (Nunavut, Canada)	79° 59′ N, 85° 56′ W, 2 m a.m.s.l.	AEROCAN (Canada)	May 2007–August 2011
Thule (North-Western Greenland)	76° 31′ N, 68° 46′ W, 225 m a.m.s.l.	AERONET (NASA/GSFC, U.S.A.)	March 2007–September 2011
Ittoqqortoomiit (Eastern Greenland)	70° 29′ N, 21° 57′ W, 68 m a.m.s.l.	AERONET (NASA/GSFC, U.S.A.)	May 2010–October 2013
Hornsund (Spisbergen, Svalbard)	77° 00′ N, 15° 34′ E, 10 m a.m.s.l.	AERONET (NASA/GSFC, U.S.A.) and Warsaw University (PAS, Poland)	March 2007–September 2012
Andenes/Andøya rocket range (Norway)	69° 18′ N, 16° 01′ E, 13 m a.m.s.l.	AERONET (NASA/GSFC, U.S.A.) and GOA, Universidad de Valladolid (Spain)	March 2002–October 2009
Sodankylä (Lapland, Northern Finland)	67° 22′ N, 26° 38′ E, 184 m a.m.s.l.	AERONET, (NASA/GSFC, U.S.A.) and FMI (Helsinki, Finland)	May 2004–September 2013
Tiksi (Northern-Central Siberia)	71° 35′ N, 128° 47′ E, 40 m a.m.s.l.	AERONET, (NASA/GSFC, U.S.A.) and IAO-SB-RAS Institute (Tomsk, Russia)	June–October of 2010, 2011 and 2012

Rocket Range, Norway) (Toledano et al. 2012); (viii) Sodankylä (Lapland, Northern Finland) by the Finnish Meteorological Institute (FMI, Helsinki, Finland) in cooperation with AERONET (NASA/GSFC, USA); and (ix) Tiksi (Northern-Central Siberia), by the IAO-SB-RAS Institute (Tomsk, Russia) in cooperation with AERONET (NASA/GSFC, USA). The geographical coordinates of the nine stations are given in Table 9.1, together with the periods during which the field measurements analyzed in the present study were conducted.

The Precision Filter Radiometer (PFR) sun-photometers (Wehrli 2000; Nyeki et al. 2012) of the Global Atmosphere Watch (GAW) network are currently employed in the Arctic at four stations: (i) Summit (Greenland), where the PMOD/WRC (**Physical Meteorological Observatory in Davos, World Radiation Centre**) (Switzerland) has used the PFR#34 model from 2001 to 2011; (ii) Ny-Ålesund (Spitsbergen, Svalbard), where the NILU (Norwegian Institute for Air Research, Norway) has employed the PFR#18 model since 2002; (iii) Sodankylä (Northern Finland) where the Finnish Meteorological Institute (FMI, Helsinki, Finland) has employed the PFR#29 and PFR#32 models from 2007 to 2011; and (iv) Kiruna (Northern Sweden), where the Swedish Meteorological and Hydrological Institute (SMHI) (Norrköping, Sweden) employed some PFR instruments since 2007. The PFR sun-photometers are equipped with a series of four narrow-band interference filters manufactured using an ion-assisted deposition technique, which are centred at the 368, 412, 500 and 862 nm peak-wavelengths (as shown in Fig. 9.1) and have their spectral transmission curves with FWHM values equal to 5 nm (Wehrli 2000). More precisely, the PFR#18 sun-photometer located at Ny-Ålesund (Spitsbergen, Svalbard) and positioned on the roof of the Sverdrup Institute, is equipped with four interference filters having peak-wavelengths of 367.7, 411.9, 500.6, 862.5 nm, while the PFR#29 and PFR#32 employed by the FMI team at Sodankylä are both equipped with interference filters having peak-wavelengths of 367.6, 411.4, 500.5, 861.6 nm. Therefore, the PFR sun-photometers do not use a narrow-band interference filter suitable for measuring the atmospheric content of water vapour. These sun-photometer models have a FOV with angular diameter of 2.5° and are equipped with a silicon photodiode employed as sensor, tilted at an angle of 3° from the optical axis to reduce inter-reflection effects, and coupled with an internal shutter, which opens for only a few seconds every 1 min during the field measurements in order to limit the effects of exposure-related degradation. A common detector head hosts the four interference filters, whose temperature is maintained at (20 ± 0.1) °C by a Peltier-type thermostatic controller, to avoid that the interference filters are subject to diurnal temperature cycling when operated outdoors. The calculations of the Ångström exponent α are made using the values of $AOT(\lambda)$ obtained at the four above-mentioned wavelengths chosen within the 440–870 nm spectral range. The geographical coordinates of the four GAW stations are given in Table 9.2, together with the field measurement periods considered in the analysis discussed below.

As a part of the POLAR-AOD network activities, the SP02 and SP022 Carter Scott models were used at Barrow (Alaska, USA) (71° 19′ N, 156° 36′ W, 8 m a. m. s. l.) by the NOAA (National Oceanic and Atmospheric Administration)/Global Monitoring Division (GMD) (Boulder, Colorado, USA) over the period from March 2000 to September 2012, in order to study the aerosol radiative properties at this remote Arctic site in Alaska. As reported by Stone (2002) and Dutton et al. (2004), the two SP02 and SP022 Carter Scott sun-photometers were equipped with: (i) an entrance quartz window, to prevent dust entering the instrument; (ii) an optical head with a collimator having an aperture diameter of 4 mm, giving an angular FOV with diameter of 5°; (iii) an UV silicon photodiode used as detector, with a

Table 9.2 List of the Arctic stations where the PFR sun-photometers of the GAW network have been regularly operated over the field measurement periods of the last decade to measure *AOT(λ)* in the present study

Sun-photometer station	Geographical coordinates and altitude	Managing institution	Measurement period
Summit (Greenland ice sheet)	72° 20′ N, 38° 45′ W, 3270 m a.m.s.l.	PMOD/WRC (Davos, Switzerland)	January 2001–October 2011
Ny-Ålesund (Spitsbergen, Svalbard)	78° 54′ N, 11° 53′ E, 5 m a.m.s.l.	NILU (Kjeller, Norway)	March 2002–September 2004, and March 2996–September 2013
Sodankylä (Lapland, Northern Finland)	67° 22′ N, 26° 38′ E, 184 m a.m.s.l.	FMI (Helsinki, Finland)	March 2007–April 2011
Kiruna (Northern Sweden)	67° 50′ N, 20° 25′ E, 424 M A.M.S.L.	SMHI (Norrköping, Sweden)	January 2007 –October 2009

detection area of 32 mm^2 and maintained at an operating temperature ranging from –50 °C to +70 °C, for which the thermal control of ±1 °C was made to correct the output voltages for any temperature-dependence effect affecting the silicon detector voltage. Regular spectral measurements of *AOT(λ)* were performed by using the two above-cited Carter Scott sun-photometers at four wavelengths, equal to 412, 500, 675 and 862 nm for the SP02 model (having FWHM values equal to 10 nm for all these four interference filters) and equal to 368, 610, 778 and 1050 nm for the SP022 model (having a FWHM value of 5 nm for the 368 nm interference filter and FWHM values of 10 nm for the interference filters centred at the 610, 778 and 1050 nm wavelengths), as shown in Fig. 9.1. Therefore, the calculations of the Ångström exponent α were made using the values of *AOT(λ)* obtained at the 412, 500, 675 and 862 nm wavelengths for the SP02 Carter Scott model, and at the 368, 610 and 778 nm wavelengths for the SP022 Carter Scott model (Tomasi et al. 2007).

In the POLAR-AOD network, three different radiometers were employed by the Alfred Wegener Institute for Polar and Marine Research (AWI) (Bremerhaven, Germany) at Ny-Ålesund (Spitsbergen, Svalbard) (78° 54′ N, 11° 53′ E, 5 m a. m. s. l.) over the period from April 2000 to September 2013, in order to measure the diurnal and nocturnal values of *AOT(λ)* in the visible and near-infrared spectral range: (a) the SP1A sun-photometer model, equipped with 17 narrow-band interference filters centred at the 351, 371, 380, 416, 443, 500, 532, 609, 675, 778, 864, 911, 947 (for PW measurements), 961 (for PW measurements), 1025, 1046, and 1062 nm wavelengths; (b) the SP2H sun-photometer model, equipped with 14 narrow-band interference filters centred at the 367, 380, 413, 441, 501, 531, 605, 673, 776, 862, 912 (for PW measurements), 949 (for PW measurements), 1023, and 1045 nm wavelengths; and (c) the STAR01 star-photometer model, equipped with 10 narrow-band interference filters centred at the 390, 441, 501, 531, 605, 673, 776, 862, 949 (for PW measurements), and 1045 nm (Herber et al. 2002). The three spectral series of interference filters have FWHM values equal to 3 nm at all the wavelengths shorter than 400 nm, and to 10 nm at longer wavelengths (see Fig. 9.1). The SP1A and SP2H sun-photometers were equipped with: (i) an entrance quartz window and an optical system having a diameter of 10 mm, giving an angular FOV diameter of 1°; (ii) a Photodiode Hamamatsu S 1337 – 66BQ used as sensor, with a detection area equal to 4 mm^2, which were maintained at an operating temperature ranging from – 40 °C to +30 °C during the field measurements, for which the thermal control was made by taking into account the temperature shift evaluated by using appropriate temperature coefficients over the whole operative temperature range. The STAR01 model was equipped with: (i) a Telescope Maksutov (180 mm/2800 mm) used as entrance optical system, having an aperture diameter of 0.5 mm and giving an angular FOV diameter of 37 s of arc (i.e of about 0.01 degrees); and (ii) an avalanche diode SQ37 used as detector, having a detection area of 0.1 mm^2 and maintained at a temperature of 30.0–30.5 °C by using a thermal control device for an external temperature ranging between – 30 °C and + 15 °C. The measurements of *AOT(λ)* were performed with the three AWI instruments described above to obtain an accuracy ranging between 5 10^{-3} and 8 10^{-3} at visible wavelengths, while the calculations of the Ångström exponent α were made using the values of *AOT(λ)* obtained over: (a) the 380–864 wavelength range for the SP1A model, (b) the 380–862 wavelength range for the SP2H model, and (c) the 390–862 wavelength range for the STAR01 model (Tomasi et al. 2007, 2012). Some examples of the best-fit procedure applied to the spectral series of *AOT(λ)* derived from the SP1A and SP2H sun-photometer measurements, and from the STAR01 star-photometer measurements performed at the AWIPEV base (Ny-Ålesund) are shown in Fig. 9.2 for some summer cloudless-sky days, in order to determine the best-fit values of the Ångström exponent α, providing evidence of the high reliability of the AWI radiometers in measuring the Ångström (1964) atmospheric turbidity parameters.

The New Portable Sun-photometer (SPM) designed by the IAO-SB-RAS Institute (Tomsk, Russia) was also employed by this research group at the Arctic station of Barentsburg (Spitsbergen, Svalbard) (78° 04′ N, 14° 13′ E, 20 m a. m. s. l.) during the April–August months of 2011 and 2012. The SPM sun-photometer design was made by inserting: (i) an entrance quartz window, to prevent dust entering the instrument; (ii) an optical entrance system having an angular FOV diameter increasing with wavelength, being equal to 2.2° at the 340 nm wavelength, 3° at the 1019 nm wavelength, 3.2° at the 1244 nm wavelength, and 4.4° at the 2134 nm wavelength; (iii) a pair of detectors, that are the Si photodiode FDUK-13 model for carrying out the direct solar irradiance measurements over the wavelength range from 340 nm to 1019 nm, and the In-Ga-As photodiode, Hamamatsu G 8373 model for performing the solar radiation measurements at the wavelengths ranging from 1244 nm to 2134 nm; (iv) a thermal control system providing a stable internal temperature of 25 °C ± 0.5 °C, to avoid variations in the output voltages due to the very hard ambient temperature conditions. This sun-photometer model was equipped with a series of twelve narrow-band interference filters having peak-wavelengths equal to 340, 379, 443, 499, 548, 676, 871, 1019, 1244, 1555, and 2134 nm, as shown in Fig. 9.1, in order to measure the spectral values of *AOT(λ)*.

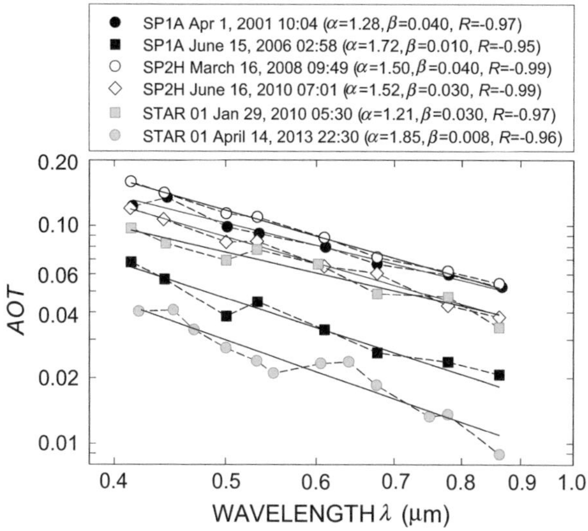

Fig. 9.2 Examples of application of the best-fit procedure based on the Ångström (1964) formula to six spectral series of *AOT(λ)* measured at the Ny-Ålesund (Koldewey/AWIPEV) station by the Alfred Wegener Institute (AWI, Germany) group: (i) measurements performed with the SP1A sun-photometer on April 1, 2001, 10:04 UTC (solid circles) and on June 15, 2006 (02:58 UTC) (solid squares); (ii) measurements performed with the SP2H sun-photometer on March 16, 2008 (09:49 UTC) (open circles) and on June 16, 2010 (07:01 UTC) (open diamonds); and (iii) measurements performed with the STAR 01 star-photometer on December 25, 2009 (20:30 UTC) (grey circles) and on January 29, 2010 (05:30 UTC) (grey squares). The best-fit values of Ångström's wavelength exponent α and atmospheric turbidity parameter β (equal to *AOT(1 μm)*) determined for the six spectral series of *AOT(λ)* are given in the legend, together with the corresponding values of regression coefficient *R*, as obtained over: (a) the 0.414–0.867 μm wavelength range for the SP1A measurements; (b) the 0.413–0.862 μm wavelength range for the SP2H measurements; and (c) the 0.414–0.861 μm wavelength range for the STAR 01 measurements

The FWHM values of the SPM sun-photometer are equal to 3 nm at the first two wavelengths, and gradually increase at the visible and near-infrared wavelengths until reaching a value of 13 nm at the 2134 nm wavelength (Sakerin et al. 2009). Using the SPM sun-photometer model, measurements of *AOT(λ)* were obtained with an accuracy ranging between 0.01 and 0.02 at all the visible and near-infrared wavelengths. The calculations of the Ångström exponent α were performed using the values of *AOT(λ)* obtained at the 443, 499, 548, 676, and 871 nm wavelengths, and therefore over the 443–871 nm spectral range in terms of the Ångström (1964) formula. The instrument was also equipped with an additional narrow-band interference filter peaked at the 941 nm wavelength to measure the atmospheric content of water vapour (PW) during each spectral scanning.

9.1.2 The Sun-Photometric Method Used to Measure the Aerosol Extinction Properties at Visible and Near-Infrared Wavelengths

The monochromatic total optical thickness $\tau_{tot}(\lambda)$ of the atmosphere is commonly calculated in terms of the well known Bouguer-Lambert–Beer law for a certain sun-photometer output voltage $V(\lambda)$ taken within a spectral channel centred at peak wavelength λ_p and for a certain apparent solar zenith angle θ_o. Thus, the monochromatic value of $\tau_{tot}(\lambda)$ can be calculated at a certain wavelength λ by using the following formula (Shaw 1976):

$$t_{tot}(l) = (1/m)\ \ln\left[F_e V_o(\lambda)\,/\,V(\lambda)\right], \tag{9.1}$$

based on the Bouguer-Lambert-Beer law, in which: (i) m is the relative optical air mass calculated as a function of solar zenith angle θ_o using a realistic model of the Arctic atmosphere (Tomasi and Petkov 2014), in which wet-air refraction and Earth/atmosphere curvature effects on the direct solar radiation beam passing through the atmosphere are properly taken into account (Thomason et al. 1983); (ii) $V(\lambda)$ is the output voltage measured by the ground-based sun-photometer (and, hence, proportional to the incoming direct solar irradiance entering the sun-photometer); (iii) $V_o(\lambda)$ is the extra-terrestrial output voltage of the sun-photometer for null m and annual mean Earth-Sun distance $D_{E\text{-}S}$; and (iv) F_e is the eccentricity correction factor equal to the square power of ratio $D_{E\text{-}S}/D$ between the annual mean Earth-Sun distance $D_{E\text{-}S}$ and the daily Earth-Sun distance D, which allows one to take into account the daily variation of $V_o(\lambda)$ occurring for the daily change in the Earth – Sun distance (Iqbal 1983). With regard to this parameter, it is worth noting that factor F_e is: (i) equal to 1.0350 on January 1; (ii) equal to 1.0351 from January 2 to 5; (iii) subsequently decreasing gradually to reach a value of 0.9666 on July 1, (iv) constant until to July 8, and (v) gradually increasing with date until reaching the value of 1.0350 on December 30 and 31.

Reliable evaluations of relative optical air mass m made for values of θ_o ranging between 0° and 87° were obtained by Tomasi and Petkov (2014) in the Arctic atmosphere, for the summer meteorological conditions observed at the Ny-Ålesund station by collecting a multi-year set of radiosounding measurements. These values are given in Table 9.3 for comparison with the values of m calculated by (i) Anderson et al. (1986) for the Subarctic Winter and Subarctic Summer models, and (ii) Tomasi et al. (1998) for the July – 75 °N Arctic atmospheric model, showing a substantial agreement over the range $0° \le \theta_o \le 75°$.

The product of relative optical air mass m by total optical thickness $\tau_{tot}(\lambda)$ of the atmosphere (see Eq. (9.1)) provides a measure of the overall extinction affecting the incoming solar radiation along the slant Sun path, and is given by the sum of various partial contributions due to Rayleigh scattering, aerosol extinction and gaseous absorption by ozone, nitrogen dioxide and N_2O_4, oxygen dimer, water vapour, molecular oxygen and carbon dioxide. Therefore, the overall atmospheric

Table 9.3 Values of relative optical air mass m of the atmosphere calculated at 30 selected values of apparent solar zenith angle θ_o over the range from 0° to 87° by (i) Anderson et al. (1986) for the Subarctic Winter and Subarctic Summer models, (ii) Tomasi et al. (1998) for the July – 75 °N Arctic model, and (iii) Tomasi and Petkov (2014) for wet-air conditions of the Arctic atmosphere measured by means of radiosounding measurements taken at Ny-Ålesund (Spitsbergen, Svalbard) (78° 54′ N, 11° 53′ E, 5 m a.m.s.l.) on several summer clear-sky days

θ_o (°)	Subarctic Winter model (Anderson et al. 1986)	Subarctic summer model (Anderson et al. 1986)	July – 75 °N Arctic model (Tomasi et al. 1998)	Arctic summer model(Tomasi and Petkov 2014)
0	1.0000	1.0000	1.0000	1.0000
10	1.0154	1.0154	1.0154	1.0154
15	1.0352	1.0352	1.0352	1.0352
20	1.0640	1.0640	1.0640	1.0640
25	1.1032	1.1032	1.1032	1.1031
30	1.1543	1.1543	1.1543	1.1543
35	1.2201	1.2201	1.2201	1.2201
40	1.3045	1.3045	1.3045	1.3044
45	1.4129	1.4127	1.4127	1.4127
50	1.5535	1.5534	1.5534	1.5533
55	1.7398	1.7395	1.7395	1.7396
60	1.9942	1.9936	1.9938	1.9936
65	2.3558	2.3551	2.3551	2.3546
68	2.6542	2.6526	2.6528	2.6521
70	2.9031	2.9011	2.9013	2.9004
72	3.2077	3.2051	3.2054	3.2038
74	3.5876	3.5837	3.5840	3.5819
75	3.8146	3.8100	3.8108	3.8078
76	4.0733	4.0676	4.0689	4.0649
77	4.3704	4.3634	4.3647	4.3599
78	4.7149	4.7055	4.7078	4.7016
79	5.1185	5.1067	5.1093	5.1016
80	5.5976	5.5825	5.5859	5.5756
81	6.1754	6.1544	6.1594	6.1454
82	6.8826	6.8545	6.8606	6.8417
83	7.7673	7.7271	7.7364	7.7096
84	8.9023	8.8430	8.8562	8.8170
85	10.4016	10.3097	10.331	10.271
86	12.4469	12.2977	12.335	12.245
87	15.4166	15.1518	15.216	15.043

extinction term $m\ \tau_{tot}(\lambda)$ given by the Bouguer-Lambert-Beer law can be correctly written as the sum of three terms, the first taking into account the extinction effects due to Rayleigh scattering produced by air molecules along the Sun path, the second including the aerosol extinction effects produced by both scattering and absorption of particulate matter, and the third due to absorption by the wings of the numerous bands of the above-cited minor gases. Therefore, the product $m\ \tau_{tot}(\lambda)$ derived from Eq. (9.1) can be written for wet-air conditions of the atmosphere as the following sum of various partial contributions:

$$\begin{aligned} m\tau_{tot}(\lambda) = {} & m\ \tau_R(\lambda) + m_a AOT(\lambda) + m_{oz} t_{oz}(\lambda) \\ & + m_{nd}\ \tau_{nd}(\lambda) + m_{od}\ \tau_{od}(\lambda) + m_{wv}\tau_{wv}(\lambda) \\ & + m_{ox}\ \tau_{ox}(\lambda) + m_{cd}\ \tau_{cd}(\lambda) \end{aligned} \tag{9.2}$$

where (i) $\tau_R(\lambda)$ is the monochromatic Rayleigh-scattering optical thickness at wavelength λ, (ii) $AOT(\lambda)$ is the monochromatic aerosol optical thickness at wavelength λ, (iii) m_a, m_{oz}, m_{nd}, m_{od}, m_{wv}, m_{ox}, and m_{cd} are the relative optical air mass functions for airborne aerosol particles, atmospheric water vapour, atmospheric ozone, nitrogen dioxide (and its dimer N_2O_4), oxygen dimer, molecular oxygen, and atmospheric carbon dioxide, respectively, and (iv) $\tau_{oz}(\lambda)$, $\tau_{nd}(\lambda)$, $\tau_{od}(\lambda)$, $\tau_{wv}(\lambda)$, $\tau_{ox}(\lambda)$, and $\tau_{cd}(\lambda)$ are the spectral values of the partial optical thicknesses generated within the main atmospheric windows by water vapour, ozone, nitrogen dioxide (and its dimer N_2O_4), oxygen dimer (O_4), molecular oxygen and carbon dioxide, respectively. Considering that the vertical profiles of oxygen and carbon dioxide molecular concentrations are similar to that of total molecular number density, it seems appropriate to use the values of relative optical air mass m given in Table 9.3 in place of those of relative optical air mass functions m_{ox} and m_{cd} considered in Eq. (9.2). In the other cases, one can use the following: (a) the values of relative optical air mass function m_a determined over the 0°–87° solar zenith angle range for background aerosol particle extinction by Tomasi and Petkov (2014) at Ny-Ålesund in different seasons and by Tomasi et al. (2015a) as summer averages at the Barrow, Eureka, and Sodankylä stations (and given in Table 9.4); and (b) the angular values of relative optical air mass functions m_{oz}, m_{nd}, m_{od}, and m_{wv} (pertinent to ozone, nitrogen dioxide (and its dimer N_2O_4), oxygen dimer O_4, and water vapour, respectively), which are given in Table 9.5, as obtained by Tomasi and Petkov (2014) through calculations made using the vertical profiles of molecular concentrations of these four minor gases collected at the Ny-Ålesund station (Spitsbergen, Svalbard) during the summer months.

In order to derive the spectral values of aerosol optical thickness $AOT(\lambda)$ from total atmospheric optical thickness $\tau_{tot}(\lambda)$ obtained from the sun-photometer measurements in terms of Eq. (9.1), it is of basic importance to determine the spectral values of $AOT(\lambda)$ according to Eq. (9.2) as the following difference:

$$\begin{aligned} AOT(\lambda) = {} & (m/m_a)\ \tau_{tot}(\lambda) - (m/m_a)\tau_R(\lambda) - (m_{oz}/m_a)\ \tau_{oz}(\lambda) \\ & - (m_{nd}/m_a)\, t_{nd}(\lambda) - (m_{od}/m_a)\, t_{od}(\lambda) \\ & - (m_{wv}/m_a)\ \tau_{wv}(\lambda) - (m/m_a)\ \tau_{ox}(\lambda) - (m/m_a)\ \tau_{cd}(\lambda), \end{aligned} \tag{9.3}$$

which states that the spectral values of $AOT(\lambda)$ measured within the main windows of the atmospheric transmission spectrum can be correctly calculated by subtracting the Rayleigh scattering optical thickness term and the various absorption optical

Table 9.4 Values of relative optical air mass m_a for aerosol particle extinction calculated at 25 selected values of apparent solar zenith angle θ_o over the range from 0° to 87° for different vertical profiles of volume aerosol extinction coefficient: (i) for an average vertical profile of background Arctic aerosol extinction coefficient and radiosounding data taken at Ny-Ålesund in summer (Tomasi and Petkov 2014), (ii) monthly average vertical profiles of background Arctic aerosol extinction coefficient and radiosounding data collected at Ny-Ålesund by Tomasi et al. (2015a), and (iii) summer average vertical profiles of background Arctic aerosol extinction coefficient and radiosounding data recorded at the Arctic sites of Barrow, Eureka, and Sodankylä

		Ny-Ålesund					
θ_o (°)	Ny-Ålesund, summer average	April, average	June–July average	October-December average	Barrow, summer average	Eureka, summer average	Sodankylä, summer average
0	1.0000	1.0000	1.0000	1.0000	1.0000	1.0000	1.0000
10	1.0154	1.0154	1.0154	1.0154	1.0154	1.0154	1.0154
20	1.0641	1.0641	1.0641	1.0641	1.0641	1.0641	1.0641
30	1.1546	1.1545	1.1545	1.1545	1.1546	1.1546	1.1546
40	1.3052	1.3049	1.3049	1.3050	1.3051	1.3051	1.3051
50	1.5552	1.5546	1.5546	1.5546	1.5551	1.5551	1.5550
55	1.7426	1.7417	1.7416	1.7417	1.7424	1.7424	1.7423
60	1.9986	1.9970	1.9969	1.9971	1.9982	1.9983	1.9981
65	2.3636	2.3608	2.3606	2.3609	2.3630	2.3631	2.3628
70	2.9186	2.9129	2.9126	2.9131	2.9173	2.9176	2.9169
72	3.2288	3.2210	3.2206	3.2213	3.2271	3.2274	3.2265
74	3.6175	3.6064	3.6058	3.6068	3.6151	3.6156	3.6143
75	3.8510	3.8375	3.8368	3.8380	3.8481	3.8487	3.8471
76	4.1179	4.1014	4.1005	4.1020	4.1144	4.1151	4.1131
77	4.4259	4.4052	4.4041	4.4061	4.4214	4.4223	4.4198
78	4.7849	4.7588	4.7574	4.7599	4.7793	4.7804	4.7772
79	5.2087	5.1750	5.1733	5.1765	5.2014	5.2029	5.1988
80	5.7162	5.6718	5.6697	5.6739	5.7065	5.7086	5.7030
81	6.3344	6.2745	6.2717	6.2776	6.3213	6.3242	6.3166
82	7.1035	7.0201	7.0165	7.0249	7.0852	7.0894	7.0786
83	8.0853	7.9647	7.9601	7.9725	8.0586	8.0653	8.0490
84	9.3804	9.1978	9.1921	9.2115	9.3395	9.3507	9.3248
85	11.163	10.870	10.864	10.897	11.097	11.117	11.073
86	13.765	13.257	13.254	13.314	13.647	13.690	13.604
87	17.893	16.917	16.934	17.061	17.657	17.765	17.571

thickness terms considered in Eq. (9.3) from the values $\tau_{tot}(\lambda)$ derived from the sun-photometer output voltages, where: (i) the ratios between the various air masses can be easily calculated using the air mass values given in Tables 9.3, 9.4 and 9.5, respectively, and (ii) the spectral values of $\tau_R(\lambda)$ can be derived from those calculated by Tomasi et al. (2010) from sets of radiosounding data collected at various Arctic sites (Cambridge Bay, Resolute, Danmarkshavn, Eureka, Alert and Ny-Ålesund) and given in Table 9.6 after their normalization to surface-level pressure $p_o = 1013.25$ hPa and surface-level absolute temperature $T_o = 273.16$ K.

Table 9.5 Values of relative optical air mass functions m_{oz} (for ozone), m_{nd} (for nitrogen dioxide and its dimer N_2O_4), m_{wv}, (for atmospheric water vapour), and m_{od} (for oxygen dimer O_4) calculated at 30 selected values of apparent solar zenith angle θ_o over the range from 0° to 87°, on the basis of the vertical profiles of molecular concentrations of these gases made by Tomasi and Petkov (2014) in summer over the Ny-Ålesund station (Spitsbergen, Svalbard) (78° 54′ N, 11° 53′ E, 5 m a.m.s.l.)

θ_o (°)	m_{oz} (for ozone)	m_{nd} (for NO_2 and N_2O_4)	m_{wv} (for atmospheric water vapour)	m_{od} (for $O_2 - O_2$)
0	1.0000	1.0000	1.0000	1.0000
10	1.0153	1.0153	1.0154	1.0154
15	1.0350	1.0350	1.0353	1.0352
20	1.0637	1.0637	1.0641	1.0640
25	1.1026	1.1025	1.1033	1.1031
30	1.1535	1.1533	1.1546	1.1543
35	1.2189	1.2186	1.2206	1.2201
40	1.3025	1.3021	1.3051	1.3044
45	1.4097	1.4091	1.4138	1.4127
50	1.5487	1.5478	1.5551	1.5533
55	1.7322	1.7308	1.7424	1.7396
60	1.9812	1.9787	1.9983	1.9936
65	2.3323	2.3280	2.3631	2.3546
68	2.6190	2.6125	2.6649	2.6521
70	2.8562	2.8476	2.9176	2.9004
72	3.1431	3.1314	3.2275	3.2038
74	3.4960	3.4795	3.6156	3.5819
75	3.7042	3.6844	3.8487	3.8078
76	3.9387	3.9147	4.1150	4.0649
77	4.2042	4.1747	4.4222	4.3599
78	4.5068	4.4703	4.7802	4.7016
79	4.8541	4.8082	5.2025	5.1016
80	5.2557	5.1969	5.7077	5.5756
81	5.7233	5.6470	6.3226	6.1454
82	6.2719	6.1710	7.0864	6.8417
83	6.9194	6.7835	8.0592	7.7096
84	7.6871	7.5003	9.3377	8.8170
85	8.5968	8.3353	11.087	10.271
86	9.6648	9.2938	13.613	12.245
87	10.885	10.358	17.535	15.043

The calculations of $\tau_R(\lambda)$ at the peak-wavelengths of the narrow-band interference filters mounted on the various sun-photometers employed at the Arctic sites can be made by interpolating in wavelength the values of $\tau_R(\lambda)$ normalized to sea-level pressure $p_o = 1013.25$ hPa and sea-level temperature $T_o = 273.16$ K, as given in Table 9.6. For this purpose, one can take into account that $\tau_R(\lambda)$ is proportional to the wavelength power term equal to $\lambda^{-\eta(\lambda)}$, with wavelength λ measured in μm,

Table 9.6 Mean values of Rayleigh-scattering optical thickness $\tau_R(\lambda)$ calculated by Tomasi et al. (2010) at 88 wavelengths from 0.34 to 2.20 μm for three 4-year sets of vertical profiles of air pressure, temperature, and moisture, defined from the radiosounding measurements performed at various Arctic sites and normalized to surface-level pressure $p_o = 1013.25$ hPa and surface-level temperature $T_o = 273.16$ K

Wavelength (μm)	Cambridge Bay (∼ 70 °N)	Resolute, Danmarkshavn (∼ 75 °N)	Eureka, alert, Ny-Ålesund (∼ 80 °N)
0.34	7.1152×10^{-1}	7.1395×10^{-1}	7.1392×10^{-1}
0.35	6.2945	6.3160	6.3157
0.36	5.5901	5.6091	5.6088
0.37	4.9863	4.9994	4.9991
0.38	4.4557	4.4710	4.4708
0.39	3.9975	4.0112	4.0110
0.40	3.5972	3.6096	3.6094
0.41	3.2459	3.2573	3.2571
0.42	2.9372	2.9473	2.9471
0.43	2.6644	2.6735	2.6734
0.44	2.4228	2.4311	2.4309
0.45	2.2081	2.2156	2.2155
0.46	2.0168	2.0236	2.0236
0.47	1.8458	1.8521	1.8521
0.48	1.6928	1.6986	1.6984
0.49	1.5554	1.5607	1.5605
0.50	1.4316	1.4365	1.4364
0.51	1.3200	1.3245	1.3244
0.52	1.2191	1.2233	1.2233
0.53	1.1277	1.1316	1.1315
0.54	1.0448	1.0484	1.0484
0.55	9.6936×10^{-2}	9.7269×10^{-2}	9.7263×10^{-2}
0.56	9.0065	9.0373	9.0369
0.57	8.3794	8.4081	8.4076
0.58	7.8061	7.8328	7.8324
0.59	7.2813	7.3061	7.3057
0.60	6.7999	6.8231	6.8227
0.61	6.3577	6.3795	6.3792
0.62	5.9511	5.9715	5.9711
0.63	5.5766	5.5956	5.5953
0.64	5.2311	5.2490	5.2487
0.65	4.9120	4.9289	4.9286
0.66	4.6171	4.6328	4.6326
0.67	4.3439	4.3587	4.3585
0.68	4.0907	4.1047	4.1045
0.69	3.8558	3.8689	3.8687
0.70	3.6375	3.6499	3.6497
0.71	3.4345	3.4462	3.4460

(continued)

Table 9.6 (continued)

Wavelength (μm)	Cambridge Bay (~ 70 °N)	Resolute, Danmarkshavn (~ 75 °N)	Eureka, alert, Ny-Ålesund (~ 80 °N)
0.72	3.2454	3.2566	3.2564
0.73	3.0692	3.0797	3.0796
0.74	2.9049	2.9149	2.9147
0.75	2.7515	2.7608	2.7607
0.76	2.6080	2.6169	2.6168
0.77	2.4738	2.4822	2.4822
0.78	2.3481	2.3562	2.3561
0.79	2.2304	2.2380	2.2379
0.80	2.1199	2.1271	2.1271
0.82	1.9188	1.9253	1.9252
0.84	1.7410	1.7469	1.7468
0.86	1.5833	1.5888	1.5887
0.88	1.4432	1.4482	1.4480
0.90	1.3182	1.3227	1.3227
0.92	1.2065	1.2107	1.2106
0.94	1.1064	1.1102	1.1102
0.96	1.0165	1.0199	1.0199
0.98	9.3552×10^{-3}	9.3872×10^{-3}	9.3866×10^{-3}
1.00	8.6247	8.6542	8.6537
1.10	5.8785	5.8987	5.8984
1.20	4.1442	4.1583	4.1581
1.30	3.0050	3.0154	3.0152
1.40	2.2320	2.2396	2.2396
1.50	1.6924	1.6982	1.6981
1.60	1.3066	1.3110	1.3109
1.70	1.0247	1.0282	1.0281
1.80	8.1486×10^{-4}	8.1765×10^{-4}	8.1760×10^{-4}
1.90	6.5614	6.5838	6.5835
2.00	5.3426	5.3609	5.3606
2.20	3.6472	3.6597	3.6594

and exponent $\eta(\lambda)$ decreasing slowly as a function of wavelength from 4.4779 at $\lambda = 0.255$ μm to 4.0017 at $\lambda = 3.75$ μm, assuming the following spectral values in the visible and near-infrared range: $\eta(0.305\ \mu m) = 4.3045$, $\eta(0.505\ \mu m) = 4.0985$, $\eta(0.810\ \mu m) = 4.0369$, $\eta(1.050\ \mu m) = 4.0217$, $\eta(2.100\ \mu m) = 4.0054$; and $\eta(3.25\ \mu m) = 4.0022$, according to the evaluations made by Tomasi et al. (2010).

In order to take into account that the vertical contents C_{oz}, C_{nd}, and C_{wv} of ozone, nitrogen dioxide and water vapour are subject to variations from one measurement day to another, it is more suitable to calculate the spectral values of optical thicknesses $\tau_{oz}(\lambda)$ (for ozone) (Inn and Tanaka 1953), $\tau_{nd}(\lambda)$ (for nitrogen dioxide and its dimer N_2O_4) (Hall Jr and Blacet 1952), and $\tau_{wv}(\lambda)$ (for water

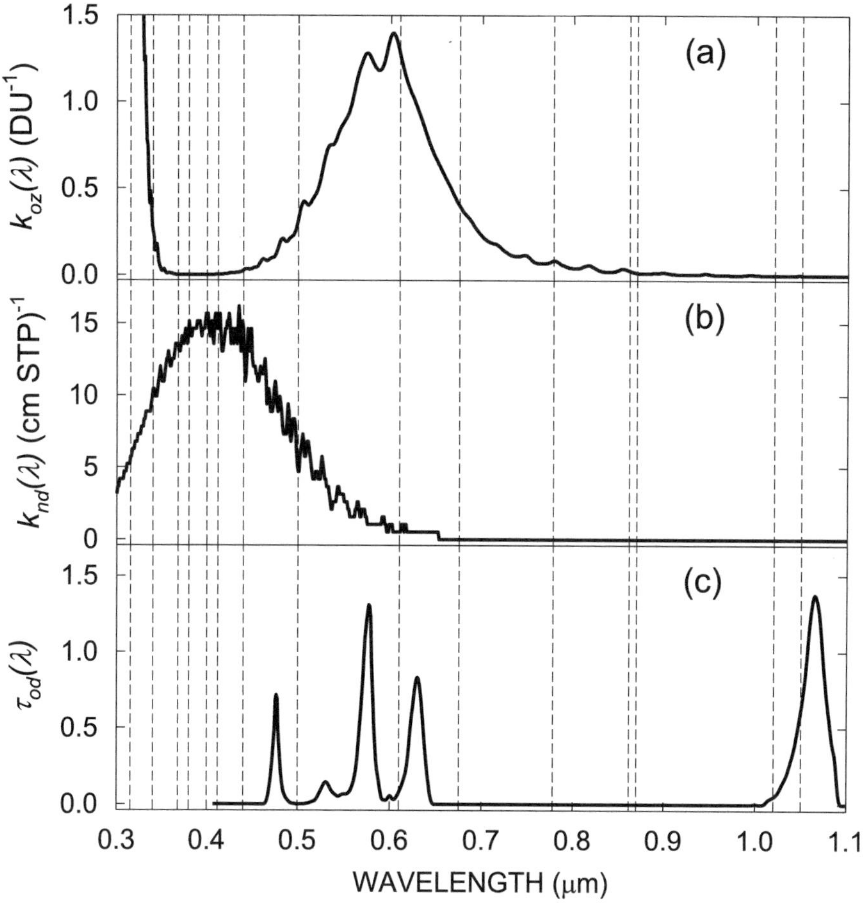

Fig. 9.3 Spectral curves over the 0.30–1.10 μm wavelength range of: (**a**) absorption coefficient $k_{oz}(\lambda)$ of ozone (O_3); (**b**) absorption coefficient $k_{nd}(\lambda)$ of nitrogen dioxide (NO_2) and its dimer (N_2O_4); and (**c**) optical thickness $\tau_{od}(\lambda)$ of oxygen dimer (O_4), as calculated according to Michalsky et al. (1999) and assuming a vertical content of O_4 molecules equal to 1.44 10^{43} molec^2 cm^{-5}, as measured by Wittrock et al. (2004) at Ny-Ålesund in summer 2002 and spring 2003. The vertical dashed lines indicate the peak-wavelengths of the narrow-band interference filters mounted on the various sun-photometers employed at the Arctic sites

vapour) as products of their respective absorption coefficients $k_{oz}(\lambda)$, $k_{nd}(\lambda)$ and $k_{wv}(\lambda)$ by their respective vertical contents measured on each day by means of the sun-photometers, or local radiosounding measurements (water vapour) or ground-based remote sensing measurements (ozone, and nitrogen dioxide) performed for instance with ground-based Brewer techniques and/or derived from satellite-borne observations (Damiani et al. 2012). The spectral curves of absorption coefficients $k_{oz}(\lambda)$ of ozone, $k_{nd}(\lambda)$ of nitrogen dioxide NO_2 and its dimer N_2O_4, and $k_{wv}(\lambda)$ of water vapour are shown in Fig. 9.3 (for ozone and nitrogen dioxide) and Fig. 9.4

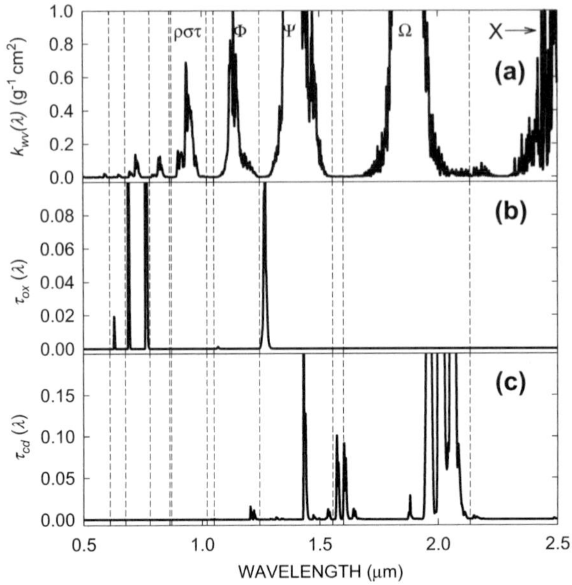

Fig. 9.4 Spectral curves over the 0.50–2.50 μm wavelength range of: (**a**) absorption coefficient $k_{wv}(\lambda)$ of water vapour in the atmospheric windows between the water vapour absorption bands labelled with letters ρστ, Φ, Ψ, Ω and X; (**b**) optical thickness $\tau_{ox}(\lambda)$ of molecular oxygen (O_2); and (**c**) optical thickness $\tau_{cd}(\lambda)$ of carbon dioxide (CO_2). All the three curves have been calculated using the MODTRAN 4 code (Kneizys et al. 1996) for standard pressure and temperature conditions of the 75 °N Arctic summer atmospheric model (Tomasi et al. 1998), in which surface-level pressure p_o is equal to 1013 hPa, surface-level temperature T_o is equal to 278.9 °K, and carbon dioxide volume concentration was assumed to be equal to 405 ppmv, according to the recent measurements performed at the Zeppelin Observatory (Spitsbergen, Svalbard). The vertical dashed lines indicate the peak-wavelengths of the narrow-band interference filters mounted on the various sun-photometers employed at the Arctic sites

for water vapour, as obtained using the MODTRAN 4 code (Berk et al. 2000) and the LOWTRAN 7 model (Kneizys et al. 1996). The spectral values of the absorption coefficients of ozone, nitrogen dioxide with its dimer, and atmospheric water vapour are given in Table 9.7 at the main window wavelengths of the sun-photometer models employed in the present study at the various Arctic stations, together with the spectral values of the oxygen dimer optical thickness $\tau_{cd}(\lambda)$. The spectral values of the optical thicknesses produced by ozone, nitrogen dioxide and water vapour were determined by assuming (i) a vertical content of ozone C_{oz} = 350 Dobson Units in the Arctic summer (June–July) atmosphere; (ii) a differential slant column of NO_2 molecules equal to 1.44 10^{43} molec^2 cm^{-5} obtained for a solar zenith angle close to 15° from the measurements performed at Ny-Ålesund (Spitsbergen, Svalbard) by Wittrock et al. (2004), yielding a vertical content of NO_2 molecules equal to about 5 10^{15} molecules cm^{-2} for clean air conditions (corresponding to 2.2 10^{-4} cm STP); and (iii) a vertical content of atmospheric water vapour equal

Table 9.7 Spectral values of the absorption coefficients of atmospheric water vapour, ozone, and nitrogen dioxide with its dimer, and oxygen dimer optical thickness at the window wavelengths of (i) the AERONET/Cimel CE-318 sun-photometers (340, 380, 440, 500, 675, 870, 1020, and 1600 nm), (ii) the GAW/PFR sun-photometers (368, 412, 500 and 862 nm), (iii) the NOAA/GMD Carter Scott sun-photometers (368, 412, 500, 610, 675, 778, 862, and 1050 nm), as obtained from the spectral data shown in Fig. 9.4 and Fig. 9.5, and (iv) some infra-red peak-wavelengths of the New Portable Sun-photometer (SPM) designed by the IAO-SB-RAS Institute (Tomsk, Russia)

Wave-length λ (μm)	Absorption coefficient of ozone $k_{oz}(\lambda)$ (D. U.)$^{-1}$ ([a])	Absorption coefficient $k_{nd}(\lambda)$ of atmo-spheric NO_2 and N_2O_4 (cm STP^{-1}) ([a])	Optical thickness of oxygen dimer $\tau_{od}(\lambda)$ (Michal-sky et al. 1999)	Absorption coefficient of water vapour $k_{wv}(\lambda)$ (g^{-1} cm^2) ([a])	Optical thickness of oxygen molecules $\tau_{ox}(\lambda)$ ([a])	Optical thickness of carbon dioxide $\tau_{cd}(\lambda)$ ([a])
0.315	1.22 10^{-3}	5.73	–	–	–	–
0.340	2.53 10^{-5}	10.43	–	–	–	–
0.368	–	13.56	–	–	–	–
0.380	–	15.13	–	–	–	–
0.400	–	15.65	–	–	–	–
0.412	2.87 10^{-7}	14.60	–	–	–	–
0.440	2.58 10^{-6}	14.61	–	–	–	–
0.500	3.17 10^{-5}	4.69	1.76 10^{-5}	6.80 10^{-5}	–	–
0.610	1.27 10^{-4}	0.52	7.80 10^{-4}	–	–	–
0.675	3.98 10^{-5}	–	–	–	–	–
0.778	8.32 10^{-6}	–	–	6.80 10^{-5}	–	–
0.862	2.58 10^{-6}	–	–	1.36 10^{-4}	1.00 10^{-4}	–
0.870	1.43 10^{-6}	–	–	–	–	–
1.020	–	–	5.49 10^{-4}	2.11 10^{-3}	–	–
1.050	–	–	6.17 10^{-3}	–	–	6.00 10^{-4}
1.244	–	–	–	4.76 10^{-4}	1.00 10^{-4}	–
1.555	–	–	–	1.09 10^{-3}	–	–
1.600	–	–	–	3.40 10^{-4}	–	5.30 10^{-2}
2.134	–	–	–	1.29 10^{-3}	–	4.00 10^{-4}

The values of absorption coefficients $k_{oz}(\lambda)$, $k_{nd}(\lambda)$ and $k_{wv}(\lambda)$ and optical thicknesses $\tau_{od}(\lambda)$,$\tau_{ox}(\lambda)$ and $\tau_{cd}(\lambda)$ have been calculated by using the transmission curves of the pass-band interference filter mounted on the sun-photometers as weighting functions at the various window wavelengths. The ozone vertical content C_{oz} was assumed to be equal to 350 Dobson Units in the Arctic atmosphere during early summer, and the NO_2 vertical content C_{nd} was assumed to be equal to 2.2 10^{-4} cm STP, as obtained for the vertical content of ~ 5 10^{15} molecules cm^{-2} measured by Wittrock et al. (2004) at Ny-Ålesund for clean air conditions in July 2002 and April 2003
[a]From calculations performed using the MODTRAN 4 code (Berk et al. 2000)

on average to 0.8 g cm^{-2} at Ny-Ålesund in June and July. The spectral curves of optical thicknesses $\tau_{od}(\lambda)$ for oxygen dimer O_4, $\tau_{ox}(\lambda)$ of molecular oxygen O_2, and $\tau_{cd}(\lambda)$ for carbon dioxide are shown in Fig. 9.3 (for oxygen dimer) and Fig. 9.4 (for molecular oxygen and carbon dioxide), such curves being calculated as follows: (i) the spectral curve of $\tau_{od}(\lambda)$ was determined according to the Michalsky et al. (1999) evaluations, assuming a vertical content of O_4 molecules close to 1.44 10^{43} molec^2 cm^{-5} (Wittrock et al. 2004), as measured at Ny-Ålesund in summer 2002 and spring 2003; (ii) the spectral curve of $\tau_{ox}(\lambda)$ relative to the molecular oxygen was calculated by assuming the O_2 molecular concentration features observed for standard pressure and temperature conditions of the Arctic atmosphere in summer (June–July); and (iii) the spectral curve of $\tau_{cd}(\lambda)$ was determined for volume concentration of CO_2 equal to 405 ppmv in the Arctic atmosphere, according to the measurements recently taken at the Zeppelin Observatory (Spitsbergen, Svalbard).

On the basis of these assumptions, the corrections defined in Eq. (9.3) have been made to derive the spectral values of $AOT(\lambda)$ from the measurements of total optical thickness $\tau_{tot}(\lambda)$ after the corrections for Rayleigh-scattering optical thickness $\tau_R(\lambda)$ (made using the data given in Table 9.6), and those made for the gaseous absorption terms described in Figs. 9.3 and 9.4, by using the evaluations of the respective air mass functions given in Tables 9.3, 9.4 and 9.5. Applying this procedure to the field measurements of $\tau_{tot}(\lambda)$ performed with the various sun-photometers, the spectral values of $AOT(\lambda)$ were determined at the various Arctic sites, showing that optical thickness $AOT(\lambda)$ is usually a smooth function of wavelength λ, which can be expressed in terms of the Ångström (1964) formula,

$$AOT\,(\lambda) = AOT\,(\lambda_o)\,\lambda^{-a}, \tag{9.4}$$

where α is the so-called Ångström's wavelength exponent, λ_o is equal to 1 μm and wavelength λ is measured in μm. Some examples of application of the best-fit procedure based on the Ångström's formula reported in Eq. (9.4) have been shown above in Fig. 9.2 for (i) a pair of spectral series of the daily mean values of $AOT(\lambda)$ measured by the AWI researchers at Ny-Ålesund with the SP1A sun-photometer on April 1, 2001, and June 15, 2006; (ii) a pair of spectral series of the daily mean values of $AOT(\lambda)$ measured at Ny-Ålesund with the SP2H sun-photometer on March 16, 2008, and June 16, 2010; and (iii) a pair of spectral series of the daily mean values of $AOT(\lambda)$ measured at Ny-Ålesund with the STAR01 star-photometer on December 25, 2009, and January 29, 2010. Actually, the analytical wavelength-dependence form defined in Eq. (9.4) can be convex or concave, varying as a function of the relative contents of the fine and accumulation/coarse particles suspended in the atmospheric column, as pointed out by O'Neill et al. (2001a, b, 2003), who demonstrated that the variations in the spectral dependence curve of $AOT(\lambda)$ described in terms of exponent α and its first spectral derivative $(d\alpha/d\lambda)$ can give a measure of the spectral interaction between the individual optical components of a bimodal size-distribution consisting of fine and accumulation/coarse particle modes, respectively. O'Neill et al. (2001b) also showed that one can exploit the spectral curvature information in the measured $AOT(\lambda)$ to achieve a direct estimate of a fine mode Ångström exponent (α_f) as well as the optical fraction of fine mode particles. The recognition that the aerosol particle size-distribution is effectively bimodal permits the extraction of optical thicknesses due to the fine particle mode and the accumulation/coarse particle mode, which give form to the overall exponent α (on this matter, see also O'Neill et al. (2003)).

In the present analysis of the sun-photometric measurements of *AOT(λ)* performed at various Arctic sites, we have considered only the values of exponent α, without analyzing the field data to evaluate the features of exponent α_f. More simply, we stated that the optical predominance of the fine particle mode occurs in general for values of α higher than 1.2, while the cases of prevailing optical extinction effects by accumulation and coarse particles are characterized by values of α lower than 0.7. The Arctic observations of *AOT(λ)* provide in general values of *AOT(λ)* lower than 0.15 in the visible. Tomasi et al. (2007) showed that the field measurements of *AOT(0.50 μm)* conducted over the 1977–2006 period exceed only rarely the value of 0.15 for the background summer aerosol conditions observed at the Arctic sites of Barrow, Alert, Summit, Ny-Ålesund, Hornsund, Sodankylä and Andenes, being higher than 0.15 only during particularly strong episodes of Arctic haze, most frequently occurring in late winter and spring, and during intense episodes of BFF smoke transport, most frequently observed during the summer months. With regard to this, Tomasi et al. (2012) found that: (i) the measurements of *AOT(0.50 μm)* recorded from 2000 to 2012 at Ny-Ålesund were estimated to exceed 0.15 in summer (June to September) only during a few cases of strong transport of BFF smoke from lower latitudes, as well in winter (December to March) only for 10% of the cases, typically associated with Arctic haze transport episodes from Europe; (ii) measurements of *AOT(0.50 μm)* > 0.15 recorded at Barrow over the same period were observed to be associated with Arctic haze transport episodes only in a few percent of cases; and (iii) daily mean background summer values of *AOT(0.50 μm)* measured at Tiksi in Siberia were found to be always lower than 0.08 during the summer 2010.

In all the cases shown in the present analysis, the best-fit values of exponent α were determined in terms of Eq. (9.4) over the spectral range $0.40 \leq \lambda \leq 0.87$ μm. It is true that our analysis of *AOT(λ)* was performed for sets of measurements conducted using different instruments, for which different network protocols are used to estimate, for instance, the spectral values of $\tau_R(\lambda)$ and the relative optical air mass m as a function of solar zenith angle, while different nominal time interval between measurements and calibration protocols have been considered for the various sun-photometers. In general, all the data-sets were cloud-screened using accurate temporal-based procedures that were developed and rigorously tested by each network group, with criteria similar to those adopted in the protocol yielding the Level 2.0 AERONET data (Smirnov et al. 2000). The NOAA/GMD data collected at the Barrow and Alert stations were further cloud-screened using the criterion of discarding the *AOT(λ)* spectral data yielding values of α lower than 0.38. This added cloud-screening criterion was applied in order to eliminate the influence of thin cirrus clouds homogeneously distributed in the sky field surrounding the Sun disk and unfortunately escaped to the temporal cloud-screening procedure.

Before presenting the evaluations of the seasonal variations affecting parameters *AOT(0.50 μm)* and α measured at the various Arctic sites, it seems useful for the reader to give a measure of the experimental errors and variability features associated with the aerosol optical characteristics, since these features vary as a function of aerosol origins and their chemical composition. On this matter, it is worth mention that: (a) the mean experimental error of *AOT(λ)* measured with the most sophisticated sun-photometers leads to an uncertainty of ~ 0.01 in the visible and near-IR estimates of *AOT*, mainly due to calibration errors (Eck et al. 1999); (b) the relative errors of the Ångström (1964) exponent α is equal to about 30% in examining the sun-photometer measurements performed at Arctic sites, and, hence, absolute errors no greater than 0.50 can be usually made when *AOT* is estimated at visible wavelengths (Mazzola et al. 2012); and (c) the spread of exponent α arising from the natural variability of the various Arctic aerosol types has been estimated by Stone (2002), Tomasi et al. (2007) and Treffeisen et al. (2007) to yield average uncertainties of ± 0.4 in the cases with Arctic background summer aerosol, and $\pm$ 0.6 in the cases presenting particular optical characteristics of the Arctic aerosol load, such as those observed during long-range transport episodes of Asian dust and BFF smoke particles.

9.1.3 Ground-Based Sun-Photometer Measurements of Aerosol Optical Thickness and Ångström's Wavelength Exponent at Various Arctic Sites

Ground-based sun-photometer measurements of *AOT(λ)* were carried out at various visible and near-infrared wavelengths during numerous experimental campaigns conducted at various Arctic sites during the past decades, providing useful information on polar aerosol optical and microphysical properties. Actually, *AOT(λ)* gives a measure of the overall aerosol extinction occurring along the vertical atmospheric path, while exponent α provides a useful information on the optical predominance of the extinction effects produced by fine, accumulation and coarse particle modes. In fact, values of α higher than 1.2 are usually observed in air masses where Aitken nuclei and very fine particles (with diameters $d < 0.24$ μm) are optically predominant, while relatively low values of $\alpha < 1.0$ are observed in all cases where accumulation mode particles (over the $0.24 \leq d \leq 2.50$ μm diameter range) and coarse mode particles (having diameters $d > 2.50$ μm) produce stronger extinction effects. The vertical profile of the aerosol volume backscattering coefficient can be determined by means of supplemental lidar measurements, which are very useful in order to identify and characterize the aerosol layers formed in the troposphere and lower stratosphere (as shown for instance by Tomasi et al. (2015a)).

Direct solar radiation measurements have been regularly conducted for cloud-free sky conditions at numerous polar sites over the past decades, using multi-spectral sun-photometers of different design. The individual measurements of the spectral values of *AOT(λ)* and the corresponding evaluations of exponent α obtained from the analysis of the field data recorded for cloudless-sky conditions were then averaged to yield daily mean values of *AOT(λ)* at the visible and near-infrared wavelengths. Since the present analysis is devoted to tropospheric aerosols, the sun-photometer measurements conducted at Arctic sites during periods presenting appreciable extinction features produced by stratospheric layers of volcanic particles were not considered in the present calculations, as made for the measurements carried out at: (i) Barrow (Alaska, USA) in May 2006 (Soufrière Hills eruption) and October 2006 (Tavurvur eruption) (Stone et al. 2014), and in July 2008 (Okmok eruption) and March 2009 (Mt. Redoubt eruption) (Tomasi et al. 2012); (ii) Ny-Ålesund (Spitsbergen, Svalbard) from mid-August to late September 2008 (Kasatochi eruption) (Hoffmann et al. 2010), (iii) Eureka PEARL (Nunavut, Canada) station from early July to early October 2009 (Sarychev eruption) (O'Neill et al. 2012), and (iv) various Scandinavian sites located at latitudes higher than 67 °N in April 2010 (Eyjafjallajokull eruption), as shown by Webley et al. (2012) who examined the WRF-Chem Model representations of the volcanic ash transport processes occurred over the Northern Scandinavian region during April 14–18, 2010, finding that the Eyjafjallajokull ash was also transported in this area at altitudes ranging between 2 and 8 km above sea-level, presenting particulate matter concentration values higher than 900 $\mu g\ m^{-3}$.

The strong extinction effects produced by the stratospheric particle layer originated from the above-mentioned volcanic eruptions caused an appreciable increase in *AOT(0.50* μm*)* and marked variations in exponent α, as shown in Fig. 9.5, which presents two multi-year series of daily mean values of *AOT(0.50* μm*)* and exponent α. The first time-series of *AOT(0.50* μm*)* and α were derived from the SP02 and SP022 Carter Scott sun-photometer measurements performed by the NOAA/GMD team at Barrow (Alaska, USA) during the 9-year period from February 2002 to October 2010, while the second time-series of *AOT(0.50* μm*)* and α were obtained over the same multi-year period from the AWI measurements conducted at the Ny-Ålesund (AWIPEV) station using the SP1A and SP2H sun-photometers. It can be easily noted in Fig. 9.5 that very sharp and marked volcanic particle extinction effects were produced by the Kasatochi and Sarychev stratospheric volcanic particle loads formed above both such strongly perturbed areas during the summer periods of 2008 and 2009, respectively.

The sets of daily mean data collected at each site and not affected by appreciable volcanic particle extinction effects were then subdivided into monthly sub-sets consisting of data measured in different years, for which the multi-year monthly mean values of *AOT(0.50* μm*)* and exponent α were determined. At each site, the corresponding relative frequency histograms were defined separately by using the daily mean values of *AOT(0.50* μm*)* and α collected during the following seasonal periods: (i) winter–spring (from December to May), when Arctic haze events are in general most frequent, and (ii) the summer–autumn (from June to October), during which relevant loads of background (BG) summer aerosol particles are in general present within the Arctic atmosphere.

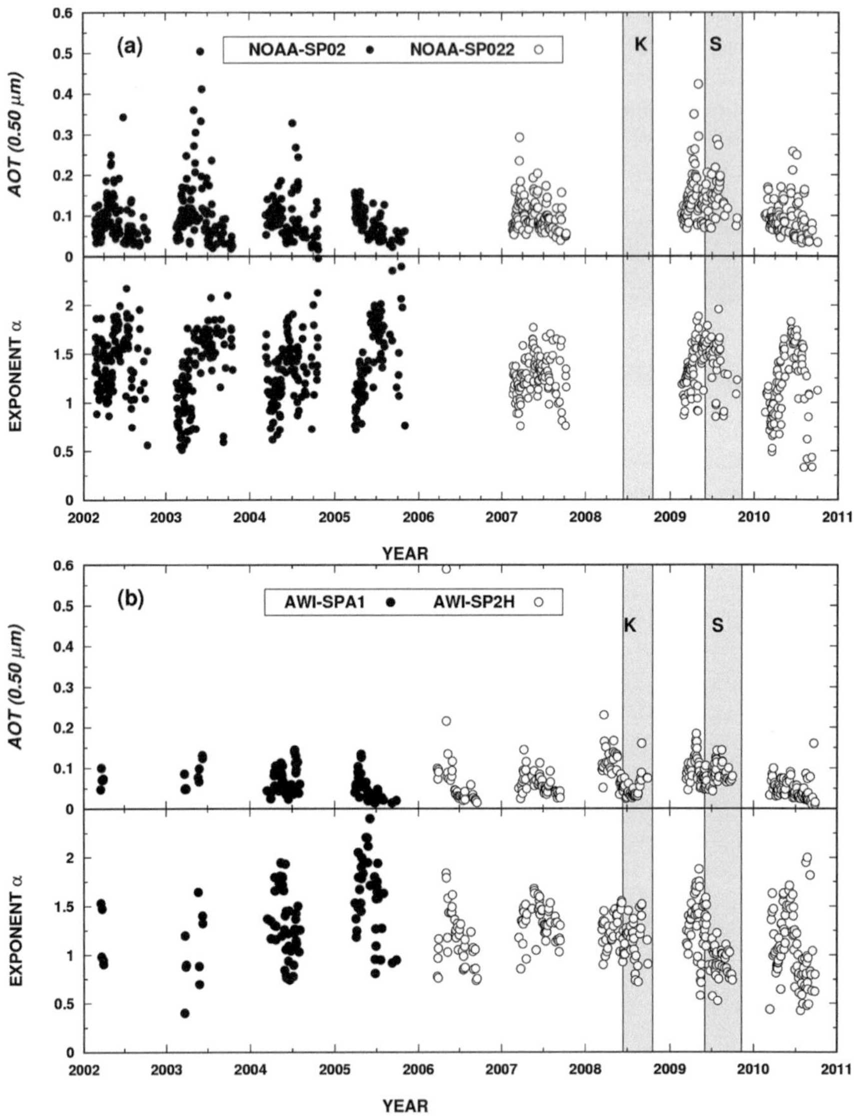

Fig. 9.5 Upper part (**a**): Time-series of the daily mean values of aerosol optical thickness *AOT(0.50 μm)* and Ångström's (1964) wavelength exponent α derived from the SP02 (solid circles) and the SP022 (open circles) sun-photometer measurements performed at Barrow (Alaska, USA) by the NOAA/GMD group during the period from February 2002 to October 2010. Lower part (**b**): As in the upper part, for the measurements performed by the AWI group at the Ny-Ålesund (AWIPEV) station (Spitsbergen, Svalbard) over the same multi-year period using the SP1A (solid circles) and the SP2H (open circles) sun-photometers. The summer periods of 2008 and 2009, during which marked volcanic particle extinction effects were produced by the Kasatochi and Sarychev eruptions above the two Arctic stations, are marked by the two grey vertical bands labelled with letters K and S, respectively

9.1.3.1 Measurements in Northern Alaska

Two multi-year sets of sun-photometer measurements conducted at Barrow (Alaska, USA) on the Arctic Ocean coast have been analysed in the present study. The first set of measurements was collected by the NOAA/GMD group using the SP02 and SP022 Carter Scott sun-photometer models during the 13-year period from March 2000 to September 2012. The technical characteristics of these two sun-photometers have been described in the Introduction, while the series of the peak-wavelengths of the narrow-band interference filters mounted on the two Carter Scott sun-photometers are given in Table 9.8. The set of spectral measurements of *AOT(λ)* was analysed to select only the measurements taken for apparently cloud-free sky conditions and then cloud-screened by applying the NOAA/GMD selection procedure to discard the cases presenting values of exponent $\alpha < 0.38$, this parameter being calculated over (i) the 0.412–0.862 μm wavelength range when examining the *AOT(λ)* measurements performed with the Carter Scott SP02 sun-photometer, and (ii) the 0.368–0.778 μm spectral range when analysing the *AOT(λ)* taken with the Carter Scott SP022 sun-photometer (Stone 2002). The second data-set collected at Barrow consists of the AERONET measurements carried out from March 2002 to September 2013 by the Brookhaven National Laboratory group (Environmental and Climate Sciences Dept., Upton, NY, USA) using the Cimel CE-318 sun-photometers equipped with narrow-band interference filters having the peak-wavelengths given in Table 9.8. The results obtained from the NOAA/GMD and AERONET measurements in terms of *AOT(0.50 μm)* and exponent α are shown in Fig. 9.6, presenting a data coverage of 94% and 78% for the overall 14-year measurement period from March 2000 to September 2013, respectively (Tomasi et al. 2015b). The NOAA/GMD monthly mean values of *AOT(0.50 μm)* were found to increase from about 0.08 in February to more than 0.12 in April and May, and then to gradually decrease until reaching values slightly lower than 0.10 in June and July, and values close to 0.07 in August and to 0.06 in September and October. The corresponding standard deviations σ_{AOT} were estimated to range between 0.03 and 0.06 in the winter-spring months, and between 0.03 and 0.12 in the summer-autumn months. Therefore, the seasonal average values of *AOT(0.50 μm)* measured by NOAA/GMD were estimated to be equal to 0.109 ± 0.050 in the winter-spring period and 0.079 ± 0.050 in the summer-autumn period, as reported in Table 9.9. The monthly mean values of exponent α determined from the NOAA/GMD measurements were found to be rather stable from February to April, varying between 1.16 and 1.18, with values of σ_α close to 0.25 during this 3-month period. In the following months, exponent α was found to increase, reaching a value of 1.42 in May and 1.55 in June, and slightly lower values equal to 1.48 and 1.30 in July and August, respectively, followed by values equal to 1.35 and 1.45 in September and October, respectively. Thus, the seasonal average values of exponent α were found to be equal to 1.24 ± 0.28 in winter-spring and 1.46 ± 0.28 in summer-autumn.

Similar results were obtained by examining the AERONET measurements, which provided monthly mean values of *AOT(0.50 μm)* increasing from about 0.13 in March to nearly 0.17 in April, and then decreasing to 0.10 in June until assuming

Table 9.8 Peak-wavelengths of the spectral channels used by the various sun-photometer models employed at the various Arctic sun-photometer stations and spectral intervals $\Delta\lambda$ used to calculate the Ångström's exponent

Sun-photometer model	Sun-photometer station	Peak wavelengths (nm) of the spectral channels	Spectral interval $\Delta\lambda$ (nm)	References
Cimel CE-318	Barrow (Alaska, USA)	380, 440, 500, 675, 870, 940 (used to measure precipitable water, *pw*), 1020	380–870	Holben et al. (1998)
Cimel CE-318	Resolute Bay (Nunavut, Canada)	380, 440, 500, 675, 870, 940 (*pw*), 1020	380–870	
Cimel CE-318, # 401	Eureka – PEARL (Nunavut, Canada)	380, 440, 500, 675, 870, 940 (*pw*), 1020	380–870	
Cimel CE-318	Thule (North-Western Greenland)	380, 440, 500, 675, 870, 940 (*pw*), 1020	380–870	
Cimel CE-318	Ittoqqortoomiit (Eastern Greenland)	380, 440, 500, 675, 870, 940 (*pw*), 1020	380–870	
Cimel CE-318	Hornsund (Spisbergen, Svalbard)	380, 440, 500, 675, 870, 940 (*pw*), 1020	380–870	
Cimel CE-318	Andenes/Andøya Rocket Range (Norway)	380, 440, 500, 675, 870, 940 (*pw*), 1020	380–870	
Cimel CE-318	Sodankylä (Lapland, Northern Finland)	380, 440, 500, 675, 870, 940 (*pw*), 1020	380–870	
Cimel CE-318	Tiksi (Northern-Central Siberia)	380, 440, 500, 675, 870, 940 (*pw*), 1020	380–870	
Prede POM-02 L	Ny-Ålesund, Rabben NIPR station (Spitsbergen, Svalbard)	315, 340, 380,400, 500, 675, 870, 940 (*pw*), 1020, 1600, 2200	380–870	Nakajima et al. (2007)

PFR#34	Summit (Greenland ice sheet)	368, 412, 500, 862	368–862	Wehrli (2000)
PFR#18	Ny-Ålesund (Spitsbergen, Svalbard), NILU	367.7, 411.9, 500.6, 862.5	367.7–862.5	
PFR#29 and PFR#32	Sodankylä (Lapland, Northern Finland), FMI	367.6, 411.4, 500.5, 861.6	367.6–861.6	
PFR	Kiruna (Northern Sweden), SMHI (Sweden)	368, 412, 500, 862	368–862	
Carter Scott SP02	Barrow (Alaska, U.S.A.), GMD/NOAA (Boulder, U.S.A); alert (Nunavut, Canada)	412, 500, 675, 862	412–862	Stone (2002)
Carter Scott SP01-A	Barrow (Alaska, U.S.A.), GMD/NOAA (Boulder, U.S.A)	367, 610, 778, 1050	367–778	
Carter Scott SP022		368, 610, 778, 1050	368–778	
SP1A sun-photometer	Ny-Ålesund (Spitsbergen, Svalbard), AWI (Germany)	371, 380, 416, 443, 500, 532, 609, 675, 778, 864, 947 (*pw*), 961 (*pw*), 1025, 1046, 1062	380–864	Herber et al. (2002)
SP2H sun-photometer		367, 380, 413, 441, 501, 531, 605, 673, 776, 862, 912 (*pw*), 949 (*pw*), 1023, 1045	380–862	
STAR01 STAR-photometer		390, 441, 501, 531, 605, 673, 776, 862, 949 (*pw*)	390–862	
New Portable Sun-photometer SPM	Barentsburg (Spitsbergen, Svalbard), IAO-SB-RAS,(Tomsk, Russia)	339, 380, 442, 500, 547, 675, 871, 941 (*pw*), 1020, 1240, 1553, 2134	380–871	Sakerin et al. (2009, 2012, 2014)

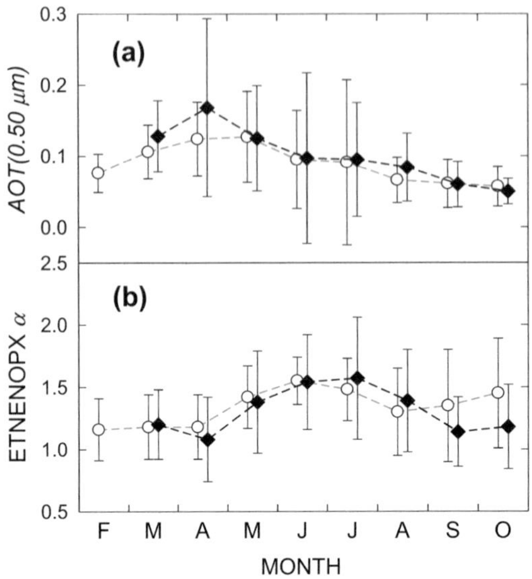

Fig. 9.6 Upper part (**a**): Time-patterns of the monthly mean values of aerosol optical thickness *AOT(0.50 μm)* given with their standard deviations (defined by the vertical bars), as obtained from the multi-year sun-photometer measurements conducted at Barrow (Alaska, USA) by: (i) the NOAA/GMD (Boulder, Colorado, USA) group using the SP02 and the SP022 sun-photometers over the period from March 2000 to September 2012 (open circles) and (ii) the Brookhaven National Laboratory (Upton, NY, USA) using the AERONET Cimel CE-318 sun-photometer in cooperation with NASA/GSFC (USA) over the period from March 2002 to September 2013 (solid diamonds). Lower part (**b**): As in the upper part for the time-patterns of the monthly mean values of wavelength exponent α given with their standard deviations (defined by the vertical bars), as obtained from the same multi-year sun-photometer measurements conducted at Barrow (Alaska, USA) by the NOAA/GMD (Boulder, Colorado, USA) from 2000 to 2012 (open circles) and by the Brookhaven National Laboratory (Upton, NY, USA) and NASA/GSFC (USA) from 2002 to 2013 (solid diamonds)

values of 0.06 and 0.05 in September and October, respectively, with σ_{AOT} ranging between 0.05 and 0.13 in the spring and summer months, and decreasing to 0.03 and 0.02 in September and October, respectively. Therefore, appreciably higher values of *AOT(0.50 μm)* were obtained from the AERONET measurements conducted at Barrow with respect to those performed by NOAA/GMD, giving average values of *AOT(0.50 μm)* equal to 0.134 $\pm$ 0.066 in winter-spring and 0.085 $\pm$ 0.055 in summer-autumn. The monthly mean values of exponent α determined from the AERONET measurements were found to describe time-patterns very similar to those derived from the NOAA/GMD measurements, equal to 1.20 in March, increasing from 1.08 in April to more than 1.50 in both June and July, and subsequently decreasing to less than 1.40 in August and varying between 1.14 and 1.18 in September and October, respectively, while σ_α was estimated to vary between 0.28 and 0.49 throughout the whole observational period. Thus, seasonal

Table 9.9 Seasonal average values of aerosol optical thickness *AOT(0.50 μm)* and Ångström wavelength exponent α obtained with their standard deviations σ_{AOT} and σ_α for the measurements sets collected during the winter–spring and summer–autumn periods at the 17 Arctic stations considered in the present study

	Winter–spring period		Summer–autumn period	
Station	$AOT(0.50\ \mu m) \pm \sigma_{AOT}$	$\alpha \pm \sigma_\alpha$	$AOT(0.50\ \mu m) \pm \sigma_{AOT}$	$\alpha \pm \sigma_\alpha$
1. Barrow (NOAA/GMD)	0.115 ± 0.050	1.24 ± 0.28	0.079 ± 0.050	1.46 ± 0.28
2. Barrow (AERONET)	0.134 ± 0.066	1.20 ± 0.36	0.085 ± 0.055	1.38 ± 0.31
3. Resolute Bay	0.117 ± 0.050	1.52 ± 0.25	0.064 ± 0.036	1.55 ± 0.26
4, Eureka-0PAL	0.093 ± 0.030	1.43 ± 0.32	0.056 ± 0.029	1.64 ± 0.24
5. Alert	0.115 ± 0.057	1.29 ± 0.30	0.066 ± 0.041	1.46 ± 0.26
6. Thule	0.093 ± 0.036	1.38 ± 0.26	0.059 ± 0.029	1.41 ± 0.26
7. Summit	0.057 ± 0.033	1.22 ± 0.29	0.052 ± 0.041	1.48 ± 0.29
8. Ittoqqortoormiit	0.068 ± 0.027	1.26 ± 0.23	0.052 ± 0.035	1.45 ± 0.29
9. Ny-Ålesund (AWI)	0.082 ± 0.031	1.32 ± 0.29	0.053 ± 0.027	1.28 ± 0.29
10. Ny-Ålesund (NILU)	0.122 ± 0.051	1.13 ± 0.33	0.066 ± 0.044	1.36 ± 0.32
11. Barentsburg	0.097 ± 0.024	1.13 ± 0.25	0.078 ± 0.025	1.29 ± 0.14
12. Hornsund	0.107 ± 0.042	1.30 ± 0.31	0.070 ± 0.032	1.31 ± 0.32
13. Sodankylä (FMI)	0.077 ± 0.040	1.24 ± 0.27	0.069 ± 0.027	1.41 ± 0.24
14. Sodankylä (AERONET)	0.069 ± 0.034	1.23 ± 0.23	0.066 ± 0.036	1.44 ± 0.28
15. Andenes	0.096 ± 0.041	1.00 ± 0.30	0.113 ± 0.035	1.19 ± 0.32
16. Kiruna	0.065 ± 0.019	1.22 ± 0.14	0.073 ± 0.027	1.45 ± 0.17
17. Tiksi	0.13 ± 0.022[a]	1.90 ± 0.10[a]	0.094 ± 0.060	1.57 ± 0.24

[a]Derived from a set consisting of four cases only

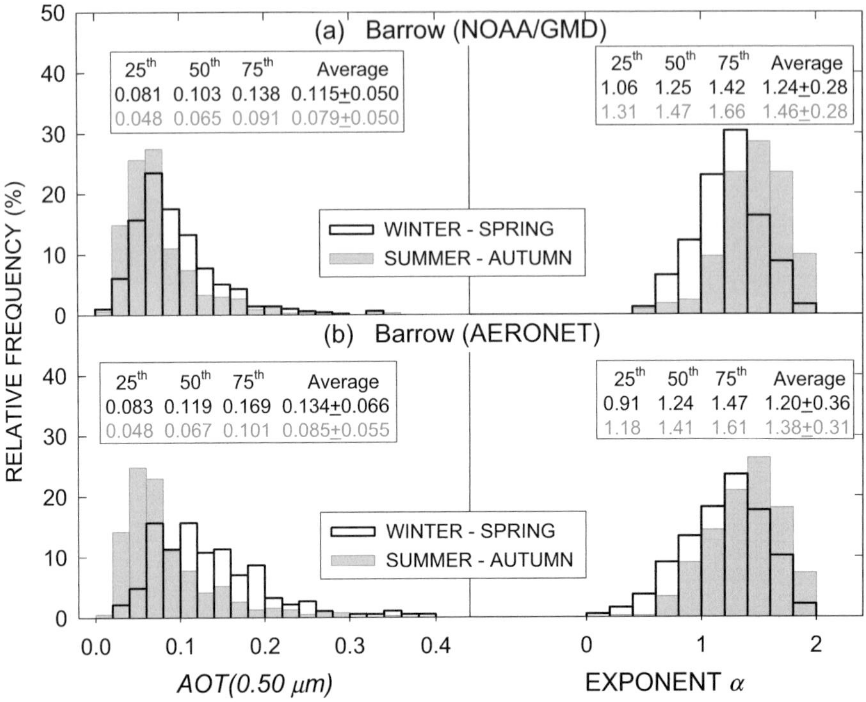

Fig. 9.7 Part (**a**): Relative frequency histograms RFHs of aerosol optical thickness *AOT(0.50 μm)* (left-hand side) and Ångström wavelength exponent α (right-hand side) obtained separately for the winter–spring (December–May) and the summer–autumn (June–October) periods from the multi-year sun-photometer measurements conducted at Barrow (Alaska, USA) by NOAA/GMD and shown in Fig. 9.6. The values of the quartiles (i.e. of the 25th, 50th, and 75th percentiles) of *AOT(0.50 μm)* and exponent α reported in the boxes have been obtained by examining the daily mean values of the two optical parameters measured at the Barrow (NOAA/GMD) station in the winter–spring (bold numbers) and summer–autumn (grey numbers) periods. Part (**b**): As in part (**a**) for the RFHs of *AOT(0.50 μm)* (left-hand side) and Ångström wavelength exponent α (right-hand side) obtained from the AERONET Cimel CE-318 sun-photometer measurements performed at Barrow (Alaska, USA) by the Brookhaven National Laboratory (Upton, NY, USA) in cooperation with NASA/GSFC (USA) over the same period from March 2002 to September 2013

average values of α equal to 1.20 ± 0.36 in winter-spring and 1.38 ± 0.31 in summer-autumn were obtained from these AERONET observations (see Table 9.9).

Figure 9.7 shows separately the relative frequency histograms of the daily mean values of *AOT(0.50 μm)* and α measured during the winter-spring and summer-autumn bi-seasonal periods at Barrow using the NOAA/GMD and AERONET sun-photometers. The frequency histograms are all normalised to yield the unit integrated area of 100% over the measured sampling intervals of *AOT(0.50 μm)* and α shown in these graphs. The three quartiles of *AOT(0.50 μm)* determined in winter-spring (corresponding to the 25th, 50th, and 75th percentiles) were found to be equal to 0.081, 0.103 and 0.138 by NOAA/GMD, and to 0.083, 0.119 and

0.169 by AERONET, while those determined in summer-autumn were estimated to be equal to 0.048, 0.065 and 0.091 by NOAA/GMD, and to 0.048, 0.067 and 0.101 by AERONET. On the basis of these results, the relative frequency histograms shown in Fig. 9.7 were found to exhibit long-tailed shapes for both the winter-spring data-sets, and higher kurtosis features for the summer-autumn data-sets. The above-mentioned long-tail characteristics could be in part due to the higher values of *AOT(0.50 μm)*, ranging mainly from 0.12 to 0.16 in April and May, which can be plausibly explained as caused by the relatively high number of Arctic haze cases observed in spring, but could also be in part due to the asymmetrical features of the frequency histograms of *AOT(0.50 μm)*. The small discrepancies found between the time-patterns of the monthly mean values of *AOT(0.50 μm)* defined for both the NOAA/GMD and AERONET data-sets as well as those between the frequency histograms of both parameters might be in part ascribed to the different observational periods of the two sun-photometers and to the different cloud screening procedures adopted by the NOAA/GMD and AERONET protocols. The different seasonal features of *AOT(0.50 μm)* and α shown in Fig. 9.6 should also arise from the variable origins of the aerosol particles, associated with the transport of continental polluted air masses in winter-spring, for the major part from Asia and to a lesser extent from North America (Hirdman et al. 2010).

Figure 9.7 shows also the relative frequency histograms of exponent α determined separately for the two seasonal periods of the year. The three quartiles of exponent α determined in winter-spring were estimated to be equal to 1.06, 1.25 and 1.42 by NOAA/GMD, and to 0.91, 1.24 and 1.47 by AERONET, while those determined in summer-autumn were estimated to be equal to 1.31, 1.47 and 1.66 by NOAA/GMD, and to 1.18, 1.41 and 1.61 by AERONET. The corresponding relative frequency histograms shown in Fig. 9.7 exhibit left-hand wings giving only a few values smaller than 0.75, probably due to the optical predominance of the coarse particle mode mainly consisting of sea-salt aerosols and/or local blown dust. Similarly, a certain fraction of values with $\alpha < 1.20$ are presumably related to Asian dust transport episodes (Di Pierro et al. 2011), which have been most frequently observed in March and April, and were often characterised by persistent extinction features typical of coarse mineral dust particles (Stone et al. 2007). As pointed out above, the bi-seasonal average values of α were found to be close to 1.20 in winter-spring for both the NOAA/GMD and AERONET measurements, and appreciably higher in summer-autumn, when average values of 1.46 and 1.38 were estimated respectively. Such a marked increase in α could be plausibly caused by the major contribution of the fine particle load provided in numerous boreal forest fire episodes by the smoke particle emissions occurring during the summer months over the coastal region around Barrow (Stohl 2006). As a result of the large variations produced by these combustion particles in the bimodal size-distribution curves consisting of fine and accumulation/coarse aerosol particles, the three quartiles of exponent α were estimated to increase by 10% to 25% on average when passing from winter-spring to summer-autumn.

9.1.3.2 Measurements in Northern Canada (Nunavut)

The results derived from the AERONET/AEROCAN measurements conducted at the Resolute Bay station (from July 2004 to October 2012) and Eureka-0PAL station (from April 2007 to September 2011) by the Environment Canada team (Downsview, North York, Ontario, Canada) and those carried out by the NOAA/GMD (Boulder, Colorado, USA) at Alert (Nunavut, Canada) (from August 2004 to September 2012) are shown in Figs. 9.8 and 9.9. They have been mainly obtained for air masses containing aerosols mainly transported from the North American and Arctic Ocean areas over these sites located in Northern Canada (Hirdman et al. 2010). The series of the peak-wavelengths of the narrow-band interference filters mounted on the Cimel CE-318 sun-photometers employed at Resolute Bay and Eureka-0PAL stations and those of the Carter Scott SP02 sun-photometer used at Alert are given in Table 9.8. Fig. 9.8 shows that: (a) the monthly mean values of $AOT(0.50\ \mu m)$ measured at Resolute Bay decrease from more than 0.15 in March to about 0.08 in May and June, and continue to more slowly decrease in the following months, until reaching values close to 0.04 in August and September, giving average seasonal values of $AOT(0.50\ \mu m)$ equal to 0.113 ± 0.046 in winter-spring and 0.053 ± 0.038 in summer-autumn (see also Table 9.9);

(b) the monthly mean values of $AOT(0.50\ \mu m)$ measured at Eureka-0PAL exhibit decreasing time-patterns from about 0.11 in April to less than 0.04 in both August and September, yielding average seasonal values of $AOT(0.50\ \mu m)$ equal to 0.093 ± 0.030 in winter-spring and 0.056 ± 0.029 in summer-autumn, as shown in Table 9.9; and (c) the monthly mean values of $AOT(0.50\ \mu m)$ measured at Alert were rather stable from March to June, ranging between 0.10 and 0.12 over such a 4-month period, and then slowly decreased to less than 0.07 in July and to 0.05 in September, providing average seasonal values of $AOT(0.50\ \mu m)$ equal to 0.115 ± 0.057 in winter-spring and 0.066 ± 0.041 in summer-autumn, as reported in Table 9.9. As pointed out by Tomasi et al. (2015b), the monthly mean values of $AOT(0.50\ \mu m)$ shown in Fig. 9.8 are rather high in March–April (varying mainly between 0.10 and 0.15, since they are presumably associated with the more frequent Arctic Haze episodes in such a period of the year) and present relatively low values in the subsequent spring and summer months, until reaching values of around 0.04–0.05 in September at all the three sites. The month-to-month differences, ranging mainly from 0.01 (as measured at Alert in late summer and autumn) to 0.05 (as observed at Resolute Bay and Alert in March–June), turn out to be comparable with the monthly mean values of standard deviation σ_{AOT}, evaluated at the three Canadian sites. Correspondingly, the monthly mean values of exponent α were estimated to: (i) slowly increase at Resolute Bay from 1.43 to 1.64 in the March–May months, and then to gradually decrease from less than 1.69 in June to 1.05 in September, yielding seasonal average values of α equal to 1.52 ± 0.25 in winter-spring and 1.55 ± 0.26 in summer-autumn (see Table 9.9); (ii) increase at the Eureka-0PAL station from 1.13 in April to more than 1.70 in both August and September, giving seasonal average values of α equal to 1.43 ± 0.32 in winter-spring and 1.64 ± 0.24 in summer-autumn; and (iii) increase at Alert from 1.08 in

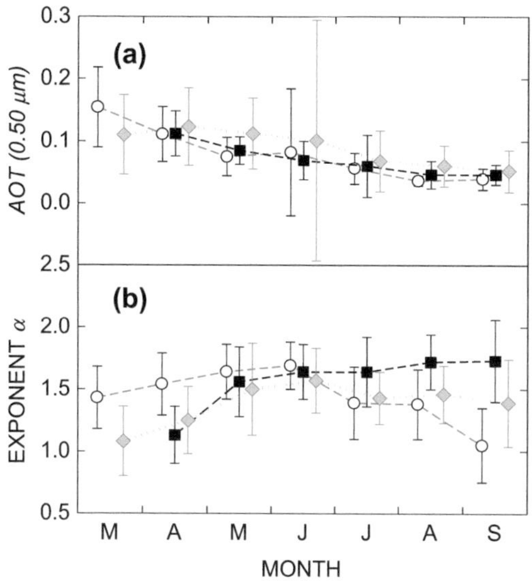

Fig. 9.8 Upper part (**a**): Time-patterns of the monthly mean values of aerosol optical thickness *AOT(0.50 μm)* given with their standard deviations (defined by vertical bars), as obtained from the multi-year sun-photometer measurements conducted at: (i) Resolute Bay (Nunavut, Canada) by Environment Canada (Ontario, Canada) with the AERONET/AEROCAN Cimel CE-318 sun-photometers over the period from July 2004 to October 2012 (open circles); (ii) Eureka-0PAL (Nunavut, Canada) by CARTEL (Sherbrooke University, Canada) with the AERONET/AEROCAN Cimel CE-318 sun-photometers over the period from April 2007 to September 2011 (solid squares); and (iii) Alert (Nunavut, Canada) by NOAA/GMD (Boulder, Colorado, USA) with the Carter Scott SP02 sun-photometers over the period from August 2004 to September 2012 (grey diamonds). Lower part (**b**): As in the upper part for the time-patterns of the monthly mean values of wavelength exponent α given with their standard deviations (defined by the vertical bars), as obtained from the same multi-year sun-photometer measurements conducted at: (i) Resolute Bay (Nunavut, Canada) by Environment Canada (Ontario, Canada) with the AERONET/AEROCAN Cimel CE-318 sun-photometers from 2004 to 2012 (open circles); (ii) Eureka-0PAL (Nunavut, Canada) by CARTEL (Sherbrooke University, Canada) with the AERONET/AEROCAN Cimel CE-318 sun-photometers from 2007 to 2011 (solid squares); and (iii) Alert (Nunavut, Canada) by NOAA/GMD (Boulder, Colorado, USA) with the Carter Scott SP02 sun-photometers from 2004 to 2012 (grey diamonds)

March to 1.57 in June, and then vary between 1.39 and 1.46 in the summer months from July to September, consequently providing seasonal average values of α equal to 1.29 ± 0.30 in winter-spring and 1.46 ± 0.26 in summer-autumn. Therefore, relatively small differences were found at the three Northern Canadian sites between the monthly mean values of α measured in winter-spring and those recorded in summer-autumn, since this parameter was found to present: (i) stable values close to 1.50 in March–June at Resolute Bay, followed by gradually decreasing values in July–September until reaching a value lower than 1.10 at the end of summer; and (ii) values increasing from around 1.10 to 1.56 in March–May, and higher than 1.60 from June to September (with $\sigma_\alpha \approx 0.30$) at the Eureka-0PAL station, and values

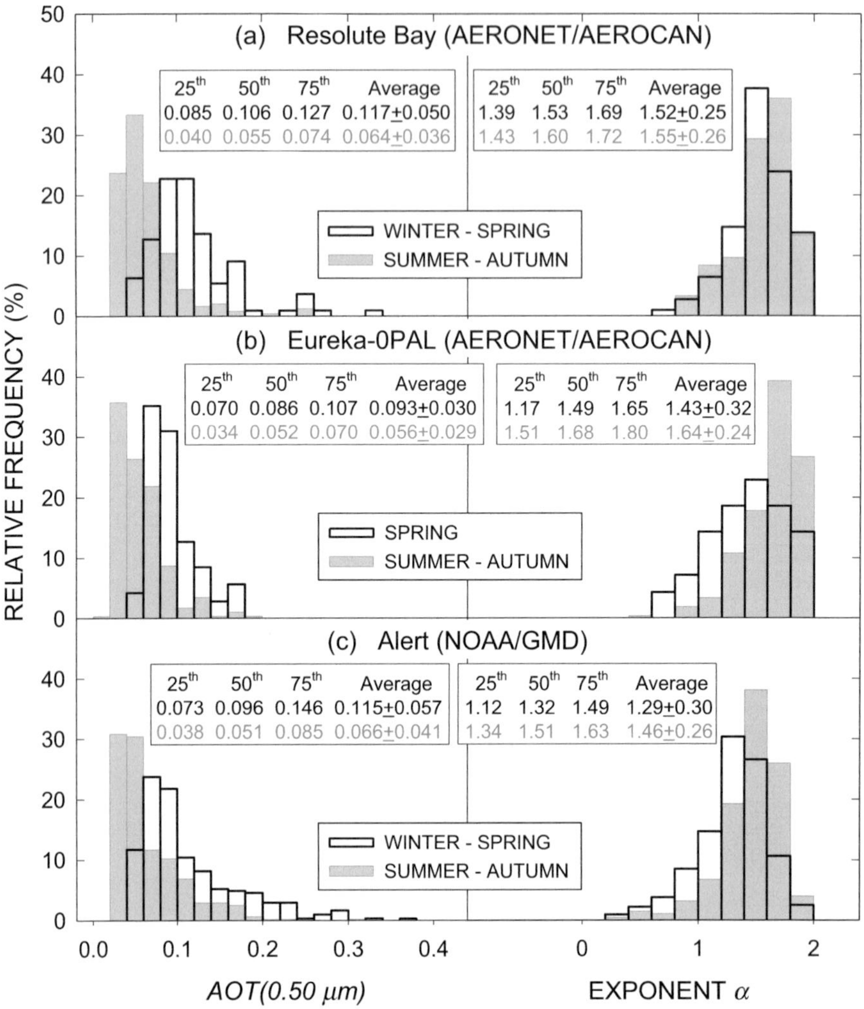

Fig. 9.9 Part (**a**): Relative frequency histograms RFHs of aerosol optical thickness *AOT(0.50 μm)* (left-hand side) and Ångström wavelength exponent α (right-hand side) obtained separately for the winter–spring (December–May) and the summer–autumn (June–October) periods from the multi-year sun-photometer measurements performed at Resolute Bay (Nunavut, Canada) by Environment Canada (Ontario, Canada) over the period from July 2004 to October 2012, using the AERONET/AEROCAN Cimel CE-318 sun-photometers. The values of the quartiles (i.e. of the 25th, 50th, and 75th percentiles) of *AOT(0.50 μm)* and exponent α reported in the boxes inserted into the graphs have been obtained by examining the daily mean values of the two optical parameters measured at this site in the winter–spring (bold numbers) and summer–autumn (grey numbers) periods. Part (**b**): As in part (**a**) for the multi-year sun-photometer measurements performed at Eureka-0PAL (Nunavut, Canada) by CARTEL (Sherbrooke University, Canada) from April 2007 to September 2011 by using the AERONET/AEROCAN Cimel CE-318 sun-photometers in the spring and summer –autumn months. Part (**c**): As in parts (**a**) and (**b**) for the multi-year sun-photometer measurements performed at Alert (Nunavut, Canada) by NOAA/GMD (Boulder, Colorado, USA) from August 2004 to September 2012 by using a Carter Scott SP02 sun-photometer in the winter-spring and summer-autumn months

increasing from 1.08 to 1.50 in spring and ranging between 1.39 and 1.57 in the summer months (also with $\sigma_\alpha \approx 0.30$) at the Alert station. The monthly mean values of α determined at Resolute Bay and Eureka-0PAL result to be appreciably higher than those measured at Barrow, presumably as a result of the weaker extinction produced by coarse sea-salt particles, and weaker contributions provided by the long-range transport of Asian dust in the spring months.

The winter-spring relative frequency histograms shown in Fig. 9.9 exhibit higher mean values of *AOT(0.50 μm)*, ranging from 0.09 to 0.12, and broader right-hand wings compared to those determined in the summer months, when narrower shape-features of the frequency histograms were obtained, yielding bi-seasonal mean values ranging between 0.06 and no more than 0.08 (Tomasi et al. 2015b). Fig. 9.9 shows that the winter-spring and summer-autumn relative frequency histograms of *AOT(0.50 μm)* are characterized by values of the three quartiles estimated to be: (a) equal to 0.085, 0.106 and 0.127, giving a winter-spring average value equal to 0.117 ± 0.050 at Resolute Bay, and equal to 0.040, 0.055 and 0.074, in summer-autumn (with an average value equal to 0.064 ± 0.036, which indicates that this parameter was subject to decrease by more than 40% from one period to the other); (b) equal to 0.070, 0.086 and 0.107 at Eureka-0PAL in winter-spring, and to 0.034, 0.052 and 0.070, in summer-autumn, and therefore decreasing by around 40% from spring to summer (although presenting appreciably lower values than those measured at Resolute Bay in both spring and summer); and (c) equal to 0.073, 0.096 and 0.146, at Alert in winter-spring, and to 0.038, 0.051 and 0.085 in summer-autumn, therefore without showing large seasonal variations with respect to those observed at the Eureka-0PAL station.

Figure 9.9 also shows that the spring and summer relative frequency histograms of the daily mean values of exponent α exhibit asymmetrical shapes: (i) in the winter-spring case they are rather wide, showing long-tailed left-hand wings and mean values varying from 1.29 (at Alert) to 1.52 (at Resolute Bay), and (ii) during summer-autumn they are characterised by mean values varying from 1.46 (at Alert) to 1.64 (at Eureka-0PAL). They present the three quartiles equal to: (a) 1.39, 1.53 and 1.69 in winter-spring and to 1.43, 1.60 and 1.72 in summer-autumn at Resolute Bay, therefore without presenting marked variations with season; (b) equal to 1.17, 1.49 and 1.65 in winter-spring, and to 1.51, 1.68 and 1.80 in summer-autumn at Eureka-0PAL, therefore presenting appreciably lower values of α than those recorded at Resolute Bay in both winter-spring and summer-autumn periods; and (c) equal to 1.12, 1.32 and 1.49 in winter-spring, and to 1.34, 1.51 and 1.63 in summer-autumn at Alert, therefore presenting appreciably lower values of α than those observed at Eureka-0PAL in both seasons.

These evaluations provide evidence that the optical extinction effects produced by fine particles predominate on those generated by the coarse particles at all the three Canadian sites considered in Fig. 9.9. The values of α determined in summer are considerably higher than those obtained in spring at both the Eureka-0PAL and Alert stations, while the increase in α observed to occur from spring to summer at Resolute Bay turns out to be more limited, due to less pronounced variations in the ratio between fine and coarse particle loads. In fact, only moderate relative increases in the coarse particle content were observed at Resolute Bay passing from spring to summer, while appreciably more marked variations in exponent α were observed to occur at Eureka and Alert passing from spring to summer, plausibly due to the more pronounced increase in the fine particle load associated with the transport of smoke particles generated by forest fire episodes taking place in the surrounding regions of North America (Stohl et al. 2006).

9.1.3.3 Measurements in Greenland

Figure 9.10 shows the monthly mean values of $AOT(0.50\ \mu m)$ derived from the multi-year sets of sun-photometer measurements conducted at: (i) Thule (located in the north-western corner of Greenland) by NASA/GSFC (USA) using an AERONET sun-photometer from March 2007 to September 2012; (ii) Summit station (in the middle of the Central Greenland ice sheet at more than 3000 m altitude above mean sea level) by the PMOD/WRC (Davos, Switzerland) group using a PFR sun-photometer from January 2001 to October 2011; and (iii) Ittoqqortoormiit (on the eastern coast of Greenland) by NASA/GSFC (USA) using an AERONET sun-photometer from May 2010 to October 2013. The peak wavelengths of the narrow-band interference filters mounted on the two Cimel CE-318 sun-photometers of the AERONET network employed at Thule and Ittoqqortoormiit, and those of the PFR#34 sun-photometer used at Summit are given in Table 9.8, together with the values of spectral intervals $\Delta\lambda$ on which the best-fit calculations of exponent α were performed. Examining the measurements performed at Thule, Tomasi et al. (2015b) found that the monthly mean values of $AOT(0.50\ \mu m)$ decreased slowly from 0.08–0.10 in the spring months to about 0.06 in June–September, with values of the standard deviation σ_{AOT} close to 0.03 in general, while exponent α was found to assume rather stable monthly mean values ranging between 1.30 and 1.40 in the spring months and between 1.35 and 1.54 in the summer months, with values of the standard deviation σ_{α} ranging between 0.21 and 0.36. Fig. 9.10 shows that: (i) the monthly mean values of $AOT(0.50\ \mu m)$ measured at Thule increase slightly from about 0.08 to 0.10 in March–April, and then decrease slowly from less than 0.10 in May to around 0.05 in September, yielding bi-seasonal average values of $AOT(0.50\ \mu m)$ equal to 0.093 $\pm$ 0.036 in winter-spring and 0.059 $\pm$ 0.029 in summer-autumn, as given in Table 9.9; (ii) the monthly mean values of $AOT(0.50\ \mu m)$ measured at Summit were found to be quite stable during the whole observational period, ranging mainly between 0.05 and 0.07 in March–August and decreasing to 0.03 in September–October, and yielding seasonal average values of $AOT(0.50\ \mu m)$ equal to 0.057 $\pm$ 0.033 in winter-spring and 0.052 $\pm$ 0.041 in summer-autumn; and (iii) the monthly mean values of $AOT(0.50\ \mu m)$ measured at Ittoqqortoormiit were estimated to slowly decrease from about 0.07 in March–May to 0.05–0.06 in early summer and to the less than 0.03 in September and October, giving seasonal average values of $AOT(0.50\ \mu m)$ equal to 0.068 $\pm$ 0.027 in winter-spring and to 0.052 $\pm$ 0.035 in summer-autumn.

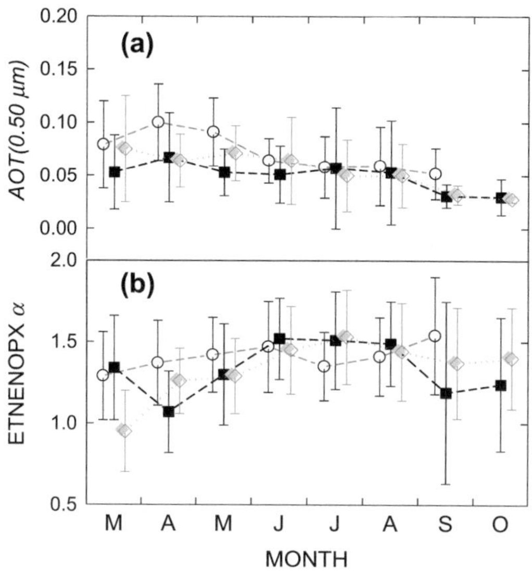

Fig. 9.10 Upper part (**a**): Time-patterns of the monthly mean values of aerosol optical thickness *AOT(0.50 μm)* given with their standard deviations (defined by vertical bars), as obtained from the multi-year sun-photometer measurements conducted at: (i) Thule (North-western Greenland) by NASA/GSFC (USA) using an AERONET Cimel CE-318 sun-photometer over the period from March 2007 to September 2012 (open circles); (ii) Summit (Central Greenland) by PMOD/WRC (Switzerland) using the PFR ≠ 34 sun-photometer of the GAW-PFR network over the period from January 2001 to October 2011 (solid squares); and (iii) Ittoqqortoormiit (Eastern Greenland) by NASA/GSFC (USA) using an AERONET Cimel CE-318 sun-photometer over the period from May 2010 to October 2013 (grey diamonds). Lower part (**b**): As in the upper part for the time-patterns of the monthly mean values of wavelength exponent α given with their standard deviations (defined by the vertical bars), as obtained from the same multi-year sun-photometer measurements conducted at: (i) Thule (North-western Greenland) by NASA/GSFC (USA) with an AERONET Cimel CE-318 sun-photometer from 2007 to 2012 (open circles); (ii) Summit (Central Greenland) by PMOD/WRC (Switzerland) with the PFR ≠ 34 sun-photometer of the GAW-PFR network from 2001 to 2011 (solid squares); and (iii) Ittoqqortoormiit (Eastern Greenland) by NASA/GSFC (USA) using an AERONET Cimel CE-318 sun-photometer over the period from May 2010 to October 2013 (grey diamonds)

Correspondingly, the monthly mean values of exponent α were evaluated: (i) to slowly increase at Thule from 1.29 in March to 1.54 in September, yielding seasonal average values equal to 1.38 ± 0.26 in winter-spring and 1.41 ± 0.26 in summer-autumn; (ii) to vary at Summit between 1.07 and 1.52 during the whole March–October period, providing seasonal average values of α equal to 1.22 ± 0.29 in winter-spring and 1.48 ± 0.29 in summer-autumn; and (iii) to increase rapidly at Ittoqqortoormiit from 0.95 in March to more than 1.50 in July, and then to maintain rather stable values ranging between 1.37 and 1.44 in August–October, giving bi-seasonal average values of α equal to 1.26 ± 0.23 in winter-spring and 1.45 ± 0.29 in summer-autumn, which are similar to those observed at Thule and Summit during both the observational periods. Examining the above results, it can be noted that rather low values of *AOT(0.50 μm)* were measured at both the coastal sites of Thule and Ittoqqortoormiit, indicating that these remote sites are not appreciably influenced by the long-range transport of particulate matter from the industrial mid-latitude regions of Europe and North America. Rather stable monthly mean values of *AOT(0.50 μm)* were also measured at the high-altitude Summit station, equal to 0.05 $\pm$ 0.03 from March to August, and to about 0.03 $\pm$ 0.01 in September and October, such low values being due to the fact that Summit is a high-altitude station likely not involved in exchange processes of air masses coming from the lower polluted atmospheric boundary layer.

Figure 9.11 shows that both the relative frequency histograms of the daily mean values of *AOT(0.50 μm)* were found to present very similar features at the three remote sites considered in the present study in Greenland, where the following values of the three quartiles were determined: (a) equal to 0.065, 0.086 and 0.112 in winter-spring at Thule, and to 0.037, 0.049 and 0.074 in summer-autumn (from June to September), these measurements presenting a decrease of about 40% from early spring to late summer, with features very similar to those shown in Fig. 9.9 for the three remote Canadian sites; (b) equal to 0.035, 0.047 and 0.062 in winter-spring at Summit, and equal to 0.029, 0.038 and 0.063 in summer-autumn, showing that a limited decrease of *AOT(0.50 μm)* no greater than 10% occurred at this high-altitude site from March to October; and (c) equal to 0.049, 0.051 and 0.081 in winter-spring at Ittoqqortoormiit, and equal to 0.033, 0.041 and 0.060 in summer-autumn (i. e. from June to October), therefore presenting atmospheric turbidity conditions that are subject to only weak changes throughout the whole period from March to October. In particular, the spring relative frequency histograms obtained from the *AOT(0.50 μm)* measurements conducted at Thule was found to assume an asymmetric shape with a mean value of 0.093, which can be reasonably explained in terms of the not rarely observed occurrence of Arctic haze events at this site. The summer relative frequency histograms of *AOT(0.50 μm)* was found to be more symmetric, presenting values of the 25th and 75th percentiles closer to the median value than in the spring relative frequency histogram, as can be seen in Fig. 9.11. The *AOT(0.50 μm)* relative frequency histograms drawn for the Ittoqqortoormiit station were found to vary only slightly when passing from the spring months to the summer-autumn period. The seasonal variations in *AOT(0.50 μm)* indicate that this optical parameter is subject to decrease appreciably from spring to late summer. On this matter, Tomasi et al. (2015b) pointed out that the seasonal variations affecting the monthly mean values of *AOT(0.50 μm)* at Ittoqqortoormiit are quite similar to those observed at Thule, since they both decrease gradually from about 0.08 in March to less than 0.04 in September and October, being σ_{AOT} equal to 0.05 in spring and considerably lower in summer, when it assumes a monthly mean value close to 0.01 in October.

Figure 9.11 shows the winter-spring and summer-autumn frequency histograms of the daily mean values of exponent α, indicating that the three quartiles of this parameter were: (a) equal to 1.20, 1.40 and 1.57 in the winter-spring months at Thule, and equal to 1.25, 1.42 and 1.58 in summer-autumn, therefore without presenting marked variations with season; (b) equal to 1.00, 1.25 and 1.41 at

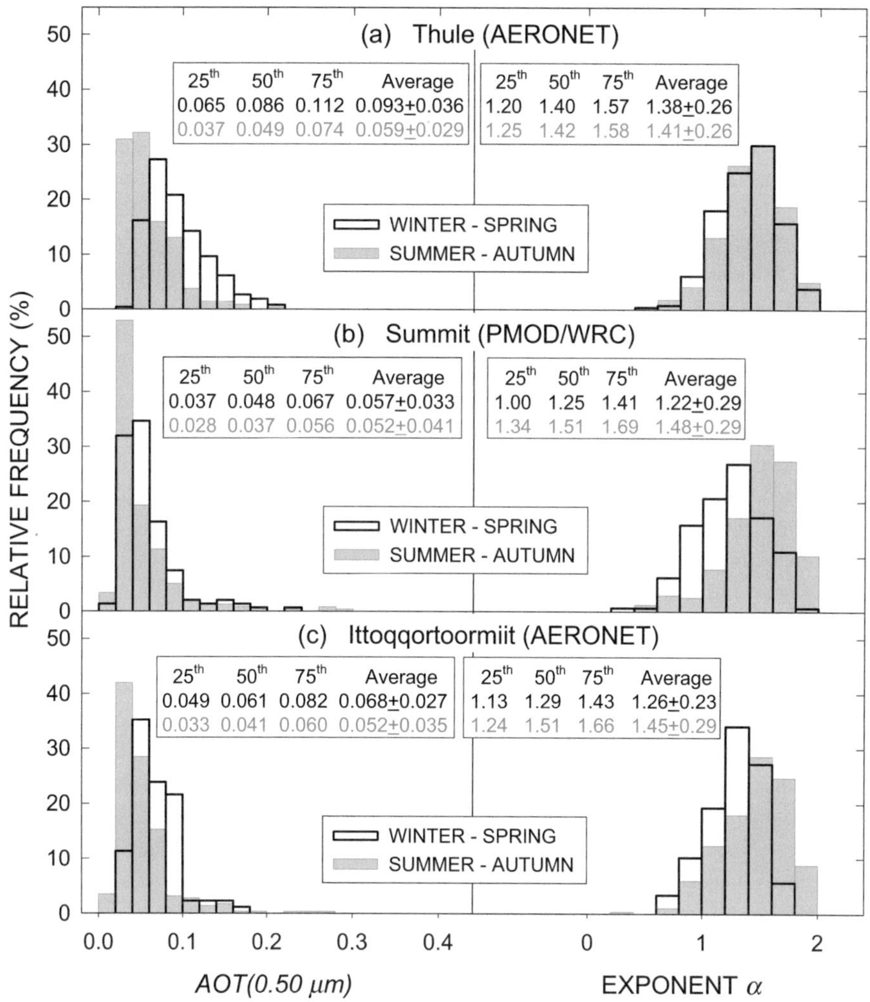

Fig. 9.11 Part (**a**): Relative frequency histograms RFHs of aerosol optical thickness *AOT(0.50 μm)* (left-hand side) and Ångström wavelength exponent α (right-hand side) obtained separately for the winter–spring (December–May) and the summer–autumn (June–October) periods from the multi-year sun-photometer measurements performed at Thule (North-western Greenland) by NASA/GSFC (USA) by using an AERONET Cimel CE-318 sun-photometer over the period from March 2007 to September 2012. The values of the quartiles (i.e. of the 25th, 50th, and 75th percentiles) of *AOT(0.50 μm)* and exponent α reported in the boxes inserted into the graphs have been obtained by examining the daily mean values of the two optical parameters measured at Thule in the winter–spring (bold numbers) and summer–autumn (grey numbers) periods. Part (**b**): As in part (a) for the multi-year sun-photometer measurements performed at Summit (Central Greenland) by PMOD/WRC (Switzerland) by using the PFR $\neq$ 34 sun-photometer of the GAW-PFR network over the period from January 2001 to October 2011. Part (**c**): As in parts (**a**) and (**b**) for the multi-year sun-photometer measurements performed at Ittoqqortoormiit (Eastern Greenland) by NASA/GSFC (USA) using an AERONET Cimel CE-318 sun-photometer over the period from May 2010 to October 2013

Summit in winter-spring, and equal to 1.34, 1.51 and 1.69 in summer-autumn, as it is typical of high-altitude aerosol extinction observations, and (c) equal to 1.13, 1.29 and 1.43 in winter-spring at Ittoqqortoormiit, and equal to 1.24, 1.51 and 1.66 in summer-autumn, when stable time-patterns of the monthly mean values of α have been measured from June to October. The seasonal relative frequency histograms of exponent α obtained at Thule exhibit very similar features in both spring and summer, presenting mean values of α equal to 1.38 in spring and 1.41 in summer-autumn, and nearly equal values of the quartiles of α, clearly indicating that no relevant seasonal changes occurred at Thule in the relative concentrations of fine and accumulation/coarse aerosol particles. The spring mean value of α was estimated to be equal to 1.27 at Ittoqqortoormiit, while the summer-autumn value was equal to 1.45, suggesting that the relative atmospheric content of fine particles was increased appreciably on average when passing from spring to summer-autumn. Such pronounced variations are probably associated with the marked extinction effects produced by the maritime accumulation/coarse mode particles mixed with relevant loads of continental particles transported mainly from Europe over this station in late spring and summer, containing in general significant mass fractions of both anthropogenic and combustion aerosol particles, which are first involved by low-level transport and uplift processes taking place outside the Arctic and then deposited by descensional motions in proximity of Greenland, as suggested by Stohl et al. (2006). Examining the measurements of $AOT(\lambda)$ and α performed at Summit, Tomasi et al. (2015b) pointed out that the spectral features of $AOT(\lambda)$ and the relatively limited seasonal variations of exponent α indicate clearly that Summit is representative of aerosol particle loads generally present in the Arctic free troposphere, which are mainly influenced by particulate matter transport processes from the European and North American sectors, and only weakly by the transport of Siberian aerosols at altitudes higher than the top-level of the atmospheric boundary layer (Hirdman et al. 2010).

The seasonal variations in the monthly mean values of parameters $AOT(0.50\ \mu m)$ and α shown in Figs. 9.10 and the frequency histograms of $AOT(0.50\ \mu m)$ and α presented in Fig. 9.11 indicate also that the optical properties of aerosol particles monitored at the three Greenland sites are closely related to the different characteristics of the aerosol particles generated in the anthropogenic/polluted regions of Europe (and more weakly in North America) and subsequently transported over Greenland, than to loads of maritime particles formed in the remote oceanic mid-latitude areas, or in the Arctic Ocean. Actually, the transport processes of anthropogenic soot aerosols are known to appreciably enhance $AOT(\lambda)$, yielding rather high values of exponent α in general (Tomasi et al. 2007; Stone et al. 2008). These long-range transport episodes may also occur within layers located in the free troposphere, as observed during airborne measurements conducted at mid-altitudes over the Arctic Ocean (Stone et al. 2010).

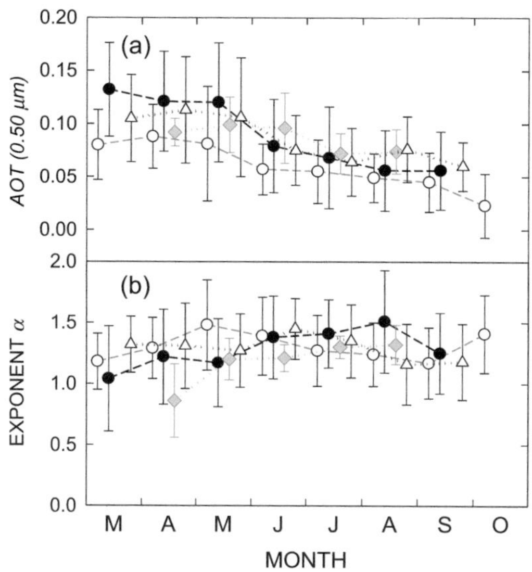

Fig. 9.12 Upper part (**a**): Time-patterns of the monthly mean values of aerosol optical thickness *AOT(0.50 μm)* given with their standard deviations (defined by vertical bars), as obtained from the multi-year sun-photometer measurements conducted at: (i) Ny-Ålesund (Spitsbergen, Svalbard) by AWI (Potsdam, Germany) using the SP1A and SP2H sun-photometers and the STAR01 star-photometer over the period from April 2000 to September 2013 (open circles); (ii) Ny-Ålesund (Spitsbergen, Svalbard) by NILU (Kjeller, Norway) using the PFR ≠ 18 sun-photometer of the GAW-PFR network over the periods from April 2002 to September 2004 and from March 2006 to September 2013 (solid circles); (iii) Barentsburg (Spitsbergen, Svalbard) by the IAO-SB-RAS (Tomsk, Russia) group with their SPM portable sun-photometer in the April–August months of 2011 and 2012 (grey diamonds); and (iv) Hornsund (Spitsbergen, Svalbard) by the Warsaw University group (PAS, Poland) in cooperation with NASA/GSFC (USA) using an AERONET Cimel CE-318 sun-photometer over the period from April 2005 to August 2013 (open triangles). Lower part (**b**): As in the upper part for the time-patterns of the monthly mean values of wavelength exponent α given with their standard deviations (defined by the vertical bars), as obtained from the same multi-year sun-photometer measurements conducted at: (i) Ny-Ålesund (Spitsbergen, Svalbard) by AWI (Potsdam, Germany) with the SP1A and SP2H sun-photometers and the STAR01 star-photometer from 2000 to 2013 (open circles); (ii) Ny-Ålesund (Spitsbergen, Svalbard) by NILU (Kjeller, Norway) with the PFR ≠ 18 sun-photometer of the GAW-PFR network from 2002 to 2004 and from 2006 to 2013 (solid circles); (iii) Barentsburg (Spitsbergen, Svalbard) by the IAO-SB-RAS (Tomsk, Russia) group with their SPM portable sun-photometer in the April–August months of 2011 and 2012 (grey diamonds); and (iv) Hornsund (Spitsbergen, Svalbard) by the Warsaw University group (PAS, Poland) in cooperation with NASA/GSFC (USA) using an AERONET Cimel CE-318 sun-photometer from 2005 to 2013 (open triangles)

9.1.3.4 Measurements in Spitsbergen (Svalbard)

The results obtained from the sun-photometer measurements carried out at Ny-Ålesund, Barentsburg and Hornsund, all located in Spitsbergen (Svalbard, Norway), are shown in Figs. 9.12 and 9.13, being derived from the measurements performed at: (i) Ny-Ålesund by the AWI (Bremerhaven, Germany) group using different sun-

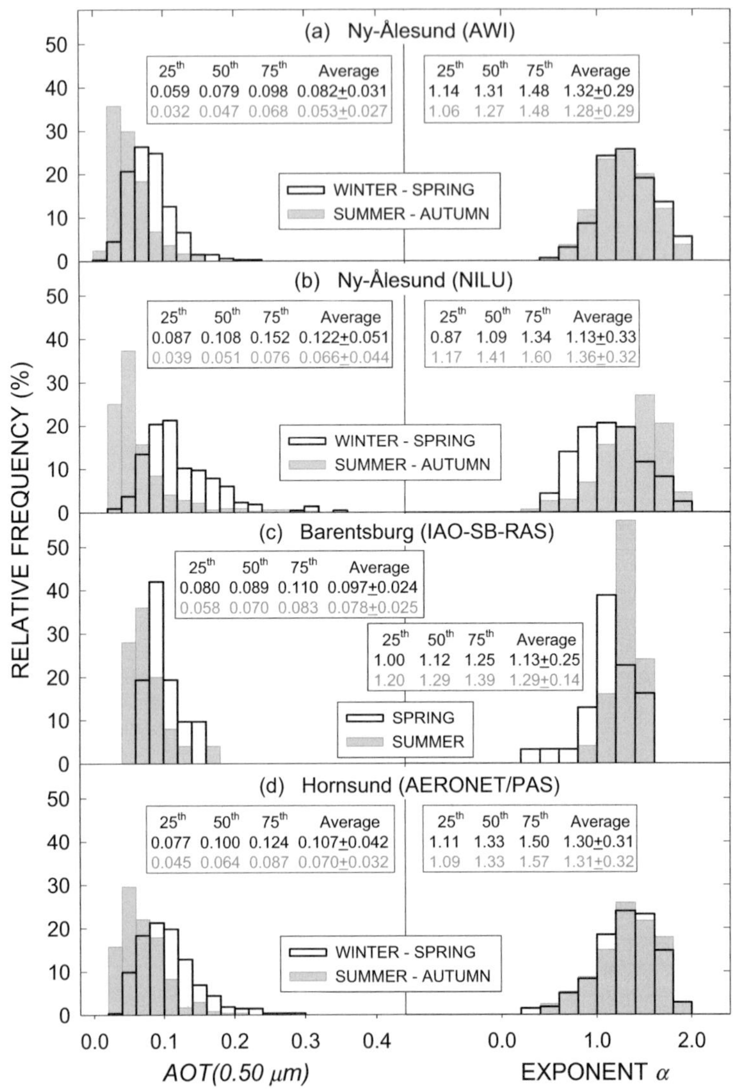

Fig. 9.13 Part (**a**): Relative frequency histograms RFHs of aerosol optical thickness *AOT (0.50 μm)* (left-hand side) and Ångström wavelength exponent α (right-hand side) obtained separately for the winter–spring (December–May) and the summer–autumn (June–October) periods from the multi-year sun-photometer measurements performed at Ny-Ålesund (Spitsbergen, Svalbard) by AWI (Potsdam, Germany) by using the SP1A and SP2H sun-photometers and the STAR01 star-photometer over the period from April 2000 to September 2013. The values of the quartiles (i.e. of the 25th, 50th, and 75th percentiles) of *AOT(0.50 μm)* and exponent α reported in the boxes inserted into the graphs have been obtained by examining the daily mean values of the two optical parameters measured at Ny- Ålesund by the AWI group in the winter–spring (bold numbers) and summer–autumn (grey numbers) periods. Part (**b**): As in part (**a**) for the multi-year sun-photometer measurements performed at Ny-Ålesund by NILU (Kjeller, Norway) by using the PFR $\neq$ 18 sun-photometer of the GAW-PFR network over the periods from April 2002

and star-photometer models from April 2000 to September 2013; (ii) Ny-Ålesund by the NILU (Kjeller, Norway) group, using some GAW-PFR sun-photometer models from March 2002 to September 2004 and from March 2006 to September 2013; (iii) Barentsburg by the IAO-SB-RAS (Tomsk, Russia) group using the SPM sun-photometer during the April–August months of 2011 and 2012; and (iv) Hornsund by the Institute of Geophysics (Warsaw University, PAS, Poland) using an AERONET network sun-photometer in cooperation with NASA/GSFC (USA) from April 2005 to August 2013. Table 9.8 provides the spectral series of the peak wavelengths of the narrow-band interference filters mounted on (i) the SP1A and SP2H sun-photometers and the STAR01 star-photometer employed by the AWI group at Ny-Ålesund, with the corresponding spectral intervals $\Delta\lambda$, over which the exponent α was calculated; (ii) the PFR#18 sun-photometer used by NILU (Kjeller, Norway) at Ny-Ålesund, for which α was calculated over the spectral range from 367.7 to 862.5 nm; (iii) the new New Portable Sun-photometer (SPM) model used by the IAO-SB-RAS (Tomsk, Russia) group at Barentsburg, for which the best-fit values of α were calculated over the 380–871 nm spectral range; and (iv) the Cimel CE-318 employed by the Warsaw University group at Hornsund, for which the interval $\Delta\lambda$ was chosen from 380 to 870 nm to calculate the best-fit values of exponent α.

The monthly mean values of *AOT(0.50 μm)* measured by AWI at Ny-Ålesund were found to vary between 0.08 and 0.09 in spring and between 0.05 and 0.06 in the summer months, subsequently decreasing to less than 0.03 in October, with σ_{AOT} lower than 0.02 on average, providing seasonal average values of *AOT(0.50 μm)* equal to 0.082 ± 0.031 in spring and 0.053 ± 0.027 in summer-autumn (see Table 9.9). The NILU monthly mean values of *AOT(0.50 μm)* were found to vary between 0.10 and 0.11 from March to May, and between 0.06 in June–July and 0.05 in September, with monthly mean values of σ_{AOT} no higher than 0.01 in all the spring and summer months, providing seasonal average values of *AOT(0.50 μm)* equal to 0.122 ± 0.051 in spring and 0.066 ± 0.044 in summer-autumn. Correspondingly, as it follows from Fig. 9.12, the monthly mean values of exponent α derived from the AWI measurements were evaluated to increase from 1.20 in March to 1.50 in May and then to decrease regularly in summer until reaching a value of about 1.20 in September and a value equal to 1.45 in October, thus yielding the seasonal average values of *AOT(0.50 μm)* equal to 1.32 ± 0.29 in spring and 1.28 ± 0.29 in summer-autumn. The monthly mean values of exponent α derived from the NILU measurements at Ny-Ålesund were quite stable during the period from March to

Fig. 9.13 (continued) to September 2004 and from March 2006 to September 2013. Part (**c**): As in parts (**a**) and (**b**) for the multi-year sun-photometer measurements performed at Barentsburg (Spitsbergen, Svalbard) by the IAO-SB-RAS (Tomsk, Russia) group using their SPM portable sun-photometer in the spring and summer months of 2011 and 2012. Part (**d**): As in parts (**a**)–(**c**) for the multi-year sun-photometer measurements performed at Hornsund (Spitsbergen, Svalbard) by the Warsaw University (PAS, Poland) group in cooperation with NASA/GSFC (USA) by employing an AERONET Cimel CE-318 sun-photometer over the period from April 2005 to August 2013

September, ranging between 1.25 in March–April and 1.50 in June–August, and giving seasonal average values of α equal to 1.13 ± 0.33 in spring and equal 1.36 ± 0.32 in summer-autumn, as reported in Table 9.9.

The relative frequency histograms of *AOT(0.50* μm*)* shown in Fig. 9.13 and derived from the AWI measurements performed at Ny-Ålesund were found to be characterized by quartiles equal to 0.059, 0.079 and 0.098 in spring and equal to 0.032, 0.047 and 0.068 in summer-autumn. The corresponding measurements conducted at Ny-Ålesund by NILU were found to provide frequency histograms of *AOT(0.50* μm*)* having quartiles equal to 0.087, 0.108 and 0.152 in spring and equal to 0.039, 0.051 and 0.076 in summer-autumn, which are therefore higher by about 10% on average than those measured by AWI. The relative frequency histograms of exponent α obtained for the measurements conducted at Ny-Ålesund and shown in Fig. 9.13 indicate that the quartiles of α derived from the AWI measurements are equal to 1.14, 1.31 and 1.48 in spring, and to 1.06, 1.27 and 1.48 in summer-autumn. The NILU measurements conducted at the same site were estimated to give the relative frequency histograms of α shown in Fig. 9.13, which present quartiles equal to 0.87, 1.09 and 1.34 in spring, and equal to 1.17, 1.41 and 1.60 in summer-autumn. These quartiles of α turn out to be appreciably higher than those derived from the AWI measurements in both spring and summer-autumn by percentages varying between 1% and 6% in spring and between 6% and 12% in summer, probably mainly due to the different measurement periods of the AWI (14-year) and NILU (8-year) sun-photometers. In fact, the comparison between the AWI and NILU results made in Fig. 9.12 shows a good agreement between the monthly mean values of *AOT(0.50* μm*)*, although the NILU values were found to be occasionally higher than those obtained by AWI (by no more than 10% on average in spring). Therefore, it is not surprising that the frequency histograms found separately for the AWI and NILU daily mean values of *AOT(0.50* μm*)* differ considerably from one season to another, being: (i) the AWI and NILU spring mean values equal to 0.082 and 0.089, respectively, with the quartiles differing by no more than 0.007 one from the other; and (ii) the AWI and NILU summer-autumn mean values equal to 0.052 and 0.059, respectively, being $\sigma_{AOT} = 0.04$ on average. The AWI monthly mean values of α were estimated to increase from less than 1.20 in March to 1.50 in May, and slowly decrease in summer-autumn until reaching values of around 1.20 in September and 1.45 in October, with $\sigma_{\alpha} = 0.10$ on average, while the NILU values were estimated to increase slowly from 1.20 in March to 1.50 in June, and then decrease to about 1.40 in September, with $\sigma_{\alpha} = 0.10$ in all months, thus presenting discrepancies no higher than 15% in general (Tomasi et al. 2015b).

The data-sets of *AOT(0.50* μm*)* and α derived from the Barentsburg measurements conducted by the IAO-SB-RAS (Tomsk, Russia) group consisted of a considerably lower number of daily measurements than those performed at Ny-Ålesund by the AWI and NILU groups. It follows from Fig. 9.12 that the monthly mean values of *AOT(0.50* μm*)* varied between 0.07 (in July and August) and 0.10 (in April and May), with σ_{AOT} lower than 0.01 on average, whereas those of α increased from about 0.90 (in April) to about 1.30 in July and August, with $\sigma_{\alpha} = 0.05$ on average, thus differing only slightly from those measured at Ny-

Ålesund by the AWI and NILU groups. The relative frequency histograms of the daily mean values of $AOT(0.50\ \mu m)$ and α were prepared giving a spring average value of $AOT(0.50\ \mu m) = 0.097 \pm 0.024$ and a summer-autumn average value of $AOT(0.50\ \mu m) = 0.078 \pm 0.025$. Correspondingly, the quartiles of $AOT(0.50\ \mu m)$ were estimated to be equal to 0.080, 0.089 and 0.110 in spring, and to 0.058, 0.070 and 0.083 in summer-autumn. Therefore, these values of $AOT(0.50\ \mu m)$ result to be appreciably higher than those measured at Ny-Ålesund with both AWI and NILU sun-photometers. The seasonal average values of α were estimated to be equal to 1.13 ± 0.25 in spring, and 1.29 ± 0.14 in summer-autumn (as shown in Fig. 9.12). The relative frequency histograms of exponent α shown in Fig. 9.13 present quartiles equal to 1.00, 1.12 and 1.25 in spring, and slightly higher values of the quartiles in summer-autumn, equal to 1.20, 1.29 and 1.39, respectively, for which the relative frequency histograms peak curve obtained at Barentsburg results to be considerably narrower than those determined at Ny-Ålesund for both AWI and NILU data-sets. The results obtained at Barentsburg from the IAO-SB-RAS (Tomsk, Russia) measurements are compared in Figs. 9.12 and 9.13 with those derived from the AERONET measurements of $AOT(0.50\ \mu m)$ and α performed at Hornsund by the Institute of Geophysics (PAS, Warsaw, Poland) over a 9-year period from April 2005 to August 2013. These results are in close agreement with those determined at Ny-Ålesund, since the monthly mean values of $AOT(0.50\ \mu m)$ were found by the PAS group to vary between 0.10 and no more than 0.12 in spring, with $\sigma_{AOT} = 0.02$ on average, and between 0.06 and 0.08 in the summer months, with σ_{AOT} lower than 0.02 in general, giving average seasonal mean values of $AOT(0.50\ \mu m)$ equal to 0.107 ± 0.042 in spring and to 0.070 ± 0.032 in summer-autumn, as reported in Table 9.9. The monthly mean values of α were estimated to be rather stable at Hornsund, mainly ranging between 1.25 and 1.35 in spring, and between 1.15 and 1.45 in summer, therefore presenting only small differences with respect to the AWI and NILU evaluations. The relative frequency histograms of $AOT(0.50\ \mu m)$ and exponent α determined at Hornsund are shown in Fig. 9.13, indicating that: (1) parameter $AOT(0.50\ \mu m)$ is characterized by quartiles equal to 0.077, 0.100 and 0.124 in spring, and equal to 0.045, 0.064 and 0.087 in summer-autumn, therefore providing seasonal average values that are appreciably higher than those measured at the other Spitsbergen sites (although with discrepancies smaller than 0.04 between the seasonal quartiles); and (2) exponent α exhibits quartiles equal to 1.11, 1.33 and 1.50 in spring, and to 1.09, 1.33 and 1.57 in summer-autumn, therefore presenting relatively small variations from one season to the other. It can be reasonably concluded that the analysis of the sun-photometer measurements conducted in the Spitsbergen area indicate that the seasonal mean values of $AOT(0.50\ \mu m)$ measured in summer are lower by 0.03 on average than those measured in spring, and the seasonal mean values of exponent α exhibit in general rather small variations, no higher than 0.04 on average.

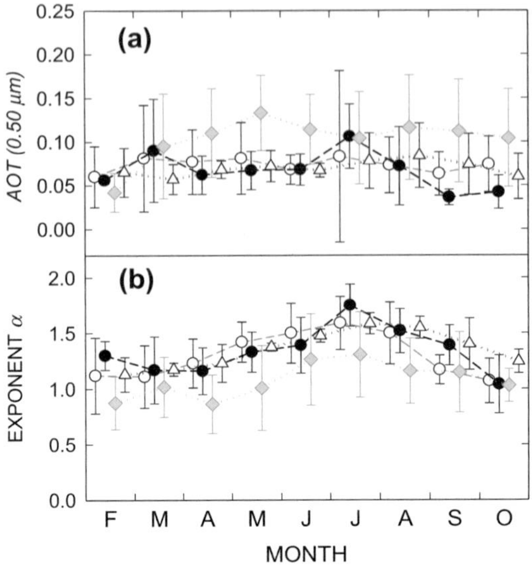

Fig. 9.14 Upper part (**a**): Time-patterns of the monthly mean values of aerosol optical thickness *AOT(0.50* μm*)* given with their standard deviations (defined by vertical bars), as obtained from the multi-year sun-photometer measurements conducted at: (i) Sodankylä (Northern Finland) by FMI (Helsinki, Finland) using the PFR $\neq$ 32 GAW sun-photometer over the period from late May 2004 to March 2013 (open circles); (ii) Sodankylä by the Arctic Research Center (FMI, Sodankylä, Finland) in cooperation with NASA/GSFC (USA) using an AERONET Cimel CE-318 sun-photometer over the period from February 2007 to November 2013 (solid circles); (iii) Andenes (Northern Norway) by the GOA (University of Valladolid, Spain) group using an AERONET Cimel CE-318 sun-photometer over the period from 2002 to 2009 (Toledano et al. 2012) (grey diamonds); and (iv) Kiruna (Northern Sweden) by the SMHI (Norrköping, Sweden) group using a GAW-PFR sun-photometer over the period from 2007 to 2009 (Toledano et al. 2012) (open triangles). Lower part (**b**): As in the upper part for the time-patterns of the monthly mean values of wavelength exponent α given with their standard deviations (defined by the vertical bars), as obtained from the same multi-year sun-photometer measurements conducted at: (i) Sodankylä (Northern Finland) by FMI (Helsinki, Finland) using the PFR $\neq$ 32 GAW sun-photometer from 2004 to March 2013 (open circles); (ii) Sodankylä by the Arctic Research Center (FMI, Sodankylä, Finland) and NASA/GSFC (USA) using an AERONET Cimel CE-318 sun-photometer from 2007 to 2013 (solid circles); (iii) Andenes (Northern Norway) by the GOA (University of Valladolid, Spain) group using an AERONET Cimel CE-318 sun-photometer from 2002 to 2009 (Toledano et al. 2012) (grey diamonds); and (iv) Kiruna (Northern Sweden) by the SMHI (Norrköping, Sweden) group using a GAW-PFR sun-photometer from 2007 to 2009 (Toledano et al. 2012) (open triangles)

9.1.3.5 Measurements in Northern Scandinavia

Figure 9.14 shows the time-patterns of the monthly mean values of *AOT(0.50* μm*)* and α derived from the sets of FMI/PFR and AERONET sun-photometer measurements carried out at Sodankylä (Northern Finland) over the periods from May 2004 to September 2013 and from February 2007 to November 2013, respectively. The spectral series of the peak wavelengths of the narrow-band interference filters mounted on: (i) the PFR#29 and PFR#32 sun-photometers employed by FMI at Sodankylä (with range $\Delta\lambda$ extending from 367.6 to 861.6 nm), and (ii) the Cimel CE-318 employed by the FMI Arctic Research Centre at Sodankylä (with $\Delta\lambda$ from 380 to 870 nm) are given in Table 9.8. The FMI/PFR monthly mean values of *AOT(0.50 μm)* were found to: (i) slowly decrease from less than 0.10 in March and April, (ii) maintain stable values ranging between 0.07 and 0.09 in May–August, and (iii) slowly decrease to 0.04 in September and 0.02 in October, when it was found to be comparable with the standard deviation σ_{AOT} obtained on these autumn days (estimated to range mainly between 0.02 and 0.04). Examining these data-sets, average values of *AOT(0.50 μm)* equal to 0.077 $\pm$ 0.040 and 0.069 $\pm$ 0.027 were obtained in the winter-spring and summer-autumn periods of the FMI sun-photometer measurements, respectively, as given in Table 9.9. The corresponding monthly mean values of exponent α were estimated to increase from 1.24 in March to more than 1.70 in July and August, and then to gradually decrease until reaching a monthly mean value of 1.17 in October, thus giving average values of α equal to 1.24 $\pm$ 0.27 in winter-spring and 1.41 $\pm$ 0.24 in summer-autumn (see Table 9.9).

Appreciably different time-patterns of the monthly mean values of *AOT(0.50 μm)* and α were obtained from the AERONET measurements conducted at Sodankylä over a shorter 7-year period (and therefore for a number of daily measurements equal to only about 30% of those made by employing the PFR sun-photometers), giving monthly mean values of *AOT(0.50 μm)* ranging between 0.05 and 0.09 in winter–spring and between 0.06 and 0.11 in June and July, and subsequently decreasing to 0.04 in September and October. Examining these measurements, average values of *AOT(0.50 μm)* equal to 0.069 $\pm$ 0.034 in winter-spring, and equal to 0.066 $\pm$ 0.036 in summer-autumn were obtained, which are very similar to those determined with the FMI/PFR sun-photometers, as shown in Table 9.9. The monthly mean values of α were estimated to vary between 1.20 and 1.30 in winter-spring, to increase to more than 1.70 in July and then to decrease to nearly 1.00 in October, providing average mean values of $\alpha = 1.23 \pm 0.23$ in winter-spring and $\alpha = 1.44 \pm 0.28$ in summer-autumn, as given in Table 9.9.

In order to provide a more complete picture of the atmospheric turbidity characteristics observed over the Northern Scandinavia region, the time-patterns of the PFR and AERONET monthly mean values of *AOT(0.50 μm)* and α determined at Sodankylä have been compared with those estimated by Toledano et al. (2012), who analysed the *AOT(0.50 μm)* and α data-sets collected at: (i) Andenes (69° 18′ N, 16° 01′ E, 380 m a.m.s.l.) in Northern Norway from January 2002 to October 2009, using the RIMA/AERONET Cimel CE-318 sun-photometer managed by the GOA group (University of Valladolid, Spain), which was equipped with the spectral series of the narrow-band interference filter peak wavelengths reported in Table 9.8; and (ii) Kiruna (67° 50′ N, 20° 25′ E, 424 m a.m.s.l.) in Northern Sweden (270 km WNW from Sodankylä) from January 2007 to October 2009, by using the GAW-PFR sun-photometer of the Swedish Meteorological and Hydrological Institute (SMHI) equipped with four narrow-band interference filters peaked at the wavelengths listed in Table 9.9 over the 368–862 nm spectral range.

The time-patterns of the monthly mean values of *AOT(0.50 μm)* and α measured at these two sites are shown in Fig. 9.18, according to the evaluations made by Toledano et al. (2012). It can be noted that the monthly mean values of *AOT(0.50 μm)* measured at Andenes were found to increase from about 0.04 in February to more than 0.13 in May, with values of σ_{AOT} ranging between 0.02 and 0.06, and then slowly decreasing during the first two summer months until reaching values close to 0.11 in September and October, the monthly mean values of σ_{AOT} being close to 0.06 throughout the whole observation period. These results were found to provide average seasonal mean values of *AOT(0.50 μm)* equal to 0.096 ± 0.041 in winter-spring and 0.113 ± 0.035 in summer-autumn at Andenes, as reported in Table 9.9. Correspondingly, the monthly mean values of parameter α were found to vary from about 0.86 to 1.02 in February–May, to subsequently increase to nearly 1.31 in July, and then decrease in late summer and autumn until reaching a value of 1.03 in October, with seasonal average values of the standard deviation σ_α varying from 0.24 to 0.38 0 in winter-spring and between 0.15 and 0.36 in summer-autumn. Analysing these measurements, average seasonal mean values of $\alpha = 1.00 \pm 0.30$ and $\alpha = 1.19 \pm 0.32$ were determined at Andenes in winter-spring and summer-autumn, respectively (as given in Table 9.9). These results indicate that the monthly mean values of α determined at Andenes are in general considerably lower than those measured at Sodankylä by means of the AERONET measurements, by about 15% on average, presumably because of the more marked extinction effects produced by the marine aerosol particles at this coastal site, the aerosol load being originated both in the off-shore and coastal regions of the Arctic Ocean and then transported toward the Andenes area, whereas Sodankylä is characterized by prevailing continental aerosol loads, presenting typically higher values of Ångström's wavelength exponent α.

Figure 9.15 presents the winter-spring and summer-autumn relative frequency histograms of *AOT(0.50 μm)* and α obtained from the measurements carried out at Sodankylä by using the PFR and AERONET sun-photometers managed by the FMI groups. The quartiles of *AOT(0.50 μm)* determined with the FMI/PFR sun-photometers are equal to 0.054, 0.066 and 0.087 in winter-spring and to 0.046, 0.065 and 0.089 in summer-autumn, without presenting marked variations with season. Correspondingly, exponent α was estimated to have quartiles equal to 1.08, 1.25 and 1.41 in winter-spring and equal to 1.23, 1.43 and 1.55 in summer-autumn, the latter being appreciably higher than those estimated in winter-spring and, therefore, indicating that a substantial increase in the fine particle mass content occur during

Fig. 9.15 (continued) sun-photometer over the period from February 2007 to November 2013. Part (**c**): As in parts (**a**) and (**b**) for the multi-year sun-photometer measurements performed at Andenes (Northern Norway) by the GOA (University of Valladolid, Spain) group using an AERONET/RIMA Cimel CE-318 sun-photometer over the 2002–2009 period (Toledano et al. 2012). Part (**d**): As in parts (**a**)–(**c**) for the multi-year sun-photometer measurements performed at Kiruna (Northern Sweden) by the SMHI (Norrköping, Sweden) group with a GAW-PFR sun-photometer from 2007 to 2009 (see also Toledano et al. 2012)

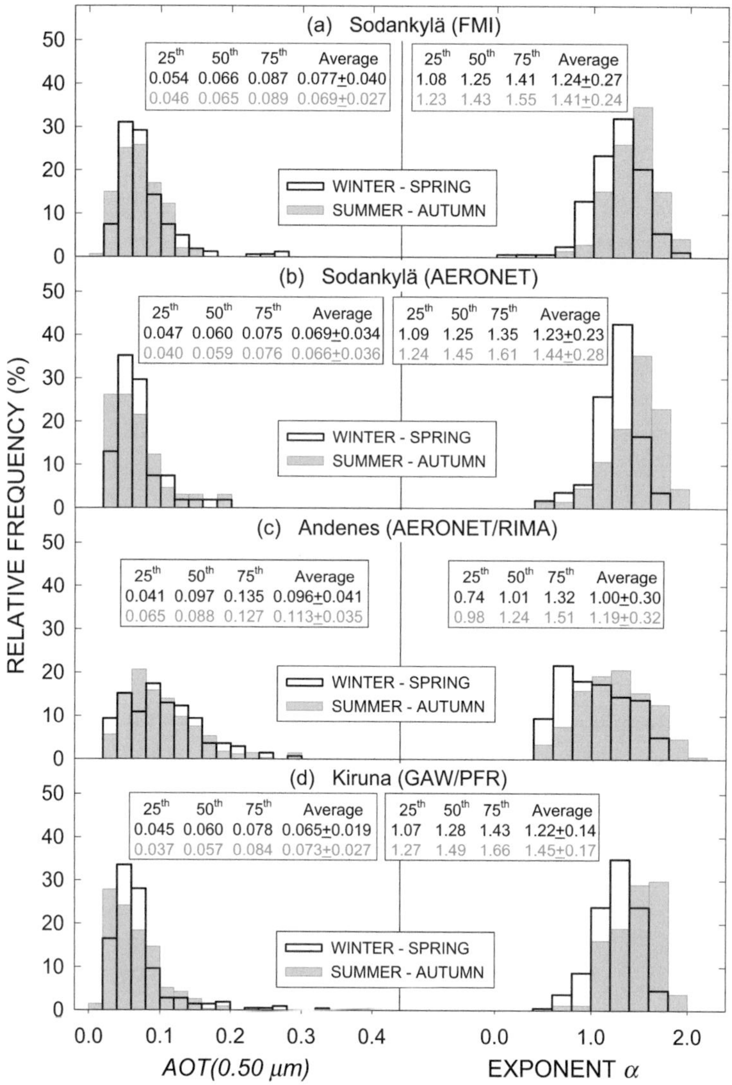

Fig. 9.15 Part (**a**): Relative frequency histograms RFHs of aerosol optical thickness *AOT (0.50 μm)* (left-hand side) and Ångström wavelength exponent α (right-hand side) obtained separately for the winter–spring (December–May) and the summer–autumn (June–October) periods from the multi-year sun-photometer measurements performed at Sodankylä (Northern Finland) by the FMI (Helsinki, Finland) group using the GAW-PFR $\neq$ 32 sun-photometer in the winter–spring and summer–autumn periods from late May 2004 to March 2013. The values of the quartiles (i.e. of the 25th, 50th, and 75th percentiles) of *AOT(0.50 μm)* and exponent α reported in the boxes inserted into the graphs have been obtained by examining the daily mean values of the two optical parameters measured at Sodankylä by the FMI group in the winter–spring (bold numbers) and summer–autumn (grey numbers) periods. Part (**b**): As in part (**a**) for the multi-year sun-photometer measurements performed at Sodankylä by the Arctic Research Centre (FMI, Finland) in cooperation with NASA/GSFC (USA) using an AERONET Cimel CE-318

the summer months, presumably caused by the increased transport of polluted air masses in summer from the Kola Peninsula, Scandinavia and Russia as well as from Britain and Eastern Europe (Aaltonen et al. 2006).

The variations in *AOT(0.50 μm)* and exponent α occurring from one season to the other were also accurately evaluated by analyzing the measurements performed at Sodankylä with the AERONET sun-photometer. The average multi-month values of *AOT(0.50 μm)* were estimated to be equal to 0.069 ± 0.034 in winter-spring and to 0.066 ± 0.036 in summer-autumn, in close agreement with the FMI/PFR measurements. Correspondingly, the average values of exponent α were found to be equal to 1.23 ± 0.23 in winter-spring and 1.44 ± 0.28 in summer-autumn, which result to agree closely with the results derived from the FMI/PFR measurements. The winter-spring and summer-autumn relative frequency histograms of *AOT(0.50 μm)* and α obtained from the AERONET measurements carried out at Sodankylä are also shown in Fig. 9.15, presenting quartiles of *AOT(0.50 μm)* equal to 0.047, 0.060 and 0.075 in winter-spring and equal to 0.040, 0.059 and 0.076 in summer-autumn, therefore showing a similar seasonal trend of *AOT(0.50 μm)* and only slightly different values of the quartiles from one period to the other. Correspondingly, the quartiles of exponent α were found to be equal to 1.09, 1.25 and 1.35 in winter-spring and to 1.24, 1.45 and 1.61 1.44 in summer-autumn, which result to substantially agree with those derived from the FMI measurements. Therefore, the AERONET measurements confirm the trend of exponent α to increase appreciably from winter-spring to summer-autumn, such a trend being mainly caused by the long-range transport of polluted aerosol from Europe and Russia, as pointed out by Aaltonen et al. (2006). On this matter, in order to explain these seasonal variations in the atmospheric turbidity conditions, it is worth mentioning that the air masses reaching the Northern Scandinavia during the year originate from the Eurasian continent and mid-latitude Atlantic Ocean regions in the 56% of the observed cases during the year, and from the Arctic Basin and Northern Atlantic Ocean in the remaining 44% of cases (Aaltonen et al. 2006).

Due to the alternation of polluted air masses from Eurasia with sea-salt particles formed over the oceanic regions, the monthly mean values of *AOT(λ)* turn out to be rather stable over the entire year, while the monthly mean values of α tend to become gradually higher in early summer, when the Arctic Basin was the principal aerosol source. Because of the efficient transport processes taking place during the year from polluted regions (characterized by the predominant content of continental aerosol particles) or oceanic areas, where the maritime aerosol content originally predominates, the FMI/PFR and AERONET Sodankylä relative frequency histograms of *AOT(0.50 μm)* did not exhibit significant differences between the seasonal mean values and the second quartiles defined in winter-spring and summer-autumn. These data present: (i) monthly mean values of *AOT(0.50 μm)* close to 0.08 in winter-spring, with their quartiles differing by no more than 0.01 one from the other, and (ii) monthly mean values of *AOT(0.50 μm)* close to 0.07 on average in summer-autumn, with differences close to 0.01 between the various quartiles. The monthly mean values of α derived from the FMI/PFR and AERONET measurements performed at Sodankylä were found to decrease by 0.10–0.20 on average when passing from winter-spring to summer-autumn. Clearer discrepancies were obtained over both the bi-seasonal periods, with FMI mean values equal to about 1.32 and 1.52 in winter-spring and summer-autumn, respectively, and AERONET mean values equal to 1.23 and 1.44 in the same two periods, providing similar values of the seasonal quartiles.

The winter-spring and summer-autumn relative frequency histograms of parameters *AOT(0.50 μm)* and α obtained from the above measurements carried out at Andenes were not shown by Toledano et al. (2012). They are presented in Fig. 9.15, as obtained for classes of *AOT(0.50 μm)* increasing in steps of 0.02 and classes of α increasing in steps of 0.20. As can be seen, they provided values of the quartiles of *AOT(0.50 μm)* equal to 0.041, 0.097 and 0.135 in winter-spring and to 0.065, 0.088 and 0.127 in summer-autumn, together with quartiles of exponent α equal to 0.74, 1.01 and 1.32 in winter-spring and equal to 0.98, 1.24 and 1.51 in summer-autumn. These evaluations made over the winter-spring and summer-autumn periods agree satisfactorily with the quartiles of these two aerosol extinction parameters determined by Rodríguez et al. (2012) for the overall set of the AERONET measurements performed at Andenes from 2002 to 2010, who found quartiles of *AOT(0.50 μm)* equal to 0.058, 0.084 and 0.122, and quartiles of exponent α equal to 1.01, 1.35 and 1.61, respectively.

The daily mean values of *AOT(0.50 μm)* and α were also measured at Kiruna (in Northern Sweden) by the SMHI group over the period from January 2007 to October 2009, by analyzing the three-year sets of measurements conducted with GAW/PFR sun-photometers. The monthly mean values of *AOT(0.50 μm)* and α were derived with their standard deviations from these measurements, after their selection made through a careful cloud-screening procedure, obtaining the following results: (i) monthly mean values of *AOT(0.50 μm)* equal to 0.065 ± 0.019 in winter-spring with quartiles equal to 0.045, 0.060 and 0.078, and 0.073 ± 0.027 in summer-autumn (see Table 9.9) with quartiles equal to 0.037, 0.057 and 0.084, and (ii) monthly mean values of exponent α equal to 1.22 ± 0.14 in winter-spring, with quartiles equal to 1.07, 1.28, and 1.43, and equal to 1.45 ± 0.17 in summer-autumn, with quartiles equal to 1.27, 1.49, and 1.66.

The comparison of the time-patterns of the monthly mean values of *AOT(0.50 μm)* and α obtained at Sodankylä from measurements conducted using two different sun-photometer models with those determined at Andenes and Kiruna (and compared in Fig. 9.14) indicates that the estimates of *AOT(0.50 μm)* and α made at Sodankylä differ considerably from those measured at Andenes, by more than 40% on average for *AOT(0.50 μm)* and about 20% for α in winter-summer, and by more than 100% for *AOT(0.50 μm)* and about 20% for α in summer-autumn, while they agree more closely with those determined at Kiruna. In fact, as can be noted in Fig. 9.14, the PFR measurements performed at Kiruna were found to provide time-patterns of the monthly mean values of *AOT(0.50 μm)*, which agree closely with those obtained at Sodankylä from the AERONET sun-photometer measurements performed in February and March, and are appreciably higher than the AERONET evaluations made by examining the measurements conducted over the whole period from April to October. These values of *AOT(0.50 μm)* are associated with monthly mean values of α that result to be considerably lower than those derived from the AERONET sun-photometer measurements conducted at Sodankylä in both winter-spring and in summer-autumn periods. The higher values of $AOT(0.50\ \mu m)$ determined at Kiruna together with lower values of α are plausibly due to the higher concentrations of fine particles observed at Kiruna (with respect to those monitored at Sodankylä) over the whole 9-month observational period.

9.1.3.6 Measurements in Northern-Central Siberia

A comparison between the winter-spring and summer-autumn average values of $AOT(0.50\ \mu m)$ and α was not performed at Tiksi, in Northern-Central Siberia (Russia), since the AERONET sun-photometer measurements were routinely conducted at this remote Siberian site only over the May–October period. Table 9.8 provides the spectral series of the peak wavelengths of the narrow-band interference filters

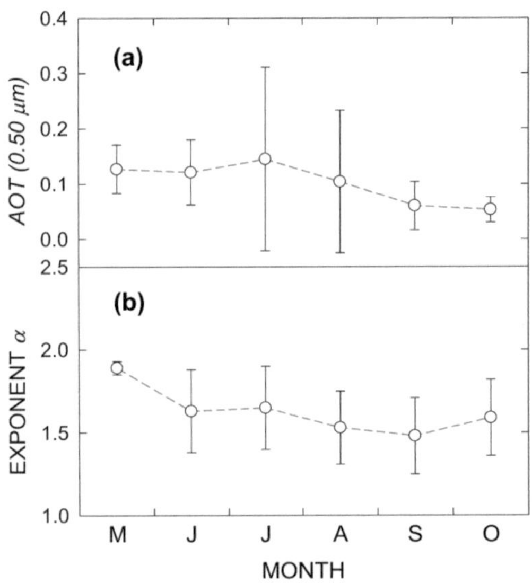

Fig. 9.16 Part (**a**): Time-patterns of the monthly mean values of (i) aerosol optical thickness $AOT(0.50\ \mu m)$ given with their standard deviations (defined by vertical bars), as obtained from the multi-year sun-photometer measurements conducted at Tiksi (North-central Siberia, Russia) by the IAO-SB-RAS (Tomsk, Russia) group in cooperation with NASA/GSFC (USA) by using an AERONET Cimel CE-318 sun-photometer in the summer–autumn months (June–October) of 2010, 2011 and 2012 (open circles). Part (**b**): As in part (**a**) for the time-patterns of the monthly mean values of wavelength exponent α given with their standard deviations (defined by the vertical bars), as obtained from the same multi-year sun-photometer measurements conducted at Tiksi (North-central Siberia, Russia) by the IAO-SB-RAS (Tomsk, Russia) group and NASA/GSFC (USA) with an AERONET Cimel CE-318 sun-photometer in the summer–autumn months of 2010, 2011 and 2012 (open circles)

mounted on the Cimel CE-318 employed at Tiksi, for which the spectral interval $\Delta\lambda$ adopted to calculate the best-fit values of exponent α was taken from 380 to 870 nm. Only four spectral series of *AOT(0.50 μm)* were measured in May at this site, from which a monthly mean value of *AOT(0.50 μm)* = 0.13 ± 0.022 was obtained, together with a monthly mean value of α equal to 1.90 ± 0.12 (as reported in Table 9.9). As can be seen in Fig. 9.16, the monthly mean values of *AOT(0.50 μm)* measured in the summer-autumn months exhibit a pronounced increase from 0.12 in May and June to nearly 0.15 in July, followed by a gradual decrease in August–October presenting a monthly mean value of 0.10 in August and values ranging between 0.05 and 0.06 in September and October. To give a measure of the dispersion of such data-set, it is useful to mention that all the above-mentioned measurements of *AOT(0.50 μm)* are associated with rather large values of standard deviation $\sigma_{AOT,}$ found to exceed 0.04 in May and June, 0.12 in July and August, and to be close to 0.04 in September and 0.02 in October. Such large variations in *AOT(0.50 μm)* were determined for monthly mean values of exponent α (found to be close to 1.50 in August–October), which have been estimated to exceed (from May to July) the monthly values of standard deviation σ_α evaluated in general to be >0.20 throughout the whole measurement period. Consequently, the summer-autumn average seasonal value of *AOT(0.50 μm)* was found to be equal to 0.094 ± 0.060, and that of α to be equal to 1.57 ± 0.24 in the same season.

On this matter, examining their sun-photometric measurements performed at Tiksi in the 2010–2013 period, Sakerin et al. (2014) obtained the following results: (i) monthly mean values of *AOT(0.50 μm)* equal to 0.127 (± 0.044) in May, 0.121 (± 0.059) in June, 0.145 (± 0.166) in July, 0.104 (± 0.129) in August, 0.060 (± 0.044) in September and 0.053 (± 0.023) in October, giving an average multi-month value of 0.093 (± 0.050), and (ii) monthly mean values of Ångström's exponent α equal to 1.891 (± 0.042) in May, 1.634 (± 0.248) in June, 1.648 (± 0.254) in July, 1.529 (± 0.224) in August, 1.481 (± 0.228) in September and 1.587 (± 0.232) in October, providing an average multi-month value of 1.57 ± 0.20. These findings fully confirm the results shown by us in Fig. 9.16, indicating that these variations were associated with: (a) seasonal mean values of the fine particle content C_f in the vertical atmospheric column, estimated by Sakerin et al. (2014) to increase from 0.017 (± 0.007) $\mu m^3/\mu m^2$ in spring to a value of 0.023 (± 0.039) $\mu m^3/\mu m^2$ in summer, and a value of 0.009 (± 0.008) $\mu m^3/\mu m^2$ in autumn, while the median fine particle radius was estimated to be equal to 0.17 (± 0.02) μm in spring, 0.19 (± 0.02) μm in summer and 0.18 (± 0.04) μm in autumn; and (b) the seasonal mean values of the coarse particle content C_c in the vertical atmospheric column, evaluated to decrease from 0.010 (± 0.011) $\mu m^3/\mu m^2$ in spring to 0.005 (± 0.006) $\mu m^3/\mu m^2$ in summer and to 0.002 (± 0.002) $\mu m^3/\mu m^2$ in autumn, while the median coarse particle radius was evaluated to slowly increase from 2.02 (± 0.51) μm in spring to 2.89 (± 0.63) μm in summer and then to decrease to 2.42 (± 0.065) μm in autumn.

Figure 9.17 shows the summer-autumn relative frequency histograms of *AOT(0.50 μm)* measured at Tiksi, which exhibits an average value equal to 0.093, with quartiles equal to 0.049, 0.074, and 0.117, respectively, which are

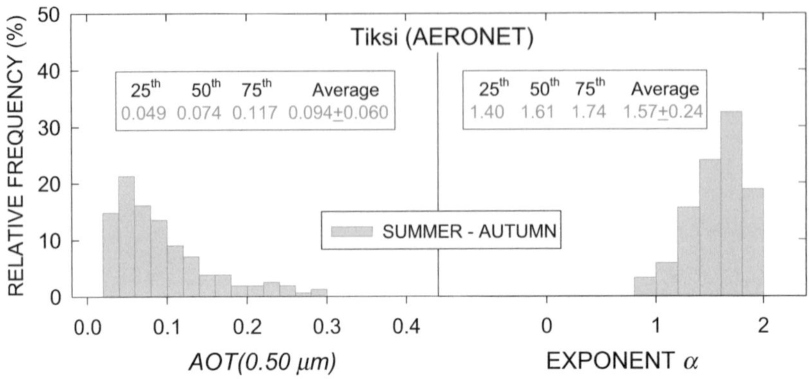

Fig. 9.17 Relative frequency histograms RFHs of aerosol optical thickness *AOT(0.50 μm)* (left-hand side) and Ångström wavelength exponent α (right-hand side) obtained examining the multi-year sun-photometer measurements performed at Tiksi (North-central Siberia, Russia) by the IAO-SB-RAS (Tomsk, Russia) group in cooperation with NASA/GSFC (USA) by using an AERONET Cimel CE-318 sun-photometer in the summer–autumn months (June–October) of 2010, 2011 and 2012. The values of the quartiles (i.e. of the 25th, 50th, and 75th percentiles) of *AOT(0.50 μm)* and exponent α reported in the boxes inserted into the graphs have been obtained by examining the daily mean values of the two optical parameters measured at this Siberian site by the IAO-SB-RAS group during the summer–autumn months of 2010, 2011 and 2012

all appreciably higher than those measured at the three Scandinavian sites during the summer season. Correspondingly, the RFH of exponent α was found to yield the average value of 1.57, showing a long-tailed left-hand wing with the 25th percentile equal to 1.39, and a long-tailed right-hand wing with the 50th and the 75th percentiles equal to 1.61 and 1.74, respectively. Such rather high values of both the first and third quartiles of α are probably due to the presence of an optically significant load of small fine particles causing an important contribution in the overall particle extinction effects. In fact, the variations in *AOT(0.50 μm)* and α observed to occur from spring to summer seem to be more plausibly due to the marked increase in the atmospheric content of fine particles, while the coarse particle concentration was estimated to decrease by about 50% from spring to summer (Sakerin et al. 2014). A plausible explanation for these spring-to-summer variations could be the increased transport of fine particles usually observed in summer as a consequence of the anthropogenic aerosol originated in the industrial regions and/or generated by the Siberian forest fires causing the long-range dispersion of large smoke particle loads, which have been estimated to reach the inner regions of Siberia and continental Asia as well as the coastal areas of the Arctic Ocean.

9.1.4 Summary on Spectral AOT Ground-Based Measurements in Arctic

The monthly mean values of *AOT(0.50 μm)* and α measured at Sodankylä with the FMI/PFR and RIMA/AERONET sun-photometers turn out to differ appreciably from those typically observed at the remote Arctic sites located at higher latitudes in Northern America, Greenland and Spitsbergen Island. Examining the results obtained in Fig. 9.6 from the measurements performed at Barrow (Alaska), the average multi-month values of *AOT(0.50 μm)* and α were estimated to vary as follows: (i) those of *AOT(0.50 μm)* between 0.115 and 0.134 in winter-spring, and between 0.079 and 0.085 in summer-autumn, and (ii) those of exponent α between 1.20 and 1.24 in winter-spring, and between 1.38 and 1.46 in summer-autumn. From the results obtained analyzing the measurements performed at Resolute Bay, Eureka – 0PAL and Alert in Nunavut (Canada) and shown in Fig. 9.8, it was estimated that: (i) the average values of *AOT(0.50 μm)* ranged between 0.093 and 0.117 in winter-spring and between 0.056 and 0.066 in summer-autumn, and (ii) the bi-seasonal average value of α between 1.29 and 1.52 in winter-spring and between 1.46 and 1.64 in summer-autumn. Analyzing the results shown in Fig. 9.10 for the sun-photometer measurements conducted at Thule, Summit and Ittoqqortoormiit in Greenland), it was found that: (i) the bi-seasonal average values of *AOT(0.50 μm)* varied between 0.057 and 0.093 in winter-spring and between 0.052 and 0.059 in summer-autumn, and (ii) the corresponding bi-seasonal average values of α ranged between 1.22 and 1.38 in winter-spring and between 1.41 and 1.48 in summer-autumn.

Examining the sun-photometer measurements conducted at Ny-Ålesund (Spitsbergen, Svalbard) and those performed at Barentsburg and Hornsund (Spitsbergen, Svalbard) and shown in Fig. 9.12 for comparison, it was estimated that: (i) the bi-seasonal average values of *AOT(0.50 μm)* varied between 0.082 and 0.122 in winter-spring and between 0.053 and 0.078 in summer-autumn over the Svalbard region, while (ii) the corresponding bi-seasonal average values of α varied between 1.13 and 1.32 in winter-spring and between 1.28 and 1.36 in summer-autumn.

The analysis of the evaluations of parameters *AOT(0.50 μm)* and α derived from the sun-photometer measurements conducted in Northern Scandinavia using the two different FMI/PFR and AERONET sun-photometer models at Sodankylä and those conducted at Andenes and Kiruna (and shown in Fig. 9.14) has provided the following results: (i) the bi-seasonal average values of *AOT(0.50 μm)* varied between 0.065 and 0.096 in winter-spring and between 0.066 and 0.113 in summer-autumn over Northern Scandinavia, while (ii) the bi-seasonal average values of α varied between 1.00 and 1.24 in winter-spring and between 1.19 and 1.45 in summer-autumn.

Finally the results shown in Fig. 9.16, as obtained from the AERONET measurements carried out at Tiksi (Siberia) indicated that the late spring value of *AOT(0.50 μm)* was close to 0.13, while that determined in summer-autumn was estimated to be close 0.094, and therefore lower by about 10%, and that of α

was close to 1.90 in May and equal to 1.57 on average in summer-autumn, thus indicating the prevailing presence of fine particles within the vertical atmospheric column.

To complete the picture of the atmospheric turbidity characteristics of the Arctic aerosol particles, it is useful to mention that the scatter-plots of the values of Ångström exponent α derived from the sun-photometer measurements analyzed in the present study and plotted versus the corresponding values of $AOT(0.50\ \mu m)$ allows also to individuate the aerosol types that are most commonly observed in the Arctic atmosphere, these aerosol types being characterized by optical properties in the visible and near-infrared that are similar to those defined by Hess et al. (1998) among the components of the most common OPAC models and those observed by Holben et al. (2001) through the AERONET network measurements taken at the high-latitude Arctic sites. Examining the AERONET Cimel CE-318 sun-photometer measurements performed at Andenes (Northern Norway) from 2002 to 2010 (as a part of the activities developed by the RIMA (Spain) network in cooperation with the Andoya Rocket Range (Andenes, Norway) group, Rodríguez et al. (2012) defined the following typical domains in terms of ranges of $AOT(0.50\ \mu m)$ and α: for the various observed aerosol types: (1) continental aerosol particles, with $AOT(0.50\ \mu m) < 0.16$ and $\alpha > 1.10$, including the 54% of the overall observed cases at Andenes, in which the extinction effects are predominantly produced by continental aerosols often mixed with a few percents of marine aerosols; (2) maritime aerosols with $AOT(0.50\ \mu m) < 0.16$ and $\alpha < 1.10$, including the 34% of the overall cases, which are mainly composed of sea salt and to a lesser extent by dust particles (Smirnov et al. 2002); (3) smoke and polluted aerosol particles with $AOT(0.50\ \mu m) \geq 0.16$ and $\alpha \geq 1.10$, observed for the 8% of the overall cases, which include an important fraction of fine particles originated by industrial activities as well as by combustion processes associated with forest fires; and (4) mixtures of particles with $AOT(0.50\ \mu m) \geq 0.16$ and $\alpha < 1.10$, observed for the 4% only of the overall examined cases, and containing Asian dust in most cases because of the relatively high mass content of coarse particles.

The scatter plots of the bi-seasonal average values of exponent α versus those of $AOT(0.50\ \mu m)$ are shown in Figs. 9.18–9.21, as obtained from the four sets of sun-photometer measurements conducted in the Arctic regions of North America, Greenland, Spitsbergen Island, and Northern Scandinavia and Siberia. The daily mean values of exponent α were estimated to vary mainly from 1.10 to 1.70 over the whole observational period of the year, and from 1.20 to 1.50 in winter-spring (with daily mean values of $AOT(0.50\ \mu m)$ ranging mainly from 0.04 to 0.12), and from 1.30 to 1.70 in summer-autumn (with daily mean values of $AOT(0.50\ \mu m)$ that are smaller than 0.08 in all cases).

These features suggest that appreciable differences exist between the aerosol extinction effects occurring in the Arctic region in winter-spring and those observed during the summer-autumn periods. In winter-spring, the median values of $AOT(0.50\ \mu m)$ vary appreciably from one Arctic site to another, due to their close dependence features on the optical characteristics of particulate matter transported from the most densely populated mid-latitude regions towards the Arctic, which

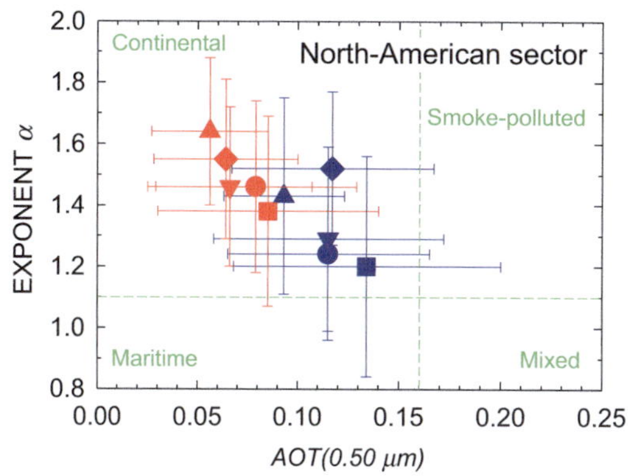

Fig. 9.18 Scatter plots of the seasonal mean values of the Ångström (1964) wavelength exponent α versus the corresponding multi-month seasonal mean values of aerosol optical thickness *AOT(0.50 μm)* determined from the sun-photometer measurements performed at the following five sites of the North American sector: (i) Barrow (NOAA/GMD) (circles), (ii) Barrow (AERONET) (squares), (iii) Resolute Bay (AERONET/AEROCAN) (diamonds), (iv) Eureka-0PAL (AERONET/AEROCAN) (up-ward triangles), and (v) Alert (NOAA/GMD) (down-ward triangles). The seasonal mean values of parameters α and *AOT(0.50 μm)* are represented using blue-color symbols for the winter–spring period and red-colour symbols for the summer–autumn period, while their standard deviations are given by the vertical and horizontal bars, respectively. The vertical and horizontal green dashed lines define the sub-domains determined by Rodríguez et al. (2012) for the continental, maritime, smoke-polluted and mixed aerosol particles

may be particularly intense in late winter and early spring. Conversely, on the summer days, when the background aerosol composition varies from one site to another as a result of the different origins and transport mechanisms of both fine and coarse mode particles moving from remote regions, the origins of aerosol particles may differ considerably from one observation day to the other. The results obtained at Barrow and shown in Fig. 9.18 provide average values and standard deviations of *AOT(0.50 μm)* and exponent α [for which the central bands of these normal distributions (limited by $\pm\, \sigma_{AOT}$ or $\pm\, \sigma_{\alpha}$, respectively) include the 68.2% of the measured data], which indicate that the two atmospheric turbidity parameters *AOT(0.50 μm)* and α were found to range at this coastal site of Alaska as follows: (i) *AOT(0.50 μm)* from 0.065 to 0.194 during the winter-spring months, and from 0.029 to 0.140 in the summer-autumn period, and (ii) α from 0.065 to 0.194 in winter-spring, and from 1.07 to 1.74 in summer-autumn. These variations in *AOT(0.50 μm)* and α from one period to the other of the year seem to be plausibly due to the relevant increase in the atmospheric content of fine particles taking place as one passes from the winter-spring to the summer-autumn periods, as a result of the transport of anthropogenic-industrial aerosols from the lower-latitude regions.

Comparing the scatter plot features of Fig. 9.18 (obtained on the basis of the average values of *AOT(0.50 μm)* and α and its standard deviations σ_{AOT} and σ_{α} given in Table 9.9) with the classification criteria adopted by Rodríguez et al. (2012) for defining the sub-domains of continental, maritime, smoke/polluted and mixed

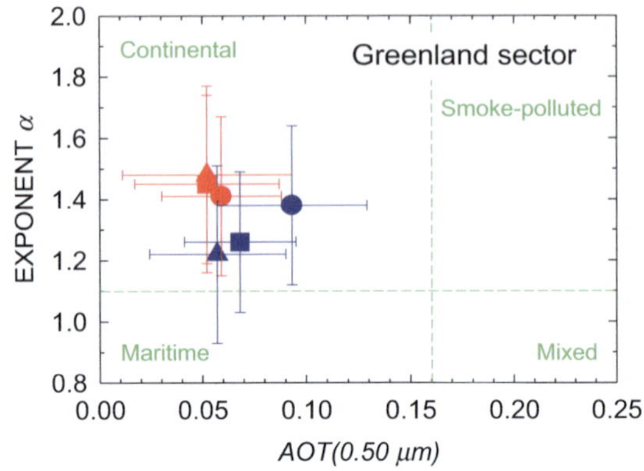

Fig. 9.19 As in Fig. 9.18 for the seasonal mean values of the Ångström (1964) wavelength exponent α versus the corresponding multi-month seasonal mean values of aerosol optical thickness *AOT(0.50 μm)* determined from the sun-photometer measurements performed at the following three sites located in Greenland: (i) Thule (AERONET) (circles), (ii) Summit (PMOD/WRC) (triangles), and (iii) Ittoqqortoormiit (AERONET) (squares). The seasonal mean values of parameters α and *AOT(0.50 μm)* are represented using blue-color symbols for the winter–spring period and red-colour symbols for the summer–autumn period, while their standard deviations are given by the vertical and horizontal dashed bars, respectively. The vertical and horizontal green dashed lines define the sub-domains determined by Rodríguez et al. (2012) for the continental, maritime, smoke-polluted and mixed aerosol particles

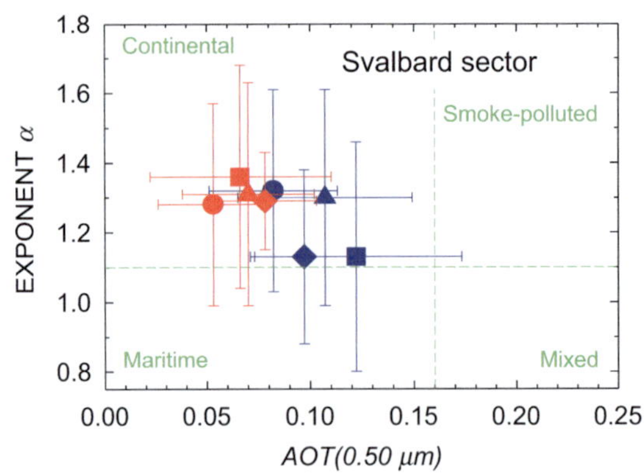

Fig. 9.20 As in Figs. 9.18 and 9.19 for the seasonal mean values of the Ångström (1964) wavelength exponent α versus the corresponding multi-month seasonal mean values of aerosol optical thickness *AOT(0.50 μm)* determined from the sun-photometer measurements performed at the following sites located in the Spitsbergen Island (Svalbard): (i) Ny-Ålesund, by using the AWI sun-photometers (circles) and the NILU/PFR sun-photometers (squares), (ii) Barentsburg, using the SPM sun-photometer model of the IAO-SB-RAS Institute (Tomsk, Russia) (diamonds), and (iii) Hornsund, using the AERONET sun-photometer employed by the Warsaw University (PAS, Poland) (up-ward triangles). The seasonal mean values of parameters α and *AOT(0.50 μm)* are represented using blue-color symbols for the winter–spring period and red-colour symbols for the summer–autumn period, while their standard deviations are given by the vertical and horizontal dashed bars, respectively. The vertical and horizontal green dashed lines define the sub-domains determined by Rodríguez et al. (2012) for the continental, maritime, smoke-polluted and mixed aerosol particles

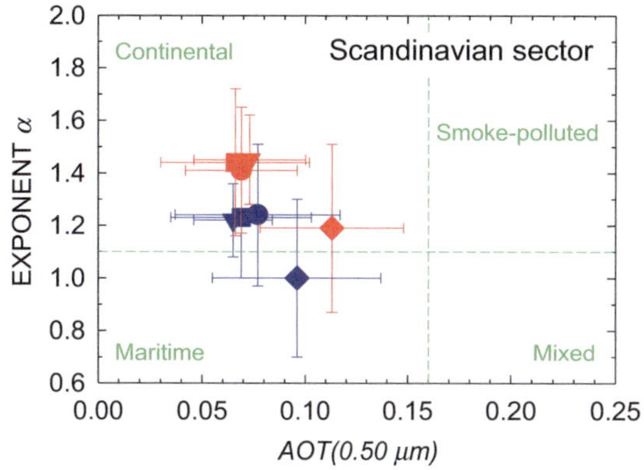

Fig. 9.21 As in Figs. 9.18, 9.19 and 9.20 for the seasonal mean values of the Ångström (1964) wavelength exponent α versus the corresponding multi-month seasonal mean values of aerosol optical thickness *AOT(0.50 μm)* determined from the sun-photometer measurements performed at the following sites of the Northern Scandinavian region: (i) Sodankylä (Finland), by using both the PFR/FMI (Helsinki, Finland) sun-photometer (circles) and the AERONET sun-photometer employed by the Arctic Research Center (FMI) (squares), (ii) Andenes (Norway), using the AERONET sun-photometer of the GOA/University of Valladolid (Spain) group (diamonds), and (iii) Kiruna (Sweden) using the PFR sun-photometer employed by SMHI (Norrköping, Sweden) (down-ward triangles). The seasonal mean values of parameters α and *AOT(0.50 μm)* are represented using blue-color symbols for the winter–spring period and red-colour symbols for the summer–autumn period, while their standard deviations are given by the vertical and horizontal dashed bars, respectively. The vertical and horizontal green dashed lines define the sub-domains determined by Rodríguez et al. (2012) for the continental, maritime, smoke-polluted and mixed aerosol particles

aerosol classes, it can be estimated that the predominant part of the aerosol cases monitored at Barrow (Alaska) exhibit the characteristics of (i) continental particles for the most part ($\sim$ 70%) of the winter-spring measurement days, and those of the maritime and smoke-polluted particle types considered by Rodríguez et al. (2012) for the remaining 30% of cases, and (ii) continental particles in summer-autumn for at least 80% of the observation days, with lower percents of maritime and smoke-polluted cases. Similarly, Fig. 9.18 shows that the sun-photometer measurements performed at the three Canadian sites and described in the present analysis were found to provide: (i) the most frequent ranges of *AOT(0.50 μm)* from 0.058 to 0.172 in winter-spring and from 0.025 to 0.107 in summer-autumn, and (ii) the most frequent ranges of α from 0.99 to 1.77 in winter-spring and from 1.20 to 1.88 in summer-autumn. These seasonal variations in *AOT(0.50 μm)* and α are plausibly due to the general increase in the atmospheric content of fine particles occurring from winter-spring to summer-autumn. According to the criteria adopted by Rodríguez et al. (2012), it can be assumed that the airborne particles monitored at the three sites of Resolute Bay, Eureka and Alert in Nunavut (Canada) generally consisted of: (i) continental particles in the winter-spring months for the major part of the cases (> 85%), and maritime aerosols in a lower percentage, and (ii) continental particles in the summer-autumn months for the major part of cases (95%), besides a few cases presenting prevailing contents of maritime and smoke-polluted aerosols.

The scatter plots of the winter-spring and summer-autumn average values of exponent α versus the corresponding average values of *AOT(0.50* μm*)* is shown in Fig. 9.19, as obtained analyzing the sun-photometer measurements conducted at the Thule, Summit, and Ittoqqortoormiit stations in Greenland. Taking into account the average values of *AOT(0.50* μm*)* and α determined in the two seasonal periods and the values of the corresponding standard deviations, and examining the scatter plot features shown in Fig. 9.19, we have estimated that: (i) *AOT(0.50* μm*)* varied mainly from less than 0.02 to more than 0.12 during the winter-spring months, and from 0.01 to 0.09 in the summer-autumn period, and (ii) exponent α ranged mainly from 0.90 to 1.60 in winter-spring and from 1.10 to 1.80 in summer-autumn. Such a decrease in *AOT(0.50* μm*)*, associated with the marked increase in α observed passing from winter-spring to summer-autumn, can be reasonably attributed to the increase in the atmospheric content of fine particles occurring during the summer-autumn months as a result of transport of anthropogenic polluted aerosols from the regions of lower latitudes. In addition, taking into account the corresponding values of σ_{AOT} and σ_{α} given in Table 9.9 for the parameters *AOT(0.50* μm*)* and α, and following the classification criteria adopted by Rodríguez et al. (2012), Fig. 9.19 clearly shows that the major part of the daily mean observations of aerosol optical properties made at Thule, Summit, and Ittoqqortoormiit were associated with: (i) continental particles in the winter-spring months (for at least 80% of the overall cases), and maritime particles ($\sim$ 15%); and (ii) continental particles in the summer-autumn months ($\sim$ 90%), with percentages of maritime particles lower than 7%, and a few percents of smoke-polluted and mixed aerosols.

Figure 9.20 shows the scatter plots of the winter-spring and summer-autumn average values of exponent α plotted versus the corresponding average values of *AOT(0.50* μm*)*, which were obtained by analyzing the sun-photometer measurements conducted at: (i) Ny-Ålesund, with both the AWI sun-photometers and the NILU/PFR sun-photometers, (ii) Barentsburg, with the SPM sun-photometer designed at the IAO-SB-RAS Institute (Tomsk, Russia) (diamonds), and (iii) Hornsund, with the AERONET sun-photometer of the Warsaw University (PAS, Poland). On the basis of the seasonal average values of *AOT(0.50* μm*)* and α and the values of the corresponding standard deviations σ_{AOT} and σ_{α} (given in Table 9.9), we estimated that: (i) *AOT(0.50* μm*)* varied mainly at the Spitsbergen (Svalbard) sites from 0.05 to 0.17 during the winter-spring period, and from 0.02 to 0.11 in summer-autumn, and (ii) exponent α varied mainly from 0.80 to 1.60 in winter-spring, and from 1.00 to 1.70 in summer-autumn. The marked of *AOT(0.50* μm*)* from winter-spring to summer-autumn and the corresponding pronounced increase of α is likely produced by the increase in the fine particle content of the atmospheric aerosol load that usually occur from winter-spring to summer-autumn in this area because of the transport of polluted aerosols from the North-European region. Examining the sun-photometer measurements conducted in the Spitsbergen Island, and using the criteria adopted by Rodríguez et al. (2012) on the basis of the average values of *AOT(0.50 μm)* and α and those of the standard deviations σ_{AOT} and σ_{α}, Fig. 9.20 provides evidence that the major part of the aerosol measurements conducted in this region are due to: (i) extinction produced in winter-spring by continental particles (for more than 85% of the observed cases), and by maritime and smoke-polluted aerosols in the residual cases, and (ii) attenuation by continental particles in summer-autumn for the major part of observations (about 90%) and only a percentage of $\sim$ 7% due to maritime particles generated over the Arctic Ocean and smoke particles ($\sim$ 3%) from forest fires and anthropogenic pollution.

The winter-spring and summer-autumn average values of exponent α were calculated by analyzing the sun-photometer measurements carried out at the sun-photometer stations located at Sodankylä in Northern Finland (where the FMI/PFR and AERONET sun-photometers were regularly employed), Andenes in Northern Norway (where the AERONET/RIMA Cimel CE-318 sun-photometer was used), and Kiruna in Northern Sweden (where the SMHI group employed from 2007 to 2009 a GAW/PFR sun-photometer model). These average values of α were plotted versus the corresponding values of of *AOT(0.50 μm)* in Fig. 9.21, showing that: (i) *AOT(0.50 μm)* varied mainly over the Northern Scandinavia from 0.03 to 0.14 during the winter-spring months, and from 0.03 to 0.15 in summer-autumn, without presenting marked differences taking place from one period to the other of the year, and (ii) exponent α varied mainly from 0.70 to 1.50 in winter-spring, and from 0.80 to 1.80 in summer-autumn, thus assuming in general appreciably higher values in summer and early autumn. The variations of both parameters *AOT(0.50 μm)* and α from winter-spring to summer-autumn are probably due to variations in the columnar content of fine particles occurring as a result of the frequent episodes of transport of polluted aerosols from the surrounding regions. Following the same criteria adopted by Rodríguez et al. (2012) in examining the scatter plot of Fig. 9.21 and using the estimates of σ_{AOT} and σ_{α} given in Table 9.9 for the sun-photometer stations located in Northern Scandinavia (Sodankylä, Andenes and Kiruna), it was evaluated that the aerosol particles monitored over Northern Europe were characterized: (i) in the winter-spring period by the predominant presence of continental particles at Sodankylä and Kiruna (for at least 85% of the measurement days) with percentages lower than 7% of maritime aerosols, and predominant contents of maritime particles at Andenes, where the transport of these particles from the Arctic Ocean is very intense; and (ii) in summer-autumn by the predominance of continental aerosols (with more than 90% of the measurement days) at Sodankylä and Kiruna, and rather low contributions of maritime aerosols, while a predominant content of maritime aerosols was achieved at Andenes in summer-autumn, evaluated to be close to 75%, with minor percentages of continental (20%) and smoke-polluted (5%) aerosols.

It was very difficult to perform a comparison between the evaluations of *AOT(0.50 μm)* and exponent α obtained in the winter-spring and summer-autumn periods at Tiksi (71° 39′ N, 128° 52′ E) (located on the shore of the Buor-Khaya Gulf of the Laptev Sea, southeast of the delta of the Lena River) since no AERONET measurements were carried out in the winter months at this site, and only a few measurements have been performed in May, giving an average value of

AOT(0.50 μm) equal to 0.13 ± 0.022 and an average value of α equal to 1.90 ± 0.10. These results suggest that *AOT(0.50 μm)* is considerably higher in May than the average value of this quantity observed in the summer–autumn months (estimated to be equal to 0.094 ± 0.060), while exponent α was found to particularly high with respect to the summer-autumn average value of 1.57 ± 0.24. These results indicate that the aerosol extinction features observed at Tiksi are predominantly produced by continental aerosols, with negligible contributions due to maritime aerosols in both winter-spring and summer-autumn periods, and not negligible contributions of forest fire particles in the summer months.

9.2 Ship-Borne Observations of Arctic Aerosols

The Arctic Ocean is the smallest of the world's oceans and its boundaries include Hudson Bay and Baffin Bay. Ship-based spectral sunphotometer measurements in the Arctic Ocean started in mid-seventies as a side product of various Soviet research expeditions. Those measurements were scarse, calibration and measurement protocol mainly unverifiable. However, that data allowed researches to have some sort of reference point or point of comparison to the island based measurements summarized in Barteneva et al. (1991), Radionov and Marshunova (1992), Radionov (2005) and Tomasi et al. (2007). Measurements listed in Table 9.10 show that aerosol optical depth at a wavelength 500 nm was generally close to 0.06–0.08 what is typical for remote oceanic areas (see summary in Smirnov et al. 2002). Elevated levels of atmospheric turbidity presented by Barteneva et al. (1991) could have been caused by El-Chichon volcanic eruption in 1983. A wide range of optical depths found by Radionov (unpublished data) could be explained by biomass burning aerosol transported from Alaska or northern parts of Canada. In the latter case a small number of measurement days did not allow us to make any definite conclusions.

In recent years atmospheric aerosol optical studies in the Arctic basin and over Spitsbergen re-emerged and hopefully will be expanded (e.g. Sakerin et al. 2015; Kabanov et al. 2017). Those measurements are better documented than before. Instrument calibration is verifiable and inter-comparison with the AERONET calibrated CIMELs showed agreement within measurement uncertainties. Overall the lower limit of aerosol optical depth is less by 0.01–0.02 than in mid-seventies. We believe that it is a result of lesser pollution in the Arctic basin. However, despite total statistics presented in Table 9.10 is still very much limited, we can anticipate the natural variability of aerosol optical depth to be within the 0.03–0.08 range (excluding forest fires and/or pollution episodes). As will be shown below this statement generally stands for the most part of data collected within the framework of the Maritime Aerosol Network.

Accurate knowledge of atmospheric aerosol optical properties is a key to the success of climate change studies (direct and indirect forcing); satellite retrieval validation; verification of global aerosol transport model simulations; and atmo-

Table 9.10 List of cruises in the Arctic Ocean

Cruise/expedition name	Cruise area	Time period	N days	τa(500 nm)	Reference
RV Professor Vize	Greenland Strait	08/1973	3	0.05-0.07	Karimova (1976)
RV Professor Vize	Greenland Sea	05-06/1976	2	0.06-0.09	Lukyanchikova and Govorushkin (1981)
Drifting station NP-22	72-76°N 160°E-135°W	05-07/1979	8	0.06-0.07	Sakunov et al. (1981)
RV Professor Vize	Arctic Ocean	05-06/1983	...	0.13-0.24	Barteneva et al. (1991)
RV Professor Vize	Greenland Sea	08/1987	3	0.06-0.22	Radionov (unpublished data)
RV Professor Zubov	Greenland Sea	06-07/1988	2	0.06-0.08	Wolgin et al. (1991)
Drifting station NP-28	~82°N 168°E	04-05/1987	2	0.21	Barteneva et al. (1991)
RV Oceania	Norway, Greenland seas	06-07/1999	6	0.04-0.14	Knobelspiesse et al. (2004)
RV Oceania	Greenland Sea	07-08/2008	9	0.06-0.09	Zielinski (unpublished data)
RV Akademik Fedorov	Arctic Ocean	08-09/2013	2	0.03-0.04	Sakerin et al. (2015)
RV Professor Khljustin	Chukchi Sea	08/2013	1	0.03	Sakerin et al. (2015)
RV Akademik Treshnikov	Barents, Kara, Laptev Seas	08-09/2015	...	...	Polkin Jr et al. (2015)
RV Akademik M.Keldysh	Barents, Kara Seas	07-10/2016	47	0.03-0.25	Kabanov et al. 2017
RV Professor Molchanov	Barents Sea	07/2017	18	0.04-0.11	Chernov et al. (2017)

spheric correction in ocean color studies. In order to improve our knowledge of aerosol optical properties over the oceans and fill the existing data gaps Aerosol Robotic Network project (Holben et al. 1998, 2001) suggested to establish Maritime Aerosol Network which would be a component of the Aerosol Robotic Network (AERONET), and would be affiliated with the AERONET calibration and data processing standards and procedures. The proposed activity would complement island–based AERONET measurements, thus extending data collection to the vast regions where no islands exist. In order to address those challenges we have developed an archival system within the AERONET browser, but specifically designed for "moving" objects –ships; developed a calibration protocol; incorporated Microtops data files into AERONET general processing, utilizing AERONET's Version 2 processing algorithm; and developed a centralized archiving and distribution – public domain web-based access.

The Microtops II sun photometer is a handheld device specifically designed to measure columnar optical depth and water vapor content (Morys et al. 2001). The Microtops II Sunphotometer has five spectral channels and can accommodate several possible filter configurations within the spectral range of 340–1020 nm. The bandwidths of the interference filters vary from 2–4 nm (UV channels) to 10 nm for visible and near-infrared channels. Detailed description of the sunphotometer and types of errors involved can be found in Morys et al. (2001), Porter et al. (2001), Ichoku et al. (2002), and Knobelspiesse et al. (2003, 2004). The instrument has built-in pressure and temperature sensors. The estimated uncertainty of the optical depth in each channel does not exceed plus or minus 0.02, which is slightly higher than the uncertainty of the AERONET field (not master) instruments as shown by Eck et al. (1999). A GPS should be connected to the sunphotometer to obtain the time of measurements and geographical position of the ship.

Microtops II instruments have shown good calibration stability over the years. The variability in calibration coefficients within a few percent over several years relative to AERONET reference CIMELs is quite acceptable. Fig. 9.22 shows the variability in calibration coefficients (extraterrestrial irradiance signal, V_0) for one particular Microtops II. Certain changes in the calibration (post-field deployment in particular) are typically associated with aerosol deposition on the optics window that occurs at sea. After window cleaning, the calibration coefficients often approximate their original (pre-deployment) values. However, for some instruments we occasionally observed filter degradation which manifests itself as a rapid changes in the calibration coefficient.

Each Microtops instrument is calibrated against an AERONET master-CIMEL sun/sky radiometer at GSFC, which was calibrated from morning Langley plot measurements on Mauna Loa. As a rule we put a master-CIMEL in a manual mode that enables it to take direct sun measurements every minute. The Microtops then takes 20–30 consecutive scans within an approximately 5–6 min interval, side-by-side with the master-CIMEL. Aerosol optical depth is retrieved by applying the AERONET processing algorithm (Version 2) to the raw data (Smirnov et al. 2004; https://aeronet.gsfc.nasa.gov/new_web/Documents/version2_table.pdf). It is highly desirable to make inter-calibration measurements in clear (with AOD at 500 nm less than 0.25) and stable atmospheric conditions to ensure accurate and stable results.

Regular measurements in various oceanic regions started in late 2006 after two pilot cruises in 2004–2005. The measurement protocol is very simple and requires an operator taking 5–6 consecutive scans (it takes slightly over a minute to complete the sequence) when the solar disk is free of clouds. It is important to note that observers are relied upon to decide when the sun is un-obscured by clouds (based on visual assessment only) and to only take measurements for those conditions. Depending on the sky conditions measurements should be repeated several times during the day. Measurement points are grouped temporally into series. If the interval between two points in a measurement sequence is more than 2 min,

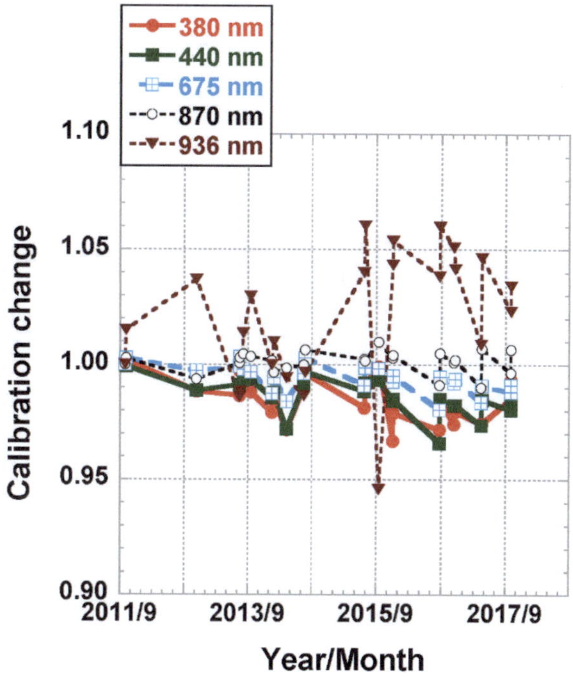

Fig. 9.22 Hand-held sunphotometer calibration history

then these points are placed into different series. A series can have one or more measurement points (typically five or more). A series is considered a single data point (an average of the measurement points in the group); and a sequence of series in a day may be used to compute the daily average.

All AODs have three data quality levels: Level 1.0 (unscreened), Level 1.5 (cloud-screened), and Level 2.0 (cloud-screened and quality assured). The Level 1.5 data series are raised to Level 2.0 (quality-assured) series after final calibration values are applied, spectral channels are evaluated for filter degradation and other possible instrumental problems or data anomalies; and manual data inspection is completed for possible cloud contaminated outliers. All data are available on the MAN web page, which is a part of the AERONET web site. A public domain web-based archive dedicated to the network activity can be found at: https://aeronet.gsfc.nasa.gov/new_web/maritime_aerosol_network.html. The current status (as of February 2018) of the network is presented in Fig. 9.23.

Establishment of the Maritime Aerosol Network allowed standardizing instrumentation, calibration, processing, and measurement protocol (Smirnov et al. 2009, 2011). Calibration has been tied to the AERONET standard and deployed processing relies on the AERONET Version 2 processing scheme. Within the framework of Maritime Aerosol Network 26 cruises were completed in the Arctic Ocean (including Hudson and Baffin Bays). Overall 952 measurement series spanning 218 days were acquired during those cruises. Table 9.11 provides cruise areas, time periods, number of measurement days and series (as defined above) and AOD ranges

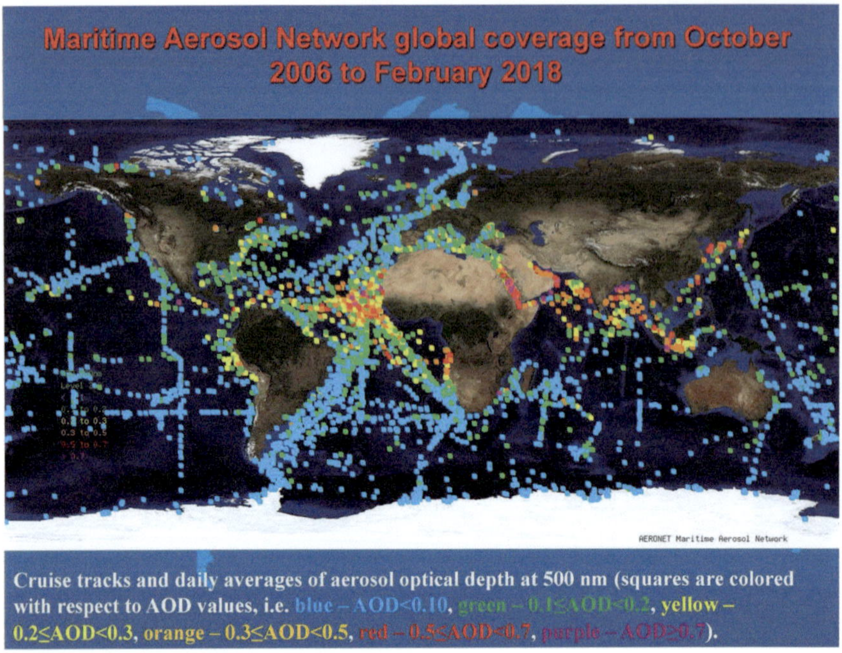

Fig. 9.23 Maritime Aerosol Network global coverage

for each cruise. Because of high uncertainty of Angstrom parameter when measure AOD is low we presented the median value for each cruise. It should be noted that majority of cruises collected data not only in the area of interest therefore we listed results only for days with data acquisition in the Arctic Ocean (including Hudson and Buffin Bays). Measurement area that corresponds to the listed in Table 9.11 cruises is shown in Fig. 9.24. Each pin on the map indicates one measurement day. As can be seen the vast majority of data were collected near Svalbard fjords and over Beaufort Sea. District clustering and group markers are presented in Fig. 9.25. Total number of days is shown in the center of each group.

From Table 9.11 one can observe a variety of optical conditions, ranging from biomass burning aerosol transported from Alaska and Northern Canada to background conditions over Beaufort Sea and near Svalbard fjords. Overall statistics is shown in Fig. 9.26. Daily AODs at 500 nm are less than 0.10 for 69% of measurement days (Fig. 9.26a). The Angstrom parameter (a general indicator of aerosol particle size) showed selectivity in spectral dependence with 79% of daily means within 0.8–1.6 range (Fig. 9.26b).

Latitudinal dependence of aerosol optical depth is presented in Fig. 9.27a. Biomass burning aerosol events are evident in the data acquired over Beaufort Sea north of 70 degrees in 2008 and 2009. Fires in northern Canada produced elevated aerosol loading over Hudson Bay in 2010, however AODs were much lower (less than 0.25) than over Beaufort Sea. Low Angstrom parameters (less than 0.5) were

Table 9.11 List of cruises in the Arctic Ocean completed within the framework of the Maritime Aerosol Network

Cruise name	Cruise area	Time period	Nd/Ns	τa(500 nm)	α(440-870)	Spectral range	PI
CCGS Louis St. Laurent	Beaufort Sea	07-08/2007	9/23	0.03–0.05	1.63	$340 \leq \lambda \leq 870$ nm	A.Proshutinsky
RV Oceania	Arctic Ocean	06-08/2007	6/62	0.03–0.07	1.26	$440 \leq \lambda \leq 870$ nm	T.Zielinski
CCGS Amundsen	Beaufort Sea	03-08/2008	44/102	0.04–0.40	1.20	$440 \leq \lambda \leq 870$ nm	P.Larouche
CCGS Amundsen	Beaufort Sea	07-08/2009	9/36	0.09–0.90	1.37	$440 \leq \lambda \leq 870$ nm	S.Belanger
RV Jan Mayen	Arctic Ocean	06/2009	4/7	0.07–0.08	1.02	$440 \leq \lambda \leq 870$ nm	C.Duarte
RV Oceania	Arctic Ocean	06-08/2009	11/102	0.05–0.18	1.45	$440 \leq \lambda \leq 870$ nm	T.Zielinski
RV Polarstern	Greenland Sea	08-09/2009	13/74	0.07–0.14	1.20	$340 \leq \lambda \leq 870$ nm	A.Macke
RV Ocean Watch	Beaufort Sea	07-08/2009	6/8	0.11–0.28	1.59	$380 \leq \lambda \leq 870$ nm	D.Covert
CCGS Amundsen	Hudson Bay	07/2010	10/22	0.04–0.23	1.72	$340 \leq \lambda \leq 870$ nm	P.Larouche
RV Oceania	Arctic Ocean	06-08/2010	8/66	0.04–0.09	0.52	$380 \leq \lambda \leq 870$ nm	T.Zielinski
RV Oceania	Arctic Ocean	06-08/2011	7/64	0.04–0.08	1.38	$440 \leq \lambda \leq 870$ nm	T.Zielinski
USCGC Healy	Arctic Ocean	08-09/2011	7/9	0.05-0.09	1.03	$440 \leq \lambda \leq 870$ nm	E.Reid
RV Oceania	Arctic Ocean	06-08/2012	7/50	0.04-0.13	1.37	$440 \leq \lambda \leq 870$ nm	T.Zielinski
RV Polarstern	Arctic Ocean	08/2012	6/8	0.05–0.21	1.24	$440 \leq \lambda \leq 870$ nm	L.Istomina
RV Oceania	Arctic Ocean	06-08/2013	2/9	0.06–0.09	1.25	$440 \leq \lambda \leq 870$ nm	T.Zielinski
CCGS Amundsen	Baffin Bay	07-08/2013	4/5	0.03–0.29	1.45	$440 \leq \lambda \leq 870$ nm	S.Belanger
RV Akademik Fedorov	Arctic Ocean	08-09/2013	2/7	0.03–0.07	1.32	$440 \leq \lambda \leq 870$ nm	L.Yurganov
RV Oceania	Arctic Ocean	06-08/2014	4/21	0.03-0.10	1.21	$440 \leq \lambda \leq 870$ nm	T.Zielinski
CCGS Amundsen	Baffin Bay	07-08/2014	7/16	0.02–0.11	1.45	$440 \leq \lambda \leq 870$ nm	S.Belanger
RV Oden	Arctic Ocean	08/2014	6/8	0.04–0.08	1.47	$440 \leq \lambda \leq 870$ nm	G. de Leeuw
RV Oceania	Arctic Ocean	06-08/2015	6/35	0.04–0.15	1.28	$440 \leq \lambda \leq 870$ nm	T.Zielinski
USCGC Healy	Arctic Ocean	08-10/2015	7/39	0.03–0.08	0.94	$380 \leq \lambda \leq 870$ nm	M.Heller
RV Oceania	Arctic Ocean	06-08/2016	9/33	0.02–0.05	0.97	$440 \leq \lambda \leq 870$ nm	T.Zielinski
RV Polarstern	Arctic Ocean	05-06/2017	8/52	0.04–0.19	1.28	$380 \leq \lambda \leq 870$ nm	S.Kinne
RV Polarstern	Arctic Ocean	06-07/2017	6/68	0.03–0.04	1.20	$380 \leq \lambda \leq 870$ nm	S.Kinne
RV Oceania	Arctic Ocean	06-08/2017	10/26	0.04–0.08	1.24	$440 \leq \lambda \leq 870$ nm	P.Makuch

Nd number of days, *Ns* number of series, α is a median value

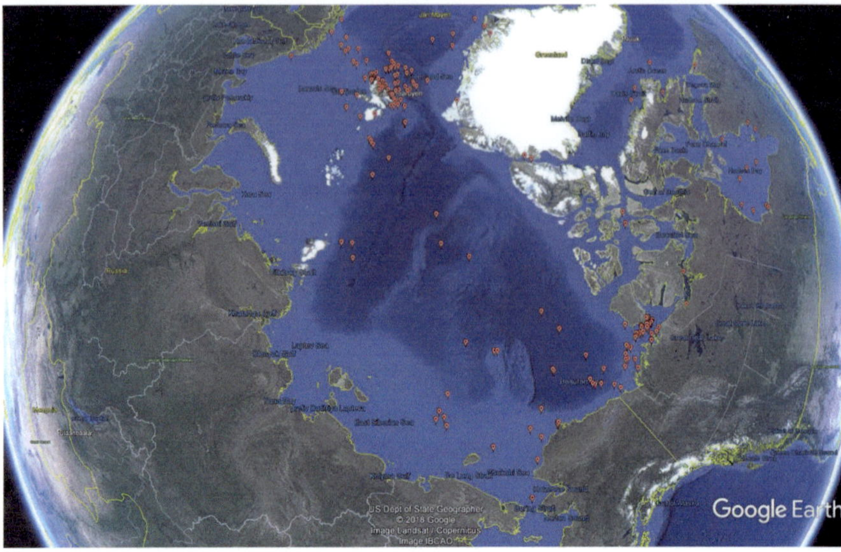

Fig. 9.24 Measurement area in the Arctic Ocean (each pin corresponds to daily averaged coordinates)

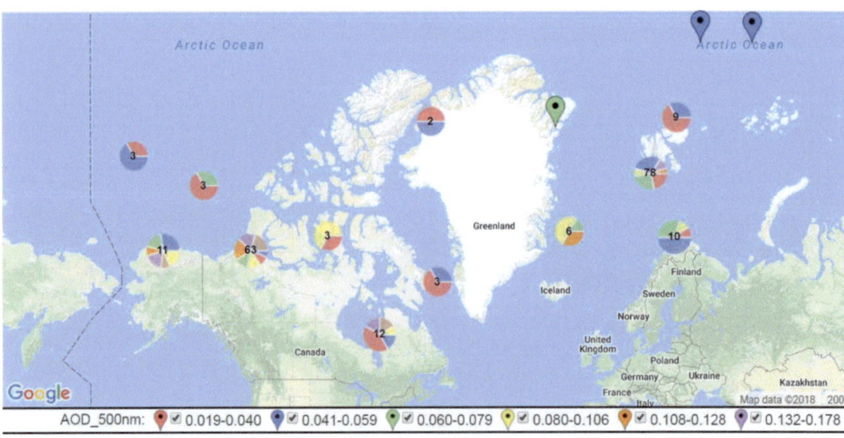

Fig. 9.25 Measurement results divided into clustering groups and colored with respect to AOD values

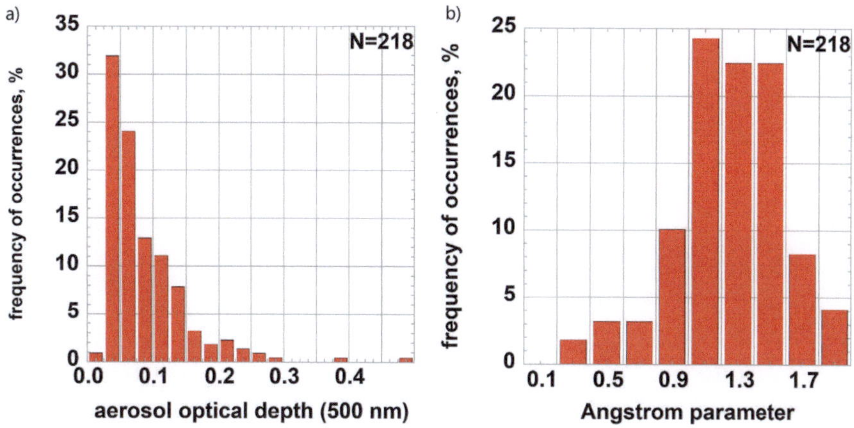

Fig. 9.26 Frequency of occurrences of daily averaged aerosol optical depth at 500 nm (**a**) and daily averaged Angstrom parameter (**b**)

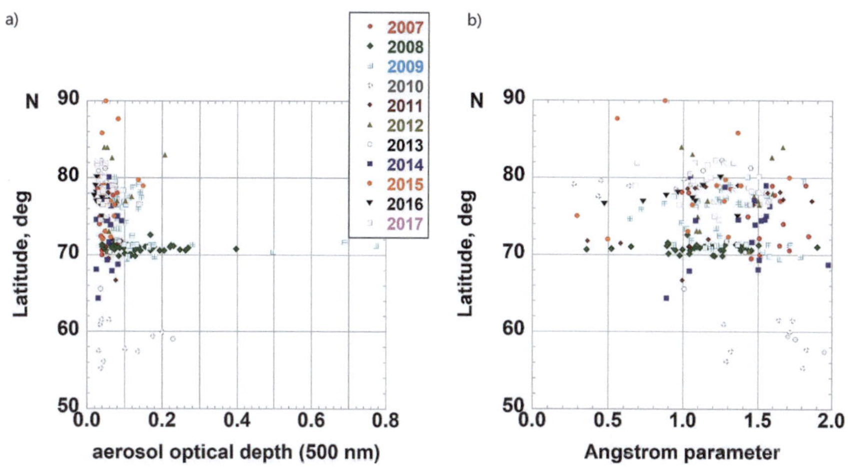

Fig. 9.27 Latitudinal dependences of aerosol optical depth (**a**) and Angstrom parameter (**b**) in the Arctic Ocean listed by year

detected north of 70 degrees (Fig. 9.27b) and possibly could be associated with pure marine or pre-foggy conditions (in rare cases we cannot exclude residual cloud contamination).

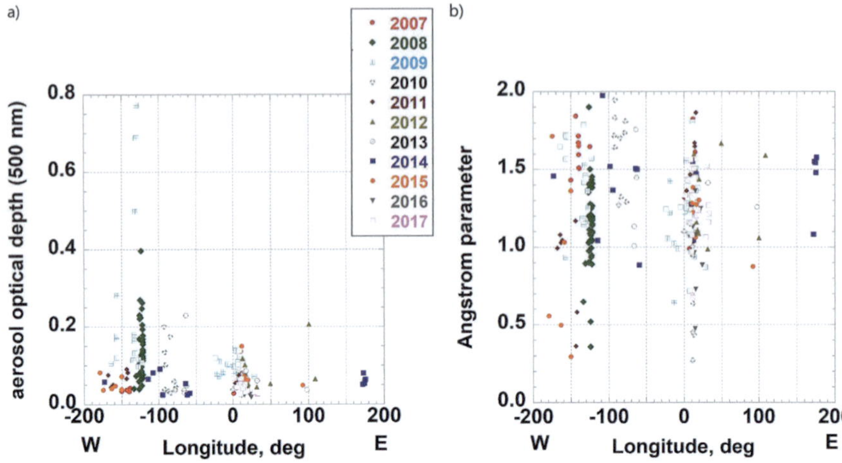

Fig. 9.28 Longitudinal dependences of aerosol optical depth (**a**) and Angstrom parameter (**b**) in the Arctic Ocean listed by year

A very nice data partition can be observed in Fig. 9.28a where daily aerosol optical depths are plotted by longitude. It shows that in Western Hemisphere variability is much higher. This is quite understandable because, as we mentioned above, of the biomass burning aerosol transport from continent. Longitudinal dependence of the Angstrom parameter did not produce any surprises with the vast majority of days having α above 0.90. We should note that low α cases occurred in Eastern and Western Hemispheres with almost equal probability (Fig. 9.28b).

9.3 Airborne and Spaceborne Remote Sensing of Arctic Aerosols

Active lidar observations and passive Sun photometer measurements are used to characterize Arctic aerosol properties from airborne observations (Tomasi et al. 2015b). While Sun photometer derive the aerosol optical depth (AOD) of the column above the aircraft, lidar have been operated in both, zenith and nadir viewing configuration and provide vertically resolved profiles of aerosol properties. By ascents and descents of the aircraft, also Sun photometer are able to obtain profile information (Russell et al. 2004).

Airborne Sun Photometer require an accurate tracking mechanism to follow the Sun while the aircraft is moving (e.g., Livingston et al. 2005; Zieger et al. 2007; Stone et al. 2010; Karol et al. 2013). Typically, the measurements cover different wavelength bands to analyse the spectral change of AOD and to invert the observations into aerosol particle number size concentrations (Shinozuka et al. 2011). On the research aircraft Polar 5 (Wesche et al. 2016), different Sun

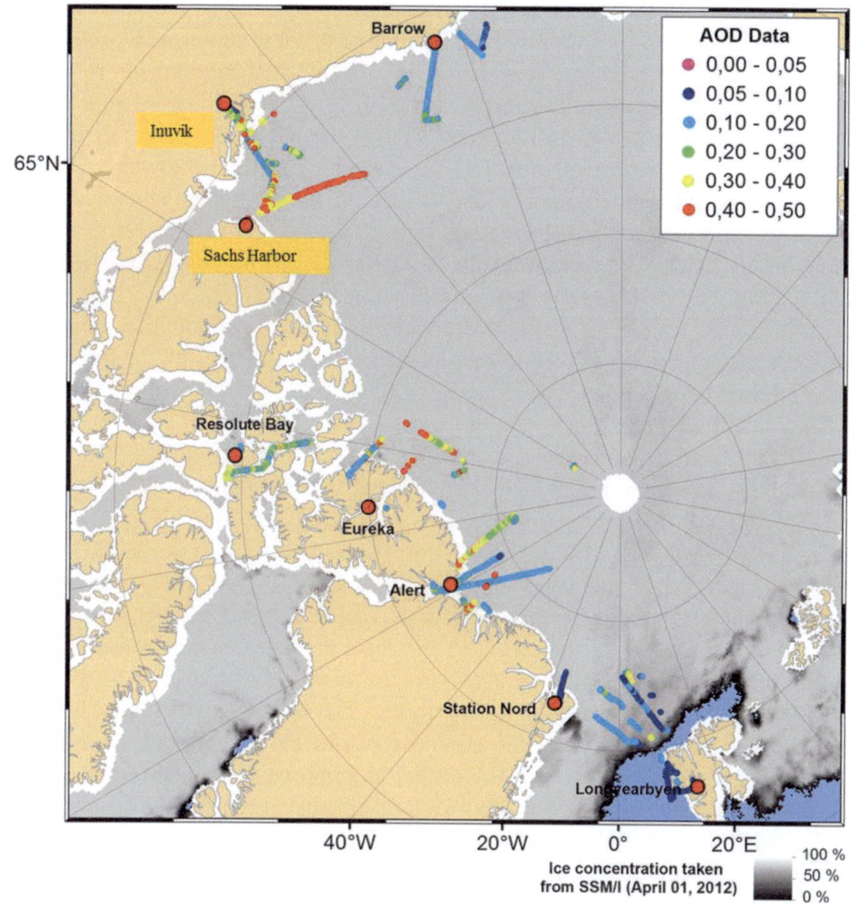

Fig. 9.29 Map of airborne Sun photometer measurements of AOD at 500 nm wavelength obtained during the PAMARCMIP campaign in 2009. The map shows the typical sea ice distribution during the observation period in Spring. (Adapted from Stone et al. 2010)

photometer with an active tracking system have been operated either installed under a quartz dome on top of the aircraft or looking across track through a cabin window made of B270 glass. To measure the direct solar radiation as well as solar radiance, the instrument optics use a sunspot which is focused via a diaphragm with a field of view of 1°. The Sun photometer typically operate a filter wheel with 8–10 selected wavelengths in the spectral range from 367 nm to 1024 nm. With knowledge of the extra-terrestrial signal, the spectral optical depth of the atmosphere as well as spectral optical depth of aerosol is derived. AOD at 500 nm wavelength measured during the Pan-Arctic Measurements and Arctic Regional Climate Model Intercomparison Project (PAMARCMIP) in 2009 over a wide Arctic area reaching from North Alaska to Svalbard are presented in Fig. 9.29 (Stone et al. 2010). The data had been recorded with the Sun photometer version looking through the cabin window which was calibrated for transmissivity in a range of view angles <25° (relative to normal view). Only valid data obtained in low altitude at about 60 m above ground providing the AOD of the full atmosphere and being not contaminated by clouds are shown. Cloud screening is an important step in the data processing to minimize an artificial enhancement of the AOD by thin clouds. Algorithms apply certain thresholds of measured irradiance and make use of the higher temporal/spatial variability of clouds compared to the rather smooth changes of aerosols properties (Stone et al. 2010). The measurements of PAMARCMIP highlight that AOD can significantly vary in the Arctic. While in general, the concentration of background aerosol is rather low, AOD increases dramatically in events of pollution long range transport (Bond et al. 2013). These spatial differences, which might be less visible in long term ground-based observations, emphasizes the need of airborne remote sensing of aerosol properties.

Different lidar techniques such as backscatter, polarisation, high spectral resolution, and dual absorption lidar have been adapted for airborne operation (Wendisch and Brenguier 2013). For aerosol and cloud remote sensing in the Arctic, the Airborne Mobile Aerosol Lidar (AMALi) instrument since 2004 is operated on the Polar 5 aircraft on a regular basis (Stachlewska et al. 2010). AMALi is a backscatter lidar instrument operating with UV (355 nm) and visible (532 nm) wavelengths with a nominal vertical resolution of 7.5 m. Linear depolarisation is measured at the 532 nm channel. For installation in the Polar 5 aircraft, AMALi can be mounted in either nadir or zenith viewing configuration. The horizontal resolution is in the range of 100 m determined by the ground speed of the aircraft and the temporal averaging over several laser pulses, needed to improve the signal-to-noise ratio. The attenuated backscatter coefficients for each channel are calculated by correcting for Rayleigh scattering. This is obtained by normalising the background corrected recorded intensities to an air density profile and a low, assumed aerosol load close to the aircraft. To minimize the effect of these assumptions, dropsonde releases from the aircraft or a nearby radiosonde launches are required.

Figure 9.30 shows two examples attenuated backscatter coefficient cross sections measured with AMALi during the Vertical Distribution of Ice in Arctic Clouds (VERDI) campaign 2012 over the Beaufort Sea (Klingebiel et al. 2015). With AMALi operating in zenith viewing configuration and Polar 5 flying in about 3 km altitude, the aerosol attenuated backscatter was normalized to the altitude range between 2100 m and 2400 m, assuming a low and homogeneous aerosol concentration at that altitude. Both cases show a strong attenuation of the lidar signal due to a layer of homogeneous boundary layer clouds with cloud top altitude at about 1000 m on 15 May 2012 (panel a) and 500 m on 5 May 2012 (panel b). Above cloud top, the low backscatter coefficients (below $0.5{\cdot}10^{-6}$ m sr^{-1}) observed on 15 May 2012 indicate a low aerosol load for this case. Contrarily, on 5 May 2012, two aerosol layers at about 1000 m and 700 m altitude can be identified above cloud top where the backscatter coefficient significantly increases up to $1.5{\cdot}10^{-6}$ m sr^{-1}. For the thicker upper aerosol layer (900–1000 m altitude), an aerosol optical depth of 0.02 ± 0.005 is estimated when an aerosol extinction

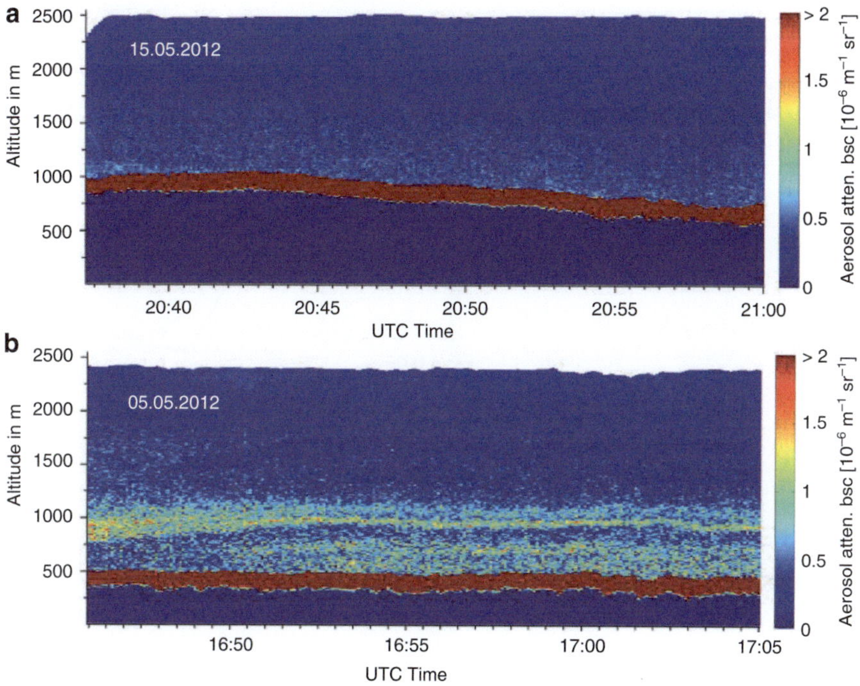

Fig. 9.30 Attenuated backscatter coefficient measured by AMALi between aircraft and the top of two boundary layer cloud during the VERDI campaign. (**a**) On 15 May 2012 the atmosphere above the cloud layer was aerosol-free. (**b**) On 05 May 2012 two aerosol layers were located above the cloud top. (Adapted from Klingebiel et al. 2015)

to backscatter ratio (also called lidar ratio) of 40 sr is assumed. Such aerosol layers located just above cloud top of boundary layer clouds have been frequently observed in the Arctic and might influence the cloud microphysical properties when aerosol particles are entrained into the cloud top (Klingebiel et al. 2015). Only airborne remote sensing observations looking from above on the cloud top can characterize these scenarios and investigate potential aerosol-cloud interactions. Ground-based lidar observations are quickly attenuated by the low-level cloud and are not able to identify aerosol layers above cloud top.

Aerosol loading across the Arctic is usually low (see Fig. 9.31). Therefore, the retrieval of spectral aerosol optical thickness using spaceborne measurements of backscattered light (especially over bright snow and ice surfaces) is a challenge. The same is true for cloud screening procedures over snow and ice (Jafariserajehlou et al. 2014). However, some results in this direction have already been reported for several optical instruments orbiting a planet (Istomina et al. 2011; Mei et al. 2013a, b, 2014; Shi et al. 2019). Look-up-tables of atmospheric reflectance for various aerosol and underlying surface models have been used in the framework of optimal estimation approaches to retrieve spectral aerosol optical thickness (Tomasi et al.

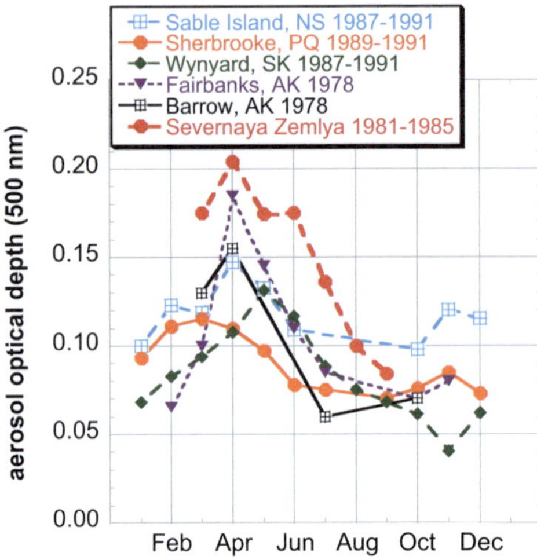

Fig. 9.31 Variation of mean monthly values of aerosol optical depth at 500 nm in Arctic air masses for sites in Canada (Sable Island, Nova Scotia; Sherbrooke, Quebec; Wynyard, Saskatchewan) (Smirnov et al. 1996), USA (Fairbanks, Barrow, Alaska) (Shaw 1982) and Russian Arctic (Severnaya Zemlya) (Barteneva et al. 1991) (Source: A. Smirnov et al. 2006)

2015b). The main problems arise due to the fact the surface contribution dominates the signal in most of cases. Therefore, small errors related to the underlying surface models (surfaces partially covered by snow and ice) lead to large uncertainties in the retrieved optical properties of Arctic aerosol. The results are more accurate in areas covered by open water and bare soil. The retrievals can become robust (especially in case of high aerosol loads) for areas covered by fresh (unpolluted) snow, where the snow reflectivity can be described by a simple two-parametric function (Kokhanovsky et al. 2019):

$$R(\lambda) = aexp\left(-\sqrt{b\alpha(\lambda)}\right), \qquad (9.5)$$

where $\alpha(\lambda)$ is the bulk ice absorption coefficient and the spectrally neutral pair (a, b) must be retrieved simultaneously with aerosol parameters (say, assuming that aerosol optical thickness obey to the Angström law). Generally, atmospheric light scattering and absorption processes reduce the underlying surface signal as detected on a satellite over bright surfaces, which can be used to retrieve aerosol properties in Arctic regions. On contrary, over ocean, the atmospheric scattering leads to the increase of the signal as detected on a satellite. Clearly, there is a region of surface reflectivity, where the influence of atmospheric effects on the reflectance of underlying surface – atmosphere system is at minimum (Seidel and Popp 2012).

In case of underlying Lambertian surface with the albedo A, the top-of-atmosphere reflectance can be presented as:

$$R = \rho + (1 - At) / (1 - Ar). \qquad (9.6)$$

Here, ρ is the top-of-atmosphere reflectance of incident solar light at $A = 0$, r is the spherical albedo of atmosphere and t is the total atmospheric transmittance (from the top-of atmosphere to the ground and back to the sensor). In case of underlying fresh (not polluted snow), one can assume the spectral dependence of A as given by Eq. (9.5) with $a = 1$ and b to be found from the measurements. Eq. (9.6) can be used to find spectral aerosol optical thickness from spectral or angular-spectral measurements of reflectance (passive remote sensing). For this one must use the channels free of gaseous absorption and assume an aerosol model (size distribution, spectral refractive index) or set of models and find the model, which best fits the measurements.

The techniques described above can be applied only during daytime measurements because they are based on analysis of backscattered sunlight. Alternatively, one can use spaceborne lidar measurements. The CALIOP (Cloud-Aerosol Lidar with Orthogonal Polarization) lidar has been flying on the CALIPSO (Cloud-Aerosol Lidar and Infrared Pathfinder Satellite Observations) satellite since June 2006 (Winker et al. 2010). CALIPSO flies in a sun-synchronous polar orbit providing coverage from 82 N to 82 S. CALIOP is nadir viewing and transmits linearly polarized laser pulses at 532 nm and 1064 nm, providing a curtain of lidar profile measurements with a footprint of 70 m (Hunt et al. 2009). The fundamental sampling resolution is 30 m vertical and 335 m horizontal. The outgoing pulses at 532 nm are linearly polarized and the receiver features two 532 nm channels sensitive to the backscattered laser return signals polarized parallel and perpendicular to the polarization of the outgoing pulse. This allows vertically-resolved discrimination of cloud from aerosols based on their depolarizing properties. Daytime solar background is measured and subtracted from each return signal profile. Nighttime profiles from the 532 nm parallel channel are calibrated by normalizing stratospheric return signals to a molecular density profile from the MERRA-2 re-analysis product (Kar et al. 2018). Calibration uncertainty from this technique is roughly 2%. The 532 nm perpendicular channel is calibrated relative to the parallel channel by inserting a pseudo-depolarizer into the beam path, placing equal amounts of light onto each of the 532 nm detectors. The two 532 nm channels are then summed to produce profiles of attenuated backscatter:

$$\beta'(r) = \left[\beta_m(r) + \beta_p(r)\right] \ T_m^2(r) T_p^2(r) \tag{9.7}$$

where $\beta_m(r)$ and $\beta_p(r)$ are molecular and particulate backscatter coefficients. T_m and T_p are atmospheric transmittances from the top of the atmosphere to range r, representing molecular and particulate attenuation respectively.

Aerosol retrievals begin with layer detection, discrimination of aerosol layers from clouds, and identification of aerosol type (Winker et al. 2009). In the Arctic, the CALIOP depolarization signals are particularly helpful in separating aerosol from thin ice clouds and diamond dust. Aerosol backscatter coefficients are retrieved within detected layers by correcting the attenuated backscatter profiles, $\beta'(r)$, for attenuation and subtracting molecular backscatter:

$$\beta_p(r) = \frac{\beta'(r)}{T_m^2(r)T_p^2(r)} - \beta_m(r) \tag{9.8}$$

Vertically resolved aerosol extinction is derived by multiplying $\beta_p(r)$ by an aerosol extinction-to-backscatter ratio, S_a, and aerosol optical depth is computed by vertically integrating the aerosol extinction profile. Details are provided in Young and Vaughan (2009) and Young et al. (2018).

During the retrieval process, horizontal averaging is performed to improve sensitivity to aerosols. The detection limit depends on the brightness of the background viewed by CALIOP. During daytime CALIOP is more sensitive over open ocean, less sensitive over surfaces covered by ice and snow, but is most sensitive at night. Detection sensitivity is a function of the vertically resolved aerosol backscatter coefficient, not aerosol extinction or AOT (Vaughan et al. 2005; Singh et al. 2005), and the fraction of AOT retrieved depends on the magnitude of AOT and the vertical distribution of the aerosol. The detection limit at night is about 0.0003 /km/sr (Winker et al. 2009), corresponding to an aerosol extinction on the order of 0.01/km (10 Mm^{-1}). To put this in perspective, this detection threshold is greater than the monthly mean aerosol extinction measured in situ at the surface at Barrow during spring, summer, and fall (Fig. 4.34; also Quinn et al. 2002). Thus, AOT reported by CALIOP in the high Arctic tends to be representative of periods of enhanced aerosol loading.

Careful validation against measurements by airborne high spectral resolution lidar (HSRL) over much of North America indicates aerosol extinction retrievals from CALIOP are accurate when aerosol is detected and the correct aerosol lidar ratio is used (Rogers et al. 2014). Comparisons with nephelometer measurements at Barrow and Alert have similarly shown that CALIOP aerosol extinction retrievals are fairly accurate when layers are detected, but detection limitations result in significant underestimates relative to monthly mean in situ measurements (di Pierro et al. 2013).

Monthly mean AOT (called also Aerosol Optical Depth (AOD)) for January and March averaged over the period 2007–2018 are shown in Figs. 9.32 and 9.33, respectively. These data are based on the Level 2 CALIOP Version 4 Aerosol Profile product. Improvements in the Version 4 algorithms have resulted in a general increase in AOT relative to Version 3 results presented in Winker et al. (2013) and di Pierro et al. (2013). Results are shown for nighttime observations only, due to the markedly lower sensitivity under daytime conditions. di Pierro et al. (2013) attempted to correct for day-night biases caused by reduced sensitivity under daytime conditions. They present results on the annual cycle of aerosol geographic and vertical distribution, and find that CALIOP reproduces the observed seasonal cycle and magnitude of surface aerosols seen in measurements at Barrow and Alert.

It is evident from Figs. 9.32 and 9.33 there is a large spatial variability in aerosol loading across the Arctic. Significant aerosol loading is seen in the north Atlantic, extending eastward into the Arctic Ocean from Svalbard to Novya Zemlya, a region which typically remains ice-free during winter. Aerosol loading is much lower in the Asian and Canadian sectors of the Arctic Ocean, which are ice covered during

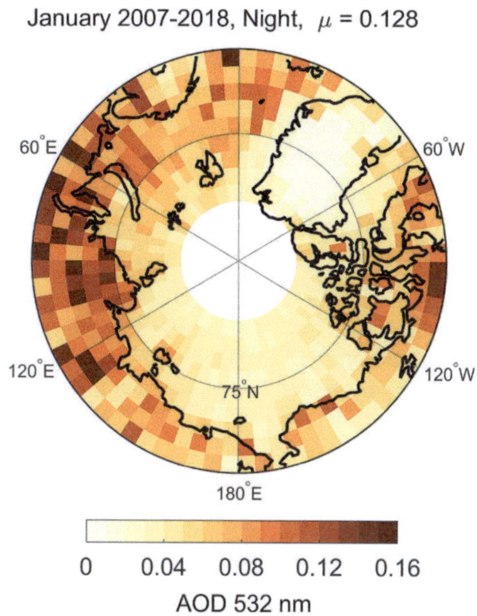

Fig. 9.32 AOD distribution, 60 N to 82 N, as detected using spaceborne lidar data for January, 2007–2018. Mean AOD = 0.128

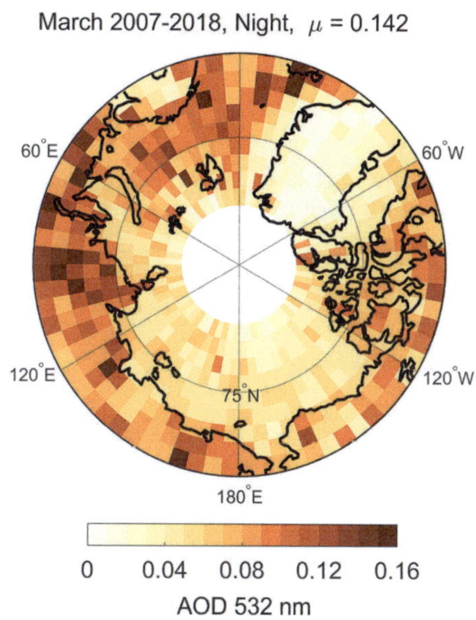

Fig. 9.33 AOD distribution, 60 N to 82 N, as detected using spaceborne lidar data for March, 2007–2018

January and March, although an increase in aerosol is seen from January to March. A persistent AOT maximum is seen over Russia west of 120E, although aerosol loading is lower over eastern Russia in January than in March. Increases seen in March may be due to Arctic haze, transport of dust from the Asian deserts, and the beginning of the boreal fire season.

CALIOP reports very low AOT above the Greenland ice sheet during both months. AOT from CALIOP is likely somewhat underestimated due to detection sensitivity limits discussed above, but Schmeisser et al. (2018) show monthly medians of 550 nm aerosol extinction at the Summit station are less than 0.002/km (2 Mm^{-1}) throughout the year, and near zero in January. If aerosol concentration was constant between the surface at Summit, with a tropopause at 10 km altitude, the median tropospheric AOT would be less than 0.015. Transport aloft, above the altitude of the Summit site, requires injection into the free troposphere. In spring and summer, dust from the Asian deserts and smoke from boreal fires can be injected at high altitudes and transported into the high Arctic, but this is rare in January and March.

Acknowledgements The authors gratefully acknowledge Richard Wagener (for the AERONET sun-photometric measurements conducted at Barrow), Ihab Abboud and Vitali Fioletov (for the AERONET sun-photometric measurements conducted at Resolute Bay and Eureka-0PAL), Norm O' Neill (for the AERONET sun-photometric measurements conducted at Eureka-0PAL), Piotr S. Sobolewski (for the AERONET sun-photometric measurements conducted at Hornsund), Rigel Kivi (for the AERONET sun-photometric measurements conducted at Sodankylä), Robert S. Stone (for the NOAA/GMD sun-photometric measurements conducted at Barrow and Alert), Stephan Nyeki (for the PMOD/WRC sun-photometric measurements conducted at Summit, in Greenland), Andreas Herber (for the AWI measurements conducted at Ny-Ålesund), Kerstin Stebel (for the NILU/PFR measurements conducted at Ny-Ålesund), Veijo Aaltonen (for the FMI/PFR sun-photometric measurements conducted at Sodankylä), and Sergey M. Sakerin (for the IAO/SB/RAS measurements conducted at Barentsburg with the SPM sun-photometer model).

B. N. Holben and A. Smirnov would like to thank Dr. Hal Maring from NASA Headquarters and Dr. Steven Platnick from EOS Project Science Office for their support of AERONET. A. Kokhanovsky thanks the European Space Agency for the support of his work via the ESRIN contracts 4000118926/16/I-NB and 4000125043 – ESA/AO/1-9101/17/I-NB EO SCIENCE FOR SOCIETY.

Manfred Wendisch and Andre Erlich are grateful for the funding provided by the Deutsche Forschungsgemeinschaft (DFG, German Research Foundation) – Project–ID 268020496 – TRR 172.

References

Aaltonen, V., Lihavainen, H., Kerminen, V.-M., Komppula, M., Hatakka, J., Eneroth, K., Kulmala, M., & Viisanen, Y. (2006). Measurements of optical properties of atmospheric aerosols in Northern Finland. *Atmospheric Chemistry and Physics, 6*(4), 1155–1164. https://doi.org/10.5194/acp-6-1155-2006.

Anderson, G. P., Clough, S. A., Kneizys, F. X., Chetwynd, J. H., & Shettle, E. P. (1986). *AFGL atmospheric constituent profiles (0–120 km)* (Environmental research papers, no. 954, AFGL-TR-86-0110). Hanscom Field: Optical Physics Division, Air Force Geophysics Laboratory, L. G, 43 pp.

Ångström, A. (1964). The parameters of atmospheric turbidity. *Tellus, 16*(1), 64–75. https://doi.org/10.1111/j.2153-3490.1964.tb00144.x.

Barteneva, O. D., Nikitinskaya, N. I., Sakunov, G. G., & Veselova, L. K. (1991). *Atmospheric transmittance in the visible and near IR spectral range* (in Russian), 224 pp., Gidrometeoizdat, St. Petersburg, Russia.

Berk, A., Anderson, G. P., Acharya, P. K., Chetwynd, J. H., Bernstein, L. S., Shettle, E. P., Matthew, M. W., & Adler-Golde, S. M. (2000). *MODTRAN 4 users's manual, 1 June 1999* (Last revised 17 April 2000) (pp. 01731–03010). Hanscom AFB: Air Force Research Laboratory, Space Vehicles Directorate.

Bond, T. C., Doherty, S. J., Fahey, D. W., Forster, P. M., Berntsen, T., DeAngelo, B. J., Flanner, M. G., Ghan, S., Kärcher, B., Koch, D., Kinne, S., Kondo, Y., Quinn, P. K., Sarofim, M. C., Schultz, M. G., Schulz, M., Venkataraman, C., Zhang, H., Zhang, S., Bellouin, N., Guttikunda, S. K., Hopke, P. K., Jacobson, M. Z., Kaiser, J. W., Klimont, Z., Lohmann, U., Schwarz, J. P., Shindell, D., Storelvmo, T., Warren, S. G., & Zender, C. S. (2013). Bounding the role of black carbon in the climate system: A scientific assessment. *Journal of Geophysical Research-Atmospheres, 118*, 5380–5552. https://doi.org/10.1002/jgrd.50171.

Chernov, D. G., Kabanov, D. M., Kozlov, V. S., Radionov, V. F., Sakerin, S. M., & Urazgildeeva, A. V. (2017). *Preliminary results of aerosol studies in the cruise of RV "Professor Molchanov" (July 2017) and comparison with other Arctic expeditions* (in Russian), XXIV Workshop "Aerosols of Siberia" (abstracts, p. 5), Tomsk, Russia.

Damiani, A., De Simone, S., Rafanelli, C., Cordero, R. R., & Laurenza, M. (2012). Three years of ground-based total ozone measurements in the Arctic: Comparison with OMI, GOME and SCIAMACHY satellite data. *Remote Sensing of Environment, 127*, 162–180. https://doi.org/10.1016/j.rse.2012.08.023.

Di Pierro, M., Jaeglé, L., & Anderson, T. L. (2011). Satellite observations of aerosol transport from East Asia to the Arctic: Three case studies. *Atmospheric Chemistry and Physics, 11*(5), 2225–2243. https://doi.org/10.5194/acp-11-2225-2011.

di Pierro, M., Jaeglé, L., Eloranta, E. W., & Sharma, S. (2013). Spatial and seasonal distribution of Arctic aerosols observed by CALIOP (2006–2012). *Atmospheric Chemistry and Physics, 13*, 7075–7095. https://doi.org/10.5194/acp-13-7075-2013.

Dutton, E., Anderson, G., Carbaugh, G., Jackson, D., Longenecker, D., Nelson, D., O'Neill, M., Stone, R., Treadwell, J., & Wendell, J. (2004). *Solar and thermal atmospheric radiation* (CMDL summary report 27) (pp. 76–96). Boulder, Colorado: Global Monitoring Division, Earth System Research Laboratory, NOAA.

Eck, T. F., Holben, B. N., Reid, J. S., Dubovik, O., Smirnov, A., O'Neill, N. T., Slutsker, I., & Kinne, S. (1999). Wavelength dependence of the optical depth of biomass burning, urban, and desert dust aerosols. *Journal of Geophysical Research, Atmospheres, 104*(D24), 31333–31349. https://doi.org/10.1029/1999JD900923.

Hall, T. C., Jr., & Blacet, F. E. (1952). Separation of the absorption spectra of NO_2 and N_2O_4 in the range of 2400–5000 Å. *The Journal of Chemical Physics, 20*(11), 1745–1749. https://doi.org/10.1063/1.1700281.

Herber, A., Thomason, L. W., Gernandt, H., Leiterer, U., Nagel, D., Schulz, K., Kaptur, J., Albrecht, T., & Notholt, J. (2002). Continuous day and night aerosol optical depth observations in the Arctic between 1991 and 1999. *Journal of Geophysical Research, Atmospheres, 107*(D10), 4097. https://doi.org/10.1029/2001JD000536.

Hess, M., Koepke, P., & Schult, I. (1998). Optical properties of aerosols and clouds: The software package OPAC. *Bulletin of the American Meteorological Society, 79*(5), 831–844. https://doi.org/10.1175/1520-0477(1998)079<0831:OPOAAC>2.0.CO;2.

Hirdman, D., Sodemann, H., Eckhardt, S., Burkhart, J. F., Jefferson, A., Mefford, T., Quinn, P. K., Sharma, S., Ström, J., & Stohl, A. (2010). Source identification of short-lived air pollutants in the Arctic using statistical analysis of measurement data and particle dispersion model output. *Atmospheric Chemistry and Physics, 10*(2), 669–693. https://doi.org/10.5194/acp-10-669-2010.

Hoffmann, A., Ritter, C., Stock, M., Maturilli, M., Eckhardt, S., Herber, A., & Neuber, R. (2010). Lidar measurements of the Kasatochi aerosol plume in august and September 2008 in Ny-Ålesund, Spitsbergen. *Journal of Geophysical Research, Atmospheres, 115*, D2. https://doi.org/10.1029/2009JD013039.

Holben, B. N., Eck, T. F., Slutsker, I., Tanré, D., Buis, J. P., Setzer, A., Vermote, E., Reagan, J. A., Kaufman, Y. J., Nakajima, T., Lavenu, F., Jankowiak, I., & Smirnov, A. (1998). AERONET – a federated instrument network and data archive for aerosol characterization. *Remote Sensing of Environment, 66*(1), 1–16. https://doi.org/10.1016/S0034-4257(98)00031-5.

Holben, B. N., Tanré, D., Smirnov, A., Eck, T. F., Slutsker, I., Abuhassan, N., Newcomb, W. W., Schafer, J. S., Chatenet, B., Lavenue, F., Kaufman, Y. J., Vande Castle, J., Setzer, A., Markham,

B., Clark, D., Frouin, R., Halthore, R., Karnieli, A., O'Neill, N. T., Pietras, C., Pinker, R. T., Voss, K., & Zibordi, G. (2001). An emerging ground-based aerosol climatology: Aerosol optical depth from AERONET. *Journal of Geophysical Research, Atmospheres, 106*(D11), 12067–12097. https://doi.org/10.1029/2001JD900014.

Hunt, W. H., et al. (2009). CALIPSO lidar description and performance assessment. *Journal of Atmospheric and Oceanic Technology, 26*, 1214–1228.

Ichoku, C., et al. (2002). Analysis of the performance characteristics of the five-channel Microtops II Sun photometer for measuring aerosol optical thickness and precipitable water vapor. *Journal of Geophysical Research, 107*(D13), 4179. https://doi.org/10.1029/2001JD001302.

Inn, E. C. Y., & Tanaka, Y. (1953). Absorption coefficient of ozone in the ultraviolet and visible regions. *Journal of the Optical Society of America, 43*(10), 870–873. https://doi.org/10.1364/JOSA.43.000870.

Iqbal, M. (1983). *An introduction to solar radiation*. Toronto: Academic, 390 pp. (See Table 1.2.1 at pages 4–5).

Istomina, L. G., von Hoyningen-Huene, W., Kokhanovsky, A. A., Schultz, E., & Burrows, J. P. (2011). Remote sensing of aerosols over snow using infrared AATSR observations. *Atmospheric Measurement Techniques, 4*, 1133–1145. https://doi.org/10.5194/amt-4-1133-2011.

Jafariserajehlou, S., Mei, L., Vountas, M., Rozanov, V., Burrows, J. P., & Hollmann, R. (2014). A cloud identification algorithm over the Arctic for use with AATSR–SLSTR measurements. *Atmospheric Measurement Techniques, 12*, 1059–1076. https://doi.org/10.5194/amt-12-1059-2019.

Kabanov, D. M., et al. (2017). Peculiarities of spatial-temporal variability of the aerosol optical depth of the atmosphere over Kara and Barents Seas in 2016. In *Proceedings of SPIE 10466, 23rd international symposium on atmospheric and ocean optics: Atmospheric physics, 104662G*. https://doi.org/10.1117/12.2284298.

Kar, J., Vaughan, M. A., Lee, K. P., Tackett, J., Avery, M., Garnier, A., Getzewich, B., Hunt, W., Josset, D., Liu, Z., Lucker, P., Magill, B., Omar, A., Pelon, J., Rogers, R., Toth, T. D., Trepte, C., Vernier, J.-P., Winker, D., & Young, S. (2018). CALIPSO lidar calibration at 532 nm: Version 4 Nighttime algorithm. *Atmospheric Measurement Techniques, 11*, 1459–1479. https://doi.org/10.5194/amt-11-1459-2018.

Karimova, G. U. (1976). Measurement results for ozone and optical characteristics of atmosphere over the North Atlantic (in Russian). *Trudy Arctic and Antarctic Institute, 327*, 209–214.

Karol, Y., Tanré, D., Goloub, P., Vervaerde, C., Balois, J. Y., Blarel, L., Podvin, T., Mortier, A., & Chaikovsky, A. (2013). Airborne sun photometer PLASMA: Concept, measurements, comparison of aerosol extinction vertical profile with lidar. *Atmospheric Measurement Techniques, 6*, 2383–2389. https://doi.org/10.5194/amt-6-2383-2013.

Klingebiel, M., de Lozar, A., Molleker, S., Weigel, R., Roth, A., Schmidt, L., Meyer, J., Ehrlich, A., Neuber, R., Wendisch, M., & Borrmann, S. (2015). Arctic low-level boundary layer clouds: In situ measurements and simulations of mono- and bimodal supercooled droplet size distributions at the top layer of liquid phase clouds. *Atmospheric Chemistry and Physics, 15*(2), 617–631. https://doi.org/10.5194/acp-15-617-2015.

Kneizys, F. X., Abreu, L. W., Anderson, G. P., Chetwind, J. H., Shettle, E. P., Berk, A., Bernstein, L. S., Robertson, D. C., Acharya, P., Rothman, L. S., Selby, J. E. A., Gallery, W. O., & Clough, S. A. (1996). The MODTRAN 2/3 report and LOWTRAN 7 model. In L. W. Abreu & G. P. Anderson (Eds.), *Contract F19628-91-C-0132*. Hanscom AFB: Phillips Laboratory, Geophysics Directorate, PL/GPOS, 261 pp.

Knobelspiesse, K. D., Pietras, C., & Fargion, G. S. (2003). Sun-pointing error correction for sea deployment of the Microtops II handheld Sun photometer. *Journal of Atmospheric and Oceanic Technology, 20*, 767–771.

Knobelspiesse, K. D., et al. (2004). Maritime aerosol optical thickness measured by handheld Sun photometers. *Remote Sensing of Environment, 93*, 87–106.

Kokhanovsky, A. A., et al. (2019). Retrieval of snow properties from the Sentinel −3 Ocean and Land Colour Instrument. *Remote Sensing, 11*(19), 2280. https://doi.org/10.3390/rs11192280.

Livingston, J. M., Schmid, B., Russell, P. B., Eilers, J. A., Kolyer, R. W., Redemann, J., Ramirez, S. R., Yee, J.-H., Swartz, W. H., Trepte, C. R., Thomason, L. W., Pitts, M. C., Avery, M. A., Randall, C. E., Lumpe, J. D., Bevilacqua, R. M., Bittner, M., Erbertseder, T., McPeters, R. D., Shetter, R. E., Browell, E. V., Kerr, J. B., & Lamb, K. (2005). Retrieval of ozone column content from airborne sun photometer measurements during SOLVE II: Comparison with coincident satellite and aircraft measurements. *Atmospheric Chemistry and Physics, 5*, 2035–2054. https://doi.org/10.5194/acp-5-2035-2005.

Lukyanchikova, N. A., & Govorushkin, A. A. (1981). Spectral characteristics of atmosphere over the northern part of the Atlantic Ocean (in Russian). *Trudy Arctic and Antarctic Institute, 370*, 160–164.

Mazzola, M., Stone, R. S., Herber, A., Tomasi, C., Lupi, A., Vitale, V., Lanconelli, C., Toledano, C., Cachorro, V. E., T, N., O'Neill, M., Shiobara, V., Aaltonen, K., Stebel, T., Zielinski, T., Petelski, J. P., Ortiz de Galisteo, B., Torres, A., Berjon, P., Goloub, Z., Li, L., Blarel, I., Abboud, E., Cuevas, M., Stock, K.-H. S., & Virkkula, A. (2012). Evaluation of sun photometer capabilities for retrievals of aerosol optical depth at high latitudes: The POLAR-AOD intercomparison campaigns. *Atmospheric Environment, 52*(C), 1–14. https://doi.org/10.1016/j.atmosenv.2011.07.042.

Mei, L. L., Xue, Y., de Leeuw, G., von Hoyningen-Huene, W., Kokhanovsky, A., Istomina, L., Guang, J., & Burrows, J. P. (2013a). Aerosol optical depth retrieval in the Arctic region using MODIS data over snow. *Remote Sensing of Environment, 128*, 234–245.

Mei, L. L., Xue, Y., Kokhanovsky, A. A., von Hoyningen-Huene, W., Istomina, L., de Leeuw, G., Burrows, J. P., Guang, J., & Jing, Y. (2013b). Aerosol optical depth retrieval over snow using AATSR data. *International Journal of Remote Sensing, 34*, 5030–5041.

Mei, L. L., Xue, Y., Kokhanovsky, A. A., von Hoyningen-Huene, W., de Leeuw, G., & Burrows, J. P. (2014). Retrieval of aerosol optical depth over land surfaces from AVHRR data. *Atmospheric Measurement Techniques, 7*, 2411–2420. https://doi.org/10.5194/amt-7-2411-2014.

Michalsky, J., Beauharnois, M., Berndt, J., Harrison, L., Kiedron, P., & Min, Q. (1999). O_2-O_2 absorption band identification based on optical depth spectra of the visible and near-infrared. *Geophysical Research Letters, 26*(11), 1581–1584. https://doi.org/10.1029/1999GL900267.

Morys, M., et al. (2001). Design, calibration, and performance of MICROTOPS II handheld ozone monitor and Sun photometer. *Journal of Geophysical Research, 106*, 14573–14582.

Nakajima, T., Yoon, S. C., Ramanathan, V., Shi, G. Y., Takemura, T., Higurashi, A., Takamura, T., Aoki, K., Sohn, B. J., Kim, S. W., Tsuruta, H., Sugimoto, N., Shimizu, A., Tanimoto, H., Sawa, Y., Lin, N. H., Lee, C. T., Goto, D., & Schutgens, N. (2007). Overview of the atmospheric brown cloud East Asian regional experiment 2005 and a study of the aerosol direct radiative forcing in East Asia. *Journal of Geophysical Research, Atmospheres, 112*, D24. https://doi.org/10.1029/2007JD009009.

Nyeki, S., Halios, C. H., Baum, W., Eleftheriadis, K., Flentje, H., Gröbner, J., Vuilleumier, L., & Wehrli, C. (2012). Ground-based aerosol optical depth trends at three high-altitude sites in Switzerland and southern Germany from 1995 to 2010. *Journal of Geophysical Research, Atmospheres, 117*(D18). https://doi.org/10.1029/2012JD017493.

O'Neill, N. T., Eck, T. F., Holben, B. N., Smirnov, A., Dubovik, O., & Royer, A. (2001a). Bimodal size distribution influences on the variation of Ångström derivatives in spectral and optical depth space. *Journal of Geophysical Research, Atmospheres, 106*(D9), 9787–9806. https://doi.org/10.1029/2000JD900245.

O'Neill, N. T., Dubovik, O., & Eck, T. F. (2001b). Modified Ångström exponent for the characterization of submicrometer aerosol. *Applied Optics, 40*(15), 2368–2375. https://doi.org/10.1364/AO.40.002368.

O'Neill, N. T., Eck, T. F., Smirnov, A., Holben, B. N., & Thulasiraman, S. (2003). Spectral discrimination of coarse and fine mode optical depth. *Journal of Geophysical Research, Atmospheres, 108*(D17), 4559. https://doi.org/10.1029/2002JD002975.

O'Neill, N. T., Perro, C., Saha, A., Lesins, G., Duck, T. J., Eloranta, E. W., Nott, G. J., Hoffmann, A., Karumudi, M. L., Ritter, C., Bourassa, A., Abboud, I., Carn, S. A., & Savastiouk, V. (2012).

Properties of Sarychev sulphate aerosols over the Arctic. *Journal of Geophysical Research, Atmospheres, 117*, D04203. https://doi.org/10.1029/2011JD016838.
Polkin, V. V. Jr, Kabanov, D. M., Radionov, V. F., & Kessel, A. S. (2015). *Spatial and temporal variability of aerosol properties in the central and western areas of Russian Arctic* (in Russian), XXII Workshop "Aerosols of Siberia" (abstracts, p. 89), Tomsk, Russia.
Porter, J. N., et al. (2001). Ship-based sunphotometer measurements using microtops Sun photometers. *Journal of Atmospheric and Oceanic Technology, 18*, 765–774.
Quinn, P. K., Miller, T. L., Bates, T. S., Ogren, J. A., Andrews, E., & Coauthors. (2002). A three-year record of simultaneously measured aerosol chemical and optical properties at Barrow, Alaska. *Journal of Geophysical Research, 107*. https://doi.org/10.1029/2001JD001248.
Radionov, V. F. (2005). *Temporal variability of the aerosol optical characteristicsof the atmosphere in the Russian Arctic (historical review)* (WMO/GAW report 162, pp. 82–85, Baltensperger, U., Barrie, L., & Wehrli, C., eds.). WMO TD N1287, Geneva.
Radionov, V. F., & Marshunova, M. S. (1992). Long-term variations in the turbidity of the Arctic atmosphere in Russia. *Atmosphere-Ocean, 30*(4), 531–549.
Rodríguez, E., Toledano, C., Cachorro, V. E., Ortiz, P., Stebel, K., Berjón, A., Blindheim, S., Gausa, M., & de Frutos, A. M. (2012). Aerosol characterization at the sub-Arctic site Andenes (69 °N, 16 °E), by the analysis of columnar optical properties. *Quarterly Journal of the Royal Meteorological Society, 138*, 471–482. https://doi.org/10.1002/qj.921.
Rogers, R. R., Vaughan, M. A., Hostetler, C. A., Burton, S. P., Ferrare, R. A., Young, S. A., Hair, J. W., Obland, M. D., Harper, D. B., Cook, A. L., & Winker, D. M. (2014). Looking through the haze: Evaluating the CALIPSO level 2 aerosol optical depth using airborne high spectral resolution lidar data. *Atmospheric Measurement Techniques, 7*, 4317–4340. https://doi.org/10.5194/amt-7-4317-2014.
Russell, P., Livingston, J., Schmid, B., Eilers, J., Kolyer, R., Redemann, J., Ramirez, S., Yee, J.-H., Swartz, W., Shetter, R., Trepte, C., Risley, A., Jr., Wenny, B., Zawodny, J., Chu, W., Pitts, M., Lumpe, J., Fromm, M., Randall, C., Hoppel, K., & Bevilacqua, R. (2004). Aerosol optical depth measurements by airborne sun photometer in SOLVE II: Comparisons to SAGE III, POAM III and airborne spectrometer measurements. *Atmospheric Chemistry and Physics, 5*, 1311–1339. https://doi.org/10.5194/acp-5-1311-2005.
Sakerin, S. M., Kabanov, D. M., Rostov, A. P., & Turchinovich, S. A. (2009). Portative solar photometer, *Pribory i Tekhnika Eksperimenta* (Instruments and Experimental Technique), 2, 181–182 (in Russian).
Sakerin, S. M., Chernov, D. G., Kabanov, D. M., Kozlov, V. S., Panchenko, M. V., Polkin, V. V., & Radionov, V. F. (2012). Preliminary results of studying the aerosol characteristics of the atmosphere in the region of Barentsburg, Spitsbergen. *Problemy Arctki i Antarktiki* (Problems of the Arctic and Antarctic), *1, 91*, 20–31 (in Russian).
Sakerin, S. M., Andreev, S. Y., Kabanov, D. M., Nikolashkin, S. V., Prakhov, A. N., Radionov, V. F., Turchinovich, J. S., Chernov, D. G., Holben, B. N., Smirnov, A., & Sorokin, M. G. (2014). On results of studies of atmospheric aerosol optical depth in Arctic regions. *Atmospheric and Oceanic Optics, 27*(6), 517–528. https://doi.org/10.1134/S1024856014060165.
Sakerin, S. M., et al. (2015). On measurements of aerosol-gas composition of the atmosphere during two expeditions in 2013 along the Northern Sea route. *Atmospheric Chemistry and Physics, 15*, 12413–12443. https://doi.org/10.5194/acp-15-12413-2015.
Sakunov, G. G., et al. (1981). Optical properties of the Arctic atmosphere (in Russian). In *Polar aerosol, cloudness and radiation* (pp. 73–89). Leningrad: Gidrometeoizdat.
Schmeisser, L., Backman, J., Ogren, J. A., Andrews, E., Asmi, E., Starkweather, S., Uttal, T., Fiebig, M., Sharma, S., Eleftheriadis, K., Vratolis, S., Bergin, M., Tunved, P., & Jefferson, A. (2018). Seasonality of aerosol optical properties in the Arctic. *Atmospheric Chemistry and Physics, 18*, 11599–11622. https://doi.org/10.5194/acp-18-11599-2018.
Seidel, F. C., & Popp, C. (2012). Critical surface albedo and its implications to aerosol remote sensing. *Atmospheric Measurement Techniques, 5*, 1653–1665. https://doi.org/10.5194/amt-5-1653-2012.

Shaw, G. E. (1976). Error analysis of multi-wavelength sun photometry. *Pure and Applied Geophysics, 114*(1), 1–14. https://doi.org/10.1007/BF00875487.

Shaw, G. E. (1982). Atmospheric turbidity in the polar regions. *Journal of Applied Meteorology, 21*, 1080–1088.

Shi, Z., T- Xing, J., Guang, Y. X., & Che, Y. (2019). Aerosol optical depth over the Arctic snow-covered regions derived from dual-viewing satellite observations. *Remote Sensing, 11*(8), 891. https://doi.org/10.3390/rs11080891.

Shinozuka, Y., Redemann, J., Livingston, J. M., Russell, P. B., Clarke, A. D., Howell, S. G., Freitag, S., O'Neill, N. T., Reid, E. A., Johnson, R., Ramachandran, S., McNaughton, C. S., Kapustin, V. N., Brekhovskikh, V., Holben, B. N., & McArthur, L. J. B. (2011). Airborne observation of aerosol optical depth during ARCTAS: Vertical profiles, inter-comparison and fine-mode fraction. *Atmospheric Chemistry and Physics, 11*, 3673–3688. https://doi.org/10.5194/acp-11-3673-2011.

Singh, U., Ismail, S., Kavaya, M., Winker, D., & Amzajerdian, F. (2005). *Space-based lidar*. In T. Fujii & T. Fukuchi (Eds.), *Laser remote sensing*. Boca Raton: CRC Press, Taylor and Francis, 888 pp.

Smirnov, A., O'Neill, N. T., Royer, A., Tarussov, A., & McArthur, B. (1996). Aerosol optical depth over Canada and the link with synoptic air mass types. *Journal of Geophysical Research, 101*(D14), 19299–19318. https://doi.org/10.1029/95JD02608.

Smirnov, A., Holben, B. N., Eck, T. F., Dubovik, O., & Slutsker, I. (2000). Cloud-screening and quality control algorithms for the AERONET database. *Remote Sensing of Environment, 73*(3), 337–349. https://doi.org/10.1016/S0034-4257(00)00109-7.

Smirnov, A., Holben, B. N., Kaufman, Y. J., Dubovik, O., Eck, T. F., Slutsker, I., Pietras, C., & Halthore, R. N. (2002). Optical properties of atmospheric aerosol in maritime environments. *Journal of Atmospheric Sciences, 59*(2), 501–523. https://doi.org/10.1175/1520-0469(2002)059<0501:OPOAAI>2.0.CO;2.

Smirnov, et al. (2004). *AERONET processing algorithms refinement*, AERONET workshop, El Arenosillo, Spain, 10–14 May 2004.

Smirnov, A., Holben, B. N., Panchenko, M. V., Sakerin, S. M., Aculinin, A. A., Chubarova, N. Y., Karner, O., Chaikovsky, A. P., Sukhinin, A. I., Pavlov, V. E., Kopelevich, O. V., Radionov, V. F., & Korotaev, G. K. (2006). *Aerosol Robotic Network activity in Russia, Moldova, Estonia, Belarus, and Ukraine – as an example of the mutually beneficial collaborative effort – results and prospective, presentation at the International Symposium on Atmospheric Radiation* (MCAP-2006). St. Petersburg, Russia, June 27–June 30, 2006.

Smirnov, A., et al. (2009). Maritime Aerosol Network as a component of Aerosol Robotic Network. *Journal of Geophysical Research, 114*, D06204. https://doi.org/10.1029/2008JD011257.

Smirnov, A., et al. (2011). Maritime aerosol network as a component of AERONET—First results and comparison with global aerosol models and satellite retrievals. *Atmospheric Measurement Techniques, 4*, 583–597. https://doi.org/10.5194/amt-4-583-2011.

Stachlewska, I. S., Neuber, R., Lampert, A., Ritter, C., & Wehrle, G. (2010). AMALi – The airborne mobile aerosol lidar for Arctic research. *Atmospheric Chemistry and Physics, 10*, 2947–2963. https://doi.org/10.5194/acp-10-2947-2010.

Stohl, A. (2006). Characteristics of atmospheric transport into the Arctic troposphere. *Journal of Geophysical Research, Atmospheres, 111*, D11306. https://doi.org/10.1029/2005JD006888.

Stohl, A., Andrews, E., Burkhart, J. F., Forster, C., Herber, A., Hoch, S. W., Kowal, D., Lunder, C., Mefford, T., Ogren, J. A., Sharma, S., Spichtinger, N., Stebel, K., Stone, R., Ström, J., Tørseth, K., Wehrli, C., & Yttri, K. E. (2006). Pan-Arctic enhancements of light absorbing aerosol concentrations due to North American boreal forest fires during summer 2004. *Journal of Geophysical Research, Atmospheres, 111*, D22214. https://doi.org/10.1029/2006JD007216.

Stone, R. S. (2002). Monitoring aerosol optical depth at Barrow, Alaska, and South Pole: Historical overview, recent results and future goals. *SIF Conference Proceedings, 80*, 123–144.

Stone, R. S., Anderson, G. P., Andrews, E., Dutton, E. G., Shettle, E. P., & Berk, A. (2007). Incursions and radiative impact of Asian dust in Northern Alaska. *Geophysical Research Letters, 34*, L14815. https://doi.org/10.1029/2007GL029878.

Stone, R. S., Anderson, G. P., Shettle, E. P., Andrews, E., Loukachine, K., Dutton, E. G., Schaaf, C., & Roman, M. O., III. (2008). Radiative impact of boreal smoke in the Arctic: Observed and modeled. *Journal of Geophysical Research, Atmospheres, 113*, D14S16. https://doi.org/10.1029/2007JD009657.

Stone, R. S., Herber, A., Vitale, V., Mazzola, M., Lupi, A., Schnell, R. C., Dutton, E. G., Liu, P. S. K., Li, S.-M., Dethloff, K., Lampert, A., Ritter, C., Stock, M., Neuber, R., & Maturilli, M. (2010). A three-dimensional characterization of Arctic aerosols from airborne Sun photometer observations: PAM-ARCMIP, April 2009. *Journal of Geophysical Research, Atmospheres, 115*, D13203. https://doi.org/10.1029/2009JD013605.

Stone, R. S., Sharma, S., Herber, A., Eleftheriadis, K., & Nelson, D. W. (2014). A characterization of Arctic aerosols on the basis of aerosol optical depth and black carbon measurements. *Elementa: Science of the Anthropocene, 2*, 27. https://doi.org/10.12952/journal.elementa.000027.

Thomason, L. W., Herman, B. M., & Reagan, J. A. (1983). The effect of atmospheric attenuators with structured vertical distributions on air mass determinations and Langley plot analyses. *Journal of the Atmospheric Sciences, 40*(7), 1851–1854. https://doi.org/10.1175/1520-0469(1983)040b1851:TEOAAWN2.0.CO;2.

Toledano, C., Cachorro, V. E., Gausa, M., Stebel, K., Aaltonen, V., Berjón, A., Ortiz de Galisteo, J. P., De Frutos, A. M., Bennouna, Y., Blindheim, S., Myhre, C. L., Zibordi, G., Wehrli, C., Kratzer, S., Hakansson, B., Carlund, T., de Leeuw, G., Herber, A., & Torres, B. (2012). Overview of sun photometer measurements of aerosol properties in Scandinavia and Svalbard. *Atmospheric Environment, 52*(C), 18–28. https://doi.org/10.1016/j.atmosenv.2011.10.022.

Tomasi, C., & Petkov, B. H. (2014). Calculations of relative optical air masses for various aerosol types and minor gases in Arctic and Antarctic atmospheres. *Journal of Geophysical Research, Atmospheres, 119*(3), 1363–1385. https://doi.org/10.1002/2013JD020600.

Tomasi, C., Vitale, V., & De Santis, L. V. (1998). Relative optical mass functions for air, water vapour, ozone and nitrogen dioxide in atmospheric models presenting different latitudinal and seasonal conditions. *Meteorology and Atmospheric Physics, 65*(1–2), 11–30. https://doi.org/10.1007/BF01030266.

Tomasi, C., Vitale, V., Lupi, A., Di Carmine, C., Campanelli, M., Herber, A., Treffeisen, R., Stone, R. S., Andrews, E., Sharma, S., Radionov, V., von Hoyningen-Huene, W., Stebel, K., Hansen, G. H., Myhre, C. L., Wehrli, C., Aaltonen, V., Lihavainen, H., Virkkula, A., Hillamo, R., Ström, J., Toledano, C., Cachorro, V. E., Ortiz, P., de Frutos, A. M., Blindheim, S., Frioud, M., Gausa, M., Zielinski, T., Petelski, T., & Yamanouchi, T. (2007). Aerosols in polar regions: a historical overview based on optical depth and in situ observations. *Journal of Geophysical Research, Atmospheres, 112*, D16. https://doi.org/10.1029/2007JD008432.

Tomasi, C., Petkov, B. H., Stone, R. S., Benedetti, E., Vitale, V., Lupi, A., Mazzola, M., Lanconelli, C., Herber, A., & von Hoyningen-Huene, W. (2010). Characterizing polar atmospheres and their effect on Rayleigh-scattering optical depth. *Journal of Geophysical Research, Atmospheres, 115*, D02205. https://doi.org/10.1029/2009JD012852.

Tomasi, C., Lupi, A., Mazzola, M., Stone, R. S., Dutton, E. G., Herber, A., Radionov, V. F., Holben, B. N., Sorokin, M. G., Sakerin, S. M., Terpugova, S. A., Sobolewski, P. S., Lanconelli, C., Petkov, B. H., Busetto, M., & Vitale, V. (2012). An update on polar aerosol optical properties using POLAR-AOD and other measurements performed during the International Polar Year. *Atmospheric Environment, 52*(C), 29–47. https://doi.org/10.1016/j.atmosenv.2012.02.055.

Tomasi, C., Petkov, B. H., Mazzola, M., Ritter, C., di Sarra, A. G., di Iorio, T., & Del Guasta, M. (2015a). Seasonal variations of the relative optical air mass function for background aerosol and thin cirrus clouds at Arctic and Antarctic sites. *Remote Sensing, 7*, 7157–7180. https://doi.org/10.3390/rs70607157.

Tomasi, C., Kokhanovsky, A. A., Lupi, A., Ritter, C., Smirnov, A., O'Neill, N. T., Stone, R. S., Holben, B. N., Nyeki, S., Wehrli, C., Stohl, A., Mazzola, M., Lanconelli, C., Vitale, V., Stebel, K., Aaltonen, V., de Leeuw, G., Rodriguez, E., Herber, A. B., Radionov, V. F., Zielinski, T., Petelski, T., Sakerin, S. M., Kabanov, D. M., Xue, Y., Mei, L., Istomina, L., Wagener, R., McArthur, B., Sobolewski, P. S., Kivi, R., Courcoux, Y., Larouche, P., Broccardo, S., & Piketh,

S. J. (2015b). Aerosol remote sensing in polar regions. *Earth-Science Reviews, 140*, 108–157. https://doi.org/10.1016/j.earscirev.2014.11.001.

Treffeisen, R., Tunved, P., Ström, J., Herber, A., Bareiss, J., Helbig, A., Stone, R. S., von Hoyningen-Huene, W., Krejci, R., Stohl, A., & Neuber, R. (2007). Arctic smoke – Aerosol characteristics during a record smoke event in the European Arctic and its radiative impact. *Atmospheric Chemistry and Physics, 7*(11), 3035–3053. https://doi.org/10.5194/acp-7-3035-2007.

Vaughan, M. A., Winker, D. M., & Powell, K. A. (2005). *CALIOP algorithm theoretical basis document, part 2: Feature detection and layer properties algorithms*. https://www-calipso.larc.nasa.gov

Webley, P. W., Steensen, T., Stuefer, M., Grell, G., Freitas, S., & Pavolonis, M. (2012). Analyzing the Eyjafjallajökull 2010 eruption using satellite remote sensing, lidar and WRF-Chem dispersion and tracking model. *Journal of Geophysical Research: Atmospheres, 117*, D00U26. https://doi.org/10.1029/2011JD016817.

Wehrli, C. (2000). Calibrations of filter radiometers for determination of atmospheric optical depths. *Metrologia, 37*, 419–422.

Wendisch, M., & Brenguier, J.-L. (Eds.). (2013). *Airborne measurements for environmental research: Methods and instruments* (Wiley series in atmospheric physics and remote sensing) (Vol. 1). Wiley.

Wesche, C., Steinhage, D., & Nixdorf, U. (2016). Polar aircraft Polar 5 and Polar 6 operated by the Alfred Wegener Institute. *Journal of Large-Scale Research Facilities, 2*, A87. https://doi.org/10.17815/jlsrf-2-153.

Winker, D. M., Vaughan, M. A., Omar, A. H., Hu, Y., Powell, K. A., Liu, Z., Hunt, W. H., & Young, S. A. (2009). Overview of the CALIPSO Mission and CALIOP data processing algorithms. *Journal of Atmospheric and Oceanic Technology, 26*, 2310–2323. https://doi.org/10.1175/2009JTECHA1281.1.

Winker, D. M., Pelon, J., Coakley, J. A., Jr., Ackerman, S. A., Charlson, R. J., Colarco, P. R., Flamant, P., Fu, Q., Hoff, R., Kittaka, C., Kubar, T. L., LeTreut, H., McCormick, M. P., Megie, G., Poole, L., Powell, K., Trepte, C., Vaughan, M. A., & Wielicki, B. A. (2010). The CALIPSO Mission: A global 3D view of aerosols and clouds. *Bulletin of the American Meteorological Society, 91*, 1211–1229. https://doi.org/10.1175/2010BAMS3009.1.

Winker, D. M., Tackett, J. L., Getzewich, B. J., Liu, Z., Vaughan, M. A., & Rogers, R. R. (2013). The global 3-D distribution of tropospheric aerosols as characterized by CALIOP. *Atmospheric Chemistry and Physics, 13*, 3345–3361. https://doi.org/10.5194/acp-13-3345-2013.

Wittrock, F., Oetjen, H., Richter, A., Fietkau, S., Medeke, T., Rozanov, A., & Burrows, J. P. (2004). MAX-DOAS measurements of atmospheric trace gases in Ny-Ålesund – Radiative transfer studies and their application. *Atmospheric Chemistry and Physics, 4*(4), 955–966. https://doi.org/10.5194/acp-4-955-2004.

Wolgin, V. M., Radionov, V. F., & Leiterer, U. (1991). Zur variabilitat der optischen eigenschaften der atmosphare in Nordatlantik. *Zeitschrift für Meteorologie, 41*, 267–272.

Young, S. A., & Vaughan, M. A. (2009). The retrieval of profiles of particulate extinction from cloud aerosol Lidar infrared pathfinder satellite observations (CALIPSO) data: Algorithm description. *Journal of Atmospheric and Oceanic Technology, 26*, 1105–1119. https://doi.org/10.1175/2008 JTECHA1221.1.

Young, S. A., Vaughan, M. A., Tackett, J. L., Garnier, A., Lambeth, J. B., & Powell, K. A. (2018). Extinction and optical depth retrievals for CALIPSO's version 4 data release. *Atmospheric Measurement Techniques, 11*, 5701–5727. https://doi.org/10.5194/amt-11-5701-2018.

Zieger, P., Ruhtz, T., Preusker, R., & Fischer, J. (2007). Dual-aureole and sun spectrometer system for airborne measurements of aerosol optical properties. *Applied Optics, 46*, 8542–8552. https://doi.org/10.1364/AO.46.008542.

Chapter 10
Radiation in the Arctic Atmosphere and Atmosphere – Cryosphere Feedbacks

Claudio Tomasi, Boyan H. Petkov, Angelo Lupi, Mauro Mazzola, Christian Lanconelli, and Ismail Gultepe

Abstract Arctic surface temperature has been increasing at a rate 2–3 times that of the global average in the last half century. Enhanced warming of the Arctic, or Arctic Amplification, is a climatic response to external forcing. Despite good results obtained by climatic models for the globe, the largest intermodel differences in surface temperature warming are found in the Arctic. The magnitude of this warming drives many different processes and determines the evolution of many climatic parameters such as clouds, sea ice extent, and land ice sheet mass. The Arctic Amplification can be attributed to the peculiar feedback processes that are triggered in the Arctic. Most of these processes include radiation interaction with the atmosphere and with the surface, all of them contributing to the radiation budget. It is then mandatory to correctly evaluate this budget both at the surface and at the top of the atmosphere and in the solar and thermal spectra. This can be done using both direct observations, from ground and from space, and model simulation via radiation transfer codes. This last approach need many observed input parameters anyhow.

In this contribution results on the evaluation of the radiation budget in the Arctic are first reviewed. Follows a detailed description of the effects of the most important

C. Tomasi · B. H. Petkov · A. Lupi · M. Mazzola (✉)
Institute of Atmospheric Sciences and Climate (ISAC), Italian National Research Council (CNR), Bologna, Italy
e-mail: c.tomasi@isac.cnr.it; b.petkov@isac.cnr.it; a.lupi@isac.cnr.it; m.mazzola@isac.cnr.it

C. Lanconelli
Joint Research Centre, Directorate D – Sustainable Resources, Unit D6 – Knowledge for Sustainable Development and Food Security, European Commission, Ispra (Varese), Italy
e-mail: Christian.LANCONELLI@ec.europa.eu

I. Gultepe
Environment and Climate Change Canada, Meteorological Research Division, Toronto, ON, Canada

Faculty of Engineering and Applied Science, Ontario Technical University, Oshawa, ON, Canada
e-mail: ismail.gultepe@canada.ca

A. Kokhanovsky, C. Tomasi (eds.), *Physics and Chemistry of the Arctic Atmosphere*, Springer Polar Sciences, https://doi.org/10.1007/978-3-030-33566-3_10

atmospheric gases (carbon dioxide, methane, ozone etc.) on both shortwave and longwave radiation ranges. The same is illustrated for aerosol loading in the Arctic, based on a large dataset of aerosol radiative properties measured by means of sun-photometers in numerous Arctic stations. Finally, the effect of the surface reflectivity characteristics on the radiation budget is illustrated by means of albedo models specific for the Arctic.

Keywords Arctic amplification · Radiation budget · Aerosols · Greenhouse gas · Albedo · Feedback processes · Reflectivity · Albedo

10.1 Introductory Remarks on the Global Energy Budget of Earth's Surface-Atmosphere System

The incoming solar radiation is the most important source of energy in the Earth's energy budget, estimated to provide an average flux currently equal to the total solar irradiance TSI = 1361 W m^{-2} at the Top-level of the atmosphere, for yearly mean Earth-Sun distance. The other sources of energy (Moon light, lightning flashes, stellar light, etc.) are very low and, hence, can be totally neglected (Kondratyev 1969). Regular measurements of TSI constant have been conducted by NASA in the seventies (Thekaekara 1973) providing an average estimate of 1353 W m^{-2}, while those of the World Radiation Center (Davos, Switzerland) performed from 1969 to 1980 gave an average value of 1367 W m^{-2} (Fröhlich and Brusa 1981). Two more recent studies significantly changed the previously accepted value of TSI into the current adopted one (Kopp and Lean (2011) and Ohmura (2012)). Considering that the solar radiation beam is intercepted by the Earth, having a cross section equal to πR_E^2 (where R_E is the Earth's radius), and that the Earth's surface is approximately equal to $4\pi R_E^2$, the mean global solar irradiance at the top-level of the atmosphere is currently estimated to be equal to 340.25 W m^{-2} (TSI divided by 4). Ohmura (2012) pointed out that the TSI observations conducted from space during the three last decades showed that the variations of TSI are periodic, being proportional to the well-known 11-year relative sunspot number with an amplitude of 0.45 W m^{-2}.

The annual global mean energy budget of the Earth was estimated by Kiehl and Trenberth (1997), who defined the schematic representation shown in Fig. 10.1. Considering that TSI was estimated over the years from 1960 to 1990 to be close to 1367 W m^{-2}, Kiehl and Trenberth (1997) assumed that the incoming solar radiation flux was equal to 342 W m^{-2}, of which: (i) 67 W m^{-2} are absorbed by the atmosphere; (ii) 77 W m^{-2} are reflected on average by clouds, aerosol particles and atmospheric air molecules toward space; (iii) 30 W m^{-2} are reflected by the surface outside the atmosphere; and (iv) 168 W m^{-2} are absorbed by the terrestrial surface. These estimates imply that: (i) the solar radiation reflected backward by the Earth is equal on average to 107 W m^{-2}, yielding a value of planetary albedo equal to 0.313, and (ii) the Earth's surface emits a long-wave radiation flux equal to 390 W m^{-2} toward the atmosphere, that corresponds to the emission of a black-

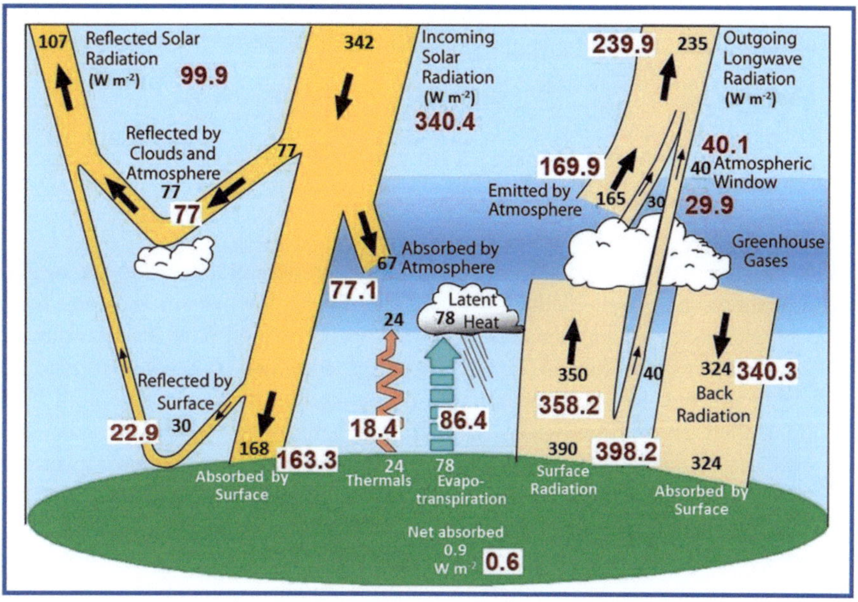

Fig. 10.1 Picture of the Earth's annual global mean energy budget based on the study of Kiehl and Trenberth (1997). The estimates (in W m^{-2}) of the incoming down-welling and up-welling radiant flux densities provided by Kiehl and Trenberth (1997) are written in black numeric characters, while those derived on the basis of a 10-year data-set collected by NASA (2016) are written in red bold numeric characters. (© American Meteorological Society. Used with permission.)

body having the temperature of 15 °C. The major part of this long-wave radiation, equal to 350 W m^{-2} (nearly 90%) is absorbed by the atmosphere, while only 40 W m^{-2} can reach the outer space on average. Thus, the atmosphere emits in turn 324 W m^{-2} toward the surface, where this energy is absorbed (causing the so-called "greenhouse effect"), and emits upward a long-wave contribution of 165 W m^{-2}, to which a further contribution of 30 W m^{-2} has to be added due to cloud emissions. Considering that 40 W m^{-2} have been emitted upward by the surface reaching the outer space, the outgoing long-wave radiation emitted by the surface-atmosphere system toward space was estimated by Kiehl and Trenberth (1997) to be of 235 W m^{-2}. The remaining fluxes capable of establishing the equilibrium conditions of the surface energy budget are those given by the turbulent exchanges of sensible heat and latent heat between the surface and the atmosphere. From conservation of water mass, the latent heat flux is equal to the global mean rate of precipitation, for which Kiehl and Trenberth (1997) estimated that this budget term is of 78 W m^{-2}. The remaining heat flux into the atmosphere from sensible heat was deduced as a residual from the global energy balance conditions at the surface, providing: (i) a net short-wave flux of 168 W m^{-2}, (ii) a net long-wave flux of 66 W m^{-2}, (iii) a latent heat flux of 78 W m^{-2}, and (iv) a sensible heat flux of 24 W m^{-2}.

Figure 10.1 reports also the most recent estimates made by NASA (2016) (printed in red bold characters), being slightly different from the above listed ones.

10.2 Some General Remarks on the Energy Budget of the Arctic Atmosphere

The energy exchange processes occurring on the Arctic surface during the various seasons shows two well distinct regimes: (i) during the winter, the long-wave atmospheric radiation and sensible heat act together as primary energy sources, while the surface radiation emission acts as the predominant sink to give form to the surface energy balance, and sensible heat is supplied on the sea-surface both from the atmosphere and the sea-water; (ii) during the summer, the exchange processes are more active and can be quite different regionally, while the long-wave incoming radiation constitutes the most important source in the energy budget, and the absorbed fraction of solar radiation is the second energy source. In these processes, the energy sinks can vary greatly depending on whether the surface is snow- and ice-covered or free of them (Ohmura 2012). Through examination of the solar global radiation data based on the International Satellite Cloud Climatology Project (ISCCP) collected during summer over an area within 500 km of the North Pole and an additional set of radiation data from the region of the North Pole provided during the French/Swiss Arctic Radiation Balance Experiment (ARBEX 2002), Raschke and Ohmura (2005) found that the mean solar global radiation for the summer season is equal to about 200 W m^{-2}. The mean solar global radiation in the central region of the Arctic Ocean was estimated to vary between 220 and 235 W m^{-2} in the June – August months. The high altitude region of the Greenland ice sheet located at altitudes higher than 2000 m a.m.s.l. shows another large area with high values of incoming global solar radiation flux in summer, exceeding 280 W m^{-2} at altitudes higher than 3000 m a.m.s.l. For instance, the 5-year average value of solar global radiation observed at the Summit GISP2 site (Central Greenland, 3203 m a.m.s.l.) was evaluated to be equal to 305 W m^{-2}. On the basis of a comparison of the solar global radiation measurements conducted over a period of several years at altitudes of 510 m, 1004 m, 1155 m and 1850 m a.m.s.l. along the mid-west slope of the Greenland ice sheet, Ohmura (2012) evaluated that the lower ablation area in summer (where the incoming flux of global solar radiation is estimated to be equal to 240 W m^{-2} on average in summer) is comparable to that observed in the Alps, except that the former is slightly more intense due to multiple reflection between the glacier surface and the sky. In the higher ablation area, at altitudes close to 1160 m a.m.s.l., the global radiation increases at a rate of 8 W m^{-2} for an altitude increase of 100 m, presenting an average value of 290 W m^{-2} at such a level. The accumulation area of the ice sheet was found to exhibit a relatively homogenous distribution of global solar radiation close to ~ 300 W m^{-2}.

Because of the presence of large areas covered by snow and/or ice, the Arctic region exhibits higher albedo characteristics than those commonly observed at mid-latitudes and considered in determining the Earth's energy budget terms shown in Fig. 10.1. The high albedo of the Arctic region plays an important role in defining the energy budget of the surface-atmosphere system during the summer months. The mean albedo of the marginal Arctic seas depends strongly on the existence or absence of sea ice. Examining the observations collected at 38 Russian North Pole drifting stations, Ohmura (2012) found that the mean albedo of the melting sea ice in summer is irrespective of latitude, being close to 0.7 in most cases, in accordance with the observations made by Persson et al. (2002) at the SHEBA tower site and those carried out during the ARBEX 2002 experiment. The mean albedo of open water in the Arctic Sea was estimated by Raschke and Ohmura (2005) to be equal to 0.09 in summer, and, hence, to be somewhat higher than at lower latitudes, primarily because of the systematically lower solar elevation angle. The albedo of the Arctic Ocean including the marginal seas was found to be mainly determined by the ice concentration. The summer albedo in the marginal Arctic sea regions was estimated to range from 0.1 to 0.5, and that measured in the central region of the multi-year ice field around North Pole was estimated to have a mean value of 0.72. On the basis of various measurements carried out over land, the following values of land albedo were determined in the Arctic by Raschke and Ohmura (2005):

(1) in the Greenland ice sheet region, the most frequent values were found to be equal to 0.82 in the accumulation area, 0.75 in the equilibrium-line area, and 0.55 in the ablation area; (2) in the areas covered by glaciers, 0.75 in the accumulation area, 0.65 in the equilibrium line area, 0.35 in the ablation area and 0.21 in the areas with nunataks and moraines; (3) in snow-covered areas, 0.70 for clean (before melt) snow, 0.40 for clean (during melt) snow, and 0.30 for dirty snow; (4) over tundra areas, 0.15; (5) over bog and muskeg areas after snow melt, 0.11; and (6) over lakes after ice melt, 0.08.

These seasonal variations of the surface reflectance characteristics occurring in the Arctic region have been also investigated by Serreze et al. (2007) over the North Polar Cap (i.e. over the Arctic regions with latitude >70 °N) during the insolation months, determining a set of monthly mean values of albedo equal to 0.71 in March, 0.66 in April, 0.64 in May, 0.54 in June, 0.45 in July, 0.48 in August, and 0.55 in September. The Arctic surfaces present a tremendous change in the energy balance and temperature development after the melt, mainly because of the strong variations in albedo causing significant regional climatic variations taking place during the summer months. An important exception in the albedo distribution of the Arctic during summer is the Greenland ice sheet, since more than 90% of the ice sheet belongs to the accumulation area that preserves high albedo during the summer period. The Greenland ice sheet exhibits a large vertical difference in the net radiation as one passes from the ablation area to the dry snow zone in summer, since it ranges from 80 W m^{-2} in the ablation area to 20 W m^{-2} at the equilibrium line, while it reaches values close to 10 W m^{-2} in the dry snow zone. This gradient determines the melt gradient on the ice sheet, and is mainly caused by the altitude variation in atmospheric long-wave radiation, favoured by the albedo change. Because of the high albedo in the accumulation area of the Greenland ice sheet, less than 60 W m^{-2} of solar radiation turns out to be absorbed. This situation is similar to that characterizing the central region of the Arctic Ocean, where the absorption affects about 65 W m^{-2}. Nevertheless, this amount is larger than the mean long-wave net radiation observed in the Arctic in summer, which varies between −30 and −35 W m^{-2} in the coastal region, and between −40 and −50 W m^{-2} in the interior of the Arctic islands and the Greenland ice sheet in summer. Consequently, there is no a negative net radiation in the Arctic in summer. The global summer hemisphere (viewed as a whole) presents a positive value of surface net radiation. The small absorption of solar radiation in the high albedo regions, however, is partially responsible for the negative annual net radiation in the central region of the Arctic Ocean, and the accumulation area of the Greenland ice sheet. Ohmura (2012) estimated that the long-wave incoming radiation is increasing in the Arctic at a rate of 4–5 W m^{-2} per decade.

Atmospheric water vapour (WV) cause important absorption effects in the energy budget of the Arctic surface – atmosphere system, since it absorbs strongly both the incoming solar radiation and the long-wave terrestrial radiation. The annual cycle of precipitable water in the North Polar Cap region was estimated by Serreze et al. (2007) to exhibit relatively low values in January and February of around 0.25 g cm^{-2}, gradually increasing values from about 0.3 g cm^{-2} in March to about 1.0 g cm^{-2} in June, a maximum of 1.3–1.4 g cm^{-2} in July, and then slowly decreasing values in the following months, until reaching the values of 0.4 g cm^{-2} in October and 0.25 g cm^{-2} in December.

In order to provide an exhaustive picture of the seasonal variations of precipitable water W in the Arctic region over the whole year, the time-patterns of the monthly mean values of precipitable water were determined by analyzing the radiosounding measurements conducted at six stations located in the North Polar Cap area at latitudes higher than 70 °N, the complete description of the procedures adopted to calculate W being given in Chap. 3 of this book, in which the analysis of 15-year sets of radiosounding measurements performed at 14 Arctic sites from 2001 to 2015 is presented.

The four stations chosen for calculating W in the Arctic atmosphere over the 2011–2015 five-year period are the following:

(1) Resolute Bay (Nunavut, Canada; 74° 42' N, 94° 58' W, 46 m a.s.l.);
(2) Ny-Ålesund (Spitsbergen, Svalbard; 78° 54' N; 11° 53′ E; 11 m a.s.l.);
(3) Ostrov Dikson (Western Siberia, Russia; 73° 30' N, 80° 24′ E, 47 m a.s.l.); and.
(4) Tiksi (Central Siberia, Russia; 71° 35' N, 128° 55′ E, 7 m a.s.l.).

Each daily value of precipitable water W was derived from one of the radiosounding measurements conducted at one of the four sites listed above, by integrating the vertical profile of absolute humidity from the surface-level to the stratospheric top-level reached by the radiosonde, and furthermore adding to this "tropospheric" value the residual content of stratospheric WV, as calculated by Tomasi et al. (2011). The time-patterns of the daily and monthly mean values of W obtained at the four stations over the five-year 2011–2015 period are shown in Fig. 10.2, which presents

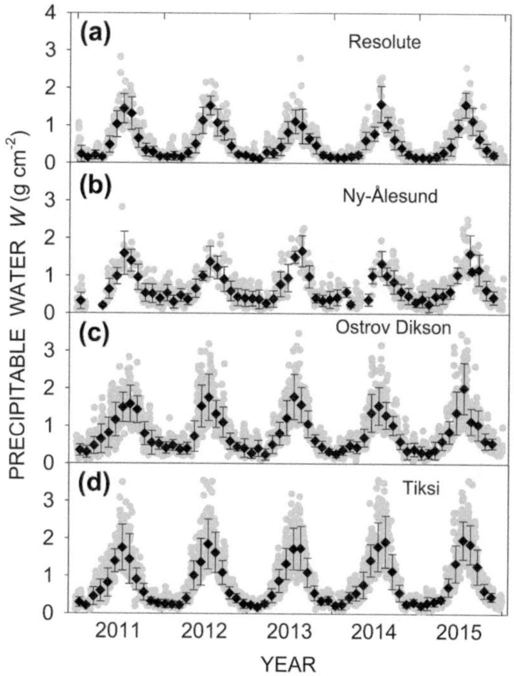

Fig. 10.2 Annual cycles of the daily values (small grey circles) and monthly mean values (solid diamonds) of precipitable water W derived from the multi-year sets of radiosounding measurements performed during the period from January 1, 2011, to December 31, 2015, at the following four Arctic sites: (**a**) Resolute; (**b**) Ny-Ålesund; (**c**) Ostrov Dikson; and (**d**) Tiksi

some examples of the annual cycles of W and gives a measure of the seasonal variations of W at the four above-selected sites of the North Polar Cap area, which agree very well with the evaluations made by Serreze et al. (2007). At Ny-Ålesund, the monthly mean values of W were estimated to be equal to a few tenths of g cm^{-2} in winter, and to increase to about 0.8 g cm^{-2} in April, followed by late spring and summer values exceeding 1 g cm^{-2} and presenting a maximum higher than 1.50 g cm^{-2} in July. Figure 10.2 also shows that W can reach daily mean values considerably higher than 2 g cm^{-2} and in a few cases higher than 3 g cm^{-2} on the warmest days of the year at the other three stations (Resolute, Ostrov Dikson and Tiksi) located at latitudes ranging between 71 °N and 75 °N.

10.3 Estimates of the Large-Scale Energy Budget of the Arctic Atmosphere

A variety of atmospheric and oceanic data were examined by Serreze et al. (2007) to define the large-scale energy budget of the Arctic atmosphere over the North Polar Cap area and the Arctic Ocean domain. This study evaluated the monthly mean values of the atmospheric energy budget terms of the Arctic region by examining the products provided by the European Centre for Medium Range Weather Forecasts

(ECMWF) during the ERA-40 study (Bromwich et al. 2002; Uppala et al. 2005; Su et al. 2006), including comparisons between the ERA-40 products and the results provided by Kalnay et al. (1996), Serreze et al. (1998, 2001, 2003), Cullather et al. (2000), Serreze and Hurst (2000) and Rogers et al. (2001) using the reanalysis conducted at the National Centers for Environmental Prediction/National Center for Atmospheric Research (NCEP/NCAR) (hereinafter more briefly referred to as NRA) in order to carry out an overall evaluation of the atmospheric energy storage and energy transport convergence, focusing on the North Polar Cap. The first step of this study was aimed at defining the annual cycles of the following four monthly mean energy budget terms of the Arctic surface-atmosphere system: (i) the time change (tendency) ($\partial A_E/\partial t$) of atmospheric energy storage A_E in the vertical column; (ii) the convergence of atmospheric energy transport F_A (represented by term $-\nabla \bullet F_A$); (iii) the net radiation at the Top-of-Atmosphere (ToA) level R_{ToA} (positive downward); and (iv) the net heat flux at the Earth's surface F_{sfc} (positive upward), being the first term related to the other three terms through the following relationship,

$$\partial A_E/\partial t = -\nabla \cdot F_A + R_{ToA} + F_{sfc}, \tag{10.1}$$

yielding the energy budget of an atmospheric column extending from the surface to the ToA-level, as obtained by Nakamura and Oort (1988) and Trenberth (1997) studies. Eq. (10.1) states that the atmospheric column gains energy in all cases where the sum of the three terms on the right-hand of the equation is positive, and the column loses energy when the right-hand term of the equation is negative. The following four set of the monthly mean evaluations of the various energy and radiation budget terms made over the year give an overall picture of the energy balance characterizing the Arctic region:

(A) *Annual cycles of the four primary atmospheric budget terms in the Arctic Ocean Domain and the North Polar Cap.*

The annual cycles of the four primary atmospheric budget terms reported in Eq. (10.1) are shown in Fig. 10.3, as obtained from the ERA-40 data-set collected over the Arctic Ocean Domain. These results indicate that: (i) the annual cycle of time change term $\partial A_E/\partial t$ exhibits the highest value (+ 25 W m^{-2}) in April and the lowest value (– 28 W m^{-2}) in September, giving a null annual mean value; (ii) the convergence of atmospheric energy transport term (i.e. of $-\nabla \bullet F_A$), is estimated to maintain rather stable values throughout the year, presenting its minimum (+ 44 W m^{-2}) in May and its maximum (+ 97 W m^{-2}) in October, giving a positive yearly average value equal to 84 W m^{-2}; (iii) the net radiation term R_{ToA} exhibits negative values during the major part of the year, presenting very low values (lower than -175 W m^{-2}) from October to February, and slightly positive values only in June and July (equal to +16 W m^{-2} and + 10 W m^{-2}, respectively), giving a yearly average value equal to -115 W m^{-2}; and (iv) the net heat flux at the surface level, F_{sfc}, exhibits its minimum (– 100 W m^{-2}) in July and a large maximum (+ 59 W m^{-2}) in November and December, with monthly mean values equal to

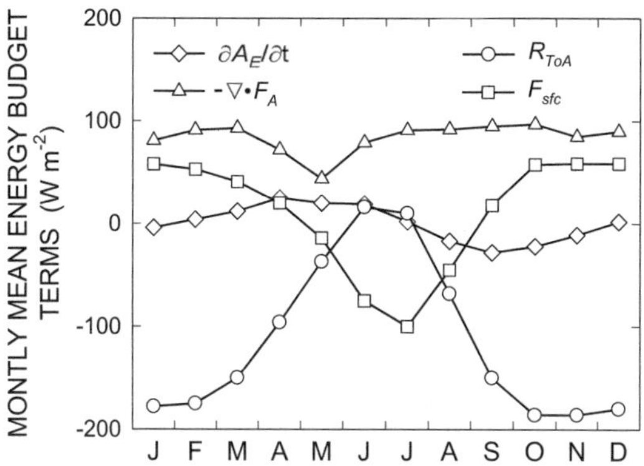

Fig. 10.3 Annual cycles of the monthly means of the four energy budget terms for the surface-atmosphere system, as determined by Serreze et al. (2007) over the Arctic Ocean Domain by examining the data-set derived from the ECMWF ERA-40 products: (1) open diamonds refer to the time change (tendency $\partial A_E/\partial t$) of atmospheric energy storage A_E in the vertical column; (2) open upward triangles refer to the convergence of atmospheric energy transport ($-\nabla \bullet F_A$); (3) open circles refer to the net radiation at the top-level of the atmosphere (R_{ToA}, positive downward); and (4) open squares refer to the net heat flux at the Earth's surface (F_{sfc}, positive upward)

+58 W m^{-2} in October and January, yielding a yearly average value equal to +11 W m^{-2}.

Similar results were derived by Serreze et al. (2007) by examining the ERA-40 data and other data collected over the whole North Polar Cap area (i.e., for latitudes >70 °N, therefore including also important land regions, such as Nunavut (in northern Canada) and Greenland). This analysis has provided: (i) monthly mean values of $\partial A_E/\partial t$ ranging between −27 W m^{-2} (in September) and + 25 W m^{-2} (in April), giving a nearly null annual mean value; (ii) monthly mean values of term $-\nabla \bullet F_A$ ranging between +66 W m^{-2} (in May) and + 114 W m^{-2} (in October), providing a yearly average value equal to +100 W m^{-2}; (iii) monthly mean values of R_{ToA} ranging between −184 W m^{-2} (in November) and + 23 W m^{-2} (in June), yielding a yearly average value equal to −110 W m^{-2}; and (iv) monthly mean values of net heat flux F_{sfc} measured at the surface, which vary between −85 W m^{-2} (in July) and + 58 W m^{-2} (in December), giving a a yearly average value of +11 W m^{-2}.

(B) *Annual cycles of the three radiation terms at the Top-of-Atmosphere level over the North Polar Cap.*

The annual cycles of the three monthly mean energy budget terms R_{ToA} (giving the net radiation flux at the ToA-level), SW_{ToA} (net short-wave radiation at the ToA-level) and LW_{ToA} (net long-wave radiation at the ToA-level) of the Arctic surface-atmosphere system determined over the North Polar Cap circular area are shown in Fig. 10.4, as obtained by Serreze et al. (2007) by examining the ERA-40 products. It can be seen that: (i) monthly mean values of net total radiation R_{ToA} are lower

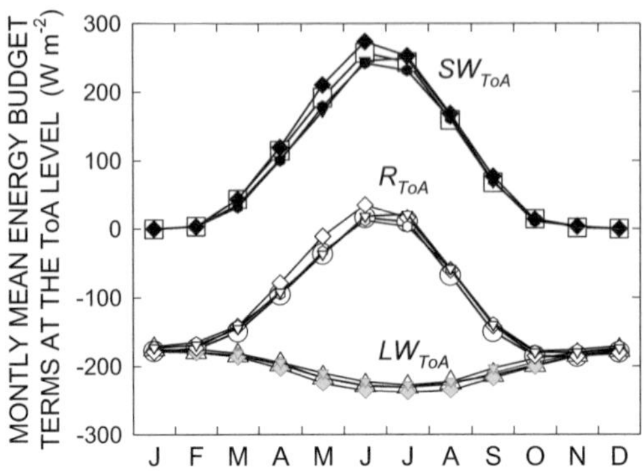

Fig. 10.4 Annual cycles of the three monthly mean energy budget terms of the surface-atmosphere system at the Top-level of the Atmosphere (ToA level) over the North Polar Cap, as determined by Serreze et al. (2007) examining the ECMWF ERA-40 products: (1) open circles refer to the net radiation at the ToA-level (R_{ToA}, positive downward); (2) open squares refer to the net short-wave radiation at the ToA-level (SW_{ToA}, positive downward); and (3) open upward triangles refer to the net long-wave radiation at the ToA-level (LW_{ToA}, positive downward), being R_{ToA} given by the difference between SW_{ToA} and LW_{ToA}). The results determined by Serreze et al. (2007) are compared with those derived by Porter et al. (2010) from: (i) the Japanese 25 Year Reanalysis Project (JRA) (diamonds); (ii) the NCEP/NCAR Reanalysis (NRA) (hexagons); and (iii) the Clouds and the Earth's Radiant Energy System (CERES) (open downward triangles). The R_{ToA} terms defined by Porter et al. (2010) are represented with open symbols, the SW_{ToA} terms with solid symbols, and the LW_{ToA} terms with grey symbols

than -180 W m^{-2} from October to December, increase gradually in the following months to reach a value equal to $+23$ W m^{-2} in June, and then decrease gradually to reach its annual minimum of -184 W m^{-2} in November, consequently providing a yearly average value equal to -110 W m^{-2}; (ii) the annual cycle of SW_{ToA} presents a null value in January and gradually increasing values from February to June, when it reaches its maximum of around 257 W m^{-2}, and then exhibits decreasing values in the following months to become null in December, which yield a yearly average value close to 91 W m^{-2}; and (iii) the annual cycle of LW_{ToA} exhibits the lowest monthly mean value in July (– 231 W m^{-2}) and the highest value in January (– 175 W m^{-2}), providing a a yearly average value of -201 W m^{-2}.

The above results determined by Serreze et al. (2007) are compared in Fig. 10.4 with those derived by Porter et al. (2010) by examining three sets of data collected during the following three projects: (a) the Japanese 25 Year Reanalysis Project (JRA) (Onogi et al. 2007); (b) the NRA (Kalnay et al. 1996); and (c) the Clouds and the Earth's Radiant Energy System (CERES) (Wielicki et al. 1996). Through these calculations, Porter et al. (2010) found that: (i) the monthly mean values of net total radiation R_{ToA} derived from the 5-year JRA data-set exhibit an annual cycle with the lowest value (– 186 W m^{-2}) in November and the highest value (+ 35 W m^{-2}) in June, providing a negative yearly average value equal to -107 W m^{-2}; (ii) the monthly mean values of net short-wave radiation at the ToA-level SW_{ToA} derived from the JRA data-set exhibit an annual cycle with null values in December and

January, and the highest value in June (equal to +273 W m^{-2}), providing a yearly average value close to +97 W m^{-2}; and (iii) the monthly mean values of net long-wave radiation at the ToA-level LW_{ToA} derived from the JRA data-set exhibit an annual cycle with its minimum of −238 W m^{-2} in June and the maximum of −179 W m^{-2} in January, providing a yearly average value of about −206 W m^{-2}.

The results derived by Porter et al. (2010) examining the NRA were found to closely agree with those derived by Serreze et al. (2007), as can be noted in Fig. 10.4. On these results, it can be pointed out that: (i) the monthly mean values of net total radiation R_{ToA} derived from the NRA data-set presents an annual cycle with the lowest value (− 178 W m^{-2}) measured in October and the highest value (+ 12 W m^{-2}) in June, giving a a yearly average value equal to −109 W m^{-2}; (ii) the monthly mean values of net short-wave radiation SW_{ToA} derived from the NRA data-set exhibit an annual cycle with null values in December and January, and the highest value in June (equal to +242 W m^{-2}), providing a yearly average value equal to +86 W m^{-2}; and (iii) the monthly mean values of net long-wave radiation LW_{ToA} derived from the NRA data-set describe an annual cycle with a minimum in July (equal to −231 W m^{-2}) and its maximum in February (equal to −175 W m^{-2}), yielding a yearly average value of nearly −200 W m^{-2}.

The results derived by Porter et al. (2010) through the analysis of the CERES data collected over the North Polar Cap from 2000 to 2005 were found to satisfactorily agree with the results obtained by Serreze et al. (2007), as shown in Fig. 10.4. In fact: (i) the monthly mean values of net total radiation R_{ToA} derived from the CERES data shows an annual cycle with the minimum (− 180 W m^{-2}) in both October and November, and the maximum (+ 21 W m^{-2}) in July, providing a yearly average value of −109 W m^{-2}; (ii) the monthly mean values of net short-wave radiation SW_{ToA} derived from the CERES data exhibit an annual cycle with null values in December and January, and the maximum higher than +248 W m^{-2} in July, giving a yearly average value of +88 W m^{-2}; and (iii) the monthly mean values of net long-wave radiation LW_{ToA} derived from the CERES data-set describe an annual cycle with a minimum lower than −227 W m^{-2} in July and a maximum of −175 W m^{-2} in both December and February, yielding a a yearly average value slightly lower than −196 W m^{-2}.

(C) *Annual cycles of the energy budget terms over the North Polar Cap area.*

The annual cycles of the three monthly mean energy budget terms concerning the convergence of atmospheric energy transport (− $\nabla \bullet F_A$), the latent heat energy, and the dry static energy are shown in Fig. 10.5, as estimated by Serreze et al. (2007) examining the ERA-40 products obtained over the North Polar Cap area. The results provided by this analysis indicate that: (i) the convergence of atmospheric energy transport term $-\nabla \bullet F_A$ is rather stable during the whole year, ranging from +66 W m^{-2} in May to about −114 W m^{-2} in October, and providing a yearly average value which results to be slightly higher than +100 W m^{-2}; (ii) the latent heat energy term exhibits rather low values throughout the whole year, ranging from +9 W m^{-2} in January to nearly +24 W m^{-2} in August, and yielding a yearly

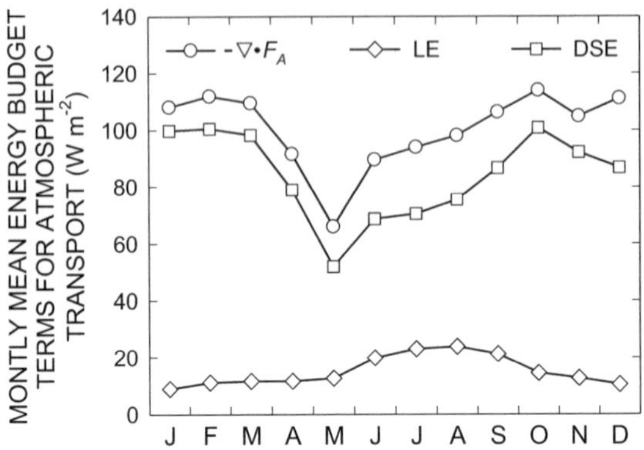

Fig. 10.5 Annual cycles of the three monthly mean energy budget terms defining the atmospheric transport in the Arctic surface-atmosphere system over the North Polar Cap, as determined by Serreze et al. (2007) examining the data derived from the ECMWF ERA-40 products: (1) open circles refer to the convergence of atmospheric energy transport ($-\nabla \bullet F_A$); (2) open diamonds refer to the latent heat energy; and (3) open squares refer to the dry static energy

average value close to +16 W m^{-2}; and (iii) the dry static energy term follows the time-patterns of term $-\nabla \bullet F_A$ throughout the year, ranging from +52 W m^{-2} in May to slightly more than +100 W m^{-2} in both February and November, thus giving a yearly average value of about +85 W m^{-2}.

(D) *Annual cycles of the five energy budget terms at the surface level over the North Polar Cap area.*

The annual cycles of the five monthly mean surface-level energy budget terms R_{sfc}, SW_{sfc}, LW_{sfc}, Q_H and Q_E of the Arctic surface-atmosphere system over the North Polar Cap area are shown in the upper part of Fig. 10.6, as obtained by Serreze et al. (2007) analyzing the ERA-40 products. The time-patterns of the monthly mean values of surface-level net radiation R_{sfc} describe a wide minimum in the summer months, presenting a monthly mean value lower than −73 W m^{-2} in July, and rather high values in the winter months, ranging between +60 W m^{-2} in December and about +56 W m^{-2} in both January and November; giving a a yearly average value of about +13 W m^{-2}. Correspondingly, the surface-level net short-wave radiation SW_{sfc} was estimated to describe a large minimum throughout the year, with a monthly mean value lower than −125 W m^{-2} in June, and nearly null values in January and December, yielding the yearly average value of about −43 W m^{-2}. The surface-level net long-wave radiation LW_{sfc} was estimated by Serreze et al. (2007) to maintain positive and rather stable monthly mean values throughout the year, presenting a minimum slightly lower than +28 W m^{-2} in July and values ranging between nearly +46 W m^{-2} in December and around +48 W m^{-2} in January–March, which provide a yearly average value equal to about +41 W m^{-2}. The turbulent sensible heat flux Q_H and the latent heat flux Q_E (both calculated at the surface level) were found to be rather stable throughout the year: (i) the term Q_H

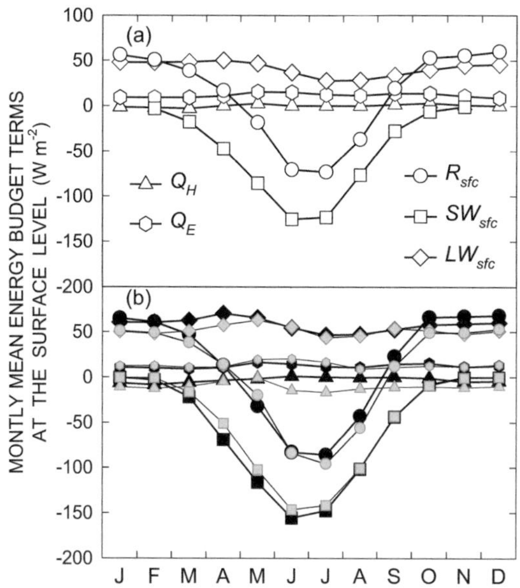

Fig. 10.6 Uppert part (**a**): Annual cycles of the five monthly mean energy budget terms of the over the North Polar Cap, as determined by Serreze et al. (2007) at the surface level by examining the data-set derived from the ECMWF ERA-40 products: (1) open circles refer to the surface-level net radiation R_{sfc} (positive downward); (2) open squares refer to the net short-wave radiation SW_{sfc} (positive downward) at the surface-level; (3) open diamonds refer to the net long-wave radiation LW_{sfs} (positive downward) at the surface-level; (4) open upward triangles refer to the turbulent sensible heat flux Q_H at the surface-level; and (5) open hexagons refer to the latent heat flux Q_E at the surface-level. Lower part (**b**): Annual cycles of the five monthly mean energy budget terms R_{sfc} (circles), SW_{sfc} (squares), LW_{sfs} (diamonds), Q_H (upward triangles) and Q_E (hexagons), as determined at the surface-level by Porter et al. (2010) for the Arctic surface-atmosphere system over the North Polar Cap by examining: (i) the JRA data (solid symbols), and (ii) the NRA data (grey symbols)

was estimated to vary between -3 W m^{-2} in March and + 3 W m^{-2} in October, giving a nearly null yearly average value, and (ii) the latent heat flux term Q_E was estimated to range between +9 W m^{-2} in both January and February and more than +15 W m^{-2} in May, and to provide a yearly average value slightly higher than +11 W m^{-2}.

The above results are compared with those derived by Porter et al. (2010) examining the surface-level data collected during the above-mentioned JRA and NRA projects, which are shown in the lower part of Fig. 10.6. Analysing the JRA data, Porter et al. (2010) found that: (i) the surface-level net radiation R_{sfc} was estimated to assume negative values from May to August, reaching a minimum of -86 W m^{-2} in July and a maximum of +68 W m^{-2} in December, yielding a yearly average value equal to +14 W m^{-2}, which is higher than the ERA-40 value determined by Serreze et al. (2007) by 1.5 W m^{-2} (i.e. by 12%); (ii) the surface-level net short-wave radiation SW_{sfc} was found to describe a large minimum during the summer months, presenting a monthly mean value lower than -155 W m^{-2} in June and nearly null values in both January and December, from which a yearly average value equal to -55.6 W m^{-2} was obtained, which is lower than the ERA-40 value by 12.8 W m^{-2} (i.e. by about 30%); (iii) the surface-level net long-wave radiation LW_{sfc} was evaluated to assume stable and positive values throughout the year, with monthly mean values higher than 70 W m^{-2} in April and lower than 47 W m^{-2} in July, yielding a yearly average value equal to $+58.0$ W m^{-2}, which is higher than the ERA-40 value by 16.6 W m^{-2} (i.e. by 40%); (iv) the turbulent sensible heat flux Q_H at the surface level was estimated by Porter et al. (2010) to range between less than -8 W m^{-2} in February and more than $+1$ W m^{-2} in June, giving a yearly average value of -3 W m^{-2}, found to be appreciably lower than the ERA-40 value determined by Serreze et al. (2007) by 3.1 W m^{-2}; and (v) the latent heat flux term Q_E at the surface level was evaluated to vary between $+9$ W m^{-2} in March and more than $+16$ W m^{-2} in May, giving a yearly average value of $+12.3$ W m^{-2}, which is higher than the ERA-40 value by 0.7 W m^{-2} only.

The results obtained by Porter et al. (2010) analyzing the surface-level NRA data-set were found to substantially agree with those derived from the JRA products (in the same study) and presented in the lower part of Fig. 10.6. In fact, it was evaluated that: (i) the monthly mean values of the surface-level net radiation R_{sfc} ranged between -96 W m^{-2} in July and + 53 W m^{-2} in December, providing a yearly average value equal to $+6.0$ W m^{-2}, which is lower than the JRA value by 8 W m^{-2}; (ii) the surface-level net short-wave radiation SW_{sfc} was found to vary between the null value in December (and a nearly null value in January) and a value lower than -146 W m^{-2} in June, thus yielding a yearly average value equal to -51.1 W m^{-2}, which results to be higher than the JRA value by 4.5 W m^{-2} only; (iii) the surface-level net long-wave radiation LW_{sfc} was estimated to maintain stable values throughout the whole year, which vary between less than 44 W m^{-2} in July and more than 62 W m^{-2} in May, for which a yearly average value equal to $+49.0$ W m^{-2} was obtained, which is lower than the JRA value by 9 W m^{-2}; (iv) the turbulent sensible heat flux Q_H measured at the surface-level was evaluated to vary between less than -16 W m^{-2} in July and less than -1 W m^{-2} in May, providing a yearly average value equal to -10.5 W m^{-2}, which is lower than the JRA value by 7.5 W m^{-2}; and (v) the latent heat flux term Q_E at the surface level was estimated to vary between $+9$ W m^{-2} in August and more than $+20$ W m^{-2} in June, yielding a yearly average value equal to $+13.5$ W m^{-2}, which is higher than the JRA value by 1.2 W m^{-2} only.

In spite of the above evaluations, considerable uncertainties still remain on the atmospheric energy budget of the Arctic atgmosphere. The differences between the monthly mean budget terms derived from the JRA and NRA data taken over the North Polar Cap were found to be as large as 23 W m^{-2}, the disparity in the annual net surface heat flux being of 9 W m^{-2}, which is equivalent to about 0.9 m of sea ice loss per year at its melting point (Serreze et al. 2007). Examining the ERA-40 products, Serreze et al. (2007) pointed out that the seasonal cycle of atmospheric energy storage is strongly modulated by the net surface flux, which is also the primary driver of seasonal changes in heat storage within the Arctic Ocean. Averaging these data over the whole Arctic Ocean domain, they also found that the July net surface flux from ERA-40 of −100 W m^{-2} (into the ocean), associated with sea ice melt and oceanic sensible heat gain, is higher than the atmospheric energy transport convergence by 91 W m^{-2}. During winter, the oceanic sensible heat loss and the sea ice growth yield an upward surface flux of 50–60 W m^{-2}, complemented with an atmospheric energy convergence of 80–90 W m^{-2} to provide a net radiation loss to space of 175–180 W m^{-2}. These findings indicate that the net surface flux has first-order impacts on the atmospheric energy budget of North Polar Cap and on that of the Arctic Ocean Domain. Because horizontal oceanic heat flux convergence and sea ice transport out of the Arctic Ocean area via Fram Strait (between Greenland and Svalbard) are fairly small, the net surface flux is the primary driver of seasonal changes in ocean heat storage. From autumn to winter, the atmosphere cools in response to the declining solar flux, and the net surface flux becomes largely positive because of sea ice growth and sensible heat loss from the ocean, adding a significant heat amount to the atmosphere. The incoming solar radiation flux increases in the spring and first summer months, so that the atmosphere gains energy. For the Arctic Ocean domain, the July net flux measured at the surface-level and derived from ERA-40 data is larger than the atmospheric transport. However, the net radiation flux at the surface-level results to be strongly negative, primarily as a result of sea ice melt and replenishment of the ocean's reservoir of sensible heat. Analyzing the evolutionary patterns of the extreme September 2005 sea ice minimum, Porter et al. (2010) obtained results that suggest a strong role of albedo feedback, in which anomalous ocean heat gain through June and July in open water areas led to further ice melt, with this ocean heat then fostering strong ice melt through August despite the fairly low solar flux. In addition, several recent studies analyzing trends and variability in the Arctic sea ice (Francis et al. 2005; Stroeve et al. 2007) suggest that the role of increased downwelling long-wave flux has become more important in recent years.

The Arctic Ocean sea ice-coverage and Northern Hemispheric snow-coverage have decreased dramatically in the last few decades (Parkinson et al. 1999). The satellite data recorded during the Earth Radiation Budget Experiment (ERBE) provided global measurements for 1985 through 1988 (Barkstrom and Smith 1986) and those collected through the analysis of the data taken with the CERES instruments from 2000 through 2004 (Wielicki et al. 1996, 1998) have shown that the effect of melting ice on the Arctic Ocean has increased the net radiation of the region from 72.5 °N to the North Pole from 2 (± 3) W m^{-2} to 9 (± 3) W m^{-2} during the 1985–2004 period, as indicated by Smith and Bess (see at http://citeseerx.ist.psu.edu/viewdoc/summary?doi=10.1.1.597.8918). The major part of this increase arises from the gradual decrease of albedo, as the ice cover is replaced by the dark ocean surface, consequently causing an increase in the solar radiation absorption. The net radiation flux increases by 0.43 W m^{-2} for every 1% decrease in ice cover (relative to the ice cover observed in 1979), only a small part of this increase of net radiation being due to changes in the outgoing long-wave radiation. Examining the above-mentioned ERBE and CERES observations made in July, when the solar declination is still high and snow and ice have had sufficient time to melt, Bess and Smith (see at https://ams.confex.com/ams/pdfpapers/102125.pdf) evaluated that the effect of melting ice on the Arctic Ocean occurring in summer has caused an average lengthening in the melt season by 10 to 17 days per decade. The maps derived from the ERBE and CERES observations quantify the ice-feedback effect on radiation, which includes also the cloud effects on the albedo characteristics at the top-level of the atmosphere, and show in particular that the solar radiation absorption was significantly decreased over northern Canada and western Russia during the last decades.

10.4 The Role of Solar Radiation in the Radiation Budget of the Arctic Atmosphere

The results obtained by Serreze et al. (2007) and those achieved by Porter et al. (2010) (which are both shown in Fig. 10.6) clearly indicate that the highest monthly mean absolute values of the net short-wave radiation reaching the surface are usually recorded in June and July over the North Polar Cap, while null monthly mean values of this quantity are recorded in January and December, in the absence of incoming solar radiation. During the rest of the year, the incoming direct solar radiation is subject to time-variable attenuation processes, due to scattering associated with air density fluctuations (Rayleigh-scattering) and aerosol particles (such as water droplets, dust particles, forest fire smokes, haze droplets) and absorption by numerous atmospheric gases, whose intensities vary with pressure and temperature conditions of the various absorbing constituents of the atmosphere.

The scattering and absorption processes produced by aerosol particles in the Arctic atmosphere are usually not intense. Examining numerous sets of direct solar irradiance reaching the surface-level at various Arctic sites, Tomasi et al. (2015) have evaluated aerosol optical thickness $AOT(0.50\ \mu m)$ values usually ranging between 0.03 and no more than 0.14, and estimates of the Ångström (1964) wavelength exponent α mainly ranging between 1.0 and 1.8, as it is typical of continental particles.

Important atmospheric absorption effects on the incoming solar irradiance are caused by numerous atmospheric constituents, such as water vapour (WV), ozone (O_3), and carbon dioxide (CO_2), whereas the atmospheric absorption produced by the main atmospheric constituents of the atmosphere (such as nitrogen and oxygen molecules) are not of primary relevance in the atmospheric radiation budget. Since the absorption of direct solar irradiance by a gaseous atmospheric species varies as a function of its molecular concentration along the sun-path and depends closely on the air pressure and temperature conditions, the calculations of the direct solar irradiance absorption characteristics have been made in the present analysis for different solar elevation angles and, hence, for different values of relative optical air mass *m* ranging from 2 to 8, with the principal purpose of evaluating the absorption effects produced at Ny-Ålesund. Less intense absorption features are produced by minor gases, such as: (i) nitrogen dioxide (NO_2) and its dimer N_2O_4, which presents a semicontinuum over the $0.30 < \lambda < 0.55$ μm range (Hall and Blacet 1952; Shaw 1976); (ii) SO_2, which exhibits a band at wavelengths $\lambda < 0.35$ μm; (iii) NOCl, ClOH and HNO_4, which show rather weak absorption bands in the UV-A spectral range; and (iv) a moderate absorption band of NO_3 characterized by semicontinuous features over the $0.40 < \lambda < 0.70$ μm range. All the absorption effects produced by these bands are rather weak and were not considered in the present study, in which only the absorption effects produced by WV, O_3 and CO_2 on the solar radiation are discussed.

10.4.1 Short-Wave Solar Radiation Absorption by Atmospheric Water Vapour

The spectrum of direct solar irradiance reaching the Earth's surface exhibits numerous absorption bands of WV over the range from 0.54 μm to more than 3.50 μm, characterized by intensities that gradually increase from the visible to the infrared. Four very weak bands of WV occupy the 0.54–0.68 μm range, of which: (i) the first band consists of a main transition at $\lambda = 0.5437$ μm, with intensity equal to 2 10^{-23} cm at 259 °K; (ii) the second band consists of a main transition only located at $\lambda = 0.5716$ μm, with intensity equal to 1 10^{-22} cm; (iii) the third band consists of three main transitions centered at the wavelengths $\lambda = 0.59175$ μm, 0.59179 μm and 0.59446 μm, with intensities equal to 3 10^{-22} cm, 3 10^{-23} cm, and 2 10^{-22} cm, respectively; and (iv) the fourth band consists of two main transitions centred at wavelengths $\lambda = 0.6316$ μm and $\lambda = 0.6516$ μm, with intensities equal to 2 10^{-23} cm and 2 10^{-22} cm, respectively. At longer wavelengths, a number of eight bands occupy the near-infrared spectral range, being commonly identified by Greek letters. They are:

(1) The band labelled with the Greek letter α occupies the range from 0.70 to 0.74 μm and exhibits its maximum at $\lambda = 0.718$ μm. It consists of five main transitions centred at the wavelengths equal to 0.6984, 0.7032, 0.7230, 0.7232, and 0.7324 μm, with intensities equal to 1 10^{-21} cm, 1 10^{-22} cm, 3 10^{-21} cm, nearly 2 10^{-23} cm, and 6 10^{-21} cm, respectively, at 259 °K.
(2) The band usually called the "0.8 μm band", which occupies the 0.796–0.847 μm range and exhibits its maximum at $\lambda = 0.810$ μm, consisting of five main transitions centred at wavelengths equal to 0.7959, 0.8059, 0.8230, 0.8237, and 0.8465 μm, with intensities equal to 1 10^{-22} cm, 6 10^{-21} cm, 6 10^{-23} cm, 1 10^{-22} cm, and 2 10^{-21} cm, respectively.

(3) The three bands (labelled with the Greek letters ρ (rho), σ (sigma) and τ (tau)) cover the 0.870–0.990 μm range, and presents their maximum at $\lambda = 0.935$ μm: (i) band ρ consists of two main transitions centred at wavelengths equal to 0.9065 μm and 0.9200 μm, with intensities equal to 2 10^{-21} cm and 4 10^{-22} cm, respectively; (ii) band σ consists of two main transitions centred at wavelengths equal to 0.9422 μm and 0.9434 μm, with intensities equal to 1 10^{-20} cm and 6 10^{-22} cm, respectively; and (iii) band τ consists of three main transitions centred at wavelengths equal to 0.9681 μm, 0.9724 μm and 1.0169 μm, with intensities equal to 2 10^{-21} cm, less than 4 10^{-23} cm and 6 10^{-23} cm, respectively.

(4) The band labelled with capital Greek letter Φ (phi) occupies the spectral range from 1.08 μm to 1.20 μm, with its maximum at $\lambda = 1.130$ μm. It consists of five main transitions centred at the wavelengths of 1.1111 μm, 1.1355 μm, 1.1413 μm, 1.1942 μm, and 1.2086 μm, with intensities equal to 3 10^{-22} cm, 8 10^{-21} cm, 1 10^{-23} cm, 7 10^{-24} cm, and 3 10^{-23} cm, respectively.

(5) The band labelled with capital Greek letter Ψ (psi), occupies the range from 1.25 μm to 1.54 μm and exhibits its maximum at $\lambda = 1.395$ μm. It consists of five main transitions centred at the wavelengths of 1.3432 μm, 1.3793 μm, 1.3887 μm, 1.4554 μm, and 1.4760 μm, with intensities equal to 1 10^{-21} cm, 1.5 10^{-19} cm, 1.5 10^{-20} cm, 1 10^{-20} cm, and 2 10^{-22} cm, respectively.

(6) The band labelled with capital Greek letter Ω (omega) occupies the spectral range from 1.69 μm to 2.08 μm, with its maximum at $\lambda = 1.870$ μm. It consists of two main transitions centred at wavelengths of 1.8758 μm and 1.9102 μm, with intensities equal to 2.2 10^{-19} cm and 7 10^{-21} cm, respectively, followed by a separated less intense transition centred at $\lambda = 2.1427$ μm and having intensity equal to 3 10^{-22} cm.

(7) The band labelled with capital Greek letter X (chi) covers the range from 2.27 μm to 2.99 μm, and exhibits its maximum at $\lambda = 2.68$ μm. It consists of two main transitions ν_2 and ν_1 centred at the 2.6625 μm and 2.7344 μm wavelengths, respectively, presenting intensities equal to 6.2 10^{-18} cm (ν_2) and 3 10^{-19} cm (ν_1).

(8) The band labelled with capital Greek letter X' covers the range from 2.99 μm to 3.57 μm, and presents its maximum at $\lambda = 3.20$ μm, being given by the main transition $2\nu_2$, centred at $\lambda = 3.1730$ μm, and having an intensity of 6.5 10^{-20} cm (Goldberg 1954; Benedict et al. 1956).

Less important absorption bands are produced by the HDO molecule in the infrared solar spectrum. This molecule presents a sequence of absorption bands centred at wavelengths of 1.1613 μm (with intensity equal to 5 10^{-24} cm), 1.5499 μm (with intensity equal to 2 10^{-24} cm), 1.5587 μm (with intensity equal to 1 10^{-23} cm), 1.8644 μm (with intensity of 3 10^{-23} cm), 1.9648 μm (with intensity of 4.6 10^{-23} cm), 2.4122 μm (with intensity of 2 10^{-23} cm), 2.4390 μm (with intensity of 2 10^{-23} cm), 2.6973 μm (with intensity of 4.4 10^{-22} cm, 3.5943 μm (with intensity of 2.3 10^{-23} cm), and 3.6715 μm (with intensity of 2.8 10^{-22} cm), all the intensities being calculated at the 259 K temperature (Benedict et al. 1956).

The above-mentioned WV bands cause very strong absorption effects on the incoming direct solar irradiance radiation at the surface-level. In order to calculate these WV absorption effects, we have determined the spectral curve of direct solar irradiance reaching the surface at the Ny-Ålesund station on a cloudless-sky day for a relative optical air mass $m = 2$ (i.e. apparent solar zenith angle $\theta_o = 60.1°$). In these calculations we have assumed the following thermodynamic and chemical composition features of the atmosphere (hereinafter called "Ny-Ålesund-spring" atmospheric model):

(a) The vertical profiles of air pressure *p(z)* (measured in hPa), air temperature *T(z)* (measured in °K), relative humidity *RH(z)* (%) and absolute humidity *q(z)* (g m^{-3}) determined by Tomasi and Petkov (2014) by averaging the 4-year data sets of radiosounding measurements of *p(z)*, *T(z)* and *RH(z)* performed by AWI (Alfred Wegener Institute, Bremerhaven, Germany) at the joint French-German AWIPEV Arctic Research Base during the spring months from 2000 to 2003. Figure 10.7 shows the vertical profiles of *T(z)*, *p(z)/T(z)*, *RH(z)*, and *q(z)* defined by Tomasi and Petkov (2014) These vertical profiles of *p(z)*, *T(z)*, and *q(z)* were completed over the height range $z > 12$ km with the monthly mean vertical profiles of these four meteorological parameters determined from 12 to 60 km altitude by Tomasi et al. (2011). The estimate of atmospheric WV content $W = 0.8$ g cm^{-2} was obtained from the radiosounding measurements conducted in April and integrated with the satellite-borne MIPAS/ENVISAT measurements of stratospheric WV content.

(b) The seasonal average spectral values of aerosol optical thickness *AOT(λ)* were calculated at Ny-Ålesund by assuming the average values of *AOT(0.50 μm)* = 0.055 and Ångström's wavelength exponent $\alpha = 1.35$, as determined by Tomasi et al. (2015) at Ny-Ålesund by analyzing the sun-photometer measurements carried out from 2000 to 2013 by the AWI group during the spring and summer months.

(c) The spectral values of Rayleigh-scattering optical thickness *ROT(λ)* were calculated at Ny-Ålesund according to the evaluations made by Tomasi et al. (2010) at this Spitsbergen site, providing values of *ROT(λ)* equal to 2.715 at $\lambda = 0.25$ μm, 1.434 10^{-1} at $\lambda = 0.50$ μm, 8.39 10^{-3} at $\lambda = 1.00$ μm, 5.35 10^{-3} at $\lambda = 2.00$ μm, and 3.34 10^{-4} at $\lambda = 4.00$ μm.

(d) The vertical profile of O_3 molecular number density was assumed according to the Tomasi and Petkov (2014) evaluations made over the altitude range from surface-level to 120 km altitude., and carefully checked by comparison with: (i) a set of balloon-borne ozonesonde measurements performed at Ny-Ålesund during the local summer months from 1978 to 1994 (Rex et al. 2000), (ii) the local Brewer measurements performed at the Dirigibile Italia station (Damiani et al. 2012), and (iii) the historical data yielding the total O_3 content in the Spitsbergen area (Vogler et al. 2006). The vertical profile of O_3 molecular number density $N_{O3}(z)$ shown in Fig. 10.7 was estimated to give a value of total O_3 content equal to 340 Dobson Units (D. U.), in agreement with the

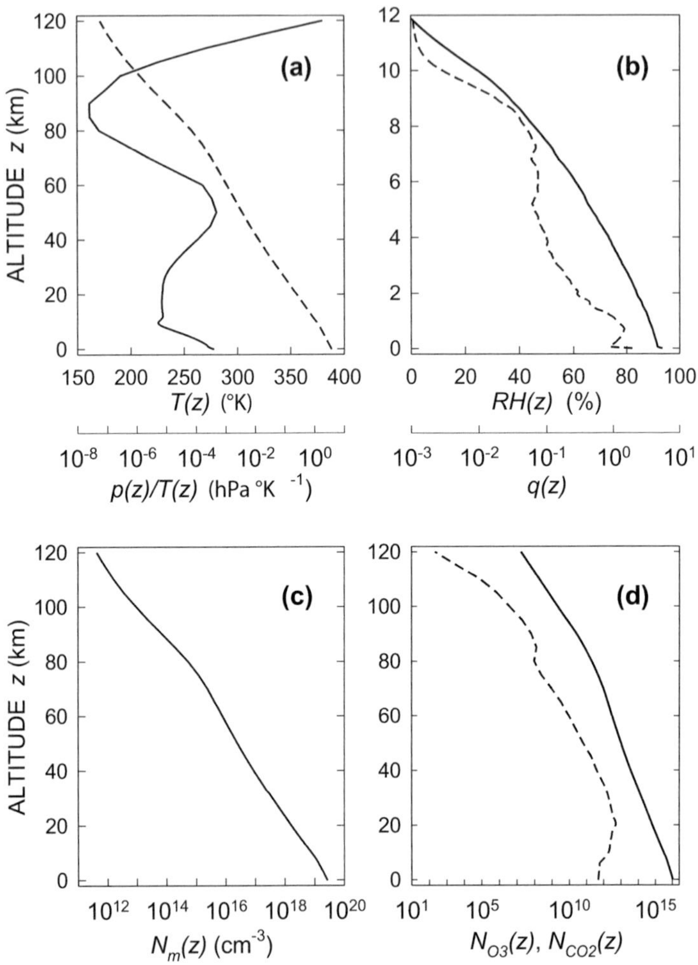

Fig. 10.7 Average vertical profiles of the meteorological and compositional parameters in the "Ny-Ålesund-spring" atmospheric model: (**a**) air temperature $T(z)$ (solid curve), and ratio $p(z)/T(z)$ (dashed curve); (**b**) relative humidity $RH(z)$ (dashed curve) and absolute humidity $q(z)$ (solid curve); (**c**) overall molecular number density $N_m(z)$; and (d) molecular number densities of ozone $N_{O3}(z)$ (dashed curve), and carbon dioxide $N_{CO2}(z)$ (solid curve)

Brewer spectrophotometer measurements recently carried out by the Italian CNR groups at Ny-Ålesund during the spring and early summer days.

(e) The vertical profile of CO_2 molecular number density $N_{CO2}(z)$ shown in Fig. 10.7 was determined by taking into account the assumptions made by Tomasi et al. (2010) on the tropospheric vertical distribution of CO_2 molecular concentration, which was assumed to maintain a stable value from the surface-level up to about 75 km altitude in the Arctic atmosphere under consideration, and then decrease gradually at upper levels until reaching a value of 40 ppmv at the 120 km height.

The spectral curves of short-wave direct solar irradiance reaching the surface were calculated for the characteristics of the Arctic atmosphere represented in terms of the above-defined "Ny-Ålesund-spring" atmospheric model by using the SBDART code (Ricchiazzi et al. 1998), for the values of relative optical air mass m ranging from 2 to 8 (i.e. for apparent solar zenith angle θ_o increasing from 60.1° to 83.2°). Three curves were first defined in the present analysis: (1) the extra-terrestrial spectral curve of direct solar irradiance $I_o(\lambda)$ defined over the 0.25–4.00 μm range; (2) the spectral curve of global (direct + diffuse) solar irradiance $I_{src}(\lambda)$ reaching the surface, as obtained for the "Ny-Ålesund-spring" atmospheric model; and (3) the spectral curve of global solar irradiance $I_{W=0}(\lambda)$ calculated at the surface-level for the "Ny-Ålesund-spring" atmospheric model, in which it was assumed that the atmospheric content of WV is null. The three curves are shown in Fig. 10.8a to provide evidence of the main absorption effects caused by the most intense WV bands and by the other atmospheric constituents. The comparison between the three curves give evidence of the high number of WV bands causing the considerable absorption features affecting the incoming solar radiation during its passage through the summer Arctic atmosphere, in spite of the relatively low values of precipitable water W measured in the summer months.

The function $\Delta I(m)$ was calculated as a function of relative optical air mass m in terms of the difference between the integrals of the spectral curves of extra-terrestrial solar irradiance $I_o(\lambda)$, and global solar irradiance $I_{srf}(\lambda)$ reaching the surface-level, both these integrals being determined over the whole short-wave range from 0.25 to 4.00 μm. Function $\Delta I(m)$ provides a measure of the attenuation effects produced on the incoming global solar irradiance by the various atmospheric constituents, through Rayleigh scattering, aerosol extinction, and absorption by WV, O_3, CO_2 and other minor gases. In Fig. 10.8a, the function $\Delta I(m)$ was calculated for relative optical air mass $m = 2$ and estimated to be equal to 167.42 W m^{-2}. Analogous calculations of $\Delta I(m)$ were also made for the values of $m = 3$ (i.e. for apparent solar zenith angles $\theta_o \approx 70.71°$), $m = 4$ ($\theta_o \approx 75.7°$), $m = 5$ ($\theta_o \approx 78.5°$), $m = 6$ ($\theta_o \approx 80.8°$), $m = 7$ ($\theta_o \approx 82.1°$) and $m = 8$ ($\theta_o \approx 83.2°$), obtaining a set of values of $\Delta I(m)$ showing that such a function gradually decreases as a function of m, as can be seen in Fig. 10.9a, until reaching a value 79.57 W m^{-2} for $m = 8$, which is equal to 47.5% of $\Delta I(m = 2)$.

The global solar irradiance function $D_{WV}(m)$ was calculated for different values of relative optical air mass m, this function being given by the difference between the integrals of the spectral curves of irradiance functions $I_{W=0}(\lambda)$ and $I_{srf}(\lambda)$ shown in Fig. 10.8a. Therefore, the global solar irradiance function $D_{WV}(m)$ provides a measure of the absorption effects produced by atmospheric WV on the direct and diffuse solar radiation reaching the surface-level, and, hence, of the overall absorption forcing produced by WV on the incoming global solar irradiance. Function $D_{WV}(m)$ was evaluated to be equal to 74.75 W m^{-2} for $m = 2$ ($\theta_o = 60.1°$) (and, hence, equal to 44.7% of the overall variation produced by the

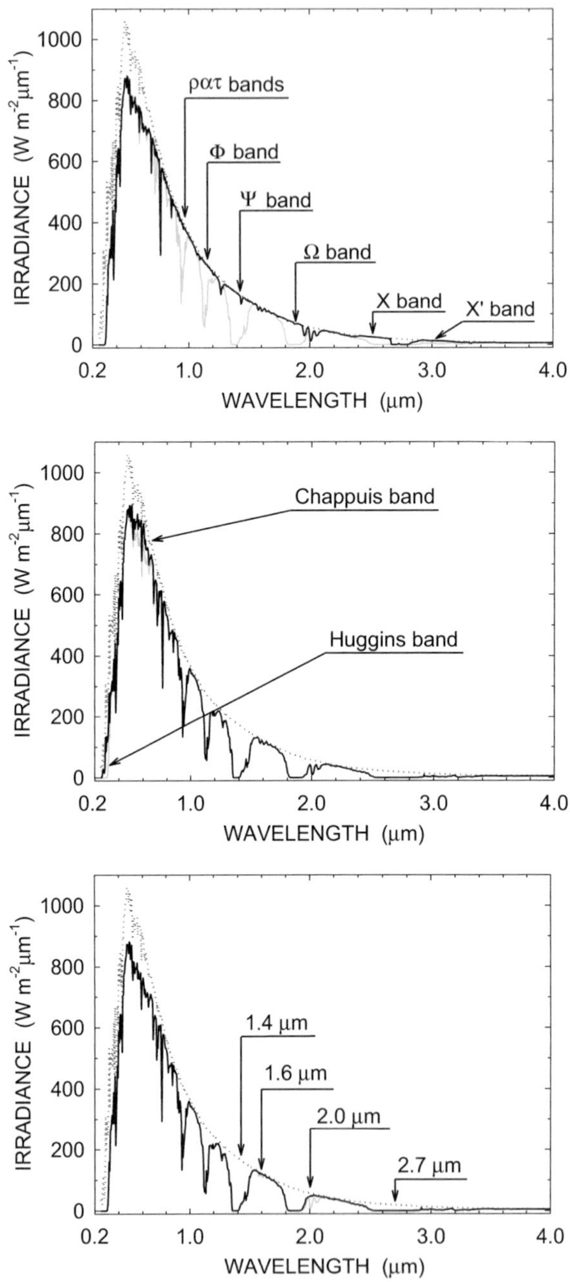

Fig. 10.8 Panel (**a**): spectral curves (calculated for relative optical air mass $m = 2$) of (i) extra-terrestrial solar irradiance $I_o(\lambda)$ (dotted curve); (ii) short-wave global solar irradiance $I_{srf}(\lambda)$ reaching the surface (grey solid curve), as calculated by using the SBDART code for the meteorological and chemical composition characteristics of the "Ny-Ålesund-spring" atmospheric model;

whole atmosphere on the incoming solar radiation). It gradually decreases as air mass m increases from 2 to 8 (as shown in Fig. 10.9a) until reaching the value of 23.25 W m^{-2} at $m = 8$. The comparison of function $D_{WV}(m)$ with $\Delta I(m)$ indicates that $D_{WV}(m)$ decreases gradually as parameter m increases, until becoming equal to 29.2% of the global solar irradiance reaching the surface for such a low solar elevation angle. Therefore, the results presented in Fig. 10.9a indicate that the strong absorption effects produced by atmospheric WV in the Arctic atmosphere during the summer months can vary between 44.7% (for $m = 2$) and 29.2% of the overall extinction effects produced by Rayleigh scattering, aerosol extinction and gaseous absorption.

10.4.2 Short-Wave Solar Radiation Absorption by Atmospheric Ozone

The overlapping of continuous absorption effects by oxygen and nitrogen with the above-mentioned absorption bands of oxygen, nitrogen and O_3 has the effect of totally preventing the incoming ultra-violet solar radiation of higher frequency from reaching the Earth's surface. The atmospheric attenuation spectrum of solar radiation in the visible spectral range is characterized by the absorption effects produced by the following weak absorption bands of minor gases: (1) two molecular oxygen bands, centred at the 0.6684 μm and 0.7621 μm wavelengths; (2) the Chappuis band produced by molecular O_3 and covering the spectral interval from 0.45 μm to 0.74 μm wavelength; (3) the absorption band due to nitrogen dioxide NO_2 and its dimer N_2O_4, characterized by a semi-continuum occupying the range from 0.25 μm to about 0.58 μm; and (4) a few weak WV absorption bands centred at the 0.54, 0.57, 0.59 and 0.64 μm wavelengths. Therefore the O_3 molecules causes very strong absorption effects on the incoming solar radiation over the whole $0.2 \leq \lambda \leq 0.3$ μm range, followed by gradually less intense absorption features

Fig. 10.8 (continued) and (iii) short-wave global solar irradiance $I_{W=0}(\lambda)$ reaching the surface-level (black solid curve), for the same atmospheric model but with a null content of WV. The arrows indicate the positions of the principal absorption bands of WV. Due to their weakness, the four WV bands occupying the 0.54–0.68 μm range, the WV band labelled with the Greek letter α (occupying the 0.70–0.74 μm range) and the so-called "0.8 μm band" are not quoted in the graph Panel (**b**): as in panel (**a**) but with the black solid curve representing the short-wave global solar irradiance $I_{O3=0}(\lambda)$ reaching the surface in the case of a null content of O_3. The arrows indicate the spectral positions of the most important absorption bands of O_3 in the ultra-violet (Huggins band over the $\lambda < 0.35$ μm range) and visible (Chappuis band over the $0.45 \leq \lambda \leq 0.74$ μm range) Panel (c): as in panel (a) but with the black solid curve representing the short-wave global solar irradiance $I_{CD=0}(\lambda)$ at the surface-level in the case of a null atmospheric content of CO_2. The arrows indicate the spectral positions of four CO_2 absorption bands

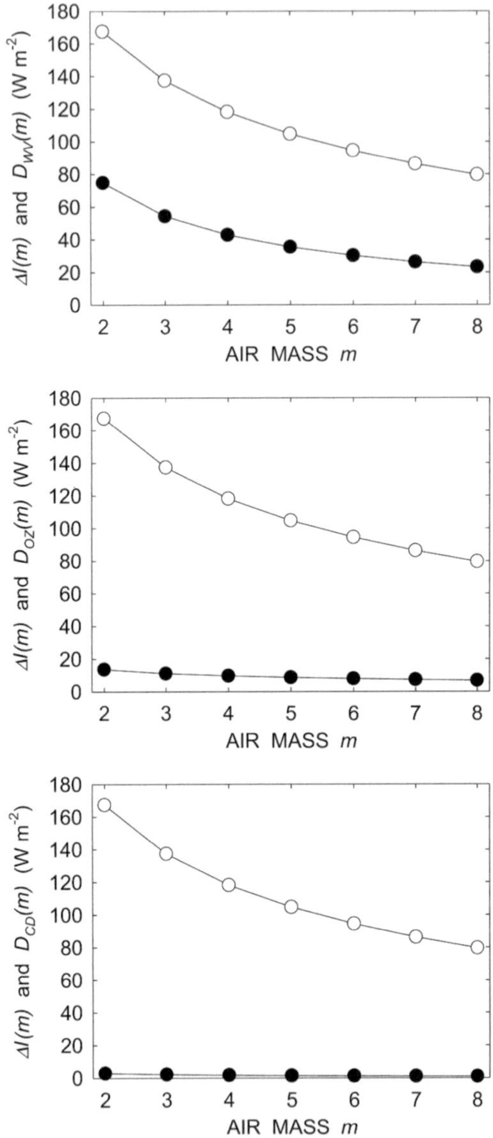

Fig. 10.9 Panel (**a**): scatter plots over the range of relative optical air mass m from 2 to 8 of the global solar irradiance function $\Delta I(m)$ (open circles) calculated as the difference between the short-wave integrals (both calculated over the 0.25–4.00 μm range) of the extra-terrestrial solar irradiance $I_o(\lambda)$ and short-wave global solar irradiance $I_{srf}(\lambda)$ reaching the surface-level, and the global solar irradiance function $D_{WV}(m)$ (solid circles) calculated as the difference between the integrals of the spectral curve of short-wave global solar irradiance $I_{W=0}(\lambda)$ reaching the surface-level determined for a null content of atmospheric WV in the "Ny-Ålesund-spring" atmosphere and the short-wave global solar irradiance $I_{srf}(\lambda)$ reaching the surface-level
Panel (**b**): as in panel (**a**) but with the solid squares representing the global solar irradiance function $D_{OZ}(m)$ calculated as the difference between the integrals of the short-wave global

at wavelengths increasing from 0.3 μm to about 0.35 μm, but they produce also relatively weak absorption effects in the visible and near-infrared (characterized by nearly continuous features over the $0.45 \le \lambda \le 0.74$ μm range), as well as in the infrared where a set of very weak absorption bands centered at wavelengths equal to 2.70, 3.28, 3.57, 4.75 and 5.75 μm are responsible for rather weak absorption effects on the incoming solar radiation.

Although the above-mentioned O_3 absorption bands are evaluated to cause a weaker impact on the short-wave radiation budget of the Earth than that produced by atmospheric WV, a similar study to that conducted in Fig. 10.8a for atmospheric WV was also carried out for atmospheric O_3. As can be seen in Fig. 10.8b, the triplet of the spectral curves of global solar irradiance were determined by assuming that the vertical atmospheric content of O_3 is equal to 340 D. U. on average during the whole year. Following a similar procedure to that adopted for WV, the function $D_{OZ}(m)$ was calculated for various entire values of relative optical air mass m, obtaining a value of $D_{OZ}(m = 2)$ equal to 13.66 W m^{-2} and decreasing values of $D_{OZ}(m)$ over the range of m from 2 to 8, until presenting a value of 6.83 W m^{-2} for $m = 8$. The estimates of $D_{OZ}(m)$ obtained for m increasing from 2 to 8 are compared in Fig. 10.9b with the corresponding evaluations of $\Delta I(m)$ made for all the attenuation effects produced by the various atmospheric constituents. This comparison indicates that $D_{OZ}(m)$ is equal to 8.2% of $\Delta I(m)$ for $m = 2$ and slowly decreases as m increases until becoming equal to nearly 8.6% of $\Delta I(m)$ for $m = 8$. The present calculations indicate also that O_3 causes overall absorption effects on the incoming solar radiation, which are equal to 18.3% of those produced by WV at $m = 2$ and equal to 29.4% of those due to WV at $m = 8$.

10.4.3 Short-Wave Radiation Absorption by Atmospheric Carbon Dioxide

The absorption bands of carbon dioxide exhibit different intensity characteristics over the solar spectrum, being present in the infrared with absorption maxima located at about 1.4, 1.6, 2.0, and 2.7 μm wavelengths. These four bands are due to groups of transitions presenting the following principal wavelengths and intensities: (i) the 1.4 μm band is due to one transition only, centered at $\lambda = 1.4341$ μm, with intensity equal to 8.6 10^{-22} cm; (ii) the 1.6 μm band consists of four main

Fig. 10.9 (continued) solar irradiance $I_{O3=0}(\lambda)$ reaching the surface-level and determined for a null atmospheric content of O_3 and the short-wave global solar irradiance $I_{srf}(\lambda)$ reaching the surface-level
Panel (**c**): as in panel (**a**) but with the solid squares representing the global solar irradiance function $D_{CD}(m)$ calculated as the difference between the integrals of the short-wave global solar irradiance $I_{CD=0}(\lambda)$ reaching the surface-level and determined for a null atmospheric content of CO_2 and the short-wave global solar irradiance $I_{srf}(\lambda)$ reaching the surface-level

transitions centred at wavelengths equal to 1.5378 μm, 1.5748 μm, 1.6057 μm (with intensity equal to 2.9 10^{-22} cm), and 1.6458 μm; (iii) the 2.0 μm band consists of six main transitions centred at wavelengths equal to 1.9486 μm, 1.9608 μm (with intensity equal to 1.6 10^{-20} cm), 2.0088 μm (with intensity equal to 3.7 10^{-20} cm), 2.0141 μm, 2.0606 μm (with intensity equal to 1.0 10^{-20} cm); and (iv) the 2.7 μm band consists of four main transitions centred at wavelengths equal to 2.6860 μm, 2.6921 μm (with intensity equal to 1.3 10^{-18} cm), 2.7678 μm (with intensity equal to 1.0 10^{-18} cm), and 2.7927 μm (Goldberg 1954; France and Dickey 1955; Migeotte et al. 1957).

Following the same procedure adopted above, the effect of the CO_2 molecules were evaluated. The results are shown in Fig. 10.8c, in which the comparison between the spectral curves of $I_{srf}(\lambda)$ and $I_{CO2\,=\,0}(\lambda)$ allow one to appreciate which are the relatively weak absorption effects of CO_2 on the incoming global solar irradiance, produced within the rather narrow spectral intervals occupied by the 1.4, 1.6, 2.0 and 2.7 μm bands of CO_2. The evaluations of functions $\Delta I(m)$ and $D_{CD}(m)$ made for all the entire values of m ranging from 2 to 8 are shown in Fig. 10.9c to give a measure of the attenuation effects produced by CO_2 absorption on the incoming global solar radiation reaching the surface. The comparison shown in Fig. 10.9c clearly indicates that the absorption effects due to CO_2 are rather weak, since $D_{CD}(m)$ was estimated to be equal to 2.92 W m^{-2} and, hence, to 1.7% of $\Delta I(m)$ for $m = 2$, and 0.95 W m^{-2} and, hence, to 1.2% of $\Delta I(m)$ for $m = 8$. These results clearly indicate that the overall absorption effects produced by CO_2 in the Arctic atmosphere on a typical summer day are rather low, since they result to be equal to only 3.9% of those caused by WV at $m = 2$ and equal to 4.1% of those produced by WV at $m = 8$.

A first consideration seems to be necessary in order to correctly compare the above evaluations of the absorption effects produced by WV, O_3 and CO_2 on the incoming global solar irradiance (and shown in Fig. 10.9a for WV, Fig. 10.9b for O_3, and Fig. 10.9c for CO_2) with the fraction of solar radiation absorbed by the atmosphere in the Earth's annual global mean energy budget defined by Kiehl and Trenberth (1997) and shown in Fig. 10.1. The evaluations of the radiative forcing terms derived from the three above-mentioned figures have to be reduced to one fourth of their values when they are compared with the Earth's annual global mean energy budget terms determined by Kiehl and Trenberth (1997), since the evaluations of the radiative forcing produced by WV, O_3 and CO_2 and shown in Fig. 10.9, have been calculated over the whole Earth's surface (approximately equal to $4\pi\ R_E^2$, being R_E the Earth's radius) and the Kiehl and Trenberth (1997) scenario refers to the Earth's cross-section approximately equal to $\pi\ R_E^2$. In addition, it is important to take into account that the calculations of the radiative forcing effects caused by WV, O_3 and CO_2 absorption effects on the incoming global short-wave solar radiation and expressed in terms of variations in the global solar irradiance function $\Delta I(m)$ and in the absorption functions $D_{WV}(m)$, $D_{OZ}(m)$, and $D_{CD}(m)$ have been made for the meteorological and chemical composition characteristics of the

Table 10.1 Daily average values of (i) incoming global short-wave solar irradiance I_o estimated outside the terrestrial atmosphere, (ii) global short-wave solar irradiance I_{srf} reaching the surface-level, and (iii) relative optical air mass m, determined on six selected days of 2017 at Ny-Ålesund, as calculated by using the SBDART code for the "Ny-Ålesund-spring" atmospheric model and the astronomical coordinates of the Sun varying from one day to another

	Daily average values		
Date	Global solar irradiance I_o (W m^{-2})	Global solar irradiance I_{srf} (W m^{-2})	Relative optical air mass m
1 April 2017	293.2	184.4	4.4
1 May 2017	422.4	292.5	3.1
1 June 2017	503.9	361.5	2.8
1 July 2017	523.4	378.0	2.7
1 August 2017	417.6	289.9	3.2
1 September 2017	347.5	228.6	3.7

"Ny-Ålesund-spring" atmospheric model, without considering that the short-wave global solar irradiance is subject to vary throughout the year since the time-period of surface insolation varies largely throughout the year, depending on the geographical coordinates of each Arctic site and the day of the year, from late winter to early autumn. To give an idea of these variations, Fig. 10.10 shows a set of time-patterns of incoming short-wave global solar irradiance I_{srf} reaching the surface, as evaluated at the Ny-Ålesund station on six selected days from April first to September first, 2017, together with the corresponding variations of the relative optical air mass m varying as a function of apparent solar zenith angle. It can be noted in Fig. 10.10 that I_{srf} increases gradually from April first to early July and then decreases slowly during the late summer and early autumn. The daily average values of I_{srf} have been evaluated in Table 10.1 to increase from about 185 W m^{-2} on April first, 2017, to reach the value of 293 W m^{-2} in early May, and further increase to 362 W m^{-2} in early June and 378 W m^{-2} in early July, whereas the corresponding daily average values of m decrease gradually from 4.4 on April first to 2.7 on July 1st. The corresponding variations of the absorption effects produced by WV, O_3 and CO_2 can be evaluated on the basis of the evaluations shown in Fig. 10.9, by taking into account the daily average values of m recorded on the various days selected above (and reported in Table 10.1). For instance, considering the values of m recorded on April first and July first, it can be noted that: (i) on April first, the daily average value of m is equal to 4.4, for which the WV absorption is responsible for an absorption effect estimated in Fig. 10.9a to be of around 40 W m^{-2}, and (ii) on July first, 2017, the daily average value of m is equal to 2.7, for which the WV absorption turns out to slightly exceed the value of 60 W m^{-2}, as shown in Fig. 10.9b.

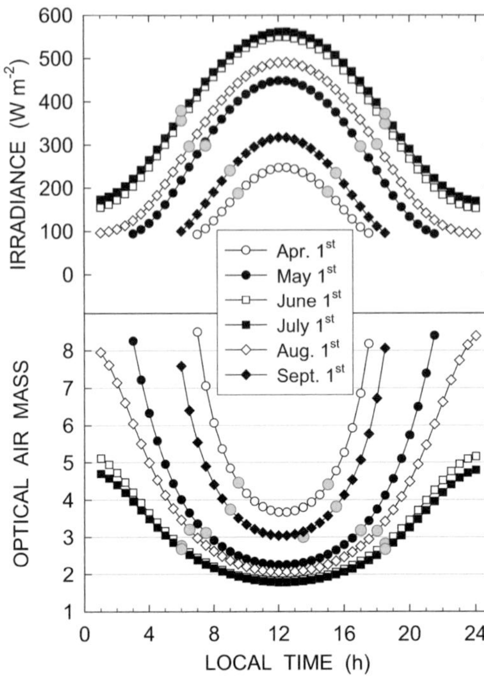

Fig. 10.10 Upper part: Diurnal variations of the incoming global short-wave solar irradiance I_{srf} at the surface calculated at Ny-Ålesund by using the SBDART code for the following days of year 2017: (**a**) April 1st (open circles), (**b**) May 1st (solid circles), (**c**) June 1st (open squares), (**d**) July 1st (solid squares), (**e**) August 1st (open diamonds), and (**f**) September 1st (solid diamonds). Lower part: Time-patterns of relative optical air mass m as a function of the local time calculated on the same days considered in the upper part. The grey symbols inserted into all the curves of both upper and lower graphs mark the values of irradiance I_{srf}, which best approximate the daily average values recorded on the six selected days (in the upper part) (see Table 10.1), and the corresponding values of relative optical air mass m (in the lower part) (Table 10.1), respectively

10.5 Absorption and Emission of Long-Wave Radiation by the Greenhouse Gases in the Arctic Atmosphere

The spectral distribution curve of the radiative energy emitted by the Earth-atmosphere system toward space approximates very well that of a black body having a temperature of 250 °K, for which the wavelength of maximum emission is close to 11 μm, as suggested by the Wien's displacement law. Taking into account that the Earth's surface-atmosphere system behaves in a similar manner to that of a grey body with average emittance ε close to 0.96, about 99% of the energy emitted by our planet results to be distributed over the 3–80 μm spectral range. This implies that the absorption by the atmospheric gases appreciably affects both the incoming solar radiation and the upwelling and down-welling fluxes of long-wave radiation observed in the terrestrial surface-atmosphere system to occur over the range $\lambda > 4$ μm. At these infrared wavelengths, the long-wave radiation is subject to intense absorption effects by some minor atmospheric gases, such as WV, CO_2 and O_3, which are responsible for important changes in the long-wave radiation balance of the surface-atmosphere system, favouring the atmospheric warming. The long-wave radiative forcing effects caused by atmospheric WV and the other most important greenhouse gases (such as CO_2, O_3, methane (CH_4) and nitrous oxide (N_2O)) are examined in the following sub-sections with the purpose of evaluating the intensities of such warming processes.

10.5.1 Water Vapour Absorption Effects on the Long-Wave Radiation Balance Terms of the Arctic Atmosphere

In the infrared range (i. e. for $\lambda > 4$ μm), two relatively weak transitions are produced by the WV at wavelengths $\lambda = 4.6272$ μm (with intensity of 2 10^{-22} cm) and $\lambda = 4.8490$ μm (with intensity of 8 10^{-23} cm at 259 °K). At longer wavelengths, the very strong vibro-rotational band (ν_2) of WV occupies the 4.88–8.70 μm range, presenting its maximum at $\lambda = 6.27$ μm, this band consisting of two important transition bands centred at $\lambda = 6.2705$ μm (with intensity of 8.3 10^{-18} cm), and 6.4232 μm (with intensity of 9 10^{-22} cm at 259 °K) (Benedict 1948; Goldberg 1954). Beside this band, there is also the HDO band centred at $\lambda = 7.1261$ μm (with intensity of 2.8 10^{-22} cm). Additional bands are due to N_2O (with transitions centred at 7.7821 μm, 8.569 μm and 16.9837 μm wavelengths) and CH_4 (with the ν_4 transition centred at the 7.6570 μm wavelength), which therefore are both important greenhouse gases to be considered in the present study because of their relevant selective absorption effects. Together with these absorption features, the vibro-rotational band ν_2 of WV defines the spectral features of the left-hand side of the 8–13 μm atmospheric window, in which the atmospheric absorption spectrum exhibits in general high transparency features and allows the large part of the terrestrial radiation emitted upward to reach the outer space. In fact, the only efficient absorption characteristics appearing within such a large atmospheric window from 8 μm to approximately 13 μm are associated with: (i) the so-called WV continuum, which has been investigated in the '70s by several authors (Bignell 1970; Burch 1970; Grassl 1974; and Tomasi et al. 1974), and (ii) the typical spectral curve of the O_3 absorption band peaked at 9.6 μm (Gerhard 1932). Therefore, the terrestrial radiation belonging to the 8–13 μm range is affected not only by selective absorption due to intense absorption lines (O_3, WV N_2O, and CH_4) but also by considerably intense wing effects due to the more distant rotational bands of WV, occupying the spectral range from 15 to more than 50 μm. These very strong absorption effects affect both the long-wave radiation emitted upward by the surface-atmosphere system and the long-wave radiation emitted downward by the terrestrial atmosphere and reaching the surface, which both contribute efficiently to regulate the atmospheric warming by the "greenhouse" gases.

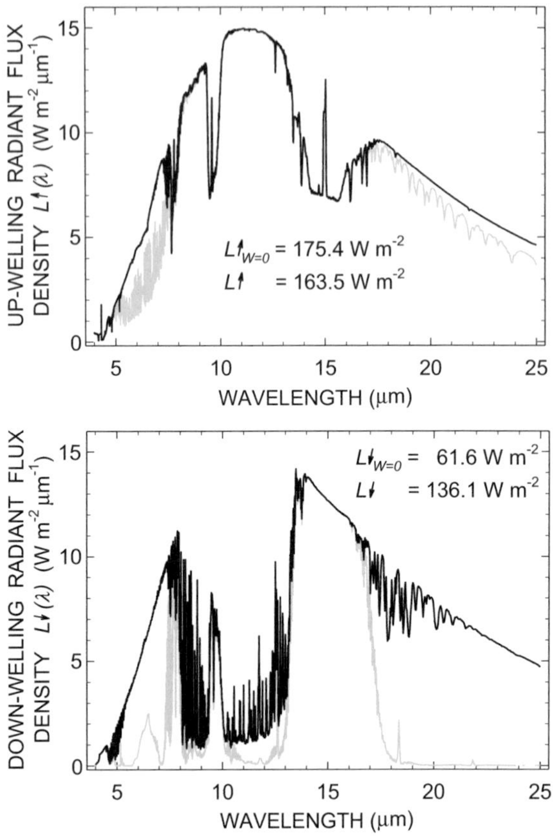

Fig. 10.11 Panel (**a**): comparison between (i) the spectral curve of up-welling radiant flux density $L\uparrow(\lambda)$ (black solid curve) of the long-wave radiation emitted upward by the Arctic surface-atmosphere system presenting cloudless-sky conditions, as calculated by using the MODTRAN 4.0 code for an atmosphere represented in terms of the "Ny-Ålesund-spring" atmospheric model having a WV content $W = 0.8$ g cm^{-2}, and (ii) the spectral curve of up-welling radiant flux density $L\uparrow(\lambda)_{W=0}$ (grey solid curve) of the long-wave radiation emitted by the "Ny-Ålesund-spring" atmospheric model assumed to have a null content of WV. Panel (**b**): as in panel (**a**) but for the down-welling radiation $L\downarrow(\lambda)$

Calculations of the up-welling radiant flux density $L\uparrow$ of the long-wave radiation emitted by the surface-atmosphere system toward space, and the down-welling radiant flux density $L\downarrow$ of the long-wave radiation emitted by the atmosphere toward the surface have been made in the present study by applying the MODTRAN 4.0 computational programme (Berk et al. 1999) to the "Ny-Ålesund-spring" atmospheric model already considered above and with: (i) volume concentration of CH_4 molecules equal to 1940.5 ppbv, and (ii) volume concentration of N_2O equal to 330 pptv, both on the basis of the measurements conducted at the Zeppelin Observatory.

The two spectral curves of (i) up-welling radiant flux density $L\uparrow(\lambda)$ of the long-wave radiation emitted by the surface-atmosphere system toward space, and (ii) down-welling radiant flux density $L\downarrow(\lambda)$ of the long-wave radiation emitted by the atmosphere toward the surface are shown in Fig. 10.11. The spectral curve of $L\uparrow(\lambda)$ is compared in Fig. 10.11a with that of up-welling radiant flux density $L\uparrow(\lambda)_{W=0}$ calculated for the same Arctic atmosphere model assumed in this case to have a null content of WV. It can be noted that the two spectral curves differ considerably over the range from 5 to about 9 μm wavelength, where strong absorption features are produced by the strong vibro-rotational band ν_2 of WV, and the whole range from 15 to 25 μm, because of the closely packed sequence of WV absorption band features as a result of the wing effects of the WV rotational bands occupying the whole range from 15 to more than 50 μm. The spectral curves of $L\uparrow(\lambda)$ and $L\uparrow(\lambda)_{W=0}$ shown in Fig. 10.11a were integrated over the 4–25 μm range to obtain the total values of up-welling radiant flux densities $L\uparrow_{W=0} = 175.4$ W m^{-2} and $L\uparrow = 163.5$ W m^{-2}, for which the difference $\Delta L\uparrow_{WV} = L\uparrow(\lambda)_{W=0} - L\uparrow(\lambda)$ due to WV absorption was estimated to be equal to 11.9 W m^{-2}. This estimate suggests that WV effects involve the 6.78% of the total up-welling radiant flux density $L\uparrow$.

As mentioned above, the spectral curve of down-welling radiant flux density $L\downarrow(\lambda)$ shown in Fig. 10.11b was calculated by applying the MODTRAN 4.0 code to the "Ny-Ålesund-spring" atmospheric model (with the contents of WV, CO_2, O_3, CH_4 and N_2O defined above). The spectral curve of $L\downarrow(\lambda)$ is compared in Fig. 10.11b with the spectral curve of down-welling radiant flux density $L\downarrow(\lambda)_{W=0}$ determined in this particular application to have a null content of WV. The comparison provides evidence of the strong variations produced by atmospheric WV absorption over the following three wavelength intervals from 5 to about 9 μm, because of strong effects due to the vibro-rotational band ν_2 of WV; within the atmospheric window from 8 to 13 μm, in which the absorption features of WV contribute efficiently to the emission of long-wave radiation toward the surface; and over the range from 15 to 25 μm, on the right-hand side of the 8–13 μm atmospheric window, where large and intense absorption effects are produced by the extended series of WV rotational bands, providing a very important contribution from 18 to 25 μm. The spectral curves of long-wave radiation flux densities $L\downarrow(\lambda)$ and $L\downarrow(\lambda)_{W=0}$ shown in Fig. 10.11b were integrated over the 4–25 μm range to determine the total values of down-welling radiant flux densities $L\downarrow$ and $L\downarrow_{W=0}$. Following such a procedure, the values of $L\downarrow = 136.1$ W m^{-2} and $L\downarrow_{W=0} = 61.6$ W m^{-2} were obtained, from which the difference $\Delta L\downarrow_{WV} = L\downarrow - L\downarrow_{W=0}$ was calculated, obtaining the value of $\Delta L\downarrow_{WV} = 74.5$ W m^{-2}, which is considerably greater than the variation in the upwelling long-wave radiation $\Delta L\uparrow_{WV}$ caused by atmospheric WV absorption and estimated to be of 11.9 W m^{-2} in the Arctic atmosphere (as shown in Fig. 10.11a). The very high value of $\Delta L\downarrow_{WV}$ determined in Fig. 10.11b indicates that atmospheric WV causes in general a strong absorption of long-wave radiation occurring in the Arctic atmosphere during summer, which is equivalent to a relative increase of 54.7% in the total greenhouse effect affecting the down-welling radiant flux density $L\downarrow$ emitted by the atmosphere toward the surface. Taking into account that higher values of precipitable water W are usually measured on the summer days at all the Arctic sites, the above evaluations of $\Delta L\downarrow_{WV}$ are expected to be twice the value of 74.5 W m^{-2} determined above for $W = 0.8$ g cm^{-2} and be higher than 150 W m^{-2} on the warmest summer days for values of W ranging between 1.6 and 1.8 g cm^{-2}, while values of $\Delta L\uparrow_{WV}$ higher than 25 W m^{-2} should be recorded on the same warmest days of summer.

10.5.2 Carbon Dioxide Absorption Effects on the Long-Wave Radiation Balance Terms in the Arctic Atmosphere

Three strong absorption bands are produced by carbon dioxide within the 4–7 μm range, centred at the wavelengths $\lambda = 4.3$ μm, $\lambda = 4.8$ μm and $\lambda = 5.2$ μm, respectively (Goldberg 1954; France and Dickey 1955; Malkmus 1963; Migeotte et al. 1957): of these bands, the 4.3 μm band (ν_3) has only one important transition centred at $\lambda = 4.2568$ μm, the 4.8 μm band ($\nu_1 + \nu_2$) consists of three main transitions centred at $\lambda = 4.6953$ μm (with intensity equal to 8.0 10^{-22} cm), $\lambda = 4.7770$ μm (with intensity equal to 2.0 10^{-21} cm), and $\lambda = 4.8150$ μm (with intensity equal to 3.2 10^{-20} cm); and the third band ($3\nu_2$) having its maximum at $\lambda = 5.2$ μm originates from only one important transition centred at $\lambda = 5.1748$ μm (with intensity equal to 2.4 10^{-21} cm). In addition, CO_2 exhibits also two relatively moderate absorption bands centred at the 9.4 μm and 10.4 μm wavelengths. However, in spite of their weakness, the absorption effects produced by these bands cannot be totally neglected since the two bands are located in the middle of the 8–13 μm atmospheric window, where the thermal radiation emitted upward by the surface-atmosphere system presents very high effects on the long-wave radiation budget of the atmosphere. They are located on both sides of the 9.6 μm ozone absorption band: (i) the relatively weak absorption band having its maximum at $\lambda = 9.4$ μm has only one important transition centred at $\lambda = 9.4003$ μm (with intensity equal to 8.6 10^{-22} cm); and (ii) the weak 10.4 μm absorption band has similarly only one important transition centred at $\lambda = 10.4058$ μm (with intensity equal to 2.7 10^{-21} cm) (Kratz et al. 1991).

On the right-hand side of the 8–13 μm atmospheric window, CO_2 exhibits a very important band for the radiative budget of the Earth, which is commonly called the 14.8 μm band. It consists of six main transitions centred at the following wavelengths: (i) 13.8729 μm (with intensity equal to 2 10^{-19} cm) (Kostkowski and Kaplan 1957); (ii) 14.9754 μm (with intensity equal to 6 10^{-19} cm), (iii) 14.9835 μm (with intensity equal to 8 10^{-18} cm), (iv) 15.0968 μm (with intensity equal to 4 10^{-20} cm), (v) 15.4197 μm (with intensity equal to 8 10^{-20} cm), and (vi) 16.1804 μm (with intensity equal to 2 10^{-19} cm) (Kaplan and Eggers 1956; Yamamoto and Sasamori 1958). For these strong absorption features, the 14.8 μm CO_2 band covers with its effects the wide range from 12.5 to 18.2 μm, defining in practice the spectral features of the right-hand side of the 8–13 μm atmospheric window. Again, all the above reported intensities refer to the temperature of 259 K.

Calculations of the up-welling and down-welling radiant flux densities have been made using the same scheme illustrated above, also for zero CO_2 value. The two spectral curves of $L{\uparrow}(\lambda)$ and $L{\uparrow}(\lambda)_{CO2\,=\,0}$ were found to differ considerably one from the other within the 12.5–18.0 μm range, in which important absorption effects are generated by the 14.8 μm band of CO_2, as a result of the complex sequence of six main transitions covering this range. Minor absorption effects can be noted in Fig. 10.12a, being produced by (i) the $3\nu_2$ and ($\nu_1 + \nu_2$) bands of CO_2 over the range from 4.5 to 5.2 μm, which both define the left-side wing of

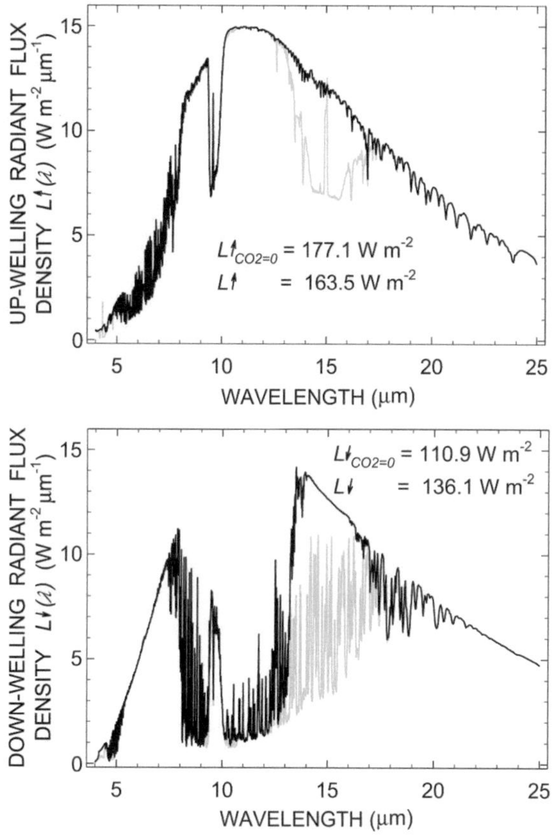

Fig. 10.12 Panel (**a**): comparison between (i) the spectral curve of up-welling radiant flux density $L\uparrow(\lambda)$ (black solid curve) of the long-wave radiation emitted upward by the Arctic surface-atmosphere system presenting cloudless-sky conditions, as calculated by using the MODTRAN 4.0 code for an atmosphere represented in terms of the "Ny-Ålesund-spring" atmospheric model having a CO_2 molecular concentration equal to 405.1 ppmv, and (ii) the spectral curve of up-welling radiant flux density $L\uparrow(\lambda)_{CO2=0}$ (grey solid curve) of long-wave radiation emitted by the "Ny-Ålesund-spring" atmospheric model assumed to have a null concentration of CO_2. Panel (**b**): as in panel (**a**) but for the down-welling radiation $L\downarrow(\lambda)$

the 8–13 atmospheric window, and (ii) the 9.4 µm and 10.4 µm secondary bands of CO_2, which are located on both wings of the 9.6 µm ozone absorption band that characterizes the spectral features of the 8–13 µm atmospheric window. The spectral curves of $L\uparrow(\lambda)$ and $L\uparrow(\lambda)_{CO2=0}$ shown in Fig. 10.12a were then integrated over the range from 4 to 25 µm to obtain the corresponding total values of up-welling radiant flux densities $L\uparrow_{CO2=0} = 177.1$ W m^{-2} and $L\uparrow = 163.5$ W m^{-2}, for which the difference $\Delta L\uparrow_{CO2} = L\uparrow_{CO2=0} - L\uparrow(\lambda)$ due to CO_2 absorption was estimated to be equal to 13.7 W m^{-2}, which is equivalent to a relative variation of 8.37% in the total up-welling radiant flux density $L\uparrow$.

The spectral curve of down-welling radiant flux densities $L\downarrow(\lambda)$ and $L\downarrow(\lambda)_{CO2=0}$ are shown in Fig. 10.11b. The comparison shows that very pronounced differences characterize the two spectral curves over the 12.5–18.0 µm range, where the 14.8 µm band of CO_2 produces important absorption features, capable of considerably "strenghtening" the right-hand side of the 8–13 µm atmospheric window, while less important effects produced by CO_2 absorption were found to occur within the 4.5–5.2 µm spectral interval (given by the $3\nu_2$ and $(\nu_1 + \nu_2)$ absorption bands of CO_2) as well as within the 9.2–10.8 µm range, because of the two minor absorption bands of CO_2 centred at wavelengths $\lambda = 9.4\ \mu m$ and $\lambda = 10.4\ \mu m$. The difference $\Delta L{\downarrow}_{CO2}$ was estimated to be equal to 25.2 W m^{-2} (as the difference between the down-welling radiant flux density $L{\downarrow}$ found to be equal to 136.1 W m^{-2} and $L{\downarrow}_{CO2\,=\,0}$ estimated to be of 110.9 W m^{-2}), which causes a relative increase of 18.5% in the down-welling radiant flux density $L{\downarrow}$ of the long-wave radiation emitted by the Arctic atmosphere toward the surface, and is appreciably higher than the decrease $\Delta L{\uparrow}_{CO2} = 13.7$ W m^{-2} in the up-welling flux of long-wave radiation emitted upward by the Arctic atmosphere caused by WV absorption. Variations of a few percent in the CO_2 molecular concentration recorded at the Arctic stations from April to September are expected to cause relatively small variations in the down-welling radiant flux density $L{\downarrow}$ that should be considerably weaker than those associated with the month-to-month changes in the Earth-Sun distance.

10.5.3 Ozone Absorption Effects on the Long-Wave Radiation Balance Terms in the Arctic Atmosphere

The molecular ozone presents a typical absorption band profile in the middle part of the 8–13 μm atmospheric window, covering the range from about 8.75 μm to 10.30 μm, with two peaks giving the overall maximum at λ close to 9.6 μm (Elsasser and Culbertson 1960). Moreover, O_3 presents also an absorption band centred at $\lambda = 14.1\ \mu m$, which is superimposed on the stronger absorption band of CO_2 presenting its maximum at the 14.8 μm wavelength. Therefore, the absorption effects of this molecule on the radiation balance of the Arctic atmosphere are expected to vary appreciably within the 8–13 μm atmospheric widow. The spectral curve of $L{\uparrow}(\lambda)$ determined for an atmospheric O_3 content equivalent to 340 D. U. is compared in Fig. 10.13a with that of up-welling radiant flux density $L{\uparrow}(\lambda)_{O3\,=\,0}$. The two spectral curves of $L{\uparrow}(\lambda)$ and $L{\uparrow}(\lambda)_{O3\,=\,0}$ were estimated to exhibit the most pronounced differences within the middle of the 8–13 μm atmospheric window, due to the strong O_3 absorption band centred at 9.6 μm, which modifies very efficiently the spectral curve of the up-welling radiant flux density. The calculations of the integrals of the spectral curves of $L{\uparrow}(\lambda)$ and $L{\uparrow}(\lambda)_{O3\,=\,0}$ shown in Fig. 10.13a provided the values of $L{\uparrow} = 163.5$ W m^{-2} and $L{\uparrow}_{O3\,=\,0} = 167.7$ W m^{-2}, respectively, yielding a value of the difference $\Delta L{\uparrow}_{O3} = L{\uparrow}(\lambda)_{O3\,=\,0} - L{\uparrow}(\lambda)$ produced by the atmospheric O_3 absorption effects which was estimated through our procedure to be equal to 4.2 W m^{-2}. This evaluation suggests that the 9.6 μm band of O_3 produces a relative increase of 2.57% in the total up-welling radiant flux density of the long-wave radiation emitted by the surface-atmosphere system toward space.

The spectral curve of the down-welling radiant flux density $L{\downarrow}(\lambda)$ of the long-wave radiation emitted by the atmosphere toward the surface determined in Fig. 10.11b is also shown in Fig. 10.13b, for being compared with the spectral curve of $L{\downarrow}(\lambda)_{O3\,=\,0}$ obtained for a null atmospheric content of O_3 (Fig. 10.13b).

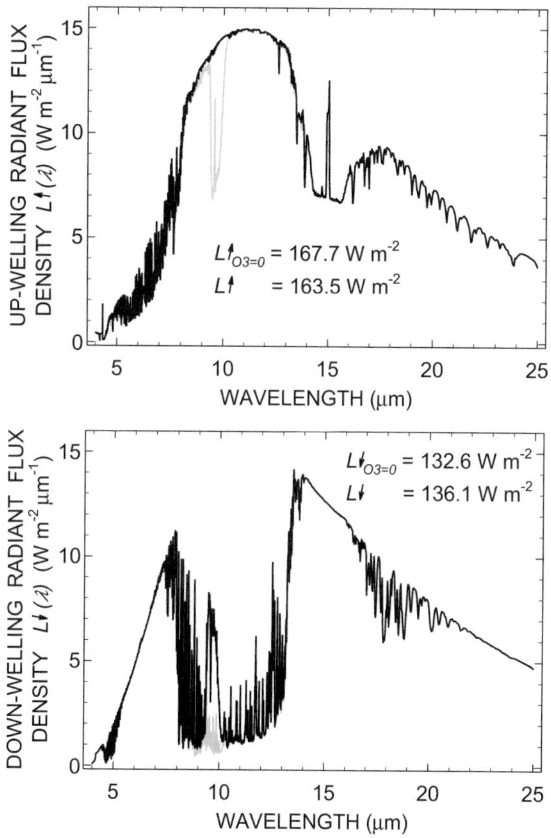

Fig. 10.13 Panel (**a**): comparison between (i) the spectral curve of up-welling radiant flux density $L\uparrow(\lambda)$ (black solid curve) of the long-wave radiation emitted upward by the Arctic surface-atmosphere system presenting cloudless-sky conditions, as calculated by using the MODTRAN 4.0 code for an atmosphere represented in terms of the "Ny-Ålesund-spring" atmospheric model having a columnar content of O_3 equivalent to 340 D. U., and (ii) the spectral curve of up-welling radiant flux density $L\uparrow(\lambda)_{O3=0}$ (grey solid curve) of the long-wave radiation emitted by the "Ny-Ålesund-spring" atmospheric model assumed to have a null columnar content of O_3. Panel (**b**): as in panel (**a**) but for the down-welling radiation $L\downarrow(\lambda)$

This comparison provides evidence of the important differences caused between the two spectral curves by the 9.6 μm band of O_3, which is totally free from external absorption bands and therefore exhibits its typical spectral absorption features in the middle of the 8–13 μm atmospheric window. The difference $\Delta L\downarrow_{O3} = L\downarrow - L\downarrow_{O3=0}$ was estimated to be equal to 3.5 W m^{-2} ($L\downarrow$ = 136.1 W m^{-2} and $L\downarrow_{O3=0}$ = 132.6 W m^{-2}), which causes a relative increase of 2.57% due to O_3 absorption in the down-welling radiant flux density $L\downarrow$ given by the long-wave radiation emitted by the Arctic atmosphere toward the surface, such an increase being appreciably lower that the decrease $\Delta L\uparrow_{O3}$ in the up-welling long-wave radiation caused by atmospheric O_3 absorption and estimated in Fig. 10.13a to be equal to 4.2 W m^{-2}. The variations in the flux densities $L\downarrow$ and $L\uparrow$ due to the changes in the atmospheric O_3 content occurring from April to September are negligible with respect to those due to the month-to-month changes in the Earth-Sun distance.

10.5.4 Methane Absorption Effects on the Long-Wave Radiation Balance Terms in the Arctic Atmosphere

Methane is a powerful greenhouse gaseous species, mainly originated from both natural sources (wetlands, such as tundra, marshes, bogs and swamps, largely present in the Arctic region; enteric fermentation of wild ruminant animals) and anthropic sources (rice paddies, domesticated ruminant animals, drilling for oil and natural gases, coal mine ventilation and degassing from coal during transport, biomass burning of forest fires, anaerobic decay of organic wastes in landfills). This molecule exhibits two strong absorption bands in the infrared, consisting of numerous overlapping peaks, of which the first is centred at $\lambda \approx 3.3$ μm (not considered in the present study), and the second at $\lambda = 7.66$ μm (Nielsen and Nielsen 1935). Due to its spectral position on the left hand-side of the 8–13 μm atmospheric window, the 7.66 μm band contributes efficiently to cause the relevant absorption of the down-welling long-wave radiation flux (emitted by the atmosphere toward the surface) as well that of the up-welling long-wave radiation flux (emitted by the surface-atmosphere system toward the outer space).

After a strong increase during the twentieth century, the atmospheric concentration of CH_4 maintained relatively stable values over the period from 1998 to 2006, during which the global average change was close to zero. The volume concentration of CH_4 was observed to increase appreciably in the Arctic region during the last 15 years, as shown by the regular measurements carried out at the Zeppelin Observatory. Considering only the observations made during the clean-air periods, the yearly average increase of CH_4 volume concentration was estimated to be of 5.2 ppbv per year. The most pronounced increase started in November/December 2005 and continued throughout the years 2007–2009, being particularly evident in the late summer–winter 2007, and summer–autumn 2009. The yearly average value of CH_4 molecular volume concentration was estimated to be equal to 1920 ppbv in 2015 at the Zeppelin Observatory, with an average increase of 11.1 ppbv per year recorded from 2013 to 2015. Therefore, it seems realistic to assume an average volume concentration of CH_4 equal to 1940.5 ppbv in 2017. Calculations of the absorption effects produced by CH_4 on the long-wave radiation budget terms of the surface-atmosphere system have been performed by us for this concentration value, by following the same procedure adopted above to calculate the absorption effects of WV, CO_2 and O_3 in the Arctic atmosphere.

The two spectral curves of $L\uparrow(\lambda)$ and $L\uparrow(\lambda)_{CH4=0}$ were found to differ appreciably one from the other only within the 7–8 μm range, where the CH_4 band centred at $\lambda = 7.66$ μm is located. The difference $\Delta L\uparrow_{CH4} = L\uparrow(\lambda)_{CH4=0} - L\uparrow(\lambda)$ caused by atmospheric CH_4 absorption effects equal to 0.79 W m^{-2} ($L\uparrow_{CH4=0} = 164.3$ W m^{-2} against that of $L\uparrow = 163.5$ W m^{-2}), which corresponds to 0.48% of the total up-welling radiant flux density $L\uparrow$.

The comparison between $L\downarrow(\lambda)$ and $L\downarrow(\lambda)_{CH4=0}$ gives evidence of the fact that marked variations in the down-welling radiant flux density $L\downarrow(\lambda)$ are caused by CH_4 absorption only within the range from 7.0 to 8.5 μm, because of the presence

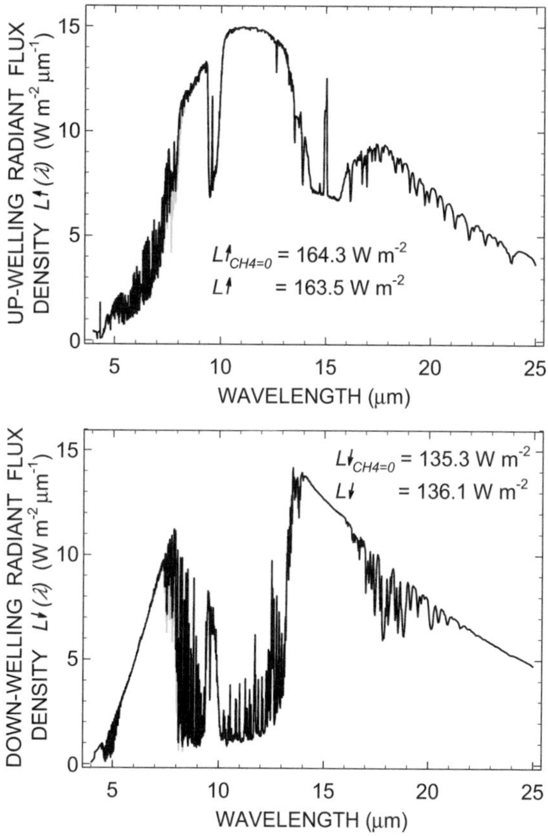

Fig. 10.14 Panel (**a**): comparison between (i) the spectral curve of up-welling radiant flux density $L\uparrow(\lambda)$ (black solid curve) of the long-wave radiation emitted upward by the Arctic surface-atmosphere system presenting cloudless-sky conditions, as calculated by using the MODTRAN 4.0 code for an atmosphere represented in terms of the "Ny-Ålesund-spring" atmospheric model having a concentration of CH_4 equal to 1940.5 ppbv, and (ii) the spectral curve of up-welling radiant flux density $L\uparrow(\lambda)_{CH4=0}$ (grey solid curve) of the long-wave radiation emitted by the "Ny-Ålesund-spring" atmospheric model assumed to have a null concentration of CH_4 molecules. Panel (**b**): as in panel (**a**) but for the down-welling radiation $L\downarrow(\lambda)$

of the 7.66 μm band of CH_4, which is capable of appreciably modifying the spectral absorption features defining the left-hand side of the 8–13 μm atmospheric window.

The spectral curves of down-welling radiant flux densities $L\downarrow(\lambda)$ and $L\downarrow(\lambda)_{CH4=0}$ are presented in Fig. 10.14b. The integral values $L\downarrow$ and $L\downarrow_{CH4=0}$ were estimated to be equal to 136.1 W m^{-2} and 135.3 W m^{-2}, respectively. The difference $\Delta L\downarrow_{CH4} = L\downarrow - L\downarrow_{CH4=0}$ was therefore found to be equal to 0.80 W m^{-2}, which indicates that CH_4 causes a relative change of only 0.59% in the down-welling radiant flux density $L\downarrow$ emitted by the Arctic atmosphere toward the surface. This estimate of $\Delta L\downarrow_{CH4}$ was therefore found to be comparable with the decrease $\Delta L\uparrow_{CH4} = 0.79$ W m^{-2} caused in the up-welling long-wave radiation by atmospheric CH_4 absorption effects (and shown in Fig. 10.14a). The variations in the flux densities $L\downarrow$ and $L\uparrow$ caused by the seasonal changes in the atmospheric concentration of CH_4 molecules recorded from April to September are negligible with respect to those taking place from spring to early autumn because of the Earth-Sun distance variations.

10.5.5 Nitrous Oxide Absorption Effects on the Long-Wave Radiation Balance Terms in the Arctic Atmosphere

Nitrous oxide is an important greenhouse gas mainly produced by oceans and tropical soils and wet forests, while weaker contributions of these are provided by dry savannas and and grasslands, and temperate soils and forests as well as by anthropic activities (through denitrification processes occurring in cultivated soils, biomass burning and coal and oil combustion processes, adipic acid combustion in industrial activities, and use in agriculture of nitrate and ammonium fertilizers). This molecule is known to exhibit three fundamental absorption bands in the infrared: (i) the band v_1 centred at $\lambda = 7.78$ μm (wavenumber $= 1285.6$ cm^{-1}), (ii) the band v_2 centred at $\lambda = 16.98$ μm (wavenumber $= 588.8$ cm^{-1}), and (iii) the band v_3 centred at $\lambda = 4.50$ μm (wavenumber $= 2223.5$ cm^{-1}) (Burch et al. 1972). Therefore, N_2O is estimated to cause relevant absorption effects on both the down-welling and the up-welling long-wave radiation fluxes.

Global volume concentration of N_2O has increased from around 270 ppbv prior to industrialization to an average global mean of 327.1 pptv in 2014. The global mean increase of the N_2O concentration in the Arctic region was estimated to be slightly stronger over the last years than that measured before 2014. At the Zeppelin Observatory, the N_2O volume concentration was found to be equal to 323.2 pptv in 2010 and to 328.1 pptv in 2015. Thus, assuming a yearly average increase of 1.1 pptv during the last years, the annual mean concentration of N_2O molecules was estimated to have reached the value of 330 pptv in 2017 at this high-latitude Arctic station.

The spectral curve of up-welling radiant flux density $L\uparrow(\lambda)$ is compared in Fig. 10.15a with that of up-welling radiant flux density $L\uparrow(\lambda)_{N2O\,=\,0}$. It can be noted in Fig. 10.15a that only small discrepancies exist between the two curves of $L\uparrow(\lambda)$ and $L\uparrow(\lambda)_{N2O\,=\,0}$, arising from the absorption effects due to the following three bands of N_2O: (i) band v_3 centred at $\lambda = 4.50$ μm, producing weak absorption features between 4 and 5 μm wavelengths; (ii) band v_1 over the range from 7.6 to 8.0 μm, being its maximum centred at $\lambda = 7.78$ μm; and (iii) band v_2 centred at $\lambda = 16.98$ μm, causing some relevant absorption effects at around 17 μm wavelength. The values of overall radiant flux density $L\uparrow_{N2O\,=\,0} = 164.4$ W m^{-2}, which is slightly higher than that of $L\uparrow = 163.5$ W m^{-2}, gives a variation of 0.93 W m^{-2} in the up-welling radiant flux density, which constitutes only the 0.57% of the total up-welling radiant flux density $L\uparrow$.

The comparison between the spectral curves of $L\downarrow(\lambda)$ and $L\downarrow(\lambda)_{N2O\,=\,0}$ shows clearly that: (i) some slight differences are due to the absorption effects of the band v_3 centred at $\lambda = 4.50$ μm, (ii) appreciable differences are generated by the band v_1 of N_2O located over the 7.6–8.0 μm range, and (iii) less important differences can be appreciated at around 17 μm wavelength, caused by the weak absorption effects of N_2O band v_2. According these evaluations, the difference $\Delta L\downarrow_{N2O} = L\downarrow - L\downarrow_{N2O\,=\,0}$ was estimated to be equal to 0.80 W m^{-2}, which indicates that N_2O causes a relative increase of 59% in the down-welling radiant flux density $L\downarrow$.

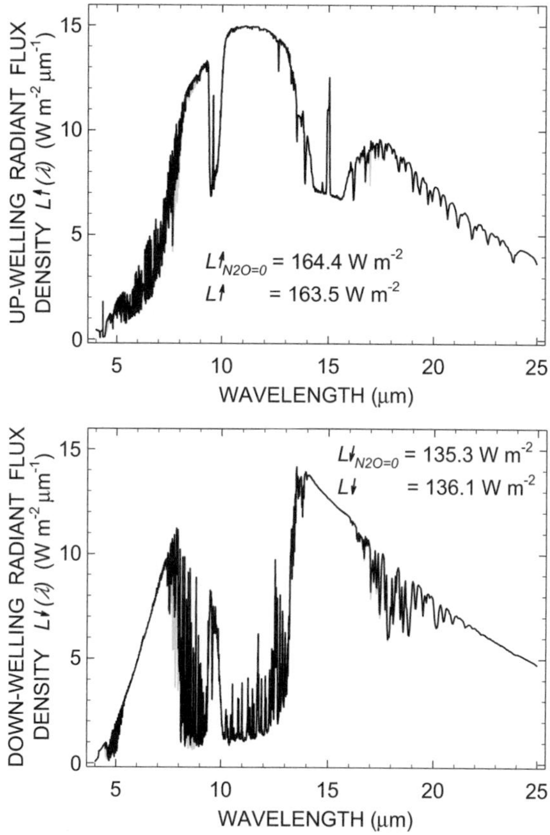

Fig. 10.15 Panel (**a**): comparison between (i) the spectral curve of up-welling radiant flux density $L\uparrow(\lambda)$ (black solid curve) of the long-wave radiation emitted upward by the Arctic surface-atmosphere system presenting cloudless-sky conditions, as calculated by using the MODTRAN 4.0 code for an atmosphere represented in terms of the "Ny-Ålesund-spring" atmospheric model having a concentration of N_2O equal to 330 pptv, and (ii) the spectral curve of up-welling radiant flux density $L\uparrow(\lambda)_{N2O=0}$ (grey solid curve) of the long-wave radiation emitted by the "Ny-Ålesund-spring" atmospheric model assumed to have a null concentration of N_2O molecules. Panel (**b**): as in panel (**a**) but for the down-welling radiation $L\downarrow(\lambda)$

Therefore, the change $\Delta L\downarrow_{N2O}$ in the down-welling long-waver radiation due to N_2O absorption is appreciably lower than the variation $\Delta L\uparrow_{N2O} = 0.93$ W m^{-2} produced by caused by N_2O absorption on the up-welling long-wave radiation (and shown in Fig. 10.15b).

Besides some less important organic gases, weak but not negligible absorption effects are also produced within the 8–13 μm atmospheric transparency window by ammonia (presenting a weak absorption band with its maximum at $\lambda = 10.55$ μm) and nitrogen dioxide (with a band centred at $\lambda = 13.3$ μm) (Tomasi and Trombetti 1985). However, due to their rather low effects in terms of radiant flux density changes, they have been not considered here in the present analysis. Taking into account as a whole of the greenhouse effects produced by WV, CO_2, O_3, CH_4 and N_2O in the Arctic atmosphere, the present calculations allowed us to evaluate that the above-mentioned five molecular species can cause an overall change in the up-welling radiant flux density of long-wave radiation equal to 30.6 W m^{-2}, which is equivalent to 18.8% of the long-wave radiation emitted by the Arctic atmosphere toward the outer space.

The above evaluations of the absorption effects produced by the five above-selected greenhouse gases on the up-welling and down-welling radiant flux densities of the long-wave radiation emitted by the Arctic atmosphere toward the surface suggest the following two final remarks:

(A) The whole change $\Delta L\uparrow$ caused by the combined absorption features of the five greenhouse gases has been estimated to be equal to 31.5 W m^{-2}, this estimate being equal to 18.8% of the total up-welling radiant flux density leaving the surface-atmosphere system at the 78 °N latitude. Compared with the overall upward flux density of 40.1 W m^{-2} produced by the atmosphere on the outgoing flux of thermal radiation in the general atmospheric model described in Fig. 10.1 by Kiehl and Trenberth (1997) and reviewed by NASA (2016), the present calculations indicate that the Arctic atmosphere generates a total up-welling radiant flux density equal to ~ 78% of that estimated over whole planetary scale, this estimate being not surprising if one considers that average temperature conditions of the atmosphere occurring over the whole planetary scale are considerably higher than those usually observed at the Arctic latitudes in summer.

(B) The overall effect produced by the five greenhouse gases on the down-welling radiant flux density reaching the surface has been estimated by us to cause a change $\Delta L\downarrow$ equal to 104.8 W m^{-2} in the long-wave radiation emitted by the Arctic atmosphere toward the surface during the early spring period, of which the predominant part of about 71% is due to WV, and about 24% to CO_2. Comparing these results with the average estimates of down-ward thermal radiation equal to 340.3 W m^{-2} on the whole planetary scale (NASA 2016), it can be concluded that WV, CO_2 and the other minor gases produce in the Arctic atmosphere a greenhouse effect contribution equivalent to about 31% of that occurring over the whole planet, plausibly because of the colder conditions characterizing the Arctic atmosphere.

10.6 Evaluations of Short-Wave Direct Aerosol Radiative Forcing Induced by Airborne Aerosols in the Arctic Atmosphere

The aerosol-induced radiative forcing can be defined as the effect produced on the surface-atmosphere system by a change in the content and optical properties of the aerosol particles suspended in the vertical atmospheric column of unit cross-section, which can be evaluated as the net radiative flux change (given by the difference between down-welling radiation and up-welling radiation) induced at the ToA-level by the aerosol particles, while the concentrations of all the other atmospheric constituents remain stable in time. Therefore, the direct aerosol radiative forcing (hereinafter referred to as DARF) term at the ToA-level gives a measure of the energy change in the climate system budget, due to a certain atmospheric aerosol load, which depends not only on the magnitude of the radiation flux change occurring at the ToA-level but also on the features of the vertical profiles of the various radiative parameters of aerosols (Hansen et al. 1997).

Because of the predominance of the short-wave effects produced by airborne aerosols over those affecting the long-wave radiation budget, the DARF effect is commonly evaluated considering only the short-wave (solar) radiation flux change, and neglecting the radiative effects produced by aerosols on both the long-wave components of solar and terrestrial radiation. These short-wave evaluations are first evaluated instantaneously at a certain number of diurnal pre-fixed hours, and subsequently calculated as diurnal averages over the 24-hour period (Bush and Valero 2003). The DARF term $\Delta F_{ToA}(t)$ can be therefore determined at the ToA-level as the difference between (i) the net radiative flux calculated for the turbid atmosphere containing a certain columnar content of aerosol particles, and (ii) the same quantity calculated in a pristine atmosphere without aerosols. Thus, the instantaneous forcing $\Delta F_{ToA}(t)$ induced at a certain time by aerosol particles can be represented in terms of the following formula:

$$\Delta F_{ToA}(t) = F_{net}(t) - F_{net}{}^{*}(t), \tag{10.2}$$

where the instantaneous ToA-level net flux $F_{net}(t)$ is given by the difference between the short-wave down-welling flux $F{\downarrow}(t)$ and the short-wave up-welling flux $F{\uparrow}(t)$, both determined at the ToA-level for an atmosphere including all its constituents, i.e.:

$$F_{net(t)} = F \downarrow (t) - F \uparrow (t). \tag{10.3}$$

In Eq. (10.2), the instantaneous net flux $F_{net^*}(t)$ at ToA-level is given for a pristine atmosphere without aerosols, i.e.:

$$F_{net}{}^{*}(t) = F{\downarrow}^{*}(t) - F{\uparrow}^{*}(t). \tag{10.4}$$

The down-welling flux $F{\downarrow}^{*}(t)$ being not altered by atmospheric aerosols, flux $F{\downarrow}^{*}(t)$ in Eq. (10.4) is equal to flux $F{\downarrow}(t)$ given in Eq. (10.3). Consequently, the instantaneous term $\Delta F_{ToA}(t)$ calculated from Eqs. (10.2), (10.3), and (10.4) is directly given by the difference,

$$\Delta F_{ToA}(t) = F{\uparrow}^{*}(t) - F \uparrow (t), \tag{10.5}$$

which shows that this DARF term can be correctly evaluated by simply subtracting the up-welling solar radiation flux (emerging from the real atmosphere with aerosols) from the up-welling solar radiation flux (emerging from the pristine atmosphere without aerosols) (Hänel et al. 1999). According to the conventional definition of radiative forcing in the atmosphere, negative values of $\Delta F_{ToA}(t)$ indicate that aerosols cause an increase in the upwelling flux of solar radiation at a certain time t and, hence, an increase in the albedo of the surface–atmosphere system, for which direct cooling effects are produced on the climate system. Conversely, positive values of $\Delta F_{ToA}(t)$ indicate that lower upwelling solar radiation fluxes are induced by aerosols at a certain time t, leading to a decrease in the overall albedo and, consequently, causing significant atmospheric warming effects (Zhao et al. 2008).

The aerosol radiative forcing ΔF_{BoA} occurring at the surface (i.e. at the Bottom-of-Atmosphere level) gives a measure of the perturbation induced by airborne aerosols in the net flux reaching the surface-level during a certain day. The instantaneous term $\Delta F_{BoA}(t)$ can be defined as the difference between the instantaneous net flux $\Phi_{net}(t)$ occurring at the surface-level in the atmosphere with aerosols and the instantaneous net flux $\Phi_{net^*}(t)$ taking place at the surface-level in the same atmosphere assumed to be without aerosols (Satheesh and Ramanathan 2000; Bush and Valero 2002, 2003), i.e.:

$$\Delta F_{BoA}(t) = \Phi_{net}(t) \quad - \Phi_{net}{}^*(t), \tag{10.6}$$

where, the net flux terms $\Phi_{net}(t)$ and $\Phi_{net^*}(t)$ at the surface-level are given by the differences between the instantaneous down-welling and up-welling fluxes of solar radiation at the surface, i.e.

$$\Phi_{net}(t) = \Phi \downarrow (t) - \Phi \uparrow (t) \tag{10.7}$$

and

$$\Phi_{net}{}^*(t) = \Phi{\downarrow}^*(t) - \Phi{\uparrow}^*(t). \tag{10.8}$$

Assuming that A is the average surface albedo over the short-wave spectral range, the instantaneous up-welling flux $\Phi\uparrow(t)$ at the surface can be expressed as the product,

$$\Phi \uparrow (t) = A \times \Phi \downarrow (t). \tag{10.9}$$

Therefore, combining Eqs. (10.7) and (10.9), the net flux Φ_{net} at the surface can be written in the following form:

$$\Phi_{net}(t) = \quad (1{-}A)\, \Phi \downarrow (t). \tag{10.10}$$

In order to evaluate the down-welling flux of solar radiation reaching the Earth's surface after its passage through the pristine atmosphere without aerosols and that passing through the turbid atmosphere (and, hence, to calculate the net flux Φ_{net}), various radiative transfer codes can be used in the studies aimed at evaluating the direct radiative effects induced by Arctic aerosol loads, such as the MODTRAN atmospheric model (Kneizys et al. 1996) or the 6S (Second Simulation of the

Satellite Signal in the Solar Spectrum) code (Vermote et al. 1997), properly applied to realistic models of the Arctic atmosphere.

The occurrences of the above-mentioned instantaneous radiative forcing effects at the ToA- and BoA-levels imply that the instantaneous aerosol thermodynamic forcing $\Delta F_{Atm}(t)$ is produced by aerosols within the atmosphere. Such a DARF term can be correctly evaluated as the difference between the instantaneous term $\Delta F_{ToA}(t)$ defined in Eq. (10.2), and the instantaneous term $\Delta F_{BoA}(t)$ defined in Eq. (10.6), according to Ramanathan et al. (2001), i.e.:

$$\Delta F_{Atm}(t) = \Delta F_{ToA}(t) - \Delta F_{BoA}(t), \tag{10.11}$$

which constitutes a change in the atmospheric energy budget that is not explicitly caused by aerosol-induced radiative effects. Unlike the DARF terms determined at the ToA- and BoA-levels, this DARF term does not modify the net energy budget of the surface-atmosphere system, but rather redistributes it internally and then affects temperature gradients and atmospheric circulation. In fact, the main contribution to $\Delta F_{Atm}(t)$ is given by the change in the amount of latent heat released by aerosol-induced changes in clouds and precipitations. Therefore, it can be expressed as the instantaneous variation in the latent heat flux passing through the atmosphere, as given by the difference between the instantaneous DARF terms $\Delta F_{ToA}(t)$ and $\Delta F_{BoA}(t)$.

As pointed out above, large uncertainties still exist on the role of columnar aerosol loading in causing climate change effects within the global circulation models (Hansen et al. 1997, 1998). In particular, large gaps remain in the knowledge of (i) the dependence of the aerosol radiative effects on the microphysical and composition characteristics of the aerosol polydispersed particles suspended in the vertical atmospheric column of unit cross-section, (ii) the vertical distribution curves of number and mass concentration and columnar aerosol radiative parameters, often characterized by multi-layered features. The same is valid for the spectral and directional characteristics of surface reflectance. With regard to this point, it is of crucial importance to take into account that surface reflectance characteristics play a fundamental role in the DARF calculations, as clearly shown by Chylek and Coakley (1974). In addition, the Lambertian reflection models commonly used to calculate the energy budget of the surface-atmosphere system do not realistically describe the reflectance characteristics of land and ocean surfaces. Thus, their use can lead to biases in the model calculations of aerosol radiative forcing, especially for high values of solar zenith angle θ_o (Ricchiazzi et al. 2005). For this reason, it seems more realistic and appropriate to use in the present study the so-called Bidirectional Reflectance Distribution Function (BRDF) models in order to represent the characteristics of surface reflectance. Using the BRDF models, it is also important to represent the spectral albedo curves of the various surfaces as the sum of two terms, related to the so-called black-sky and white-sky albedo concepts proposed by Lewis (1995).

10.6.1 The Surface Reflectance and Albedo Characteristics in the Arctic Region

The instantaneous DARF terms $\Delta F_{ToA}(t)$ and $\Delta F_{BoA}(t)$ induced by columnar aerosol are both subject to vary as a function of the spectral and spatial features of the underlying surface reflectance (Tomasi et al. 2013). Therefore, the knowledge of the spectral and spatial characteristics of the surface reflectance is of basic importance when the DARF effects have to be calculated at both the ToA- and BoA-levels. Chylek and Coakley (1974) and Coakley and Chylek (1975) used the two-stream approximation procedure in order to simulate the radiative transfer processes occurring in a plane atmosphere containing a layer of aerosol particles with well-defined absorptance and reflectance characteristics, showing that the cooling or warming effects induced by an aerosol layer suspended in the atmosphere depend on both the surface albedo characteristics and the absorptance and reflectance properties of the airborne aerosol particles. This implies that the DARF effects induced at the ToA-level by a certain particle load can change sign by passing from cooling to warming effects, as the airborne aerosol particles are transported from oceanic areas, typically presenting relatively low surface albedo conditions, to the Arctic regions covered by snow fields and glaciers (such as those located in the interior part of Greenland). The Chylek and Coakley (1974) calculations also evidenced that the DARF effect at the ToA-level tends to change sign from cooling to warming for a certain surface albedo value, as the particulate matter absorptance increases with respect to reflectance, until exceeding a critical value of the absorptance/reflectance ratio.

In order to improve the representation of the spectral and geometrical dependence features of the BRDF parameters, more realistic approaches were implemented through numerous tests and specific applications aimed at determining more suitable models of bidirectional reflectance for evaluating the DARF effects over surfaces presenting different reflectance characteristics (Vermote et al. 1997; Wanner et al. 1997). In these applications, it is useful to consider the bidirectional reflectance factor $R(\lambda, \theta_o, \varphi_o, \vartheta, \varphi)$, commonly used to represent the ratio between the real surface reflectance and the BRDF reflectance of an ideal Lambertian reflector, and generally assumed to depend not only on wavelength λ but also on the four angular coordinates of the Sun surface–external viewer system. Varying as a function of such angular parameters, factor $R(\lambda, \theta_o, \varphi_o, \vartheta, \varphi)$ provides the ratio between (i) the upwelling irradiance $F\uparrow(\vartheta, \varphi)$ reflected by the surface in a certain upwelling direction, and (ii) the incident flux of a collimated incoming solar radiation beam having direction defined by solar zenith angle θ_o and azimuth angle φ_o. The plane perpendicular to the surface, containing both the Sun and the ground-level reference spot, defines the principal plane of reflection. Function $R(\lambda, \theta_o, \varphi_o, \vartheta, \varphi)$ used to represent the surface reflectance is commonly assumed to exhibit a cylindrical symmetry with respect to the principal plane of reflection. Thus, its mathematical representation is in general made considering only the difference φ between the two angles φ_o and φ.

In order to obtain precise calculations of the DARF effects at the ToA-level for each geometrical configuration of the surface-atmosphere system defined by the angular parameters θ_o, ϑ and φ, it was decided to use a set of BRDF models for representing the surface reflectance properties of the Arctic surfaces in terms of the spectral and angular dependence features of the following three reflectance functions:

(1) The *spectral directional hemispherical reflectance* (black-sky albedo) $R_{bs}(\lambda, \theta_o)$, obtained through integration of the BRDF function $R(\lambda, \theta_o, \phi_o = 0^\circ, \vartheta, \varphi)$ over the upward solid angle equal to 2π, to represent the spectral curve of surface reflectance as a function of solar zenith angle θ_o (Román et al. 2010). The function is considered to be valid in the ideal case in which the diffuse component of the global solar radiation field is assumed to be null. It is worth pointing out that this spectral function provides the spectral curve of the so-called black-sky albedo for $\theta_o = 0°$, expressed in terms of the following analytical form,

$$R_{bs}\left(\lambda, \theta_o = 0^\circ\right) = \frac{1}{\pi} \int_0^{2\pi} \int_0^{\pi/2} R\left(\lambda, \theta_o, \varphi_o, \vartheta, \phi\right) \cos \vartheta \, \sin \vartheta \, d\vartheta \, d\phi \tag{10.12}$$

(2) In the approximation of an isotropic diffuse-sky illumination and for an azimuth independent bidirectional reflectance, the *spectral bi-hemispherical reflectance* (white-sky albedo) $R_{ws}(\lambda)$ can be calculated by integrating the black-sky albedo $R_{bs}(\lambda, \theta_o)$ over the zenith angular range. Therefore, function $R_{ws}(\lambda)$ represents the spectral curve of surface albedo relative to the diffuse component of incoming solar radiation:

$$R_{ws}(\lambda) = 2 \int_0^{\pi/2} R_{bs}(\lambda, \theta_o) \cos\theta_o \sin\theta_o \, d\theta_o \tag{10.13}$$

Bearing in mind that the white-sky albedo is given by the integral of the black-sky albedo $R_{bs}(\lambda, \theta_o)$ over the entire intervals of the two down-welling polar angles, and that $R_{bs}(\lambda, \theta_o)$ is given in Eq. (10.12) as the double hemispherical integral of the bidirectional reflectance factor R over the entire ranges of the two upwelling polar angles, it is evident that $R_{ws}(\lambda)$ does not depend on the geometrical configuration of the "Sun-surface-external viewer" system. In cases of isotropic surface reflectance conditions, the white-sky albedo can be assumed to be that of an equivalent Lambertian reflector. Its features can be better defined by employing satellite data derived from the observations of the Sun-synchronous multispectral sensors mounted on the Terra polar platform, such as the MISR (Diner et al. 1998; Bothwell et al. 2002) and MODIS (Christopher and Zhang 2002) satellite-borne sensors.

(3) The spectral curve of *surface albedo* $R_L(\lambda, \theta_o)$,which is obtained as the weighted average of the spectral surface reflectance contributions associated with the black-sky and white-sky albedo conditions of the solar radiation field, respectively (Lewis and Barnsley 1994; Lucht et al. 2000). This approximate function can be calculated in terms of the analytical form of the Lewis (1995) function defined by the following equation,

$$R_L(\lambda, \theta_o) = R_{bs}(\theta_o)\ [1 - D\downarrow(\lambda)] + R_{ws}(\theta_o)\, D\downarrow(\lambda)\,, \tag{10.14}$$

where $D\downarrow(\lambda)$ is the spectral curve of the diffuse fraction of down-welling (global) solar radiation $I\downarrow(\lambda)$ reaching the surface, which can be calculated as a function of solar zenith angle θ_o using the 6S code for each atmospheric content of aerosol particles.

10.6.2 Use of BRDF Non-Lambertian Surface Reflectance Models in the Arctic Region

In order to calculate the DARF effects induced by aerosols in the Arctic region, a set of BRDF non-Lambertian surface reflectance models were defined for different reflectance features yielding values of albedo increasing gradually from less than 0.05 (for ocean water surfaces) to more than 0.9 (for ice-covered surfaces). In the present study, only six surface reflectance models were considered, of which only the OS1 model was used to describe the ocean surfaces, only the VS1 model was adopted to describe the vegetation-covered surfaces, and four polar surface (PS) models were employed to represent the surface reflectance characteristics of the snow- and ice –covered surfaces, each model being defined with specific spectral and angular characteristics of its surface reflectance:

(1) The OS1 model was based on a BRDF surface reflectance representation of an oceanic area described in terms of the OCEAN hyperspectral model (Morel 1988) and developed using the OCEAN sub-routine given in the 6S code. This model takes also into account the characteristics defined by the whitecaps model of Koepke (1984) and includes: (i) the improved modeling features of spectral reflectance of whitecaps that have been proposed by Kokhanovsky (2004)), (ii) the sun glint reflectance effects (Cox and Munk 1954) and (iii) the effects due to Fresnel's reflection (Born and Wolf 1975). The OCEAN subroutine was used to calculate the BRDF function curves for all the triplets of angular coordinates θ_o, η and φ considered in the OS1 model by assuming: (i) the wind speed $V_w = 2$ m s^{-1}; (ii) the sea-water pigment concentration $C_p = 0.1$ mg m^{-3}, this assumption being made considering that variations in C_p of more than four orders of magnitude can cause only relatively small changes in the surface

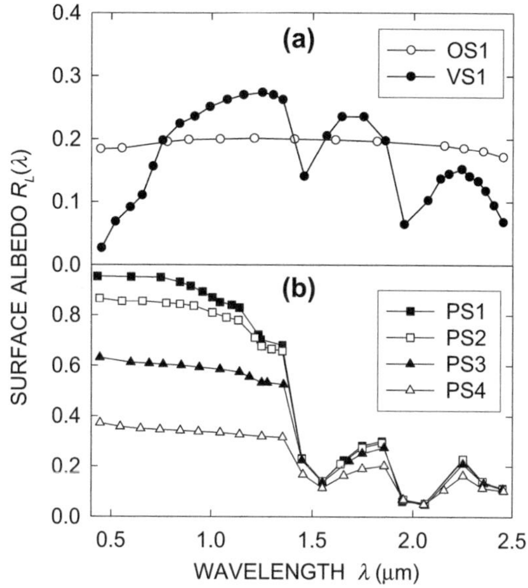

Fig. 10.16 Upper part (**a**): spectral curves of surface albedo $R_L(\lambda)$ (Lewis and Barnsley 1994), as defined in Eq. (10.14) over the 0.40–2.50 μm range for the BRDF surface reflectance models OS1 (for an oceanic surface) and VS1 (for a vegetation-covered surface). Lower part (**b**): as in the upper part but for the four PS (polar surface) reflectance models defined by Tomasi et al. (2013). All the six reflectance models were determined for solar zenith angle $\theta_o = 60°$

reflectance leading to relative variations of reflectance not exceeding 10%; and (iii) sea-water salt concentration $C_s = 34.3$ ppt, this value being assumed by taking into account that an increase in C_s from 0 to 48 ppt is estimated to induce surface reflectance changes much smaller than 1%. The corresponding spectral curve of surface albedo $R_L(\lambda, \theta_o = 60°)$ is shown in Fig. 10.16, giving the spectral values of this parameter reported in Table 10.2 for the OS1 model and pertaining to 21 selected wavelengths over the spectral range from 0.40 to 2.50 μm. Table 10.3 provides the values of reflectance parameters $R_{bs}(\theta_o = 0°)$ and R_{ws}, and those of broadband albedo $A(\theta_o)$ calculated for nine increasing values of θ_o taken in steps of 10° over the 0°–80° range for model OS1, the values of $A(\theta_o)$ being obtained through integration of $R_L(\lambda, \theta_o)$ over the 0.40–2.50 μm range. Table 10.4 provides the spectral values of white-sky albedo $R_{ws}(\lambda)$ defined in Eq. (10.4) for 21 wavelengths selected over the 0.40–2.50 μm spectral range, while the spectral curve of white-sky albedo $R_{ws}(\lambda)$ defined in Eq. (10.2) for the BRDF surface reflectance model OS1 is shown in Fig. 10.17 over the 0.40–2.50 μm spectral range.

(2) The VS1 model was based on a BRDF reflectance model describing the characteristics of a vegetated surface, as derived using the Kuusk (1994, 1995) canopy reflectance model, which considers the diffuse and specular reflection of short-wave radiation on leaves, the canopy hot spot, and non-lambertian soil. The Kuusk (1994, 1995) model utilizes the PROSPECT code of Jacquemoud (1993) for simulating the chlorophyll absorption features, and the Nilson and Kuusk (1989) algorithm for representing the reflectance anisotropy characteristics of a one-layer canopy coverage. It was employed in the 6S code to evaluate the reflectance properties of vegetated surfaces.

Table 10.2 Monochromatic values of surface albedo $R_L(\theta_o, \lambda)$ defined in Eq. (10.4c) and shown in Fig. 10.16, as obtained at 21 wavelengths selected over the 0.40–2.50 μm spectral range for the six BRDF surface reflectance models used in the present

	BDRF surface reflectance models					
Wave-length λ (μm)	OS1	VS1	PS1	PS2	PS3	PS4
0.45	0.184	0.034	0.957	0.866	0.633	0.377
0.55	0.186	0.077	0.956	0.859	0.620	0.362
0.65	0.190	0.108	0.955	0.855	0.609	0.350
0.75	0.194	0.200	0.956	0.852	0.607	0.348
0.85	0.197	0.228	0.935	0.845	0.601	0.343
0.95	0.199	0.244	0.903	0.830	0.594	0.338
1.05	0.200	0.262	0.857	0.803	0.588	0.338
1.15	0.201	0.271	0.828	0.780	0.574	0.327
1.25	0.201	0.278	0.702	0.678	0.535	0.319
1.35	0.200	0.262	0.680	0.660	0.527	0.318
1.45	0.200	0.138	0.227	0.226	0.211	0.167
1.55	0.199	0.199	0.134	0.134	0.129	0.112
1.65	0.198	0.241	0.220	0.219	0.206	0.167
1.75	0.197	0.233	0.280	0.278	0.255	0.193
1.85	0.196	0.195	0.302	0.299	0.273	0.204
1.95	0.194	0.059	0.068	0.068	0.068	0.068
2.05	0.192	0.096	0.050	0.050	0.051	0.054
2.15	0.189	0.143	0.126	0.125	0.121	0.109
2.25	0.185	0.150	0.229	0.227	0.211	0.167
2.35	0.179	0.122	0.139	0.138	0.134	0.119
2.45	0.171	0.067	0.113	0.113	0.111	0.105

In order to characterize the physical, optical and biological properties of such a vegetation cover, the angular BRDF values were calculated for the following characteristics: (i) chlorophyll content$CAB = 100\ \mu g\ cm^{-2}$; (ii) leaf water equivalent thickness$C_w = = 0.04$ cm; (iii) effective number N_e of elementary layers inside a leaf equal to 1.09; (iv) ratio $C_n = 0.9$, this parameter being calculated as the ratio between the refractive indices of the leaf surface wax and internal material; (v) weight $S_1 = 0.213$ of the first Price (1990) function for the soil reflectance; (vi) Leaf Area Index $U_L = 0.1$; (vii) leaf angle distribution with elliptical eccentricity $E_e = 0.972$; (viii) leaf distribution with modal inclination $Q_m = 10.7°$; and (ix) relative leaf size $S_L = 0.1$ with respect to the canopy depth.

The spectral curve of surface albedo $R_L(\lambda, \theta_o = 60°)$ determined for the VS1 model is shown in Fig. 10.17, while the spectral values of $R_L(\lambda, \theta_o = 60°)$ defined for the VS1 model are reported in Table 10.2 for 21 selected wavelengths over the spectral range from 0.40 to 2.50 μm. Table 10.3 provides the values of reflectance parameters $R_{bs}(\theta_o = 0°)$ and R_{ws}, and those of broadband albedo $A(\theta_o)$ calculated for nine increasing values of θ_o taken

Table 10.3 Values of directional-hemispherical reflectance $R_{bs}(0°)$ (black-sky albedo) at null zenith illumination angle, bi-hemispherical reflectance R_{ws} (white-sky albedo) and broadband albedo $A(\theta_o)$ calculated in terms of Eq. (10.4d) for nine values of solar zenith angle $A(\theta_o)$, as obtained for the six BRDF surface reflectance models used in the present study and the spectral characteristics of the pure continental aerosol model. The values of $A(80°)$ given for the OS1 model is not reported in the table since the OS1 model is affected by very high uncertainties for values of $\theta_o > 75°$

	BRDF surface reflectance models					
Reflectance Parameters	OS1	VS1	PS1	PS2	PS3	PS4
$R_{bs}(0°)$	0.026	0.134	0.824	0.720	0.461	0.214
R_{ws}	0.070	0.153	0.847	0.761	0.536	0.296
$A(0°)(m = 1.000)$	0.030	0.135	0.827	0.726	0.472	0.223
$A(10°)(m = 1.0148)$	0.030	0.134	0.827	0.727	0.475	0.226
$A(20°)(m = 1.0634)$	0.030	0.133	0.830	0.732	0.483	0.234
$A(30°)(m = 1.1536)$	0.035	0.139	0.834	0.739	0.496	0.249
$A(40°)(m = 1.3037)$	0.048	0.141	0.840	0.749	0.514	0.269
$A(50°)(m = 1.5526)$	0.083	0.149	0.847	0.761	0.537	0.296
$A(60°)(m = 1.9928)$	0.193	0.155	0.854	0.775	0.564	0.329
$A(70°)(m = 2.8999)$	0.601	0.175	0.862	0.789	0.591	0.365
$A(80°)(m = 5.5803)$	–	0.202	0.865	0.796	0.608	0.390

in steps of 10° over the 0°–80° range for the VS1 model, the values of $A(\theta_o)$ being obtained through integration of $R_L(\lambda, \theta_o)$ over the 0.40–2.50 μm range. Table 10.4 gives the spectral values of white-sky albedo $R_{ws}(\lambda)$ defined in Eq. (10.4) for the VS1 model at 21 wavelengths ranging from 0.40 to 2.50 μm.

(3) The four PS surface reflectance models have been determined by Tomasi et al. (2013). The main parameters of the PS models are the size-distributions of the snow and black carbon grains, represented by means of log-normal size-distribution functions, and the concentration of dust and/or black carbon particles deposited on the snow surface (Wiscombe and Warren 1980; Warren and Wiscombe 1980). In modeling such surface reflectance features, it was taken into account that the effects produced on reflectance by a volume concentration of black carbon particles equal to 1 ppb are comparable with those produced by a volume concentration of dust particles equal to 100 ppm. Both types of this particulate matter have been observed to cause a relevant reduction of the surface albedo over the $\lambda < 1$ μm range of the solar radiation spectrum, where soot particles produce a rather flat spectral curve of reflectance, and dust particles cause an appreciable decrease in albedo at the visible wavelengths.

Snow grains with typical sizes varying between 50 and 500 μm were found to produce only weak variations of reflectance in the visible part of the solar spectrum. These features agree in general with the recent results found by Kokhanovsky and Breon (2012), who have investigated the anisotropic characteristics of the snow reflectance, and defined a semi-empirical spectral model based on detailed representations of the dependence of reflectance on view zenith angle and azimuthal

Table 10.4 Monochromatic values of white-sky albedo $R_{ws}(\lambda)$ defined in Eq. (10.4b) and shown in Fig. 10.13b, as obtained at 21 wavelengths selected over the 0.40–2.50 μm spectral range for the six BRDF surface reflectance models used in the present

	BRDF surface reflectance models					
Wavelength λ (μm)	OS1	VS1	PS1	PS2	PS4	PS4
0.45	0.087	0.034	0.954	0.857	0.612	0.351
0.55	0.071	0.076	0.953	0.847	0.592	0.330
0.65	0.066	0.106	0.950	0.840	0.578	0.316
0.75	0.065	0.201	0.950	0.836	0.574	0.312
0.85	0.065	0.229	0.928	0.828	0.566	0.305
0.95	0.065	0.244	0.891	0.810	0.558	0.300
1.05	0.065	0.262	0.840	0.781	0.551	0.299
1.15	0.064	0.270	0.809	0.756	0.536	0.289
1.25	0.064	0.277	0.671	0.646	0.495	0.280
1.35	0.064	0.260	0.648	0.626	0.487	0.279
1.45	0.064	0.135	0.193	0.192	0.179	0.140
1.55	0.063	0.197	0.110	0.110	0.105	0.091
1.65	0.063	0.239	0.187	0.186	0.175	0.139
1.75	0.062	0.231	0.242	0.240	0.219	0.163
1.85	0.062	0.192	0.263	0.260	0.236	0.172
1.95	0.061	0.058	0.054	0.054	0.054	0.054
2.05	0.061	0.094	0.039	0.039	0.040	0.043
2.15	0.060	0.141	0.102	0.102	0.099	0.088
2.25	0.059	0.148	0.194	0.192	0.178	0.139
2.35	0.058	0.120	0.113	0.113	0.109	0.097
2.45	0.055	0.066	0.091	0.091	0.090	0.085

reflectance variations. Thus, various simulations of surface albedo were made by assuming that different additional concentrations of soot particles are present (having log-normal size-distribution curves centred at radius $r = 0.050\ \mu$m) together with the main component of the surface layer constituted by snow grains having a mean radius of 100 μm. In order to determine the spectral values of single scattering albedo $\omega(\lambda)$ and asymmetry factor $g(\lambda)$ of the snow grains, use was made of the model of fractal grains defined by Kokhanovsky and Zege (2004) and by Kokhanovsky et al. (2005, 2011), which allowed us to obtain: (i) evaluations of the asymmetry parameters close to the *in situ* measurements of these parameters; and (ii) results similar to those reported in the ground measurements of snow angular reflectance.

Using these data, the spectral values of single scattering albedo $\omega(\lambda)$ and asymmetry factor $g(\lambda)$ of such snow grain size-distributions were calculated at 217 selected wavelengths over the 0.40–2.50 μm range, according to the Warren and Wiscombe (1980) and Warren (1984) estimates. The two parts of the complex refractive index of snow grains were determined by taking into account the optical properties of black carbon particles, which have been calculated at the same 217 wavelengths by applying an interpolation procedure in wavelength for the 11 monochromatic values provided by the 6S parameterization for the aerosol soot component.

Following this procedure, values of the real part $n(\lambda)$ ranging between 1.75 and 1.90 were obtained, together with values of the imaginary part $\chi(\lambda)$ varying between 0.43 and 0.57 over the entire solar spectrum. The direct and diffuse solar radiation components were then calculated following the semi-empirical parameterization method of Wiscombe and Warren (1980), based on the use of the above-calculated values of optical parameters $\omega(\lambda)$ and $g(\lambda)$ of the bi-modal size-distribution consisting of soot particles and snow grains. Four models were obtained by following the above procedure, in which: (i) the optical properties of ice and snow measured in the Arctic were taken into account, and (ii) the following increasing values of the soot volume concentration C^* (i.e. of the relative content of soot particles inside the snow particle, measured in ppm) were assumed: (1) $C^* = 2\ 10^{-3}$ ppm in the PS1 model representing almost pure snow; (2) $C^* = 4\ 10^{-2}$ ppm in the PS2 model used to represent slightly contaminated snow; (3) $C^* = 4\ 10^{-1}$ ppm in the PS3 model used to represent a type of considerably contaminated snow (according to the measurements performed by Grenfell and Maycut (1977)), and (4) $C^* = 2$ ppm in the PS4 model here adopted to represent the heavy black-carbon contaminated snow.

The spectral curves of surface albedo $R_L(\lambda, \theta_o = 60°)$ determined for the four above-described PS models are shown in Fig. 10.17, while the monochromatic values of $R_L(\lambda, \theta_o = 60°)$ obtained for the four PS models and for the OS1 and VS1 models are given in Table 10.2 at the 21 above-selected wavelengths from 0.40 to 2.50 μm, and the corresponding values of the broadband albedo $A(\theta_o)$ determined at

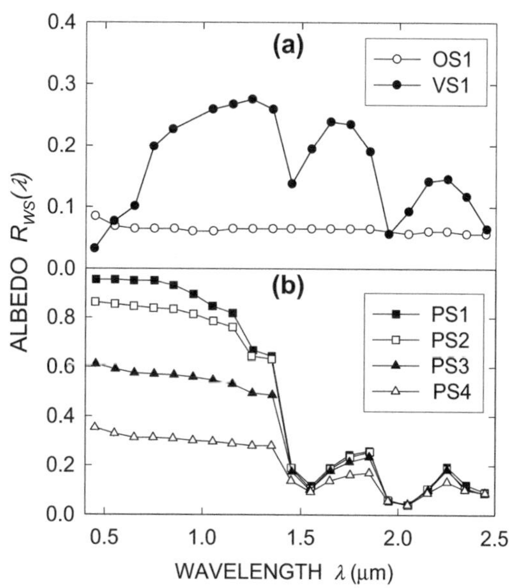

Fig. 10.17 Upper part (**a**): Spectral curves of white-sky albedo $R_{ws}(\lambda)$ defined in Eq. (10.2) over the 0.40–2.50 μm spectral range for the BRDF surface reflectance models OS1 (open circles) and VS1 (solid circles). Lower part (**b**): As in the upper part but for the four Polar Surface reflectance models determined by Tomasi et al. (2013) and labelled PS1 (open triangles), PS2 (solid triangles), PS3 (opend diamonds), and PS4 (solid diamonds). All the six reflectance models were determined for solar zenith angle $\theta_o = 60°$

nine increasing values of θ_o from 0° to 80° are reported in Table 10.3 for all the four PS models (and the OS1 and VS1 models), being all calculated through integration of $R_L(\lambda, \theta_o)$ over the 0.40–2.50 μm range, by assuming that the atmospheric aerosol content consists of the pure continental aerosol particles represented in terms of the M-8 model defined by Tomasi et al. (2013), giving (i) an aerosol optical thickness equal to 0.10 at the 0.55 μm wavelength, (ii) Ångström wavelength exponent equal to 1.15 to represent the background optical properties of the Arctic aerosol, (iii) single scattering albedo equal to 0.852 at the visible wavelength $\lambda = 0.55$ μm, and (iv) mass density equal to 2.161 g cm^{-3}. Thus, broadband albedo is given by the ratio between: (1) the integral over the 0.40–2.50 μm range of the product of surface albedo $R_L(\lambda, \theta_o)$ by the incoming solar irradiance $I{\downarrow}(\lambda, \theta_o)$ at the surface, and (2) the integral over the 0.40–2.50 μm range of the incoming solar irradiance $I{\downarrow}(\lambda, \theta_o)$ at the surface. The solar irradiance $I{\downarrow}(\lambda, \theta_o)$ reaching the surface-level was assumed to be attenuated by the M-8 aerosol particles, which consist of 67.29% dust-like (DL) particles, 27.88% water-soluble (WS) particles, 0.96% soot (SO) particles, and 3.87% liquid water (LW) particles, being the DL, WS and SO components defined in the 6S code.

The snow surface reflectance was found to decrease gradually as the wavelength increases beyond 1 μm, presenting throughout the middle infrared the typical features of the water absorption bands labelled with the Greek capital letters Ψ (over the 1.25–1.54 μm spectral range) and Ω (over the 1.69–2.08 μm range). The broadband white-sky albedo of snow-covered surfaces is given at each wavelength by the ratio $R_{ws}(\lambda) = I_{srf}{\uparrow}(\lambda)/I_{srf}{\downarrow}(\lambda)$ between the upwelling short-wave global (solar) irradiance $I_{srf}{\uparrow}(\lambda)$ reflected up-ward by the snow-covered surface and the down-ward short-wave global (solar) irradiance $I_{srf}{\downarrow}(\lambda)$ reaching the surface, both the irradiance terms being calculated over the semi-hemispherical space. Broadband white-sky albedo R_{ws} of snow-covered surfaces is estimated at visible and near-infrared wavelengths to vary usually from 0.30 for extremely polluted snow cover conditions (such those of the PS4 model) to more than 0.85 for clean snow cover conditions (such as those represented by the PS1 model).

10.6.3 Calculation of the Diurnally Averaged Aerosol Forcing

Some examples of the large variations of the incoming short-wave global solar irradiance calculated on various spring and summer days at Ny-Ålesund of 2017 are shown in Fig. 10.10. Correspondingly, the DARF effects induced by aerosols are subject to vary from one hour to another of the same day, from one day to another and from one season to the other. To reduce the effects of solar radiation variability during a day, the so-called "diurnally averaged aerosol forcing" (hereinafter referred to as ΔDF) is calculated as the integral of instantaneous radiative forcing $\Delta F(t)$ at the ToA- or the BoA-levels over the sunlit period (from sunrise to sunset) and divided by the 24-hour period (Bush and Valero 2003; Xu et al. 2003; Kim et al. 2005), as made on the basis of the following two equations,

$$\Delta DF_{ToA} = \int_{sunrise}^{sunset} \Delta F_{ToA}(t)\ dt\ /\ 24\ hr \tag{10.15}$$

and

$$\Delta DF_{BoA} = \int_{sunrise}^{sunset} \Delta F_{BoA}(t)\ dt\ /\ 24\ hr, \tag{10.16}$$

determined for the DARF effects at the ToA- and BoA-levels, respectively, and considering that the sunlit period varies as a function of latitude and season in both Eqs. (10.15) and (10.16). The evaluations of ΔDF_{ToA} and ΔDF_{BoA} obtained from field measurements performed at an Arctic site may be subject to significant variations from one season to another, as shown in the examples of Fig. 10.10.

The absolute magnitude of the DARF terms at the ToA-level closely depends on the fraction of solar radiation entering the atmosphere, but also on the amount of particulate matter present within the atmospheric column and the optical characteristics of airborne aerosols. An appropriate parameter for defining the columnar particle amount is the aerosol optical thickness *AOT(λ)*, which provides a measure of the monochromatic extinction of solar radiation occurring along the vertical atmospheric path. Thus, in cases where *AOT(λ)* increases, the radiative forcing is expected to increase through features closely related to the optical characteristics of airborne aerosols. The absolute magnitude of the ToA-level DARF effect is given by the amount of up-welling solar radiation reflected back by the surface-atmosphere system towards space, as a result of aerosol scattering processes. This quantity depends closely on the number of aerosol particles present in the atmospheric column and extinguishing the incoming solar radiation, which is approximately proportional to *AOT(λ)*. New calculations of the DARF effects are presented in the following section, as obtained from field measurements of *AOT(λ)* and aerosol optical characteristics measured in the Arctic region, by paying particular attention in order to define: (i) the aerosol radiative parameters measured within the vertical atmospheric column by means of ground-based remote sensing techniques and *in situ* sampling measurements, and (ii) the spectral and directional characteristics of surface reflectance models adopted to to represent the surface albedo characteristics of the various Arctic areas.

10.6.4 Calculations of the DARF Effects at the ToA- and BoA-Levels of the Arctic Atmosphere

Regular and precise measurements of *AOT(λ)* were performed at several visible and near-infrared wavelengths during the POLAR-AOD campaigns conducted at various Arctic sites (Tomasi et al. 2007), from which measurements of the Ångström

(1964) exponent α were obtained, together with additional measurements of the aerosol physical, chemical, and optical properties. The DARF effects induced by various Arctic aerosol types were evaluated from these measurements by following the procedure defined by Tomasi et al. (2014). The main purpose was to determine the daily time-patterns of the instantaneous DARF terms and calculate the diurnally averaged DARF effects for various surface albedo models. The analysis of experimental data was made through the six following steps:

(1) *Analysis of field data to determine the columnar aerosol extinction parameters*

The daily mean values of *AOT(λ)* at visible and near-infrared wavelengths were determined by examining the daily sets of sun-photometer measurements performed by the various POLAR-AOD research groups over the past 10 years (Tomasi et al. 2007, 2012). The most significant estimates of *AOT(0.50 μm)* and exponent α were carried out at the following five polar sites:

(i) Barrow, in northern Alaska, where measurements of *AOT(λ)* and exponent α were performed by NOAA/GMD (Boulder, Colorado, USA) using the SP02 and SP022 Carter Scott sun-photometers from February 2002 to October 2010 (Tomasi et al. 2012). Four different types of atmospheric turbidity conditions were studied during the 2002–2010 period by determining the daily mean values of *AOT(0.50 μm)* and *α(0.368–0.862 μm)* for: (a) Arctic Haze (AH) usually observed in the January – May period and giving average values of *AOT(0.50 μm)* = 0.116 and $\alpha = 1.28$; (b) background summer aerosol, with average values of *AOT(0.50 μm)* = 0.078 and $\alpha = 1.40$; (c) Asian Dust (AD), observed during some transport episodes monitored in spring 2002, for which a peak value of *AOT(0.50 μm)* = 0.20 was measured by Stone et al. (2007), with a mean value of $\alpha = 0.80$, clearly due to prevailing extinction by coarse particles; and (d) Boreal Forest Fire smoke (hereinafter referred to as BFF), for which the average value of *AOT(0.50 μm)* = 0.30 was measured in July 2004, with an average value of $\alpha = 1.20$, which clearly indicates that the most pronounced optical effects were produced by fine particles in such episodes (Tomasi et al. 2007).

(ii) Ny-Ålesund, where measurements of *AOT(λ)* and α were performed by AWI (Bremerhaven, Germany) using the SP1A and SP2H sun-photometers and the STAR01 star-photometer from May 2006 to September 2010 (Herber et al. 2002). Three different sets of atmospheric turbidity conditions were examined for: (a) AH conditions in winter and spring, with average values of *AOT(0.50 μm)* = 0.080 and α(0.380–0.864 μm) = 1.25; (b) Arctic Dense Summer (ADS) aerosol, with average values of *AOT(0.50 μm)* = 0.12 and $\alpha = 1.00$; and (c) background summer aerosol, with average values of *AOT(0.50 μm)* = 0.041 and $\alpha = 1.20$.

(iii) Summit where measurements of *AOT(λ)* and α were performed for background summer aerosol extinction features since 2002 by PMOD/WRC (Davos, Switzerland) using the PFR sun-photometers, obtaining the average values of *AOT(0.50 μm)* = 0.039 ± 0.020 and $\alpha = 1.48 \pm 0.10$.

(iv) Sodankylä, in northern Finland, where measurements of *AOT*(λ) and α were performed by FMI (Helsinki, Finland) during various campaigns from 2004 to 2012, using a pair of PFR sun-photometers and finding average values of *AOT*(0.50 μm) = 0.066 and α = 1.25 for AH episodes observed in winter and early spring, and of *AOT*(0.50 μm) = 0.060 and α = 1.42 for background summer aerosol extinction characteristics.

(v) Tiksi, where measurements of *AOT*(λ) and α were taken in summer 2010 with the AERONET Cimel CE-318 sun-photometer of NASA/GSFC (Holben et al. 1998), obtaining the seasonal average values of *AOT* (0.50 μm) = 0.085 and α = 1.60 for background (BG) summer aerosol.

(2) *Determination of the columnar aerosol refractive index*

Due to the great variety of both natural and anthropogenic particle sources in the Arctic region (Hirdman et al. 2010), the different aerosol types observed at the five above-mentioned Arctic sites were assumed to consist of linear combinations of a fine/accumulation particle mode and an accumulation/coarse particle mode. The columnar aerosol number contents of such two particle modes have been then varied to fit the average median values of *AOT*(0.50 μm) and exponent α(0.40–0.87 μm) given in Table 10.5. According to the above-mentioned measurements of particulate chemical composition, the fine/accumulation particles were assumed to consist mainly of water-soluble, soot, sea salt, and insoluble substances, and the accumulation/coarse particles of variable mass fractions of water-soluble, sea salt, insoluble matter, and mineral dust transported from the remote mid-latitude regions of the Northern hemisphere. For these mixed characteristics of the multimodal size-distribution curves, it was assumed that the fine/accumulation and accumulation/coarse particles consist of the mass mixing percentages given in Table 10.6. Thus, parameters $n(\lambda)$ and $\chi(\lambda)$ were calculated in terms of the mass composition percentages assumed by Hess et al. (1998) for the OPAC components calculated at RH = 50%, which exhibit chemical composition features varying from one particle size-class to another.

On the basis of these assumptions, the monochromatic values of $n(\lambda)$ and $\chi(\lambda)$ were separately determined for the fine/accumulation and accumulation/coarse particle modes, by calculating the weighted averages of such values for the best-fit evaluations of the number concentrations of fine/accumulation and accumulation/coarse particles, yielding the median values of *AOT*(0.50 μm) and α(0.40–0.87 μm) given in Table 10.5. The overall values of $n(0.50\ \mu m)$ and $\chi(0.50\ \mu m)$ obtained as linear combinations of such optical parameters calculated separately for the fine/accumulation and the accumulation/coarse particle fractions given in Table 10.6 are provided in Table 10.7. The overall spectral curves of $n(\lambda)$ and $\chi(\lambda)$ are shown in: (a) Fig. 10.18 for the four Arctic aerosol models defined at Barrow (AH, AD, BFF particles, and BG summer aerosol); (b) Fig. 10.19 for the three Arctic aerosol models defined at Ny-Ålesund (AH, ADS aerosol, and BG summer aerosol); and (c) Fig. 10.20 for the BG summer aerosol model determined at Summit, the

Table 10.5 Daily median values of aerosol optical thickness *AOT(0.50 μm)* and Ångström's (1964) exponent α determined over the 0.40–0.87 μm range from the sun-photometer measurements performed at five Arctic sites during the POLAR-AOD campaigns conducted for different atmospheric turbidity conditions

Arctic site	Geographical coordinates and altitude	Aerosol type and season	Average median value of *AOT(0.50 μm)*	Average median value of exponent α*(0.40–0.87 μm)*
Barrow (Alaska)	71° 19' N, 156° 36' W, 8 m a.m.s.l.	Arctic haze (winter – Early spring)	0.116	1.28
		Asian dust	0.200	0.80
		Boreal Forest fire smoke	0.300	1.20
		Background summer aerosol	0.078	1.40
Ny-Ålesund (Spitsbergen, Svalbard)	78° 54' N, 11° 53′ E, 5 m a.m.s.l.	Arctic haze (winter – Early spring)	0.080	1.29
		Arctic dense summer aerosol	0.120	1.00
		Background summer aerosol	0.041	1.20
Summit (Greenland ice sheet)	72° 20' N, 38° 45' W, 3270 m a.m.s.l.	Background summer aerosol	0.039	1.48
Sodankylä (Finland)/	67° 22' N, 26° 38′ E, 184 m a.m.s.l.	Arctic haze (winter – Early spring)	0.066	1.25
		Background summer aerosol	0.060	1.42
Tiksi (Northern-Central Siberia)	71° 35' N, 128° 47′ E, 40 m a.m.s.l.	Background summer aerosol	0.085	1.60

AH and BG summer aerosol models determined at Sodankylä (Finland), and the BG summer aerosol determined at Tiksi. Figures 10.18, 10.19, and 10.20 show that $n(\lambda)$ varies in the visible between about 1.40 (for the AH model observed at Sodankylä, mainly consisting of polluted aerosol) and about 1.53 (for the AD model determined at Barrow), while $\chi(\lambda)$ was correspondingly estimated to range in the visible between 3.7 10^{-3} (for the Ny-Ålesund ADS aerosol model) and 6.3 10^{-2} (for the Barrow AD model). Particularly low values of $\chi(\lambda)$ were estimated at visible wavelengths for the BG summer aerosol monitored at Ny-Ålesund (see Fig. 10.19) and for the AH case observed at Sodankylä (see Fig. 10.20).

Table 10.6 Mass mixing percentages of fine and coarse particle loads defined in terms of the OPAC (relative humidity RH = 50%) components (Hess et al. 1998), as assumed at the various POLAR-AOD Arctic sites for the different atmospheric turbidity conditions defined by Tomasi et al. (2007, 2012)

Arctic site	Aerosol type and season	Mass mixing percentages separately assumed for the fine/accumulatyion and the accumulation/coarse particles defined by Hess et al. (1998) to determine the composition of the two aerosol components					
		Water-soluble (fine)	Sea salt (accum.)	Sea salt (coarse)	Insoluble (accum.)	Soot (fine)	Mineral (coarse)
Barrow (Alaska)	Arctic haze (winter – early spring):						
	Fine/accum. particle fraction	41.3%	34.9%	–	21.8%	2.0%	–
	Accum./coarse particle fraction	4.1%	–	57.6%	38.3%	–	–
	Asian dust:						
	Fine/accum. particle fraction	24.0%	10.0%	–	66.0%	–	–
	Accum./coarse particle fraction	6.0%	–	4.0%	30.0%	–	60.0%
	Boral Forest fire smoke:						
	Fine/accum. particle fraction	34.0%	29.0%	–	35.0%	2.0%	–
	Accum./coarse particle fraction	29.0%	–	35.0%	35.0%	1.0%	–
	Background summer aerosol						
	Fine/accum. particle fraction	35.0%	17.8%	–	46.0%	1.2%	–
	Accum./coarse particle fraction	1.3%	–	70.9%	27.8%	–	–
Ny-Ålesund (Spitsbergen, Svalbard)	Arctic haze (winter – early spring):						
	Fine/accum. particle fraction	98.0%	–	–	–	2.0%	–
	Accum./coarse particle fraction	–	–	90.9%	9.1%	–	–
	Arctic dense summer aerosol						
	Fine/accum. particle fraction	35.0%	10.0%	–	54.0%	1.0%	–
	Accum./coarse particle fraction	5.0%	–	95.0%	–	–	–
	Background summer aerosol						
	Fine/accum. particle fraction	98.0%	–	–	–	2.0%	–
	Accum./coarse particle fraction	–	–	50.0%	50%	–	–

(continued)

Table 10.6 (continued)

Arctic site	Aerosol type and season	Mass mixing percentages separately assumed for the fine/accumulatyion and the accumulation/coarse particles defined by Hess et al. (1998) to determine the composition of the two aerosol components					
		Water-soluble (fine)	Sea salt (accum.)	Sea salt (coarse)	Insoluble (accum.)	Soot (fine)	Mineral (coarse)
Summit (Greenland)	Background summer aerosol						
	Fine/accum. particle fraction	98.0%	–	–	–	2.0%	–
	Accum./coarse particle fraction	–	–	50.0%	50.0%	–	–
Sodankylä (Finland)	Arctic haze (winter – early spring):						
	Arctic (AR) OPAC model with RH = 50%	40.6%	54.4%	–	3.0%	2.0%	–
	Background summer aerosol						
	Fine/accum. particle fraction	98.0%	–	–	–	2.0%	–
	Accum./coarse particle fraction	–	–	50.0%	50.0%	–	–
Tiksi (Siberia)	Background summer aerosol						
	Fine/accum. particle fraction	98.0%	–			2.0%	–
	Accum./coarse particle fraction	–	–	50.0%	50.0%		–

Table 10.7 Daily mean values of the real part $n(0.50\ \mu m)$ and imaginary part $\chi(0.50\ \mu m)$ of the complex refractive index, the volume extinction coefficient $\beta_{ext}(0.50\ \mu m)$ (obtained for the overall columnar particle number contents derived for the median values of $AOT(\lambda)$ given in Table 10.5), and the average values of columnar aerosol single scattering albedo $\omega(0.55\ \mu m)$ determined for the various Arctic aerosol types listed in Table 10.5 at the various Arctic sites, and for the radiative characteristics of columnar aerosol particles defined in Table 10.6

Aerosol type	Measurement site	Average values of columnar aerosol refractive index at the 0.50 μm wavelength		Average columnar volume extinction coefficient $\beta_{ext}(0.50\ \mu m)$(km^{-1})	Columnar aerosol single scattering albedo $\omega(0.55\ \mu m)$
		Real part $n(0.50\ \mu m)$	Imaginary part $\chi(0.50\ \mu m)$		
Background summer aerosol	Barrow	1.444	$9.2\ 10^{-3}$	$7.44\ 10^{-3}$	0.978
	Ny-Ålesund	1.461	$7.9\ 10^{-3}$	$3.92\ 10^{-3}$	0.966
	Summit	1.449	$8.8\ 10^{-3}$	$3.27\ 10^{-3}$	0.969
	Sodankylä	1.452	$8.6\ 10^{-3}$	$5.71\ 10^{-3}$	0.965
	Tiksi	1.444	$9.2\ 10^{-3}$	$8.04\ 10^{-3}$	0.977
Arctic haze (winter – Early spring)	Barrow	1.399	$3.3\ 10^{-3}$	$1.11\ 10^{-2}$	0.840
	Ny-Ålesund	1.424	$7.3\ 10^{-3}$	$7.63\ 10^{-3}$	0.949
	Sodankylä	1.399	$3.9\ 10^{-3}$	$6.26\ 10^{-3}$	0.840
Arctic dense summer aerosol	Ny-Ålesund	1.437	$3.7\ 10^{-3}$	$1.16\ 10^{-2}$	0.852
Asian dust	Barrow	1.527	$6.3\ 10^{-3}$	$1.99\ 10^{-2}$	0.858
Boreal Forest fire smoke	Barrow	1.469	$2.5\ 10^{-3}$	$2.89\ 10^{-2}$	0.758

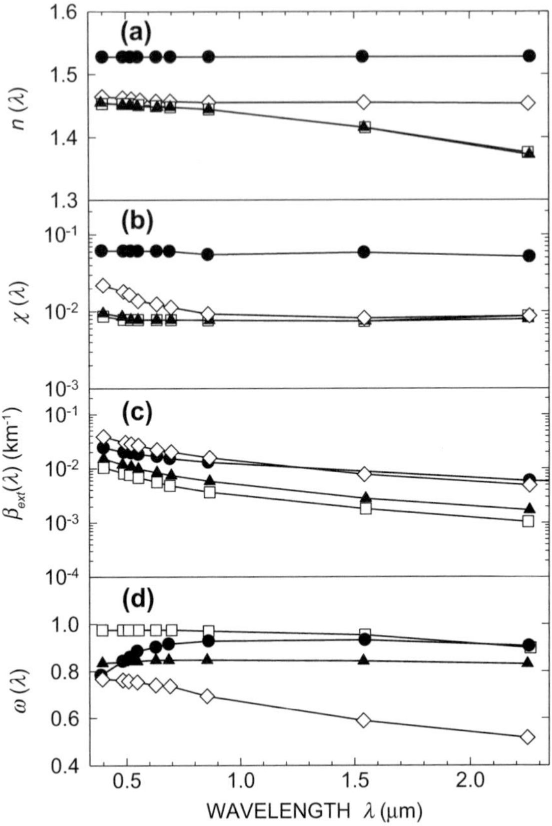

Fig. 10.18 Spectral curves of the real part $n(\lambda)$ and imaginary part $\chi(\lambda)$ of the aerosol refractive index (shown in parts (**a**) and (**b**), respectively), volume extinction coefficient $\beta_{ext}(\lambda)$ (part (**c**)), and columnar aerosol single scattering albedo $\omega(\lambda)$ (part (**d**)), determined over the 0.45–2.50 μm range for the following aerosol size-distribution models defined on the basis of the sun-photometer measurements performed at Barrow (Alaska, USA) for: (i) Arctic haze (solid triangles), (ii) Asian dust (solid circles), (iii) particles from boreal forest fires (open diamonds), and (iv) background summer aerosol (open squares). The corresponding size-distribution curves are shown in Fig. 10.31, while the linear combinations of fine and coarse particle modes are based on the use of the mass fractions given in Table 10.5 for the various OPAC components

(3) *Determination of the size-distribution curves of columnar aerosol*

The eleven multimodal size-distribution curves of the BG summer aerosol, and ADS, AH, AD, and BFF smoke particles observed at the five above-mentioned Arctic sites were determined by using the same procedure adopted in the previous step to define the values of $n(\lambda)$ and $\chi(\lambda)$: in this case, the best-fit linear combinations of the fine/accumulation and accumulation/coarse particle modes and their columnar number concentrations were made to vary until fitting the median values of $AOT(0.50\ \mu m)$ and $\alpha(0.40–0.87\ \mu m)$ given in Table 10.5. The

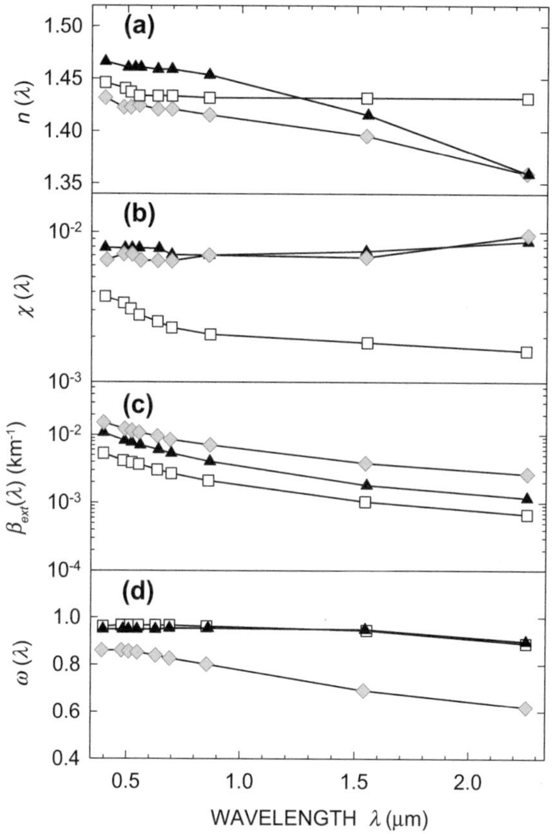

Fig. 10.19 Spectral curves of the real part $n(\lambda)$ and imaginary part $\chi(\lambda)$ of the particulate matter refractive index (shown in parts (**a**) and (**b**), respectively), volume extinction coefficient $\beta_{ext}(\lambda)$ (part (**c**)), and columnar aerosol single scattering albedo $\omega(\lambda)$ (part (**d**)), determined over the 0.45–2.50 μm range for the following aerosol size-distribution models defined on the basis of the sun-photometer measurements performed at Ny-Ålesund for: (i) Arctic haze (solid triangles), (ii) Arctic Dense Summer aerosol (grey diamonds), and (iii) background summer aerosol (open squares). The corresponding size-distribution curves are shown in Fig. 10.21, while the linear combinations of fine and coarse particle modes are based on the use of the mass fractions given in Table 10.5 for the various OPAC components

composition of the fine/accumulation particles was assumed to be given by (i) most significant mass fractions of water-soluble (W-S) and insoluble substances, and sea salt, together with small fractions of soot substances in the AH and BG summer aerosol cases; (ii) a prevailing mass fraction of insoluble substances and mineral dust in the AD case; and (iii) relevant mass fractions of water-soluble and insoluble substances, and a relatively low fraction of small sea-salt aerosol, together with a rather marked fraction of soot substances in the BFF smoke case. The accumulation/coarse particles were assumed to consist mainly of water-soluble,

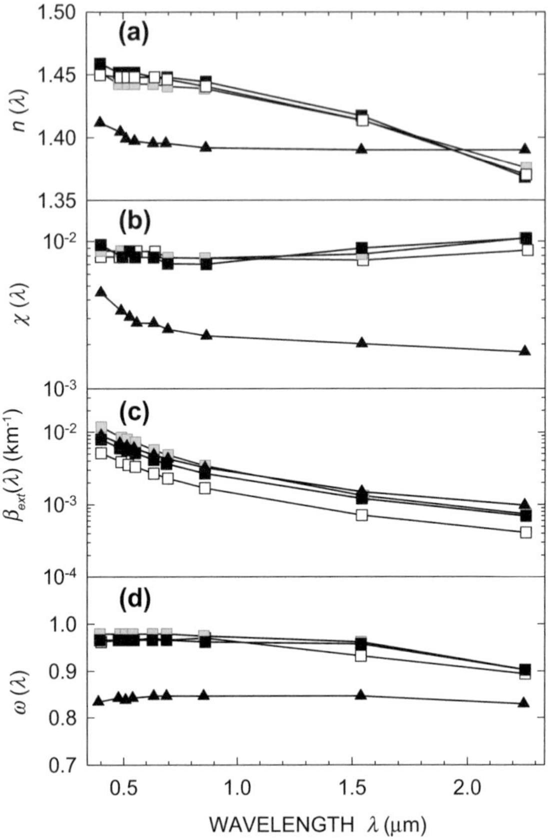

Fig. 10.20 Spectral curves of the real part $n(\lambda)$ and imaginary part $\chi(\lambda)$ of the particulate matter refractive index (shown in parts (**a**) and (**b**), respectively), volume extinction coefficient $\beta_{ext}(\lambda)$ (part (**c**)), and columnar aerosol single scattering albedo $\omega(\lambda)$ (part (**d**)), determined over the 0.45–2.50 μm range for the following aerosol size-distribution models defined on the basis of the sun-photometer measurements performed at: (i) Summit for background summer aerosol (open squares); (ii) Sodankylä for Arctic haze (solid triangles); (iii) Sodankylä for background summer aerosol (solid squares); and (iv) Tiksi for background summer aerosol (grey squares). The corresponding size-distribution curves are shown in Fig. 10.21, while the linear combinations of fine and coarse particle modes are based on the use of the mass fractions given in Table 10.5 for the various OPAC components

sea-salt coarse, and insoluble particles. For the mass mixing percentages given in Table 10.6 for the various OPAC components of the fine and coarse particle modes, the overall particle size-distribution curves were found to consist in general of 2–4 modes of fine/accumulation particles, and of 1–4 modes of accumulation/coarse particles, which are in part superimposed, as can be seen in Fig. 10.21 for the Arctic BG aerosol cases.

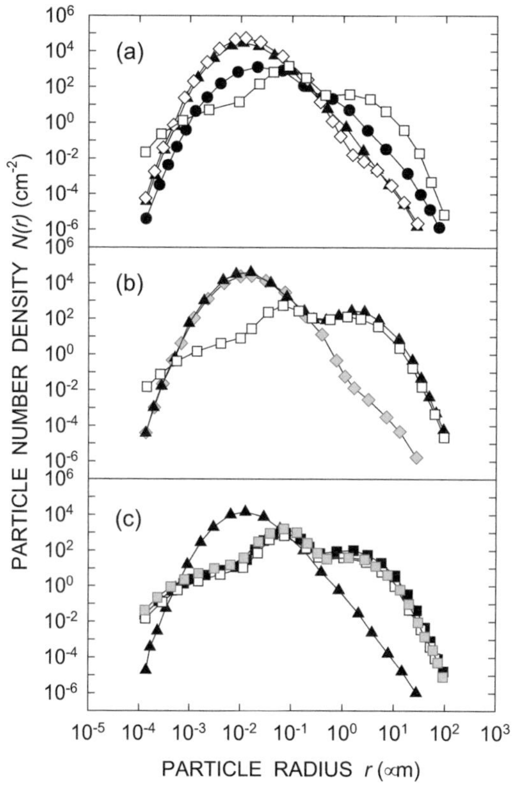

Fig. 10.21 Multimodal size-distribution curves of columnar particle number density $N(r) = dN/d(\ln r)$ in the vertical atmospheric column of unit cross section (and therefore measured in cm^{-2}) as a function of particle radius. The curves of $N(r)$ have been obtained as linear combinations of the OPAC components defined for relative humidity RH = 50% to best-fit the atmospheric turbidity parameters given in Table 10.4 and weighted using the mass percentages given in Table 10.5, to represent: (**a**) the following four Arctic aerosol types monitored at Barrow: (i) Arctic haze (solid triangles), (ii) Asian dust (solid circles), (iii) particles from boreal forest fires (open diamonds), and (iv) background summer aerosol (open squares); (**b**) the following three Arctic aerosol types monitored at Ny-Ålesund: (i) Arctic haze (solid triangles), (ii) Arctic Dense Summer aerosol (grey diamonds), and (iii) background summer aerosol (open squares); and (**c**) the four aerosol types monitored at (i) Summit for background summer aerosol (open squares), (ii) Sodankylä for Arctic haze (solid triangles) and background summer aerosol (solid squares), and (iii) Tiksi for background summer aerosol (grey squares)

In particular, Fig. 10.21 shows the size-distribution curves of columnar total particle number density $N(r)$ determined for (i) the AH case observed at Barrow, (ii) the BFF smoke case observed at Barrow, (iii) the ADS aerosol case monitored at Ny-Ålesund, and (iv) the AH case measured at Sodankylä. They mainly consist of fine particle modes, with the volume fractions of fine particles prevailing over those of coarse particles by at least one order of magnitude. In the AD case monitored at Barrow, the size-distribution curve shown in Fig. 10.21 exhibits clearly bimodal features, with a largely predominant contribution of coarse particles. In the other cases, pertaining to the AH case observed at Ny-Ålesund, and the five BG summer aerosol cases monitored at Barrow, Ny-Ålesund, Summit, Sodankylä, and Tiksi, the size-distribution curves of $N(r)$ were found to show substantially bimodal characteristics, with a predominant volume content of coarse particles in the atmospheric column, which is higher than that of overall fine particles by several orders of magnitude.

(4) *Determination of the columnar aerosol single-scattering albedo*

The spectral values of volume coefficients $\beta_{sca}(\lambda)$, $\beta_{abs}(\lambda)$, and $\beta_{ext}(\lambda)$ were calculated for all the Arctic aerosol types considered in the present analysis, with the main purpose of determining the spectral characteristics of columnar single scattering albedo $\omega(\lambda)$ for the various Arctic aerosol types. Figures 10.18, 10.19, and 10.20 show the spectral curves of $\beta_{ext}(\lambda)$ calculated for the overall columnar number contents of aerosol particles equal to those derived from the median values of $AOT(\lambda)$ given in Table 10.5. They present similar spectral patterns for all the aerosol types monitored at the five Arctic sites, except for the AD case observed at Barrow, which was found to exhibit more slowly decreasing features, due to the significant extinction effects produced by the coarse particles mobilized in the desert regions.

The spectral values of average $\omega(\lambda)$ were calculated along the vertical atmospheric path for all the multimodal size-distribution curves defined at the Arctic sites and shown in Figs. 10.18, 10.19, and 10.20 for the 11 above-defined Arctic aerosol models, as determined for the spectral values of $n(\lambda)$ and $\chi(\lambda)$ shown in the upper parts of such graphs. The values of $\omega(\lambda)$ were estimated to vary at visible wavelengths between about 0.76 (for the BFF smoke case monitored at Barrow) and nearly 0.98 (for the BG summer aerosol cases observed at Barrow and Tiksi). Single scattering albedo $\omega(\lambda)$ exhibits slowly decreasing spectral patterns from the visible to the near-infrared, except for the AD case observed at Barrow, in which $\omega(\lambda)$ was estimated to increase from less than 0.8 to more than 0.9 over the 0.40–1.00 range. Relatively low values of $\omega(\lambda)$ were found at: (i) Barrow, for BFF particles (slightly lower than 0.80 at visible wavelengths, as can be seen in Fig. 10.18) because of the high content of soot matter generated by the forest fires), (ii) Ny-Ålesund, for the ADS aerosol, in which a certain fraction of anthropogenic-industrial particles was found, being transported from Northern Europe over long distances (presenting values of $\omega(\lambda)$ close to 0.85 at visible wavelengths, as shown in Fig. 10.19), and (iii) Sodankylä, for AH mainly transported from Northern Europe, in which values of $\omega(\lambda)$ ranging between 0.82 and 0.84 were determined at visible wavelengths, as can be noted in Fig. 10.20.

(5) *Determination of the local surface albedo models*

The surface reflectance characteristics over the Arctic Ocean are known to vary considerably with the local seasonal climatic conditions: ocean surfaces are usually frosted in the winter months, presenting high-reflectance characteristics for both ice and snow coverages, while the ice-free sea surface usually observed near the coast in late spring and summer presents relatively low surface reflectance characteristics, which can vary considerably as a function of wind velocity at sea level and wave motion. The ocean surface reflectance characteristics observed in summer at Barrow, Ny-Ålesund, Sodankylä, and Tiksi on clear-sky days of boreal late spring and summer were assumed to be most realistically represented by the OS1 surface albedo model defined by Tomasi et al. (2013) for a wind velocity of 2 m/s at sea level, giving values of white-sky albedo $R_{ws} = 0.069$, black-sky albedo $R_{bs}(\theta_o = 0°) = 0.026$, and broadband albedo $A(\theta_o = 60°) = 0.193$. Conversely, the land areas can present very different surface reflectance features in the Arctic, since they depend on the type of vegetation coverage and on the snow precipitation regime, therefore varying with season and latitude. In the Arctic region, there are two main types of vegetation coverage that are most commonly referred to as "taiga" and "tundra". Taiga covers extended areas of coniferous forests, mostly consisting of pines, spruces, and larches, which grow in inland Canada and Alaska in North America, some low-land/coastal areas of Iceland, most of the Scandinavian peninsula, and a large area of Russia, from Karelia to the eastern Pacific Ocean coast, thus including northern Siberia. Tundra vegetation consists of dwarf shrubs, sedges and grasses, mosses, lichens, and sometimes scattered trees, because the tree growth is hindered by low temperatures and short growing seasons. For this reason, Arctic tundra occurs north of the taiga belt, in remote areas where the subsoil is permafrost, or permanently frozen soil. Examining the sets of MCD43C3 products derived from the MODIS Level 3.0 surface albedo data recorded over land, within all the seven MODIS spectral bands, the average white-sky albedo maps were determined in the Arctic region during the local summer period (July 2009) (Tomasi et al. 2014), showing values of R_{ws} mainly ranging between 0.15 and 0.30 at Barrow and Sodankylä, and equal to around 0.60 at Ny-Ålesund and around 0.80 at Summit, and close to 0.40 at Tiksi. Therefore, the taiga surface reflectance characteristics near Barrow, Sodankylä, and Tiksi can be realistically represented in summer by using the VS1 surface albedo model of Tomasi et al. (2013), which yields average values of $R_{ws} = 0.153$, $R_{bs}(\theta_o = 0°) = 0.134$, and $A(\theta_o = 60°) = 0.155$. Conversely, the surface albedo conditions observed at Summit can be well represented in summer by using the PS1 surface albedo model, and those over the snow-free land areas surrounding Ny-Ålesund by using model PS4.

The analysis of the MODIS Level 3.0 surface albedo data recorded over land in the Arctic region, during early April of 2009, provided a white-sky albedo map presenting typical spring conditions in the Arctic and showing that the sun-photometer stations are surrounded by areas covered by snow fields and glaciers in that season, with rather high surface albedo conditions, varying mainly between 0.50 and 0.75. Raschke and Ohmura (2005) estimated that the surface albedo observed in summer (June – August) over the Arctic areas covered by glaciers ranges: (i) between 0.57 and 0.85 in the accumulation areas (with a most frequent value of 0.75 in general, and an average value of 0.82 on the Greenland ice sheet); (ii) between 0.37 and 0.77 in the equilibrium line area (with a most frequent value of 0.65 in general, and an average value of 0.75 on the Greenland ice sheet); and (iii) between 0.11 and 0.70 in the ablation area (with a most frequent value of 0.35 in general, and an average value of 0.55 on the Greenland ice sheet). Therefore, the spectral curves of surface albedo $R_L(\lambda)$ and white-sky albedo $R_{ws}(\lambda)$ shown in Figs. 10.16 and 10.17, respectively, for the PS1 and PS2 surface reflectance models appear to be suitable for providing realistic representations of the surface albedo conditions of the glaciers and snow-covered fields during the spring months. For instance, the measurements of surface short-wave albedo conducted by Muscari et al. (2014) at the Thule Air Base (in north-western Greenland) clearly show that surface albedo ranges between 0.6 and 0.7 during the March – May period, when the surface is covered by snow, and decreases to less than 0.10 in June – August when the surface is not covered by snow.

The nunataks and moraines exhibit values of surface albedo ranging mainly between 0.09 and 0.28, with a most frequent value equal to 0.21, so that model PS4 appears to be the most realistic among the PS models shown in Figs. 10.16 and 10.17. On this matter, Ohmura (2012) pointed out that the clean snow-covered areas present surface albedo values ranging between 0.65 and 0.85 in summer before melt, with a most frequent value of 0.70. These evaluations suggest that the PS2 and PS3 surface reflectance models can be more confidently used in these cases. Conversely, Ohmura (2012) pointed out that snow-covered areas exhibit surface albedo ranging between 0.3 and 0.6 during melt (with a most frequent value of 0.4) and between 0.2 and 0.4 when the snow is dirty (with a most frequent value of 0.3), suggesting that linear combinations of the PS3 and PS4 models can be appropriately used to realistically represent the clean snow areas during melt, and the PS4 model is more suitable for the surfaces covered by dirty snow. With regard to these applications, it is useful to mention that the PS3 model defined by Tomasi et al. (2013) for snow-covered surfaces provides average values of $R_{ws} = 0.536$, $R_{bs}(\theta_o = 0°) = 0.461$, and $A(\theta_o = 60°) = 0.564$, which are appreciably lower than those obtained with models PS1 and PS2. Tundra not covered by fresh snow was estimated by Ohmura (2012) to present surface albedo conditions ranging between 0.07 and 0.22, with a most frequent value of 0.15, which can be well represented by using the VS1 surface reflectance model shown in Figs. 10.16 and 10.17.

Models PS1 and PS2 were deemed to be more appropriate for representing the land surface reflectance features observed near Barrow in winter-spring, over both ice-covered sea and snow-covered land areas. In addition, as pointed out above, model PS1 is suitable to represent the inner parts of Greenland, which are covered by fresh-snow fields and glaciers throughout the year, including the high-altitude area of Summit. Model PS2 was also chosen to represent the surface albedo characteristics of the snow fields surrounding the Sodankylä station in winter and early spring. Model PS3 was reliably employed in summer to simulate the land area characteristics around Ny-Ålesund and, together with model PS4, to represent the mixed ice-covered and ice-free areas in Spitsbergen, not far from Ny-Ålesund, since it yields average values of $R_{ws} = 0.296$, $R_{bs}(\theta_o = 0°) = 0.214$, and $A(\theta_o = 60°) = 0.329$. The spectral curves of white-sky albedo $R_{ws}(\lambda)$ determined for models PS1, PS2, PS3, and PS4 are shown in Fig. 10.17, to give an idea of the surface albedo characteristics that can be reliably used to calculate the instantaneous DARF values at the ToA- and BoA-levels. The calculations of the instantaneous DARF terms at the Arctic sites were made using the spectral curves of surface albedo $R_L(\lambda, \theta_o)$ calculated at the various solar zenith angles for the four PS models shown in Fig. 10.17. The most representative PS model was chosen at each site and for each aerosol type, to simulate the characteristics of white-sky albedo R_{ws} derived from the MCD43C3 products, as indicated in Table 10.4 for the various aerosol types.

(6) *Calculations of the daily time-patterns of instantaneous DARF terms and diurnally averaged DARF effects*

The Arctic region exhibits extensive areas characterized by high-reflectance features, which strongly reflect the incoming solar radiation. For the rather low solar elevation angles usually observed at the Arctic sites, aerosols are expected to induce in general significant direct climatic effects as well as indirect effects by changing the microphysical properties of clouds (Stone 2002). These effects can be very strong during the poleward transport episodes of aerosols often involving particles of industrial and urban origins, desert dust, and biomass-burning particulate matter (Shaw 1995), and specifically: (i) anthropogenic polluted aerosol transported over long distances toward the Arctic from the densely populated mid-latitude areas; (ii) biomass-burning and forest fire smokes transported from the mid-latitude and temperate polar regions; and (iii) dust mobilized in the mid-latitude desert and agricultural regions.

The mean time-patterns of instantaneous DARF terms were determined at various Arctic sites, for all the aerosol types listed in Table 10.5, by using the 6S radiative transfer code for (i) the median values of $AOT(0.50\ \mu m)$ and $\alpha(0.40–0.87\ \mu m)$ given in Table 10.5, which were assumed to be stable during the sunlit period of clear-sky days; (ii) the values of solar zenith angle θ_o calculated at the various hours of the day for the longitude and latitude coordinates of the various stations, as listed in Table 10.5; (iii) the multimodal size-distribution curves determined as linear combinations of the OPAC components defined for relative humidity $RH = 50\%$ in Table 10.6, according to the calculations made at step (3) for the various aerosol types; (iv) the optical characteristics of the multimodal size-distribution curves determined at step (2); and (v) the surface albedo models chosen at step (5) for the various Arctic sites.

All the time-patterns of instantaneous DARF terms $\Delta F_{ToA}(t)$, $\Delta F_{BoA}(t)$, and $\Delta F_{Atm}(t)$ determined above were then integrated over the sunlit periods recorded at the various Arctic sites to calculate the corresponding diurnally averaged values of the DARF terms ΔDF_{ToA}, ΔDF_{BoA}, and ΔDF_{Atm}. The results are shown in Table 10.8 for all the Arctic aerosol types considered in the present study. The first five cases considered in Table 10.8 pertain to the BG summer aerosol measurements

Table 10.8 Daily mean values of the diurnal average DARF terms ΔDF_{ToA}, ΔDF_{BoA} and ΔDF_{Atm} obtained for different aerosol types, surface albedo models and values of aerosol optical thickness *AOT(0.55 μm)*, as obtained for the Arctic aerosol types defined in Table 10.5 by analyzing the sun-photometer measurements performed at the various Arctic sites and taking into account the radiative characteristics of columnar aerosol particles given in Table 10.7

Polar aerosol type	Measurement site	Surface albedo model	ΔDF_{ToA} (W m^{-2})	ΔDF_{BoA} (W m^{-2})	ΔDF_{Atm} (W m^{-2})	*AOT(0.55 μm)*
BG summer aerosol	Barrow	OS1	−9.2	−1.9	−7.3	0.068
		VS1	−4.2	−6.5	+ 2.3	
BG summer aerosol	Ny-Ålesund	OS1	−7.2	+ 1.5	−8.6	0.037
		PS4	+ 1.0	−2.3	+ 3.4	
BG summer aerosol	Summit	PS1	+ 2.2	−0.2	+ 2.5	0.034
BG summer aerosol	Sodankylä	OS1	−6.5	−1.6	−4.9	0.052
		VS1	−3.0	−5.5	+ 2.4	
		PS2	+ 3.0	−0.4	+ 3.4	
BG summer aerosol	Tiksi	OS1	−10.3	−1.6	−8.7	0.073
		VS1	−5.6	−9.0	+ 3.3	
Arctic haze	Barrow	OS1	−5.5	+ 6.5	−12.0	0.103
		PS1	+ 9.0	+ 0.0	+ 8.9	
		PS2	+ 8.8	+ 0.1	+ 8.7	
Arctic haze	Ny-Ålesund	OS1	−5.3	−0.6	−4.7	0.071
		PS3	+ 1.0	−0.4	+ 1.4	
Arctic haze	Sodankylä	OS1	−7.7	−1.1	−6.5	0.059
		VS1	−3.5	−5.3	+ 1.8	
		PS2	+ 2.2	−0.3	+ 2.5	
Arctic dense summer aerosol	Ny-Ålesund	OS1	−10.4	+ 4.2	−14.6	0.109
		PS4	+ 13.0	−5.2	+ 18.2	
Asian dust	Barrow	OS1	−9.3	+ 16.0	−25.3	0.222
		PS2	+ 18.5	+ 0.3	+ 18.2	
Boreal Forest fire smoke	Barrow	OS1	−32.9	−6.7	−26.2	0.268
		VS1	−20.2	−28.6	+ 8.3	

conducted at Barrow, Summit, Ny-Ålesund, Sodankylä and Tiksi for values of *AOT(0.55 μm)* ranging between 0.034 (Summit) and 0.073 (Tiksi). The values of ΔDF_{ToA} obtained for oceanic surface albedo conditions (OS1 model) were estimated to vary between −6.5 W m^{-2} (Sodankylä) and −10.3 W m^{-2} (Tiksi), while: (1) those obtained for VS1 surface albedo conditions were estimated to vary between −3.0 W m^{-2} (Sodankylä) and −5.6 W m^{-2} (Tiksi), and (2) those determined for the high albedo conditions of snow-covered PS surfaces were estimated to vary between +1.0 W m^{-2} (Ny-Ålesund) and + 3.0 W m^{-2} (Sodankylä). The values of ΔDF_{BoA} determined for the low surface albedo conditions of the OS1 model were estimated to vary between −1.9 W m^{-2} (Barrow) and + 1.5 W m^{-2}

(Ny-Ålesund), while: (a) those obtained for the VS1 surface albedo conditions were estimated to vary between −5.5 W m^{-2} (Sodankylä) and −9.0 W m^{-2} (Tiksi), and (b) those evaluated for the high albedo conditions of the snow-covered PS surfaces were estimated to vary between −2.3 W m^{-2} (Ny-Ålesund) and −0.2 W m^{-2} (Summit). The values of ΔDF_{Atm} determined for the low surface albedo (OS1 model) conditions were estimated to vary between −8.7 W m^{-2} (Tiksi) and −4.9 W m^{-2} (Sodankylä), while: (1) those obtained for the VS1 surface albedo conditions were estimated to vary between +2.3 W m^{-2} (Barrow) and + 3.3 W m^{-2} (Tiksi), and (2) those evaluated for the high albedo conditions of the snow-covered surfaces were estimated to vary between +2.5 W m^{-2} (Summit) and + 3.4 W m^{-2} (Ny-Ålesund and Sodankylä).

The three AH cases were measured at Barrow, Ny-Ålesund and Sodankylä, for values of $AOT(0.55\ \mu m)$ ranging between 0.059 (Sodankylä) and 0.103 (Barrow), providing: (1) values of ΔDF_{ToA} obtained for the OS1 surface albedo conditions varying between −5.3 W m^{-2} (Ny-Ålesund) and −7.7 W m^{-2} (Sodankylä); (2) a value of ΔDF_{ToA} equal to −3.5 W m^{-2} (Sodankylä) for the VS1 surface albedo conditions; and (3) values of ΔDF_{ToA} calculated for ice- and snow-covered PS surfaces varying between +1.0 (Ny-Ålesund) and + 9.0 W m^{-2} (Barrow). The values of ΔDF_{BoA} determined for the OS1 low surface albedo conditions were estimated to vary between −1.1 W m^{-2} (Sodankylä) and + 6.5 W m^{-2} (Barrow), while: (1) the case pertaining to the VS1 surface albedo conditions assumed at Sodankylä was estimated to be equal to −5.3 W m^{-2}; and (2) those evaluated for the high albedo conditions of the snow-covered PS surfaces varied between −0.4 W m^{-2} (Ny-Ålesund) and + 0.1 W m^{-2} (Barrow). The values of ΔDF_{Atm} determined for the low OS1 surface albedo conditions were estimated to vary between −4.7 W m^{-2} (Ny-Ålesund) and −12.0 W m^{-2} (Barrow), while: (1) that considered at Sodankylä for the VS1 surface albedo conditions was found to be equal to +1.8 W m^{-2}; and (2) those calculated for the high albedo conditions of the snow-covered PS surfaces were estimated to vary between +1.4 W m^{-2} (Ny-Ålesund) and + 8.9 W m^{-2} (Barrow).

The ADS aerosol case observed at Ny-Ålesund for $AOT(0.55\ \mu m) = 0.109$ was estimated to yield: (1) values of ΔDF_{ToA} ranging between −10. 4 W m^{-2} (for the OS1 surface albedo conditions) and + 13.0 W m^{-2} (for the PS4 surface); (2) values of ΔDF_{BoA} ranging between −5.2 W m^{-2} (for the PS4 surface) and + 4.2 W m^{-2} (for the OS1 surface); and (3) values of ΔDF_{Atm} ranging between −14.6 W m^{-2} (for the OS1 surface) and + 18.2 W m^{-2} (for the PS4 surface). The AD case observed at Barrow (for $AOT(0.55\ \mu m) = 0.222$) was estimated to yield: (1) values of ΔDF_{ToA} ranging between −9.3 W m^{-2} (for the OS1 surface) and + 18.5 W m^{-2} (for PS2 surface albedo conditions); (2) values of ΔDF_{BoA} ranging from +0.3 W m^{-2} (for the PS2 surface) to +16.0 W m^{-2} (for the OS1 surface); and (3) values of ΔDF_{Atm} varying between −25.3 W m^{-2} (for the OS1 surface) and + 18.2 W m^{-2} (for the PS2 surface). The BFF smoke case observed at Barrow (for $AOT(0.55\ \mu m) = 0.268$) was estimated to give: (1) values of ΔDF_{ToA} ranging between −32.9 W m^{-2} (for the OS1 surface) and −20.2 W m^{-2} (for the VS1 surface); (2) values of ΔDF_{BoA}

Table 10.9 Daily mean values of the diurnal average DARF efficiencies E_{ToA}, E_{BoA} and E_{Atm} obtained for different aerosol types, surface albedo models and values of aerosol optical thickness *AOT(0.55 μm)*, as obtained for the Arctic aerosol types defined in Table 10.5 by analyzing the sun-photometer measurements performed at the various Arctic sites and taking into account the radiative characteristics of columnar aerosol particles given in Table 10.7

Polar aerosol type	Measurement site	Surface albedo model	E_{ToA} (W m^{-2})	E_{BoA} (W m^{-2})	E_{Atm} (W m^{-2})
BG summer aerosol	Barrow	OS1	−118.0	−23.8	−94.2
		VS1	−53.6	−82.9	+ 29.3
BG summer aerosol	Ny-Ålesund	OS1	−174.6	+ 36.3	−210.9
		PS4	+ 23.5	−56.6	+ 82.0
BG summer aerosol	Summit	PS1	+ 57.3	−6.3	+ 63.5
BG summer aerosol	Sodankylä	OS1	−108.6	−26.9	−81.7
		VS1	−50.0	−90.8	+ 40.8
		PS2	+ 50.3	−6.5	+ 56.9
BG summer aerosol	Tiksi	OS1	−121.5	−18.7	+ 102.8
		VS1	−66.3	−105.3	+ 39.0
Arctic haze	Barrow	OS1	−47.1	+ 55.9	−103.0
		PS1	+ 77.2	+ 0.2	+ 77.1
		PS2	+ 76.2	+ 0.9	+ 75.3
Arctic haze	Ny-Ålesund	OS1	−66.2	−7.3	−58.9
		PS3	+ 13.1	−4.5	+ 17.6
Arctic haze	Sodankylä	OS1	−116.2	−17.3	−99.0
		VS1	−52.9	−79.6	+ 26.7
		PS2	+ 33.1	−4.5	+ 37.7
Arctic dense summer aerosol	Ny-Ålesund	OS1	−86.7	+ 35.3	−122.0
		PS4	+ 108.0	−43.1	+ 151.2
Asian dust	Barrow	OS1	−46.5	+ 80.0	−126.5
		PS2	+ 92.3	+ 1.3	+ 91.0
Boreal Forest fire smoke	Barrow	OS1	−109.8	−22.3	−87.5
		VS1	−67.5	−95.3	+ 27.8

ranging between −28.6 W m^{-2} (for the VS1 surface) and −6.7 W m^{-2} (for the OS1 surface); and (3) values of ΔDF_{Atm} varying between −26.2 W m^{-2} (for the OS1 surface) and + 8.3 W m^{-2} (for the VS1 surface) (Table 10.8). The corresponding DARF efficiencies are reported in Table 10.9.

10.7 Concluding Remarks

The analysis of the ERA-40 products conducted by Serreze et al. (2007) provided evidence of the annual cycles described by the main various atmospheric budget terms over the Arctic Ocean Domain. It can be seen in Fig. 10.3 that: (i) the convergence of atmospheric energy transport maintains nearly stable conditions throughout the year, presenting monthly mean values ranging mainly between 50 and 100 W m^{-2}, with the higher values recorded in March and October, and the lowest value in May; (ii) the net radiation at the ToA-level shows the most negative monthly mean values in January and December (in the absence of incoming solar radiation) and nearly null values in June and July, when the outgoing solar radiation (reflected by the surface-atmosphere system) fully compensates the outgoing long-wave radiation emitted by the surface atmosphere system toward the outer space; and (iii) the neat heat flux at the terrestrial surface yields the highest (positive) monthly mean values (close to +50 W m^{-2}) in January and December and the lowest (negative) values (close to around −100 W m^{-2}) in July, when the exchange of latent heat from the surface to the atmosphere is most intense. The calculations of Serreze et al. (2007) over the North Polar Cap area (for latitudes >70 °N) were made by analyzing the ERA-40 products to determine the monthly mean values of: (i) the net short-wave radiation SW_{ToA} at the ToA-level, (ii) the net long-wave radiation LW_{ToA} at the ToA-level, and (iii) the net radiation R_{ToA} at the ToA-level, all calculated as differences between the SW_{ToA} and LW_{ToA} fluxes. As can be seen in Fig. 10.4, the monthly mean values of SW_{ToA} reach a maximum largely exceeding the value of 250 W m^{-2} in June and July, while null values are recorded in January and December. Conversely, the monthly mean values of LW_{ToA} are rather stable throughout the year, and compensate the effects produced by the SW_{ToA} contributions, providing nearly null values of R_{ToA} in June and July. The time-patterns of these annual cycles have been compared in Fig. 10.4 with those determined by Porter et al. (2010), who analyzed the data-sets provided by the JRA, the NRA project, and the CERES programme, respectively. The comparison shows that a substantial agreement exists between the Serreze et al. (2007) and Porter et al. (2010) evaluations of the annual cycles of the SW_{ToA}, LW_{ToA} and R_{ToA} monthly mean budget terms.

The time-patterns of the monthly mean terms of (i) the convergence of atmospheric energy transport (giving a yearly average value of +100.4 W m^{-2}), (ii) latent heat energy (giving a yearly average value of +15.7 W m^{-2}), and (iii) dry static energy (giving a yearly average value of +84.7 W m^{-2}) were shown in Fig. 10.5, as obtained by Serreze et al. (2007) over the North Polar Cap area, providing evidence that the time-patterns of convergence of atmospheric energy transport and dry static energy assume rather stable values in the winter months and describe both a minimum in spring (May) followed by a gradually increasing trend in the summer months. Conversely, the time-patterns of the monthly mean terms of latent heat energy appear to be quite stable during the year, with valucs exceeding the yearly average during the June – October months.

The calculations made by Serreze et al. (2007) over the North Polar Cap for (i) the net short-wave radiation SW_{sfc} at the surface-level, (ii) the net long-wave radiation LW_{sfc} at the surface-level, and (iii) the net radiation R_{sfc} at the surface-level were shown in Fig. 10.6, indicating that the most intense monthly mean values of SW_{sfc} are produced by the incoming solar radiation in June and July, while the monthly mean values of LW_{sfc} maintain almost stable values throughout the year, leading to time-patterns of net radiation R_{sfc} at the surface-level slightly exceeding the June – July value of 70 W m^{-2} for the down-welling radiation. The time-patterns of the monthly mean values of turbulent sensible heat flux at the surface-level, and latent heat flux at the surface-level are also shown in Fig. 10.6, according to the Serreze et al. (2007) calculations, presenting nearly stable time-patterns over the whole year. These annual cycles have been compared in Fig. 10.6 with those determined by Porter et al. (2010), who analyzed the data-sets provided by the JRA, NRA and CERES projects and obtained evaluations that closely agree with those of Serreze et al. (2007).

A further contribution to the knowledge of the atmospheric absorption processes affecting the incoming solar radiation at Arctic latitudes was given by the evaluations made in Figs. 10.8, 10.9, 10.10, 10.11, 10.12, 10.13, 10.14, and 10.15 of the fractions of short-wave solar radiation absorbed by atmospheric WV, O_3 and CO_2 for different solar elevation angles, giving a measure of the important role of atmospheric WV in absorbing the incoming global solar radiation at the surface-level. In addition, accurate evaluations of the absorption effects produced on the up-welling and down-welling fluxes of long-wave radiation observed in the Arctic atmosphere during the summer months was separately made for the atmospheric contents of WV and O_3 measured at Ny-Ålesund and the volume concentrations of CO_2, CH_4 and N_2O currently recorded at the Zeppelin Observatory. The analysis of these calculations of the long-wave radiation budgert terms indicated that: (a) fluxes of up-welling long-wave radiation equal to 11.9 W m^{-2} at the ToA-level and of down-welling long-wave radiation equal to 74.5 W m^{-2} at the surface-level result to be absorbed on average by atmospheric WV present in the Arctic atmosphere during the summer months; (b) fluxes of up-welling long-wave radiation equal to 13.7 W m^{-2} at the ToA-level and of down-welling long-wave radiation equal to 25.2 W m^{-2} at the surface-level are currently absorbed on average by atmospheric CO_2 in the Arctic region; (c) fluxes of up-welling long-wave radiation equal to 4.2 W m^{-2} at the ToA-level and of down-welling long-wave radiation equal to 3.5 W m^{-2} at the surface-level are absorbed on average by atmospheric O_3; (d) fluxes of up-welling long-wave radiation equal to 0.79 W m^{-2} at the ToA-level and of down-welling long-wave radiation equal to 0.80 W m^{-2} at the surface-level are currently absorbed on average by atmospheric CH_4; and (e) fluxes of up-welling long-wave radiation equal to 0.93 W m^{-2} at the ToA-level and of down-welling long-wave radiation equal to 0.80 W m^{-2} at the surface-level are currently absorbed by atmospheric N_2O.

On the basis of the above calculations, the overall absorption effects produced all together by the five greenhouse gases result to be equal to 31.5 W m^{-2} on the up-welling long-wave radiation reaching the ToA-level and equal to 104.8 W m^{-2} on the down-welling long-wave radiation at the surface-level. Therefore, the absorption effects produced by the five greenhouse gases result to be comparable with the direct aerosol-induced radiative forcing (DARF) terms ΔDF_{ToA} (determined at the ToA-level), ΔDF_{BoA} (at the surface-level), and ΔDF_{Atm} (within the atmosphere), as produced by aerosol particles at the Arctic sites of Barrow, Ny-Ålesund, Summit,

Sodankylä and Tiksi, presenting the optical characteristics carefully described in the present study.

In fact, the diurnally averaged values of the DARF forcing term ΔDF_{ToA} at the ToA-level were found to vary largely as a function of the aerosol type and surface reflectance characteristics presenting the following features:

- Term ΔDF_{ToA} induced by BG summer aerosol was evaluated to vary mainly between -10 W m^{-2} and + 3 W m^{-2}, presenting the lowest value of -10.3 W m^{-2} at Tiksi (over the oceanic OS1 surface), and the highest value equal to $+3.0$ W m^{-2} at Sodankylä (over the snow-covered PS2 surface).
- Term ΔDF_{ToA} induced by AH particles was estimated to vary between -7.7 W m^{-2} at Sodankylä, (over the oceanic OS1 surface) and + 9.0 W m^{-2} at Barrow (over the ice-covered PS1 surface).
- Term ΔDF_{ToA} induced by ADS aerosol at the Ny-Ålesund station was estimated to range from -10.4 W m^{-2} (over the oceanic OS1 surface) to $+13.0$ W m^{-2} (over the snow-covered PS4 surface).
- Term ΔDF_{ToA} induced by AD particles at Barrow was evaluated to vary between -9.3 W m^{-2} (over the oceanic OS1 surface) and + 18.5 W m^{-2} (over the ice-covered PS2 surface).
- Term ΔDF_{ToA} induced by BFF smoke particles observed at Barrow was evaluated to be strongly negative (and, hence, causing intense absorption effects) and ranging between -32.9 W m^{-2} (over the oceanic OS1 surface) and -20.2 W m^{-2} (over the vegetation-covered VS1 surface).

Therefore, the radiative forcing term ΔDF_{ToA} induced by airborne aerosols in the Arctic atmosphere was estimated to vary in general between -32 W m^{-2} and + 15 W m^{-2} for cloudless-sky conditions and different surface-albedo conditions, while the radiative forcing caused by atmospheric WV absorption on the up-welling long-wave radiation at the ToA-level was estimated to be on average equal to -11.9 W m^{-2} on summer days characterized by cloudless-sky conditions, and, hence, comparable with those induced on average by the various Arctic aerosol loads. On this matter, it is worth mentioning that opposite radiative forcing effects are produced by purely scattering aerosols (such as those mainly consisting of nss-sulphates and other water-soluble substances) and by strongly absorbing aerosol particles such as those containing high percentages of BC and other combustion substances). Yang et al. (2018) estimated that the mean radiative forcing effects caused at the ToA-level of the Arctic atmosphere by anthropogenic sulfate are negative (i. e. causing cooling effects) and equivalent to about one third of the positive radiative forcing effects caused by black carbon particles. On the basis of Arctic climate sensitivity factors, Yang et al. (2018) estimated that about -0.19 K of the Arctic surface temperature cooling is produced on average by anthropogenic sulfate, with about one fourth of such radiative forcing arising from the East-Asian sources of this particulate matter component.

The diurnally averaged values of the DARF forcing term ΔDF_{BoA} induced at the surface-level were estimated to vary for the various aerosol types and surface reflectance characteristics, as follows:

- Term ΔDF_{BoA} induced by BG summer aerosol was estimated to vary between -9.0 W m^{-2} at Tiksi (over the VS1 surface) and + 1.5 W m^{-2} at Ny-Ålesund (over the OS1 surface).
- Term ΔDF_{BoA} induced by AH particles was estimated to vary between -5.3 W m^{-2} at Sodankylä (over the VS1 surface) and + 6.5 W m^{-2} at Barrow (over the OS1 surface).
- Term ΔDF_{BoA} induced by ADS aerosol at Ny-Ålesund was evaluated to vary between -5.2 W m^{-2} (over the snow-covered PS4 surface) and + 4.2 W m^{-2} (over the OS1 surface).
- Term ΔDF_{BoA} induced by AD particles at Barrow was evaluated to vary between +0.3 W m^{-2} (over the ice-covered PS2 surface) and + 16.0 W m^{-2} (over the OS1 surface).
- Term ΔDF_{BoA} induced by BFF smoke particles at Barrow were estimated to range between -28.6 W m^{-2} (over the VS1 surface) and -6.7 W m^{-2} (over the OS1 surface).

Therefore, the radiative forcing term ΔDF_{BoA} induced at the surface-level by the Arctic aerosols were evaluated to vary for the most part of the herein considered cases between -5 W m^{-2} and + 6 W m^{-2} for BG aerosol and AH conditions over the OS1, VS1, PS2 and PS4 surfaces, while the radiative forcing effects caused by atmospheric WV absorption on the down-welling long-wave radiation at the surface-level were evaluated to be on average equal to -74.5 W m^{-2} on summer days presenting cloudless-sky conditions. These results indicate that considerably stronger effects are in general caused on the radiative budget of the Arctic atmosphere-surface system by the atmospheric WV absorption of incoming solar radiation than those induced by the various types of Arctic aerosol particles.

The diurnally averaged values of the DARF forcing term ΔDF_{Atm} induced by aerosols within the atmosphere were estimated to vary largely as a function of aerosol type and surface reflectance features, as follows:

- Term ΔDF_{Atm} induced by BG summer aerosol was estimated to vary between -8.7 W m^{-2} at Tiksi (over the OS1 surface) and + 3.4 W m^{-2} at Ny-Ålesund (over the snow-covered PS4 surface) and Sodankylä (over the ice-covered PS2 surface).
- Term ΔDF_{Atm} induced by AH particles was estimated to vary between -12.0 W m^{-2} at Barrow (over the OS1 surface) and + 8.9 W m^{-2} at Barrow (over the ice-covered PS1 surface).
- Term ΔDF_{Atm} induced by the ADS aerosol at Ny-Ålesund was evaluated to range between -14.6 W m^{-2} (over the OS1 surface) and + 18.2 W m^{-2} (over the snow-covered PS4 surface).
- Term ΔDF_{Atm} induced by the AD particles at Barrow was evaluated to vary between -25.3 W m^{-2} (over the OS1 surface) and + 18.2 W m^{-2} (over the ice-covered PS2 surface).
- Term ΔDF_{Atm} induced by the BFF smoke particles at Barrow were estimated to vary between -26.2 W m^{-2} (over the OS1 surface) and + 8.3 W m^{-2} (over the VS1 surface).

On the basis of these evaluations, the radiative forcing term ΔDF_{Atm} induced by Arctic aerosols within the atmosphere were estimated to vary most frequently: (i) between -12 W m^{-2} and + 18 W m^{-2} for BG aerosol and AH particles over oceanic surfaces or surfaces covered by vegetation or snow, and (ii) between -25 W m^{-2} and + 18 W m^{-2} on clear-sky days characterized by long-range transport of AD or BFF smoke plumes, while: (a) the radiative forcing caused by atmospheric WV absorption on the down-welling long-wave radiation crossing the Arctic atmosphere was evaluated to be on average equal to -74.5 W m^{-2} and, hence, to cause considerably higher effects on the radiative budget of the Arctic atmosphere-surface system than those due to aerosols, and (b) the radiative forcing caused by the absorption of incoming global solar irradiance by atmospheric WV was estimated to be equal to more than -60 W m^{-2} for a daily average value of the apparent solar elevation angle equal to 21.6°, giving a value of relative optical air mass $m = 2.7$.

References

Ångström, A. (1964). The parameters of atmospheric turbidity. *Tellus, 16*(1), 64–75. https://doi.org/10.1111/j.2153-3490.1964.tb00144.x.

Barkstrom, B. R., & Smith, G. L. (1986). The earth radiation budget experiment: Science and implementation. *Review of Geophysics, 24*(2), 379–390. https://doi.org/10.1029/RG024i002p00379.

Benedict, W. S. (1948). New bands in the vibration – Rotation spectrum of water vapor. *Physical Review, 74*, 1246–1247.

Benedict, W. S., Gailar, N., & Plyler, E. K. (1956). Rotation-vibration spectra of deuterated water vapor. *The Journal of Chemical Physics, 24*(6), 1139–1165. https://doi.org/10.1063/1.1742731.

Berk, A., Anderson, G. P., Acharya, P. K., Chetwynd, J. H., Bernstein, L. S., Shettle, E. P., Matthew, M. W., & Adler-Golden, S. M. (1999). MODTRAN4 user's manual. Air Force Research Laboratory, Space Vehicles Directorate, Air Force Materiel Command, Hanscom AFB, MA 01731-3010, 1 June 1999. Last revised 17 April 2000, 94 pp.

Bess, T. D., & Smith, G. L. Interannual variability of Arctic radiation balance in July. Available at https://ams.confex.com/ams/pdfpapers/102125.pdf

Bignell, K. J. (1970). The water-vapour infra-red continuum. *Quarterly Journal of the Royal Meteorological Society, 96*(409), 390–403. https://doi.org/10.1002/qj.49709640904.

Born, M., & Wolf, E. (1975). *Principles of optics, electromagnetic theory of propagation, interference and direction of light* (5th edn). Oxford: Pergamon Press.

Bothwell, G. W., Hansen, E. G., Vargo, R. E., & Miller, K. C. (2002). The multiangle imaging spectro-radiometer science data system, its products, tools, and performance. *IEEE Transactions on Geoscience and Remote Sensing, 40*, 1467–1476. https://doi.org/10.1109/TGRS.2002.801152.

Bromwich, D. H., Wang, S.-H., & Monaghan, A. J. (2002). *ERA-40 Representation of the Arctic Atmospheric Moisture Budget* (ERA-40 Project Report Series 3). Workshop on Re-Analysis, 5–9 November, 2001, European Center for Medium Range Weather Forecasts, pp. 287–297.

Burch D. E. (1970). *Investigation of the absorption of infrared radiation by atmospheric gases* (Semi-annual technical report, Contract No. F 19628-69-C-0263). Newport Beach: Philco-Ford Corporation.

Burch, D. E., Gryvnak, D. A., & Pembrook, J. D. (1972). *Infrared absorption bands of nitrous oxide* (AFCRL-72-0387 (U-5037) Report, Contract No. F19628-69-C-0263, Project No. 5130, 87 pp). Newport Beach: Philco-Ford Corporation, Aeronutronic Division.

Bush, B. C., & Valero, F. P. J. (2002). Spectral aerosol radiative forcing at the surface during the Indian Ocean experiment (INDOEX). *Journal of Geophysical Research, Atmospheres, 107*, 8003. https://doi.org/10.1029/2000JD000020.

Bush, B. C., & Valero, F. P. J. (2003). Surface aerosol radiative forcing at Gosan during the ACE-Asia campaign. *Journal of Geophysical Research, Atmospheres, 108*, 8660. https://doi.org/10.1029/2002JD003233.

Christopher, S. A., & Zhang, J. (2002). Shortwave aerosol radiative forcing from MODIS and CERES observations over the oceans. *Geophysical Research Letters, 29*(18), 1859. https://doi.org/10.1029/2002GL014803.

Chylek, P., & Coakley, J. A., Jr. (1974). Aerosols and climate. *Science, 183*, 75–77. https://doi.org/10.1126/science.183.4120.75.

Coakley, J. A., Jr., & Chylek, P. (1975). The two-stream approximation in radiative transfer: Including the angle of the incident radiation. *Journal of the Atmospheric Sciences, 32*(2), 409–418. https://doi.org/10.1175/1520-0469(1975)032<0409:TTSAIR>2.0.CO;2.

Cox, C., & Munk, W. (1954). Statistics of the sea surface derived from sun glitter. *Journal of Marine Research, 13*, 198–227.

Cullather, R. I., Bromwich, D. H., & Serreze, M. C. (2000). The atmospheric hydrologic cycle over the Arctic basin from reanalyses. Part I: Comparison with observations and previous studies. *Journal of Climate, 13*, 923–937. https://doi.org/10.1175/1520-0442(2000)013<0923:TAHCOT>2.0.CO;2.

Damiani, A., De Simone, S., Rafanelli, C., Cordero, R. R., & Laurenza, M. (2012). Three years of ground-based total ozone measurements in the Arctic: Comparison with OMI, GOME and SCIAMACHY satellite data. *Remote Sensing of Environment, 127*, 162–180. https://doi.org/10.1016/j.rse.2012.08.023.

Diner, D. J., Beckert, J. C., Reilly, T. H., Bruegge, C. J., Conel, J. E., Kahn, R. A., Martonchik, J. V., Ackerman, T. P., Davies, R., Gerstl, S. A. W., Gordon, H. R., Muller, J.-P., Myneni, R. B., Sellers, P. J., Pinty, B., & Verstraete, M. M. (1998). Multi-angle imaging spectro radiometer (MISR) instrument description and experiment overview. *IEEE Transactions on Geoscience and Remote Sensing, 36*, 1072–1087. https://doi.org/10.1109/36.700992.

Elsasser, W. M., & Culbetson, M. F. (1960). *Atmospheric radiation tables*. Meteorological monographs published by the American Meteorological Society, Vol. 4(23), August 1960, Boston, MA (USA), 43 pp.

France, W. L., & Dickey, F. P. (1955). Fine structure of the 2.7 micron carbon dioxide rotation-vibration band. *The Journal of Chemical Physics, 23*(3), 471–474. https://doi.org/10.1063/1.1742012.

Francis, J. A., Hunter, E., Key, J. R., & Wang, X. (2005). Clues to variability in Arctic minimum sea ice extent. *Geophysical Research Letters, 32*, L21501. https://doi.org/10.1029/2005GL024376.

Fröhlich, C., & Brusa, R. W. (1981). Solar radiation and its variation in time. *Solar Physics, 74*(1), 209–215. https://doi.org/10.1007/BF00151291.

Gerhard, S. L. (1932). The infrared absorption spectrum and the molecular structure of ozone. *Physical Review, 42*(5), 622–631.

Goldberg, L. (1954). Absorption spectrum of the atmosphere. In G. P. Kuiper (Ed.), *The Earth as a planet* (pp. 484–490). Chicago: Chicago University Press.

Grassl, H. (1974). Influence of different absorbers in the window region on radiative cooling (and on surface temperature determination). *Beiträge zur Physik der Atmosphäre (Contributions to Atmospheric Physics), 47*(1), 1–13.

Grenfell, T. C., & Maycut, G. A. (1977). The optical properties of ice and snow in the Arctic Basin. *Journal of Glaciology, 18*(78), 445–463. https://doi.org/10.1017/S0022143000021122.

Hall, T. C., Jr., & Blacet, F. E. (1952). Separation of the absorption spectrum of NO_2 and N_2O_4 in range of 2400–5000 Å. *The Journal of Chemical Physics, 20*(11), 1745–1749. https://doi.org/10.1063/1.1700281.

Hänel, G., Adam, W., Bundke, U., Komguem, L., & Leiterer, U. (1999). Optical properties of boundary layer particles, columnar absorption and direct radiative forcing by particles in the solar spectral region. *Journal of Aerosol Science, 30*(1), S171–S172.

Hansen, J., Sato, M., & Ruedy, R. (1997). Radiative forcing and climate response. *Journal of Geophysical Research, Atmospheres, 102*(D6), 6831–6864. https://doi.org/10.1029/96JD03436.

Hansen, J. E., Sato, M., Lacis, A., Ruedy, R., Tegen, I., & Matthews, E. (1998). Climate forcings in the industrial era. *Proceedings of the National Academy of Sciences (PNAS), USA, 95*(22), 12753–12758. https://doi.org/10.1073/pnas.95.22.12753.

Herber, A., Thomason, L. W., Gernandt, H., Leiterer, U., Nagel, D., Schulz, K.-H., Kaptur, J., Albrecht, T., & Notholt, J. (2002). Continuous day and night aerosol optical depth observations in the Arctic between 1991 and 1999. *Journal of Geophysical Research, Atmospheres, 107*(D10, 4097), AAC 6-1–AAC 6-13. https://doi.org/10.1029/2001JD000536.

Hess, M., Koepke, P., & Schult, I. (1998). Optical properties of aerosols and clouds: The software package OPAC. *Bulletin of the American Meteorological Society, Volume, 79*(5), 831–844. https://doi.org/10.1175/1520-0477(1998)079<0831:OPOAAC>2.0.CO;2.

Hirdman, D., Sodemann, H., Eckhardt, S., Burkhart, J. F., Jefferson, A., Mefford, T., Quinn, P. K., Sharma, S., Ström, J., & Stohl, A. (2010). Source identification of short-lived air pollutants in the Arctic using statistical analysis of measurement data and particle dispersion model output. *Atmospheric Chemistry and Physics, 10*, 669–693. https://doi.org/10.5194/acp-10-669-2010.

Holben, B. N., Eck, T. F., Slutsker, I., Tanré, D., Buis, J. P., Setzer, A., Vermote, E., Reagan, J. A., Kaufman, Y. J., Nakajima, T., Lavenu, F., Jankowiak, I., & Smirnov, A. (1998). AERONET – A federated instrument network and data archive for aerosol characterization. *Remote Sensing of Environment, 66*, 1–16. S0034-4257(98)00031-5.

Jacquemoud, S. (1993). Inversion of the PROSPECT + SAIL canopy reflectance model from AVISIR equivalent spectra – Theorectical study. *Remote Sensing of Environment, 44*(2–3), 281–292.

Kalnay, E., Kanamitsu, M., Kistler, R., Collins, W., Deaven, D., Gandin, L., Iredell, M., Saha, S., White, G., Woollen, J., Zhu, Y., Leetmaa, A., Reynolds, R., Chelliah, M., Ebisuzaki, W., Higgins, W., Janowiak, J., Mo, K. C., Ropelewski, C., Wang, J., Jenne, R., & Joseph, D. (1996). The NCEP/NCAR 40-year reanalysis project. *Bulletin of the American Meteorological Society, 77*(3), 437–471. https://doi.org/10.1175/1520-0477(1996)077<0437:TNYRP>2.0.CO;2.

Kaplan, L. D., & Eggers, D. F. (1956). Intensity and line-width of the 15-micron CO_2 band, determined by a curve-of-growth method. *The Journal of Chemical Physics, 25*(5), 876–883. https://doi.org/10.1063/1.1743135.

Kiehl, J., & Trenberth, K. (1997). Earth's annual global mean energy budget. *Bulletin of the American Meteorological Society, 78*(2), 197–206. https://doi.org/10.1175/1520-0477(1997)078<0197:EAGMEB>2.0.CO;2.

Kim, S.-W., Jefferson, A., Yoon, S.-C., Dutton, E. G., Ogren, J. A., Valero, F. P. J., Kim, J., & Holben, B. N. (2005). Comparisons of aerosol optical depth and surface short-wave irradiance and their effect on the aerosol surface radiative forcing estimation. *Journal of Geophysical Research, Atmospheres, 110*, D07204. https://doi.org/10.1029/2004JD004989.

Kneizys, F. X., Abreu, L. W., Anderson, G. P., Chetwynd, J. H., Shettle, E. P., Berk, A., Bernstein, L. S., Robertson, D. C., Acharya, P., Rothman, L. S., Selby, J. E. A., Gallery, W. O., & Clough, S. A. (1996). The MODTRAN 2/3 report and LOWTRAN 7 model (L. W. Abreu, & G. P. Anderson, eds.). Contract F19628-91-C.0132, Phillips Laboratory, Geophysics Directorate, PL/GPOS, Hanscom AFB, MA., 261 pp.

Koepke, P. (1984). Effective reflectance of oceanic whitecaps. *Applied Optics, 23*(11), 1816–1824. https://doi.org/10.1364/AO.23.001816.

Kokhanovsky, A. A. (2004). Spectral reflectance of whitecaps. *Journal of Geophysical Research, Oceans, 109*, C05021. https://doi.org/10.1029/2003JC002177.

Kokhanovsky, A. A., & Breon, F.-M. (2012). Validation of an analytical snow BRDF model using PARASOL multi-angular and multispectral observations. *IEEE Geoscience and Remote Sensing Letters, 9*(5), 928–932. https://doi.org/10.1109/LGRS.2012.2185775.

Kokhanovsky, A. A., & Zege, E. P. (2004). Scattering optics of snow. *Applied Optics, 43*(7), 1589–1602. https://doi.org/10.1364/AO.43.001589.

Kokhanovsky, A. A., Aoki, T., Hachikubo, A., Hori, M., & Zege, E. P. (2005). Reflective properties of natural snow: Approximate asymptotic theory versus *in situ* measurements. *IEEE Transactions on Geosciences and Remote Sensing, 43*(7), 1529–1535. https://doi.org/10.1109/TGRS.2005.848414.

Kokhanovsky, A. A., Rozanov, V. V., Aoki, T., Odermatt, D., Brockmann, C., Krüger, O., Bouvet, M., Drusch, M., & Hori, M. (2011). Sizing snow grains using backscattered solar light. *International Journal of Remote Sensing, 32*(22), 6975–7008. https://doi.org/10.1080/01431161.2011.560621.

Kondratyev, K. Y. (1969). *Radiation in the atmosphere*, Chapter 1: Radiant energy, the main concepts and definitions (pp. 1–48); Chapter 3: Radiation absorption in the atmosphere (pp. 85–139); Chapter 4: Scattering of radiation in the atmosphere (pp. 161–216). New York/London: Academic, 912 pp.

Kopp, G., & Lean, J. L. (2011). A new, lower value of total solar irradiance: Evidence and climate significance. *Geophysical Research Letters, 38*, L01706. https://doi.org/10.1029/2010GL045777.

Kostkowski, J. K., & Kaplan, L. D. (1957). Absolute intensities of the 721 and 742 cm^{-1} bands of CO_2. *The Journal of Chemical Physics, 26*, 1252–1253. https://doi.org/10.1063/1.1743501.

Kratz, D. P., Gao, B.-C., & Kiehl, J. T. (1991). Study of the radiative effects of the 9.4- and 10.4-micron bands of carbon dioxide. *Journal of Geophysical Research, Atmospheres, 96*, D5, 9021–9026. https://doi.org/10.1029/89JD01004.

Kuusk, A. (1994). A multispectral canopy reflectance model. *Remote Sensing of Environment, 50*(2), 75–82. https://doi.org/10.1016/0034-4257(94)90035-3.

Kuusk, A. (1995). A fast, invertible canopy reflectance model. *Remote Sensing of Environment, 51*(3), 342–350. https://doi.org/10.1016/0034-4257(94)00059-V.

Lewis, P. (1995). On the implementation of linear kernel-driven BRDF models. In: *Proceedings of the annual conference of Remote Sensing Society* '95, "Remote Sensing in Action", Southampton, UK, 11–14 September, 1995, pp. 333–340.

Lewis, P., & Barnsley, M. J. (1994). Influence of the sky radiance distribution on various formulations of the Earth surface albedo. In *Proceedings of the sixth international symposium on physical measurements and signatures in remote sensing*, Val d'Isere (France), January 17–21, 1994, pp. 707–715.

Lucht, W., Schaaf, C. B., & Strahler, A. H. (2000). An algorithm for the retrieval of albedo from space using semiempirical BRDF models. *IEEE Transactions on Geoscience and Remote Sensing, 38*(2 Part 2), 977–998. https://doi.org/10.1109/36.841980.

Malkmus, W. (1963). Infrared emissivity of carbon dioxide (4.3-μ band). *Journal of the Optical Society of America, 53*(8), 951–961. https://doi.org/10.1364/JOSA.53.000951.

Migeotte, M., Niven, L., & Swensson, J. (1957). *The solar spectrum from 2.8 to 23.7 Microns,* final report – Phase a (Part 1.1.1, Air Force Cambridge Research Center, Contract No. AF 61(514)-432). Liège: Institute d'Astrophysique de l'Université de Liège.

Morel, A. (1988). Optical modeling of the upper ocean in relation to its biogenous matter content (Case I waters). *Journal of Geophysical Research, Oceans, 93*(C9), 10749–10768. https://doi.org/10.1029/JC093iC09p10749.

Muscari, G., Di Biagio, C., di Sarra, A., Cacciani, M., Ascanius, S. E., Bertagnolio, P. P., Cesaroni, C., de Zafra, R. L., Eriksen, P., Fiocco, G., Fiorucci, I., & Fuà, D. (2014). Observations of surface radiation and stratospheric processes at Thule Air Base, Greenland, during the IPY. *Annals of Geophysics, 57*(3), SS0323. https://doi.org/10.4401/ag-6382.

Nakamura, N., & Oort, A. H. (1988). Atmospheric heat budgets of the polar regions. *Journal of Geophysical Research, Atmospheres, 93*(D8), 9510–9524. https://doi.org/10.1029/JD093iD08p09510.

NASA (2016), *The NASA Earth's energy budget poster*. See: https://science-edu.larc.nasa.gov/energy_budget/

Nielsen, A. H., & Nielsen, H. H. (1935). The infrared absorption bands of methane. *Physical Review, 48*(11), 864–867. https://doi.org/10.1103/PhysRev.48.864.

Nilson, T., & Kuusk, A. (1989). A reflectance model for the homogeneous plant canopy and its inversion. *Remote Sensing of Environment, 27*, 157–167. https://doi.org/10.1016/0034-4257(89)90015-1.

Ohmura, A. (2012). Present status and variations in the Arctic energy balance. *Polar Science, 6*, 5–13. https://doi.org/10.1016/j.polar.2012.03.003.

Onogi, K., Tsutsui, J., Koide, H., Sakamoto, M., Kobayashi, S., Hatsushika, H., Matsumoto, T., Yamazaki, N., Kamahori, H., Takahashi, K., Kadokura, S., Wada, K., Kato, K., Oyama, R., Ose, T., Mannoji, N., & Taira, R. (2007). The JRA-25 reanalysis. *Journal of the Meteorological Society of Japan, Series II, 85*(3), 369–432. https://doi.org/10.2151/jmsj.85.369.

Parkinson, C. L., Cavalieri, D. J., Gloersen, P., Zwally, H. J., & Comiso, J. C. (1999). Arctic Sea ice extents, areas, and trends, 1978–1996. *Journal of Geophysical Research, Oceans, 104*(C9), 20837–20856. https://doi.org/10.1029/1999JC900082.

Persson, P. O., Fairall, C. W., Andreas, E. L., Guest, P. S., & Perovich, D. K. (2002). Measurements near the atmospheric surface flux group tower at SHEBA: Near-surface conditions and surface energy budget. *Journal of Geophysical Research, Oceans, 107*, C10. https://doi.org/10.1029/2000JC000705.

Porter, D. F., Cassano, J. J., Serreze, M. C., & Kindig, D. N. (2010). New estimates of the large-scale Arctic atmospheric energy budget. *Journal of Geophysical Research, Atmospheres, 115*, D08108. https://doi.org/10.1029/2009JD012653.

Ramanathan, V., Crutzen, P. J., Lelieveld, J., Mitra, A. P., Althausen, D., Anderson, J., Andreae, M. O., Cantrell, W., Cass, G. R., Chung, C. E., Clarke, A. D., Coakley, J. A., Collins, W. D., Conant, W. C., Dulac, F., Heintzenberg, J., Heymsfield, A. J., Holben, B., Howell, S., Hudson, J., Jayaraman, A., Kiehl, J. T., Krishnamurti, T. N., Lubin, D., McFarquhar, G., Novakov, T., Ogren, J. A., Podgorny, I. A., Prather, K., Priestley, K., Prospero, J. M., Quinn, P. K., Rajeev, K., Rasch, P., Rupert, S., Sadourny, R., Satheesh, S. K., Shaw, G. E., Sheridan, P., & Valero, F. P. J. (2001). Indian Ocean experiment: An integrated analysis of the climate forcing and effects of the great Indo-Asian haze. *Journal of Geophysical Research, Atmospheres, 106*(D22), 28371–28398. https://doi.org/10.1029/2001JD900133.

Raschke, E., & Ohmura, A. (2005). Radiation budget of the climate system (Chapter 4). In: M. Hantel (Ed.), *Observed global climate. Group V: Geophysics, Landolt-Börnstein Numerical and Functional Relationships in Science and Technology* (New series, Vol. 6, pp. 4.1–4.42). Berlin/Heidelberg/New York: Springer.

Rex, M., Dethloff, K., Handorf, D., Herber, A., Lehmann, R., Neuber, R., Notholt, J., Rinke, A., von der Gathen, P., Weisheim, A., & Gernandt, H. (2000). Arctic and Antarctic ozone layer observations: Chemical and dynamical aspects of variability and long-term changes in the polar stratosphere. *Polar Research, 19*(2), 193–204. https://doi.org/10.1111/j.1751-8369.2000.tb00343.x.

Ricchiazzi, P., Yang, S., Gautier, C., & Sowle, D. (1998). SBDART: A research and teaching software tool for plane-parallel radiative transfer in the Earth's atmosphere. *Bulletin of the American Meteorological Society, 79*(10), 2101–2114. https://doi.org/10.1175/1520-0477(1998)079<2101:SARATS>2.0.CO;2.

Ricchiazzi, P., O'Hirok, W., & Gautier, C. (2005). The effect of non-lambertian surface reflectance on aerosol radiative forcing. *Fifteenth ARM Science Team Meeting Proceedings, Daytona Beach, Florida, March, 14–18*. 8 pp.

Rogers, A. N., Bromwich, D. H., Sinclair, E. N., & Cullather, R. I. (2001). The atmospheric hydrologic cycle over the Arctic basin from reanalyses. Part II: Interannual variability. *Journal of Climate, 14*, 2414–2429. https://doi.org/10.1175/1520-0442(2001) 014<2414:TAHCOT>2.0.CO;2.

Román, M. O., Schaaf, C. B., Lewis, P., Gao, F., Anderson, G. P., Privette, J. L., Strahler, A. H., Woodcock, C. E., & Barnsley, M. (2010). Assessing the coupling between surface albedo derived from MODIS and the fraction of diffuse skylight over spatially-characterized landscapes. *Remote Sensing of Environment, 114*(4), 738–760. https://doi.org/10.1016/j.rse.2009.11.014.

Satheesh, S. K., & Ramanathan, V. (2000). Large differences in tropical aerosol forcing at the top of the atmosphere and Earth's surface. *Nature, 405*, 60–63. https://doi.org/10.1038/35011039.

Serreze, M. C., & Hurst, C. M. (2000). Representation of mean Arctic precipitation from NCEP-NCAR and ERA reanalyses. *Journal of Climate, 13*, 182–201. https://doi.org/10.1175/1520-0442(2000)013<0182:ROMAPF>2.0.CO;2.

Serreze, M. C., Key, J. R., Box, J. E., Maslanik, J. A., & Steffen, K. (1998). A new monthly climatology of global radiation for the Arctic and comparisons with NCEP-NCAR reanalysis and ISCCP-C2 fields. *Journal of Climate, 11*, 121–136. https://doi.org/10.1175/1520-0442(1998)011<0121:ANMCOG>2.0.CO;2.

Serreze, M. C., Lynch, A. H., & Clark, M. P. (2001). The Arctic frontal zone as seen in the NCEP/NCAR reanalysis. *Journal of Climate, 14*, 1550–1567. https://doi.org/10.1175/1520-0442(2001)014<1550:TAFZAS>2.0.CO;2.

Serreze, M. C., Bromwich, D. H., Clark, M. P., Etringer, A. J., Zhang, T., & Lammers, R. (2003). The large-scale hydroclimatology of the terrestrial Arctic drainage. *Journal of Geophysical Research, 108*(D2), 8160. https://doi.org/10.1029/2001JD000919.

Serreze, M. C., Barrett, A. P., Slater, A. G., Steele, M., Zhang, J., & Trenberth, K. E. (2007). The large-scale energy budget of the Arctic. *Journal of Geophysical Research, Atmospheres, 112*, D11122. https://doi.org/10.1029/2006JD008230.

Shaw, G. E. (1976). Nitrogen dioxide - optical absorption in the visible, *Journal of Geophysical Research. Oceans and Atmospheres, 81*, 5791–5792. https://doi.org/10.1029/JC081i033p05791.

Shaw, G. E. (1995). The arctic haze phenomenon. *Bulletin of the American Meteorological Society, 76*, 2403–2413. https://doi.org/10.1175/1520-0477(1995)076<2403:TAHP>2.0.CO;2.

Smith and Bess. See at http://citeseerx.ist.psu.edu/viewdoc/summary?doi=10.1.1.597.8918

Smith, G. L., & Bess, T. D. *Arctic radiation budget changes*. Available at http://citeseerx.ist.psu.edu/viewdoc/summary?doi=10.1.1.597.8918

Stone, R. S. (2002). Monitoring aerosol optical depth at Barrow, Alaska, and South Pole. Historical overview, recent results and future goals. *SIF Conference Proceedings, 80*, 123–144.

Stone, R. S., Anderson, G. P., Andrews, E., Dutton, E. G., Shettle, E. P., & Berk, A. (2007). Incursions and radiative impact of Asian dust in northern Alaska. *Geophysical Research Letters, 34*, L14815. https://doi.org/10.1029/2007GL029878.

Stroeve, J., Holland, M. M., Meier, W., Scambos, T., & Serreze, M. (2007). Arctic Sea ice decline: Faster than forecast. *Geophysical Research Letters, 34*, L09501. https://doi.org/10.1029/2007GL029703.

Su, F., Adam, J. C., Trenberth, K. E., & Lettenmaier, D. P. (2006). Evaluation of surface water fluxes of the pan-Arctic land region with a land surface model and ERA-40 reanalysis. *Journal of Geophysical Research, Atmospheres, Volume, 111*, D05110. https://doi.org/10.1029/2005JD006387.

Thekaekara, M. P. (1973). Solar energy outside the Earth's atmosphere. *Solar Energy, 14*(2), 109–127. https://doi.org/10.1016/0038-092X(73)90028-5.

Tomasi, C., & Petkov, B. H. (2014). Calculations of relative optical air masses for various aerosol types and minor gases in Arctic and Antarctic atmospheres. *Journal of Geophysical Research, Atmospheres, 119*(3), 1363–1385. https://doi.org/10.1002/2013JD020600.

Tomasi, C., & Trombetti, F. (1985). Absorption and emission by minor atmospheric gases in the radiation balance of the earth. *La Rivista del Nuovo Cimento*, Vol. 8, Series 3, no. 2, 89 pp. (Monograph).

Tomasi, C., Guzzi, R., & Vittori, O. (1974). A search for the *e*-effect in the atmospheric water vapor continuum. *Journal of the Atmospheric Sciences, 31*(1), 255–260. https://doi.org/10.1175/1520-0469(1974)031<0255:ASFTEI>2.0.CO;2.

Tomasi, C., Vitale, V., Lupi, A., Di Carmine, C., Campanelli, M., Herber, A., Treffeisen, R., Stone, R. S., Andrews, E., Sharma, S., Radionov, V., von Hoyningen-Huene, W., Stebel, K., Hansen, G. H., Myhre, C. L., Wehrli, C., Aaltonen, V., Lihavainen, H., Virkkula, A., Hillamo, R., Ström, J., Toledano, C., Cachorro, V., Ortiz, P., de Frutos, A., Blindheim, S., Frioud, M., Gausa, M., Zielinski, T., Petelski, T., & Yamanouchi, T. (2007). Aerosols in polar regions: A historical overview based on optical depth and in situ observations. *Journal of Geophysical Research, Atmospheres, 12*, D16205. https://doi.org/10.1029/2007JD008432.

Tomasi, C., Petkov, B., Stone, R. S., Benedetti, E., Vitale, V., Lupi, A., Mazzola, M., Lanconelli, C., Herber, A., & von Hoyningen-Huene, W. (2010). Characterizing polar atmospheres and their effect on Rayleigh-scattering optical depth. *Journal of Geophysical Research, Atmospheres, 115*, D2. https://doi.org/10.10129/2009JD012852.

Tomasi, C., Petkov, B., Dinelli, B. M., Castelli, E., Arnone, E., & Papandrea, E. (2011). Monthly mean vertical profiles of pressure, temperature, and water vapour volume mixing ratio in the polar stratosphere and low mesosphere from a multi-year set of MIPAS-ENVISAT limb-scanning measurements. *Journal of Atmospheric and Solar-Terrestrial Physics, 73*(16), 2237–2271. https://doi.org/10.1016/j.jastp.2011.06.018.

Tomasi, C., Lupi, A., Mazzola, M., Stone, R. S., Dutton, E. G., Herber, A., Radionov, V. F., Holben, B. N., Sorokin, M. G., Sakerin, S. M., Terpugova, S. A., Sobolesky, P. S., Lanconelli, C., Petkov, B. H., Busetto, M., & Vitale, V. (2012). An update on polar aerosol optical properties using POLAR-AOD and other measurements performed during the International Polar Year. *Atmospheric Environment, 52, C*, 29–47. https://doi.org/10.1016/j.atmosenv.2012.02.055.

Tomasi, C., Lanconelli, C., Lupi, A., & Mazzola, M. (2013). Dependence of direct aerosol radiative forcing on the optical properties of atmospheric aerosol and underlying surface (Chapter 11). In A. A. Kokhanovsky (Ed.), *Light scattering reviews, volume 8, radiative transfer and light scattering* (pp. 505–626). Berlin: Springer Praxis Books/Environmental Sciences. https://doi.org/10.1007/978-3-642-32106-1.

Tomasi, C., Lanconelli, C., Lupi, A., & Mazzola, M. (2014). Diurnally averaged direct aerosol-induced radiative forcing from cloud-free sky field measurements performed during seven regional experiments (Chapter 8). In A. A. Kokhanovsky (Ed.), *Light scattering reviews, volume 9, Radiative Transfer and Light Scattering* (pp. 297–425). Berlin: Springer Praxis Books/Environmental Sciences. https://doi.org/10.1007/978-3-642-37985-7_8.

Tomasi, C., Kokhanovsky, A. A., Lupi, A., Ritter, C., Smirnov, A., O'Neill, N. T., Stone, R. S., Holben, B. N., Nyeki, S., Wehrli, C., Stohl, A., Mazzola, M., Lanconelli, C., Vitale, V., Stebel, K., Aaltonen, V., de Leeuw, G., Rodriguez, E., Herber, A. B., Radionov, V. F., Zielinski, T., Petelski, T., Sakerin, S. M., Kabanov, D. M., Xue, Y., Mei, L., Istomina, L., Wagener, R., McArthur, B., Sobolewski, P. S., Kivi, R., Courcoux, Y., Larouche, P., Broccardo, S., & Piketh, S. J. (2015). Aerosol remote sensing in polar regions. *Earth-Science Reviews, 140*, 108–157. https://doi.org/10.1016/j.earscirev.2014.11.001.

Trenberth, K. E. (1997). Using atmospheric budgets as a constraint on surface fluxes. *Journal of Climate, 10*, 2796–2809. https://doi.org/10.1175/1520-0442(1997)010<2796:UABAAC>2.0.CO;2.

Uppala, S. M., Allberg, P. W. K., Simmons, A. J., Andrae, U., Dacostabechtold, V., Fiorino, M., Gibson, J. K., Haseler, J., Hernandez, A., Kelly, G. A., Li, X., Onogi, K., Saarinen, S., Sokka, N., Allan, R. P., Andersson, E., Arpe, K., Balmaseda, M. A., Beljaars, A. C. M., Vandeberg, L., Bidlot, J., Bormann, N., Caires, S., Chevallier, F., Dethof, A., Dragosavac, M., Fisher, M., Fuentes, M., Hagemann, S., Olm, E. H., Hoskins, B. J., Isaksen, L., Janssen, P. A. E. M., Jenne, R., McNally, A. P., Mahfouf, J.-F., Morcrette, J.-J., Rayner, N. A., Saunders, R. W., Simon, P., Sterl, A., Trenberth, K. E., Untch, A., Vasiljevic, D., Viterbo, P., & Woollen, J. (2005). The ERA-40 reanalysis. *Quarterly Journal of the Royal Meteorological Society, 131*(612), 2961–3012. https://doi.org/10.1256/qj.04.176.

Vermote, E. F., Tanré, D., Deuzé, J. L., Herman, M., & Morcrette, J.-J. (1997). Second simulation of the satellite signal in the solar spectrum (6S): An overview. *IEEE Transactions on Geoscience and Remote Sensing, 35*(3), 675–686. https://doi.org/10.1109/36.581987.

Vogler, C., Brönnimann, S., & Hansen, G. (2006). Re-evaluation of the 1950–1962 total ozone record from Longyearbyen, Svalbard. *Atmospheric Chemistry and Physics, 6*(12), 4763–4773. https://doi.org/10.5194/acp-6-4763-2006.

Wanner, W., Strahler, A. H., Hu, B., Lewis, P., Muller, J.-P., Li, X., Schaaf, C. L. B., & Barnsley, M. J. (1997). Global retrieval of bidirectional reflectance and albedo over land from EOS MODIS and MISR data: Theory and algorithm. *Journal of Geophysical Research, Atmospheres, 102*(D14), 17143–17161. https://doi.org/10.1029/96JD03295.

Warren, S. G. (1984). Optical constants of ice from the ultraviolet to the microwave. *Applied Optics, 23*(8), 1206–1225. https://doi.org/10.1364/AO.23.001206.

Warren, S. G., & Wiscombe, W. J. (1980). A model for the spectral albedo of snow. II: Snow containing atmospheric aerosols. *Journal of the Atmospheric Sciences, 37*(12), 2734–2745. https://doi.org/10.1175/1520-0469(1980)037<2734:AMFTSA>2.0.CO;2.

Wielicki, B. A., Barkstrom, B. R., Harrison, E. F., Lee, R. B., III, Louis Smith, G., & Cooper, J. E. (1996). Clouds and the Earth's radiant energy system (CERES): An earth observing system experiment. *Bulletin of the American Meteorological Society, 77*(5), 853–868. https://doi.org/10.1175/1520-0477(1996)077<0853:CATERE>2.0.CO;2.

Wielicki, B. A., Barkstrom, B. R., Baum, B. A., Charlock, T. P., Green, R. N., Kratz, D. P., Lee, R. B., III, Minnis, P., Smith, G. L., Wong, T., Young, D. F., Cess, R. D., Coakley, J. A., Jr., Crommelynck, D. H., Donner, L., Kandel, R., King, M. D., Miller, J., Ramanathan, V., Randall, D. A., Stowe, L. L., & Welch, R. M. (1998). Clouds and the Earth's Radiant Energy System (CERES): Algorithm overview. *IEEE Transactions on Geoscience and Remote Sensing, 36*(4), 1127–1141. https://doi.org/10.1109/36.701020.

Wiscombe, W. J., & Warren, S. G. (1980). A model for the spectral albedo of snow. I: Pure snow. *Journal of the Atmospheric Sciences, 37*(12), 2712–2733. https://doi.org/10.1175/1520-0469(1980)037<2712:AMFTSA>2.0.CO;2.

Xu, J., Bergin, M. H., Greenwald, R., & Russell, P. B. (2003). Direct aerosol radiative forcing in the Yangtze delta region of China: Observation and model estimation. *Journal of Geophysical Research, Atmospheres, 108*, 4060. https://doi.org/10.1029/2002JD002550.

Yamamoto, G., & Sasamori, T. (1958). Calculation of the absorption of the 15 micron carbon dioxide band. *Scientific Report Tōhoku University, Series 5, Geophysics, 10*, 37–57.

Yang, Y., Wang, H., Smith, S. J., Easter, R. C., & Rasch, P. J. (2018). Sulfate aerosol in the Arctic: Source attribution and radiative forcing. *Journal of Geophysical Research, Atmospheres, 123*, 1899–1918. https://doi.org/10.1002/2017JD027298.

Zhao, T. X.-P., Yu, H., Laszlo, I., Chin, M., & Conant, W. C. (2008). Derivation of component aerosol direct radiative forcing at the top of atmosphere for clear-sky oceans. *Journal of Quantitative Spectroscopy and Radiative Transfer, 109*(7), 1162–1186. https://doi.org/10.1016/j.jqsrt.2007.10.006.

Chapter 11
Climate Change in the Arctic

Torben Koenigk, Jeff Key, and Timo Vihma

Abstract Observations over the last decades showed large changes in the Arctic regions with a strong warming of the Arctic, which is about twice that of the global mean warming. The largest warming rates with up to 10 K since 1980 are reached near the surface in the Chukchi Sea in autumn and in the Barents Sea in winter. Changes in Arctic climate are a result of complex interactions between the cryosphere, atmosphere, and ocean and different processes contribute to the amplified warming signal such as the ice albedo feedback, changes in clouds and water vapour, enhanced meridional energy transport in the atmosphere and in the ocean, vertical mixing in Arctic winter inversions and temperature feedbacks.

The observed warming is concurrent with a large reduction of the sea ice cover particularly in summer and autumn. The impact of Arctic amplification and sea ice retreat on the atmospheric circulation is still discussed. Positive winter sea level pressure trends along the Siberian Arctic coast have been linked to negative winter temperature trends over Central Asia.

Ocean heat and freshwater transports into and out of the Arctic undergo changes as well with potentially strong consequences for deep water formation in the North Atlantic Ocean and the entire large scale oceanic circulation.

Climate projections indicate that the Arctic will continue to warm faster than the rest of the world in the twenty-first century. Whether summer sea ice is going to melt completely depends on the future emission scenario.

T. Koenigk (✉)
Swedish Meteorological and Hydrological Institute, Norrköping, Sweden
e-mail: torben.koenigk@smhi.se

J. Key
National Oceanic and Atmospheric Administration, Madison, WI, USA
e-mail: jeff.key@noaa.gov

T. Vihma
Finnish Meteorological Institute, Helsinki, Finland
e-mail: timo.vihma@fmi.fi

A. Kokhanovsky, C. Tomasi (eds.), *Physics and Chemistry of the Arctic Atmosphere*,
Springer Polar Sciences, https://doi.org/10.1007/978-3-030-33566-3_11

This chapter will review the state of knowledge of mechanisms of the observed changes, the potential consequences of future Arctic warming for sea ice, ocean and atmosphere, and uncertainties due to emission scenarios, model shortcomings and natural variability.

Keywords Arctic temperature amplification · Arctic climate change · Natural variability · Future climate scenarios · Arctic observations · Climate modelling · Ocean- sea ice - atmosphere interactions · Arctic - lower latitude interactions · Climate processes

11.1 Introduction

The Arctic plays an important role in the global climate system. Snow and ice cover have a large influence on both the local and remote climate. Small changes in the sea ice cover may have a strong impact on the large-scale atmospheric circulation, which affects the local climate but could also affect remote regions. The Arctic Ocean is a large source of freshwater for the global oceans. The export of freshwater from the Arctic alters the deep water formation in the North Atlantic and affects thus the entire global ocean circulation system.

Observations over the last few decades showed an ongoing climate change in the Arctic regions with a strong warming of the Arctic, which is about twice of the global mean warming. The observed warming is concurrent with a large reduction of the sea ice cover, which has accelerated since 2000 with recent minima in September of 2007 and 2012.

Arctic warming and sea ice reduction have large impacts on the Arctic ecosystem and social-economic activities in the Arctic. Changes in Arctic ice cover affect ice algae and phytoplankton. Consequently, fish stocks move further to the north, affecting fishery and coastal communities. With the reduction of sea ice, new shipping routes through the Arctic open and the exploitation of nature resources such as oil and gas is becoming more economically feasible. This puts the Arctic environment at risk, and actions need be taken to prevent Arctic pollution. Further, increased shipping in the tough Arctic environment requires rescue plans for people and nature in case of accidents.

11.2 Observed Changes in Recent Decades

11.2.1 Atmosphere

Observations indicate that Arctic near-surface temperature trends have been about twice the rate of the global mean warming over the last few decades (ACIA 2005; Stocker et al. 2013; Richter-Menge and Jeffries 2011). The largest warming rates

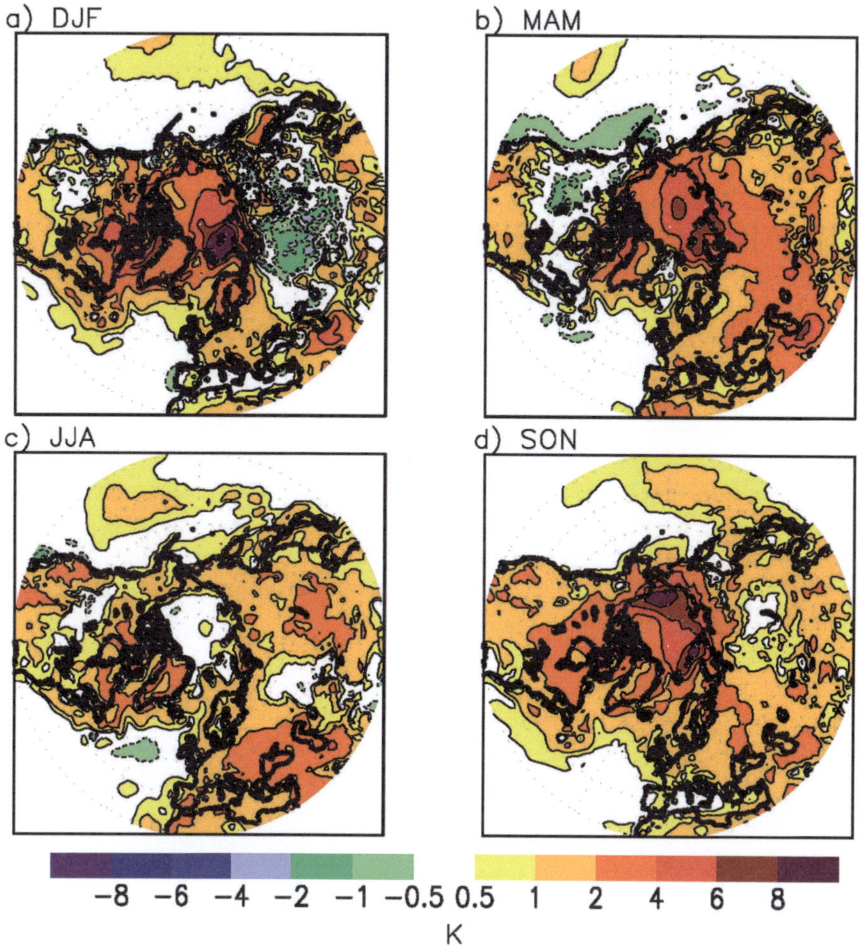

Fig. 11.1 Seasonal two meter air temperature trends calculated from ERA-Interim reanalysis data (Dee et al. 2011) between 1980 and 2017

with up to 10 K between 1980 and 2017 are reached in the Chukchi Sea in autumn and in the Barents Sea in winter (Fig. 11.1). In contrast, surface warming in summer is small and no amplification of surface temperature can be seen.

The warming, particularly in winter but to a lesser degree also in autumn and spring, is especially pronounced at the surface and is strongly reduced with height. However, even above the surface, the warming is amplified compared to lower latitudes (Cohen et al. 2014). In summer, the strongest warming occurs above the surface at 900–950 hPa height.

The winter Arctic is often dominated by a near surface temperature inversion. The warmest temperatures typically occur at about 850–900 hPa while the surface is on average approximately 6 K colder than the air above (Medeiros et al. 2011; Koenigk et al. 2013).

The observed strong warming at the surface in winter leads to a reduction of the winter inversion weather situations in the Arctic. This has impacts on cloud and precipitation formation. A reduced winter inversion allows for more formation of clouds further up in the atmosphere replacing low level clouds. On the other hand, more open water regions lead to increased formation of low-level clouds. While there is rather large agreement that precipitation in the warming Arctic has increased, uncertainties in cloud trends are still large.

Different processes have been suggested to contribute to the amplified warming signal. The traditional explanation is the ice albedo feedback (Serreze et al. 2009; Screen and Simmonds 2010a, b; Letterly et al. 2018), where an initial global warming signal leads to reduced sea ice, which in turn leads to additional warming due to reduced reflectivity of solar radiation at the surface. However, other studies suggested that changes in clouds and water vapour (Graversen and Wang 2009; Liu et al. 2008, 2009; Liu et al. 2018), enhanced meridional energy transport in the atmosphere (Graversen et al. 2008) and in the ocean (Spielhagen et al. 2011; Koenigk and Brodeau 2014), vertical mixing in Arctic winter inversions (Bintanja et al. 2011), and temperature feedbacks (Pithan and Mauritsen 2014) might play a similarly large role as the sea ice albedo feedback.

The change patterns in the atmospheric circulation in recent decades vary strongly throughout the year (Fig. 11.2) and most trends in the period 1980–2017 are rather small compared to the natural variability. The trends are of similar amplitude in all seasons. The most robust pattern across seasons is a positive SLP trend over parts of the North Pacific. Also, the positive trend along the Siberian Arctic coast in winter raised attention, and there is an ongoing discussion if this trend can be linked to Arctic sea ice reduction. More details on the potential impact of the Arctic amplification on lower latitudes is provided in Sect. 11.3.

11.2.2 Sea Ice

The rapid Arctic warming is accompanied by a strong reduction of Arctic sea ice. Satellite data, which can reliably estimate the Arctic sea ice extent, provides a record starting in 1979 (Fig. 11.3). Since then, the sea ice extent has been decreased in all seasons (Stroeve et al. 2012; Devasthale et al. 2013; Meier et al. 2014). The trend is most negative in late summer (Fig. 11.3), and the average minimum ice extent in the last decade has been more than 30% below the average of the 1980s and 1990s. The lowest minimum sea ice extent was observed in September 2012 with an ice extent of about 3.6 million km^2, approximately 50% below the average of the 80s and 90s. Spatially, the summer sea ice reduction is most pronounced in the Beaufort and East Siberian Seas.

The maximum winter sea ice extent decreased from around 16.5 million km^2 in the 80s and 90s to around 14.5 million km^2 in the recent decade. A reduction of

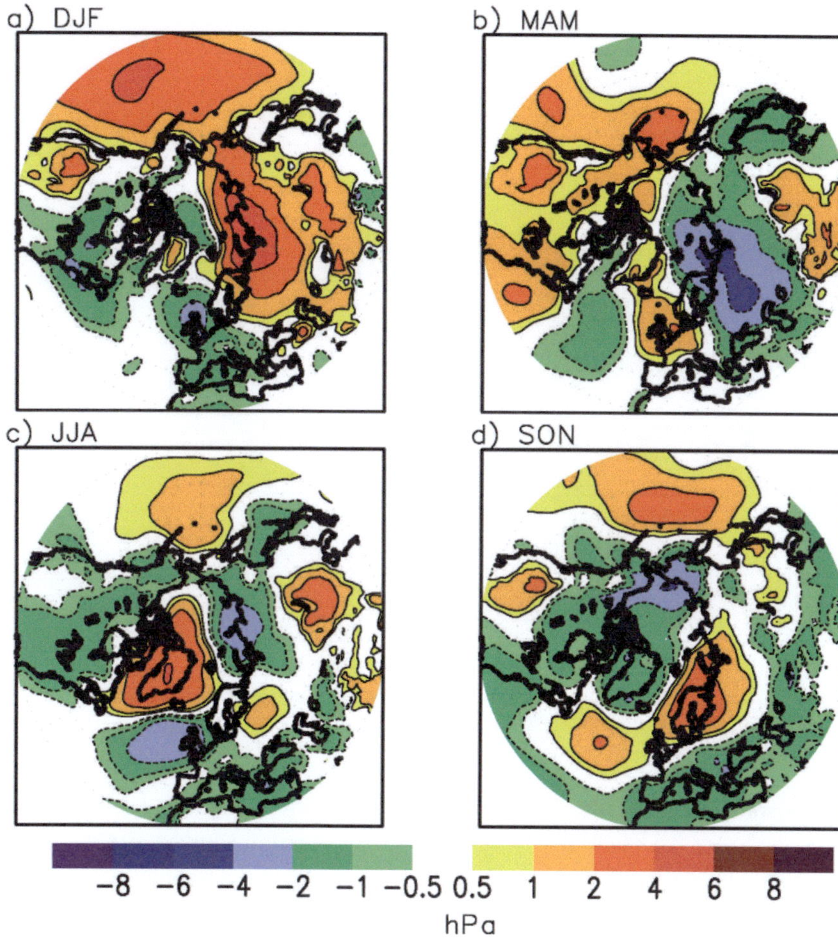

Fig. 11.2 As Fig. 11.1 but for SLP

sea ice has a direct impact on temperature changes in the Arctic. In the cold season, sea ice isolates the warm ocean (around freezing temperature) from the very cold atmosphere (on average around −30 °C). Where sea ice disappears, a large amount of heat is released from the ocean to the atmosphere, leading to local warming rates of more than 20 °C.

Sea ice thickness measurements are still difficult and no reliable, long term Arctic-wide observations are available. However, in-situ measurements, estimates from satellites and model simulations indicate that sea ice has become much thinner in recent decades. The most commonly used comprehensive ice volume data set is the Pan-Arctic Ice Ocean Modeling and Assimilation System (PIOMAS) data set (Schweiger et al. 2011). It indicates that the total sea ice volume in September shrank by more than 50% since the 1980s and 1990s. In contrast to sea ice extent,

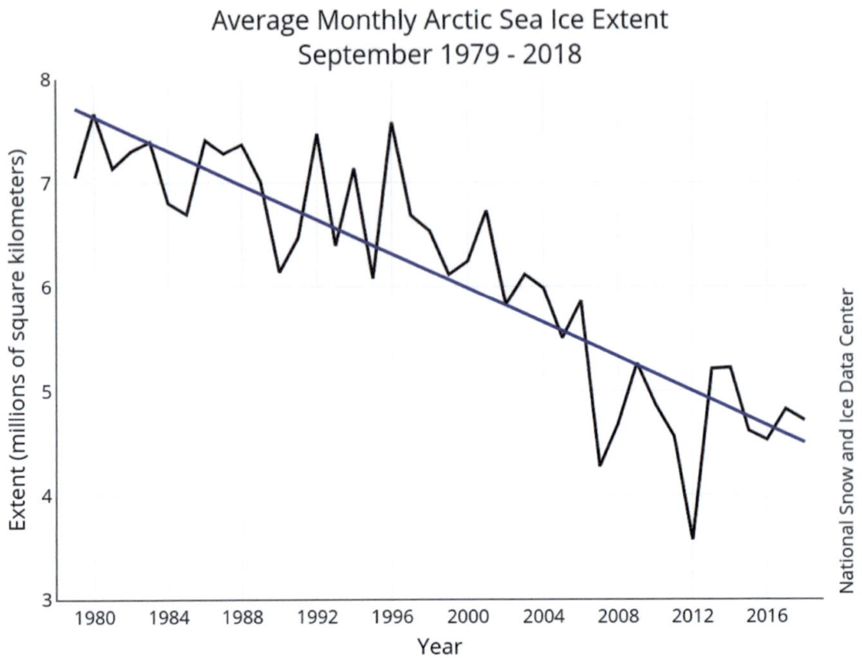

Fig. 11.3 Arctic sea ice extent in September between 1979 and 2018. (National Snow and Ice Data Center, NSIDC; Fetterer et al. 2017)

sea ice volume reduction is rather uniformly occurring throughout the entire year. Thus, the lesser ice extent decrease in winter compared to summer is misleading: the remaining winter sea ice is thinning at the same rate as the summer sea ice. This also has as a consequence that the amount of old, multi-year sea ice in the Arctic is reduced. Large parts of the Arctic Ocean are now covered by ice that is one year and younger.

11.2.3 Ocean

The ocean circulation in the Arctic in layers near the surface is characterized by a pronounced Beaufort Gyre, transpolar currents from the Siberian coast towards Fram Strait and southerly flows through Fram Strait as well as inflowing waters through the Barents Sea Opening and further along the Siberian coast.

The Arctic Ocean is a large source of freshwater and the Arctic freshwater export has an important impact on the formation of deep water in the North Atlantic Ocean and thus potentially the entire thermohaline circulation (Dickson et al. 1988; Häkkinen 1999; Haak et al. 2003; Koenigk et al. 2006).

Observations indicate an increase in the discharge in river runoff and precipitation in the second half of the twentieth century (Peterson et al. 2002; McClelland et al. 2006). Furthermore, sea ice is expected to melt at least partly. All this led to freshening of the Arctic Ocean. Although, observations are rare, it seems that most of the additional freshwater input into the Arctic in recent decades have been stored in the Arctic, particularly in the Canadian Arctic and Beaufort Gyre, and were not exported into the North Atlantic Ocean yet. As one main reason for this, a strengthening of the anticyclonic wind pattern has been discussed. Such a strengthening would cause a convergence of fresh surface water in the Beaufort Gyre. A release of this freshwater, which is now several times the freshwater amount that has been exported into the North Atlantic during the Great Salinity Anomaly (Dickson et al. 1988) could have large consequences for the global thermohaline circulation.

Observations also suggest an increase in the oceanic heat transport into the Arctic (Polyakov et al. 2005) and indicate a potential for accelerated sea ice melt. Hereby, the Barents Sea plays a key role for the heat transport since most of the heat coming from the south has to pass the rather small and shallow Barents Sea. Observations of the temperature of the inflowing Atlantic water show a positive trend over recent decades (Skagseth et al. 2008. Smedsrud et al. 2013). Schlichtholz (2016) showed, based on proxies, that the temperature of the Atlantic water reaching the Arctic has never been as high as today during the last 2000 years.

The additional heat that arrives in the Barents Sea is mainly passed to the atmosphere, thus contributing to Arctic warming and affecting atmospheric stability and circulation (Koenigk and Brodeau 2014). The rest of the additional heat is used to reduce the sea ice in the Barents Sea or is mixed into the Arctic waters below the cold, fresh surface layer, thus affecting the oceanic stratification of the Arctic Ocean.

11.2.4 Cloud-Ice Interactions, Forcings, and Feedbacks

Observational assessments have shown that trends in Arctic cloud and sea ice cover are linked. Liu et al. (2012) employed an equilibrium feedback assessment to quantify this relationship using satellite-derived sea ice concentration and cloud cover in the Arctic over the period 2000–2010. They found that a 1% decrease in sea ice concentration leads to an increase in cloud cover of approximately 0.4%, and that 22–34% of the cloud cover variance can be explained by the variability in sea ice extent. This suggests that a further decline in sea ice cover will result in an even cloudier Arctic.

Cloud feedbacks are the main factor behind the uncertainty in patterns of warming, primarily through their indirect influence on meridional transport between the tropics and the Arctic (Bonan et al. 2018). The radiative effects of Arctic clouds on the surface energy budget - cloud “forcing” - can also result in warming (or cooling) through their direct effect on sea ice growth and melt. Similarly, changes in cloud cover influence the ice albedo and snow albedo feedbacks, which have large-scale impacts on warming through changes in the absorption of solar radiation at the surface.

For example, Liu and Key (2014) examined the record minimum Arctic sea ice extent in September 2012 and the "rebound" of the ice cover the following year. During the 2012–2013 winter the sea ice cover quickly returned, carrying through to the summer when the ice extent was 48% greater than the same time in 2012. Most of this rebound was in the Chukchi and Beaufort Seas, areas experiencing the greatest recent decline in sea ice. While factors such as ice dynamics, heat transport, and solar insolation may have contributed, it was demonstrated that below-average Arctic cloud cover in January–February 2013 resulted in a strongly negative surface radiation budget, which cooled the surface and allowed for greater ice growth. More thick ice was observed in March 2013 relative to March 2012 in the western Arctic Ocean, and the areas of ice growth estimated from the negative cloud cover anomaly that were advected from winter to summer corresponded well with the September ice concentration anomaly pattern. Therefore, decreased wintertime cloud cover appears to have played an important role in the return of the sea ice cover the following summer, providing a partial explanation for large year-to-year variations.

Letterly et al. (2016) extended this work to cover the period 1983–2013. Using satellite data and two climate reanalyses it was shown that cloud forcing anomalies in the East Siberian and Kara Seas precondition the ice pack and, as a result, explain 25% of the variance in late summer sea ice concentration. This suggests that winter cloud forcing anomalies in this area have predictive capabilities for summer sea ice anomalies across much of the central and Eurasian Arctic.

Changes in Arctic climate are a result of complex interactions between the cryosphere, atmosphere, and ocean. With regard to climate warming, more cryosphere feedbacks are positive and result in further warming than are negative, resulting in a reduced rate of warming or cooling (Callaghan et al. 2011). Both the ice albedo feedback over the ocean and the snow albedo feedback over land have played an important role in Arctic warming. Recent declines in sea ice and snow extent have led to an increase in the absorption of solar energy at the surface, which has resulted in additional surface heating and a further decline in snow and ice. These are the ice and snow albedo feedbacks. Letterly et al. (2018) used satellite data covering the period 1982–2015 to show that the positive trend in solar absorption over the Arctic Ocean is more than double the trend over Arctic land, and that the magnitude of the ice–albedo feedback is four times that of the snow–albedo feedback in summer. The timing of the high-to-low albedo transition is important for the future, as it has shifted closer to the summer solstice over ocean when solar insolation is highest, but further away from the summer solstice toward earlier in the year over land. The implication of this shift is that decreasing sea ice cover, not changes in terrestrial snow cover, has been the dominant radiative feedback mechanism in recent decades and will continue to play an important role in future Arctic climate change.

11.3 Effects of Arctic Changes on Mid-Latitude Weather and Climate

In Sect. 11.2, rapid changes in Arctic climate in the last decades were discussed. Sea ice has been substantially reduced in extent and thickness, particularly since the early 2000s. At the same time, several mid-latitude regions experienced a number of cold winters and parts of Eurasia and North America show negative winter temperature trends since the 1980s.

Effects of Arctic changes on mid-latitude weather and climate is a challenging study topic, which has recently received considerable attention (Cohen et al. 2014; Walsh 2014; Vihma 2014, 2017; Gao et al. 2015; Overland et al. 2015) and perspective papers (Overland et al. 2016; Francis 2017; Screen 2017). Studies based on both observational data and model simulations have indicated a connection between variations of late summer Arctic sea ice extent and winter mid-latitude conditions (Petoukhov and Semenov 2010; Francis et al. 2009; Francis and Vavrus 2012; Yang and Christensen 2012; Overland and Wang 2010; Hopsch et al. 2012; Garcia-Serrano and Frankkignoul 2014; Mori et al. 2014). Most of these studies found that a reduction or negative anomaly in late summer sea ice extent leads to winter atmospheric circulation anomalies resembling the negative phase of the North Atlantic Oscillation (NAO) and thus to more frequent cold winters in mid-latitudes of Eurasia. On the contrary, there are also studies suggesting that Arctic warming or sea ice decline sometimes favors the positive phase of NAO (Orsolini et al. 2012; Screen et al. 2014). Contrasting results have also been obtained for the Arctic effects on the Polar front jet stream (Francis and Vavrus 2012; Barnes 2013).

Despite these controversies, advances been made in understanding the multi-faceted effects of Arctic changes. We can presently recognize several dynamical mechanisms that transmit the effects of Arctic changes on mid-latitude weather and climate, six of them introduced below. Some of them have might have contributed to the recent cold winters in mid-latitudes.

11.3.1 Dynamical Mechanisms

First, due to Arctic Amplification, wintertime cold-air outbreaks originating from the Arctic have become warmer (Serreze et al. 2011; Screen 2014). This has opposed the recent occurrence of cold mid-latitude winters. However, contrary to general trend in mid-latitudes, in East Asia wintertime cold-air outbreaks have become colder (Kim et al. 2014).

Second, due to Arctic Amplification, the north-south geopotential height gradients have become weaker, in particular in mid- and upper-troposphere. This has decreased the westerly winds according to the thermal wind law (Francis and Vavrus 2012, 2015), and resulted in a wavier jet stream (Vavrus et al. 2017), which has slowed down the eastward motion of weather patterns, increasing the likelihood of extremes related to persistent weather patterns (Overland et al. 2015; Coumou et al. 2015). However, the tropical amplification of climate warming in the upper troposphere tends to oppose these effects (McGraw and Barnes 2016). The net effect of Arctic and tropical amplifications is unclear and may depend on the season and region. This is seen, among others, in regionally different observed and predicted changes in blocking occurrence. According to Hoskins and Woollings (2015), a general decrease in blocking occurrence is expected under warming climate but, according to Masato et al. (2013) and Dunn-Sigouin and Son (2013), blocking may increase, e.g., in summer in western Russia when the European blocking shifts eastward. Further, Mori et al. (2014) show that Arctic sea ice decline has favored Eurasian blocking, and Hanna et al. (2016) associate recent strong events of Greenland Blocking to Arctic Amplification.

Third, several studies suggest that Arctic changes have triggered planetary wave trains responsible for remote effects in mid-latitudes. For example, spring sea ice anomalies in the Arctic have been associated with anomalies in summer precipitation in China (Zhao et al. 2004; Wu et al. 2009, 2013; Guo et al. 2014; Uotila et al. 2014) and Europe. In the case of Europe, the Sea of Okhotsk, Hudson Bay and Labrador Sea have been named as the most influential regions of sea ice decline (Mesquita et al. 2011; Wu et al. 2013). The summertime effects may be particularly strong, if the planetary waves resonate with quasi-stationary thermal and orographic forcing. According to Petoukhov et al. (2013) and Coumou et al. (2014, 2015), trapping of planetary waves typically with zonal wave numbers 6, 7, or 8 have contribution to heat waves in Europe (Summer 2003), Russia (2010), and the USA (2011), as well as the floods in Pakistan (2010).

Fourth, the Arctic sea ice decline and changes in Eurasian snow cover in autumn have probably contributed to Eurasian winter weather via stratosphere-troposphere interactions (e.g. Cohen et al. 2012; Jaiser et al. 2013; Garcia-Serrano and Frankignoul 2014). They argue that reduced autumn sea ice and thus more humid air affects precipitation patterns which lead to enhanced autumn snow cover over Siberia. Cohen (2007) showed that this can affect the Siberian High and changes planetary wave fluxes into the stratosphere. Also, a more direct link between sea ice and stratospheric circulation through generating anomalous Rossby waves (Honda et al. 2009; Peings and Magnusdottir 2014) or by preconditioning static stability and baroclinicity (Rinke et al. 2013) has been suggested. All these suggested processes might slow down the Polar Vortex, which in turn feeds back to the troposphere leading to a negative NAO/AO signal. Inoue et al. (2012) and Koenigk et al. (2016) suggested that the meridional temperature gradient due to ice variations in the Barents Sea is the main reason for the observed SLP anomalies and the connected "warm Arctic – cold continent" temperature patterns in recent years. Earlier studies (Magnusdottir et al. 2004; Deser et al. 2004; Alexander et al. 2004; Koenigk et al. 2006) also supported a link via the meridional temperature gradient caused

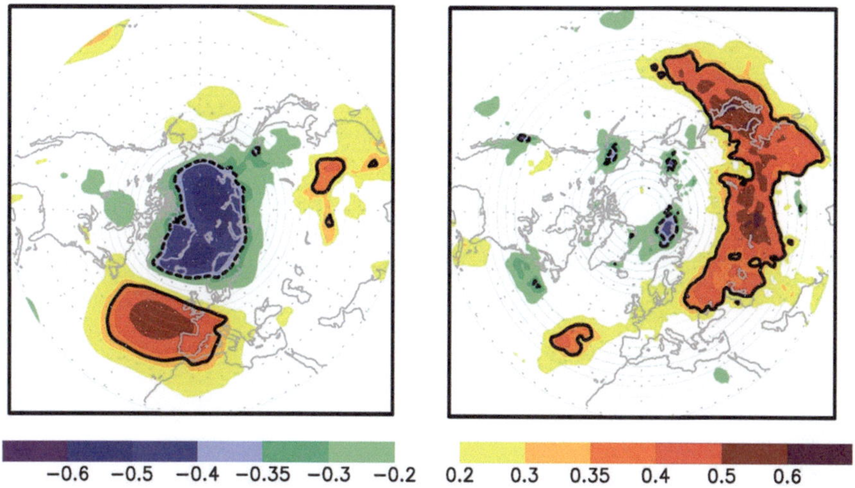

Fig. 11.4 Correlation between Barents-Kara Seas ice area in November and next winters SLP (hPa, left) and T2 m (K, right) in the period 1980–2013 in ERA-interim atmosphere data and OSISAF sea ice data

by sea ice anomalies leading to a baroclinic adjustment to the introduced mass gradient anomaly.

Fifth, combined effects of the Arctic and Pacific have been identified, so that the atmospheric sensitivity to sea ice loss depends on SST anomalies related to the Pacific Decadal Oscillation (PDO). For the same pattern and amount of sea-ice loss, warming is larger during the negative PDO phase (Screen and Francis 2016).

Sixth, sea ice in the Barents Sea has been identified as the region with the strongest impact on NAO and Eurasian winters (Inoue et al. 2012; Koenigk et al. 2016). Figure 11.4 shows that November sea ice anomalies in the Barents-Kara Seas are significantly correlated to a NAO-like SLP pattern and widespread surface air temperature anomalies over Eurasia in the following winter. Negative sea ice anomalies would be followed by a negative NAO and a cooling over large parts of Eurasia. It is an unsolved question whether this atmospheric response to sea ice anomalies is linear. Liptak and Strong (2014) showed, performing sensitivity experiments with the atmosphere model CAM, a similar response of the NAO to both negative and positive sea ice anomalies in the Barents Sea. Koenigk et al. (2016) performed a composite analysis for observed low and high ice cases in autumn Barents-Kara Seas ice area and found a very strong NAO- like response after low ice conditions while the response after high ice conditions was not significant. Also, an earlier study by Petoukhov and Semenov (2010) showed a dependency of the response on the size of the ice anomaly, thus indicating a non-linear response of the atmosphere to ice anomalies.

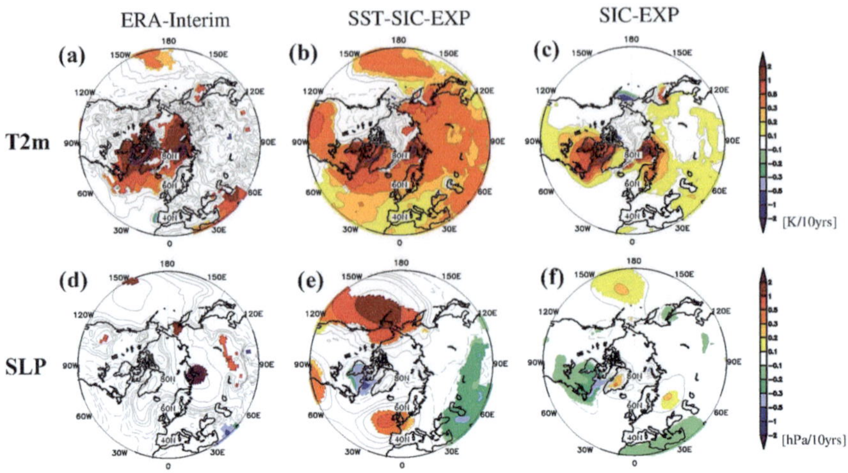

Fig. 11.5 (**a–c**) Mean observed 2 m temperature trends in DJF from 1982 to 2014. (**a**) is for ERA-Interim, (**b–c**) are for the simulated multi-model ensemble mean trends in SST-SIC-EXP and SIC-EXP, respectively. (**d–f**) as in (**a–c**), but for the SLP. Shading is applied for the statistical significance by a Student's t-test with 95% confidence level. (From Ogawa et al. 2018, Fig. 1)

11.3.2 Recent Model Experiments

The largest existing multi-model ensemble study to analyse the potential impact of sea ice on lower-latitudes has been performed in the GREENICE project (https://greenice.w.uib.no/). Six different Atmosphere General Circulation Models (AGCM) have been used to perform two sets of ensemble experiments. In the first, observed daily varying sea ice concentration and sea surface temperature (SST) between 1982 and 2014 have been prescribed. In the second experiment, the varying SST has been replaced by the climatological annual cycle over 1982–2014. The results indicate that the impact of sea ice trend and variations on mid-latitudes is low (Ogawa et al. 2018). Figure 11.5 shows that the observed negative trends over Eurasia are almost nowhere significant and neither the simulations with full SST and sea ice forcing nor the runs with only varying sea ice forcing show any mid-latitude cooling. Further, Ogawa et al. (2018) showed that the Siberian temperature trends are extremely strongly varying among ensemble members and that the observed trends are well captured by these variations.

Also results from long coupled climate model simulations by Koenigk and Brodeau (2017), highlighted the large natural variability on timescales of up to 30 years. For selected 30-year periods, both strong cooling trends and strong warming trends of similar size of the observed one have been found. Similarly, correlation patterns between autumn Barents-Kara Seas ice area and winter SLP are heavily varying depending on the period. 30-year periods are found where low sea ice in the Barents/Kara Seas is followed by negative NAO-like SLP patterns and other 30-year periods with the opposite correlation between ice NAO (Fig. 11.6).

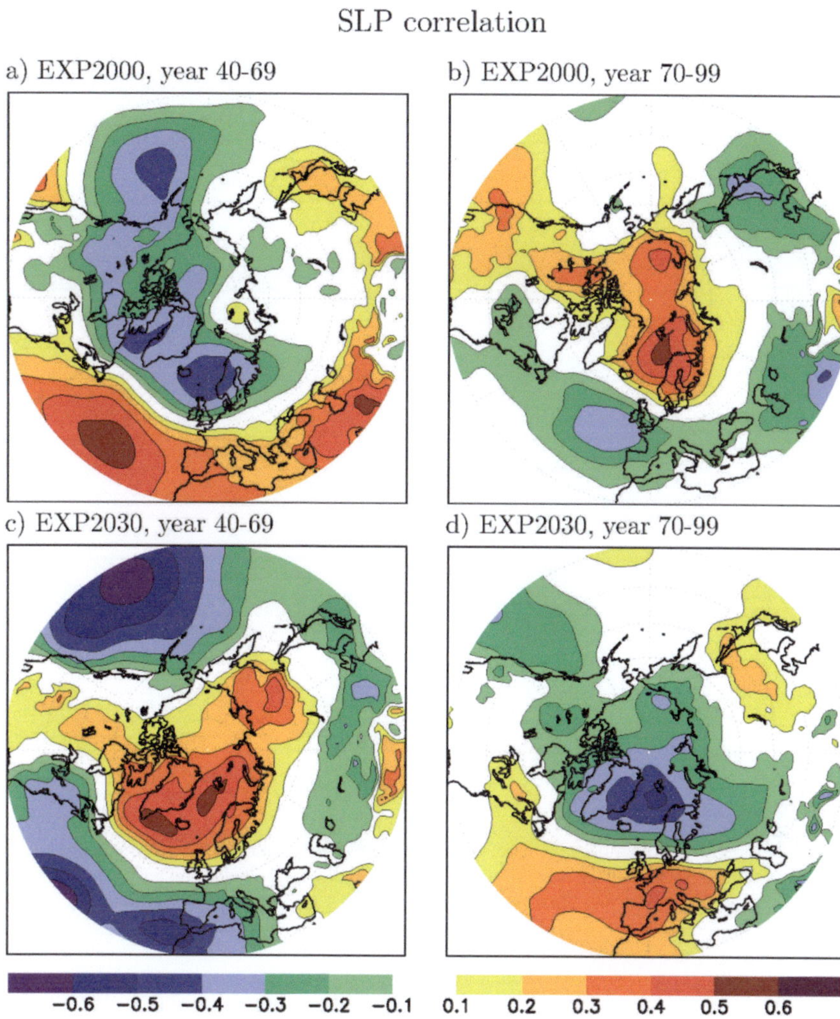

Fig. 11.6 Correlation between sea ice area in the Barents-Kara Seas (70°N–82°N, 15°E–100°E) in November and SLP in the following winter (DJF) in EXP2000 (**a**, **b**) and EXP2030 (**c**, **d**). Correlations are shown for two selected 30-year time periods: years 40–69 (**a**, **c**) and years 70–99 (**b**, **d**). (From Koenigk and Brodeau 2017, Fig. 13)

An update of the observational time series until winter 2017 indicates a substantial reduction of the negative winter trends over Siberia due to mainly warm winters in the winters 2014–2017 (Fig. 11.7). This is caused by a return of the NAO-index towards normal or above normal values – the trend in the winter NAO-index from 1980–2017 is zero – and a return of the Pacific Decadal Oscillation (PDO) index to normal values. The sea ice extent in the Barents-Kara Seas stays low even after

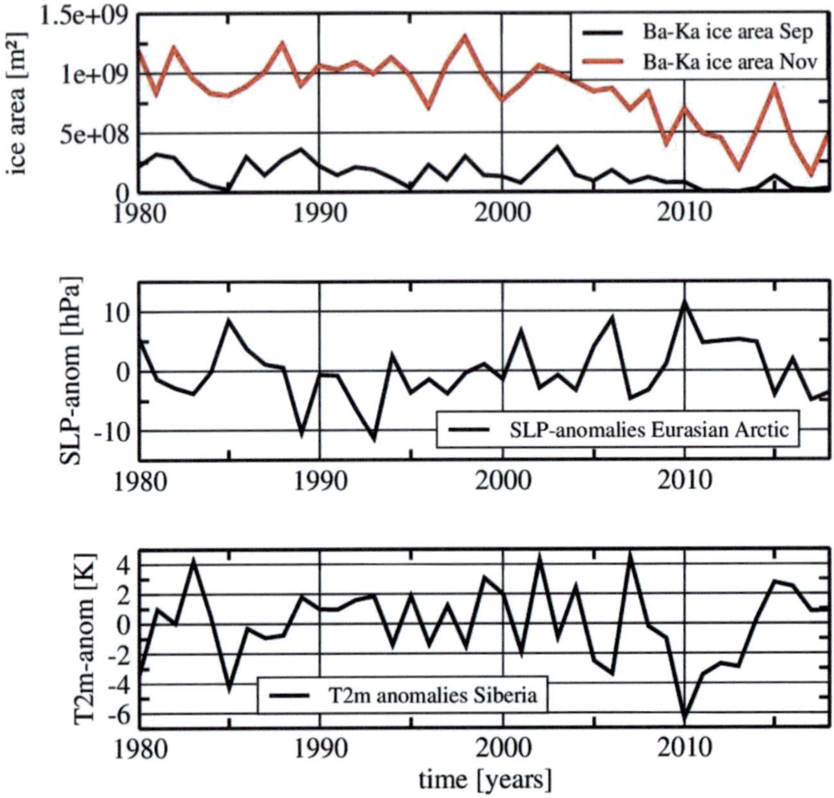

Fig. 11.7 First row: November and September Sea ice area in the Barents and Kara Seas, averaged over 70°N–82°N, 15°E–100°E. Second row: DJF SLP-anomalies, averaged over a box along the Eurasian Arctic coast with the strongest trend in 1980–2013 (65°N–80°N, 30°E–120°E). Third row: DJF T2 m-anomalies, averaged over a box in Central Eurasia with the strongest negative trend 1980–2013 (50°N–60°N, 60°E–100°E)

2013, which indicates that the linkage to the NAO and to Siberian cooling might be lower than anticipated by many studies a few years ago.

The fact that extending the time series by a few years leads to strong changes in the trends, shows clearly the high natural variability, and that time series are still too short to make final statements on the linkage between Arctic sea ice and mid-latitude climate and weather.

Even if the linkage between low sea ice and mean winter temperature is less pronounced, it is still possible that reduced sea ice could affect the occurrence of weather extremes. While much focus in recent years was on cold episodes in winter, also a number of very warm winter periods occurred in many northern hemispheric mid-latitude areas. These have been generally linked to global warming while it is possible that even these warm extremes are particularly extreme due to a reduced meridional temperature gradient caused by Arctic warming and related wavier and slower moving Rossby waves as discussed by Francis and Vavrus (2012). Warm and cold periods might cancel out each other in the seasonal means.

11.3.3 Discussion

On the basis of numerous studies, it is evident that the effects of a certain Arctic forcing depend on the region, season, and the state of the atmosphere and ocean. Accordingly, Arctic forcing may reinforce or weaken existing weather patterns. Small signal-to-noise ratios and multiple forcing factors acting simultaneously in a chaotic dynamical system makes the Arctic effects on mid-latitudes inconsistent, episodic, non-linear and hard to distinguish from other effects (Overland et al. 2016). This is one of the reasons why different studies have yielded different and controversial results. The other reasons include (a) different study periods and regions, (b) application of different (sometimes ill-suited) metrics to quantify the linkages, and (c) errors and uncertainties in reanalyses and climate models. Further, some studies have focused on the effects of Arctic Amplification, whereas some have only focused on the effects of Arctic sea ice decline.

As Shepherd (2016) formulated: "The question is not whether Arctic changes are affecting mid-latitudes but rather how and by how much." We need to better understand the role of the Arctic forcing compared to other forcings and natural variability. The Arctic forcing always acts in concert with other forcings on mid-latitude weather and climate. These include SST variations in the tropical and northern Atlantic (Davini et al. 2015; Hall et al. 2015; Yeager and Robson 2017) and Pacific (Screen and Francis 2016), which may be partly responsible for the recent cooling over the northern continents (Kosaka and Xie 2013; Peings and Magnusdottir 2014). As the recent Arctic Amplification only started in late 1990s, the observed time series are still short, which makes it challenging to distinguished between forced effects and natural variability (Barnes 2013; Screen et al. 2014; Sun et al. 2016; Koenigk and Brodeau 2017; McCusker et al. 2016; Ogawa et al. 2018). Further, dynamical mechanisms originating from the Arctic Amplification partly compensate for each other. The most direct effect - warmer cold-air outbreaks originating from the Arctic - favours mid-latitude warming, enhancing the effects of greenhouse forcing (Screen et al. 2015), whereas many of the other effects described above often tend to generate mid-latitude cooling, opposing (but not dominating) the global warming. In addition to the temperature trends, an important aspect is that the Arctic Amplification probably favors more persistent weather patterns in mid-latitudes, increasing the risk of floods, droughts, as well as long-lasting heat waves and cold spells.

11.4 Projected Changes until the End of the Twenty-First Century

Climate projections indicate that the Arctic will continue to warm faster than the rest of the world in the twenty-first century. The projected changes in atmosphere, ocean and sea ice are closely linked, and it is often difficult to distinguish between causes and effects. In this section, we will focus on the simulated mean changes of Arctic key variables in global and regional future projections.

11.4.1 Projected Changes in Sea Ice

The sea ice conditions strongly govern the exchanges of heat, freshwater and momentum between ocean and atmosphere. Thus, sea ice changes can have a huge impact on the atmosphere above. At the same time, the sea ice is itself very sensitive to changes in atmosphere conditions.

Satellite-based retrievals of sea ice extent are now available since 1979, covering a period of 40 years (Meier et al. 2014). The northern hemispheric ice extent showed a strong negative trend during this period, particularly during late summer. Future climate model projections indicate a continuation of this negative sea ice trend in the twenty-first century (Stocker et al. 2013). However, whether or not sea ice will completely melt in late summer in the twenty-first century depends on the emission scenario and on the model. While most global CMIP5 models indicate a total summer sea ice loss under the strong RCP8.5 emission scenario, around 1/3 of all CMIP5 models indicate an almost total loss of late summer sea ice loss using the RCP4.5 emission scenario (Stroeve et al. 2012). Under the low emission scenario RCP2.6, almost all models show a survival of Arctic summer sea ice.

Sea ice reduction does not only vary between different emission scenarios but also between different models (using the same emission scenario) and due to natural variations. For the RCP8.5, September sea ice disappears between year 2020 and later than year 2100 dependent on the model. Massonnet et al. (2012) selected those models, which agree best with observed historical ice concentrations and trends and found that the most likely date of first Arctic sea ice loss under RCP8.5 would be between year 2040 and 2060. However, there is no guarantee that models with high agreement of historical sea ice conditions to observations are most reliable for the future.

Generally, many models show a smaller sea ice reduction in recent decades than the observations. Thus, a few studies made the attempt to correct the models for this potential underestimation, e.g., Wang and Overland (2012) estimated that sea ice loss is likely to happen in the 2030s taking into account the potential underestimation of the observed sea ice trends in the models.

Despite the underestimation of the CMIP5-model mean trend, the observed trend is still inside the range of sea ice trends of the CMIP5 models. Thus, it cannot be ruled out that natural variations caused the apparent difference between simulated and observed summer sea ice trend. Ding et al. (2017) concluded that 30–50% of the observed summer sea ice decline since 1979 could be due to natural variations.

The natural variation of sea ice trends is commonly estimated by comparing the results from different simulations with the same model. The CESM large ensemble (Deser et al. 2010) consists of 40 ensemble members and shows large spread across different ensemble members. The year when an ice-free September is reached for the first time varies between 2032 and 2053 in this ensemble for the RCP8.5 emission scenario (Jahn et al. 2016).

Spatially, summer sea ice disappears first in the Beaufort Sea and along the Siberian shelf seas; sea ice disappears last along the north coast of Greenland and the north of the Canadian Archipelago. In winter, sea ice extent will be reduced but at a somewhat slower rate as in summer. In contrast to summer, sea ice melt is largest along the North Atlantic Arctic ice edges, particularly in the Barents Sea. The Barents Sea will thus be the first Arctic region where sea ice disappears year around. Koenigk and Brodeau (2014), using the EC-Earth climate model, estimated that the Barents Sea will be ice free year-around at around year 2030.

The year to year variations of Arctic sea ice extent might increase in future when sea ice becomes thinner and thus more vulnerable to atmospheric circulation anomalies. Dynamical regional downscalings of global climate model simulations with the regional coupled atmosphere-ocean-sea ice model RCAO show a number of 2007-like rapid sea ice reduction events throughout the first half of the twenty-first century with partial recovery thereafter (Koenigk et al. 2011). Anomalously warm temperatures in the winter before the summer event and anomalous summer circulation situations that advect warm air into the Arctic have been identified as the main causes for rapid reduction events (Döscher and Koenigk 2013; Paquin et al. 2013). The warm winter conditions reduce sea ice formation, and thus sea ice thickness is thinner than normal in the beginning of the melting seasons and can more easily be melted during summer. The summer atmospheric circulation anomalies, which lead to sea ice reduction events can vary, but they have in common that they favour enhanced meridional flow and thus advection of warm air into the Arctic during the summer.

11.4.2 Projected Changes in the Atmosphere

Climate model simulations of the future indicate an accelerated climate change in the Arctic in the twenty-first century but the spread between different models is large (e.g. Holland and Bitz 2003). At the decadal time scale, this spread is mainly due to a large inherent variability in the Arctic system, while at longer time scales different model physics may be the cause of this large uncertainty (Sorteberg et al. 2005).

11.4.2.1 Turbulent Heat Fluxes

Changes in the sea ice cover have strong consequences for sensible and latent heat fluxes at the surface in the Arctic. The most extreme local changes in the sensible heat fluxes occur in the North Atlantic Arctic section and can reach up to 35 W m^{-2} under the RCP4.5 emission scenario (Koenigk et al. 2013). The primary reason for the changes is related to the retreating sea-ice edge in the northern North Atlantic and is dominated by winter conditions. In the RCP8.5 emission scenario, there is also a widespread but smaller increase in the sensible heat flux over the entire sea-ice area of about 5–15 W m^{-2}, also dominated by winter conditions. This is likely the result of thinner winter ice and thus more heat being conducted to the surface from the underlying ocean.

The changes in the latent heat flux across the different emission scenarios exhibit a small change maximum east of Iceland and a large area of increased fluxes centered in the Barents Sea, extending into the Kara Sea. Maximum changes reach about 35–40 W m^{-2}, but do not increase in magnitude with emission scenario; instead the area of the maxima increases. Over the Greenland Sea, between these two maxima, there is a slight decrease in the upward flux, which remains unchanged with emission scenario. Thus, the pattern of change is less markedly tied to the retreating ice edge in latent than in sensible heat flux. Like for the sensible heat flux, the latent heat flux also exhibit a widespread increase in the annual average upward flux over the central Arctic Ocean; however, unlike for the sensible heat flux, the latent heat flux in summer shows a widespread decrease of about 5–10 W m^{-2} over the central Arctic Ocean sea ice.

11.4.2.2 Temperature

The warming until 2081–2100 compared to the period 1986–2005 varies from 2.2 °C in RCP2.6 until 8.3 °C in RCP8.5 averaged across all CMIP5-models. As comparison, same values for the global mean warming are 1.0 °C and 3.7 °C. However, the spread across models is large and 5 and 95% confidence interval ranges from −0.5 °C to 5.0 °C in RCP2.6 and 5.2 °C to 11.4 °C in RCP8.5.

The strongest warming in the Arctic occurs in autumn and winter and is linked to the decline of sea ice, which allows for large vertical surface heat fluxes from the relatively warm ocean (around freezing level) to the cold atmosphere above. While the warming in autumn is relatively uniformly distributed over the entire Arctic Ocean, the winter warming is most pronounced in the Barents Sea region since here winter sea ice declines strongest (Chapman and Walsh 2007; Koenigk et al. 2013). Figure 11.8 shows the results for the EC-Earth model. The warming is by far the smallest in summer and does not exceed one Kelvin over most of the Arctic Ocean. This is caused by the fact that the surface stays near 0 °C in summer until almost all sea ice has been melted. Only in RCP8.5, some Arctic Ocean regions warm up due to the earlier onset of the sea ice melt period.

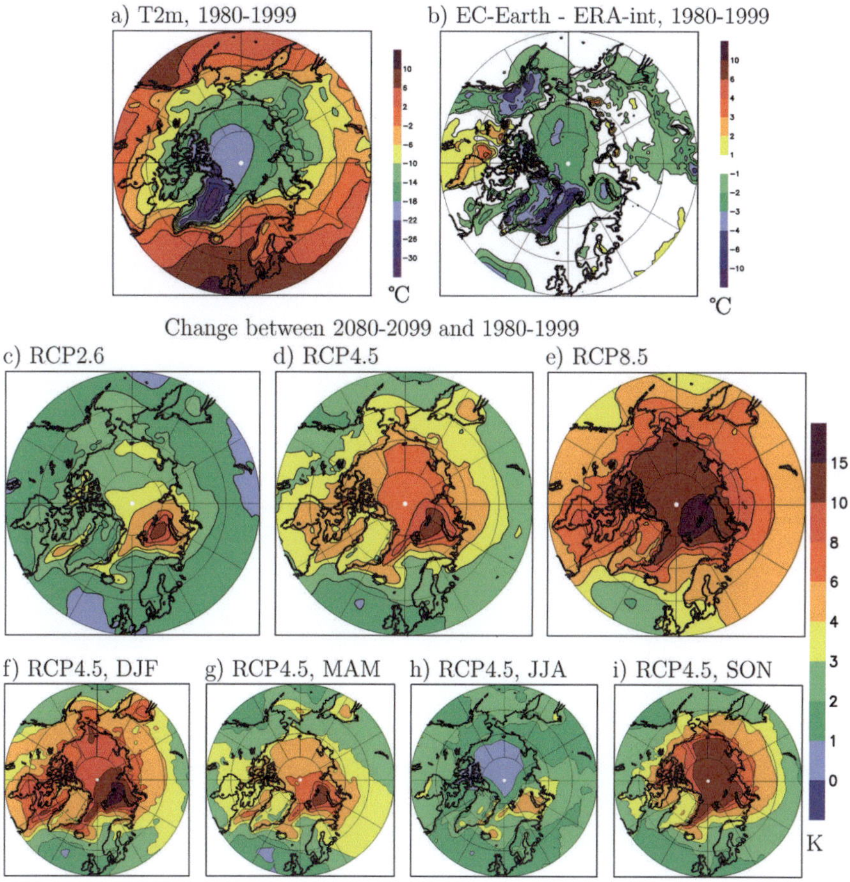

Fig. 11.8 (**a**) Annual mean two meter air temperature in °C in EC-Earth, averaged over 1980–1999 and (**b**) difference to ERA-Interim reanalysis. (**c–e**) Change in annual mean two meter air temperature between 2080–2099 and 1980–1999 in RCP2.6 (**c**), RCP4.5 (**d**) and RCP8.5 (**e**). (**f–i**) Seasonal mean two meter air temperature changes between 2080–2099 and 1980–1999 in RCP4.5. Shown are ensemble means. Note that there is only one RCP2.6 simulation. (From Koenigk et al. 2013, Fig. 7)

The zonal mean, vertical winter temperature distribution in the Arctic in the twentieth century is dominated by a strong surface near inversion with warmest average temperatures at about 850 to 900 hPa (Fig. 11.9). At the surface, the temperature is up to 6 K colder than the air aloft, primarily due to the effect of the negative net radiation at the surface. Thus, the lower atmosphere is sometimes very stable stratified. In summer, temperature inversion is much weaker and often not existent; the temperature decrease with height is, however, relatively small compared to the lower latitudes.

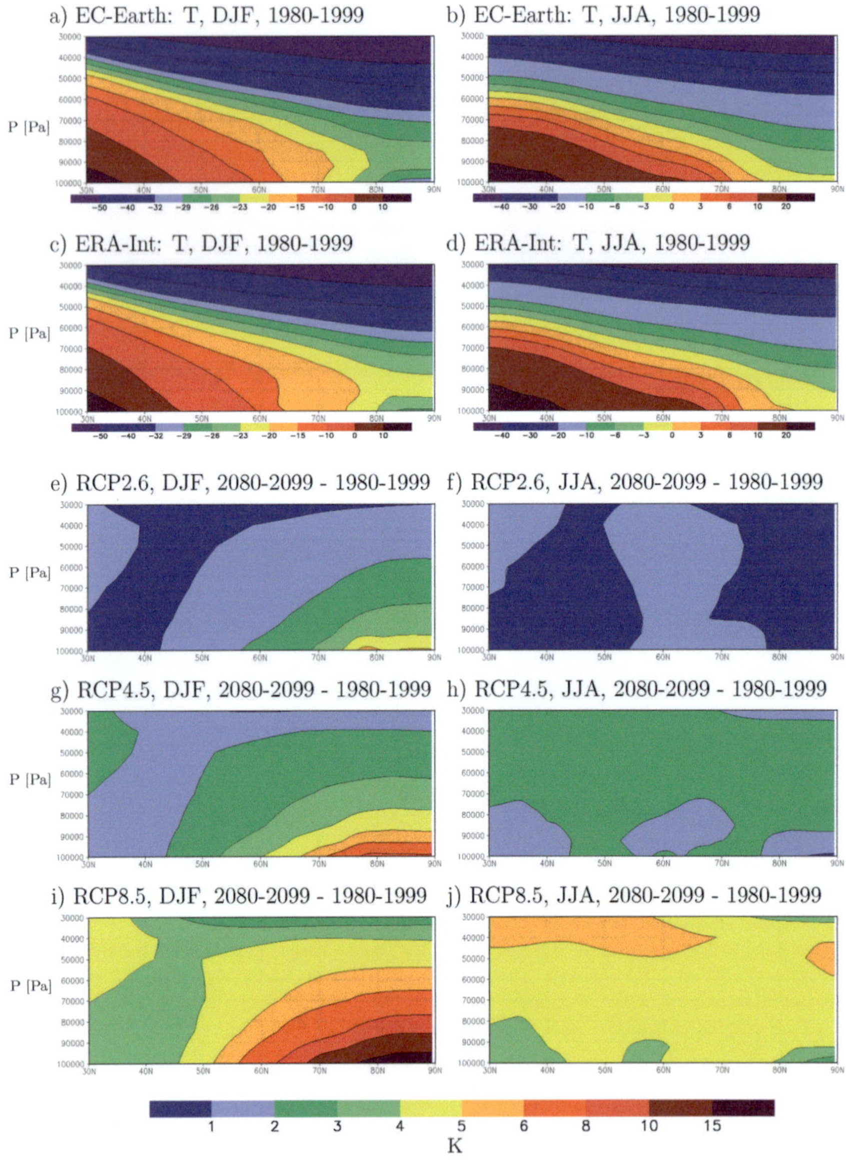

Fig. 11.9 Vertical temperature distribution (in °C) as a function of latitude in winter (DJF, **a**) and summer (JJA, **b**) averaged over 1980–1999 in the twentieth century simulations with EC-Earth and ERA-Interim reanalysis (**c**, **d**). (**e**) and (**f**) Changes in winter and summer in RCP2.6 between 2080–2099 and 1980–1999. (**g**) and (**h**) Same as (**e**) and (**f**) but for RCP4.5. (**i**) and (**j**) Same as (**e**) and (**f**) but for RCP8.5. Shown are ensemble means except for RCP2.6. (From Koenigk et al. 2013, Fig. 9)

The simulated future changes differ distinctively between summer and winter. In summer, the atmosphere is relatively uniformly warmed by about 0–2 K in RCP2.6, 1–3 K in RCP4.5 and 3–6 K in RCP8.5. The warming near the surface is slightly reduced north of 80°N because of the melting sea ice and the cold ocean.

In winter, the warming near the surface is strongly amplified in high northern latitudes compared to lower latitudes. The temperature amplification decreases with increasing height, and above 600 hPa, no amplification can be found anymore. Thus, the Arctic atmosphere becomes less stable during winter and the winter temperature inversion totally disappears at the end of the twenty-first century in RCP8.5.

Regional downscalings of four different global model future projections in the CORDEX (Co-Ordinated Regional Downscaling Experiment)-context reveal the results from the global models (Koenigk et al. 2015).

The diurnal temperature rate shows a decreasing trend in reanalysis data since 1980 and according to both global and regional future model projections, this negative trend will continue in the future (Koenigk et al. 2013; Rinke and Dethloff 2008). Again, the change is most pronounced over future ice-free oceans in winter.

11.4.2.3 Sea Level Pressure and Atmospheric Circulation

Most CMIP5 models indicate a reduction of sea level pressure (SLP) in the Arctic in winter and summer (Stocker et al. 2013) until the end of the twenty-first century. This SLP reduction increases and is becoming more robust in the higher emission scenarios. From the North Atlantic towards Southern Europe, models tend to show an increase of SLP but the response is not robust across models. Regional model future projections reproduce most of the results from the driving global models (Koenigk et al. 2015). Figure 11.10 shows the SLP response to greenhouse gas warming in the EC-Earth model. The SLP change pattern indicates a slightly northward shift of the storm track over the North Atlantic and an extension of the storm track into the Arctic. The zonal mean winds at the surface are somewhat increased at around 60°N. In the upper troposphere, a stronger increase of zonal winds and a slight poleward and upward shift of the tropospheric jets are simulated by the models. The dynamical mechanisms behind these shifts are not yet fully understood but might be linked to increasing upper tropospheric temperature gradients.

11.4.2.4 Cloudiness and Precipitation

Observations of clouds and precipitation are very uncertain in the Arctic. Cloud cover is one of the largest uncertainties in model predictions of Arctic climate. Cloud amounts in global climate models and atmospheric reanalyses vary widely and may have large biases. Liu and Key (2016) examined the performance of five atmospheric reanalysis products in depicting monthly mean Arctic cloud amount anomalies (the National Aeronautics and Space Administration (NASA) Modern-Era Retrospective

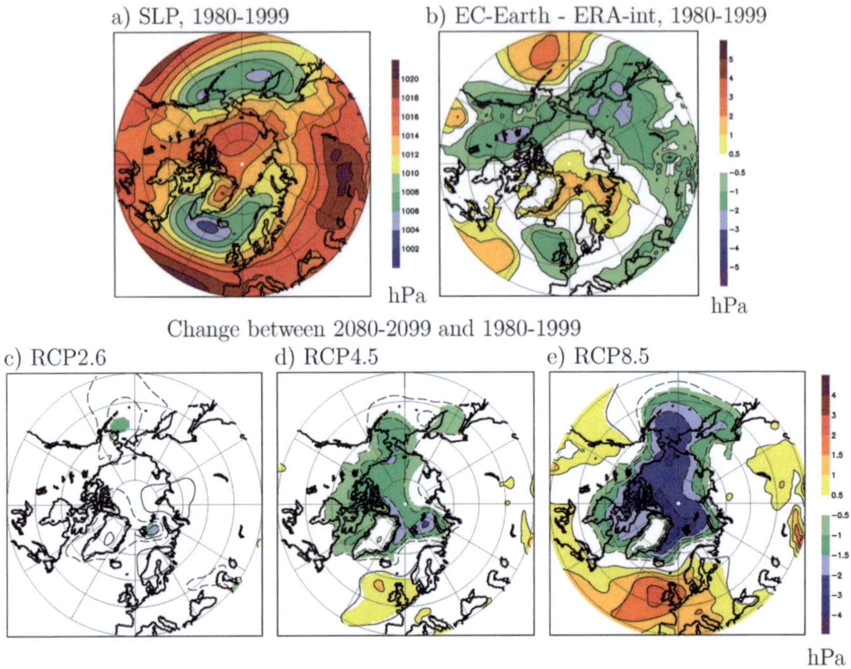

Fig. 11.10 (**a**) Annual mean SLP in hPa in EC-Earth, averaged over 1980–1999 and (**b**) difference to ERA-Interim reanalysis. c-e) Change in annual mean SLP between 2080–2099 and 1980–1999 in RCP2.6 (**c**), RCP4.5 (**d**) and RCP8.5 (**e**). Shown are ensemble means. Note that there is only one RCP2.6 simulation. The ensemble mean SLP-change is significant at the 95% significance level in all colored areas. (From Koenigk et al. 2013, Fig. 10)

Analysis for Research and Applications (MERRA and MERRA-2), the European Centre for Medium-Range Weather Forecasts (ECMWF) interim reanalysis, and the National Centers for Environmental Prediction (NCEP)/National Center for Atmospheric Research (NCAR) Reanalysis-1, NCEP/U.S. Department of Energy (DOE) Reanalysis-2). Reanalysis cloud anomalies were compared to satellite products from the Moderate Resolution Imaging Spectroradiometer (MODIS) satellite observations and the Cloud–Aerosol Lidar and Infrared Pathfinder Satellite Observation (CALIPSO). It was shown that all five reanalysis products exhibit biases in the mean cloud amount, especially in winter, but that the reanalyses do exhibit some capability for depicting the monthly mean cloud amount anomalies. There were no significant differences in the overall performance of reanalysis products. They all perform best in July, August, and September and worst in November, December, and January. All reanalysis datasets perform better over land than over ocean.

The CMIP5 multi-model mean indicates an increase of both total cloudiness and precipitation in the Arctic in the future. The increase of total cloud cover is rather small, on the order of a few percent in simulations based on the RCP2.6 and the

RCP4.5 emission scenarios, but up to 10% under the RCP8.5 emission scenario. Inter-model spread is large, and the response depends on the region, season, and the cloud type. While the total amount of clouds increases in large parts of the year, the amount of low clouds is reduced in regions where the winter temperature inversion disappears.

Models agree largely on an increase of precipitation in the entire Arctic and in all seasons in the future. The increase reaches more than 50% until the end of the twenty-first century (Stocker et al. 2013). In absolute values, the increase is largest in areas where sea ice disappears, which highlights the importance of increased vertical sensible and latent heat fluxes from the ocean to the atmosphere in these regions for the precipitation. Generally, the precipitation change in the Arctic increases linearly with the Arctic temperature change (Koenigk et al. 2015).

Regional downscaling of a number of global model scenarios with the regional atmosphere model RCA (Koenigk et al. 2015) indicated a stronger precipitation increase in summer compared to the global models. It remains unclear if this is caused by higher resolution or caused by different physical parameterizations in the regional RCA-model.

11.4.3 Projected Changes in the Arctic Ocean

The warming of the Arctic, increased river runoff and precipitation and the enhanced exposure to winds due to sea ice decline have strong effects on the Arctic Ocean. Despite the potentially large impact, literature on future Arctic Ocean changes is still limited. Here, we will mainly focus on results from CMIP5 simulations with the EC-Earth model (Koenigk et al. 2013; Koenigk and Brodeau 2014).

11.4.3.1 Ocean Circulation

In the future, the reduction of sea ice leads to an increased wind stress on the ocean and to a strengthening of the surface flow. In RCP2.6 and RCP4.5, the ocean currents are generally strengthened compared to the twentieth century but the patterns stay similar. The inflow through the Barents Sea Opening into the Arctic is strengthened, particularly in RCP8.5 due to increased south-westerly winds over the Nordic Seas. Also, the circulation pattern in the Arctic Basin changes substantially in RCP8.5. The size of the Beaufort Gyre is reduced and is displaced towards Chukchi Sea. This leads to strongly reduced velocities in the transpolar drift.

11.4.3.2 Freshwater and Heat in the Arctic Ocean

The twentieth century Arctic Ocean is characterized by a cold and fresh upper layer and warmer and saltier water masses in deeper layers according to the World Ocean

Atlas 2009 (Antonov et al. 2010; Locarnini et al. 2010). The northward flowing warm and salty Atlantic water masses are cooled when they reach the Arctic, are getting denser and sink down near the ice edge to a depth of around 500 m and spread around the Arctic Ocean in this depth and form a warm and salty layer. Within the twenty-first century, enhanced oceanic heat inflow into the Arctic from the Atlantic strengthens this warm layer.

In the low emission scenario RCP2.6, the cold surface layer persists during the entire twenty-first century while in the high emission scenario RCP8.5, the surface layer is warmed by up to 3 K at the end of the twenty-first century and is getting much warmer than the deep ocean. The reduced sea ice area allows stronger wind mixing and a penetration of the warm surface waters almost down to the warm intermediate layer.

In the twenty-first century, river runoff into the Arctic will likely increase due to more precipitation in the catchment areas of the large Arctic rivers (Koenigk et al. 2007, 2013, Fig. 11.11). Precipitation over the Arctic Ocean basin itself will also grow faster than the evaporation. Both, the increase in river runoff and the growing positive difference between precipitation and evaporation lead to more freshwater input into the Arctic. Consequently, the Arctic Ocean is getting fresher. The largest reduction of salinity in the twenty-first century occurs near the surface - by up to 1.5 psu in the upper 100 m. The salinity differences between different RCPs are not as pronounced as for temperature but we see a clear tendency to more freshening in the higher emission scenarios.

More input of freshwater into the Arctic could be compensated by increased outflow of freshwater out of the Arctic. However, model results indicate that the change in the total Arctic freshwater export is relatively small and first increasing in the second half of the twenty-first century (Fig. 11.11). Investigating the individual straits, a strong increase of the total freshwater export (solid + liquid freshwater) through the Canadian Archipelago is evident, while the total exports through the Fram Strait and the Barents Sea are slightly reduced. The transport through Bering Strait shows a slight increase.

The freshwater transports are dominated by the liquid freshwater transport except for Fram Strait. In Fram Strait, the ice export dominates the total export and its variability. In the twenty-first century, the decrease in sea ice leads to strongly decreased Fram Strait ice exports, which is compensated by an increase in the liquid export. As the increase in freshwater input into the Arctic exceeds the increase in the

Fig. 11.11 (continued) and the sum of all (black) in RCP4.5 (solid) and RCP8.5 (dashed) in EC-Earth. Ensemble means are shown. (**b**) Annual mean P-E (blue) and river runoff (red) in m^3/s in RCP4.5 (solid) and RCP8.5 (dashed). Ensemble means are shown. (**c**) Annual mean liquid (solid) and solid (dashed) fresh water transports in m^3/s through Fram Strait in ensemble means of twentieth century simulations (black), RCP4.5 (blue) and RCP8.5 (red). (**d**) Annual mean Arctic freshwater content in m^3 (same area as in Fig. 16) in ensemble means of twentieth century simulations (black), RCP4.5 (blue) and RCP8.5 (red). As reference salinity 34.9 psu has been used. (From Koenigk et al. 2013, Fig. 17)

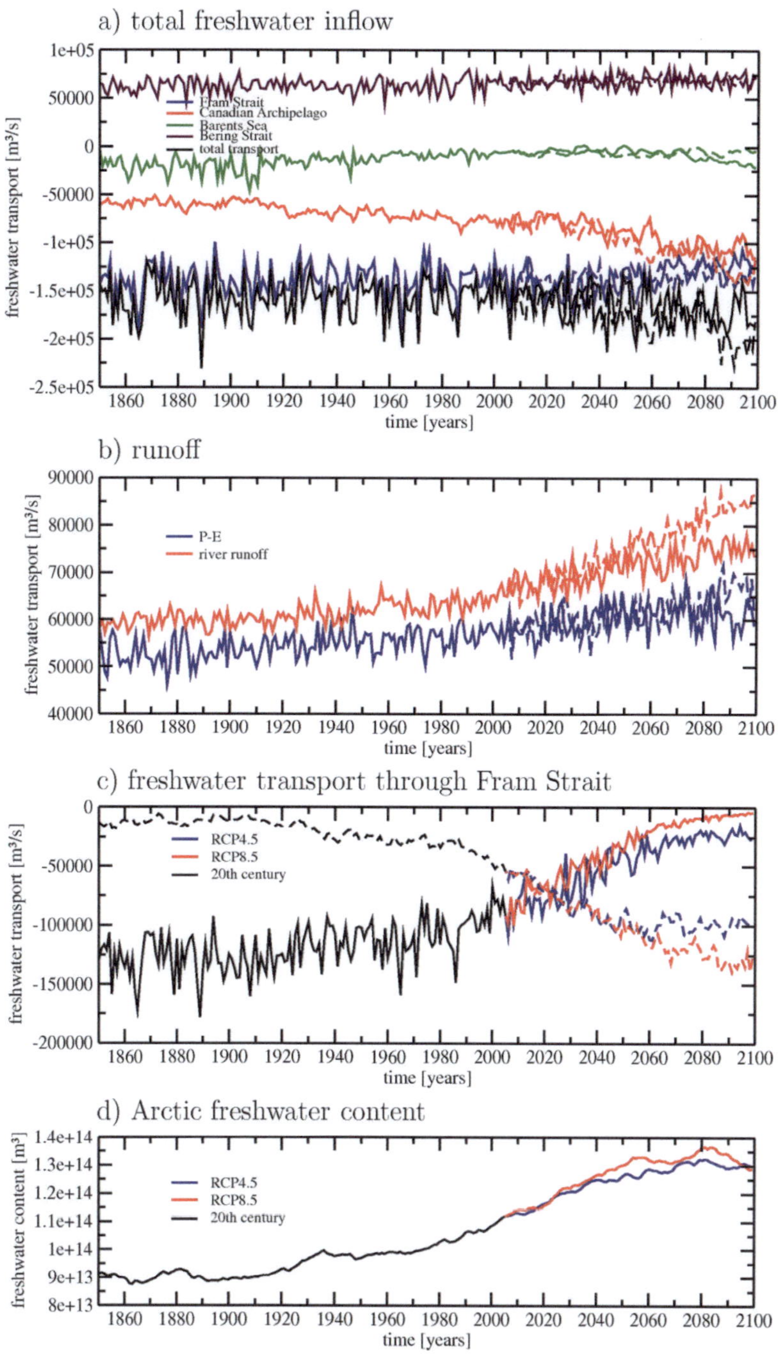

Fig. 11.11 (**a**) Annual mean total (liquid + solid) freshwater inputs into the Arctic in m^3/s through Fram Strait (blue), Canadian Archipelago (red), Barents Sea (green) and Bering Strait (maroon)

exports in the twenty-first century simulations, the fresh water content in the Arctic is increased and the Arctic Ocean is getting fresher.

Observations show an increasing ocean heat transport into the Arctic, particularly through the Barents Sea Opening (Skagseth et al. 2008; Smedsrud et al. 2013) in recent decades. This increase of ocean heat transport is projected to continue in the twenty-first century (Koenigk and Brodeau 2014), particularly through the Barents Sea Opening. The strong increase of heat transport in the Barents Sea is due both to the temperature increase of the transported water masses and increased northward volume transport.

The changes in ocean heat transport strongly contribute to the reduced sea ice cover in the Barents and Kara Sea region and govern sea ice variations at decadal scale time periods. However, most of the additional heat is passed to the atmosphere and contributes thus to the warming of the Arctic atmosphere and the Arctic amplification.

11.5 Conclusions

The Arctic climate is in rapid transition. There are few places in the world where the ongoing climate change is as obvious as in the Arctic:

- Sea ice area and volume have been reduced by more than 30% and 50%, respectively, in the last 30 years.
- Temperature, particularly at the surface, increases much faster than the global mean. Locally, temperature increased by almost 10 °C since 1980.
- Atmospheric stratification has changed, affecting cloud formation and precipitation.
- Increased precipitation, more river runoff into the Arctic, and melting land and sea ice led to a freshening of the Arctic Ocean.
- Arctic warming, the reduction of sea ice and related changes in the temperature difference between high and low latitudes might affect circulation and occurrence of extremes in lower latitudes. However, many uncertainties remain on the governing processes and the amplitude of this connection.
- Changes in sea ice and cloud cover are linked such that 22–34% of the cloud cover variance can be explained by the variability in sea ice extent. A further decline in sea ice cover will result in an even cloudier Arctic.
- The magnitude of the Arctic ice–albedo feedback (ocean) is four times that of the snow–albedo feedback (land) in summer. Decreasing sea ice cover, not changes in terrestrial snow cover, has been the dominant radiative feedback mechanism in recent decades and will continue to play an important role in future Arctic climate change.

Future climate scenarios show a continuous warming of the Arctic with the risk of total loss of late summer Arctic sea ice. This would lead to a completely different

Arctic climate, a “blue Arctic”, where ocean and atmosphere would interact much more directly compared to the past.

Acknowledgements Torben Koenigk has been supported by the NordForsk-funded Nordic Centre of Excellence project (award 76654) Arctic Climate Predictions: Pathways to Resilient, Sustainable Societies (ARCPATH) together with the JPI-CLimate-Belmont Forum project 407, InterDec. The research on cloud-ice interactions and snow- and ice-albedo feedbacks described herein was supported by the US National Oceanic and Atmospheric Administration (NOAA) Climate Data Records program. Thanks are due to Aaron Letterly, Yinghui Liu, Xuanji Wang, and Richard Dworak for their contributions to those projects. The views, opinions, and findings contained in this report are those of the author(s) and should not be construed as an official National Oceanic and Atmospheric Administration or US Government position, policy, or decision.

Timo Vihma has been supported by the Academy of Finland (contract 317999).

References

ACIA. (2005). *Arctic climate impact assessment* (p. 1042). Cambridge, UK: Cambridge University Press.

Alexander, M., Bhatt, U., Walsh, J., Timlin, M., Miller, J., & Scott, J. (2004). The atmospheric response to realistic Arctic Sea ice anomalies in an AGCM during winter. *Journal of Climate, 17*, 890–905.

Antonov, J. I., Seidov, D., Boyer, T. P., Locarnini, R. A., Mishonov, A. V., Garcia, H. E., Baranova, O. K., Zweng, M. M., Johnson, D. R. (2010). World Ocean Atlas 2009, Volume 2: Salinity. S. Levitus, Ed., NOAA Atlas NESDIS 69, U.S. Government Printing Office, Washington, DC, 184 pp.

Barnes, E. A. (2013). Revisiting the evidence linking Arctic amplification to extreme weather in midlatitudes. *Geophysical Research Letters, 40*, 4734–4739.

Bintanja, R., Graversen, R. G., & Hazeleger, W. (2011). Arctic winter warming amplified by the thermal inversion and consequent low infrared cooling to space. *Nature Geoscience Letters*. https://doi.org/10.1038/NGEO1285.

Bonan, D. B., Armour, K. C., Roe, G. H., Siler, N., & Feldl, N. (2018). Sources of uncertainty in the Meridional pattern of climate change. *Geophysical Research Letters, 45*(17), 9131–9140. https://doi.org/10.1029/2018GL079429.

Callaghan, T., Johansson, M., Key, J., Prowse, T., Ananicheva, M., & Klepikov, A. (2011). Feedbacks and interactions: From the Arctic cryosphere to the climate system. *Ambio, 40*, 75–86. https://doi.org/10.1007/s13280-011-0215-8.

Chapman, W. L., & Walsh, J. E. (2007). Simulations of Arctic temperature and pressure by global coupled models. *Journal of Climate, 20*, 609–632. https://doi.org/10.1175/JCLI4026.1.

Cohen, J., Barlow, M., Kushner, P. J., & Saito, K. (2007). Stratosphere-troposphere coupling and links with Eurasian land surface variability. *Journal of Climate, 20*, 5335–5343.

Cohen, J. L., Furtado, J. C., Barlow, M. A., Alexeev, V. A., & Cherry, J. E. (2012). Arctic warming, increasing snow cover and widespread boreal winter cooling. *Environmental Research Letters, 7*. https://doi.org/10.1088/1748-9326/7/1/014007.

Cohen, J., Screen, J. A., Furtado, J. C., Barlow, M., Whittleston, D., Coumou, D., et al. (2014). Recent Arctic amplification and extreme mid-latitude weather. *Nature Geoscience, 7*, 627–637. https://doi.org/10.1038/NGEO2234.

Coumou, D., Petoukhov, V., Rahmstorf, S., Petri, S., & Schellnguber, H. J. (2014). Quasi-resonant circulation regimes and hemispheric synchronization of extreme weather in boreal summer. *Proceedings of the National Academy of Sciences of the United States of America*, 12331–12336.

Coumou, D., Lehmann, J., & Beckmann, J. (2015). The weakening summer circulation in the northern hemisphere midlatitudes. *Science, 348*, 324–327. https://doi.org/10.1126/science.1261768.

Davini, P., von Hardenberg, J., & Corti, S. (2015). Tropical origin for the impacts of the Atlantic multidecadal variability on the euro-Atlantic climate. *Environmental Research Letters, 10*, 094010. https://doi.org/10.1088/1748-9326/10/9/094010.

Dee, D. P., Uppala, S. M., Simmons, A. J., Berrisford, P., Poli, P., Kobayashi, S., et al. (2011). The ERA-interim reanalysis: Configuration and performance of the data assimilation system. *The Quarterly Journal of the Royal Meteorological Society, 137*, 553–597. https://doi.org/10.1002/qj.828.

Deser, C., Magnusdottir, G., Saravanan, R., & Philips, A. (2004). The effects of North Atlantic SST and sea ice anomalies on the winter circulation in CCM3. Part II: Direct and indirect components of the response. *Journal of Climate, 17*, 2160–2176.

Deser, C., Tomas, R., Alexander, M., & Lawrence, D. (2010). The seasonal atmospheric response to projected Arctic Sea ice loss in the late twenty-first century. *Journal of Climate, 23*, 333–351. https://doi.org/10.1175/2009JCLI3053.1.

Devasthale, A., Sedlar, J., Koenigk, T., & Fetzer, E. J. (2013). The thermodynamic state of the Arctic atmosphere observed by AIRS: Comparisons during the record minimum sea ice extentso f 2007 and 2012. *Atmospheric Chemistry and Physics, 13*, 7441–7450. https://doi.org/10.5194/acp-13-7441-2013.

Dickson, R., Meincke, J., & Malmberg, S. A. (1988). The "great salinity anomaly" in the northern North Atlantic, 1968-1982. *Progress in Oceanography, 20*, 103–151.

Ding, Q., Schweiger, A., L'Heureux, M., Battisti, D. S., Po-Chedley, S., Johnson, N. C., Blanchard-Wrigglesworth, E., Harnos, K., Zhang, Q., Eastman, R., & Steig, E. J. (2017). Influence of high-latitude atmospheric circulation changes on summertime Arctic Sea ice. *Nature Climate Change*. https://doi.org/10.1038/NCLIMATE3241.

Döscher, R., & Koenigk, T. (2013). Arctic rapid sea ice loss events in regional coupled climate scenario experiments. *Ocean Science*, os-2012-65. https://doi.org/10.5194/os-9-217-2013.

Dunn-Sigouin, E., & Son, S. W. (2013). Northern hemisphere blocking frequency and duration in the CMIP5 models. *Journal of Geophysical Research – Atmospheres, 118*, 1179–1188. https://doi.org/10.1002/jgrd.50143.

Fetterer, F., Knowles, K., Meier, W., Savoie, M., Windnagel, A. K. (2017). updated daily. Sea Ice Index, Version 3.0. Boulder, Colorado USA. NSIDC: National Snow and Ice Data Center. doi: https://doi.org/10.7265/N5K072F8. Accessed September 2018.

Francis, J. A. (2017). Why are Arctic linkages to extreme weather still up in the air? *The Bulletin of the American Meteorological Society, 98*, 2551–2557.

Francis, J. A., & Vavrus, S. J. (2012). Evidence linking Arctic amplification to extreme weather in midlatitudes. *Geophysical Research Letters, 39*, L06801.

Francis, J. A., & Vavrus, S. J. (2015). Evidence for a wavier jet stream in response to rapid Arctic warming. *Environmental Research Letters, 10*. https://doi.org/10.1088/1748-9326/10/1/014005.

Francis, J. A., Chan, W., Leathers, D. J., Miller, J. R., & Veron, D. E. (2009). Winter northern hemisphere weather patterns remember summer Arctic Sea-ice extent. *Geophysical Research Letters, 36*(L07503). https://doi.org/10.1029/2009GL037274.

Gao, Y., Sun, Y., Li, F., He, S., Sandven, S., Yan, Q., Zhang, Z., Lohmann, K., Keenlyside, N., Furevik, T., & Suo, L. (2015). Arctic Sea ice and Eurasian climate: A review. *Advances in Atmospheric Sciences, 32*, 92–114.

Garcia-Serrano, J., & Frankkignoul, C. (2014). High predictability of the winter euro-Atlantic climate from cryospheric variability. *Nature Geoscience*. https://doi.org/10.1038/NGEO2118.

Graversen, R. G., & Wang, M. (2009). Polar amplification in a coupled climate model with locked albedo. *Climate Dynamics, 33*, 629–643. https://doi.org/10.1007/s00382-009-0535-6.

Graversen, R. G., Mauritsen, T., Tjernström, T., Källén, E., & Svensson, G. (2008). Vertical structure of recent Arctic warming. *Nature, 541*. https://doi.org/10.1038/nature06502.

Guo, D., Gao, Y., Bethke, I., Gong, D., Johannessen, O. M., & Wang, H. (2014). Mechanism on how the spring Arctic Sea ice impacts the east Asian summer monsoon. *Theoretical and Applied Climatology, 115*, 107–119. https://doi.org/10.1007/s00704-013-0872-6.

Haak, H., Jungclaus, J., Mikolajewicz, U., & Latif, M. (2003). Formation and propagation of great salinity anomalies. *Geophysical Research Letters, 30*(9), 26/1–26/4.

Häkkinen, S. (1999). A simulation of thermohaline effects of a great salinity anomaly. *Journal of Climate*, (6), 1781–1795.

Hall, R., Erdélyi, R., Hanna, E., Jones, J. M., & Scaife, A. A. (2015). Drivers of North Atlantic polar front jet stream variability. *International Journal of Climatology, 35*, 1697–1720. https://doi.org/10.1002/joc.4121.

Hanna, E., Cropper, T. E., Hall, R. J., & Cappelen, J. (2016). Greenland blocking index 1851–2015: A regional climate change signal. *International Journal of Climatology, 36*, 4847–4861.

Holland, M. M., & Bitz, C. M. (2003). Polar amplification of climate change in coupled models. *Climate Dynamics, 21*, 221–232.

Honda, M., Inoue, J., & Yamane, S. (2009). Influence of low Arctic Sea-ice minima on anomalously cold Eurasian winters. *Geophysical Research Letters, 36*(L08707). https://doi.org/10.1029/2008GL037079.

Hopsch, S., Cohen, J., & Dethloff, K. (2012). Analysis of a link between fall Arctic Sea ice concentration and atmospheric patterns in the following winter. *Tellus A, 64*(18624). https://doi.org/10.3402/tellusa.v64i0.18264.

Hoskins, B., & Woollings, T. (2015). Persistent extratropical regimes and climate extremes. *Current Climate Change Reports*. https://doi.org/10.1007/s40641-015-0020-8.

Inoue, J., Masatake, H. E., & Koutarou, T. (2012). The role of Barents Sea ice in the wintertime cyclone track and emergence of a warm-Arctic cold-Siberian anomaly. *Journal of Climate, 25*(7), 2561–2568.

Jahn, A., Kay, J. E., Holland, M. M., & Hall, D. M. (2016). How predictable is the timing of a summer ice-free Arctic? *Geophysical Research Letters, 43*(17). https://doi.org/10.1002/2016GL070067.

Jaiser, R., Dethloff, K., & Handorf, D. (2013). Stratospheric response to Arctic Sea ice retreat and associated planetary wave propagation changes. *Tellus A, 65*, 19375. https://doi.org/10.3402/tellusa.v65i0.19375.

Kim, B. M., Son, S. W., Min, S. K., Jeong, J. H., Kim, S. J., Zhang, X., Shim, T., & Yoon, Y. H. (2014). Weakening of the stratospheric polar vortex by Arctic Sea-ice loss. *Nature Communications, 5*, 4646. https://doi.org/10.1038/ncomms5646.

Koenigk, T., & Brodeau, L. (2014). Ocean heat transport into the Arctic in the twentieth and twenty-first century in EC-earth. *Climate Dynamics, 42*, 3101–3120. https://doi.org/10.1007/s00382-013-1821-x.

Koenigk, T., & Brodeau, L. (2017). Arctic climate and its interaction with lower latitudes under different levels of anthropogenic warming in a global coupled climate model. *Climate Dynamics, 49*, 471–492. https://doi.org/10.1007/s00382-016-3354-6.

Koenigk, T., Mikolajewicz, U., Haak, H., & Jungclaus, J. (2006). Variability of Fram Strait sea ice export: Causes, impacts and feedbacks in a coupled climate model. *Climate Dynamics, 26*, 17–34. https://doi.org/10.1007/s00382-005-0060-1.

Koenigk, T., Mikolajewicz, U., Haak, H., & Jungclaus, J. (2007). Arctic freshwater export in the 20th and 21st century. *Journal of Geophysical Research, 112*. https://doi.org/10.1029/2006JG000274.

Koenigk, T., Döscher, R., & Nikulin, G. (2011). Arctic future scenario experiments with a coupled regional climate model. *Tellus, 63A*(1), 69–86. https://doi.org/10.1111/j.1600-0870.2010.00474.x.

Koenigk, T., Brodeau, L., Graversen, R. G., Karlsson, J., Svensson, G., Tjernström, M., Willen, U., & Wyser, K. (2013). Arctic climate change in 21st century CMIP5 simulations with EC-earth. *Climate Dynamics, 40*, 2720–2742. https://doi.org/10.1007/s00382-012-1505-y.

Koenigk, T., Berg, P., & Döscher, R. (2015). Arctic climate change in an ensemble of regional CORDEX simulations. *Polar Research, 34*, 24603. https://doi.org/10.3402/polar.v34.24603.

Koenigk, T., Caian, M., Nikulin, G., & Schimanke, S. (2016). Regional Arctic Sea ice variations as predictor for winter climate conditions. *Climate Dynamics, 46*, 317–337. https://doi.org/10.1007/s00382-015-2586-1.

Kosaka, Y., & Xie. (2013). Recent global-warming hiatus tied to equatorial Pacific surface cooling. *Nature, 501*, 403–407. https://doi.org/10.1038/nature12534.

Letterly, A., Key, J., & Liu, Y. (2016). The influence of winter cloud on summer sea ice in the Arctic, 1983–2013. *Journal of Geophysical Research – Atmospheres, 121*. https://doi.org/10.1002/2015JD024316.

Letterly, A., Key, J., & Liu, Y. (2018). Arctic climate: Changes in sea ice extent outweigh changes in snow cover. *The Cryosphere, 12*, 3373–3382. https://doi.org/10.5194/tc-12-3373-2018.

Liptak, J., & Strong, C. (2014). The winter atmospheric response to sea ice anomalies in the Barents Sea. *Journal of Climate, 27*, 914–924. https://doi.org/10.1175/JCLI-D-13-00186.1.

Liu, Y., & Key, J. (2014). Less winter cloud aids summer 2013 Arctic Sea ice return from 2012 minimum. *Environmental Research Letters, 9*, 044002. https://doi.org/10.1088/1748-9326/9/4/044002.

Liu, Y., & Key, J. (2016). Assessment of Arctic cloud cover anomalies in atmospheric reanalysis products using satellite data. *Journal of Climate, 29*, 6065–6083. https://doi.org/10.1175/JCLI-D-15-0861.1.

Liu, Y., Key, J. R., & Wang, X. (2008). The influence of changes in cloud cover on recent surface temperature trends in the Arctic. *Journal of Climate, 21*, 705–715.

Liu, Y., Key, J., & Wang, X. (2009). Influence of changes in sea ice concentration and cloud cover on recent Arctic surface temperature trends. *Geophysical Research Letters, 36*, L20710. https://doi.org/10.1029/2009GL040708.

Liu, Y., Key, J. R., Liu, Z., Wang, X., & Vavrus, S. J. (2012). A cloudier Arctic expected with diminishing sea ice. *Geophysical Research Letters, 39*, L05705. https://doi.org/10.1029/2012GL051251.

Liu, Y., Key, J., Vavrus, S., & Woods, C. (2018). Time evolution of cloud response to moisture intrusions into the Arctic during winter. *Journal of Climate, 31*(22), 9389–9405. https://doi.org/10.1175/JCLI-D-17-0896.1.

Locarnini, R. A., Mishonov, A. V., Antonov, J. I., Boyer, T. P., Garcia, H. E., Baranova, O. K., Zweng, M. M., Johnson, D. R. (2010). World Ocean Atlas 2009, Volume 1: Temperature. S. Levitus, Ed., NOAA Atlas NESDIS 69, U.S. Government Printing Office, Washington, DC, 184 pp.

Magnusdottir, G., Deser, C., & Saravanan, R. (2004). The effects of North Atlantic SST and sea ice anomalies on the winter circulation in CCM3. Part I: Main features and storm track characteristics of the response. *Journal of Climate, 17*(5), 857–876.

Masato, G., Hoskins, B. J., & Woollings, T. (2013). Winter and summer northern hemisphere blocking in CMIP5 models. *Journal of Climate, 26*, 7044–7059.

Massonnet, F., Fichefet, T., Goosse, H., Bitz, C. M., Philippon-Berthier, G., Holland, M. M., & Barriat, P. Y. (2012). Constraining projections of summer Arctic Sea ice. *The Cryosphere, 6*, 1383–1394. https://doi.org/10.5194/tcd-6-1383-2012.

McClelland, J., Dery, S., Peterson, B., & Holmes, R. M. (2006). A pan-Arctic evaluation of changes in river discharge during the latter half of the 20th century. *Geophysical Research Letters, 33*, L06715. https://doi.org/10.1029/2006GL025753.

McCusker, K. E., Fyfe, K. C., & Sigmond, M. (2016). Twenty-five winters of unexpected Eurasian cooling unlikely due to Arctic Sea-ice loss. *Nature Geoscience, 9*, 838–842. https://doi.org/10.1038/NGEO2820.

McGraw, M., & Barnes, E. A. (2016). Seasonal sensitivity of the Eddy-driven jet to tropospheric heating in an idealized AGCM. *Journal of Climate, 29*. https://doi.org/10.1175/JCLI-D-15-0723.1.

Medeiros, D. C., Tomas, R. A., & Kay, J. E. (2011). Arctic inversion strength in climate models. *Journal of Climate, 24*, 4733–4740.

Meier, W., Hovelsrud, G., van Oort, B., Key, J., Kovacs, K., Michel, C., Haas, C., Granskog, M., Gerland, S., Perovich, D., Makshtas, A., & Reist, J. (2014). Arctic Sea ice in transformation:

A review of recent observed changes and impacts on biology and human activity. *Reviews of Geophysics, 51*. https://doi.org/10.1002/2013RG000431.
Mesquita, M. D. S., Hodges, K. I., Atkinson, D. A., & Bader, J. (2011). Sea-ice anomalies in the Sea of Okhotsk and the relationship with storm tracks in the northern hemisphere during winter. *Tellus A, 63*, 312–323.
Mori, M., Watanabe, M., Shiogama, H., Inoue, J., & Kimoto, M. (2014). Robust Arctic Sea-ice influence on the frequent Eurasian cold winters in past decades. *Nature Geoscience, 7*, 869–873. https://doi.org/10.1038/NGEO2277.
Ogawa, F., Keenlyside, N., Gao, Y., Koenigk, T., Yang, S., Suo, L., et al. (2018). Evaluating impacts of recent Arctic Sea ice loss on the northern hemisphere winter climate change. *Geophysical Research Letters, 45*. https://doi.org/10.1002/2017GL076502.
Orsolini, Y. J., Senan, R., Benestad, R. E., & Melsom, A. (2012). Autumn atmospheric response to the 2007 low Arctic Sea ice extent in coupled ocean–atmosphere hindcasts. *Climate Dynamics, 38*, 2437–2448.
Overland, J. E., & Wang, M. (2010). Large-scale atmospheric circulation changes are associated with the recent loss of Arctic Sea ice. *Tellus, 62*. https://doi.org/10.1111/j.1600-0870.2009.00421.x.
Overland, J., Francis, J., Hall, R., Hanna, E., Kim, S. J., & Vihma, T. (2015). The melting Arctic and Midlatitude weather patterns: Are they connected? *Journal of Climate, 28*, 7917–7932. https://doi.org/10.1175/JCLI-D-14-00822.1.
Overland, J. E., Dethloff, K., Francis, J. A., Hall, R. J., Hanna, E., Kim, S. J., Screen, J. A., Shepherd, T. G., & Vihma, T. (2016). The melting Arctic and Midlatitude weather patterns: Forced Chaos and a way forward. *Nature Climate Change*. https://doi.org/10.1038/NCLIMATE3121.
Paquin, J. P., Döscher, R., Sushama, L., & Koenigk, T. (2013). Causes and consequences of mid-21st century rapid ice loss events simulated by the Rossby Centre regional Atmosphere-Ocean model. *Tellus A, 2013*(65), 19110. https://doi.org/10.3402/tellusa.v65i0.19110.
Peings, Y., & Magnusdottir. (2014). Response of the wintertime Northern Hemisphere atmospheric circulation to current and projected Arctic sea ice decline: a numerical study with CAM5. *Journal of Climate*. https://doi.org/10.1175/JCLI-D-13-00272.1.
Peterson, B., Holmes, R., McClelland, J., Vorosmarty, C., Lammers, R., Shiklmanov, A., Shiklomanov, I., & Rahmstorf, S. (2002). Increasing river discharge to the Arctic Ocean. *Science, 298*, 2171–2173.
Petoukhov, V., & Semenov, V. A. (2010). A link between reduced Barents-Kara Sea ice and cold winter extremes over northern continents. *Journal of Geophysical Research, 115*(D21111). https://doi.org/10.1029/2009JD013568.
Petoukhov, V., Rahmstorf, S., Petri, S., & Schellnhuber, H. J. (2013). Quasiresonant amplification of planetary waves and recent northern hemisphere weather extremes. *Proceedings of the National Academy of Sciences, 110*, 5336–5341.
Pithan, F., & Mauritsen, T. (2014). Arctic amplification dominated by temperature feedbacks in contemporary climate models. *Nature Geoscience, 7*. https://doi.org/10.1038/NGEO2071.
Polyakov, I. V., Beszczynska, A., Carmack, E. C., Dmitrenko, I. A., Fahrbach, E., Frolov, E. A., et al. (2005). One more step toward a warmer Arctic. *Geophysical Research Letters, 32*, L17605. https://doi.org/10.1029/2005GL023740.
Richter-Menge, J., & Jeffries, M. (2011). The Arctic, in "State of the climate in 2010". *Bulletin of the American Meteorological Society, 92*(6), S143–S160.
Rinke, A., & Dethloff, K. (2008). Simulated circum-Arctic climate changes by the end of the 21st century. *Global and Planetary Change, 62*(2008), 173–186.
Rinke, A., Dethloff, K., Dorn, W., Handorf, D., & Moore, J. C. (2013). Simulated Arctic atmospheric feedbacks associated with late summer sea ice anomalies. *Journal of Geophysical Research: Atmospheres, 118*, 7698–7714. https://doi.org/10.1002/jgrd.50584.
Schlichtholz, P. (2016). Empirical relationships between summertime oceanic heat anomalies in the Nordic seas and large-scale atmospheric circulation in the following winter. *Climate Dynamics, 47*, 1735. https://doi.org/10.1007/s00382-015-2930-5.

Schweiger, A., Lindsay, R., Zhang, J., Steele, M., & Stern, H. (2011). Uncertainty in modeled arctic sea ice volume. *Journal of Geophysical Research*. https://doi.org/10.1029/2011JC007084.

Screen, J. A. (2014). Arctic amplification decreases temperature variance in northern mid- to high-latitudes. *Nature Climate Change, 4*, 577–582.

Screen, J. A. (2017). Climate science: Far-flung effects of Arctic warming. *Nature Geoscience, 10*, 253–254.

Screen, J. A., & Francis, J. A. (2016). Contribution of sea-ice loss to Arctic amplification is regulated by Pacific Ocean decadal variability. *Nature Climate Change*. https://doi.org/10.1038/NCLIMATE3011.

Screen, J. A., & Simmonds, I. (2010a). Increasing fall-winter energy loss from the Arctic Ocean and its role in Arctic temperature amplification. *Geophysical Research Letters, 37*(L16707). https://doi.org/10.1029/2010GL044136.

Screen, J. A., & Simmonds, I. (2010b). The central role of diminishing sea ice in recent Arctic temperature amplification. *Nature, 464*. https://doi.org/10.1038/nature09051.

Screen, J. A., Deser, C., Simmonds, I., & Tomas, R. (2014). Atmospheric impacts of Arctic Sea-ice loss, 1979–2009: Separating forced change from atmospheric internal variability. *Climate Dynamics*. https://doi.org/10.1007/s00382-013-1830-9.

Screen, J. A., Deser, C., & Sun, L. (2015). Reduced risk of North American cold extremes due to continued Arctic Sea ice loss. *Bulletin of the American Meteorological Society*, 1489–1503. https://doi.org/10.1175/BAMS-D-14-00185.1.

Serreze, M. C., Barrett, A. P., Stroeve, J. C., Kindig, D. N., & Holland, M. M. (2009). The emergence of surface-based Arctic amplification. *The Cryosphere, 3*, 11–19.

Serreze, M. C., Barrett, A. P., & Cassano, J. J. (2011). Circulation and surface controls on the lower tropospheric temperature field of the Arctic. *Journal of Geophysical Research, 116*, D07104.

Shepherd, T. G. (2016). Effects of a warming Arctic. *Science, 353*, 989–990.

Skagseth, Ø., Furevik, T., Ingvaldsen, R., Loeng, H., Mork, K. A., Orvik, K. A., & Ozhigin, V. (2008). Chapter 2: Volume and heat transports to the Arctic Ocean via the Norwegian and Barents seas. In B. Dickson, J. Meincke, & P. Rhines (Eds.), *Arctic-Subarctic Ocean fluxes: Defining the role of Nordic seas in climate*. Berlin: Springer.

Smedsrud, L. H., Esau, I., Ingvaldsen, R. B., Eldevik, T., Haugan, P. M., Li, C., et al. (2013). The role of the Barents Sea in the Arctic climate system. *Reviews of Geophysics, 51*, 415–449. https://doi.org/10.1002/rog.20017.

Sorteberg, A., Furevik, T., Drange, H., & Kvamstø, N. G. (2005). Effects of simulated natural variability on Arctic temperature projections. *Geophysical Research Letters, 32*(L18708). https://doi.org/10.1029/2005GL023404.

Spielhagen, R. F., Werner, K., Aagaard Sörensen, S., Zamelczyk, K., Kandiano, E., Budeus, G., Marchitto, T. M., & Hald, M. (2011). Enhanced modern heat transfer to the Arctic by warm Atlantic water. *Science, 331*, 450–454. https://doi.org/10.1126/science.1197397.

Stocker, T., et al. (2013). Climate change 2013: The physical science basis. In *Contribution of working group I to the fifth assessment report of the intergovernmental panel on climate change*. Cambridge, UK/New York: Cambridge University Press.

Stroeve, J. C., Kattsov, V., Barrett, A., Serreze, M., Pavlova, T., Holland, M., & Meier, W. M. (2012). Trends in Arctic Sea ice extent from CMIP5, CMIP3 and observations. *Geophysical Research Letters, 39*(16). https://doi.org/10.1029/2012GL052676.

Sun, L., Perlwitz, J., & Hoerling, M. (2016). What caused the recent "warm Arctic, cold continents" trend pattern in winter temperatures? *Geophysical Research Letters, 43*, 5345–5352. https://doi.org/10.1002/2016GL069024.

Uotila, P., Karpechko, A., & Vihma, T. (2014). Links between Arctic sea ice and extreme summer precipitation in China: An alternative view. *Advances in Polar Science, 25*, 222–233. https://doi.org/10.13679/j.advps.2014.4.00222.

Vavrus, S. J., Wang, F., Martin, J. E., Francis, J. A., Peings, Y., & Cattiaux, J. (2017). Changes in north American atmospheric circulation and extreme weather: Influence of Arctic amplification and northern hemisphere snow cover. *Journal of Climate, 30*, 4317–4333.

Vihma, T. (2014). Effects of Arctic Sea ice decline on weather and climate: A review. *Surveys in Geophysics, 35*, 1175–1214. https://doi.org/10.1007/s10712-014-9284-0.

Vihma, T. (2017). Weather extremes linked to interaction of the Arctic and mid-latitudes. In S.-Y. Wang et al. (Eds.), *Climate extremes: Mechanisms and potential prediction* (Geophysical monograph series, 226) (pp. 39–49). American Geophysical Union.

Walsh, J. E. (2014). Intensified warming of the Arctic: Causes and impacts on middle latitudes. *Global and Planetary Change, 117*, 52–63. https://doi.org/10.1016/j.gloplacha.2014.03.003.

Wang, M., & Overland, J. (2012). A sea ice free summer Arctic within 30 years: An update from CMIP5 models. *Geophysical Research Letters, 39*, L18501. https://doi.org/10.1029/2012GL052868.

Wu, B., Zhang, R., Wang, B., & D'Arrigo, R. (2009). On the association between spring Arctic Sea ice concentration and Chinese summer rainfall. *Geophysical Research Letters, 36*, L09501.

Wu, B., Zhang, R., D'Arrigo, R., & Su, J. (2013). On the relationship between winter sea ice and summer atmospheric circulation over Eurasia. *Journal of Climate, 26*, 5523–5536.

Yang, S., & Christensen, J. H. (2012). Arctic Sea ice reduction and European cold winters in CMIP5 climate change experiments. *Geophysical Research Letters, 39*(L20707). https://doi.org/10.1029/2012GL053333.

Yeager, S. G., & Robson, J. I. (2017). Recent progress in understanding and predicting Atlantic decadal climate variability. *Current Climate Change Reports, 3*, 112–127.

Zhao, P., Zhang, X., Zhou, X., Ikeda, M., & Yin, Y. (2004). The sea ice extent anomaly in the North Pacific and its impact on the east Asian summer monsoon rainfall. *Journal of Climate, 17*, 3434–3447.

Index

A. Kokhanovsky, C. Tomasi (eds.), *Physics and Chemistry of the Arctic Atmosphere*, Springer Polar Sciences, https://doi.org/10.1007/978-3-030-33566-3

B

C

MIX
Papier aus verantwortungsvollen Quellen
Paper from responsible sources
FSC® C105338

If you have any concerns about our products,
you can contact us on
ProductSafety@springernature.com

In case Publisher is established outside the EU,
the EU authorized representative is:
Springer Nature Customer Service Center GmbH
Europaplatz 3, 69115 Heidelberg, Germany

Printed by Libri Plureos GmbH
in Hamburg, Germany